Plant Biodiversity and Genetic Resources

Plant Biodiversity and Genetic Resources

Editors

Andreas W. Ebert

Johannes M. M. Engels

MDPI • Basel • Beijing • Wuhan • Barcelona • Belgrade • Manchester • Tokyo • Cluj • Tianjin

Editors
Andreas W. Ebert
World Vegetable Center
Taiwan

Johannes M. M. Engels
Bioversity International
Italy

Editorial Office
MDPI
St. Alban-Anlage 66
4052 Basel, Switzerland

This is a reprint of articles from the Special Issue published online in the open access journal *Plants* (ISSN 2223-7747) (available at: https://www.mdpi.com/journal/plants/special_issues/plant_biodiversity_genetic_resources).

For citation purposes, cite each article independently as indicated on the article page online and as indicated below:

LastName, A.A.; LastName, B.B.; LastName, C.C. Article Title. *Journal Name* **Year**, *Volume Number*, Page Range.

ISBN 978-3-0365-0894-8 (Hbk)
ISBN 978-3-0365-0895-5 (PDF)

Cover image courtesy of Andreas W. Ebert.

Contents

About the Editors

Andreas W. Ebert graduated in May 1976 from the University of Hohenheim, Stuttgart, Germany with a diploma degree in Agricultural Sciences. In February 1980, he obtained a Ph.D. with distinction (magna cum laude) from the same university for his work on "hormonal aspects of crop regulation in apple" at the Institute for Fruit and Vegetable Production and Viticulture. Thereafter, he held a postdoctoral position at the Glasshouse Crops Research Institute in Littlehampton, UK, where he conducted research on assimilate translocation in tomatoes. At a later stage, he worked on the somatic embryogenesis of coconut at Wye College, University of London.

He has dedicated his entire professional life to agricultural research for development in the tropics and subtropics. He worked for the German Agency for International Cooperation (GIZ) from 1981 to 2002 in various development projects in Brazil, the Philippines, West Africa, Malawi, and China. From 2002 to 2008, he joined CATIE in Costa Rica as Team Leader of an interdisciplinary plant genetic resources group, directed CATIE's genebank, and served as Professor at its Graduate School. From 2008 until his retirement in December 2015, he held the position of Genebank Manager and Global Theme Leader—Germplasm (germplasm conservation, utilization, and gene discovery) at the World Vegetable Center's headquarters in Taiwan.

Currently, he is a freelance consultant in tropical agriculture and horticulture and the management of plant genetic resources and genebanks. He is passionate about the conservation and sustainable use of plant genetic resources, especially landraces, farmers' varieties, and crop wild relatives—key resources for crop improvement and adaptation to climate change. His publications are accessible at: https://www.researchgate.net/profile/Andreas_Ebert2.

Johannes M. M. Engels, Honorary Research Fellow, Alliance of Bioversity International and CIAT, 00053 Rome/Maccarese, Italy, and freelance consultant Upon completion of his M.Sc. at Wageningen University, The Netherlands with majors in plant breeding, genetics, and pedagogics and didactics, in 1974 he joined the German Agency for International Cooperation (GIZ) to establish and operate a regional genebank for Central America at CATIE, Turrialba, Costa Rica. During his stay of almost six years, he conducted field work on cacao genetic resource management aspects and defended his Ph.D. thesis at Wageningen University in 1986.

In 1981, he accepted the leadership of a German-funded genebank project in Ethiopia. In 1985, he played a key role in organizing an international conference on plant genetic resources (PGR) in Addis Ababa and subsequently published a book on the same topic. Linking PGR conservation management activities with economic development became a major focus of his work. In 1988, he joined the International Board for Plant Genetic Resources (IBPGR), based in Rome, and accepted responsibilities as the regional coordinator for South and Southeast Asia, based at the National Bureau for Plant Genetic Resources (NBPGR) in New Delhi, India. During his three-year stay, he played an active role in strengthening national PGR activities and programs in the various countries of the region.

In 1991, he moved to the IBPGR headquarters at FAO in Rome to coordinate research activities worldwide, with a focus on germplasm and genebank management. The genebank focus was gradually widened to also include in situ conservation aspects and the integration of conservation activities into agricultural and economic development. Networking of researchers worldwide was an essential element of the research approach that IBPGR, and subsequently IPGRI, took as a research institute "without walls" and as an honest broker. In leading positions, he was able to influence directions, approaches, and scope of research activities, including the PGR program of CGIAR. The addition of policy research allowed more direct linkages of PGR conservation and development in national, regional, and global contexts. His subsequent involvement in several regional and crop networks facilitated the implementation of research and policy results. His publications are accessible at: https://www.researchgate.net/profile/Johannes_Engels

Preface to "Plant Biodiversity and Genetic Resources"

The foundation of our global food supply is based on the result of thousands of years of crop selection and improvements made to wild and semi-domesticated species, crop wild relatives, and landraces, which have given rise to present-day cultivated crop varieties. In addition, in many parts of the world people continue to collect and use food from plants that still exist in the wild. The "wild" genes of crop wild relatives and landraces are an important resource to improve our crops with respect to agronomic characteristics such as phenology, growing seasons, sensitivity to inputs (i.e., fertilizer and water), resistance to diseases and insect pests, and tolerance to heat, drought, and salinity. The availability of such genetic diversity is critical for plant breeding, especially with climate change. Moreover, the genetic diversity within and between species gives rise to a multitude of characteristics that enable plants, animals, and microbes to fulfil different roles in the environment and to adapt to steadily changing conditions, as this diversity will ensure the continued functioning of ecosystems and the provision of ecosystem services.

Current over-reliance on three major staple crops, i.e., maize, rice, and wheat, has inherent agronomic, ecological, nutritional, and economic risks and is unsustainable in the long run. The wider use of underutilized minor or orphan crops provides more options to build temporal and spatial heterogeneity into uniform cropping systems, thus helping to maintain and improve the efficiency and resilience of agroecosystems, enhance dietary diversity, and combat malnutrition. Production systems and the underlying genetic resources, including crop wild relatives that are (still) found in cultivated, abandoned, and protected lands and especially in natural ecosystems such as forests (ranging from tropical to temperate), are severely threatened due to drastic land-use changes, overexploitation of resources, and manmade and natural disasters. Climate change is already affecting the distribution of plants and associated species, their population sizes, and life cycles. Efficient adaptation strategies for a changing climate require, among other measures, the effective and rational conservation and sustainable utilization of the remaining (especially agricultural) biodiversity, both in nature and farmers' fields (in situ) as well as in genebanks and botanic gardens (ex situ), and easy, facilitated access to genetic resources of crops and their wild relatives by plant breeders. Appropriate compensation of farmers and others who contribute to such conservation efforts is needed for such services as well as a share of the benefits that derive from the use of these resources.

To develop and grow "climate-smart" crop varieties for sustainable production systems, farmers and plant breeders worldwide are in dire need of access to a wide range of traits and genes, often found in plant genetic resources located far away from major production areas. This raises a multitude of policy issues and concerns regarding access and benefit sharing, ownership, intellectual property rights, and patents imposed on PGRFA and breeding lines, as well as implications of transgenic crops for biodiversity and sustainable agriculture.

Andreas W. Ebert, Johannes M. M. Engels
Editors

plants

MDPI

Editorial

Plant Biodiversity and Genetic Resources Matter!

Andreas W. Ebert [1,*,†] and Johannes M. M. Engels [2]

1 World Vegetable Center, 60 Yi-Min Liao, Shanhua, Tainan 74151, Taiwan
2 Alliance of Bioversity International and CIAT, Maccarese, 00054 Rome, Italy; j.engels@cgiar.org
* Correspondence: ebert.andreas6@gmail.com
† Current address: Oderstr. 3, 73529 Schwäbisch Gmünd, Germany.

Received: 1 December 2020; Accepted: 3 December 2020; Published: 4 December 2020

Abstract: Plant biodiversity is the foundation of our present-day food supply (including functional food and medicine) and offers humankind multiple other benefits in terms of ecosystem functions and resilience to climate change, as well as other perturbations. This Special Issue on 'Plant Biodiversity and Genetic Resources' comprises 32 papers covering a wide array of aspects from the definition and identification of hotspots of wild and domesticated plant biodiversity to the specifics of conservation of genetic resources of crop genepools, including breeding and research materials, landraces and crop wild relatives which collectively are the pillars of modern plant breeding, as well as of localized breeding efforts by farmers and farming communities. The integration of genomics and phenomics into germplasm and genebank management enhances the value of crop germplasm conserved ex situ, and is likely to increase its utilization in plant breeding, but presents major challenges for data management and the sharing of this information with potential users. Furthermore, also a better integration of in situ and ex situ conservation efforts will contribute to a more effective conservation and certainly to a more sustainable and efficient utilization. Other aspects such as policy, access and benefit-sharing that directly impact the use of plant biodiversity and genetic resources, as well as balanced nutrition and enhanced resilience of production systems that depend on their increased use, are also being treated. The editorial concludes with six key messages on plant biodiversity, genetic erosion, genetic resources and plant breeding, agricultural diversification, conservation of agrobiodiversity, and the evolving role and importance of genebanks.

Keywords: genetic erosion; agrobiodiversity; crop wild relatives; ex situ and in situ conservation; plant breeding; diversification; climate change; sustainability; food and nutrition security; policy

---

1. Introduction

The foundation of the current global food supply is based on thousands of years of phenotypic selection of plant species and genotypes with favorable traits for cultivation and human nutrition by early farmers. This selection process eventually led to the domestication of crop species and the development of landraces that were well-adapted to local conditions. The rediscovery of Mendel's laws of trait inheritance in 1900 and Darwin's concept of natural selection provided the basis of modern plant breeding [1], giving rise to modern crop varieties. The development of high-yielding, uniform commercial crop cultivars since the early 1900s, especially for temperate crops, and later during the Green Revolution also for tropical and subtropical crops, led to global annual productivity gains of 1.0% for wheat, 0.8% for rice, 0.7% for maize, 0.6% for millets, and 0.5% for sorghum [2]. However, the wide dissemination of these high-yielding cultivars caused a drastic replacement and progressive elimination of the locally adapted landraces, in particular in tropical and subtropical countries. Moreover, the often-narrow genetic base of modern cultivars made them more vulnerable to the incidence of new races of diseases or insect pests. As such, breeders gradually started to explore landraces and especially crop wild relatives (CWR; the wild species from which modern crops derived and their close relatives) with useful traits, such as pathogen and insect pest resistance as well as tolerance to drought, salinity, and low or high temperature, in order to introgress the respective genes and/or quantitative trait loci (QTL) into modern crop varieties to make them more resilient.

It is crucial to conserve CWR and landraces if we wish to retain sufficient genetic diversity for current and future plant breeding programs [3]. This can be achieved through in situ conservation of CWR in their natural habitats or through the continuous use of landraces by farmers on their own farm, and thus maintaining the evolutionary processes that result in adaptation. However, due to the alarming rates of genetic erosion in nature, it is critically important to complement in situ conservation with ex situ conservation in genebanks—either as dried seeds with low seed moisture content (3–7%) at freezing temperatures (amenable for orthodox seeds) [4], or in field genebanks, through in vitro culture or cryopreservation (rapid freezing and storage at extremely low temperatures) [5]. The latter concerns crops that do not produce orthodox seeds with desiccation and low temperature tolerance (recalcitrant seeds; e.g. avocado), or do not produce seeds at all as is the case for bananas, or are generally clonally propagated, as seeds are not true-to-type as for instance potato [5]. It should be noted that ex situ approaches 'freeze' the genetic diversity, as no (or very limited) possibilities for adaptive evolutionary processes take place.

Production systems and the underpinning genetic resources, including crop wild relatives, which are (still) found on both cultivated and protected land, and especially in natural ecosystems such as forests, are severely threatened due to drastic land-use changes, over-exploitation of resources and human-made and natural disasters. Climate change is already significantly affecting the distribution of plants and associated species, their population sizes, and life cycles [6]. Currently, two in five plant species are estimated to be threatened with extinction [7]. The major threats to plants in situ, as assessed for the Red List of Threatened Species by the International Union for Conservation of Nature (IUCN), are as follows: agriculture and aquaculture (32.8%), biological resource use (21.1%), modification of natural systems (10.8%), residential and commercial developments (10.5%), invasive species, genes, and diseases (6.5%), pollution (5.4%), and climate change (4.1%). Yet, all genetic diversity is far from known to humankind, for instance only in 2019 close to 2000 new species of plants were discovered and described for the first time from Asia (36%), South America (34%), Africa (12%), Oceania (8%), Europe (5%), and North America (5%) [7]. Among those newly discovered species, there are also species of high potential relevance for agriculture, such as six species of *Allium*, described by scientists in Turkey, as well as wild relatives of cassava (*Manihot esculenta*), yams (*Dioscorea* spp.) and sweet potato (*Ipomoea* spp.) encountered in Brazil. If we are serious about slowing down the on-going genetic erosion and saving those species which are threatened or close to extinction, as well as those currently still being discovered by science, whose potential is still unknown, it is mandatory to conserve the genetic diversity found especially in biodiversity hotspots of wild and domesticated biodiversity [8]. This will benefit current and future generations.

The current over-reliance on a handful of major staple crops has inherent agronomic, ecological, nutritional, and economic risks, and is unsustainable in the long run. The wider cultivation of today's underutilized minor crops [9] as well as intercropping [10] provide more options to build much needed heterogeneity into increasingly uniform cropping systems. This, in turn, will help to cushion the negative impact of climate change and to maintain and enhance evolutionary processes, efficiency and resilience of agroecosystems, as well as enhancing dietary diversity and combating malnutrition [11]. Efficient adaptation strategies for agriculture in a dynamically changing climate require, among other measures, effective and rational conservation and sustainable utilization of agricultural biodiversity, both in situ as well as in genebanks, and facilitated access to genetic resources of crops and their wild relatives by plant breeders and other users [12]. International access and exchange of genetic resources relies on the safe movement of germplasm across borders [13], with germplasm health requirements clearly differing from one country to another, and this germplasm movement raises a multitude of policy issues and concerns regarding access and benefit-sharing, ownership, intellectual property rights and patents imposed on plant genetic resources for food and agriculture (PGRFA) and breeding lines [14], as well as implications of transgenic crops for biodiversity and sustainable agriculture [15].

We encouraged contributions to this Special Issue on plant biodiversity and genetic resources in the form of original research and review papers with the aim to disseminate and promote knowledge on a wide array of aspects in this important field, which underpins sustainable agriculture and food and nutrition security in a globalized world, facing multiple challenges. This Special Issue comprises 32 papers covering a wide array of aspects from the definition and identification of hotspots of wild and domesticated plant biodiversity [8] to the specifics of the conservation of crop germplasm and their wild relatives in situ and ex situ, including germplasm and genebank management and genebank phenomics, to enhance the value and utilization of crop germplasm in breeding. The various modes of ex situ germplasm conservation in the form of orthodox seeds, as in vitro collections and/or in liquid nitrogen, have been described [5], and special attention has been drawn to the importance of seed physiology for the successful long-term conservation of orthodox seeds [4]. One paper assessed the specific case of seed germination after 30 years of storage in permafrost [16].

Attention has also been drawn to the challenges of some crops and crop wild relatives that present specific hurdles for successful, long-term ex situ conservation, such as forage germplasm (grasses, herbaceous legumes, and fodder trees) [17], wild bananas presenting varying levels of seed desiccation tolerance [18], wild food plants [19,20], and crop wild relatives [3]. As crop wild relatives are mostly found in the wild and are rather difficult to be conserved ex situ, the merits of payment for ecosystem services for in situ conservation have been analyzed by Tyak et al. [21] for these germplasm resources, which are increasingly coming into the focus of plant breeders, given their broad genetic diversity, harboring many (new) traits of interest to build resistance to diseases and insect pests as well as tolerance to abiotic stresses into resilient modern crop cultivars. With respect to germplasm management, an example of the use of single nucleotide polymorphisms (SNP) markers for the identification of duplicate accessions of *Brassica oleracea* in genebank collections has been presented [22], while another paper highlights the importance of information management to assist in germplasm and genebank management and to enhance the use of germplasm [23]. Genebank phenomics is a rather novel approach in modern genebanking, and Nguyen and Norton [24] shed light on digital phenotyping methods that enable capturing traits during annual seed regeneration events to enrich genebank phenotypic datasets, thus adding value to crop collections and increasing their usefulness for the identification of traits of interest for breeding.

The international germplasm collections hosted by 11 CGIAR Centers include over 760,000 accessions of crops, forages, and trees [14], and constitute a major proportion of the international germplasm exchange. Over the last ten years, the CGIAR genebanks distributed over 1.1 million PGRFA samples to recipients in 163 countries. Therefore, the CGIAR Germplasm Health Units play a major role in safe global germplasm movement and exchange and the prevention of the trans-boundary spread of pests and diseases [13]. Halewood et al. [14] elucidated the state of international and national access and benefit-sharing laws, noting some unresolved tensions regarding access and benefit sharing and digital genomic sequence information that have the potential to undermine international cooperation to conserve, share and use germplasm for use in food and agriculture.

Several papers highlight the importance of vegetables as well as traditional, underutilized and wild food plants for food and nutrition security in general [19,25], in pilot studies in Kenya [20], and on atolls in the South Pacific [26], including specific crops, such as Hairy Stork's Bill (*Erodium crassifolium*) [27], and the sister of the common pomegranate (*Punica protopunica*), the latter also having interesting medicinal properties [28]. A number of papers focus on the genetic diversity of specific crops or specific traits in a range of food crops for the benefit of plant breeding, such as genome-wide association mapping for stripe rust resistance in spring wheat [29], diversity studies for drought and heat stress in maize landraces [30], nitrogen fixation and water use efficiency in common bean landraces and cultivars in Honduras [31], species identification of Katsouni pea on Greek Islands [32], wild potato germplasm evaluation for starch content and nitrogen utilization efficiency [33], diversity, population structure and marker-trait association for 100-seed weight in a safflower (*Carthamus tinctorius*) germplasm panel [34], the composition of Cypriot grapevine varieties [35], species assignment, genetic diversity and phylogeographic relationships of wild germplasm of macadamia [36], genetic diversity and population structure of *Rhododendron rex* subsp. *rex* [37], and genetic distinctiveness of a Sicilian manna ash (*Fraxinus angustifolia*) collection [38]. A further paper looks at the genetic distinctiveness of cogongrass (*Imperata cylindrica*), an invasive species in the Southern United States [39].

Rosero et al. [40] advocate a dual strategy of breeding for drought tolerance in staple crops and introducing drought-tolerant, underutilized species in cropping systems to enhance their resilience to drought. In the context of public breeding programs in 18 developing countries, Galluzzi et al. [41] analyze the specific role of genetic resources in breeding for climate change. The importance of vegetable genetic resources for nutrition security and vegetable breeding is highlighted in a paper by Ebert [25].

2. Key Messages

2.1. Plant Biodiversity

Plant biodiversity encompasses the ecosystem level, species diversity, and the genetic diversity within species. While biodiversity hotspots are mainly defined by species richness, agrobiodiversity hotspots refer to the centers of origin and diversity of (major) crops and their wild relatives, harboring genetic resources that are of high and increasing relevance for plant breeding. Pironon et al. [8] observed spatial congruence between biodiversity and agrobiodiversity hotspots and proposed a unifying concept, taking into consideration not only species richness but also the multiple benefits plants offer to humankind in terms of ecosystem and agro-ecosystem functions, the inter- and intra-specific diversity of the chemical properties of plants for human nutrition and medicinal use, as well as the provision of gene sources for plant breeding. Rather than considering only major global food crops, agrobiodiversity conservation and use strategies should explicitly include the huge number of edible plants, which are currently underutilized and/or only of regional or often local importance, including wild food plants harvested in the wild that provide significant food and other benefits to humanity [19,20].

2.2. Genetic Erosion

As noted by Antonelli et al. [7], two out of five plant species are estimated to be threatened with extinction. The major threats to plants are of anthropogenic nature and encompass agriculture, overexploitation of biological resources in the wild, modification, fragmentation, and destruction of natural ecosystems, rapidly expanding residential and commercial developments, pollution, and climate change. Van de Wouw et al. [42] highlighted three major evolutionary bottlenecks which led to an enormous reduction in the genetic diversity of cultivated plants. A first major drop in genetic diversity occurred during the development steps from the wild ancestors into the domesticated forms (domestication bottleneck), followed by a dispersal bottleneck when only a small population or even a single plant may have been the foundation of the introduced crop, far away from the area of domestication (e.g., introduction of soybean to North America and the introduction of coffee, traceable to one single plant, to South America). Finally, the modernization bottleneck is due to the success of modern plant breeding and the wide dissemination and adoption of high-yielding cultivars that led and continue to lead to a drastic replacement of landraces. If we are not trying to halt the ongoing processes of dramatic genetic erosion and habitat destruction and fragmentation, we are threatening the web that is sustaining our lives.

Several papers in this Special Issue have addressed genetic erosion aspects and provided suggestions and experiences regarding how this erosion can be stemmed for wild and cultivated biodiversity [8], CWR [3,21], wild food plants [19,20], vegetables [25], and forages [17].

2.3. Genetic Resources and Plant Breeding

It is crucial to conserve landraces and CWR of crops if we wish to retain sufficient genetic diversity for current and future plant breeding programs, and to develop resilient cultivars that can withstand the multiple biotic and abiotic challenges that are exacerbated by climate change [3,25]. However, public sector breeders in developing countries are still confronted with obstacles such as accessing germplasm across national borders and the lack of appropriate technologies and skills to exploit sets of germplasm accessions composed of landraces and CWR [41]. It has been proposed that the commonly used estimated breeding value (EBV) of a parent in a cross could be expanded from individuals to species and populations, and in this case go beyond mere heritability values to additionally include crossing considerations (e.g., ploidy, mating system), evolutionary factors (e.g., phylogenetic relationship), and ecological factors (e.g., environment the species is thriving in) [8]. This multi-pronged approach of an expanded EBV could be used to produce a ranking of species for an individual breeding program, based on the traits required for a desired phenotype. Employing this EBV evaluation tool for characterizing the utility variation across species (and populations) could help define priority areas for in situ and ex situ conservation according to specific breeding targets.

Examples of conservation activities that impact on the use of the genetic resources that have been covered in this Special Issue include the following: easy access to genetic resources increases the capacity of breeders to respond to climate change and the availability of appropriate technologies [41]; access to traditional knowledge on the use of wild plant species [27]; a systematic association-mapping of wheat varieties with SNP markers was successfully used to associate adult plant stripe rust resistance with specific rust races, and results can be used in marker-assisted selection [29]; the analysis of a local genetic panel of manna ash with a continental dataset allowed conclusions on the presence of a possible glacial refuge, and thus facilitates the collecting and use of more genetic diversity [38]; the systematic characterization of ancient grape germplasm in Cyprus allowed the discovery of so far unnoticed genetic diversity [35]; literature searches and conducting field surveys allowed the identification of unknown wild food plants in Kenya [20]; fact sheets promoted the use of traditional food plants in the South Pacific [26]; the exploitation of the local genetic diversity of traditional pea landraces in Greece is fundamental for conservation practices and crop improvement through breeding strategies [32]; the evaluation of maize landrace accessions under heat and drought stresses resulted in invaluable sources of genes/alleles for adaptation breeding [30]; the review of recent efforts that build evidence of the importance of wild food plants in selected countries, while providing examples of cross-sectoral cooperation and multi-stakeholder approaches, contributes to enhancing their sustainable use [19]; the advances in conventional and molecular breeding for the drought tolerance of conventional staple crops, and the introduction of drought-tolerant neglected and underutilized species into existing production systems has the potential to enhance the resilience of agricultural production under conditions of water scarcity [40]; the utilization of advanced phenotyping tools, coupled with high-throughput genotyping, will accelerate the use of genetic resources and fast-track the development of more resilient food crops for the future [24]; and genomics-assisted breeding is increasingly facilitating the introgression of favorable genes and quantitative trait loci from wild species into cultigens, and will lead to a wider use of crop wild relatives in the development of resilient cultivars [25].

2.4. Agricultural Diversification

Relying on only a handful crops and on the enormous advances in plant breeding to feed the global population has greatly improved (global) food security, but has also contributed to malnutrition and has left farmers vulnerable to climate change. A meta-analysis of over 5000 original studies led Tamburini et al. [43] to the conclusion that agricultural "diversification enhances biodiversity, pollination, pest control, nutrient cycling, soil fertility, and water regulation without compromising crop yields". Multispecies cropping systems constitute practical applications of ecological principles based on biodiversity, plant interactions, and other natural regulation mechanisms [44], and may lead to higher productivity, yield stability, ecological sustainability, and resilience to disruptions caused by climate change and other natural events. Diversification with a wider use of neglected and undervalued fruit and vegetable crops and semi-domesticated species, either intercropped with main staples in cereal-based systems or as stand-alone crops, would also contribute to the increased availability (and consumption) of nutritious food and, thus, help combat the triple burden of malnutrition [25]. However, as underutilized crops and semi-wild plant species rarely draw the attention of plant breeders, they often do not meet market standards and the expectation of consumers, and thus might slowly disappear with the advances in the breeding of the major food crops. Therefore, special attention should be given to the conservation and research of underutilized, threatened crop species and their associated genetic diversity by genebank curators. In addition, breeding efforts in these minor crops should be promoted, so that they will keep their place in farms, contributing to more sustainable production systems and benefitting consumers with diverse, often nutrient-dense food.

2.5. Conservation of Agrobiodiversity

The choice of the conservation approach, i.e., in situ and ex situ, as well as specific methods of the latter, will depend on the purpose of the conservation. In case we would like to maintain evolutionary processes that allow targeted genetic resources to evolve, we have to opt for in situ approaches. In cases where farmers are still actively depending on and managing crop diversity for their production, on-farm conservation would apply. Where long-term conservation is important, combined with easy access to diversity, genebank and other ex situ approaches are preferred.

About 80% of plant species produce viable seeds that are desiccation-tolerant and can easily withstand long-term low temperature storage; this presents the most convenient form of ex situ storage of germplasm of most domesticated crops and their close wild relatives. Special attention has been paid to the importance of seed physiology for the successful long-term conservation of orthodox seeds [4]. For other crops that do not produce orthodox seeds with desiccation and low temperature tolerance, or do not produce seeds at all, or are generally clonally propagated as seeds are not true-to-type, different conservation options are available. Those crops can be safely conserved in field genebanks, through in vitro culture or cryopreservation [5]. Crop wild relatives of the secondary and tertiary genepool are mostly found in the wild and are rather difficult to conserve ex situ. These resources are best maintained in situ, in their natural habitat, usually in protected and undisturbed areas where they can continue to evolve [3]. Payment for ecosystem services may be a useful tool to support the in situ conservation of CWR [21].

The availability of comprehensive information on the composition of genebank collections is an important prerequisite to further develop such collections and thus, to increase their utility for breeders and other users [23]. The on-going and increasingly accelerating genetic erosion, as well as the fact that a significant amount of 'new' genetic diversity is still being discovered every year, highlight the importance and urgency of conserving and systematically studying the genetic diversity found, especially in biodiversity hotspots of wild and domesticated biodiversity [8]. This will help in achieving the food and nutrition security of a still growing global population, and will secure other benefits provided by the plant kingdom, such as environmental services and the functional ingredients of plants.

2.6. The Evolving Role and Importance of Genebanks

Genetic resources conserved in genebanks are often considered to be sub-utilized for breeding purposes. This is at least partly due to the fact that most genebanks can offer users only limited, often incomplete passport and basic characterization data, while evaluation data are mostly missing. Characterization data are generally obtained during germplasm regeneration events and mostly focus on morphological descriptors with a high degree of heritability. The limited availability of genotypic and phenotypic information on the majority of genebank accessions is a major reason why breeders are hesitant to use genebank accessions in their breeding programs. A strategic three-step approach has been proposed by McCouch et al. [45] to enhance the mining of genetic resources conserved in genebanks. Such an approach combines genomics and phenomics with efficient information management to enhance the value of available germplasm, thus making it user-friendly for breeders. Enormous progress has been made on the molecular front with the availability of modern cost-effective high throughput molecular tools and advanced statistical analysis that enables the genomic selection of traits of interest. Although many QTL have been identified for biotic and abiotic stresses, their practical use is still limited due to the lack of high throughput phenotyping and molecular platforms [46]. Nguyen and Norton [24] suggested the use of digital high-throughput phenotyping methods that can capture traits of interest to breeders during annual seed regeneration events to enrich phenotypic datasets of genebank accessions. However, high initial investment costs, data capture and standardization, quality assurance, and data analysis are major technical challenges related to genebank phenotyping that restrict the number of genebanks which can adopt and effectively run a high-throughput phenotyping platform.

Finally, to make comprehensive information on germplasm accessions available for global users, there is a clear need for a cooperative platform for data collection, analysis and sharing. The International Plant Phenotyping Network brings plant phenotyping centers together and facilitates the sharing of technical advances in high-throughput phenotyping equipment, data capture and standardization for a range of crop phenotypes [47]. In order to facilitate the use of conserved germplasm, genebanks should consider offering this material in a suitable form to potential users. Examples of facilitating the use of germplasm are as follows: the splitting of heterogeneous accessions into uniform lines [48]; the generation of core, minicore [49] and trait-specific collections [50], as well as single seed descents [51]. In addition, the conservation and distribution of genetic stocks to breeders might facilitate the use of genetic resources and thus increase the availability of more genetic diversity by breeders. For an overview of the complex set of operations in a typical seed genebank, please see Hay and Sershen [52] (p. 4, Figure 1).

As reported by Halewood et al. [14], there is widespread concern, especially among developing countries, regarding the governance of digital genomic sequence information, leading to international disagreement over access and benefit sharing regulations. However, despite the enormous enthusiasm regarding the importance of dematerialized digital genomic sequences for the potential engineering of new plants, the conservation of accessions in physical form will remain important as genotyping needs the complement of phenotyping to be of relevance for plant breeders. Future research may show the value of physical collections in ways that no one can anticipate today [53].

Author Contributions: A.W.E. and J.M.M.E. conceived of, drafted, and edited this paper equally. All authors have read and agreed to the published version of the manuscript.

Funding: This research received no external funding.

Conflicts of Interest: The authors declare no conflict of interest.

References

1. Hallauer, A.R. Evolution of Plant Breeding. *Crop Breed. Appl. Biotechnol.* **2011**, *11*, 197–206. [CrossRef]
2. Evenson, R.E.; Gollin, D. Assessing the Impact of the Green Revolution, 1960 to 2000. *Science* **2003**, *300*, 758–762. [CrossRef] [PubMed]
3. Engels, J.M.M.; Thormann, I. Main Challenges and Actions Needed to Improve Conservation and Sustainable Use of Our Crop Wild Relatives. *Plants* **2020**, *9*, 968. [CrossRef] [PubMed]
4. Whitehouse, K.J.; Hay, F.R.; Lusty, C. Why Seed Physiology Is Important for Genebanking. *Plants* **2020**, *9*, 584. [CrossRef]
5. Panis, B.; Nagel, M.; Van den Houwe, I. Challenges and prospects for the conservation of crop genetic resources in field genebanks, in in vitro collection and/or in liquid nitrogen. *Plants* **2020**, *9*, 1634. [CrossRef]
6. Bellard, C.; Bertelsmeier, C.; Leadley, P.; Thuiller, W.; Courchamp, F. Impacts of climate change on the future of biodiversity. *Ecol. Lett.* **2012**, *15*, 365–377. [CrossRef]

7. Antonelli, A.; Smith, R.J.; Fry, C.; Simmonds, M.S.J.; Kersey, P.J.; Pritchard, H.W.; Abbo, M.S.; Acedo, C.; Adams, J.; Ainsworth, A.M.; et al. *State of the World's Plants and Fungi*; Research Report; Royal Botanic Gardens (Kew); Sfumato Foundation: London, UK, 2020.
8. Pironon, S.; Borrell, J.S.; Ondo, I.; Douglas, R.; Phillips, C.; Khoury, C.K.; Kantar, M.B.; Fumia, N.; Soto Gomez, M.; Viruel, J.; et al. Toward Unifying Global Hotspots of Wild and Domesticated Biodiversity. *Plants* **2020**, *9*, 1128. [CrossRef]
9. Dempewolf, H.; Bordoni, P.; Rieseberg, L.H.; Engels, J.M.M. Food security: Crop species diversity. *Science* **2010**, *328*, 169–170. [CrossRef]
10. Bybee-Finley, K.; Ryan, M.R. Advancing intercropping research and practices in industrialized agricultural landscapes. *Agriculture* **2018**, *8*, 80. [CrossRef]
11. Ebert, A.W. Potential of underutilized traditional vegetables and legume crops to contribute to food and nutritional security, income and more sustainable production systems. *Sustainability* **2014**, *6*, 319–335. [CrossRef]
12. Dempewolf, H.; Baute, G.; Anderson, J.; Kilian, B.; Smith, C.; Guarino, L. Past and future use of wild relatives in crop breeding. *Crop Sci.* **2017**, *57*, 1070–1082. [CrossRef]
13. Kumar, P.L.; Cuvero, M.; Kreuze, J.F.; Muller, G.; Kulkarni, G.; Kumar, S.G.; Massart, S.; Mezzelema, M.; Alakonya, A.; Muchugi, A.; et al. Phytosanitary Interventions for Safe Global Germplasm Exchange and Prevention of Transboundary Pest Spread: The Role of CGIAR Germplasm Health Units. *Plants* **2020**. submitted.
14. Halewood, M.; Jamora, N.; Noriega, I.L.; Anglin, N.L.; Wenzl, P.; Payne, T.; Ndjiondjop, M.N.; Guarino, L.; Kumar, P.; Yazbek, M.; et al. Germplasm acquisition and distribution by CGIAR genebanks. *Plants* **2020**, *9*, 1296. [CrossRef] [PubMed]
15. Jacobsen, S.E.; Sørensen, M.; Pedersen, S.M.; Weiner, J. Feeding the world: Genetically modified crops versus agricultural biodiversity. *Agron. Sustain. Dev.* **2013**, *33*, 651–662. [CrossRef]
16. Solberg, S.Ø.; Brodal, G.; von Bothmer, R.; Meen, E.; Yndgaard, F.; Andreasen, C.; Asdal, Å. Seed Germination after 30 Years Storage in Permafrost. *Plants* **2020**, *9*, 579. [CrossRef]
17. Hanson, J.; Ellis, R.H. Progress and Challenges in Ex Situ Conservation of Forage Germplasm: Grasses, Herbaceous Legumes and Fodder Trees. *Plants* **2020**, *9*, 446. [CrossRef]
18. Kallow, S.; Longin, K.; Sleziak, N.F.; Janssens, S.B.; Vandelook, F.; Dickie, J.; Swennen, R.; Paofa, J.; Carpentier, S.; Panis, B. Challenges for Ex Situ Conservation of Wild Bananas: Seeds Collected in Papua New Guinea Have Variable Levels of Desiccation Tolerance. *Plants* **2020**, *9*, 1243. [CrossRef]
19. Borelli, T.; Hunter, D.; Powell, B.; Ulian, T.; Mattana, E.; Termote, C.; Pawera, L.; Beltrame, D.; Penafiel, D.; Tan, A.; et al. Born to Eat Wild: An Integrated Conservation Approach to Secure Wild Food Plants for Food Security and Nutrition. *Plants* **2020**, *9*, 1299. [CrossRef]
20. Mutie, F.M.; Rono, P.C.; Kathambi, V.; Hu, G.W.; Wang, Q.F. Conservation of Wild Food Plants and Their Potential for Combatting Food Insecurity in Kenya as Exemplified by the Drylands of Kitui County. *Plants* **2020**, *9*, 1017. [CrossRef]
21. Tyack, N.; Dempewolf, H.; Khoury, C.K. The Potential of Payment for Ecosystem Services for Crop Wild Relative Conservation. *Plants* **2020**, *9*, 1305. [CrossRef]
22. Palmé, A.E.; Hagenblad, J.; Solberg, S.Ø.; Aloisi, K.; Artemyeva, A. SNP Markers and Evaluation of Duplicate Holdings of *Brassica oleracea* in Two European Genebanks. *Plants* **2020**, *9*, 925. [CrossRef] [PubMed]
23. Weise, S.; Lohwasser, U.; Oppermann, M. Document or Lose It—On the Importance of Information Management for Genetic Resources Conservation in Genebanks. *Plants* **2020**, *9*, 1050. [CrossRef] [PubMed]
24. Nguyen, G.N.; Norton, S.L. Genebank phenomics: A strategic approach to enhance value and utilization of crop germplasm. *Plants* **2020**, *9*, 817. [CrossRef] [PubMed]
25. Ebert, A.W. The role of vegetable genetic resources in nutrition security and vegetable breeding. *Plants* **2020**, *9*, 736. [CrossRef] [PubMed]
26. Lyons, G.; Dean, G.; Tongaiaba, R.; Halavatau, S.; Nakabuta, K.; Lonalona, M.; Susumu, G. Macro-and Micronutrients from Traditional Food Plants Could Improve Nutrition and Reduce Non-Communicable Diseases of Islanders on Atolls in the South Pacific. *Plants* **2020**, *9*, 942. [CrossRef] [PubMed]
27. Cohen, S.; Koltai, H.; Selvaraj, G.; Mazuz, M.; Segoli, M.; Bustan, A.; Guy, O. Assessment of the Nutritional and Medicinal Potential of Tubers from Hairy Stork's-Bill (*Erodium crassifolium* L'Hér), a Wild Plant Species Inhabiting Arid Southeast Mediterranean Regions. *Plants* **2020**, *9*, 1069. [CrossRef]

28. Guerrero-Solano, J.A.; Jaramillo-Morales, O.A.; Jiménez-Cabrera, T.; Urrutia-Hernández, T.A.; Chehue-Romero, A.; Olvera-Hernández, E.G.; Bautista, M. *Punica protopunica* Balf., the Forgotten Sister of the Common Pomegranate (*Punica granatum* L.): Features and Medicinal Properties—A Review. *Plants* **2020**, *9*, 1214. [CrossRef]
29. Habib, M.; Awan, F.S.; Sadia, B.; Zia, M.A. Genome-Wide Association Mapping for Stripe Rust Resistance in Pakistani Spring Wheat Genotypes. *Plants* **2020**, *9*, 1056. [CrossRef]
30. Nelimor, C.; Badu-Apraku, B.; Tetteh, A.Y.; N'guetta, A.S. Assessment of genetic diversity for drought, heat and combined drought and heat stress tolerance in early maturing maize landraces. *Plants* **2019**, *8*, 518. [CrossRef]
31. Wilker, J.; Humphries, S.; Rosas-Sotomayor, J.C.; Gómez Cerna, M.; Torkamaneh, D.; Edwards, M.; Navabi, A.; Pauls, K.P. Genetic Diversity, Nitrogen Fixation, and Water Use Efficiency in a Panel of Honduran Common Bean (*Phaseolus vulgaris* L.) Landraces and Modern Genotypes. *Plants* **2020**, *9*, 1238. [CrossRef]
32. Stavridou, E.; Lagiotis, G.; Karapetsi, L.; Osathanunkul, M.; Madesis, P. DNA Fingerprinting and Species Identification Uncovers the Genetic Diversity of Katsouni Pea in the Greek Islands Amorgos and Schinoussa. *Plants* **2020**, *9*, 479. [CrossRef] [PubMed]
33. Bachmann-Pfabe, S.; Dehmer, K.J. Evaluation of Wild Potato Germplasm for Tuber Starch Content and Nitrogen Utilization Efficiency. *Plants* **2020**, *9*, 833. [CrossRef] [PubMed]
34. Ali, F.; Nadeem, M.A.; Barut, M.; Habyarimana, E.; Chaudhary, H.J.; Khalil, I.H.; Alsaleh, A.; Hatipoğlu, R.; Karaköy, T.; Kurt, C.; et al. Genetic Diversity, Population Structure and Marker-Trait Association for 100-Seed Weight in International Safflower Panel Using SilicoDArT Marker Information. *Plants* **2020**, *9*, 652. [CrossRef] [PubMed]
35. Grigoriou, A.; Tsaniklidis, G.; Hagidimitriou, M.; Nikoloudakis, N. The Cypriot Indigenous Grapevine Germplasm Is a Multi-Clonal Varietal Mixture. *Plants* **2020**, *9*, 1034. [CrossRef] [PubMed]
36. Mai, T.; Alam, M.; Hardner, C.; Henry, R.; Topp, B. Genetic Structure of Wild Germplasm of Macadamia: Species Assignment, Diversity and Phylogeographic Relationships. *Plants* **2020**, *9*, 714. [CrossRef]
37. Zhang, X.; Liu, Y.H.; Wang, Y.H.; Shen, S.K. Genetic Diversity and Population Structure of *Rhododendron rex* Subsp. *rex* Inferred from Microsatellite Markers and Chloroplast DNA Sequences. *Plants* **2020**, *9*, 338. [CrossRef]
38. Abbate, L.; Mercati, F.; Di Noto, G.; Heuertz, M.; Carimi, F.; del Bosco, S.F.; Schicchi, R. Genetic Distinctiveness Highlights the Conservation Value of a Sicilian Manna Ash Germplasm Collection Assigned to *Fraxinus angustifolia* (Oleaceae). *Plants* **2020**, *9*, 1035. [CrossRef]
39. Lucardi, R.D.; Wallace, L.E.; Ervin, G.N. Patterns of Genetic Diversity in Highly Invasive Species: Cogongrass (*Imperata cylindrica*) Expansion in the Invaded Range of the Southern United States (US). *Plants* **2020**, *9*, 423. [CrossRef]
40. Rosero, A.; Berdugo-Cely, J.A.; Šamajová, O.; Šamaj, J.; Cerkal, R. A Dual Strategy of Breeding for Drought Tolerance and Introducing Drought-Tolerant, Underutilized Crops into Production Systems to Enhance Their Resilience to Water Deficiency. *Plants* **2020**, *9*, 1263. [CrossRef]
41. Galluzzi, G.; Seyoum, A.; Halewood, M.; López Noriega, I.; Welch, E.W. The Role of Genetic Resources in Breeding for Climate Change: The Case of Public Breeding Programmes in Eighteen Developing Countries. *Plants* **2020**, *9*, 1129. [CrossRef]
42. Van De Wouw, M.; Kik, C.; Van Hintum, T.; Van Treuren, R.; Visser, B. Genetic erosion in crops: Concept, research results and challenges. *Plant Genet. Resour.* **2010**, *8*, 1–15. [CrossRef]
43. Tamburini, G.; Bommarco, R.; Wanger, T.C.; Kremen, C.; van der Heijden, M.G.A.; Liebman, M.; Hallin, S. Agricultural diversification promotes multiple ecosystem services without compromising yield. *Sci. Adv.* **2020**, *6*, eaba1715. [CrossRef] [PubMed]
44. Malézieux, E.; Crozat, Y.; Dupraz, C.; Laurans, M.; Makowski, D.; Ozier-Lafontaine, H.; Rapidel, B.; De Tourdonnet, S.; Valantin-Morison, M. Mixing plant species in cropping systems: Concepts, tools and models. A review. *Agron. Sustain. Dev.* **2009**, *29*, 43–62. [CrossRef]
45. McCouch, S.R.; McNally, K.L.; Wang, W.; Sackville, H.R. Genomics of gene banks: A case study in rice. *Am. J. Bot.* **2012**, *99*, 407–423. [CrossRef]
46. Tadesse, W.; Sanchez-Garcia, M.; Assefa, S.G.; Amri, A.; Bishaw, Z.; Ogbonnaya, F.C.; Baum, M. Genetic gains in wheat breeding and its role in feeding the world. *Crop Breed. Genet. Genom.* **2019**, *1*, e190005.

47. Pieruschka, R.; Schurr, U. Plant phenotyping: Past, present, and future. *Plant Phenom.* **2019**, *2019*, 7507131. [CrossRef]
48. Van Hintum, T.J.L.; Sackville Hamilton, N.R.; Engels, J.M.M.; van Treuren, R. Accession Management Strategies: Splitting and Lumping. In *Managing Plant Genetic Diversity*; Engels, J.M.M., Rao Ramanatha, V., Brown, A.H.D., Jackson, M.T., Eds.; CABI Publishing: Wallingford, UK, 2002; pp. 113–120.
49. Schafleitner, R.; Nair, R.M.; Rathore, A.; Wang, Y.W.; Lin, C.Y.; Chu, S.H.; Lin, P.Y.; Chang, J.C.; Ebert, A.W. The AVRDC–The World Vegetable Center mungbean (*Vigna radiata*) core and mini core collections. *BMC Genom.* **2015**, *16*, 344. [CrossRef]
50. Vetriventhan, M.; Upadhyaya, H.D. Diversity and trait-specific sources for productivity and nutritional traits in the global proso millet (*Panicum miliaceum* L.) germplasm collection. *Crop J.* **2018**, *6*, 451–463. [CrossRef]
51. Saxena, K.; Saxena, R.K.; Varshney, R.K. Use of immature seed germination and single seed descent for rapid genetic gains in pigeonpea. *Plant Breed.* **2017**, *136*, 954–957. [CrossRef]
52. Hay, F.R.; Sershen, N. *New Technologies to Improve the Ex Situ Conservation of Plant Genetic Resources*; Burleigh Dodds Series in Agricultural Science; Burleigh Dodds Science Publishing Limited: Sawston, Cambridge, UK, 2021; pp. 1–32. [CrossRef]
53. Gollin, D. Conserving genetic resources for agriculture: Economic implications of emerging science. *Food Secur.* **2020**, *12*, 919–927. [CrossRef]

plants

MDPI

Review

Toward Unifying Global Hotspots of Wild and Domesticated Biodiversity

Samuel Pironon [1,*], James S. Borrell [1], Ian Ondo [1], Ruben Douglas [1], Charlotte Phillips [2], Colin K. Khoury [3,4], Michael B. Kantar [5], Nathan Fumia [5], Marybel Soto Gomez [6,7], Juan Viruel [1], Rafael Govaerts [1], Félix Forest [1] and Alexandre Antonelli [1,8]

1. Royal Botanic Gardens, Kew, Richmond TW93AQ, UK; j.borrell@kew.org (J.S.B.); i.ondo@kew.org (I.O.); r.douglas@kew.org (R.D.); j.viruel@kew.org (J.V.); r.govaerts@kew.org (R.G.); f.forest@kew.org (F.F.); a.antonelli@kew.org (A.A.)
2. Royal Botanic Gardens, Kew, Wakehurst Place TW93AE, UK; c.phillips@kew.org
3. International Center for Tropical Agriculture (CIAT), Cali 6713, Colombia; c.khoury@cgiar.org
4. Department of Biology, Saint Louis University, St. Louis, MO 63103, USA
5. Department of Tropical Plant and Soil Science, University of Hawaii at Manoa, Honolulu, HI 96822, USA; mbkantar@hawaii.edu (M.B.K.); nfumia@hawaii.edu (N.F.)
6. Department of Botany, University of British Columbia, Vancouver, BC V6T1Z4, Canada; marybel.soto@gmail.com
7. UBC Botanical Garden and Centre for Plant Research, University of British Columbia, Vancouver, BC V6T1Z4, Canada
8. Gothenburg Global Biodiversity Centre, Department of Biological and Environmental Sciences, University of Gothenburg, 40530 Göteborg, Sweden

* Correspondence: s.pironon@kew.org

Received: 17 July 2020; Accepted: 27 August 2020; Published: 31 August 2020

Abstract: Global biodiversity hotspots are areas containing high levels of species richness, endemism and threat. Similarly, regions of agriculturally relevant diversity have been identified where many domesticated plants and animals originated, and co-occurred with their wild ancestors and relatives. The agro-biodiversity in these regions has, likewise, often been considered threatened. Biodiversity and agro-biodiversity hotspots partly overlap, but their geographic intricacies have rarely been investigated together. Here we review the history of these two concepts and explore their geographic relationship by analysing global distribution and human use data for all plants, and for major crops and associated wild relatives. We highlight a geographic continuum between agro-biodiversity hotspots that contain high richness in species that are intensively used and well known by humanity (i.e., major crops and most viewed species on Wikipedia) and biodiversity hotspots encompassing species that are less heavily used and documented (i.e., crop wild relatives and species lacking information on Wikipedia). Our contribution highlights the key considerations needed for further developing a unifying concept of agro-biodiversity hotspots that encompasses multiple facets of diversity (including genetic and phylogenetic) and the linkage with overall biodiversity. This integration will ultimately enhance our understanding of the geography of human-plant interactions and help guide the preservation of nature and its contributions to people.

Keywords: agro-biodiversity; breeding; centres of origin; conservation; crop wild relatives; domestication; geographic distribution; phylogenetic diversity; useful plants; Vavilov centres

1. Introduction

Biogeographers and conservation biologists have long been interested in identifying and characterizing geographic regions containing a higher concentration of biodiversity and derived natural resources than surrounding areas, ranging from within- and among-species diversity through to ecosystem services [1–3], at different spatial, temporal and taxonomic scales [4]. Centres (also known as hotspots) and peripheries (coldspots) of plant diversity have been shown to be unevenly distributed and to play a fundamental role in shaping ecosystems and delivering associated benefits to humans and other species [5,6]. Mapping efforts contribute to a better fundamental understanding of both biodiversity (e.g., species extinction, diversification and co-existence) [7], and the interaction between people and nature, including the resulting socio-economic benefits and threats [8–11]. This is a particularly urgent endeavour in the current context of a rapidly growing global human population with increasing consumer expectations, posing a serious threat for both plant diversity and its long-term contributions to people [12]. Factors, such as land use and climate change, pollution, direct exploitation of species, and biological invasions have direct and indirect impacts on plant diversity [13,14], potentially undermining current and unrealised plant-based adaptive solutions [15,16] and traditional knowledge associated with plant uses [17]. Documenting the distribution of plant diversity and its uses are, therefore, critical steps towards developing the transformative changes required to achieve socio-economic sustainability, while preserving life on Earth.

Conservationists, constrained by finite resources, have used the concept of biodiversity hotspots to advocate for the allocation of international efforts and resources in regions of the world containing exceptionally diverse, unique and threatened biodiversity [18]. However, biodiversity hotspots often fail to capture the multi-faceted nature of biodiversity. For example, hotspot designations may consider a narrow range of organisms [19], and miss non-terrestrial habitats [3], phylogenetic and functional diversity (but see [20]). The seminal definition of a biodiversity hotspot, which was based solely on plants from tropical forests, highlights this shortcoming [21]. An additional limitation of biodiversity hotspots, as currently defined, is that they rarely consider anthropogenic interactions as anything other than a threat [22]. However, the sustainable and responsible use of nature, as demonstrated by traditional small-scale livelihoods and indigenous communities around the world, presents opportunities for resolving the current biodiversity crisis and global challenges facing humanity [12,15,23]. Finally, although often considered implicitly [13,18], nature's contributions to people are rarely considered when setting global conservation priorities, except for a few regulating ecosystem services [2,3,24].

Agro-biodiversity is a sub-component of biodiversity that accounts for the variety of life that contributes to food and agriculture [25]. In the broadest sense, it can be broken down into two components: (a) Planned agro-biodiversity, which refers to the diversity within and across species (domesticated or undomesticated) that are used by people, and (b) associated agro-biodiversity, which refers to species that surround and/or enhance planned agro-biodiversity [26]. In order to identify species and forms of high potential for improving and sustaining agriculture, research efforts have generally focused on mapping centres of origin and diversity of major domesticated plant species (mostly food crops) and associated wild relatives [27,28]. While the distribution of plants of highest importance for commodity production and human nutrition is now widely studied [6,29], much uncertainty remains about the large fraction of neglected and underutilized species that contribute to a wide array of provisioning, support, regulation and cultural services [30]. Moreover, although recent international treaties incentivized the preservation of plant genetic resources in the face of major global challenges [30], the distribution of useful plant genetic diversity is mainly studied through proxies (e.g., taxonomy, geographic distribution and environmental data) [28]. Likewise, the drivers and spatial patterns of decline in agro-biodiversity remain poorly understood [15,17].

Here we explore the relationship between hotspots of wild plant diversity and regions containing high levels of agro-biodiversity by (i) reviewing the history of the two concepts, (ii) characterizing the degree to which they overlap in space and the human and biological drivers of these geographic (in-) congruencies, and (iii) considering how to better integrate the multi-faceted aspects of useful plant diversity, including genetic and phylogenetic diversity. Finally, we propose a new general framework and discuss future avenues for obtaining an improved understanding and preservation of global hotspots of useful plant diversity [31,32].

2. The History of Diversity Hotspots

2.1. Biodiversity Hotspots

The distribution of life on Earth has long been investigated by naturalists aiming to understand, preserve and exploit the natural world [33–35]. These influential observations were followed by a more systematic documentation of the potential distribution of plant diversity globally based on the compilation of data from regional checklists and pioneering modelling efforts [36–38]. It was only in 1988 that British environmentalist Norman Myers (1934–2019) coined the term "hotspot" to define 10 tropical forest areas considered to be both irreplaceable (i.e., due to high concentrations of endemic plant species) and vulnerable (i.e., due to high rates of deforestation) [21]. Eight hotspots were added subsequently, including four in Mediterranean regions [39]. Later, the definition of a hotspot underwent a major update by considering strict quantitative criteria to designate areas containing at least 0.5% of the world's flora, but less than 30% of its original vegetation—resulting in the addition of seven additional hotspots [40]. A recent major update incorporated new data and took account of both terrestrial vertebrates and plants to delineate a total of 35 global biodiversity hotspots (Figure 1) [18]. Besides their importance for biodiversity, hotspots are home to more than two billion people and encompass some of humanity's highest population growth, as well as poverty rates [22].

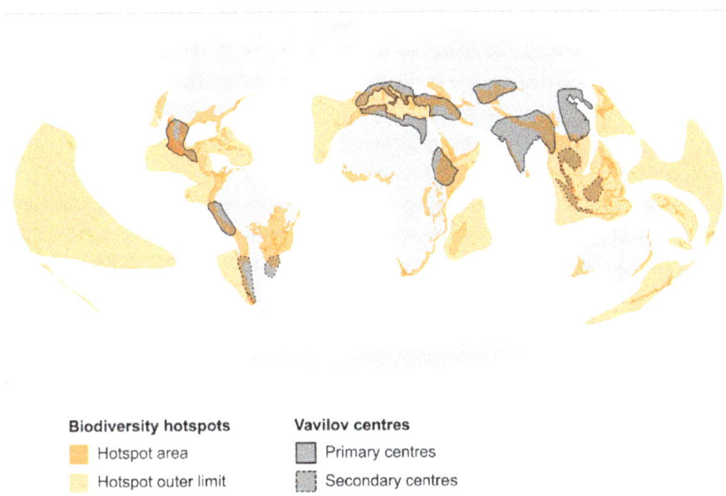

Figure 1. Global distribution of biodiversity hotspots and Vavilov centres. Biodiversity hotspots were first defined by Myers in 1988 [21], and now comprise 35 regions of high species richness, endemism and threat, as last updated in 2011 by Mittermeier and colleagues [18]. Islands constituting biodiversity hotspots are highlighted by outer hotspot limits. Vavilov first defined centres of origin of cultivated species and wild relatives in 1924; he provided an update in 1935, comprising eight primary and three secondary centres [27].

Since the definition of the hotspot concept by Myers, several other approaches have been explored to define and refine important biodiversity areas. Alongside terrestrial plants and vertebrates, other taxa have been considered to date, including marine mammals [41], phytoplankton [42] and soil invertebrates [43]. Conservationists have also argued for the protection of remaining wilderness, considered to be the most ecologically intact areas of the world, as these have been shown to experience substantially fewer threats, and thus, contribute more to the persistence of biodiversity [44]. Whereas, early work focused on species richness, recent studies have also advocated for the consideration of functional traits and evolutionary history, which may permit a better understanding of nature's contribution to both people and nature itself [45]. Incorporating knowledge across various disciplines is of particular importance as the multiple facets of biodiversity that they represent often do not overlap in space [20,46,47]. Spatial congruencies between biodiversity and direct measures of ecosystem services, such as carbon and water, are also increasingly being investigated [2,3].

Plants have been central to the definition of biodiversity hotspots since the pioneering work of Myers. Regions of high plant diversity have been investigated in multiple ways in the last decades, mainly by attempting to obtain finer continuous maps of (rare) plant species richness, as opposed to considering categorical hotspots. Barthlott and colleagues produced an influential estimate at the scale of ecoregions [5,48,49]. In contrast, researchers at the Royal Botanic Gardens, Kew gathered information at a semi-administrative scale [50,51], and adapted and expanded the concept of Important Plant Areas to the tropics [52], a programme that remains under development [53]. More recently, vascular plant species richness has been interpolated globally based on ~1000 local estimates [54], extrapolated based on the relationship between the occurrences of ~200,000 species and their environment [3], or investigated through the examination of commonness-rarity patterns [55]. Hotspots of high plant richness and endemism include Mesoamerica, the tropical Andes, the Amazon, Brazil's Atlantic rainforest, Central Africa, the western Ghats, South-East Asia, and many islands (e.g., Madagascar, New Guinea), mountainous regions (e.g., Alps, Caucasus) and Mediterranean areas (e.g., Cape floristic region). Despite their international recognition, most of these regions have become increasingly depleted under ongoing human pressures, whereas richness may increase at their periphery through repeated introductions in gardens and disturbed habitats [56]. These processes are leading to a global loss of diversity and increasing biotic homogenisation [57]. In this context, mapping priority areas and taxa for conservation remain, at least, as crucial and urgent as when hotspots were initially identified more than three decades ago.

2.2. *Agro-Biodiversity Hotspots*

Starting some 150 years ago, global botanical, geographic, linguistic, and archaeological evidence were combined to identify the geographic origins of crops, including distinctions among Old versus New World species [58]. These works were largely built on developments in plant systematics (e.g., Linnaeus (1707–1787), Alefeld (1820–1872), de Candolle (1806–1893)), phytogeography (e.g., Willdenow (1765–1812), von Humboldt (1769–1859), Wegener (1880–1930)), and evolution by natural and artificial selection (e.g., Darwin (1809–1882)). A number of these scientists were extensive travellers, whose contributions to their fields were catalysed by their voyages. However, none travelled as much as Nikolaï I. Vavilov (1887–1943). Informed by previous phytogeographic research and the rediscovery of Gregor J. Mendel's primary works in genetics, the Russian agricultural scientist pursued genetic variation in crops and their wild relatives, exploring five continents over several decades. Through his field experiences, Vavilov came to propose a set of independent "centres of origin" of cultivated food plants around the world, based fundamentally on where he saw a maximum concentration of diversity of traditional varieties of a wide range of crops, along with their wild relatives. Vavilov initially proposed three centres of origin of forms (1924), progressing to as many as eight primary centres, and including several sub-centres (Figure 1). These putative centres, which later in his tragically curtailed life he called hearths of origin, included Mesoamerica; parts of the Andes, Chile and Brazil-Paraguay; the Mediterranean; the Near East; Ethiopia; Central Asia; India; China; and Indo-Malaysia [27].

Since Vavilov, the regions of origin and diversity of different crops have been debated, investigated and refined, benefiting from an expanding body of archaeological, linguistic, genetic, and taxonomic information [6,59–66]. "Centres of diversity" came to be the preferred term over "centres of origin", to account for the difficulty in assigning an exact place of origin for most crops, and due to the understanding that high concentrations of crop varieties and related wild species are not in every case located where crops were initially domesticated [62]. "Regions" (also "megacentres" per Zhukovsky and "non-centres" per Harlan) rather than "centres" became preferred to reflect the large size of many of these geographic areas, and again, the difficulty of pinpointing exact locations where crops were domesticated [6,67]. At the present time, multidisciplinary evidence supports the identification of ca. 24–28 different areas around the world where crop domestication occurred independently, mostly beginning in the early to middle Holocene (approximately 11,700–6000 years ago), and in a few cases more recently [66,68,69]. Not all of the identified areas would be considered by most researchers as a "centre" or "region" of origin or diversity, as only a limited number of crops were domesticated in some of these.

There was an acceleration in the movement of crop plants across the globe between 1500–1700, as they were introduced to colonizing countries, their colonies and other regions with emerging export-oriented production [9,70]. Agricultural development and globalisation have made a number of crop species available to consumers worldwide, but in turn, increased homogeneity in global agriculture [71,72]. Added to the geographic decoupling of agricultural production and consumption [73,74], this homogenisation has deteriorated the connection between crops and their geographic origin [6]. Nevertheless, these areas of origin continue to hold foremost importance with regard to crop genetic diversity as their crops diversified for thousands of years under natural and human selection, including via further introgression with wild relatives [75,76]. During Vavilov's voyages a century ago, it was already apparent that the diversity of crops that people grew and consumed was changing as a result of globalisation. Major efforts commenced, particularly during the 1970s and 1980s, to collect traditional crop landraces and wild relatives for safeguarding in genebanks, and to also support in situ conservation, often in collaboration with subsistence agriculturalists [77]. Such efforts continue today, often linked to seed banking and germplasm collections, based on the recognition of persisting gaps in conservation of agro-biodiversity [28,78].

The history of agro-biodiversity was primarily written by colonial powers and white male explorers, conferring little or no room to traditional knowledge holders [79]. Moving ahead in tackling the challenges of mapping, understanding, protecting and further exploring the potential of crop plants and associated wild relatives, it is crucial that benefits of this work are shared in equitable ways [80]. In particular, access must be ensured in low-income countries to new or neglected crops, especially those that offer climate resilience, nutritious contents and other desirable traits.

3. From Biodiversity to Agro-Biodiversity Hotspots: A Geographic Continuum

The geographic distribution of biodiversity and agro-diversity hotspots has long been investigated, but rarely together. Very little is known about their spatial intricacies despite recent calls for their integration [31]. Several areas of high plant diversity do not include primary regions of agro-biodiversity (e.g., California, Caribbean, Brazilian Cerrado, South Africa, Madagascar, Pacific islands) and a few agro-biodiversity regions are not recognized as diversity hotspots (e.g., large parts of Eastern Asia and India) (Figure 1). Although human activities are altering this pattern [12,14], global plant species richness generally decreases with increasing latitude [5,54], reflecting past environmental changes, land configuration and the evolutionary histories of species [81,82]. On the other hand, the history of global human migrations, civilisations, economy and cultural preferences have been profoundly intertwined with the distribution and availability of natural resources (species richness, abundance and properties) to shape regions of agro-biodiversity [9]. Here, we present a new set of analyses to explore, illustrate and discuss the spatial congruence between biodiversity and agro-biodiversity hotspots, and its relationship with two important processes—human selection of species and species evolutionary history.

3.1. *From Popularity to Anonymity*

Vavilov provided an early spatial representation of the origins of cultivated plants by mapping the distribution of major food species. However, by restricting (justifiably) his focus mainly to selected food plants encountered in his field experiences, Vavilov was unable to produce a comprehensive assessment of the distribution of all plant species selected for use by humans across the world. Identifying which plant species are most selected by humans for a broader range of uses in addition to food, and characterizing their geographic distribution is, thus, key for defining agro-biodiversity hotspots and for relating their variation to the wider context of biodiversity.

In our currently globalized world and big data era, it is now possible to investigate human preferences for species more extensively. Here we assessed the popularity of most vascular plants on the free online encyclopaedia Wikipedia (www.wikipedia.org), using search data as a proxy for cultural preference, knowledge of plant species, use by humans and domestication intensity. As Wikipedia can be edited by anyone at any given time, it cannot be considered a reliable source of information without critical evaluation. However, our analysis remains independent from the quality of Wikipedia articles as it only examines the interaction between users and the web platform by quantifying numbers of page views. Species names and occupied geographic regions were retrieved for 339,924 species from the World Checklist of Vascular Plants (WCVP) [51]. Wikipedia uses the taxonomic backbone of the WCVP, so no additional name matching was performed. Geographic regions were retrieved at the national or sub-national level (finest level three) of the World Geographical Scheme for Recording Plant Distribution (WGSRPD), which was developed by the International Working Group on Taxonomic Databases for Plant Sciences [83]. We retrieved species popularity on Wikipedia as measured by the number of page views over the period 1 January 2016 to 1 January 2020 using the R package pageviews. This information relates to English Wikipedia pages only as it is the language with the highest number of pages overall and often the source of translations into other languages. We ranked these data according to three categories: (1) The 1000 most popular species; this includes plants used as food, medicine, timber, ornamentals and cosmetics (Supplementary Materials), (2) the remaining species covered by Wikipedia (i.e., those with an available page), (3) the remaining species not documented in Wikipedia (i.e., those without a page).

The global richness distribution of the 1000 most popular species is strikingly similar to the Vavilov primary centres with particularly high richness in subtropical regions of the Northern hemisphere (Figure 2a). There are also expansions towards temperate areas (i.e., Eastern North America, Europe and Central Asia), while the Northern Andes, Eastern Africa and the Indo-Malayan regions have relatively low crop richness at a global scale, but high richness within their respective continents. Richness generally increases towards the tropics for species that are documented in Wikipedia but fall outside of the 1000 most popular category, with particularly high concentrations in the Mediterranean and subtropical regions that are characterized by high plant species endemism but low richness in major crops (e.g., Western North America, South Africa, Australia) [5] (Figure 2b). In contrast, richness in species not documented in Wikipedia tends to follow a latitudinal gradient similar to that observed for total plant diversity and most similar to biodiversity hotspots (Figure 2c) [18,84]. These findings highlight the relationship between the spatial structure of biodiversity and agro-biodiversity, and people's knowledge, perception and use of nature. Indeed, we observe the existence of a geographic continuum between popular plant diversity that may be more intensively used by humanity (based on the existence of a Wikipedia page; 2A) and anonymous plant diversity that may be less heavily used (based on the lack of a Wikipedia page; 2C). One artefactual limitation in our assessment of the distribution of plant popularity is the general over-representation of English-speaking regions (e.g., United States, Canada, Australia) given that our data extraction came from English Wikipedia pages only. This also likely explains relatively low values of popular species richness in regions, such as the Andes or Ethiopia (Figure 2a), although those are still visible at the continental scale.

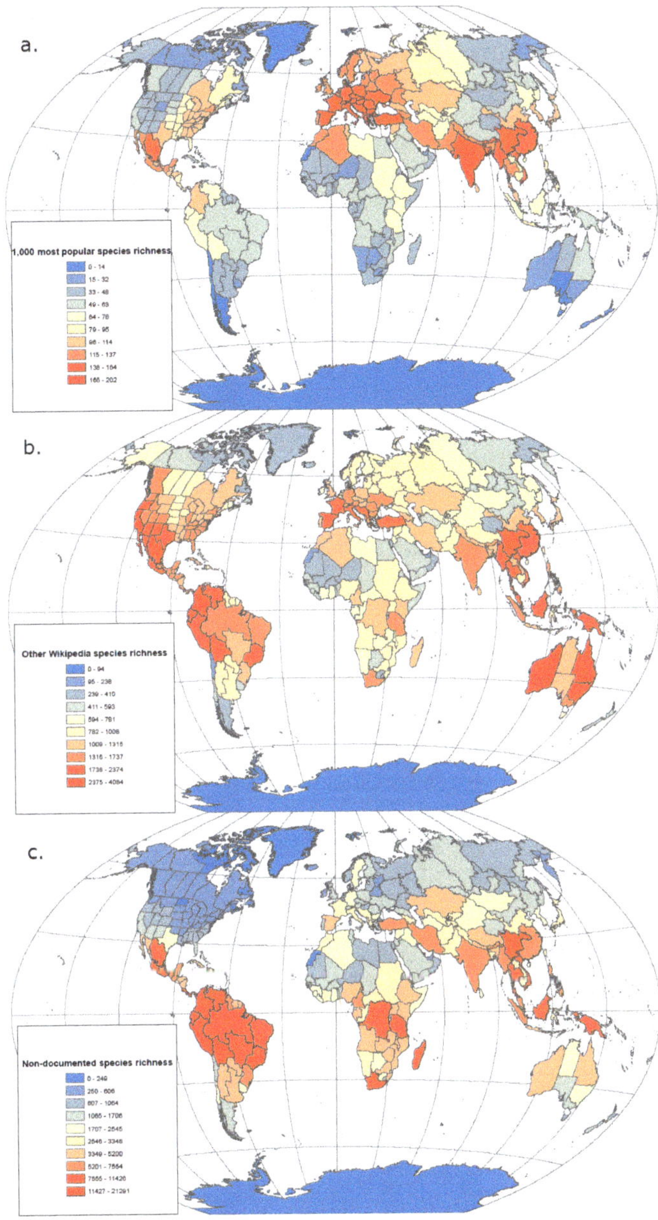

Figure 2. Global distribution of the species richness of plants ranging from popular to anonymous based on English Wikipedia page views. Global distribution of (**a**) the 1000 most popular plant species on Wikipedia; (**b**) 49,019 species documented in Wikipedia, excluding the 1000 most popular ones; (**c**) 280,905 species not documented in Wikipedia. Popularity was measured as the number of views of the Wikipedia webpage of each species. Native distribution data was retrieved from the World Checklist of Vascular Plants at the national or sub-national level of the World Geographical Scheme for Recording Plant Distribution (WGSRPD) [51].

3.2. From Domesticates to Wild Relatives

Alongside major crops, Vavilov was also interested in documenting, mapping, collecting, using, and preserving wild ancestors and closely related species [27]. By considering less known and undomesticated (or less domesticated) parent and sister species, Vavilov directly connected his definition of regions of origin/diversity with wild plant diversity. Recent studies illustrate this link between biodiversity and agro-biodiversity hotspots; these assessed the distribution of the closest and more distant relatives of major crops and found high species richness in plant diversity hotspots, such as the Brazilian Cerrado and Atlantic Forest or South-East Asia [28,85].

Here we assessed the current distribution of 222 major international crops and 2,731 of their wild relative species using comprehensive lists from the USDA ARS GRIN-Global Taxonomy and geographic data from the World Checklist of Vascular Plants, again retrieved at the national and sub-national level (level three) of the WGSRPD [51]. Crop wild relatives are classified across three gene pools based on both their relatedness (using phylogenetics and systematics) and crossing ability with the crop [86,87]: gene pool one comprises the most closely related (even conspecific) wild species that are generally fully interfertile with the crop; gene pool two includes more distant relatives that may be crossed to the crop with more difficulty; and gene pool three typically contains the most distantly related and least compatible species within the genus to which the crop belongs (sometimes including other genera). Here, we assess changes in geographic patterns across a gradient from cultivated species to their closest wild relatives to their more distant wild relatives, by mapping richness for crops and each associated gene pool separately. When more than one species was identified in a gene pool for a given crop, we merged their distribution to assign the same weight to each crop, thus, avoiding overrepresentation of genera with many wild relatives. Geographic data were not available at the infra-specific level (i.e., sub-species, varieties, forms) for all taxa, and so we performed analyses at the species level.

Geographic patterns in major crop species richness strongly overlap with the Vavilov centres (Figure 3a) and are also similar to the 1000 most popular plant species on Wikipedia (Figure 2a). Given that gene pool one is composed of the closest crop wild relatives, including progenitors and/or wild types of the crop species, the distribution of gene pool one species richness is very similar to that of the crops (Figure 3b). Slight increases and decreases are respectively observed inside and outside Europe, which may be explained by more extensive documentation of European crop wild relatives compared to those that occur in other regions [88]. Gene pools two and three provide a more diffuse representation of the Vavilov centres: species richness decreases in most of the Vavilov centres, but increases in surrounding regions, particularly (but not exclusively) towards the tropics and plant diversity hotspots (Figure 3c,d). Although centres of crop diversity remain visible when mapping secondary and tertiary gene pools, the geographic signal diminishes as we move further away from the most explored and popular branches of the tree of life. In contrast, areas of high biodiversity start to emerge, which is even more striking when species richness mapping does not account for the over-representation of genera with many wild relatives [28,85]. This reinforces the existence of a geographic continuum between agro-biodiversity (i.e., widely used and cultivated crops) and biodiversity (i.e., non-domesticated and less used sister species of crops) related to the strength of the interaction between humans and plants, but also plant evolutionary history.

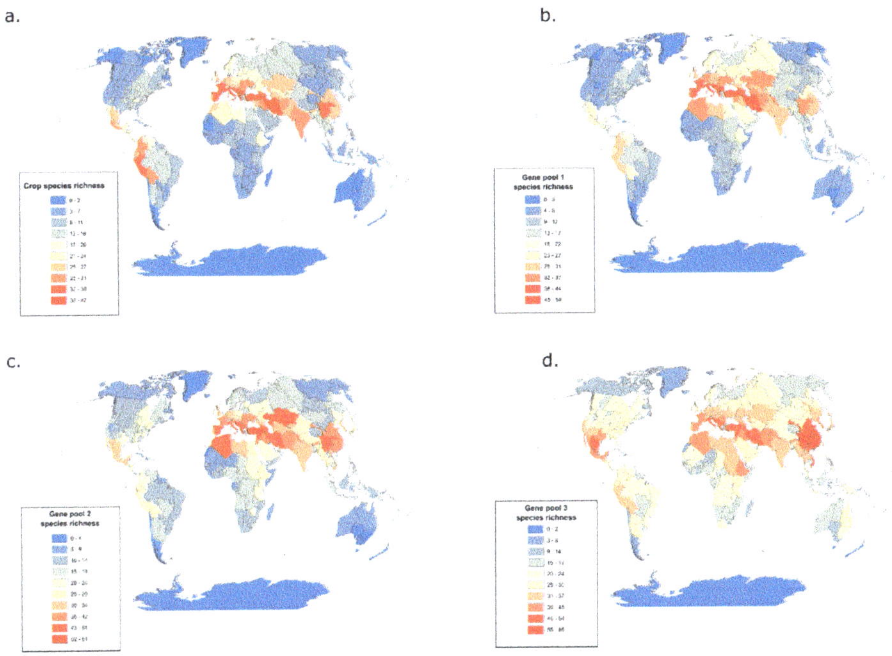

Figure 3. Global distribution of the richness of major crops and their wild relatives. Global distribution of (**a**) 222 major crops, (**b**) 361 wild relative species in gene pool one, (**c**) 1040 wild relative species in gene pool two and (**d**) 2358 wild relative species in gene pool three; see text for gene pool characterisation. Species identities and gene pool classifications were retrieved from the USDA ARS GRIN Global Taxonomy. Distribution data was retrieved from the World Checklist of Vascular Plants at level three of the World Geographical Scheme for Recording Plant Distribution (WGSRPD) [51]. When more than one wild relative species was identified for a crop in a gene pool, we merged their distribution to assign the same weight to each crop and avoid genera with many wild relatives to be overrepresented.

4. Integrating Genetic and Phylogenetic Diversity into Agro-Biodiversity Hotspots

The designation of hotspots of plant species richness and rarity has provided a way to focus attention on the intrinsic value of biodiversity at a global scale. However, the contributions of plants to livelihoods are rarely considered in this context, despite the predominance of utilitarian arguments in conservation. Going beyond species counts (i.e., taxonomic diversity) to describe and understand structural and/or chemical properties of species and their diversity is of critical importance. The wide range of plant properties cannot currently be quantified across all species but is often related to genetic variation within and among species [89]. Characterizing genetic and phylogenetic diversity and their geographic distribution may, therefore, provide a useful framework for identifying other currently unknown forms of diversity and associated usages, and for preserving plant genetic resources.

4.1. Hotspots of Phylogenetic Diversity

Some species are the sole remaining representative of ancient lineages, while others are part of recent and rapid radiations (i.e., increases in species richness related to elevated speciation rate) that may comprise hundreds of closely related species. Therefore, conservation biologists now account for the fact that the evolutionary histories of species are not equivalent when setting priorities [90]. To address this inconsistency, phylogenetic diversity (PD) was proposed as an approach based on evolutionary information [91]. The PD of a given area is equal to the sum of all the branches on a phylogenetic tree linking the set of species that occur in this area. Areas containing high PD will, therefore, reflect higher concentrations in distantly related species. While previous studies have cast doubt on the ability of PD to provide a different answer to species richness for prioritisation [92,93], the decoupling of biodiversity patterns, based on taxon richness and evolutionary history, have since been clearly demonstrated (e.g., Reference [94]). Phylogenetic diversity and associated metrics have been widely used to explore biodiversity patterns among and within biodiversity hotspots [95–97].

One of the most important characteristics of PD is its potential to act as a surrogate for feature diversity (i.e., the diversity of characters or traits of species), which encompasses the qualities of plants that are beneficial for humans. While the relationship between PD and feature diversity remains contentious [45,98], it is nevertheless an attribute that is of particular importance for the identification of areas rich in crop wild relatives and species with unexplored uses for humans. Assuming that PD is a suitable surrogate for feature diversity, maintaining PD would not only help retain the evolutionary potential of species, but also maximise the potential unanticipated benefits that biodiversity may have in the future for humans (i.e., biodiversity option values), particularly in the face of global change [99,100]. In the context of agro-biodiversity hotspots and the identification of new sources of plant properties, PD as a metric of choice has a key role to play (Figure 4a). Ultimately, however, the identification of hotspots of agro-biodiversity would be best served by the integration of various metrics capturing the multiple facets of biodiversity [101], and also by considering the human dimension for better understanding the portion, intensity and modes of use of biodiversity by humanity.

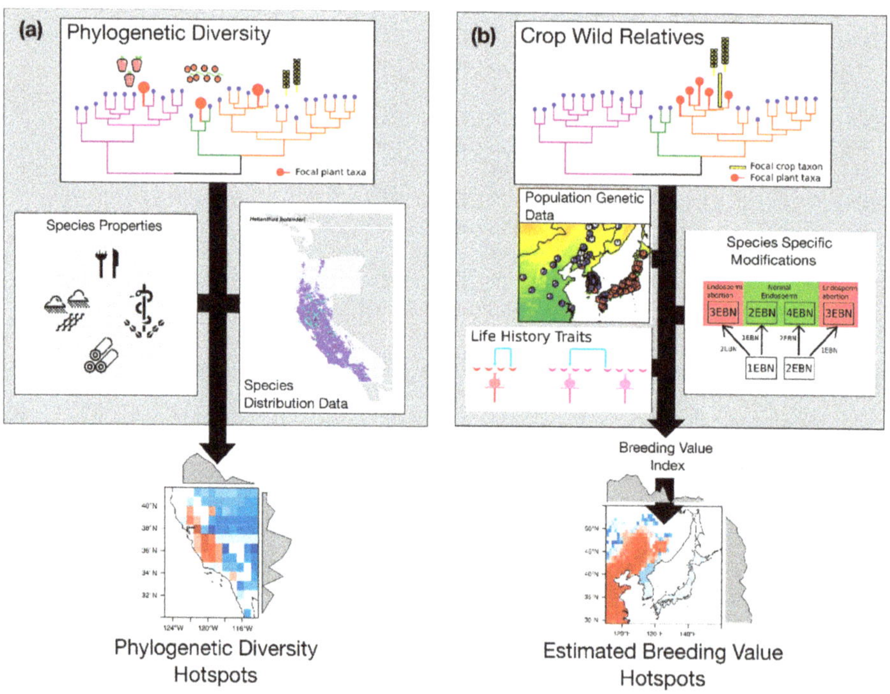

Figure 4. A proposed general framework for the further inclusion of genetic information into the mapping of agro-biodiversity hotspots. (**a**) A relationship between species evolutionary history and their physical or chemical properties (i.e., features or phenotype) would use phylogenetic diversity as a proxy for feature diversity, the latter being less readily quantifiable across wide ranges of species, regions and features. Combined with species distribution data, phylogenetic diversity could ultimately identifies hotspots of feature diversity and priority areas for the conservation of species' contribution to people; (**b**) By considering phylogenetic information together with population genomics, cytology, life

history and/or ecological data, an Estimated Breeding Value (EBV) could be computed for crop wild relative species and/or populations. Combined with distribution data, hotspots of EBV could then be mapped to identify areas containing high concentrations in valuable wild gene sources for preservation and crop improvement.

4.2. Hotspots of Breeding Value

Genetic diversity represents the raw material that humans have relied upon for millennia for the maintenance and improvement of crops. Plant breeding is a long-term process [102] aimed at enhancing traits of interest (e.g., yield, quality, disease tolerance, abiotic stress tolerance) using extant variation [103]. There is tremendous genetic and phenotypic diversity in crop wild relatives distributed across the plant tree of life [69,104]. Identifying the top priority branches (from species to populations) that will generate the largest changes in trait values, while having the closest form to current crops is of great interest to plant breeders and agriculturalists. Populations that exist at the edges of distribution ranges may have great utility for breeding as these often occur in more isolated and extreme ecological conditions and display high levels of genetic and phenotypic differentiation [1,66,78,105]. Identification of potentially useful crop wild relatives is generally based on heuristic approaches (e.g., place X has a similar environment to place Y, so translocation is expected to lead to positive results), or by using large-scale germplasm collections to search for specific traits of interest [106,107]. However, these approaches, as currently applied, may not fully explore available plant genetic diversity (e.g., Reference [108]) and are often not available for crops that are less economically important.

There have been recent efforts to identify the geographic regions where crop wild relative species are concentrated [28]. Characterisation of the distribution of populations that are most compatible with existing crops and exhibit phenotypes of interest has somewhat lagged behind. Nevertheless, there are now concerted efforts to identify and incorporate these taxa into breeding programmes for producing viable new cultivars by using integrative approaches that leverage large amounts of data from phylogenetics, population genomics, cytology, life history, ecological niches, and predictions of future environmental conditions that crops may experience [78,109–111].

Building on historic advancements, it may be possible to incorporate the data currently used in heuristic approaches for crop wild relative identification into a more general framework to help the decision-making within breeding programmes. Such a framework could potentially be modelled from the Estimated Breeding Value (EBV), a common breeding programme metric that could expedite selection of donor species. An EBV is the potential of an individual as a genetic parent, considering the heritability of a given trait under selection [103]. Typically, EBVs are obtained from narrow breeding populations of a single species by multiplying the narrow-sense heritability (calculated either by variance decomposition or parent-offspring regression) by the difference between the parent performance and population mean, which provides an estimate of how the progeny of a specific parent will perform relative to an average parent. We propose that EBVs could also be calculated for crop wild relative species and populations, as long as the phenotype of interest is clearly defined. Applied to species and populations instead of individuals, the EBV could go beyond heritability to additionally incorporate biological factors (e.g., ploidy, mating system), evolutionary factors (e.g., phylogenetic relationship), and ecological factors (e.g., species environmental niche) in a hierarchical way for prioritizing species that may be of greater potential use to plant breeding. Functionally, this proposed use of EBV would produce a ranking of species for an individual breeding programme, based on the desired phenotype by crossing interaction. Ultimately, characterizing this variation in utility across species (and populations) may help identify priority areas for in- and ex-situ conservation related to specific breeding targets (Figure 4b).

5. Towards a Unified Concept of Agro-Biodiversity Hotspot

Hotspots of both biodiversity and agro-biodiversity have long relied on counting numbers of species (i.e., species richness; Figure 5a) and assessing threats (Figure 5b). While biodiversity scientists have mainly focused on numbers of rare species (including many narrowly distributed taxa) for conservation, agronomists have been interested in diversity within gene pools (i.e., numbers of domesticated species and wild relatives) [27,28,85] and within crops (e.g., numbers of landraces) [112]. Although taxon counts remain extremely useful, new approaches are now proposed to account for the multi-faceted nature of (agro-)biodiversity, such as functional diversity accounting for the diversity in species chemical properties and eco-/agri-system functions [113], phylogenetic diversity as a potential proxy for functional and property diversity (Figure 5c) or for identifying gene sources for breeding programmes (Figure 5d). As highlighted by their seminal definition from N. Myers and recent publications [114], biodiversity hotspots are also deeply related to the distribution of a wide range of threats, many of which are shared with agro-biodiversity (e.g., land use and climate change, pollution, biological invasions, over-harvesting), but less formally included in the geographic assessments of the latter (Figure 5b) [12,16]. Although the different facets and threats of agro-biodiversity are not all expected to overlap geographically, our paper proposes to assess them jointly rather than separately.

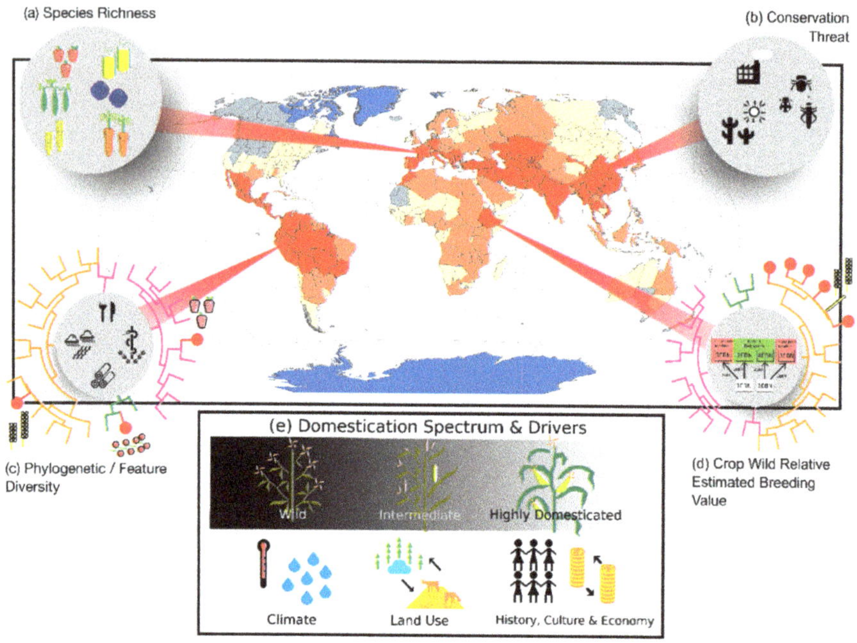

Figure 5. The conceptual framework for the identification of agro-biodiversity hotspots, including (**a**) plant species richness (applicable to infra-specific levels as well); (**b**) threats; (**c**) species (or infra-specific taxa) evolutionary and features diversity; (**d**) crop wild relatives estimated breeding value; and accounting for (**e**) the geographic continuum between hotspots of wild species diversity and regions containing high concentrations in major crops (i.e., highly domesticated species), and its environmental and human drivers. The map does not provide a new estimate of the distribution of agro-biodiversity hotspots, but rather illustrates a combination of the potential two ends of the domestication spectrum: Major crop species richness (Figure 3a) and species richness undocumented in Wikipedia (Figure 2c). High to low species richness is represented from red to blue.

Primary regions of agro-biodiversity have focused on relatively few important crops selected by researchers, whereas they have mainly explored wild relatives of these domesticated species, which are not always considered for other uses than crop improvement. Given the existence of a wide domestication spectrum, ranging from major global crops (i.e., highly domesticated species) to those harvested in the wild [16], we believe that further work on regions of agro-biodiversity should expand the focus to better include the long list of plants that provide food and other cultural benefits to humanity (Figure 5e). This will be made effective through the documentation of the tremendous diversity of neglected and under-utilized species of the world (with more than 30,000 useful plant species known to date [115]). Many of these species occur naturally in low-income countries, including already established biodiversity hotspots, which are also often home to large human populations and cultural diversity [22,116]. Understanding the drivers of the distributions of nature's contribution to people across the domestication spectrum (from climate and land use to socio-economic factors; Figure 5e) is also fundamental to define hotspots and design conservation and development efforts to sustain socio-environmental sustainability.

The recognition of biodiversity hotspots and agro-biodiversity (Vavilov) centres have played important roles to raise public awareness, foster research and attract political action to preserve and use natural resources sustainably. Given the urgency and magnitude of the global challenges outlined by the United Nations' Sustainable Development Goals and the recent report on biodiversity loss by the intergovernmental science-policy platform on biodiversity and ecosystem Services (IPBES) [12], it is more important than ever to refine, integrate and disseminate such powerful concepts. Failing to protect hotspots of natural resources, and especially agro-biodiversity, would have damaging consequences on nature and human livelihoods, both at those centres and in their peripheries. Our paper calls for the further development and integration of a range of commonly used and more recently proposed indices, while accounting for the key interaction with biodiversity, into the agro-biodiversity hotspot concept.

Supplementary Materials: The following are available online at http://www.mdpi.com/2223-7747/9/9/1128/s1.

Author Contributions: Conceptualisation, S.P.; data analysis, J.S.B., I.O., S.P.; data curation, C.P., C.K.K., R.G.; writing—original draft preparation, C.K.K., F.F., M.B.K., N.F., S.P.; writing—review and editing, J.S.B., C.P., C.K.K., M.B.K., M.S.G., J.V., F.F., A.A., S.P.; visualisation, I.O., R.D., N.F., M.B.K., S.P.; supervision, A.A., S.P. All authors have read and agreed to the published version of the manuscript.

Funding: S.P., R.D. and I.O. received funding from Norway's International Climate and Forest Initiative (NICFI). J.B. received funding from [BB/S014896/1] GCRF BBSRC Landscape genomic-environment diversity data to model existing and novel agri-systems under climate change to enhance food security in Ethiopia. A.A. received funding from the Swedish Research Council, the Swedish Foundation for Strategic Research, the Knut and Alice Wallenberg Foundation, and the Royal Botanic Gardens, Kew.

Acknowledgments: We thank Justin Moat and Amanda Cooper for their help with designing the maps; Nicholas Black, Alan Paton, and Robert Turner for their assistance with data from the World Checklist of Vascular Plants; and four anonymous reviewers for constructive suggestions that helped improve this manuscript.

Conflicts of Interest: The authors declare no conflict of interest.

References

1. Pironon, S.; Papuga, G.; Villellas, J.; Angert, A.L.; García, M.B.; Thompson, J.D. Geographic variation in genetic and demographic performance: New insights from an old biogeographical paradigm: The centre-periphery hypothesis. *Biol. Rev.* **2017**, *92*, 1877–1909. [CrossRef] [PubMed]
2. Chaplin-Kramer, R.; Sharp, R.P.; Weil, C.; Bennett, E.M.; Pascual, U.; Arkema, K.K.; Brauman, K.A.; Bryant, B.P.; Guerry, A.D.; Haddad, N.M.; et al. Global modeling of nature's contributions to people. *Science* **2019**, *366*, 255–258. [CrossRef] [PubMed]
3. Jung, M.; Arnell, A.; de Lamo, X.; García-Rangel, S.; Lewis, M.; Mark, J.; Merow, C.; Miles, L.; Ondo, I.; Pironon, S.; et al. Areas of global importance for terrestrial biodiversity, carbon, and water. *bioRxiv* **2020**. [CrossRef]
4. Jetz, W.; McGeoch, M.A.; Guralnick, R.; Ferrier, S.; Beck, J.; Costello, M.J.; Fernandez, M.; Geller, G.N.; Keil, P.; Merow, C.; et al. Essential biodiversity variables for mapping and monitoring species populations. *Nat. Ecol. Evol.* **2019**, *3*, 539–551. [CrossRef]
5. Kier, G.; Kreft, H.; Lee, T.M.; Jetz, W.; Ibisch, P.L.; Nowicki, C.; Mutke, J.; Barthlott, W. A global assessment of endemism and species richness across island and mainland regions. *Proc. Natl. Acad. Sci. USA* **2009**, *106*, 9322–9327. [CrossRef]

6. Khoury, C.K.; Achicanoy, H.A.; Bjorkman, A.D.; Navarro-Racines, C.; Guarino, L.; Flores-Palacios, X.; Engels, J.M.M.; Wiersema, J.H.; Dempewolf, H.; Sotelo, S.; et al. Origins of food crops connect countries worldwide. *Proc. R. Soc. B Biol. Sci.* **2016**, *283*, 20160792. [CrossRef]
7. Lomolino, M.V.; Riddle, B.R.; Brown, J.H. *Biogeography*; Sinauer Associates: Sunderland, UK, 2006; ISBN 978-0-87893-062-3.
8. Harlan, J.R.; Hutchinson, J.B.; Clark, G.; Jope, E.M.; Riley, R. Plant and animal distribution in relation to domestication. *Philos. Trans. R. Soc. Lond. B Biol. Sci.* **1976**, *275*, 13–25. [CrossRef]
9. Diamond, J. Evolution, consequences and future of plant and animal domestication. *Nature* **2002**, *418*, 700–707. [CrossRef]
10. Soga, M.; Gaston, K.J. The ecology of human–nature interactions. *Proc. R. Soc. B Biol. Sci.* **2020**, *287*, 20191882. [CrossRef]
11. Hartel, T.; Nita, A.; Rozylowicz, L. Understanding human–nature connections through value networks: The case of ancient wood-pastures of Central Romania. *Sustain. Sci.* **2020**, *15*, 1357–1367. [CrossRef]
12. Díaz, S.; Settele, J.; Brondízio, E.S.; Ngo, H.T.; Agard, J.; Arneth, A.; Balvanera, P.; Brauman, K.A.; Butchart, S.H.M.; Chan, K.M.A.; et al. Pervasive human-driven decline of life on Earth points to the need for transformative change. *Science* **2019**, *366*. [CrossRef] [PubMed]
13. Corlett, R.T. Plant diversity in a changing world: Status, trends, and conservation needs. *Plant Divers.* **2016**, *38*, 10–16. [CrossRef] [PubMed]
14. Newbold, T.; Hudson, L.N.; Arnell, A.P.; Contu, S.; Palma, A.D.; Ferrier, S.; Hill, S.L.L.; Hoskins, A.J.; Lysenko, I.; Phillips, H.R.P.; et al. Has land use pushed terrestrial biodiversity beyond the planetary boundary? A global assessment. *Science* **2016**, *353*, 288–291. [CrossRef] [PubMed]
15. Pironon, S.; Etherington, T.R.; Borrell, J.S.; Kühn, N.; Macias-Fauria, M.; Ondo, I.; Tovar, C.; Wilkin, P.; Willis, K.J. Potential adaptive strategies for 29 sub-Saharan crops under future climate change. *Nat. Clim. Chang.* **2019**, *9*, 758–763. [CrossRef]
16. Borrell, J.S.; Dodsworth, S.; Forest, F.; Pérez-Escobar, O.A.; Lee, M.A.; Mattana, E.; Stevenson, P.C.; Howes, M.-J.R.; Pritchard, H.W.; Ballesteros, D.; et al. The climatic challenge: Which plants will people use in the next century? *Environ. Exp. Bot.* **2020**, *170*, 103872. [CrossRef]
17. Cámara-Leret, R.; Fortuna, M.A.; Bascompte, J. Indigenous knowledge networks in the face of global change. *Proc. Natl. Acad. Sci. USA* **2019**, 201821843. [CrossRef]
18. Mittermeier, R.A.; Turner, W.R.; Larsen, F.W.; Brooks, T.M.; Gascon, C. Global biodiversity conservation: The critical role of hotspots. In *Biodiversity Hotspots*; Zachos, F.E., Habel, J.C., Eds.; Springer: Berlin/Heidelberg, Germany, 2011; pp. 3–22. ISBN 978-3-642-20991-8.
19. Jenkins, C.N.; Pimm, S.L.; Joppa, L.N. Global patterns of terrestrial vertebrate diversity and conservation. *Proc. Natl. Acad. Sci. USA* **2013**, *110*, E2602–E2610. [CrossRef]
20. Pollock, L.J.; Thuiller, W.; Jetz, W. Large conservation gains possible for global biodiversity facets. *Nature* **2017**, *546*, 141–144. [CrossRef]
21. Myers, N. Threatened biotas: "Hot spots" in tropical forests. *Environmentalist* **1988**, *8*, 187–208. [CrossRef]
22. Cincotta, R.P.; Wisnewski, J.; Engelman, R. Human population in the biodiversity hotspots. *Nature* **2000**, *404*, 990–992. [CrossRef]
23. Renard, D.; Tilman, D. National food production stabilized by crop diversity. *Nature* **2019**, *571*, 257–260. [CrossRef] [PubMed]
24. Naidoo, R.; Balmford, A.; Costanza, R.; Fisher, B.; Green, R.E.; Lehner, B.; Malcolm, T.R.; Ricketts, T.H. Global mapping of ecosystem services and conservation priorities. *Proc. Natl. Acad. Sci. USA* **2008**, *105*, 9495–9500. [CrossRef] [PubMed]
25. Hammer, K.; Arrowsmith, N.; Gladis, T. Agrobiodiversity with emphasis on plant genetic resources. *Naturwissenschaften* **2003**, *90*, 241–250. [CrossRef]
26. Jackson, L.E.; Pascual, U.; Hodgkin, T. Utilizing and conserving agrobiodiversity in agricultural landscapes. *Agric. Ecosyst. Environ.* **2007**, *121*, 196–210. [CrossRef]
27. Vavilov, N.I.; Vavylov, M.I.; Vavílov, N.Í.; Dorofeev, V.F. *Origin and Geography of Cultivated Plants*; Cambridge University Press: Cambridge, UK, 1992; ISBN 978-0-521-40427-3.
28. Castañeda-Álvarez, N.P.; Khoury, C.K.; Achicanoy, H.A.; Bernau, V.; Dempewolf, H.; Eastwood, R.J.; Guarino, L.; Harker, R.H.; Jarvis, A.; Maxted, N.; et al. Global conservation priorities for crop wild relatives. *Nat. Plants* **2016**, *2*, 16022. [CrossRef]

29. Monfreda, C.; Ramankutty, N.; Foley, J.A. Farming the planet: (2) Geographic distribution of crop areas, yields, physiological types, and net primary production in the year 2000. *Glob. Biogeochem. Cycles* **2008**, *22*, GB1022. [CrossRef]
30. Khoury, C.K.; Amariles, D.; Soto, J.S.; Diaz, M.V.; Sotelo, S.; Sosa, C.C.; Ramírez-Villegas, J.; Achicanoy, H.A.; Velásquez-Tibatá, J.; Guarino, L.; et al. Comprehensiveness of conservation of useful wild plants: An operational indicator for biodiversity and sustainable development targets. *Ecol. Indic.* **2019**, *98*, 420–429. [CrossRef]
31. Attwood, S.J.; Park, S.E.; Marshall, P.; Fanshawe, J.H.; Gaisberger, H. *An Argument for Integrating Wild and Agricultural Biodiversity Conservation*; Handbooks Online; Routledge: London, UK, 2017; ISBN 978-0-415-74692-2.
32. Pilling, D.; Bélanger, J.; Hoffmann, I. Declining biodiversity for food and agriculture needs urgent global action. *Nat. Food* **2020**, *1*, 144–147. [CrossRef]
33. de Candolle, A. *Géographie Botanique Raisonnée ou Exposition des Faits Principaux et des Lois Concernant la Distribution Géographique des Plantes de l'Époque Actuelle*; Masson, V., Ed.; Nabu Press: London, UK, 1855.
34. Wallace, A.R. *The Geographical Distribution of Animals*; Harper and Brothers: New York, NY, USA, 1876.
35. Cailleux, A. *Biogéographie Mondiale*; Presses Universitaires de France: Paris, France, 1953.
36. Wulff, E.W. Versuch einer einteilung der vegetation der erde in pflanzengeographische gebiete auf grund der artenzahl. *Repertorium Specierum Novarum Regni Vegetabilis* **1935**, *12*, 57–83.
37. Malyshev, L.I. Quantitative analysis of flora: Spatial diversity, the level of specific richness and representativity of sampling areas. *Bot. Zhurnal* **1975**, *60*, 1537–1550.
38. Frodin, D.G. *Guide to Standard Floras of the World*; Cambridge University Press: Cambridge, UK, 1984; ISBN 978-1-139-42865-1.
39. Myers, N. The biodiversity challenge: Expanded hot-spots analysis. *Environmentalist* **1990**, *10*, 243–256. [CrossRef] [PubMed]
40. Myers, N.; Mittermeier, R.A.; Mittermeier, C.G.; da Fonseca, G.A.B.; Kent, J. Biodiversity hotspots for conservation priorities. *Nature* **2000**, *403*, 853–858. [CrossRef] [PubMed]
41. Davidson, A.D.; Boyer, A.G.; Kim, H.; Pompa-Mansilla, S.; Hamilton, M.J.; Costa, D.P.; Ceballos, G.; Brown, J.H. Drivers and hotspots of extinction risk in marine mammals. *Proc. Natl. Acad. Sci. USA* **2012**, *109*, 3395–3400. [CrossRef] [PubMed]
42. Righetti, D.; Vogt, M.; Gruber, N.; Psomas, A.; Zimmermann, N.E. Global pattern of phytoplankton diversity driven by temperature and environmental variability. *Sci. Adv.* **2019**, *5*, eaau6253. [CrossRef] [PubMed]
43. van den Hoogen, J.; Geisen, S.; Routh, D.; Ferris, H.; Traunspurger, W.; Wardle, D.A.; de Goede, R.G.M.; Adams, B.J.; Ahmad, W.; Andriuzzi, W.S.; et al. Soil nematode abundance and functional group composition at a global scale. *Nature* **2019**, *572*, 194–198. [CrossRef] [PubMed]
44. Di Marco, M.; Ferrier, S.; Harwood, T.D.; Hoskins, A.J.; Watson, J.E.M. Wilderness areas halve the extinction risk of terrestrial biodiversity. *Nature* **2019**, *573*, 582–585. [CrossRef]
45. Tucker, C.M.; Davies, T.J.; Cadotte, M.W.; Pearse, W.D. On the relationship between phylogenetic diversity and trait diversity. *Ecology* **2018**, *99*, 1473–1479. [CrossRef]
46. Devictor, V.; Mouillot, D.; Meynard, C.; Jiguet, F.; Thuiller, W.; Mouquet, N. Spatial mismatch and congruence between taxonomic, phylogenetic and functional diversity: The need for integrative conservation strategies in a changing world: Spatial mismatch between diversity facets. *Ecol. Lett.* **2010**, *13*, 1030–1040. [CrossRef]
47. Azevedo, J.A.R.; Guedes, T.B.; Nogueira, C.D.C.; Passos, P.; Sawaya, R.J.; Prudente, A.L.C.; Barbo, F.E.; Strüssmann, C.; Franco, F.L.; Arzamendia, V.; et al. Museums and cradles of diversity are geographically coincident for narrowly distributed Neotropical snakes. *Ecography* **2020**, *43*, 328–339. [CrossRef]
48. Barthlott, W.; Lauer, W.; Placke, A. Global distribution of species diversity in vascular plants: Towards a world map of phytodiversity (globale verteilung der artenvielfalt höherer pflanzen: Vorarbeiten zu einer weltkarte der phytodiversität). *Erdkunde* **1996**, *50*, 317–327. [CrossRef]
49. Kier, G.; Mutke, J.; Dinerstein, E.; Ricketts, T.H.; Küper, W.; Kreft, H.; Barthlott, W. Global patterns of plant diversity and floristic knowledge. *J. Biogeogr.* **2005**, *32*, 1107–1116. [CrossRef]
50. Pimm, S.L.; Jenkins, C.N.; Abell, R.; Brooks, T.M.; Gittleman, J.L.; Joppa, L.N.; Raven, P.H.; Roberts, C.M.; Sexton, J.O. The biodiversity of species and their rates of extinction, distribution, and protection. *Science* **2014**, *344*. [CrossRef] [PubMed]

51. WCVP. World Checklist of Vascular Plants, Version 2.0. Facilitated by the Royal Botanic Gardens, Kew. 2020. Published on the Internet. Available online: http://wcvp.science.kew.org/ (accessed on 12 April 2020).
52. Darbyshire, I.; Anderson, S.; Asatryan, A.; Byfield, A.; Cheek, M.; Clubbe, C.; Ghrabi, Z.; Harris, T.; Heatubun, C.D.; Kalema, J.; et al. Important plant areas: Revised selection criteria for a global approach to plant conservation. *Biodivers. Conserv.* **2017**, *26*, 1767–1800. [CrossRef]
53. Tropical Important Plant Areas. Available online: https://www.kew.org/science/our-science/projects/tropical-important-plant-areas (accessed on 28 August 2020).
54. Kreft, H.; Jetz, W. Global patterns and determinants of vascular plant diversity. *Proc. Natl. Acad. Sci. USA* **2007**, *104*, 5925–5930. [CrossRef] [PubMed]
55. Enquist, B.J.; Feng, X.; Boyle, B.; Maitner, B.; Newman, E.A.; Jørgensen, P.M.; Roehrdanz, P.R.; Thiers, B.M.; Burger, J.R.; Corlett, R.T.; et al. The commonness of rarity: Global and future distribution of rarity across land plants. *Sci. Adv.* **2019**, *5*, eaaz0414. [CrossRef]
56. van Kleunen, M.; Xu, X.; Yang, Q.; Maurel, N.; Zhang, Z.; Dawson, W.; Essl, F.; Kreft, H.; Pergl, J.; Pyšek, P.; et al. Economic use of plants is key to their naturalization success. *Nat. Commun.* **2020**, *11*, 3201. [CrossRef]
57. Olden, J.D.; LeRoy Poff, N.; Douglas, M.R.; Douglas, M.E.; Fausch, K.D. Ecological and evolutionary consequences of biotic homogenization. *Trends Ecol. Evol.* **2004**, *19*, 18–24. [CrossRef]
58. de Candolle, A. *Origin of Cultivated Plants*; D. Appleton: New York, NY, USA, 1885.
59. Zhukovsky, P.M. New centres of origin and new gene centres of cultivated plants including specifically endemic microcentres of species closely allied to cultivated species. *Bot. J. (Russian Bot. Z.)* **1968**, *53*, 430–460.
60. Sinskaya, E.N. *Historical Geography of Cultivated Floras (at the Dawn of Agriculture)*; Kolos: Leningrad, Russia, 1969.
61. Harlan, J.R. Agricultural origins: Centers and noncenters. *Science* **1971**, *174*, 468–474. [CrossRef]
62. Zeven, A.C.; de Wet, J.M.J. *Dictionary of Cultivated Plants and Their Centres of Diversity: Excluding Most Ornamentals, Forest Trees and Lower Plants*; CAPD: Wageningen, The Netherlands, 1982.
63. Hawkes, J.G. *The Diversity of Crop Plants*; Harvard University Press: Cambridge, MA, USA, 1983.
64. Doebley, J.F.; Gaut, B.S.; Smith, B.D. The molecular genetics of crop domestication. *Cell* **2006**, *127*, 1309–1321. [CrossRef]
65. Price, T.D.; Bar-Yosef, O. The origins of agriculture: New data, new ideas: An Introduction to supplement 4. *Curr. Anthropol.* **2011**, *52*, S163–S174. [CrossRef]
66. Larson, G.; Piperno, D.R.; Allaby, R.G.; Purugganan, M.D.; Andersson, L.; Arroyo-Kalin, M.; Barton, L.; Vigueira, C.C.; Denham, T.; Dobney, K.; et al. Current perspectives and the future of domestication studies. *Proc. Natl. Acad. Sci. USA* **2014**, *111*, 6139–6146. [CrossRef] [PubMed]
67. Chomicki, G.; Renner, S.S. Watermelon origin solved with molecular phylogenetics including Linnaean material: Another example of museomics. *New Phytol.* **2015**, *205*, 526–532. [CrossRef] [PubMed]
68. Purugganan, M.D.; Fuller, D.Q. The nature of selection during plant domestication. *Nature* **2009**, *457*, 843–848. [CrossRef] [PubMed]
69. Meyer, R.S.; DuVal, A.E.; Jensen, H.R. Patterns and processes in crop domestication: An historical review and quantitative analysis of 203 global food crops: Tansley review. *New Phytol.* **2012**, *196*, 29–48. [CrossRef]
70. Romão, R.L. Northeast Brazil: A secondary center of diversity for watermelon (*Citrullus lanatus*). *Genet. Resour. Crop Evol.* **2000**, *47*, 207–213. [CrossRef]
71. Kearney, J. Food consumption trends and drivers. *Philos. Trans. R. Soc. B Biol. Sci.* **2010**, *365*, 2793–2807. [CrossRef]
72. Khoury, C.K.; Bjorkman, A.D.; Dempewolf, H.; Ramirez-Villegas, J.; Guarino, L.; Jarvis, A.; Rieseberg, L.H.; Struik, P.C. Increasing homogeneity in global food supplies and the implications for food security. *Proc. Natl. Acad. Sci. USA* **2014**, *111*, 4001–4006. [CrossRef]
73. Fader, M.; Gerten, D.; Krause, M.; Lucht, W.; Cramer, W. Spatial decoupling of agricultural production and consumption: Quantifying dependences of countries on food imports due to domestic land and water constraints. *Environ. Res. Lett.* **2013**, *8*, 014046. [CrossRef]
74. MacDonald, G.K.; Brauman, K.A.; Sun, S.; Carlson, K.M.; Cassidy, E.S.; Gerber, J.S.; West, P.C. Rethinking Agricultural Trade Relationships in an Era of Globalization. *BioScience* **2015**, *65*, 275–289. [CrossRef]
75. Badr, A.; M, K.; Sch, R.; Rabey, H.E.; Effgen, S.; Ibrahim, H.H.; Pozzi, C.; Rohde, W.; Salamini, F. On the origin and domestication history of barley (*Hordeum vulgare*). *Mol. Biol. Evol.* **2000**, *17*, 499–510. [CrossRef] [PubMed]

76. Acosta-Gallegos, J.A.; Kelly, J.D.; Gepts, P. Prebreeding in common bean and use of genetic diversity from wild germplasm. *Crop Sci.* **2007**, *47*, S-44–S-59. [CrossRef]
77. Gepts, P. Plant genetic resources conservation and utilization: The Accomplishments and future of a societal insurance policy. *Crop Sci.* **2006**, *46*, 2278–2292. [CrossRef]
78. Dempewolf, H.; Baute, G.; Anderson, J.; Kilian, B.; Smith, C.; Guarino, L. Past and Future use of wild relatives in crop breeding. *Crop Sci.* **2017**, *57*, 1070–1082. [CrossRef]
79. Das, S.; Lowe, M. Nature read in black and white: Decolonial approaches to interpreting natural history collections. *J. Nat. Sci. Collect.* **2018**, *6*, 4–14.
80. Antonelli, A.; Smith, R.J.; Simmonds, M.S.J. Unlocking the properties of plants and fungi for sustainable development. *Nat. Plants* **2019**, *5*, 1100–1102. [CrossRef]
81. Morueta-Holme, N.; Enquist, B.J.; McGill, B.J.; Boyle, B.; Jørgensen, P.M.; Ott, J.E.; Peet, R.K.; Šímová, I.; Sloat, L.L.; Thiers, B.; et al. Habitat area and climate stability determine geographical variation in plant species range sizes. *Ecol. Lett.* **2013**, *16*, 1446–1454. [CrossRef]
82. Pontarp, M.; Bunnefeld, L.; Cabral, J.S.; Etienne, R.S.; Fritz, S.A.; Gillespie, R.; Graham, C.H.; Hagen, O.; Hartig, F.; Huang, S.; et al. The latitudinal diversity gradient: Novel understanding through mechanistic eco-evolutionary models. *Trends Ecol. Evol.* **2019**, *34*, 211–223. [CrossRef]
83. Brummitt, R.K. *World Geographical Scheme for Recording Plant Distributions, Edition 2*; Biodiversity Information Standards (TDWG): Geneva, Switzerland, 2001.
84. Brummitt, N.; Araújo, A.C.; Harris, T. Areas of plant diversity—What do we know? *Plants People Planet* **2020**. [CrossRef]
85. Vincent, H.; Amri, A.; Castañeda-Álvarez, N.P.; Dempewolf, H.; Dulloo, E.; Guarino, L.; Hole, D.; Mba, C.; Toledo, A.; Maxted, N. Modeling of crop wild relative species identifies areas globally for in situ conservation. *Commun. Biol.* **2019**, *2*, 1–8. [CrossRef]
86. Harlan, J.R.; de Wet, J.M.J. Toward a rational classification of cultivated plants. *Taxon* **1971**, *20*, 509–517. [CrossRef]
87. Maxted, N.; Ford-Lloyd, B.V.; Jury, S.; Kell, S.; Scholten, M. Towards a definition of a crop wild relative. *Biodivers. Conserv.* **2006**, *15*, 2673–2685. [CrossRef]
88. Vincent, H.; Wiersema, J.; Kell, S.; Fielder, H.; Dobbie, S.; Castañeda-Álvarez, N.P.; Guarino, L.; Eastwood, R.; León, B.; Maxted, N. A prioritized crop wild relative inventory to help underpin global food security. *Biol. Conserv.* **2013**, *167*, 265–275. [CrossRef]
89. Morrell, P.L.; Buckler, E.S.; Ross-Ibarra, J. Crop genomics: Advances and applications. *Nat. Rev. Genet.* **2012**, *13*, 85–96. [CrossRef]
90. Winter, M.; Devictor, V.; Schweiger, O. Phylogenetic diversity and nature conservation: Where are we? *Trends Ecol. Evol.* **2013**, *28*, 199–204. [CrossRef] [PubMed]
91. Faith, D.P. Conservation evaluation and phylogenetic diversity. *Biol. Conserv.* **1992**, *61*, 1–10. [CrossRef]
92. Rodrigues, A.S.L.; Brooks, T.M.; Gaston, K.J. Phylogeny and conservation. In *Phylogeny and Conservation*; Cambridge University Press: Cambridge, UK, 2005; pp. 101–119.
93. Brooks, T.M.; Mittermeier, R.A.; da Fonseca, G.A.B.; Gerlach, J.; Hoffmann, M.; Lamoreux, J.F.; Mittermeier, C.G.; Pilgrim, J.D.; Rodrigues, A.S.L. Global biodiversity conservation priorities. *Science* **2006**, *313*, 58–61. [CrossRef]
94. Forest, F.; Grenyer, R.; Rouget, M.; Davies, T.J.; Cowling, R.M.; Faith, D.P.; Balmford, A.; Manning, J.C.; Procheş, Ş.; van der Bank, M.; et al. Preserving the evolutionary potential of floras in biodiversity hotspots. *Nature* **2007**, *445*, 757–760. [CrossRef]
95. Sechrest, W.; Brooks, T.M.; da Fonseca, G.A.B.; Konstant, W.R.; Mittermeier, R.A.; Purvis, A.; Rylands, A.B.; Gittleman, J.L. Hotspots and the conservation of evolutionary history. *Proc. Natl. Acad. Sci. USA* **2002**, *99*, 2067–2071. [CrossRef]
96. Costion, C.M.; Edwards, W.; Ford, A.J.; Metcalfe, D.J.; Cross, H.B.; Harrington, M.G.; Richardson, J.E.; Hilbert, D.W.; Lowe, A.J.; Crayn, D.M. Using phylogenetic diversity to identify ancient rain forest refugia and diversification zones in a biodiversity hotspot. *Divers. Distrib.* **2015**, *21*, 279–289. [CrossRef]
97. Buerki, S.; Callmander, M.W.; Bachman, S.; Moat, J.; Labat, J.-N.; Forest, F. Incorporating evolutionary history into conservation planning in biodiversity hotspots. *Philos. Trans. R. Soc. B Biol. Sci.* **2015**, *370*, 20140014. [CrossRef]

98. Mazel, F.; Pennell, M.W.; Cadotte, M.W.; Diaz, S.; Dalla Riva, G.V.; Grenyer, R.; Leprieur, F.; Mooers, A.O.; Mouillot, D.; Tucker, C.M.; et al. Prioritizing phylogenetic diversity captures functional diversity unreliably. *Nat. Commun.* **2018**, *9*, 2888. [CrossRef] [PubMed]
99. Larsen, F.W.; Turner, W.R.; Brooks, T.M. Conserving critical sites for biodiversity provides disproportionate benefits to people. *PLoS ONE* **2012**, *7*. [CrossRef]
100. Gascon, C.; Brooks, T.M.; Contreras-MacBeath, T.; Heard, N.; Konstant, W.; Lamoreux, J.; Launay, F.; Maunder, M.; Mittermeier, R.A.; Molur, S.; et al. The importance and benefits of species. *Curr. Biol.* **2015**, *25*, R431–R438. [CrossRef] [PubMed]
101. Swenson, N.G. The role of evolutionary processes in producing biodiversity patterns, and the interrelationships between taxonomic, functional and phylogenetic biodiversity. *Am. J. Bot.* **2011**, *98*, 472–480. [CrossRef] [PubMed]
102. Pardey, P.G.; Chan-Kang, C.; Dehmer, S.P.; Beddow, J.M. Agricultural R&D is on the move. *Nat. News* **2016**, *537*, 301–303. [CrossRef]
103. Bernardo, R. *Essentials of Plant Breeding*; Stemma Press: Woodbury, MN, USA, 2014.
104. Hajjar, R.; Hodgkin, T. The use of wild relatives in crop improvement: A survey of developments over the last 20 years. *Euphytica* **2007**, *156*, 1–13. [CrossRef]
105. Nakazato, T.; Warren, D.L.; Moyle, L.C. Ecological and geographic modes of species divergence in wild tomatoes. *Am. J. Bot.* **2010**, *97*, 680–693. [CrossRef]
106. Jarquin, D.; Specht, J.; Lorenz, A. Prospects of genomic prediction in the USDA soybean germplasm collection: Historical data creates robust models for enhancing selection of accessions. *G3 Genes Genomes Genet.* **2016**, *6*, 2329–2341. [CrossRef]
107. Stocks, J.J.; Metheringham, C.L.; Plumb, W.J.; Lee, S.J.; Kelly, L.J.; Nichols, R.A.; Buggs, R.J.A. Genomic basis of European ash tree resistance to ash dieback fungus. *Nat. Ecol. Evol.* **2019**, *3*, 1686–1696. [CrossRef]
108. Wambugu, P.W.; Ndjiondjop, M.-N.; Henry, R.J. Role of genomics in promoting the utilization of plant genetic resources in genebanks. *Brief. Funct. Genom.* **2018**, *17*, 198–206. [CrossRef] [PubMed]
109. Crop Wild Relative—GRIN-Global Web v 1.10.6.2. Available online: https://npgsweb.ars-grin.gov/gringlobal/taxon/taxonomysearchcwr.aspx (accessed on 17 July 2020).
110. Miller, R.E.; Khoury, C.K. The gene pool concept applied to crop wild relatives: An evolutionary perspective. In *North American Crop Wild Relatives, Volume 1: Conservation Strategies*; Greene, S.L., Williams, K.A., Khoury, C.K., Kantar, M.B., Marek, L.F., Eds.; Springer International Publishing: Cham, Switzerland, 2018; pp. 167–188, ISBN 978-3-319-95101-0.
111. Viruel, J.; Kantar, M.B.; Gargiulo, R.; Hesketh-Prichard, P.; Leong, N.; Cockel, C.; Forest, F.; Gravendeel, B.; Perez-Barrales, R.; Leitch, I.J.; et al. Crop wild phylorelatives (CWPs): Phylogenetic distance, cytogenetic compatibility and breeding system data enable estimation of crop wild relative gene pool classification. *Bot. J. Linn. Soc.* 2020; in press.
112. Pacicco, L.; Bodesmo, M.; Torricelli, R.; Negri, V. A methodological approach to identify agro-biodiversity hotspots for priority in situ conservation of plant genetic resources. *PLoS ONE* **2018**, *13*, e0197709. [CrossRef] [PubMed]
113. Wood, S.A.; Karp, D.S.; DeClerck, F.; Kremen, C.; Naeem, S.; Palm, C.A. Functional traits in agriculture: Agrobiodiversity and ecosystem services. *Trends Ecol. Evol.* **2015**, *30*, 531–539. [CrossRef] [PubMed]
114. Bowler, D.E.; Bjorkman, A.D.; Dornelas, M.; Myers-Smith, I.H.; Navarro, L.M.; Niamir, A.; Supp, S.R.; Waldock, C.; Winter, M.; Vellend, M.; et al. Mapping human pressures on biodiversity across the planet uncovers anthropogenic threat complexes. *People Nat.* **2020**, *2*, 380–394. [CrossRef]
115. Royal Botanic Gardens, Kew. *State of the World's Plants*; Royal Botanic Gardens: Kew, UK, 2016; ISBN 978-1-84246-628-5.
116. Loh, J.; Harmon, D. A global index of biocultural diversity. *Ecol. Indic.* **2005**, *5*, 231–241. [CrossRef]

plants

MDPI

Review

Why Seed Physiology Is Important for Genebanking

Katherine J. Whitehouse [1], Fiona R. Hay [2,*] and Charlotte Lusty [3]

1 Australian Grains Genebank, Agriculture Victoria Research, Departments of Jobs, Precincts and Regions, Private Bag 260, Horsham, Victoria 3401, Australia; katherine.whitehouse@agriculture.vic.gov.au
2 Department of Agroecology, Aarhus University, Forsøgsvej 1, 4200 Slagelse, Denmark
3 Global Crop Diversity Trust, Platz Der Vereinten Nationen 7, 53113 Bonn, Germany; charlotte.lusty@croptrust.org
* Correspondence: fiona.hay@agro.au.dk; Tel.: +45-87-15-60-00

Received: 1 April 2020; Accepted: 20 April 2020; Published: 2 May 2020

Abstract: Genebank management is a field in its own right; it is multifaceted, requiring a diverse set of skills and knowledge. Seed physiology is one area that is critical to the successful operation of seed genebanks, requiring understanding of seed quality during development and maturation, seed dormancy and germination, and seed longevity in storage of the target species. Careful management of the workflow between these activities, as seeds move from harvest to storage, and the recording and management of all relevant associated data, is key to ensuring the effective conservation of plant genetic resources. This review will discuss various aspects of seed physiology that genebank managers should be aware of, to ensure appropriate decisions are made about the handling and management of their seed collections.

Keywords: agrobiodiversity; genebank; genebank management; plant genetic resources; seed physiology; seed quality management

1. Introduction

Storing seeds in genebanks is the most effective way of conserving and sharing most of our existing agrobiodiversity (that of orthodox species). It is also a relatively simple activity; seeds are dried and stored at low temperature. However, within that simple statement lies a whole series of operations, which, if not carefully followed and controlled, risk the loss of the agrobiodiversity that genebanks are meant to be conserving. Effective management of the individual operations and of the whole workflow, especially when managing a crop genebank with thousands of accessions, requires some understanding of seeds and how they might be impacted in terms of their quality and viability during the different production and processing steps. In this review, we discuss various aspects of seed physiology that we consider particularly relevant to genebanks, including understanding whether a seed is viable or just dormant and how dormancy might be overcome; the acquisition and loss of the ability to tolerate desiccation during seed development; seed ageing leading to loss of viability; understanding how long seed lots will remain above a threshold viability level based on genebank viability monitoring data, modelling, and comparative longevity experiments; how seed longevity might be affected by various pre- and post-harvest factors; and the statistics of analysing seed germination/viability data. The review is directed primarily towards genebanks that conserve agrobiodiversity; the science of seed banking to conserve species diversity has been reviewed elsewhere [1,2]. Some wild species seed banks invest significant resources into researching the seeds of the species they conserve. In contrast, many agricultural genebanks do not have a mandate to do much research and, in the interests of efficiency in genebank operations, avoid collecting data that could be classified as non-essential to managing their collections. We nonetheless touch upon various areas where genebanks could collect data that would be of interest and use to the network of genebanks and to the scientific community. In relation to this, we also encourage renewed interaction between genebanks and the seed science and testing community.

2. Understanding Why a Seed Lot Is Showing No or Low Germination

Air-dry seeds of some orthodox species are predicted to survive for very long periods at sub-zero temperatures, typical of long-term genebank storage conditions (−20 °C) [3]. However, many genebanks regenerate [regeneration is used to describe the process of planting seeds to produce more seeds for genebank storage, be that as a result of declining viability, where the process might also be called 'rejuvenation', or to increase the number of seeds available ('multiplication')] their material more frequently than longevity predictions would suggest is necessary (unpublished data). This may be because insufficient seeds are being placed into storage [4], because seeds are being distributed at higher rates than expected or a viability monitoring result indicates low viability (i.e., below the acceptable threshold), or simply for reassurance that the seeds in storage, because they are 'fresh', must be viable. Indeed, some accessions may be regenerated frequently because it is difficult to produce seeds in large quantities in one season (because of, for example, a small plot size, poor plant establishment, or low fecundity) and/or because the initial germination result is poor. There are two fundamental reasons why a seed lot might show no or low germination, either in the initial germination test or in a viability monitoring germination test, including seed dormancy and/or non-optimal germination conditions or low seed viability. Indeed, the most important first consideration is to understand whether or not the result is an accurate reflection of the viability of the seed lot.

2.1. *Viability versus Germinability*

Ultimately, it is only when a seed germinates that we can be sure that the genetic information it contains can be expressed or exploited, be that for crop production or for scientific research. Thus, a germination test is considered the ultimate method for monitoring 'viability', even though it requires knowledge of dormancy-breaking treatments and optimum germination conditions. However, there are other ways in which the viability of a seed lot can be assessed. At the end of a germination test, any seeds that have not germinated could be assessed via a 'cut-test' to see if they are dead seeds (soft, mouldy); seeds that have not taken up moisture (hard-seededness, a trait that is particular prevalent in seeds from species in families such as the Fabaceae and Malvaceae); empty seeds; or seeds that still have firm, normal-coloured tissues. Other viability tests would be made on a separate sample of seeds. The International Seed Testing Association has a whole chapter of their *International Rules for Seed Testing* dedicated to the use of the tetrazolium test to rapidly assess seed lot viability, based on the living parts of the viable seed staining red, with optimized procedures for specific species or genera. The method usually involves imbibition of water, seed sectioning, and exposure to a solution of 2,3,5-triphenyl tetrazolium chloride or bromide for a specified period of time [5]. How to interpret the results (staining pattern) is also described [5]. The tetrazolium test is often criticised for being somewhat subjective; therefore, it should only be used instead of a germination test when it is not possible to remove seed dormancy. The species and genera included in the *International Rules for Seed Testing* tetrazolium chapter [5] cover the most important crops and wild species conserved by genebanks. Fluorescein diacetate has also been used as a vital stain for seeds, in particular for orchid seeds, which are very small and would thus be difficult to evaluate using tetrazolium [6,7], but also for larger seeds [8]. The methodology is similar to the tetrazolium test, except viability is assessed by observation under a fluorescence microscope, making this method less attractive than the tetrazolium test, which generally does not require any specialised equipment.

If the viability test result is consistent with the germination result, the two test results together can be considered reliable. Indeed, this is how the tetrazolium viability test is used in some genebanks, particularly those working on wild species, where the dormancy mechanism of different and diverse species may not always be known. When the two results are comparable, then the dormancy-breaking treatment(s) and/or germination test conditions are considered appropriate. If the viability result is higher than the germination result, further dormancy-breaking treatment(s) and/or germination conditions may be tried, if there are sufficient seeds available. It is also possible that the viability result is less than the germination result, in which case the tetrazolium test conditions may not be optimal. Some genebanks systematically use the tetrazolium test for viability monitoring, perhaps because it can ultimately be more straightforward than lots of different, multi-step germination tests. This is not recommended, except perhaps under special circumstances.

2.2. Seed Dormancy and Germination Behaviour

A non-dormant seed will readily germinate given appropriate environmental conditions, most importantly, moisture and temperature. The optimum temperature range for germination can vary between species, but is likely to reflect the ambient temperature in the habitat where the species naturally occurs or is typically planted, and at the time of year when the plant is expected to germinate [9,10]. In contrast, dormancy mechanisms can be multifaceted and diverse, although still climate-related [9,10]. Species-specific information on germination protocols and dormancy breaking treatments are widely available, again particularly for crop species and their wild relatives [5,10–13]. It is, however, important to be aware that, in some species, dormancy may be released or induced during storage, which may mean that the germination procedure initially used may have to be adapted following storage [14–16].

Where dormancy is suspected, because a cut test indicates that seeds are probably still viable at the end of a germination test, further germination tests with different dormancy-breaking treatments should be performed. The same literature as described in the previous section will help guide what treatments should be prioritized for testing, for example, if the 'problem' species is closely related (same genus) and comes from a similar environment as a species for which dormancy behaviour is documented. Complex, multi-factorial experiments are unlikely to be conducted in genebank laboratories (Table 1), unless there is a plan to publish the findings, in which case a more structured experimental design that can be appropriately analysed (see Section 4 below) might be better received. Nonetheless, it is helpful to share any such 'new' data on the germination requirements of previously unstudied species—precisely because it is useful information for other genebanks and for end-users. Indeed, most genebanks give germination advice when they distribute seed samples, if the information is available.

2.3. Desiccation Tolerance

Not all plant species produce seeds that tolerate the drying procedure, which is one of the steps in preparing seeds for genebank storage, and seeds of some other species, even if desiccation-tolerant, do not survive for long periods under low-temperature storage. However, the proportion of species that show this non-orthodox seed storage behaviour is low (8%, [17]), and of course, species that are already effectively managed and conserved in genebanks are clearly orthodox. However, seeds of orthodox species can be desiccation-intolerant if they are harvested too early or, conversely, if germination has commenced, for example, if the seeds are prone to vivipary. Seeds acquire physiological traits, including desiccation tolerance, during development. The relative timing of the acquisition of desiccation tolerance varies between species, but is normally acquired by or around the time that seeds reach mass maturity, when the seeds stop accumulating dry matter, and thus reach their maximum dry weight ([18] and references therein). Because dry weight will be reflected in the fresh weight of the seeds and other physical characteristics of the seed, it is perhaps unlikely that seeds of an orthodox species will be harvested before desiccation tolerance is acquired, unless the species flowers and fruits asynchronously. For species with such indeterminate flowering, it may be difficult to avoid harvesting seeds that have yet to acquire desiccation tolerance, and this may be a reason that a seed lot shows poor germination in the initial viability test. It may be possible to remove the desiccation intolerant seeds in the seed lot, by sorting based on density, size, and/or appearance. Alternatively, slow drying the whole seed lot or pre-sorting the seeds according to likely maturity and then slow drying the least mature seeds may allow maturation events to continue in the least mature seeds [19].

Table 1. Examples of how genebank operations might differ if resources are limited and there is a need for high levels of efficiency compared with what could be done if resources are available, which would be more sound and/or beneficial from a scientific perspective.

Optimum Strategy from a Genebank Management Perspective	More Scientifically Sound or of Interest from a Scientific Perspective
Only one sample of each accession in the active and base collections.	To compare the physiological response of seed lots produced in different crop seasons or environments; to have seeds of different ages to test at the same time (e.g., to understand seed longevity).
Once-over harvesting strategy.	Harvest seeds as they reach maturity.
Stop monitoring seeds once the seed lot that represents an accession has been replaced.	Continue monitoring viability to collect more data to inform seed longevity. This is better, not just because there are more data, but also because more of the data will cover the range where viability is expected to decline faster, enabling more robust model fitting.
Only consider one or a few different dormancy-breaking treatments at a time.	Factorial dormancy breaking/germination experiment, with different treatments and/or germination temperatures and treatment combinations. This should be a priority for 'new' species where there is little information on dormancy and germination requirements.
Initial viability test to confirm initial seed quality is sufficient.	Initial seed storage experiment to estimate initial seed storage potential, for setting seed lot-based monitoring intervals, and/or for confirming that the ranking of seed lots for longevity based on experimental storage corresponds with the ranking in genebank storage.
Minimal viability monitoring tests, e.g., only test a subset from each harvest season.	Monitor the viability of all samples at frequent intervals to get more data on relative seed longevity of different samples and of the same samples in different storage environments (e.g., medium- vs. long-term storage).
Only score for germination once or twice during a viability monitoring test.	Regular scoring of germination during a germination viability monitoring test to get information on speed of germination (vigour measures) and how vigour declines as seeds age.

Vivipary is a characteristic of some species or varieties whose seeds have little or no dormancy as the seeds mature on the plant. If the seeds are exposed to germination-promoting conditions, in particular, high moisture owing to, for example, late rain, then the seeds may start to germinate while still on the plant [20]. Vivipary is common in tomatoes and peppers, as well as some cereals (e.g., temperate rice). Upon germination, orthodox seeds lose the ability to tolerate desiccation. Imbibition of water alters the physical properties of the membrane lipid bilayer, making the cellular membranes susceptible to injury upon subsequent desiccation [21–23]. Furthermore, as metabolism is reinstated upon imbibition, subsequent desiccation can lead to the initiation of further lipid peroxidation (see below) in the dehydrated tissues, which can continue upon re-imbibition [21]. Hence, any seeds showing on-plant germination will have to be discarded; indeed, it may not be possible to identify all the seeds that have started to germinate and thus lost desiccation tolerance, meaning that the entire seed lot may have to be rejected.

2.4. Seed Ageing

Seed ageing is inevitable, regardless of the conditions under which the seeds are stored, though of course, the aim of genebanking is to slow down the rate of ageing so as to preserve the genetic integrity of the seeds and to make sure the genetic information is available for use. Seeds are hygroscopic, meaning they exchange water with their surroundings until they reach equilibrium. The tendency of water to move into the tissues from the external environment is dependent upon the relative humidity (RH) of the atmosphere and the moisture content of the seed. It also depends on the chemical composition (oil content), size, and seed coat properties. Seeds with a higher oil content have a lower moisture content at a specific RH compared with lower oil content seeds. How a seed interacts with water can be explained by sorption isotherms, the shape of which reflects the availability of water within the seeds to support different chemical reactions [24–27]. After the attainment of mass maturity, that is, the end of seed-filling when seeds have reached maximum dry weight, seeds equilibrate with the ambient environmental conditions, usually by losing moisture, unless conditions are very humid. Nonetheless, they may still be harvested at a moisture content conducive to high rates of ageing reactions. This is why it is critical that seeds are dried as quickly as possible following harvest, in order to reduce the rate of ageing reactions (unless slow drying may allow the continuation of the acquisition of desiccation tolerance, as discussed above).

The type and rate of ageing reactions depend on seed moisture content, temperature, and oxygen availability [28]. For example, respiration will only occur in the presence of oxygen and at a high enough RH that the cytoplasm allows molecular motion. As a result, seeds can be classified as either undergoing wet or dry ageing, determined by the viscosity of the cytoplasm (low viscosity at RH ≥ca.75% (wet ageing) and highly viscous/glassy state at RH <ca.75% (dry ageing)). The difference in the types of reactions that occur during these two states differentially impacts hormonal regulation and signalling pathways involved in longevity [29]. The major contributors to seed deterioration, resulting in loss of membrane integrity, reduced energy metabolism, protein carbonylation, impairment in RNA and protein synthesis, and RNA degradation, are lipid peroxidation and free radical accumulation [30]. DNA damage inhibits effective transcription and replication [31], key processes that are activated during the imbibition stage of germination [32,33]. In this stage, repair mechanisms are activated, in order to restore a functional physiological state [32,33]. However, if damage has accumulated above a certain threshold, seeds will begin to show a delay in germination, as it takes longer to repair all the damage. This explains why seed lot vigour declines and the proportion of abnormal seedlings increases, before complete loss of viability [20], although the temporal pattern from the origin of deterioration to death, either within an individual seed or a seed lot, is still not fully understood [29]. It is important to note however, that if vigour and/or abnormal germination was used to indicate the extent to which a seed lot has already aged, there may be an expectation that accessions be regenerated at an earlier stage of the ageing process. This could lead to more frequent regeneration, which would not be encouraged owing to the inherent costs and risks. Thus, better understanding of the vigour of accessions and how it changes during storage is again perhaps something that is more desirable from a scientific perspective and/or of interest to end-users, rather than being something that genebanks should aspire to assess routinely (Table 1).

3. Understanding How Long a Seed Lot Will Continue to Show Good Germination and the Factors That Influence That Period

Thresholds for the acceptable percentage viability of a seed lot have been defined in various standards [34,35]. Understanding how long a seed lot is likely to maintain viability above the threshold value will improve the efficiency of genebank operations. For example, in some cases, monitoring intervals could be extended beyond 5 or 10 years, which are the 'default' intervals for species for which evidence of seed longevity is lacking (but which are nonetheless 'expected' to be short- or long-lived, respectively, in long-term genebank storage, at −18 °C, in hermetically closed containers [34]). This will save use of seeds, resources, and particularly staff time. Conversely, expectations regarding seed longevity may also highlight where there might be problems owing to biology, perhaps, or where interventions might be appropriate in the seed harvesting–processing–storage chain.

3.1. Historical Viability Monitoring Data

A number of genebanks have now published historical viability monitoring data for seeds in medium- and/or long-term storage, including the United States Department of Agriculture [36]; the National Agrobiodiversity Center of South Korea [37]; the Centre for Genetic Resources, the Netherlands (CGN [38,39]); the International Rice Research Institute (IRRI [40]); and the International Livestock Research Institute (ILRI [41,42]). Some of these publications highlight the problems of analysing such data, including changes in protocols and/or storage conditions, failure to overcome dormancy, and censoring. Testing of seed lots in genebank storage often ceases once the seed lot is 'replaced' by newly harvested seeds following a cycle of regeneration (literally or metaphorically in that a seed lot may simply cease to be actively managed; Table 1). Thus, the data are censored and most of the data collected have values between 100% and whatever the viability standard might be, typically 75% or 85%, and higher if seed lots are primarily replaced owing to low seed quantity. This makes the data difficult to analyse using methods such as generalized linear modelling (e.g., probit analysis) of germination percentage data, which might otherwise be used (see below).

Nonetheless, these papers in general both confirm the effectiveness of seed genebanks as a means of conserving agrobiodiversity, and perhaps flag accessions that are likely to have shorter seed longevity, for example, accessions of carrot, parsnip, and onion [36] or temperate varieties of rice [40], or accessions that show unexpected behaviour in response to genebank storage, as reported for wheat and barley [39]. In the case of the analysis of the data from the IRRI genebank, it was not deemed necessary to change the monitoring interval, largely because there appeared to be a trend of declining longevity of the seeds being placed into storage, for seeds harvested between 1992 and 2001 (the last year's harvest covered by the analysis was 2002 [40]). On the other hand, CGN concluded that they could delay the first monitoring test to 25 years after harvest (seeds in long-term storage at −20 °C [38]). There are also plans to adjust retest intervals, according to genus at the ILRI genebank, to one-third of the time predicted for viability to fall to the viability standard, where those predictions are reliable (Y. Woldemariam and J. Hanson, pers. comm.). The revised intervals will be incorporated into the genebank management software used at ILRI.

From personal experience, downloading, sorting, and analysis of genebank monitoring data is not a straightforward exercise, in part because of the length of time over which the data have been collected and entered (or not) into the database—or different databases—by different staff, with a range of associated errors that accrue over time. In future, this process should not constitute a 'major exercise'; rather, genebank managers should be prepared to 'regularly' analyse data or even have some sort of analysis continuously running in the background, within the genebank management software, though the basis of that analysis (algorithms to handle the data, analytical approach) is not known; machine learning may help in this process [43].

3.2. Seed Longevity Predictions

Orthodox seeds are not simply defined as those seeds that tolerate desiccation; they are those seeds whose longevity systematically increases in response to a reduction in moisture and temperature [44]. The only longevity model currently available to describe these relations is the Ellis-Roberts viability equations (VEs) [45], which are based on fitting negative cumulative normal distributions to the proportion of viable seeds within samples of seeds stored under different constant moisture and temperature conditions (usually achieved by equilibrating seeds in a particular environment and then sealing them inside air-tight containers, such as aluminium foil bags). The equations have species-specific parameters that quantify the inherent longevity of seeds of a particular species and how longevity changes as moisture content changes. Two further parameters describe the effect of changes in temperature on seed longevity, but the temperature response is thought to be similar across all species [46]. To date, the species-specific parameters have been solved for approximately 70 species, mostly agricultural crops [13]. This means that genebank managers can predict the longevity of seed lots of these species, based on the initial germination test result, for example, using the seed viability constants menu of the Seed Information Database [13]. This is one of the reasons that the initial germination test, ideally made at the time when the seed lot is placed into genebank storage, is so important, not just to confirm the seeds are above the viability threshold, but also to have some indication of how long they might stay above that viability threshold. For example, the predicted length of time it takes for the viability of rice (*Oryza sativa* L.) seeds in long-term storage (6.1% moisture content, −20 °C) to fall to 85% is 255 years when the initial viability is 95%, but 103 years if initial viability is 90% (estimated using the Seed Information Database [13]).

Of course, the VEs are not infallible; there is error associated with any prediction, based on the errors associated with the parameter estimates caused by random (and perhaps experimental) errors in the original data [47], though prediction errors have not usually been calculated. Furthermore, the initial viability result is only an estimate of the true viability of a seed lot [48,49]. Of more concern, the species-specific parameters may not be as robust as previously thought [49,50]. The VEs are also critiqued because the cumulative normal distribution has a particular well-defined shape. In fact, deviances from the basic symmetrical sigmoid curve can be accommodated, and in many cases, attributed to a biological response. For example, viability may be more or less maintained above a threshold value that is less than 100% [51] or there may be a dormancy-breaking response at the start of storage, before viability declines [52]. It would also be possible to describe sub-populations within the seed lot that are responding to storage differently, provided there are sufficient data.

Despite the concerns around using the VEs to make predictions of longevity, they are still extremely useful, not least for emphasising the importance of drying; even seemingly small declines in moisture content, over a certain range, result in significant increases in longevity [27]. The low moisture content limit to this range varies between species, largely owing to differences in seed oil content; it also shifts to a higher moisture content as the storage temperature is reduced [53]. Importantly, drying below this moisture content does not result in either further improvement in longevity or decline in longevity [27,53,54]. As some of the original work in this area found that the low moisture content limit was reached when seeds were equilibrated to a rather low humidity (approximately 10% at 20 °C for seeds subsequently aged at 65 °C) [55], the first official Genebank Standards recommended drying seeds for genebank storage at 10–25 °C and 10–15% RH [56]. Many genebanks installed drying rooms set to these conditions to efficiently dry seeds. Some genebanks check the moisture content of seeds before packing. The official seed testing method to determine seed moisture content is to weigh a sample of seeds before and after oven-drying (at 103 or 130 °C, depending on species [5]). This is a destructive test and may require far more seeds than a genebank would like to use. A more efficient, non-destructive method is to simply check whether the equilibrium relative humidity (eRH) of the seeds is equal or close to the RH of the environment in which they have been dried, using a suitable hygrometer [57]. If they are in equilibrium, the seeds will not be able to dry further, and are ready to be packed; it is not necessary to know the actual moisture content of the seeds.

The VEs also illustrate why, if different samples of seeds from the same seed lot are stored at the same moisture content (because they are hermetically packed at the same time) at different temperatures, it is not necessary to monitor the viability of both samples from the start. The seeds at the higher temperature will lose viability faster than those at the lower temperature. Thus, for example, it would not be necessary to test the seeds in long-term storage (at the lower temperature) until the viability of the seeds in medium-term storage has declined to the viability standard (Figure 1) [48]. This could save resources if both samples are currently being tested, though of course, it means there would be less viability monitoring data for modelling longevity. Similarly, if samples of seeds are packed for safety duplication (Box 1) in high quality moisture-proof containers at the same time as the samples intended for long-term storage (packed similarly), provided the seeds are stored at the same temperature in the two locations or at a lower temperature in the safety duplicate location than the originating genebank, then it is not necessary to test the seeds that have been sent as the safety duplicate. Moreover, there should not be any real need to include additional samples for retrieval and testing, which would only serve to add extra work load to genebank staff, who may already be challenged to carry out the requisite number of monitoring tests of seeds in the active and base collections.

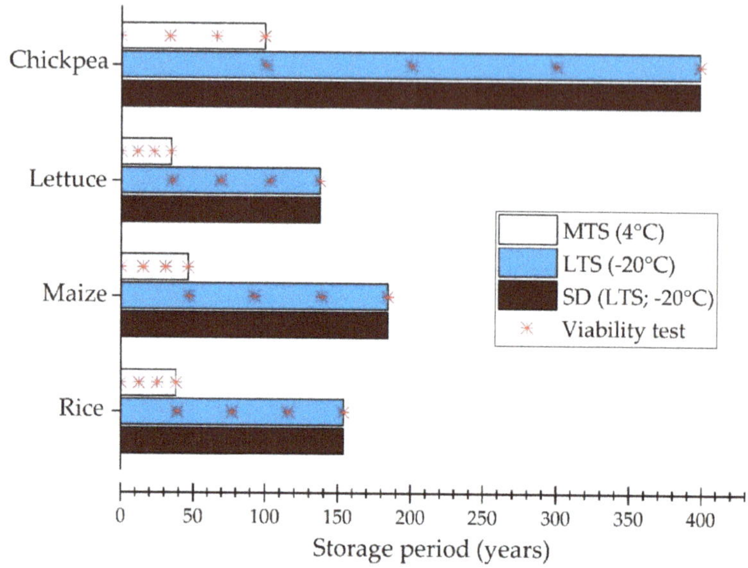

Figure 1. Predicted time for viability to fall from 92% to 85% for seeds of different crops stored in the medium-term store (MTS), the long-term store (LTS), or as a safety duplicate (SD) under LTS conditions in another location. Predications made using the Seed Viability Constants tool of the Seed Information Database [13], based on drying seeds to equilibrium with 15% relative humidity at 15 °C. The red stars indicate the timing of viability monitoring tests, at intervals of one-third of the time predicted for viability to fall to 85%.

It is often asked whether moving seeds between temperatures, for example, from the medium-term storage environment to the laboratory, to take a sample for distribution or for viability monitoring, has an effect on the longevity of the seeds. This may happen many times in the 'lifetime' of a genebank seed lot. The evidence, again from applying the VEs, is that there is no effect of moving the seeds per se; the viability is affected only by as much as would be expected from the brief period spent at the higher temperature where the seeds are allowed to equilibrate and are sampled [58].

Box 1. Safety duplication and the Svalbard Global Seed Vault.

The Genebank Standards [34] recommend, as a safety measure in case of natural disasters, that a sample of all original seeds collected or seed accessions only held by that genebank should be duplicated in another location. Ideally, this location should be in a different country or even continent i.e., somewhere that is not at risk to the same natural and/or human-caused catastrophes, and preferably one where there is no socio-political uncertainty and environmental risk. This "black-box" collection is not active therefore it is the responsibility of the depositor to ensure the sample size is sufficient (enough to carry out at least three regenerations) and of a high quality (>85% germination). If the storage conditions are the same at both locations then the rate of loss in viability of the black-box collection should equal that of the sample in the original genebank, the viability of which is monitored.

The Svalbard Global Seed Vault (SGSV) on Svalbard, Norway, opened in 2008 and offers the long-term storage of safety duplicates from the world's 1700 national and international genebanks. The vault currently holds more than 1,173,000 samples (www.seedvault.no) and represents the world's largest collection of crop diversity. It is strategically located in a highly secure zone and is the ultimate insurance policy against loss of plant agrobiodiversity.

3.3. Comparative Seed Longevity Studies

A more robust method for assessing the storage potential of different seed lots would be to carry out a seed storage experiment (SSE) rather than just an initial germination test (Table 1). This involves storing samples of seeds at a relatively high temperature and moisture content (e.g., 45 °C after equilibrating seeds at 60% RH and 20 °C) and removing a sample at regular intervals for germination testing [59,60]. After probit analysis of the germination data, the seed lots can be ranked according to the time for viability to decrease to a specific viability [61,62] and/or categorized into longevity categories [63]. If the slopes of the survival curves can be constrained to be the same for all of the seed lots (potentially possible if the seeds are of the same crop), then the theoretical initial viability can be inserted into the VEs to make more precise predictions of longevity. Monitoring intervals for the seeds in genebank storage can then be adjusted on a seed lot basis.

3.4. Maturity at Harvest

Seed longevity continues to increase after the seeds have acquired desiccation tolerance, in the late maturation phase of seed development [64–67]. However, the time when 'final' or 'maximum achievable' longevity is reached and how long it is maintained thereafter will vary between species, varieties, accessions, and seasons. This is because the longevity of maturing seeds still in the field is highly dependent on both the genotype and the environment, and the interaction of the two. This makes it virtually impossible to consistently harvest different accessions when they have peak longevity.

In the late maturation phase of seed development, the seeds no longer have a vascular connection with the mother plant. The moisture content of the seeds, and thus the availability of water to support different types of chemical reactions, will depend on the ambient conditions (humidity and temperature). The drying that occurs in situ is considered important for stimulating the improvement in seed longevity [52,65], which is why it is often recommended to wait for seeds to dry in the field before harvest [18]. This is easy for shatter-resistant crops; for crops and species that readily disperse their seeds after the vascular connection with the mother plant is broken, delaying the harvest may result in loss of seeds. For such species, bagging inflorescences before they enter the late maturation phase may be necessary [54].

If seeds that have dried on the plant subsequently take up moisture before they have been harvested, depending on the moisture content they attain, longevity may in fact continue to improve, either while the seeds have a high moisture content, provided the seeds do not germinate, and/or upon redrying [68]. If an intermittent moisture content is reached, ageing may occur at a rather fast rate until either the seeds dry again or moisture content increases further. If the moisture content increases to a level where respiration can occur, it is possible that the damage accumulated during ageing will be repaired and seed quality restored/maintained [69]. If the seeds have suffered a substantial amount of damage, they may not able to reach the same longevity as was previously attained. It is ultimately the net changes in seed quality (improvement vs. deterioration) that will determine the potential longevity of the seeds when they are harvested. Developmental events that result in improvements in longevity may also continue ex planta if seeds are harvested prematurely and held at conditions similar to what they might naturally encounter in planta [52,60,70], or upon rehydration if seeds were dried too quickly for maximum quality to be attained [71].

As yet, there is no tool to assess how far seeds have progressed through maturation processes before they are harvested, not least because physical changes that occur during maturation vary so widely between species and varieties [72,73]. In temperate and dry climates, the most reliable method is to see whether seeds are in equilibrium with ambient conditions, by placing a sample of seeds in a portable hygrometer and comparing the seed equilibrium relative humidity with the ambient relative humidity.

3.5. Post-Harvest Handling

The post-harvest environment and seed processing operations also affect seed quality and subsequent longevity [61,74]. If seeds have already dried to equilibrium with ambient conditions and reached a relatively low moisture level (<85% equilibrium relative humidity), seeds should be dried as soon as possible to minimize ageing [75]. In semi-arid and arid climates, drying, if necessary, may be done outside under well-ventilated conditions, but avoiding risks such as insect predation or over-heating in direct sunlight. In such environments, care should also be taken not to allow drying to progress too far, as cracking may occur if the seeds need to be threshed. Otherwise, the seeds should be transferred to a controlled drying environment as soon as possible.

As discussed previously, if seeds are harvested with a high moisture content and are likely to be metabolically active, seeds should be dried under conditions that optimally stimulate late maturation phase metabolism. For example, in rice and soybean, the quality of seeds harvested, when still at a high moisture content, was improved by drying for an initial period at a high temperature (45 °C) [50,52,76,77]. It is thought that the loss in moisture is a critical factor controlling the maturation process, by inducing the stress response and other protective mechanisms [78], which significantly increase seed quality. However, drying at such a relatively high temperature is in conflict with the current Genebank Standards that recommend drying mature seeds for long-term storage at a low temperature (5–20 °C) and relative humidity (10–25%) [34]. This recommendation was driven by the requirement for a single, simple, and safe procedure for diverse species in all locations worldwide, and assumed that seeds were already comparatively dry (for example, seeds dried using heated air [79], but requiring further drying to a low moisture content for long-term storage). This was determined by combining the seed viability equation with equations describing the effect of environment on seed drying rate and seed temperature in constant-temperature heated-air dryers in contrasting species [80]. Although the "safe" temperature limit for drying seeds varies between species, high temperatures are usually avoided to reduce the risk of seed deterioration, especially when seeds have a high moisture content and during the later stages of drying when evaporative cooling will no longer suppress the temperature within the seeds [79–81]. Despite this, the recommended low temperature and low humidity conditions for post-harvest seed drying are neither species-specific nor dependent upon initial moisture content [34].

3.6. Length of Time Before Storage

Most genebanks clean seed lots before packing for storage to remove, for example, off-types, immature, diseased, and damaged seeds. This is often a largely manual process or involves the use of fairly basic equipment such as graded sieves or blowing machines. It is important that the seeds do not sit for too long in an uncontrolled environment while cleaning occurs, as this might mean that they take up moisture and rates of ageing increase. Redrying may be necessary before packing, which should also be done in a controlled environment. Most drying and processing environments, even though they are controlled, are maintained at temperatures that are comfortable for genebank staff, but that are thus higher than desirable from the point of view of maintaining the quality of the seeds. Therefore, it is important that the seeds are packed and transferred to proper genebank storage as soon as possible.

4. Statistics of Seed Testing

Genebanks, as a rule, do not conduct experiments on genebank accessions as part of routine operations (Table 1), but they nonetheless do gather data that may be relevant to analyse, for example, to understand the diversity of the material they conserve, to create core-sets based on the diversity represented, or to identify potential gaps in a collection. Some of these data may be suitable for methods of analysis that are typical within the agricultural sciences. However, viability monitoring and other germination data cannot be treated and analysed in the same way [82]. This is because seed germination is a binary response; a seed will either germinate or not germinate, and a sample of seeds will show a germination result that falls between 0% and 100%.

4.1. Comparing Two Germination/Viability Results

There are a number of reasons why two germination or viability results might be compared, the most obvious being to compare the results of a germination test with that of a viability test (see above), or the results of a pair of germination tests made using different pre-treatments and/or test conditions. Testing a viability monitoring result against a previous monitoring result or the initial test result would not be common, because genebanks tend to 'accept' a viability result and not worry whether or not it is significantly different from any other value. Indeed, genebanks do not even routinely compare a viability monitoring result against the viability threshold value, even though the germination result is only ever an estimate of the whole seed lot viability (because not all the seeds are tested). One test that could be used for comparing two germination results is the χ^2 test to test whether the probability of success (scored as viable or germinated) is the same for seeds from two different treatments (e.g., stored or not, viability or germination). The algorithm for the χ^2-test is likely to be based on maximum likelihood estimation, in which case the test is the same as fitting a single-factor binary logistic regression, that is, a generalized linear model (GLM) with logit link function (the link function transforms the response variable, for example, the proportion of germinating seeds, to the linear 'predictor' variable that is fitted in the analysis) and binomial error distribution based on the number of seeds tested.

4.2. Analysis of a Factorial Germination Experiment

It may be that a dormancy-germination experiment is multi-factorial, in which case the obvious extension to the χ^2 test is to fit a GLM with different parameters estimating the effect of the different factors included in the experiment. Different link functions may be used to relate the response variable to the predictor [83], but the most common is probably the logit link function, that is, binary logistic regression analysis, for example, [84]. A further extension of this method is to fit a generalized linear mixed model (GLMM) to take into account fixed (factor-related) and random effects (e.g., if seeds are sown across multiple test units such as Petri dishes or rolled paper towels) [82,85]. It is rarely appropriate to use analysis of variance (ANOVA), because the nature of germination data is likely to violate the assumptions of the analysis; that is, that the errors after fitting the model follow the same, normal distribution across treatment groups [82].

4.3. Analysing a Series of Germination Results

Although not routine in most genebanks, recording the progress of germination of a sample of seeds in a germination test over some days or weeks (or in some cases, even longer) is common in studies related to determining optimum dormancy-breaking/germination requirements and understanding seed vigour (speed of germination). Germination progress data are also collected to assess seed lot behaviour in response to stress (e.g., water stress) or, conversely, to stimulants, for example, [86,87]. By convention, many analyses of such behaviour have neither taken the binomial error distribution of germination behaviour into account nor used independent samples, e.g., [88]. If independent samples are used, for scoring just once and then discarding, again, a GLM or GLMM might be used to analyse the data; if samples are repeatedly scored for germination, time-to-event model fitting would be more appropriate [83]. Germination data for a series of samples from a seed lot in storage can be analysed by fitting a GLM. By convention, this has been by probit analysis (i.e., fitting a GLM with a probit link function) and is the basis of the Ellis-Roberts viability equations (see above).

5. Concluding Remarks

Genebank managers often come from diverse backgrounds, with different scientific training and experience. On the other hand, to be a genebank manager requires a diverse set of skills and acquaintance with, if not knowledge of, a broad range of topics. As such, it can be difficult to find an ideal person to fill a senior genebank management position, unless there is an applicant who is moving from another genebank or has been an understudy for the role. Consequently, it may be likely, or even inevitable, that some expertise gets overlooked and/or is difficult to find compared with a few decades ago, when for example, plant taxonomy and seed science were perhaps more likely to be covered in botany and plant science degrees. As such, incoming genebank managers may be on a steep learning curve and be limited by a lack of understanding of some aspect(s) of genebank science. It is key to the management of a genebank that staff understand the reasons behind patterns of germination shown by individual crops and seed lots; are primed to make efforts to increase the efficiency of operations; and are able to communicate this to donors, reviewers, and the public. If not the head of the genebank, then other genebank staff should have training in seed science and technology and/or access to seed physiology experts, and deliberate strategic investments should be made in capacity building in this area.

Seed handling has been a focus of recent quality management audits and external reviews that have been conducted under the CGIAR Genebank Platform [89]. Standard operating procedures (SOPs) have been drafted and audited for key operations at each of the 11 CGIAR genebanks. In addition to document audits, expert reviewers have visited each genebank to see the procedures in practice and to validate individual SOPs. A large number of the reviewers' observations and recommendations concerned seed processing, handling, and multiplication of seed lots and related data management. It is clear that seed quality management practice is a highly dynamic and influential aspect of genebank management, where a lack of oversight at any point in the long history of a permanent genebank can profoundly impact efficiency and long-term conservation. Continual expert input and review, and research and experimentation, as well as innovation and automation on various areas of seed management, clearly play major roles in the sustainability of major seed collections.

In this review, we have discussed various aspects of seed physiology that are particularly relevant for seed genebanks if appropriate decisions about the handling of seed germplasm are to be made. Some of this may seem like common sense to those that are trained in seed science, but in fact, many of the observations relate to questions that do get asked by genebank staff at all management levels. Further, there should be more interaction between the seed science/testing community and genebanks. This would help genebank staff stay abreast of new scientific developments in seed physiology and, in particular, testing methods, and thereby help them to continue to improve the efficiency and effectiveness of genebanks operations. It is also important for seed scientists to be aware of areas on which to focus their research to support the work of the genebanks in conserving and making available plant genetic resources.

Author Contributions: F.R.H. and K.J.W. conceived the article; K.J.W., F.R.H., and C.L. wrote the article. All authors have read and agreed to the published version of the manuscript.

Funding: The writing of this article was partially funded through the CGIAR Genebank Platform, as part of the Seed Quality Management activity of the Conservation Module of the Platform.

Acknowledgments: We thank the editors of this special issue, Andreas Ebert and Jan Engels, for inviting us to write this article and for providing guidance on the content.

Conflicts of Interest: The authors declare no conflict of interest.

References

1. Smith, R.D.; Dickie, J.B.; Linington, S.H.; Pritchard, H.W.; Probert, R.J. (Eds.) *Seed Conservation: Turning Science into Practice*; Royal Botanic Gardens Kew: Richmond, UK, 2003.
2. Hay, F.R.; Probert, R.J. Advances in seed conservation of wild plant species: A review of recent research. *Cons. Phys.* **2013**, *1*, cot030. [CrossRef] [PubMed]
3. Hong, T.D.; Linington, S.; Ellis, R.H. Seed storage behaviour: A compendium. In *Handbook for Genebanks No. 4*; International Plant Genetic Resources: Rome, Italy, 1996.
4. Sackville Hamilton, N.R.; Chorlton, K.H. Regeneration of accessions in seed collections: A decision guide. In *Handbook for Genebanks No. 5*; International Plant Genetic Resources Institute: Rome, Italy, 1997.
5. ISTA. *International Rules for Seed Testing*; International Seed Testing Association: Basserdorf, Switzerland, 2020.
6. Pritchard, H.W. Determination of orchid seed viability using fluorescein diacetate. *Plant Cell Environ.* **1985**, *8*, 727–730.
7. Wood, C.B.; Pritchard, H.W. Determination of intra-specific variation in orchid seed viability using fluorescein diacetate. *Seed Sci. Technol.* **2004**, *32*, 629–635. [CrossRef]

8. Noland, T.L.; Mohammed, G.H. Fluorescein diacetate as a viability stain for tree roots and seeds. *New For.* **1997**, *14*, 221–232. [CrossRef]
9. Baskin, C.C.; Baskin, J.M. When breaking seed dormancy is a problem. Try a move-along experiment. *NPJ* **2003**, *4*, 17–21. [CrossRef]
10. Baskin, C.C.; Baskin, J.M. *Seeds, Ecology, Biogeography, and Evolution of Dormancy and Germination*, 2nd ed.; Academic Press: San Diego, CA, USA; London, UK; Waltham, MA, USA, 2014.
11. Ellis, R.H.; Hong, T.D.; Roberts, E.H. *Handbook of Seed Technology for Genebanks. Volume I. Principles and Methodology*; International Board for Plant Genetic Resources: Rome, Italy, 1985.
12. Ellis, R.H.; Hong, T.D.; Roberts, E.H. *Handbook of Seed Technology for Genebanks. Volume II. Compendium of Specific Germination Information and Test Recommendations*; International Board for Plant Genetic Resources: Rome, Italy, 1985.
13. Royal Botanic Gardens Kew: Seed Information Database (SID). Version 7.1. Available online: http://data.kew.org/sid/ (accessed on 10 March 2020).
14. Crawford, A.D.; Steadman, K.J.; Plummer, J.A.; Cochrane, A.; Probert, R.J. Analysis of seed-bank data confirms suitability of international seed-storage standards for the Australian flora. *Aus. J. Bot.* **2007**, *55*, 18–29. [CrossRef]
15. Pérez-García, F.; Gómez-Campo, C.; Ellis, R.H. Successful long-term ultra-dry storage of seed of 15 species of Brassicaceae in a genebank: Variation in ability to germinate over 40 years and dormancy. *Seed Sci. Technol.* **2009**, *37*, 640–649. [CrossRef]
16. González-Benito, M.E.; Pérez-García, F.; Tejeda, G.; Gómez-Campo, C. Effect of gaseous environment and water content on seed viability of four Brassicaceae species after 36 years. *Seed Sci. Technol.* **2011**, *39*, 443–451. [CrossRef]
17. Wyse, S.V.; Dickie, J.B. Predicting the global incidence of seed desiccation sensitivity. *J. Ecol.* **2017**, *105*, 1082–1093. [CrossRef]
18. Ellis, R.H. Temporal patterns of seed quality development, decline, and timing of maximum quality during seed development and maturation. *Seed Sci. Res.* **2019**, *29*, 135–142. [CrossRef]
19. Hay, F.R.; Probert, R.J. Seed maturity and the effects of different drying conditions on desiccation tolerance and seed longevity in foxglove (*Digitalis purpurea* L.). *Ann. Bot.* **1995**, *76*, 639–647. [CrossRef]
20. Bewley, J.D.; Bradford, J.K.; Hilhorst, H.W.M.; Nonogaki, H. *Seeds. Physiology of development, Germination and Dormancy*, 3rd ed.; Springer: New York, NY, USA, 2013.
21. Leprince, O.; Deltour, R.; Thorpe, P.C.; Atherton, N.M.; Hendry, G.A.F. The role of free radicals and radical processing systems in loss of desiccation tolerance in germinating maize (*Zea mays* L.). *New Phytol.* **1990**, *116*, 573–580. [CrossRef]
22. Buitink, J.; Leprince, O.; Hoekstra, F.A. Dehydration-induced redistribution of amphiphilic molecules between cytoplasm and lipids is associated with desiccation tolerance in seeds. *Plant Physiol.* **2000**, *124*, 1413–1425. [CrossRef] [PubMed]
23. Hoekstra, F.A.; Golovina, E.A.; Buitink, J. Mechanisms of plant desiccation tolerance. *Trends Plant Sci.* **2001**, *6*, 431–438. [CrossRef]
24. Vertucci, C.W.; Leopold, A.C. Bound water in soybean seed and its relation to respiration and imbibitional damage. *Plant Physiol.* **1984**, *75*, 114–117. [CrossRef]
25. Vertucci, C.W.; Leopold, A.C. Physiological activities associated with hydration level in seeds. In *Membranes, Metabolism and Dry Organisms*; Leopold, A.C., Ed.; Cornell University: Ithaca, NY, USA, 1986; pp. 35–49.
26. Vertucci, C.W. The effects of low water contents on physiological activities of seeds. *Physiol. Plant.* **1989**, *77*, 172–179. [CrossRef]
27. Roberts, E.H.; Ellis, R.H. Water and seed survival. *Ann. Bot.* **1989**, *63*, 39–52. [CrossRef]
28. Ellis, R.H.; Hong, T.D. Seed longevity—Moisture content relationships in hermetic and open storage. *Seed Sci. Technol.* **2007**, *35*, 423–431. [CrossRef]
29. Zinsmeister, J.; Leprince, O.; Buitink, J. Molecular and environmental factors regulating seed longevity. *Biochem. J.* **2020**, *477*, 305–323. [CrossRef]
30. Corbineau, F. Markers of seed quality: From present to future. *Seed Sci. Res.* **2012**, *22*, 61–68. [CrossRef]
31. Waterworth, W.M.; Bray, C.M.; West, C.E. The importance of safeguarding genome integrity in germination and seed longevity. *J. Exp. Bot.* **2015**, *66*, 3549–3558. [CrossRef] [PubMed]

32. Waterworth, W.; Masnavi, G.; Bhardwaj, R.M.; Jiang, Q.; Bray, C.M.; West, C.E. A plant DNA ligase is an important determinant of seed longevity. *Plant J.* **2010**, *63*, 848–860. [CrossRef] [PubMed]
33. Dona, M.; Balestrazzi, A.; Mondoni, A.; Rossi, G.; Ventura, L.; Buttafava, A.; Macavei, A.; Sabatini, M.E.; Valassi, A.; Cabonera, D. DNA profiling, telomere analysis and antioxidant properties as tools for monitoring *ex situ* seed longevity. *Ann. Bot.* **2013**, *111*, 987–998. [CrossRef] [PubMed]
34. FAO. *Genebank Standards for Plant Genetic Resources for Food and Agriculture, Rev Ed.*; FAO: Rome, Italy, 2014.
35. Royal Botanic Gardens Kew: Seed Conservation Standards for 'MSB Partnership Collections'. Final Version February 2015. Updated November 2019. Available online: http://brahmsonline.kew.org/Content/Projects/msbp/resources/Training/MSBP-Seed-Conservation-Standards.pdf (accessed on 10 March 2020).
36. Walters, C.; Wheeler, L.M.; Grotenhuis, J.M. Longevity of seeds stored in a Genebank: Species characteristics. *Seed Sci. Res.* **2005**, *15*, 1–20. [CrossRef]
37. Lee, H.S.; Jeon, Y.A.; Lee, Y.Y.; Lee, S.Y.; Kim, Y.G. Comparison of seed viability among 42 species stored in a genebank. *Korean J. Crop. Sci.* **2013**, *58*, 432–438. [CrossRef]
38. Van Treuren, R.; de Groot, E.C.; van Hintum, T.J.L. Preservation of seed viability during 25 years of storage under standard genebank conditions. *Genet. Resour. Crop Evol.* **2013**, *60*, 1407–1421. [CrossRef]
39. Van Treuren, R.; Bas, N.; Kodde, J.; Groot, S.P.C.; Kik, C. Rapid loss of seed viability in *ex situ* conserved wheat and barley at 4 °C as compared to −20 °C storage. *Conserv. Physiol.* **2018**, *6*, coy033. [CrossRef]
40. Hay, F.R.; de Guzman, F.; Sackville Hamilton, N.R. Viability monitoring intervals for genebank samples of *Oryza sativa*. *Seed Sci. Technol.* **2015**, *43*, 218–237. [CrossRef]
41. Ellis, R.H.; Nasehzadeh, M.; Hanson, J.; Woldemariam, Y. Medium-term seed storage of 50 genera of forage legumes and evidence-based genebank monitoring intervals. *Genet. Resour. Crop Evol.* **2018**, *65*, 207–623. [CrossRef]
42. Ellis, R.H.; Nasehzadeh, M.; Hanson, J.; Ndiwa, N.; Woldemariam, Y. Medium-term seed storage of diverse genera of forage grasses, evidence-based genebank monitoring intervals, and regeneration standards. *Genet. Resour. Crop. Evol.* **2019**, *66*, 723–734. [CrossRef]
43. Brüning, R.S.; Oppermann, M.; Willner, E.; Börner, A.; Nagel, M. Application of probit analysis and neural networks to predict seed longevity. In Proceedings of the ISTA Seed Symposium of the 32nd ISTA Congress, Hyderabad, India, 26–28 June 2019; pp. 2–16.
44. Roberts, E.H. Predicting the storage life of seeds. *Seed Sci. Technol.* **1973**, *1*, 499–514.
45. Ellis, R.H.; Roberts, E.H. Improved equations for the prediction of seed longevity. *Ann. Bot.* **1980**, *45*, 13–30. [CrossRef]
46. Dickie, J.B.; Ellis, R.H.; Kraak, H.L.; Ryder, K.; Tompsett, P.B. Temperature and seed storage longevity. *Ann. Bot.* **1990**, *65*, 197–204. [CrossRef]
47. Hay, F.R.; Mead, A.; Manger, K.; Wilson, F.J. One-step analysis of seed storage data and the longevity of *Arabidopsis thaliana* seeds. *J. Exp. Bot.* **2003**, *54*, 737–746. [CrossRef]
48. Hay, F.R.; Whitehouse, K.J. Rethinking the approach to viability monitoring in seed genebanks. *Conserv. Physiol.* **2017**, *5*. [CrossRef]
49. Lee, J.S.; Velasco-Punzalan, M.; Pacleb, M.; Valdez, R.; Kretzschmar, T.; McNally, K.L.; Ismail, A.M.; Sta Cruz, P.C.; Sackville Hamilton, N.R.; Hay, F.R. Variation in seed longevity among diverse Indica rice varieties. *Ann. Bot.* **2019**, *124*, 447–460. [CrossRef]
50. Whitehouse, K.J.; Hay, F.R.; Ellis, R.H. Improvement in rice seed storage longevity from high-temperature drying is a consistent positive function of harvest moisture content above a critical value. *Seed Sci. Res.* **2018**, *28*, 332–339. [CrossRef]
51. Butler, L.H.; Hay, F.R.; Ellis, R.H.; Smith, R.D.; Murray, T.B. Priming and re-drying improve the survival of mature seeds of *Digitalis purpurea* during storage. *Ann. Bot.* **2009**, *103*, 1261–1270. [CrossRef]
52. Whitehouse, K.J.; Hay, F.R.; Ellis, R.H. Increases in desiccation-phase developing rice seeds: Response to high-temperature drying depends on harvest moisture content. *Ann. Bot.* **2015**, *116*, 247–259. [CrossRef]
53. Ellis, R.H.; Hong, T.D. Temperature sensitivity of the low-moisture-content limit to negative seed longevity–moisture content relationships in hermetic storage. *Ann. Bot.* **2006**, *97*, 785–791. [CrossRef]
54. Kameswara Rao, N.; Dulloo, M.E.; Engels, J.M.M. A review of factors that influence the production of quality seed for long-term conservation in genebanks. *Genet. Resour. Crop. Evol.* **2017**, *64*, 1061–1074. [CrossRef]
55. Ellis, R.H.; Hong, T.D.; Roberts, E.H.; Tao, K.-L. Low moisture content limits to relations between seed longevity and moisture. *Ann. Bot.* **1990**, *65*, 493–504. [CrossRef]

56. FAO/IPGRI. *Genebank Standards*; Food and Agriculture Organization of the United Nations, Rome, International Plant Genetic Resources Institute: Rome, Italy, 1994.
57. Gold, K.; Manger, K. Measuring Seed Moisture Status Using a Hygrometer. Millennium Seed Bank Partnership Technical Information Sheet 5. Royal Botanic Gardens Kew, UK. 2014. Available online: http://brahmsonline.kew.org/Content/Projects/msbp/resources/Training/05-eRH-moisture-measurement.pdf (accessed on 15 April 2020).
58. Hung, L.Q.; Hong, T.D.; Ellis, R.H. Constant, fluctuating and effective temperature and seed longevity: A tomato (*Lycopersicon esculentum* Mill.) exemplar. *Ann. Bot.* **2001**, *88*, 465–470. [CrossRef]
59. Newton, R.; Hay, F.R.; Probert, R. A Protocol for Comparative Seed Longevity Testing. Millennium Seed Bank Partnership Technical Information Sheet 1. Royal Botanic Gardens Kew, UK. 2014. Available online: http://brahmsonline.kew.org/Content/Projects/msbp/resources/Training/01-Comparative-longevity.pdf (accessed on 15 April 2020).
60. Hay, F.R.; Valdes, R.; Lee, J.S.; Sta Cruz, P.C. Seed longevity phenotyping: Recommendations on research methodology. *J. Exp. Bot.* **2019**, *70*, 425–434. [CrossRef] [PubMed]
61. Probert, R.J.; Adams, J.; Coneybeer, J.; Crawford, A.; Hay, F. Seed quality for conservation is critically affected by pre-storage factors. *Aus. J. Bot.* **2007**, *55*, 326–335. [CrossRef]
62. Merritt, D.J.; Martyn, A.J.; Ainsley, P.; Young, R.E.; Seed, L.U.; Thorpe, M.; Hay, F.R.; Commander, L.E.; Shackelford, N.; Offord, C.A.; et al. A continental-scale study of seed lifespan in storage reveals seed, plant, and environmental traits associated with longevity. *Bio. Conserv.* **2014**, *23*, 1081–1104. [CrossRef]
63. Mondoni, A.; Probert, R.J.; Rossi, G.; Vegini, E.; Hay, F.R. Seeds of alpine plants are short lived: Implications for long-term conservation. *Ann. Bot.* **2011**, *107*, 171–179. [CrossRef]
64. Hay, F.R.; Smith, R.D.; Ellis, R.H.; Butler, L.H. Developmental changes in the germinability, desiccation tolerance, hardseededness, and longevity of individual seeds of *Trifolium ambiguum*. *Ann. Bot.* **2010**, *105*, 1035–1052. [CrossRef]
65. Leprince, O.; Pellizzaro, A.; Berriri, S.; Buitink, J. Late seed maturation: Drying without drying. *J. Exp. Bot.* **2017**, *68*, 827–841.
66. Pereira Lima, J.J.; Buitink, J.; Lalanne, D.; Rossi, R.F.; Pelletier, S.; da Silva, E.A.A.; Leprince, O. Molecular characterisation of the acquisition of longevity during seed maturation in soybean. *PLoS ONE* **2017**, *12*. [CrossRef]
67. Basso, D.P.; Hoshino-Bezerra, A.A.; Sartori, M.M.P.; Buitink, J.; Leprince, O.; de Silva, E.E.A. Late seed maturation improves the preservation of seedling emergence during storage in soybean. *J. Seed Sci.* **2014**, *40*, 185–192. [CrossRef]
68. Ellis, R.H.; Yadav, G. Effect of simulated rainfall during wheat seed development and maturation on subsequent seed longevity is reversible. *Seed Sci. Res.* **2016**, *26*, 67–76. [CrossRef]
69. Ellis, R.H.; Pieta Filho, C. The development of seed quality in spring and winter cultivars of barley and wheat. *Seed Sci. Res.* **1992**, *2*, 9–15. [CrossRef]
70. Hong, T.D.; Gedebo, A.; Ellis, R.H. Accumulation of sugars during the onset and development of desiccation tolerance in immature seeds of Norway maple (*Acer platanoides* L.) stored moist. *Seed Sci. Res.* **2000**, *10*, 147–152. [CrossRef]
71. Butler, L.H.; Hay, F.R.; Ellis, R.H.; Smith, R.D. Post-abscission pre-dispersal seeds of *Digitalis purpurea* L. remain in a developmental state that is not terminated by desiccation *ex planta*. *Ann. Bot.* **2009**, *103*, 785–794. [CrossRef]
72. Hay, F.R.; Smith, R.D. Seed maturity: When to collect seeds from wild plants. In *Seed Conservation. Turning Science into Practice*; Smith, R.D., Dickie, J.B., Linington, S.H., Pritchard, H.W., Probert, R.J., Eds.; The Royal Botanic Gardens Kew: Richmond, UK, 2003; pp. 97–133.
73. Hay, F.R.; Timple, S.; van Duijn, B. Can chlorophyll fluorescence be used to determine the optimal time to harvest rice seeds for long-term genebank storage? *Seed Sci. Res.* **2015**, *25*, 321–334. [CrossRef]
74. Hay, F.R.; Klin, J.; Probert, R.J. Can a post-harvest ripening treatment extend the longevity of *Rhododendron* L. seeds? *Sci. Hortic.* **2006**, *111*, 80–83. [CrossRef]
75. Hay, F.R.; Probert, R.J. Collecting and handling seeds in the field. In *Collecting Plant Genetic Diversity: Technical Guidelines–2011 Update*; Guarino, L., Goldberg, E., Eds.; Bioversity International: Rome, Italy, 2011.
76. Whitehouse, K.J.; Hay, F.R.; Ellis, R.H. High-temperature stress during drying improves subsequent rice (*Oryza sativa* L.) seed longevity. *Seed Sci. Res.* **2017**, *27*, 281–291. [CrossRef]

77. Whitehouse, K.J.; Owoborode, O.F.; Adebayo, O.O.; Oyatomi, O.A.; Olaniyan, A.B.; Abberton, M.T.; Hay, F.R. Further evidence that the genebank standards for drying orthodox seeds may not be optimal for subsequent seed longevity. *Biopreserv. Biobank.* **2018**, *16*, 327–336. [CrossRef]
78. Radawan, A.; Hara, M.; Kleinwächter Selmar, D. Dehydrin expression in seeds and maturation drying: A paradigm change. *Plant Biol.* **2014**, *16*, 853–855. [CrossRef]
79. Nellist, M.E. Safe drying temperatures for seed grain. In *Seed Production*; Hebblewaite, P.D., Ed.; Butterworth: London, UK, 1980; pp. 371–388.
80. Cromarty, A.S.; Ellis, R.H.; Roberts, E.H. *The Design of Seed Storage Facilities for Genetic Conservation*; International Board for Plant Genetic Resources: Rome, Italy, 1982.
81. McDonald, M.B.; Copeland, L. *Seed Production: Principles and Practices*; Chapman & Hall: New York, NY, USA, 1997.
82. Sileshi, G.W. A critique of current trends in the statistical analysis of seed germination and viability data. *Seed Sci. Res.* **2012**, *22*, 145–159. [CrossRef]
83. Hay, F.R.; Mead, A.; Bloomberg, M. Modelling seed germination in response to continuous variables: Use and limitations of probit analysis and alternative approaches. *Seed Sci. Res.* **2014**, *24*, 165–186. [CrossRef]
84. Newton, R.J.; Hay, F.R.; Ellis, R.H. Temporal patterns of seed germination in early spring-flowering temperate woodland geophytes are modified by warming. *Ann. Bot.* **2020**. [CrossRef] [PubMed]
85. Gianinetti, A. Basic features of the analysis of germination data with generalized linear mixed models. *Data* **2020**, *5*, 6. [CrossRef]
86. Gianinetti, A.; Cohn, M.A. Seed dormancy in red rice. XII: Population-based analysis of dry- afterripening with a hydrotime model. *Seed Sci. Res.* **2007**, *17*, 253–271. [CrossRef]
87. Bradford, K.J.; Benech-Arnold, R.L.; Côme, D.; Corbineau, F. Quantifying the sensitivity of barley seed germination to oxygen, abscisic acid, and gibberellin using a population-based threshold model. *J. Exp. Bot.* **2008**, *59*, 335–347. [CrossRef]
88. Joosen, R.V.L.; Kodde, J.; Willems, L.A.J.; Ligerink, W.; van der Plas, L.H.W.; Hilhorst, H.W.M. GERMINATOR: A software package for high-throughput scoring and curve fitting of Arabidopsis seed germination. *Plant J.* **2010**, *62*, 148–159. [CrossRef]
89. Genebank Reviews. Available online: https://www.genebanks.org/resources/genebank-reviews/ (accessed on 15 April 2020).

plants

MDPI

Review

Challenges and Prospects for the Conservation of Crop Genetic Resources in Field Genebanks, in In Vitro Collections and/or in Liquid Nitrogen

Bart Panis [1,2,*,†], **Manuela Nagel** [3,†] **and Ines Van den houwe** [1,2,†]

1 Alliance of Bioversity International and CIAT, c/o KU Leuven, Willem de Croylaan 42, P.O. Box 2455, 3001 Leuven, Belgium; i.vandenhouwe@cgiar.org
2 Department of Biosystems, KU Leuven, Willem de Croylaan 42, 3001 Leuven, Belgium
3 Leibniz Institute of Plant Genetics and Crop Plant Research (IPK), OT Gatersleben, Corrensstrasse 3, D-06466 Seeland, Germany; nagel@ipk-gatersleben.de
* Correspondence: b.panis@cgiar.org
† All authors contributed equally.

Received: 15 October 2020; Accepted: 18 November 2020; Published: 24 November 2020

Abstract: The conservation of crop genetic resources, including their wild relatives, is of utmost importance for the future of mankind. Most crops produce orthodox seeds and can, therefore, be stored in seed genebanks. However, this is not an option for crops and species that produce recalcitrant (non-storable) seeds such as cacao, coffee and avocado, for crops that do not produce seeds at all; therefore, they are inevitably vegetatively propagated such as bananas, or crops that are predominantly clonally propagated as their seeds are not true to type, such as potato, cassava and many fruit trees. Field, in vitro and cryopreserved collections provide an alternative in such cases. In this paper, an overview is given on how to manage and setup a field, in vitro and cryopreserved collections, as well as advantages and associated problems taking into account the practical, financial and safety issues in the long-term. In addition, the need for identification of unique accessions and elimination of duplicates is discussed. The different conservation methods are illustrated with practical examples and experiences from national and international genebanks. Finally, the importance of establishing safe and long-term conservation methods and associated backup possibilities is highlighted in the frame of the global COVID-19 pandemic.

Keywords: clonal crops; collection management; cryobiotechnology; cryopreservation; field collections; field maintenance; germplasm storage; in vitro conservation; recalcitrant seeds

1. Introduction

In the course of crop domestication, many plants have been selected for quantity and/or quality of their seed, while some have been cultivated for their roots, tubers, fruits, stems and leaves. Plant genetic resources for food and agriculture (PGRFA) are of strategic importance to ensure sustainable crop production [1], nutritious food and food security for humans and to enhance economic prosperity of the present and future generations. They comprise the sum of genes, gene combinations or genotypes which serve as a reservoir for direct use in food production systems and for breeding new varieties [2].

Since the beginning of agriculture, selection of plants and seeds for sowing, growing, harvest and storage gave rise to locally adapted varieties, so-called "landraces", that reveal specific variations of morphological and yield characteristics and quality traits. In the mid-19th century, the rediscovery of Gregor Mendel's work and the introduction of breeding schemes resulted in the development of high-yielding and more stress-tolerant varieties leading to higher crop yields. This laid the foundation for the green revolution taking place in the middle of the last century bringing about increased agricultural production to feed the exponentially growing world population. However, the expansion of industrial mono-cropping with the replacement of landraces by modern breeding varieties has caused the loss of 75% of plant genetic diversity, with more than 90% of crop varieties having disappeared from farmers' fields [3]. In this era of global environmental problems, climate change, and booming population growth, it is of paramount importance that the remaining crop genetic resources are kept available to sustain the agricultural production systems, to feed the world population a healthy diet and to tackle future demanding challenges [4].

Since the 16th century, botanical gardens have collected and preserved a variation of more than 80,000 plant species in about 3400 gardens all over the world [5]. They primarily have an interest in conserving the widest possible plant diversity and crop wild relatives that can be an important source material for breeders. PGRFA are conserved ex situ in specialized repositories, often termed genebanks that have been established since the mid-20th century. In contrast to most botanical gardens, genebanks focus on both intra- and inter-specific crop diversity. There are more than 17,000 national, regional and international institutes and organizations dealing with the conservation and sustainable use of PGRFA [5]. Currently, 711 gene banks and 16 international/regional centers in 90 countries retain more than 5.4 million accessions from over 7051 genera, mainly focusing their conservation efforts on crop species, including landraces and crop wild relatives, breeding materials and cultivars [6].

Most of the major food crops produce orthodox seeds that tolerate intense dehydration and low temperatures, thus seed storage under dry and cool conditions is naturally the most widely adopted method for long-term ex situ conservation at relatively low costs. About 45% of the accessions stored as seeds are cereals, i.e., wheat (*Triticum aestivum*), triticale (*Triticum secale*), rice (*Oryza sativa*), oat (*Avena sativa*), rye (Secale cereale), barley (*Hordeum vulgare*), maize (*Zea mays*) and sorghum (*Sorghum bicolor*), followed by food legumes (15%), forages (9%) and vegetables (7%) [5].

In contrast, a large number of food crops are not storable through seeds and thus need different conservation approaches [7]. This category of plants consists of important species that produce desiccation sensitive, recalcitrant or intermediate seeds, such as coconut (*Cocos nucifera*), cacao (*Theobroma cacao*), avocado (*Persea americana*) and citrus (*Citrus* spp.) and species that are seedless such as edible banana (*Musa* spp.) and garlic (*Allium sativum*). Species including yucca (*Yucca* sp.) and bamboo (*Bambuseae* sp.) that have long life cycles and take years or decades to reproduce also fall into this category. Other species that produce orthodox seeds but require the conservation of particular gene combinations or genotypes, such as root and tuber crops, notably potato (*Solanum tuberosum*), cassava (*Manihot esculenta*), yam (*Dioscorea* spp.), taro (*Colocasia esculenta*) and several fruit and nut trees are also included. These crops are propagated vegetatively, and each genotype needs to be maintained as a clone.

Two main ex situ conservation approaches can be distinguished for these crops: the conservation of plants in field genebanks and the maintenance of propagules in tissue culture, either (i) as active growing cultures in short- and medium-term storage (i.e., in vitro storage), or (ii) in frozen state at ultra-low temperature in liquid nitrogen for long-term storage (cryopreservation). These approaches, the challenges entailed, and prospects offered to secure crop diversity ex situ will be discussed in detail in this chapter and are depicted in Figure 1.

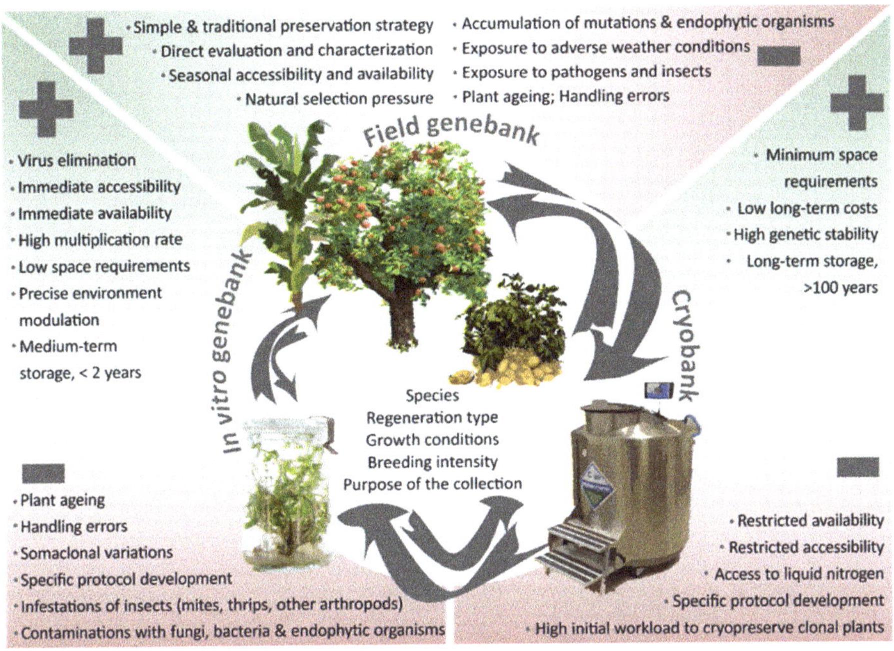

Figure 1. Pros and cons of storing crop genetic resources in field genebanks, in vitro and through cryopreservation. Arrows indicate the potential source material and target approach to maintain plant genetic resources for food and agriculture (PGRFA) and can be specific for each plant species.

2. Field Genebanks

Clonally propagated crops and their wild relatives belong to some 34 plant families, including herbs, shrubs, trees and vines [8]. Agronomically important genotypes that were selected over centuries for their specific properties can only be conserved in a vegetative mode. Consequently, the simplest and most traditional way to establish a collection is to gather specific genotypes from farmers' fields, gardens or in the wild and then grow them in the field genebanks where they continue to grow when maintained appropriately. Even under the highest standards of management, germplasm maintained in the field can deteriorate due to a wide variety of climate conditions, ageing of the plants, diseases and pests, hence the need for timely regeneration. For example, depending on the rootstock and orchard conditions, apple (*Malus* sp.) trees may need to be repropagated periodically after 25–50 years [9].

Some of the largest collections were established at the beginning of the 20th century and are based on the efforts of passionate geneticists and plant explorers such as Nikolai I. Vavilov, Frank N. Meyer [10] and Hans Stubbe among others [11]. In the 1970s, the International Board for Plant Genetic Resources (IBPGR) promoted and sponsored numerous collection missions. To preserve clonal material, 23 field genebanks of nine major crops were established worldwide [12]. As such, approximately 400,000 accessions are currently held in field genebanks of international, regional and national authorities. A recent study commissioned by the Alliance of Bioversity International and the International Center for Tropical Agriculture (CIAT), the International Potato Center (CIP) and the Global Crop Diversity Trust [13] showed that of the 20 institutions surveyed, the vast majority of the clonal plant material is kept in the field. Worldwide, major genebanks maintain potato (98,285 accessions), apple (59,922), cassava (36,529), citrus (36,410), sweet potato (*Ipomoea batatas*, 35,478), coffee (*Coffea* spp., 30,483) and cacao (23,107). The largest field genebanks are located in the USA (potato, sweet potato and apple), Japan (apple, citrus and sweet potato), Russia (potato and apple) and Brazil (citrus and coffee). In addition, there are numerous smaller collections, such as for grape vine (*Vitis vinifera* L.), garlic

(Figure 2, see Box 1), Jerusalem artichoke (*Helianthus tuberosus* L.), and Andean root crops, which are of high value as luxury foods and condiments or are of regional or religious significance.

2.1. Management of Field Genebanks

The specific needs of a crop with respect to growth requirements and its multiplication cycle ultimately determine the field conditions and the structural design of the field genebanks. Propagules of some annual crops, such as potato, shallot or yams, are cultivated by up to 10 plants in the field and require good agricultural practices including crop rotation. After harvest, these propagules must be kept under suitable storage conditions until the next growing season [14]. By contrast, woody crops, such as apple, pear, coffee and grapevine are grown at the same locations for many years, often with two plants per accession. Some of these crops require budding or grafting to rootstock resistant to root nematodes, layering or rooting stem cuttings. For example, the US Department of Agriculture—Agricultural Research Service (USDA-ARS) National Plant Germplasm System (NPGS) maintains 5004 apple accessions in the field and 1603 *M.* × *domestica* seed accessions in Geneva, NY, USA. Due to the large number, about 3100 field accessions are only represented by a single tree [15]. By comparison, the JKI Dresden-Pillnitz, but also universities, governmental institutes, communes, non-governmental organizations and private individuals are responsible for the maintenance of fruit collections in Germany. These collections include landraces that partly date back to the 12th century and some of these are at risk. Therefore, the German Fruit Genebank was launched in 2007 and consists of six fruit-specific networks (apple, cherry and plum (both *Prunus* spp.), berries (*Rubus* spp.), strawberry (*Fragaria* spp.), pear). For apple, an expert group selected 743 unique accessions that were duplicated and are represented by two trees each in at least two sites [16]. Although, establishing networks requires additional efforts, the responsibility for maintaining such valuable resources is shared and the partners mutually benefit from the concerted expertise and the security status of the collections.

Depending on the collection strategy and the breeding programs, both annual and perennial crops can vary in proportions of landraces, breeding lines and cultivars and are used as different types of donors for plant breeding. CIP, for example, maintains the largest collection of 4487 potatoes landraces at Huancayo, Peru at a 3200 m elevation. [17]. To increase the productivity and, hence, farmer's incomes in Africa, Asia and Latin America, CIP and partners developed training guides for the positive selection of propagules of landraces that have no visible symptoms of diseases or abiotic stress [18]. By comparison, at IPK (Leibniz Institute of Plant Genetics and Crop Plant Research), in addition to 3300 landraces, also 2900 wild potato accessions are maintained as orthodox seeds. As some potato species are self-incompatible, seeds are propagated as a population in the greenhouse and are screened for resistances, to late blight (*Phytophthora infestans*) on tubers or pale potato cyst nematodes (*Globodera pallida*) [19]. Selected donors are further used to develop introgression lines that can be later used by the breeding industry. In other words, breeding intensity and objectives determine the composition and have a strong impact on the management system of the collections.

Figure 2. *Allium* field collection at IPK, Germany.

Box 1. Field collection of Allium genetic resources at IPK, Germany.

The IPK maintains one of the largest *Allium* collections worldwide and comprises the Taxonomic Reference Collection of 1300 accessions and 287 species and the *Allium* Crop Collection covering 1400 accessions of 76 species (Figure 2). About 2100 accessions are permanently maintained in the field because of (i) their inability to form seeds, i.e., in garlic and shallots (*Allium cepa* var. *aggregatum*), (ii) the presence of heterozygous seeds, i.e., in onion (*Allium cepa* var. *cepa*), or (iii) the traditional breeding strategy as clonal varieties as in the case of some ornamental *Allium* hybrids [20].

Allium accessions, except shallot and onion, are maintained as a permanent crop at the same site and 3 to 6 plants are planted within a 2 m^2 plot. These accessions require permanent management including regular identification, weeding, phytosanitary treatments and seed harvest to avoid the establishment of different accessions or hybrids within the same plots. The soil quality and the continental climate in Central Germany, including mild winters, reduce the risk of pest and diseases and provide optimum growth conditions for a longer term. However, each 5 or 6 years, the Allium gardens are replanted to overcome the problem of soil exhaustion [21].

Special attention must be given to 425 garlic accessions and 82 shallot accessions that are partially replanted every year. In autumn, to avoid infections of nematodes and wireworms, cloves are treated with insecticides and planted in the field. At the end of July of the following year, accessions develop cloves (lateral bulbs) and in some cases bolt and develop bulbils which dry up. These can be used for germplasm distribution. For introduction into tissue culture or cryopreservation, bulbs and bulbils require a further after-ripening period of up to 2 months until the physiological dormancy [22] is broken. During this period, storage of the bulbs and, in the case of garlic also the bulbils, is conducted at between 4 and 10 °C [23].

Various strategies have been developed to support the field genetic resources of garlic and shallot. About 30 years ago, in vitro slow-growth maintenance at a temperature between 2 and 10 °C was initiated for about 700 accessions. As such, cultures can be kept for up to 12 months without subculturing. However, after many subculture cycles, the accumulation of microbes delimitates the in vitro storage. Therefore, the number of in vitro accessions was reduced to about 25 accessions. Nowadays, the introduction of garlic and shallot accessions into in vitro conditions is only carried out as preparatory step for cryopreservation. Since the PVS3 vitrification approach has been successfully developed for garlic and shallot [24], more than 210 accessions (Table 1) were successfully cryopreserved. To protect further the allelic diversity of garlic, the Research Institute of Crop Production (CRI) in the Czech Republic, the Research Institute of Horticulture (RIH) in Poland and the IPK established a European Core Collection within the frame of "A European Genebank Integrated System" (AEGIS) [25].

2.2. *Advantage of Field Collections: Characterization and Evaluation*

The propagation of clonal plants in the field is the conventional method for the preservation of genetic material and traditional knowledge of the farming system [26]. In areas where plants are historically grown, the cultivation can be carried out by local farmers. The sites are usually well-established with pest control and forecasting models used [27]. Their exposure to natural conditions allows a limited selection according to environmental conditions and a competition between plant propagules [8]. A tremendous benefit of holding field collections is that images and voucher specimens can be immediately assessed and made available online [27–29].

Descriptors have been elaborated for a wide range of crop species [27,29,30]. The descriptors and the list of Multi-Crop Passport Descriptors provide an international format that facilitates comparisons among and within collections [14]. The data often follow FAIR—findable, accessible, interoperable and reusable—principles [29] and well-characterized material can be further distributed and evaluated and made available to breeding programs. At IPK in 2019, of 3300 maintained potato cultivars, about 450 accessions were propagated in the field and of these 1027 sub-samples were distributed. A main advantage is that the material is immediately available for evaluation and distribution. For example, at a field-based germplasm collection in New Zealand genomic DNA was extracted and screened for the presence of TG689 and 57R haplotypes linked to the *H1* gene. These haplotypes act as potential predictors for the resistance against a pathotype of potato cyst nematode (*Globodera rostochiensis*), an economically important pest [31]. Similarly, banana germplasm was screened for the resistance against the *Fusarium oxysporum* f. sp. *cubense* tropical race 4 (*Foc TR4*) that has seriously threatened global banana production. From 129 evaluated accessions, 10 were highly resistant to the virus [32]. The readily available data and material can thus have a strong impact on the material selected for breeding.

2.3. *Problems Associated with Field Collections*

In field collections, plants are exposed to a natural environment, which can include unfavorable climatic conditions, such as drought, heat or frost, similar to crop production fields. Additionally, pests and diseases can threaten the material, especially the less adapted or susceptible accessions. One of the important apple diseases, fire blight, caused by the bacterium *Erwinia amylovora*, can severely damage or even eradicate susceptible apple accession [33,34]. In 2013, the *Allium* field collection at IPK was challenged with a serious infestation of larval stage click beetles (*Elateridae*), also known as wireworms. Of 2150 accessions, 52 accessions were completely lost and 73 could be recovered by replanting and duplication. Furthermore, depending on the plant species and their life cycles, clonal collections must be rejuvenated periodically. The Tropical Agricultural Research and Higher Education Center (CATIE) in Costa Rica hosts one of the largest collections of *Coffea arabica* L. and maintains plants that were planted in the 1970s. In the year 2000, a project to rejuvenate the collection was initiated and is still ongoing [14,35]. Compared to reproduction by seeds, vegetatively propagated populations are also susceptible to the accumulation of viruses, bacteria and fungi that might be similar to Muller's ratchet [8]. As a consequence, the maintenance of field collections requires substantial efforts in terms of agricultural measures such as manual selection, evaluation, mechanical weed control, rejuvenation and specific disease control. Therefore, the number of replicates is often limited to between 5 and 10 for cassava, 10 and 12 for sweet potato, 2 and 9 for garlic, 1 and 3 for trees and shrubs and 3 and 20 for bananas [14]. However, spatially separated replicate plants (i.e., trees and vines) can aid in minimizing loss of accessions by pests or mismanagement. However, even provided everything is handled with the utmost care and state-of-the-art technologies, human errors can easily happen and accessions can be mixed up. A genotyping study of 250 accessions of the CIP potato collection showed that only about 80% of the collection was still comparable with voucher mother plants which arrived at CIP about 30 years before [36]. At Arabidopsis stock centers, it was estimated that about 3 to 14% of the materials are misidentified with most errors caused by incorrect labeling [37]. There is also the possibility that genetic lines segregate and spontaneous mutations occur. Especially under environmental stress, whole-genome and whole-methylome sequencing revealed that the Arabidopsis lineages accumulate 100% more genetic mutations and epigenetic modifications under stress compared to non-stress conditions [38]. This underlines the importance of cautious evaluation, involving trained staff, and the application of complementary preservation methods such as in vitro storage and cryopreservation.

3. In Vitro Collections

In vitro culture (or tissue culture) of plants is a biotechnological technique in which plant parts are isolated from in vivo plants, disinfected to free the explants from bacteria and fungi and transferred onto well-defined and sterile tissue culture media that provides the plant tissue with the necessary nutrients for growth and multiplication. In a relatively small space, the environment can be controlled precisely and plant growth can be easily observed and manipulated. In vitro approaches are commonly used for large-scale micro-propagation, reproduction purposes including embryo rescue, ploidy manipulations, protoplast fusions and somatic embryogenesis and are appropriate tools for short- and mid-term storage of plant genetic resources. Although the feasibility of using in vitro culture methods for plant genetic resources conservation was already known in the 1970s, it was only in the 1980s that the International Board for Plant Genetic Resources (IBPGR) established a working group of specialists to investigate the critical aspects of in vitro plant conservation [39,40]. Since then, in vitro collections have been setup for many vegetatively propagated crops. Additionally, Genebank Standards for PGRFA maintained in vitro were developed [1], forming the benchmark for establishing standard operating procedures and quality management systems to ensure effective, safe and efficient conservation of these genetic resources. Nowadays, in vitro collections for PGRFA comprise potato (9700 accessions), cassava (8700), sweet potato (6400), yam (3200), banana (2000) and taro (1200) [40–42]. The largest collections are maintained by international organizations such as Bioversity International, the International Center for Tropical Agriculture (CIAT), CIP, the International Institute of Tropical Agriculture (IITA) and in national institutes such as Brazilian Agricultural Research Corporation (EMBRAPA) in Brazil, CRI in Czech Republic, the IPK in Germany and the USDA-NPGS in the USA [13,42].

3.1. Setting Up In Vitro Collections

At tissue culture facilities, plant tissues are maintained in specific and aseptic growth conditions. To culture explants in vitro, typically tissue culture media are used that contain water, macro and micro nutrients (salts), a gelling agent, plant growth regulators (plant hormones) and sugars. The supplement of sugars as source of energy and building blocks is essential because, unlike in vivo plants, in vitro plants lack the ability to photosynthesize effectively. Explants are kept in a culture room with controlled temperature and light regimes. Since tissue culture material continuously grows and undergoes ageing, plants regularly need to be trimmed, divided in separate propagules and transferred to new culture media. Explant survival and vitality of in vitro plants depend on media composition, temperature and light intensity and further factors such as light spectrum, vessel type and size, number of explants and gas exchange rate [43–45], all of which can affect the growth of the plants.

The goals of applying slow-growth conditions in plant collections is to reduce frequency of subculture events that are labor (and thus cost) intensive and to minimize the risk of loss of accessions due to handling errors and genetic instability induced by the tissue culture environment. Under optimal growth conditions, subculture frequencies range from one to three months, whereas at slow growth conditions, the subculture period can be from one to two years. Three so-called Medium Term Storage (MTS) approaches can be distinguished: physical growth limitation, chemical growth limitation and nutrient limitation [40]; all aimed to reduce the metabolic activity of the plantlets. Physical growth limitation is often achieved by lowering temperature and often combined by applying low light intensities. Cold-tolerant species such as garlic, potato and most *Mentha* species can be kept viable at temperatures from 0 to 5 °C and 2 to 4 μmol m^{-2} s^{-1} for up to 24 months [46], whereas cold sensitive species such as many tropical plants should be stored at relatively high temperatures. Examples are *Musa* [47] (Figure 3, see Box 2) and sweet potato [48] both requiring a temperature of 15 °C or higher;

for pineapple this is 21 °C [49]. Such higher storage temperatures result in shorter sub-culture intervals (six to twelve months). Nutritional growth limitation decreases the supply of carbon and inorganic nutrients whereas chemical growth limitation involves the application of osmotically active agents such as mannitol, sorbitol, polyethylene glycol (PEG) or growth retardants such as abscisic acid (ABA) and hydrazides. In potato, for example, extended periods in MTS can be achieved by different protocols which either apply Murashige and Skoog (MS) medium, sucrose and the growth retardant ancymidol at 6 °C resulting in storage periods of 12 months [50]; MS and mannitol at 6 °C for storage periods of 16 months [51] or MS, sucrose, mannitol at 6 °C for storage periods of 30 months [52]. This shows that often combinations and variations of different slow-growth approaches need to be adapted for different species. An overview of growth limiting protocols is summarized by Chauhan, Singh and Quraishi [53].

3.2. Advantages of In Vitro Collection

An important benefit of in vitro collections, compared to field collections, is that the plant material is free of most pests and diseases. An exception is in the case of viruses, which can easily be transmitted through tissue culture, often symptomless. As international quarantine restrictions are very stringent for some species, a prerequisite for germplasm to be internationally distributed is that the sample must be healthy and free of harmful pathogens. Several in vitro tissue culture techniques, such as thermotherapy, chemotherapy and meristem tip culture, can be applied to eradicate the viruses [42]. In addition, other eradication methods such as electrotherapy [54] and cryotherapy [55] have been effective in several crops including grape, potato, sweet potato and banana. Cryotherapy using vitrification proved to eliminate effectively Cucumber Mosaic Virus (CMV) and Banana Streak Virus (BSV) from in vitro meristematic tissues of the dessert banana cv. Williams (AAA, Cavendish subgroup) with 30% and 90% of the regenerated plants being CMV- and BSV-free, respectively [56]. Often, combinations of eradication techniques are used to increase their effectiveness.

In contrast to field gene banks, plant materials kept at MTS conditions, also called in vitro "active" collections can be delivered on a year-round basis. For most crops, the availability of samples from a field gene bank is bound to the development stage of the plant and season. In garlic, for example, cloves and bulbils follow a seasonal development and are only available and highly viable between July and February in the northern hemisphere. Moreover, the supply of material from the field can be limited because of the often poor plant propagation rates. For instance, banana has an annual field multiplication rate through suckering as low as five to 20 suckers per year depending on the clone, age of the plant and climatic and culture conditions [57,58]. Furthermore, the international and inter-regional movement of propagules from the field such as suckers of banana, corm pieces or small corms of cocoyam or taro, tubers of oca, ulluco or other plant parts such as wood cuttings of apple or blueberry, involves the risk of transmission of harmful pest and diseases. Strict quarantine measure must thus be taken. For crops with insect- or mite-transmitted viruses, it is therefore useful to maintain virus-free stocks in screenhouses although this implies higher cost for making materials available to users. In general, it is recommended that vegetative material is distributed to requesters as tissue cultures, derived from conventional vegetative propagation material, and indexed free from pathogens. Moreover, tissue culture using multiplication-inducing growth regulators, often cytokinins, allows rapid and mass propagation of plants. Hence, hundreds of propagules can be produced within a few months.

3.3. Problems Associated with In Vitro Collections

The artificial tissue culture environment can be stressful for plant cells and thus poses a challenge when plant germplasm needs to be maintained for extended periods of time. This can result in so-called somaclonal variations that are changes in the DNA sequences that are not derived from recombination [59]. It can have epigenetic origin which reflects the adaptation process of cells to a different environment [60]. Different factors have an effect on the genetic mutation rate, among them exposure of the chromosomal DNA to different chemicals present in the culture medium and the survival of the variants in a non-selective tissue culture environment. Although the mutagenic activity of plant growth regulators is debated, it is generally accepted that stimulated rapid disorganised growth induces somaclonal variation [59]. This has been shown for thiadiazuron (TDZ) in bananas [61] and the synthetic auxin 2,4-dichlorophenoxyacetic acid (2,4-D) in *Curcuma* aromatic plants [62]. Bairu, et al. [63] observed that, with increasing subculture events, the frequency of somaclonal variation of micropropagated bananas increased, sometimes as high as 72%. However, in potato, random DNA methylation changes occurred only in individual samples at a very low frequency for 3 of 469 markers [64]. Furthermore, no differences in genetic and epigenetic changes have been reported for in vitro propagated pea clones that have been maintained for more than 24 years [65], indicating that somaclonal variation is strongly species-dependent. A general principle is that, the more disorganized the cultured tissues, the higher chance of mutations [59,66]. Therefore, for slow-growth storage, organized tissue systems, such as shoot cultures, are preferred over non organized tissues like callus and suspension cells and the in vitro storage is preferably conducted on hormone-free culture media, as has been used for *Mentha* species [67].

Other constraints in tissue culture collections are the occurrence of cellular ageing and senescence during prolonged cultivation. In eight-year-old peach palm (*Bactris gasipaes*) cultures, processes of plant cell death and senescence were visible through nuclear condensation, cell degradation and the development of large intercellular spaces [68]. The effect of cellular ageing or senescence may appear in parallel with slow growing endophytic microbes that can accumulate over time. Bacteria or fungi are known to colonize almost all healthy plant tissues without necessarily damaging the host or eliciting any defense responses [69]. For example, the in vitro collection of *Allium* species maintained at slow-growth conditions suffers from the presence of endophytes and the storage duration of individual clones is considerably shortened [70]. At the banana collections at the Bioversity Genebank, the tissue cultures are regularly tested for the presence of slow growing (endophytic) microorganisms using a broad spectrum bacterial growth medium. When this method was first applied, the presence of cryptic contaminants in 5% of the stored germplasm was revealed [71]. Although some endopyhtic strains are known to show plant growth promoting behavior such as PsJN in potato [72], the majority is detrimental for in vitro cultures in case they accumulate.

Aside from the challenges that are related to intrinsic processes linked to in vitro culture, external factors can also result in a complete loss of cultures, such as malfunctions in air-conditioning and lighting systems. Moreover, human errors such as accidental microbial contaminations, physical mixing of accession samples, but also documentation errors, e.g., mislabeling, or misidentification, are even more likely to cause serious problems in the operation of an in vitro genebank [73]. Furthermore, mites, thrips and other small arthropods can cause extensive fungal contaminations in tissue cultures and are difficult to eradicate [14]. Therefore, quality management systems including barcode labelling, cleaning management and regular monitoring of the stored materials should be implemented as standard procedures at slow-growth storage facilities [1]. An extra safety measure is duplicating the collection, either in vitro or in cryopreservation, and preferably in another distant location to ensure that the duplicate is properly secured. For example, the *Mentha* collection at IPK consists of two subsets that are maintained in different culture rooms one kept at 2 °C and one at 10 °C. The CIAT genebank in Colombia sends its duplicates of its cassava in vitro clones off site to CIP, Peru where they are maintained at 23 °C and under a controlled photoperiod. For *Musa* at the Bioveristy genebank, maintaining 70% of its in vitro clones in a cryopreserved base collection, a cryopreserved sample is safely duplicated in IRD France.

Figure 3. Banana in vitro collection at the International Transit Centre (ITC), Belgium.

Box 2. Collection of bananas at the Bioversity Genebank, Belgium.

The international banana collection, also known as the Bioversity International Transit Centre (ITC), Belgium, was founded in 1985. It is the largest repository of *Musa* species in the world holding more than 1600 banana accessions sourced from 38 countries. The genebank, located in a non-banana growing country, does not manage a field collection of its own, but has close links with national and regional field collections around the world, serving as a back-up repository for their accessions. The ITC conserves the widest diversity of cultivated sweet and starchy bananas (75%) belonging to 17 genome groups and 52 subgroups, a representation of the wild genepool with specimens of 34 species (16%), and a range of high yielding and disease-resistant advanced varieties (9%). As bananas are vegetatively propagated with seed fertility limited to the wild forms, ex situ conservation of banana species in the form of in vitro cultures is the most suitable approach when the material should remain available for regeneration, multiplication and international distribution.

Germplasm in the collection is stored as multiple shoot clusters obtained from individual shoot tips. Since the early days, low temperature, combined with light limitation, has been successfully applied as growth-retarding factors to reduce the frequency of subculturing considerably. In the storage room, the temperature is 16 °C and light intensity 25 μMol/m^2/s (Figure 3). It was demonstrated that under the conditions applied, storage duration is nearly one year on average for the total diversity maintained. However, large differences in transfer interval, ranging between 3 and 22 months, occur among the different genomic groups and even within the same subgroup [47].

To keep the collection alive and in good condition, the active in vitro collection requires close monitoring not only to assess the viability and need for re-culturing of accessions, but also to keep the collection free from contamination [71] which may interfere with the storage and use of the germplasm. In addition, samples of accessions maintained for more than 10 years in vitro are evaluated in the greenhouse from where rejuvenated (renewed) in vitro stocks are established. To keep the levels of somaclonal variation of the stored germplasm as low as possible, older accessions are also systematically grown in the field for an integrity check and in parallel to the field detection of changes in morphology, cytological and molecular (SSR) characterization is performed [74] to confirm the identity of the accession.

Each accession in the active collection has a number of 20 replicate cultures maintained on one single multiplication-inducing growth medium, ensuring safe storage of the clone and direct availability for distribution and use. Over the years, the gene bank has played a crucial role in the international exchange of banana germplasm.

Box 2. *Cont.*

A system for the safe movement of germplasm is in place with substantial efforts going into the testing and sanitizing of the collection from banana viruses [75]. Each day, three to four clean in vitro samples of the stored banana accessions are distributed to users somewhere in the world to underpin research, breeding and development activities. Depending on the purposes and needs of the germplasm user, the required growth stage of the material may differ: if no tissue culture capacity is available, in vitro rooted plants for direct soil acclimation can be distributed whereas users may need shoot cultures in the multiplication phase for in vitro studies or as virus-free stock for further in vitro propagation. In any case, it is recommended that recipients are provided with detailed handling instructions, the composition of the growth medium and specifications of the growth conditions. Since molecular techniques have become increasingly important in biodiversity studies, scientists and breeders have a growing interest in accessing the DNA of the widest range of genetic diversity present in collections. To meet this changing need, the Bioversity banana genebank established a collection of freeze-dried leaf tissues isolated from greenhouse plants stored at −20 °C. Leaf samples of some 850 accessions are readily available to customers, forming a low-cost alternative for DNA-banking and for distribution of living plant material that is time consuming to prepare, requires phytosanitary checks, special packaging, and fast shipping to assure arrival in good conditions.

The first report on successful shoot tip cryopreservation of *Musa* species was published in 1996 [76]. Sucrose precultured meristem clumps belonging to seven different genomic groups were rapidly frozen in liquid nitrogen, from which the shoots were regenerated. However, regeneration rates were low (ranging between 0% and 50%) and variable depending on the cultivar. A new protocol, i.e., droplet vitrification, was therefore needed (see Section 4.1.2) [77] resulting in higher and more cultivar-independent regeneration rates (between 50% and 95%). Since then and thanks to funding provided by the World Bank and the Gatsby Foundation, routine cryopreservation of clean and field characterized banana accessions kept in the in vitro collection was established. Such secure funding sources proved to be important since cryopreservation of *Musa* species is a very labor intensive and costly process, especially because hundreds of tiny meristems need to be excised under the binocular microscope. Using the droplet vitrification protocol, one skilled technical staff member can cryopreserve about 40 to 50 *Musa* accessions per year.

The ITC is a successful example of how complementary conservation approaches can be used to increase the security of collections. With samples of 1175 accessions also maintained in liquid nitrogen (Figure 4), the cryopreserved base collection serves as source for replacing materials in case the accession samples in slow growth are lost due to accidental contamination or genetic variations. This cryopreserved collection is "backed up" (duplicated) in Cryotanks held at the Institut de Recherche pour le Development (IRD), Montpellier, France.

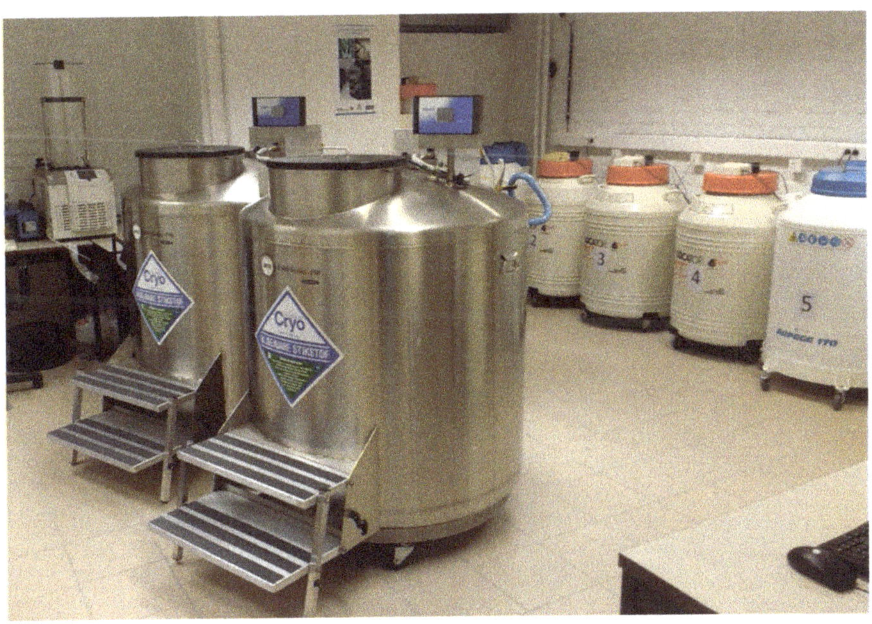

Figure 4. Cryopreservation facilities at the ITC.

4. Cryopreserved Collections

Cryopreservation (or storage of biological material at ultra-low temperatures) is the obvious solution to the above-mentioned limitations, since at these conditions metabolic, physical and chemical alterations are unlikely to occur, even after hundreds of years of storage. Usually, cryogenic storage takes place in liquid nitrogen (−196°) or its vapor phase (between −140 and −180 °C) (Figure 4). The main hurdle associated with cryopreservation is the formation of lethal ice crystals. Complete drying plant tissues, thus preventing the formation of ice crystals is not an option since the presence of water is inevitably linked with life. The only way to avoid ice crystal formation of a watery solution is by making use of the physical phase called "vitrification", i.e., the solidification of a liquid forming an amorphous "or glassy" structure. All cryopreservation procedures developed for biological materials are based on optimizing the chance for vitrification. To attain this, two conditions must be met: (i) application of ultra-rapid cooling rates, limiting the time period that an ice crystal can form before all molecules are immobilized by the ultra-low temperature, and (ii) concentrating the cell solution resulting in relatively more molecules in interference with the organization of water molecules turning into ice crystals.

4.1. Setting-Up Cryopreserved Collections

Since the first report by Akira Sakai in 1965 [78] on the survival of plant tissues exposed to liquid nitrogen, a wide variety of plant cryopreservation protocols have been established, among them dormant bud cryopreservation, classical (slow) freezing, encapsulation-dehydration, and a range of vitrifications solution-based protocols (for an overview, see [79]). Currently, dormant bud cryopreservation and droplet vitrification are commonly applied (Table 1).

Table 1. Cryopreservation methods used in world's largest crop genebanks that use cryopreservation for storage of their vegetatively propagated germplasm.

Institute	Country	Crop	Cryopreservation Method	Number of Accessions	Ref
AFOCEL	France	Elm	• Dormant bud freezing	440	[80]
Bioversity International, Leuven	Belgium	Banana	• Droplet vitrification	1100	Panis, personal communication, 2020
Crop Research Institute, Prague	Czech Republic	garlic	• Droplet vitrification	157	[81]
International Center for Tropical Agriculture (CIAT), Cali	Colombia	cassava	• Droplet vitrification • Encapsulation/dehydration	480	[80]
International Institute of Tropical Agriculture (IITA), Ibadan	Nigeria	Yam	• Droplet vitrification	27	[82]
International Potato Center (CIP), Lima	Peru	Potato	• Straw vitrification • Droplet vitrification	3264 (Situation 14 October 2020)	[83]
Julius Kühn-Institut (JKI), Institut für Züchtungsforschung an Obst, Dresden	Germany	Strawberry	• Vitrification	194	[84]
Leibniz Institute of Plant Genetics and Crop Plant Research (IPK), Gatersleben	Germany	Potato,	• Droplet freezing • Droplet vitrification	1818	Nagel, personal communication, 2020
Leibniz Institute of Plant Genetics and Crop Plant Research (IPK), Gatersleben	Germany	Mint	• Droplet vitrification	157	Nagel, personal communication, 2020
Leibniz Institute of Plant Genetics and Crop Plant Research (IPK), Gatersleben	Germany	Garlic and shallot	• Droplet vitrification	213	Nagel, personal communication, 2020
National Agrobiodiversity Center (NAAS), RDA, Suwon	South Korea	Garlic	• Droplet vitrification	1158	[85]
National Institute of Agrobiological Sciences (NIAS), Tsukuba	Japan	Mulberry	• Dormant bud freezing	1236	[80]
Tissue Culture and Cryopreservation Unit, NBPGR, Delhi	India	Mulberry	• Dormant bud freezing	329	[80]
USDA-ARS, Ford Collins and Corvallis	USA	Citrus	• Droplet vitrification	451	[86]
USDA-ARS, Ford Collins and Corvallis	USA	Apple	• Dormant bud freezing	2155	[87]

4.1.1. Dormant Bud Cryopreservation

The number of crops or plants species that can be cryopreserved through dormant bud cryopreservation is limited since two requirements must be met: (i) the species produces buds that go into a dormant phase, usually induced in winter by a prolonged period of low temperature and/or photoperiod [88], before being prepared for cryostorage, and (ii) buds recovered from cryopreservation should respond to bud grafting.

A typical protocol for apple consists of (i) collecting dormant field material in mid-winter, (ii) air-dehydrating the twigs at −5 °C to 25–30% moisture content, (iii) subsequently applying slow freezing at 1 °C h^{-1} to −30 °C and holding this temperature for one hour, before (iv) the twigs are plunged into liquid nitrogen for storage [89]. After rewarming, buds are grafted onto a suitable rootstock. A comparative advantage of this protocol over the other cryopreservation protocols is that, during the whole procedure, no in vitro culture phase is involved; the material for conservation is transferred, after a proper treatment, from the field to the liquid nitrogen tank and, at the time of recovering, back from the tank to the field. This model saves much time [90] and resources and reduces the risks of contamination relative to classic shoot tip-based cryopreservation methods. To date, only apple, pear (*Pyrus* sp.) [91] and sour cherry (*Prunus cerasus*) [92] have been prepared for conservation with this method. An adapted dormant bud cryopreservation protocol where cryopreserved dormant buds are not grafted but directly planted out is developed for *Vaccinium* [87]. Alternatively, buds can be recovered in vitro as in the case of mulberry (*Morus* sp.) [93] and currants (*Ribes* sp.) [94].

4.1.2. Droplet Vitrification

Droplet vitrification can be considered as a "generic" cryopreservation protocol for hydrated tissues, such as in vitro cultures [77,95], as opposed to dry tissues such as seeds and dormant buds. Because of its resulting high post-thaw regeneration rates, relative user-friendliness, and applicability to many plants species, it is now the widest applied protocol for cryopreserved germplasm collections. In short, 1 to 2 mm meristem tips are excised, precultured (or not) on a sucrose medium and treated with two highly concentrated liquids called loading solution (LS) and plant vitrification solution (PVS2 or PVS3) leading to a more concentrated (and thus vitrifiable) cell solution. The dehydrated meristem tips are transferred onto a small strip of aluminum foil and directly plunged in liquid nitrogen for storage. Rapid rewarming takes place in a third liquid, called recovery solution (RS) after which the meristems are transferred onto an in vitro medium for plant recovery. Since its first report in 2005 (See Box 2 [77]), this cryopreservation protocols has now been successfully applied to 111 plant species according the publications reported in Web of Science (Situation 1 July 2020) among which some very important staple foods are found, such as potato [96–98], taro [99], cassava and yam [82], sweet potato [100], and also ornamentals (*Pelargonium* spp. [101], *Nandina domestica* [102]), fruit trees (*Actinidia chinensis* (kiwi fruit)) [103], medicinal plants (*Byrsonima intermedia*) [104] and conifers (*Sequoia sempervirens* (redwood) [105]). For most of these crops/species, cryocollections are being established.

A more recent development in plant cryopreservation is the establishment of two cryo-plate vitrification methods: the vitrification cryo-plate (V Cryo-plate [106]) and dehydration cryo-plate (D Cryo-plate [81,107]). These methods follow the same principles as the droplet vitrification method; small volumes (droplets) cool down more rapidly to the temperature of liquid nitrogen compared to big volumes. The main difference is that, with these recent methods, meristems are enclosed in tiny drops of calcium alginate placed on the aluminum plate before being dehydrated and subsequently plunged in liquid nitrogen. Some advantages of the V Cryo-plate and D Cryo-plate method are (1) simple procedure to store plates in liquid nitrogen, (2) processing large numbers at a time and (3) secured high cooling and heating speed [108]. Post-cryopreservation regeneration rates of these methods are not significantly different from the droplet vitrification method [81], therefore, the choice of method to use is personal preference.

Routine cryopreservation of crop collections began only a few decades ago. Currently, about 18 genebanks have cryopreserved crop collections [13,67,81,98,109]. It is estimated that about 100,000 unique accessions of vegetatively propagated and recalcitrant seed crops potentially need long-term conservation through cryopreservation while currently only about 10,000 accessions are cryopreserved [13].

4.2. Advantages of a Cryopreserved Collection: Safety Backup for Clonally Propagated Crops

Cryopreservation is a cost- and labor-efficient conservation method that insures genetic stability over time [70], it may also be used for establishing a backup. For orthodox seeds such a safety backup facility already exists—the Svalbard Global Seed Vault (SGSV) in Norway. This facility is built by the Norwegian government and operated by the "Global Crop Diversity Trust" and NordGen, to store safety duplicates of national and international seed collections [110] to protect the diversity from irreversible loss due to natural and human-caused disasters. The SGSV consists of chambers maintained at −18 °C, dug into a mountain in the permafrost on Spitzbergen island at the arctic circle. Currently more than one million seed samples are conserved there by many national and international research institutes and genebanks.

Clonal crops can be duplicated in the field or in vitro but the ultimate backup may include a cryopreserved duplication of the conserved accessions. An additional advantage of the cryopreserved backup compared to the global seed vault is that the cryopreservation backup would serve for hundreds of years and does not need to be regenerated after a few decades of storage, depending on the species after 50 to 100 years, as is the case for seeds. Periodic regeneration of cryopreserved tissues is, however, necessary to evaluate successful conservation.

In 2017, a feasibility study concluded that a safety backup facility is currently required to accommodate 5000–10,000 accessions arising from ongoing cryopreservation activities [13] and that this facility should be expanded in a later phase to host all unique clonally propagated crop accessions. Such facility should operate according to the same policies and principles that govern the SGSV.

4.3. Problems Associated with Cryopreserved Collections

The range of crops represented in "cryobanks" is still rather restricted, and more than 90% of these accessions are composed of a few crops such as potatoes, cassava, apple, bananas and plantains, mulberry, garlic and strawberry. The main reasons why cryopreservation for the long-term conservation of vegetatively propagated crops is not applied more widely and on a larger scale are reviewed in detail by [13]. These challenges are linked to (i) protocol development: for many plant species efficient cryopreservation protocols are not available yet, (ii) problems with implementation of existing cryopreservation protocols such as genotype-specific responses and insufficient supply of healthy plant material, and (iii) challenges related to cryobanking capacities such as insufficient funding, lack of skilled personnel with knowledge on plant genetic resources [111] and lack of equipment/infrastructure.

5. Identification of Unique Accessions and Elimination of Duplicates

The elimination of duplicates in clonal collections has been an ongoing task since their establishment because (i) core collections need to be set up that contain the widest possible genetic diversity within the smallest number of accessions, and (ii) costs associated with maintaining clonal collections are high. CIP initially gathered more than 15,000 accessions of native potato cultivars from Latin America. Morphological characteristics and electrophoretic bands were used to identify duplicates and to reduce the collections to about 3500 accessions [17]. A reduction of 13% of the duplicates found in white clover resulted in a per accession saving of USD 500 regeneration cost in the field [112]. Additionally, other collections were screened for their redundancies. In eight Dutch apple collections, molecular fingerprinting revealed that 32% of the accessions were duplicates [113], a similar ratio to the duplicate proportion has been found in a barley seed collection [114]. In the US NPGS, about 12.5% of 1910 apple accessions were considered to be identical [115] comparable to the redundancy of a natural population of cacao [116]. Current next-generation sequencing technologies allow an increase in the number of studied alleles from about 10 SSR markers to several thousand SNPs and elucidate the relationship and ancestries between accessions and species resulting in a better identification of duplicates. The elimination of valuable genotypes has thus to be carefully balanced against the cost of the active preservation.

6. Costs Associated with the Different Conservation Methods

The aim, costs and safety considerations of a collection are important arguments in defining the necessary conservation approaches. To overcome the major drawbacks of field conservation, in vitro and especially cryopreservation are an option to secure clonal crop collections. The introduction of the genetic resources into in vitro conditions or cryopreservation requires well-equipped laboratories and trained personnel. It also requires the development of crop–specific growth media, the optimization of growth conditions, and the development of cryopreservation protocols. In 2000, the maintenance costs for one cassava accession at CIAT, Colombia, was estimated to be USD 7, USD 25, and USD 43 in the field, in vitro and in cryo conditions, respectively [14]. For garlic, the cost to maintain nine garlic plants (one accession) in the field and replant them annually is EUR 47, about 81% of this amount is due to labor costs. The introduction of one garlic accession into cryopreservation is about EUR 363, of which 52% is labor costs. The type of plant material has a strong impact on the cost of cryopreservation. When bulbils/inflorescences can be used, about EUR 318 is required whereas when cryopreservation is carried out with in vitro plants, the costs can increase to EUR 433 and up to EUR 557. The total depends on the number of in vitro subcultures required to achieve the appropriate number of plants for cryopreservation. Subsequent maintenance of such cryopreserved accessions, however, is rather low since this only demands regular filling up of the cryotanks with liquid nitrogen. Therefore, the break-even point when costs of field conservation are equal to that of cryopreservation is achieved after 8 to 13 years [21]. This estimate is very similar to the calculation made for bananas where the cumulative costs between in vitro and in cryo equate after 15 years. In general, the in vitro initiation of a banana accession at the ITC Leuven, Belgium, costs USD 861. The cost for cryopreserving such an accession is USD 1296. The costs accumulate differentially and are assumed in-perpetuity to reach USD 3686 and USD 1431, respectively. Nowadays, only few in vitro collections, such as the 1200 garlic accessions in Korea [85], about 150 mint accessions at IPK [67] and 150 citrus accessions in NPGS are completely preserved through cryopreservation [117]. However, it has to be considered that the material is only immediately available for distribution, characterization and evaluation when kept either in the field or in vitro.

7. Conclusions

The objective of a genebank is to conserve crop genetic resources. The advantage of a field genebank is that characterization and evaluation can be performed on mature plants at limited additional cost. Yet, the method has significant restrictions regarding its security, costs, and sustainability [14]. In this respect, in vitro conservation offers advantages over field conservation and is recognized as an invaluable complementary approach to secure the diversity of the collection. With the potential that tissue culture protocols can be developed for practically any plant species, the technique has been successfully applied over the past decades for short- to medium-term conservation of a wide range of crop species [118]. While offering the possibility of disease elimination and rapid clonal propagation of healthy plants, it is a suitable method for exchanging germplasm accommodating national and international phytosanitary requirements and regulations. However, the possible occurrence of genetic instability as a result of enhanced selection pressure under in vitro conditions compared to in vivo may be an obstacle to its use for long-term preservation of plant germplasm [49]. The chance that somatic mutations occur in tissues maintained at the ultra-low storage temperature of liquid nitrogen (−196 °C) is very slim since all metabolic processes are suspended [119]. In this regard, cryopreservation is the best option for unlimited and secure conservation of regenerative tissues in the long term. Nevertheless, the latter also requires adequate infrastructure, reliable electricity provision and well-trained staff.

The global COVID-19 pandemic has reminded us that particularly clonal ex situ collections in the field and in vitro that require continuous maintenance are vulnerable and insufficiently secured. "Essential" activities for maintenance of the germplasm in gene banks were temporarily interrupted for concerns out of human safety and health reasons, hence putting valuable genetic resources at risk of loss. While the initial costs of developing a crop- or species-specific cryopreservation protocol [120] and labor input for processing the samples in liquid nitrogen are high, once established, the maintenance of a cryopreserved collection requires very little input of resources including human intervention, making it probably the most cost-effective and secure option for long-term conservation.

Author Contributions: All authors contribute equally to this manuscript. All authors have read and agreed to the published version of the manuscript.

Funding: This research received no external funding.

Acknowledgments: B.P. and I.V.d.h. gratefully acknowledge the Gene Bank CGIAR Research Programme and the CGIAR Research Programme on Roots, Tubers and Bananas (RTB), and the Directorate-general Development Cooperation and Humanitarian, Belgium (DGD) for financial support of the project "Safeguarding vegetatively-propagated crop diversity to nourish people now and in the future". In addition, many thanks to Yu-Chun Liao and Maurizio Lambardi for critical proof-reading the text.

Conflicts of Interest: The authors declare no conflict of interest.

References

1. *Genebank Standards for Plant Genetic Resources for Food and Agriculture*; Food and Agriculture Organization of the United Nations: Rome, Italy, 2014.
2. Gepts, P.; Hancock, J. The future of plant breeding. *Crop Sci.* **2006**, *46*, 1630–1634. [CrossRef]
3. *Building on Gender, Agrobiodiversity and Local Knowledge*; Food and Agriculture Organization of the United Nations: Rome, Italy, 2004.
4. Thrupp, L.A. Linking agricultural biodiversity and food security: The valuable role of agrobiodiversity for sustainable agriculture. *Int. Aff.* **2000**, *76*, 265–281. [CrossRef] [PubMed]
5. *The State of the World's Biodiversity for Food and Agriculture*; Food and Agriculture Organization of the United Nations: Rome, Italy, 2019.
6. *WIEWS—World Information and Early Warning System on Plant Genetic Resources for Food and Agriculture*; Food and Agriculture Orgnization of the United Nations: Rome, Italy, 2020.
7. Rajasekharan, P.E.; Sahijram, L. In vitro conservation of plant germplasm. In *Plant Biology and Biotechnology*; Bahadur, B., Venkat Rajam, M., Sahijram, L., Krishnamurthy, K., Eds.; Springer: New Delhi, India, 2015. [CrossRef]
8. McKey, D.; Elias, M.; Pujol, B.; Duputie, A. The evolutionary ecology of clonally propagated domesticated plants. *New Phytol.* **2010**, *186*, 318–332. [CrossRef] [PubMed]
9. Bramel, P.J.; Volk, G.M. *A Global Strategy for the Conservation and Use of Apple Genetic Resources*; Global Crop Diversity Trust: Bonn, Germany, 2019.
10. Damania, A.B. History, achievements, and current status of genetic resources conservation. *Agron. J.* **2008**, *100*, S27–S39. [CrossRef]
11. Müntz, K.; Wobus, U. *Das Institut Gatersleben und Seine Geschichte Genetik und Kulturpflanzenforschung in Drei Politischen Systemen*; Springer spectrum: Berlin/Heidelberg, Germany, 2013; p. 459. [CrossRef]
12. Hawkes, J.G.; Maxted, N.; Ford-Lloyd, B.V. *The Ex Situ Conservation of Plant Genetic Resources*; Springer: Dordrecht, The Netherlands, 2000.
13. Acker, J.P.; Adkins, S.; Alves, A.; Horna, D.; Toll, J. *Feasibility Study for a Safety Backup Cryopreservation Facility*; Bioversity International: Maccarese-Stazione, Italy, 2017.
14. Reed, B.M.; Engelmann, F.; Dulloo, M.E.; Engels, J.M.M. *Technical Guidelines for the Management of Field and In Vitro Germplasm Collections*; International Plant Genetics Resources Institute: Rome, Italy, 2004.
15. Volk, G.; Henk, A. Historic american apple cultivars: Identification and availability. *J. Am. Soc. Hortic. Sci.* **2016**, *141*, 292–301. [CrossRef]
16. Höfer, M.; Flachowsky, H.; Hanke, M.V. German Fruit Genebank–looking back 10 years after launching a national network for sustainable preservation of fruit genetic resources. *J. Cultiv. Plants* **2019**, *71*. [CrossRef]

17. Huaman, S.; Salas, A.; Gomez, R.; Panta, A.; Toledo, J. Conservation of potato genetic resources at CIP. In *Potato, Global Research & Development*; Khurana, S.M.P., Shekhawat, G.S., Singh, B.P., Pandey, S.K., Eds.; Indian Potato Association: Shimla, India, 2000; Volume 1.
18. Forbes, G.A.; Charkowski, A.; Andrade-Piedra, J.; Parker, M.L.; Schulte-Geldermann, E. Potato Seed Systems. In *The Potato Crop: Its Agricultural, Nutritional and Social Contribution to Humankind*; Campos, H., Ortiz, O., Eds.; Springer International Publishing: Cham, Switzerland, 2020; pp. 431–447. [CrossRef]
19. Bachmann-Pfabe, S.; Hammann, T.; Kruse, J.; Dehmer, K.J. Screening of wild potato genetic resources for combined resistance to late blight on tubers and pale potato cyst nematodes. *Euphytica* **2019**, *215*, 48. [CrossRef]
20. Keller, E.R.J.; Kik, C. Allium genetic resources. *Compend. Plant Genome* **2018**. [CrossRef]
21. Keller, E.R.J.; Zanke, C.D.; Senula, A.; Breuing, A.; Hardeweg, B.; Winkelmann, T. Comparing costs for different conservation strategies of garlic (Allium sativum L.) germplasm in genebanks. *Genet. Resour. Crop Evol.* **2013**, *60*, 913–926. [CrossRef]
22. Takagi, H. Garlic Allium sativum L. In *Onions and Allied Crops*; Brewster, J.L., Rabinowitch, H.D., Eds.; CRC Press: Boca Raton, FL, USA, 1990; Volume 3, pp. 109–157.
23. Sharma, K.; Lee, Y.R.; Park, S.W.; Nile, S.H. Importance of growth hormones and temperature for physiological regulation of dormancy and sprouting in onions. *Food Rev. Int.* **2016**, *32*, 233–255. [CrossRef]
24. Senula, A.; Nagel, M. Cryopreservation of plant shoot tips of potato, mint, garlic, and shallot using Plant Vitrification Solution 3. In *Cryopreservation and Freeze-Drying Protocols*; Humana: New York, NY, USA, 2020; Volume 2180, pp. 647–661.
25. Keller, E.R.J.; Zanke, C.D.; Blattner, F.R.; Kik, C.; Stavelikova, H.; Zamecnik, J.; Esnault, F.; Kotlinska, T.; Solberg, S.; Miccolis, V. EURALLIVEG: Establishment of a european core collection by cryopreservation and virus elimination in garlic. *Acta Hortic.* **2012**, *969*, 319–327. [CrossRef]
26. Engels, J.M.M.; Visser, L. *A Guide to Effective Management of Germplasm Collections*; International Plant Genetic Resources Institute: Rome, Italy, 2003.
27. Volk, G.M.; Chao, C.T.; Norelli, J.; Brown, S.K.; Fazio, G.; Peace, C.; McFerson, J.; Zhong, G.Y.; Bretting, P. The vulnerability of US apple (Malus) genetic resources. *Genet. Resour. Crop Evol.* **2015**, *62*, 765–794. [CrossRef]
28. Herbarium, WU. Available online: https://herbarium.univie.ac.at (accessed on 20 November 2020).
29. Wilkinson, M.D.; Dumontier, M.; Aalbersberg, I.J.; Appleton, G.; Axton, M.; Baak, A.; Blomberg, N.; Boiten, J.-W.; da Silva Santos, L.B.; Bourne, P.E.; et al. The FAIR guiding principles for scientific data management and stewardship. *Sci. Data* **2016**, *3*, 160018. [CrossRef] [PubMed]
30. The Alliance of Bioversity International and CIAT. Available online: www.bioversityinternational.org (accessed on 20 November 2020).
31. Meiyalaghan, S.; Paget, M.; Thompson, S.; Thomson, S.; Baldwin, S.; Anderson, J.; Genet, R.; Lewthwaite, S. High resolution DNA melting markers for identification of H1-linked resistance to potato cyst nematode. *Mol. Breed.* **2018**, *38*. [CrossRef]
32. Zuo, C.; Deng, G.; Li, B.; Huo, H.; Li, C.; Hu, C.; Kuang, R.; Yang, Q.; Dong, T.; Sheng, O.; et al. Germplasm screening of Musa spp. for resistance to Fusarium oxysporum f. sp. cubense tropical race 4 (Foc TR4). *Eur. J. Plant Pathol.* **2018**, *151*, 723–734. [CrossRef]
33. Peil, A.; Emeriewen, O.F.; Khan, A.; Kostick, S.; Malnoy, M. Status of fire blight resistance breeding in Malus. *J. Plant Pathol.* **2020**. [CrossRef]
34. Höfer, M. Cryopreservation of winter-dormant apple buds: Establishment of a duplicate collection of Malus germplasm. *Plant Cell Tissue Organ Cult.* **2015**, *121*, 647–656. [CrossRef]
35. Bramel, P.; Krishnan, S.; Horna, D.; Lainoff, B.; Montagnon, C. *Global Conservation Strategy for Coffee Genetic Resources*; Crop Trust: Bonn, Germany, 2017.
36. Ellis, D.; Chavez, O.; Coombs, J.; Soto, J.; Gomez, R.; Douches, D.; Panta, A.; Silvestre, R.; Anglin, N.L. Genetic identity in genebanks: Application of the SolCAP 12K SNP array in fingerprinting and diversity analysis in the global in trust potato collection. *Genome* **2018**, *61*, 523–537. [CrossRef]
37. Anastasio, A.E.; Platt, A.; Horton, M.; Grotewold, E.; Scholl, R.; Borevitz, J.O.; Nordborg, M.; Bergelson, J. Source verification of mis-identified Arabidopsis thaliana accessions. *Plant J.* **2011**, *67*, 554–566. [CrossRef]

38. Jiang, C.; Mithani, A.; Belfield, E.J.; Mott, R.; Hurst, L.D.; Harberd, N.P. Environmentally responsive genome-wide accumulation of de novo *Arabidopsis thaliana* mutations and epimutations. *Genome Res.* **2014**, *24*, 1821–1829. [CrossRef]
39. Withers, L.A. *Tissue Culture Storage for Genetic Conservation*; IBPGR: Rome, Italy, 1980.
40. Benson, E.E.; Harding, K.; Debouck, D.; Dumet, D.; Escobar, R.; Mafla, G.; Panis, B.; Panta, A.; Tay, D.; Vandenhouwe, I.; et al. *Refinement and Standardization of Storage Procedures for Clonal Crops = Global Public Goods Phase 2: Part 1. Project Landscape and General Status of Clonal Crop In Vitro Conservation Technologies*; System-wide Genetic Resources Programme: Rome, Italy, 2011.
41. Genesys. Available online: https://www.genesys-pgr.org/ (accessed on 20 November 2020).
42. Bamberg, J.B.; Martin, M.W.; Abad, J.; Jenderek, M.M.; Tanner, J.; Donnelly, D.J.; Nassar, A.M.K.; Veilleux, R.E.; Novy, R.G. In vitro technology at the US Potato Genebank. *In Vitro Cell Dev. Plant* **2016**, *52*, 213–225. [CrossRef]
43. Kumar, N.; Reddy, M. In vitro plant propagation: A review. *J. For. Environ. Sci.* **2011**, *27*, 61–72.
44. Chen, C.; Chen, J.-J. Measurement of gas exchange rates in plant tissue culture vessels. *Plant Cell Tissue Organ Cult.* **2002**, *71*, 103–109. [CrossRef]
45. Ramírez-Mosqueda, M.A.; Iglesias-Andreu, L.G.; Luna-Sánchez, I.J. Light quality affects growth and development of in vitro plantlet of Vanilla planifolia Jacks. *S. Afr. J. Bot.* **2017**, *109*, 288–293. [CrossRef]
46. Keller, E.R.J.; Senula, A.; Leunufna, S.; Grube, M. Slow growth storage and cryopreservation—Tools to facilitate germplasm maintenance of vegetatively propagated crops in living plant collections. *Int. J. Refrig.* **2006**, *29*, 411–417. [CrossRef]
47. Vandenhouwe, I.; Desmet, K.; Dumontcel, H.T.; Swennen, R. Variability in storage potential of banana shoot cultures under medium-term storage-conditions. *Plant Cell Tissue Organ Cult.* **1995**, *42*, 269–274. [CrossRef]
48. Jackson, D.M.; Harrison, H.F.; Jarret, R.L.; Wadl, P.A. Phenotypic variation in leaf morphology of the USDA, ARS Sweetpotato (Ipomoea batatas) germplasm collection. *Hortscience* **2020**, *55*, 465–475. [CrossRef]
49. da Silva, R.L.; Ferreira, C.F.; da Silva Ledo, C.A.; de Souza, E.H.; da Silva, P.H.; de Carvalho Costa, M.A.P.; Souza, F.V.D. Viability and genetic stability of pineapple germplasm after 10 years of in vitro conservation. *Plant Cell Tissue Organ Cult. (PCTOC)* **2016**, *127*, 123–133. [CrossRef]
50. Sarkar, D.; Chakrabarti, S.K.; Naik, P.S. Slow-growth conservation of potato microplants: Efficacy of ancymidol for long-term storage in vitro. *Euphytica* **2001**, *117*, 133–142. [CrossRef]
51. Sarkar, D.; Kaushik, S.; Naik, P.S. Minimal growth conservation of potato microplants: Silver thiosulfate reduces ethylene-induced growth abnormalities during prolonged storage in vitro. *Plant Cell Rep.* **1999**, *18*, 897–903. [CrossRef]
52. Sarkar, D.; Naik, P.S. Factors affecting minimal growth conservation of potato microplants in vitro. *Euphytica* **1998**, *102*, 275–280. [CrossRef]
53. Chauhan, R.; Singh, V.; Quraishi, A. In vitro conservation through slow-growth storage. In *Synthetic Seeds: Germplasm Regeneration, Preservation and Prospects*; Faisal, M., Alatar, A.A., Eds.; Springer International Publishing: Cham, Swizerland, 2019; pp. 397–416. [CrossRef]
54. Guta, I.C.; Buciumenau, E.C.; Tataru, L.D.; Oprescu, B.; Topala, C.M. New approach of electrotherapy for grapevine virus elimination. *Acta Hortic.* **2019**, *1242*, 697–702. [CrossRef]
55. Wang, Q.C.; Valkonen, J.P.T. Cryotherapy of shoot tips: Novel pathogen eradication method. *Trends Plant Sci.* **2009**, *14*, 119–122. [CrossRef] [PubMed]
56. Helliot, B.; Panis, B.; Poumay, Y.; Swennen, R.; Lepoivre, P.; Frison, E. Cryopreservation for the elimination of cucumber mosaic and banana streak viruses from banana (Musa spp.). *Plant Cell Rep.* **2002**, *20*, 1117–1122. [CrossRef]
57. Singh, D.H.; Uma, S.; Selvarajan, R.; Karihaloo, J. *Micropropagation for Production of Quality Banana Planting Material in Asia-Pacific*; Asia-Pacific Consortium on Agricultural Biotechnology (APCoAB): New Delhi, India, 2011; p. 92.
58. Bhende, S.; Kurien, S. Sucker production in banana. *J. Trop. Agric.* **2015**, *53*, 97–106.
59. Bairu, M.W.; Aremu, A.O.; van Staden, J. Somaclonal variation in plants: Causes and detection methods. *Plant Growth Regul.* **2011**, *63*, 147–173. [CrossRef]
60. Miguel, C.; Marum, L. An epigenetic view of plant cells cultured in vitro: Somaclonal variation and beyond. *J. Exp. Bot.* **2011**, *62*, 3713–3725. [CrossRef]

61. Roels, S.; Escalona, M.; Cejas, I.; Noceda, C.; Rodriguez, R.; Canal, M.J.; Sandoval, J.; Debergh, P. Optimization of plantain (Musa AAB) micropropagation by temporary immersion system. *Plant Cell Tissue Organ Cult.* **2005**, *82*, 57–66. [CrossRef]
62. Mohanty, S.; Panda, M.K.; Subudhi, E.; Nayak, S. Plant regeneration from callus culture of Curcuma aromatica and in vitro detection of somaclonal variation through cytophotometric analysis. *Biol. Plant.* **2008**, *52*, 783–786. [CrossRef]
63. Bairu, M.W.; Fennell, C.W.; van Staden, J. The effect of plant growth regulators on somaclonal variation in Cavendish banana (Musa AAA cv. 'Zelig'). *Sci. Hortic. Amst.* **2006**, *108*, 347–351. [CrossRef]
64. Kaczmarczyk, A.; Houben, A.; Keller, E.R.J.; Mette, M.F. Influence of cryopreservation on the cytosine methylation state of potato genomic DNA. *Cryoletters* **2010**, *31*, 380–391. [PubMed]
65. Smykal, P.; Valledor, L.; Rodriguez, R.; Griga, M. Assessment of genetic and epigenetic stability in long-term in vitro shoot culture of pea (*Pisum sativum* L.). *Plant Cell Rep.* **2007**, *26*, 1985–1998. [CrossRef]
66. Skirvin, R.M.; Mcpheeters, K.D.; Norton, M. Sources and frequency of somaclonal variation. *Hortscience* **1994**, *29*, 1232–1237. [CrossRef]
67. Senula, A.; Büchner, D.; Keller, E.R.; Nagel, M. An improved cryopreservation protocol for mentha spp. Based on Pvs3 as the cryoprotectant. *Cryo Lett.* **2018**, *39*, 345–353.
68. Graner, E.M.; Brondani, G.E.; de Almeida, C.V.; Batagin-Piotto, K.D.; de Almeida, M. Study of senescence in old cultures of the *Bactris gasipaes* Kunth in vitro. *Plant Cell Tissue Organ Cult. (PCTOC)* **2015**, *120*, 1169–1189. [CrossRef]
69. Wilson, D. Endophyte: The evolution of a term, and clarification of its use and definition. *Oikos* **1995**, *73*, 274–276. [CrossRef]
70. Keller, E.R.J.; Panis, B.; Engelmann, F. In vitro storage and cryopreservation as substantial complements in concerted actions to better maintain and use crop germplasm. *Acta Hortic.* **2012**, 35–50. [CrossRef]
71. Van den houwe, I.; Swennen, R. Characterization and control of bacterial contaminants in in vitro cultures of banana (Mus spp.). In Proceedings of the International Symposium on Methods and Markers for Quality Assurance in Micropopagation, Cork, Ireland, 24–27 August 1999; pp. 69–79.
72. Sheibani-Tezerji, R.; Rattei, T.; Sessitsch, A.; Trognitz, F.; Mitter, B. Transcriptome profiling of the endophyte *Burkholderia phytofirmans* PsJN indicates sensing of the plant environment and drought stress. *mBio* **2015**, *6*, e00621-15. [CrossRef]
73. Jarret, R.L.; Florkowski, W.J. In vitro active vs field genebank maintenance of sweet-potato germplasm—major costs and considerations. *Hortscience* **1990**, *25*, 141–146. [CrossRef]
74. Christelova, P.; de Langhe, E.; Hribova, E.; Cizkova, J.; Sardos, J.; Husakova, M.; van den Houwe, I.; Sutanto, A.; Kepler, A.K.; Swennen, R.; et al. Molecular and cytological characterization of the global Musa germplasm collection provides insights into the treasure of banana diversity. *Biodivers. Conserv.* **2017**, *26*, 801–824. [CrossRef]
75. Thomas, J.E. *MusaNet Technical Guidelines for the Safe Movement of Musa Germplasm*, 3rd ed.; Bioversity International: Rome, Italy, 2015.
76. Panis, B.; Totte, N.; VanNimmen, K.; Withers, L.A.; Swennen, R. Cryopreservation of banana (Musa spp.) meristem cultures after preculture on sucrose. *Plant Sci.* **1996**, *121*, 95–106. [CrossRef]
77. Panis, B.; Piette, B.; Swennen, R. Droplet vitrification of apical meristems: A cryopreservation protocol applicable to all Musaceae. *Plant Sci.* **2005**, *168*, 45–55. [CrossRef]
78. Sakai, A. Survival of plant tissue at super-low temperature III. relation between effective prefreezing temperatures and the degree of front hardiness. *Plant Physiol.* **1965**, *40*, 882–887. [CrossRef]
79. Panis, B. Sixty years of plant cryopreservation: From freezing hardy mulberry twigs to establishing reference crop collections for future generations. In Proceedings of the III International Symposium on Plant Cryopreservation, Bangkok, Thailand, 26–28 March 2018; pp. 1–8.
80. Niino, T.; Arizaga, M.V. Cryopreservation for preservation of potato genetic resources. *Breed Sci.* **2015**, *65*, 41–52. [CrossRef] [PubMed]
81. Wang, M.R.; Lambardi, M.; Engelmann, F.; Pathirana, R.; Panis, B.; Volk, G.M.; Wang, Q.C. Advances in cryopreservation of in vitro-derived propagules: Technologies and explant sources. *Plant Cell Tissue Org.* **2020**. [CrossRef]
82. Dumet, D.; Diebiru, E.; Adeyemi, A.; Akinyemi, O.; Gueye, B.; Franco, J. Cryopreservation for the 'in perpetuity' conservation of yam and cassava genetic resources. *Cryoletters* **2013**, *34*, 107–118.

83. CIP. Cryopreservation. Available online: https://cipotato.org/genebankcip/process/cryopreservation/ (accessed on 14 October 2020).
84. Hofer, M.; Hanke, M.V. Cryopreservation of fruit germplasm. *In Vitro Cell Dev. Plant* **2017**, *53*, 372–381. [CrossRef]
85. Kim, H.H.; Popova, E.; Shin, D.J.; Yi, J.Y.; Kim, C.H.; Lee, J.S.; Yoon, M.K.; Engelmann, F. Cryobanking of Korean Allium Germplasm Collections: Results from a 10 Year Experience. *Cryoletters* **2012**, *33*, 45–57.
86. Volk, G.M.; Jenderek, M.M.; Walters, C.; Bonnart, R.; Shepherd, A.; Skogerboe, D.; Hall, B.D.; Moreland, B.; Krueger, R.; Polek, M. Implementation of Citrus shoot tip cryopreservation in the USDA-ARS National Plant Germplasm System. In Proceedings of the III International Symposium on Plant Cryopreservation, Bangkok, Thailand, 26–28 March 2018.
87. Jenderek, M.M.; Tanner, J.D.; Ambruzs, B.D.; West, M.; Postman, J.D.; Hummer, K.E. Twig pre-harvest temperature significantly influences effective cryopreservation of Vaccinium dormant buds. *Cryobiology* **2017**, *74*, 154–159. [CrossRef]
88. Maurya, J.P.; Bhalerao, R.P. Photoperiod- and temperature-mediated control of growth cessation and dormancy in trees: A molecular perspective. *Ann. Bot. Lond.* **2017**, *120*, 351–360. [CrossRef]
89. Stushnoff, C. Cryopreservation of apple genetic-resources. *Can. J. Plant Sci.* **1987**, *67*, 1151–1154. [CrossRef]
90. Lambardi, M.; Benelli, C.; de Carlo, A.; Ozudogru, E.A.; Previati, A.; Ellis, D. Cryopreservation of ancient apple cultivars of veneto: A Comparison between PVS2-vitrification and dormant-bud techniques. *Int. Symp. Cryopreserv. Hortic. Species* **2011**, *908*, 191–198. [CrossRef]
91. Guyader, A.; Guisnel, R.; Simmoneau, F.; Rocand, B.; Le Bras, C.; Grapin, A.; Chatelet, P.; Dussert, S.; Engelmann, F.; Feugey, L.; et al. First results on cryopreservation by dormant bud technique of a set of Malus and Pyrus cultivars from the INRA biological resources centre. In Proceedings of the Cryopreservation of crop species in Europe: COST Action 871: CryoPlanet: Proceedings of the final meeting, Angers, France, 8–11 February 2011.
92. Towill, L.E.; Forsline, P.L. Cryopreservation of sour cherry (Prunus cerasus L.) using a dormant vegetative bud method. *Cryoletters* **1999**, *20*, 215–222.
93. Fukui, K.; Shirata, K.; Niino, T.; Kashif, I.M. Cryopreservation of Mulberry Winter Buds in Japan. *Int. Symp. Cryopreserv. Hortic. Species* **2011**, *908*, 483–488. [CrossRef]
94. Rantala, S.; Kaseva, J.; Karhu, S.; Vetelainen, M.; Uosukainen, M.; Haggman, H. Cryopreservation of Ribes nigrum (L.) dormant buds: Recovery via in vitro culture to the field. *Plant Cell Tissue Organ Cult.* **2019**, *138*, 109–119. [CrossRef]
95. Panis, B.; Piette, B.; Andre, E.; van den Houwe, I.; Swennen, R. Droplet vitrification: The first generic cryopreservation protocol for organized plant tissues? *Int. Symp. Cryopreserv. Hortic. Species* **2011**, *908*, 157–163. [CrossRef]
96. Folgado, R.; Panis, B.; Sergeant, K.; Renaut, J.; Swennen, R.; Hausman, J.F. Unravelling the effect of sucrose and cold pretreatment on cryopreservation of potato through sugar analysis and proteomics. *Cryobiology* **2015**, *71*, 432–441. [CrossRef]
97. Panta, A.; Panis, B.; Ynouye, C.; Swennen, R.; Roca, W.; Tay, D.; Ellis, D. Improved cryopreservation method for the long-term conservation of the world potato germplasm collection. *Plant Cell Tissue Organ Cult.* **2015**, *120*, 117–125. [CrossRef]
98. Kopnick, C.; Grube, M.; Stock, J.; Senula, A.; Mock, H.P.; Nagel, M. Changes of soluble sugars and ATP content during DMSO droplet freezing and PVS3 droplet vitrification of potato shoot tips. *Cryobiology* **2018**, *85*, 79–86. [CrossRef]
99. Sant, R.; Panis, B.; Taylor, M.; Tyagi, A. Cryopreservation of shoot-tips by droplet vitrification applicable to all taro (Colocasia esculenta var. esculenta) accessions. *Plant Cell Tissue Organ Cult.* **2008**, *92*, 107–111. [CrossRef]
100. Vollmer, R.; Panta, A.; Tay, D.; Roca, W.; Ellis, D. Effect of sucrose preculture and PVS2 exposure on the cryopreservation of sweet potato shoot tips [Ipomoea batatas (L.) Lam.] Using the PVS2 Droplet Vitrification. *Ii Int. Symp. Plant Cryopreserv.* **2014**, *1039*, 265–271. [CrossRef]
101. Gallard, A.; Panis, B.; Dorion, N.; Swennen, R.; Grapin, A. Cryopreservation of pelargonium apices by droplet-vitrification. *Cryoletters* **2008**, *29*, 243–251. [PubMed]

102. Ozudogru, A.; da Silva, D.P.C.; Kaya, E.; Dradi, G.; Paiva, R.; Lambardi, M. In vitro Conservation and Cryopreservation of Nandina domestica, an Outdoor Ornamental Shrub. *Not. Bot. Horti Agrobo.* **2013**, *41*, 638–645. [CrossRef]
103. Mathew, L.; Burritt, D.J.; McLachlan, A.; Pathirana, R. Combined pre-treatments enhance antioxidant metabolism and improve survival of cryopreserved kiwifruit shoot tips. *Plant Cell Tissue Organ Cult.* **2019**, *138*, 193–205. [CrossRef]
104. Silva, L.C.; Paiva, R.; Swennen, R.; Andre, E.; Panis, B. Shoot-tip cryopreservation by droplet vitrification of Byrsonima intermedia A. Juss.: A woody tropical and medicinal plant species from Brazilian cerrado. *Cryo Lett.* **2013**, *34*, 338–348.
105. Ozudogru, E.A.; Kirdok, E.; Kaya, E.; Capuana, M.; Benelli, C.; Engelmann, F. Cryopreservation of Redwood (Sequoia Sempervirens (D. Don.) Endl.) in vitro buds using vitrification-based techniques. *Cryoletters* **2011**, *32*, 99–110.
106. Yamamoto, S.; Rafique, T.; Priyantha, W.S.; Fukui, K.; Matsumoto, T.; Niino, T. Development of a cryopreservation procedure using aluminium cryo-plates. *Cryo Lett.* **2011**, *32*, 256–265.
107. Niino, T.; Yamamoto, S.I.; Fukui, K.; Castillo Martinez, C.R.; Arizaga, M.V.; Matsumoto, T.; Engelmann, F. Dehydration improves cryopreservation of mat rush (Juncus decipiens Nakai) basal stem buds on cryo-plates. *Cryo Lett.* **2013**, *34*, 549–560.
108. Tanaka, D.; Sakuma, Y.; Yamamoto; Valle Arizaga, M.; Niino, T.; Matsumoto, T. Development of −80 °C storage for Allium shoot tips using D cryo-plate method. *Plant Cell Tissue Organ Cult. (PCTOC)* **2020**. [CrossRef]
109. Ruta, C.; Lambardi, M.; Ozudogru, E.A. Biobanking of vegetable genetic resources by in vitro conservation and cryopreservation. *Biodiv. Conserv.* **2020**, *29*, 3495–3532. [CrossRef]
110. Asdal, A.; Guarino, L. The svalbard global seed vault: 10 Years-1 million samples. *Biopreserv. Biobank.* **2018**, *16*, 391–392. [CrossRef] [PubMed]
111. Volk, G.; Namuth-Covert, D.; Byrne, P. Training in plant genetic resources management: A way forward. *Crop Sci.* **2019**, *59*. [CrossRef]
112. Greene, S.L.; Pederson, G.A. Eliminating duplicates in germplasm collections: A white clover example. *Crop Sci.* **1996**, *36*, 1398–1400. [CrossRef]
113. van Treuren, R.; Kemp, H.; Ernsting, G.; Jongejans, B.; Houtman, H.; Visser, L. Microsatellite genotyping of apple (Malus x domestica Borkh.) genetic resources in the Netherlands: Application in collection management and variety identification. *Genet. Resour. Crop Evol.* **2010**, *57*, 853–865. [CrossRef]
114. Milner, S.G.; Jost, M.; Taketa, S.; Mazon, E.R.; Himmelbach, A.; Oppermann, M.; Weise, S.; Knupffer, H.; Basterrechea, M.; Konig, P.; et al. Genebank genomics highlights the diversity of a global barley collection. *Nat. Genet.* **2019**, *51*, 319–326. [CrossRef]
115. Gross, B.L.; Volk, G.M.; Richards, C.M.; Forsline, P.L.; Fazio, G.; Chao, C.T. Identification of "duplicate" accessions within the USDA-ARS national plant germplasm system malus collection. *J. Am. Soc. Hortic. Sci.* **2012**, *137*, 333–342. [CrossRef]
116. Zhang, D.; Arevalo-Gardini, E.; Mischke, S.; Zuniga-Cernades, L.; Barreto-Chavez, A.; del Aguila, J.A. Genetic diversity and structure of managed and semi-natural populations of cocoa (*Theobroma cacao*) in the Huallaga and Ucayali Valleys of Peru. *Ann. Bot. Lond.* **2006**, *98*, 647–655. [CrossRef]
117. Volk, G.M.; Bonnart, R.; Shepherd, A.; Yin, Z.F.; Lee, R.; Polek, M.; Krueger, R. Citrus cryopreservation: Viability of diverse taxa and histological observations. *Plant Cell Tissue Organ Cult.* **2017**, *128*, 327–334. [CrossRef]
118. Ashmore, S.E. *Status Report on the Development and Application of In Vitro Techniques for the Conservation and Use of Plant Genetic Resources*; International Plant Genetic Resources Institute: Rome, Italy, 1997.
119. Kaviani, B. Conservation of plant genetic resources by cryopreservation. *Aust. J. Crop Sci.* **2011**, *5*, 778–800.
120. Pence, V.C. Tissue cryopreservation for plant conservation: Potential and challenges. *Int. J. Plant Sci.* **2014**, *175*, 40–45. [CrossRef]

plants

MDPI

Article

Seed Germination after 30 Years Storage in Permafrost

Svein Øivind Solberg [1,*], Guro Brodal [2], Roland von Bothmer [3], Eivind Meen [4], Flemming Yndgaard [5], Christian Andreasen [6] and Åsmund Asdal [5]

1 Faculty of Applied Ecology, Agricultural Sciences and Biotechnology, Inland Norway University of Applied Sciences, NO-2418 Elverum, Norway
2 Norwegian Institute of Bioeconomy Research, NIBIO, P.O. Box 115, NO-1431 Ås, Norway; guro.brodal@nibio.no
3 The Faculty of Landscape Planning, Horticulture and Agricultural Sciences has Six Departments, Swedish University of Agricultural Sciences, SE-230 53 Alnarp, Sweden; roland.von.bothmer@slu.se
4 Kimen Seed Laboratory, P.O. Box 164, NO-1431 Ås, Norway; eivind.meen@kimen.no
5 Nordic Genetic Resource Center, Box 41, SE-230 53 Alnarp, Sweden; flemming@yndgaard.se (F.Y.); asmund.asdal@nordgen.org (A.A.)
6 Department of Plant and Environmental Sciences, Copenhagen University, Højbakkegaard Allé 13, DK2630 Taastrup, Denmark; can@plen.ku.dk
* Correspondence: svein.solberg@inn.no

Received: 13 April 2020; Accepted: 29 April 2020; Published: 2 May 2020

Abstract: More than 30 years ago, the Nordic Gene Bank established a long-term experiment on seeds stored under permafrost conditions in an abandoned mine corridor in Svalbard, as a tool to monitor storage life under these conditions. The study included seeds from 16 Nordic agricultural and horticultural crops, each represented by two or three cultivars (altogether 38 accessions). All seeds were ultra-dried to 3–5% moisture before being sealed in glass tubes. Germination tests were performed in accordance with the International Seed Testing Association (ISTA) protocols. At the initiation of the experiment, the samples showed good germination with the median value at 92%. The overall picture remained stable over the first twenty to twenty-five years. However, the variation became larger over time and at 30 years, the median value had dropped to 80%. At the lower end, with a high drop in germination, we found rye, wheat, and English ryegrass. At the upper end, we found Kentucky bluegrass and cucumber. The lowest germination was found in samples with the highest initial seed moisture levels. Pre-storage conditions are likely to be of major importance for longevity.

Keywords: ex situ conservation; germination; longevity; plant genetic resources; seed storage

1. Introduction

Most food plants produce seeds that can be stored under low temperature and moisture conditions. Much of our knowledge on seed longevity is based on artificial ageing experiments, where seeds are exposed to suboptimal conditions of elevated temperature and moisture for some weeks, and storage life is predicted based on the seed moisture content, storage temperature, and seed lot characters [1,2]. Such calculations have predicted that high-quality seeds could survive ideal conditions for hundreds of years or more [3,4], which was good news for many gene banks, but might be unrealistic as such studies have rarely been confirmed in long-term storage studies. Long-term seed storage is crucial for ex situ gene bank conservation [5–7]. Gene banks maintain crop diversity and facilitate the utilization of seeds for breeding, research, education, and other purposes [8–11]. Safety back-ups are kept, ideally at a second location, to spread the risks [12,13]. The Global Seed Vault at the Arctic Archipelago of Svalbard was opened in 2008 and is facilitating such back-up collections with a world outreach [14]. However, more than twenty years before this, the Nordic Gene Bank (NGB) started a small seed storage facility in an abandoned coalmine corridor in Svalbard. Different options had been considered, such as inland ice caves in Greenland or mountain caves in Jotunheimen, Norway, but in the end, a coalmine in the permafrost in Svalbard was chosen due to its good logistics, despite its remote location [15]. The temperature of the Global Seed Vault in Svalbard *was* −18 °C compared to the −3.5 °C present in the abandoned coalmine corridor. The rock temperature was stable, which meant that it was independent of an external energy supply. As a tool to monitor storage life under these permafrost conditions, a long-term seed storage experiment was initiated. The experiment started in 1986 and included samples for germination monitoring until 2086, thus it was termed "the 100 year experiment". Important crops for Nordic agriculture and horticulture were included. The investigation is still ongoing, and in this paper, we summarize the results after the first 30 years.

2. Results and Discussion

2.1. Overall Patterns

The overall picture of seed germination development over storage time across all accessions (species and cultivars) is illustrated by boxplots (Figure 1). A lower and upper percentile defines the box, in which 75% of the observations were found. A marked line denotes the median germination value, and the whiskers and small circles show observations away from the box. At the initiation of the experiment in 1986 (year 0 = y0), most of the seed lots showed an excellent germination ability. The box-range was from 83% to 95%, and the median was at 92% germination. The overall picture was relatively stable over the first twenty to twenty-five years, but the variation increased over time as some seed lots showed a reduced germination. After 30 years, the median value across all the lots was 80%, but with outlier samples below 40%, and some of the lots in the 50–70% range.

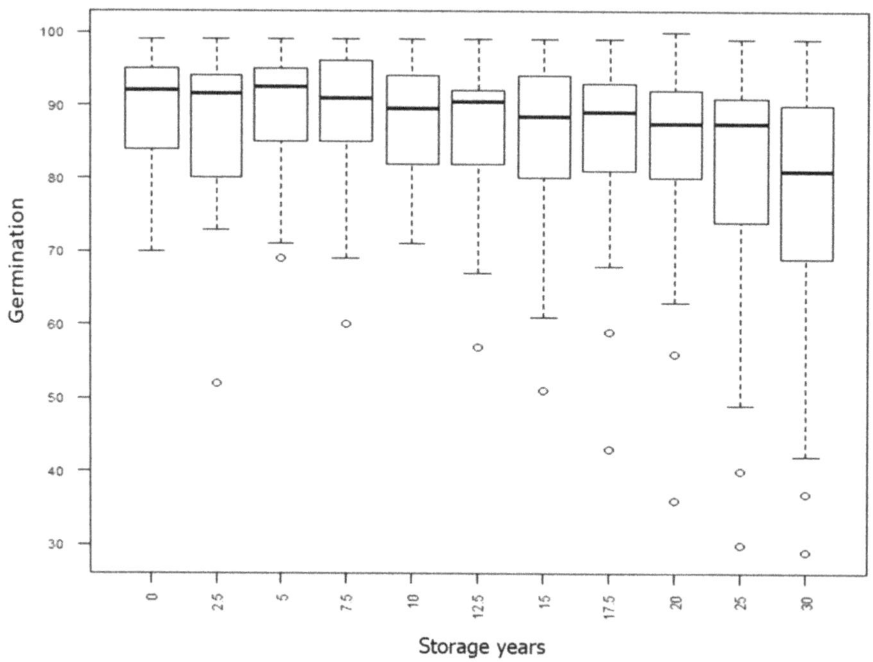

Figure 1. Boxplots showing the germination percentages across species and cultivars throughout the first thirty years (year 0, year 2.5, year 5, year 7.5, year 10, year 12.5, year 15, year 17.5, year 20, year 25, and year 30).

Figure 2 shows a dendrogram of a cluster analysis of the germination results of the accessions. The dissimilarity values of the fusion level values of the dendrogram indicated the cutting level six clusters to be correct. The largest cluster, cluster 1, contained 19 lots. It contained both normal and lots with seed-borne pathogens (i-prefix lots). Also, cluster 2 contained both normal and seed-borne pathogens, as did cluster 6. In cluster 1 we found Barley 1, Ryegrass 1, Ryegrass 2, Timothy 2, Bluegrass 2, Redclover 2, Rape 1, Rape 2, Onion 1, Onion 2, Carrot 1, Carrot 2, Cauliflower 2, i1 Wheat, i2 Wheat, i Barley, i Meadow_fescue, i Onion, and i Cabbage. In cluster two, we found Barley 2, Bluegrass 1, Beet 2, Lettuce, Cabbage, Cucumber 1, Cucumber 2, Cauliflower 1, and i-Timothy. In cluster three, we found Wheat 1, Timothy 1, Redclover 1, and i Carrot. In cluster 4, we found Wheat 2. In cluster 5, we found Rye 1 and Rye 2. In cluster 6, we found Beet 1, i Lettuce, and i Beet. Lots with seed-borne pathogens (i-prefix lots) were spread over the clusters, showing that such pathogens do not explain much of the seed longevity. Furthermore, seed lots of the same species were only partly in the same clusters; for example, the two cucumber (*Cucumis sativus* L.) lots and the two rye (*Secale cereale* L.) lots. This also shows that interspecies differences are not of importance for explaining seed longevity.

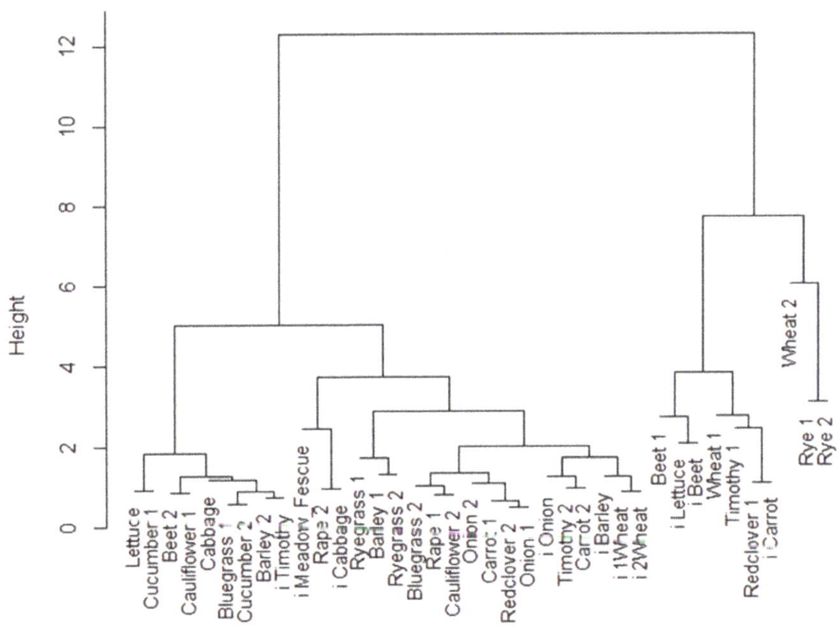

Figure 2. The dendrogram of a cluster analysis of the germination results of the seed lots.

The seeds of all the species remained viable after 30 years in permafrost under the given conditions (dried to 3–5% moisture content and stored in sealed glass ampoules). Crops, but also cultivars, within the same crops showed different results. Similar patterns have also been observed in other long-term experiments, both under ambient [16,17] or −18 °C conditions [18–21]. The cold storage of seeds should provide improved longevity compared to ambient storage [22]. Another factor of importance is seed maturation [23,24].

For instance, the potential seed longevity of barley was found to be best if it was harvested a week or so after grain filling was completed [25]. The same has been found in tomatos (*Solanum lycopersicum* L. var. *lycopersicum*) [26]. Furthermore, weather, seed coat damages, diseases, and pests may influence seed storage life [24,27]. In our experiment, the samples were all from newly harvested seeds, but we do not have data on the weather or other pre-harvest conditions. We only assume that the seed lots selected for the experiment were of a high quality.

2.2. Crop-Wise Results

Our results show good performance for many of the vegetables, while the picture is more varied for cereals and forage species (Figure S1). Table 1 gives the germination result for each seed lot. According to the FAO's gene bank standard [13], "regeneration shall be carried out when the viability drops below 85 percent of the initial viability or when the remaining seed quantity is less than what is required for three sowings of a representative population of the accession." Seed lots with the highest loss in germination, here defined as a loss in 15% or more over the 30 years, were found in both of the rye lots, two of the three English ryegrass (*Lolium perenne* L.) lots (Ryegrass 1, Ryegrass 2), two out of four wheat (*Triticum aestivum* L.) lots (Wheat 1, Wheat 2), one of the three timothy (*Phleum pratense* L.) lots (Timothy 1), one of the three barley (*Hordeum vulgare* L.) lots (Barley 1), and in the one seed lot of meadow fescue (*Schedonorus pratensis* (Huds.) P. Beauv.). The literature shows that especially rye, but also some forage grasses, can be relatively short-lived [28–31]. Intermediate storage performance, with a 5–15% loss in germination over the 30-year period, was found in two of the three lettuce (*Lactuca sativa* L.) lots (i Lettuce, Lettuce 2), two of the three carrot (*Daucus carota* subsp. *sativus* (Hoffm.) Schübl & G. Martens) lots (Carrot 2, i-Carrot), one of the four wheat lots (i1-Wheat), one of the three barley lots (Barley 2), one of the three timothy lots (Timothy 2), one of the two cauliflower (*Brassica oleracea* L. var. *botrytis*) lots (Cauliflower 2), one of the two oilseed rape (*Brassica napus* L.) lots (Rape 1), and one out of the three red clover (*Trifolium pratense* L.) lots (Redclover 1). The most long-lived, with a loss in germination less than 5% after 30 years of storage, were found in all of the three beet (*Beta vulgaris* L.) lots, all of the three onion (*Allium cepa* L.) lots, both of the cucumber lots, both of the Kentucky bluegrass (*Poa pratensis* L.) seed lots, the only cabbage (*Brassica oleracea* L.) seed lot, and in the last of the red clover, cauliflower, barley, wheat, oilseed rape, carrot, and lettuce seed lots. Other studies have also shown that beet seeds as well as cucumber seeds have retained a high germination level over time, but, in contrast to our study, onions are generally found to be short-lived [17–19,29–31]. Our results are a little surprising as we found that onion, and to some extent lettuce, showed no decline in germination over the 30-year period.

Table 1. Germination percentages for all the seed lots (values below 70% in bold), and loss in germination (Δ germ) over the first 30 years, calculated by averaging the germination percentage of the three first test occasions (year 0, year 2.5, and year 5) minus the last test occasion (year 30).

	Storage Years											
Seed Lot ID	**0**	**2.5**	**5**	**7.5**	**10**	**12.5**	**15**	**17.5**	**20**	**25**	**30**	**Δ Germ**
Barley 1	95	96	96	96	95	91	95	95	79	90	76	−20%
Barley 2	95	94	96	97	94	91	97	94	93	92	86	−9%
i Barley	90	89	94	94	92	82	89	88	85	85	88	−3%
Wheat 1	75	75	83	83	80	85	77	76	71	**58**	**57**	−21%
Wheat 2	70	**52**	89	89	82	86	79	76	**64**	**40**	**37**	−33%
i1 Wheat	89	85	89	89	88	84	87	87	87	82	80	−8%
i2 Wheat	90	83	91	93	89	87	85	89	84	83	87	−1%
Rye 1	76	74	84	84	74	**57**	**51**	**43**	**36**	**30**	**29**	−49%
Rye 2	81	78	87	83	74	**67**	**61**	**59**	**56**	**49**	**48**	−34%
Ryegrass 1	99	96	96	96	95	93	93	88	85	**74**	**59**	−38%
Ryegrass 2	96	95	97	95	96	93	94	92	94	87	**69**	−27%
Timothy 1	73	77	77	**69**	79	71	73	**68**	**63**	**63**	**42**	−34%
Timothy 2	90	91	93	89	90	92	88	92	85	81	78	−13%
i Timothy	94	95	92	96	92	94	95	92	94	91	92	−2%
Bluegrass 1	92	95	95	97	97	96	95	94	92	93	92	−2%
Bluegrass 2	94	91	89	96	93	91	94	90	92	89	88	−3%
iMeadow fescue	92	90	89	86	88	90	87	81	81	**69**	**54**	−36%
Red clover 1	75	74	76	80	75	74	**68**	76	**68**	73	**68**	−7%
Red clover 2	93	94	94	92	89	91	89	89	90	91	90	−4%

Table 1. *Cont.*

	Storage Years											
Seed Lot ID	**0**	**2.5**	**5**	**7.5**	**10**	**12.5**	**15**	**17.5**	**20**	**25**	**30**	**Δ Germ**
Beet 1	79	79	71	**60**	76	72	75	70	80	70	78	0%
Beet 2	97	97	97	97	97	98	94	95	95	98	97	0%
i Beet	78	78	69	77	81	81	80	82	82	79	77	0%
Rape 1	95	94	95	96	90	91	90	90	92	91	87	−8%
Rape 2	84	84	83	87	85	84	84	82	88	82	81	−3%
Onion 1	92	92	92	93	90	89	89	88	91	90	91	−1%
Onion 2	89	94	92	92	91	94	92	94	90	92	89	−3%
i Onion	92	93	87	87	88	90	85	91	87	88	88	−3%
Lettuce 1	98	95	98	98	98	92	99	98	98	99	99	0%
Lettuce 2	97	93	96	90	94	97	93	93	92	92	90	−5%
i Lettuce	84	80	78	80	76	71	78	77	73	75	74	−7%
Cucumber 1	99	99	99	99	99	99	99	99	100	98	98	−1%
Cucumber 2	93	94	94	95	97	92	96	95	92	92	90	−4%
Carrot 1	92	90	93	90	89	91	86	89	88	90	91	−1%
Carrot 2	91	93	94	86	88	90	85	89	90	89	81	−12%
i Carrot	76	73	72	74	71	75	**67**	78	**65**	71	**68**	−6%
Cauliflower 1	95	93	94	98	98	98	98	94	96	94	91	−3%
Cauliflower 2	95	92	94	95	94	92	90	92	93	90	79	−15%
i Cabbage	87	82	85	85	86	82	86	82	80	76	80	−5%

2.3. *Moisture Measurements*

For all samples, the variation between the highest and lowest moisture content over the ten years was between 0.3% and 0.8% (Table 2). The data showed that two lots had a higher moisture content exceeding 5% in the initial test; these were the samples Wheat 2 'Solid', with 6.3% humidity, and Rye 1 'Petkus II', with 5.3% humidity. For wheat, one seed lot showed a low decline, and two seed lots showed a steep decline in germination. Different pre-harvest conditions or genetic factors may explain a steep decline in germination. We saw an effect of drying the seeds to lower than 5% internal moisture content before packing. The two samples with the highest internal humidity (Wheat 2 and Rye 1) were among the ones that showed the most significant drop in germination in our experiment. The current FAO standards [13], which are used by most gene banks, recommend drying for three months at 15 °C and 15% RH. According to our experience, this would give a seed moisture content exceeding 5%, thus decreasing the longevity.

Table 2. Seed moisture content (in %) in the different lots over the first ten years of the experiment.

	Storage Years					
Seed Lot	**0**	**2.5**	**5.0**	**7.5**	**10**	**Max/Min**
Barley 1	5.0	4.7	4.6	4.7	4.8	5.0/4.6
Barley 2	4.6	4.6	4.3	4.3	4.7	4.7/4.3
Wheat 1	4.0	4.3	4.3	4.3	4.3	4.3/4.0
Wheat 2	6.3	5.8	5.5	5.8	5.8	6.3/5.5
Rye 1	5.3	5.1	4.9	5.1	5.2	5.3/4.9
Rye 2	4.9	4.7	4.4	4.7	4.8	4.9/4.4
Beet 1	3.7	4.3	4.3	3.7	4.1	4.3/3.7
Beet 2	3.5	3.7	3.6	3.9	3.5	3.9/3.5
Cucumber 1	3.0	3.5	3.6	2.8	2.9	3.6/2.8
Cucumber 1	2.8	3.3	2.7	2.8	2.7	3.3/2.7

3. Conclusions

The study has so far revealed valuable results concerning the longevity of seeds after 30 years in permafrost. Nine out of the 38 seed lots showed a germination loss exceeding 15%, which is the level recommended by the FAO for carrying out regeneration. Rye and ryegrass in particular showed a rapid decline, while many of the vegetables showed a low decline in germination. The results are relevant for the seeds in the first Nordic back-up collection stored in the abandoned coalmine. A given storage condition is an essential characteristic for comparing results with other experiments. Here, seeds were stored in permafrost with a stable sub-zero temperature. We had no reference material at −18 °C conditions. Despite the limited number of samples and the lack of a −18 °C control, the observations add knowledge about the longevity of seeds.

4. Materials and Methods

4.1. Seed Samples and Seed Storage

In total, 38 seed lots covering 16 crops were included in the study (Table 3). Each species was represented by two to four seed lots, except for cabbage and meadow fescue, which had only one. The seed lot ID was given by crop name and code. A number after the crop name (1 or 2) refers to seed lots where we have no information on seed-borne diseases. A prefix "i" letter applies to seed lots where we know that seed-borne diseases were present.

Table 3. Overview of the examined crops and sample information. A prefix number refers to the samples with no information on seed-borne diseases, while a prefix i-letter refers to the samples where we detected seed-borne diseases to be present at the start of the experiment.

Crop/Species	Sample ID and Cultivars (Country of Origin)
Barley (*Hordeum vulgare*)	1 = Inga Abed (DNK), 2 = Tunga (NOR), i = Bamse (NOR)
Wheat (*Triticum aestivum*)	1 = Vakka (FIN), 2 = Solid (SWE), i1 = Runar (NOR), i2 = Line 79 CBW A72 (CAN)
Rye (*Secale cereale*)	1 = Petkus II (DNK), 2 = Voima (DNK)
English ryegrass (*Lolium perenne*)	1 = Pippin (DNK), 2 = Riikka (FIN)
Timothy (*Phleum pratense*)	1 = Tammisto (FIN), 2 = Bodin (NOR), I = Forus (NOR)
Kentucky bluegrass (*Poa pratensis*)	1 = Hankkijan Kyösti (FIN), 2 = Annika (DNK)
Meadow fescue (*Schedonorus pratensis*)	I = Salten (NOR)
Red clover (*Trifolium pratense*)	1 = Jokioinen (FIN), 2 = Molstad (NOR)
Beet (*Beta vulgaris*)	1 = 311 N typ (SWE), 2 = 70500 (DNK)^2, i = Hilma (GBR)
Oilseed rape (*Brassica napus*)	1 = Jupiter (SWE), 2 = Linrama (DNK)
Onion (*Allium cepa*)	1 = Hamund (SWE), 2 = Owa (DNK), i = Laskala (NOR)
Lettuce (*Lactuca sativa*)	1 = Attraktion Sana (DNK), 2 = Hilro (SWE), i = Attractie (NLD)
Cucumber (*Cucumis sativus*)	1 = Langelands gigant (DNK), 2 = Rhensk Druv (SWE)
Carrot (*Daucus carota*)	1 = Nantes Fancy (DNK), 2 = Regulus (SWE), I = Forto Nantes (NLD)
Cauliflower (*B. oleracea* v. *botrytis*)	1 = Svavit (SWE), 2 = Pari (DNK)
Cabbage (*Brassica oleracea*)	I = Trønder Lunde (NOR)

The lots were from different cultivars and origins. No cultivation details are available except that the seed lots should be of good quality. Initial germination tests were conducted (y0). Before storage, all the seeds were dried to 3–5% internal moisture content using a Munters (Kista, Sweden) dehumidifier adjusted to 10% RH in a room at 25 °C. After that, they were placed in a freezer overnight before they were sealed in glass ampoules. Each seed lot was divided into sub-samples of 2 × 500 seeds for each withdrawal. The sub-samples were boxed and labelled with the date for withdrawal, transported to Longyearbyen, Svalbard, and placed in the abandoned coalmine corridor. The temperature inside the coalmine was measured as −3.5 °C (± 0.2 °C).

4.2. Germination Studies

The first set of samples was tested in December 1986 (year 0), with a plan to have the last tests in 2086 (year 100). Since the start of the experiment, 11 boxes have been retrieved. The moisture content of the seed was monitored from year 0 to year 10 (Table 4). The germination tests and the moisture content tests were carried out at the Kimen Seed Laboratory (Ås, Norway). The testing followed the International Seed Testing Association (ISTA) rules [32,33]. The details of the germination conditions for the different species in the study are shown in Table 5. In the case of oilseed rape and onion, the number of days to final count was, however, larger than the number of days in the ISTA protocols. In accordance with the protocols, we germinated 4x100 or 8x50 seeds per sample and each replicate was compared to the mean. If a large variation was detected, the samples were re-tested. Different types of filter paper (Seedburo Equipment Co, Des Plaines, IL, USA) were used as substrates for the germination tests.

Table 4. Dates of the retrieval of the samples over the first 30 years, and types of tests conducted.

Year	Date	Germination	Moisture
0	1986, December	X	X
2.5	1989, June	X	X
5	1991, December	X	X
7.5	1994, June	X	X
10	1996, December	X	X
12.5	1999, June	X	
15	2001, December	X	
17.5	2004, June	X	
20	2006, December	X	
25	2011, December	X	
30	2016, December	X	

Table 5. Germination conditions for the different crops included in the experiment.

Crop/Species	Germination Substrate	Temperature (°C)	Final Count (Day Number)
Barley	BP	20	7
Wheat	BP	20	8
Rye	BP	20	7
English ryegrass	TP	20<=>30	14
Timothy	TP	15<=>25	10
Kentucky bluegrass	TP	15<=>25	28
Meadow fescue	TP	20<=>30	14
Red clover	TP	20	10
Beet	PP	20	14
Oilseed rape	TP	20	10 [3]
Onion	TP	15	14 [4]
Lettuce	TP	20-light	7
Cucumber	TP	20<=>30	8
Carrot	TP	20	14
Cauliflower, cabbage	TP	20<=>30	10

For cereals, in-between paper methods (BP) were used; the seeds were placed on one moist paper with a second paper on top, and thereafter rolled and placed vertically in plastic. Inside the roll, the seeds germinated, and the seedlings developed. For grasses, red clover, and vegetables (except for beet), a Jacobsen apparatus was used [34], which is a plate where circular pieces of wet filter paper and seeds are placed and kept moist by a wick (TP). The seeds of beet were germinated between humidified pleated filter paper. The germination temperature varied from alternating between +20 °C for 16 h and +30 °C for 8 h to constant ambient temperatures. Lettuce required light for germination, but for the other species light was not applied. Tests of actual moisture content were made along with seed germination tests in year 0 and the first 10 years (Table 2).

4.3. Data Analyses

R software was used for statistical examination. One value was missing for year 5 for red clover (cultivar Jokioinen). It was replaced with the average of the four nearest neighbor values: 76.0%. The short descriptor names of the eleven test results are year 0, year 2.5, year 5, year 7.5, year 10, year 12.5, year 15, year 17.5, year 20, year 25, and year 30, where year 0 represents the results of the trial that started in 1986. Summary statistics and boxplots were used to overview the data. The R function 'time series' was used to illustrate the fluctuation of the results over the 11 test occasions over the first 30 years. We calculated the loss in germination (Δ germ) by averaging the germination percentage of the three first test occasions (y0, y2.5, and y5) minus the test occasion at y30. We then categorized the samples into three: the most short-lived group (with Δ germ of 15% or more, which is at a level where regeneration should have been carried out), the intermediate group (Δ germ 5% to 15%), and the most long-lived (Δ germ less than 5%, or with less than 95% germination loss over the 30 years).

Supplementary Materials: The following are available online at http://www.mdpi.com/2223-7747/9/5/579/s1, Figure S1: Illustration of germination for each seed lot included in the trial.

Author Contributions: Conceptualization, F.Y.; Methodology, G.B., F.Y., and E.M.; Formal analysis, F.Y.; Investigation, S.Ø.S. and F.Y.; Resources, Å.A.; Data curation, F.Y.; Writing—Original draft preparation, S.Ø.S.; Writing—Review and editing, C.A., Å.A., G.B., and R.v.B.; Visualization, F.Y.; Supervision, Å.A. and R.v.B.; Project administration, Å.A.; Funding acquisition, S.Ø.S. and Å.A. All authors have read and agreed to the published version of the manuscript.

Funding: We also thank the Nordic Joint Committee for Food and Agricultural Research for network funds to publish the results from the first 30 years of the long-term experiment (Grant number SLU 202100-2817).

Acknowledgments: We would like to thank Store Norske Spitsbergen Kulkompani for the seed storage and for organizing transport. A special thanks to the technical staff at the Kimen Seed Laboratory/Norwegian State Seed Testing Station for the exact and laborious germination and moisture tests, and to H. Tangerås who administered, and was responsible for, the laboratory work for most of the 30 years. We would also like to honor the Norwegian Ministry of Agriculture and Food and the Nordic Gene Bank/NordGen staff involved in this long-term seed trial.

Conflicts of Interest: The authors declare no conflict of interest. The funders had no role in the design of the study, in the collection, analysis, or interpretation of data, in the writing of the manuscript, or in the decision to publish the results.

References

1. Ellis, R.H.; Roberts, E.H. Improved equations for the prediction of seed longevity. *Ann. Bot.* **1980**, *45*, 13–30. [CrossRef]
2. Priestley, D.A. *Seed Aging: Implications for Seed Storage and Persistence in the Soil*; Cornell University Press: Ithaca, NY, USA, 1986.
3. Roos, E.E. Precepts of successful seed storage. In *Physiology of Seed Deterioration, Special Publication no. 11*; McDonald, M.B., Nelson, C., Eds.; Crop Science Society of America: Madison, WI, USA, 1986; pp. 1–25.
4. Ellis, R.H. The longevity of seeds. *HortScience* **1991**, *26*, 1119–1125. [CrossRef]
5. Frankel, O.; Hawkes, J. *Crop Genetic Resources for Today and Tomorrow*; Cambridge University Press: Cambridge, UK; New York, NY, USA, 1975; 492p.
6. Pistorius, R. *Scientists, Plants and Politics – A History of the Plant Genetic Resources Movement*; International Board for Plant Genetic Resources: Rome, Italy, 1997; 134p.
7. FAO. *The Second Report on the State of the World's Plant Genetic Resources for Food and Agriculture*; Food and Agriculture Organization of the United Nations: Rome, Italy, 2010.
8. Plucknett, D.L.; Smith, N.J.H.; Williams, J.T.; Anishetty, N.M. *Gene Banks and the World's Food*; Princeton University Press: Princeton, NJ, USA, 1987.
9. Harlan, J.R. *The Living Fields. Our Agricultural Heritage*; Cambridge University Press: Cambridge, UK; New York, NY, USA, 1995; 273p.

10. Walters, C. Principles for preserving germplasm in genebanks. In *Ex Situ Plant Conservation: Supporting Species Survival in the Wild*; Guerrant, E., Havens, K., Maunder, M., Eds.; Island Press: Covelo, CA, USA, 2004; pp. 442–453.
11. Fowler, C.; Hodgkin, T. Plant genetic resources for food and agriculture: Assessing global availability. *Annu. Rev. Environ. Resour.* **2004**, *29*, 143–179. [CrossRef]
12. Frankel, O. Genetic resources: The founding years. Part IV: After twenty years. *Diversity* **1987**, *3*, 25–27.
13. FAO. *Genebank Standards for Plant Genetic Resources for Food and Agriculture*; Food and Agriculture Organization of the United Nations: Rome, Italy, 2014.
14. Westengen, O.T.; Jeppson, S.; Guarino, L. Global *ex-situ* crop diversity conservation and the Svalbard Global Seed Vault: Assessing the current status. *PLoS ONE* **2013**, *8*, e64146. [CrossRef] [PubMed]
15. Yndgaard, F. Genebank security storage in permafrost. *Plant Genet. Resour. Newsl.* **1985**, *62*, 2–7.
16. Priestley, D.A.; Cullinan, V.I.; Wolfe, J. Differences in seed longevity at the species level. *Plant Cell Environ.* **1985**, *8*, 557–562. [CrossRef]
17. Nagel, M.; Börner, A. The longevity of crop seeds stored under ambient conditions. *Seed Sci. Res.* **2010**, *20*, 1–12. [CrossRef]
18. Rincker, C.M. Germination of forage crop seeds after 20 years of subfreezing storage. *Crop Sci.* **1983**, *23*, 229–231. [CrossRef]
19. Roos, E.E.; Davidson, D.A. Record longevities of vegetable seeds in storage. *HortScience* **1986**, *27*, 393–396. [CrossRef]
20. Walters, C.; Wheeler, L.; Grotenhuis, J.M. Longevity of seeds stored in a genebank: Species characteristics. *Seed Sci. Res.* **1992**, *15*, 1–20. [CrossRef]
21. Desheva, G. The longevity of crop seeds stored under long-term condition in the national gene bank of Bulgaria. *Agriculture (Polnohospodárstvo)* **2016**, *62*, 90–100. [CrossRef]
22. Cromarty, A.S.; Ellis, R.H.; Roberts, E.H. *The Design of Seed Storage Facilities for Genetic Conservation*; International Board for Plant Genetic Resources: Rome, Italy, 1982; 96p.
23. Austin, R.B. Effect of environment before harvesting on viability. In *Viability of Seeds*; Roberts, E.H., Ed.; Chapman and Hall: London, UK, 1972; pp. 114–149.
24. Justice, O.L.; Bass, L.N. *Principles and Practices of Seed Storage*; USDA Handbook on Agriculture No. 506; USDA: Washington, DC, USA, 1978; 289p.
25. Pieta Filho, C.; Ellis, R.H. The development of seed quality in spring barley in four environments. I. Germination and longevity. *Seed Sci. Res.* **1991**, *1*, 163–177. [CrossRef]
26. Dias, D.C.F.S.; Ribeiro, F.P.; Dias, L.A.S.; Silva, D.J.H.; Vidigal, D.S. Tomato seed quality in relation to fruit maturation and post-harvest storage. *Seed Sci. Technol.* **2006**, *34*, 691–699. [CrossRef]
27. Rao, N.K.; Dulloo, M.E.; Engels, J.M.M. A review of factors that influence the production of quality seed for long-term conservation in genebanks. *Genet. Resour. Crop Evol.* **2017**, *64*, 1061–1074.
28. Nagel, M.; Arif, M.A.R.; Rosenhauer, M.; Börner, A. *Longevity of Seeds—Intraspecific Differences in the Gatersleben Genebank Collections*; Tagung der Pflanzenzüchter und Saatgutkaufleute Österreichs, Lehr- und Forschungsanstalt für Landwirtschaft Raumberg-Gumpenstein: Irdning, Austria, 2010; Volume 60, pp. 179–181.
29. Steiner, A.; Ruckenbauer, P. Germination of 110-year-old cereal and weed seeds, the Vienna Sample of 1877. Verification of effective ultra-dry storage at ambient temperature. *Seed Sci. Res.* **1995**, *5*, 195–199. [CrossRef]
30. Probert, R.; Adams, J.; Coneybeer., J.; Crawford, A.; Hay, F. Seed quality for conservation is critically affected by pre-storage factors. *Australian J. Bot.* **2007**, *55*, 326–335. [CrossRef]
31. Nagel, M.; Vogel, H.; Landjeva, S.; Buck-Sorlin, G.; Lohwasser, U.; Scholz, U.; Börner, A. Seed conservation in *ex situ* genebanks-genetic studies on longevity in barley. *Euphytica* **2009**, *170*, 5–14. [CrossRef]
32. ISTA. International rules for seed testing. *Seed Sci. Technol.* **1985**, *13*, 299–355.
33. ISTA. *International Rules for Seed Testing*; International Seed Testing Association: Basserdorf, Switzerland, 2017.
34. ISTA. International rules for seed testing, Chapter 5. In *The Germination Tests*; International Seed Testing Association: Basserdorf, Switzerland, 2011; p. 16.

plants

MDPI

Review

Progress and Challenges in Ex Situ Conservation of Forage Germplasm: Grasses, Herbaceous Legumes and Fodder Trees

Jean Hanson [1,*] and Richard H. Ellis [2]

1 International Livestock Research Institute, P.O. Box 5689, Addis Ababa, Ethiopia
2 School of Agriculture, Policy and Development, University of Reading, PO Box 237, Reading RG6 6AR, UK; r.h.ellis@reading.ac.uk
* Correspondence: jeanhanson2010@gmail.com

Received: 19 February 2020; Accepted: 12 March 2020; Published: 2 April 2020

Abstract: Forages provide an important livestock feed resource globally, particularly for millions of smallholder farmers, and have important roles in natural resource management and carbon sequestration, reducing soil erosion and mitigating the effects of climate change. Forage germplasm remains the basis for the selection and development of new, higher-yielding and better adaptedgenotypes to meet the increasing demand for livestock feed. Rapid rates of genetic erosion of forage diversity due to land-use change from natural pastures and rangelands to crop production to meet the food security requirements of a growing global population, together with pressures from a changing climate, highlight the necessity for ex situ seed conservation of forage genetic resources to provide germplasm for use by future generations. Whilst many forage species have orthodox seeds, the diverse range of genera and species which provide forage is a challenge in terms of the wide scope of information and understanding on conservation methods that genebank managers require—particularly for tropical forages, many of which are comparatively under-researched. We review the challenges to the conservation of tropical forage species by seed in ex situ genebanks and provide information on optimum methods for their management.

Keywords: genebanks; forage germplasm; grasses; legumes; seed storage; conservation; seed longevity; seed germination; monitoring; regeneration

1. Introduction

The term 'forage' encompasses many different types of plants used for livestock feed, including grasses, herbaceous and tree legumes, as well as other non-leguminous forbs and trees varying greatly in plant habit and adaptation. Forages are an important livestock feed resource globally, particularly for millions of smallholder farmers who rely on natural pastures and grasslands as the basis of their sustainable livestock systems, allowing ruminants to convert feed that cannot be used directly by humans into milk and meat to provide essential nutrients that are required for human health, growth and cognitive development [1]; these animals also provide fibre and skin for clothing and footwear. Forages are effective in maintaining the natural resource base [2]. They stabilise the soil, provide ground cover and windbreaks to prevent or reduce soil erosion, increase soil carbon content by strong rooting and decomposition of leaf litter, and, through symbiotic nitrogen fixation by rhizobia with legumes, they capture nitrogen from the atmosphere to the soil. There has been much research aimed at capturing these benefits, and there are now many alternative ways of introducing sown forages into the farming system [2].

Most forage diversity originates in natural grasslands, one of the largest and most important natural ecosystems [3] covering over 52 million square kilometers globally. Grasslands are home to over a thousand forage species with actual or potential use to support livestock production systems in tropical and sub-tropical areas, although only about 70 species have been commercially developed, mostly through the direct selection of germplasm accessions or by plant breeding in few important species, as feeds in these regions (Table 1). Rapid rates of genetic erosion of forage diversity are occurring in grasslands, mainly due to land-use change converting natural pasture and marginal areas to crop production to meet the food security requirements of a growing global population. This erosion together with pressures from a changing climate, resulting in extreme weather events (such as drought, floods and cyclones), has a negative effect on natural habitats where forage diversity is still found and highlights the necessity for ex situ conservation of forage genetic resources to provide diverse forage germplasm for the use of current and future generations. Most forages in use today are wild species or genotypes selected from wild populations and safeguarding sources of forage germplasm is particularly important because with few forage breeding programmes [4]—and most of those focusing mainly on temperate grassland species, including alfalfa, ryegrasses and clovers—forage germplasm remains the basis for the future selection and development of the new feeds which are particularly needed for tropical and sub-tropical regions.

Table 1. Common tropical and sub-tropical forage species and climate zone suitability for cultivation.

Forage Species	Climatic Zone of Cultivation				
	Arid	**Semi-Arid**	**Sub-Humid**	**Humid**	**Highland**
	Herbaceous legumes				
Aeschynomene americana			√		
Alysicarpus glumaceus		√			
Arachis pintoi			√	√	
Calopogonium mucunoides			√	√	
Centrosema acutifolium			√	√	
Centrosema brasilianum		√			
Centrosema macrocarpum				√	
Centrosema pascuorum		√	√		
Centrosema pubescens			√	√	
Chamaecrista rotundifolia		√	√		
Clitoria ternatea	√	√	√		
Desmodium intortum			√		√
Desmodium uncinatum			√		
Lablab purpureus	√	√	√		
Macroptilium atropurpureum		√	√		
Macrotyloma axillare			√		√
Medicago sativa					√
Neonotonia wightii		√	√		
Pueraria phaseoloides			√	√	
Rhynchosia minima		√			
Stylosanthes fruticosa	√	√			
Stylosanthes guianensis		√	√	√	
Stylosanthes hamata	√	√	√	√	
Stylosanthes scabra		√	√	√	
Stylosanthes seabrana			√	√	
Teramnus labialis			√		
Trifolium decorum					√
Trifolium rueppellianum					√
Trifolium pratense					√
Trifolium quartinianum					√
Trifolium repens					√
Trifolium semipilosum					√
Trifolium steudneri					√
Trifolium tembense					√
Vicia villosa					√
Vicia sativa					√
Vigna unguiculata	√	√	√		
Zornia glabra		√	√		
Zornia latifolia		√	√		

Table 1. *Cont.*

Forage Species	Climatic Zone of Cultivation				
	Arid	Semi-Arid	Sub-Humid	Humid	Highland
Fodder trees					
Cajanus cajan		√	√		
Calliandra calothyrsus			√		
Cytisus proliferus					√
Desmanthus virgatus			√	√	
Faidherbia albida		√			
Gliricidia sepium		√	√	√	
Leucaena diversifolia					√
Leucaena leucocephala		√	√	√	
Leucaena pallida					√
Leucaena revoluta					√
Sesbania sesban		√	√		√
Grasses					
Avena sativa					√
Bothriochloa pertusa			√		
Brachiaria decumbens			√	√	
Cenchrus ciliaris		√	√		
Chloris gayana		√	√		
Cynodon dactylon		√	√		
Melinis minutiflora			√		
Panicum coloratum		√	√		√
Panicum maximum			√	√	
Paspalum dilatatum			√		
Paspalum plicatulum			√		√
Pennisetum clandestinum					√
Pennisetum purpureum			√	√	
Setaria sphacelata			√	√	
Sorghum almum		√	√		
Urochloa mosambicensis			√	√	

Zone definitions; Arid zone-100–500 mm rainfall, 0–180 growing days per annum; Semi-arid zone-600–1000 mm rainfall, 0–180 growing days per annum; Sub-humid zone-1000–1500 mm rainfall, 180–270 growing days per annum; Humid zone- >1500 mm rainfall, >270 growing days per annum; Highland zone- >1500 m altitude.

Forages are not as well represented as food crops in ex situ collections with almost 182,000 accessions representing over 1000 species of grasses, legumes and fodder trees maintained in 80 national and international genebanks registered in Genesys (www.genesys-pgr.org) compared to the approximately 7.4 million plant accessions stored in around 1750 genebanks globally [5]. These accessions registered in Genesys were collected from a wide range of sites across tropical and temperate regions (Figure 1) and contain a large amount of diversity with potential use in forage breeding. However, despite the diversity available in ex situ collections, the germplasm from these genebanks remains underutilised in crop and forage breeding programmes. Current genomic technologies that can be used to screen large collections efficiently offer opportunities for increased identification of useful genes and the use of germplasm [5].

Figure 1. Map of forage accession collection sites. Data accessed through Genesys Global Portal on Plant Genetic Resources, www.genesys-pgr.org, 12 February 2020.

2. Management of Forage Germplasm

Seed genebank protocols have been developed with both crop and wild species in mind, but the focus of most genebanks is very much crop genetic resources conservation with an acknowledgement of the priority to include crop wild relatives in collections. Whilst the knowledge base and dominant thinking was around conservation of crop plants, the same principles have been applied successfully to the ex situ conservation of wild plants [6,7]. The principle difference, and source of problems in implementing standard protocols, is that seeds from wild populations, crop wild relatives and forages tend not to show the high intra-seed lot uniformity of crop seeds [7]. Seed of forage species show considerable variability within and between seed lots and, similarly, within and between accessions of a species that make it difficult to develop inclusive protocols.

In the case of forage species, the information available on seed handling from seed collection through storage to germination, multiplication and regeneration is less than for the major staple food crops, and many of the problems facing forage plant genebank managers are similar to those facing those managing wild species genebanks:

- A wide range of diverse species;
- High variability amongst and within each species;
- Comparatively little relevant published information for many species;
- and limited practical experience with some species.

Nonetheless, it is wrong to imply that little is known about seeds of forages: Table 2 summarises examples of selected sources of information on protocols and advice on seed production, dormancy, germination, and survival in genebank storage for various genera of forage grasses and forage legumes. The International Seed Testing Association (ISTA) has included germination test methods for many forage species in the recent rules for seed testing [8]. Detailed information on seed longevity of forages is also included in several publications [9–12]. Even for species or genera in which advice cannot be found, such information sources are useful to indicate how to handle seed of similar species or genera. Nonetheless, some of this information and advice was developed for commercial seed production and may need to be interpreted and adapted for use by genebanks to support the ex situ conservation of forage genetic resources. Some of the particular difficulties in handling forage seed accessions in genebanks stem from the tendency of forage species to show substantial within-accession variation in the timing of flowering, seed development and maturation; they reflect plant diversity in general. As a result, few generalisations are possible when discussing the production, harvesting, and pre-storage processing of seeds of forage species. These diverse species can be determinate or indeterminate in their development; annual, biennial, or perennial; flowers can be self-fertile, require cross-pollination or are apomictic, or a mixture of the three; and flowering may often continue for several weeks [13–19]. Moreover, the spike in most grasses is fragile, with abscission at the rachis or rachilla, and thus, seeds shed to the ground as they mature [20]. Once mature, the seeds also differ in dormancy, germination requirements and in survival during storage [21].

Table 2. Examples of information sources (reference numbers shown) on methods of seed production; dormancy (and dormancy-breaking methods), germination (germination testing and promotion for accession monitoring), and seed survival in genebanks (seed survival periods of different genera of forage grasses and forage legumes in genebanks).

Family/Genus	Seed Production	Dormancy/ Germination	Seed Survival in Genebanks
Fabaceae			
Acacia		[8,21]	[9,11]
Aeschynomene	[13]		[11]
Albizia			[11]
Alysicarpus	[13,15]		[11]
Argyrolobium			[11]
Astragalus	[15,17]	[8,21]	
Cajanus			[11]
Calopogonium	[13]	[8,21]	[11]
Canavalia			[11]
Cassia	[13]		[11]
Centrosema	[13,17,19]	[8,21]	[11]
Chamaecrista			[11]
Clitoria			[11]
Coronilla	[15,17]	[8,21]	
Crotalaria	[15]	[8,21]	[11]
Desmanthus			[11]
Desmodium	[13]	[8,21]	[11]
Entada			[11]
Erythrina			[11]
Faidherbia			[11]
Galactia			[11]
Gliricidia			[11]
Glycine		[8,21]	[9,11]
Indigofera	[15]	[21]	[11]
Lablab	[13]	[8,21]	[11]
Lathyrus	[15]	[8,21]	[9,11]
Lespedeza	[15,17,19]	[8,21]	[9]
Leucaena	[13,17]	[21]	[11]
Lotononis	[13]	[8,21]	[11]
Lotus	[15,17,19]	[8,21]	[9,11]
Lupinus	[15]	[8,21]	[9,11]
Macroptilium	[17,19]	[8,21]	[11]
Macrotyloma			[11]
Medicago	[15,17,19]	[8,21]	[9,11]
Melilotus	[15,17,19]	[8,21]	[9,11]
Mucuna		[8,21]	[11]
Neonotonia			[11]
Onobrychis	[15,17,19]	[8,21]	[9]
Ornithopus	[17,19]	[8,21]	[11]
Phaseolus		[8,21]	[9,11]
Pisum	[15]	[8,21]	[9,11]
Prosopis			[11]
Pseudarthria			[11]
Psophocarpus		[8,21]	[11]
Pueraria	[15]	[8,21]	
Rhynchosia			[11]
Senna			[9,11]
Sesbania		[21]	[11]
Stizolobium	[15]		
Stylosanthes	[13,17,19]	[8,21]	[11]
Tephrosia		[8,21]	[11]
Teramnus			[11]
Trifolium	[13,15,17,19]	[8,21]	[9–11]
Vicia	[15,17,19]	[8,21]	[9,11]
Vigna	[15]	[8,21]	[9,11]
Zornia	[13]		[11]

Table 2. *Cont.*

Family/Genus	Seed Production	Dormancy/ Germination	Seed Survival in Genebanks
Poaceae			
Aegilops		[21]	
Agropyron	[15,17,19]	[8,21]	[9,12]
Agrostis	[15,17]	[8,21]	[9,10]
Alopecurus	[17,19]	[8]	
Andropogon	[13,15,17,19]	[8,21]	[12]
Arrhenatherum	[15,17,19]	[8]	
Aristida		[21]	[12]
Avena		[8,21]	[9,12]
Axonopus	[15]	[8]	
Bothriochloa	[13]	[8,21]	[12]
Bouteloua	[15,17]	[8,21]	
Brachiaria	[13,17,19]	[8,21]	[12]
Bromus	[15,17,19]	[8,21]	[9,12]
Buchloe	[15]	[8]	
Cenchrus	[13,17]	[8]	[12]
Chloris	[13,15,19]	[8,21]	[12]
Cymbogon		[21]	
Cynodon	[15]	[8,21]	[12]
Dactylis	[15,17,19]	[8,21]	[9,10,12]
Digitaria		[8,21]	[12]
Echinochloa		[8,21]	[9,12]
Eleusine		[8,21]	[9,12]
Elymus	[15,17,19]	[8]	[9,12]
Eragrostis	[13,15,17]	[8,21]	[9,12]
Eremochloa	[15]		
Festuca	[15,19]	[8,21]	[9,10,12]
Hordeum	[15]	[8,21]	[9,12]
Hyparrhenia	[13]		
Lolium	[15,17]	[8,21]	[9,12]
Melinis	[17]	[8]	[12]
Oryzopsis	[15]	[8,21]	
Panicum	[13,15,17]	[8,21]	[9,12]
Paspalum	[13,15,17]	[8,21]	[12]
Pennisetum	[13,15,17]	[8,21]	[9,12]
Phalaris	[15,17]	[8,21]	[12]
Phleum	[15,17,19]	[8,21]	[9,10,12]
Poa	[15,17]	[8,21]	[9,10]
Setaria	[13,17,19]	[8,21]	[9,12]
Sorghastrum	[15]	[8,21]	
Sorghum	[17,19]	[8,21]	[9,12]
Sporobolus	[15]		[9]
Stipa	[15]	[21]	
Themeda		[21]	
Tripsacum	[15]		
x*Triticosecale*		[8,21]	[12]
Urochloa	[13,17]	[8]	[12]

3. Germplasm Collecting

One major difference between cultivated and wild forages and the major food crops is the collection location and strategy. Wild forages are found in grasslands and natural pastures, either with small numbers of plants distributed over a large area or as large areas of single species in open grasslands, compared to crop germplasm, which is found as landraces in farmers' fields. Whilst crop landraces have more uniform maturity and little seed shattering, forages tend to vary within accessions in the timing of seed maturity, and in some grasses, seed shattering occurs as seeds ripen or pods dehisce in some forage legumes. This results in some seeds being collected whilst they are immature and unable to survive the drying and storage process, resulting in poor quality or small numbers of seeds being stored [6]. In some species, a post-harvest ripening treatment can be applied, keeping seeds in high humidity to facilitate ripening [22]. With only small numbers of seeds being collected for individual accessions, seed multiplication is usually required before conservation compared to crop landraces, which can often be collected in the required quantity of seeds to go straight for conservation, safety duplication and distribution. In some cases where the species covers large areas, such as in native grasslands, random collection strategies can be applied similar to those used for crops. Some forages must be collected as cuttings because at collection time there are few ripe seeds due to shattering, as reported in the germplasm of *Panicum coloratum* [23], or the species is a poor seeder and only a small percentage of the florets result in caryopses, as in Napier grass (*Pennisetum purpureum*) [24].

4. Multiplication and Regeneration

Regeneration and multiplication are especially challenging for forages; it can be a slow process for perennial species and accessions represented by few seeds. As explained above, the original sample size of many accessions is small, and most need to be multiplied at the outset to provide sufficient seeds for long term storage. In the absence of information on the amount of outcrossing in many forage species, each accession should be assumed to be fully outcrossing and thus isolated to ensure the maintenance of genetic integrity. In addition, there is a high risk of loss of diversity and changes in genetic integrity due to the small sample size available to use for the regeneration of accessions with few seeds [25]. Accessions of slow-growing and maturing fodder trees often remain in the field for several years before they are capable of flowering to produce enough seeds to meet storage, distribution and safety duplication needs. This results in limited availability of accessions of some species. General guidelines for regeneration procedures for forages have been developed for grasses and forage legumes [26,27].

Protocols for large-scale seed production and multiplication and the basic problems inherent in forage seed multiplication, collection, and regeneration are similar but not identical. Many forages are recognised as crops, but they tend to be selected for vegetative rather than reproductive traits. Thus selection for long periods of vigorous vegetative growth by forage plant breeders to benefit livestock production has had to be tempered by the requirement for the plants to ultimately flower and produce sufficient seeds if new varieties are to be multiplied and disseminated [28]. Moreover, many forages have not been subject to plant breeding, which has focused on *Medicago* species, clovers, ryegrasses, oats, *Brachiaria* grass, Guinea grass and Napier grass. These species have been used in crossing programmes and named varieties are commercially available from seed companies, whilst other forage species are selections from the wild and are naturally adapted to a wide range of contrasting habitats. For range grassland species, for example, to persist requires seeds able to survive in the soil seedbank for considerable periods after shedding, and/or to survive passage through the animal gut. Such characteristics require traits of (considerable) seed dormancy and/or hardseededness, which hinder the promotion of prompt germination when accessions are sown for testing, regeneration, or utilisation. Hence the promotion of seed germination can require considerable intervention, including the use of growth promoters, such as potassium nitrate and gibberellic acid, as well as seed coat removal in grasses and testa scarification in legumes and/or long test periods [21,29].

The major consequence of these inherent traits of forages for genebank regeneration is that as seed harvest time approaches, the seeds are at different stages of development and maturation on any one day, with the likelihood of seed being shed from plants increasing as the harvest is delayed, whilst other seeds ripen. Shedding not only reduces the numbers of seed harvested but may also cause subsequent confusion in the gene bank. Seeds of grass accessions may appear dormant to some, whereas the problem is one of "empty seeds" whereby seed were shed, but the firm seed covering structures may give the impression that the seed remains present, e.g., in *Brachiaria humidicola* [30]. Three different methods are widely used to identify and quantify empty seeds within accessions [29]: dissection (particularly of seeds that fail to germinate when tested); X-radiography; and (rarely used by genebanks to identify empty seeds but routinely used for seed cleaning) the use of a seed blower to separate by density the empty seed fraction. X-radiography can be combined with a subsequent germination test on the same seed sample and so save valuable seeds and provide a permanent image for records [29]. "Empty seeds" are not expected in forage legumes, but X-ray images can identify poorly-developed seeds and also those with insect damage.

The decision as to when to harvest forage seeds is, therefore, a compromise between the quality and uniformity of the seed and maximising their number. Consequently, many genebanks undertake the time- and labour-consuming daily sequential harvesting of ripe seeds: first, to maximise the number of seeds harvested; and second, to reduce selection and change in genetic integrity of the accession which would occur if only early- or late- flowering and maturing plants were selected.

In terms of when to harvest seed, there is no single point in developmental time across all species and all environments where seed quality (subsequent seed storage longevity) is maximal [31]. Nonetheless, harvesting when seeds are at, or almost at, harvest maturity (i.e., their moisture content is close to equilibrium with ambient relative humidity) is a reasonable solution to this dilemma because although maximal seed quality may have been attained prior to this time net seed deterioration is rarely observed on any scale before harvest maturity [31]. This tallies with forage seed production practices where the harvest is when the earliest to develop seeds or pods change colour as they mature visually, to, for example, yellow or brown, or shortly before they would be shed; and similarly in wild species more generally [6,32]. Whilst the lack of uniformity in seed maturation in forages means that the seed crop varies in the timing of harvest maturity, where direct harvesting is not possible, forage seed producers use a variety of seed harvesting approaches to reduce the consequences of variation in developmental timing (Table 3). Combing, bagging, and suction tend to be associated with multiple, sequential harvests of seed from plants where seed maturity date varies considerably—and these approaches are all used by forage plant genebanks.

Table 3. Seed harvesting techniques for accessions which show considerable plant-to-plant variation in the timing of seed maturity.

Detail	Swathing	Swathing and Sweating	Desiccation	Combing	Bagging	Suction
Description	Cutting the seed crop and leaving in a loose swath to dry for later mechanical harvesting	Heaping or sheathing seed heads for about 3 days before threshing to raise temperature	Spraying chemical desiccant on the crop to accelerate the drying of less-mature seed	Running a hand over the seed heads or shaking seed heads gently into a bag	Covering seed heads or pods with a net, cloth or paper bag	Vacuuming mature seeds from the plant or ground under the plant
Advantages	Maturation of later-developing seeds as moisture uptake ends and swath dries; capture of more mature seeds in the swath if they shatter; low labour requirement	Aids maturation and abscission [33]	Enables direct harvesting of dry seeds; low labour requirement	Only mature seeds are collected	Captures seeds that shatter or dehisce	Immature seeds remain on the plant to mature and ripe seeds are not lost by shattering
Disadvantages	Some seeds will be lost from the swath; specialist equipment	Immature seeds may also abscise; prolonged treatment may damage seeds; specialist equipment	Cost and risk of desiccant; specialist equipment	Labour intensive; repeated, frequent combing required	Labour intensive; fungal contamination in bags during rains	Labour intensive; weed seeds and debris can be collected from the ground
Examples	*Stylosanthes* species; temperate grasses	*Panicum maximum*; tropical grasses	*Vicia faba*; *Avena sativa*	*Vicia villosa*	*Brachiaria* species	*Stylosanthes* species; *Buchloe dactyloides* [17]

5. Seed Longevity

Monitoring the viability of seed of accessions in large collections regularly is costly in terms of seeds, labour and other resources, and it is important, therefore, to prioritise monitoring those accessions with brief longevity [34]. Assessing the seed storage longevity of wild species in genebanks is complex and difficult to predict as it is influenced by genotype, production environment, post-harvest handling and processing and storage conditions [7], and until recently, there was limited information on the longevity of many species [6]. Lack of data for evidence-based longevity predictions for forages led to species being classed as good or poor storers [35]. However, recent analysis of over 30 years of germination monitoring data from seeds stored in the medium-term store (seeds stored at circa 8 °C with 5% moisture content) in the genebank at ILRI allowed the calculation of seed longevity under these conditions for a range of forage genera and identified many, but not all, forage grass and legume species as having long-lived orthodox seeds [11,12]. Variation was observed in seed longevity among genera and between species of the same genus and indicated that some forage seeds have minimal dormancy and a longevity comparable to seeds of the major food crops, whilst other species showed high levels of hardseededness or dormancy that required dormancy-breaking procedures to be used in the germination test protocols.

Advice on breaking seed dormancy and promoting germination for each of 58 plant families is already published [21]. Typical successful dormancy-breaking procedures suggested therein for grasses include a germination test environment of a constant 16–21 °C or alternating temperature of 23/9 °C (12 h/12 h) for temperate accessions, or a constant 21–26 °C or alternating temperature of 33/19 °C (12 h/12 h) for tropical accessions, and/or 10^{-3} M potassium nitrate, removal of seed-covering structures, and pre-chilling at 2–6 °C for up to 8 weeks. One caution is that although (white) light can promote seed germination, the germination of some grasses can be inhibited by high light intensity: for example, in the temperate grasses *Bromus mollis* and *Bromus sterilis* [36] and the tropical grasses *Echinochloa turnerana*, *Panicum maximum* and *Brachiaria humidicola* [37]. This can be avoided by limiting the period of exposure to light each day, as well as the dose. Typical successful dormancy-breaking/germination promoting procedures suggested for legumes [21] are seed scarification and a germination test environment of a constant 11–16 °C or alternating temperature of 23/9 °C (12 h/12 h) for temperate accessions, or a constant 21–26 °C or alternating temperature of 33/19 °C (12 h/12 h) for tropical accessions. One caution is that very dry legume seeds can be sensitive to a rapid uptake of water and so may benefit from initial hydration in a moist atmosphere at 100% relative humidity. A further source of readily-available information on characteristics of seed of diverse species relevant to genebanks is the Seed Information Database (Royal Botanic Gardens Kew, http://data.kew.org/sid/) [38]. This online compilation of various types of seed information from a wide range of sources, including for seed storage behavior [39], is provided to support seed genebank operations globally by the Millennium Seed Bank Project.

Monitoring seed viability in a large collection of many individual accessions using germination tests is labour intensive, time-consuming, and depletes the amount of seeds available for each accession. Hence monitoring should be kept to a minimum, but it is nonetheless essential to determine when germination declines and thus regeneration becomes necessary to avoid the loss of that accession from the genebank. The rational determination of accession monitoring interval is important for efficient, low-cost, but effective genebank management. Analysis of historic genebank data has been applied to provide evidence-based estimates of the different monitoring intervals necessary for diverse forage species [11,12] and indicated that 15-year monitoring intervals are suitable for many long-lived forage seeds with high viability at the time of entering storage. Even in medium-term seed stores, high-quality seed of some forages need only infrequent monitoring (Table 4).

There is a virtuous circle amongst high initial accession viability, excellent storage environment, monitoring period, and regeneration period for the genetic integrity of accessions. The first two support infrequent monitoring and yet more infrequent regeneration, whilst loss in viability during storage, accession depletion through frequent monitoring, and frequent regeneration each put the genetic integrity of accessions at risk [24,40]. Moreover, accession monitoring and regeneration are expensive undertakings. It is essential to retain much larger samples for regeneration than for distribution to users if minor alleles at a low frequency in a genetically heterogeneous accession are to be conserved through cycles of storage and regeneration (infrequent if seed viability is maintained long term). One analysis suggests that, if feasible, large samples of 500 seeds be grown out to regenerate the most original seed source to maintain allelic richness within an accession containing many alleles at low frequency [24].

Table 4. List of forage legume species with long-lived seeds under medium-term storage (circa 8 °C with 5% moisture content) at ILRI for which monitoring intervals of 15 years for high-quality seed accessions under such storage conditions are proposed.

Genus	Specific Epithet of the Species
Acacia	*angustissima, boliviana*
Alysicarpus	*ferrugineus, glumaceus, longifolius, monilifer, ovalifolius, rugosus, vaginalis*
Calapogonium	*mucunoides*
Chamaecrista	*mimosoides, nigricans, pilosa, rotundifolia*
Desmanthus	*acuminatus, covillei, leptophyllus, pubescens, tatuhyensis, virgatus*
Desmodium	*adscendens, barbatum, cinereum, dichotomum, discolor, distortum, incanum, intortum, salicifolium, sandwicense, tortuosum, uncinatum, velutinum*
Indigofera	*arrecta, brevicalyx, colutea, cryptantha, hirsuta, hochstetteri, spicata, suffruticosa*
Leucaena	*diversifolia, leucocephala, pallida, pulverulenta, shannonii, trichandra, trichodes*
Lupinus	*albus, angustifolius, luteus, mutabilis*
Macroptilium	*atropurpureum, bracteatum, lathyroides*
Macrotyloma	*africanum, axillare, daltonii, uniflorum*
Medicago	*lupulina, minima, polymorpha, sativa, scutellata, truncatula*
Melilotus	*albus, officinalis*
Pisum	*sativum*
Stylosanthes	*calcicola, capitata, fruticosa, guianensis, hamata, humilis, scabra*
Tephrosia	*bracteolata, noctiflora, pumila, purpurea, villosa*
Teramnus	*labialis, repens, uncinatus*
Trifolium	*baccarinii, bilineatum, burchellianum, cryptopodium, decorum, masaiense, mattirolianum, multinerve, polystachyum, pratense, quartinianum, repens, resupinatum, rueppellianum, semipilosum, simense, steudneri, tembense*
Vigna	*luteola, oblongifolia, parkeri, peduncularis, racemosa, radiata, unguiculata, vexillata*
Zornia	*glabra, latifolia*

6. Cost Efficiency of Managing Wild Species

The biological differences between crop species on the one hand and forage and wild species on the other are reflected in the costs of management and conservation in genebanks. There is limited data available on the actual costs of conservation other than from the genebanks of the centres of the Consultative Group for International Agricultural Research (CGIAR) [41,42]. Using these data as an example, the cost of conservation and management per sample in CIAT in 2000 was more than double for forages than for beans [41]. Costs also differ between locations even for the conservation of the same species. Studies in 2006 and 2009 calculated the costs of acquisition, characterisation, safety duplication, medium and long term storage, germination and seed health monitoring, regeneration, seed processing, information management, distribution and general management and concluded that the cost of forage germplasm conserved in CIAT was $315 per accession, whilst in ILRI these estimates ranged from $125–242 per accession, depending on the type of forage [42]. Such cost comparisons must account for the costs of operating in specific locations worldwide, as well as those incurred by different processes and procedures, and each Centre's prioritisation of species and activities. The reproductive biology of the species, which determines the type of conservation and regeneration procedures, and the quality of management attained in the genebank, contributed most to the variation in the costs of genebank operations [42]. Crops with orthodox seeds had the lowest costs of conservation, whilst vegetatively propagated crops, trees and wild relatives had higher costs.

The size of the collection also contributes to the costs of management. There is a marginal cost to adding an additional accession into a seed genebank with sufficient space to accommodate the seeds, but the benefit to cost ratio is high. The basic questions that genebank managers ask are what is the value of adding additional accessions, which germplasm should genebanks collect and store, and when is the collection large enough and contains the genetic traits demanded by users. The probability of finding accessions of value for users depends on the gene frequency of the trait that is being sought, how many accessions are tested and whether accessions are selected at random or selected based on prior knowledge on where to look for specific genes. Economic principles can be applied to modelling the value of accessions [43]. Sampling more accessions provides a higher probability of finding genes in low frequencies within the accessions but comes at a cost. For rare genes with a probability of 0.01 of presence in accessions, sampling 200 seeds gives an 87% chance of the gene being present in the sample. There is a marginal or incremental probability of finding the gene in the population, and the probability reduces as more accessions are added because if the gene is rare, the marginal value of adding an accession is low because it is difficult to find the gene. If the gene is common, the marginal value of adding accessions is also low because it has probably already been located. The marginal value of adding more accessions is high for intermediate ranges of gene scarcity [43,44]. Economically this supports keeping collections of wild species and forages relatively small, in terms of the number of accessions within a single species, because of their lower probability for use and high costs of conservation resulting from their reproductive biology and conservation challenges.

Cost–benefit analysis could be used to determine if the benefits from using the germplasm in forage development that has been, or could, in future, be realised, justify the conservation costs. However, this would require future germplasm utilisation to be quantified, which is very difficult to do. The value of the germplasm within genebanks lies in the future—when it will be used to transfer useful traits to forage cultivars. The option value of having the germplasm available to respond to as yet unidentified future needs, such as changing feed needs due to climate change or intensifying livestock production systems, and the existence value that society derives from knowing that something exists and will be available for future needs may be more powerful economic drivers of conservation of forage diversity than the cost–benefit analyses that can be done on its actual use [45]. In reality, most genebanks have fixed budgets, and so the management priority is to apply those resources to the best effect.

One example is the relative cost and risk of collecting forages and conserving ex situ in seed or field genebanks or maintaining them in situ in conservation areas within the extensive grasslands where they originated and continue to evolve and adapt. In situ conservation is a good alternative for forage species that are well protected in the wild, provided there is no evidence of a future threat or of historic genetic erosion to these forage genotypes. A good example of this is in sub-Saharan Africa, where many of the indigenous forages are found in well-protected rangelands in national parks [46]. On the other hand, ex situ conservation of forage germplasm is an essential precaution where over-grazing and other poor land management practices occur; it also makes the germplasm more accessible to users.

7. Policy Issues

The legal framework for crop collections is the multilateral system of the International Treaty on Plant Genetic Resources for Food and Agriculture (ITPGRFA) [47]. This covers the major crops where countries have high interdependence on germplasm for food security, allowing them to be freely exchanged under a standard material transfer agreement (SMTA) to be used for food and agriculture with benefit-sharing provisions based on their commercial use. Germplasm flows through facilitated access under the ITPGRFA from the genebanks of the CGIAR Centres are well documented [48,49] with 4 million samples of germplasm made available under the SMTA from 2007–2016 to national partners and breeders, representing 93% of reported global distributions under the multilateral system [49].

The list of crops covered by the ITPGRFA includes some temperate forages and crop wild relatives that are also used as forages, but not many of the common and important tropical forage species in Table 1 of this review. Expanding collections with new germplasm will need to be done under bilateral agreements on access and benefit-sharing under the terms of the Nagoya Protocol. Reaching bilateral agreements on access and benefit-sharing under the Nagoya Protocol will be complicated and, in some cases, may not be possible in the short term as countries reform their national laws on access [50].

Whilst the economic cost–benefit analysis of conserving large ex situ collections of forages combined with the option to conserve forages in secure conservation areas in national parks indicates that there are other more cost-effective options for the conservation of forage diversity in many cases, the forage germplasm currently held in ex situ collections may well be the more accessible source for research and forage development because it avoids the need to collect and to negotiate conditions of access and benefit-sharing to new germplasm. The latter may take time, and so it is important that the germplasm already collected is properly managed and safeguarded in ex situ genebanks and continues to be made available to users.

8. Conclusions

Although collections in genebanks are vulnerable to loss of diversity [51] and the conservation of forage in ex situ collections has many challenges as outlined in this review, seed storage remains the most cost-effective and efficient method for their conservation and sustainable use for the immediate future. Ex situ conservation of forage germplasm must be linked to use in forage development to realise the potential benefits from the high costs of conservation. Unlike crops, where genebank accessions are used in breeding programmes to combine with other genotypes, many forage accessions are selected for direct use based on their adaptation, productivity and nutritional quality and released as cultivars due to the limited number of forage breeding programmes [4] and time and cost required to screen large numbers of accessions for useful traits [5]. To use wild species as forages, the challenges faced in seed dormancy, germination and harvesting for seed production of these species will need to be addressed methodically and research carried out to improve ease of cultivation and support adoption.

There are fundamental questions on which germplasm genebanks should collect and store, optimum size of the collection and how much diversity is sufficient to meet user needs remain open for forages. The answer depends on what types of forages are required for future sustainable livestock systems. Climate and environmental issues are shaping the perception of consuming livestock products in western diets, and people are being urged to reduce meat consumption. Yet in the developing world, many children survive on diets deficient in proteins and animal products and require milk and meat in their diets for cognitive development [52]. Livestock production systems are intensifying in the tropics with an increase in crossbred dairy cows with higher feed requirements. Natural pastureland is reducing in area as grassland and forest are converted for cropping, whilst climates are tending to dry and warm, pushing forage production to marginal land or planted as part of intensified crop-livestock production systems. The intensification of dairy systems through the use of higher quality feed in East Africa is predicted to increase milk yields without increasing greenhouse gas emissions, addressing both productivity and environmental issues [53]. This intensification is leading to the production of more crop residues and by-products for feed and, as a result, the need for higher quality protein-rich forages to supplement these lower quality residues. Additionally, in an effort to reduce greenhouse gas emissions, forage producers could reduce the use of inorganic nitrogen fertilisers replacing them with forage legumes, capable of symbiotic nitrogen fixation, and use more perennial forage shrubs and trees to reduce carbon loss from soils from cultivation. Hence, it is likely that the forages of the future will be developed from accessions of forage legumes and fodder tree species from ex situ collections to meet these needs. Given the anticipated continued demand for forage germplasm, challenges in their management will be reduced through research to support their conservation and availability for forage development.

Author Contributions: Both authors contributed to this paper's conceptualization; writing—original draft preparation, J.H.; writing—review and editing, J.H. and R.H.E. All authors have read and agreed to the published version of the manuscript.

Funding: This research received no external funding.

Conflicts of Interest: The authors declare no conflict of interest.

References

1. Thorning, T.K.; Raben, A.; Tholstrup, T.; Soedamah-Muthu, S.S.; Givens, I.; Astrup, A. Milk and dairy products: Good or bad for human health? An assessment of the totality of scientific evidence. *Food Nutr. Res.* **2016**, *60*, 32527. [CrossRef] [PubMed]
2. Peters, M.; Horne, P.; Schmidt, A.; Holmann, F.; Kerridge, P.C.; Tarawali, S.A.; Schultze-Kraft, R.; Lascano, C.E.; Argel, P.; Stür, W.; et al. *The Role of Forages in Reducing Poverty and Degradation of Natural Resources in tropical Production Systems*; AgREN Network Paper 117; ODI Agricultural and Extension Network: London, UK, 2001; pp. 1–12.
3. Suttie, J.M.; Reynolds, S.G.; Batello, C. *Grasslands of the World*; Plant Protection and Production Series 34; Food and Agriculture Organization of the United Nations: Rome, Italy, 2005.
4. Sandhu, J.S.; Kumar, D.; Yadav, V.K.; Singh, T.; Sah, R.P.; Radhakrishna, A. Recent trends in breeding of tropical grass and forage species. In Proceedings of the 23rd International Grassland Congress, New Delhi, India, 20–24 November 2015.
5. Yu, X.; Li, X.; Guo, T.; Zhu, C.; Wu, Y.; Mitchell, S.E.; Roozeboom, K.L.; Wang, D.; Wang, M.L.; Pederson, G.A.; et al. Genomic prediction contributing to a promising global strategy to turbocharge gene banks. *Nat. Plants* **2016**, *2*, 16150. [CrossRef] [PubMed]
6. Hay, F.R.; Probert, R.J. Advances in seed conservation of wild plant species: A review of recent research. *Conserv. Physiol.* **2013**, *1*. [CrossRef] [PubMed]
7. Walters, C. Genebanking seeds from natural populations. *Nat. Areas J.* **2015**, *35*, 98–105. [CrossRef]
8. International Seed Testing Association. *International Rules for Seed Testing 2018*; ISTA: Bassersdorf, Switzerland, 2018.
9. Walters, C.; Wheeler, L.M.; Grotenhuis, J.M. Longevity of seeds stored in a genebank: Species characteristics. *Seed Sci. Res.* **2005**, *15*, 1–20. [CrossRef]
10. Van Treuren, R.; de Groot, E.C.; van Hintum, T.J.L. Preservation of seed viability during 25 years of storage under standard genebank conditions. *Genet. Resour. Crop Evol.* **2013**, *60*, 1407–1421. [CrossRef]
11. Ellis, R.H.; Nasehzadeh, M.; Hanson, J.; Woldemariam, Y. Medium-term seed storage of 50 genera of forage legumes and evidence-based genebank monitoring intervals. *Genet. Resour. Crop Evol.* **2018**, *65*, 607–623. [CrossRef]
12. Ellis, R.H.; Nasehzadeh, M.; Hanson, J.; Ndiwa, N.; Woldemariam, Y. Medium-term seed storage of diverse genera of forages grasses, evidence-based genebank monitoring intervals, and regeneration standards. *Genet. Resour. Crop Evol.* **2019**, *66*, 723–734. [CrossRef]
13. Humphreys, L.R.; Riveros, F. *Tropical Pasture Seed Production*; FAO Plant Production and Protection Paper 8; Food and Agriculture Organization of the United Nations: Rome, Italy, 1986; pp. 1–203.
14. Fairey, D.T.; Hampton, J.G. (Eds.) *Forage Seed Production. 1. Temperate Species*; CABI: Wallingford, UK, 1997; pp. 1–420.
15. McDonald, M.B.; Copeland, L. *Seed Production Principles and Practices*; Chapman and Hall: New York, NY, USA, 1997; pp. 1–749.
16. Sackville Hamilton, N.R.; Chorlton, K.H. *Regeneration of Accessions in Seed Collections: A Decision Guide*; Handbook for genebanks No. 5; International Plant Genetic Resources Institute: Rome, Italy, 1997; pp. 1–75.
17. Kelly, A.F.; George, R.A.T. *Encyclopaedia of Seed Production of World Crops*; Wiley: Chichester, UK, 1998; pp. 1–403.
18. Loch, D.S.; Ferguson, J.E. (Eds.) *Forage Seed Production. Volume 2. Tropical and Subtropical Species*; CABI: Wallingford, UK, 1999; pp. 1–479.
19. George, R.A.T. *Agricultural Seed Production*; CABI: Wallingford, UK, 2011; pp. 1–204.
20. McWilliam, J.R. The development and significance of seed retention in grasses. In *Seed Production*; Hebblethwaite, P.D., Ed.; Butterworths: London, UK, 1980; pp. 51–60.

21. Ellis, R.H.; Hong, T.D.; Roberts, E.H. Compendium of Specific Germination Information and Test Recommendations. In *Handbook of Seed Technology for Genebanks*; International Board for Plant Genetic Resources: Rome, Italy, 1985; Volume II, pp. 211–667.
22. Hay, F.R.; Probert, R.J. Collecting and handling seeds in the field. In *Collecting Plant Genetic Diversity: Technical Guidelines—2011 Update*; Guarino, L., Ramanatha Rao, V., Goldberg, E., Eds.; Bioversity International: Rome, Italy, 2011; pp. 1–8.
23. Tomás, A.; Berone, G.; Dreher, N.; Barrios, C.; Pisani, M. Variation in seed shattering in a germplasm collection of Panicum coloratum L. var. makarikariensis Goossens. In Proceedings of the 7th International Herbage Seed Conference, Dallas, TX, USA, 11–13 April 2010; pp. 26–32.
24. Singh, B.P.; Singh, H.P.; Obeng, E. Elephant grass. In *Biofuel Crops: Production, Physiology and Genetics*; Singh, B.P., Ed.; CAB International: Fort Valley, GA, USA, 2013; pp. 271–291.
25. Richards, C.M.; Lockwood, D.R.; Volk, G.M.; Walters, C. Modeling demographics and genetic diversity in ex situ collections during seed storage and regeneration. *Crop Sci.* **2010**, *50*, 2440–2447. [CrossRef]
26. Hanson, J.; Schultze-Krafte, R. *Regeneration Guidelines: Forage Grasses*; International Livestock Research Institute: Addis Ababa, Ethiopia, 2009; pp. 1–7.
27. Hanson, J.; Amri, A.; Street, K.; Shehadeh, A.; Rukhkyan, N.; Snowball, R. *Regeneration Guidelines: Forage Legumes*; International Livestock Research Institute: Addis Ababa, Ethiopia, 2009; pp. 1–8.
28. Griffiths, D.J.; Lewis, J.; Bean, E.W. Problems of Breeding for Seed Production in Grasses. In *Seed Production*; Hebblethwaite, P.D., Ed.; Butterworths: London, UK, 1980; pp. 37–49.
29. Ellis, R.H.; Hong, T.D.; Roberts, E.H. Principles and Methodology. In *Handbook of Seed Technology for Genebanks*; International Board for Plant Genetic Resources: Rome, Italy, 1985; Volume I, pp. 1–210.
30. Goedert, C.O. Seed Dormancy of Tropical Forage Grasses and Implications for the Conservation of Genetic Resources. Ph.D. Thesis, University of Reading, Reading, UK, 1984; pp. 1–190.
31. Ellis, R.H. Temporal patterns of seed quality development, decline, and timing of maximum quality during seed development and maturation. *Seed Sci. Res.* **2019**, *29*, 135–142. [CrossRef]
32. Hay, F.R.; Smith, R.D. Seed maturity: When to collect seeds from wild plants. In *Seed Conservation: Turning Science into Practice*; Smith, R.D., Dickie, J.B., Linington, S.H., Pritchard, H.W., Probert, R.J., Eds.; Royal Botanic Gardens: London, UK, 2003; pp. 97–133.
33. Hopkinson, J.M.; English, B.H.; Harty, R.L. Sweating of panicoid tropical pasture grass seeds. *Seed Sci. Technol.* **2003**, *31*, 367–377. [CrossRef]
34. Hay, F.R.; Whitehouse, K.J. Rethinking the approach to viability monitoring in seed genebanks. *Conserv. Physiol.* **2017**, *5*, cox009. [CrossRef] [PubMed]
35. Chin, H.F.; Hanson, J. Seed quality: Seed storage. In *Forage Seed Production Volume II Tropical and Subtropical Species*; Loch, D.S., Ferguson, J., Eds.; CABI International: Cambridge, UK, 1999; pp. 303–315.
36. Ellis, R.H.; Hong, T.D.; Roberts, E.H. The response of seeds of *Bromus sterilis* L. and *Bromus mollis* L. to white light of varying photon flux density and photoperiod. *New Phytol.* **1986**, *104*, 485–496. [CrossRef]
37. Ellis, R.H.; Hong, T.D.; Roberts, E.H. Quantal response of seed germination in *Brachiaria humidicola*, *Echinochloa turnerana*, *Eragrostis tef* and *Panicum maximum* to photon dose for the low energy reaction and the high irradiance reaction. *J. Exp. Bot.* **1986**, *37*, 742–753. [CrossRef]
38. Royal Botanic Gardens Kew Seed Information Database (SID) Version 7.1. Available online: http://data.kew.org/sid/ (accessed on 11 November 2019).
39. Hong, T.D.; Linington, S.; Ellis, R.H. *Seed Storage Behaviour: A Compendium*; International Plant Genetic Resources Institute: Rome, Italy, 1996; pp. 1–656.
40. International Board for Plant Genetic Resources. *Report of IBPGR Working Group on Engineering, Design and Cost Aspects of Long-Term Seed Storage Facilities*; IBPGR: Rome, Italy, 1976.
41. Koo, B.; Pardey, P.G.; Wright, B.D. *Saving Seeds. The Economics of Conserving Crop Genetic Resources ex situ in the Future Harvest Centres of the CGIAR*; Centre for Agriculture and Bioscience International (CABI) Publishing: Wallingford, UK, 2004; pp. 1–232.
42. Horna, D.; Smale, M. *Final Report on Evaluating Cost-Effectiveness of Collection Management: Ex-situ Conservation of Plant Genetic Resources in the CG System*; IPGRI: Rome, Italy, 2010; pp. 1–136.
43. Simpson, R.D.; Sedjo, R.A. The Value of Genetic Resources for Use in Agricultural Improvement. In *Agricultural Values of Plant Genetic Resources*; Evenson, R.E., Gollin, D., Santaniello, V., Eds.; CABI Publishing: Wallingford, UK, 1998; pp. 55–66.

44. Zohrabian, A.; Traxler, G.; Caudill, S.; Smale, M. *The Marginal Value of an Accession*; Biotechnology and Genetic Resources Policies, Brief 9; Bioversity International: Rome, Italy, 2003; pp. 1–2.
45. Hanson, J.; Schultze-Kraft, R.; Peters, M.; Wenzl, P.; Amri, A.; Shehadeh, A.; Yazbek, M. Forage diversity conservation and use. In *The Impact of Research at the International Livestock Research Institute, 1975–2018*; McIntire, J., Grace, D., Eds.; CABI Publishing: Wallingford, UK, 2020; Chapter 12. (in press)
46. Reid, R.S.; Serneels, S.; Nyabenge, M.; Hanson, J. The changing face of pastoral systems in grass dominated ecosystems of Eastern Africa. In *Grasslands of the World. Plant Protection and Production Series 34*; Suttie, J.M., Reynolds, S.G., Batello, C., Eds.; Food and Agriculture Organization of the United Nations: Rome, Italy, 2005; pp. 19–76.
47. FAO. *International Treaty on Plant Genetic Resources for Food and Agriculture*; Food and Agriculture Organization of the United Nations: Rome, Italy, 2009.
48. Galluzzi, G.; Halewood, M.; Noriega, I.L.; Vernooy, R. Twenty-five years of international exchanges of plant genetic resources facilitated by the CGIAR genebanks: A case study on global interdependence. *Biodivers. Conserv.* **2016**, *25*, 1421–1446. [CrossRef]
49. Noriega, I.; Halewood, M.; Abberton, M.; Amri, A.; Angarawai, I.; Anglin, N.; Blümmel, M.; Bouman, B.; Campos, H.; Costich, D.; et al. CGIAR Operations under the Plant Treaty Framework. *Crop Sci.* **2019**, *59*, 819–832. [CrossRef]
50. Brink, M.; van Hintum, T. Genebank Operation in the Arena of Access and Benefit-Sharing Policies. *Front. Plant Sci.* **2020**, *10*, 1712. [CrossRef] [PubMed]
51. Fu, Y.B. The vulnerability of plant genetic resources conserved ex situ. *Crop Sci.* **2017**, *57*, 1–15. [CrossRef]
52. Balehegn, M.; Mekuriaw, Z.; Miller, L.; Mckune, S.; Adesogan, A.T. Animal-sourced foods for improved cognitive development. *Anim. Front.* **2019**, *9*, 50–57. [CrossRef] [PubMed]
53. Brandt, P.; Yesuf, G.; Herold, M.; Rufino, M.C. Intensification of dairy production can increase the GHG mitigation potential of the land use sector in East Africa. *Glob. Chang. Biol.* **2019**, 1–18. [CrossRef] [PubMed]

plants

MDPI

Article

Challenges for Ex Situ Conservation of Wild Bananas: Seeds Collected in Papua New Guinea Have Variable Levels of Desiccation Tolerance

Simon Kallow [1,2,3,*], Kevin Longin [2], Natalia Fanega Sleziak [4], Steven B. Janssens [3,5], Filip Vandelook [3], John Dickie [1], Rony Swennen [2,4,6], Janet Paofa [7], Sebastien Carpentier [2,4] and Bart Panis [2,4]

1 Royal Botanic Gardens Kew, Millennium Seed Bank, Wakehurst, Ardingly, Sussex RH17 6TN, UK; j.dickie@kew.org
2 Department of Biosystems, Katholieke Universiteit Leuven, Willem de Croylaan 42, 3001 Leuven, Belgium; kevin.longin@kuleuven.be (K.L.); rony.swennen@kuleuven.be (R.S.); sebastien.carpentier@kuleuven.be (S.C.); bart.panis@kuleuven.be (B.P.)
3 Meise Botanic Garden, Nieuwelaan 38, 1860 Meise, Belgium; steven.janssens@plantentuinmeise.be (S.B.J.); filip.vandelook@plantentuinmeise.be (F.V.)
4 Bioversity International, Willem de Croylaan 42, 3001 Leuven, Belgium; natalia.fanegasleziak@kuleuven.be
5 Plant Conservation and Population Biology, Katholieke Universiteit Leuven, Kasteelpark Arenberg 31, 3001 Leuven, Belgium
6 International Institute of Tropical Agriculture, c/o Nelson Mandela African Institution of Science and Technology, Arusha 0206, Tanzania
7 National Agricultural Research Institute, Laloki 121, Papua New Guinea; janet.paofa@nari.org.pg
* Correspondence: S.Kallow@Kew.org

Received: 30 July 2020; Accepted: 17 September 2020; Published: 21 September 2020

Abstract: Ex situ seed conservation of banana crop wild relatives (*Musa* spp. L.), is constrained by critical knowledge gaps in their storage and germination behaviour. Additionally, challenges in collecting seeds from wild populations impact the quality of seed collections. It is, therefore, crucial to evaluate the viability of seeds from such collecting missions in order to improve the value of future seed collections. We evaluate the seed viability of 37 accessions of seven *Musa* species, collected from wild populations in Papua New Guinea, during two collecting missions. Seeds from one mission had already been stored in conventional storage (dried for four months at 15% relative humidity, 20 °C and stored for two months at 15% relative humdity, −20 °C), so a post-storage test was carried out. Seeds from the second mission were assessed freshly extracted and following desiccation. We used embryo rescue techniques to overcome the barrier of germinating in vivo *Musa* seeds. Seeds from the first mission had low viability (19 ± 27% mean and standard deviation) after storage for two months at 15% relative humidity and −20 °C. *Musa balbisiana* Colla seeds had significantly higher post-storage germination than other species ($p < 0.01$). Desiccation reduced germination of the seeds from the second collecting mission, from 84 ± 22% (at 16.7 ± 2.4% moisture content) to 36 ± 30% (at 2.4 ± 0.8% moisture content). There was considerable variation between and (to a lesser extent) within accessions, a proportion of individual seeds of all but one species (*Musa ingens* N.W.Simmonds) survived desiccation and sub-zero temperature storage. We identified that seeds from the basal end of the infructescence were less likely to be viable after storage ($p < 0.001$); and made morphological observations that identify seeds and infructescences with higher viability in relation to their developmental maturity. We highlight the need for research into seed eco-physiology of crop wild relatives in order to improve future collecting missions.

Keywords: banana; crop wild relatives; ex situ conservation; desiccation tolerance; *Musa*; Papua New Guinea; plant genetic resources; seed conservation; seed storage behaviour

1. Introduction

Crop wild relatives (CWRs) possess genetic material useful for improving crops in an increasingly challenging context [1–3]. They comprise a large untapped genepool of alleles potentially useful for breeding [4]. Examples from banana CWRs include improved drought tolerance [5] and resistance to several diseases [6,7]. At the same time, many CWRs are threatened with extinction [8,9] making their conservation imperative for both biodiversity and food security [10,11]. Effective plant conservation employs complementary in situ and ex situ strategies [12,13]. Such an approach is notably encouraged for CWR conservation [14,15]. Accordingly, ambitious efforts to collect and conserve many CWRs ex situ have recently been made [16–20]. However, CWRs, and banana CWRs (*Musa* spp. L.) in particular, are poorly represented in ex situ collections [21].

Banana CWRs are diploid wild species whose fruits contain many dark coloured seeds. Edible bananas, selected to avoid seeds in the fruit pulp, can be diploid, triploid or tetraploid. There are around 80 species in the genus *Musa* [22,23], and over 1000 edible banana cultivars [24,25]. The management of banana germplasm is co-ordinated in a global network of 31 collections containing over 6600 accessions of in vitro or field plants [26]. Only 163 of these accessions are CWRs. Moreover, of these, 122 are of the two most important banana CWRs (*Musa acuminata* Colla and *M. balbisiana* Colla), the other 41 accessions include 33 *Musa* species. Additionally, there are 131 *Musa* seed collections of only seven species stored in seed banks [27]. This means many species are only represented by a single genotype and for many wild banana species, no accessions exist. The diversity of banana CWRs in ex situ conservation is therefore highly constrained and expansion of the inter and intra-specific diversity of the collection is clearly needed.

Musa are pioneers or early successional tall herbs of tropical to subtropical rainforests. Native distribution area ranges from Southeast Asia to Pacific regions [22]. Papua New Guinea (PNG), the world's most floristically diverse island [28], is an important centre for both wild banana and cultivar diversity [29–32]. Sixteen wild *Musa* taxa occur in PNG [33]. Several collecting missions have been made in PNG to characterise and collect both cultivar and CWR germplasm [34–38]. These include seed collections, two of which we evaluate here.

Ex situ conservation using seeds can be a highly effective way of conserving the genetic diversity of plant populations [39,40]. This is useful for further conservation activities, phenotyping and breeding. Furthermore, conservation using seed is a relatively cost effective method of ex situ conservation [41]. In order to make high quality seed collections of wild species, understanding of seed development and storage behaviour are crucial [42]. Seed storage behaviour can classically be categorised into three groups. The majority of seeds are easily dried (to 2–5% MC) and stored at sub-zero temperatures, these are *orthodox* seeds [43]. Secondly, *recalcitrant* seeds do not survive drying to below 20–30% moisture content and are sensitive to low temperatures [43]. Finally, seeds that do not fit well into these binary categories are often called *intermediate,* and show partial sensitivity to drying and cold storage in particular [44]. Seeds of recalcitrant and intermediate species should be stored cryogenically, whereas orthodox seeds may be stored conventionally (at −20 °C) following desiccation [45].

For wild species, and especially banana CWRs, critical knowledge gaps exist in how best to collect, store and germinate their seeds. For *Musa*, only six species have been assessed for their storage behaviour, results of which are inconclusive [46–52]. Additionally, germination of seeds is notoriously inconsistent and dormancy poorly understood [53–55]. Embryo rescue techniques are therefore commonly used to germinate seeds in breeding programmes [56]. Together, these critical knowledge gaps hamper storage and access to banana genetic material [54].

Substantial challenges associated with collecting seeds from wild species impact the quality of seed collections [57]. Non-uniform seed development across a population, low seed numbers and sub-optimal post-harvest handling may be problematic [57–60]. Post-harvest handling is difficult because it is often not possible to control the temperature and humidity of seeds on collecting missions, e.g., whilst in a vehicle or when moving from place to place. Furthermore, there are significant practical challenges in collecting seed material from populations of wild species, the location of which may be remote, inaccessible and previously unknown. Evaluation of material from actual collecting missions can provide useful concrete evidence of these particular challenges, and lessons can be learnt to improve the quality of collections in the future.

In this study, we make use of seeds from two recent collecting missions to PNG. Seeds from one mission were already stored in Meise Botanic Gardens seed bank (called 'batch 1', and described by Eyland et al. [38]); the others were collected during the course of this investigation (called 'batch 2'). By evaluating seed viability of these collections, we address some of the issues and knowledge gaps described, by answering the following questions: (1) What is the viability of *Musa* seeds stored in Meise Botanic Gardens seed bank (for two months at 15% relative humidity (RH), −20 °C)? (2) Do seeds of some *Musa* species have higher viability after storage than others? (3) Do seeds from different parts of the infructescence have higher viability after storage than others? (4) How does desiccation affect seed viability? (5) Is it possible to predict storage behaviour of *Musa* seeds based on their physical properties? (6) Does seed maturity affect viability during dry storage? We use in vitro embryo rescue techniques to quantify viability. This provides the most reliable estimate of viability and removes dormancy constrains that limit germination in *Musa* seeds [61–63].

2. Results

2.1. Viability Evaluation of Seeds Stored in the Seed Bank

2.1.1. Overall Viability

The post-storage viability of batch 1 seeds (already stored in Meise Botanic Gardens seed bank) was markedly low with considerable variance between the accessions (Figure 1A). Across all accessions and hands, germination was on average 19 ± 27% (mean and standard deviation used hereafter; empty seeds are excluded in the percentages). In the present study, we use the term *accession* to mean a seed collection from a single individual plant including all the fruits of an infructescence. The term *bunch* refers to an infructescence. A bunch can be subdivided into *hands,* these are groups of fruits from the former clusters of flowers subtended by one bract [64]. Embryos that showed no reaction were 73 ± 29%. Other embryonic reactions, callus formation and embryo darkening without further outgrowth, were minimal (respectively, 0.2 ± 1.2%, and 3 ± 9%). Microbial contamination of the sample was 4 ± 16%. Overall, 24 ± 23% of seeds contained no identifiable embryos.

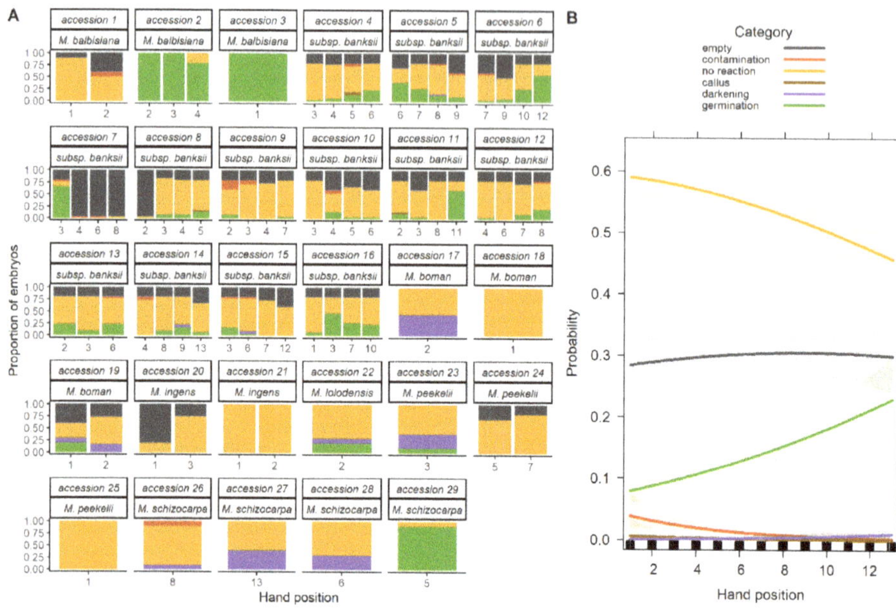

Figure 1. (**A**) Germination responses of embryos rescued from 29 accessions of *Musa* species following drying for 4 months at 15% relative humidity 20 °C and storage for 2 months at 15% relative humidity −20 °C. 'Hand position' refers to the position in the infructescence of the hand from which seeds were collected, with '1' being closest to the basal end of the bunch ($n = 23 \pm 10$ seeds). (**B**) Predicted probability of five embryo rescue outcomes of *Musa acuminata* subsp. *banksii* seeds extracted from different hand positions in the infructescence. Probabilities based on the multinomial logistic regression model of the response of seeds from 50 hands (representing 13 accessions; $n = 30$ seeds for each hand). Shaded areas are 95% standard errors of the estimated regression coefficients.

2.1.2. Effect of Species

Musa balbisiana Colla seeds (accessions #2 and #3), showed significantly higher germination than other species after storage ($p < 0.01$), in parallel with less embryos showing no reaction. This is demonstrated by the multinomial logistic regression (MLR) model (Table S1A), whereby the log odds of germination against an increase in no reaction is 0.865, but for all other species, log odds are negative. This is despite one of the three *M. balbisiana* accessions (accession #1) having no viability. One *M. schizocarpa* accession also showed high viability (accession #29, 90%), in contrast to the other three (accessions #26–28).

2.1.3. Effect of Position in the Infructescence

Musa acuminata subsp. *banksii* seeds from hands with a higher number, i.e., that were more recently pollinated, had embryos that were significantly more likely to germinate after storage ($p < 0.001$, Figure 1B, Table S1B); again this was concurrent with a reduced likelihood of no reaction.

2.2. *Effect of Desiccation*

2.2.1. Overall Effect of Desiccation

Seeds of batch 2 were tested before and after desiccation. Before desiccation (16.7 ± 2.4% moisture content (MC)), germination was on average 84 ± 22% (Figure 2A). After seven days desiccation (to 2.4 ± 0.8% MC), germination decreased to 36 ± 30%. As embryos dried, their percentage without any sign of germination increased from 3 ± 7% to 55 ± 27%. Again, there was considerable variance between accessions. Accession #37, *M. balbisiana*, was excluded from the analysis as the initial MC was an outlier compared to all the others (35% MC).

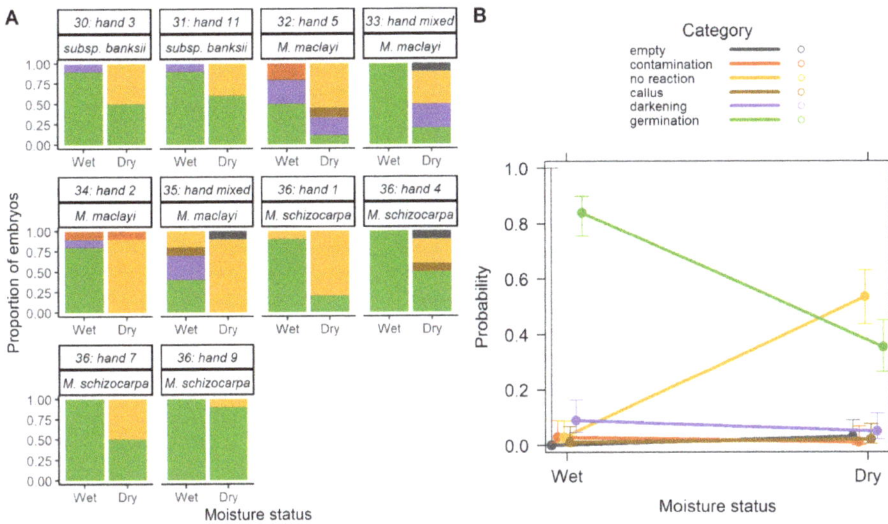

Figure 2. (**A**) Embryo rescue outcomes of *Musa* seeds (batch 2) before desiccation at 16.7 ± 2.4% moisture content ('Wet'), and after desiccation for seven days in a desiccator to 2.4 ± 0.8% moisture content ('Dry'). Accession and hand numbers are included above each chart. Seeds were germinated using embryo rescue and results recorded 28 days after transfer to the growth medium ($n = 10$). (**B**) Predicted probability of embryo rescue results according to the moisture status of seeds. Plot is on predicted values of the multinomial logistic regression model coefficients in Table S1C, data in Figure 2A. 95% standard errors shown.

2.2.2. Desiccation Tolerance and Species

Viability after desiccation differed by species. The *M. acuminata* subsp. *banksii* and *M. schizocarpa* accessions showed the highest germination after desiccation (55 ± 7%, 54 ± 29%, respectively). *M. maclayi* accessions had the lowest viability after desiccation (3 ± 6%). The MLR model based on drying as a factor showed a significant effect on germination in relation to no reaction ($p < 0.001$, Figure 2B, Table S1C). Additionally, seeds in wet condition were also more likely to darken and be contaminated ($p < 0.001$, $p < 0.01$, respectively). In the model, there is clear interchange of embryos that germinate with those that show no reaction (Figure 2B).

2.3. *Prediction of Seed Storage Behaviour*

Predicted seed storage behaviour (using the method of Hong and Ellis [65] and Ellis et al. [66]), identified that accessions straddled both intermediate and orthodox categories (Figure 3). Nine of the 11 accessions were predicted to be intermediate and two orthodox. The accessions predicted to be intermediate exceeded the threshold for weight rather than moisture content (apart from accession #37, *M. balbisiana*, previously identified as an outlier with high MC). There was no correlation between predicted seed storage behaviour and post-desiccation germination. Only seeds from batch 2 were used, as moisture content measurement of seeds before desiccation is required by the model.

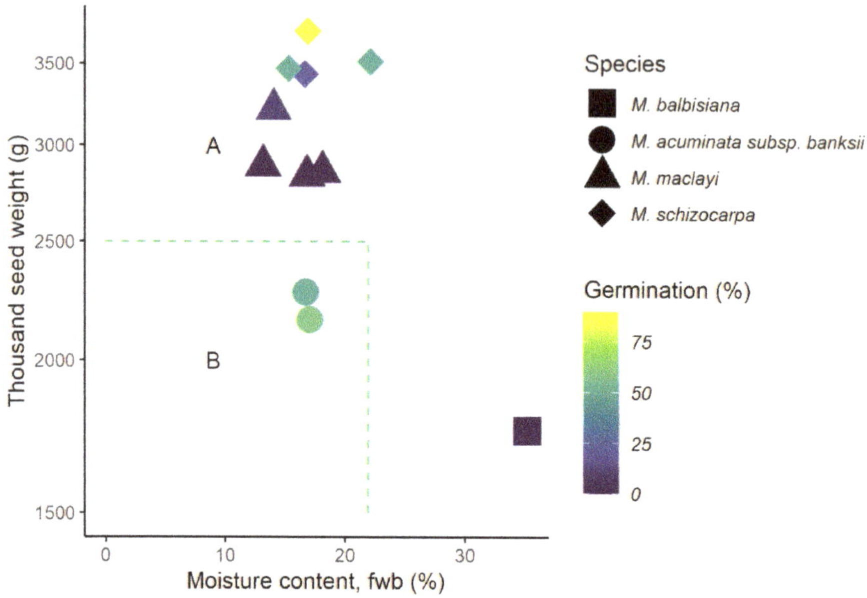

Figure 3. Predicted storage of behaviour of *Musa* accessions (batch 2) using the diagnostic key of Hong and Ellis [65] and Ellis et al. [66]. Area A includes accessions predicted to have intermediate storage behaviour, accessions in area B are predicted to have orthodox storage behaviour. Accessions are coloured according to the germination percentage of seeds after seven days desiccation to 2.4 ± 0.8% moisture content. Moisture content is calculated on the fresh weight basis ('fwb'). Seeds were germinated using embryo rescues and assessed 28 days after transfer to growth medium.

2.4. *Dry storage and Maturity*

2.4.1. Effect on Viability

Two *M. acuminata* subsp. *banksii* accessions were selected from batch 1. Accession #4 was from a less mature and accession #11 from a more mature bunch according to observations in the field. Seeds were tested before and during dry storage (contained within a paper bag, stored in a humidity controlled room at 15%RH, 20 °C). Seeds had a mean moisture content of 11.2 ± 2.7% before drying (Figure 4A). After seven days drying, moisture content reduced to 6.6 ± 2.5%. Moisture content remained about the same with further drying time, so that after six months moisture content was 6.5 ± 2.35%. Differences in moisture content between the accessions were not statistically significant.

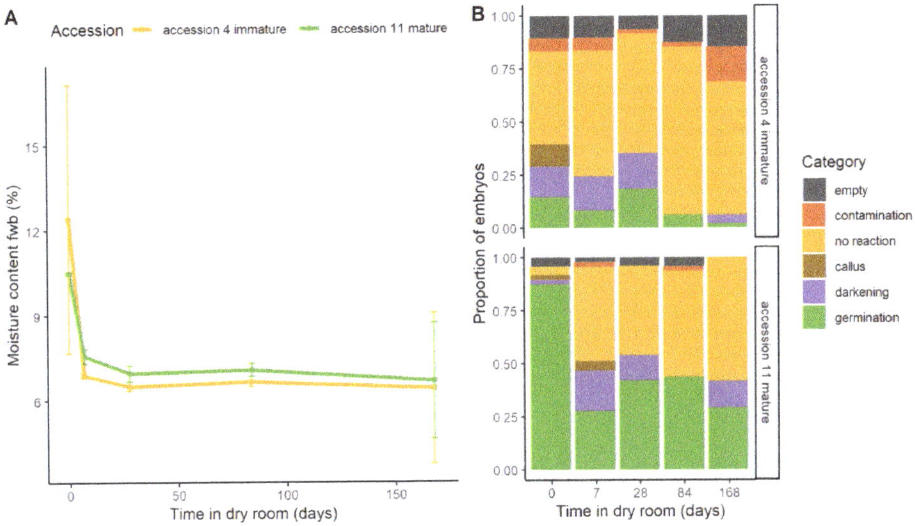

Figure 4. (**A**) Moisture content during dry storage, calculated on fresh weight basis ('fwb'). Seeds from a mature and an immature accession of *Musa acuminata* subsp. *banksii* were used. Seeds were dry stored at 15%RH, 20 °C for up to 168 days. (**B**) The effect of dry storage on embryo rescue outcomes. Outcomes recorded 28 days after the transfer of each embryo to growth medium (n = 48).

The mature seeds (accession #11) had much higher initial germination rates than the immature (accession #4), 91.3% and 16.3%, respectively (Figure 4B). Embryos from both accessions reduced in germination after seven days drying, to 28.3% for the mature accession and to 9.1% for the immature accession. For the mature seeds, this level of germination remained about the same with further drying time, so that after 6 months of drying, germination was 29.2%. Embryos from the immature accession continued to reduce in germination with further drying time, so that after 6 months dry storage, germination was negligible (2.4%). The proportion of embryos that germinated was notably reduced and exchanged to a correspondingly larger proportion of embryos that displayed no germination reaction, and to a lesser extent, darkening. Contamination also increased with time of drying for the immature seeds.

2.4.2. Effect on Morphology

Observations from magnified images of the selected *M. acuminata* subsp. *banksii* accessions showed apparent under-developed seed coats in the less mature seeds. The different layers of the seed-coat integuments are evident rather than fused, they are also lighter in colour (Figure 5). During drying, these layers were observed to separate. Additionally, there is a noticeably greater effect of desiccation on the structure of the endosperm and shape of the embryo. Less mature seeds display increased airspaces in the endosperm on desiccation. Embryos of the less mature seeds show greater loss of structure during desiccation.

Figure 5. Photographs of immature and mature bunches and their seeds of two *Musa acuminata* subsp. *banksii* accessions. Seed images taken before and after 1 month drying in a dry room (15% relative humidity, 20 °C). Seed images taken on a Keyence VHX5000 at 150 x magnification.

3. Discussion

3.1. Key Findings

This assessment of seed storability of banana CWR seeds from PNG collecting missions illustrates some of the challenges involved in making high quality collections of wild species for ex situ conservation. In particular, this assessment demonstrates some of the difficulties involved in making seed collections of wild species and how critical knowledge gaps impact the value of such collections. Our evaluation shows substantial loss of seed viability during seed banking which can be attributed to variable levels of desiccation tolerance. There was considerable variation between accessions, and some species (*M. balbisiana* and to a lesser extent *M. schizocarpa* and *M. acuminata* subsp. *banksii*) maintained higher viability during storage compared to others.

3.2. Desiccation Sensitivity

The low germination rates (19%) of seeds that were stored in the seed bank (batch 1), suggests a problem with maintaining the viability of collections in conventional storage (15%RH, −20 °C). However, as this was a viability assessment of seeds already stored, it is not possible to draw specific conclusions as to why viability is low: seeds may have had low initial viability or lost viability during transport, for example. By testing batch 2 seeds both before and after desiccation, it is clear that desiccation sensitivity is a major contributor to loss in viability. On average, these seeds reduced germination from 84% to 36% during rapid desiccation (from 17% to 2.4% moisture content). We therefore surmise that loss of viability is primarily a result of sensitivity to rapid desiccation. Further research is needed to fully understand whether loss in viability is caused by desiccation per se, or whether speed of desiccation is an important contributing factor.

3.3. Seed Storage Behaviour

Variation in our results, with respect to seed storage classification, is in line with previous studies. For instance, several studies demonstrate desiccation sensitivity where seeds lose viability at 6% MC [46] or, for extracted embryos, to 10–15%MC [50,51,67]. Other studies found that seeds tolerate drying, but do not state to what moisture content [47,48].

High viability of a few accessions stored under very low moisture and sub-zero temperature in the present study, suggests that (at least for *M. balbisiana*) orthodox storage class is likely. For others, our results suggest that intermediate storage classification may be appropriate, as significant proportions of seeds lost viability on desiccation and freezing. *Musa* storage behaviour is, therefore, at the threshold between orthodox and intermediate storage classes, as illustrated by results of the predictive model (Figure 3). It should be noted, however, that for all species (apart from *M. ingens*), a proportion of seeds survived desiccation, or even desiccation and sub-zero storage. Storage class was, therefore, variable within an accession, and orthodox behaviour of at least a small proportion of seeds was possible, if rare. Whilst storage classification is helpful, a continuum of storage behaviour is known to exist [68], even within the same genus or species, depending on when and where seeds were collected [45,69–72]. In the present study, a continuation in desiccation tolerance is evident within the same accession and even from fruits in the same hand.

3.4. Variation between Infructescences

3.4.1. Species and Climate

Differences in post-storage viability were greater between-infructescences (from different maternal plants), than within-infructescences (from hands of the same plant, Figure 1A). This may be related to differences in seed storage behaviour at the species level, to the maturity level of the whole infructescence or perhaps the microclimate [73]. Viability levels were consequently strongly linked to the fruit-bearing plant.

In batch 1, *M. balbisiana* seeds showed significantly higher post-storage viability than the other species. This species is characterized by a wide, yet often introduced, distribution across the tropics and subtropics. Notably, *M. balbisiana* is not considered a native species to PNG [74,75], but rather has its native distribution in the more seasonal subtropical Northern Indo-Burmese region [76]. By contrast, the other wild banana species studied are native to the Equatorial wet to moist ecoregions of PNG. As such, *M. balbisiana*, has also shown to have high leaf wax content that contributes to drought tolerance [5] and is therefore probably better adapted to seasonal changes in precipitation and temperature than the *Musa* species native to PNG [33,74,75]. This then suggests that within the whole genus, there may well be a range of desiccation tolerance levels possibly according to species distributions. It should, however, be caveated that our observation is based on only a small number of samples and a wider survey should be carried out for further conclusions. Nonetheless, it is well known that there is a correlation between the bioclimatic distribution of a plant and seed storage behaviour: higher annual precipitation is positively correlated with recalcitrance [77]. Interestingly, differences in precipitation in the native region of the *Musa* species examined here are in fact greater when only the precipitation in the driest quarter of the year is considered, rather than for annual precipitation (see Figure S1). We therefore suggest that the precipitation in the driest quarter might possibly have a stronger correlation to seed storage behaviour for *Musa* than annual precipitation, as the possible impact of a dry season may be masked in seemingly high precipitation regions. We therefore propose that *Musa* seeds collected from species adapted to more pronounced dry seasons may have better desiccation tolerance and therefore better survive storage. However, further research is required in this area.

3.4.2. Seed Maturity

We identified physical properties that were seemingly linked to the level of seed maturity at the time of harvest. Larger fruits with softer pulp texture and seeds with a more powdery endosperm were considered to be more mature. Seeds from the bunch categorised as more mature in the field had greater embryo rescue germination percentages, both before and after dry storage (15%RH, 20 °C), compared to the seeds identified from the less mature bunch. In the laboratory, it was observed that the less mature seeds had higher initial moisture content that reduced to a greater extent, and an under-developed seed coat: a light brown inner integument as opposed to dark brown to black, that was less well fused with the outer integument (Figure 5). The small sample size notwithstanding, the importance of seed maturity for desiccation tolerance is consistent with current understanding of the development of desiccation tolerance during late seed maturation [73,78,79]. Desiccation tolerance is acquired at 'mass maturity' after maximum dry weight is achieved and the vascular connection between the maternal plant and seed is terminated [80]. Following this, seed moisture content equilibrates with the environment prior to dispersal. Often this is described as the 'point of natural dispersal' [60]. The difficulty for improving the quality of seed collections is how to translate theory into practice, particularly for seeds that are contained within large pulpy fruits like bananas.

Regarding seeds that were collected from field collections, Simmonds [47] found that, for maximum in vivo germination, *M. balbisiana* seeds should be collected 'mature'. Unfortunately, he did not define what 'mature' meant in this instance. However, he detected a window of six weeks whereby high germination can be achieved (>80%), four weeks before and two weeks after maturity. Additionally, he found that fruit of *M. acuminata* should be collected green or yellow (rather than black or rotten) to achieve high germination. Furthermore, Uma et al. [81] found that at 70% maturity (full maturity being 110 days after (self-)pollination) 'Pisang Jajee' (a *M. acuminata* genotype) embryos were discernible and endosperms had converted from a liquid to semi-solid state; this also coincided with thickening of the integuments. They also found that seeds, in order to germinate, should be at least 90% mature, and immature embryos were more likely to produce calluses.

Collecting mature seeds during collecting missions is much more challenging than from field germplasm collections. Collectors must access bunches before they are consumed and seeds are dispersed by birds and mammals [82–84]. Humans also harvest wild bananas for food, construction and artistry [37,85]. It is therefore important to be able to identify fruits that contain seeds that are mature enough to be desiccation tolerant, without knowing flowering times, whilst they may not have yet attracted frugivores. Based on our results, we suggest that seeds should have powdery endosperms and well-formed integuments with fused layers, without which many seeds will be lost during storage; however, clearer definitions should be developed for collectors.

It may also be possible to improve desiccation tolerance and longevity of seeds by using a treatment that mimics late maturation on the plant, as has been shown for other species [78,86–88]. Indeed, in one study [89], Simmonds found that seeds from 'ripe' and 'over-ripe' bananas that were dried in the fruit at a temperature similar to what may be found on the plant (in an oven at 45 °C), germinated better than seeds that were not dried in this manner. Assessing and furthering seed maturity whilst avoiding dispersal is clearly a key factor in improving the quality of future banana seed collecting.

3.5. Variation within Infructescences

Heterogeneity of maturity within an infructescence has been highlighted as a cause for variable desiccation tolerance within seed accessions of other wild species [57,60]. We observed a small but significant within-infructescence effect, in that seeds from the male bud end of the infructescence (seeds from flowers that were more recently pollenated) were around 15% more likely to germinate (post-storage) than the peduncle basal end (Figure 1B). This, perhaps surprising, effect may be caused be caused by variation in seed-vigour or seed-aging, discussed below.

It is well known for other species, that there are differences in physical and physiological properties in seed-vigour within the same infructescence [90]. For species of temperate regions variation is often correlated with seasonality [91–93]. In tropical species, the effect of climate on seed properties is not well known. However, pollen, seed set and germination success of banana seeds (during breeding programmes) have been found to correlate with climatic conditions [94,95]. Alternatively, seed-vigour, including the ability to tolerate stress, deteriorates according to temperature and moisture [71]. When seeds are kept in the fruit for relatively long periods of time, for example, during collecting missions, seed-aging can occur. This can potentially influence the ability of seeds to withstand the stress of desiccation later on. As seeds from basal hands are produced first, when they are harvested they are already in a more advanced state: fruit may soften quicker and have higher moisture content, and the exocarp may be rotting. This all means that aging is more likely to occur if they are then kept in the fruit during the remainder of the trip and until they are transported to the laboratory (see Figure S2 for a photograph of the fruits of batch 1 after transit to Belgium). This could explain why the older seeds within a bunch may display lower desiccation tolerance. However, as this effect was relatively small compared to the overall maternal effect, it seems that the within-bunch maturity appears at least to have less of an effect than the maturity of the whole bunch.

3.6. Limitations and Assumptions

3.6.1. Embryo Rescue

We used embryo rescue techniques to estimate viability in the present study. Whilst this is the best current method for estimating *Musa* seed viability, there are limitations and assumptions that should be stated. Firstly, the purpose of a viability measure is to estimate the proportion of seeds that are capable of developing into seedlings or plants [42]. Embryo rescue 'short-cuts' some of the constraints that could limit this process of in vivo germination. For instance, if an embryo germinates in vitro, it does not necessarily mean that it is capable of developing into a seedling or plant. For this to happen the embryo must also push off the seed micropyle cap and develop roots that can access the soil. We accounted for this in our analysis by categorising separately embryos that did not develop fully formed shoots, but rather formed calluses or showed no growth but darkening of the embryo. Secondly, embryo rescue evaluation, in our method, is at 28 days; however, it is possible that germination may be slow and only is evident after this period. According to the literature, 28 days should normally be enough time [61,62,96], but it is an assumption that, at this point in time, the germination process is concluded for all embryos. Finally, the conclusions of this study are based on the assumption that embryos showing no germination reaction are in fact dead. However, it is possible that desiccation does not kill the embryo, but rather causes a deep level of physiological dormancy that is not removed by excision from the rest of the seed. To account for this, we carried out tetrazolium tests on embryos that showed no germination reaction on embryo rescue. These embryos showed no staining. This indicates that embryo rescue produces the maximum measure of viability.

3.6.2. Conservation and Research Material

Whilst the benefit of using seeds from collecting missions allows results to be impactful for future missions, limitations are also introduced by using such material. One of the main limitations we faced was the limited availability of seeds for research. This inevitably constrains the interpretation of results (hence the large amount of deviation) because sample sizes and replicates were small. Seed numbers were limited for two reasons. One, because it is difficult to access seed material in suitable time periods from third parties, despite relevant treaties [97]. Two, because there are conflicting demands for material. There is an expectation and requirement for seeds to be placed into storage 'for conservation'. This may conflict with availability of adequate material for research into how best to store and germinate seeds. Our results highlight the need for seed collecting for research purposes in addition to, and ideally prior to, collecting missions whose primary purpose is conservation. In practice, as here, these two processes often run concurrently.

4. Materials and Methods

4.1. Study Region

The study region was between Latitude 2° to 8° South, and Longitude 141° to 151° East. Seeds were collected in the Papua New Guinean provinces of Morobe, Madang and Sandaun on the island of New Guinea, and the province of West New Britain on the island of New Britain (Figure 6). These locations are in the tropical and subtropical moist broadleaf forest biomes [98]. Mean annual precipitation and mean annual temperature, at the collecting locations are 2695 ± 562 mm and 24.9 ± 1.9 °C, respectively (averages for years 1970–2000) [99].

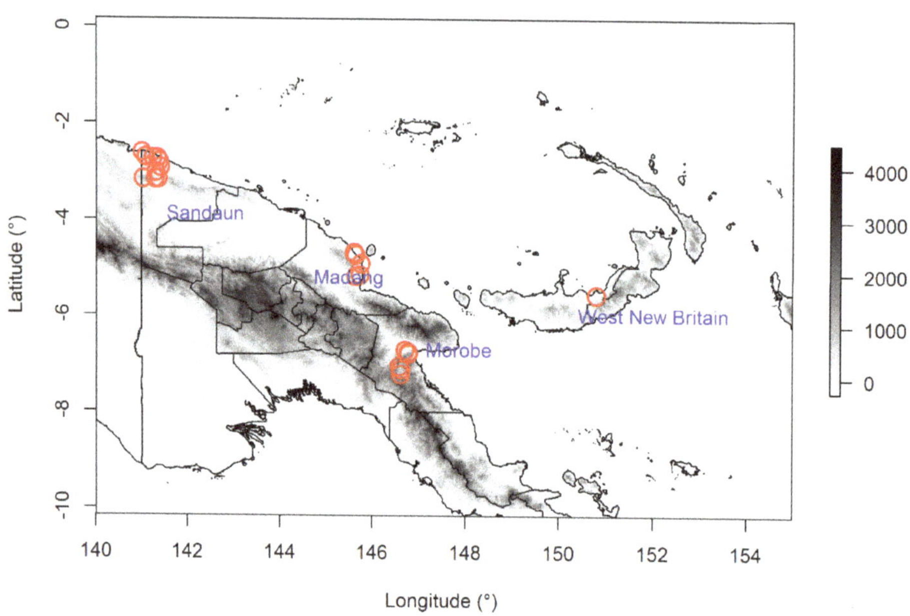

Figure 6. Collecting locations of seeds used in this study (red circles) and relevant province names. Map is shaded according to elevation (meters above sea level, data from: http://srtm.csi.cgiar.org).

4.2. Plant Material

4.2.1. Accessions

Overall, 37 *Musa* seed accessions were used in this study. Accessions were from a total of seven species: *Musa balbisiana* Colla, *M. acuminata* subsp. *banksii* (F. Muell.) N.W. Simmonds, *M. boman* Argent, *M. ingens* N.W. Simmonds, *M. lolodensis* Cheesman, *M. peekelii* Lauterb, *M. schizocarpa* N.W. Simmonds (Table 1).

Table 1. Details of seed collections.

Batch	Accession	Species	Province	Latitude	Longitude	Date Collected
1	1	*M. balbisiana*	Morobe	S 07°03′23″	E 146°34′56″	14/05/2019
1	2	*M. balbisiana*	Morobe	S 07°07′35″	E 146°36′57″	14/05/2019
1	3	*M. balbisiana*	Madang	S 04°41′49″	E 145°36′49″	17/05/2019
1	4	*M. acuminata* subsp .*banksii*	Morobe	S 06°43′34″	E 146°42′40″	14/05/2019
1	5	*M. acuminata* subsp .*banksii*	Morobe	S 06°43′34″	E 146°42′40″	14/05/2019
1	6	*M. acuminata* subsp .*banksii*	Morobe	S 07°13′42″	E 146°36′32″	14/05/2019
1	7	*M. acuminata* subsp .*banksii*	Morobe	S 07°13′42″	E 146°36′32″	14/05/2019
1	8	*M. acuminata* subsp .*banksii*	Morobe	S 06°47′14″	E 146°47′00″	15/05/2019
1	9	*M. acuminata* subsp .*banksii*	Morobe	S 06°46′04″	E 146°46′46″	15/05/2019
1	10	*M. acuminata* subsp .*banksii*	Madang	S 04°41′31″	E 145°36′60″	17/05/2019
1	11	*M. acuminata* subsp .*banksii*	Sandaun	S 02°43′58″	E 141°15′20″	20/05/2019
1	12	*M. acuminata* subsp .*banksii*	Sandaun	S 03°09′53″	E 141°21′57″	21/05/2019
1	13	*M. acuminata* subsp .*banksii*	Sandaun	S 03°09′51″	E 141°18′18″	21/05/2019
1	14	*M. acuminata* subsp .*banksii*	Sandaun	S 02°55′56″	E 141°25′09″	21/05/2019
1	15	*M. acuminata* subsp .*banksii*	Sandaun	S 02°42′17″	E 141°05′36″	22/05/2019
1	16	*M. acuminata* subsp .*banksii*	Sandaun	S 02°42′46″	E 141°05′45″	22/05/2019
1	17	*M. boman*	Sandaun	S 03°01′46″	E 141°19′18″	21/05/2019
1	18	*M. boman*	Sandaun	S 02°48′36″	E 141°23′44″	21/05/2019
1	19	*M. boman*	Sandaun	S 03°09′38″	E 141°02′22″	21/05/2019
1	20	*M. ingens*	Morobe	S 06°48′03″	E 146°46′24″	15/05/2019
1	21	*M. ingens*	Morobe	S 06°47′14″	E 146°47′00″	15/05/2019
1	22	*M. lolodensis*	Sandaun	S 03°01′46″	E 141°19′18″	21/05/2019
1	23	*M. peekelii*	Madang	S 05°09′44″	E 145°44′51″	16/05/2019
1	24	*M. peekelii*	Madang	S 04°55′07″	E 145°45′51″	18/05/2019
1	25	*M. peekelii*	Madang	S 04°55′07″	E 145°45′51″	18/05/2019
1	26	*M. schizocarpa*	Madang	S 04°44′17″	E 145°38′60″	17/05/2019
1	27	*M. schizocarpa*	Sandaun	S 02°44′19″	E 141°21′11″	19/05/2019
1	28	*M. schizocarpa*	Sandaun	S 02°43′58″	E 141°15′20″	20/05/2019
1	29	*M. schizocarpa*	Sandaun	S 02°37′02″	E 141°00′52″	22/05/2019
2	30	*M. acuminata* subsp .*banksii*	Madang	S 05°11′58″	E 145°39′30″	16/10/2019
2	31	*M. acuminata* subsp .*banksii*	Madang	S 05°11′58″	E 145°39′30″	16/10/2019
2	32	*M. maclayi*	West New Britain	S 05°33′54″	E 150°47′29″	03/10/2019
2	33	*M. maclayi*	West New Britain	S 05°33′54″	E 150°47′29″	03/10/2019
2	34	*M. maclayi*	West New Britain	S 05°33′54″	E 150°47′29″	03/10/2019
2	35	*M. maclayi*	West New Britain	S 05°33′54″	E 150°47′29″	03/10/2019
2	36	*M. schizocarpa*	Madang	S 05°11′58″	E 145°39′30″	16/10/2019
2	37	*M. balbisiana*	Madang	S 05°11′58″	E 145°39′30″	15/10/2018

4.2.2. Seed Batches

Seeds were collected during two field missions to Papua New Guinea. Batch 1 was collected in May 2019, at the end of the wet season, and included 29 accessions, described by Eyland et al. [38]. Batch 2 was collected in October 2019, at the start of the wet season and contained eight accessions (Table 1).

4.3. Seed Collection, Field Evaluation and Transportation

Seeds were collected from wild populations that occurred either in primary or secondary forests. At the time of collecting, seed maturity was assessed by dissecting approximately 10 seeds per bunch and examining the embryos and endosperms. Seeds were considered mature when embryos were capitate in shape (mushroom-like) and endosperms were powdery as opposed to wet or milky. Only bunches with seemingly mature seeds were collected (although, some bunches proved to be not completely mature, see results). Each bunch was photographed on site. Hands were removed from the bunch and numbered according to position, with 1 being at the basal end, i.e., they were produced first. Hands were placed in paper bags, which were then placed in cardboard boxes for storage during the remaining field mission. Accessions were then transported to Belgium for extraction. Transportation was initiated within one week of the end of the two-week collecting mission and took approximately one week to complete by aeroplane. During shipping, temperatures were greater than 0 °C and less than 25 °C. Fruits were therefore received within four weeks of collecting in the field.

4.4. Seed Processing

4.4.1. Extraction

Seeds were extracted by peeling the epicarp and crumbling or squashing the endocarp and removing seeds by hand. Excess fruit pulp was removed by washing in running water if necessary. In case fruits were hard, they were soaked in water for 24 h prior to seed extraction. It took a week to extract all the seeds from batch 1, and one day for the seeds of batch 2. Seeds were maintained under ambient laboratory conditions (approximately 60–80% relative humidity, 20 °C) for a maximum of seven days whilst all extractions were completed, this also allowed removal of excess water gained during washing. Moisture content of a subset of three accessions of batch 1 seeds and all accessions of batch 2 seeds was measured after extractions were completed (see Section 4.4.2 for method).

4.4.2. Moisture Content Measurement

Moisture content (MC) was calculated on a fresh weight basis (FWB) using the formula:

$$MC(\%) = \frac{(fresh\ weight - dry\ weight)}{fresh\ weight} \times 100$$

Seeds were weighed in plastic boats, dried at 70 °C for three days, and re-weighed. The MC of seeds was then calculated. Seeds were dried whole, as seeds coats were previously assessed as water permeable (our own data not shown and see [100]). Our own previous results also showed that embryo moisture content was 2% higher than whole seeds for non-desiccated seeds (at 10%MC) and 3%MC higher after desiccation (to 3%MC). Whole seeds were used here because accurately measuring the moisture content of embryos requires many samples that were not available because of their small size. Three replicates of 10 seeds were used to assess moisture content unless otherwise stated.

4.4.3. Storage

For storage in the seed bank at Meise Botanic Gardens, Belgium, seeds were further dried for four months at 15%RH and 20 °C, and then placed in cold storage at 15%RH and −20 °C sealed in aluminium envelopes. Seeds were in cold storage for two months prior to viability evaluation. The moisture content of seeds was taken prior to transfer to cold storage.

4.5. Viability Evaluation of Seeds Stored in the Seed Bank

We used embryo rescue techniques to evaluate viability [61–63]. This is the most effective measure of *Musa* seed viability compared to whole seed germination [61,62,96] and the tetrazolium chloride test [101] (Simon Kallow, pers. obs.).

We evaluated the viability of batch 1 seed accessions that had been stored in the Meise Botanic Gardens cold storage for two months. No pre-storage viability evaluation had been made. The MC of a subset of three accessions was assessed on removal from storage.

For embryo rescue, seeds were sterilised by soaking them in 96% ethanol for 3 min, followed by 20 min in 1% NaOCl (diluted commercial bleach 5%), containing 1 drop of detergent per 100 mL. Seeds were then rinsed three times in sterile water. Embryos were extracted from seeds using a sterile forceps and scalpel by making an incision in the seed coat next to the micropyle with the scalpel and by manipulating the seed with scalpel and forceps until the testa split open exposing the endosperm and embryo; embryos were then removed by careful manipulation. Embryos were transferred onto autoclaved half MS medium [102] in tubes with the haustorium in contact with medium and the embryonic axis upwards. All procedures were carried out in a laminar flow cabinet. Tubes containing embryos were incubated in the dark at 27 °C for 14 days, after which they were put in a growth chamber for an additional 14 days (24 h photo-period, 27 °C, 50 μE m^{-2} s^{-1} illumination provided by 36 W Osram cool-white fluorescent tubes). Six possible observations were recorded after 28 days: empty (no embryo, identified during excision), contamination, no reaction, callus formation, darkening and germination. *Musa* embryos are regarded as non-dormant when cultured in vitro [61–63,96,103], so seeds showing no reaction during this period were considered dead. Embryos that form calluses or that darken are considered alive but unlikely to regenerate into seedlings. An average of 23 ± 10 seeds from an average of 3 hands were tested for 29 accessions. Seed availability for this evaluation was highly constrained.

4.6. *Effect of Desiccation*

Following the results of the viability evaluation of stored seeds (Section 2.1), batch 2 was collected. We assessed the effect of desiccation on the eight accessions included, using embryo rescue (as previously described). This was done before desiccation and then after seven days of enforced desiccation. Seeds were desiccated by placing them on plastic boats suspended over silica gel sealed in a desiccator. The environment in the desiccator was approximately 2.4% RH and 20 °C. Ten seeds per accession were used both before and after extraction. Moisture content was measured for each accession before and after desiccation using ten seeds.

4.7. *Prediction of Seed Storage Behaviour*

Seed mass and initial moisture content was used to predict seed storage behaviour according to the model of Hong and Ellis [65] and Ellis et al. [66]. For this, seeds with a 1000 seed weight of less than 2500 g and a moisture content of less than 22% are predicted to have orthodox storage behaviour. Seeds with higher mass and moisture content (>4000 g and >40%MC) are predicted to be recalcitrant. Seeds in between these limits are predicted to be intermediate. Only seeds in batch 2 were used for this prediction as fresh seed moisture content is a requirement. Seeds were weighed within seven days of extraction whilst maintained in ambient conditions (60–80%RH and 20 °C) to remove excess moisture gained during extraction. Five replicates of 50 seeds were weighed and the mean of this was used to calculate 1000 seed weight for each accession. Moisture content was measured as described above. Moisture content and mass for each accession was then plotted in a scatter chart.

4.8. *Survival during Dry Storage*

4.8.1. Effect of Maturity

Two *M. acuminata* subsp. *banksii* accessions from batch 1 were selected from a seemingly more mature (accession #11) and a less mature bunch (accession #4). Maturity level was identified during collecting. The seemingly mature bunch had darker fruit colour and softer pulp texture, as well as more powdery endosperm compared to the less mature bunch (Figure 5). Subsample seeds were selected across all hands and mixed, so that the seeds used reflected the entire accession. For these selected accessions, embryo rescue was carried out after extraction as described above, and then after 1 week, 1, 3 and 6 months of dry storage in a paper bag stored in a dry room (15%RH, 20 °C). Forty eight seeds were used from each accession at each time point. The MC of seeds of each accession at each time point described was measured.

4.8.2. Effect on Morphology

Seeds from the selected *M. acuminata* subsp. *banksii* accessions (#4 and #11) were dissected and photographed at the same time points and conditions described above. A digital microscope (Keyence VHX5000) was used at 150–200 x magnification. Ten seeds were used per accession, condition and time point.

4.9. Statistical Analysis

Counts from the categorised outcomes of the embryo rescue tests were transformed into lists where each embryo's reaction was a nominal outcome. These data were then used to build multinomial logistic regression (MLR) models to analyse the log-odds of the embryo rescue outcome category using the *nnet* R package [104]. Maximum models were reduced by comparing Akaike information criterion (AIC) and carrying out the likelihood ratio test. Effects plots were produced from the models by predicting and then plotting data using the effects R package [105]. Statistics were carried out in R v 3.6.2 [106].

5. Conclusions

The aim of this study was to assess the viability of banana seeds collected during two collecting missions in order to inform ex situ conservation of banana CWRs. (1) We found that in general *Musa* seeds collected in PNG and stored in the seed bank had low viability. (2) There was considerable variation between accessions, *Musa balbisiana* seeds had significantly higher post-storage viability than other species. (3) Variation within accessions, according to the position in the infructescence, was significant, with seeds of *M. acuminata* subsp. *banksii* from the basal end having lower viability after storage than from the male bud end. (4) Freshly extracted seeds lost much of their viability during desiccation. (5) Predictions of seed storage behaviour based on physical properties indicate that *Musa* seeds are at the threshold of orthodox and intermediate classification; this is in keeping with our embryo rescue results. (6) *M. acuminata* subsp. *banksii* seeds, identified in the field as more mature, had higher viability before and during dry storage than less mature seeds, but this level was also reduced after dry storage.

This assessment of seed viability demonstrates the importance of advancing understanding of the seed storage behaviour of CWRs. In particular, we show how the ecology and adaption of species and the development of their seeds in time effects the viability of seeds collected for storage.

Supplementary Materials: The following are available online at http://www.mdpi.com/2223-7747/9/9/1243/s1, Figure S1: (A) Annual precipitation, and (B) precipitation of the driest quarter (three-month period), across the native distribution of the species evaluated in this study. Data extracted from WorldClim v2.0 based on occurrence records of species (data compiled by A. Mertens). Green dots displays means. Figure S2: Batch 1 fruits during processing after arrival at Meise Botanic Gardens, Belgium. Fruits are separated according to hand position on the bunch. Table S1: Multinomial logistic regression coefficients in log-odds (logits) and standard deviations in parentheses. (A) Embryo rescue outcome of batch 1 post storage. (B) Embryo rescue outcome of post-storage *Musa acuminata* subsp. *banksii* accessions in batch 1 according to hand position (1 being at the basal peduncle end of the infructescence). (C) Embryo rescue outcome of seeds from batch 2 after drying in a desiccator for seven days, a factor with two levels ('Wet' and 'Dry'). Embryo rescue outcome categorised after 28 days. Stars designate significance levels for p values * ≤ 0.05, ** ≤ 0.01, *** ≤ 0.001.

Author Contributions: Conceptualization, S.K., K.L., S.B.J., S.C. and B.P.; methodology, S.K., K.L., S.B.J., F.V., J.P., and B.P.; software, S.K.; validation, K.L. and B.P.; formal analysis, S.K.; investigation, S.K., K.L., and N.F.S.; resources, F.V., J.P., S.C. and B.P.; data curation, S.K., K.L., N.F.S., S.J.B., F.V. and B.P.; writing—original draft preparation, S.K.; writing—review and editing, S.K., S.B.J., F.V., J.D., R.S., S.C. and B.P.; visualization, S.K.; supervision, S.J.B., J.D., R.S. and B.P.; project administration, S.K., S.B.J., S.C. and B.P.; funding acquisition, S.B.J., J.D., R.S., S.C. and B.P. All authors have read and agreed to the published version of the manuscript.

Funding: This work was funded as a sub-grant from the University of Queensland from the Bill & Melinda Gates Foundation project 'BBTV mitigation: Community management in Nigeria, and screening wild banana progenitors for resistance' [OPP1130226]. In addition, this study received funding from the Genebank CGIAR Research Program, from Research Foundation-Flanders (FWO) (No. G0 D9318 N) and from CROP TRUST (GS15024). The collection mission in PNG was funded by the Global TRUST foundation project "Crop wild Relatives Evaluation of drought tolerance in wild bananas from Papua New Guinea, grant no: GS15024. The authors thank all donors who supported this work also through their contributions to the CGIAR Fund (http://www.cgiar.org/funders), and in particular to the CGIAR Research Program Roots, Tubers and Bananas (RTB-CRP).

Acknowledgments: We gratefully acknowledge all those who contributed to the making of these seed collections: Joel Pilon, Ian Nabo, Billy Pitalai, Godfried Savi, Philip Daur, Elasanty Yabu, Samson Itau, Joel Lapiu, Tony Kunou, Steven Kambase, Joe Guaf, Nemothy Sinoksor, Julie Sardos, Gabriel Sachter-Smith and David Eyland. Thank you also to Arne Mertens for the data to plot Figure S1 and Richard Ellis for helpful advice.

Conflicts of Interest: The authors declare no conflict of interest.

References

1. Redden, R.; Yadaz, S.S.; Maxted, N.; Dulloo, E.M.; Guarino, L.; Smith, P.E. *Crop Wild Relatives and Climate Change*; Redden, R., Yadaz, S.S., Maxted, N., Ehsan Dulloo, M., Guarino, L., Smith, P., Eds.; Wiley-Blackwell: Hoboken, NJ, USA, 2015; p. 400.
2. Guarino, L.; Lobell, D.B. A walk on the wild side. *Nat. Clim. Chang.* **2011**, *1*, 374–375. [CrossRef]
3. Hajjar, R.; Hodgkin, T. The use of wild relatives in crop improvement: A survey of developments over the last 20 years. *Euphytica* **2007**, *156*, 1–13. [CrossRef]
4. Dempewolf, H.; Baute, G.; Anderson, J.; Kilian, B.; Smith, C.; Guarino, L. Past and future use of wild relatives in crop breeding. *Crop Sci.* **2017**, *57*, 1070–1082. [CrossRef]
5. Sampangi-Ramaiah, M.H.; Ravishankar, K.V.; Shivashankar, K.S.; Roy, T.K.; Rekha, A.; Hunashikatti, L.R. Developmental changes in the composition of leaf cuticular wax of banana influenced by wax biosynthesis gene expression: A case study in *Musa acuminata* and *Musa balbisiana*. *Acta Physiol. Plant.* **2019**, *41*, 9. [CrossRef]
6. Vuylsteke, D.R.; Swennen, R.L.; Ortiz, R. Development and performance of black sigatoka resistant tetraploid hybrids of plantain (*Musa* spp. AAB group). *Euphytica* **1993**, *65*, 33–42. [CrossRef]
7. Parac, E.P.; Lalusin, A.G.; Pangga, I.B.; Cruz, F.C.S. Characteristics of selected Hybrids of Abaca (*Musa textilis* Nee) with resistance to bunchy top. *Philipp. Agric. Sci.* **2020**, *103*, 1–12.
8. Jarvis, A.; Lane, A.; Hijmans, R.J. The effect of climate change on crop wild relatives. *Agric. Ecosyst. Environ.* **2008**, *126*, 13–23. [CrossRef]
9. Ford-Lloyd, B.V.; Schmidt, M.; Armstrong, S.J.; Barazani, O.; Engels, J.; Hadas, R.; Hammer, K.; Kell, S.P.; Kang, D.M.; Khoshbakht, K.; et al. Crop wild relatives: Undervalued, underutilized and under threat? *Bioscience* **2011**, *61*, 559–565. [CrossRef]
10. Schatz, G.E. Plants on the IUCN Red List: Setting priorities to inform conservation. *Trends Plant Sci.* **2009**, *14*, 638–642. [CrossRef]
11. UN General Assembly. *Transforming Our World: The 2030 Agenda for Sustainable Development*; United Nations: New York, NY, USA, 2015.
12. Secretariat of the Convention on Biological Diversity. Global Strategy for Plant Conservation; 2010. Available online: https://www.bgci.org/our-work/policy-and-advocacy/the-global-strategy-for-plant-conservation/ (accessed on 18 September 2020).
13. Cohen, J.I.; Williams, J.T.; Plucknett, D.L.; Shands, H. Ex situ conservation of plant genetic resources: Global developments and environmental concerns. *Science* **1991**, *253*, 866–872. [CrossRef]
14. FAO. *Second Global Plan of Action for Plant Genetic Resources for Food and Agriculture*; Commission on Genetic Resources for Food and Agriculture, Food and Agriculture Organization of the United Nations: Rome, Italy, 2012.
15. Maxted, N.; Dulloo, E.; Ford-Lloyd, B.V.; Iriondo, J.M.; Jarvis, A. Gap analysis: A tool for complementary genetic conservation assessment. *Divers. Distrib.* **2008**, *14*, 1018–1030. [CrossRef]
16. Global Crop Diversity Trust. *A global Rescue: Safeguarding the World's Crop Wild Relatives*; Global Crop Diversity Trust: Bonn, Germany, 2019.
17. Pradheep, K.; Singh, M.; Sultan, S.M.; Singh, K.; Parimalan, R.; Ahlawat, S.P. Diversity in wild relatives of wheat: An expedition collection from cold-arid Indian Himalayas. *Genet. Resour. Crop Evol.* **2019**, *66*, 275–285. [CrossRef]
18. Phillips, J.; Kyratzis, A.; Christoudoulou, C.; Kell, S.; Maxted, N. Development of a national crop wild relative conservation strategy for Cyprus. *Genet. Resour. Crop Evol.* **2014**, *61*, 817–827. [CrossRef]

19. Dempewolf, H.; Eastwood, R.J.; Guarino, L.; Khoury, C.K.; Muller, J.V.; Toll, J. Adapting agriculture to climate change: A global initiative to collect, conserve, and use crop wild relatives. *Agroecol. Sustain. Food Syst.* **2014**, *38*, 369–377. [CrossRef]
20. Holubec, V.; Hauptvogel, P.; Paprstein, F.; Podyma, W.; Sevcikoa, M.; Vymyslicky, T. Results of projects on collecting, mapping, monitoring, and conserving of plant genetic resources 1990–2008. *Czech J. Genet. Plant Breed.* **2010**, *46*, S2–S8. [CrossRef]
21. Castañeda-Álvarez, N.P.; Khoury, C.K.; Achicanoy, H.A.; Bernau, V.; Dempewolf, H.; Eastwood, R.J.; Guarino, L.; Harker, R.H.; Jarvis, A.; Maxted, N.; et al. Global conservation priorities for crop wild relatives. *Nat. Plants* **2016**, *2*, 16022. [CrossRef]
22. POWO. *Plants of the World Online*; Royal Botanic Gardens: Kew, UK, 2019.
23. Häkkinen, M.; Väre, H. Typification and check-list of *Musa* L. names (Musaceae) with nomenclatural notes. *Adansonia* **2008**, *30*, 63–112.
24. FAO. Banana Facts and Figures. Available online: http://www.fao.org/economic/est/est-commodities/bananas/bananafacts/en/#.XcWis-hKg2x (accessed on 8 November 2019).
25. Daniells, J.; Jenny, C.; Karamura, D.; Tomekpe, K. *Musalogue: Diversity in the Genus Musa*; International Network for the Improvement of Banana and Plantain: Montpellier, France, 2001.
26. Ruas, M.; Guignon, V.; Sempere, G.; Sardos, J.; Hueber, Y.; Duvergey, H.; Andrieu, A.; Chase, R.; Jenny, C.; Hazekamp, T.; et al. MGIS: Managing Banana (*Musa* spp.) genetic resources information and high-throughput genotyping data. *Database* **2017**, *2017*. [CrossRef]
27. Data Providers and the Crop Trust. Genesys Global Portal on Plant Genetic Resources. 2020. Available online: https://www.genesys-pgr.org/ (accessed on 17 September 2020).
28. Cámara-Leret, R.; Frodin, D.G.; Adema, F.; Al, E. New Guinea has the world's richest island flora. *Nature* **2020**. [CrossRef]
29. Perrier, X.; De Langhe, E.; Donohue, M.; Lentfer, C.; Vrydaghs, L.; Bakry, F.; Carreel, F.; Hippolyte, I.; Horry, J.P.; Jenny, C.; et al. Multidisciplinary perspectives on banana (*Musa* spp.) domestication. *Proc. Natl. Acad. Sci. USA* **2011**, *108*, 11311–11318. [CrossRef]
30. Perrier, X.; Bakry, F.; Carreel, F.; Jenny, C.; Horry, J.P.; Lebot, V.; Hippolyte, I. Combining biological approaches to shed light on the evolution of edible bananas. *Ethnobot. Res. Appl.* **2009**, *7*, 199–216. [CrossRef]
31. Christelová, P.; De Langhe, E.; Hřibová, E.; Čížková, J.; Sardos, J.; Hušáková, M.; Van den houwe, I.; Sutanto, A.; Kepler, A.K.; Swennen, R.; et al. Molecular and cytological characterization of the global *Musa* germplasm collection provides insights into the treasure of banana diversity. *Biodivers. Conserv.* **2017**, *26*, 801–824. [CrossRef]
32. Sardos, J.; Perrier, X.; Dolezel, J.; Hribova, E.; Christelova, P.; Van den Houwe, I.; Kilian, A.; Roux, N. DArT whole genome profiling provides insights on the evolution and taxonomy of edible banana (*Musa* spp.). *Ann. Bot.* **2016**, *118*, 1269–1278. [CrossRef] [PubMed]
33. Govaerts, R.; Häkkinen, M. *World Checklist of Musaceae*; Royal Botanic Gardens, Kew: Kew, UK, 2006.
34. Arnaud, E.; Horry, J.P. *Musalogue: A Catalogue of Musa Germplasm: Papua New Guinea Collecting Missions 1988–1989*; International Network for the Improvement of Banana and Plantain: Montpellier, France, 1997; p. 127.
35. Sachter-Smith, G.; Paufa, J.; Rauka, G.; Sardos, J.; Janssens, S. *Bananas of the Autonomous Region of Bougainville: A Catalog of Banana Diversity Seen on the Islands of Bougainville and Buka, Papua New Guinea*. 2017, p. 97. Available online: http://www.musalit.org/seeMore.php?id=17006 (accessed on 17 September 2020).
36. Sardos, J.; Paofa, J.; Janssens, S.; Sachter-Smith, G.; Rauka, G.; Roux, N. Banana Collecting Mission in the Autonomous Region of Bougainville (AROB), Papua New Guinea. 2017. Available online: https://www.researchgate.net/publication/318259511 (accessed on 17 September 2020).
37. Sardos, J.; Christelova, P.; Cizkova, J.; Paofa, J.; Sachter-Smith, G.L.; Janssens, S.B.; Rauka, G.; Ruas, M.; Daniells, J.W.; Dolezel, J.; et al. Collection of new diversity of wild and cultivated bananas (*Musa* spp.) in the Autonomous Region of Bougainville, Papua New Guinea. *Genet. Resour. Crop Evol.* **2018**, *65*, 2267–2286. [CrossRef]
38. Eyland, D.; Breton, C.; Sardos, J.; Kallow, S.; Panis, B.; Swennen, R.; Paofa, J.; Tardieu, F.; Welcker, C.; Janssens, S.B.; et al. Filling the gaps in gene banks: Collecting, characterizing and phenotyping wild banana relatives of Papua new Guinea. *Crop Sci.* **2020**. [CrossRef]

39. Gargiulo, R.; Saubin, M.; Rizzuto, G.; West, B.; Fay, M.F.; Kallow, S.; Trivedi, C. Genetic diversity in British populations of *Taxus baccata* L.: Is the seedbank collection representative of the genetic variation in the wild? *Biol. Conserv.* **2019**, *233*, 289–297. [CrossRef]
40. Hoban, S.; Kallow, S.; Trivedi, C. Implementing a new approach to effective conservation of genetic diversity, with ash (*Fraxinus excelsior*) in the UK as a case study. *Biol. Conserv.* **2018**, *225*, 10–21. [CrossRef]
41. Li, D.Z.; Pritchard, H.W. The science and economics of ex situ plant conservation. *Trends Plant Sci.* **2009**, *14*, 614–621. [CrossRef]
42. Whitehouse, K.J.; Hay, F.R.; Lusty, C. Why seed physiology is important for genebanking. *Plants* **2020**, *9*, 584. [CrossRef]
43. Roberts, E.H. Predicting the storage life of seeds. *Seed Sci. Technol.* **1973**, *1*, 499–514.
44. Ellis, R.H.; Hong, T.D.; Roberts, E.H. An intermediate category of seed storage behaviour. 1, Coffee. *J. Exp. Bot.* **1990**, *41*, 1167–1174. [CrossRef]
45. Walters, C. Orthodoxy, recalcitrance and in-between: Describing variation in seed storage characteristics using threshold responses to water loss. *Planta* **2015**, *242*, 397–406. [CrossRef] [PubMed]
46. Chin, H.F. Germination and Storage of Banana Seeds. In *Proceedings of the Workshop on New Frontiers in Resistance Breeding for Nematode, Fusarium and Sigatoka, Kuala Lumpur, Malaysia, 2–5th October 1995. Frison, E., Horry, J.P., Waele, D.D., Eds.*; International Network for the Improvement of Banana and Plantain: Montpellier, France, 1996; pp. 218–229.
47. Simmonds, N.W. The germination of banana seeds. *Trop. Agric.* **1952**, *29*, 35–49.
48. Stotzky, G.; Cox, E.A. Seed germination studies in *Musa*. II. Alternating temperature requirement for the germination of *Musa balbisiana*. *Am. J. Bot.* **1962**, *49*, 763–770. [CrossRef]
49. Boyce, K.G. Report of the seed storage committee 1986–1989. *Seed Sci. Technol.* **1989**, *17*, 135–142.
50. Chin, H.F.; Krishnapillay, B. Cryogenic storage of some horticultural species. In Proceedings of the IV International Symposium on Seed Research in Horticulture, Angers, France, 5–9 September 1988; pp. 107–112.
51. Darjo, P.; Bakry, F. Conservation and germination of banana seeds. *Fruits* **1990**, *45*, 103–113.
52. Nagano, S.; Mori, G.; Oda, M. Seed germinability in *Musa velutina* Wendl. & Drude is markedly lowered by 1 week in dry-storage. *J. Hortic. Sci. Biotechnol.* **2009**, *84*, 325–328.
53. Fortescue, J.A.; Turner, D. Reproductive biology. In *Banana Breeding: Progress and Challenges*; Pillay, M., Tenkouano, A., Eds.; Taylor & Francis: Boca Raton, FL, USA, 2011; pp. 145–179.
54. Laliberté, B. *Global Strategy for The Conservation and Use of Musa (Banana) Genetic Resources: A Consultative Document Prepared by The Global Musa Genetic Resources Network (MusaNet)*; Bioversity International: Rome, Italy, 2016; p. lxiv + 146.
55. Ellis, R.H.; Hong, T.D.; Roberts, E.H. *Handbook of Seed Technology for Genebanks*; International Board for Plant Genetic Resources (IBPGR): Rome, Italy, 1985; Volume 2, p. 456.
56. Batte, M.; Swennen, R.; Uwimana, B.; Akech, V.; Brown, A.; Tumuhimbise, R.; Hovmalm, H.P.; Geleta, M.; Ortiz, R. Crossbreeding East African Highland Bananas: Lessons learnt relevant to the botany of the crop after 21 years of genetic enhancement. *Front. Plant Sci.* **2019**, *10*. [CrossRef]
57. Hay, F.R.; Probert, R.J. Advances in seed conservation of wild plant species: A review of recent research. *Conserv. Physiol.* **2013**, *1*. [CrossRef]
58. Rao, N.K.; Dulloo, M.E.; Engels, J.M.M. A review of factors that influence the production of quality seed for long-term conservation in genebanks. *Genet. Resour. Crop Evol.* **2017**, *64*, 1061–1074. [CrossRef]
59. Probert, R.; Adams, J.; Coneybeer, J.; Crawford, A.; Hay, F. Seed quality for conservation is critically affected by pre-storage factors. *Aust. J. Bot.* **2007**, *55*, 326–335. [CrossRef]
60. Hay, F.; Probert, R. Collecting and handling seeds in the field. In *Collecting Plant Genetic Diversity: Technical Guidelines–2011 Update*; Guarino, L., Ramanatha Rao, V., Goldberg, E., Eds.; Bioiversity International: Rome, Italy, 2011.
61. Pancholi, N.; Wetten, A.; Caligari, P.D.S. Germination of *Musa velutina* seeds: Comparison of *in vivo* and *in vitro* systems. *In Vitro Cell. Dev. Biol. Plant* **1995**, *31*, 127–130. [CrossRef]
62. Afele, J.C.; De Langhe, E. Increasing in vitro germination of *Musa balbisiana* seed. *Plant Cell Tissue Organ Cult.* **1991**, *27*, 33–36. [CrossRef]
63. Cox, E.A.; Stotzky, G.; Goos, R.D. *In vitro* culture of *Musa balbisiana* Colla embryos. *Nature* **1960**, *185*, 403–404. [CrossRef]
64. IPGRI. *Descriptors for Banana (Musa spp.)*; IPGRI: Rome, Italy, 1996.

65. Hong, T.D.; Ellis, R.H. *Ex situ* biodiversity conservation by seed storage: Multiple-criteria keys to estimate seed storage behaviour. *Seed Sci. Technol.* **1996**, *25*, 157–161.
66. Ellis, R.H.; Mai-Hong, T.; Hong, T.D.; Tan, T.T.; Xuan-Chuong, N.D.; Hung, L.Q.; Ngoc-Tam, B.; Le-Tam, V.T. Comparative analysis by protocol and key of seed storage behaviour of sixty Vietnamese tree species. *Seed Sci. Technol.* **2007**, *35*, 460–476. [CrossRef]
67. Abdelnouresquivel, A.; Mora, A.; Villalobos, V. Cryopreservation of zygotic embryos of *Musa acuminata* (AA) and *M. balbisiana* (BB). *Cryo-Lett.* **1992**, *13*, 159–164.
68. Berjak, P.; Pammenter, N.W. Progress in the understanding and manipulation of desiccation-sensitive (recalcitrant) seeds. *Basic Appl. Asp. Seed Biol.* **1997**, *30*, 689–703.
69. Ellis, R.H.; Hong, T.D.; Roberts, E.H. An intermediate category of seed storage behaviour. 2, effects of provenance, immaturity and imbibition on desiccation-tolerance in coffee. *J. Exp. Bot.* **1991**, *42*, 653–657. [CrossRef]
70. Dussert, S.; Chabrillange, N.; Engelmann, F.; Anthony, F.; Louarn, J.; Hamon, S. Relationship between seed desiccation sensitivity, seed water content at maturity and climatic characteristics of native environments of nine Coffea L. species. *Seed Sci. Res.* **2000**, *10*, 293–300. [CrossRef]
71. Walters, C. Understanding the mechanisms and kinetics of seed aging. *Seed Sci. Res.* **1998**, *8*, 223–244. [CrossRef]
72. Berjak, P.; Pammenter, N.W. From Avicennia to Zizania: Seed recalcitrance in perspective. *Ann. Bot.* **2008**, *101*, 213–228. [CrossRef] [PubMed]
73. Ellis, R.H. Temporal patterns of seed quality development, decline, and timing of maximum quality during seed development and maturation. *Seed Sci. Res.* **2019**, *29*, 135–142. [CrossRef]
74. De Langhe, E.; Perrier, X.; Donohue, M.; Denham, T. The original banana split: Multi-disciplinary implication of the generation of African and Pacific plantains in island southeast Asia. *Ethnobot. Res. Appl.* **2015**, *14*, 299–312. [CrossRef]
75. Argent, G.C.G. The wild bananas of Papua New Guinea. *Notes Royal Bot. Gard. Edinb.* **1976**, *35*, 77–114.
76. Janssens, S.B.; Vandelook, F.; De Langhe, E.; Verstraete, B.; Smets, E.; Vandenhouwe, I.; Swennen, R. Evolutionary dynamics and biogeography of Musaceae reveal a correlation between the diversification of the banana family and the geological and climatic history of Southeast Asia. *New Phytol.* **2016**, *210*, 1453–1465. [CrossRef]
77. Tweddle, J.C.; Dickie, J.B.; Baskin, C.C.; Baskin, J.M. Ecological aspects of seed desiccation sensitivity. *J. Ecol.* **2003**, *91*, 294–304. [CrossRef]
78. Hay, F.R.; Probert, R.J. Seed maturity and the effects of different drying conditions on desiccation tolerance and seed longevity in foxglove (*Digitalis purpurea* L.). *Ann. Bot.* **1995**, *76*, 639–647. [CrossRef]
79. Hong, T.D.; Ellis, R.H. *A Protocol to Determine Seed Storage Behaviour*; International Plant Genetic Resources Institute: Rome, Italy, 1996.
80. Ellis, R.; Pieta Filho, C. The development of seed quality in spring and winter cultivars of barley and wheat. *Seed Sci. Res.* **1992**, *2*, 9–15. [CrossRef]
81. Uma, S.; Lakshmi, S.; Saraswathi, M.S.; Akbar, A.; Mustaffa, M.M. Embryo rescue and plant regeneration in banana (*Musa* spp.). *Plant Cell Tissue Organ Cult.* **2010**, *105*, 105–111. [CrossRef]
82. Marod, D.; Pinyo, P.; Duengkae, P.; Hiroshi, T. The role of wild banana (Musa acuminata Colla) on wildlife diversity in mixed deciduous forest, Kanchanaburi Province, Western Thailand. *Kasetsart J. Nat. Sci.* **2010**, *44*, 35–43.
83. Meng, L.Z.; Gao, X.X.; Chen, J.; Martin, K. Spatial and temporal effects on seed dispersal and seed predation of *Musa acuminata* in southern Yunnan, China. *Integr. Zool.* **2012**, *7*, 30–40. [CrossRef] [PubMed]
84. Kulavmode, A.R.; Hegde, V.M.; Kambale, A.A. A note on seed dispersal of rock banana *Ensete superbum* by Asian palm civet *Paradoxurus hermaphroditus* at Sinhgad Fort, Maharashtra, India. *J. Bombay Nat. Hist. Soc.* **2015**, *112*, 25–26. [CrossRef]
85. Kumar, K.P.S.; Bhowmik, D.; Duraivel, S.; Umadevi, M. Tradition and medicinal uses of banana. *J. Pharmacogn. Phytochem.* **2012**, *1*, 51–63.
86. Whitehouse, K.J.; Hay, F.R.; Ellis, R.H. Increases in the longevity of desiccation-phase developing rice seeds: Response to high-temperature drying depends on harvest moisture content. *Ann. Bot.* **2015**, *116*, 247–259. [CrossRef] [PubMed]

87. Whitehouse, K.J.; Hay, F.R.; Ellis, R.H. High-temperature stress during drying improves subsequent rice (*Oryza sativa* L.) seed longevity. *Seed Sci. Res.* **2017**, *27*, 281–291. [CrossRef]
88. Whitehouse, K.J.; Hay, F.R.; Ellis, R.H. Improvement in rice seed storage longevity from high-temperature drying is a consistent positive function of harvest moisture content above a critical value. *Seed Sci. Res.* **2018**, *28*, 332–339. [CrossRef]
89. Simmonds, N.W. Experiments on the germination of banana seeds. *Trop. Agric.* **1959**, *36*, 259–273.
90. Michaels, H.J.; Benner, B.; Hartgerink, A.P.; Lee, T.D.; Rice, S.; Willson, M.F.; Bertin, R.I. Seed size variation: Magnitude, distribution, and ecological correlates. *Evol. Ecol.* **1988**, *2*, 157–166. [CrossRef]
91. Cavers, P.B.; Steel, M.G. Patterns of change in seed weight over time on individual plants. *Am. Nat.* **1984**, *124*, 324–335. [CrossRef]
92. Boe, A.; Bortnem, R.; Johnson, P.J. Changes in weight and germinability of black medic seed over a growing season, with a new seed predator. In Proceedings of the 101st Annual Meeting of South-Dakota-Academy-of-Science, Univ Sioux Falls, Sioux Falls, SD, USA, 8–9 April 2016; pp. 105–117.
93. Susko, D.J.; Lovett-Doust, L. Patterns of seed mass variation and their effects on seedling traits in *Alliaria petiolata* (Brassicaceae). *Am. J. Bot.* **2000**, *87*, 56–66. [CrossRef] [PubMed]
94. Ortiz, R.; Vuylsteke, D. Factors influencing seed set in triploid *Musa* spp. and production of euploid hybrids. *Ann. Bot.* **1995**, *75*, 151–155. [CrossRef]
95. Ssebuliba, R.; Makumbi, D.; Pillay, M. Patterns of seed set in East African Highland Banana (*Musa* spp.) hybrids. *J. New Seeds* **2009**, *10*, 160–170. [CrossRef]
96. Asif, M.; Mak, C.; Othman, R. In vitro zygotic embryo culture of wild *Musa acuminata* ssp.*malaccensis* and factors affecting germination and seedling growth. *Plant Cell Tissue Organ Cult.* **2001**, *67*, 267–270. [CrossRef]
97. FAO. *International Treaty on Plant Genetic Resources for Food and Agriculture*; Food and Agriculture Organization of the United Nations: Rome, Italy, 2009.
98. Olson, D.M.; Dinerstein, E.; Wikramanayake, E.D.; Burgess, N.D.; Powell, G.V.N.; Underwood, E.C.; D'Amico, J.A.; Itoua, I.; Strand, H.E.; Morrison, J.C.; et al. Terrestrial ecoregions of the worlds: A new map of life on Earth. *Bioscience* **2001**, *51*, 933–938. [CrossRef]
99. Fick, S.E.; Hijmans, R.J. Worldclim 2: New 1-km spatial resolution climate surfaces for global land areas. *Int. J. Clim.* **2017**, *25*, 1965–1978. [CrossRef]
100. Puteh, A.B.; Aris, E.M.; Sinniah, U.R.; Rahman, M.; Mohamad, R.B.; Abdullah, N.A.P. Seed anatomy, moisture content and scarification influence on imbibition in wild banana (*Musa acuminata* Colla) ecotypes. *Afr. J. Biotechnol.* **2011**, *10*, 14373–14379.
101. Leist, N.; Kr ämer, S. *Supplements 2011 to ISTA Working Sheets on Tetrazolium Testing*; ISTA: Bassersdorf, Switzerland, 2011; Volume 1.
102. Murashige, T.; Skoog, F. A revised medium for rapid growth and bio assays with tobacco tissue cultures. *Physiol. Plant.* **1962**, *15*, 473–497. [CrossRef]
103. Vineesh, P.S.; Skaria, R.; Mukunthakumar, S.; Padmesh, P.; Decruse, S.W. Seed germination and cryostorage of *Musa acuminata* subsp. *burmannica* from Western ghats. *S. Afr. J. Bot.* **2015**, *100*, 158–163. [CrossRef]
104. Ripley, B.; Venables, W. *Modern Applied Statistics with S*, 4th ed.; Springer: New York, NY, USA, 2002.
105. Fox, J.; Hong, J. Effect displays in R for multinomial and proportional-odds logit models: Extensions to the effects package. *J. Stat. Softw.* **2009**, *32*, 1–24. [CrossRef]
106. R Core Team. *R: A Language and Environment for Statistical Computing*; R Foundation for Statistical Computing: Vienna, Austria, 2019.

plants

MDPI

Review

Born to Eat Wild: An Integrated Conservation Approach to Secure Wild Food Plants for Food Security and Nutrition

Teresa Borelli [1,*], Danny Hunter [1], Bronwen Powell [2], Tiziana Ulian [3], Efisio Mattana [3], Céline Termote [1], Lukas Pawera [4,5], Daniela Beltrame [6], Daniela Penafiel [7,8], Ayfer Tan [9], Mary Taylor [10] and Johannes Engels [1]

1 Alliance of Bioversity International and CIAT, via dei Tre Denari 472/a, 00054 Rome, Italy; D.Hunter@cgiar.org (D.H.); C.Termote@cgiar.org (C.T.); j.engels@cgiar.org (J.E.)
2 Center for International Forestry Research, Penn State University, State College, PA 16802, USA; bxp15@psu.edu
3 Royal Botanic Gardens Kew, Wakehurst, Ardingly, West Sussex RH17 6TN, UK; t.ulian@kew.org (T.U.); E.Mattana@kew.org (E.M.)
4 Faculty of Tropical AgriSciences, Czech University of Life Sciences Prague, Kamýcká 129, 16500 Praha-Suchdol, Czech Republic; paweralukas@gmail.com
5 The Indigenous Partnership for Agrobiodiversity and Food Sovereignty, c/o Alliance of Bioversity International and CIAT, Via dei Tre Denari 472/a, 00054 Rome, Italy
6 Biodiversity for Food and Nutrition Project, Ministry of the Environment, Brasília-DF 70068-900, Brazil; dani.moura.oliveira@gmail.com
7 Escuela Superior Politécnica del Litoral, Centro de Investigaciones Rurales–FCSH, Campus Gustavo Galindo-km. 30.5 vía Perimetral, Guayaquil 090112, Ecuador; ddpenafi@espol.edu.ec
8 Faculty of Medicine, Universidad de Especialidades Espíritu Santo, Samborondon 091650, Ecuador
9 Aegean Agricultural Research Institute, Menemen, Izmir P.O. Box 9 35661, Turkey; ayfer_tan@yahoo.com
10 Environmental Studies, University of the Sunshine Coast, Maroochydore, QLD 4556, Australia; maryt@oxalis.plus.com
* Correspondence: t.borelli@cgiar.org

Received: 31 July 2020; Accepted: 24 September 2020; Published: 1 October 2020

Abstract: Overlooked in national reports and in conservation programs, wild food plants (WFPs) have been a vital component of food and nutrition security for centuries. Recently, several countries have reported on the widespread and regular consumption of WFPs, particularly by rural and indigenous communities but also in urban contexts. They are reported as critical for livelihood resilience and for providing essential micronutrients to people enduring food shortages or other emergency situations. However, threats derived from changes in land use and climate, overexploitation and urbanization are reducing the availability of these biological resources in the wild and contributing to the loss of traditional knowledge associated with their use. Meanwhile, few policy measures are in place explicitly targeting their conservation and sustainable use. This can be partially attributed to a lack of scientific evidence and awareness among policymakers and relevant stakeholders of the untapped potential of WFPs, accompanied by market and non-market barriers limiting their use. This paper reviews recent efforts being undertaken in several countries to build evidence of the importance of WFPs, while providing examples of cross-sectoral cooperation and multi-stakeholder approaches that are contributing to advance their conservation and sustainable use. An integrated conservation approach is proposed contributing to secure their availability for future generations.

Keywords: wild food plants; food security; nutrition data; multi-sectoral collaboration; policy; conservation

1. Introduction

The practice of consuming wild food plants (WFPs) is as old as human prehistory. Early humans obtained their food by hunting, fishing and gathering these plants, or parts of plants (e.g., stems, roots, flowers, fruits, leaves, buds, and seeds), that were safe for human consumption. It was not until 10,000 years BC that people started settling into more permanent homesteads and domesticating plant species (mostly carbohydrate-rich staples) while maintaining some hunter-gatherer activities and collecting WFPs from the wild [1,2]. This still holds true for some traditional horticultural societies today (e.g., the Machiguenga in South America) [3]. All of the plants we now call domestic crops were once WFPs, altered by human manipulation to achieve domestication by selecting more favorable plant traits. With plant domestication and farming came also the development of weeds; that is, unwanted plant species in cultivated fields, and many of the WFPs consumed today include relatives of today's crops.

Today, the term "wild" is mostly taken to indicate species that grow spontaneously in self-sustaining populations outside cultivated areas, in field margins, forests, woodland, grassland, and wetlands (e.g., paddy fields), independently of human activity [4]. However, the distinction between "wild" and "cultivated" or "domesticated" is not so clear-cut and many WFPs fall somewhere in between these extremes depending on the degree of human intervention and management. For example, they can grow spontaneously in areas that are or have been themselves cultivated [4,5], or, as in the case of the "quelites" greens in Mesoamerica (e.g., the genus *Amaranthus, Chenopodium, Porophyllum, Portulaca, Crotalaria,* and *Anoda*), they have become the focus of systematic in situ management practices such as "selective harvesting" and "let standing", with important repercussions on plant communities [6]. Another known management practice is that of "encouraging growing" recorded by Cruz-Garcia [7] in the Peruvian Amazon along the deforestation border. Surveys revealed that, out of 30 wild food plant species identified, 20 are actively managed by local farmers and that most are transplanted from the forest to their agricultural fields for easy access. From these, 57% of the species are classified as weeds, yet are perceived by farmers to play a role in food security, particularly with increasing deforestation and reduced availability of food plants [7].

In this review paper, the term "wild food plants" is extended to all those food plants (herbs and spices included) that are also semi-domesticated, in addition to economically important non-timber forest food products, such as açaí berries and Brazil nuts [8]. As they are often wild relatives of domesticated species, WFPs have potential for domestication and can provide a pool of genetic resources for hybridization and selective breeding [9].

2. The Importance of Wild Food Plants Today

WFPs continue to play a vital role in the subsistence of many human populations particularly when the availability of food crops is scarce, when household budgets are insufficient to buy enough food or when access to markets is challenging [5,8,10–16]. Wild foods are also integral to traditional food systems and have nutritional and cultural value for many indigenous peoples [4,5,17,18]. Deeply connected to their land, indigenous peoples, who represent 5% of the global population [19], are often the sole custodians of rich and diverse knowledge relating to plant uses and traditional food systems and to local food biodiversity existing within the ecosystems they inhabit [18]. Traditional communities also have better ecological knowledge about local environments and their customary users, making monitoring and regulating of natural resources easier [20].

Although the caloric contribution of WFPs to people's diets is generally low compared to staple foods [21], these species contribute to diet diversification in many geographical settings where otherwise monotonous diets may prevail [22–26]. Wild foods (both plants and non) provided between 1% and 19% of the iron consumed, between 5% and 45% of the calcium and between 0% and 31% of the vitamin A equivalents (RAE) in the diets of women and children in studies from Benin, Tanzania, and the Philippines [21]. These neglected biological resources have, in fact, been shown to contain equally, if not higher amounts, of nutrients than more widely available commercial crops [5,27–29], and, if properly assessed and managed, could be introduced in national food and nutrition security and sovereignty strategies that focus on nutrient adequacy rather than quantity of staples, while being culturally acceptable.

WFPs could also be central to efforts directed at empowering local market actors as well as reducing the distance between consumers and producers and the overreliance on globalized value chains. Although, recent research by Kinnunen et al. [30] highlights the unfeasibility of localizing production for important global staples such as rice, maize and temperate cereals, there is increasing evidence that the local trade of minor crops, traditional varieties, and WFPs has potential to empower communities and increase livelihoods in rural areas, particularly of women and youth [31,32]. Meanwhile, the COVID-19 crisis has revealed the vulnerability of our global food systems to disease-related disruptions and shocks [33–35]. For example, the imposed travel restrictions on people and goods as a result of the lockdowns are causing logistical bottlenecks in food supply chains [36]. Given the national and international trade restrictions, long supply chains are struggling to cope with the rise in food demand for non-perishable food supplies [37], while short supply chains are suffering due to the closing of informal and local open-air markets [38], where the majority of the world's population still obtains fruits, horticultural, and other perishable products [37,39]. At the same time, the pandemic has opened up opportunities for a new food system paradigm that supports local self-sufficiency and domestic agricultural production and sees home and community gardens, traditional agroecosystems, and farmers' markets as essential services [38,40]. With food shortages affecting specialized, high value horticultural crops [41], people are turning to traditional vegetables and WFPs as a sustainable source of food, vitamins and nutrients [42], not to mention for herbal ingredients, traditional medicine formulations or new biopharmaceuticals [38,43,44].

This paper builds mainly on the authors' own efforts being undertaken in several countries to provide evidence of the role of WFPs in supporting nutrition and livelihood security. This paper also provides examples of cross-sectoral cooperation and multi-stakeholder approaches that are contributing to the better conservation and use of WFPs, including by fostering linkages between in and ex situ conservation. In the case of WFPs, "use" includes the various practices and activities involved in (i) domesticating wild species; (ii) the management of wild species and their habitats in and around production systems to promote the delivery of ecosystem services; and (iii) the introduction of wild species into production and consumption systems, for example by creating demand for the species, and regulating their harvesting in the wild. Lastly, details will be provided of a proposed integrated conservation approach that focuses on local interventions based on traditional food systems.

2.1. Diversity (Geographical Use) and Contribution to Diets

The use of WFPs in many countries is confirmed by national contributions to the recent "State of the World's Biodiversity for Food and Agriculture"—"SOWBFA"—of the Food and Agriculture Organization of the United Nations [45]. Of 91 countries reporting information for compiling the report, 69 nations reported a total of 1955 wild plant species that contribute to food security and nutrition in their respective countries, as well as making diets healthier and more diverse. However, as the examples provided by the authors and recently published papers [46] demonstrate, the number is probably much higher and these species remain largely unreported in national statistics, as does the actual contribution of these biological resources to national economies in many parts of the world [47]. Table S1 in the Supplementary materials lists the wide range of plant families that encompass the edible wild and semi-cultivated plant species researched by the authors and mentioned in the text as contributing to food and nutrition security. The list, as the review carried out by the authors, is by no means exhaustive and could undoubtedly include more.

2.1.1. Africa

As part of the MGU Useful Plants Project (UPP) managed by the Royal Botanic Gardens, Kew, UK, institutional partners working alongside local communities in Botswana, Kenya, Mali, South Africa, and Mexico identified 615 species used for food across the five countries [48]. Information on seed conservation, propagation and traditional uses has been published for 48 of them and is now available on the internet [49]. In Africa, the species included the baobab (*Adansonia digitata)*, the mongongo tree (*Schinziophyton rautanenii*), and the morama bean (*Tylosema esculentum*) [49]. Research undertaken by Bioversity International in the early 1990s has documented 210 African leafy vegetables in Kenya alone [50]. These are wild or semi-domesticated species that are grown mostly for household consumption or traded informally, but which have seen a revival particularly in urban and peri-urban areas [51]. In Western Kenya, between 23 and 42 African leafy vegetables continue to be consumed by local communities depending on the district. Eleven species, including amaranth (*Amaranthus* spp.), spider plant (*Cleome gynandra*), and African nightshades (*Solanum* spp.) were selected for further research as part of the African Leafy Vegetables program from 1996 to 2004 (Bioversity International and EIARD, 2013; Gotor and Irungu, 2010) [51,52] as well as for the Biodiversity for Food and Nutrition Project [28,53]. In addition to filling the nutrient gap, a cost of diet study carried out in Eastern Baringo, Kenya, has shown that wild plant species, especially vegetables, are able to significantly reduce (by 30–70%) the cost of a nutritious meal for women and children aged 6 to 23 months in hypothetically-modeled lowest cost nutritious diets [54].

2.1.2. South America

In Ecuador, one of the top seven mega diverse countries in the world, wild edible fruits and plants collected from a diverse range of habitats play a fundamental role in traditional diets, particularly for the indigenous communities living in forest areas. Studies in the country by Penafiel et al. [55,56], documenting local knowledge on the use of WFPs among the Andean indigenous communities of Guasaganda (Cotopaxi) and the Andean Kichwa mothers of Arosemena Tola (Napo), recorded the culinary use of 49 and 10 WPFs, respectively. Brazil also contains vast amounts of wild food plant diversity [57]. Some of this diversity is of national and regional relevance, e.g., Brazil nut (*Bertholletia excelsa*) and açaí (*Euterpe oleracea*), but most is of local value and its potential nutritional and economic value remains unexplored and unexploited [58]. The "Plants for the Future (PPF) Initiative", a prioritization exercise undertaken by the Ministry of the Environment that set out to explore the wealth of Brazil's plant biodiversity, has identified a considerable number of wild species of nutritional value and market potential. Across the country's five eco-regions, out of 78 native undervalued edible plant species, 49 are found exclusively in the wild (mostly fruits and nuts) [28]. Mostly found in forest areas, the species are managed by family farmers or harvested from the wild by local communities using traditional practices. The link between local communities and nature is such that the Brazilian ministries of Agriculture, Environment, and Social Development have coined the term "sociobiodiversity" to describe these traditionally managed biodiversity-derived goods that are sold in local markets, provide incomes and improve the livelihoods of traditional communities, while protecting biodiversity and the environment.

2.1.3. The Mediterranean

In the Mediterranean, WFPs are still common in traditional cuisine and are widely consumed locally [59,60]. In their compendium of gathered Mediterranean food plants, Rivera et al. [61] identified approximately 2300 different WFPs and fungi taxa in this region alone, of which 1000 are strictly used locally. As part of the Biodiversity for Food and Nutrition (BFN) project, Turkey prioritized 42 wild edible plants for further research [28] out of hundreds of known species [59,62,63], while across Morocco, Nassif and Tanji [64] compiled a list of 246 wild plant species used as food. While many WFPs are only used occasionally or in small regional areas, some are central to Moroccan diets and culinary traditions. Aromatic herbs such as thyme (*Thymus* spp.), mint (*Mentha* spp.), and sage (*Salvia* spp.) are the most widely consumed wild plants; however, they contribute little to the diet in terms of energy (kcal) and nutrients because they are used as condiments [65]. Wild leafy vegetables, on the other hand, are a seasonally important component of Moroccan diets, particularly in rural Morocco where 86% of households reported consuming wild leafy vegetables (WLVs) on a regular basis [66]. Some of the most commonly consumed WLVs (many of which are also consumed in Turkey and other Mediterranean countries) include mallow (*Malva* spp.), purslane (*Portulaca oleracea)*, goosefoots (*Chenopodium* spp.), docks and sorrels (*Rumex* spp.), fennel (*Foeniculum* sp. cf *F. vulgare*), golden thistle (*Scolymus hispanicus)*, and watercress (*Nasturtium officinale)*. They are commonly served as a cooked salad or side dish, eaten in moderate portion sizes (approximately 50 g per meal). Argan oil (*Sideroxylon spinosum*), capers (*Capparis spinosa* and *C. decidua*), acorns (*Quercus* spp.), and the fruits of the strawberry tree (*Arbutus unedo)* as well as jujube (*Ziziphus jujube*), mulberry (*Morus* spp.), and blackberry (*Rubus* spp.) are other commonly consumed plant products in this region [64,65,67].

2.1.4. Asia Pacific

The consumption of WFPs and food trees makes a significant contribution to human health in the Pacific region [68]. In "Food Plants of Papua New Guinea, A Compendium" [69], Bruce French produces a list of food plants, many of which are sourced from the wild, including root crops and staples, legumes, green leafy and other vegetables, nuts, fruits and what are categorized as "minor foods and flavorings". For example, the kernel of wild *karuka* (*Pandanus brosimos*), endemic to Papua New Guinea (PNG), is eaten by about one-third of the rural population [70], particularly by communities living at high altitudes. When the fruit matures, villagers migrate to high altitudes to harvest the fruit and extract the nuts. Nuts have not been recorded in the main highland markets, but it is possible that they are sold in some high-altitude locations [71]. PNG and surrounding region are also one of the few places in the world where communities obtain the majority of their carbohydrate staple from a wild food plant: sago [72,73].

In Niue (Polynesia), the traditional processing of wild arrowroot (*Tacca* spp.) is still an ongoing practice. Starch processed from the root is a local delicacy used to make local puddings and breads [74]. Thaman [75] lists 60 WFPs used in Fiji, noting that these plants play an important role as emergency or famine foods when extreme climatic events disrupt cultivation. Among these are wild marine seaweeds such as sea grapes (*Caulerpa racemosa*), known as nama, and other edible seaweeds that are still widely consumed. "The Guide to the Common Edible and Medicinal Sea Plants of the Pacific Islands" provides an insight into marine WFPs and the benefits that can be gained from their use [76].

In the mosaic tropical landscape of West Sumatra, Indonesia, composed predominantly of rice fields, home gardens, cacao agroforestry, and forests, with the help of local communities, the Food, Agrobiodiversity and Diet (FAD) project has identified 85 WFPs [77]. In this region, WFPs are consumed less than in the past, and the FAD project aimed to raise knowledge and awareness of wild foods by organizing workshops, traditional food competitions, and sharing community materials such as illustrations, posters, and community guidebooks, on food plants for nutrition and health [78].

2.2. *Income Generation*

In many parts of the world, WFPs are not only harvested for subsistence [79–81]. Gathered in excess, they are sold in local markets to generate income, thereby contributing to the household economies of gatherers and collectors, usually women, or to bolster the incomes of migrants and unemployed moving from rural to urban areas [82]. For example, in the Chimanimani communities living in the Trans-Frontier Conservation Area in Mozambique, the fruits of *Uapaca kirkiana* and *Strychnos madagascariensis* are sold for a reasonable profit and represent an important source of income outside the maize harvest season (March to May) [83]. In their review paper, Hickey et al. [84] showed that 50% of almost 8000 households sampled in forested areas of 24 developing countries across Asia, Africa, and Latin America derived their income from wild food collection. The study also highlighted that the sale of plant foods contributed 2.3% to total household income across the study sites on three continents, the proportion increasing to 2.8% in Africa and Latin America, particularly in poorer households.

In parts of Turkey, where WFPs are central to traditional cooking, wild edibles are sold unprocessed in local markets and processed (e.g., pickled, canned, or frozen) in district markets or supermarkets via wholesalers and middlemen [85]. In 2012, in the Pacific Island States of Fiji, Samoa, and Tonga the yearly production and revenue from the harvesting and sales of the seaweed *Caulerpa racemosa* was valued at USD 266,492 [86]. However, the true extent of this revenue is not always available. For example, a recent European assessment established the value of collected non-wood forest products, mainly food plants, at € 19.5 billion with value per hectare rising to € 77.8, and ten times above the official European estimates [87]. Many markets for WFPs are informal, and market players may hold back information because of illicit harvesting in local conservation areas [82]. Data about geographic and temporal distribution, production cost, quantity harvested, and price is also often limited. Increased profits can often lead to overexploitation of WFPs and negative outcomes for the entire community [20]. To avert this possibility, participatory research is key to establish sustainable management guidelines and harvest rates, and to monitor the ecological impacts of increased use [83].

2.3. Threats to WFPs

Despite the realization of the potential use of WFPs in food security and poverty reduction strategies, the SOWBFA, along with other recent global reports [46], warn us that this precious diversity is fast disappearing, particularly in forest habitats [88,89]. Land use changes (e.g., conversion to agriculture, change in agricultural practices and infrastructure development), habitat destruction (resulting from timber harvesting, fuelwood collection, grazing, and forest fires) and overharvesting collectively account for 62% of the threats reported to WFPs in SOWBFA, which mostly grow beyond the limit of protected areas [45,90,91]. The SOWBFA used the Sampled Red List Index for Plants of the International Union for Conservation on Nature (IUCN) [92] to determine that, of a total 822 WFP species considered across 7 different classes, 73% are currently at low risk of extinction (Figure 1), with some classes highly threatened in the wild (e.g., WFPs that are derived from conifers and cycads). However, the IUCN Red List Index for Plants includes global conservation assessments for only one third (31%) of known WFPs. Local assessments for many WFPs that are currently excluded from the IUCN assessment paint a very different story indicating the need to consider community perceptions when ascribing risk class (Table 1). Furthermore, an assessment of the comprehensiveness of conservation of 1587 WFP taxa (including cereals, fruit, and nuts), carried out by the International Center for Tropical Agriculture (CIAT) as part of a larger study to identify conservation gaps for useful plants, shows that only 3.3% of WFPs are sufficiently conserved ex situ, i.e., in gene banks or in other living plant repositories, while 89.1% require urgent off-site conservation measures given the impending threats to their existence [93]. Their continued use in diets, when accompanied by careful sustainable management by the communities consuming them, and protection of WFP habitats, on the other hand, seems to have ensured their momentary conservation in situ, in the natural habitats in which they grow. Of the WFP taxa analyzed 42.1% are sufficiently conserved, 46.7% deserve medium priority and 11.1% require stepping up conservation measures [93]. Nonetheless, Khoury et al. [93] caution against the overreliance on protected areas for the long-term conservation of these species. Rapidly warming temperatures and habitat destruction can alter the species' geographic distribution, driving them across the artificially designated boundaries of many protected areas in pursuit of favourable growing conditions [94].

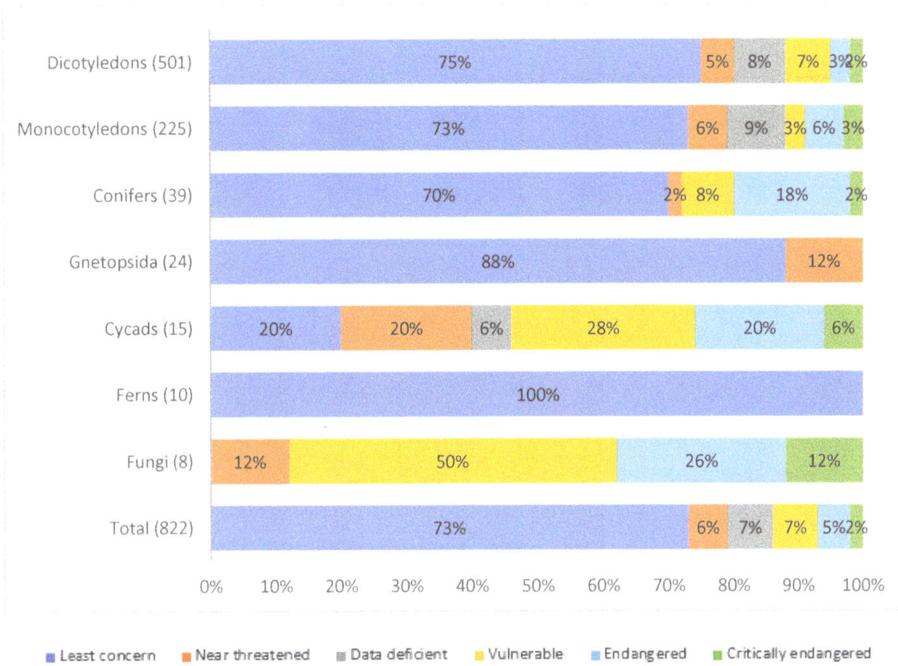

Figure 1. Number of WFPs and fungi on the IUCN Red List of Threatened Species classified by class and risk category Source: IUCN Red List 2017. Adapted from FAO [45].

Given that many WFPs grow in agricultural systems (as weeds, in hedge rows, as wild trees in agroforestry systems, and in small forest patches [5,21]), agricultural change, including intensification, more pesticide use and removal of trees can threaten the existence of these biological resources [12,14,95]. Food production systems that pollute the environment by using large quantities of fertilizers, pesticides, and herbicides, are also major causes of biodiversity loss [45,88,96]. Applying chemical herbicides in rice fields or agroforestry plots, for example, leads to the reduced availability of WFPs in West Sumatra, Indonesia [77]. WFPs that survive aerial spraying with herbicides are contaminated by these harmful substances, making them unfit for human consumption, while pesticides wipe out many of the pollinators needed for plant reproduction.

Overharvesting can also be an important pressure on non-timber forest products, including wild foods [97]. This is the case for Morocco and Turkey. Morocco is the twelfth global exporter of medicinal and aromatic plants, a trade that places extensive harvesting pressure on many of the species traditionally used as herbs [98]. A rapid vulnerability assessment carried out by Lamrani-Alaoui and Hassikou identified six species that grow across wide areas of Morocco's government owned forests (*Thymus satureioides, Lavandula dentata, Origanum compactum, Origanum elongatum, Salvia rosmarinus* and *Fraxinus dimorpha*) as needing urgent conservation, restoration, and sustainable management measures [98].

Table 1. Local threat assessments carried out in partnership with local communities have identified increasing dangers to the survival of nutritious and locally important WFPs. A selection from the authors' project sites is provided along with suggested measures for conservation and sustainable use.

Country	Species Name	Local Name	Edible Use	Main Nutritional Benefits	Habitat	Threat Status (IUCN, Community to Other)	Threats/Suggestion for Conservation	Photo
Brazil	*Astrocaryum aculeatum*	Tucumã	Fruit pulp	Rich in vitamin A as well as lauric, myristic and oleic acid [99]	Amazon rainforest	No IUCN assessment	Habitat loss—deforestation/ Preserve natural habitats	 Credit: J. Camillo
	Euterpe edulis	Jussara	Fruit pulp consumed as puree, palm heart (discouraged)	The fruit is rich in antioxidants [99]	Dense shady forest (Atlantic forest)	No IUCN assessment, listed as Vulnerable in the Red Book of Brazilian Flora [100]	Habitat loss—deforestation, overharvesting of palm heart/ Preserve natural habitats	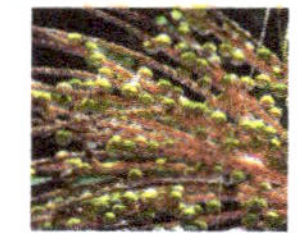 Credit: A. Popovkin
	Butia eriospatha	Butiá	Fruit pulp, seed	Good source of fiber, potassium, and vitamin C (equivalent to levels found in oranges) [99]	Highland mixed shady forests (Araucaria forest), around 800–900 m elevation	IUCN—Vulnerable	Habitat loss—deforestation/ Preserve natural habitats	 Credit: G. Lopes
	Dipteryx alata	Baru nut	Fruit/Nut	High in fiber; the nut is high in quality protein [99]	Tropical savannah (Cerrado)	IUCN—Vulnerable	Habitat loss—deforestation/ Preserve natural habitats	 Credit: J. Camillo
	Hancornia speciosa	Mangaba	Fruit	Excellent source of vitamin C, folates and a good source of carotenoids and vitamin E [99]	Scrublands (Caatinga) and barren lands in central Brazil	No IUCN assessment	Habitat loss—deforestation/ Preserve natural habitats	 Credit: J. Camillo

Table 1. *Cont.*

Country	Species Name	Local Name	Edible Use	Main Nutritional Benefits	Habitat	Threat Status (IUCN, Community to Other)	Threats/Suggestion for Conservation	Photo
Ecuador	*Vasconcellea microcarpa (Carica microcarpa)*	Col de monte	Leaves	N/A	Forest	IUCN—Least concern	Deforestation/Nutrition education needed	 Credit: X. Scheldeman
	Pouteria multiflora	Logma	Fruit	N/A	Forest	No IUCN assessment	Deforestation/Nutrition education needed	 Credit: IKIAM
	Hypolepis hostilis	Garabato yuyo	Leafy green vegetable (fern)	N/A	Forest	No IUCN assessment	Loss of traditional food culture /Use as complementary food for infants	 Credit: IKIAM
	Plukenetia volubilis	Sachainchi	Nut	Good source of lipids, proteins, and essential amino acids (e.g., cysteine, tyrosine, threonine, and tryptophan), vitamin E and polyphenols [101]	Home garden	No IUCN assessment	Loss of traditional food culture /Use as complementary food for infants	 Credit: IKIAM
Fiji/ Samoa	*Caulerpa racemosa*	Nama, Limu	Sea vegetable	Contains proteins, fiber, minerals, vitamins, polyunsaturated fatty acids, and bioactive anti-oxidants [102]	Near reefs, in shallow waters	No IUCN assessment	Unsustainable harvesting, storm surges, cyclones	 Credit: N.Hobgood

Table 1. *Cont.*

Country	Species Name	Local Name	Edible Use	Main Nutritional Benefits	Habitat	Threat Status (IUCN, Community to Other)	Threats/Suggestion for Conservation	Photo
Kenya	*Cleome gynandra*	Spider plant, Ofsaga, saga, liSaga, lisaka	Leaves used as vegetables [103]	High in β-carotene, folic acid, vitamin C, calcium and a good source of vitamin E, iron [104]	Roadsides, field margins, semi-domesticated	No IUCN assessment	No organized collecting missions	 Credit: BFN Kenya
	Amaranthus tortuosus	Amaranth, Ekichabo, Dodo	Leaves used as vegetables and seed crushed for flour	Good source of proteins, fibers, calcium, iron, riboflavin, niacin and vitamin C and an excellent source of lysine [104]	Roadsides, field margins, semi-domesticated	No IUCN assessment	No organized collecting missions	 Credit: BFN Kenya
	Chorchorus olitorius	Jute mallow, murere	Leaves used as vegetables	High levels of β-carotene, vitamin C, folic acid, calcium and iron [104]	Roadsides, field margins, semi-domesticated	No IUCN assessment	No organized collecting missions	 Credit: C. Kerr
Morocco	*Nasturtium officinale*	Watercress, Gernounch	Leafy vegetable	Rich in vitamin K, vitamin A, vitamin C, vitamin B6, manganese, calcium, and folate [105]	Springs, river edges, irrigation canals	IUCN—Least Concern	Paving of irrigation canals may decrease community access, changing diet and preference and leading to decreased use	 Credit: M. Lavin
	Malva sylvestris	Mallow, Tibi, Khobiza, Bakola houra	Leafy vegetable	Strong antioxidant properties, rich in phenols, flavonoids, carotenoids, and tocopherols, alpha-linolenic acid and minerals [106]	Fields, field margins, along irrigation canals and roads	IUCN—Least Concern	Changing diet and preferences may lead to decreased use	 Credit: B. Powell

Table 1. *Cont.*

Country	Species Name	Local Name	Edible Use	Main Nutritional Benefits	Habitat	Threat Status (IUCN, Community to Other)	Threats/Suggestion for Conservation	Photo
	Sideroxylon spinosum	Argan	Edible oil	Good source of linoleic and oleic fatty acids. Rich source of tocopherol (vitamin E) [107]	Dry forests from the Atlantic coast to 800 m elevation	IUCN criteria at national level—Vulnerable [108]	Social-ecological systems change driven by commodification and globalization	 Credit: B. Powell
Niue	*Tacca leontopetaloides*	Kai Niue	Root starch	Good source of carbohydrate, also contains vitamin C, fat, and protein [109]	Uncultivated land	IUCN Least Concern	General lack of information	 Credit: B. Dupont
Papua New Guinea	*Pandanus brosimos*	Karuka	Fruit (boiled) & extracted nut	Good source of protein and oil especially for highland communities [71]	Forest, high altitudes	No IUCN assessment	No known threat, but general lack of information	 Credit: Green Dean
Turkey	*Scolymus hispanicus*	Golden thistle, Şevketi bostan	Roots and young leaves	Rich in dietary fiber, magnesium and calcium [105,110]	Disturbed habitats and fallow fields	No IUCN assessment	Overharvesting/ domestication programs initiated	 Credit: BFN Turkey
	Eremurus spectabilis	Foxtail lily, Çiriş otu	Shoots, buds and young leaves	Rich in antioxidants and minerals [111]. High in vitamin C [112]	Dry and stony grazed hillsides	No IUCN assessment	Overharvesting/ domestication programs initiated	 Credit: K.D. Zinnert

Table 1. *Cont.*

Country	Species Name	Local Name	Edible Use	Main Nutritional Benefits	Habitat	Threat Status (IUCN, Community to Other)	Threats/Suggestion for Conservation	Photo
West Sumatra, Indonesia	*Elateriospermum tapos*	Tapuih	Seeds consumed raw or fermented	Rich in protein and unsaturated fatty acids [113]	Forest/agroforest	No IUCN assessment	Perceived as rare by local communities/Preserve forest and multi-strata agroforests	 Credit: L. Pawera
	Mangifera foetida	Ambacam, Bacang	Fruits consumed raw or cooked	Rich in vitamins A and C [114]	Forest, agroforest, homegardens	IUCN -Least Concern	Perceived as rare by local communities/Preserve forests and multi-strata agroforests	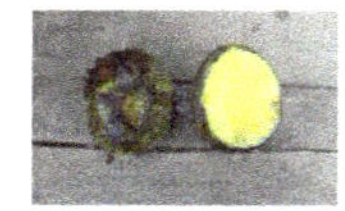 Credit: L. Pawera
	Diplazium esculentum	Pakis, Pahu	Young shoots as a vegetable, cooked	Rich in vitamin B9 (folate) [114]	Forests, wetlands	IUCN—Least concern	Relatively common/Preserve forests and wetlands	 Credit: L. Pawera
	Ipomoea aquatica	Kangkung air, Kangkuang liar	Leaves and stems as a vegetable, cooked	Rich in Iron and provitamin A [114]	Rivers, ponds, rice fields	No IUCN assessment	Threatened by overuse of herbicides/Reduce the use of herbicides and keep clean water bodies	 Credit: L. Pawera

Exacerbating these problems in the different geographies are the uncertain effects of climate change, which in many countries is expected to lead to increased variability in seasonality, temperatures and precipitations and increased incidence of hurricanes and wildfires [89]. Climate change is also predicted to severely impact cultivated plants, affecting crop production in specific geographic locations [115], stripping nutrients from staple crops [46,116] and making WFPs all the more important for food and nutrition security. Although generally highly adaptable and often more drought tolerant than cultivated crops, WFPs, as many useful plants, are also not fully resistant to climate change [116]. In the past, many WFPs survived major climatic fluctuations, but thematic studies on the implications of future climate change suggest important impacts on the ability of wild species to survive. This includes WFPs, particularly in tropical regions where economies are already fragile and capacity may be inadequate to protect these species effectively [94,116]. One key impact that could threaten WFP use is the likely shifts in both WFP geographic ranges and phenological changes in ripening times. This could create mismatches with traditional knowledge and practices of the communities that traditionally harvest them [117].

At present, there are very few formal systematic efforts that support and regulate the conservation and sustainable use of WFPs [118]. A sample survey of some of the most recent National Biodiversity Strategies and Action Plans (NBSAPs) submitted to the CBD as part of the reporting requirements of member states (e.g., Chile, Morocco, and Portugal), shows that rarely do these strategies refer specifically to WFPs or, if they do, are very vague in terms of the measures needed to protect them. Actions are mostly limited to ex situ conservation measures [46], while no concrete activities are put forward to support their conservation via sustainable use [45]. Furthermore, appropriate and effective governance mechanisms are seldom in place to safeguard the rights of indigenous people and local communities to sustainably manage and benefit from the use of WFPs (and prevent their over-exploitation by others) [119].

The use of wild species, however, is explicitly recognized as useful for improving food and nutrition security in several international agreements, strategies and action plans: in the 2030 Agenda for Sustainable Development (SDG2, Target 2.5), the International Treaty on Plant Genetic Resources for Food and Agriculture (International Treaty), the Second Global Plan of Action for Plant Genetic Resources for Food and Agriculture (Second GPA), and in the Global Strategy for Plant Conservation of the Convention on Biological Diversity (CBD). The CBD, the main international agreement aimed at conserving biological diversity, accords explicit recognition to sustainable use for the long-term conservation of ecosystems, species and genes, which must continue to be used, but "in a way and at a rate that does not lead to the long-term decline of biological diversity" [120]. Intrinsic in the term "sustainable use" is that it generates benefits (e.g., nutritional, cultural, and financial) for the custodians and users of these wild species. These benefits encourage people to continue conserving these biological resources and the habitats in which they grow or live. However, the real challenge is to ensure this sustainability is maintained given the rising demands on global resources imposed by population growth and economic development, combined with the uncertain effects of climate change mentioned above.

3. Barriers to the Greater Use of WFPs

The disregard of WFPs for food security and nutrition can be partly attributed to a lack of evidence and awareness among policymakers and other stakeholders of the importance of wild foods to diets, livelihoods, and food security, coupled with a number of market and non-market barriers limiting their untapped potential.

Underpinning the lack of recognition for WFPs is also limited or short-term research and extension funding to support the exploration of non-conventional, traditional and indigenous food resources. Many of these barriers were summarized by Heywood [4] and are still very much valid today:

- lack of information about the extent of their use and importance in rural economies;
- lack of information, especially statistics, concerning the economic value of WFPs;

- lack of reliable methods for measuring their contribution to farm households and the rural economy;
- lack of information on the sustainability of current harvest levels;
- poorly developed infrastructure and markets for WFPs, with the exception of small number of products (e.g., Açaí berries);
- unevenness of supply;
- lack of quality standards;
- general lack of storage and processing technology;
- availability of substitutes;
- policies and research mostly favoring commodity crops and commercial agriculture.

Like other neglected and underutilized species, additional barriers to the promotion of WFPs in food production and consumption patterns include: limited and fragmented data of the nutritional importance of these species; fragmented data on the quality and nutritional impacts of WFPs on household nutrition [121]; and knowledge gaps on the species' biology and ecology to develop domestication and management strategies [45,46].

Unfavorable and disabling national policies, coupled with the many stakeholders and interests involved, represent an additional obstacle to greater recognition for WFPs. The main policy barriers were identified and summarized by the Strategic Framework for Underutilized Plant Species [122], of which WFPs are part of. These are provided in Table 2 below.

Table 2. Barriers that hinder the improvement of national policy frameworks towards supporting WFPs.

Awareness	Focus	Financial Support	External Pressures
No adequate data	Mismatch with national priorities	No international financial or donor support	International trade favor R&D on conventional crops
Lack of priority in education and information systems	Limited capacity (institutional, research) to work with WFPs	Weak economies for investing in R&D for WFPs	International R&D priorities influence national priorities

Further contributing to the demise of WFPs, is the low recognition of value and perception of these foods as being "women's food" [66] "food for the poor" or "famine foods" to be harvested only when staple crops fail, as well as lack of institutional capacity to mainstream this diversity into national production and consumption patterns [28]. On the other hand, in some regions, such as West Sumatra, communities perceive WFPs positively, but the main barrier to their greater use is their reduced availability caused by land degradation and agriculture intensification [77]. In many places, traditional wild leafy vegetables are disappearing from local diets due to changing dietary patterns and preferences driven by globalization and increasing market integration [123]. Wild leafy vegetables (WLV) and wild food plants in general are undervalued and seen as "un-modern" in Morocco, Turkey [28,66], and many other parts of the world. This lack of value places the role of WFPs in the diet at risk, although it may ease pressures on overharvesting. In Brazil and Kenya, changing dietary patterns and lifestyles has reduced the diversity and availability of wild fruit and vegetables in market settings, which focus instead on a limited number of exotic crops [124]. This has led to people consuming sub-optimal diets that are increasingly unhealthy, unsustainable, and inequitable for many populations [125].

3.1. Contribution to Nutrition and Diets

The contribution of wild food biodiversity to diets and nutrition is simultaneously limited by a severe lack of food composition data for many neglected and underutilized cultivated and wild foods [126] as well as by a lack of accurate botanical identification for many foods recorded in dietary records or food composition tables [45,127,128]. Nutrient composition data indicates the presence and quantity of nutrients (e.g., energy, proteins, minerals, and vitamins) as well as the compounds that can impact the bioavailability of nutrients within a food. These data are combined with dietary records of the foods consumed to assess whether individuals or groups are meeting their dietary requirements [129]. Nutrient composition data do not exist for many WFPs, and when they do there may be high variation in nutrient composition for a given species across space and time [130]. The few WFPs that have nutrient composition data and that are included in local food composition tables are often identified by local names. This hinders the use of these data to fill nutrition gaps in other locations where the same species might be present and used but is identified by a different local name. Many data sets lump all wild foods into a single food group (e.g., wild greens). For example, in analyzing data on wild harvests in 24 developing countries across Africa, Asia, and Latin America, Hickey et al. [84] found that only a small percentage (0.9%) of the collected mushrooms were identified by species, the rest was reported nonspecifically as "mushrooms".

In some cases, when WFPs are lost from the diet they may be replaced by similar cultivated species, but in other cases they are not. Anecdotal evidence from Morocco suggests that when people stop or reduce the consumption of WLVs in their diet, these are not replaced with cultivated alternatives, leading to a reduced consumption of any leafy vegetable and fruit and vegetables in general. This is particularly worrying given global recommendations [131] to consume at least 5 servings of fresh fruit and vegetables (including berries, green leafy and cruciferous vegetables, and legumes) per day as a protective measure against cardiovascular diseases and type II diabetes [132–136].

Practical challenges also exist in measuring wild food consumption and contribution to the diet relative to other foods [5,137]. Although in recent years, several investigations have tried to assess the role of wild food biodiversity and the contribution of forests and agroforestry systems to human dietary intakes [13,14], the real dietary contribution of wild food plants, berries, fruit, nuts, and mushrooms harvested within and around people's homesteads and in forested areas remains poorly understood. Geographical variations exist regarding the proportion of WFPs consumed. While in in the global North WFPs mostly have cultural and recreational value [138], in some low-income countries they significantly enrich people's diets [119]. Rowland et al. [13] found that the collection of forest foods represents a regular livelihood strategy for many households and that forest dependent communities living in specific sites in Brazil, Cameroon, and Ethiopia derive as much as 80–96% of wild fruits and vegetables from the forest. In some areas, the nutritional contribution of fruits and vegetables is such to cover 50% and above the minimum dietary recommendation for these food groups [13]. Differences in consumption might also vary by ethnicity. For example, in documenting the traditional food systems of Western Sumatra, Pawera et al. [77] found that different ethnic communities living in the same environment have different knowledge and uses for the same WFPs. Seasonal fluctuations in WFP occurrence and therefore consumption by local communities might also not be adequately captured with a single survey [139]. Other challenges include cultural and language barriers and perceived power imbalances during questionnaire administration that can alter the surveys' accuracy and reliability [137]. There is a huge body of research that only lists the edible species known to community members but neglects to quantify the use of WFPs in local recipes nor is their use standardized in nutrition surveys [121].

3.2. Gathering Grounds, Collection Practices and Use

An additional knowledge gap is represented by the lack of information on traditional gathering grounds and the sustainability of collection practices. In the SOWBFA, the ecosystem origin reported for WFPs is either from forests (>25%) or unknown (>45%) [45]. An often-overlooked practice is urban and peri-urban foraging for WFPs. In their cross-continental study of urban foraging spanning India, South Africa, Sweden, and the US, Shackleton et al. [140] found that urban foraging is a widespread custom that is practiced independently of wealth and social status and is driven by different motivations varying in time and place. Wooded areas on public land, local lake beds, and other urban habitats harbor nutritionally rich greens and fruits. Even spontaneous vegetation growing in alleyways was reportedly used by Indian residents for food and culinary use [140]. Aside from direct consumption and small-scale trade, other benefits include "improved physical and psychological health, sense of place, increased ecological knowledge, stronger connections with nature, food, income or cash saving, and a source of pride" [140]. The important cultural ecosystem services offered by these plants are apparent in a study of WFP gathering and consumption trends across Spain [141]. The authors observe that WFPs continue to be used in areas with deep-rooted culinary traditions and in some instances have become gourmet ingredients for chefs. Schulp et al. [142] also suggest that the cultural benefits of wild foods in the European Union might exceed their income and food benefits and observe that wild mushroom and food plant collecting are generally highest in Southern European countries where gastronomic identity is strongest.

4. An Integrated Approach for Conserving and Sustainably Using WFPs

With the gradual disappearance of WFPs from nature and diets, the question is how to effectively promote their sustainable use and simultaneously conserve them for food security and nutrition. Because they exist on a continuum of human management, from truly wild to semi-domesticated [7], and because the germplasm and other plant material (e.g., tissue, embryos etc.) of some species may not be suitable for ex situ conservation [143], both in situ and ex situ conservation should be combined for optimal results [144–146] (Figure 2). In situ conservation strategies can complement ex situ conservation and allow WFPs to continue to evolve adaptive traits in their natural environments while benefiting those who need them most, particularly in areas where high diversity, rural poverty and malnutrition coexist.

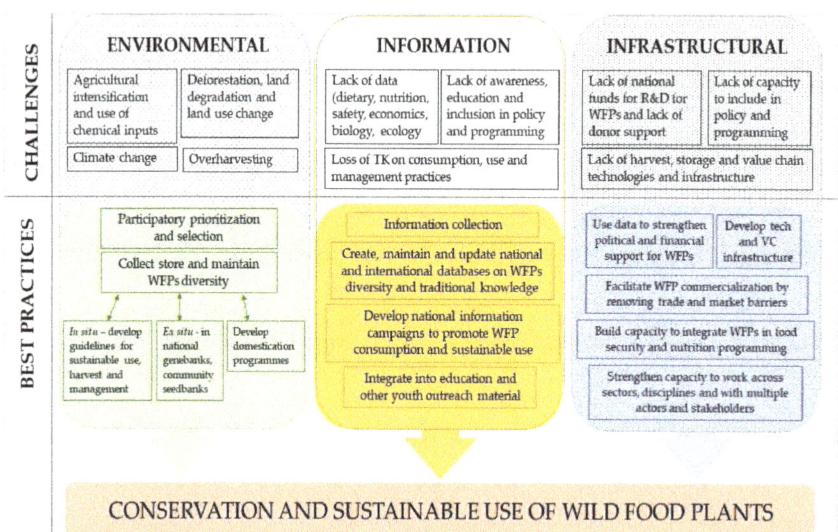

Figure 2. Proposed best practices for the long-term co-creation of conservation and sustainable use of WFPs help overcome many of the challenges identified.

Above, we have identified an array of threats to WFPs including: land use changes, deforestation and degradation; agricultural change, intensification and chemical input use; overharvest or unsustainable harvesting; loss of traditional management practices that communities used to promote the production of wild food plants (for example, pruning and burning); and climate change. We also identified a range of barriers that are contributing to the loss of use and value for WFPs, such as, lack of information (diet, nutrition, safety economics, and ecological); lack of harvest, storage and value chain tech and infrastructure; and lack of awareness, education and inclusion in policy and programming.

In the subsequent sections of this paper we propose a set of best practice actions that can be taken to support sustainable use and conservation of WFPs. This set of actions laid out in Figure 2 will act to overcome or mitigate against many of the threats and barriers identified.

The proposed set of best practice actions includes: the collection of information (identify the diversity of WFPs that are present in a given environment, information on nutrient composition and contribution to diet, economic importance, and ecological studies to determine sustainable offtake); (ii) prioritize the species with greatest potential to fill nutrition gaps, greatest need in terms of conservation, greatest cultural importance; (iii) protect species that are vulnerable through ex situ conservation; (iv) promote the use and management of WFPs in natural environments (in situ) (including sustainable management and collection guidelines where needed); (v) develop domestication programs where necessary and possible to avoid overexploitation in the wild; (vi) build local capacity to improve storage, processing, value chains, and markets (and all related technology and infrastructure); (vii) integrate WFP into programming and education and other youth outreach so as to raise awareness; (viii) develop and strengthen policies that support the conservation and sustainable use of WFPs; and (ix) and build donor commitment to funding efforts to support sustainable use and conservation of WFPs.

Each community and each WFP are unique, and will require a different set of actions, possibly occurring in a different order. Successful implementation of the set of best practice actions best suited to any given context will require working in a coordinated fashion across disciplines and sectors at the local, regional, and international level, and is largely dependent on the close and active participation of the national and local stakeholders. Due to the limits of time-bound projects (e.g., capacity, resources), it is rare for a single project or intervention to cover all elements or actions needed for a comprehensive and integrated approach. Below we present examples of best practice actions that we believe have successfully helped to further the conservation and sustainable use of WFPs.

4.1. Identify and Prioritize

The identification of WFPs to include in conservation and sustainable use strategies will almost invariably require close collaboration with indigenous and local communities who are the main users and custodians of this diversity. As opposed to extractive methods, participatory research approaches that integrate traditional and scientific knowledge are the most appropriate to collect information on WFPs while maximizing benefits for the communities involved [147]. Prior to the intervention, the community should be aware and agree on every aspect of the research process so that the methods, the analysis and the purposes of the data collection are clearly understood [148]. Ethnobotanical surveys and free-listing exercises are the most commonly used methods to complement scientific ecogeographic assessments. In the majority of the studies discussed in this paper, focus group discussions conducted with knowledgeable key informants were able to help fill knowledge gaps in WFP availability and use. Useful tools for assessing the potential of WFPs to fill seasonal food insecurity gaps, and low dietary diversity characterized by low fruit and vegetable consumption, are seasonal calendars, such as the one shown in Figure 3 developed by BFN Brazil to investigate flowering and fruiting seasons for wild fruit species. Data can then be transformed into an accessible and understandable tool to assist communities and decision makers adopt healthier diets based on local biodiversity [149].

Market surveys are also useful to understand what WFPs might be available for consumption within a community. A notable example is represented by BFN Turkey, which undertook market surveys and key informant interviews with 2334 local wild plant gatherers, sellers and consumers of WFPs to capture the diversity of WFPs still being used across three ecogeographic regions [28,150]. Documenting the use of wild plants in this participatory way has several benefits that include: (i) facilitating knowledge transmission from elders to younger generations and between community members; (ii) stimulating local innovation without undermining cultural traditions and local governance mechanisms, and (iii) ensuring that the community can use this diversity to address its own questions, challenges and needs [147]. In Western Kenya, for example, biodiversity surveys and dietary health assessments were followed by a series of participatory workshops that enabled five communities to gain and share knowledge on available wild and cultivated biodiversity, discuss options on ways to use this diversity to improve malnutrition within their communities, rank and prioritize the most suitable species and develop their own community action plans (CAP) towards this end. In collaboration with the ministries of Agriculture and Health, training was then provided to assist with the integration of the chosen species—mostly African leafy vegetables and legumes—into sustainable production systems and diets. One year into CAP implementation, mean dietary diversity scores and the percentage of children meeting minimum dietary diversity had significantly increased in all the households in the sublocation, irrespective of participation in the scheme, indicating the adoption of these best practices by neighboring households. The dietary diversity scores of women from participating households had also significantly increased [151].

Figure 3. Research into the flowering and fruiting period of wild fruits and greens within a given geography can be used to develop an adaptable tool for informed decision making at both community and government level. Credit: BFN Brazil.

With all probability, surveys will reveal a long list of species that could be the focus of further research and promotion in food and nutrition strategies. Realistically, limited resources will often require a prioritization exercise that reduces the list to a manageable number. Since the intent is to ensure that WFPs are safeguarded and sustainably consumed, again community participation in the prioritization process is key, for example, to single out species that could be conserved in seed saving facilities, domesticated, or included in breeding programs, or to identify WFPs with the potential to contribute to nutrition, climate-change resilience and other aspects of community well-being. In the earlier example from Turkey, the BFN team developed an ad hoc sustainability index to reduce an initial sample size of 43 species, mostly WFPs, to three target species—foxtail lily (*Eremurus spectabilis*), golden thistle (*Scolymus hispanicus*), and einkorn wheat (*Triticum monococcum*)—which have since been the object of domestication research as well as post-harvest handling and value chain analysis [150].

4.2. The Nutritional Importance of WFPs and Associated Traditional Knowledge

Understanding the nutrient content and health properties of WFPs (e.g., compositional data) as well as their contribution to diets will also greatly assist in the prioritization process. Compositional data is key to national nutritional planning and for developing locally fitting dietary guidelines. Seldom, however, does nutrition information appear in national food composition tables and databases, and if data does exist it is either scattered across different sources in institutional databases, in academic journals and grey literature, or is outdated and/or incomplete making data compilation a daunting task [28,150]. An additional hurdle is the standardization of food composition values. Common component names are often expressed inaccurately (e.g., vitamin A: retinol activity equivalents vs. retinol equivalents) or differ in terms of units, denominators, significant figures, maximum decimal places, and conversion factors [152]. Of further importance, is the documentation and protection of traditional knowledge related to consumption and preparation of WFPs, available largely in the recollections of elderly users, i.e., rural, indigenous, and forest-dependent communities, including local farmers, and city migrants. Some information may be available in national floras, in herbaria and in ethnological studies of local human ethnical groups, but additional botanical, culinary, nutritional, cultural research is required to fill this knowledge gap. As explained in Section 4.1, to avoid issues of misuse and abuse of this information, it is important that respondents are always adequately informed about data use, that sources are acknowledged, and that the data is made available in public databases. Biological knowledge on individual species is also frequently lacking but particularly essential for both in situ and ex situ conservation.

One of the most recent and comprehensive attempts to fill the evidence gap in food composition data is provided by the GEF-supported Biodiversity for Food and Nutrition Project (BFN). Led by Brazil, Kenya, Sri Lanka, and Turkey, and implemented by Bioversity International with support from the UN Environment Programme (UNEP) and the Food and Agriculture Organization of the United Nations (FAO), the project has generated food composition data for 185 plant species, many of them wild, particularly in Brazil and Turkey [28,150,153]. Because of the high costs associated with food composition analysis, the four countries carried out literature reviews prior to the project to identify information gaps and narrow down the list of potentially interesting species to a practicable sample size for analysis and to select the species with the greatest potential for conservation, domestication/management, promotion and marketing. Following the literature review and identified data gaps, food composition analysis was carried out for those species and nutrients for which information was missing or incomplete. Examples of the high nutrient content of WFPs was demonstrated as part of the BFN project in Brazil and Turkey [28]. Similar results were obtained in Indonesia by reviewing the country's food composition data [114] in which wild leafy vegetables are reported to contain higher amounts of limiting micronutrients than more commonly consumed greens (Figure 4).

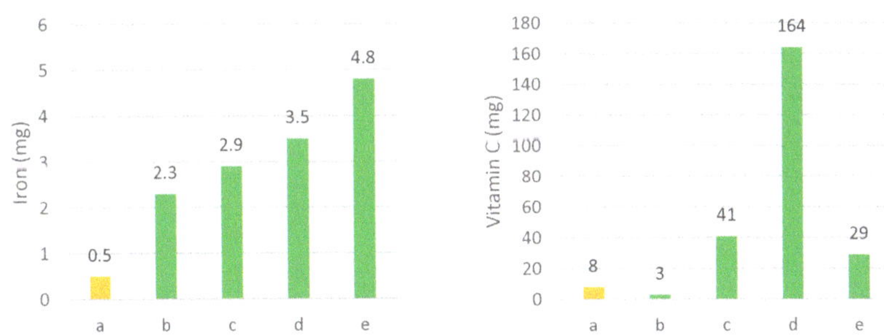

(a) (b)

Figure 4. Four wild leafy vegetables from West Sumatra are compared to common lettuce (*Lactuca sativa*) in terms of (**a**) iron content (mg) and (**b**) vitamin C content (mg). In the graphs the letters stand for a lettuce; b vegetable fern "pakis" (*Diplazium esculentum*); c nightshade "leunca/ranti" (*Solanum americanum*), d sweet leaf "katuk/nasi-nasi" (*Sauropus androgynus*), and e water mimosa "komen" (*Neptunia prostrata*). Values are expressed per 100g of fresh, raw ingredient. *Source*: Indonesian Food Composition Data [114].

Species selection and prioritization, literature reviews, and generating food composition data is only the first step of a comprehensive and integrated conservation approach.

4.3. Collecting, Storing and Maintaining WFP Diversity

Once the species have been identified and prioritized, consideration will need to be given to safeguarding the species for future use, either through ex situ or in situ conservation strategies. Particularly for WFPs, ex situ measures are envisioned as a support to their propagation and reintroduction for habitat restoration [154]. In both cases, to be effective, conservation should involve a wide range of stakeholders working together both in the public and private sectors, across the agricultural and environmental domains [145].

In many cases, governments have established national plant genetic resources programs and seed saving facilities (e.g., gene banks) for ex situ conservation.

However, seed and planting material produced by these "formal" facilities are often inaccessible to smallholder farmers due to strict regulations limiting exchange, little involvement of community actors in the governance, and management of these services [146], as well as imbalances in seed availability, access, and quality for smallholders [155]. However, alternatives do exist. The MGU Useful Plants project, for example, worked closely with communities in Botswana, Kenya, Mali, South Africa, and Mexico to select useful indigenous plants for which high-quality seed collections were established. Seed lots were also banked in the five countries as well as being duplicated and tested at the Millennium Seed Bank in Kew [48,154,156]. Research on seed germination helped support plant propagation activities. The propagules were then planted in community gardens while facilities were established or improved at the local level to facilitate conservation of the prioritized species. Training and knowledge on seed conservation in seed conservation, plant propagation, and planting activities were also provided [49]. This form of conserving WFPs, which takes place in situ, either "on farm" or "in the wild" in natural habitats or protected areas, provides greater opportunity for the involvement of local communities. Once hotspots of WFP occurrence are identified, farmers and indigenous communities living within and around those habitats and protected areas should be involved in conservation activities with due recognition given to their roles and rights in managing WFPs. Further guidance on the establishment of sites for active in situ conservation (i.e., where populations are actively monitored and maintained) of WFPs can be found in the "Voluntary Guidelines on the Conservation and Sustainable Use of Crop Wild Relatives and Wild Food Plants" [145].

Midway between these two conservation approaches are community seed banks or gene banks, which are community-maintained facilities that preserve seeds and other planting materials for local use [146]. These are a collective forms of crop conservation that provide farmers with access to seed, planting material, and traditional knowledge that may otherwise be lost [147]. They also foster community engagement and strengthen the understanding of farmers' and community' intellectual property [157]. By documenting and storing biodiversity and associated traditional knowledge, the seed banks also raise awareness of unique biodiversity in a given area. The community-based organizations (CBO) operating in Vihiga County, Western Kenya, have now established their own community seed bank for African indigenous vegetables and legumes [158]. Creating markets for the seeds and planting material can create additional conservation incentives. Such is the case for the communities in Botswana engaged in the MGU UPP who collect the edible seeds of *Tylosema esculentum* (Burch.) A. Schreib (morama bean) for conservation and cultivation, consumption, sale, and processing into numerous marketable food products. Likewise, the Tsetseng community, through their community trust, have become leading innovators in marketable morama products [159].

4.4. Domestication Programmes and Guidelines for Sustainable Collection

Depending on conservation status and extent of utilization of WFPs, domestication programs may be required to facilitate cultivation of these wild species and thus to ease the pressure on wild populations and rebuild and restore the genetic diversity that has been lost. A successful example is provided by Turkey in its quest to reduce overexploitation of golden thistle (*Scolymus hispanicus* L.). Golden thistle is a flowering plant that is widely consumed across Turkey and is traditionally collected from the wild for its roots and immature leaves that are sold in local markets [62,160]. Selected by the BFN Project as one of the target species for potential commercialization, breeding, and domestication efforts were undertaken by the Aegean Agricultural Research Institute and the University of Anadolu in collaboration with 37 farmers to select, characterize, and evaluate the species [161]. Nurseries established following initial selection of the hardiest plants produced an average yield of 3.9 tons/ha and up to a maximum of 7 tons/ha. A cultivar called "Sari" was registered and seeds distributed to farmers in the İzmir province. Golden thistle is now cultivated on an area of 100 ha^{-1} [161]. To complement seed distribution, guidelines for the sustainable production of golden thistle were also produced to assist farmers in addressing critical aspects such as climate and soil conditions, plant management, harvest, and seed production (Figure 5).

Figure 5. From left to right, top to bottom. BFN Turkey work with farmers to test domesticated golden thistle (*Scolymus hispanicus* L.); the sustainable production guidelines; and harvested golden thistle roots ready for sale Credit: BFN Turkey.

4.5. Strengthening Policies in Support of WFP Conservation and Sustainable Use

Once baseline data has been gathered, guidelines exist to assist countries in preparing a National Plan for the Conservation and Sustainable Use of Wild Food Plants and crop wild relatives, including setting up a monitoring plan for WFP diversity [145]. The scope of the action plan, its application and effectiveness will very much depend on the national context, the existing policy framework and institutional arrangements, the range of stakeholders involved and their interrelationships, as well as on the resources available. Guidelines are also provided for undertaking preparatory work towards this end [145]. Suggestions are made on how to promote wider use of crop wild relatives and WFPs, but few examples are given to show countries what is practical or actionable. One possibility, which has shown great promise in the BFN Project, is to link producers and collectors to institutional or private sector markets enabling them to benefit from the authorized trade of WFPs. Brazil has used its two largest public procurement programs, the Food Procurement Program (PAA) and the National School Feeding Program (PNAE), to stimulate engagement by family farmers and wild plant collectors (known as extractivists) in sustainable agriculture and the management of Brazilian food diversity, including many WFPs. Both PNAE and PAA are in fact obliged to buy a proportion of the food they distribute from family farmers and pay a 30% bonus for organic or agroecological produce, preferring suppliers from indigenous and traditional communities [162]. PAA also supports activities aimed at the conservation, production, storage, and distribution of local or traditional seed varieties (Beltrame et al., 2020) [28]. Working closely with government actors and using the nutritional data generated as part of the BFN project, BFN Brazil was also able to promote the publication of Ordinance N° 284/2018, which officially recognizes the nutritional and sociocultural value of over 100 plant species native to the Amazon, Caatinga, Atlantic Forest, and Cerrado biomes. This has boosted the market value of native biodiversity including WFPs, with ministries now referring to the list in the "Sociobiodiversity Ordinance" to purchase biodiversity and farmers and collectors eager to join the scheme. In order to do so, however, producers must adhere to procurement regulations, and follow training and guidelines for organic production and the sustainable management of these resources in the wild (Figure 6). Similar linkages were fostered in the other three BFN countries, increasing structured demand for African leafy vegetables in Kenya, for WFPs in Turkey and for native fruits such as wood apple (*Limonia acidissima*) in Sri Lanka including via private sector linkages (see next section).

Figure 6. Best practices for the sustainable harvesting and management of pequi (*Caryocar brasiliense* Cambess), common to Brazil's Cerrado region. The guidelines are produced by the Ministry of Agriculture, Livestock and Supply (MAPA) in support of producers/extractivists intending to take part in the public procurement schemes.

4.6. Raising Public Awareness of the Importance of WFPs

Raising public awareness of the important contribution WFPs can make to diets and livelihoods is another effective way to secure research and policy investments targeting their conservation and use and creating a mutually reinforcing virtuous cycle [163]. This is probably the area in which countries invested in protecting WFPs let lose their creativity and excel in finding ingenious, innovative, and culturally acceptable ways of communicating the importance of WFPs to different target groups. Naturally, collaborating and partnering with the broadest range of stakeholders, e.g., farmer groups, NGOs, private sector enterprises, schools, the media, and ministries will ensure that there is clear and cohesive messaging that is able to reach the widest possible audience.

4.6.1. Youth

As future consumers and protectors of biodiversity, youth are an important target audience for WFP messaging. Awareness raising campaigns can take advantage of recurring activities such as biodiversity festivals or food fairs to organize nature walks or competitions for younger participants [28,150] or join forces with relevant ministries (e.g., Environment, Agriculture, and Education) to introduce messaging around WFP conservation and use in curricular activities and courses (Figure 7). For older students, WFPs offer an interesting opportunity for "greening" vocational training, particularly in the food and beverage sector. In Turkey, to raise the profile of WFPs, the BFN project partnered with the Halim Foçali Vocational School organizing a series of lectures and hands-on activities for 16 student chefs, who were trained to recognize and collect local edible species and use them in their cooking classes. Future plans for the institute include the establishment of an herb garden on the school premises where WFPs will be grown and harvested for use in cooking courses. Interest in the program from the National Education Directorate of Foça has led to plans for extending the training to other schools and officially include traditional WFPs in the school curriculum [150]. School gardens are also an effective way of promoting greater interest in biodiversity and can act as important conduit for improving nutrition, well-being and education of schoolchildren and their families [164], as well as acting as conservation networks for tree genetic resources [165] and reviving traditional food systems and culture [164].

Figure 7. A nutrition education booklet from Ecuador that includes WFPs as a food group. On the left, the cover depicts the forest as an alternative source of foods, mainly fruits, while on the right, five food groups are shown along with a list of 13 WFPs (mostly aromatic plants) used for preparing hot beverages.

4.6.2. Communities and Households

As previously mentioned, it will be important to ensure that the main users of this diversity the communities that continue collecting and maintaining WFPs are aware of the species' nutritional and sociocultural importance. Seasonal food availability booklets and calendars, such as the one shown in Figure 5, and simple, locally appropriate picture posters (Figure 8) can serve the dual purpose of revitalizing the use of WFPs and imparting basic nutrition information derived from national nutrition guidelines. Translated into local languages, these tools can be used by government extension workers and NGO practitioners to offer an overview of local diet quality and consumption patterns derived from baseline assessments and provide recommendations on how WFPs and other local agrobiodiversity can fill existing nutrient gaps. To avoid issues of overharvesting, it will be important, that the above information is complemented by easy-to-understand guidelines on the sustainable collection and management of these species, as shown by the informative brochures that accompanied the revival of WFPs in Turkey (Figure 9).

Figure 8. Community poster in local language developed in West Sumatra, Indonesia, as part of the "Food, Agrobiodiversity, and Diet project" explaining the health benefits of local food plants that are rich in protein, vitamin A, vitamin C, and iron, and Mandailing children learning about local foods. Credit: Lukas Pawera.

Figure 9. Foragers' guide to edible wild plants and illustrations taken from "A children's guide to the collection of wild edibles", produced by BFN Turkey to complement activities aimed at raising the profile of Turkish WFPs. Credit: BFN Turkey.

4.6.3. Policymakers

Policymakers and key change agents who can support the conservation and use of WFPs are to be found within the following sectors: nutrition, health, agriculture, forestry, education, environment, trade, planning, poverty reduction, food security, rural development, economy, and finance at national, regional, and international levels. Whatever their background, for effective decision-making to occur, policymakers need access to timely, independent and reliable information, in a simple and useful form, accompanied by the cost implications of the research, indicating whether it is feasible and affordable [166]. As demonstrated by the endorsement of the "Sociobiodiversity Ordinance" in Brazil, for example, nutrition evidence generated via food composition analysis was critical for expanding the list of sociobiodiversity species to include previously neglected WFPs, and for subsequent policy uptake by national programs dealing with food and nutrition security.

The recognition of WFPs as important elements of healthy diets and rural resilience has thus resulted in increased federal funding (approximately US$6 Million per year) for public procurement programs to purchase sociobiodiversity products directly from family farmers and provided an indication of the untapped market potential of WFPs in institutional markets [167]. The increased appreciation of the role WFPs play in rural diets is also leading to investigations into the affordability of diets that include WFPs [126]. As mentioned earlier, the study carried out in Eastern Baringo, Kenya, has shown that wild plant species, especially vegetables, are able to significantly reduce (by 30–70%) the cost of nutritious meal for vulnerable groups [54]. The tool, which provides an insight into the affordability of nutritious foods, offers a useful entry point for policy discussion around the types of commodities and delivery channels that are likely to achieve nutritional outcomes particularly for the most vulnerable segments of the population [168].

4.6.4. Broader Audiences

Recent interest in food and gastronomy programs worldwide has acted as the perfect jumping board for WFPs, particularly in developing countries. Many of the approaches adopted by BFN project countries have extensively been described [28,150,153], and broadly involve communities partnering with celebrity chefs, gastronomists, or taking advantage of existing food festivals to organize information and hands-on events on WFP collection, transformation, and cooking (Figure 10). Innovative approaches for reaching out to broader audiences are described in detail by Gee and Lee (2020) who look at emerging youth-led innovations that can be productively applied to the conservation and sustainable use of food biodiversity, including WFPs. The realms of social media and mobile technology are rapidly evolving, and via mobile apps consumers are now able to (i) find local crops in season and plan grocery purchases, (ii) identify plants through a global photo database, (iii) learn about wild edible plants (Wild Edibles and Foraging Flashcard Lite), (iv) and even trace fresh crops back to farms using blockchain technology. On the production side, a growing number of applications, including in developing countries, offer "smart phone farmers" unprecedented access to crop, field, and market information, which could easily be extended to incorporate WFPs. Gee and Lee [169] also explore the benefits of creating conservation networks for biodiversity through international movements such as via "Campesina" and Slow Food, which can connect different actors who are motivated to improve global and community-based food systems using food biodiversity.

Figure 10. Front covers of recipe books developed as part of the WFP-focused projects in Brazil, Ecuador and Kenya. Credit: BFN Brazil, IKIAM and BFN Kenya.

5. Conclusions

While WFPs contribute to the diets and livelihoods of millions of people worldwide at the local level, there is still much that we do not yet fully understand about them and thus their role is not fully appreciated. This makes it a challenge when it comes to decisions and actions that might support more effective national and international conservation, sustainable management, and useful strategies for WFPs. Some of these actions are summarized in Table S2. While there are an increasing number of publications outlining the importance of WFPs, usually at a local level, there is largely a scarcity of data and information at a national level, and conservation assessments are still limited. This fails to convey the full contribution that WFPs make to food security and nutrition and the overall importance of these biological resources to national economies in many parts of the world. Furthermore, while we increasingly learn more about some of the threats which impact WFPs, we still know so little about their biology and ecology as well as the dynamics of their use and how climate change is impacting them now and in the future. The integrated conservation approach described in this paper is intended to guide stakeholders in creating plans and strategies to ensure that WFPs are used sustainably and are conserved for generations to come.

In this review we survey the contribution of WFPs to food security, nutrition, and livelihoods in a variety of geographical settings, many of which have benefited from the availability of donor-funded projects and therefore the dedicated attention of researchers and their organizations. It is by no means a comprehensive review. However, the limited cases and examples it highlights clearly demonstrate that the contribution of WFPs to food security, nutrition, and livelihoods is significant. With increased development attention and research investments, including a more effective enabling policy environment, the role of WFPs could be strengthened in the future.

A greater understanding and appreciation, especially by decision-makers, of the nutritional value of WFPs and their contribution to food security and nutrition could see the enhanced inclusion of WFPs in important national nutrition policy instruments such as dietary guidelines, development plans, or in nutrition education and school curricula. Greater use should also go hand in hand with increased research and investments targeting existing biological and ecological knowledge gaps on WFPs, such as plant demographic studies to calculate sustainable harvest levels in the wild or studies on seed biology and ecology to ensure they are adequately conserved ex situ. If WFPs were provided with greater policy recognition and support, especially through policy incentives and the development of innovative market-based demand approaches (with clear benefits arising to custodians), it would help create longer-term economic viability. This, in turn, could help greatly in better linking the conservation of WFPs and their sustainable traditional management and use, something which is currently missing in most national Plant Genetic Resources conservation strategies and action plans.

Supplementary Materials: The following are available online at http://www.mdpi.com/2223-7747/9/10/1299/s1, Table S1: Plant families that include WFPs and semi-cultivated species that are known to contribute to food and nutrition security; Table S2: Summary of actions that can be undertaken across the four pillars by the main stakeholders involved in WFP conservation and use.

Author Contributions: Conceptualization, T.B. and D.H.; investigation, T.U., E.M., D.B., C.T., D.P., L.P., B.P., and A.T.; writing—original draft, T.B.; and writing—review and editing, T.B., B.P., T.U., E.M., D.H., D.B., C.T., D.P., L.P., A.T.; M.T., and J.E. All authors have read and agreed to the published version of the manuscript.

Funding: Overall support for the Biodiversity for Food and Nutrition (BFN) Project was provided by the Global Environment Facility (GEF Project ID 3808). Co-funding and implementation support were received from the UN Environment Programme; the Food and Agriculture Organization of the United Nations; Bioversity International and the governments of Brazil, Kenya, Sri Lanka, and Turkey. Additional funding was received by the Australian Centre for International Agriculture Research for work in Kenya (HORT2014/100, GP2017/007 and GP2018/101), as well as the CGIAR Research Program on Agriculture for Nutrition and Health (A4NH). For the Food, Agrobiodiversity, and Diet (FAD) Project in West Sumatra, Indonesia, the authors would like to thank the Neys-van Hoogstraten Foundation (Project IN305), and ALFABET mobility for supporting Lukas Pawera under the Erasmus Mundus Action 2 Programme. For work in Kenya, the support of Biovision Foundation Switzerland and A4NH CRP are gratefully acknowledged. The "MGU—Useful Plants Project" was funded by MGU, a kind and generous philanthropist based in Spain. E.M. is supported by the Kew Future Leaders Fellowship—Diversity and Livelihoods, of the Royal Botanic Gardens, Kew, UK.

Acknowledgments: The authors would also like to gratefully acknowledge the many colleagues and institutions in participating countries that contributed to the work described in this paper.

Conflicts of Interest: The authors declare no conflict of interest.

Abbreviations

BFN	Biodiversity for Food and Nutrition Project
CAP	Community action plans
CBD	Convention on Biological Diversity
CBO	Community-based organization
CIAT	International Center for Tropical Agriculture now part of the Alliance of Bioversity International and CIAT
EIARD	European Initiative for Agricultural Research for Development
FAD	Food, Agrobiodiversity and Diet Project
FAO	Food and Agriculture Organization of the United Nations
GEF	Global Environment Facility
GPA	Global Plan of Action for Plant Genetic Resources for Food and Agriculture of the FAO
IKIAM	Universidad Regional Amazónica - Amazon Regional University (Ecuador)
IUCN	International Union for Conservation of Nature
MAPA	Ministério da Agricultura, Pecuária e Abastecimento - Ministry of Agriculture, Livestock and Supply (Brazil)
NBSAPs	National Biodiversity Strategies and Action Plan (of the CBD)
NGO	Non-governmental organization
PAA	Programa de Aquisição de Alimentos - Food Procurement Program (Brazil)
PNAE	Programa nacional de alimentação escolar - National School Feeding Program (Brazil)
PNG	Papua New Guinea
PPF	Plantas Paro o Futuro (Plants for the Future Initiative – Brazil)
RAE	Retinol Activity Equivalents
R&D	Research and Development
SDG	Sustainable Development Goal

SOWBFA	State of the World's Biodiversity for Food and Agriculture of the FAO
UNEP	UN Environment Programme
UPP	Useful Plants Project, Kew
WFPs	Wild food plants
WLVs	Wild leafy vegetables

References

1. Zohary, D.; Hopf, M.; Weiss, E. *Domestication of Plants in the Old World. The Origin and Spread of Domesticated Plants in South-West Asia, Europe, and the Mediterranean Basin*, 4th ed.; Oxford University Press: New York, NY, USA, 2012. [CrossRef]
2. Sowunmi, M.A. The beginnings of agriculture in West Africa: Botanical evidence. *Curr. Anthropol.* **1985**, *26*, 127–129. [CrossRef]
3. Johnson, A.; Behrens, C.A. Nutritional criteria in machiguenga food production decisions: A linear-programming analysis. *Hum. Ecol.* **1982**, *10*, 167–189. [CrossRef]
4. Heywood, V.H. *Use and Potential of Wild Plants in Farm Households*; FAO Farm System Management Series; FAO: Rome, Italy, 1999; Volume 15.
5. Bharucha, Z.; Pretty, J. The roles and values of wild foods in agricultural systems. *Philos. Trans. R. Soc. B Biol. Sci.* **2010**, *365*, 2913–2926. [CrossRef] [PubMed]
6. Casas, A.; Otero-Arnaiz, A.; Perez-Negron, E.; Valiente-Banuet, A. In situ management and domestication of plants in Mesoamerica. *Ann. Bot.* **2007**, *100*, 1101–1115. [CrossRef]
7. Cruz-Garcia, G.S. Management and motivations to manage "wild" food plants. A case study in a mestizo village in the amazon deforestation frontier. *Front. Ecol. Evol.* **2017**, *5*, 127. [CrossRef]
8. Heywood, V.H. Overview of Agricultural Biodiversity and Its Contribution to Nutrition and Health. In *Diversifying Food and Diets: Using Agricultural Biodiversity to Improve Nutrition and Health*; Fanzo, J., Hunter, D., Borelli, T., Mattei, F., Eds.; Routledge: London, UK, 2013; pp. 35–67.
9. Commission on Genetic Resources for Food and Agriculture—FAO. *Biodiversity for Food and Agriculture—Revised Draft—Needs and Possible Actions CGRFA/NFP-BFA-1/18/2*; FAO: Rome, Italy, 2018.
10. Asprilla-Perea, J.; Díaz-Puente, J.M. Importance of wild foods to household food security in tropical forest areas. *Food Secur.* **2019**, *11*, 15–22. [CrossRef]
11. Bioversity International. *Mainstreaming Agrobiodiversity in Sustainable Food Systems: Scientific Foundations for an Agrobiodiversity Index*, 1st ed.; Bioversity International: Rome, Italy, 2017.
12. Broegaard, R.B.; Rasmussen, L.V.; Dawson, N.; Mertz, O.; Vongvisouk, T.; Grogan, K. Wild food collection and nutrition under commercial agriculture expansion in agriculture-forest landscapes. *For. Policy Econ.* **2017**. [CrossRef]
13. Rowland, D.; Ickowitz, A.; Powell, B.; Nasi, R.; Sunderland, T. Forest foods and healthy diets: Quantifying the contributions. *Environ. Conserv.* **2017**, *44*, 102–114. [CrossRef]
14. Ickowitz, A.; Powell, B.; Rowland, D.; Jones, A.; Sunderland, T. Agricultural intensification, dietary diversity, and markets in the global food security narrative. *Glob. Food Sec.* **2019**, *20*, 9–16. [CrossRef]
15. Carvalho, A.M.; Barata, A.M. The Consumption of Wild Edible Plants. In *Wild Plants, Mushrooms and Nuts*; John Wiley & Sons, Ltd.: Chichester, UK, 2016; pp. 159–198. [CrossRef]
16. Ojelel, S.; Mucunguzi, P.; Katuura, E.; Kakudidi, E.K.; Namaganda, M.; Kalema, J. Wild edible plants used by communities in and around selected forest reserves of Teso-Karamoja Region, Uganda. *J. Ethnobiol. Ethnomed.* **2019**, *15*, 3. [CrossRef]
17. Kuhnlein, H.V.; Erasmus, B.; Spigelski, D. *Indigenous Peoples' Food Systems: The Many Dimensions of Culture, Diversity and Environment for Nutrition and Health*; Food and Agriculture Organization of the United Nations: Rome, Italy, 2009.
18. Kuhnlein, H.V. Holding on to Agrobiodiversity: Human Nutrition and Health of Indigenous Peoples. In *Routledge Handbook of Agricultural Biodiversity*; Hunter, D., Guarino, L., Spillane, C., McKeown, P.C., Eds.; Routledge: London, UK, 2017; p. 692.
19. United Nations. *State of the World's Indigenous Peoples*; United Nations: New York, NY, USA, 2009.
20. Angelsen, A. Policies for reduced deforestation and their impact on agricultural production. *Proc. Natl. Acad. Sci. USA* **2010**, *107*, 19639–19644. [CrossRef] [PubMed]

21. Powell, B.; Thilsted, S.H.; Ickowitz, A.; Termote, C.; Sunderland, T.; Herforth, A. Improving diets with wild and cultivated biodiversity from across the landscape. *Food Secur.* **2015**, *7*, 535–554. [CrossRef]
22. Aryal, K.P.; Poudel, S.; Chaudhary, R.P.; Chettri, N.; Chaudhary, P.; Ning, W.; Kotru, R. Diversity and use of wild and non-cultivated edible plants in the western Himalaya. *J. Ethnobiol. Ethnomed.* **2018**, *14*, 10. [CrossRef] [PubMed]
23. Boedecker, J.; Termote, C.; Assogbadjo, A.E.; Van Damme, P.; Lachat, C. Dietary contribution of wild edible plants to women's diets in the buffer zone around the Lama forest, Benin—An underutilized potential. *Food Secur.* **2014**, *6*, 833–849. [CrossRef]
24. Ju, Y.; Zhuo, J.; Liu, B.; Long, C. Eating from the wild: Diversity of wild edible plants used by Tibetans in Shangri-La Region, Yunnan, China. *J. Ethnobiol. Ethnomed.* **2013**, *9*, 28. [CrossRef]
25. Shumsky, S.A.; Hickey, G.M.; Pelletier, B.; Johns, T. Understanding the contribution of wild edible plants to rural social-ecological resilience in Semi-Arid Kenya. *Ecol. Soc.* **2014**, *19*, art34. [CrossRef]
26. Smith, E.; Ahmed, S.; Dupuis, V.; Running Crane, M.; Eggers, M.; Pierre, M.; Flagg, K.; Byker Shanks, C. Contribution of wild foods to diet, food security, and cultural values amidst climate change. *J. Agric. Food Syst. Community Dev.* **2019**, 1–24. [CrossRef]
27. Fernández-Ruiz, V.; Morales, P.; Ruiz-Rodríguez, B.M.; Isasa, E.T. Nutrients and Bioactive Compounds in Wild Fruits Through Different Continents. In *Wild Plants, Mushrooms and Nuts*; John Wiley & Sons, Ltd.: Chichester, UK, 2016; pp. 263–314. [CrossRef]
28. Hunter, D.; Borelli, T.; Beltrame, D.M.O.; Oliveira, C.N.S.; Coradin, L.; Wasike, V.W.; Wasilwa, L.; Mwai, J.; Manjella, A.; Samarasinghe, G.W.L.; et al. The potential of neglected and underutilized species for improving diets and nutrition. *Planta* **2019**, *250*, 709–729. [CrossRef]
29. Morales, P.; Herrera, P.G.; González, M.C.M.; Hurtado, M.C.; de Cortes Sánchez Mata, M. Wild Greens As Source of Nutritive and Bioactive Compounds Over the World. In *Wild Plants, Mushrooms and Nuts*; John Wiley & Sons, Ltd.: Chichester, UK, 2016; pp. 199–261. [CrossRef]
30. Kinnunen, P.; Guillaume, J.H.A.; Taka, M.; D'Odorico, P.; Siebert, S.; Puma, M.J.; Jalava, M.; Kummu, M. Local food crop production can fulfil demand for less than one-third of the population. *Nat. Food* **2020**, *1*, 229–237. [CrossRef]
31. Padulosi, S.; Mal, B.; King, O.; Gotor, E. Minor millets as a central element for sustainably enhanced incomes, empowerment, and nutrition in rural India. *Sustainability* **2015**, *7*, 8904–8933. [CrossRef]
32. Shackleton, S.; Paumgarten, F.; Kassa, H.; Husselman, M.; Zida, M. Opportunities for enhancing poor women's socioeconomic empowerment in the value chains of three African Non-Timber Forest Products (NTFPs). *Int. For. Rev.* **2011**, *13*, 136–151. [CrossRef]
33. Torero, M. Without food, there can be no exit from the pandemic. *Nature* **2020**, *580*, 588–589. [CrossRef] [PubMed]
34. HLPE. *Food Security and Nutrition: Building a Global Narrative towards 2030*; HLPE: Rome, Italy, 2020.
35. Béné, C. Resilience of local food systems and links to food security—A review of some important concepts in the context of COVID-19 and other shocks. *Food Secur.* **2020**. [CrossRef] [PubMed]
36. Fernandes, N. Economic effects of coronavirus outbreak (COVID-19) on the world economy. *SSRN Electron. J.* **2020**. [CrossRef]
37. United Nations. *Policy Brief: The Impact of COVID-19 on Food Security and Nutrition*; United Nations: New York, NY, USA, 2020; p. 23.
38. Vandebroek, I.; Pieroni, A.; Stepp, J.R.; Hanazaki, N.; Ladio, A.; Alves, R.R.N.; Picking, D.; Delgoda, R.; Maroyi, A.; van Andel, T.; et al. Reshaping the future of ethnobiology research after the COVID-19 pandemic. *Nat. Plants* **2020**, *6*, 723–730. [CrossRef]
39. Cappelli, A.; Cini, E. Will the COVID-19 pandemic make us reconsider the relevance of short food supply chains and local productions? *Trends Food Sci. Technol.* **2020**, *99*, 566–567. [CrossRef]
40. IPES-Food. *COVID-19 and the Crisis in Food Systems: Symptoms, Causes, and Potential Solutions*; IPES-Food: Brussels, Belgium, 2020.
41. Poppick, L. The Effects of COVID-19 Will Ripple through Food Systems. Available online: https://www.scientificamerican.com/article/the-effects-of-covid-19-will-ripple-through-food-systems/ (accessed on 13 July 2020).

42. Mururia, D.; Mwale, A. Demand for Indigenous Vegetables Soar as Residents Grapple with COVID-19 Economic Shocks. Available online: https://www.kenyanews.go.ke/demand-for-indigenous-vegetables-soar-as-residents-grapple-with-covid-19-economic-shocks/ (accessed on 13 July 2020).
43. Giuliano, G. Coronavirus: From Wild Tobacco New Perspectives in the Treatment of COVID-19. Available online: https://www.enea.it/en/news-enea/news/coronavirus-from-wild-tobacco-new-perspectives-in-the-treatment-of-covid-19 (accessed on 13 July 2020).
44. Timoshyna, A.; Ling, X.; Zhang, K. COVID-19—The Role of Wild Plants in Health Treatment and Why Sustainability of Their Trade Matters. Available online: https://www.traffic.org/news/covid-19-the-role-of-wild-plants-in-health-treatment/ (accessed on 13 July 2020).
45. FAO. *The State of the World's Biodiversity for Food and Agriculture*; FAO: Rome, Italy, 2019. [CrossRef]
46. Ulian, T.; Diazgranados, M.; Pironon, S.; Padulosi, S.; Davies, L.; Howes, M.-J.; Borrell, J.; Ondo, I.; Perez Escobar, O.A.; Sharrock, S.; et al. Unlocking plant and fungal resources to support food security and promote sustainable agriculture. *Plants People Planet* **2020**. [CrossRef]
47. Zulu, D.; Ellis, R.H.; Culham, A. Collection, consumption, and sale of Lusala (*Dioscorea hirtiflora*)—A wild yam—By rural households in southern province, Zambia. *Econ. Bot.* **2019**, *73*, 47–63. [CrossRef]
48. Ulian, T.; Sacandé, M.; Hudson, A.; Mattana, E. Conservation of indigenous plants to support community livelihoods: The MGU—Useful plants project. *J. Environ. Plan. Manag.* **2017**, *60*, 668–683. [CrossRef]
49. Ulian, T.; Flores, C.; Lira, R.; Mamatsharaga, A.; Mogotsi, K.K. *Wild Plants for a Sustainable Future: 110 Multipurpose Species*, 1st ed.; Ulian, T., Flores, C., Lira, R., Mamatsharaga, A., Mogotsi, K.K., Muthoka, P., Ngwako, S., Nyamongo, D.O., Omondi, W., Sanogo, A.K., et al., Eds.; Kew Publishing: Kew, UK, 2019.
50. Maundu, P.M.; Njiro, E.I.; Chweya, J.A.; Imungi, J.K.; Seme, E.N. Kenya. In *The Biodiversity of Traditional Leafy Vegetables*; Chweya, J.A., Eyzaguirre, P.B., Eds.; International Plant Genetic Resources Institute: Rome, Italy, 1999; pp. 51–84.
51. Gotor, E.; Irungu, C. The impact of bioversity international's African leafy vegetables programme in Kenya. *Impact Assess. Proj. Apprais.* **2010**, *28*, 41–55. [CrossRef]
52. Bioversity International; European Initiative for Agricultural Research for Development. *African Leafy Vegetables Come out of the Shade*; Bioversity International: Rome, Italy, 2013; p. 6.
53. Borelli, T.; Gee, E.; Hunter, D. Neglected no more: Reframing the food systems narrative using agricultural biodiversity. In *Biodiversity, Food and Nutrition. A New Agenda for Sustainable Food Systems*; Hunter, D., Borelli, T., Gee, E., Eds.; Routledge: Oxford, UK, 2020.
54. Sarfo, J.; Keding, G.B.; Boedecker, J.; Pawelzik, E.; Termote, C. The impact of local agrobiodiversity and food interventions on cost, nutritional adequacy and affordability of women and children's diet in northern Kenya: A modeling exercise. *Front. Nutr.* **2020**. [CrossRef] [PubMed]
55. Penafiel, D.; Lachat, C.; Espinel, R.; Van Damme, P.; Kolsteren, P. A systematic review on the contributions of edible plant and animal biodiversity to human diets. *Ecohealth* **2011**, *8*, 381–399. [CrossRef] [PubMed]
56. Penafiel, D.; Vanhove, W.; Espinel, R.L.; Van Damme, P. Food biodiversity includes both locally cultivated and wild food species in Guasaganda, Central Ecuador. *J. Ethn. Foods* **2019**, *6*, 25. [CrossRef]
57. Jacob, M.C.M.; Araújo de Medeiros, M.F.; Albuquerque, U.P. Biodiverse food plants in the Semiarid Region of Brazil have unknown potential: A systematic review. *PLoS ONE* **2020**, *15*, e0230936. [CrossRef] [PubMed]
58. Baldermann, S.; Blagojević, L.; Frede, K.; Klopsch, R.; Neugart, S.; Neumann, A.; Ngwene, B.; Norkeweit, J.; Schröter, D.; Schröter, A.; et al. Are neglected plants the food for the future? *CRC Crit. Rev. Plant Sci.* **2016**, *35*, 106–119. [CrossRef]
59. Dogan, Y.; Baslar, S.; Ay, G.; Mert, H.H. The use of wild edible plants in western and central Anatolia (Turkey). *Econ. Bot.* **2004**, *58*, 684–690. [CrossRef]
60. Taylor, J.; Sarkis, L.; Hani, N.; Abulaila, K.; Ulian, T. Conservation and Sustainable Use of Wild Edible Plants in the Eastern Mediterranean Region. In *Role of Mediterranean Forests in the Paris Agreement, Proceedings of the Sixth Mediterranean Forest Week, Brummana, Lebanon, 1–5 April 2019*; Mohanna, C., Ed.; Forêt Méditerranéenne: Brummana, Lebanon, 2019; pp. 293–300.
61. Rivera, D.; Obón, C.; Heinrich, M.; Inocencio, C.; Verde, A.; Fajardo, J. Gathered Mediterranean Food Plants—Ethnobotanical Investigations and Historical Development. In *Local Mediterranean Food Plants and Nutraceuticals*; Heinrich, M., Müller, W.E., Galli, C., Eds.; KARGER: Basel, Switzerland, 2006; pp. 18–74. [CrossRef]

62. Dogan, Y. Traditionally Used Wild Edible Greens in the Aegean Region of Turkey. *Acta Soc. Bot. Pol.* **2012**, *81*, 329–342. [CrossRef]
63. Dogan, Y.; Ugulu, I.; Durkan, N. Wild edible plants sold in the local markets of Izmir, Turkey. *Pak. J. Bot.* **2013**, *45* (Suppl. 1), 177–184.
64. Nassif, F.; Tanji, A. Gathered food plants in Morocco: The long forgotten species in ethnobotanical research. *Life Sci. Leafl.* **2013**, *3*, 17–54.
65. Hadjichambis, A.C.; Paraskeva-Hadjichambi, D.; Della, A.; Elena Giusti, M.; De Pasquale, C.; Lenzarini, C.; Censorii, E.; Reyes Gonzales-Tejero, M.; Patricia Sanchez-Rojas, C.; Ramiro-Gutierrez, J.M.; et al. Wild and semi-domesticated food plant consumption in seven Circum-Mediterranean areas. *Int. J. Food Sci. Nutr.* **2008**, *59*, 383–414. [CrossRef] [PubMed]
66. Powell, B.; Ouarghidi, A.; Johns, T.; Ibn Tattou, M.; Eyzaguirre, P. Wild leafy vegetable use and knowledge across multiple sites in Morocco: A case study for transmission of local knowledge? *J. Ethnobiol. Ethnomed.* **2014**, *10*. [CrossRef] [PubMed]
67. Ellatifi, M. The situation of non-wood forest products in Morocco. In *Proceedings of the Joint FAO/ECE/ILO Committee on Forest Technology, Management and Training Seminar Proceedings on Harvesting of Non-Wood Forest Products, Menemen-Izmir, Turkey, 2–8 October 2000*; Food and Agricultural Organization of The United Nations: Menemen, Turkey, 2000.
68. Thomson, L.; Doran, J.; Clarke, B. *Trees for Life in Oceania: Conservation and Utilisation of Genetic Diversity*; ACIAR Mono.; Australian Centre for International Agricultural Research: Canberra, Australia, 2018.
69. French, B. *Food Plants of Papua New Guinea*; Australia and Pacific Science Foundation: Sheffield, Tasmania, 1986.
70. Powell, J.M. Ethnobotany. In *New Guinea Vegetation*; Paijmans, K., Ed.; Elsevier Scientific Publishing: Amsterdam, The Netherlands, 1976; pp. 106–183.
71. Bourke, M.R.; Allen, B. Nuts. In *Food and Agriculture in Papua New Guinea*; Bourke, M., Harwood, T., Eds.; ANU E Press: Canberra, Australia; The Australian National University: Canberra, Australia, 2009; pp. 215–222.
72. Townsend, P.K. Sago production in a New Guinea economy. *Hum. Ecol.* **1974**, *2*, 217–236. [CrossRef]
73. Roscoe, P. The hunters and gatherers of New Guinea. *Curr. Anthropol.* **2002**, *43*, 153–162. [CrossRef]
74. Government of Niue. *The State of Niue's Biodiversity for Food and Agriculture*; Niue Country Report; FAO: Rome, Italy, 2019.
75. Thaman, R.R. The Evolution of the Fiji Food System. In *Food and Nutrition in Fiji: A Historical View*; Jansen, A.A.J., Parkinson, S., Robertson, A.F.S., Eds.; University of the South Pacific: Suva, Fiji, 1990; pp. 23–109.
76. Novaczek, I. *A Guide to the Common and Edible Medicinal Sea Plants of the Pacific Islands*; Secretariat of the Pacific Community: Suva, Fiji, 2001.
77. Pawera, L.; Khomsan, A.; Zuhud, E.A.M.; Hunter, D.; Ickowitz, A.; Polesny, Z. Wild food plants and trends in their use: From knowledge and perceptions to drivers of change in West Sumatra, Indonesia. *Foods* **2020**, *9*, 1240. [CrossRef] [PubMed]
78. Pawera, L.; Lipoeto, N.I.; Khomsan, A.; Zuhud, E.A.D. *Food Plants of Minang and Mandailing Communities in Pasaman District, West Sumatra. A Community Guidebook on Biodiversity for Nutrition and Health*; Swisscontact: Jakarta, Indonesia, 2018.
79. Sen, A. *Poverty and Famines: An Essay on Entitlement and Deprivation*; Clarendon Press: Oxford, UK, 1981.
80. Shackleton, S.; Cocks, M.; Dold, T.; Kaschula, S.; Mbata, K.; Mickels-Kokwe, G.; von Maltitz, G. Contribution of Non-Wood Forest Products to Livelihoods and Poverty Alleviation. In *The Dry Forests and Woodlands of Africa. Managing for Products and Services*; Chidumayo, E.N., Gumbo, D.J., Eds.; Routledge: Abingdon, UK, 2010; pp. 93–129.
81. FAO. *The State of the World's Forests. Forest Pathways to Sustainable Development*; Food and Agriculture Organization of the United Nations: Rome, Italy, 2018. [CrossRef]
82. Petersen, L.M.; Moll, E.J.; Collins, R.; Hockings, M.T. Development of a compendium of local, wild-harvested species used in the informal economy trade, Cape Town, South Africa. *Ecol. Soc.* **2012**, *17*. [CrossRef]
83. Hudson, A.; Milliken, W.; Timberlake, J.; Giovannini, P.; Fijamo, V.; Massunde, J.; Chipanga, H.; Nivunga, M.; Ulian, T. Natural plant resources for sustainable development: Insights from community use in the Chimanimani trans-frontier conservation area, Mozambique. *Hum. Ecol.* **2020**, *48*, 55–67. [CrossRef]

84. Hickey, G.M.; Pouliot, M.; Smith-Hall, C.; Wunder, S.; Nielsen, M.R. Quantifying the economic contribution of wild food harvests to rural livelihoods: A global-comparative analysis. *Food Policy* **2016**, *62*, 122–132. [CrossRef]
85. Karabak, S. Economic and socio-cultural importance of edible wild species. *ANADOLU J. AARI* **2017**, *27*, 26–38.
86. Morris, C.; Bala, S.; South, G.R.; Lako, J.; Lober, M.; Simos, T. Supply chain and marketing of sea grapes, *Caulerpa racemosa* (Forsskål) J. Agardh (Chlorophyta: Caulerpaceae) in Fiji, Samoa and Tonga. *J. Appl. Phycol.* **2014**, *26*, 783–789. [CrossRef]
87. Lovrić, M.; Da Re, R.; Vidale, E.; Prokofieva, I.; Wong, J.; Pettenella, D.; Verkerk, P.J.; Mavsar, R. Non-wood forest products in Europe—A quantitative overview. *For. Policy Econ.* **2020**, *116*, 102175. [CrossRef]
88. Díaz, S.; Settele, J.; Eduardo, B.; Ngo, H.T.; Guèze, M.; Agard, J.; Arneth, A.; Balvanera, P.; Brauman, K.; Butchart, S.; et al. *The Global Assessment Report on Report on Biodiversity and Ecosystem Services. Summary for Policymakers*; IPBES: Bonn, Germany, 2019. [CrossRef]
89. Masson-Delmotte, V.; Zhai, P.; Pörtner, H.-O.; Roberts, D.; Skea, J.; Calvo, E.; Priyadarshi, B.; Shukla, R.; Ferrat, M.; Haughey, E.; et al. *Climate Change and Land: IPCC Report*; IPCC: Geneva, Switzerland, 2019.
90. Royal Botanic Gardens, Kew. *State of the World's Plants*; Royal Botanic Gardens, Kew: London, UK, 2016.
91. Sunderland, T.C.H. Food Security: Why Is Biodiversity Important? *Int. For. Rev.* **2011**, *13*, 265–274. [CrossRef]
92. Kew. *Plants under Pressure—A Global Assessment*; Royal Botanic Gardens: Kew, UK; Natural History Museum: London, UK, 2016.
93. Khoury, C.K.; Amariles, D.; Soto, J.S.; Diaz, M.V.; Sotelo, S.; Sosa, C.C.; Ramírez-Villegas, J.; Achicanoy, H.A.; Velásquez-Tibatá, J.; Guarino, L.; et al. Comprehensiveness of conservation of useful wild plants: An operational indicator for biodiversity and sustainable development targets. *Ecol. Indic.* **2019**, *98*, 420–429. [CrossRef]
94. Jarvis, A.; Upadhyaya, H.; Gowda, C.; Aggarwal, P.K.; Fugisaka, S.; Anderson, B. *Climate Change and Its Effect on Conservation and Use of Plant Genetic Resources for Food and Agriculture and Associated Biodiversity for Food Security*; Thematic Study SoW Report PGRFA; FAO: Rome Italy, 2008; p. 26.
95. Cooper, M.; Zvoleff, A.; Gonzalez-Roglich, M.; Tusiime, F.; Musumba, M.; Noon, M.; Alele, P.; Nyiratuza, M. Geographic factors predict wild food and nonfood NTFP collection by households across four African countries. *For. Policy Econ.* **2018**, *96*, 38–53. [CrossRef] [PubMed]
96. Willett, W.; Rockström, J.; Loken, B.; Springmann, M.; Lang, T.; Vermeulen, S.; Garnett, T.; Tilman, D.; DeClerck, F.; Wood, A.; et al. Food in the anthropocene: The EAT–Lancet Commission on healthy diets from sustainable food systems. *Lancet* **2019**, *393*, 447–492. [CrossRef]
97. Belcher, B.; Ruíz-Pérez, M.; Achdiawan, R. Global patterns and trends in the use and management of commercial NTFPs: Implications for livelihoods and conservation. *World Dev.* **2005**, *33*, 1435–1452. [CrossRef]
98. Lamrani-Alaoui, M.; Hassikou, R. Rapid risk assessment to harvesting of wild medicinal and aromatic plant species in Morocco for conservation and sustainable management purposes. *Biodivers. Conserv.* **2018**, *27*, 2729–2745. [CrossRef]
99. Government of Brazil; BFN Project. SiBBr—Information Platform on Brazilian Biodiversity. Available online: https://ferramentas.sibbr.gov.br/ficha/bin/view/FN/ (accessed on 29 July 2020).
100. Martinelli, G.; Moraes, M.A. *Livro Vermelho da Flora do Brasil*, 1st ed.; Instituto de Pesquisas Jardim Botânico do Rio de Janeiro: Rio de Janeiro, Brazil, 2013.
101. Kim, D.-S.; Joo, N. Nutritional composition of *Sacha inchi* (*Plukenetia volubilis* L.) as affected by different cooking methods. *Int. J. Food Prop.* **2019**, *22*, 1235–1241. [CrossRef]
102. de Gaillande, C.; Payri, C.; Remoissenet, G.; Zubia, M. Caulerpa consumption, nutritional value and farming in the Indo-Pacific Region. *J. Appl. Phycol.* **2017**, *29*, 2249–2266. [CrossRef]
103. Chweya, J.A.; Mnzava, N.A. *Cat's Whiskers. Cleome Gynandra L. Promoting the Conservation and Use of Underutilized and Neglected Crops*; International Plant Genetic Resources Institute: Rome, Italy, 1997.
104. FAO/Government of Kenya. *Kenya Food Composition Tables*; Food and Agricultural Organization of the United Nations: Nairobi, Kenya, 2018.
105. BFN Project. BFN Species Database. Available online: http://www.b4fn.org/resources/species-database/ (accessed on 29 July 2020).

106. Barros, L.; Carvalho, A.M.; Ferreira, I.C.F.R. Leaves, flowers, immature fruits and leafy flowered stems of *Malva sylvestris*: A comparative study of the nutraceutical potential and composition. *Food Chem. Toxicol.* **2010**, *48*, 1466–1472. [CrossRef]
107. El Abbassi, A.; Khalid, N.; Zbakh, H.; Ahmad, A. Physicochemical characteristics, nutritional properties, and health benefits of Argan oil: A review. *Crit. Rev. Food Sci. Nutr.* **2014**, *54*, 1401–1414. [CrossRef]
108. Garzuglia, M. *Threatened, Endangered and Vulnerable Tree Species: A Comparison between FRA 2005 and the IUCN Red List*; Forest Resources Assessment (FRA) Working Paper Series; FRA WP 106/E; FAO: Rome, Italy, 2006.
109. Wardah; Sambas, E.N.; Ridwan; Ariani, D. Starch product of wild plants species Jalawure (*Tacca leontopetaloides* L.) Kuntze as the source of food security in the south coastal west Java. *IOP Conf. Ser. Mater. Sci. Eng.* **2017**, *193*, 012035. [CrossRef]
110. García-Herrera, P.; Sánchez-Mata, M.C.; Cámara, M.; Fernández-Ruiz, V.; Díez-Marqués, C.; Molina, M.; Tardío, J. Nutrient composition of six wild edible Mediterranean Asteraceae plants of dietary interest. *J. Food Compos. Anal.* **2014**, *34*, 163–170. [CrossRef]
111. Tosun, M.; Tosun1, M.; Ercisli, S.; Ozer, H.; Turan, M.; Polat, T.; Ozturk, E.; Padem, H.; Kilicgun, H. Chemical Composition and Antioxidant Activity of Foxtail Lily (Eremurus Spectabilis). *Acta Sci. Pol. Hortorum Cultus* **2012**, *11*, 145–153.
112. Cinar, A.; Tugrul Ay, S.; Ayas, F.; Karabak, S.; Guzelsoy, N.; Ucurum, O. Foxtail Lilly (*Eremurus spectabilis* M. Bieb.) as priority species of biodiversity for food and nutrition project of Turkey. *ANADOLU J. AARI* **2017**, *27*, 69–73.
113. Choonhahirun, A. Proximate composition and functional properties of Pra (Elateriospermun Tapos Blume) seed flour. *Afr. J. Biotechnol.* **2010**, *9*, 5946–5949.
114. Indonesian Food Composition Data. Available online: https://www.panganku.org/en-EN/tentang_kami (accessed on 14 July 2020).
115. Rojas, M.; Lambert, F.; Ramirez-Villegas, J.; Challinor, A.J. Emergence of robust precipitation changes across crop production areas in the 21st century. *Proc. Natl. Acad. Sci. USA* **2019**, *116*, 6673–6678. [CrossRef]
116. Borrell, J.S.; Dodsworth, S.; Forest, F.; Pérez-Escobar, O.A. The climatic challenge: Which plants will people use in the next century? *J. Exp. Bot.* **2020**, *170*. [CrossRef]
117. Anderson, D.; Ford, J.D.; Way, R.G. The impacts of climate and social changes on cloudberry (*Bakeapple*) picking: A case study from southeastern Labrador. *Hum. Ecol.* **2018**, *46*, 849–863. [CrossRef]
118. FAO. *Sixteenth Regular Session of the Commission on Genetic Resources for Food and Agriculture—CGRFA-16/17/Report Rev.1*; FAO: Rome, Italy, 2017.
119. Ingram, V.; Vinceti, B.; van Vliet, N. Wild Plant and Animal Genetic Resources. In *Routledge Handbook of Agricultural Biodiversity*; Hunter, D., Guarino, L., Spillane, C., McKeown, P.C., Eds.; Routledge: Abingdon, UK, 2017; pp. 65–85. [CrossRef]
120. United Nations. *Convention on Biological Diversity*; United Nations: Rio de Janeiro, Brazil, 1992; p. 28.
121. Tata Ngome, P.I.; Shackleton, C.; Degrande, A.; Tieguhong, J.C. Addressing constraints in promoting wild edible plants' utilization in household nutrition: Case of the Congo basin forest area. *Agric. Food Secur.* **2017**, *6*, 1–10. [CrossRef]
122. Jaenicke, H.; Höschle-Zeledon, I. (Eds.) *Strategic Framework for Underutilized Plant Species Research and Development, with Special Reference to Asia and the Pacific, and to Sub-Saharan Africa*; International Centre for Underutilised Crops, Colombo, Sri Lanka and Global Facilitation Unit for Underutilized Species: Rome, Italy, 2006.
123. Hawkes, C. Uneven dietary development: Linking the policies and processes of globalization with the nutrition transition, obesity and diet-related chronic diseases. *Glob. Health* **2006**, *2*. [CrossRef]
124. Biodiversity, Food and Nutrition. *A New Agenda for Sustainable Food Systems*, 1st ed.; Hunter, D., Borelli, T., Gee, E., Eds.; Routledge: Oxford, UK, 2020.
125. Fanzo, J.; Davis, C. Can diets be healthy, sustainable, and equitable? *Curr. Obes. Rep.* **2019**, *8*, 495–503. [CrossRef] [PubMed]
126. Termote, C.; Raneri, J.; Deptford, A.; Cogill, B. Assessing the potential of wild foods to reduce the cost of a nutritionally adequate diet: An example from eastern Baringo District, Kenya. *Food Nutr. Bull.* **2014**, *35*, 458–479. [CrossRef] [PubMed]
127. Nesbitt, M.; McBurney, R.P.H.; Broin, M.; Beentje, H.J. Linking biodiversity, food and nutrition: The importance of plant identification and nomenclature. *J. Food Compos. Anal.* **2010**, *23*, 486–498. [CrossRef]

128. Kennedy, G.; Lee, W.T.K.; Termote, C.; Charrondière, R.; Tung, J.Y.A. *Guidelines on Assessing Biodiverse Foods in Dietary Intake Surveys*; Food and Agricultural Organization of the United Nations: Rome, Italy, 2017.
129. Elmadfa, I.; Meyer, A.L. Importance of food composition data to nutrition and public health. *Eur. J. Clin. Nutr.* **2010**, *64* (Suppl. S3), S4–S7. [CrossRef] [PubMed]
130. Msuya, J.M.; Mamiro, P.; Weinberger, K. Iron, Zinc and β-Carotene nutrient potential of non-cultivated. Indigenous vegetables in Tanzania. *Acta Hortic.* **2009**, *806*, 217–222. [CrossRef]
131. World Health Organization (WHO); Food and Agriculture Organization of the United Nations (FAO). *Diet, Nutrition and the Prevention of Chronic Diseases: Report of a Joint WHO/FAO Expert Consultation*; WHO Technical Report Series 916; World Health Organization: Geneva, Switzerland, 2003.
132. Afshin, A.; Sur, P.J.; Fay, K.A.; Cornaby, L.; Ferrara, G.; Salama, J.S.; Mullany, E.C.; Abate, K.H.; Abbafati, C.; Abebe, Z.; et al. Health effects of dietary risks in 195 countries, 1990–2017: A systematic analysis for the global burden of disease study 2017. *Lancet* **2019**, *393*, 1958–1972. [CrossRef]
133. Boeing, H.; Bechthold, A.; Bub, A.; Ellinger, S.; Haller, D.; Kroke, A.; Leschik-Bonnet, E.; Müller, M.J.; Oberritter, H.; Schulze, M.; et al. Critical review: Vegetables and fruit in the prevention of chronic diseases. *Eur. J. Nutr.* **2012**, *51*, 637–663. [CrossRef]
134. Hall, J.N.; Moore, S.; Harper, S.B.; Lynch, J.W. Global variability in fruit and vegetable consumption. *Am. J. Prev. Med.* **2009**, *36*, 402–409.e5. [CrossRef]
135. Slavin, J.L.; Lloyd, B. Health benefits of fruits and vegetables. *Adv. Nutr.* **2012**, *3*, 506–516. [CrossRef]
136. Micha, R.; Khatibzadeh, S.; Shi, P.; Andrews, K.G.; Engell, R.E.; Mozaffarian, D. Global, regional and national consumption of major food groups in 1990 and 2010: A systematic analysis including 266 country-specific nutrition surveys worldwide. *BMJ Open* **2015**, *5*, e008705. [CrossRef]
137. de Bruyn, J.; Msuya, J.; Ferguson, E. Evaluating pictorial charts as a means of collecting participant-recorded data on household dietary diversity in low-literacy communities in Tanzania. *Br. J. Nutr.* **2019**, *122*, 1432–1440. [CrossRef] [PubMed]
138. Łuczaj, Ł.; Pieroni, A.; Tardío, J.; Pardo-de-Santayana, M.; Sõukand, R.; Svanberg, I.; Kalle, R. Wild food plant use in 21st century Europe: The disappearance of old traditions and the search for new cuisines involving wild edibles. *Acta Soc. Bot. Pol.* **2012**, *81*, 359–370. [CrossRef]
139. Hattersley, L.; Cogill, B.; Hunter, D.; Kennedy, G. Evidence for the Role of Biodiversity in Supporting Healthy, Diverse Diets and Nutrition. In *Biodiversity, Food and Nutrition: A New Agenda for Sustainable Food Systems*; Hunter, D., Borelli, T., Gee, E., Eds.; Routledge: Abingdon, UK, 2020.
140. Shackleton, C.M.; Hurley, P.T.; Dahlberg, A.C.; Emery, M.R.; Nagendra, H. Urban foraging: A ubiquitous human practice overlooked by urban planners, policy, and research. *Sustainability* **2017**, *9*, 1884. [CrossRef]
141. Reyes-García, V.; Menendez-Baceta, G.; Aceituno-Mata, L.; Acosta-Naranjo, R.; Calvet-Mir, L.; Domínguez, P.; Garnatje, T.; Gómez-Baggethun, E.; Molina-Bustamante, M.; Molina, M.; et al. From famine foods to delicatessen: Interpreting trends in the use of wild edible plants through cultural ecosystem services. *Ecol. Econ.* **2015**, *120*, 303–311. [CrossRef]
142. Schulp, C.J.E.; Thuiller, W.; Verburg, P.H. Wild food in Europe: A synthesis of knowledge and data of terrestrial wild food as an ecosystem service. *Ecol. Econ.* **2014**, *105*, 292–305. [CrossRef]
143. Li, D.-Z.; Pritchard, H.W. The science and economics of ex situ plant conservation. *Trends Plant Sci.* **2009**, *14*, 614–621. [CrossRef]
144. Commission on Genetic Resources for Food and Agriculture—FAO. *The Second Global Plan of Action for Plant Genetic Resources for Food and Agriculture*; Food and Agricultural Organization of the United Nations: Rome, Italy, 2012.
145. FAO. *Voluntary Guidelines for the Conservation and Sustainable Use of Crop Wild Relatives and Wild Food Plants*; Food and Agriculture Organization of the United Nations: Rome, Italy, 2017.
146. Vernooy, R.; Bessette, G.; Otieno, G. (Eds.) *Resilient Seed Systems: Handbook*, 2nd ed.; Bioversity International: Rome, Italy, 2019.
147. PAR. *Assessing Agrobiodiversity: A Compendium of Methods*; Mijatović, D., Hodgkin, T., Eds.; Platform for Agrobiodiversity Research: Rome, Italy, 2018.
148. Secretariat of the Convention on Biological Diversity (SCBD). *Nagoya Protocol on Access to Genetic Resources and the Fair and Equitable Sharing of Benefits Arising from Their Utilization to the Convention on Biological Diversity*; United Nations: Montreal, QC, Canada, 2011; p. 25.

149. Bioversity International. *Seasonal Calendar of Fruits and Vegetables for a Diversified Diet in Mandla and Dindori Districts, Madhya Pradesh, India*; Bioversity International: Rome, Italy; Action for Social Advancement: Madhya Pradesh, India, 2018; p. 1.
150. Gee, E.; Borelli, T.; Beltrame, D.M.O.; Oliveira, C.N.S.; Coradin, L.; Wasike, V.; Manjella, A.; Samarasinghe, G.; Güner, B.; Tan, A.; et al. The ABC of Mainstreaming Biodiversity for Food and Nutrition. Concepts, Theory and Practice. In *Biodiversity Food and Nutrition. A New Agenda for Sustainable Food Systems*; Hunter, D., Borelli, T., Gee, E., Eds.; Routledge: Abingdon, UK, 2020.
151. Boedecker, J.; Odhiambo Odour, F.; Lachat, C.; Van Damme, P.; Kennedy, G.; Termote, C. Participatory farm diversification and nutrition education increase dietary diversity in Western Kenya. *Matern. Child Nutr.* **2019**, *15*. [CrossRef]
152. FAO/INFOODS. *FAO/INFOODS Guidelines for Checking Food Composition Data Prior to the Publication of a User Table/Database—Version 1.0*; FAO: Rome, Italy, 2012.
153. Borelli, T.; Hunter, D.; Padulosi, S.; Amaya, N.; Meldrum, G.; de Oliveira Beltrame, D.M.; Samarasinghe, G.; Wasike, V.W.; Güner, B.; Tan, A.; et al. Local solutions for sustainable food systems: The contribution of orphan crops and wild edible species. *Agronomy* **2020**, *10*, 231. [CrossRef]
154. Rodríguez-Arévalo, I.; Mattana, E.; García, L.; Liu, U.; Lira, R.; Dávila, P.; Hudson, A.; Pritchard, H.W.; Ulian, T. Conserving seeds of useful wild plants in Mexico: Main issues and recommendations. *Genet. Resour. Crop Evol.* **2017**, *64*, 1141–1190. [CrossRef]
155. Sperling, L.; McGuire, S. Fatal gaps in seed security strategy. *Food Secur.* **2012**, *4*, 569–579. [CrossRef]
156. Mattana, E.; Sacande, M.; Abdul Sanogo, K.; Lira, R.; Gomez-Barreiro, P.; Rogledi, M.; Ulian, T. Thermal requirements for seed germination of underutilized Lippia species. *S. Afr. J. Bot.* **2017**, *109*, 223–230. [CrossRef]
157. McCune, L.M. The protection of indigenous peoples' seed rights during ethnobotanical research. *Ethnobiol. Lett.* **2018**, *9*, 67–75. [CrossRef]
158. Aluso, L.; Nzuki, R. From the Field: How On-Farm Biodiversity Is Transforming Community Livelihoods in Kenya. Available online: https://alliancebioversityciat.org/news_and_blogs/on-farm-biodiversity-transforming-communitys-livelihoods-in-resilient-landscapes/ (accessed on 29 July 2020).
159. Mogotsi, K.K.; Ulian, T. Conserving indigenous food plants in Botswana: The case of *Morama* bean. *Samara* **2013**, *24*.
160. Tan, A.; Adanacıoglu, N.; Karabak, S.; Aykas, L.; Tas, N.; Taylan, T. Biodiversity for food and nutrition: Edible wild plant species of Aegean Region of Turkey. *ANADOLU J. AARI* **2017**, *27*, 1–8.
161. Karik, U. The effect of different harvest dates on the yield and quality of golden thistle (*Scolymus hispanicus* L.). *Turk. J. Feild Crops* **2019**, *24*, 230–236. [CrossRef]
162. Beltrame, D.M.D.O.; Oliveira, C.N.S.; Borelli, T.; Santiago, R.D.A.C.; Tronco Monego, E.; de Rosso, V.V.; Coradin, L.; Hunter, D. Diversifying institutional food procurement—Opportunities and barriers for integrating biodiversity for food and nutrition in Brazil. *Raízes* **2016**, *36*, 55–69. [CrossRef]
163. Bacchetta, L.; Visioli, F.; Cappelli, G.; Caruso, E.; Martin, G.; Nemeth, E.; Bacchetta, G.; Bedini, G.; Wezel, A.; van Asseldonk, T.; et al. A manifesto for the valorization of wild edible plants. *J. Ethnopharmacol.* **2016**, *191*, 180–187. [CrossRef]
164. Gonsalves, J.; Hunter, D.; Lauridsen, N. School Gardens: Multiple Functions and Multiple Outcomes. In *Agrobiodiversity, School Gardens and Healthy Diets. Promoting Biodiversity, Food and Sustainable Nutrition*; Hunter, D., Monville-Oro, E., Burgos, B., Rogel, C.N., Calub, B., Gonsalves, J., Lauridsen, N., Eds.; Routledge: Abingdon, UK, 2020; pp. 1–32.
165. Grazioli, F.; Turdieva, M.; Kettle, C.J. The Role of School Gardens as Conservation Networks for Tree Genetic Resources. In *Agrobiodiversity, School Gardens and Healthy Diets: Promoting Biodiversity, Food and Sustainable Nutrition*; Hunter, D., Monville-Oro, E., Burgos, B., Rogel, C.N., Calub, B.M., Gonsalves, J., Lauridsen, N., Eds.; Routledge: Abingdon, UK, 2020; p. 302.
166. Sutcliffe, S.; Court, J. *Evidence-Based Policymaking: What Is It? How Does It Work? What Relevance for Developing Countries*; Overseas Development Institute: London, UK, 2005.
167. Beltrame, D.; Borelli, T.; Oliveira, C.; Coradin, L.; Hunter, D. Biodiversity for Food and Nutrition: Promoting Food Security and Nutrition through Institutional Markets in Brazil. In *Public Food Procurement for Sustainable Food Systems and Healthy Diets*; Swensson, L., Tartanac, F., Hunter, D., Schneider, S., Eds.; Food and Agricultural Organization of the United Nations: Rome, Italy, 2020.

168. Baldi, G.; Martini, E.; Catharina, M.; Muslimatun, S.; Fahmida, U.; Jahari, A.B.; Hardinsyah; Frega, R.; Geniez, P.; Grede, N.; et al. Cost of the Diet (CoD) tool: First results from Indonesia and applications for policy discussion on food and nutrition security. *Food Nutr. Bull.* **2013**, *34* (Suppl. S2), S35–S42. [CrossRef]
169. Gee, E.; Lee, H. Start Me up! Food Biodiversity and Youth-Led Innovations. In *Biodiversity, Food and Nutrition. A New Agenda for Sustainable Food Systems*; Hunter, D., Borelli, T., Gee, E., Eds.; Routledge: Abingdon, UK, 2020; pp. 255–274.

plants

MDPI

Article

Conservation of Wild Food Plants and Their Potential for Combatting Food Insecurity in Kenya as Exemplified by the Drylands of Kitui County

Fredrick Munyao Mutie [1,2,3], Peninah Cheptoo Rono [1,2,3], Vivian Kathambi [4], Guang-Wan Hu [1,2,*] and Qing-Feng Wang [1,2]

1 CAS Key Laboratory of Plant Germplasm Enhancement and Speciality Agriculture, Wuhan Botanical Gen, Chinese Academy of Sciences, Wuhan 430074, China; munyao@wbgcas.cn (F.M.M.); peninah@wbgcas.cn (P.C.R.); qfwang@wbgcas.cn (Q.-F.W.)

2 Sino-Africa Joint Research Center, Chinese Academy of Sciences, Wuhan 430074, China

3 University of Chinese Academy of Sciences, Beijing 100049, China

4 National Museums of Kenya, East African Herbarium, P.O. Box 40658, Nairobi 00100, Kenya; vivian@wbgcas.cn

* Correspondence: guangwanhu@wbgcas.cn

Received: 22 June 2020; Accepted: 8 August 2020; Published: 12 August 2020

Abstract: Wild food plants are important resources for people living in dry areas of Kenya. A botanical inventory of vascular plants of Kitui county was compiled from specimens collected during field investigations in Kitui county, at the East African (EA) herbarium and from literature reporting on plants of Kitui county. To obtain an inventory of wild edible plants found in Kitui county, literature reporting on wild edible plants of Kenya were searched and combined with the use reports obtained from field surveys in Kitui county. A total of 199 wild plants found in Kitui county have the potential of being utilized as foods in different ways. Plant species growing either as trees or shrubs (83 species) and herbs (36 species) are the dominant life forms while the best represented plant families are Leguminosae (25 species) and Malvaceae (17 species). Fruits (124 reports) and leaves (56 reports) are the common plant parts collected for food. Fruits (120 species) and vegetables (44 species) are the common wild food types in Kitui county. Further studies on species distribution are necessary to address conservation concerns that may threaten such plants.

Keywords: climate change; conservation; crop wild relatives (CWR); drylands; Kitui county; wild food plants

1. Introduction

Arid and semi-arid areas cover about 80% of the Kenyan land mass and are characterized by a hot and dry climate and soils of poor agricultural potential [1]. The lives of people living in dry areas of Kenya are thus constrained by frequent droughts, which in turn trigger further challenges such as poor grazing resources and poor water quality. This in turn results in poverty and human to human conflicts over scarce resources. These challenges coupled with poor veterinary services make pastoralism an increasingly difficult source of livelihood for dryland communities [2]. Climate manifestation in East Africa has also proved to be difficult where warming is likely to lead to dryness in some areas and a higher precipitation in others [3], meaning that local populations are forced to cope with climatic uncertainties [4,5]. Livelihood diversification is the main way to cope with drought [6]. There is evidence of livelihood diversification in dry areas of Kenya to cope with changing climatic conditions such as the extraction of gums and resins by some communities as a source of income [7,8] while wild plants are also reported as important sources of traditional foods [9–11].

Many foodstuffs consumed in tropical Africa are derived from wild plants [9]. Those plants are utilized in different ways such as fruits, vegetables, cereals, roots, and tubers [11]. Plant derived foods such as fruits and nuts provide many nutritional benefits to the body [12]. Fruits can improve the nutrition of poor people who may suffer from deficiencies in vitamins, minerals, and other macronutrients. Many fruits are also important sources of vitamins A and C which may be lacking in the diet. For example, vitamin C which is found in significant quantities in many fruits is essential for protecting body cells and improves the adsorption of nonheme iron from plant-based foods. As a result of low intake of vitamin A, an estimated 50 million children in Africa are at risk of its deficiency, making it the third greatest health problem in the continent, preceded by malaria and HIV/AIDS [13]. Traditionally, children used to eat wild foods such as fruits and nuts during herding, which served them with nutritional benefits [12]. Although foods from wild plants may not at the present time form a major part of the diet of the local communities in Kenya as exemplified by the life of Dorobo people [14], traditional foods are still culturally accepted and are an integral part of the diet of local inhabitants [11]. For example, among the Dorobo people, plant species such as *Grewia tephrodermis*, *Vangueria madagascariensis*, *Vigna frutescens*, and *Vatovaea pseudolablab* are reported to have been served as staple foods for many years [14]. Furthermore, some traditional local vegetable species such as leaf amaranth are sold in the local markets in Kenya [15]. The local people have the knowledge on preparation and production of traditional foods, which require minimal additional inputs which are affordable to many, including the poor people [15]. Some wild foods also have medicinal properties to the human body and can be processed through various methods such as boiling, fermentation, and sun drying by the local people [11]. In some societies, some traditional foods from wild plants might be considered to be of no or low commercial value hence their collection is mostly meant for local consumption [11]. However, many wild indigenous fruits are sold locally in Kenyan markets such as fruits of *Adansonia digitata* which is also processed by coloring the seed pulp to make a snack. Its products are also in global demand for novel foods, pharmaceuticals, and cosmetics where the European Union, United States, Japan, and South Africa are reportedly potential markets [13].

The exploitation of native flora can be a buffer against periodical famines which are becoming prevalent in tropical areas [9]. About 60% of the Kenyan population face starvation due to lack of physical and economic access to adequate calories [16]. Kenya is endowed with diverse plant species which are estimated to comprise about 6293 indigenous vascular plants [17]. These include an estimated 800 food plants [10] some of which are underutilized food plants [16] such as *Amaranthus* spp. (leaf amaranth), *Solanum americanum* (African nightshade), *Cleome gynandra* (spider plant), *Cucumis dipsaceus* (Hedgehog cucumber), *Commelina forskaolei* (Rat's ear), and *Cucurbita* spp. (pumpkin leaves) which can all be utilized as green leafy vegetables [15]. Such plant species were relied upon in the past as sources of vitamins, minerals, and proteins by rural societies [16]. Despite their importance, ethnobotanical knowledge of traditional wild foods is declining in Kenya [10,11]. Women, children, and herders play important roles during collection of wild edible plants [14,18] and can therefore be considered as important custodians of such knowledge in Kenya. In northern Kenya for example, collection of gums is mostly done by married women in an effort to provide an additional income for their households [8], perhaps adopting the roles of single parenthood especially the widowed. In addition, some wild food plants considered to be of minor significance are gathered by little children and are at times used as diet supplements and emergency foods [9,10]. During collection of some wild edible plants in Kenya such as *Ficus* fruits (figs), *Vangueria* fruits, *Craibia laurentii* nuts, and *Maerua kirkii* nuts, children accompany their mothers to help in gathering while collection of some species such as tubers of *Cyphia glandulosa* is reportedly done by children as they go on with their duties [14]. Collection of wild edible plants is mostly done by poor and illiterate people, where such activities have perhaps been normalized as survival strategies during the dry periods of the year when there are insufficient resources available for human survival [8]. For example, among the hunter-gatherer communities in Kenya, wild foods may comprise the main diet of the day at certain times such as during famines [10].

Introduction of exotic vegetables has diverted the focus on indigenous vegetables in Kenya [16]. Recognizing the value of wild food plants can be useful in conservation of germplasm for the future generations [9] as well as buffering against famine in the changing climatic conditions [18]. The need for the recognition of the nutritional value of traditional foods has resulted in campaigns for them to be incorporated into the rural and urban diets [16]. In spite of this, few studies in Kenya [9,14,19–22] have focused on documentation of wild food plants at local levels. In Kitui county, studies have mostly focused on documentation of medicinal plants [23–25] with little attention given to wild edible plants. Local utilization and acceptance of underutilized vegetable species has been reported in Kitui county where cowpeas are the most popular vegetable species [15]. The overall aim of this study is to highlight the potential of wild edible plants in Kitui county as resources which can be utilized in combatting food insecurity and famine by the rural dryland communities. The study is based on the assumption that the plant species documented as wild food plants elsewhere in Kenya but currently not yet used as such in Kitui county, have the potential to also be adopted and utilized as food plants in this region as well.

1.1. Study Area

Kitui county is a tract of land located at 0°10′ S and 39°0′ E, between Athi and Tana rivers occupying an area of 30,496.4 km^2 [26,27] (Figure 1). The area is mainly inhabited by Kamba people while Tharaka people are found in the North of Tana River [26]. Kitui Kamba also interact with Oromo and Somali ethnic groups during droughts when the latter two move seeking pastures [2]. The area experiences infrequent rain and lacks permanent waters except in the Athi and Tana rivers, hence water scarcity is a major problem during the dry season. The area also lacks fertile soils hence chronic droughts and famines are major adversities to the people of Kitui [2,26,28]. As a result, the inhabitants rely heavily on forest resources especially in wetter zones near hills [2]. According to the 2009 national census, the population of Kitui county was 1,012,709 with 531,427 females and 481,282 males with a population density of 44 persons per square kilometer [27].

Figure 1. Map of Kenya showing the location of Kitui county.

Kitui county is largely a low plateau rising from 300 m above the sea level through various inselbergs reaching to an altitude of about 1638 m above the sea level [28]. The highest altitudes reach about 1800 m above the sea level. The climate of Kitui county varies from arid to semi-arid with a minimum mean annual temperature varying from 14 to 22 °C and a maximum mean annual temperature ranging from 26 to 34 °C. There are two rainy seasons where the long rains start from March and end in June while the short rains fall from October to December with a mean annual rainfall of 250–1050 mm [27]. Low rainfall amounts are experienced in the extremely hot lowlands while higher rainfall amounts are experienced on the hilltops [27–29]. As a result, the highlands are wetter and highly populated while the dry lowlands are sparsely populated [28]. Considering the amount of rainfall received in the drylands [20] and their elevational ranges above the sea level [30], Kitui county is a typical dryland region.

The vegetation of Kitui county is characterized by low, stunted, dense thorn bushes with thick undergrowth and occasional baobab trees. Much of the area lacks vegetation except on the hills [29] where scrublands and wooded bushlands are found [31] with *Drypetes*, *Combretum*, *Vepris*, and *Croton* as the dominant species [32]. In the dry areas, the dominant vegetation is composed of *Acacia* and *Commiphora* bushlands and woodlands. The vegetation on humid and cooler hills varies and mostly include *Terminalia brownii* and *Acacia polyacantha* while exotic trees such as *Grevillea robusta*, *Cupressus* spp., *Eucalyptus* spp., and *Pinus* spp. are planted on some slopes and mountains [28]. The uncultivated and intact lands are composed of dry bushes [26]. There are several hilltops containing a high diversity of plant and animal species [32]. Such highlands reportedly provide a link between coastal forests and the Kenya highland forests, resulting in the presence of unique species adapted to each individual highland [18,32]. The plant diversity in Kitui county is high and is used for traditional foods, teas, medicines among other uses by the local communities [18,28].

1.2. Food Plants of Kitui County

The people of Kitui practice mixed farming which involves growing a variety of crops and keeping livestock [18,28] where cattle are kept as a security against famine [33]. In 2009, 80% of the county population was reported to rely on agriculture for economic income [27]. Kitui county is one of the regions which has a high diversity of local foods in Kenya including cultivated food crops. Some vegetable species used by the local communities include African nightshade, cowpeas (*Vigna unguiculata*), *Commelina forskaolii*, leaf amaranth, spider plant, *Cucumis dipsaceus*, and pumpkin leaves [15]. The main food crops cultivated include millet, sorghum (*Sorghum bicolor*), lablab beans (*Lablab purpureus*), pigeon peas (*Cajanus cajan*), cowpeas, maize (*Zea mays*), and green grams (*Vigna* spp.) while mangoes (*Mangifera indica*) are among the cultivated fruits. In some cases, global vegetables such as tomatoes and other green leafy vegetables are also cultivated in addition to the aforementioned traditional vegetable species. Wild fruits eaten include *Adansonia digitata* (baobab), *Grewia villosa*, *Vitex doniana* (Black plum), *Lannea alata*, *Uvaria schefleri*, *Berchemia discolor*, *Azanza garckeana*, *Tamarindus indica* (Indian date), *Vangueria madagascariensis* (Spanish tamarind), and *Cordia monoica* (Sandpaper saucer-berry) [2,15,18,28]. Goods sold in Kitui local markets during the colonial period include cultivated grain crops, sugarcane, and unspecified vegetables [34].

2. Results and Discussion

2.1. Diversity of Edible Plants

A total of 199 plant species in 52 families and 114 genera currently growing in Kitui county have been documented as wild food plants in different parts of Kenya (Table 1). Some of the common wild edible fruits reported during the field work are shown (Figure 2). Leguminosae is the best represented plant family (25 species in 13 genera) followed by Malvaceae (17 species in six genera). Previous studies have reported Leguminosae to be the largest plant family in the flora of various parts of Kitui county [25,32]. It is also the largest plant family in the flora of Kenya [17]. In addition, it has been recorded to comprise most of traditional food plants utilized elsewhere in Kenya [14]. Legumes are important food plants in poor rural African communities where they provide proteins, essential amino acids, macronutrients, minerals, and vitamins. African legumes are also tolerant to drought and are therefore strategic food sources in arid areas especially under the current climatic fluctuations. Despite this, African legumes are poorly studied, and some important economic species are still obtained from the wild [35].

Table 1. Number of genera and species of wild edible plants in Kitui county by family.

Family	Genus	Species	Family	Genus	Species
Leguminosae	13	25	Convolvulaceae	1	2
Malvaceae	6	17	Ebenaceae	2	2
Rubiaceae	6	11	Euphorbiaceae	2	2
Anacardiaceae	3	10	Poaceae	1	2
Cucurbitaceae	7	9	Portulacaceae	1	2
Lamiaceae	4	9	Salvadoraceae	2	2
Burseraceae	2	7	Bignoniaceae	1	1
Moraceae	2	7	Campanulaceae	1	1
Amaranthaceae	3	6	Cannabaceae	1	1
Capparaceae	3	6	Cleomaceae	1	1
Rhamnaceae	3	6	Clusiaceae	1	1
Apocynaceae	5	5	Combretaceae	1	1
Rutaceae	3	5	Menispermaceae	1	1
Verbenaceae	2	5	Nymphaeaceae	1	1
Zygophyllaceae	1	5	Olacaceae	1	1
Annonaceae	2	4	Opiliaceae	1	1
Phyllanthaceae	3	4	Pedaliaceae	1	1
Boraginaceae	1	3	Polygonaceae	1	1
Commelinaceae	1	3	Salicaceae	1	1
Compositae	3	3	Santalaceae	1	1
Cyperaceae	2	3	Sapotaceae	1	1
Loganiaceae	1	3	Solanaceae	1	1
Oleaceae	2	3	Xanthorrhoeaceae	1	1
Sapindaceae	3	3	Geraniaceae	1	1
Vitaceae	2	3	Putranjivaceae	1	1
Arecaceae	2	2	Talinaceae	1	1

Figure 2. Some of the wild edible fruits encountered during field study in Kitui county: (**A**) *Balanites aegyptiaca* (L.) Delile (Zygophyllaceae); (**B**) *Tamarindus indica* L. (Leguminosae); (**C**) *Berchemia discolor* (Klotzsch) Hemsl. (Rhamnaceae); (**D**) *Vangueria madagascariensis* J.F.Gmel. (Rubiaceae); (**E**) *Cordia sinensis* Lam. (Boraginaceae); (**F**) *Grewia villosa* Willd. (Malvaceae); (**G**) *Commiphora edulis* (Klotzsch) Engl. (Burseraceae); (**H**) *Lannea schweinfurthii* Engl. (Anacardiaceae); (**I**) *Cynanchum hastifolium* K.Schum. (Apocynaceae); (**J**) *Uvaria scheffleri* Diels (Annonaceae).

2.2. Growth Habit

Wild edible plant species growing as either shrubs or trees are the dominant life forms (77) recorded followed by herbs (32) and trees (21) (Figure 3). A recent ethnobotanical survey in Kitui county reported shrubs and trees to be the frequent medicinal plants reported by herbalists [25]. The vegetation of Kitui is also characterized by bushlands and woodlands composed of low, stunted thorn bushes and under-growths [28,29]. Such vegetation types are likely to be dominated by shrubs or trees.

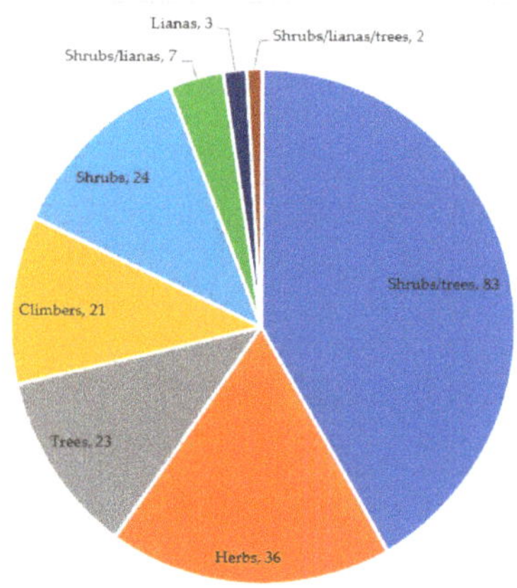

Figure 3. Growth habits of the wild edible plants in Kitui county.

2.3. Plant Parts Used

Fruits comprise the majority of plant parts utilized as food (124 reports), followed by leaves (56 reports) while roots and barks are also frequently reported. Other plant parts such as flowers and galls are sparingly reported. A single plant may have different parts collected for food; hence such species are represented by more than one report (Figure 4).

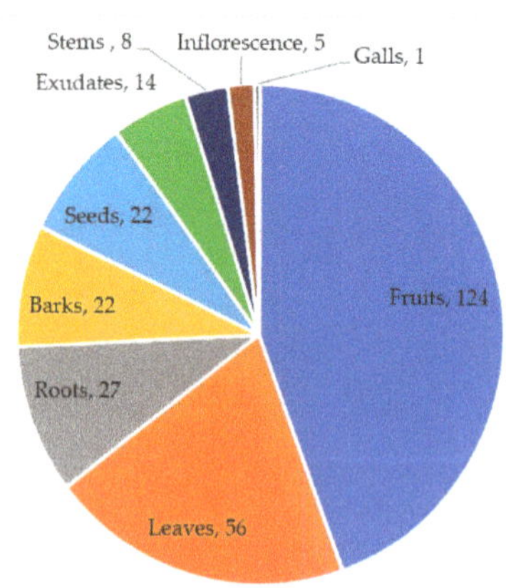

Figure 4. Number of species by parts of plants eaten (numbers represent reports per category).

2.4. Food Types Obtained from Wild Edible Plants

Foods obtained from wild edible plants reported in Kitui county fall into different categories, where the best represented food types are fruits (120 species), vegetables (44 species), and beverages (28 species) (Table 2).

Table 2. Classification of food types and the number of species in each type.

Food Types	Number of Species	Specific Food Type	Number of Species
Fruits	120	Eaten raw	119
		Cooked	1
Vegetables	44	Green vegetables	43
		Tuber vegetables	1
Beverages	28	Tisanes	22
		Beers	4
		Wines	1
		Coffee substitutes	1
Food additives	22	Flavoring agents	17
		Sweeteners	2
		Fermenting agents	2
		Water clarifiers	2
		Milk curdlers	1
		Meat tenderizers	2
Starch foods	21	Tubers	21
Seed foods	22	Other seeds	13
		Pulses	5
		Cereals	2
		Pseudo-cereals	2
Gums and resins	13	Eaten raw	13
Others	24	Leaves chewed raw	8
		Barks chewed raw	7
		Roots chewed raw	4
		Inflorescence eaten raw	5
		Edible cotyledon/embryo	3
		Internal juice of fruit drunk	1
		Galls	1
		Stem pith chewed raw	1

Fruits are reported to be eaten raw, cooked, or used in preparation of beverages such as wine and beer. They are also used as food additives such as flavoring agents in foods and soups or as fermenting agents in preparation of local brews. Fruits are among the frequently utilized wild edible plant parts in rural areas of Kenya [9,22]. Consumption of fresh fruits is beneficial to the body since they provide the body with resources such as mineral salts, vitamins A and C, carbohydrates, natural sugars, and water [12]. Some fruits are also consumed as snacks in some rural parts of Kenya [9] while in some regions, some wild fruits are considered to be of little nutritional value and therefore consumed as supplementary and emergency foods [9,28]. Utilization of wild fruits is at times constrained by some fruit plants being widely dispersed in their natural habitats making it difficult to gather enough while other plants produce small fruits which may also be unpalatable [9]. In Kitui county, some wild and cultivated fruits complement each other, where the ripening seasons alternate successively, maintaining a continued supply of fruits to the local communities. The dependence on wild fruits is reported to be higher in drier lowland areas of Kitui where cultivated fruit species are few [28]. Such areas also experience low amounts of rainfall [27–29], making wild fruits an important part of the local diet. In addition, fruits of *Adansonia adansonia*, *Vitex doniana*, *Azanza garckeana*, *Tamarindus indica*, and *Vangueria madagascariensis* are sold at the local markets of Kitui [18]. During the field study at Mutomo subcounty, fruits of *Berchemia discolor* were also reported to be collected and sold at Mutomo market.

Leaves are mostly utilized as green vegetables and as food additives in preparation of tisanes while in some cases, sour leaves are chewed raw. Germinating seeds are also eaten raw or cooked as vegetables [36]. Leafy vegetables are major contributors to local diets of rural populations and are also abundant in local markets. It is likely that they provide similar nutritional composition as cultivated vegetables such as vitamins and minerals and are also of medicinal value to the body [9,12,16]. In Kitui county, deficiencies of vitamin A and zinc are reportedly widespread [15] hence leafy vegetables can play an important role in the diet of the local inhabitants. An advantage of picking wild vegetables is that they provide an opportunity to pick a variety of different plant taxa which in turn offers a diversity in the dietary composition compared to cultivated green vegetables. A single diet of wild vegetables may comprise of different plant taxa thus ensuring maximum nutritional benefits to the body [21]. However, some vegetable plant species bear small leaves, while others are bitter. In addition, nutritional composition and palatability of vegetables vary with season [21]. Combination of such characters mean that wild vegetables require skills and time in their preparation which might result in their avoidance [16]. Similar to cultivated vegetables that are mixed together during times of scarcity, wild vegetables are also mixed to gather enough [9]. Wild vegetables are also mixed with cultivated vegetables to improve the taste [21]. Preparation of wild vegetables may involve boiling to wash them before cooking begins, probably as a way of dealing with bitterness in some vegetable species such as *Solanum americanum* which may contain toxic alkaloids. Bitter tasting or toxic populations of wild vegetables can also be avoided during the time of picking in the wild [9,21]. Leafy vegetables can be obtained from natural habitats such as forests and in disturbed places including farmlands [21]. Cowpeas are the main vegetables in Kitui county while other vegetables are underutilized [15] or used in the absence of cowpeas [28]. For example, *Commelina africana*, a wild vegetable which grows in the farm and in the wild sprouts earlier after the rains, providing an early vegetable before maturation of cowpeas [28]. Some species of wild vegetables such as *Amaranthus graecizans*, *Solanum americanum*, and *Cleome gynandra* occur naturally including in disturbed habitats although their occurrence depends on the right season which coincides with rains [9,21]. Other leafy vegetables such as *Oxygonium sinuatum*, *Commelina africana*, and *C. benghalensis* may occur as invasive weeds in cultivations [9]. Such adaptations to the local environments make indigenous vegetables suitable candidate species for combatting food insecurity by poor people living in dry areas of Kenya [16]. Some wild vegetable plants, which are utilized in other parts of Kenya grow naturally as weeds and may be underutilized by the local communities in Kitui county. Nutritional education and cooking demonstration of underutilized wild leafy vegetables was reported to result to an increase in their utilization in Kitui county [15]. According to Ichikawa et al. [37], major food plants might be shared between different communities while minor food plants may vary from one community to another. Local people are cautious with trying cultivation of vegetables they are not familiar with [15]. To enhance diversification of the ways of obtaining foods by the local communities, it is therefore important to create awareness on the utilization of local food plants not known or less prioritized by the local communities.

Exudates comprise of gums, resins, and wines tapped from plants. Gums and resins are produced by plants throughout the season including during the dry periods of the year. Although most of such exudates are collected from plants in arid and semi-arid areas that are of poor agricultural potential, only small quantities are meant for domestic consumption and much of the material is collected for sale [8]. However, gums and resins are still locally eaten during food scarcity and also have medicinal benefits [38]. Their collection is mostly carried out during the dry season by women and children in poor communities or by opportunists interested in income generation. Harvesting of gums and resins is a viable alternative for strengthening livelihood diversification in the drylands [38] especially during the dry periods of the year when other sources of livelihood such as dependence on livestock resources are constrained by insufficient pasture [8]. Despite the important role played by gums and resins in the lives of the local communities, collectors encounter various challenges such as poor harvesting methods, contamination of the collected materials, and improper post-harvest handling techniques resulting in overall reduction of the quality of the end products and hence low prices in the market.

These coupled with the poor markets where the collectors mostly sell to the local shops and further complicated by the presence of local agents and opportunist buyers makes income generation from gums and resins unsustainable. The potential of gums and resins in alleviating poverty in dry areas of Kenya are hence underutilized [8]. Development of better markets for gums and resins would be an important step towards maximizing the benefits of such products to the local communities in dry areas of Kenya including in Kitui county.

Barks are reported to be used in preparation of tisanes or as food additives such as flavorings. Raw roots are eaten as starch foods, cooked as vegetables, or used as food additives. Roots and tubers are important sources of energy since they are rich in starch. Freshly harvested roots also contain a large water content [12]. Roots and stems from some plant species are sweet and succulent hence they are chewed raw to quench thirst. In some instances, roots serve as immediate sources of food especially during grazing when the herders have little or no time to cook. Such methods of utilizing wild plants are important attributes for people to survive in dry areas [20]. Some root tubers are cooked to reduce the poisonous compounds that may be present while others are prepared through drying and pounding before consumption [9,20,36]. For example, the roots of *Thilachium africanum* are poisonous but edible when cooked [39]. Stems of some plant species such as *Albizia amara* are used as food additives, which is boiled in soup and also used as a meat tenderizer. Other plant parts reported are flowers which are eaten raw or picked with leafy parts and prepared together as vegetables, while the internal parts of galls are eaten raw. Wild edible seeds reported include pulses (seeds from legumes), cereals (seeds from grasses), pseudo-cereals (non-grass seeds that serve a similar purpose as cereals), and other seeds which are prepared through boiling, roasting, or eaten raw. Cereals and pseudo-cereals are ground into flour which is made into other dishes such as porridges. Legume seeds are important sources of proteins, iron, niacin, and vitamins hence are used as meat substitutes while other seeds are sources unsaturated fats, vitamins, and minerals such as phosphorus, calcium, and fluorine [12].

Beverages such as beer, wine, and tisanes are also prepared from wild plant parts. Tisanes include both infusions and decoctions taken as beverages, bitter teas, teas with essential oils, stimulant teas, and medicinal teas [40]. Utilization of herbal teas is dated back to the medieval medicine when they were used for therapeutic purposes [41]. Therapeutic classification of herbal teas in Kenya was also done by Ichikawa [14] who referred to them as narcotics and herbal medicines. However, Maundu et al. [36] treated some of them as foods. In this study, infusions and decoctions prepared and taken in place of caffeinated drinks and herbal additives added into caffeinated drinks are categorized as tisanes. They are prepared from the leaves and barks while in some instances, seeds are used. Tisanes are mostly prepared from plant families (Verbanaceae, Lamiaceae, Rutaceae, Burseraceae, and Anacardiaceae) comprising of aromatic, glandular, or resinous and oil producing species [39,42,43] although some species in Leguminosae, Sapindaceae, and Rhamnaceae families are also used. According to Maundu et al. [36], plants with essential oils such as *Ocimum* species are used for flavoring tea. These plants serve both as flavoring agents and as substitutes for caffeinated teas. For example, during a field survey at Mutomo subcounty in Kitui, leaves of *Zanthoxylum chalybeum* were reported by the local residents to add a good flavor to caffeinated teas while the bark of *Acacia nilotica* was said to be used as a substitute for caffeinated teas. Stem bark of *Acacia nilotica* is also reported to be a stimulant [44]. Further studies are needed to determine the role of the reported plant species in preparation of tisanes since such preparations may be categorized as flavorings, teas, coffee substitutes, and as herbal medicines. Some plant species are used in making beverages such as herbal beer and wine. These include the fruits of *Hyphaene compressa* which contain a liquid that is brewed into beer and the fruits of *Cordia sinensis* and *Balanites rotundifolia* which are used in preparation of local brews [36]. Palm exudate liquid, tapped from the vascular bundles of *Phoenix reclinata*, is also drunk as wine [45].

Food additives are also obtained from edible plants. These include spices, herbs, and seasonings which are of small nutritive value hence consumed in small amounts to stimulate appetite by enhancing flavor [12].

An inventory of the wild edible plants in Kitui county is provided in Table 3. Those plant species which at present are already used as wild food plants are marked with an asterisk (*) and a number sign (#).

Table 3. An inventory of wild edible plant species occurring in Kitui county. Information given under 'presence in Kitui county' refers to voucher specimens collected during field work by the authors (designated as SAJIT-Mutie MU), specimens at the East African (EA) herbarium or in publications citing the presence of the species in Kitui county. The plant use information refers to records of use of the plants for anywhere in Kenya, not necessarily in Kitui county, unless the name of the species is accompanied by an Asterix (*) (indicating the plant was cited during a field survey as edible) or by a number sign (#) indicating the plant use in Kitui county was obtained from literature.

Family and Plant Species	Kamba Name	Growth Habit	Presence in Kitui County	Part Used; Use in Kenya	Source of Information
Amaranthaceae					
# *Amaranthus dubius* Mart. ex Thell.	W'oa, telele, terere	Herb	Lind and Agnew 5642 (EA)	Leaves; eaten as a vegetable.	[15,36]
Amaranthus graecizans L.	W'oa, telele, terere	Herb	Gilbert 70 (EA)	Leaves; eaten as a vegetable.	[36]
Amaranthus hybridus L.	W'oa, telele, terere	Herb	Gilbert 98 (EA)	Leaves; eaten as a vegetable.	[21,36,42]
Amaranthus spinosus L.		Herb	Gilbert 97 (EA)	Leaves; eaten as a vegetable.	[36]
Digera muricata (L.) Mart.	Walange	Herb	Someren 2024 (EA)	Leaves, flowers; leaves eaten as a vegetable. Flower nectar is edible.	[36]
* *Aerva lanata* (L.) Juss.	-	Herb	SAJIT-Mutie MU0268 (EA)	Leaves; eaten as a vegetable.	[36]
Anacardiaceae					
# *Lannea alata* (Engl.) Engl.	Kitungu, ndungu, mukolya	Shrub or tree	Kuchar 14876 (EA)	Fruits, bark; ripe fruits edible. Bark used in tea.	[18,36,39]
* *Lannea schweinfurthii* Engl.	Muasi, kyuasi	Shrub or tree	SAJIT-Mutie MU0271 (EA)	Fruits, bark; ripe fruits edible. Bark used for making tea.	[18,39]
# *Lannea triphylla* (Hochst. ex A. Rich.) Engl.	Muthaalwa, kithaalwa, kithaala, nzaala	Shrub or tree	[36]	Fruits, bark, root; ripe fruits are edible. Sweet succulent roots and bark chewed raw to quench thirst. Bark used in tea.	[18–20,36,39]
# *Searsia natalensis* (Bernh. ex C.Krauss) F.A.Barkley	Kitheu, mutheu	Shrub or tree	Evans 167 (EA)	Fruits, bark, roots, leaves; ripe fruits are edible. Bark used in tea. Roots boiled in soup. Young shoots chewed raw.	[18,36,39]
* *Lannea rivae* Sacleux	Kithaalwa, muthaalwa, kithaala, kithaalua kya kiima	Shrub or tree	SAJIT-Mutie MU0305 (EA, HIB)	Fruits, bark; ripe fruits are edible. Bark is sweet and chewed raw.	[36,39]
Lannea schimperi (Hochst. ex A.Rich.) Engl.	kithoona, kithauna, nthoona	Shrub or tree	[36]	Fruits, bark; ripe fruits are edible. Bark used in tea.	[36]
Searsia tenuinervis (Engl.) Moffett	Kitheu	Shrub or tree	[36]	Fruits, leaves; ripe fruits are edible. Young shoots and leaves chewed raw.	[36,39]
Searsia quartiniana (A.Rich.) A.J.Mill.	Mutheu	Shrub or tree	[36]	Ripe fruits; edible.	[18,36]
Searsia pyroides (Burch.) Moffett	Kitheu, mutheu, mutheu munene	Shrub or tree	[36]	Ripe fruits; edible.	[36,39]
# *Sclerocarya birrea* (A.Rich.) Hochst.	Muua, muuw'a, mauw'a	Tree	Bogdan AB4379 (EA)	Fruits, seeds; ripe fruits are edible. Internal seed contents eaten raw.	[18,36,39,46]

Table 3. *Cont.*

Family and Plant Species	Kamba Name	Growth Habit	Presence in Kitui County	Part Used; Use in Kenya	Source of Information
Annonaceae					
#* *Uvaria scheffleri* Diels	Mukukuma	Shrub or liana	SAJIT-Mutie MU0290 (EA)	Ripe fruits; edible.	[18,19,36,39]
# *Annona senegalensis* Pers.	Makulo, mutomoko 'wild custard apple, wild soursop'	Shrub or tree	[36]	Bark, fruits; ripe fruits are edible. Bark chewed raw.	[39,47]
# *Uvaria acuminata* Oliv.	-	Shrub or liana	Mwachala et al., 476 (EA)	Ripe fruits; edible.	[36,39]
Uvaria lucida Bojer ex Benth.	-	Shrub or liana	Mbonge 14 (EA)	Ripe fruits; edible.	[36]
Apocynaceae					
* *Saba comorensis* (Bojer ex A.DC.) Pichon	Kilia, kiongwa, kyongoa, mongoa	Liana	SAJIT-Mutie MU0278 (EA)	Ripe fruits; edible.	[22,36,39]
* *Cynanchum hastifolium* K.Schum.	-	Climber	SAJIT-Mutie MU0261 (EA)	Unripe fruits; edible.	[19]
# *Carissa spinarum* L.	Mukawa, mutote, ngawa, ndote, nzunu	Shrub	[48]	Flowers, fruits, roots; flowers and ripe fruits are edible. Boiled roots eaten as vegetables and used as a flavor in soup.	[18–20,36,39,47]
Acokanthera schimperi (A.DC.) Schweinf.	Kivai	Shrub or tree	[36]	Ripe fruits; edible.	[18,36,39]
* *Pentarrhinum insipidum* E.Mey.	-	Climber	SAJIT-Mutie MU0139 (EA)	Fruits, leaves; leaves eaten as a vegetable. Ripe fruits are edible.	[20,36]
Arecaceae					
Phoenix reclinata Jacq.	Makindu 'wild date palm'	Tree	[36,48]	Fruits, stem; ripe fruits are edible. Wine is tapped from stem.	[36,39]
# *Hyphaene compressa* H.Wendl.	Mukoma, ilala	Tree	[36]	Fruits, leaves; seedling embryo is edible. Fruit pulp eaten raw. Juice from immature fruits drunk fresh or used to make beer.	[36,39,47]
Bignoniaceae					
# *Kigelia africana* (Lam.) Benth.	Kiatine, muatine 'sausage tree'	Shrub or tree	[48]	Fruits; used for fermenting traditional beer.	[18,39]
# *Cordia monoica* Roxb.	Muthii, kithei, nthei	Shrub or tree	[48]	Ripe fruits; edible.	[18,19,36,39,47]
#* *Cordia sinensis* Lam.	Muthea, kithea, muthei-munini, kithia	Shrub or tree	SAJIT-Mutie MU0292 (EA)	Exudate, roots, fruits; roots eaten raw. Ripe fruits are edible. Fruit pulp used for brewing local beer. Produces an edible gum.	[18,19,36,39]
Cordia crenata Delile	-	Shrub or tree	Kirika et al., GBK2/10/2005 (EA)	Ripe fruits; edible.	[39]
Burseraceae					
#* *Commiphora baluensis* Engl.	Itula, mutula	Tree	SAJIT-Mutie MU0254 (EA)	Bark; used in making tea.	[18,19,39]
* *Commiphora edulis* (Klotzsch) Engl.	Kyoa kika	Shrub or tree	SAJIT-Mutie MU0193 (EA)	Fruits; seed aril from ripe fruit is edible.	[19]
# *Boswellia neglecta* S.Moore	Kinondo	Shrub or tree	Bally B1612 (EA)	Exudate, bark; resin from bark chewed. Bark used for making tea.	[18,36,39,49]

Table 3. *Cont.*

Family and Plant Species	Kamba Name	Growth Habit	Presence in Kitui County	Part Used; Use in Kenya	Source of Information
Commiphora campestris Engl	-	Tree	[48]	Exudate; produces an edible resin.	[49]
* *Commiphora africana* (A.Rich.) Endl.	Kitungu, mutungu, itula	Shrub or tree	Kuchar 15067 (EA)	Exudate, roots, bark; produces an edible gum. Roots of young plants chewed raw to quench thirst. Bark used for making tea.	[18,36,39]
Commiphora rostrata Engl.	Inywamanzi	Shrub or tree	[48]	Bark, leaves, stem; leaves chewed raw or cooked to add flavor in foods. Bark used for tea. Stem pith and bark of young plants chewed raw to quench thirst.	[18,20,36,39]
Commiphora schimperi (O.Bergman) Engl.	Mutungu	Shrub or tree	[36,48]	Exudate, roots, bark; produces an edible resin. Roots chewed to quench thirst. Inner red bark boiled in tea.	[19,36,39]
Campanulaceae					
# *Cyphia glandulifera* Hochst. ex A.Rich.	Ngomo	Herb	[36]	Roots, leaves; leaves eaten as a vegetable. Tubers eaten raw.	[14,19]
Cannabaceae					
Trema orientalis (L.) Blume	-	Shrub or tree	[18]	Ripe fruits; edible.	[39]
Capparaceae					
# *Boscia coriacea* Graells	Isivu	Shrub or tree	SAJIT-Mutie MU0122 (EA)	Fruits, seeds; fruits are edible. Seeds edible when boiled.	[18,36,39,47]
# *Maerua decumbens* (Brongn.) DeWolf	Kinatha, munatha	Herb or shrub	Kuchar 15244 (EA)	Roots, fruits, seeds; ripe fruits eaten raw or cooked. Seeds edible when boiled. Root bark chewed raw. Roots added to water as a sweetener.	[18,36,39,47]
# *Maerua denhardtiorum* Gilg	Itembokambola	Shrub	Kuchar 14991 (EA)	Ripe fruits; edible.	[18,19,39,47]
Maerua kirkii F. White	Ivovotwe	Shrub or tree	Kimani 86 (EA)	Nuts; boiled and eaten.	[19]
Thilachium africanum Lour.	Mutunguu	Shrub or tree	Greenway 9228 (EA)	Roots; cooked and eaten.	[39]
Thilachium thomasii Gilg	Kitungulu	Shrub or tree	Spjut and Muchai 4655 (EA)	Roots, fruits; ripe fruits are edible. Tubers eaten or cooked and the resultant liquid drunk or used for making tea. Peeled roots used as flocculants in water.	[36,39]
Cleomaceae					
# *Cleome gynandra* L.	Mwianzo, mukakai, sake, mwaanzo, ithea-utuku	Herb	Hucks 341 (EA)	Leaves; eaten as a vegetable.	[19,21,36,42]
Clusiaceae					
# *Garcinia livingstonei* T.Anderson	Mukanga, kikangakanywa, ngangakanywa	Tree	Adamson B6084 (EA)	Ripe fruits; edible.	[19,36,39,47]
Combretaceae					
# *Terminalia brownii* Fresen.	Muuku, kiuku	Shrub or tree	Gillett 19774 (EA)	Fruits; eaten by children.	[18,39,47]

Table 3. *Cont.*

Family and Plant Species	Kamba Name	Growth Habit	Presence in Kitui County	Part Used; Use in Kenya	Source of Information
Commelinaceae					
# *Commelina africana* L.	Kikowe	Herb	[36]	Leaves; eaten as a vegetable.	[20,28,36]
Commelina benghalensis L.	Itula	Herb	[36]	Leaves; eaten as a vegetable.	[36]
# *Commelina forskaolii* Vahl	Kikowe, kikoe	Herb	[36]	Leaves; eaten as a vegetable.	[15,36]
Compositae					
* *Launaea cornuta* (Hochst. ex Oliv. and Hiern) C.Jeffrey	Uthunga, muthunga	Herb	SAJIT-Mutie MU0209 (EA)	Leaves; eaten as a vegetable.	[22,36]
Cyanthillium cinereum (L.) H.Rob.	–	Herb	Kuchar 15163 (EA)	Leaves; eaten as a vegetable.	[36]
Galinsoga parviflora Cav.	–	Herb	Sheldrick TNP/E/109 (EA)	Leaves; eaten as a vegetable.	[36]
Convolvulaceae					
Ipomoea lapathifolia Hallier f.	Nzola, kinzola	Herb	Ossent 441A (EA)	Roots; tubers eaten raw.	[36]
Ipomoea mombassana Vatke	Ukwai wa nthi, wimbia, musele, uthui	Climber	Napper 1591 (EA)	Leaves; eaten as a vegetable.	[36]
Cucurbitaceae					
# *Momordica spinosa* Chiov.	–	Liana	Kuchar 14829 (EA)	Ripe fruits; edible.	[39,47]
* *Coccinia grandis* (L.) Voigt	Kimuya, kimowe, imore, imondiu	Climber	SAJIT-Mutie MU0176 (HIB)	Leaves, fruits; leaves eaten as a vegetable. Ripe fruits eaten raw or ground into flour and used to make porridge.	[36]
#* *Cucumis dipsaceus* Ehrenb. ex Spach	Kikungi, kyambatwa	Climber	SAJIT-Mutie MU0145 (HIB)	Leaves; eaten as a vegetable.	[36]
* *Kedrostis pseudogijef* C. Jeffrey	Mukauw'u	Climber	SAJIT-Mutie MU0212 (EA)	Leaves, fruits; leaves eaten as a vegetable. Ripe fruits are edible.	[36]
Kedrostis gijef C. Jeffrey	Witulu	Climber	Kuchar 1503 (EA)	Leaves, fruits; leaves eaten as a vegetable. Ripe fruits are edible.	[36]
* *Momordica rostrata* A. Zimm.	Kiongoa, kyongoa	Climber	SAJIT-Mutie MU0149 (HIB)	Leaves, seeds, fruits; ripe fruits are edible. Roasted seeds are edible. Leaves eaten as a vegetable.	[36]
# *Lagenaria siceraria* (Molina) Standl.	Ungu, kikuu, yungu	Climber	[36]	Leaves, fruits, seeds; young fruits edible when cooked. Seeds roasted and eaten. Leaves eaten as a vegetable.	[28,36]
# *Citrullus lanatus* (Thunb.) Matsum. and Nakai	Itikitiki	Climber	[28]	Fruits, leaves, seeds; ripe fruits are edible. Dry seeds ground into flour, mixed with sorghum flour and used to make porridge. Leaves eaten as a vegetable.	[28,36]
Peponium vogelii	–	Climber	Kimani 80 (EA)	Ripe fruits; edible.	[19]
Cyperaceae					
Cyperus blysmoides Hochst. ex C.B.Clarke	–	Herb	Edwards 23 (EA)	Roots; bulbs and stem bases eaten raw.	[20,28]
Cyperus rotundus L.	–	Herb	Porter 51 (EA)	Roots; stem bases are edible.	[36]
Kyllinga alba	–	Herb	Kuchar 8848 (EA)	Root bulbs; edible.	[19]

Table 3. *Cont.*

Family and Plant Species	Kamba Name	Growth Habit	Presence in Kitui County	Part Used; Use in Kenya	Source of Information
Ebenaceae					
#* *Diospyros mespiliformis* Hochst. ex A.DC.	Mukongoo 'African ebony'	Tree	SAJIT-Mutie MU0179 (HIB)	Ripe fruits; edible.	[18,36,39,47]
Euclea divinorum Hiern	Mukinyai, mukinyai, mukuthi, nginyai	Shrub or tree	Kuchar 15097 (EA)	Fruits, bark; ripe fruits are edible. Bark added to soup as an appetizer.	[18,36,39]
Euphorbiaceae					
#* *Croton dichogamus* Pax	Mwalula, muthiani	Shrub or tree	SAJIT-Mutie MU0245 (EA)	Bark; used as a flavor in soup.	[18,20,39]
Euphorbia schefleri Pax	–	Shrub or tree	SAJIT-Mutie MU0188 (HIB)	Stems; smoke from wood used as a meat tenderizer.	[39]
Geraniaceae					
Pelargonium quinquelobatum Hochst. ex Rich.	–	Herb	Muasya 2459 (EA)	Stems; eaten raw.	[19]
Lamiaceae					
#* *Vitex payos* (Lour.) Merr.	Kimuu, muu	Shrub or tree	SAJIT-Mutie MU0286 (EA, HIB)	Ripe fruits; edible.	[36,39,47]
* *Vitex strickeri* Vatke and Hildebr.	Mwalika	Shrub or liana	SAJIT-Mutie MU0264 (EA)	Ripe fruits; edible.	[39]
* *Hoslundia opposita* Vahl	Musovi, musovasovi	Shrub	SAJIT-Mutie MU0244 (EA)	Fruits, leaves, stems; ripe fruits are edible. Leaves and stems used in tea.	[19,36,39]
Ocimum basilicum L.	Mutaa	Herb	[48]	Leaves; used for flavoring tea.	[18,36]
Ocimum kilimandscharicum Gürke	Wenye	Herb or shrub	Brilloe B303 (EA)	Leaves; used for flavoring tea.	[36,39]
Ocimum gratissimum L.	Mukandu	Shrub	Mbonge 6 (EA)	Leaves; used for flavoring tea.	[36,39]
* *Premna oligotricha* Baker	Mukaakaa	Shrub	SAJIT-Mutie MU0183 (EA)	Ripe fruits; edible.	[19]
# *Premna resinosa* (Hochst.) Schauer		Shrub	Kirika et al., NMK455 (EA)	Ripe fruits; edible.	[19,47]
# *Vitex doniana* Sweet	Kimuu 'Black plum, vitex'	Tree	[18]	Ripe fruits; edible.	[18,36,39]
Leguminosae					
#* *Acacia nilotica* (L.) Delile	Musemei, musemeli	Tree	SAJIT-Mutie MU0224 (EA)	Bark, fruit; bark and pods boiled with sugar to make tea. Pods eaten during famine.	[18,36,39]
Acacia reficiens Wawra	–	Shrub or tree	Ament and Magogo 418 (EA)	Sweet inner bark; chewed raw.	[19]
#* *Acacia senegal* (L.) Willd.	King'olola	Shrub or tree	SAJIT-Mutie MU0122 (HIB)	Exudate; produces an edible gum.	[18,19,36,49]
# *Acacia seyal* Delile	Kinyua, kisewa	Shrub or tree	Robertson 4288 (EA)	Exudate, bark; produces an edible gum. Bark chewed raw or ground into powder to make tea.	[18,36,39,49]
Acacia gerrardii Benth.	Munina, kithi, muthii	Shrub or tree	[18]	Bark; used to make soup.	[18,39]
Acacia drepanolobium Sjostedt	Kiunga, muuga	Shrub or tree	[36]	Galls, fruits; inner flesh of the galls is edible. Young fruits are edible.	[36,39]
Acacia hockii De Wild.	Muuga, kinyua 'white thorn'	Shrub or tree	Gardner 1088 (EA)	Exudate, bark; produces an edible gum. Inner bark chewed raw to quench thirst.	[36,39]

Table 3. *Cont.*

Family and Plant Species	Kamba Name	Growth Habit	Presence in Kitui County	Part Used; Use in Kenya	Source of Information
* *Acacia tortilis* (Forssk.) Hayne	Mwaa, kilaa, mulaa, muaa, ulaa	Tree	Sangai 935 (EA)	Exudate, fruits; produces an edible gum. Ripe pods eaten or ground into flour which is mixed with tea or blood.	[18,36,39]
Albizia amara (Roxb.) B.Boivin	Mwowa, muundua, kiundua, muundua	Tree	[36,48]	Exudate, stems; produces an edible gum. Dried stems used as an additive in food or soup and as a meat tenderizer.	[36,39]
Bauhinia thonningii Schum.	Mukolokolo	Shrub or tree	[36,48]	Fruits, leaves; dry fruit pulp is edible. Young sour shoots used in porridge or chewed raw.	[18,36]
Crotalaria brevidens var. *parviflora* (Baker f.) Polhill	Kamusuusuu	Herb	[50]	Leaves; eaten as a vegetable.	[36]
Eriosema shirense Baker f.	Ng'athu	Herb	[36]	Roots tubers; edible.	[36]
Craibia laurentii De Wild.	–	Tree	Mwachala et al., 487 (EA)	Seeds; beans eaten after boiling for several hours.	[19]
# *Vigna membranacea* A.Rich.	Ithookwe	Climber	Gillett 19475 (EA)	Roots, leaves; leaves eaten as a vegetable. Roots eaten raw or roasted.	[19,20,36]
Vigna frutescens A.Rich.	–	Climber	Bally B1536 (EA)	Root tubers; eaten raw or roasted.	[19,20,36]
Vigna praecox Verdc.	–	Climber	SAJIT-Mutie MU0115 (HIB)	Roots; boiled or roasted and eaten.	[19]
#* *Tamarindus indica* L.	Kithumula, muthumula, kikwasu, nthumula, nzumula, ngwasu	Tree	SAJIT-Mutie MU0208 (EA)	Fruits, leaves, seeds; fruit pulp eaten raw or used as a flavor in porridge or beer. Young leaves chewed raw or cooked as a vegetable. Seeds fried and eaten.	[18,22,36,39]
Tylosema fassoglensis (Schweinf.) Torre and Hillc.	Ivole	Climber	Hucks and Hucks 217 (EA)	Seeds, pods; seeds eaten raw, roasted or used as a coffee substitute. Unripe pods eaten raw.	[36]
# *Vatovaea pseudolablab* (Harms) J.B.Gillett	Kilukyo	Shrub or liana	[36]	Roots, leaves, flowers, pods, seeds; tubers cooked or roasted for food or eaten raw to quench thirst. Seeds eaten raw or cooked. Roots ground into flour and used for making porridge. Immature leaves, flowers and pods cooked as vegetables.	[19,20,36,39]
# *Cajanus cajan* (L.) Millsp.	Nzuu	Shrub	[36]	Seeds; cooked and eaten.	[22,36,51]
Lablab purpureus (L.) Sweet	Mbumbu, ngiima, nzavi	Climber	[36]	Seeds, leaves; beans cooked and eaten. Leaves eaten as a vegetable.	[19,36]
* *Vigna vexillata* (L.) A.Rich.	–	Climber	SAJIT-Mutie MU0257 (EA, HIB)	Roots; chewed raw to quench thirst.	[20]
# *Vigna unguiculata* (L.) Walp.	Nzooko, nthooko	Climber	[36]	Leaves; eaten as a vegetable.	[22,28]

Table 3. *Cont.*

Family and Plant Species	Kamba Name	Growth Habit	Presence in Kitui County	Part Used; Use in Kenya	Source of Information
Ormocarpum kirkii S.Moore	Muthingii	Shrub or tree	[48]	Leaves; eaten as a vegetable.	[18,22,39]
Albizia anthelmintica Brongn.	Mwowa, kyalundathi, kyowa kisamba	Shrub or tree	SAJIT-Mutie MU0194 (EA)	Leaves; eaten as a vegetable.	[22,39]
Loganiaceae					
#* *Strychnos decussata* (Pappe) Gilg	Mutolongwe	Shrub or tree	SAJIT-Mutie MU0109 (EA)	Ripe fruits; edible.	[39,47]
* *Strychnos henningsii* Gilg	Muteta	Shrub or tree	SAJIT-Mutie MU0200 (EA)	Roots, stems, bark, fruits; roots, stems and bark added to soup as a flavor. Fruits used for flavoring beer.	[18,36,39]
* *Strychnos spinosa* Lam.	Kyae, kimee, mae	Shrub or tree	SAJIT-Mutie MU0162 (EA)	Ripe fruits; edible.	[36,39]
Malvaceae					
#* *Azanza garckeana* (F.Hoffm.) Exell and Hillc.	Kitotoo, Mutoo	Tree	SAJIT-Mutie MU0289 (EA, HIB)	Ripe fruits; edible.	[18,36,39]
#* *Grewia tephrodermis* K.Schum.	Mulawa, kikalwa, ngalwa, ilawa	Shrub or tree	SAJIT-Mutie MU0220 (EA)	Ripe fruits; edible.	[18,19,36,39,47]
#* *Grewia villosa* Willd.	Muvu	Shrub	SAJIT-Mutie MU0206 (EA)	Ripe fruits; edible.	[18,19,36,39,47]
# *Grewia mollis* Juss.	–	Shrub or tree	Thomas 671 (EA)	Ripe fruits; edible.	[39,47]
* *Grewia arborea* (Forssk.) Lam.	Nguni	Shrub or tree	SAJIT-Mutie MU0321 (EA)	Ripe fruits; edible.	[39]
* *Grewia forbesii* Harv. ex Mast.	Mutalenda	Shrub, liana, tree	SAJIT-Mutie MU0270 (HIB)	Ripe fruits; edible.	[36,39]
Grewia lilacina K.Schum.	–	Shrub	Kirika et al., NMK462 (EA)	Fruits; edible.	[19]
Grewia similis K.Schum.	Mutuva	Shrub or liana	Edwards 681 (EA)	Ripe fruits; edible.	[19,36,39]
* *Grewia tembensis* Fresen.	Mutuva, nduva	Shrub	SAJIT-Mutie MU0242 (EA)	Ripe fruits; edible.	[19,36,39]
# *Grewia tenax* (Forssk.) Fiori	–	Shrub	Kirika et al., NMK457 (EA)	Ripe fruits; edible.	[19,36,39,47]
Grewia trichocarpa Hochst. ex A.Rich.	–	Shrub or tree	Lind and Agnew 5656 (EA)	Ripe fruits; edible.	[19,39]
Hibiscus greenwayi Baker f.	–	Shrub	[52]	Leaves, stems; young leaves eaten raw. Sweet stems chewed raw.	[19]
# *Adansonia digitata* L.	Kiamba, muamba	Tree	Bally 11691 (EA)	Roots, leaves, seeds; Root tips eaten during famine. Roots of germinating seeds are edible. Young leaves eaten as a vegetable. Roasted seeds are edible. Seed pulp eaten raw or boiled and the juice used as a sauce or added to porridge.	[18,20,36,39,47]
# *Corchorus olitorius* L.	–	Herb	[15]	Leaves; eaten as a vegetable.	[15,22]
* *Corchorus trilocularis* L.	–	Herb	SAJIT-Mutie MU0134 (EA)	Leaves; eaten as a vegetable.	[36]
* *Corchorus tridens* L.	–	Herb	SAJIT-Mutie MU0133 (HIB)	Leaves; eaten as a vegetable.	[22]
Sterculia stenocarpa H.J.P.Winkl.	–	Shrub or tree	Joana 7411 (EA)	Fruits; edible.	[19]

Table 3. *Cont.*

Family and Plant Species	Kamba Name	Growth Habit	Presence in Kitui County	Part Used; Use in Kenya	Source of Information
Moraceae					
* *Dorstenia hildebrandtii* var. *schlechteri* (Engl.) Hijman	–	Herb	SAJIT-Mutie MU0281 (EA, HIB)	Roots; eaten raw.	[19]
Ficus capreifolia Delile	–	Shrub or tree	Adamson 19716 (EA)	Ripe fruits; edible.	[39]
* *Ficus glumosa* Delile	Kionywe	Shrub or tree	SAJIT-Mutie MU0259 (EA, HIB)	Ripe fruits; edible.	[19,39]
Ficus populifolia Vahl	–	Shrub or tree	Gillett 18574 (EA)	Ripe fruits; edible.	[39]
#* *Ficus sycomorus* L.	Mukuyu	Tree	SAJIT-Mutie MU0202 (EA)	Ripe fruits; figs eaten or dried and made into flour which is mixed with maize flour for making porridge.	[18,19,36,47]
Ficus sur Forssk	–	Tree	[48]	Ripe fruits; edible.	[19,39,48]
Ficus vasta Forssk.	Mumbu, mukuyu	Tree	[48]	Ripe fruits; edible.	[39,48]
Menispermaceae					
Chasmanthera dependens Hochst.	Uswe	Liana	SAJIT-Mutie MU0039 (EA)	Roots, stems; roots boiled in milk as a drink for a child. Stems are edible.	[19,20,39]
Nymphaeaceae					
* *Nymphaea nouchali* var. *caerulea* (Savigny) Verdc.	–	Herb	SAJIT-Mutie MU0186 (EA)	Roots, flowers, fruits, seeds; edible.	[20,36]
Olacaceae					
# *Ximenia americana* L.	Kitula, mutula	Shrub or tree	[36]	Fruits, bark; ripe fruits are edible. Root bark used for tea.	[18,19,36,39,47]
Jasminum abyssinicum Hochst. ex DC.	Mukaksu	Climber	SAJIT-Mutie MU0154 (HIB)	Roots; roots boiled in broth or soup.	[20,39]
Olea europaea L.	Muthata, molialundi	Shrub or tree	[18]	Ripe fruits; edible.	[18,39]
Olea capensis L.	'Elgon Olive, East African Olive'	Tree	[18]	Ripe fruits; edible.	[39]
Opiliaceae					
# *Opilia campestris* Engl.	Kiburuburu, mubrubru	Shrub	[18]	Ripe fruits; edible.	[18,19,39,47]
Pedaliaceae					
* *Sesamum calycinum* Welw.	Luta	Herb	SAJIT-Mutie MU0081 (EA)	Leaves; eaten as a vegetable.	[36]
Phyllanthaceae					
Antidesma venosum E.Mey. ex Tul.	Mukala, kitelanthia, kitolanthia	Shrub or tree	[36]	Ripe fruits; edible.	[36,39]
Bridelia scleroneura Müll.Arg.	–	Shrub or tree	Bally 1567 [42]	Ripe fruits; edible.	[39]
#* *Bridelia taitensis* Vatke and Pax ex Pax	Yathia, muandi, mwaanzia	Shrub or tree	SAJIT-Mutie MU0039 (EA)	Ripe fruits; edible.	[18,36,39,47]
# *Flueggea virosa* (Roxb. ex Willd.) Royle	Mukuluu, mukururu	Shrub	[48]	Ripe fruits; edible.	[18,19,36,39]
Poaceae					
Dactyloctenium aegyptium (L.) Willd.	Ukuku	Herb	[36]	Roots, seeds; rhizomes chewed raw. Grains ground into flour or chewed raw.	[36]
Dactyloctenium giganteum B.S.Fisher and Schweick.	Ukuku	Herb	[36]	Seeds; grains ground into flour for making porridge.	[36]
Polygonaceae					
Oxygonum sinuatum (Hochst. and Steud ex Meisn.) Dammer	Song'e	Herb	Bally 13179 (EA)	Leaves; eaten as a vegetable or chewed raw.	[19,36]

Table 3. *Cont.*

Family and Plant Species	Kamba Name	Growth Habit	Presence in Kitui County	Part Used; Use in Kenya	Source of Information
Portulacaceae					
Portulaca oleracea L.	Kamama, kamumama, kinyukwi	Herb	[36]	Leaves, seeds; leaves and slender stems eaten raw or cooked as a vegetable. Seeds ground into flour for making porridge.	[36]
* *Portulaca quadrifida* L.	Kenyinyia, kamumama	Herb	SAJIT-Mutie MU0121 (EA)	Leaves, seeds; leaves and slender stems eaten raw or cooked as a vegetable. Seeds ground into flour for making porridge.	[36]
Putranjivaceae					
Drypetes gerrardii Hutch.	-	Tree	Burry 4 (EA)	Fruits; eaten raw.	[19]
Rhamnaceae					
* *Berchemia discolor* (Klotzsch) Hemsl.	Kisanawa, kisaaya, nzaaya, nzanawa	Shrub or tree	SAJIT-Mutie MU0293 (EA)	Fruits, exudate; ripe fruits are edible. Produces an edible gum.	[18,19,36,39]
# *Ziziphus mucronata* Willd.	Kitola usuu, kitolousuu, muae	Shrub or tree	[36,48]	Bark, fruits; ripe fruits are edible. Bark used in tea.	[18,19,36,39]
# *Scutia myrtina* (Burm.f.) Kurz	Mtanda mboo, kitumbuu, mbombo	Shrub or tree	[36]	Roots, fruits; ripe fruits are edible. Roots used in soup.	[18,19,36,39,47]
Ziziphus abyssinica Hochst. ex A.Rich.	Muae, kitolousuu	Shrub, liana or tree	[53]	Ripe fruits; edible.	[19,39]
Ziziphus pubescens Oliv.	-	Shrub or tree	[42]	Ripe fruits; edible.	[36,39]
Ziziphus jujuba Mill.	-	Shrub or tree	[36]	Ripe fruits; edible and made into flour.	[22,36,39]
Rubiaceae					
Canthium glaucum Hiern	-	Shrub or tree	[36]	Ripe fruits; edible.	[36]
Pavetta gardeniifolia Hochst. ex A.Rich.	-	Shrub	[54]	Fruits; edible.	[19]
#* *Rothmannia urcelliformis* (Hiern) Bullock ex Robyns	Mutendeluka	Shrub or tree	SAJIT-Mutie MU0164 (HIB)	Ripe fruits; edible.	[18]
#* *Vangueria madagascariensis* J.F.Gmel.	Kikomoa, mukomoa	Shrub or tree	SAJIT-Mutie MU0280 (EA, HIB)	Ripe fruits; edible and used for flavoring beer.	[18,19,36]
Rothmannia fischeri (K.Schum.) Bullock ex Oberm.	Muendeluka	Shrub or tree	Owino and Mathenge 214 (EA)	Ripe fruits; edible.	[39]
#* *Tennantia sennii* (Chiov.) Verdc. and Bridson	Kisilingu	Shrub	SAJIT-Mutie MU0207 (EA)	Ripe fruits; edible.	[39,47]
# *Vangueria infausta* Burch.	Kikomoa, mukomoa, muteleli	Shrub or tree	Joana B1142 (EA)	Ripe fruits; edible.	[36,39,47]
Vangueria volkensii K.Schum.	Kikomoa, mukomoa	Shrub or tree	Gibbons OX635 (EA)	Ripe fruits; edible.	[36,39]
#* *Vangueria schumanniana* (Robyns) Lantz	Mukomole, kitootoo, ngomole, ndootoo	Shrub	Napper 1590 (EA)	Fruits, stems; ripe fruits are edible. Stems smoked and inserted into gourds of milk to induce good flavor in milk.	[36,39,47]
Vangueria apiculata K.Schum.	Kikomoa, mukomoa	Shrub or tree	[52]	Fruits; edible and used for flavoring beer.	[36,39]
* *Meyna tetraphylla* (Schweinf. ex Hiern) Robyns	Kitotoo, kitootoo, kakomoa, kitolousuu	Shrub or tree	Bally 1636 [42]	Ripe fruits; edible.	[36,39]

Table 3. *Cont.*

Family and Plant Species	Kamba Name	Growth Habit	Presence in Kitui County	Part Used; Use in Kenya	Source of Information
Rutaceae					
#* *Zanthoxylum chalybeum* Engl.	Mukenea, mukanu	Shrub or tree	SAJIT-Mutie MU0317 (EA)	Bark, leaves, fruit; bark and fruits used as food spices. Leaves and fruits used in flavoring tea. Bark used in making or flavoring tea.	[18,36,39]
Vepris glomerata Engl.	–	Shrub or tree	Trapnell 2406; Thomas 673 [42]	Ripe fruits; edible.	[39]
* *Vepris simplicifolia* (Engl.) Mziray	Mutuyu	Shrub or tree	SAJIT-Mutie MU0234 (EA, HIB)	Fruits; edible.	[19]
Harrisonia abyssinica Oliv.	Mkiliulu	Shrub or tree	Mutie MU0185 (HIB)	Ripe fruits; edible.	[18,39]
* *Zanthoxylum holtzianum* (Engl.) P.G. Waterman	–	Shrub or tree	SAJIT-Mutie MU0180 (HIB)	Ripe fruits; edible.	[22]
Salicaceae					
Flacourtia indica (Burm.f.) Merr.	Kiathani, kikathani	Shrub or tree	Festo and Luke 2291 (EA)	Ripe fruits; edible.	[36,39]
Salvadoraceae					
# *Dobera glabra* (Forssk.) Juss. ex Poir.	Kisiu, kithio, kikaitha	Shrub or tree	Edwards EAH 12315 [42]	Exudate, fruits, seeds; produces an edible gum. Ripe fruits are edible. Boiled seeds are edible.	[18,36,39,47]
# *Salvadora persica* L.	Mukayau	Shrub or tree	Pearce 405 (EA)	Ripe fruits; edible.	[18,19,36,39,47]
Santalaceae					
Osyris lanceolata Hochst. and Steud.	Kithawa	Shrub or tree	Birch 59/13 (EA)	Ripe fruits; edible.	[39]
Sapindaceae					
Allophylus africanus P.Beauv.	–	Shrub or tree	SAJIT-Mutie MU0165 (HIB)	Fruits; edible.	[19]
# *Pappea capensis* Eckl. and Zeyh.	Kyuua, kiva, mba	Shrub or tree	[18,36]	Fruits, bark; ripe and unripe fruits are edible. Dry inner bark used for tea.	[18,36,39,47]
Haplocoelum foliolosum (Hiern) Bullock	Mukumu, mukumi	Shrub or tree	[54]	Ripe fruits; edible.	[18,39]
Sapotaceae					
# *Manilkara mochisia* (Baker) Dubard	Kinako, kisaa	Tree	[36]	Ripe fruits; edible.	[18,36,47]
Solanaceae					
# *Solanum americanum* Mill.	Kitulu	Herb	[36]	Leaves; eaten as a vegetable.	[21,22,36]
Talinaceae					
* *Talinum portulacifolium* (Forssk.) Asch. ex Schweinf.	–	Herb	SAJIT-Mutie MU0014 (EA)	Leaves; eaten raw.	[19]
Verbenaceae					
Lantana camara L.	Kitavisi, mukiti, musomolo	Shrub	[36]	Ripe fruits; edible.	[36,39]
Lantana humuliformis Verdc.	–	Shrub	Kuchar 14908 (EA)	Ripe fruits; edible.	[36]
Lantana ukambensis (Vatke) Verdc.	–	Herb	Napier 1567 (EA)	Ripe fruits; edible.	[36]
Lippia javanica (Burm.f.) Spreng.	Muthiethi	Shrub	[18]	Fruits, leaves; ripe fruits are edible. Leaves used for tea.	[18,36]
Lippia kituiensis Vatke	Muthiethi, muthiiti, muthyeti	Shrub	[36]	Fruits, leaves; ripe fruits edible. Leaves used for tea.	[36,39]

Table 3. *Cont.*

Family and Plant Species	Kamba Name	Growth Habit	Presence in Kitui County	Part Used; Use in Kenya	Source of Information
Vitaceae					
Cissus aphyllantha Gilg	Mwelengwa	Shrub or liana	SAJIT-Mutie MU0247 (EA)	Ripe fruits; edible.	[39]
Cissus rotundifolia Vahl	Itulu	Shrub	SAJIT-Mutie MU0128 (EA)	Ripe fruits; edible.	[39]
#*Cyphostemma adenocaule* (Steud. ex A.Rich.) Desc. ex Wild and R.B.Drumm.	-	Climber	SAJIT-Mutie MU0143 (EA)	Leaves; eaten as a vegetable.	[22]
Xanthorrhoeaceae					
* *Aloe secundiflora* Engl.	Kiluma	Herb	SAJIT-Mutie MU0191 (EA)	Roots, flowers, peduncle; roots used to ferment traditional beer. Flower nectar is edible. Sweet base of inflorescence is chewed raw.	[36]
Zygophyllaceae					
#* *Balanites aegyptiaca* (L.) Delile	Kilului, kiluluwi, mulului	Tree	SAJIT-Mutie MU0196 (EA)	Exudate, fruits, leaves, seeds; produces an edible gum. Ripe fruits are edible. Leaves and tender shoots eaten as a vegetable. Inner part of a seed is edible when boiled.	[18,36,39,47]
Balanites glabra Mildbr. and Schltr.	Kilului	Shrub or tree	[48]	Ripe fruits; edible.	[39]
Balanites pedicellaris Mildbr. and Schltr.	-	Shrub or tree	[36]	Seeds, fruits; ripe fruits are edible. Inner part of the seed cooked and eaten.	[36,39]
Balanites rotundifolia (Tiegh.) Blatt.	Kilului	Shrub or tree	[36]	Fruit, seeds; fruit pulp is edible and used to make local brew. Inner part of seed is edible when cooked.	[36,39]
Balanites wilsoniama Dawe and Sprague	Kivuw'a	Tree	[32]	Ripe fruits; edible.	[36]

2.5. Potential of Crop Wild Relatives (CWR) in Kitui County

Crop wild relatives (CWR) form an important part of gene pool for the improvement of cultivated crops [55]. The genetic relationship between many of the tropical CWR and the cultivated crops is unknown [56]. In Kenya for example, wild sorghum populations are reportedly widespread in various habitats such as in protected areas, roadsides, and farmlands. Such resources are regarded as weeds in farmers' fields and are facing the risk of genetic contamination through pollen-mediated crop-wild introgression [57]. The negligence of CWR and land races from the notion that they will remain to be readily available in the wild is causing their degradation [58]. Some of the wild plants utilized as wild foods in Kitui county that have cultivated relatives in the area include *Amaranthus* species such as *A. dubius* [15]. The Amaranthaceae family also exhibits the highest diversity of species used as traditional vegetables in Kenya [51], hence such group of plants form an important gene pool for future improvement of cultivated members. The leaves of *Vigna membranacea* (traditional vegetable) are reported to taste similar as cultivated *V. unguiculata*, a species composed of various subspecies and several cultivars in Kenya [36]. *Vigna unguiculata* is also the second most popular grain legume in Kenya after beans, and it is estimated that 85% of the area under its cultivation in Kenya lies in arid and semi-arid areas [51]. In Tharaka for example, an arid area adjacent to Kitui, cowpeas are cultivated by about 80% of the households [59]. Pigeon pea (*Cajanus cajan*) is also an important crop in dry areas although its diversity is limited to only one species [51,59]. *Cajanus cajan* is regarded as an indigenous plant in Kenya [43], hence wild forms might form an important resource base for improvement of cultivated members especially in dry areas such as Kitui county. Other important cultivated plant species with wild forms in Kitui include *Lagenaria siceraria* and *Citrullus lanatus* [28,36,51]. *Solanum americanum* is also a vegetable species growing in the wild and cultivated in Kitui county [15]. Some of its wild forms are bitter tasting and hence avoided during vegetable collection [9,21]. Such forms might be neglected leading to their possible disappearance in the wild. Although the socio-economic importance of CWR is well known, their conservation has not been systematically addressed and their current extinction levels might result in serious social and economic problems if threats facing them are not adequately addressed [58]. In Kenya, the decline of plant genetic resources is at its peak following the effects of global warming, increased population, and desertification [51]. Conservation efforts of such critically important group of plants is therefore vital if they are to be relied upon in the future [60]. Since it is evident that drylands of Kenya harbor wild plants with a potential to combat food insecurity as exemplified by Kitui county, collection of CWR and other food plants' germplasm and its conservation are important steps towards ensuring maximum benefits from such resources.

2.6. Conservation of Natural Habitats in Kitui County

Availability of wild food plants depends on the ecology of a given area and the history of its deforestation [9]. In Kenya, there is an ongoing loss of wild food species and the traditional knowledge associated with them especially in areas of high agricultural potential, where much of the original vegetation has been cleared for agriculture and infrastructure [10]. Domestication of some wild vegetables is however reportedly ongoing in some regions where vegetable plants such as *Cleome gynandra*, once introduced continue to self-reseed in subsequent years [9] hence becoming a long-term source of leafy vegetables. Other vegetables species under domestication in Kenya include *Amaranthus* spp., *Solanum americamum*, *Basella alba*, and *Sesamum angustifolium* which may also be spared in the farmland during cultivation of weeds [10]. Many of the food plants occur in natural forests while some are preserved by the local inhabitants in their farmlands. In Kitui county, fruit plants such as *Tamarindus indica* and *Balanites aegyptiaca* are preserved in farmlands for their medicinal uses [25]. Wild fruit trees are also left standing when other plants are being cleared for farmlands or charcoal. Wild food plants of Kitui county are threatened by the local communities who cut them for charcoal, thus also leading to loss of indigenous knowledge associated with them [28]. According to Mutie et al. [25], some medicinal plants in Kitui county such as *Strychnos henningsii* and *Vepris simplicifolia* which are also reported as food plants are decreasing in the wild as a result of human activities. Wild food plants are most important to the communities who reside in dry areas, which are more vulnerable to droughts [9]. Such areas are mostly inhabited by pastoral groups whose major threat to plant diversity is overstocking [10]. Diversity of wild edible plants is also reported to be richer in savanna zones compared to other forests zones [14]. It is also in the drier regions where the vegetation has been conserved to the greatest extent in some regions of Kenya [9]. The hills of Kitui are perceived by the local people to harbor important medicinal and food plant species [18,23]. In addition, high plant diversity and species endemism are reported in the hills of Kitui [32]. Such hills are vital ecosystems for adaptation towards the changing climatic conditions through provision of important ecosystems goods such as wild foods [18]. Mutomo hill plant sanctuary, one of the hills in Kitui county has been recently reported as a potentially important area for conservation of medicinal plants [25]. Conservation of important plants including wild food plants in other hills of Kitui county needs assessment and prioritization through community awareness so as to ease pressure exerted on wild plant populations by the local communities.

3. Materials and Methods

A botanical inventory of vascular plants of Kitui county was first compiled from data collected during three different botanical surveys in various parts of Kitui county between May 2018 and February 2019 by the Sino-Africa Joint Investigation Team (SAJIT). These include an ethnobotanical survey of medicinal plants carried out in Mutomo subcounty [25], which included citations of wild edible plants by the respondents. Further floral surveys were carried out in Endau hills, Mutitu hills, and Mui basin where the local people cited the wild plants used as food whenever they encountered them. Where possible, specimens at the EA herbarium in Kenya were checked to obtain plant species previously collected from Kitui county. Voucher specimens reported in this study have been deposited at Hubei Institute of Botany (HIB) herbarium in China and at the EA.

The data was supplemented by other data obtained from various literature such as published articles, conference proceedings, botanical survey reports, and the monographs of the Flora of Tropical East Africa reporting on plants of Kitui county (voucher materials are represented by specimen numbers seen in literature or by references citing the presence of the reported plant species in Kitui county). This yielded a plant checklist totaling to 931 vascular plant species, the most comprehensive checklist of the region to date (unpublished results). To obtain an inventory of wild edible plants of Kitui county, literature reporting on wild edible plants of Kenya was searched from various sources and combined with the use reports obtained from field surveys. Data were searched using key words 'plants, flora, edible plants, wild fruits, fruits, livelihood diversification in drylands, vegetable plants, nuts and seeds, useful plants, edible tubers, wild teas.' The key words were combined with 'Kitui' and 'Kenya', each at a time in order to determine the area of data collection. To exhaust the information gathered, if a plant species was found to be edible in Kenya and not yet recorded in Kitui county, another search category was initiated ('Kitui county' plus 'scientific name the plant'). The indigenous plant species and their growth habits were determined based on the local monographs of Kenyan flora [39,43] and the monographs of the Flora of Tropical East Africa [42]. All plant name synonymies were resolved using The Plant List database (http://www.theplantlist.org/). The plant species were then classified into parts utilized for food and into different food categories according to Cook [61]. The data were entered and analyzed in Microsoft Excel 2016.

4. Conclusions

Although further circumspection is needed before the potential adoption of these plants for food, this study nevertheless presents wild plants as important sources of food for the local communities living in dry areas of Kenya such as Kitui county. Investigation of herbarium materials and further botanical surveys are still necessary to determine the undetected food plants. The local communities have incorporated some conservation measures in their farmlands. There is still a need to sensitize them further on the need of preserving natural habitats and involve them in collecting of germplasm of edible wild plants and their relatives for ex-situ conservation. Disturbed places play an important role in human nutrition through provision of green vegetables, some of which may occur as invasive weeds. Prioritizing on proper harvesting, storage, and marketing of wild foods produced seasonally in large quantities might be an important step in maximizing the nutritional benefits of dryland communities. Lastly, understanding the distribution of ethnobotanical knowledge among individuals and the role of age, gender, and the level of education are important factors in conservation of wild edible plants in dryland areas of Kenya in general.

Author Contributions: Investigation and writing: F.M.M.; Investigation: V.K. and P.C.R.; Conceptualization and supervision: G.-W.H. and Q.-F.W. All authors have read and agreed to the published version of the manuscript.

Funding: This research project was supported by the National Natural Science Foundation of China, 31970211 and the Sino-Africa Joint Research Centre, CAS (SAJC201614).

Acknowledgments: The authors thank Kitui county government and the Kenya Forest Service for issuing the permits and to the National Museums of Kenya for allowing them to access the East African (EA) herbarium.

Conflicts of Interest: The authors declare no conflict of interest.

References

1. Kigomo, N.B. State of Forest genetic Resources in Kenya. In *Forest Genetic Resources Working Papers, Proceedings of the Sub-Regional Workshop FAO/IPGRI/ICRAF on the Conservation, Management, Sustainable Utilisation and Enhancement of Forest Genetic Resources in Sahelian and North-Sudanian Africa*; FAO: Rome, Italy, 2001; p. 27.
2. Eriksen, S.; Lind, J. Adaptation as a political process: Adjusting to drought and conflict in Kenya's drylands. *Environ. Manag.* **2009**, *43*, 817–835. [CrossRef] [PubMed]
3. Boko, M.; Niang, I.; Nyong, A.; Vogel, C.; Githeko, A.; Medany, M.; Osman-Elasha, B.; Tabo, R.; Yanda, P. Africa. In *Climate Change 2007: Impacts, Adaptation and Vulnerability, Contribution of Working Group II to the Fourth Assessment Report of the Intergovernmental Panel on Climate Change*; Parry, M.L., Canziani, O.F., Palutikof, J.P., van der Linden, P.J., Hanson, C.E., Eds.; Cambridge University Press: Cambridge, UK, 2007; pp. 433–467.
4. Hulme, M.; Doherty, R.; Ngara, T.; New, M.; Lister, D. African climate change: 1900–2100. *Clim. Res.* **2001**, *17*, 145–168. [CrossRef]
5. Owuor, B.; Eriksen, S.; Mauta, W. Adapting to climate change in a dryland mountain environment in Kenya. *Mt. Res. Dev.* **2005**, *25*, 310–316. [CrossRef]
6. Scoones, I. *Hazards and Opportunities: Farming Livelihoods in Dryland Africa. Lessons from Zimbabwe*; Zed Books: London, UK, 1996.
7. Gachathi, F.N.; Eriksen, S. Gums and resins: The potential for supporting sustainable adaptation in Kenya's drylands. *Clim. Dev.* **2011**, *3*, 59–70. [CrossRef]
8. Vellema, W.; Mujawamariya, G.; D'Haese, M. Gum arabic collection in northern Kenya: Unexploited resources, underdeveloped markets. *Afr. Focus* **2014**, *27*, 69–86. [CrossRef]
9. Johns, T.; Kokwaro, J.O. Food plants of the Luo of Siaya District, Kenya. *Econ. Bot.* **1991**, *45*, 103–113. [CrossRef]
10. Maundu, P. Utilization and conservation status of wild food plants in Kenya. In *Biodiversity of African Plants*; Springer: Dordrecht, The Netherlands, 1996; pp. 678–683. [CrossRef]
11. Kinyuru, J.N.; Konyole, S.O.; Kenji, G.M.; Onyango, C.A.; Owino, V.O.; Owuor, B.O.; Estambale, B.B.; Friis, H.; Roos, N. Identification of traditional foods with public health potential for complementary feeding in Western Kenya. *J. Food Res.* **2012**, *1*, 148–158. [CrossRef]
12. Nyambo, A.; Nyomora, A.; Ruffo, C.; Tengnas, B. *Fruits and Nuts: Species with Potential for Tanzania*; World Agroforestry Centre, Eastern and Central Africa Regional Programme: Nairobi, Kenya, 2005.
13. Mbora, A.; Jamnadass, R.; Lillesø, J.B. *Growing High Priority Fruits and Nuts in Kenya: Uses and Management*; World Agroforestry Centre: Nairobi, Kenya, 2008.
14. Ichikawa, M. The Utilization of Wild Food Plants by the Suiei Dorobo in Northern Kenya. *J. Anthropol. Soc. Nippon* **1980**, *88*, 25–48. [CrossRef]
15. Kariuki, L.; Maundu, P.; Morimoto, Y. *Some intervention strategies for promoting underutilized species: Case of local vegetables in Kitui District, Kenya. II International Symposium on Underutilized Plant Species: Crops for the Future-Beyond Food Security*; International Society for Horticultural Science: Kuala Lumpur, Malaysia, 2011. [CrossRef]
16. Okeno, J.A.; Chebet, D.K.; Mathenge, P.W. Status of indigenous vegetable utilization in Kenya. In *XXVI International Horticultural Congress: Horticultural Science in Emerging Economies, Issues and Constraints*; ISHS: Leuven, Belgium, 2002; Volume 621, pp. 95–100. [CrossRef]
17. Zhou, Y.; Liu, B.; Mbuni, Y.; Yan, X.; Mwachala, G. Vascular flora of Kenya based on the Flora of Tropical East Africa. *PhytoKeys* **2017**, *126*, 113–126. [CrossRef]
18. Gachathi, F.N. A vital habitat to local livelihood and drought coping strategies in the drylands: The case of Endau Hilltop forest, Kitui District, Kenya. In *Proceedings of the Two Workshops*; Eriksen, S., Owuor, B., Nyukuri, E., Eds.; World Agroforestry Centre: Nairobi, Kenya; KEFRI Research Centre: Kitui, Kenya, 2005; pp. 81–102.
19. Ichikawa, M. A Preliminary Report on the Ethnobotany of the Suiei Dorobo in Northern Kenya. *Afr. Study Monogr.* **1987**, *7*, 1–52. [CrossRef]
20. Kabuye, C.H. Edible roots from wild plants in arid and semi-arid Kenya. *J. Arid Environ.* **1986**, *11*, 65–74. [CrossRef]

21. Marshall, F. Agriculture and use of wild and weedy greens by the Piik AP Oom Okiek of Kenya. *Econ. Bot.* **2001**, *55*, 32–46. [CrossRef]
22. Rajat, J.; Jefwa, J.; Mwafaida, J. Survey on Indigenous Food Plants of Kaya Kauma and Kaya Tsolokero in Kilifi County Kenya. *J. Life Sci.* **2017**, *11*, 82–90. [CrossRef]
23. Njoroge, G.N.; Kaibui, I.M.; Njenga, P.K.; Odhiambo, P.O. Utilization of priority traditional medicinal plants and local people's knowledge on their conservation status in arid lands of Kenya (Mwingi District). *J. Ethnobiol. Ethnomedicine* **2010**, *6*, 1–8. [CrossRef] [PubMed]
24. Kisangau, D.P.; Kauti, M.; Mwobobia, R.; Kanui, T.; Musimba, N. Traditional Knowledge on Use of Medicinal Plants in Kitui County, Kenya. *Int. J. Ethnobiol. Ethnomedicine* **2017**, *4*, 1–10.
25. Mutie, F.M.; Gao, L.-L.; Kathambi, V.; Rono, P.C.; Musili, P.M.; Ngugi, G.; Hu, G.-W.; Wang, Q.-F. An Ethnobotanical Survey of a Dryland Botanical Garden and Its Environs in Kenya: The Mutomo Hill Plant Sanctuary. *Evid. Based Complement. Altern. Med.* **2020**, *2020*, 1–22. [CrossRef]
26. Dundas, C. History of Kitui. *J. R. Anthropol. Inst. G. B. Irel.* **1913**, *43*, 480–549. [CrossRef]
27. County, G.K. *County Integrated Development Plan 2013–2017: Planning for Sustainable Socio-Economic Growth and Development*; County Government of Kitui: Kitui, Kenya, 2014; p. 600.
28. Bélair, C.; Ichikawa, K.; Wong, B.Y.L.; Mulongoy, K.J. (Eds.) *Sustainable Use of Biological Diversity in Socio-Ecological Production Landscapes, Background to the 'Satoyama Initiative for the Benefit of Biodiversity and Human Well-Being'*; Technical Series; Secretariat of the Convention on Biological Diversity: Montreal, QC, Canada, 2010; pp. 67–72.
29. Saggerson, E.P. *Geology of the South Kitui Area: Degree Sheet 53, SW Quarter, with Colored Geological Map*; Geological Survey of Kenya: Nairobi, Kenya, 1957; Volume 37.
30. Grace, O.M.; Malombe, I.; Davis, S.D.; Pearce, T.; Simmonds, M. Identifying priority species in dryland Kenya for conservation in the Millennium Seed Bank Project. In *Proceedings of the Systematics and Conservation of African Plants*; Royal Botanic Gardens: London, UK, 2010; pp. 749–758.
31. Lind, E.M.; Morrison, M.E.; Hamilton, A. *East African Vegetation*; Longman Group Limited: London, UK, 1974.
32. Malonza, P.K.; Muasya, A.M.; Lange, C.; Webala, P.; Mulwa, R.K.; Wasonga, D.V.; Mwachala, G.; Malombe, I.; Muasya, J.; Kirika, P.; et al. *Biodiversity Assessment in Dryland Hilltops of Kitui and Mwingi Districts*; National Museums of Kenya: Nairobi, Kenya, 2006.
33. Tignor, R.L. Kamba political protest: The destocking controversy of 1938. *Afr. Hist. Stud.* **1971**, *4*, 237–251. [CrossRef]
34. Stanner, W. The Kitui Kamba Market, 1938–1939. *Ethnology* **1969**, *8*, 125–138. [CrossRef]
35. Duodu, K.G.; Apea-Bah, F.B. African legumes: Nutritional and health-promoting attributes. In *Gluten-Free Ancient Grains*; Woodhead Publishing: Cambridge, UK, 2017; pp. 223–269.
36. Maundu, M.; Ngugi, W.; Kabuye, H. *Traditional Food Plants of Kenya*; National Museums of Kenya: Nairobi, Kenya, 1999.
37. Ichikawa, M.; Kimura, D.; Terashima, H. AFlora: A database of traditional plant use in tropical Africa. *Syst. Geogr. Plants* **2001**, *71*, 759–764. [CrossRef]
38. Chretin, M.; Chikamai, B.; Loktari, P.E.; Ngichili, J.; Loupa, N.; Odee, D.; Lesueur, D. The current situation and prospects for gum Arabic in Kenya: A promising sector for pastoralists living in arid lands. *Int. For. Rev.* **2008**, *10*, 14–22. [CrossRef]
39. Beentje, H.; Adamson, J.; Bhanderi, D. *Kenya Trees, Shrubs, and Lianas*; National Museums of Kenya: Nairobi, Kenya, 1994.
40. Van Wyk, B.-E.; Gorelik, B. The history and ethnobotany of Cape herbal teas. *S. Afr. J. Bot.* **2017**, *170*, 13–38. [CrossRef]
41. Gill, M. Speciality and herbal teas. In *Tea*; Springer: Dordrecht, The Netherlands, 1992; pp. 513–534. [CrossRef]
42. Milne-Redhead, E.; Turrill, W.B. (Eds.) *Flora of Tropical East Africa*; Crown Agents: London, UK, 1974; pp. 1952–2012.
43. Agnew, A.D.Q. *Upland Kenya Wild Flowers and Ferns: A Flora of the Flowers, Ferns, Grasses, and Sedges of Highland Kenya*; Nature Kenya-The East Africa Natural History Society: Nairobi, Kenya, 2013.
44. Muthee, J.K.; Gakuya, D.W.; Mbaria, J.M.; Kareru, P.G.; Mulei, C.M.; Njonge, F.K. Ethnobotanical study of anthelmintic and other medicinal plants traditionally used in Loitoktok District of Kenya. *J. Ethnopharmacol.* **2011**, *135*, 15–21. [CrossRef] [PubMed]

45. Kinnaird, F. Competition for a forest palm: Use of *Phoenix reclinata* by human and nonhuman primates. *Conserv. Biol.* **1992**, *6*, 101–107. [CrossRef]
46. Muok, B.O.; Owuor, B.; Dawson, I.; Were, J. The potential of indigenous fruit trees: Results of a survey in Kitui District, Kenya. *Agrofor. Today* **2000**, *12*, 13–16.
47. Simitu, P.J. Consumption and Conservation of Drylands' Indigenous Fruit Trees for Rural Livelihood Improvement in Mwingi District, Kenya. Master's Thesis, Kenyatta University, Kenya Drive, Nairobi, Kenya, 2011.
48. Hayashi, I.; Gachathi, F. A check list of the plant species at the site of the Kenya and Japan Social Forestry Training Project, Kitui, Kenya. *Veg. Sci.* **1998**, *15*, 71–77. [CrossRef]
49. Chikamai, B.; Gachathi, F.N. Gum and Resin Resources in Isiolo District, Kenya: Ethnobotanical and Reconnaissance Survey. *East Afr. Agric. For. J.* **1994**, *59*, 345–351. [CrossRef]
50. Polhill, R.M. Miscellaneous notes on African species of Crotalaria, L.: II. *Kew Bull.* **1968**, *22*, 169–348. [CrossRef]
51. Wambugu, P.W.; Muthamia, Z.K. Country report on the state of plant genetic resources for food and agriculture: Kenya. In *The Second Report on the State of the World's Plant Genetic Resources for Food and Agriculture*; FAO: Rome, Italy, 2009; p. 26.
52. Bally, P.R.O. The Mutomo hill plant sanctuary, Kenya. *Biol. Conserv.* **1968**, *1*, 90–91. [CrossRef]
53. Verdcourt, B. An enumeration of the Rhamnaceae in the East African Herbarium with the description of a new species of Lasiodiscus. *Bull. du Jard. Bot. l'État a Bruxelles* **1957**, *27*, 351–363. [CrossRef]
54. Mistry, J.; Berardi, A. *Savannas and Dry Forests: Linking People with Nature*; Ashgate Publishing, Ltd.: Farnham, UK, 2006.
55. Tanksley, S.D.; McCouch, S.R. Seed banks and molecular maps: Unlocking genetic potential from the wild. *Science* **1997**, *277*, 1063–1066. [CrossRef] [PubMed]
56. Maxted, N.; Ford-Lloyd, B.V.; Jury, S.; Kell, S.; Scholten, M. Towards a definition of a crop wild relative. *Biodivers. Conserv.* **2006**, *15*, 2673–2685. [CrossRef]
57. Okeno, J.A.; Mutegi, E.; de Villiers, S.; Wolt, J.D.; Misra, M.K. Morphological Variation in the Wild-Weedy Complex of Sorghum bicolor In Situ in Western Kenya: Preliminary Evidence of Crop-to-Wild Gene Flow? *Int. J. Plant Sci.* **2012**, *173*, 507–515. [CrossRef]
58. Maxted, N.; Brehm, J.M.; Kell, S. *Resource Book for the Preparation of National Plans for Conservation of Crop Wild Relatives and Landraces*; FAO: Rome, Italy, 2013; p. 463.
59. Labeyrie, V.; Rono, B.; Leclerc, C. How social organization shapes crop diversity: An ecological anthropology approach among Tharaka farmers of Mount Kenya. *Agric. Hum. Values* **2013**, *31*, 97–107. [CrossRef]
60. Maxted, N.; Scholten, M.; Codd, R.; Ford-Lloyd, B. Creation and use of a national inventory of crop wild relatives. *Biol. Conserv.* **2007**, *140*, 142–159. [CrossRef]
61. Cook, F.E.M. *Economic Botany Data Collection Standard*; Royal Botanic Gardens: London, UK, 1995; p. 146.

plants

MDPI

Review

Main Challenges and Actions Needed to Improve Conservation and Sustainable Use of Our Crop Wild Relatives

Johannes M. M. Engels [1,*] and Imke Thormann [2]

1 Alliance of Bioversity International and CIAT, Maccarese, 00054 Rome, Italy
2 Federal Office for Agriculture and Food, 53179 Bonn, Germany; imke.thormann@gmail.com
* Correspondence: j.engels@cgiar.org

Received: 5 July 2020; Accepted: 27 July 2020; Published: 30 July 2020

Abstract: Crop wild relatives (CWR, plural CWRs) are those wild species that are regarded as the ancestors of our cultivated crops. It was only at the end of the last century that they were accorded a high priority for their conservation and, thus, for many genebanks, they are a new and somewhat unknown set of plant genetic resources for food and agriculture. After defining and characterizing CWR and their general threat status, providing an assessment of biological peculiarities of CWR with respect to conservation management, illustrating the need for prioritization and addressing the importance of data and information, we made a detailed assessment of specific aspects of CWRs of direct relevance for their conservation and use. This assessment was complemented by an overview of the current status of CWRs conservation and use, including facts and figures on the in situ conservation, on the ex situ conservation in genebanks and botanic gardens, as well as of the advantages of a combination of in situ and ex situ conservation, the so-called complementary conservation approach. In addition, a brief assessment of the situation with respect to the use of CWRs was made. From these assessments we derived the needs for action in order to achieve a more effective and efficient conservation and use, specifically with respect to the documentation of CWRs, their in situ and ex situ, as well as their complementarity conservation, and how synergies between these components can be obtained. The review was concluded with suggestions on how use can be strengthened, as well as the conservation system at large at the local, national, and regional/international level. Finally, based on the foregoing assessments, a number of recommendations were elaborated on how CWRs can be better conserved and used in order to exploit their potential benefits more effectively.

Keywords: crop wild relatives; biological features; conservation; use; local; national and global efforts; policy; genetic diversity; gene donors; pre-breeding; breeding; cross-sectoral collaboration

1. Introduction

Today's cultivated crop plants have undergone more or less drastic changes since their first cultivation and domestication. The first signs of domesticating wild plant (and animal) species date back 10,500 years in Western Asia and domestication has since then been practiced in different parts of the world by different groups of people on new species [1]. The duration and intensity of this domestication process have been very variable from one crop to the other [2]. The one thing that all crops have in common is that they originated from (one or more) wild and naturally occurring species. For a number of crops, the domestication process is well known, based on archaeological finds and (experimental) research. In general, this process started with gathering in particular wild grasses and leguminous species, followed by their cultivation closer to the homestead and gradually undergoing transformation from wild into domesticated taxa [3–5]. In some instances, crops are the result of natural or man-made hybrids between two wild ancestor species (e.g., banana: *Musa acuminata* and

M. balbisiana); in other cases, the wild relative is a subspecies of the cultivated crop (e.g., *Vitis vinifera*) or there is no difference between the wild and the domesticated species (e.g., the olive tree, *Olea europea* which has wild, weedy, and cultivated forms, and many forage crops), which are just two different forms of the same species. For other crops, the domestication process is much less known or even completely obscure, including which wild species might have been involved as ancestor(s) of the crop in question (e.g., *Triticum spelta*, spelt). For some crops, the domestication process is still ongoing, especially in local fruit trees [6]. Possibly the most important consequence of the domestication process is that the genetic diversity available in the crop genepool (in the narrow sense) is usually much smaller than that in the related wild species [7,8]. In this paper, we focus on the wild species that are related to our crops, i.e., the crop wild relatives (CWRs). They have in different ways contributed (genetically) to the domestication process and thus can be regarded as the ancestral species or progenitors of our present crops, and they are a valuable resource of genetic diversity and traits for plant breeding.

It has taken several years after the global initiation of systematic collecting and conserving threatened landraces of our crops, somewhere in the 1960/70's, until CWRs were systematically included, both at the national and international level. In 1975, a global collecting program of threatened landraces and CWRs was initiated under the coordination of the International Board for Plant Genetic Resources (IBPGR) and approximately 220,000 samples were collected during more than 1000 collecting missions in more than 130 countries, largely before 1995. The collected materials were sent to and subsequently stored in selected national and regional/international genebanks around the world [9,10]. The inclusion of CWRs in collecting efforts was triggered by the observed genetic erosion, as well as by the apparent need to include more genetic diversity for the advancement of breeding programs of major crops (e.g., potato), triggered by the success of using CWRs in breeding programs, such as the tomato, for specific traits [11]. Due to breeding programs in need for more diversity, the first 'push' for CWR conservation came from the international CGIAR research centers, as well as some (international) breeding companies in the 1970/80's [12].

Only during the past few decades, significant successes of transferring traits from CWRs into cultivated crops have been reported, mostly to overcome biotic stresses, such as pests and diseases, as well as abiotic stresses, such as drought tolerance [8,13]. More recently, adaptability to changing environmental conditions, in particular those caused by climate change, has also become important. Only gradually, CWRs became a priority for the more advanced national plant genetic resources centers for food and agriculture (PGRFA), such as in the USA, UK, Germany, The Netherlands, and Australia. Possibly the biggest 'push' for the conservation of CWRs was the advancement of molecular biology and genetic tools and techniques that greatly facilitate the transfer of traits, genes, and alleles from one species to another, almost independent of how closely they are related to each other.

The above-mentioned developments certainly had an important impact on the increasing (political) conservation priorities accorded to CWRs since the late 1980's/early 1990's. This has been reflected by the inclusion of CWRs in the text of the Convention on Biological Diversity (CBD) [14] and, in 2010, in the AICHI Biodiversity Targets, in particular Target 13, as well as in target 9, of its Global Strategy for Plant Conservation, where CWRs and wild food plants were accorded a high priority for conservation [15]. In almost half of the 18 priority activities of the Second Global Plan of Action (GPA II), adopted in 2011 by the Food and Agricultural Organization of the United Nations (FAO) Member Countries, it makes (again, like in the first GPA agreed upon in 1996) a special reference to CWRs and wild food plants, highlighting the need to strengthen their conservation and sustainable use [16]. More recently, CWRs have been included in the United Nations' Sustainable Development Goals (SDG) [17]. The recent Global Assessment Report on Biodiversity and Ecosystem Services, published in 2019 by the United Nations' Intergovernmental Science-Policy Platform on Biodiversity and Ecosystem Services (IPBES) [18], mentions CWRs explicitly as species that are important for long-term food security, helping render ecosystems more resilient to stressors including climate change, pests and pathogens, and that lack effective protection. The report highlights the decreasing number of CWRs and mentions that many hotspots of agrobiodiversity and CWRs are under threat or not formally protected.

In response to this increasing visibility and importance of CWRs in global and international political agendas since the early 1990's, numerous projects, tools, and guidelines have been initiated and developed at local/national, regional, and global levels. Examples for the latter are the voluntary guidelines for the conservation of CWRs and wild food plants at the national level [19] or the interactive toolkit for CWR conservation planning [20].

Besides the more political framework facilitating conservation, technical and managerial considerations are also important in order to effectively include CWR species in routine conservation programs. As treated in the following sections, a number of specific requirements can be identified that determine the ability of genebanks, in particular, to cope more effectively with CWR conservation. Especially, the availability of adequate knowledge and experience to manage this very variable and sometimes extremely difficult category of genetic resources is one of the main hurdles to overcome.

It has been a long and is yet a continuous struggle to get CWRs as a high priority on, in particular, local and national conservation agendas [21,22]. Reasons for this are limited financial resources available to many conservation and use programs; the lack of technological resources to effectively exploit these resources; an increasing debate on access to and availability of PGRFA; the sometimes severe technical challenges, which the conservation of CWRs' can cause to genebanks, due to biological peculiarities of CWRs; as well as the relatively low priorities these resources have for local people. Against this backdrop, the paper investigates the reasons for these constraints, focusing on difficulties, opportunities and synergies that characterize the conservation and use of CWRs. Furthermore, due to the biological peculiarities of CWRs, there is a need for a strong collaboration between actors operating at different levels, especially between local/national and international, as well as between different sectors, such as agriculture and environment.

2. Definition and Classification of CWRs

A 'simple' and broad definition of a CWR is that all wild species belonging to the same genus (and that coincides in most cases with the same genepool) of a given crop are treated as a crop wild relative [23]. A narrower definition refers to the genepool concept developed by Harlan and de Wet [24]. They used the easiness of crossing a given wild relative with the crop species in question as the basis for their classification. When a CWR species crosses easily with the related crop, the species is defined as a genepool I species (GP1a = cultivated form of the crop and GP1b = wild or weedy form of the crop). Wild relatives from whom genes can be transferred to the crop, but with difficulties using conventional breeding techniques, are included in genepool II. Those wild relatives that cannot be crossed with a given crop and where gene transfer is only possible using sophisticated techniques, such as embryo rescue, somatic fusion or genetic engineering, are defined as genepool III species. Although this classification is very 'utility driven' and from a plant breeding perspective, it makes good practical sense, as crossing barriers are a major limiting factor for the use of CWRs in conventional plant breeding.

However, for the majority of crop complexes, particularly those from tropical areas, too little information on crossability is available to use the genepool concept. Therefore, an alternative concept has been proposed by Maxted et al. [23], based on the existing taxonomic hierarchy to define to which of four recognized taxon groups a given species belongs. Taxon group TG1a corresponds to the crop, CWRs in TG1b correspond to the same species as the crop, CWRs of TG2 are in the same series or section as the crop, TG3 is the same subgenus as the crop, and CWRs of TG4 are those in the same genus. Thus, without detailed information on the reproductive isolation, this concept can be used to establish the degree of relationship between a CWR and a crop [23].

The number of CWR species account for about 21% of the world's flora [19,25], assuming that any species belonging to the same genus as a given crop is a CWR. On that basis, it has been estimated that there are 50,000 to 60,000 CWR and wild food plant species worldwide [19]. For Europe, Kell et al. [26]

found that 17,495 (8624 of them endemic), out of approximately 20,590 species, or 85% of the European flora, comprise crop and CWR species. Maxted et al. [23] argued that a more targeted list of globally important CWR species could be obtained by focusing on the crop genepool GP1b or on taxon groups TG1b and TG2, containing the closest CWR species. By applying this to genera that contain major and minor food crops, as defined by Groombridge and Jenkins [27], that the resulting 77 genera contain 10,739 CWR species that are congeneric to these genera, and of these 221 are very close wild relatives and 471 close wild relatives [25]. Thus, as a working estimate, there would be, globally, around 700 closely related CWR species (i.e., less than 0.26% of the world flora), which are of a high value in terms of their potential use in plant breeding programs and would deserve the highest priority to conserve the genetic diversity of major and minor food crops [21,28].

Vincent et al. [29] used the genepool and taxon group concepts to estimate CWR relatedness for 173 priority crops included in Groombridge and Jenkins [27] and the Annex 1 of the International Treaty for PGRFA. Additional taxa more remotely related to crops were added if they had useful traits for crop improvement. The inventory contains 1667 taxa, belonging to 1392 species in 108 genera and 37 families. It also includes ancillary data, such as their regional and national occurrence, seed storage behavior, and herbaria, housing major collections of CWRs. This inventory is available online as searchable resource, called the Harlan and de Wet inventory, and is actively maintained [30]. This list can be regarded as the most comprehensive one, based on clear criteria. A number of other global priority lists, typically developed in the context of specific projects, are less comprehensive, have less well defined or complex criteria, and have not been used as widely as the list by Vincent et al. [29]. Two African regional checklists [31,32] and several national checklists and inventories have also been developed and are available on the CWR global portal [33].

3. General Threat Status of CWR

Since the successes of the so-called Green Revolution in the sixties and seventies of the last century, with the breeding of high-yielding varieties of a number of important food crops worldwide, in particular by the CGIAR research centers, a vast replacement of traditional varieties of these crops by the newly bred varieties resulted in a significant loss of genetic diversity and triggered a systematic collecting and conservation of in particular landraces in the newly established genebanks. The Green Revolution also impacted on the agricultural production systems through the promotion of fertilizers and the use of pesticides, leading to a much more intensive agriculture. This development impacted also indirectly on CWRs, especially those that grew in cultivated fields, on field margins and along roadsides. Consequently, they were included in the global collecting efforts coordinated by IBPGR [9]. The authors reported that 25% of the collecting missions were dedicated to CWRs. About 60,000, or 27%, of the 220,000 collected samples were CWRs, mostly forages, including forage shrubs and trees (53.2%), followed by wild cereals (10.4%), wild legumes (9.4%), wild vegetables (7.6%), and wild root and tuber species (7.6%).

As for other wild plant species, the genetic diversity of CWRs continues to be eroded by global threats, such as: changing land use; climate change and natural calamities, becoming possibly the biggest threat through different specific impacts on CWRs; changes in agricultural practices; over-exploitation or excessive use; nitrogen deposition; and invasive species. Other factors include overgrazing and desertification; agricultural subsidies, such as that of biofuel crops, maize, and rubber; the development of aquaculture; reclamation of wasteland; pollution; and others [22].

Specific examples of global threats leading to genetic erosion of CWR species have been presented by [22,34] and [9]. The latter authors noted much fewer publications on genetic erosion of wild plants and CWRs, compared to those on crop species. Jarvis et al. [35] predicted a loss of almost half of the current geographic ranges of CWRs of peanuts in South America, cowpeas in Africa, and wild potatoes in Central and South America. They also projected that between 16% and 22% of these species would go extinct by 2055. Lira et al. [36] concluded from model studies in Mexico that eight of the wild Cucurbitaceae taxa will not survive under accepted climate change models. Erosion of traditional crops and their wild relatives is greatest in cereals, followed by vegetables, fruits, and nuts and food legumes [15]. As part of the GPA II implementation assessment for the period 2012–2014, 32 countries reported to FAO to have conducted more than 5200 PGRFA surveys, covering 1823 species (predominantly wild). Of these, 56.3% were rated as threatened, i.e., they were no longer cultivated or did no longer occur in situ in most of their previous areas of cultivation or occurrence [22].

The most commonly applied means of assessing threats to wild taxa are The IUCN Red List of Threatened Species criteria [37], including for CWRs [38]. Some countries, e.g., Germany, have their own system for assessing endangerment status at national level [39]. IUCN has started to place some focus on CWR threat assessments. Their Plants for People Initiative, for example, included the assessment of high priority CWRs. CWRs are flagged within the IUCN Species Information System. The IUCN Red List of Threatened Species version 2017-2 included 760 CWR assessments [40]. The IUCN Red List status was assessed for 572 CWR species in Europe, and 11.5% of these species were classified as threatened (categories 'vulnerable', 'endangered', or 'critically endangered') and 26 species were reported as 'near threatened' [41]. Bolivia established a red list of CWRs using the IUCN criteria [42]. Maxted et al. [28] reported that the loss of genetic diversity within CWR species is likely to be much greater than the loss of species. Most of the species that are able to survive the threats they are exposed to will suffer some genetic erosion or loss of genetic diversity. The increasing impact of climate change is likely to impose heavy selection pressure on CWR populations. This could easily lead to a loss of genetic diversity and, consequently, species may not be able to adapt as readily and quickly to changing conditions as before. Thus, this vital diversity that is required to underpin food security might not be any more available to breeders [28].

Genetic erosion occurs also in genebanks due to intercrossing with other accessions during regeneration, selection, genetic drift, and shift because of unsuitable growing conditions, loss of viability in storage, or also due to human errors during cultivation. As CWRs are difficult to grow, genebanks might tend to wait as long as possible with regenerating them and, thus, seeds might lose their viability and thus cause genetic erosion [43]. The lack of knowledge about the biology of CWR species, the absence of a good infrastructure for their cultivation, and other factors, such as adequate funding for conservation, might well contribute further to genetic erosion, in particular within accessions [9].

4. An Assessment of Peculiarities of CWRs with Respect to Conservation Management

4.1. Biological Peculiarities

CWR species possess characteristics that allow them to survive in nature. Such characteristics are, in many instances, not suitable for cultivation. As CWRs are most valued and valuable as reservoirs of new genetic diversity and traits required by plant breeders, this diversity is evolving in nature while being exposed and adapting to (changing) environmental conditions. Storage in a genebank would not allow such adaptation processes to take place while being conserved. This means that one has to consider where to conserve the CWR, i.e., in their natural habitat (i.e., in situ), in a genebank or botanic garden (i.e., ex situ), and/or a combination of the two. Both the GPA II [16] and the CBD [15] regard in situ conservation as the strategy of choice for CWRs, backed by ex situ.

With respect to in situ conservation, the obvious advantages compared to ex situ conservation are that CWRs can be conserved dynamically, providing for ongoing evolution and for a wider coverage of their genetic diversity. However, a number of preconditions to achieve this are presently not met, including lack of biological information on the species themselves, their taxonomy, distribution, and threat status.

With respect to ex situ conservation, one should realize that crop species have lost most or all of the 'wild' characteristics during the domestication process. Typical examples are shattering, day length sensitivity, variable and non-determined flowering period, fragile ears (in the case of cereals), etc., which CWRs do possess. Thus, their management in an ex situ condition might be very difficult and requires ample experience. Many wild species have a limited distribution area, compared to most crops, and are an integral part of 'their' natural ecosystem. Their adaptability might be limited and, thus, also their ability to adapt to new environments (i.e., in particular, those of a genebank setting) might be low. Consequently, their optimum ecological conditions should be known when growing them outside their distribution area, in order to produce healthy and vigorous seeds/planting materials for subsequent storage. Furthermore, their biological reproduction 'system' should be known to ensure an effective reproduction, especially in case pollinators are required.

Storage behavior of CWR seeds might be unknown as seed biological aspects are unknown and, thus, require testing to ensure optimum storage conditions; standard viability seed testing methods might not function properly and/or more advanced viability tests might be used; collected seeds might be very variable in quality, i.e., not uniform in their maturation status and, thus, with variable longevity expectations; seeds might have dormancy and/or could possess hard seeds, whereas no treatments are (yet) known; typically only small samples have been collected and, thus, there is in general a need for (immediate) multiplication before storage; possible presence of pest and disease in or on the material (vegetative material, non-orthodox seeds, and/or orthodox seeds) could have implications for outgrowing in the field or greenhouse, for viability testing, storage, and distribution [44].

The lack of knowledge and information on the existence, distribution, and genetic diversity patterns of CWRs make their adequate collecting difficult. This includes the application of the best possible sampling strategy, including the number of plants per population (if there would be such an option to decide), the number of populations for a defined area, or even the entire distribution area of a given CWR, the right timing of the collecting mission, etc. (for details of these and other collecting aspects see [45–47]). This general lack of information is certainly one of the main reasons why CWR genetic diversity is sub-optimally represented in ex situ collections.

Notwithstanding the high importance accorded to in situ conservation of CWR, in particular in protected areas [21], the effectiveness is reported to be more uncertain than in genebanks. At the same time it should be noted that the main rationale for in situ conservation is based on the likelihood that continued exposure to changing selective forces will generate and favor new genetic variation and, thus, there is an increased chance that rare alleles that may be of value to future agriculture are maintained [48].

In addition, considering the rather huge numbers of CWR species reported (50,000–60,000 species), the need to conserve adequate representation of selected populations for each CWR species is creating big challenges for an efficient conservation of CWR diversity [28].

4.2. *Managerial Responsibility- and Awareness-Related Issues*

It should be realized when establishing priorities for CWR conservation that their natural distribution does not follow, in most instances, national borders. Consequently, consultations with neighboring countries could facilitate comprehensive and effective conservation of the entire CWR genepool. In addition, information on the spread and possible distribution patterns of the genetic diversity within a given CWR genepool will be very helpful to identify possible sites for in situ conservation and/or to apply the most efficient sampling strategy when collecting.

According to the CBD, the CWR occurrences are under the sovereignty of the countries in which they grow. Therefore, in situ conservation of these species has, necessarily, to include a strong national component and any regional or global in situ conservation approach should be based on and/or aim to integrate or complement such national and local in situ actions. CWR in situ conservation cannot be centralized at national or international level, as is possible with ex situ conservation in genebanks.

According to FAO [21,22], in many countries, CWRs do 'fall between the cracks' of the responsibilities of the environmental and agricultural sectors. This makes it difficult to decide which organizational entity should be the 'logical' institution to assume the conservation responsibility in a given country. Constraints related to this decision are the fact that CWRs are still a not sufficiently known genetic resource, that they have been knowingly or unknowingly included in nature protection measures without specific management or monitoring activities [28,48], and that they have been maintained by botanic gardens or genebanks without communication with other stakeholders.

Due to the disadvantaged position of CWRs compared to the domesticated genetic resources in most countries, the public awareness on CWRs is, in general, very low; there is no or only a weak political lobby within institutes and countries and, thus, a low priority to apply or provide funding for their conservation. Furthermore, there is a need for training and capacity building; skills such as taxonomy are limited and dwindling, creating dependencies on other organizations and countries. Especially in (remote) rural areas, there is a big need for better awareness and appreciation of CWRs, their diversity, and their role in breeding and adaptation to climate change for sustainable agriculture in order to stand any change of creating sustainable conservation initiatives.

The establishment and operation of in situ conservation sites can present administrative, logistical, and legal problems. For instance, CWR species that occur in 'disturbed' habitats, such as road-sides and field margins, as well as abandoned agricultural areas, will most likely not be 'included' in a protected area [28] and, thus, will require either their 'own' in situ conservation efforts, for instance, as part of an on-farm management scheme, and/or should be included in ex situ conservation. However, in many instances, their existence might not be known to the national PGRFA programs and/or the local authorities or conservation projects and, thus, are not on anybody's radar.

When considering the conservation of CWRs in protected areas, it should be noted that this type of in situ conservation is likely passive, meaning that CWR populations located in protected areas are not being actively managed and monitored, as most of the protected areas that harbor CWR species do not have specific CWR management plans [25]. Active and effective conservation of CWR populations located in protected areas could be achieved by expanding the management plans by including specific actions targeted to CWR [16]. Furthermore, climate change might lead to pronounced range contractions or range shifts for many CWRs. This led Aguirre-Gutiérrez et al. [49] to investigate the impact of climate change on CWRs and to combine this with monitoring programs, as well as collecting of CWRs for backing up in ex situ conditions. They conclude that in situ conservation measures, when ignoring the effects of climate change, will not be effective for many CWR species and that large-scale ex situ conservation actions are needed to safeguard CWRs.

CWRs can create problems for genebanks to manage them in routine operations, in particular, when specific required species information is lacking. For instance, to regenerate or multiply CWR accessions in the field or green or screen house, a genebank manager has to cultivate these wild species and, therefore, has to find answers to manage characteristics, such as a possible low germination rate, the unknown reproductive biology of the species, possibility of small sample sizes, shattering, non-homogenous ripening, etc., in order to meet the agreed standards for genebanks [50,51]. The lack of knowledge, experience, and facilities to adequately manage CWRs in genebanks is widely recognized. Thus, many genebanks will have to seek collaboration with other scientists in the country or with other genebanks that have more expertise in conserving CWRs. One option could be participation in a regional CWR network, through which the coordination of activities with neighboring countries could be achieved, sharing of responsibilities could be obtained, etc. The European Cooperative Program for Plant Genetic Resources (ECPGR) and its virtual European genebank, AEGIS, is an example of such a regional network [52]. At the same time, it should be noted that the conservation of CWRs only ex situ would not be feasible because of the sheer number of species and the need to sample and conserve eco-geographically and genetically diverse populations for each species in a dynamic way [28].

4.3. The Need for Prioritization of CWR Taxa

Considering the large numbers of species that are classified as CWRs, the usually limited financial resources for conservation, and the fact that many CWR species are not well known and in most cases lack critical information, there is a strong need to set clear priorities for their effective conservation. Possible prioritization criteria for CWRs should address aspects such as:

1. the degree of threat of the species;

2. their genetic closeness to the crop species;
3. the demand for specific traits/species by the (potential) users (and thus their economic potential);
4. the distribution area (uniqueness, incl. endemism; centre of origin/diversity) and occurrence of a given species;
5. the conservation status of a given species, including in other (neighboring) countries of the distribution area;
6. the (physical as well as legal) availability; and
7. the international legal and policy instruments vis-à-vis the national legal framework.

These criteria are based on priority-setting criteria that have been used and reported in [53–56]. When countries need to prioritize CWR species they will select a number of these criteria in accordance with their national context. The choice and assigned importance of criteria are therefore likely to vary between countries, while the most commonly included criteria are the economic importance of the related crop, the genetic closeness to the crop, and the threat status of the CWR.

Whereas priority-setting is a 'standard requirement' in conservation, both for in situ as well as for ex situ approaches, there are some specific impediments to the prioritization process of CWRs. Possibly the most important factor is the lack of information/knowledge on the species themselves (see also the following section). Another important constraint is that CWRs are typically not 'directly' used and, thus, not part of a traditional 'food system' (and consequently of a traditional knowledge system) or of an agricultural production system and, thus, their intrinsic value is often not recognized.

4.4. Availability of and Access to Data and Information

Availability of and access to data and information about CWRs, i.e., their occurrences, distribution, and threat status, their taxonomy, biological characteristics, ecological requirements, habitats, uses and genetic and phenotypic characterization and evaluation, are essential for the planning and implementation of effective conservation and use of CWRs. Existing information is yet mostly scattered, held in different formats (including non-digital) by very disparate entities, many outside the PGR community, and often not readily available. In hardly any data source, CWRs are flagged or tagged as such. Accessing this information is, therefore, resource intensive and time consuming, even more so as comparing datasets is often very difficult due to the variety of standards, formats, and data management models used [26,57–59]. However, quite some progress in proposing descriptors and data collection formats has been made in the past few years, e.g., [26,60–64]. In addition, data are often incomplete and new and/or more data need to be generated or collected. For example, data about occurrences of CWR populations are usually derived from databases of ex situ genebank accessions and herbaria specimen records. These most often do not reflect a comprehensive picture of the species' distribution, can include very old records, and do not include data about the population status of the recorded occurrence. Field surveys and collecting require solid taxonomic knowledge of the local flora, which can be difficult to source. A global database or catalogue that collects into one place data about CWR inventories, occurrences, distribution, and in situ conservation actions currently does not exist.

5. The Current CWR Conservation and Use Status

5.1. Facts and Figures on CWR Conservation

5.1.1. In Situ Conservation

Whereas the CBD [14], as well as the GPA [16], recognize the importance of CWR in situ conservation and regard ex situ conservation as a complementary conservation effort, the progress of CWR in situ conservation remains slow and difficult. In the second State of the World (SOW II) report, it is noted that in situ conservation is often envisaged to take place in protected areas or habitats and can be targeted at the species or at the ecosystem in which they occur [21]. However, the report also noted that in situ conservation of wild species of agricultural importance occurs mainly as an unplanned result of efforts to protect particular habitats or charismatic species. Furthermore, existing in situ protected areas do not always meet the required management standards to maintain CWR populations and their genetic diversity long-term [65]. Whereas the number of protected areas globally has increased considerably and the total area covered by protection expanded from 13 in 1996 to 20.3 million square kilometers in 2020, covering 15.2% of the terrestrial surface [66], it should also be mentioned that, in general, areas with the greatest diversity, for instance within centers of origin and/or diversity of our crops, have received significantly less protection than the global average [21].

Several countries informed as part of the SOW II report [21] the establishment of protected areas for CWRs, e.g., Armenia (CWRs of cereals), Ethiopia (wild populations of *Coffea arabica*), Mexico (*Zea perennis* and *Z. diploperennis*, CWR species of maize), China (86 in situ conservation sites for CWRs of different crops), Turkey (protected areas for CWRs of cereals and legumes), and Syria (protected areas for CWRs of cereals, legumes, and fruit trees). Hunter and Heywood [55] reported the establishment of a citrus wild relatives' gene sanctuary in northeast India in 1981. A similar genetic reserve for wild relatives, including relatives of lychee, longan, and citrus, was established in Vietnam. They also mentioned that certain wild species of mangoes and other wild relatives are known to occur in biosphere reserves, national parks, and other reserves in India, Indonesia, Singapore, the Philippines, Thailand, and Sri Lanka, but little targeted in situ conservation has been undertaken. In Europe, the first CWR genetic reserves were designated in 2019, when, in Germany, a network of genetic reserves for four wild celery species was established [67–69]. As of February 2020, the network included 15 genetic reserves and more are in the process of being established.

The aforementioned summary assessment of GPA II [22] noted an increased attention to CWRs in the context of in situ conservation and management. Overall, 14.2% of the over 15,000 in situ conservation sites that were listed in 20 country reports had management plans addressing CWRs and wild food plants. A total of 78 activities on in situ conservation and management were implemented with institutional support in 19 countries. A total of 16 countries reported an estimated total of 2141 CWRs, including species from primary and secondary genepools, as well as species previously used for breeding but belonging to the tertiary genepools, and wild food plants, actively conserved in in situ areas. The average per country is amounting to 134 CWR species with a maximum of 840 species in one country. However, the overall developments, with respect to the implementation of the in situ conservation priority activities of GPA II, were limited in scope and the reporting countries rated their achievements with respect to this priority activity as the lowest across all the 18 priority areas that make up the Second GPA [22].

Vincent et al. [65] assessed 167 of the most important food crops for improving food security and income generation and identified 1425 priority CWR species related to these crops. They modeled the distributions of 791 of these priority CWRs as the basis for the identification of 150 sites for in situ conservation. Individual CWR species, in general, were found to be well represented in current protected areas; only 35 (2.5%) of the studied species, related to 28 crops, were distributed exclusively outside of protected areas. If a threshold of 50% or more of the potential genetic diversity of a CWR, based on ecogeographic land characterization diversity [70], occurring within protected areas, is considered adequate for genetic conservation, then 112 of the assessed CWR species are under-conserved, while 91% of CWRs are well represented within existing protected areas. Effectively conserving the top 10 CWR sites inside protected areas and the top 10 sites outside protected areas as defined in the pragmatic scenario, would only require active management of ~2000 km^2 globally and would protect 475 CWR species, and 1257 unique CWR/adaptive scenario combinations. Vincent et al. [65] propose to manage these as a global in situ conservation network.

As any other wild species, most of the CWRs might not have any direct economic or nutritional relevance to local communities and, thus, might not be of interest to them. In fact, some of them might even be weedy and constitute a nuisance to local farmers. Therefore, CWRs might not be very attractive for inclusion in a local 'on farm' conservation program [15], and in case their distribution area is not part of a protected area setting, local communities will not be interested in participating in a conservation activity if no benefits/funding will be provided. Only in cases where the CWR species occur in a protected area (targeted or 'by chance': [15]), their conservation might be easier and more sustainable as long as some sort of a monitoring system does exist.

In some cases, however, CWRs play a known and appreciated role in local and, typically, traditional cropping systems and, thus, will be valued by local farmers or communities. Consequently, conservation approaches might be easier and could directly involve the local people, as long as benefits will be generated through such activities. Examples of such situations include the regular re-domestication of *Dioscorea cayensis* subsp. *rotundata* in Benin; the use of *Dioscorea* spp. in West African countries by facilitating the introgression between wild and domesticated yams, as this is an important improvement strategy; the use of *Ensete ventricosum* in Ethiopia for regular incorporation of 'wild' seedlings into the fields of the cultivated crop; or the selection of wild walnut genotypes for cultivation in Kyrgyzstan [21]. From a crop evolutionary perspective and more related to traditional agricultural production systems, tolerating CWR species which are weeds, especially in field-borders, as pollinators of the cultivated material and, thus, assumingly increasing the genetic diversity of the crop for subsequent selection, is another example. However, also the opposite can be true that CWR play a detrimental role in farmers' field, for instance, as noxious weeds.

5.1.2. Ex Situ Conservation

Traditionally, ex situ conservation is the main approach that countries have taken to conserve CWRs. Genebanks play an important role in the overall conservation of CWR germplasm; in fact, they (should) provide a link between in situ conservation and the users' communities at various levels. This role is essential as they typically are specialized in long-term conservation, distributing or exchanging requested materials, characterizing and evaluating the stored accessions, keeping detailed information records on the individual accessions and, in some instances, conducting pre-breeding activities to facilitate the use.

Genesys, the largest global database on ex situ conserved germplasm accessions, provided data (as of 11.01.2020) for 4,097,112 accessions, of which only 12% are classified as wild material, thus possibly also including some non-CWR species [71]. The European Search Catalogue for Plant Genetic Resources (EURISCO) [72] contains data for 2,023,530 accessions. Among those, 12.15% are reported as wild. According to Ford-Lloyd et al. [34], the 1095 CWR species reported in EURISCO, at the time the research was undertaken, only represented 6% of the 17,495 CWR species found in Europe. This means that 94% of European CWR species are not conserved in ex situ collections.

The SOW II report [21] provides an average percentage of wild species, predominantly CWRs, for each of the 11 major crop groups, varying from 4% (food legumes and fiber crops) to 35% (forages) and 46% (industrial and ornamental plants). The overall mean for the almost 7 million reported accessions of wild plants is 10%, most of them being CWRs.

For a number of reasons, many CWRs are represented by a small number of accessions per species in the collections, both in genebanks and in botanic gardens. As an example, of the 1076 global priority CWR taxa identified in a study about global CWR conservation priorities [73], 'only' 763 or 70.9% are included in genebanks; among those, 257 taxa are represented by less than 10 accessions each. Over 95% of the taxa examined were found to be insufficiently represented in genebank collections with respect to their full range of geographic and ecological variation in their native distribution area. In many instances one would find just few accessions per taxon, e.g., only 5.4% of the CWR taxa in EURISCO are represented by 10 or more accessions, whereas 90.5% of the CWR taxa have less than 5 accessions.

Due to the already mentioned difficulty to collect adequately sized numbers of seeds/plants per population, many of the accessions consist of (too) small quantities of seeds and are genetically poorly sampled [74]. In addition, the stored seed samples have frequently a low(er) viability due to the difficulties to grow them out for regeneration purposes [75]. Another aspect, related to lack of information/knowledge, concerns taxonomic identification of the CWR, including to which crop genepool they belong. This will directly impact on the priority-setting and possible subsequent conservation, both in situ and ex situ, as well as on their use.

In a study of ex situ holdings of 23 selected genepools of the major crops included in Annex I of the International Treaty, i.e., those materials that countries agreed to form the backbone of the multilateral system of the International Treaty, the authors calculated an average non-weighted percentage of CWR accessions in genebank collections (without the international collections held by the CGIAR genebanks) of the selected genepool worldwide of 9.6%, ranging from 0% for coconuts (there are no CWRs known) to 33% for grass pea (a little bred crop) (Figure 1). The total global holdings considered in the study of the selected genepools (without the collections held by the CGIAR genebanks) were 3,149,371 accessions [76].

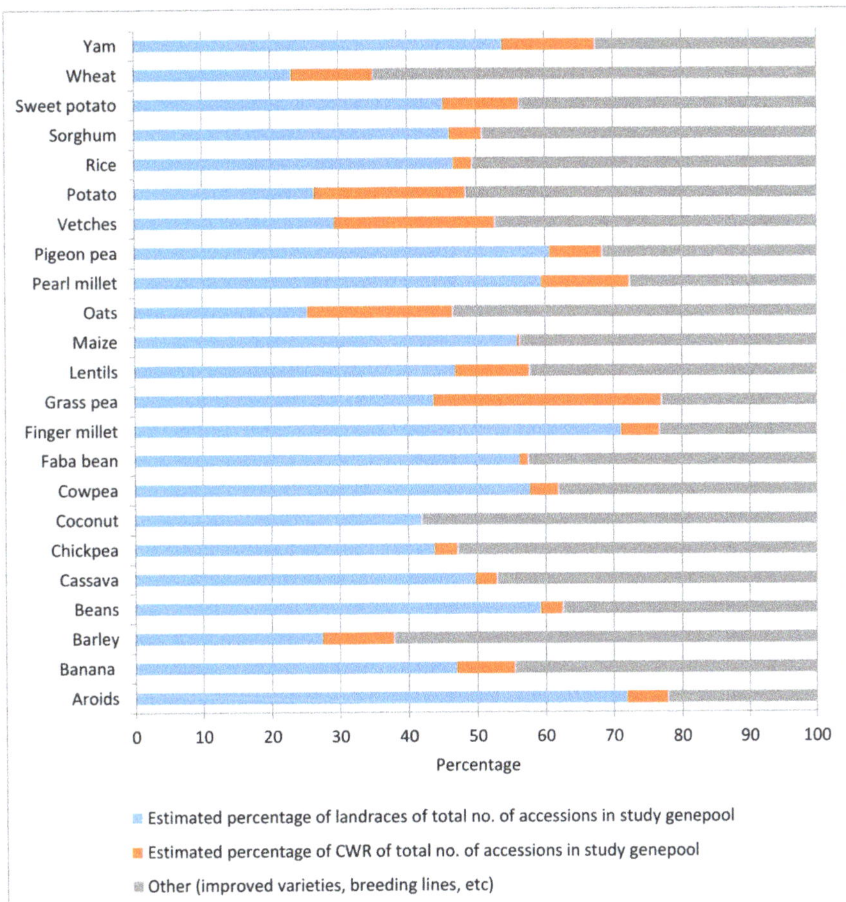

Figure 1. Percentages of CWR and landrace accessions in genebank collections of 23 selected crop genepools.

When looking at the primary genepool, 242 of the 1667 CWR taxa included in the Harlan and de Wet CWRs inventory were found to be under-represented in ex situ collections and the countries identified as the highest priority for further germplasm collecting are China, Mexico, and Brazil [29]. Khoury et al. [77] used gap analysis to assess the degree of representation of *Cucurbita* CWR taxa in conservation in situ, as well as ex situ in genebanks and botanic gardens. For the *Cucurbita* genus, including 16 CWR and six cultivated species, the authors established detailed taxon-related ex situ, as well as in situ (i.e., protected areas) conservation priorities and suggested further in situ protected areas that would cover the greatest amount of populations of the largest number of taxa. Khoury et al. [77] concluded that 68.8% of wild *Cucurbita* taxa were assessed as high or medium priority for further collecting for ex situ conservation and 81.3% had a high or medium priority for further protection in situ, including all of the progenitors of the cultivated species. Furthermore, four taxa were listed as having very few accessions and, thus, very limited diversity is available for crop breeding. Khoury et al. [77] suggested that these figures might be considered as 'typical' for the CWRs at large.

Besides their conservation in situ and in genebanks, botanic gardens have also been collecting and storing CWR materials in their collections, as demonstrated by the PlantSearch database, which is an information platform for 1155 botanic gardens that collectively maintain plant, seed, or tissue collections of 589,526 taxa [78]. The database reveals that botanic gardens maintain at least 30% of all known plant species in their own collections, including that more than 41% of species assessed are globally threatened. Many of these wild species are CWRs. Almost one-third (315, or 28.6%) of the 1076 aforementioned global priority CWR taxa are maintained by botanic gardens [79].

A recent major effort of collecting new CWR samples was made by the project "Adapting Agriculture to Climate Change" [80], which focuses on the wild relatives of 29 crops included in Annex 1 of the International Treaty; over 4500 new CWR samples were collected for ex situ storage, evaluated for useful traits, and enhanced or pre-bred for use in crop improvement programs.

5.1.3. Complementary Conservation

As already noted above, both the CBD [15] and GPA II [16] refer to the need to complement in situ conservation efforts with ex situ measures. Genebanks have recognized strengths in facilitating easy and targeted access to specific material (which is problematic for in situ conserved material) and to allow secure and long-term conservation as part of the conservation and use continuum. Especially when environmental change is too rapid for evolutionary change and adaptation, or migration, it can be easily understood how and why ex situ measures would complement or even replace in situ conservation and thus provide for the most effective approach [22,81]. Such a complementary approach requires that in situ and ex situ conservation measures have to be carefully planned and combined, thus securing a holistic combination of the two, which capitalizes on strengths and avoids weaknesses of one or the other. This will require a good understanding of the (seed) biology of the species, their threat status, priorities assigned to the individual CWR species, and other aspects; an assignment of clear responsibilities, including, for instance, to the agricultural and environmental sectors; if applicable, to link conservation and development; adequate and comprehensive information management; facilitation of adequate coordination with other stakeholders and countries; the verification of clear ownership rights over areas where the to-be-conserved CWRs occur; support of public awareness on the importance of CWR conservation; and, where necessary, to ensure the engagement of the broader public.

As an example, Hunter and Changtragoon [82] conclude, on the basis of regional project experiences, that for wild relatives of tropical fruit trees, any conservation strategy should contain elements of both in situ and ex situ conservation and should have a focus on conservation, both inside and outside protected areas. It should also ensure coordination of planning and implementation, institutionalize the practice of wild relative conservation, promote public awareness and understanding, create a suitable policy environment, and highlight the many benefits derived from their sustainable conservation and use. In situ approaches seem feasible for conserving wild relatives of tropical fruit trees, but experiences with targeted species and actions inside and outside protected areas appear to be relatively few. Consequently, wild relatives of tropical fruit trees remain a largely under-conserved natural resource, both ex situ and in situ, and are continuously under threat in their natural habitat from neglect and over-harvesting [82]. Vincent et al. [65] note the generally accepted requirement for complementary conservation, i.e., to also cover in situ conserved materials in genebanks, a process that has started recently. They further see a particular need to develop CWR in situ activities that enable the conservation of geographically partitioned genetic diversity which retains potential for local environmental-evolutionary adaptation.

5.2. Facts and Figures on CWR Use

The term 'use' needs to be applied in its widest sense for CWRs. The traditional understanding is the use of genetic diversity in plant breeding by crossing cultivated material, usually advanced varieties with CWRs and through a strong selection to obtain genotypes, with the traits that have been transferred from the CWR species. Furthermore, CWRs are an important target for research on crop evolution and are, indirectly, an important component of research on the origin and spread of agriculture. With the increasing focus of conserving CWR in situ (including on-farm), the 'direct' use of CWRs by local communities and farmers has now also received some more attention. Another dimension of 'using' CWRs is their not well understood and accepted role in and contributions to the evolution of crops and plants at large. Through the overall conservation efforts of the flora (and fauna) in natural habitats and protected areas, of which CWRs are an integral part, they contribute to a healthier environment, healthy ecosystems, and the provision of ecosystem services. However, this latter aspect is not part of the focus of this paper. Furthermore, the appreciation of the economic value of CWRs and their contribution to the global economy is an aspect that would fall under the term 'use'.

In tropical zones, wild fruit harvested from forests contribute significantly to the total income and to sustainable nutritious diets of many rural households, apart from contributing substantially to important ecosystem services [29]. Wild relatives and wild-growing semi-domesticated species of tropical fruit trees also provide services to domesticated fruit trees in terms of resistance to extreme abiotic and biotic stresses through their high levels of genetic diversity [82].

More widely applied is the use of CWRs in pre-breeding and breeding programs and in research, in particular in countries with strong breeding companies, where facilities and technologies, as well as funding, are available to exploit these 'difficult' resources. Today, climate change is causing dramatic changes that are being experienced around the globe, especially global warming and the related increase of severe erratic weather conditions. These changes have a significant impact on agricultural production systems that need to be addressed as well. To allow crops to cope with and/or to adapt to more extreme weather conditions, including heat, drought, flooding, and increased salinity, there is a strong need for more genetic diversity than currently available for most crops from which plant breeders can select specific traits and resistance genes to 'equip' new varieties to cope with these changing conditions. In particular, the use of CWRs, as a known source of traits for introgression into the crops, has proven to offer such solutions, especially to overcome biotic stresses [8]. As CWRs do possess a much wider array of traits and allelic diversity, as well as 'new' genetic variation compared to our modern crops, they are an important asset to be included in the breeding pools of our plant breeders and, thus, to be accorded a high priority in their conservation and research and management activities that facilitate their use by plant breeders, worldwide [73,83,84].

'Historical' examples of CWRs in plant breeding include the use of wild *Aegilops*, *Secale*, *Haynaldia*, and *Agropyron* species in wheat breeding [85], the introduction of resistance to late blight, which is caused by *Phytophthora infestans* and is found in the wild potato *Solanum demissum* [86], as well as other disease resistances and tolerances from different potato CWRs [87]. Resistance against stem rust caused by *Puccinia graminis* subsp. *graminis* derived from the wild wheat *Aegilops tauschii* [88], in another example. In the early 1970's, resistance to the grassy stunt virus was found in wild *Oryza nivara* and now this gene can be found in almost all material bred by the International Rice Research Institute in the Philippines [34]. Maxted and Kell [25] reviewed the use of CWR in crop improvement in 291 papers reporting the identification and transfer of useful traits from 185 CWR taxa into 29 crop species. Wheat and rice accounted for almost 84% of the transfers and 56% of the inter-specific trait transfers related to pest and disease resistances.

The above historical examples demonstrate the past focus on trying to identify traits of interest through phenotypic characterization and evaluation [28]. Whereas the inclusion of genetic diversity from the wild genepool in breeding activities was difficult [21], the advancements in molecular genetics and the related tools allow a much more 'targeted' use of CWRs. Through the possibility of transferring specific parts of the genome, i.e., traits, genes, and/or alleles into the genetic background of improved breeding materials, the hesitation of using CWRs is fading and, thus, their importance for breeding is increasing. According to Ford-Lloyd et al. [34], genomic-based resources, map-based cloning, analysis of quantitative trait loci, gene isolation, and genetic modification are increasingly significant to exploit the potential of CWRs. Genomic databases containing information on genes associated with adaptive characters must increasingly be linked to web-enabled databases of ex situ conserved CWR germplasm, such as EURISCO [72]. Furthermore, predictive characterization, Focused Identification of Germplasm Strategy (FIGS) [28] and eco-geographical filtering method [89] are other promising approaches to facilitate the use of CWRs in breeding.

The number of CWR genomes sequenced has grown significantly over the past decade and in 2016 the number of crop genomes sequenced was 'only' about three times higher than that of sequenced CWR species, which were about 40 [90]. For example, Bertioli et al. [91] sequenced two wild peanut species (*Arachis ipaensis* and *A. duranensis*). Peanut is an important food source for many farmers in the developing world. The CWR genome sequences will provide breeders with new tools for enhancing the crop, and for developing new varieties more resistant to pests, diseases or with improved abiotic tolerance traits. It is hoped that this positive trend of more CWRs to be sequenced continues and thus, allows a better exploitation of the important traits that CWRs harbor, including quantitatively inherited traits.

A study carried out by PwC [92] assigned an indicative value of $42 billion to the CWRs of 29 major food crops, with a potential to reach a value of $120 billion in the future. All these 29 crops are included in Annex 1 of the International Treaty on PGRFA. Pimentel et al. [93] reported an estimated value of $115 billion that CWRs contributed toward increased crop yields per year worldwide. In addition to their economic value, CWRs are also being valued for their not so well-known contributions to ecosystem services [34]. Tyack and Dempewolf [94] have reviewed past economic values of CWRs, including the previously cited studies, and propose an improved conceptual model for understanding the economic value of CWRs under climate change, expanding it from the focus of gross production to including a series of other values and costs.

6. What Needs to Be Done to Conserve and Use CWRs More Effectively?

From the information, facts and figures presented above, it is apparent that further concerted assessment and conservation efforts are required in order to keep these valuable resources and the traits therein available and accessible to the users, now and in the future. In this section, we summarize findings and identify actions for efficient conservation and sustainable use of priority CWRs. Important aspects that require attention to underpin the conservation efforts are presented.

6.1. Documentation

Documentation and availability of CWR data are the basis for the assessments of conservation and threat status, conservation planning, and monitoring, but are yet insufficient to provide more precise assessments and concrete figures about status and trends of CWR diversity. In recent years, tools and descriptors have been developed to support CWR data collection and management (see Supplementary Materials Table S1). The Secretariat of the International Treaty is currently developing a globally agreed descriptor list for CWR data exchange as a further step towards harmonizing CWR data recording and exchange and facilitating the development of national and global CWR databases. Based on these standards and tools, all relevant data at national level required for CWR conservation planning and management should be brought together in an accessible as well as standardized format into national CWR databases or portals. Furthermore, the development of a global CWR data portal, analogue to Genesys, the global hub for ex situ data, should be considered. National CWR databases could then provide data to this global resource. Such a global portal would allow reaching a better understanding of global CWR distribution and conservation status. It would serve as an important tool for sharing information and supporting more effective planning, conservation, and monitoring at the national and international levels, as well as international collaboration in CWR conservation.

An increased recognition among the actors within the environmental sector responsible for nature protection and protected area management that CWRs constitute a group of very valuable PGRFA, would possibly support flagging and data recording in their respective databases and monitoring activities, and integration of CWR conservation aspects into existing nature protection networks and activities.

6.2. In Situ Conservation

As each country is responsible for the conservation of the natural resources within its territory, CWR conservation is logically and mainly addressed at national level. To secure these resources effectively and long-term, systematic and coordinated conservation is essential, as well as integrating in situ and ex situ measures. In most occasions, however, CWR in situ conservation has been carried out within the framework of projects, which are limited in time, hardly ever running for more than five years. A more stable organizational and financial basis for CWR conservation at the national level is therefore required in most countries. This can be supported and facilitated by developing a national strategic action plan for CWR conservation.

There is no single method for planning CWR conservation or for developing such a strategic plan, as related factors, such as financial and human resources, availability and quality of baseline data, the range, role and responsibility of relevant stakeholders, or the commitment of national governments, vary between countries. Nevertheless, a series of steps and decisions in the conservation planning process are likely to be common in most situations. These include the development of a CWR checklist, prioritization of CWRs, development of an inventory of the priority CWRs, threat assessments, gap and diversity analyses, and the identification of priority sites and actions for in situ and ex situ conservation [56].

The development of a national CWR organizational plan and an efficient coordination mechanism are important to facilitate coordination and collaboration. These measures require and will greatly benefit from the establishment or provision of a nation-wide information platform that facilitates the routine operations, allows the necessary coordination, and enables adequate reporting. The collaboration between the various important stakeholders at the local, provincial, and national levels is a prerequisite for effective and sustainable conservation operations. At the national level, adequate coordination between, in particular, the ministries of agriculture and environment and their implementation bodies is critically important to facilitate the identification and management of protected areas that target or include CWRs and to allow the participation of key stakeholders in the planning and implementation of projects and activities, including the support of research and awareness creation. Considering the specialized skills and facilities required for efficient and effective conservation of the CWR genepools, close collaboration with neighboring countries, possibly in the context of a regional network, seems to be very important to allow an adequate conservation of the total genetic diversity range of a given CWR species.

6.3. Ex Situ Conservation

Targeted and adequate collecting of highly threatened and prioritized CWR materials from their natural distribution areas, as well as of populations that are requested for research and use, is a critically important step to avoid genetic erosion and to facilitate use. A close collaboration with local communities and their conservation activities is important, as well as coordination with botanical gardens and other ex situ conservation programs. During collecting, it is important that an adequate number of populations of targeted CWR species is sampled and that the samples are of an adequate size. To ensure effective conservation for each collected CWR species, specific conservation standards need to be used; where necessary, further research might be required. One such research area is on seed biological aspects (see, for instance, [44,51] and/or the application of already developed advanced methods, e.g., on germination testing, using potential markers as volatile compound [95,96], changes in methylation [97,98], or DNA and RNA integrity [99,100]). The morphological and/or molecular characterization as well as further evaluation of conserved samples will be an essential step to facilitate their use, where applicable this should be done in collaboration with neighboring countries. One other example could be the application of cryopreservation of embryos, cells, tissues, or seeds as a long-term conservation method, especially for CWRs that cannot be conserved in the form of orthodox seeds.

A national CWR priority list provides the foundation for targeted collecting of threatened populations and for the development of complementary conservation efforts that reflect the long-term conservation needs, the biology of the species, the needs of users, accessibility to specific materials, and the requirement of exchanging/distributing germplasm. Well planned characterization and evaluation of prioritized accessions will increase our required knowledge and understanding of the genetic diversity aspects of the CWRs and thus enable and facilitate effective conservation as well as the targeted and sustainable use of conserved material.

6.4. Complementary Conservation Approaches

When planning CWR conservation approaches, a number of considerations will be important to take into account, especially when realizing that in general limited information is available about these resources. Furthermore, different infrastructures and technologies are needed to collect, conserve and monitor the material under conservation. In addition, geographical, technological, scientific as well as political/legal aspects will have to be considered and should complement each other well. As mentioned before, complementary conservation is not a 'method', but rather a conceptual framework that helps with the systematic planning of conservation efforts for a given species and under specific 'local conditions'. An example of such a framework is provided in [101]. So far, little practical experience can be reported. The approach should lead to practical and efficient, long-lasting, and cost-effective conservation activities for a given species. Examples of such pragmatic approaches would be to include populations of CWR species conserved in situ also in ex situ storage as a safety back-up and to facilitate their access for use. In case species cannot be (safely) conserved in situ, for instance, due to financial or administrative constraints or when the species is highly threatened, attempts should be made to conserve the threatened species ex situ in a genebank.

As use might be regarded as the ultimate goal of a conservation effort, it seems obvious to involve the users (primarily breeders) also in a prioritization and conservation planning exercise. Thus, the requirements of possible users of conserved germplasm can be duly reflected in the conservation approach, including specific aspects such as that the conserved materials can be shared easily with users in an appropriate form and quantity.

The very fact that only limited practical experience has been made with complementary conservation, the fact that the best possible combinations will vary from place to place and species to species, means that it will require more research to allow optimal solutions for effective and efficient conservation and sustainable use of individual CWR species to be identified. The development of a generic decision tree and supporting guidelines could be an important contribution to a more comprehensive, effective, and efficient complementary conservation of CWRs, at the various levels.

6.5. Supporting Use

Concerted efforts that facilitate the use of conserved CWR germplasm, either in in situ or ex situ conditions, are needed to enable a more effective and increased use of the often-unique genetic diversity contained in these threatened resources. Such efforts can be very diverse and include for example better management practices in a genebank or protected area, with respect to the representation of genetic diversity (as populations and/or as pure lines, etc.), ensuring an adequate coverage of the genetic diversity that exists within a species in the collection, and very importantly increasing the level of characterization and evaluation of individual accessions (both morphological and molecular), providing much more information on the CWRs conserved in genebanks and improving the availability and accessibility of data.

6.6. Strengthening the Conservation System

In the context of this paper, the national approach is possibly the most relevant one, but with the clear understanding that the 'real action' will have to be undertaken 'on the ground' at the local level and, whenever possible, for both in situ and ex situ approaches. However, when considering the many difficulties to ensure an effective and secured conservation of these species, it is obvious that many of the less well-endowed local genebanks and botanic gardens will require support to implement such conservation activities adequately, in order to contribute to a sustainable and long-term safeguarding of CWR.

6.6.1. National Level

There are a number of steps that need to be addressed at the national level to achieve effective, efficient and long-lasting conservation of CWR. The FAO published voluntary guidelines on the conservation of CWRs and wild food plants that provide an overview of all relevant steps that should be considered while planning and implementing conservation activities [19]. Some of these steps are mentioned in the following list:

1. Establishment of a comprehensive picture of the national botanic diversity;
2. Elaborating a national CWR checklist and inventory, e.g., [38,102] and, in parallel, ensuring an adequate integration of CWR conservation with broader national ecosystem, habitat and species conservation plans;
3. Prioritization of CWR taxa/diversity;
4. Eco-geographic and genetic diversity analysis of the priority CWR taxa;
5. Identification of threats to priority CWR taxa and important CWR areas;
6. Gap analysis and establishment of CWR conservation goals;
7. Development of in situ/ex situ CWR national conservation actions [50,103], in accordance with the other forms of conservation, mentioned in point 2 above;
8. Identification of key national CWR protected areas based on gap analysis, on the CWR inventory and occurrence data, the threat status as well as of CWRs under-represented in genebanks;
9. Establishment of national CWR genetic reserves as well as of targeted CWR ex situ collections; and
10. Elaboration of concrete suggestions on how to strengthen utilization, research and education.

A helpful website in preparing and implementing CWR checklists and inventories, as well as conservation strategies, might be the 'CWR Global Portal', established and updated by Bioversity International (now called the Alliance of Bioversity International and CIAT) [104]. It provides access to the Interactive toolkit for CWR conservation planning [56]. Guidelines and tools that can support national CWR documentation, prioritization, conservation planning, and implementation are summarized in Supplementary Materials Table S1.

A close collaboration between the national PGRFA program and those concerned with protected areas in a given country will be indispensable to avoid mistakes, to ensure that the best possible management approaches are being used, and that the existing strengths spread over people and institutions are being combined for successful implementation of in situ conservation. This collaboration can also address concerns that typically only a limited number of CWR species is included in protected areas.

6.6.2. Local Level

The national CWR conservation approach will obviously have to address and include the local level actors' roles and responsibilities. However, often there is very limited published information on specific aspects at the local level that could be included in the planning and implementation processes [55]. A number of obvious aspects can be listed, including the involvement (and active engagement) of all relevant stakeholders in the preparation of management plans for target species. This is a crucial prerequisite when the CWRs are part of a protected area that can no longer be used, for instance, for collecting fresh fruits by the local communities in the neighborhood of such an area. Maxted and Kell [25] included the way to involve local communities in their report as a research question. They also propose an interesting approach in promoting CWR in situ conservation in less formally designated protected areas such as Indigenous and Community Conserved Areas (ICCAs). For the latter, see IUCN [105]. ICCAs are areas where indigenous peoples and local communities have conserved, for millennia, natural environments and species for economic (as well as cultural, spiritual, and aesthetic) reasons, independent of more formal conservation sector interventions. Brooks et al. [106] note that the establishment of genetic or other kinds of reserves for CWRs in areas not yet under protection in times of rapidly rising human population, climate change, and ecosystem instability is a complex goal, which necessitates a carefully researched strategic approach. Sites competing for reserve status would need to be assessed and prioritized for their longer-term sustainability, in terms of the predicted impact of climate change on the site and the economic development plans associated with local communities as well as at the national level [107].

6.6.3. Global Level

Dilemmas with CWRs: Distribution areas of CWR species (at least those of the major food crops) in the tropics/subtropics are, to a large extent, located in countries with limited financial and/or technological resources, limited conservation programs, limited legal frameworks, few breeding program, and which can derive little direct benefits from CWR conservation (especially for local communities). In contrast, interest in these species is largely found in 'the North' where financial and technological resources are ample, knowledge is advanced, and where most of the breeding happens. Access to these species, however, is often limited and thus their use in breeding and research for global benefit difficult. Possibly, the only real solution would be to agree within the framework of the existing global instruments, in particular, the FAO Commission on Genetic Resources for Food and Agriculture and the International Treaty, to accord a high(er) priority to the conservation and sustainable use of these threatened resources, to study them more extensively, and to make the diversity freely available as foreseen by these instruments. A mechanism to enable the badly needed global coordination and facilitation of the frequently complex conservation activities, as well as to provide a platform for identifying and prioritizing research activities on CWRs, would be an important help in effectively and efficiently conserving and sustainably utilizing CWRs.

7. Conclusions

CWRs have been identified as threatened resources that are understudied, not properly conserved, and that possess a tremendous potential for the breeding of our crops. The latter is particularly important because of climate change, which calls for the urgent development of better adapted crops and varieties for the changing growing conditions in our vulnerable production systems. The protection of the environment is yet another important consideration that can be achieved, or at least important contributions can be made through the increase of crops and varieties that require less harmful inputs and provide still stable and high production levels.

In the above text, we distilled a number of actions that are recommended to be implemented at the various levels, whenever possible, in a timely and collaborative manner. Whereas a number of these recommendations can be implemented by individual countries, others will require agreement and coordination at the global level, where possible, using existing mechanisms and instruments.

7.1. Documentation

1. Collating, creating, and sharing more information and knowledge on CWR species, in particular, by stimulating and conducting more research.
2. Establishing national databases and inventories to enable better coordination and implementation of CWR conservation.
3. Developing a global data portal/platform for the exchange and provision of CWR data and information, including tools and guidelines that will facilitate a better coordinated and more efficient conservation, worldwide.

7.2. In Situ Conservation

1. Facilitating and encouraging the inclusion of CWRs in national and local conservation agendas and ensuring that they are being given an adequate priority supported possibly by a longer-term financial and organizational structure.
2. Complementing the management and monitoring of CWR in situ conservation sites and genetic reserves with ex situ conservation efforts of the priority species.
3. Identifying existing and novel mechanisms to finance and govern the proposed global coordination and facilitation of CWR in situ conservation should be of high priority. The proposed global network could play an important role in setting standards, sharing experiences, and providing the platform for monitoring and coordination and, thus, to provide a fundamental basis for ensuring our future food security.
4. Increasing the awareness and recognition among actors, especially within the environmental sector, about CWRs as important group of wild species that need to be conserved.

7.3. Ex Situ Conservation

1. Ensuring adequate ex situ conservation of threatened national priority CWRs.
2. Ensuring adequate ex situ conservation of a globally agreed list of priority CWRs (e.g., [29]) through national/regional/international genebanks, in particular those that already have global or regional conservation responsibilities for the corresponding crop genepools.
3. The identification and/or application of new methods to assess the viability of seeds, not requiring seed germination tests, could address current difficulties with viability tests and with small seed samples.
4. Large-scale research on CWR seed biology can lead to methods allowing for long-term storage of seeds of these species in genebanks. One such specific research area is the use of cryopreservation for long-term conservation.

7.4. Complementary Conservation and Collaboration

1. Development of a generic decision tree on complementary conservation approaches that can be applied to individual CWR species. Supporting guidelines should be developed to facilitate the application of the decision tree and the subsequent implementation of the conservation efforts, using gained experiences with individual species and cases as a basis.
2. Ensuring ready access to the genetic resources and related information, both from in situ as well as ex situ conservation within the framework of existing legal instruments.
3. Facilitating and coordinating phenotypic and molecular characterization of the priority CWRs to provide a basis for pre-breeding and breeding activities through the involvement of conservation, research, and breeding stakeholders.

4. Facilitating/strengthening the collaboration between stakeholders for more effective and efficient conservation, research and use of CWRs as well as to facilitate the transfer of technologies at the local, national, regional, and global levels.

7.5. Conservation System

1. Increasing awareness on the importance of and threat to CWRs, including through the active involvement of botanic gardens to 'demonstrate' this genetic wealth and the relationship between the CWRs and crop species.
2. Facilitating the training of staff on skills that strengthen the implementation of the above activity areas.
3. Providing a more stable organizational and financial basis for CWR conservation at national level.

Supplementary Materials: The following are available online at http://www.mdpi.com/2223-7747/9/8/968/s1, Table S1: Guidelines and tools for CWR conservation, including references [108–112].

Author Contributions: Specifically, conceptualization, J.M.M.E.; methodology, J.M.M.E. and I.T.; formal analysis, J.M.M.E. and I.T.; investigation, J.M.M.E. and I.T.; writing-original draft preparation, J.M.M.E.; writing-review and editing, J.M.M.E. and I.T.; visualization, I.T.; supervision, J.M.M.E. All authors have read and agreed to the published version of the manuscript.

Funding: This research received no external funding.

Conflicts of Interest: The authors declare no conflict of interest.

References

1. Willcox, G. The beginnings of cereal cultivation and domestication in Southwest Asia. In *A Companion to the Archaeology of the Ancient Near East*, 1st ed.; Potts, D.T., Ed.; Blackwell Publishing Ltd.: Oxford, UK, 2012; pp. 163–180.
2. Purugganan, M.D. Evolutionary Insights into the Nature of Plant Domestication. *Curr. Boil.* **2019**, *29*, R705–R714. [CrossRef] [PubMed]
3. Purugganan, M.D.; Fuller, D.Q. The nature of selection during plant domestication. *Nature* **2009**, *457*, 843–848. [CrossRef] [PubMed]
4. White, C.E. The Emergence and Intensification of Cultivation Practices at the Pre-Pottery Neolithic Site of El-Hemmeh, Jordan: An Archaeobotanical Study. Ph.D. Dissertation, Boston University, Boston, MA, USA, 2013; 240p. Available online: https://hdl.handle.net/2144/12888 (accessed on 18 June 2020).
5. Weide, A.; Riehl, S.; Zeidi, M.; Conard, N.J. A systematic review of wild grass exploitation in relation to emerging cereal cultivation throughout the Epipalaeolithic and aceramic Neolithic of the Fertile Crescent. *PLoS ONE* **2018**, *13*, e0189811. [CrossRef] [PubMed]
6. Gepts, P. Domestication of Plants. In *Encyclopedia of Agriculture and Food Systems*; van Alfen, N.K., Ed.; Elsevier: San Diego, CA, USA, 2014; Volume 2, pp. 474–486.
7. Tanksley, S.D.; McCouch, S.R. Seed banks and molecular maps: Unlocking genetic potential from the wild. *Science* **1997**, *277*, 1063–1066. [CrossRef] [PubMed]
8. Dempewolf, H.; Baute, G.; Anderson, J.; Kilian, B.; Smith, C.; Guarino, L. Past and Future Use of Wild Relatives in Crop Breeding. *Crop Sci.* **2017**, *57*, 1070–1082. [CrossRef]
9. Thormann, I.; Engels, J.M.M. Genetic diversity and erosion—A global perspective. In *Genetic Diversity and Erosion in Plants—Indicators and Prevention*; Ahuja, M.R., Jain, S.M., Eds.; Springer: Berlin, Germany, 2015; Chapter 10; Volume 1, pp. 263–294.
10. Thormann, I.; Fiorino, E.; Halewood, M.; Engels, J.M.M. Plant genetic resources collections and associated information as baseline resource for genetic diversity studies—An assessment of the IBPGR supported collections. *Genet. Resour. Crop Evol.* **2015**, *62*, 1279–1293. [CrossRef]
11. Rick, C.M.; Chetelat, R. Utilization of related wild species for tomato improvement, First International Symposium on Solanaceae for Fresh Market. *Acta Hortic.* **1995**, *412*, 21–38. [CrossRef]
12. Hoyt, E. *Conserving the Wild Relatives of Crops*; IBPGR: Rome, Italy, 1988; 45p.

13. Mammadov, J.; Buyyarapu, R.; Guttikonda, S.K.; Parliament, K.; Abdurakhmonov, I.Y.; Kumpatla, S.P. Wild Relatives of Maize, Rice, Cotton, and Soybean: Treasure Troves for Tolerance to Biotic and Abiotic Stresses. *Front. Plant Sci.* **2018**, *9*, 886. [CrossRef]
14. CBD. *Global Strategy for Plant Conservation*; Secretariat of the Convention on Biological Diversity: Montreal, QC, Canada, 2002; 48p.
15. CBD. *Notification: Strengthening the In Situ Conservation of Plant Genetic Resources for Food and Agriculture through Incorporation of Crop Wild Relatives under Areas Important for Biodiversity in Protected Area Networks and Other Effective Area-Based Conservation Measures (Aichi Biodiversity Targets 7, 11, 12 and 13 and Global Strategy for Plant Conservation Targets 5, 6, 7 and 9)*; CBD Secretariat: Montreal, QC, Canada, 2015; 11p.
16. FAO. *Second Global Plan of Action for Plant Genetic Resources for Food and Agriculture. Commission on Genetic Resources for Food and Agriculture*; Food and Agriculture Organization of the United Nations: Rome, Italy, 2012; 96p, Available online: http://www.fao.org/3/i2624e/i2624e00.pdf (accessed on 18 June 2020).
17. United Nations. *Transforming Our World: The 2030 Agenda for Sustainable Development*; General Assembly, Seventieth Session; Agenda Items 15 and 116, A/RES/70/1; United Nations: New York, NY, USA, 2015; 35p.
18. IPBES. *Global Assessment Report on Biodiversity and Ecosystem Services*; Intergovernmental Science-Policy Platform on Biodiversity and Ecosystem Services: Bonn, Germany, 2019.
19. FAO. *Voluntary Guidelines for the Conservation and Sustainable Use of Crop Wild Relatives and Wild Food Plants*; Food and Agriculture Organization of the United Nations: Rome, Italy, 2017; 106p.
20. Brehm, J.M.; Kell, S.; Thormann, I.; Gaisberger, H.; Dulloo, M.E.; Maxted, N. New tools for crop wild relative conservation planning. *Plant Genet. Resour.* **2019**, *17*, 208–212. [CrossRef]
21. FAO. *The Second Report on the State of the World's Plant Genetic Resources for Food and Agriculture*; Food and Agriculture Organization of the United Nations: Rome, Italy, 2010; 399p.
22. FAO. *Assessment of the Implementation of the Second Global Plan of Action for Plant Genetic Resources for Food and Agriculture 2012–2014*; Food and Agriculture Organization of the United Nations: Rome, Italy, 2017; 76p.
23. Maxted, N.; Ford-Lloyd, B.V.; Jury, S.L.; Kell, S.P.; Scholten, M.A. Towards a definition of a crop wild relative. *Biodivers. Conserv.* **2006**, *15*, 2673–2685. [CrossRef]
24. Harlan, J.R.; de Wet, J.M.J. Towards a rational classification of cultivated plants. *Taxon* **1971**, *20*, 509–517. [CrossRef]
25. Maxted, N.; Kell, S.P. *Establishment of a Global Network for the In Situ Conservation of Crop Wild Relatives: Status and Needs Background Study Paper No. 39*; FAO Commission on Genetic Resources for Food and Agriculture: Rome, Italy, 2009; 224p.
26. Kell, S.P.; Knüpffer, H.; Jury, S.L.; Ford-Lloyd, B.V.; Maxted, N. Crops and wild relatives of the Euro-Mediterranean region: Making and using a conservation catalogue. In *Crop Wild Relative Conservation and Use*; Maxted, N., Ford-Lloyd, B.V., Kell, S.P., Iriondo, J.M., Dulloo, M.E., Turok, J., Eds.; CAB International: Wallingford, UK, 2008; pp. 69–109.
27. Groombridge, B.; Jenkins, M.D. *World Atlas of Biodiversity*; Prepared by the UNEP World Conservation Monitoring Centre; University of California Press: Berkeley, CA, USA, 2002.
28. Maxted, N.; Kell, S.P.; Toledo, A.; Dulloo, E.M.; Heywood, V.; Hodgkin, T.; Hunter, D.; Guarino, L.; Jarvis, A.; Ford-Lloyd, B.V. A Global Approach to Crop Wild Relative Conservation: Securing the Gene Pool for Food and Agriculture. *Kew Bull.* **2010**, *65*, 561–576. [CrossRef]
29. Vincent, H.; Wiersema, J.; Dobbie, S.; Kell, S.P.; Fielder, H.; Castañeda-Álvarez, N.P.; Eastwood, R.P.; Guarino, L.; Maxted, N. A prioritized crop wild relative inventory to help underpin global food security. *Biol. Conserv.* **2013**, *167*, 265–275. [CrossRef]
30. The Harlan and de Wet Crop Wild Relatives Inventory. Available online: https://www.cwrdiversity.org/checklist/ (accessed on 30 June 2020).
31. Lala, S.; Amri, A.; Maxted, N. Towards the conservation of crop wild relative diversity in North Africa: Checklist, prioritization and inventory. *Genet. Resour. Crop Evol.* **2018**, *65*, 113–124. [CrossRef]
32. Allen, E.; Gaisberger, H.; Brehm, J.M.; Maxted, N.; Thormann, I.; Lupupa, T.; Dulloo, M.E.; Kell, S.P. A crop wild relative inventory for Southern Africa: A first step in linking conservation and use of valuable wild populations for enhancing food security. *Plant Genet. Resour.* **2019**, *17*, 128–139. [CrossRef]
33. CWR Checklists, Strategies, Action Plans. Available online: http://www.cropwildrelatives.org/cwr-strategies/ (accessed on 1 July 2020).

34. Ford-Lloyd, B.V.; Schmidt, M.; Armstrong, S.J.; Barazani, O.; Engels, J.; Hadas, R.; Hammer, K.; Kell, S.P.; Kang, D.; Khoshbakht, K.; et al. Crop Wild Relatives—Undervalued, Underutilized and under Threat? *BioScience* **2011**, *61*, 559–565. [CrossRef]
35. Jarvis, A.; Lane, A.; Hijmans, R.J. The Effect of Climate Change on Crop Wild Relatives. *Agric. Ecosyst. Environ.* **2008**, *126*, 13–23. [CrossRef]
36. Lira, R.; Téllez, O.; Dávila, P. The effects of climate change on the geographic distribution of Mexican wild relatives of domesticated Cucurbitaceae. *Genet. Resour. Crop Evol.* **2009**, *56*, 691–703. [CrossRef]
37. IUCN. *IUCN Red List Categories and Criteria: Version 3.1*, 2nd ed.; IUCN Species Survival Commission, IUCN: Gland, Switzerland; Cambridge, UK, 2012; 32p.
38. Brehm, J.M.; Mitchell, M.; Maxted, N.; Ford-Lloyd, B.V. Martins-Loução, M.A. IUCN Red Listing of Crop Wild Relatives: Is a National Approach as Difficult as Some Think? In *Crop Wild Relative Conservation and Use*; Maxted, N., Ford-Lloyd, B.V., Kell, S.P., Iriondo, J.M., Dulloo, M.E., Turok, J., Eds.; CAB International: Wallingford, UK, 2008.
39. Ludwig, G.; Haupt, H.; Gruttke, H.; Binot-Hafke, M. Methodik der Gefährdungsanalyse für Rote Listen. In *Rote Liste gefährdeter Tiere, Pflanzen und Pilze Deutschlands. Band 1: Wirbeltiere*; Haupt, H., Ludwig, G., Gruttke, H., Binot-Hafke, M., Otto, C., Pauly, A., Eds.; Landwirtschaftsverlag: Münster, Germany, 2009; Volume 70, pp. 19–71.
40. FAO. *Study on the Linkages between Protected Areas and the Conservation of Biodiversity for Food and Agriculture. Thematic Study for the State of the World's Biodiversity for Food and Agriculture*; Commission on Genetic Resources for Food and Agriculture, Food and Agriculture Organization of the United Nations: Rome, Italy, 2020; 22p.
41. Bilz, M.; Kell, S.P.; Maxted, N.; Lansdown, R.V. *European Red List of Vascular Plants*; Publications Office of the European Union: Brussels, Belgium, 2011; 130p.
42. Mora, A.; Zapata-Ferrufino, B.; Hunter, D.; Navarro, G.; Galeano, G.; Apaza, K.S.; Baudoin, M.J.; Dulloo, M.E.; Cuellar, S.; Libro Rojo de Parientes Silvestres de Bolivia; et al. Ministerio de Medio Ambiente y Agua, Viceministerio de Medio Ambiente Biodiversidad Y Cambios Climáticos (VMABCC)/Biodiversity International. 2009. Available online: http://www.cropwildrelatives.org/fileadmin/documents/Red%20List_Bolivia_optim.pdf (accessed on 18 June 2020).
43. Engels, J.M.M. (Bioversity International, Rome, Italy). Personal Communication, 2020.
44. Rao, K.N.; Dulloo, M.E.; Engels, J.M.M. A review of factors that influence the production of quality seed for long-term conservation in genebanks. *Genet. Resour. Crop Evol.* **2017**, *64*, 1061–1074.
45. Engels, J.M.M. An introduction to plant germplasm exploration and collecting: Planning, methods and procedures, follow-up. In *Collecting Plant Genetic Diversity: Technical Guidelines*; Guarino, L., Rao, R.V., Goldberg, E., Eds.; Bioversity International: Rome, Italy, 2011.
46. Hoban, S.; Schlarbaum, S. Optimal sampling of seeds from plant populations for *ex-situ* conservation of genetic biodiversity, considering realistic population structure. *Biol. Conserv.* **2014**, *177*, 90–99. [CrossRef]
47. Hoban, S.; Strand, A. Ex situ seed collections will benefit from considering spatial sampling design and species' reproductive biology. *Biol. Conserv.* **2015**, *187*, 182–191. [CrossRef]
48. Bellon, M.R.; Dulloo, E.; Sardos, J.; Thormann, I.; Burdon, J.J. In situ conservation—Harnessing natural and human-derived evolutionary forces to ensure future crop adaptation. *Evol. Appl.* **2017**, *10*, 965–977. [CrossRef]
49. Aguirre-Gutiérrez, J.; van Treuren, R.; Hoekstra, R.; van Hintum, T.J.L. Crop wild relatives range shifts and conservation in Europe under climate change. *Divers. Distrib.* **2017**, *23*, 739–750. [CrossRef]
50. FAO. *Genebank Standards for Plant Genetic Resources for Food and Agriculture*; Food and Agriculture Organization of the United Nations: Rome, Italy, 2014; 166p.
51. Hanson, J.; Ellis, R.H. Progress and Challenges in Ex Situ Conservation of Forage Germplasm: Grasses, Herbaceous Legumes and Fodder Trees. *Plants* **2020**, *9*, 446. [CrossRef]
52. Engels, J.M.M.; Maggioni, L. Managing germplasm in a virtual European genebank (AEGIS) through networking. In *Theorien der Lebendsammlung. Pflanzen, Mikroben und Tiere als Biofakte in Genbanken. (Lebenswissenschaften im Dialog 25)*; Karafyllis, N.C., Ed.; Euro. Verlag: Freiburg/München, Germany, 2018; pp. 169–197.
53. Ford-Lloyd, B.; Kell, S.P.; Maxted, N. Establishing Conservation Priorities for Crop Wild Relatives. In *Crop Wild Relative Conservation and Use*; Maxted, N., Ford-Lloyd, B., Kell, S.P., Iriondo, J.M., Dulloo, M.E., Turok, J., Eds.; CAB International: Wallingford, UK, 2008; pp. 110–119.

54. Akparov, Z.I.; Aronsson, M.; Asdal, A.; Avagyan, A.; Bartha, B.; Benediková, D.; Berishvili, T.; Bocci, R.; Bullinska-Radomska, Z.; Cop, J.; et al. Current and future threats and opportunities facing European crop wild relative and landrace diversity. In *Agrobiodiversity Conservation: Securing the Diversity of Crop Wild Relatives and Landraces*; Maxted, N., Dulloo, E.M., Ford-Lloyd, B.V., Frese, L., Iriondo, J., Pinheiro-de-Carvalho, M.A.A., Eds.; CABI: Wallingford, UK, 2011; pp. 333–354.
55. Hunter, D.; Heywood, V. (Eds.) *Crop Wild Relatives. A Manual of in situ Conservation*; Routledge: London, UK, 2011; 414p.
56. Brehm, J.M.; Kell, S.; Thormann, I.; Gaisberger, H.; Dulloo, E.; Maxted, N. *Interactive Toolkit for Crop Wild Relative Conservation Planning Version 1.0*; University of Birmingham: Birmingham, UK; Bioversity International: Rome, Italy, 2017. Available online: http://www.cropwildrelatives.org/conservation-toolkit/ (accessed on 18 June 2020).
57. Thormann, I.; Jarvis, D.I.; Dearing, J.A.; Hodgkin, T. Internationally available information sources for the development of in situ conservation strategies for wild species useful for food and agriculture. *Plant Genet. Resour. Newsl.* **1999**, *118*, 38–50.
58. Thormann, I.; Lane, A.; Durah, K.; Dulloo, M.E.; Gaiji, S. Crop wild relative information: Developing a tool for its management and use. In *Crop Wild Relative Conservation and Use*; Maxted, N., Ford-Lloyd, B.V., Kell, S.P., Iriondo, J., Dulloo, M.E., Turok, T., Eds.; CAB International: Wallingford, UK, 2008; pp. 504–512.
59. Thormann, I. Published Sources of Information on Wild Plant Species. In *Collecting Plant Genetic Diversity: Technical Guidelines*; Guarino, L., Rao, V.R., Goldberg, E., Eds.; Bioversity International: Rome, Italy, 2011; 10p, ISBN 978-92-9043-922-6. Available online: http://cropgenebank.sgrp.cgiar.org/index.php?option=com_content&view=article&id=657 (accessed on 18 June 2020).
60. Thormann, I.; Alercia, A.; Dulloo, M.E. *Core Descriptors for In Situ Conservation of Crop Wild Relatives v.1*; Bioversity International: Rome, Italy, 2013; 28p.
61. Thormann, I.; Kell, S.; Brehm, J.M.; Dulloo, M.E.; Maxted, N. *CWR Checklist and Inventory Data Template v.1*; Harvard Dataverse: Cambridge, MA, USA, 2017; Available online: https://dataverse.harvard.edu/dataset.xhtml?persistentId=10.7910/DVN/B8YOQL. (accessed on 18 June 2020). [CrossRef]
62. Brehm, J.M.; Kell, S.; Thormann, I.; Gaisberger, H.; Dulloo, E.; Maxted, N. *Occurrence Data Collation Template v*; Harvard Dataverse: Cambridge, MA, USA, 2017; Available online: https://dataverse.harvard.edu/dataset.xhtml?persistentId=10.7910/DVN/5B9IV5. (accessed on 18 June 2020). [CrossRef]
63. Thormann-Wiersema, J.H.; León, B. The GRIN Taxonomy Crop Wild Relative Inventory. In *Enhancing Crop Genepool Use: Capturing Wild Relative and Landrace Diversity for Crop Improvement*; Maxted, N., Dulloo, M.E., Ford-Lloyd, B.V., Eds.; CAB International: Wallingford, UK, 2016; pp. 453–458.
64. Bioversity International, University of Birmingham. *Crop Wild Relative Checklist and Inventory Descriptors, v.1*; Bioversity International: Rome, Italy, 2017; 26p.
65. Vincent, H.; Amri, A.; Castañeda-Álvarez, N.P.; Dempewolf, H.; Dulloo, E.M.; Guarino, L.; Hole, D.; Mba, C.; Toledo, A.; Maxted, N. Modeling of crop wild relative species identifies areas globally for in situ conservation. *Commun. Biol.* **2019**, *2*, 1–8. [CrossRef]
66. UNEP-WCMC, IUCN and NGS. *Protected Planet Live Report 2020*; UNEP-WCMC: Cambridge, UK; IUCN: Gland, Switzerland; NGS: Washington, DC, USA, 2020.
67. Frese, L.; Bönisch, M.; Herden, T.; Zander, M.; Friesen, N. In-situ-Erhaltung von Wildselleriearten. *Nat. Landsch.* **2018**, *50*, 155–163.
68. Bönisch, M.; Frese, L. Designation of Genetic Reserves for Wild Celery Species in Germany. Crop Wild Relative Issue 12; ISSN 1742-3694 (Online). In Press. Available online: http://farmerspride.eu/.
69. Thormann, I. The German Network of Genetic Reserves. Crop Wild Relative Issue 12; ISSN 1742-3694 (Online). In Press. Available online: http://farmerspride.eu/.
70. Parra-Quijano, M.; Iriondo, J.M.; Torres, E. Ecogeographical land characterization maps as a tool for assessing plant adaptation and their implications in agrobiodiversity studies. *Genet. Resour. Crop Evol.* **2011**, *59*, 205–217. [CrossRef]
71. Genesys Is an Online Platform Where You Can Find Information about Plant Genetic Resources for Food and Agriculture (PGRFA) Conserved in Genebanks Worldwide. Available online: https://www.genesys-pgr.org/ (accessed on 1 July 2020).
72. Eurisco. Finding Seeds for the Future. Available online: http://eurisco.ecpgr.org (accessed on 1 July 2020).

73. Castañeda-Álvarez, N.P.; Khoury, C.K.; Achicanoy, H.A.; Bernau, V.; Dempewolf, H.; Eastwood, R.J.; Guarino, L.; Harker, R.H.; Jarvis, A.; Maxted, N.; et al. Global conservation priorities for crop wild relatives. *Nat. Plants* **2016**, *2*, 16–22. [CrossRef]
74. Heywood, V.H. The role of botanic gardens in ex situ conservation of agrobiodiversity. In *Implementation of the Global Plan of Action in Europe—Conservation and Sustainable Utilization of Plant Genetic Resources for Food and Agriculture, Proceedings of the European Symposium, Braunschweig, Germany, 30 June–3 July 1988*; Gass, T., Frese, L., Begemann, F., Lipman, E., Eds.; International Plant Genetic Resources Institute: Rome, Italy, 1999; pp. 102–107.
75. Dempewolf, H. (Global Crop Diversity Trust, Bonn, Germany). Personal communication, 2020.
76. Engels, J.; Thormann, I. *Final Report of a Consultancy "Increasing Climate Resilience for Poor Farmers: The role of National Plant Genetic Resource Collections"*; Kreditanstalt für Wiederaufbau (KfW) and Global Crop Diversity Trust: Frankfurt/Bonn, Germany, 2017; 129p, Unpublished, manuscript in preparation.
77. Khoury, C.K.; Carver, D.; Kates, H.R.; Achicanoy, H.A.; van Zonneveld, M.; Thomas, E.; Heinitz, C.; Jarret, R.; Labate, J.A.; Reitsma, K.; et al. Distributions, conservation status, and abiotic stress tolerance potential of wild cucurbits (*Cucurbita* L.). *Plants People Planet* **2019**, *2*, 269–283. [CrossRef]
78. Welcome to PlantSearch! Botanic Gardens Conservation. International. Available online: https://tools.bgci.org/plant_search.php (accessed on 30 June 2020).
79. Meyer, A.; Barton, N. Botanic Gardens Are Important Contributors to Crop Wild Relative Preservation. *Crop Sci.* **2019**, *59*, 2404–2412. [CrossRef]
80. Dempewolf, H.; Eastwood, R.J.; Guarino, L.; Khoury, C.K.; Muller, J.V.; Toll, J. Adapting agriculture to climate change: A global initiative to collect, conserve, and use crop wild relatives. *Agroecol. Sustain. Food Syst.* **2014**, *38*, 369–377. [CrossRef]
81. Engels, J.M.M. Complementary strategies for improved conservation and use of plant genetic resources. In *Towards Sustainable National Plant Genetic Resources Programmes—Policy, Planning and Conservation Issues*; Engels, J.M.M., Vodouhe, R., Thompson, J., Zannou, A., Hehne, E., Grum, M., Eds.; IPGRI: Rome, Italy, 2000; pp. 69–77.
82. Hunter, D.; Changtragoon, S. Good practices for conservation and sustainable use of crop wild relatives of tropical fruit tree diversity. In *Tropical Fruit Tree Diversity. Good Practices for In Situ and On-Farm Conservation*; Sthapit, B., Lamers, H., Rao, R.V., Bailey, A., Eds.; Earthscan from Routledge: London, UK, 2016; pp. 83–96.
83. Jha, U.C.; Bohra, A.; Singh, N.P. Heat stress in crop plants: Its nature, impacts and integrated breeding strategies to improve heat tolerance. *Plant Breed.* **2014**, *133*, 679–701. [CrossRef]
84. Dulloo, M.E.; Fiorino, E.; Thormann, I. Research on Conservation and Use of Crop Wild Relatives. In *Crop Wild Relatives and Climate Change*; Chapter 7; Redden, R., Yadav, S.S., Maxted, N., Dulloo, M.E., Guarino, L., Smith, P., Eds.; Wiley-Blackwell: Hoboken, NJ, USA, 2015; pp. 108–129.
85. Vavilov, N.I. The origin, variation, immunity and breeding of cultivated plants. *Chron. Bot.* **1951**, *13*, 1–54. [CrossRef]
86. Prescott-Allen, C.; Prescott-Allen, R. *The First Resource: Wild Species in the North American Economy*; Yale University Press: New Haven, CT, USA, 1986.
87. Engels, J.M.M. Genetische bronnen, hun conservering en aanwending in de aardappelveredeling. In *Genetic Resources, Conservation and Use in the Potato Breeding*; Agricultural University Wageningen: Wageningen, The Netherlands, 1974; 83p.
88. Kilian, B.; Martin, W.; Salamini, F. Genetic diversity, evolution and domestication of wheat and barley in the Fertile Crescent. In *Evolution in Action*; Glaubrecht, M., Ed.; Springer: Berlin/Heidelberg, Germany, 2010; pp. 137–166.
89. Thormann, I.; Parra-Quijano, M.; Endresen, D.T.F.; Rubio-Teso, M.L.; Iriondo, M.J.; Maxted, N. *Predictive Characterization of Crop Wild Relatives and Landraces. Technical Guidelines, Version 1*; Bioversity International: Rome, Italy, 2014; 40p.
90. Willis, K.J. *State of the World's Plants Report—2017*; Royal Botanic Gardens: Kew, UK, 2017; 96p.
91. Bertioli, D.J.; Cannon, S.B.; Froenicke, L.; Huang, G.; Farmer, A.D.; Cannon, E.K.; Liu, X.; Gao, N.; Clevenger, J.; Dash, S.; et al. The genome sequences of *Arachis duranensis* and *Arachis ipaensis*, the diploid ancestors of cultivated peanut. *Nat. Genet.* **2016**, *48*, 438–446. [CrossRef]

92. PricewaterhouseCoopers PwC, 2013. Crop Wild Relatives: A Valuable Resource for Crop Development. www.pwc.co.uk/valuations. Available online: https://pwc.blogs.com/files/pwc-seed-bank-analysis-for-msb-0713.pdf (accessed on 15 January 2020).
93. Pimentel, D.; Wilson, C.; McCullum, C.; Huang, R.; Dwen, P.; Flack, J.; Tran, Q.; Saltman, T.; Cliff, B. Economic and environmental benefits of biodiversity. *Bioscience* **1997**, *47*, 747–757. [CrossRef]
94. Tyack, N.; Dempewolf, H. The Economics of Crop Wild Relatives under Climate Change. In *Crop Wild Relatives and Climate Change*, 1st ed.; Redden, R., Yadav, S.S., Maxted, N., Dulloo, M.E., Guarino, L., Smith, P., Eds.; Wiley Online Library John Wiley & Sons, Inc.: Hoboken, NJ, USA, 2015.
95. Mira, S.; Hill, L.M.; González-Benito, M.E.; Ibáñez, M.A.; Walters, C. Volatile emission in dry seeds as a way to probe chemical reactions during initial asymptomatic deterioration. *J. Exp. Bot.* **2016**, *67*, 1783–1793. [CrossRef]
96. Colville, L.; Bradley, E.L.; Lloyd, A.S.; Pritchard, H.W.; Castle, L.; Kranner, I. Volatile fingerprints of seeds of four species indicate the involvement of alcoholic fermentation, lipid peroxidation, and Maillard reactions in seed deterioration during ageing and desiccation stress. *J. Exp. Bot.* **2012**, *63*, 6519–6530. [CrossRef] [PubMed]
97. Michalak, M.; Plitta-Michalak, B.P.; Naskręt-Barciszewska, M.; Barciszewski, J.; Bujarska-Borkowska, B.; Chmielarz, P. Global 5-methylcytosine alterations in DNA during ageing of *Quercus robur* seeds. *Ann. Bot.* **2015**, *116*, 369–376. [CrossRef]
98. Mira, S.; Pirredda, M.; Martín-Sánchez, M.; Marchessi, J.; Martín, C. DNA methylation and integrity in aged seeds and regenerated plants. *Seed Sci. Res.* **2020**, 1–9. [CrossRef]
99. Kranner, I.; Chen, H.; Pritchard, H.W.; Pearce, S.R.; Birtic, S. Inter-nucleosomal DNA fragmentation and loss of RNA integrity during seed ageing. *Plant Growth Regul.* **2011**, *63*, 63–72. [CrossRef]
100. Fleming, M.B.; Patterson, E.L.; Reeves, P.A.; Richards, C.M.; Gaines, T.A.; Walters, C. Exploring the fate of mRNA in aging seeds: Protection, destruction, or slow decay? *J. Exp. Bot.* **2018**, *69*, 4309–4321. [CrossRef]
101. Dulloo, M.E.; Ramanatha, V.R.; Engelmann, F.; Engels, J.M.M. Complementary conservation strategy for coconuts. In *Coconuts Genetic Resources*; Batugal, P., Ramanatha, V.R., Oliver, J., Eds.; International Plant Genetic Resources Institute-Regional Office for Asia, the Pacific, and Oceania (IPGRI-APO): Serdang, Malaysia, 2005; 18p.
102. Teso, M.L.R.; Torres, M.E.; Parra-Quijano, M.; Iriondo, J.M. Prioritization of crop wild relatives in Spain. *Crop Wild Relat.* **2012**, *18*, 18–22.
103. van Treuren, R. *PGR Management in the 21st Century. Crop Wild Relatives: Climate Change and Niche Modeling*; Wageningen University and Research, Centre for Genetic Resources: Wageningen, The Netherlands, 2017; Available online: https://edepot.wur.nl/441411 (accessed on 2 May 2020).
104. Crop Wild Relatives Global Portal. Available online: http://www.cropwildrelatives.org/ (accessed on 1 July 2020).
105. Indigenous and Community Conserved Areas: A Bold New Frontier for Conservation. Available online: https://www.iucn.org/content/indigenous-and-community-conserved-areas-a-bold-new-frontier-conservation (accessed on 1 July 2020).
106. Brooks, T.M.; Mittermeier, R.A.; da Fonseca, G.A.B.; Gerlach, J.; Hoffmann, M.; Lamoreux, J.F.; Mittermeier, C.G.; Pilgrim, J.D.; Rodrigues, A.S.L. Global biodiversity conservation priorities. *Science* **2006**, *313*, 58–61. [CrossRef]
107. Khoury, C.K.; Amariles, D.; Soto, J.S.; Diaz, M.V.; Sotelo, S.; Sosa, C.; Ramírez-Villegas, J.; Achicanoy, H.A.; Velásquez-Tibatá, J.; Guarino, L.; et al. Comprehensiveness of conservation of useful wild plants: An operational indicator for biodiversity and sustainable development targets. *Ecol. Indic.* **2019**, *98*, 420–429. [CrossRef]
108. Dulloo, M.E.; Magos Brehm, J.; Kell, S.; Thormann, I.; Maxted, N. *Template for the Preparation of a National Strategic Action Plan for the Conservation and Sustainable Use of Crop Wild Relatives*; Harvard Dataverse: Cambridge, MA, USA, 2017; Volume 1, 23p, Available online: https://doi.org/10.7910/DVN/QH9XWB (accessed on 18 June 2020). [CrossRef]
109. Brehm, J.M.; Kell, S.; Thormann, I.; Maxted, N.; Dulloo, E. Template for the preparation of a technical background document for a national strategic action plan for the conservation and sustainable use of crop Wild Relatives. *Harv. Dataverse* **2017**, 23. [CrossRef]

110. Álvarez, N.P.C.; Vincent, H.A.; Kell, S.P.; Eastwood, R.J.; Maxted, N. Ecogeographic surveys. In *Collecting Plant Genetic Diversity: Technical Guidelines*; Guarino, L., Rao, V.R., Goldberg, E., Eds.; Bioversity International: Rome, Italy, 2011; 23p.
111. Parra-Quijano, M.; Torres, E.; Iriondo, J.M.; López, F. *CAPFITOGEN Tools User Manual, Version 2.0*; International Treaty on Plant Genetic Resources for Food and Agriculture, Food and Agriculture Organization of the United Nations: Rome, Italy, 2016; 260p.
112. Centro Internacional de Agricultura Tropical—CIAT. A Global Database for the Distributions of Crop Wild Relatives. Version 1.12. Available online: https://doi.org/10.15468/jyrthk (accessed on 28 June 2020).

plants

MDPI

Review

The Potential of Payment for Ecosystem Services for Crop Wild Relative Conservation

Nicholas Tyack [1,*], Hannes Dempewolf [2] and Colin K. Khoury [3]

1 Graduate Institute of International and Development Studies, 1202 Geneva, Switzerland
2 Global Crop Diversity Trust, 53113 Bonn, Germany; Hannes.dempewolf@croptrust.org*
3 International Center for Tropical Agriculture (CIAT), 6713 Cali, Colombia; c.khoury@cgiar.org
* Correspondence: nicholas.tyack@graduateinstitute.ch

Received: 28 August 2020; Accepted: 25 September 2020; Published: 2 October 2020

Abstract: Crop wild relatives (CWR) have proven to be very valuable in agricultural breeding programs but remain a relatively under-utilized and under-protected resource. CWR have provided resistance to pests and diseases, abiotic stress tolerance, quality improvements and yield increases with the annual contribution of these traits to agriculture estimated at USD 115 billion globally and are considered to possess many valuable traits that have not yet been explored. The use of the genetic diversity found in CWR for breeding provides much-needed resilience to modern agricultural systems and has great potential to help sustainably increase agricultural production to feed a growing world population in the face of climate change and other stresses. A number of CWR taxa are at risk, however, necessitating coordinated local, national, regional and global efforts to preserve the genetic diversity of these plants through complementary in situ and ex situ conservation efforts. We discuss the absence of adequate institutional frameworks to incentivize CWR conservation services and propose payment for ecosystem services (PES) as an under-explored mechanism for financing these efforts. Such mechanisms could serve as a potentially powerful tool for enhancing the long-term protection of CWR.

Keywords: crop wild relatives; payment for ecosystem services; payment for environmental services; agrobiodiversity conservation; climate change; agricultural adaptation

---

1. Introduction

Crop wild relatives (CWR) are wild or weedy plants that are the progenitors or close relatives of crops including those that readily cross with cultivated taxa as well as more distantly related species within the same or related genera [1]. CWR constitute genetic resources with demonstrated value for plant breeding due to their useful traits, which are not easily found in crops due to domestication bottlenecks and the subsequent narrowing of the cultivated gene pool through breeding efforts [2]. Further, CWR are often important cultural resources being directly harvested for food, spice, medicine, ceremony or other uses [3] and provide various ecosystem services within their natural habitats.

Introgression of CWR traits has led to improvements in an ever-growing list of crops including enhanced resistance to pests and diseases, tolerance to abiotic stresses such as drought, heat and salinity, yield increases, quality improvements and other desired characteristics [1,4]. Maxted and Kell identify the transfer of useful traits from 185 CWR taxa to 29 crop species [5]. Estimates of the value of these contributions have ranged from USD 8 million (USD 17.6 million in 2020) per year for increased sugar content and improved taste from a single wild tomato species [6] to USD 267–384 million (USD 310–445 million in 2020) per year for the wild sunflower gene pool [7]. Pimentel et al. estimate the contributions of CWR genetic material to increased crop yields at approximately USD 20 billion per year (USD 32.4 billion in 2020) in the United States and USD 115 billion (USD 186.3 billion in 2020) per year worldwide [8].

However, with more than a fifth of plant species worldwide threatened with extinction [9] amidst what has been called the "sixth great extinction event in Earth's history" [10], CWR are facing threats from development and other forms of habitat modification, the industrialization of agriculture, invasive species, pollution, overharvesting, overgrazing and climate change. A recent European Red List assessment of Vascular Plants found that 11.5% of assessed CWR species in Europe were threatened with another 4.5% at Near Threatened status [11]. Jarvis et al. estimated that 16–22% of peanut, potato and cowpea wild relative species are likely to become extinct by 2055 [12] while Ureta et al. predicted that climate change is likely to lead to severe reductions in the potential distributions of maize wild relatives [13].

Up to double the amount of food that is presently produced may be required by 2050 to feed a global population of around 9 billion [14]. At the same time, normal growing season temperatures are expected to exceed the most extreme seasonal temperatures recorded from 1900 to 2006 by the end of the 21st century, which will likely have severe effects on the cultivation of many crops [15,16]. For example, rice grain yield can decline by up to 10% for each 1 °C increase in growing season minimum temperature in the dry season [17]. With rapidly evolving techniques in molecular biology, it is increasingly feasible to access and utilize genetic material from CWR including distant relatives and the expanded use of CWR with these tools provides a potential pathway to contribute to meeting these challenges through the development of crop varieties that are more resilient and productive [1,2,18]. For example, a gene from the Asian wild rice species *Oryza rufipogon* Griff. has been shown to significantly increase rice yields [19]. The incorporation of CWR derived genetic diversity into elite gene pools has been a key tool for plant breeders for many decades with hundreds of different taxa that have been used in this way especially to introduce disease tolerance traits into domesticated crops. In more recent years the value of crop wild relatives to also address abiotic stress tolerances, including many of relevance to climate change, is becoming more widely recognized and many breeding programs around the world are using these genetic resources in pursuit of that goal. As such, the range of stakeholders involved is also broadening beyond the public breeding programs in universities and national and international agricultural research programs. Several private companies are adding pre-breeders to their staff who often engage in pre-competitive public-private partnerships to utilize CWR and other types of genetic resources [2].

However, the users of CWR diversity for the most part remain far removed from where CWR are found in nature and CWR remain a largely unrecognized group within the field of conservation policy and the ecosystem services literature [20]. The continued lack of sufficient investment in CWR conservation may lead to permanent gaps in the pool of wild genetic resources available to crop breeders. Their extirpation will also have negative impacts locally including the loss of their contributions as wild-harvested plants and the disappearance of the ecosystem services they provide.

At the root of these conservation deficiencies is a lack of, or inadequacies in, institutions and payment systems by which the beneficiaries of CWR conservation services could compensate those who can supply them. Adaptation of payment for ecosystem services (PES) mechanisms to CWR offers a potentially useful tool for correcting this failure and enabling the creation of a missing institutional framework for the conservation of wild and weedy genetic diversity to support future agricultural research and crop improvement efforts. While early PES schemes have encountered challenges in implementation, the mechanism has been shown to be effective in strengthening ecosystem service provision [21] and well-designed PES programs can offer a low-cost and efficient solution for the mitigation of market failures associated with ecosystem service provision such as those associated with the carbon fixation services provided by forests or water filtration services provided by wetlands and riparian buffers [22]. The use of PES over the past years has provided potential PES practitioners with a number of lessons that can help improve program design [23]. Initial research investigating the potential for implementing PES for CWR in fact already exists in the case of several CWR taxa in Zambia where competitive tenders have been held for farmer conservation of CWR in field margins [24] as well as experiences from Latin America, which provide insights for the design of payment for agrobiodiversity conservation service programs more broadly [25]. Importantly, the successful conservation of CWR requires a combination of action on different geographic (local, national, global) and social (individual, market, societal) levels spanning in situ, ex situ and on-farm conservation [26]. If designed well, PES instruments provide the flexibility to further CWR conservation across these different dimensions.

We describe here the ways in which economic benefits flow from the conservation of CWR, discuss the absence of adequate institutional frameworks to incentivize CWR conservation services and discuss the potential of payment for ecosystem services (PES) as a tool for the conservation and sustainable use of CWR genetic diversity.

2. CWR and Ecosystem Services

Crop wild relative populations provide a number of ecosystem services (Table 1), which include direct contributions locally as well as cultural and genetic resource services more widely. With regard to agriculture and plant breeding, the conservation of these populations provides the important "supporting" ecosystem service of plant genetic diversity [27]. CWR germplasm, as a tangible material product resulting from ecosystem processes, is an "ecosystem good" [28] that is collected from wild, weedy or human managed habitats, deposited in gene banks or other repositories for ex situ conservation, undergoes a process of pre-breeding and breeding and finally results in the introduction of beneficial traits into crop varieties (provisioning service). In traditional farming systems, wild relatives of numerous crops also provide genetic diversity "spontaneously" to crops in the field through the natural gene flow; for example, wild and cultivated populations of cowpea often overlap and there is evidence of substantial hybridization between the two in the field [29]. The Millennium Ecosystem Assessment has also considered genetic resources as primarily a provisioning good/service [30] while other authors see genetic diversity as providing mainly a supporting service to agriculture [27]. The Economics of Ecosystems and Biodiversity (TEEB) clarifies that the collection of useful genetic resources from nature implies a provisioning service whereas the maintenance of genetic diversity, e.g., through the in situ conservation of CWR, should be considered to be a supporting or "habitat" service [31].

Table 1. List of ecosystem services provided by CWR.

Ecosystem Service	Examples
Supporting service	The conservation of CWR populations maintains genetic diversity and allows for the continuing evolution of the gene pool as a resource for future crop improvement, providing an important supporting service to help meet future demand for improved crop varieties and resilient agricultural systems [32].
Regulating services	CWR can regulate certain ecosystem processes such as pest and disease control, pollination efficiency, nutrient cycling, decomposition, erosion control and carbon sequestration [27].
Provisioning services	CWR provide a provisioning service of genetic resources when CWR germplasm is collected from wild populations and used by plant breeders to develop improved varieties. In addition, some CWR are harvested and directly used for food, spice, medicine, ceremony or other purposes [28].
Cultural services	CWR are a part of the world's natural and cultural heritage with potential for ecotourism (e.g., wild coffee forests in Ethiopia) [33].

When CWR and other forms of agrobiodiversity are conserved and used for crop improvement, agricultural system resilience may be increased through a greater resistance to pests and diseases among other factors. Changing conditions and the outbreak of new pests and diseases can cause significant losses in productivity as occurred during the 1970–1971 Southern corn leaf blight outbreak, which led to the loss of almost 710 million bushels of the US maize crop [34]. Such epidemics can turn into catastrophes such as the Irish Potato Famine, an outbreak of the late blight disease associated with the death or displacement of 25% of the Irish population [35].

The use of CWR to breed more resistant or resilient crop varieties can help to avert persistent crop failures stemming from a genetic deficiency; for example, one of the very few potato cultivars immune to the most recent and virulent strains of late blight is Sarpo Mira with its durable, broad-spectrum resistance coming from genes that originated in the wild potato species *Solanum demissum* Lindl. [36]. The use of CWR can provide insurance value to agriculture by minimizing the risks posed by climate change, droughts and pests and diseases to the genetically homogenous monocultures of modern industrial agriculture. Though valuable traits can also be found in traditional crop varieties and other sources, in some cases the unique traits found in the genomes of CWR are essentially a non-substitutable good as shown also by the provision of cytoplasmic male sterility from sunflower CWR [37].

CWR germplasm is a mostly renewable and non-rival resource in that it can be multiplied and shared at a relatively low cost (for most crops) though it is possible to exclude others from using it. As with other forms of plant genetic resources, CWR thus constitute an imperfect public good with both public and private characteristics. Importantly, various publicly-held collections of CWR of a list of globally important crops are required to be shared freely by the International Treaty on Plant Genetic Resources for Food and Agriculture (ITPGRFA) [38], which came into force in 2004 and strengthened the public and non-excludable aspects of these resources. The ITPGRFA is a major international agreement that governs access to the CWR species of many agricultural, horticultural and forage crops. Parties are governed by the Treaty's Multilateral System of access and benefit sharing, which specifies that publicly-held CWR materials included in Annex 1 of the ITPGRFA in countries that have ratified the Treaty are non-excludable. They must be made available to those who request them for the purpose of research and breeding, rendering these species a global public good whose benefits can be enjoyed by all although payment is required to the Treaty's Benefit Sharing Fund if the accessed materials are used to breed a commercialized variety that itself is available without restrictions for further research and breeding. Under this Multilateral System, plant breeders and other researchers around the world have access to CWR germplasm held in the gene banks of the CGIAR and other international collections as well as the national gene banks of the countries that have ratified and implemented the Treaty (though a number of countries possessing significant CWR resources have not yet done so including China, Israel and Mexico). Non-public collections can also be placed under the umbrella of the Treaty. CWR not under the purview of the Multilateral System of the ITPGRFA fall under the governance of the Nagoya Protocol on Access and Benefit Sharing under the Convention on Biological Diversity [39] and are subject to bilateral agreement negotiations.

Land containing CWR populations maintains two important ecosystem services related to the production of the ecosystem good of CWR germplasm. First, the CWR habitat essentially serves as a natural repository for these resources. Second, it also serves as an incubator for the creation of novel allelic variation and genotypes, some of which may be useful to breeders. When the CWR habitat is destroyed, both of these services are lost as well. CWR populations often contain higher allelic diversity than cultivated diversity as a result of genetic bottlenecks and continued evolution, as has been demonstrated for soybean, and a range of CWR populations maintained across diverse topographies with different microclimates can provide both a valuable resource of genetic diversity and also incubate new traits and alleles of value [40]. Conversely, changes in the CWR habitat can destroy these resources.

In situ genetic reserves are thus an important conservation strategy, complementary to gene banks and other ex situ repositories (and vice versa) and systems employing both in situ and ex situ conservation are generally considered to be the most robust in terms of successful long-term conservation [41,42]. A population of CWR conserved in situ typically contains more genetic diversity than a single accession that is collected and stored in a gene bank, which is a subset meant to represent that population and provides a storage service associated with the conservation of genetic diversity. The incubator service refers to the maintenance of evolutionary relationships and processes in populations preserved in situ. Continued evolution of CWR populations has value for adapting crops to changes in pests and disease pressures [43]. As ecosystems change, for example through climatic shift and the subsequent migration of species, populations of CWR may adapt and these adaptations may benefit future breeding efforts. As an example, CWR living in dry ecosystems that become even drier due to climate change may respond to selection for increased drought tolerance and become particularly useful in the context of crop improvement for this trait even on short time scales [44]. It is important to note that CWR occur not only in wild ecosystems but also in disturbed and in human managed environments including in and around agricultural fields. While these may be considered as locally useful resources by farmers, wild-harvesters and others, they also may be regarded as agricultural nuisances, for example in wild pumpkins (*Cucurbita* L.), where populations close to agricultural fields in the United States and Mexico may be purposely extirpated because the gene flow can introduce bitterness into the fruit of cucurbit crops [45].

Ex situ conservation provides the opportunity to conserve a great diversity of samples in a small area and is thus indispensable for maintaining CWR genetic resources and facilitating their use for crop improvement. However, each sample in a gene bank only represents a single snapshot in time of a limited amount of the genetic diversity of the wild population. The in situ conservation of a CWR population coupled with ex situ protection thus provides the range of conditions under which wild populations may continue to evolve and also be readily available for agricultural research.

3. The Rationale for Payment for Ecosystem Services for CWR Conservation

Crop wild relatives have demonstrated economic and cultural value [46] and potentially even greater future value but major gaps have been recognized in their conservation, both in situ and ex situ. However, in spite of this demonstrated economic value and the threatened status of many CWR species, with over 70% of CWR taxa identified as high priority for further collecting [47], strong institutional frameworks to support the conservation of CWR do not exist. In the absence of such a framework, the owners of land serving as a habitat for CWR populations are not compensated for conserving CWR and thus have no incentive to do so. In fact, they may very well have various incentives to convert CWR habitat into farmland or develop it for other purposes. Indeed, it is not private actors (individuals or firms) who have implemented most of the past CWR conservation projects but rather the financing for most past major initiatives to conserve CWR in situ to date have typically come from internationally funded projects (e.g., by the Global Environmental Facility).

A key issue associated with the protection of CWR populations in situ is that many of the agricultural benefits arising from CWR conservation are separated both temporally and geographically from those who have the ability to provide them. For example, CWR might remain in a gene bank for several years or even decades before it is selected for breeding, after which it typically takes ten years for the effort to result in a released variety that provides economic benefits.

In this paper, we explore payment for ecosystem service (PES) instruments as incentive mechanisms to provide compensation for the provision of CWR conservation services. Such market instruments have the potential to help bridge the gap between those willing to pay for CWR conservation and those able to conserve them and create stronger institutional frameworks for CWR conservation.

4. Payment for Ecosystem Services Mechanisms for CWR Conservation

The adoption of payment for ecosystem services (PES) to the CWR context could incentivize the provision of CWR conservation services, bridging the gap between the beneficiaries and providers of these services. The core idea of PES is that "external (ecosystem service) beneficiaries make direct, contractual and conditional payments to local landholders and users in return for adopting practices that secure ecosystem conservation and restoration" [48].

Although payment for ecosystem services has been hailed as arguably "the most promising innovation in conservation since [the enactment of the Convention of Biological Diversity at] Rio 1992" [48], the mechanism has not been widely applied to agrobiodiversity conservation, particularly wild agrobiodiversity [49]. Noting the lack of research in this area, Narloch et al. [49] also proposed "payments for agrobiodiversity conservation services" (PACS) as a PES-like solution to the loss of landraces and other local crop varieties, which they define as an "economic instrument to tackle market, intervention and global appropriation failures associated with the public good characteristics of agrobiodiversity conservation services through the use of (monetary or in kind) reward mechanisms in order to increase the private benefits from local (plant and animal genetic resources)".

Though PACS is designed to incentivize the on-farm conservation of under-utilized and endangered crop varieties, PES instruments are also promising in the context of mitigating the market failure associated with the under-conservation of CWR by increasing the private benefits of conserving CWR and could fund both in situ and ex situ conservation activities.

4.1. Who Will Pay for CWR Conservation?

Large scale CWR conservation efforts are slowly increasing at an international and various national levels as seen in the examples of the Sierra de Manantlan Biosphere reserve in Mexico, the Ammiad Project in Israel, the Erebuni Reserve in Armenia and the Global Environment Facility project "In situ conservation of crop wild relatives through enhanced information management and field application", which developed national CWR conservation strategies in Armenia, Bolivia, Madagascar, Sri Lanka and Uzbekistan [7]. Given the value of CWR for adapting agriculture to climate change, the focus of a major CWR-related project over the past decade [50], PES mechanisms for a complementary system of in situ and ex situ conservation of CWR species may also be attractive investment targets for the proposed USD 100 billion Green Climate Fund, which envisions significant investment in climate change adaptation and mitigation measures [51].

There may also be under-explored demand for CWR conservation in the private sector. For example, the drug company Merck & Co. paid an upfront fee of USD 1 million to Costa Rica's National Institute of Biodiversity to help conserve rainforest biodiversity in exchange for the rights to use samples of the plants, insects and microorganisms collected through this program to create new pharmaceutical products [52]. Investing in payments for CWR conservation services could provide agricultural sector companies with germplasm that might be difficult to access otherwise. A PES for CWR conservation program funded by a private agricultural firm could be seen as an investment in the long-term sustainability of the industry and could also be an opportunity for green marketing and corporate social responsibility programs [53,54]. Given that firms are under increasing pressure from stakeholders to reduce their impacts on ecosystems and biodiversity [55], companies that invest in PES for CWR conservation could gain a competitive advantage by advertising their activities through sustainability labelling programs.

Individual governments, companies or organizations may fund PES programs for CWR conservation unilaterally or could contribute to a CWR conservation fund that aims to conserve the wild genetic resources of a given crop gene pool or set of crops, providing streamlined access to contributors. The creation of CWR endowments could also aid in the sustainability and permanence of such programs. The interest from such funds would be used to pay those safeguarding the plants on a frequent basis contingent on the persistence of the CWR populations or for the maintenance of CWR within a protected area. Such programs could be arranged through already existing access and benefit sharing agreements, either bilateral (Nagoya Protocol) or multilateral (ITPGRFA) and could contribute new resources to existing funds, such as the ITPGRFA's Benefit Sharing Fund.

A key challenge facing PES for CWR programs is that the benefits of CWR conservation are typically distant both spatially and temporally from the conservation activities (with the exception of farmers who manage CWR on-farm). In addition, given that CWR and plant genetic resources held as part of the Multilateral System as a whole act as an important global public good, the benefits of these conservation activities are highly diffused. Thus, it is likely that the majority of funds made available for CWR conservation through PES would come either from international institutions (such as the Benefit Sharing Fund of the ITPGRFA, GEF or the Green Climate Fund) or from national governments.

4.2. Who Will Provide CWR Conservation Services?

Crop wild relative conservation differs from the protection of cultivated crop diversity in that CWR taxa are wild species that generally do not require farmers for their persistence. That said, many CWR taxa can be weedy and can benefit from disturbance caused by humans; for example, on roadsides or on the sides of agricultural fields including CWR of maize as managed within the Sierra de Manantlan Biosphere Reserve. These plant taxa thus constitute a particular class of wild and weedy biodiversity that calls for specific forms of conservation.

While PES for CWR conservation will in some way require different designs from past PES programs, many similarities exist and lessons can be taken from past experiences with PES and PACS more specifically. Depending on the taxa, PES for in situ or on-farm CWR conservation might target farmers (as in Zambia for wild millet, cowpea and sorghum taxa [24]), private landowners, forest managers or conservationists working in the context of protected areas. In some circumstances, successful CWR conservation will require communities to collaborate who do not typically work together (e.g., conservationists/ecologists and agricultural scientists and crop breeders) to further collaborate and PES offers a tool for bridging such gaps similar to the REDD+ (Reducing Emissions from Deforestation and Forest Degradation in Developing Countries) program, which has brought together climate scientists and those working in forestry management and biodiversity conservation.

4.3. Towards Designing a CWR PES Conservation Portfolio

In this section, we describe a portfolio of CWR conservation actions that fit into the PES framework. These programs could also be coupled with already existing PES schemes by "bundling" CWR with other ecosystem services such as carbon fixation (e.g., REDD+) or water purification [56]. Though this portfolio focuses on in situ conservation, gene banks and other ex situ conservation repository actors are important to long-term CWR conservation and accessibility for use; thus, PES programs for in situ CWR conservation would best include integrated aspects with the ex situ conservation community. A PES scheme for CWR conservation could take many forms including:

Within preserves. CWR can be conserved through the creation of new preserves, the addition of new land containing CWR populations to existing reserves or by providing new funding to the budgets of existing reserves to support CWR conservation programs.

On farms or other highly human managed environments. Property owners could be compensated for protecting CWR populations on their lands. One manner by which to implement such work could be through the inclusion of CWR within agri-environmental payment schemes or other, already existing, PES programs such as REDD+. Many countries have agri-environmental payment programs designed to preserve the provision of environmental public goods such as biodiversity, cultural heritage and scenery. These programs theoretically have the potential to correct market failures associated with these goods [57]. Such programs could, as an example, provide compensation to farmers for conserving CWR in field margins [24] or for setting aside larger portions of their cropland as non-agricultural conservation lands. Owners and managers of roadsides could also be compensated for the protection of CWR within their mowing and herbicide use activities.

Use in landscaping, forage programs and plantations. PES could be used to fund the use of CWR in landscaping, forage programs and plantations of edible and medicinal species. These three strategies could be particularly useful for expanding the distributions of threatened CWR and may enable long-term conservation without further funding if the target species becomes sufficiently popular.

Use in restoration projects. CWR could be prioritized or subsidized for use in restoration activities in their native ranges through PES funding mechanisms.

The creation of new protected areas for CWR conservation may emphasize those that contain several important CWR populations following an optimal reserve design [58,59] as exemplified by the Sierra de Manantlan Biosphere Reserve, with a core area preserving the CWR habitat surrounded by transition and buffer zone containing settlements. Sometimes sites need not be very large, as pioneered by the plant micro-reserve initiative in Valencia, Spain [60] and the Ammiad Project in Israel, a one-hectare site that nonetheless has thus far been successful in conserving an important population of wild wheat [31]. Creating new preserves may be expensive but is sometimes necessary and feasible in areas where the presence of CWR populations overlap with high levels of other ecosystem services and species that are targeted for conservation.

Contracting with farmers and other local people, on the other hand, could help decrease the costs of conserving a CWR population by eliminating the need to purchase the land and turn it into a protected area, allow the transfer of the payments if the CWR population shifts due to climate change and enable those administering the fund to shift the payments to other populations if the original goes extinct. For example, research conducted in Zambia to determine how much farmers would have to be paid to conserve CWR in field margins estimated conservation costs at between USD 23–91 per hectare per year [24]. Payment could be monetary or in kind. CWR conservation may also be compatible with agroforestry activities such as shade-grown coffee or cacao, allowing local people to continue economic activities in the area and farmers that take part in these programs may have the opportunity to command a premium for their products through eco-labelling, gaining multiple benefits from their involvement. As CWR are present in a wide range of habitats, this type of PES scheme for in situ conservation of CWR could also include the management of roadside CWR populations, plans for sustainable management plans of harvested populations of edible CWR and the conservation of CWR populations present in agricultural landscapes by farmers. In any such strategy, payments should be contingent on the quality of the conservation effort (generally, whether the population continues to persist, whether it increases or decreases in size, etc.) to incentivize the effective conservation of the CWR populations. Although it may prove cheaper than the creation of new CWR preserves, this strategy may be difficult to successfully design and implement due to complications with land tenure, questions about who to pay, payment structure and local culture. The long-term sustainability of the strategy will most likely be contingent upon whether or not payments continue.

4.4. Prioritizing CWR for Conservation

Given limited budgets, it is important to prioritize the most important resources to conserve. For CWR, the potential PES practitioner must prioritize between crop gene pools, individual species and intraspecific populations. Crop gene pools may be prioritized, among other factors, by the economic value of the crop to which they are related, the value of the crop for food security or other cultural values and/or contributions made by the crop to development and poverty alleviation. Another tool that can be used for prioritization is the so-called Weitzman approach. The Weitzman theorem uses diversity, risk status and conservation cost indices to construct a priority portfolio of conservation targets to maximize the diversity that can be conserved with any given quantity of funding [61,62]. The geographic prioritization of CWR for conservation is possible as their native distributions are known to be concentrated in primary regions of diversity around the world. Previous efforts to map the ranges of over 1000 CWR related to 81 globally important crops distinguished areas of the Mediterranean, Near East and southern Europe, South America, Southeast and East Asia and Mesoamerica as particularly CWR rich, with up to 84 taxa potentially overlapping in a 25 km^2 area in Turkey, one of the top global hotspots for CWR diversity (an online tool has been developed that allows conservation practitioners to identify where CWR taxon richness is the highest [63]). A related gap analysis methodology has combined conservation and native distribution data on CWR to map under-collected (ex situ) and under-protected (in situ) areas, identifying priority populations for conservation [64,65].

Another important consideration is the potential usefulness of CWR species and populations to crop improvement efforts, which has been described for the CWR of many crops [66]. Consulting with breeders of the crop of interest to further determine which CWR species and populations they are most interested in may provide further context for selecting conservation targets. A maximum diversity approach might alternatively be adopted in which CWR are selected based on genetic diversity and genetic uniqueness since specific trait values are often challenging to measure in CWR [2].

5. Assessing the Effectiveness of PES for CWR

The success of PES mechanisms designed to conserve CWR-related ecosystem services may be assessed according to three main criteria: ecological effectiveness, economic efficiency and social equity [47,67]. This section discusses strategies for maximizing these aspects of PES programs for CWR conservation.

5.1. Ecological Effectiveness

The ecological effectiveness of PES schemes for CWR conservation refers to the efficacy and sustainability of programs in contributing to the preservation of the targeted CWR genetic diversity in the medium-term. A major concern of any PES scheme for the in situ conservation of CWR is how effective its payments are in preserving the ecological functions of CWR habitats and the evolutionary potential of the CWR populations. Practitioners designing a PES program for CWR conservation should assess whether or not the CWR population itself can be expected to survive indefinitely into the future; i.e., what is the risk that the population will go extinct, eliminating the benefits the payments for ecosystem services were meant to achieve. If the population is likely to go extinct in the medium-term even with conservation action, it may be a priority for collecting for ex situ conservation but may not be a wise choice for a PES program for in situ CWR protection. Methods from plant conservation biology can be utilized to set and measure the progress of populations, while studies have also predicted the effects of climate change on the range of specific CWR species into the future using methodologies that could be adapted to determine which potential preserve areas are more likely than others to remain as a suitable habitat for CWR species in future climates [10,13,68]. Last, the management of in situ CWR populations conserved through the PES mechanism should include measures to reduce the risks posed by development, livestock, crop introgression and other threats. Genetic erosion may also occur through the selection for desired traits in CWR that are used in landscaping, plantations and in forage programs and care should be taken to prevent the excessive narrowing of the conserved genetic diversity.

5.2. Economic Efficiency

The term economic efficiency in the context of PES for CWR conservation refers to the use of project resources such that the net economic benefit resulting from the PES program is maximized. Economic efficiency on a larger scale in this context would imply that the limited funds available for CWR conservation globally are spent in ways that maximize the economic benefits flowing from these projects (through crop improvement and other uses). Though it is difficult to predict which CWR populations will end up being most useful, the prioritization techniques discussed in Section 4.4 may indicate means by which to maximize the value of the genetic material conserved while the cost effectiveness of PES schemes for CWR conservation can be enhanced by preserving the most unique populations and/or by preserving overlapping populations of many taxa of interest. In the case of multiple habitats containing similar amounts of CWR genetic diversity, costs can be cut by selecting the habitat that is the cheapest to conserve. This holds true for the other elements of the PES programs as well, as long as the quality of the service provided remains as high as with less expensive providers. Employing a conservation auction in which potential service providers reveal their cost structures through the process of bidding may be useful in driving down the costs of conservation by helping to select the lowest-cost project partners [49]. Furthermore, it should be noted that complementary funding for the characterization and evaluation of CWR germplasm and its use in pre-breeding activities is essential for increasing the economic benefits flowing from CWR conservation.

5.3. Social Equity

Finally, social equity should be a key consideration for the design of PES schemes for CWR conservation. Though many authors have emphasized economic efficiency as the primary goal of PES schemes, incorporating social equity concerns is important to the success of PES mechanisms designed for CWR conservation so as to avoid the so-called "PES curse" of negative social impacts [47,69]. There may be more overlap between social equity concerns and the ecological effectiveness and economic efficiency aspects of PES than previously considered, at least for CWR.

The history of CWR conservation contains several examples of conservation programs that were designed to be both ecologically effective and socially equitable. In addition, it should be noted that even if CWR are not cultivated species, on-farm conservation techniques still are essential for the successful conservation of many CWR taxa, as shown by recent research conducted by Fagandini Ruiz et al. [70] on quinoa wild relatives in southern Peru around Lake Titicaca. Six quinoa CWR were found to be present both on permanent native meadows and cultivated land with fallow cycles and plot borders [70]. Other examples of how farmers and local communities have been included in CWR conservation efforts are numerous. For example, the Potato Park in Peru, or Parque de la Papa, is a biocultural heritage area that preserves wild relatives and landrace varieties of potato as well as other Andean crops like quinoa and oca. The park is maintained by six local Quechua communities and its management uses customary laws and institutions to aid in the conservation and sustainable use of the area's natural resources [71]. The Global Environment Facility's CWR Project developed a management plan for wild yams in Madagascar to allow the sustainable harvest and management for these CWR instead of simply cutting off access to the plants that locals had harvested, eaten and sold for centuries. The Sierra de Manantlan Biosphere Reserve in Mexico, with its focus on people as an integral part of the ecosystem, combines the in situ conservation of maize wild relatives and landraces with the development of local agrarian communities, ecotourism and sustainable forest management. The disease-resistant maize wild relative *Zea diploperennis* (Iltis et al.) is preserved within the core zones of the preserve with strict protection, along with *Zea perennis* (Hitchc. Reeves and Mangels) and subspecies *Zea mays* spp. *parviglumis* (H.H. Iltis and Doebley), yet around 40,000 people live in the buffer zone [72]. A project funded by the ITPGRFA's Benefit Sharing Fund is helping to train local farmers and their families in the conservation of a maize wild relative in Nicaragua's Apacunca Genetic Reserve and the area surrounding it as part of a wider package of development activities, seeking to involve communities in the recovery, conservation and use of teosinte (a maize CWR) while ensuring that they receive some benefit as well. Thus, local communities have played a central role in many past CWR conservation projects.

Projects inclusive of social equity considerations tend to be internally originated and driven, owned by the community, fully backed by local practice and culture and strongly supported by other stakeholders [7]. Those designing PES schemes for CWR conservation should recognize the synergies between social equity and the ecological effectiveness and economic efficiency of CWR preservation mechanisms while at the same time acknowledging the potential tradeoffs between these goals. It might not always be possible to use these strategies but social equity considerations should at least be considered during PES mechanism design. Participatory approaches present opportunities for CWR conservation to bring added benefits through the social and economic empowerment of often-marginalized groups by sharing the benefits of the program with those who live nearby; for example, through their involvement in the planting of CWR species or in the maintenance of plantations of edible CWR and in training and job creation in ecotourism activities centered on CWR. They may also help to tap into local ethnobotanical knowledge through the engagement of local parabotanists who may be better suited to identify, manage and educate others about the CWR resources of a particular area. For example, local rural communities were found to have a detailed knowledge of the utility of the flora in the Sierra de Manantlan, using more than 500 of the plant species present in the area [73].

6. Conclusions

Payment for ecosystem services (PES) has been shown to be an effective mechanism mitigating market failures associated with the provision of ecosystem services such as water filtration, carbon fixation and a number of other economically and culturally valuable functions provided by the natural world. In this article, we argue that PES may offer a useful tool for ensuring the conservation of priority, at-risk populations of CWR of important crops and may assist CWR conservation efforts at a local, national, regional and global scale.

Currently, adequate institutional frameworks to support the conservation of CWR worldwide do not exist, in part due to insufficient incentives for providing CWR conservation services. Payment for ecosystem services could be a promising tool for solving this problem by directly linking payments from public and private beneficiaries of CWR conservation services to their suppliers, bridging substantial spatial and temporal gaps. PES in particular offers a flexible mechanism for advancing CWR conservation that may prove successful in a broad range of situations and scenarios in developing and developed countries alike.

The loss of CWR genetic diversity results in the irreversible destruction of resources of significant importance to the sustainability and resilience of future agriculture, to local food and cultural security and to the provision of local ecosystem services. To ensure that this diversity is present and available when needed, it is necessary that investments be made in the conservation of CWR. The payment for ecosystem services (PES) mechanism has the potential to aid in this goal by providing economic incentives for the maintenance of CWR resources.

Author Contributions: Conceptualization, N.T. and H.D.; investigation, N.T., H.D. and C.K.K.; writing—original draft preparation, N.T.; writing—review and editing, N.T., H.D. and C.K.K.; funding acquisition, N.T. All authors have read and agreed to the published version of the manuscript.

Funding: The initial research resulting in this article was conducted in 2011–2012 with financial support from the Fulbright Program.

Acknowledgments: Many thanks to Luigi Guarino, Ehsan Dulloo, Bo Cutter, Devra Jarvis, Adam Drucker, Nanete Neves, Peter Tyack, Lisen Runsten, Vijay Kolinjivadi and Chelsea Smith for providing comments on previous draft versions. Thanks also to the Fulbright Program for funding and to the Global Crop Diversity Trust for hosting this research.

Conflicts of Interest: The authors declare no conflict of interest and the funders had no role in the conceptualization or writing of the manuscript.

References

1. Dempewolf, H.; Baute, G.; Anderson, J.; Kilian, B.; Smith, C.; Guarino, L. Past and future use of wild relatives in crop breeding. *Crop Sci.* **2017**, *57*, 1070–1082. [CrossRef]
2. Tanksley, S.D.; McCouch, S.R. Seed Banks and Molecular Maps: Unlocking Genetic Potential from the Wild. *Science* **1997**, *277*, 1063–1066. [CrossRef] [PubMed]
3. Khoury, C.K.; Carver, D.; Barchenger, D.W.; Barboza, G.; van Zonneweld, M.; Jarret, R.; Bohs, L.; Kantar, M.B.; Uchanski, M.; Mercer, K.; et al. Modeled distributions and conservation status of the wild relatives of chile peppers (*Capsicum* L.). *Divers. Distrib.* **2019**, *26*, 209–225. [CrossRef]
4. Hajjar, R.; Hodgkin, T. The use of wild relatives in crop improvement: A survey of developments over the past 20 years. *Euphytica* **2007**, *156*, 1–13. [CrossRef]
5. Maxted, N.; Kell, S.P. *Establishment of a Global Network for the In Situ Conservation of Crop Wild Relatives: Status and Needs*; FAO Commission on Genetic Resources for Food and Agriculture: Rome, Italy, 2009; p. 266.
6. Iltis, H.H. *Serendipity in the Exploration of Biodiversity. What Good are Weedy Tomatoes?* Wilson, E.O., Ed.; Biodiversity, National Academy Press: Washington, DC, USA, 1988; pp. 98–105.
7. Hunter, D.; Heywood, V. *Crop Wild Relatives: A Manual of In Situ Conservation*, 1st ed.; Earthscan: London, UK, 2011.
8. Pimentel, D.; Wilson, C.; McCullum, C.; Huang, R.; Dwen, P.; Flack, J.; Tran, Q.; Saltman, T.; Cliff, B. Economic and environmental benefits of biodiversity. *BioScience* **1997**, *47*, 747–757. [CrossRef]
9. Brummitt, N.; Bachman, S. *Plants under pressure a global assessment. In The First Report of the IUCN Sampled Red List Index for Plants*; Royal Botanic Gardens: Kew, UK, 2015.
10. Canadell, J.; Noble, I. Challenges of a changing Earth. *Trends Ecol. Evol.* **2001**, *16*, 664–666. [CrossRef]

11. Bilz, M.; Kell, S.P.; Maxted, N.; Lansdown, R.V. *European Red List of Vascular Plants*; Publications Office of the European Union: Luxembourg, 2011.
12. Jarvis, A.; Lane, A.; Hijmans, R.J. The effect of climate change on crop wild relatives. *Agric. Ecosyst. Environ.* **2008**, *126*, 13–23. [CrossRef]
13. Ureta, C.; Martínez-Meyer, E.; Perales, H.R.; Álvarez-Buylla, E.R. Projecting the effects of climate change on the distribution of maize races and their wild relatives in Mexico. *Glob. Chang. Biol.* **2011**, *18*, 1073–1082. [CrossRef]
14. Godfray, H.C.J.; Beddington, J.R.; Crute, I.R.; Haddad, L.; Lawrence, D.; Muir, J.F.; Pretty, J.; Robinson, S.; Thomas, S.M.; Toulmin, C. Food Security: The Challenge of Feeding 9 Billion People. *Science* **2010**, *327*, 812–818. [CrossRef]
15. Battisti, D.S.; Naylor, R.L. Historical Warnings of Future Food Insecurity with Unprecedented Seasonal Heat. *Science* **2009**, *323*, 240–244. [CrossRef]
16. Lobell, D.B.; Burke, M.B.; Tebaldi, C.; Mastrandrea, M.D.; Falcon, W.P.; Naylor, R.L. Prioritizing Climate Change Adaptation Needs for Food Security in 2030. *Science* **2008**, *319*, 607–610. [CrossRef] [PubMed]
17. Peng, S.; Huang, J.; Sheehy, J.E.; Laza, R.C.; Visperas, R.M.; Zhong, X.; Centeno, G.S.; Khush, G.S.; Cassman, K.G. Rice yields decline with higher night temperature from global warming. *Proc. Natl. Acad. Sci. USA* **2004**, *101*, 9971–9975. [CrossRef] [PubMed]
18. Zamir, D. Improving plant breeding with exotic genetic libraries. *Nat. Genet.* **2001**, *2*, 983–989. [CrossRef] [PubMed]
19. Thalapati, S.; Batchu, A.K.; Neelamraju, S.; Ramanan, R. *Os11Gsk* gene from a wild rice, *Oryza rufipogon* improves yield in rice. *Funct. Integr. Genet.* **2012**, *12*, 277–289. [CrossRef] [PubMed]
20. Ford-Lloyd, B.V.; Schmidt, M.; Armstrong, S.J.; Barazani, O.; Engels, J.; Hadas, R.; Hammer, K.; Kell, S.P.; Kang, D.; Khoshbakht, K.; et al. Crop Wild Relatives—Undervalued, Underutilized and under Threat? *BioScience* **2011**, *61*, 559–565. [CrossRef]
21. Jayachandran, S.; de Laat, J.; Lambin, E.F.; Stanton, C.Y.; Audy, R.; Thomas, N.E. Cash for carbon: A randomized trial of payments for ecosystem services to reduce deforestation. *Science* **2017**, *357*, 267–273. [CrossRef]
22. Börner, J.; Baylis, K.; Corbera, E.; Ezzine-de-Blas, D.; Honey-Rosés, J.; Persson, U.M.; Wunder, S. The effectiveness of payments for environmental services. *World Dev.* **2017**, *96*, 359–374. [CrossRef]
23. Wunder, S.; Borner, J.; Ezzine-de-Blas, D.; Feder, S.; Pagiola, S. Payments for environmental services: Past performance and pending potentials. *Annu. Rev. Resour. Econ.* **2020**, *12*. [CrossRef]
24. Wainwright, W.; Drucker, A.G.; Maxted, N.; Brehm, J.M.; Ng'uni, D.; Moran, D. Estimating in situ conservation costs of Zambian crop wild relatives under alternative conservation goals. *Land Use Policy* **2019**, *81*, 632–643. [CrossRef]
25. Drucker, A.G.; Ramirez, M. Payments for agrobiodiversity conservation services: An overview of Latin American experiences, lessons learned and upscaling challenges. *Land Use Policy* **2020**, *99*, 104810. [CrossRef]
26. Louafi, S.; Bazile, D.; Noyer, J.L. Conserving and cultivating agricultural genetic diversity: Transcending established divides. In *Cultivating Biodiversity to Transform Agriculture*; Etienne, H., Ed.; Springer [Allemagne]: Berlin/Heidelberg, Germany, 2013; pp. 181–230.
27. Zhang, W.; Ricketts, T.H.; Kremen, C.; Carney, K.; Swinton, S.M. Ecosystem services and dis-services to agriculture. *Ecol. Econ.* **2007**, *64*, 253–260. [CrossRef]
28. Brown, T.C.; Bergstrom, J.C.; Loomis, J.B. Defining, Valuing and Providing Ecosystem Goods and Services. *Nat. Resour. J.* **2007**, *47*, 329–376.
29. *Millennium Ecosystem Assessment (MEA) Ecosystems and Human Well-being: Biodiversity Synthesis*; Island Press: Washington DC, USA, 2005.
30. TEEB. *The Economics of Ecosystems and Biodiversity Ecological and Economic Foundations*; Kumar, P., Ed.; Earthscan: London, UK; Washington, DC, USA, 2010.
31. Anikster, Y.; Feldman, M.; Horovitz, A. The Ammiad experiment. In *Plant Genetic Conservation: The in Situ Approach*; Maxted, N., Ford-Lloyd, B.V., Hawkes, J.G., Eds.; Chapman and Hall: London, UK, 1997; pp. 239–253.
32. Takahashi, R.; Todo, Y. Coffee Certification and Forest Quality: Evidence from a Wild Coffee Forest in Ethiopia. *World Dev.* **2017**, *92*, 158–166. [CrossRef]

33. Rawal, K.M. Natural hybridization among wild, weedy and cultivated *Vigna unguiculata* (L.) Walp. *Euphytica* **1975**, *24*, 699–707. [CrossRef]
34. Tatum, L.A. The Southern Corn Leaf Blight Epidemic. *Science* **1971**, *19*, 1113–1116. [CrossRef]
35. Fraser, E.D.G. Social Vulnerability and Ecological Fragility: Building Bridges between Social and Natural Sciences using the Irish Potato Famine as a Case Study. *Conserv. Ecol.* **2003**, *7*, 9. [CrossRef]
36. Kim, H.J.; Lee, H.R.; Jo, K.R.; Mortazavian, S.M.; Huigen, D.J.; Evenhuis, B.; Kessel, G.; Visser, R.G.; Jacobsen, E.; Vossen, J.H. Broad spectrum late blight resistance in potato differential set plants MaR8 and MaR9 is conferred by multiple stacked R genes. *Theor. Appl. Genet.* **2012**, *124*, 923–935. [CrossRef]
37. Seiler, G.J.; Qi, L.L.; Marek, L.F. Utilization of sunflower crop wild relatives for cultivated sunflower improvement. *Crop Sci.* **2017**, *57*, 1083–1101. [CrossRef]
38. International Treaty on Plant Genetic Resources for Food and Agriculture. Available online: http://www.fao.org/plant-treaty/en (accessed on 29 September 2020).
39. The Nagoya Protocol on Access and Benefit-Sharing. Available online: https://www.cbd.int/abs (accessed on 29 September 2020).
40. Nawaz, M.A.; Lin, X.; Chan, T.; Ham, J.; Shin, T.; Ercisli, S.; Golokhvast, K.S.; Lam, H.; Chung, G. Korean Wild Soybeans (*Glycine soja* Sieb & Zucc.): Geographic Distribution and Germplasm Conservation. *Agronomy* **2020**, *10*, 214.
41. Del Rio, A.H.; Bamberg, J.B.; Huaman, Z.; Salas, A.; Vega, S.E. Assessing changes in the genetic diversity of potato gene banks. 2. In situ vs ex situ. *Theor. Appl. Genet.* **1997**, *95*, 199–204. [CrossRef]
42. Greene, S.L.; Kisha, T.J.; Yu, L.-X.; Parra-Quijano, M. Conserving Plants in Gene Banks and Nature: Investigating Complementarity with Trifolium thompsonii Morton. *PLoS ONE* **2014**, *9*, e105145. [CrossRef] [PubMed]
43. Thompson, J.N.; Burdon, J.J. Gene-for-gene coevolution between plants and parasites. *Nature* **1992**, *360*, 121–125. [CrossRef]
44. Nevo, E.; Fu, Y.B.; Pavlicek, T.; Khalifa, S.; Tavasi, M.; Beiles, A. Evolution of wild cereals during 28 years of global warming in Israel. *Proc.Natl.Acad.Sci.USA* **2012**, *109*, 2412–2415. [CrossRef]
45. Khoury, C.K.; Carver, D.; Kates, H.R.; Achicanoy, H.A.; van Zonneveld, M.; Thomas, E.; Heinitz, C.; Jarret, R.; Labate, J.A.; Reitsma, K.; et al. Distributions, conservation status, and abiotic stress tolerance potential of wild cucurbits (*Cucurbita* L.). *Plants People Planet* **2019**, *2*, 269–283. [CrossRef]
46. Tyack, N.; Dempewolf, H. The economics of crop wild relatives under climate change. In *Crop Wild Relatives and Climate Change*; Redden, R., Yadav, S.S., Maxted, N., Dulloo, M.E., Guarino, L., Smith, P., Eds.; John Wiley & Sons: Hoboken, NJ, USA, 2015; pp. 281–291. [CrossRef]
47. Castañeda-Álvarez, N.P.; Khoury, C.K.; Achicanoy, H.A.; Bernau, V.; Dempewolf, H.; Eastwood, R.J.; Guarino, L.; Harker, R.H.; Jarvis, A.; Maxted, N.; et al. Global conservation priorities for crop wild relatives. *Nat Plants.* **2016**, *2*, 16022. [CrossRef]
48. Wunder, S. *Payments for Environmental Services: Some Nuts and Bolts*; CIFOR Occasional Paper No. 42; Center for International Forestry Research: Jakarta, Indonesia, 2005.
49. Narloch, U.; Drucker, A.G.; Pascual, U. Payments for agrobiodiversity conservation services for sustained on-farm utilization of plant and animal genetic resources. *Ecol. Econ.* **2011**, *70*, 1837–1845. [CrossRef]
50. Dempewolf, H.; Eastwood, R.J.; Guarino, L.; Khoury, C.K.; Müller, J.V.; Toll, J. Adapting agriculture to climate change: A global initiative to collect, conserve and use crop wild relatives. *Agroecol. Sustain. Food Syst.* **2014**, *38*, 369–377. [CrossRef]
51. Bodansky, D. The Copenhagen Climate Change Conference: A Postmortem. *Am. J. Int. Law* **2010**, *104*, 230–240. [CrossRef]
52. Roberts, L. Chemical prospecting: Hope for vanishing ecosystems? *Science* **1992**, *256*, 1142–1143. [CrossRef]
53. Elkington, J. Towards the Sustainable Corporation: Win-Win-Win Business Strategies for Sustainable Development. *Calif. Manag. Rev.* **1994**, *36*, 90–100. [CrossRef]
54. McDaniel, S.W.; Rylander, D.H. Strategic Green Marketing. *J. Consum. Mark.* **1993**, *10*, 4–10. [CrossRef]
55. Houdet, J.; Trommetter, M.; Weber, J. Understanding changes in business strategies regarding biodiversity and ecosystem services. *Ecol. Econ.* **2012**, *73*, 37–46. [CrossRef]
56. Raudsepp-Hearne, C.; Peterson, G.D.; Bennett, E.M. Ecosystem service bundles for analyzing tradeoffs in diverse landscapes. *PNAS* **2010**, *107*, 5242–5247. [CrossRef] [PubMed]

57. Hasund, K.P. Indicator-based agri-environmental payments: A payment-by- result model for public goods with a Swedish application. *Land Use Policy* **2012**, *30*, 223–233. [CrossRef]
58. Batisse, M. Developing and focusing the biosphere reserve concept. *Nat. Resour.* **1986**, *22*, 1–10.
59. Cox, G.W. *Conservation Ecology*; W.C. Brown: Dubuque, IA, USA, 1993.
60. Laguna, E. The plant micro-reserve initiative in the Valencian Community (Spain) and its use to conserve populations of crop wild relatives. *Crop Wild Relat.* **2004**, *2*, 10–13.
61. Weitzman, M.L. The Noah's ark problem. *Econometrica* **1998**, *66*, 1279–1298. [CrossRef]
62. Metrick, A.; Weitzman, M.L. Conflicts and choices in biodiversity preservation. *J. Econ. Perspect.* **1998**, *12*, 21–34. [CrossRef]
63. Crop Wild Relatives Global Atlas. Available online: www.cwrdiversity.org/distribution-map (accessed on 29 September 2020).
64. Ramirez-Villegas, J.; Khoury, C.; Jarvis, A.; Debouck, D.G.; Guarino, L. A Gap Analysis Methodology for Collecting Crop Genepools: A Case Study with Phaseolus Beans. *PLoS ONE* **2010**, *5*, e13497. [CrossRef]
65. Khoury, C.K.; Amariles, D.; Soto, J.S.; Diaz, M.V.; Sotelo, S.; Sosa, C.C.; Ramírez-Villegas, J.; Achicanoy, H.A.; Velásquez-Tibatá, J.; Guarino, L.; et al. Comprehensiveness of conservation of useful wild plants: An operational indicator for biodiversity and sustainable development targets. *Ecol. Indic.* **2019**, *98*, 420–429. [CrossRef]
66. Miller, R.E.; Khoury, C.K. The Gene Pool Concept Applied to Crop Wild Relatives: An Evolutionary Perspective. In *North American Crop Wild Relatives: Volume 1: Conservation Strategies*; Greene, S.L., Williams, K.A., Khoury, C.K., Kantar, M.B., Marek, L.F., Eds.; Springer International Publishing AG: Switzerland, 2018; pp. 167–188. [CrossRef]
67. FAO. *Payments for Ecosystem Services and Food Security*; FAo: Rome, Italy, 2011.
68. Lira, R.; Tellez, O.; Davila, P. The effects of climate change on the geographic distribution of Mexican wild relatives of domesticated Cucurbitaceae. *Genet. Resour. Crop Evol.* **2008**, *56*, 691–703. [CrossRef]
69. Pascual, U.; Muradian, R.; Rodríguez, L.C.; Duraiappah, A. Exploring the links between equity and efficiency in payments for environmental services: A conceptual approach. *Ecol. Econ.* **2010**, *69*, 1237–1244. [CrossRef]
70. Fagandini Ruiz, F.; Bazile, D.; Drucker, A.G.; Tapia, M.; Chura, E. Geographical distribution of quinoa crop wild relatives in the Peruvian Andes: A participatory mapping initiative. *Environ. Dev. Sustain.* **2020**, 1–22. [CrossRef]
71. Argumedo, A. *Collective Trademarks and Biocultural Heritage—Towards New Indications of Distinction for Indigenous Peoples in the Potato Park, Peru*; International Institute for Environment and Development: London, UK, 2013.
72. Louette, D.; Smale, M. *Genetic Diversity and Maize Seed Management in a Traditional Mexican Community: Implications for In Situ Conservation of Maize*; NRG Paper 96-03; CIMMYT: Mexico, D.F., Mexico, 1996.
73. Benz, B.F.; Santana, F.; Piñeda, R.; Cevallos, J.; Robles, L.; De Niz, D. Characterization of mestizo plant use in the Sierra de Manantlán, Jalisco-Colima, Mexico. *J. Ethnobiol.* **1994**, *14*, 23–41.

plants

Article

SNP Markers and Evaluation of Duplicate Holdings of *Brassica oleracea* in Two European Genebanks

Anna E. Palmé [1], Jenny Hagenblad [2], Svein Øivind Solberg [3,*], Karolina Aloisi [1] and Anna Artemyeva [4]

1 Nordic Genetic Resource Centre, Smedjevägen 3, SE-230 53 Alnarp, Sweden; anna.palme@nordgen.org (A.E.P.); karolina.aloisi@nordgen.org (K.A.)
2 Department of Physics, Chemistry and Biology, Linköping University, SE-581 83 Linköping, Sweden; jenny.hagenblad@liu.se
3 Faculty of Applied Ecology, Agricultural Sciences and Biotechnology, Inland Norway University of Applied Sciences, NO-2418 Elverum, Norway
4 N. I. Vavilov Institute of Plant Genetic Resources (VIR), 42-44, B. Morskaya Street, 190000 St. Petersburg, Russia; akme11@yandex.ru
* Correspondence: svein.solberg@inn.no; Tel.: +46-735-401-516

Received: 17 June 2020; Accepted: 20 July 2020; Published: 22 July 2020

Abstract: Around the world, there are more than 1500 genebanks storing plant genetic resources to be used in breeding and research. Such resources are essential for future food security, but many genebanks experience backlogs in their conservation work, often combined with limited budgets. Therefore, avoiding duplicate holdings is on the agenda. A process of coordination has started, aiming at sharing the responsibility of maintaining the unique accessions while allowing access according to the international treaty for plant genetic resources. Identifying duplicate holdings based on passport data has been one component of this. In the past, and especially in vegetables, different selections within the same varieties were common and the naming practices of cultivars/selections were flexible. Here, we examined 10 accession pairs/groups of cabbage (*Brassica oleracea* var. *capitata*) with similar names maintained in the Russian and Nordic genebanks. The accessions were analyzed for 11 morphological traits and with a SNP (Single Nucleotide Polymorphism) array developed for *B. napus*. Both proved to be useful tools for understanding the genetic structure among the accessions and for identifying duplicates, and a subset of 500 SNP markers are suggested for future *Brassica oleracea* genetic characterization. Within five out of 10 pairs/groups, we detected clear genetic differences among the accessions, and three of these were confirmed by significant differences in one or several morphological traits. In one case, a white cabbage and a red cabbage had similar accession names. The study highlights the necessity to be careful when identifying duplicate accessions based solely on the name, especially in older cross-pollinated species such as cabbage.

Keywords: *Brassica oleracea*; conservation; diversity; genebank; plant genetic resources; SNP

1. Introduction

A report from the Food and Agriculture Organization of the United Nations (FAO) indicates that up to 70% of the 7.4 million accessions around the world might be duplicate holdings [1]. At the same time, genebanks are struggling with inadequate resources and backlogs in regeneration, characterization, and documentation [2]. Taking a bird's-eye view, duplication is not an efficient conservation approach. At a local level, each collection holder aims to have a large and influential collection. Requesting and maintaining accessions from other genebanks (duplication) has been one way to do this. The European Genebank Integrated System (AEGIS) has managed to involve institutions in more than 30 countries in an action for coordination and collaboration on plant genetic resource conservation [3]. The main idea is to share responsibility by establishing and operating a European Collection of unique and important germplasm and to increase the conservation efficiency and quality while facilitating the use of the genetic resources [4].

A critical step in this initiative has been the selection of accessions (generally seeds, conserved in genebanks). There are several challenges in such selections, but one is how to handle accessions with the same or similar names [5]. Same or similar names could be due to the duplication of accessions among genebanks, but can also indicate different selections (enterprises' selections) and/or a flexible naming practice of seed material in the past [6,7]. Official variety lists (cultivar lists) and control came in the mid-twentieth century [8,9], and plant breeders' rights came after the ratification of the International Union for the Protection of New Varieties of Plants, which was launched in 1969 [10]. In 1969, local companies' selections were still listed under similar names in Scandinavia, but almost all of them were removed from the national variety lists between 1970 and 1980 [11]. All this has resulted in a large number of older varieties with the same or similar names. An important question is whether accessions of such varieties should be regarded as duplicates when efforts are made to increase the efficiency in genebank conservation.

Recent developments such as using passport data with digital object identifiers (DOI) on accessions and transactions [12], large-scale morphological characterization and phenotyping [13], and molecular studies with next-generation sequencing platforms [14–18] have all improved the possibilities to identify duplicates. Regardless of the approach, proper data and transparent genebank information systems are needed to facilitate duplication assessments [19,20]. *Brassica oleracea* L. (2n = 18) comprises many important crops, including cauliflower, broccoli, and cabbages as well as wild species and subspecies which are cross-compatible with the cultivars [21,22]. The issue of duplicate genebank holdings of *B. oleracea* has been raised [23], and genetic diversity has been investigated using, for example, AFLP (Amplified Fragment Length Polymorphism) markers [24–26] and microsatellites [27]. In these studies, substantial diversity within accessions was observed, but so was a clear differentiation among accessions. A high-density SNP genotyping array for *Brassica napus* and its ancestral diploid species has been developed [28], and in this study we test it on *B. oleracea*.

The main objectives of this study were: (1) to examine the suitability of this Illumina Infinium SNP array to study the genetic diversity and structure in cabbage; (2) to evaluate potential duplicates among genebank accessions with similar names by using morphological traits and the mentioned genetic markers. If the SNP array could be used successfully, we wanted to identify a sub-set of SNP markers to be used for the future screenings of a larger number of accessions in a process of identifying duplicates and incorrectly labelled accessions at a European or global scale.

2. Results

The *Brassica* Working Group of the European Cooperative Programme for Plant Genetic Resources (ECPGR) has prioritized AEGIS, which means there is an ongoing process of searching for potential duplicates. Based on the cabbage passport data from the N. I. Vavilov Institute of Plant Genetic Resources in St. Petersburg (VIR) and the Nordic Genetic Resource Centre (NGB), we were able to identify 40 pairs, triplets, and groups (hereafter termed groups) based on the "accession name" or "donor name". Here, we present the results for 10 such groups, with a total of 27 accessions (Table 1).

2.1. Morphological Diversity

Two types of morphological descriptors were analyzed: continuous descriptors, such as plant height and leaf length, and categorical descriptors, such as leaf color and head shape. Many of the continuous, numeric descriptors were positively correlated (arrows pointing to the right in Figure 1). The two first principal components represented 42% and 23% of the total variation. Accessions with the same or similar names grouped together, but only to a certain extent.

Table 1. Overview of the accessions included in the study, sorted into groups based on accession name. For each accession, a code is given. Information is provided on the accession number, genebank holding institute, acquisition year, and donor institute. Acquisition year indicates the year when the accession entered the genebank holding institute, and the donor institute is the organization from which the material was sent.

Code	Accession Name	Accession Number	Gene Bank	Acquisition Year, Donor Institute
Amager Tall group				
A1	Amager Hög	NGB11705	NGB	1995, Olson & Sons AB, Sweden
A3	Amager Høj Grøn Grami	NGB1875	NGB	1980, Dæhnfeldt A/S, Denmark
A4	Grami	K2537	VIR	1988, Unknown, Denmark
A6	Amager Høj, Grøn, Toftø 67	NGB1873	NGB	1980, FDB Frø, Denmark
A7	Amager Tall Resistent	K2475	VIR	1980, Unknown, Denmark
Amager Short pair				
A8	Amager Kurzstunkiger Orginal	K1485	VIR	1935, Ohlsens Enke, Denmark
A9	Amager L NF Orginal	K2248	VIR	1967, Unknown, Norway
Amager Winter pair				
A8	Amager Kurzstunkiger Orginal	K1485	VIR	1935, Ohlsens Enke, Denmark
A9	Amager L NF Orginal	K2248	VIR	1967, Unknown, Norway
Blåtopp group				
B1	Amager Faales Blatopp	K1181	VIR	1930, Norsk Frø, Norway
B2	Blatopp	K2250	VIR	1967, Unknown, Norway
B3	Blatopp Kvithamar	K2243	VIR	1967, Unknown, Norway
B6	Blåtopp Kvithamar	NGB4555	NGB	1984, Unknown, Norway
Kissendrup pair				
K1	Kissendrup	K111	VIR	1935, Unknown, Denmark
K2	Kissendrup Tagenshus	NGB1996	NGB	1980, Hansens Amagerfrø, Denamrk
Langendijker group				
L1	Langendijk Summer	K181	VIR	1959, Unknown, Denmark
L2	Langendijker Sommer Debut	NGB1997	NGB	1980, Ohlsens Enke, Denmark
L3	Langendijker Sommer Lanso	NGB1998	NGB	1980, Hansens Amagerfrø, Denmark
Ruhm v Enkhizen group				
R1	Ruhm von Enkhizen	NGB11718	NGB	1996, Hansens Amagerfrø, Denmark
R2	Ruhm von Enkhizen Haba	NGB1888	NGB	1980, Hansens Amagerfrø, Denmark
R3	Ruhm von Enkhizen, B Hunderup	NGB2431	NGB	1982, Unknown, Denmark
Jåtunsalgets pair				
J1	Jatunsalgets Vinterkål Berbes	K2139	VIR	1959, Unknown, Norway
J2	Jåtunsalgets Vinterkål	NGB5007	NGB	1983, Unknown, Norway
Loke pair				
LO1	Loke	K2489	VIR	1982, Unknown, Sweden
LO2	Loke	NGB12050	NGB	1997, Unknown, Denmark
Stavanger Torv pair				
ST1	Stavanger Torv	K2175	VIR	1961, Unknown, Norway
ST2	Stavanger Torg	NGB8515	NGB	1990, NLH, Norway

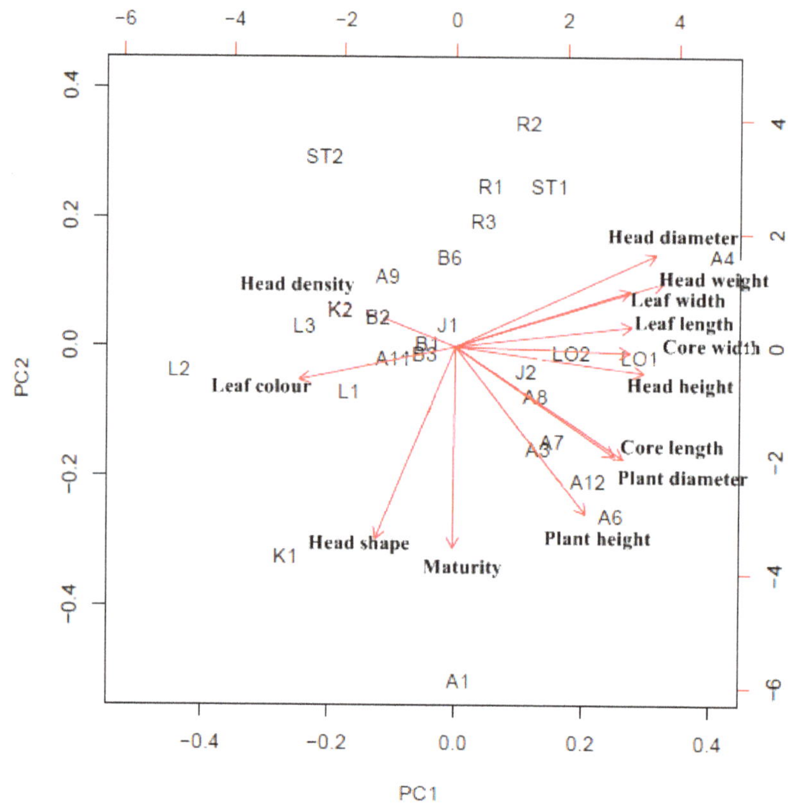

Figure 1. Morphological data: Principal component analysis (PCA) biplot where the descriptors of importance are given as arrows and where the length of an arrow is a measure of the descriptor's variance and the angle between arrows is a measure of the correlation between descriptors, with a small angle expressing a high correlation. PC1 and PC2 explain 42% and 23% of the total variation. Accession names are abbreviated as codes; see Table 1.

Within the Amager Tall group, A1 (Amager Hög) and A4 (Grami) did not cluster with A3 (Amager Høj Grøn Grami), A6 (Amager Høj, Grøn, Toftø 67), and A7 (Amager Tall Resistent). The biplot furthermore indicated that the two Amager Winter accessions (A11 and A12), the two Stavanger Torg accessions (ST1 and ST2), and the two Kissendrup accessions (K1 and K2) did not cluster within their respective group.

The two Stavanger Torv accessions were early maturing and needed around 120 days to maturation, while most of the other accessions needed 140 days or more (Figure 2). There was also a large variation in the time to maturity and other traits within many of the accessions. The results from the Tukey multiple comparisons of means (Table S1) showed that A1 differed significantly from A4, A6, and A7 in the time to maturity (all $p < 0.05$). Furthermore, A4 differed from A1 and A6 in the leaf lamina length and plant height, and A4 from A1 in head weight (all $p < 0.05$). A11 and A12 differed in plant height and, notably, also in leaf lamina color, where A11 plants were purple while A12 plants were green. Significant differences were also detected among the Langendijker Summer accessions. The L2 and L3 plants were purple, while L1 had a mixture of purple and green leaves. In addition, L1 differed from L2 in core length ($p < 0.05$). The Stavanger Torg accessions ST1 and ST2 differed in head height and head density, while the Kissendrup accessions differed in the time to maturity (all $p < 0.05$).

No clear differentiation was detected among the accessions within the Blåtopp, Ruhm von Enkhizen, Loke, or Jåtunsalgets Vinterkål groups (Table S1).

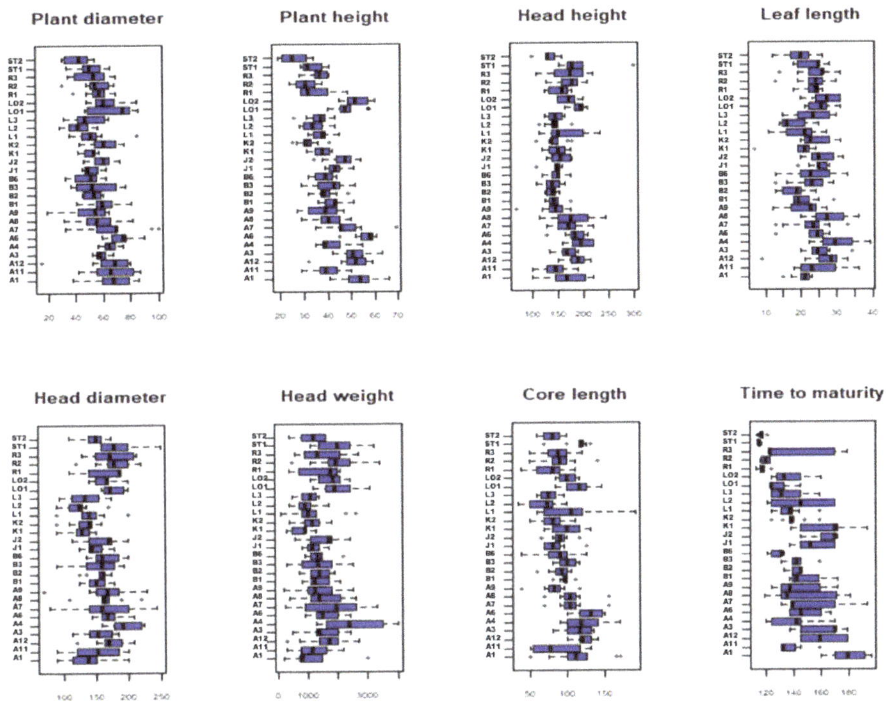

Figure 2. Boxplots describing the variation in the continuous morphological descriptors. Accession names are abbreviated as codes; see Table 1.

2.2. Marker Efficiency and Accession Diversity

Among the 5965 markers, 3969 failed to amplify in all the genotyped cabbage plants. Of these, mapping data was available for 3750 markers, 99% of which were located on the A genome in *B. napus*, and therefore were not expected to be found in *B. oleracea*. The largest proportion of failed markers was found on the *B. napus* chromosome A04 (0.540), with the lowest on chromosome C06 (0.002). Duplicate samples showed a high consistency across runs. Each re-genotyped individual differed in only two markers. Individual 135 differed in Bn-A01-p16331424 and Bn-scaff_15712_6-p1025930, and individual 136 in Bn-scaff_15877_1-p926737 and Bn-scaff_16553_1-p6743.

The highest average number of alleles across all the loci was found in accession A4 (1.8) and the lowest in accession K1 (1.2) (Table 2). The same accessions had the highest and the lowest observed heterozygosity and the lowest genetic diversity, calculated as Nei's h (expected heterozygosity under Hardy–Weinberg Equilibrium, HWE) (Table 2). In some cases, accessions with similar names had similar levels of within-accession diversity—for example, ST1 and ST2—but in many cases, different levels of diversity were observed among the accessions within a group (Table 2). The genetic diversity of the accessions from NGB did not differ significantly from those from VIR (*t*-test, $p = 0.734$). Genetic diversity was not significantly correlated with acquisition year for the full data ($p = 0.177$) nor for accessions from NGB ($p = 0.687$), but was positively correlated with the acquisition year for VIR accessions ($c = 0.553$, $p < 0.05$).

Table 2. Genetic diversity in individual accessions as measured by SNP markers.

Code	Average No Alleles	Nei's h	Observed Heterozygosity
A1	1.1	0.06	0.06
A3	1.4	0.14	0.15
A4	1.8	0.27	0.30
A6	1.5	0.16	0.17
A7	1.3	0.14	0.12
A8	1.5	0.21	0.22
A9	1.3	0.13	0.14
A11	1.6	0.17	0.15
A12	1.4	0.14	0.15
B1	1.4	0.13	0.15
B2	1.4	0.14	0.18
B3	1.3	0.12	0.13
B6	1.4	0.13	0.14
K1	1.2	0.06	0.06
K2	1.4	0.14	0.15
L1	1.4	0.16	0.15
L2	1.4	0.13	0.10
L3	1.5	0.18	0.17
R1	1.5	0.17	0.19
R2	1.5	0.17	0.16
R3	1.4	0.14	0.13
J1	1.3	0.10	0.10
J2	1.4	0.14	0.15
LO1	1.6	0.19	0.22
LO2	1.6	0.19	0.18
ST1	1.3	0.11	0.14
ST2	1.3	0.12	0.13

2.3. Accession Comparisons

All the pairwise F_{ST} values were significantly different from 0 ($p < 0.001$ for all comparisons), indicating genetic differentiation among all accessions. In general, the average pairwise F_{ST} values were lower between pairs of accessions belonging to the same group (average 0.193) than between pairs of accessions belonging to different groups (average 0.270) (*t*-test, $p < 0.001$). The lowest F_{ST} value (0.087) was found between accessions B6 and B3 (Table 3), both from the Blåtopp group and with a similar morphology (Figure 1, Table S1). The same was true for the Loke pair (LO1 and LO2), with a very low F_{ST} between the accessions (Table 3) and no significant morphological differences (Table S1). The highest F_{ST} values within groups were found between the pairs A8 and A9 (0.326) and B2 and B3 (0.321, Table 3), accessions showing little morphological differentiation (Figure 1, Table S1). The highest F_{ST} value overall was found between the accessions A1 and K1 (0.564, Table S2), the two accessions with the lowest level of genetic diversity. The pattern was reflected when looking at the average number of pairwise shared alleles, where accessions A1 and K1 shared few alleles with many of the other accessions. The lowest number of shared alleles was found between the accessions K1 and A11, and the highest between A1 and A4 (Table S3).

Table 3. Pairwise F_{ST} values within groups based on all markers (full dataset) and from subsets of 10 markers (subsample average). Subsample values that do not include the pairwise F_{ST} value for the full dataset are highlighted in bold.

Accession 1	Accession 2	Full Dataset F_{ST}	Subsample Average F_{ST}	$F_{ST} - 1$ SE	$F_{ST} + 1$ SE
A1	A3	0.216	0.218	0.215	0.222
A1	A4	0.184	0.184	0.182	0.186
A1	A6	0.252	0.243	0.240	**0.247**
A1	A7	0.304	0.309	**0.305**	0.314
A3	A4	0.103	0.104	0.103	0.105
A3	A6	0.149	0.149	0.147	0.151
A3	A7	0.175	0.173	0.171	0.176
A4	A6	0.110	0.111	0.109	0.112
A4	A7	0.132	0.132	0.131	0.134
A6	A7	0.166	0.169	0.166	0.171
A8	A9	0.326	0.327	0.324	0.331
A11	A12	0.144	0.143	0.141	0.145
B1	B2	0.291	0.294	0.290	0.298
B1	B3	0.228	0.230	0.226	0.234
B1	B6	0.218	0.216	0.213	0.220
B2	B3	0.321	0.324	0.320	0.328
B2	B6	0.304	0.309	**0.305**	0.313
B3	B6	0.087	0.085	0.084	**0.087**
K1	K2	0.253	0.249	0.245	**0.253**
L1	L2	0.188	0.188	0.185	0.190
L1	L3	0.160	0.162	0.160	0.164
L3	L2	0.117	0.117	0.115	0.119
R1	R2	0.111	0.110	0.108	0.111
R1	R3	0.144	0.149	**0.147**	0.151
R2	R3	0.129	0.129	0.127	0.131
J1	J2	0.202	0.201	0.198	0.205
LO1	LO2	0.095	0.095	0.094	0.097
ST1	ST2	0.300	0.294	0.290	**0.298**

A STRUCTURE analysis showed equally high support for two and three clusters (Figure S1). At K = 2, the first cluster contained A1, A3, A6, A7, A12, B1, B3, B6, J1, and J2 and the second contained A8, K1, K2, L1, L2, and L3. The remaining accessions showed some level mixed clustering, with similar degrees of mixture for R1, R2, R3 and ST1 and ST2 (Figure 3). At K = 3, the cluster consisting of accessions A8, K1, K2, L1, L2, and L3 remained intact. A second cluster consisted of some of the accessions showing mixed clustering at K = 2: R1, R2, R3, and ST1. The third cluster contained accessions A1, A3, A6, A7, A12, B1, B3, B6, J1, and J2, with the remaining accessions showing various degrees of mixed clustering (Figure 3b). In general, accessions within the same group tended to belong to the same clusters (Figure 3). For example, all the Kissendrup, Langendijk, Ruhm von Enkhuizen, Jatunsalgets, and Loke accessions clustered within the same group. The most notable exception was accession A8, Amager Kurzstrunkiger Original. Based on the accession name, A8 was expected to cluster with the other Amager accessions, but instead the accession clearly clustered with the Kissendrup and Langendijker accessions (Figure 3).

The PCA based on the allele frequencies in the different accessions (Figure 4) supported the clustering found with STRUCTURE (Figure 3) to a large degree and the morphology-based clustering to a lesser degree (Figure 1). One cluster contained A8, K1, K2, L1, L2, and L3, (upper right); one cluster contained accessions R1, R2, R3, and ST1 (lower center); and one cluster contained accessions A1, A3, A6, A7, A9, A12, B1, B3, B6, J1, LO1, J2, and LO2 (upper left). Accessions A4, B2, A11, and ST2 were located between the latter two clusters, substantiating the structure analysis at K = 3. There was no evidence of clustering according to the genebank origin. The VIR accessions acquired at an earlier date tended to be located less centrally in the PCA ($c = -0.540$, $p = 0.056$) than the NGB accessions. An individual-based PCA showed a good agreement with the accession average-based

PCA. Most individuals of each accession clustered together, but two exceptions were found in accession L3 and accession A4 (Figure S2).

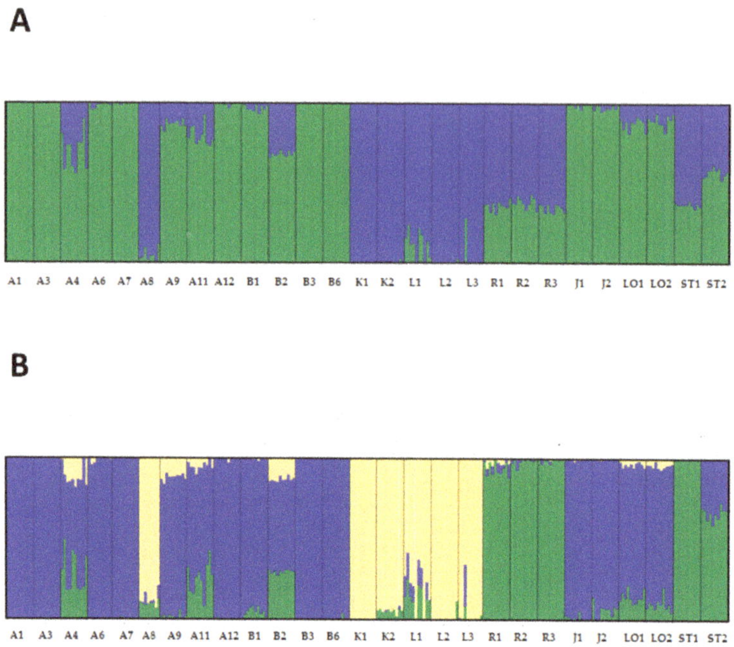

Figure 3. STRUCTURE analysis based on the full SNP dataset, assuming (**A**) two genetic clusters (K = 2); (**B**) three genetic clusters (K = 3). Different colours symbolize the different genetic clusters identified in each individual.

2.4. Genotypic and Morphological Comparisons

In the accession-level PCA based on the SNP data, the first principal component (PC1) separated most of the purple-leafed accessions from most of the green-leafed accessions. The green-leafed accession A8, however, clustered among the purple-leafed accessions, while the purple-leafed accession A11 clustered among the green-leafed (Figure 4). The accession L1, with purple/green leaves, clustered among the purple accessions. Neither head density nor head shape showed any clustering in the accession-level PCA (data not shown).

2.5. Limiting the Number of SNP Markers

Subsets of the data were analyzed to determine whether a more limited number of markers could be used to capture a similar amount of information as the full dataset. Already, with as few as 10 randomly chosen markers, the F_{ST} values obtained were in most cases similar to those calculated from the full dataset (Table 3). Increasing the number of markers to 50, 100, and 500 markers, respectively, reduced the variance of the F_{ST} estimates (Figure S3) from an average of 0.088 for 10 markers to 0.011 for 500 markers. The mean F_{ST} values did, however, not change consistently in any given direction, and could either increase or decrease with an increasing number of markers. The mean F_{ST}, however, always changed less than 0.01.

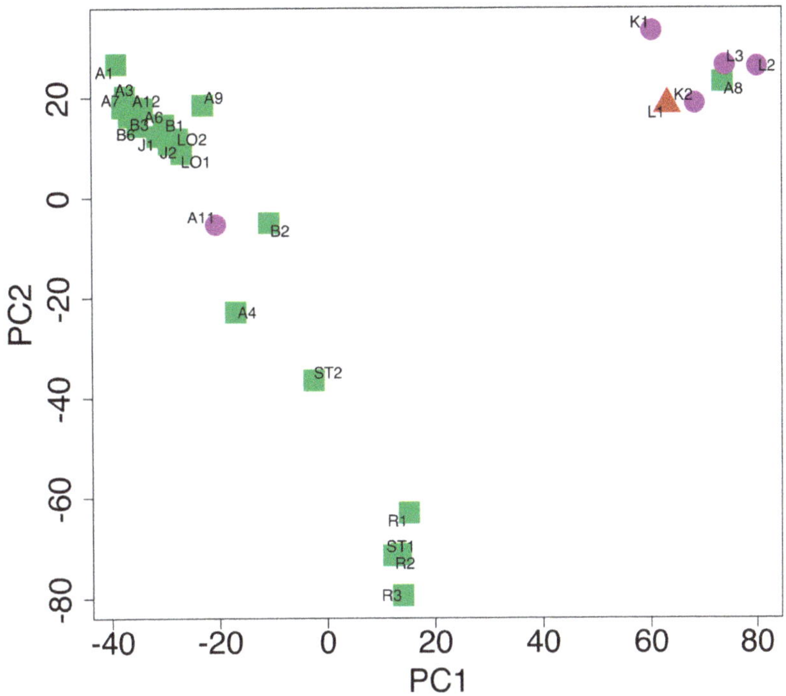

Figure 4. SNPs markers data: Accession-level principal component analysis (PCA) of the SNP markers, where the use of color refers to the leaf color of the cabbage accessions (purple circles = purple leaves, green squares = green leaves, reddish-brown triangles = purple/green leaves). PC1 and PC2 explain 15% and 10% of the total variation, respectively.

Individual-level PCA clustering according to accession could be obtained with a limited number of markers. Already, when subsampling the 20 markers with the alleles providing the largest segregation along PC1 and PC2, a reasonable clustering according to accession could be obtained (Figure S4a). The same number of markers used on the accession-level PCA could only separate accessions along PC1 (Figure S4b).

A single subsample for each 10, 50, 100, and 500 markers was used to investigate whether the same clustering could be obtained in a STRUCTURE analysis as with the full dataset. All subsets indicated K = 2 to be the level of structuring best explaining the data. With only 10 markers, very limited power was obtained to identify structuring, although the clustering of the Jatunsalgets Vinterkål group (J1 and J2) and the Ruhm group (R1, R2, and R3) could be discerned. Surprisingly, as few as 50 markers yielded a structuring similar to that obtained for the full dataset, as well as from 500 markers (Figure S5). A STRUCTURE analysis of 100 randomly chosen markers, however, showed that a lower number of markers was not sufficient to reliably replicate the results of the full dataset.

To explore the efficiency of 50 markers to capture the same structure at K = 2 as the full dataset, an additional 9 datasets of 50 randomly chosen markers were generated. Comparisons of the STRUCTURE analysis of the 10 sets of 50 markers showed that some but not all clusters were reliably identified in each subset (Figure S6). In particular, the clustering of the Jatunsalgets Vinterkål (J1 and J2) and the Loke (LO1 and LO2) groups varied. An analysis with the software CLUMPP showed that the clustering among the 10 sets was not very consistent (H = 0.752).

A subset of markers was chosen with the aim to provide a good resolution for discriminating between accessions. The markers were chosen to provide as high a discriminatory power as possible along the first two principal components in the accession-level PCA (Figure 4). The markers were further evaluated to show a high level of genetic diversity (h > 0.3) and not found to be in high linkage disequilibrium (D′ < 0.25) with each other. In total, 500 markers were chosen (Table S4).

3. Discussion

3.1. Discrimination Power and the Number of Markers

The array used in this study for genotyping *B. oleracea* was originally developed for *B. napus* [28]. Not surprisingly, the array showed the greatest efficiency for markers located on the "C" genome in *B. napus*, with less than 1% of the markers on most chromosomes failing to amplify. Nevertheless, as many as 50% of the markers mapping to the "A" genome in *B. napus* were able to successfully amplify in our *B. oleracea*. Further studies are needed to discern the mapping location of these markers in *B. oleracea* and to evaluate the amount of cross-amplification between the "A" and the "C" genome.

Although the cost of genotyping and sequencing is becoming ever lower, using a limited number of markers while retaining sufficient discriminatory power is still of interest. We found that a random sample of only a few percent of the amplifying markers would have provided us with an overall picture of reasonable similarity to the one obtained with the full dataset. In most cases, the F_{ST} value between two accessions in the same group could be estimated with reasonable accuracy with as few as 50 or 100 markers, similar to what was reported by Willing et al. [29].

A STRUCTURE analysis further suggested that as few as 50 randomly chosen markers could often capture a large part of the genetic structuring in the data, although the clustering of some accessions was inconsistent and the single analysis of 100 random markers failed to replicate the clustering obtained with the full dataset. In addition, parameters such as the amount of gene flow and the evenness of sampling have been shown to influence the number of markers needed to discriminate between groups of differently related individuals [30]. The minimum number of markers needed will vary from organism to organism and from case to case, but may be surprisingly low. Choosing the 20 most discriminatory markers for individual-level PCA resulted in an outcome with a high similarity to that of the full dataset. However, these 20 markers might not be the most discriminatory in another sample of cabbage accessions. For this reason, we recommend a subset of at least 500 SNP markers for duplication assessments in cabbage in order to get a robust, detailed result. A list of 500 SNP markers with a high discriminating power, high level of genetic diversity, and low linkage disequilibrium with each other can be found in Table S4).

3.2. Duplication Assessment and Genetic Similarity

This study clearly demonstrates that the same or similar names do not necessarily mean a duplicate holding. Both the STRUCTURE and PCA analyses detected genetic differences among accessions grouped by accession name. In the STRUCTURE analysis, five out of the 10 groups showed genetic differences within the group: A4 vs. A1, A3, A6, and A7 (Amager Tall group); A8 vs. A9 (Amager Short group); A11 vs. A12 (Amager Winter group); B2 vs. B1, B3, and B6 (Blåtopp group); and ST1 vs. ST2 (Stavanger Torv group, Figure 3). A similar pattern could be seen in the PCA based on SNP markers (Figure 4). Three of the five cases were supported by significant differences in one or several morphological traits (Table S1). The difference between the two Amager Winter accessions was obvious from the morphology, where A11 (Amager Winter) was identified as a red cabbage type, while A12 (Amager Vinter Gefion) was a white cabbage. The differences between the two Stavanger Torv accessions (ST1 and ST2) were less obvious, both being early maturing white cabbage types but with significant differences in two head characters. Within the Amager Tall group, morphological data also confirmed that accessions were different (Table S1), while no clear differences in morphology were observed in the Amager Short group or the Blåtopp group.

Genetic similarity can be used as a criterion to identify and handle duplicate holdings in genebanks [23,24]. In clonal and highly inbred material, it can be a relatively easy task to determine whether accessions are duplicates, but in open pollinated crops such as cabbage, the genetic structure is more complex. Even accessions that have the same origin—for example, accessions donated from the same seed lot to different genebanks—are expected to diverge over time [27]. In these species, the overall pattern of genetic diversity needs to be taken into consideration.

In instances where several lines of evidence (e.g., low F_{ST} values, a large proportion of shared alleles, morphologic similarity, and common clustering in the STRUCTURE analysis and PCA) suggest a high similarity compared to the average, accessions can be considered duplicates and one of the accessions can be removed from the genebank holdings with a minimum loss of genetic diversity (or bulked) [23]. Examples from our dataset that could be bulked, removed, or given lower priority in the conservation could be B3 and B6 and one of the LO accessions. Our study has shown that using accession names alone is not a good strategy to reduce duplicate holdings, as the same or similar names does not mean identical genetic composition. A combined method using both accession names and other passport data as a first step and then marker evaluations as a second step would be a better approach. Alternatively, morphological evaluations could be used or a more extensive passport data evaluation trying to trace the transactions of accessions between genebanks, for example, by using donor accession numbers or other relevant information. The ECPGR Brassica group has established an online tool for identifying duplicate holdings based on accession names and other passport data. This is a useful first step that could be taken into a next step with an extensive evaluation of the potential duplicates with the developed marker set.

3.3. Cultivation History and Naming Practices

There can be several explanations as to why accessions with the same or similar names are genetically and/or morphologically different. Minor differences could be explained by breeding history and naming practices. As mentioned, selections within a cultivar were common in the 19th and first part of the 20th century, and a cultivar could have many and complicated names [7]. One example is "Jatunsalgets Vinterkål Berbes St. Orginal" (J1) or "Jåtunsalgets Vinterkål" (J2). These two accessions are morphologically close and cluster together both in the STRUCTURE analysis and the SNP-based PCA, but the accessions are not identical, either morphologically or genetically. Jåtunsalget was a small Norwegian seed enterprise with only these two accessions recorded in the ECPGR *Brassica* database [31]. "Jåtunsalgets Vinterkål" was listed on the Norwegian variety list in 1979 [32], however the pedigree was "Jåtun Amager x en Hollansk sort i 1929", which means "a selection of Amager crossed with a Dutch cultivar in the year 1929". Most likely, "Jåtun Vinterkål" was marketed already in the 1930s but was listed much later. The oldest (and most original) accession is J1, acquired by VIR in 1953, while J2 (from NGB) entered the Nordic Genebank more than 20 years later from unknown sources [32].

A more complicated example is the Amager varieties. The ECPGR *Brassica* database [31] shows 102 records with "Amager" in their accession name. Amager is a geographical area and a village just outside Copenhagen that hosted both seed enterprises and an extensive vegetable production. We divided the Amager accessions in our study into three sub-groups based on naming; one of them was the Amager Tall sub-group, where the Danish word "Høj" or the Swedish "Hög" both mean "Tall" (with 21 accessions in the ECPGR database). Our study included five such accessions where A4 (accession name "Grami") was genetically clearly different from the other four. In most selections, there is a second name—e.g., "Grön Grami", "Grön Toftø", or "Resistent"—describing further selection properties or enterprises' names. We hypothesized that A4 ("Grami" from VIR) would be similar to A3 ("Amager Høj Grøn Grami" from NGB), but this was not the case. In retrospect, we should maybe not have grouped A4 with the Amager Tall accessions, as the only link to Amager was through the name "Grami", which was used also in the name of A3.

Within the remaining two Amager sub-groups, genetic differences were also detected. Amager Winter is commonly known to be a white cabbage (green leaf laminas), marketed, amongst others, by A. Hansen Amagerfrø in Denmark in the mid-20th century [7]. In the ECPGR *Brassica* database [31], there are six accessions fitting this name, and we included two of these.

Test cultivations showed that A11 (Amager Winter, K192) was a red cabbage (purple leaf laminas). A11 is from an unknown source in Denmark, acquired by VIR in 1969, and the accession has so far been through at least six regeneration cycles at the VIR experimental field. The accession was listed as purple at the time of entry (as accession number K192 in the VIR catalogue). Certainly, there have been red Amager cabbages traded. According to the Nordic Genetic Resource Center cultivar database [32], a red cabbage cultivar/selection named "Amager" was released in 1959 and was bred by Østergård frøavl (Denmark, breeders name Stenballe P 59 68). Other red Amager cabbage cultivars/selections were "Amager 304", bred by A. Hansen Amagerfrø (breeders name Tagenhus P 59 69, released in 1959); "Holdbar Amager", bred by L. Dæhnfeldt (breeders name Toftø S 1960, released in 1960); and "Amager Caro" and "Amager Rega", both released in 1974 by Ohsens Enke (Denmark). A11 could not be one of the latter, as it was acquired by VIR already in 1969, but it could be one of the earlier developed red "Amager" cultivars. What is certain is that A11 is not a duplicate of A12, which has a similar name (Amager Vinter Gefion, NGB1879) but is a genetically different white cabbage. Although A11 is different from the remaining accessions in the "Amager" group in the PCA, it does show a higher similarity to the "Amager" group than to other red cabbages in the PCA (Figure 4). This, together with the results of the STRUCTURE analysis (Figure 3), tentatively suggests that the accession is the result of breeding the purple color into an Amager background. It is hard to know if the polymorphism observed in accession L1 (compared to L2 and L3; Figure 3) is due to gene flow from accessions with a different leaf color, as there are both green and purple plants in this accession, or if there are other explanations.

The genetic data showed clear differences between the Amager Shor pair (Amager Kurzstunkiger Orginal, A8) and (Amager L NF Orginal, A9). Based on passport data, we know they are from different seed companies (one in Denmark and one in Norway) and were included in the collection at VIR in different years (1935 and 1967, respectively). Accessions with the name "Amager Kurztunkig" are found in in Germany, Poland, the Check Republic, and Belarus [31], most likely duplicated with the original accession from VIR included in this study. The prefix "Kurz" is a German word and means "Lav" (in Scandinavian) or short in English. Our reason for pairing A8 and A9 was the prefix "Kurz" in A8 and the abbreviation L ("Lav"?) in A9. Regarding plant height, they were both short, and they had quite similar morphological characteristics, but were not very close in the morphological PCA (Figure 1). Most likely, A9 is an Amager selection from Norsk Frø (NF). From Norway, a cultivar with a similar name (Amager L1 Orginal, not included in this study) is known, with the pedigree "Jåtun Amager x Jåtunsalgets vinterkål 1932" [32]. The cultivar was marketed from 1933 onward, but it was approved as late as in 1961. Amager Kurzstunkiger came to VIR in 1935 from a Danish enterprise but with a German accession name. Certainly, A8 and A9 have a different history and, as demonstrated, they are genetically different.

3.4. The Effect of Genebank Conservation

Changes in genetic composition, major or minor, may take place during field regeneration [33–37]. Our study was not designed to track changes from generation to generation, but some of the differences we observe are probably the result of this process. Minor changes are expected during regeneration in genebanks, especially if a low number of plants are used [38] or if insufficient isolation is used during flowering.

Genebanks use a standard number of plants, usually 20 to 50 individuals, and net cage isolation with pollinators to reduce the risks of genetic changes during regeneration. The FAO genebank standards do not specify the number of plants [39]. At VIR, regeneration takes place typically every 5–7 years, meaning that material acquired in the 1930s has been through at least 10 regeneration cycles. This is expected to result in an increase in differentiation among accessions and a loss of genetic diversity within accessions. The heterozygosity is predicted to decrease each generation in proportion to the population size [40]. For example, with 10 regeneration cycles and a population size of 20, a 22% decrease in heterozygosity is expected on average ($H_t = (1-1/2N)\ H_t-1$; N=diploid population size).

We found that accessions acquired a long time ago tended to have lower genetic diversity than more recent additions, a pattern that is in agreement with the loss of genetic diversity from genetic drift. Additionally, more recently acquired accessions tended to cluster more centrally in the PCA plots, which could also be the result of genetic drift acting to differentiate older accessions.

Other factors such as selection and gene flow could also affect genebank accessions. Some selection from local conditions—both those connected to the environment at the regeneration site, such as climate and soil conditions, and those linked to cultivation practices, such as harvesting time and methods—is expected. In addition, if accessions are not completely isolated, gene flow will occur from other cultivated genebank accessions of the same species, from cultivated fields in the area, and from weeds. Bees and flies are the main pollinators of cabbages [41], and to avoid unwanted gene flow through cross-accession pollination, isolation is crucial. Contrary to genetic drift, external gene flow is expected to increase diversity within the accession and decrease divergence among accessions cultivated at the same time.

Van Hintum et al. [24] demonstrated that the genetic changes caused by regeneration within an accession were of similar magnitude to differences among genebank accessions of cabbages with the same or similar names. Therefore, they questioned the rationale behind conserving a large number of accessions with the same or similar names. Our study supports the occurrence of genetic change during regeneration and similarity among some accessions with similar names. At the same time, however, we have shown that similar names do not always imply the same genetic material.

3.5. Implications for Genebank Conservation

All genebanks have limited budgets and need to adapt their operations to their economic frame. One way to adapt is to remove or pool/bulk duplicates and thus make funds available for the high-quality conservation of the remaining unique accessions. This approach has been used in many species, including *Brassica* crops [23,42].

AEGIS has suggested a roadmap for how to handle duplicate holdings at the European level, identifying the most appropriate accessions based on passport data [4]. However, the decision to remove duplicates is the responsibility of individual genebanks. Our study clearly shows that relying exclusively on accession name when identifying duplicates can be risky, especially with old cross-pollinating cultivars with complex breeding history and naming practices. We find that in five out of the 10 groups, accessions with same or similar names have clear genetic differences. In most cases, such differences were corroborated by significant differences in one or several morphological traits.

Additional passport data such as accession numbers, donor institute, donor accession number, etc. can help pinpoint the origin of the accession and the time of split from other accessions. If such data is available, the chance of correctly identifying duplicates based solely on documentation increases. For recently acquired material, for example—modern cultivars—this can be a safe approach. However, for older cultivars and landraces this is often difficult. Documentation is often missing and, as discussed above, accessions can have diverged substantially from a common origin via complex breeding histories and regeneration in genebanks.

Using detailed morphological characterization has been suggested [43], as has combined morphological and molecular characterization, and molecular characterization alone [18,44]. Our current findings lend support to the need for characterization before deciding to remove or bulk accessions. However, a major challenge is the costs of such characterizations. Most genebanks are underfunded and have backlogs in regeneration and viability monitoring. For tracking future duplications, the introduction of DOI on the accession level [12] would make transactions between genebanks easier, but it cannot capture what is already duplicated. International collaboration and genotyping using the next-generation sequencing could be a cost-effective way forward [14–18]. Here, proper information on the accession level could go hand-in-hand with the facilitation of the use of the germplasm as genetic information is catalogued, linked to the accessions, and made available for the users.

4. Materials and Methods

4.1. Plant Material, Cultivation, and Morphological Characterization

In Europe, there are 35 *Brassica* collections, located in 24 countries and with more than 11,000 *B. oleracea* accessions [45]. In total, 980 cabbage accessions are maintained at VIR and 189 at NGB. An initial study [46] characterized six groups for morphological traits, and some differences within the groups could be detected. In this study, another 10 groups (Table 1) were examined to see if there were differences among accessions within a group. For each group 10 plants per accession were planted. These plants were randomized, and each of the 10 plants was characterized. The study consisted of 10 such randomized pair/triplet characterizations. The planting distance was 50 cm between the plants. The work was done at Alnarp, Sweden (55°′ N, 13°′ E); the soil was loamy clay and the fertilization was 100 kg ha-1 PROMAGNA 11-5-18™ (Yara, Norway) at planting and 30 kg ha-1 YaraMila 22-0-12™ (Yara, Norway) one month after planting. Plants were irrigated and biological control and fungicides were applied. Plants were evaluated just before harvesting. SI units were used for plant leaf, head, and core size parameters and UPOV (International Union for the Protection of New Varieties of Plants) [47] descriptors for leaf color, head shape, and head density. Details are provided in Table S1. Principal Component Analysis (PCA) bi-plots were used for an overview of the data and characters. An ANOVA was performed for each numeric character and included data from all the individuals in that group. If the ANOVA indicated significant differences among accessions, a Tukey multiple comparison of means [48] was used to identify accessions that differed from each other. $\chi2$ statistics were used for categorical characters.

4.2. DNA Extraction

As far as we know, the selected accessions have not previously been included in any molecular studies. DNA extraction was conducted on the same 10 plants per accession that were morphologically characterized. Leaf samples (2 cm^2) were collected from the plants cultivated in the field, placed in 2 ml Eppendorf tubes, immediately frozen in liquid nitrogen, and subsequently freeze-dried overnight in a LyoLab 3000 (Heto Lab Equipment). The freeze-dried material was powdered in a mixer mill (Merck Retsch MM 300) using steel beads; hereafter, 600 µl of CTAB buffer was added to each powdered sample (0.1 M Tris; pH 8.0, 0.01 M EDTA, 0.7 M NaCl, 1% CTAB, and 1% β-mercaptoethanol) according to Doyle and Doyle [49]. The samples were incubated in a thermomixer (Eppendorf) at 600 rpm and 60 °C for 60 min. DNA was extracted by adding one volume of chloroform/isoamylalcohol (24:1), then they were mixed and centrifuged for 20 min at 13,200 rpm. The supernatants were transferred to new tubes and 5 µl of RNAse (1 mg/ml) was added and incubated in the thermomixer (600 rpm, 37 °C for 30 min). Cold isopropanol (0.8 V) was added, followed by mixing and centrifugation (10 min at 13,200 rpm). The DNA pellet was cleaned in 500 µl of wash-buffer (76% ethanol, 0.2 M sodium acetate) for 20 min at room temperature with subsequent centrifugation for 5 min in 13,200 rpm, followed by 500 µl of rinse-buffer (76% ethanol, 0.01 M ammonium acetate) and mixing and centrifugation for 5 min at 13,200 rpm. The samples were left to dry in room temperature for 1 h and were then re-suspended in 50 µl of ddH_2O. The DNA concentration and ratio (at 260 nm and 280 nm) were determined using an Eppendorf BioSpectrometer. The DNA concentrations of the samples were adjusted for further analysis.

4.3. SNP Analysis and Statistics

Array genotyping was performed by TraitGenetics GmbH (Gatersleben, Germany) and with a 15 K Illumina Infinium array that contains a subset of markers from the Brassica 60K array [28]. The array had previously been tested for *B. oleracea* and had a total of 13,714 SNPs. An initial run was carried out with 92 individuals, and a second with 178 individuals. Ten individuals were analyzed from each accession in order to capture the within-accession variation. Cabbage accessions are expected to harbour substantial within-accession variation, and therefore more individuals per accession are needed. By analyzing 10 individuals per accession, we gained an adequate picture of the within-accession variation and at the same time we were able to include many accessions in the study. The first run with 92 individuals was a test run, and since that was successful the second run with 178 individuals was performed in the same way to increase the number of individuals analyzed. After the merging of the two runs, failed and invariant markers were removed, as were markers failing in more than 50% of the individuals. The remaining 5965 markers were used for further analysis. Of these, 68% (4090 markers) had less than 10% missing data. After the removal of the above markers, individuals with more than 40% failed markers were removed (2 individuals + 2 controls). Of the remaining individuals, all had less than 10% missing data.

Two individuals of the accession K2248 were genotyped in both runs. The second genotyping of both individuals had a lower success rate and was removed from the downstream analysis. In total, 266 individuals were kept for further analyses and were analyzed with 5965 markers.

Deviations from the HWE (Hardy-Weinberg Equilibrium) were tested using a $\chi 2$ test with and without Bonferroni correction. All the accessions had less than 10% of the markers deviating from the HWE before the Bonferroni correction. No marker deviated significantly from the HWE in any accession after the Bonferroni correction, and hence no marker was removed for this reason.

Wright's F_{ST} [50] and Nei's h were estimated according to Nei [51] using purpose-written perl scripts. For the F_{ST} values, significance was determined by permutation tests (1000 permutations). Subsets of the markers were analyzed to evaluate if the F_{ST} values between pairs of accessions from the same group could have been estimated with similar accuracy with a more limited number of markers. Subsets of 500, 100, 50, and 10 markers were randomly drawn from the dataset and used for calculating the F_{ST} values. This was repeated 1000 times for each number of markers, and the average F_{ST} values and standard error for the 1000 replicates were calculated.

PCA of the genetic data was carried out using R v 3.2.4 [52] and the *prcomp* command. For an accession-level PCA, the allele frequencies for each allele at each locus were treated as independent variables, while in the individual-level PCA the number of copies of each allele at each locus was used.

The software STRUCTURE (v 2.3.4) [53,54] was used to explore the data for genetic structuring. The software was run with a burn-in length of 20,000 iterations, followed by 50,000 iterations for estimating the parameters, with non-amplifying markers treated as missing data. Each analysis using the admixture model was repeated 10 times for each number of clusters (K = 1 to 10), until the likelihood values for the runs no longer improved. The number of clusters observed in the dataset was evaluated by calculating the ΔK according to Evanno et al. [55]. CLUMPP v 1.1.1 [56] was used to compare the results of individual runs and to calculate the similarity coefficients, H, and the average matrix of ancestry. In CLUMPP, the Full- Search, Greedy, and LargeKGreedy algorithms were used for comparing runs with $K < 4$, $4 \leq K \leq 6$, and $K > 6$, respectively. The graphical presentation of the results was obtained using DISTRUCT v 1.1 [57]. STRUCTURE was also used to analyze the subsets of randomly chosen markers (10, 50, 100, and 500 markers, respectively), and the repeatability of the STRUCTURE analysis of 50 randomly chosen markers was analyzed using CLUMPP.

5. Conclusions

Our study is a contribution to AEGIS and the work to avoid unwanted duplication holdings. We tested the SNP markers developed for *B. napus* and found that many of these genetic markers (nearly 6000) were suitable for an analysis of the genetic structure and duplicate identification of *B. oleracea*. Of these, a subset of 500 markers are recommended for a future large-scale analysis of *B. oleracea* var. *capitata*. Both the genetic SNP data and the morphological data demonstrate the complex relationships among old cabbage cultivars and show that similar accession names do not necessarily mean that accessions are genetically or morphologically similar. This emphasizes that in the case of old cultivars of cross-pollinating species such as cabbage, extra care should be taken when identifying duplicates.

Supplementary Materials: The following are available online at http://www.mdpi.com/2223-7747/9/8/925/s1: Figure S1: Data for determining optimal number of clusters (K) in the STRUCTURE analysis of the full SNP dataset.

Figure S2: Results of the individual level principal component analysis (PCA) based on SNP markers. Figure S3: F_{ST} values between different accessions based on 1000 subsamples of each 10, 50, 100 and 500 SNP markers respectively. Figure S4: Results from PCA when subsampling 20 SNP markers: A) individual level analysis and B) accession-level analysis. Figure S5: Subsampling of SNP markers for STUCTURE analysis assuming two genetic clusters (K=2). Figure S6: Comparison of STRUCTURE analyses of 10 subsets of 50 randomly chosen SNP markers. Table S1: Morphological descriptors with means and standard deviations. Table S2: Pairwise F_{ST} values among all investigated accessions. Table S3: Average number of pairwise shared SNP alleles among all investigated accessions. Table S4: Marker subset suggested for the identification of duplicate accessions (Excel file).

Author Contributions: Conceptualization, A.A. and S.Ø.S.; methodology, J.H., K.A., and A.E.P.; formal analysis, K.A., J.H., and A.E.P.; investigation, A.E.P.; writing—original draft preparation, A.P., J.H., and S.Ø.S.; writing—review and editing, all; visualization, J.H., S.Ø.S., and A.E.P.; project administration, S.Ø.S. and A.A.; funding acquisition, S.Ø.S. and A.A. All authors have read and agreed to the published version of the manuscript.

Funding: This research was supported by the Nordic countries (Norway, Sweden, Denmark, Finland, and Iceland) through the Nordic Counsel of Ministers.

Acknowledgments: We like to thank Flemming Yndgaard and Jerker Niss at Nordic Genetic Resource Center for assistance in the visualization of morphological data and for field cultivation, respectively.

Conflicts of Interest: The authors declare no conflict of interest. The funders had no role in the design of the study; in the collection, analysis, or interpretation of data; in the writing of the manuscript; or in the decision to publish the results.

References

1. FAO. *The Second Report on the State of the World's Plant Genetic Resources for Food and Agriculture*; Food and Agriculture Organization of the United Nations: Rome, Italy, 2010.
2. Fu, Y.B. The Vulnerability of Plant Genetic Resources Conserved Ex Situ. *Crop Sci.* **2017**, *57*, 2314–2328. [CrossRef]
3. ECPGR. A European Genebank Integrated System AEGIS. Available online: http://www.ecpgr.cgiar.org/aegis/about-aegis/ (accessed on 11 June 2020).
4. Engels, J.M.M.; Maggioni, L. *Benefits of Establishing and Operating a European Collection of Unique and Important Germplasm*; ECPGR Secretariat, Bioversity International: Maccarese, Rome, Italy, 2015.
5. Germeier, C.U.; Frese, L.; Bücken, S. Concepts and data models for treatment of duplicate groups and sharing of responsibilities in genetic resources information systems. *Genet. Resour. Crop Evol.* **2003**, *50*, 693–705. [CrossRef]
6. Vilmorin-Andrieux, M.M. *The Vegetable Garden*; Ten Speed Press: Berkeley, CA, USA, 1885.
7. Börjesson, A. *Sorter Av Köksväxter: Svenske Priskuranter Fra 1800-Talet till Ca 1930*; NordGen Publication Series 2015:01; Nordic Genetic Resource Centre: Alnarp, Sweden, 2015.
8. Lamm, R.; Tometorp, G.; Åvall, H. Klassificerande försök med köksväxter. *Medd. Statens Trädgårdsförsök* **1945**, *26*, 165–202. (In Swedish)
9. Mackay, I.; Horwell, A.; Garner, J.; White, J.; McKee, J.; Philpott, H. Reanalyses of the historical series of UK variety trials to quantify the contributions of genetic and environmental factors to trends and variability in yield over time. *Theor. Appl. Genet.* **2011**, *122*, 225–238. [CrossRef]
10. UPOV. The International Union for the Protection of New Varieties of Plants. Available online: http://www.upov.int (accessed on 1 April 2020).
11. Solberg, S.Ø.; Breian, L. Commercial cultivars and farmers' access to crop diversity: A case study from the Nordic region. *Agric. Food Sci.* **2015**, *24*, 150–163. [CrossRef]
12. Alercia, A.; López, F.M.; Sackville Hamilton, N.R.; Marsella, M. *Digital Object Identifiers for Food Crops-Descriptors and Guidelines of the Global Information System*; Food and Agriculture Organization of the United Nations: Rome, Italy, 2018.
13. Zamir, D. Where Have All the Crop Phenotypes Gone? *PLoS Biol.* **2013**, *11*, e1001595. [CrossRef]
14. McCouch, S.R.; McNally, K.L.; Wang, W.; Sackville Hamilton, R. Genomics of gene banks: A case study in rice. *Am. J. Bot.* **2012**, *99*, 407–423. [CrossRef]
15. Poland, J.A.; Rife, T.W. Genotyping-by-Sequencing for Plant Breeding and Genetics. *Plant Genome* **2012**, *5*, 92–102. [CrossRef]

16. Huang, Y.F.; Poland, J.A.; Wight, C.P.; Jackson, E.W.; Tinker, N.A. Using genotyping-by-sequencing (GBS) for genomic discovery in cultivated oat. *PLoS ONE* **2014**, *9*, e102448. [CrossRef]
17. Anglin, N.L.; Amri, A.; Kehel, Z.; Ellis, D. A case of need: Linking traits to Genebank accessions. *Biopreservation Biobanking* **2018**, *16*, 337–349. [CrossRef]
18. Singh, N.; Wu, S.; Raupp, W.J.; Sehgal, S.; Arora, S.; Tiwari, V.; Vikram, P.; Singh, S.; Chhuneja, P.; Gill, B.S.; et al. Efficient curation of genebanks using next generation sequencing reveals substantial duplication of germplasm accessions. *Sci. Rep.* **2019**, *9*, 650. [CrossRef] [PubMed]
19. McCouch, S.; Baute, G.J.; Bradeen, J.; Bramel, P.; Bretting, P.K.; Buckler, E.; Burke, J.M.; Charest, D.; Cloutier, S.; Cole, G.; et al. Agriculture: Feeding the future. *Nature* **2013**, *499*, 23–24. [CrossRef] [PubMed]
20. Van Treuren, R.; van Hintum, T.J.L. Next-generation genebanking: Plant genetic resources management and utilization in the sequencing era. *Plant Genet. Resour.* **2014**, *12*, 298–307. [CrossRef]
21. Bothmer, R.V.; Gustafsson, M.; Snogerup, S. Brassica sect. Brassica (Brassicaceae) II. Inter- and intraspecific crosses with cultivars of *B. oleracea*. *Genet. Resour. Crop Evol.* **1995**, *42*, 165–178. [CrossRef]
22. Nagaharu, U. Genome analysis in Brassica with special reference to the experimental formation of *B. napus* and peculiar mode of fertilization. *Japan J. Bot.* **1935**, *7*, 389–452.
23. Hintum, T.J.L.; van Sackville Hamilton, N.R.S.; Engels, J.M.M.; Treuren, R.V. Accession Management Strategies: Splitting and Lumping. In *Managing Plant Genetic Diversity*; Engels, J.M.M., Ramanatha, R.V., Brown, A.H.D., Jackson, T., Eds.; CABI Publishing: Wallingford, UK, 2002; pp. 113–120.
24. Van Hintum, T.J.; van De Wiel, C.C.M.; Visser, D.L.; Van Treuren, R.; Vosman, B. The distribution of genetic diversity in a *Brassica oleracea* genebank collection related to the effects on diversity of regeneration, as measured with AFLPs. *Theor. Appl. Genet.* **2007**, *114*, 777–786. [CrossRef]
25. Christensen, S.; von Bothmer, R.; Poulsen, G.; Maggioni, L.; Phillip, M.; Andersen, B.A.; Jørgensen, R.B. AFLP analysis of genetic diversity in leafy kale (*Brassica oleracea* L. convar. *acephala* (DC.) Alef.) landraces, cultivars and wild populations in Europe. *Genet. Resour. Crop Evol.* **2011**, *58*, 657–666.
26. Faltusova, Z.; Kucera, L.; Ovesna, J. Genetic diversity of *Brassica oleracea* var. *capitata* gene bank accessions assessed by AFLP. *Elect. J. Biotech.* **2011**, *14*. [CrossRef]
27. Izzah, N.K.; Lee, J.; Perumal, S.; Park, J.Y.; Ahn, K.; Fu, D.; Kim, G.B.; Nam, Y.W.; Yang, T.J. Microsatellite-based analysis of genetic diversity in 91 commercial *Brassica oleracea* L. cultivars belonging to six varietal groups. *Genet. Resour. Crop Evol.* **2013**, *60*, 1967–1986. [CrossRef]
28. Clarke, W.E.; Higgins, E.E.; Plieske, J.; Wieseke, R.; Sidebottom, C.; Khedikar, Y.; Batley, J.; Edwards, D.; Meng, J.; Li, R.; et al. A high-density SNP genotyping array for *Brassica napus* and its ancestral diploid species based on optimised selection of single-locus markers in the allotetraploid genome. *Theor. Appl. Genet.* **2016**, *129*, 1887–1899. [CrossRef]
29. Willing, E.M.; Dreyer, C.; Van Oosterhout, C. Estimates of genetic differentiation measured by F_{ST} do not necessarily require large sample sizes when using many SNP markers. *PLoS ONE* **2012**, *7*, e42649. [CrossRef] [PubMed]
30. Nelson, M.F.; Anderson, N.O. How many marker loci are necessary? Analysis of dominant marker data sets using two popular population genetic algorithms. *Ecol. Evol.* **2013**, *3*, 3455–3470. [CrossRef] [PubMed]
31. Menting, F.; Bas, N. The ECPGR Brassica Database. Available online: http://ecpgr.cgn.wur.nl/Brasedb/ (accessed on 15 November 2017).
32. Nordic Genetic Resource Center. Seed Information Database. Available online: https://nordgen.org/ (accessed on 15 November 2019).
33. Soleri, D.; Smith, S.E. Morphological and phenological comparisons of two hopi maize varieties conserved in situ and ex situ. *Econ. Bot.* **1995**, *49*, 56–77. [CrossRef]
34. Gomez, O.J.; Blair, M.W.; Frankow-Lindberg, B.E.; Gullberg, U. Comparative study of common bean (*Phaseolus vulgaris* L.) landraces conserved ex situ in genebanks and in situ by farmers. *Genet. Res. Crop Evol.* **2005**, *52*, 371–380. [CrossRef]
35. Negri, V.; Tiranti, B. Effectiveness of in situ and ex situ conservation of crop diversity. What a *Phaseolus vulgaris* L. landrace case study can tell us. *Genetica* **2010**, *138*, 985–998. [CrossRef]
36. Hagenblad, J.; Zie, J.; Leino, M.W. Exploring the population genetics of genebank and historical landrace varieties. *Genet. Resour. Crop Evol.* **2012**, *59*, 1185–1199. [CrossRef]
37. Solberg, S.Ø.; Yndgaard, F.; Palmé, A. Morphological and phenological consequences of ex situ conservation of natural populations of red clover (*Trifolium pratense* L.). *Plant Genet Resour.* **2015**, *15*, 97–108. [CrossRef]

38. Ellstrand, N.C.; Elam, D.R. Population genetic consequences of small population size: Implications for plant conservation. *Annu. Rev. Ecol. Syst.* **1993**, *24*, 217–242. [CrossRef]
39. FAO. *Genebank Standards for Plant Genetic Resources for Food and Agriculture*; Food and Agriculture Organization of the United Nations: Rome, Italy, 2014.
40. Wright, S. Evolution in Mendelian Populations. *Genetics* **1931**, *16*, 97–159.
41. Bohart, G.E.; Todd, F.E. Pollination of Seed Crops by Insects. In *Seeds. The Yearbook of Agriculture*; US Government Printing Office: Washington, DC, USA, 1961; pp. 240–246.
42. Cruz, V.; Nason, J.; Luhman, R.; Marek, L.; Shoemaker, R.; Brummer, E.; Gardner, C. Analysis of bulked and redundant accessions of *Brassica* germplasm using assignment tests of microsatellite markers. *Euphytica* **2006**, *152*, 339–349. [CrossRef]
43. Diederichsen, A. Duplication assessments in Nordic *Avena sativa* accessions at the Canadian national genebank. *Genet. Resour. Crop Evol.* **2009**, *56*, 587–597. [CrossRef]
44. Lund, B.; Ortiz, R.; Skovgaard, I.M.; Waugh, R.; Andersen, S.B. Analysis of potential duplicates in barley gene bank collections using re-sampling of microsatellite data. *Theor. Appl. Genet.* **2003**, *106*, 1129–1138. [CrossRef] [PubMed]
45. Branca, F.; Bas, N.; Artemyeva, A.; De Haro, A.; Maggioni, L. Activities of the Brassica Working Group of the European Cooperative Programme for Plant Genetic Resources (ECPGR). *Acta Hort.* **2013**, *1005*, 149–155. [CrossRef]
46. Solberg, S.Ø.; Artemyeva, A.; Yndgaard, F.; Dorre, M.; Niss, J.; Burleigh, S. Duplication assessments in *Brassica* vegetable accessions. *Plant Genet. Resour.* **2017**, *16*, 201–208. [CrossRef]
47. UPOV. *UPOV Guidelines for the Conduct of Tests for Distinctness, Uniformity and Stability TG/48/7*; International Union for the Protection of New Varieties of Plants: Geneva, Switzerland, 2004.
48. Crawley, M.J. *The R Book*; John Wiley & Sons Ltd.: Chichester, UK, 2009.
49. Doyle, J.J.; Doyle, J.L. Isolation of plant DNA from fresh tissue. *Focus* **1990**, *12*, 13–15.
50. Wright, S. The genetical structure of populations. *Ann. Eugen.* **1951**, *15*, 323–354. [CrossRef]
51. Nei, M. Analysis of gene diversity in subdivided populations. *Proc. Natl. Acad. Sci. USA* **1973**, *70*, 3321–3323. [CrossRef]
52. R Core Team. *R: A Language and Environment for Statistical Computing*; R Foundation for Statistical Computing: Vienna, Austria; Available online: https://www.R-project.org/ (accessed on 11 June 2020).
53. Pritchard, J.K.; Stephens, M.; Donnelly, P. Inference of population structure using multilocus genotype data. *Genetics* **2000**, *155*, 945–959.
54. Falush, D.; Stephens, M.; Pritchard, J.K. Inference of population structure using multilocus genotype data: Linked loci and correlated allele frequencies. *Genetics* **2003**, *164*, 1567–1587.
55. Evanno, G.; Regnaut, S.; Goudet, J. Detecting the number of clusters of individuals using the software STRUCTURE: A simulation study. *Mol. Ecol.* **2005**, *14*, 2611–2620. [CrossRef]
56. Jakobsson, M.; Rosenberg, N.A. CLUMPP: A cluster matching and permutation program for dealing with label switching and multimodality in analysis of population structure. *Bioinformatics* **2007**, *23*, 1801–1806. [CrossRef] [PubMed]
57. Rosenberg, N.A. DISTRUCT: A program for the graphical display of population structure. *Mol. Ecol. Notes* **2004**, *4*, 137–138. [CrossRef]

plants

MDPI

Review

Document or Lose It—On the Importance of Information Management for Genetic Resources Conservation in Genebanks

Stephan Weise *, Ulrike Lohwasser and Markus Oppermann

Leibniz Institute of Plant Genetics and Crop Plant Research (IPK) Gatersleben, Corrensstr. 3, 06466 Seeland, Germany; lohwasse@ipk-gatersleben.de (U.L.); opperman@ipk-gatersleben.de (M.O.)

* Correspondence: weise@ipk-gatersleben.de

Received: 15 July 2020; Accepted: 14 August 2020; Published: 18 August 2020

Abstract: Genebanks play an important role in the long-term conservation of plant genetic resources and are complementary to the conservation of diversity in farmers' fields and in nature. In this context, documentation plays a critical role. Without well-structured documentation, it is not possible to make statements about the value of a resource, especially with regard to its potential for breeding and research. In particular, comprehensive information management is a prerequisite for the further development of genebank collections. This requires detailed information about the composition of a collection, thus allowing statements about which species and/or regions of origin are under-represented. This task is of strategic importance, especially due to the threats to crop plants and their wild relatives caused by advancing climate change. Both the actual conservation management and the fulfilment of legal obligations depend on information. Hence, documentation units have been established in almost all genebanks worldwide. They all face the challenge that knowledge about genebank accessions must be permanently managed and passed on across generations. International standards such as Multi-Crop Passport Descriptors (MCPD) have been established for the exchange of data between genebanks, and allow the operation of international information systems, such as the World Information and Early Warning System on Plant Genetic Resources for Food and Agriculture (WIEWS), the European Search Catalogue for Plant Genetic Resources (EURISCO) or Genesys.

Keywords: documentation; genebank; plant genetic resources; agricultural biodiversity

1. Introduction

For many centuries, humans have been taking advantage of the plant world and adapting it to their needs. The resulting diversity of useful plants is the main source of food for humans and animals [1]. In addition to the nutritional aspect, plants also supply raw materials for the chemical and pharmaceutical industries and are renewable energy sources [2,3]. Global biodiversity is severely threatened by human intervention, not the least against the background of advancing climate change [4]. This also applies to the diversity of crops.

Genebanks play an important role in the long-term conservation of these plant genetic resources [5]. They complement the conservation of diversity in farmers' fields and in nature. There are about 1800 genebanks worldwide, more than 600 of them in Europe [6]. The important genetic diversity stored in genebanks can provide new impulses for research and breeding, e.g., by introducing new alleles into existing breeding stocks [7], which have a morphologically and physiologically beneficial effect on the characteristics of the plants. Besides maintenance and regeneration, an important task of genebanks is therefore the phenotypic characterisation of accessions [8].

In this context, documentation plays a crucial role [9,10]. Imagine a supermarket with shelves where all the goods are unlabelled. In addition, only a few persons know what is on which shelf.

There are no records of where the goods come from, how old they are and for how much they have to be priced. It is obvious that such a store cannot work. The same holds true for genebank holdings. Without having as much information as possible available in a well-structured way, it is not possible to make informed statements about the value of a resource, especially with regard to its breeding and research potential. The wealth of information covers many areas, from data necessary to optimally manage collections over genebank basic data (passport data) to phenotypic and comprehensive genetic data. One of the greatest challenges of genebanks, apart from the conservation of accessions, is the management of these data [11,12].

This article aims to provide an overview of the relevant systems and structures of plant genetic resources documentation in genebanks. The development of documentation, the current situation and international cooperation as well as the specification of data for long-term and sustainable conservation management are described. In addition, the current needs for integrative information management based on several requirements for future genebank documentation are described.

2. Information Required

For the long-term conservation and use of plant genetic resources, it is necessary to document a large amount of information at various levels, especially for the identification and characterisation of accessions.

2.1. Basic Data

Basic information on plant genetic resources is contained in the passport data. They serve in particular to identify the material and contain information such as the accession number, the scientific name and information on the origin and acquisition of the material (year of acquisition, donor, collecting mission, location of collecting). Ideally, these data are following the standard of the Multi-Crop Passport Descriptors (MCPD) [13,14], which has been developed as a uniform, global format. This corresponds with the recommendations of the Genebank Standards [15] of the Food and Agriculture Organization of the United Nations (FAO), which were developed by the FAO Commission on Genetic Resources for Food and Agriculture (http://www.fao.org/cgrfa/).

The geographical origin of a genebank accession (particularly in connection with environmental data) can provide information on possible adaptations to biotic/abiotic stress factors. Such data should be complemented by information on the type of material (biostatus, e.g., wild form, landrace, etc.). Other important data include phenotypic characterisation of the individual accession, including morphological and agronomical traits.

This basic information helps to identify the individual samples of the genetic resource and to estimate its value, especially its potential for breeding, but also for research.

2.2. Stable and Unique Identifiers

Genebanks have been existing for many decades. This long period of time implies that the conditions at genebanks have been and continue to change. On the one hand, changes result from technological progress. This means that both the type of storage and the data itself must be adapted. On the other hand, social, political and economic changes occur, which lead to organisational changes in genebanks. This means that the description and use of plant genetic resources are also subject to constant change and may result in different identifiers for an accession over the time. In addition, the exchange between genebanks is common in order to establish conservation security through targeted safety duplication and to complement collections in the individual countries. Moreover, material is supplied to researchers and breeders, but also to other users.

Before the introduction of information systems spanning several collections, a local, sometimes temporary, unambiguous identification of an accession was sufficient. However, changing identifiers lead to chains of identifiers of an accession and make it more difficult to trace transferred material. Even within a collection, unambiguity could not always be guaranteed, for example, in the case of multiple accession numbers for the same material in the course of time. Identical identifiers in aggregating information systems also pose a problem [16]. This led directly to the consideration of introducing a system that would provide unique and stable identifiers for genebank accessions [17]. Of the various approaches that can be considered for this purpose, digital object identifiers (DOIs) appear to be the most common. The idea of the DOI dates back to the 1990s [18]. A DOI is a unique and permanent digital identifier of a (digital) object. For the description of the object, metadata are associated with it. The name resolution of a DOI is performed using a handle system such as doi.org. The DOI system was introduced in 2000 and is maintained by the International DOI Foundation (IDF) [19]. The core is a central database. It contains the URL under which the referenced object is currently available. The organisation that has registered the DOI is responsible for updating the database entry whenever the metadata are changed. Not the least due to the support by the Secretariat of the International Treaty on Plant Genetic Resources for Food and Agriculture (ITPGRFA), DOIs have established themselves as a quasi-standard for plant genetic resources material [20]. An undeniable advantage of DOIs is their high acceptance in the scientific community. However, it should be critically noted that accessions of plant genetic resources are not non-modifiable digital objects. Data describing accessions are generally subject to changes and additions.

Moreover, it is an indispensable task of the documentation units of the genebanks to map the historically used identifiers of the accessions to these new, unique identifiers.

2.3. Data for Collection Development

What is needed to recognise which parts of a collection are over- or under-represented? First and foremost, the composition of a collection naturally depends on its exact purpose. In addition, as comprehensive information as possible on individual accessions is a further essential prerequisite for the further development of genebank collections. With the help of the botanical determinations of accessions, a genebank collection can be examined to determine the extent to which the gene pool of a genus is completely represented or which species/subspecies are missing. In addition, the within species/subspecies diversity should be adequately represented. By adding geographical information, it is possible to specify more precisely from which regions material is missing and should be collected. Such detailed information on the composition of a collection is necessary to perform gap analyses [21–23]. In addition, precise geographical data can be used to draw conclusions about ecological conditions. This task is of strategic importance, especially because of the threat posed to crops and their wild relatives by progressing climate change [24].

The comparison of data on accessions also allows statements to be made about potential duplicates within a collection. Nevertheless, this task is not trivial. A reliable statement can only be made by the joint analysis of phenotypic, genotypic and passport data in combination with comparative cultivations [25,26].

By analysing combined data from different genebanks, it is also possible to check which accessions are maintained in other collections and, if necessary, could be obtained from there [27,28]. Such comparisons are an important means of expanding a collection in a targeted manner through exchange with other genebanks. This can also contribute to making the genebank work more effectively. One approach that pursues this goal is the European Cooperative Programme for Plant Genetic Resources (ECPGR) initiative, A European Genebank Integrated System (AEGIS, https://www.ecpgr.cgiar.org/aegis/) [6,29]. AEGIS aims to identify which accession in different genebanks is the most appropriate accession (MAA). In this case, genebanks participating in AEGIS undertake to assume responsibility for conservation and to maintain this accession according to uniform standards in the long term.

2.4. Securing the Legal Status of Acquisitions

There are international agreements governing the conservation and sustainable use of plant genetic resources as well as access and benefit sharing (ABS), in particular, the Convention on Biological

Diversity (CBD, https://www.cbd.int/), which entered into force in 1993, and its supplementary agreement, the Nagoya Protocol of 2014 (Nagoya Protocol on Access to Genetic Resources and the Fair and Equitable Sharing of Benefits Arising from Their Utilization to the Convention on Biological Diversity, https://www.cbd.int/abs/about/). To this end, it is essential to document the origin of genebank accessions, as well as the time of inclusion in a collection and existing collecting permits.

In accordance with the CBD, the International Treaty on Plant Genetic Resources for Food and Agriculture (ITPGRFA, http://www.fao.org/plant-treaty/) came into force in 2004 and the signatory states committed themselves to conserve, characterise and evaluate plant genetic resources and to ensure their sustainable use. The main component of the Treaty is the Multilateral System of Access and Benefit Sharing (MLS), which regulates facilitated access to plant genetic resources that have been included in the MLS and the equitable sharing of the resulting benefits. A Standard Material Transfer Agreement (SMTA, http://www.fao.org/plant-treaty/areas-of-work/the-multilateral-system/the-smta/en/) was created for this purpose. The reporting obligations attached to the SMTA are usually part of the tasks of the documentation units of the genebanks.

2.5. Material Management Data

The actual conservation management in the genebanks is also dependent on information, e.g., on storage conditions and location, and the quality and quantity of seeds and other plant propagules at accession level, in order to enable efficient and effective genebank management. This includes information such as germinability or storage quantity as well as information on health tests that have been carried out or are pending. In addition, information on the regeneration of the individual accessions must be documented as well as how and where the material is stored (e.g., active or base collection, cold store, shelf, safety duplication sites) [15]. Depending on size and information technology equipment, the solutions of the individual genebanks differ [12,30].

2.6. Data to Protect Against Losses

How is it generally attempted to protect plant genetic resources in genebanks from loss? A fundamental idea of genebanks is the protection against losses of accessions or entire collections, e.g., in case of war or catastrophe. The FAO Genebank Standards [15] recommend to realise this by intentional duplication of unique accessions in different genebanks of geographically distant areas. In recent years, the Global Seed Vault (https://www.croptrust.org/our-work/svalbard-global-seed-vault/) in Svalbard was introduced as an additional safety backup system [31–33]. That this has its justification is evident in the example of the genebank of the International Center for Agricultural Research in the Dry Areas (ICARDA), which was reconstructed from security samples in Morocco and Lebanon (https://www.seedvault.no/news/withdrawal-of-icarda-aleppo-seeds-accomplished/). However, such a mechanism has not yet been established for the associated data. It also does not make sense to save data as a black box, as is possible with the physical material. In the case of loss of material, information about where safety duplicates are located, but also who did receive this material (both donors and recipients of material samples), are the keys to finding the material again. Unfortunately, this information is primarily in the genebank, which could be lost. Usually only the basic passport data are shared along with safety duplicates, in some cases not even this. In addition, not all genebanks have a strategy and the ability to secure data which goes beyond this. Similar to the idea of the Global Seed Vault, a global data safe or corresponding, connected repositories would be one way to meet this challenge. Of course, this must not result in losing reference to the physical material. The International Nucleotide Sequence Database Collaboration (INSDC, http://www.insdc.org/) shows that such an approach can work [34,35]. The INSDC dates back to the 1980s and ensures the continuous synchronisation of DNA and RNA sequence data from the three major international systems, GenBank (https://www.ncbi.nlm.nih.gov/Genbank/), the DNA Data Bank of Japan (DDBJ, https://www.ddbj.nig.ac.jp/) and the European Nucleotide Archive (ENA, https://www.ebi.ac.uk/ena).

3. Information Management for Genetic Resources Conservation

3.1. Documentation Development

The main intention of genebanks is to conserve collections of plant genetic resources for posterity. This means that the documentation of the material must also be ensured across generations. In order to meet the resulting requirements, documentation units have been established in almost all genebanks worldwide. They all face the challenge that the knowledge about the genebank material must be continuously managed, supplemented and passed on across technical and personnel generations. The knowledge must not be tied to individual persons alone. Some genebanks are many decades old, which means that there have been several changes of personnel during this period and possibly also one or more predecessor institutions. Ideally, however, there should still be complete documentation of all accessions of plant genetic resources acquired, maintained and distributed, covering the entire period since the foundation of the genebank.

Historically, the documentation structures and methods of the management of genebanks in the first decades were influenced by experiences from botanical gardens and focused on breeding activities. Accordingly, there was a collection-oriented documentation of the origins, often in the form of collecting mission reports, and a practice-oriented documentation of the collections as inventory books and cultivation documentation. The taxonomic classification of the accessions was an important ordering criterion. Many genebanks regularly created a catalogue, a so-called Index Seminum, to make their holdings publicly accessible. Depending on the size and focus of the genebank, more or less complex index card systems and registries were established. Archives with correspondence and collecting mission reports supplemented these, but often only with weak linkage to collection management. These document holdings were difficult to search and exploit. Due to their limited storage capacity, the first electronic systems were therefore established parallel to paper documentation in order to facilitate searchability, filtering and presentation of the information. With the advancing technical development of databases and spreadsheet programs, different IT solutions were created depending on size, focus and organisational form (centralised or decentralised) of the genebank, which increasingly replaced paper documentation [36–40].

Many larger genebanks have implemented in-house genebank information systems in recent years, e.g., GENIS [41], GBIS [42] or GBIMS [43]. In addition, there was also cooperation between genebanks, sometimes across national borders. For example, the management system SESTO (https://sesto.nordgen.org/) was developed for the joint use of the Nordic countries and was operated by the Nordic Genetic Resource Centre (NordGen) in Alnarp, Sweden. SESTO was also used by the genebanks of the Baltic countries for documentation purposes. However, even today there are still a large number of smaller collections that do not have these capabilities and manage data with simpler means, such as MS Excel lists [44]. For those, the freely available system, GRIN-Global (https://www.grin-global.org/) [45], is increasingly establishing itself as an alternative to proprietary systems, e.g., as seen in Barata et al. [30]. This system was originally developed for the Germplasm Resource Information Network (GRIN) of the United States Department of Agriculture (USDA) and has been made available as an open source version jointly by the Global Crop Diversity Trust, Bioversity International and the USDA's Agricultural Research Service since 2011. GRIN-Global enables to manage phenotypic data in addition to the basic passport data. Furthermore, the system allows the maintenance of material management data, e.g., germination rates or existing storage quantities. The data can be curated via a corresponding interface. Furthermore, GRIN-Global has an online search and ordering system. This represents a major step forward for structured and sustainable documentation. The Genebank Information System (GBIS) [42] of the German Federal ex situ Genebank for Agricultural and Horticultural Crop Species is following a similar line. Just like GRIN-Global, GBIS allows to manage different types of data (passport and phenotypic data, management data, plant health tests, germination rates, orders, etc.). An online search and ordering system is also publicly available. However, in contrast to GRIN-Global, GBIS is explicitly designed for use in a single genebank. For this purpose, the system is fully integrated into the specific work processes. In this context, GRIN-Global and GBIS shall serve as examples of two different philosophies in the development of information systems in genebanks.

3.2. Current Situation

In order to provide an overview of the current situation of information management in genebanks, the authors conducted an ad hoc survey among the National Inventory Focal Points of the European Search Catalogue for Plant Genetic Resources (EURISCO) network (see below). They were asked whether they could provide information on (1) how the genebanks in their respective countries manage their data and (2) which data are managed in addition to basic passport data. From 40 persons contacted, 30 replied. Even though this survey is not representative and certainly reflects only a part of the overall situation, it provides a basic overview. A more extensive survey, aiming in particular at perspective approaches to the future development of the documentation systems in the individual countries, would be a logical consequence, but was not feasible in the context of this article.

As expected, the information management in genebanks is very diverse. Since the data of different domains are often managed in different systems, there were multiple answers. Forty percent of the respondents stated that individual information systems are used in the genebanks. Beginning in July 2020, the Nordic and Baltic countries started to use a joint system based on GRIN-Global (Nordic Baltic Genebanks Information System (NBIS), https://www.nordic-baltic-genebanks.org/gringlobal/). Thus, 28% in total use the GRIN-Global system, while only 3% apply other systems that are used in different genebanks. More than 40% use MS Access or FoxPro for data management. The most widespread use is MS Excel (59%). Thirty-four percent still document in paper form (Figure 1).

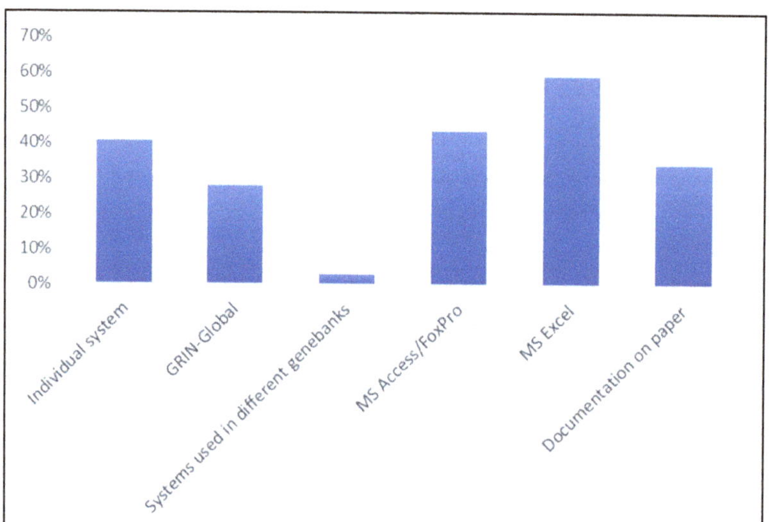

Figure 1. Overview of the management systems used in various genebanks. Different systems are often used depending on the type of data. Therefore, multiple answers were possible.

All respondents indicated that they manage passport data. This was to be expected because this is the basic information for genebanks. In addition, it was stated that more than 90% also hold phenotypic data and information on seed stocks. Sixty-nine percent also manage data on seed orders.

3.3. International Collaboration

Despite ever better IT support through genebank information systems, these remained largely isolated from each other. Since the 1980s, a start has been made on compiling data on accessions of one or more crop species obtained in genebanks in a region or even worldwide into databases, the so-called Central Crop Databases (CCDBs) [46]. Two of the earliest Central Crop Databases are the European Barley Database [47] and the European Prunus Database [48]. The CCDBs have strengthened the cooperation between genebanks and have been made possible through networking genebanks and collections. In addition, the CCDBs aimed to make genebank material more accessible to users and to identify possible duplicates between the individual collections. However, these goals could only be achieved to a limited extent, particularly due to the limited availability of these databases and their low data quality or lack of such data [49]. Another major challenge was the long lack of uniform standards for the description and exchange of passport data. In 1997, a first draft of the Multi-Crop Passport Descriptors (MCPD) was presented [50], which was subsequently developed into a globally accepted standard [13,14].

Aggregating platforms and databases for a cross-genebank search for suitable accessions have also been and are being developed. MCPD and Darwin Core [51,52] have been established for the exchange of passport data between genebanks and these platforms. It is only through such standards that it is possible to feed and operate international information systems such as the World Information and Early Warning System on Plant Genetic Resources for Food and Agriculture (WIEWS, http://www.fao.org/wiews/), EURISCO (http://eurisco.ecpgr.org/), Genesys (https://www.genesys-pgr.org/) or the Global Biodiversity Information Facility (GBIF, https://www.gbif.org/), which combine the information as homogeneously as possible and make it available beyond the boundaries of the individual genebank collection.

While the intention of WIEWS is to provide periodic, country-driven assessments of the plant genetic resources conservation status for FAO, the European Search Catalogue for Plant Genetic Resources (EURISCO) provides detailed accession-specific information on the majority of European collections [53]. Of the approximately 600 European collections of plant genetic resources, more than 400 provide their accession-level data to EURISCO. Information on the geographical location of the genebanks can be found on the EURISCO website. The development of EURISCO started in 1999. For this purpose, a network of national inventories covering 43 countries was successively established. These national inventories bring together the data from the respective collections of their countries and then make them available to the EURISCO information system in a coordinated manner. EURISCO currently documents more than two million genebank accessions, the data of which are regularly updated. The MCPD standard plays a key role here.

Another approach is the Genesys information system, which has been developed since 2008 and is based on the System-wide Information Network for Genetic Resources (SINGER) system. SINGER was an integrated system for the management and exchange of data on plant genetic resources held in Consultative Group on International Agricultural Research (CGIAR) genebanks (https://cgiar.org/). This system was originally developed in 1994 and was put on a new technological basis with Genesys. In addition, EURISCO as the European hub and the US GRIN system as the North American hub regularly feed their data into Genesys as well. A source of additional Australian data is the University of Queensland's Crop Trait Mining Informatics Platform [54]. The aim of the platform is to make information available through the existing Genesys system to support the development and use of plant genetic resources. Both EURISCO and Genesys comprise passport data as well as phenotypic data.

The above-mentioned GBIF is both an international network and an infrastructure designed to make data on global biodiversity freely and permanently available. For this purpose, standards and tools for the exchange of information are provided. With its all-encompassing approach to aggregating data from all areas of biodiversity, GBIF holds a special position. In the plant sector, this network mainly includes data from natural history collections, such as botanical gardens, herbaria and other biodiversity databases, but also includes data on plant genetic resources, which are made available either via aggregators, such as EURISCO, or directly through the genebanks.

The systems just described have made a significant contribution to standardising the documentation of plant genetic resources and improving international cooperation. They make it easier to find information on accessions maintained in a large number of genebanks. In addition, they provide positive impulses for the coordination of the conservation of collections. In particular, the European approach AEGIS is being promoted as a virtual European genebank and aims to define the efficient conservation of genebank material using uniform standards.

In general, networks at the regional and/or international level play an outstanding role in developing the above approaches. They contribute to bundling existing strengths and resources in order to master common challenges. In this context, the European Cooperative Programme for Plant Genetic Resources (ECPGR, https://www.ecpgr.cgiar.org/) should be mentioned, under which a large number of Central Crop Databases as well as the EURISCO system were developed. From the field of natural science collections, the Distributed System of Scientific Collections (DiSSCo, https://www.dissco.eu/) approach should be mentioned here. DiSSCo aims to curate and make accessible the holdings of European natural science collections according to uniform criteria. This approach could provide additional impulses to advance the international networking of genebanks and their documentation.

In the context of international cooperation, reference should be made here once again to the International Treaty on Plant Genetic Resources for Food and Agriculture (ITPGRFA, see above). Article 17 of the Treaty contains the vision for the development of a global information system. Since 2015, a Global Information System (GLIS) for plant genetic resources on the basis of existing information systems has been underway. The goal of the GLIS is to provide a global entry point for knowledge and information to support the conservation, management and use of plant genetic resources (http://www.fao.org/plant-treaty/areas-of-work/global-information-system/). Besides the GLIS DOI portal (https://ssl.fao.org/glis/) of the Treaty Secretariat, WIEWS, Genesys, EURISCO and GRIN-Global are major components of GLIS.

Successful international cooperation, also in the field of documentation, supports the fulfilment of the Sustainable Development Goals (SDG, http://www.fao.org/sustainable-development-goals/) developed by 193 member states of the United Nations.

Finally, it is worth mentioning the international DivSeek network (https://divseekintl.org/), which is a worldwide collaboration to support the creation, integration and exchange of data on plant genetic resources. A number of working groups have been established for this purpose.

4. Challenges

As already mentioned, genebanks manage more than just passport data and material management data on accessions. For plant genetic resources, however, there are no generally accepted standards for capturing and exchanging data other than passport data [55]. This is a particular challenge for phenotypic and genotypic data.

Phenotypic data are collected in genebanks for various reasons. On the one hand, they provide important information for a better exploitation of the collection material. On the other hand, they support the management of collections, for example, they serve to ensure the quality of seed multiplication. Many genebanks collect a range of phenotypic and agronomic traits during each multiplication cycle to characterise the material, but also to detect potential mixing or swapping of material. This is particularly important for cross-pollinated species. Initiatives to harmonise the collection of phenotypic data in the field of genebanks have been underway since the late 1970s. The IPGRI/Bioversity descriptor lists developed for different crop species are a good example of this, e.g., [56–58]. However, it never was possible to achieve general acceptance. In many genebanks, the phenotyping of the material is based on such lists, but they have often been further developed and adapted to the respective practice. All data collected in this way are of limited comparability. In addition, there is a large number of scientific experiments, not carried out by the genebanks themselves, in which phenotypic data are collected from genebank material. The European information system EURISCO (see above) also collects phenotypic data [53]. Due to the problems just described, it was decided here not to standardise the data itself, but only the exchange format. This is a minimum consensus format that only contains the fields that are absolutely necessary. The idea behind this was to first gather a critical mass of data from which it is worthwhile to start a discussion with providers and users about harmonising traits and methods [59]. In this context, there are current approaches that aim to improve the comparability and traceability of phenotypic data, e.g., by recommendations for more extensive documentation of metadata such as Minimum Information About a Plant Phenotyping Experiment (MIAPPE) [55,60,61]. Mapping of traits and methods to ontology terms, such as Crop Ontology [62,63], is also promising.

Platforms for combining and analysing the data from the different domains are also essential, and can be adapted to the needs of projects or communities. A promising and already successfully used approach is the platform Germinate, developed by the James Hutton Institute, which is used in various projects to represent their data [64].

In addition to phenotypic data on genebank collections, genotypic data are increasingly coming into focus. Genotypic data can help to better exploit the treasures stored in genebanks [65,66]. Against this background, reference should be made here to pilot projects, which aim to exploit entire collections on a molecular level. For example, the BRIDGE project carried out the genotyping of the entire barley collection with more than 20,000 accessions of the German Federal ex situ Genebank for Agricultural and Horticultural Crop Species [67]. The analysis of such genome-wide genotyping-by-sequencing data provides the basis for gene annotation, marker-assisted selection or a better understanding of the population structures of globally domesticated crops, to name a few. In addition, molecular data can also be used to improve the curation of genebank collections, for example, with regard to potential duplication [68].

Moreover, such approaches can contribute significantly to the long-term development of traditional genebanks into biodigital resource centres. This means integrated centres which, in addition to the actual plant genetic resources, also provide a large amount of associated information from various data domains and thus enable better and more targeted access to the material [26,69]. According to this objective and based on first experiences, e.g., from the above-mentioned BRIDGE project, a number of further research projects with international participation are currently taking place. Examples are the Activated GEnebank NeTwork (AGENT, https://www.agent-project.eu/) and Intelligent Collections of Food Legumes Genetic Resources for European Agrofood Systems (INCREASE, https://increase-h2020.eu/) projects.

However, new challenges arise from this thoroughly gratifying development. Single-seed descent (SSD) lines must be generated for genotyping. However, depending on the type of accession, these SSD lines reflect only a selection from the original accessions. This is particularly problematic in the case of populations. Strictly speaking, the SSD lines would have to be preserved as precision collections in addition to the original accessions. The challenge for the documentation is that the data do not explicitly represent the SSD lines either. For example, it is doubtful that an SSD line that is derived from a landrace can still be called a landrace, since the character of a landrace is greatly lost through the selection of a single-seed descent. Conversely, the data obtained by genotyping an SSD line cannot be transferred to the entire accession. The same applies to phenotypic data that is not necessarily valid for both the original accession and the SSD line created from it. If, for capacity reasons, not all SSD lines can be preserved, assigning the data collected to the original accessions without comment should be avoided.

5. Conclusions

In this article, an overview was given of the most important topics, systems and structures of the documentation of plant genetic resources in genebanks. Furthermore, existing challenges were described.

In the context of the conservation of plant genetic resources, structured documentation of the material is essential. This includes a variety of basic information, such as origin or legal status, but also data on conservation management, such as storage quantities, germinability, etc. In addition, phenotypic and agronomic characterisation play an important role. Only through this it is possible to exploit the potential of the conserved material for research and breeding.

Over the past decades, there have been significant efforts to intensify cooperation between genebanks in order to conserve plant biodiversity in the best possible way. To this end, networks have been established, such as the European Cooperative Programme for Plant Genetic Resources. Such networks have been an important basis for the development of international aggregator systems such as EURISCO or Genesys, which serve as central entry points for the search for plant genetic resources.

These information systems are currently limited to passport data and phenotypic data. In terms of a sustainable use of genebank collections, it seems promising to continuously expand the interactions between the globally available information systems and to enrich the existing information on genebank accessions with additional data from other domains. In this context, a number of pilot projects have been successfully carried out in the recent past, which aimed at the molecular exploitation of genebank collections. These promising approaches must be continued in the future.

In order to meet the requirements of the modern research landscape, which is characterised by constant diversification and big data, the information systems of the genebanks must be able to connect with other data sources and to make their data available in a network. The linkage with each other and with other data domains leads to new knowledge about plant genetic resources, which ultimately improves their usability and the reputation of the genebanks.

Author Contributions: Coordinated the drafting, S.W.; conceptualization and writing, S.W., U.L. and M.O. All authors have read and agreed to the published version of the manuscript.

Funding: This research received no external funding.

Acknowledgments: We would like to thank the National Inventory Focal Points of the EURISCO network for providing data on the information management situation in their countries through an ad hoc survey. We also thank Helmut Knüpffer for providing information on the history of genebank documentation.

Conflicts of Interest: The authors declare no conflict of interest.

References

1. Grusak, M.A.; DellaPenna, D. Improving the Nutrient Composition of Plants to Enhance Human Nutrition and Health. *Annu. Rev. Plant. Physiol. Plant. Mol. Biol.* **1999**, *50*, 133–161. [CrossRef] [PubMed]
2. Metzger, J.O.; Bornscheuer, U. Lipids as renewable resources: Current state of chemical and biotechnological conversion and diversification. *Appl. Microbiol. Biotechnol.* **2006**, *71*, 13–22. [CrossRef] [PubMed]
3. Tilman, D.; Hill, J.; Lehman, C. Carbon-Negative Biofuels from Low-Input High-Diversity Grassland Biomass. *Science* **2006**, *314*, 1598–1600. [CrossRef] [PubMed]
4. Tollefson, J. Humans are driving one million species to extinction. *Nature* **2019**, *569*, 171. [CrossRef] [PubMed]
5. Hoisington, D.; Khairallah, M.; Reeves, T.; Ribaut, J.-M.; Skovmand, B.; Taba, S.; Warburton, M. Plant genetic resources: What can they contribute toward increased crop productivity? *Proc. Natl. Acad. Sci. USA* **1999**, *96*, 5937–5943. [CrossRef]
6. Engels, J.M.M.; Maggioni, L. AEGIS: A regionally based approach to PGR conservation. In *Agrobiodiversity Conservation: Securing the Diversity of Crop Wild Relatives and Landraces*; Maxted, N., Dulloo, M.E., Ford-Lloyd, B.V., Frese, L., Iriondo, J.M., de Pinheiro Carvalho, M.A.A., Eds.; CABI: Wallingford, UK, 2012; pp. 321–326. [CrossRef]
7. Keilwagen, J.; Kilian, B.; Özkan, H.; Babben, S.; Perovic, D.; Mayer, K.F.X.; Walther, A.; Poskar, C.H.; Ordon, F.; Eversole, K.; et al. Separating the wheat from the chaff—a strategy to utilize plant genetic resources from ex situ genebanks. *Sci. Rep.* **2014**, *4*, 5231. [CrossRef]
8. Tanksley, S.D.; McCouch, S.R. Seed Banks and Molecular Maps: Unlocking Genetic Potential from the Wild. *Science* **1997**, *277*, 1063–1066. [CrossRef]

9. Díez, M.J.; De la Rosa, L.; Martín, I.; Guasch, L.; Cartea, M.E.; Mallor, C.; Casals, J.; Simó, J.; Rivera, A.; Anastasio, G.; et al. Plant Genebanks: Present Situation and Proposals for Their Improvement. the Case of the Spanish Network. *Front. Plant. Sci.* **2018**, *9*, 1794. [CrossRef]
10. Fowler, C.; Hodgkin, T. Plant Genetic Resources for Food and Agriculture: Assessing Global Availability. *Annu. Rev. Environ. Resour.* **2004**, *29*, 143–179. [CrossRef]
11. FAO. *The State of the World's Plant Genetic Resources for Food and Agriculture*; Food and Agriculture Organization of the United Nations: Rome, Italy, 1997.
12. FAO. *The Second Report on the State of the World's Plant. Genetic Resources for Food and Agriculture*; Commission on Genetic Resources for Food and Agriculture; Food and Agriculture Organization of the United Nations: Rome, Italy, 2010.
13. Alercia, A.; Diulgheroff, S.; Metz, T. *FAO/IPGRI Multi-Crop. Passport Descriptors (MCPD)*; Food and Agriculture Organization of the United Nations (FAO): Rome, Italy; International Plant Genetic Resources Instit. (IPGRI): Rome, Italy, 2001.
14. Alercia, A.; Diulgheroff, S.; Mackay, M. *FAO/Bioversity Multi-Crop. Passport Descriptors, V.2.1 (MCPD V.2.1)*; Food and Agriculture Organization of the United Nations (FAO): Rome, Italy; Bioversity International: Rome, Italy, 2015.
15. FAO. *Genebank Standards for Plant. Genetic Resources for Food and Agriculture*; Commission on Genetic Resources for Food and Agriculture; Food and Agriculture Organization of the United Nations: Rome, Italy, 2014.
16. Richards, K.; White, R.; Nicolson, N.; Pyle, R. *A Beginner's Guide to Persistent Identifiers*; GBIF Secretariat: Copenhagen, Denmark, 2011. [CrossRef]
17. Garrity, G.M.; Thompson, L.M.; Ussery, D.W.; Paskin, N.; Baker, D.; Desmeth, P.; Schindel, D.E.; Ong, P.S. Studies on Monitoring and Tracking Genetic Resources. *Stand. Genom. Sci.* **2009**, *1*, 78–86.
18. Davidson, L.A.D.; Kimberly. Digital Object Identifiers: Promise and Problems for Scholarly Publishing. *J. Electron. Publ.* **1998**, *4*. [CrossRef]
19. Paskin, N. Digital Object Identifier (DOI®) System. In *Encyclopedia of Library and Information Sciences*, 3rd ed.; Bates, M.J., Maack, M.N., Eds.; CRC Press: Boca Raton, FL, USA, 2009; pp. 1586–1592. [CrossRef]
20. Alercia, A.; López, F.M.; Sackville Hamilton, N.R.; Marsella, M. *Digital Object Identifiers for Food Crops—Descriptors and Guidelines of the Global Information System*; Food and Agriculture Organization of the United Nations: Rome, Italy, 2018.
21. Jennings, M.D. Gap analysis: Concepts, methods, and recent results. *Landsc. Ecol.* **2000**, *15*, 5–20. [CrossRef]
22. Margules, C.R.; Pressey, R.L. Systematic conservation planning. *Nature* **2000**, *405*, 243–253. [CrossRef]
23. Maxted, N.; Dulloo, E.V.; Ford-Lloyd, B.; Iriondo, J.M.; Jarvis, A. Gap analysis: A tool for complementary genetic conservation assessment. *Divers. Distrib.* **2008**, *14*, 1018–1030. [CrossRef]
24. Aguirre-Gutiérrez, J.; van Treuren, R.; Hoekstra, R.; van Hintum, T.J.L. Crop wild relatives range shifts and conservation in Europe under climate change. *Divers. Distrib.* **2017**, *23*, 739–750. [CrossRef]
25. Diederichsen, A. Duplication assessments in Nordic Avena sativa accessions at the Canadian national genebank. *Genet. Resour. Crop. Evol.* **2009**, *56*, 587–597. [CrossRef]
26. Mascher, M.; Schreiber, M.; Scholz, U.; Graner, A.; Reif, J.C.; Stein, N. Genebank genomics bridges the gap between the conservation of crop diversity and plant breeding. *Nat. Genet.* **2019**, *51*, 1076–1081. [CrossRef]
27. Van Hintum, T.J.L.; Visser, D.L. Duplication within and between germplasm collections. *Genet. Resour. Crop. Evol.* **1995**, *42*, 135–145. [CrossRef]
28. Van Treuren, R.; van Hintum, T.J.L.; van de Wiel, C.C.M. Marker-assisted optimization of an expert-based strategy for the acquisition of modern lettuce varieties to improve a genebank collection. *Genet. Resour. Crop. Evol.* **2008**, *55*, 319–330. [CrossRef]
29. Engels, J.M.M.; Maggioni, L. Managing germplasm in a virtual European genebank (AEGIS) through networking. In *Theorien der Lebendsammlung: Pflanzen, Mikroben und Tiere als Biofakte in Genbanken*; Karafyllis, N.C., Ed.; Verlag Karl Alber: Freiburg, Germany, 2018; Volume 25, pp. 169–197.
30. Barata, A.M.R.F.; Oliveira, J.; Lima, J.M.; Nobrega, H.; Pinheiro de Carvalho, M.A.A.; Dias, S. Implementation of a PGR global documentation system in Portugal. In *Enhancing Crop Genepool Use: Capturing Wild Relative and Landrace Diversity for Crop Improvement*; Maxted, N., Dulloo, M.E., Ford-Lloyd, B.V., Eds.; CABI: Wallingford, Oxfordshire, UK, 2016; pp. 441–452. [CrossRef]

31. Qvenild, M. Svalbard Global Seed Vault: A 'Noah's Ark' for the world's seeds. *Dev. Pract.* **2008**, *18*, 110–116. [CrossRef]
32. Fowler, C. The Svalbard Seed Vault and Crop Security. *BioScience* **2008**, *58*, 190–191. [CrossRef]
33. Westengen, O.T.; Jeppson, S.; Guarino, L. Global Ex-Situ Crop Diversity Conservation and the Svalbard Global Seed Vault: Assessing the Current Status. *PLoS ONE* **2013**, *8*, e64146. [CrossRef] [PubMed]
34. Cochrane, G.; Karsch-Mizrachi, I.; Nakamura, Y.; Collaboration, o.b.o.t.I.N.S.D. The International Nucleotide Sequence Database Collaboration. *Nucleic Acids Res.* **2010**, *39*, D15–D18. [CrossRef] [PubMed]
35. Karsch-Mizrachi, I.; Nakamura, Y.; Cochrane, G.; International Nucleotide Sequence Database Collaboration. The International Nucleotide Sequence Database Collaboration. *Nucleic Acids Res.* **2011**, *40*, D33–D37. [CrossRef]
36. Seidewitz, L. *Thesaurus for the International Standardisation of Gene Bank Documentation. Part, I. Cereals (incl. Maize, Rice and Sorghum)*; Genebank, institut fur Pflanzenbau und Saatgutgutforschung der Forschungsanstalt fur Landwirtschaft: Braunschweig, Germany, 1974.
37. April, J. An information system to service the US National Germplasm System. In *Technical Aspects of Information Management and Means of Communication in Plant Genetic Resources Work for a Future Utilization of Genetic Material in Plant Breeding, Braunschweig, Germany, 7–8 November 1978*; Dambroth, M., Ed.; FAL: Braunschweig, Germany, 1979; pp. 73–78.
38. Blixt, S. Germplasm Documentation: The Future. In Proceedings of the FAO/UNEP/IBPGR Technical Conference on Crop Genetic Resources, Rome, Italy, 6–10 April 1981.
39. Knüpffer, H. Computer in Genbanken—Eine Übersicht. *Die Kult.* **1983**, *31*, 77–143. [CrossRef]
40. Konopka, J.; Hanson, J. *Information Handling Systems for Genebank Management*; International Board for Plant Genetic Resources: Rome, Italy, 1985.
41. Menting, F.B.J.; van Veller, M.G.P.; van Hintum, T.J.L. *Genis Data Dictionary*; Centre For Genetic Resources (GCN): Wageningen, DC, USA, 2007.
42. Oppermann, M.; Weise, S.; Dittmann, C.; Knüpffer, H. GBIS: The information system of the German Genebank. *Database* **2015**, *2015*, bav021. [CrossRef]
43. Agrawal, R.C.; Behera, D.; Saxena, S. Genebank Information Management System (GBIMS). *Comput. Electron. Agric.* **2007**, *59*, 90–96. [CrossRef]
44. Khoury, C.; Laliberté, B.; Guarino, L. Trends in ex situ conservation of plant genetic resources: A review of global crop and regional conservation strategies. *Genet. Resour. Crop. Evol.* **2010**, *57*, 625–639. [CrossRef]
45. Postman, J.; Hummer, K.; Ayala-Silva, T.; Bretting, P.; Franko, T.; Kinard, G.; Bohning, M.; Emberland, G.; Sinnott, Q.; Mackay, M.; et al. GRIN-Global: An international project to develop a global plant genebank information management system. *Acta Hortic.* **2010**, *859*, 49–55. [CrossRef]
46. Gass, T.; Lipman, E.; Maggioni, L. The role of Central Crop Databases in the European Cooperative Programme for Crop Genetic Resources Networks (ECP/GR). In *Central Crop Databases: Tools for Plant Genetic Resources Management*; Lipman, E., Jongen, M.W.M., van Hintum, Th.J.L., Gass, T., Maggioni, L., Eds.; European Cooperative Programme for Crop Genetic Resources Networks (ECP/GR): Rome, Italy, 1997; pp. 22–29.
47. Knüpffer, H. The European barley database of the ECP/GR: An introduction. *Die Kult.* **1988**, *36*, 135–162. [CrossRef]
48. Perret, P.M. The ECP/GR Prunus Working Group: A collaborative action programme for Prunus genetic resources in Europe. *Acta Hortic.* **1988**, *224*, 19–32. [CrossRef]
49. Van Hintum, T.J.L. Central Crop Databases—An overview. In *Central Crop Databases: Tools for Plant Genetic Resources Management*; Lipman, E., Jongen, M.W.M., van Hintum, Th.J.L., Gass, T., Maggioni, L., Eds.; European Cooperative Programme for Crop Genetic Resources Networks (ECP/GR): Rome, Italy, 1997; pp. 18–21.
50. Hazekamp, T.; Serwinski, J.; Alercia, A. Multicrop passport descriptors. In *Central Crop Databases: Tools for Plant Genetic Resources Management*; Lipman, E., Jongen, M.W.M., van Hintum, Th.J.L., Gass, T., Maggioni, L., Eds.; European Cooperative Programme for Crop Genetic Resources Networks (ECP/GR): Rome, Italy, 1997; pp. 40–44.
51. Wieczorek, J.; Bloom, D.; Guralnick, R.; Blum, S.; Döring, M.; Giovanni, R.; Robertson, T.; Vieglais, D. Darwin Core: An Evolving Community-Developed Biodiversity Data Standard. *PLoS ONE* **2012**, *7*, e29715. [CrossRef] [PubMed]

52. Endresen, D.; Knüpffer, H. The Darwin Core extension for genebanks opens up new opportunities for sharing genebank datasets. *Biodivers. Inform.* **2012**, *8*, 12–29. [CrossRef]
53. Weise, S.; Oppermann, M.; Maggioni, L.; van Hintum, T.; Knüpffer, H. EURISCO: The European search catalogue for plant genetic resources. *Nucleic Acids Res.* **2017**, *45*, D1003–D1008. [CrossRef]
54. Mackay, M.; Taylor, M. *Crop Trait Mining Informatics Platform (Crop TMIP)*; The University of Queensland: Gatton, Australia, 2014. [CrossRef]
55. Krajewski, P.; Chen, D.; Ćwiek, H.; van Dijk, A.D.J.; Fiorani, F.; Kersey, P.; Klukas, C.; Lange, M.; Markiewicz, A.; Nap, J.P.; et al. Towards recommendations for metadata and data handling in plant phenotyping. *J. Exp. Bot.* **2015**, *66*, 5417–5427. [CrossRef]
56. International Board for Plant Genetic Resources. *Plum Descriptors*; Commission of the European Communities (CEC): Rome, Italy, 1984.
57. Williams, P.H. *Descriptors for Brassica and Raphanus*; International Board for Plant Genetic Resources: Rome, Italy, 1990.
58. International Plant Genetic Resources Institution. *Descriptors for Allium (Allium spp.)*; European Cooperative Programme for Crop Genetic Resources Networks (ECP/GR): Wellesbourne, UK, 2001.
59. Maggioni, L.; van Hintum, T.J.L.; Lipman, E. Tailoring the Documentation of Plant Genetic Resources in Europe to the Needs of the User. In Proceedings of the Workshop of the ECPGR Documentation and Information Working Group, Prague, Czech Republic, 20–22 May 2014.
60. Ćwiek-Kupczyńska, H.; Altmann, T.; Arend, D.; Arnaud, E.; Chen, D.; Cornut, G.; Fiorani, F.; Frohmberg, W.; Junker, A.; Klukas, C.; et al. Measures for interoperability of phenotypic data: Minimum information requirements and formatting. *Plant. Methods* **2016**, *12*, 44. [CrossRef]
61. Papoutsoglou, E.A.; Faria, D.; Arend, D.; Arnaud, E.; Athanasiadis, I.N.; Chaves, I.; Coppens, F.; Cornut, G.; Costa, B.V.; Ćwiek-Kupczyńska, H.; et al. Enabling reusability of plant phenomic datasets with MIAPPE 1.1. *New Phytol.* **2020**, *227*, 260–273. [CrossRef]
62. Shrestha, R.; Arnaud, E.; Mauleon, R.; Senger, M.; Davenport, G.F.; Hancock, D.; Morrison, N.; Bruskiewich, R.; McLaren, G. Multifunctional crop trait ontology for breeders' data: Field book, annotation, data discovery and semantic enrichment of the literature. *Aob Plants* **2010**, *2010*, plq008. [CrossRef]
63. Shrestha, R.; Matteis, L.; Skofic, M.; Portugal, A.; McLaren, G.; Hyman, G.; Arnaud, E. Bridging the phenotypic and genetic data useful for integrated breeding through a data annotation using the Crop Ontology developed by the crop communities of practice. *Front. Physiol.* **2012**, *3*, 326. [CrossRef]
64. Raubach, S.; Kilian, B.; Dreher, K.; Amri, A.; Bassi, F.M.; Boukar, O.; Cook, D.; Cruickshank, A.; Fatokun, C.; El Haddad, N.; et al. From bits to bites: Advancement of the germinate platform to support genetic resources collections and pre-breeding informatics for crop wild relatives. *Crop. Sci.* **2020**. [CrossRef]
65. Van Treuren, R.; van Hintum, T.J.L. Next-generation genebanking: Plant genetic resources management and utilization in the sequencing era. *Plant. Genet. Resour.* **2014**, *12*, 298–307. [CrossRef]
66. Wambugu, P.W.; Ndjiondjop, M.-N.; Henry, R.J. Role of genomics in promoting the utilization of plant genetic resources in genebanks. *Brief. Funct. Genom.* **2018**, *17*, 198–206. [CrossRef] [PubMed]
67. Milner, S.G.; Jost, M.; Taketa, S.; Mazón, E.R.; Himmelbach, A.; Oppermann, M.; Weise, S.; Knüpffer, H.; Basterrechea, M.; König, P.; et al. Genebank genomics highlights the diversity of a global barley collection. *Nat. Genet.* **2019**, *51*, 319–326. [CrossRef] [PubMed]
68. Singh, N.; Wu, S.; Raupp, W.J.; Sehgal, S.; Arora, S.; Tiwari, V.; Vikram, P.; Singh, S.; Chhuneja, P.; Gill, B.S.; et al. Efficient curation of genebanks using next generation sequencing reveals substantial duplication of germplasm accessions. *Sci. Rep.* **2019**, *9*, 650. [CrossRef]
69. González, M.Y.; Philipp, N.; Schulthess, A.W.; Weise, S.; Zhao, Y.; Börner, A.; Oppermann, M.; Graner, A.; Reif, J.C. Unlocking historical phenotypic data from an ex situ collection to enhance the informed utilization of genetic resources of barley (Hordeum sp.). *Theor. Appl. Genet.* **2018**, *131*, 2009–2019. [CrossRef]

plants

MDPI

Review

Genebank Phenomics: A Strategic Approach to Enhance Value and Utilization of Crop Germplasm

Giao N. Nguyen * and Sally L. Norton

Australian Grains Genebank, Agriculture Victoria, 110 Natimuk Road, Horsham 3400, Australia; sally.norton@agriculture.vic.gov.au

* Correspondence: giao.nguyen@agriculture.vic.gov.au; Tel.: +61-3-4344-3315

Received: 3 June 2020; Accepted: 26 June 2020; Published: 29 June 2020

Abstract: Genetically diverse plant germplasm stored in ex-situ genebanks are excellent resources for breeding new high yielding and sustainable crop varieties to ensure future food security. Novel alleles have been discovered through routine genebank activities such as seed regeneration and characterization, with subsequent utilization providing significant genetic gains and improvements for the selection of favorable traits, including yield, biotic, and abiotic resistance. Although some genebanks have implemented cost-effective genotyping technologies through advances in DNA technology, the adoption of modern phenotyping is lagging. The introduction of advanced phenotyping technologies in recent decades has provided genebank scientists with time and cost-effective screening tools to obtain valuable phenotypic data for more traits on large germplasm collections during routine activities. The utilization of these phenotyping tools, coupled with high-throughput genotyping, will accelerate the use of genetic resources and fast-track the development of more resilient food crops for the future. In this review, we highlight current digital phenotyping methods that can capture traits during annual seed regeneration to enrich genebank phenotypic datasets. Next, we describe strategies for the collection and use of phenotypic data of specific traits for downstream research using high-throughput phenotyping technology. Finally, we examine the challenges and future perspectives of genebank phenomics.

Keywords: high-throughput phenotyping; statistical modelling; phenotypic breeding; genomic selection

1. Introduction

The global population is forecasted to reach 9.6 billion people by 2050, with the current 1.3% annual growth rate of crop productivity required to increase to 2.4% to meet the expected food security needs [1]. Concomitantly, climate change is impacting global food and biofuel production chains through rising temperatures, increasing carbon dioxide concentrations, unpredictable rainfall patterns, and soil degradation [2]. Extreme weather conditions, such as drought and heat that occurs during critical crop growth phases, reduce yield and production of major grain crops, and can trigger the emergence of new pests and diseases that can cause further production losses [2,3]. To produce enough food to meet this increasing demand under current and future growing conditions, it is critical that plant breeders develop new high yielding, environmentally resilient, and sustainable crop varieties.

Crop improvement through breeding relies heavily on genebanks worldwide to provide genetically diverse material that contains genes and alleles that govern desirable agronomic traits [4]. Novel alleles discovered from genebank genetic resources underpin the selection of, and enhance the genetic gain in breeding programs for favorable traits such as high yield, abiotic, and biotic stress adaptation [5–7]. However, under current practices, most genebanks can only offer end-users limited passport and basic characterization data based on morphological traits guided by standard international descriptors [8]. Only a small number of accessions have agronomic and quality trait data available [9]. There are around 1750 individual genebanks worldwide that preserve approximately 7.4 million accessions of agricultural genetic materials [10]. However, only 10% of these accessions are used for breeding purposes, partly due to poor phenotypic and genotypic characterization or lack of evaluation for agronomic traits [11] or because they are not publicly available [5].

Li et al. [12] pointed out two major limitations preventing the exploitation of genebank genetic resources for breeding programs: 1) time and available resources for thorough characterization of accessions at a large scale; and 2) identifying and introducing the allelic variance into elite breeding materials. This missing characterization data makes searching for an accession with specific desirable agronomic traits from within the millions of accessions held in genebanks, like 'finding the proverbial needle in a haystack' [13]. Therefore, to improve the utilization of germplasm, genebanks are increasingly required to move beyond providing basic passport data that defines only the identity and origin of the genetic resources, to thoroughly catalogue and make publicly available additional information for accessions such as agronomic, physiological, and genetic traits that meets the specific needs of end-users [14–16].

McCouch et al. [17] proposed a strategic three-step approach to effectively mine genebank genetic resources that combines genomics and phenomics with efficient database management to enhance the value of available germplasm that is readily available to breeders. Although the use of genomics by genebanks have advanced due to the development of DNA technology and next-generation genome sequencing [18,19], genebank phenomics still lag in the valorizing of available plant genetic resources [20,21]. The lack of robust, cost-efficient phenotyping tools and systematic collection of phenotypic data of accessions are currently a bottleneck, restricting the exploration and utilization of genebank genetic resources for downstream research and breeding [22]. A vast amount of useful agronomic and physiological information from genebank seed regeneration trials are not systematically recorded, contributing to the underutilization of germplasm [5]. Since standard genebank characterization practices can be expensive and time-consuming, a strategic cost-effective approach for simultaneously collecting multiple phenotypic trait data from genebank accessions during routine annual seed regenerations is essential to efficiently collect this valuable data and provide it to end-users [7,23,24]. This phenotypic data can be readily available for use in combination with genotypic information in subsequent genomic studies and breeding purposes [19,25]. High-throughput phenotyping (HTP) using sensors and imagers is a promising, efficient, and cost-effective approach to collect phenotypic data for multiple traits across large scale trials, that can then be used together with genomic data for accurate selection in breeding [26–28]. This approach has been successfully applied for genomic selection in wheat using various sensor-derived representations of agronomic and adaptive traits [29,30].

Although there are numerous excellent reviews on genebank mining using genomic approaches [13,18,31,32], only a handful of literature has addressed the exploitation of plant genetic resources using a phenomic approach is available and even do not fully cover genebank management practice as a whole [17,33–35]. In this manuscript, we examine: (i) current HTP methods that can be applied to phenotype accessions to leverage genebanks' phenotypic dataset; (ii) compatible crop traits that can be phenotyped by HTP technology and catalogued together with passport data in a genebank database; and (iii) data management strategies to effectively exploit these phenotypic data for future use. Finally, we discuss the challenges and future perspectives of genebank phenomics. Although there are numerous HTP methods, we limit our discussions to those that are more applicable to characterize and evaluate genebank germplasm in accordance with the international crop descriptors.

2. Phenomics to Unlock the Genetic Potential of Genebank Germplasm

2.1. Plant Phenomics and Its Potential Applications for Plant Genetic Resources Research

Plant phenomics is a multidisciplinary field that enables the systematic and comprehensive research and development of robust HTP tools and methods for data capture, processing, handling, and meta-analysis of phenotypic properties, growth, the performance of crops, and their environments [22,36]. The foundation of plant phenomics is the advent of HTP technology, which contrasts with more conventional arduous manual and destructive phenotyping, as it uses sensor- or image-based instruments to non-destructively simultaneously measure morphological, agronomic, and physiological characteristics of crops on a large scale across time and space. HTP technology is fundamentally based on principles of interaction between plant cellular components and natural light spectra between 400–2500 nm [37]. By capturing and analyzing these interactions proximally or remotely, important morphological, agronomic and physiological properties can be derived such as crop growth status, phenology, water and nutrient content, and yield potential [38]. The HTP approach has been widely used for decades in agriculture and plant science research with promising outcomes [39,40].

Various HTP platforms that use a combination of multiple sensors have been developed over many years that are suitable for plant science research in controlled and field conditions [27,39]. In the controlled environment, the system such as the automated Scanalyzer 3D imaging platform developed by LemnaTec GmbH (Aachen, Germany) has been effectively used to phenotype various crops and traits [41,42]. In the field, many HTP platforms are currently deployed such as the Field Scanalyzer gantry type [43]; manned- [44] or unmanned ground vehicles (UGV) [45]; and manned- [29] or unmanned aerial vehicles (UAV) [46,47]. These platforms are equipped with multiple sensor types and can be used to capture various crop traits at the same time. These sensor technologies will continue to advance over time and are likely to become less expensive, and hence more affordable for use in plant science applications.

Sensors developed for HTP can be broadly classified into either active or passive sensors that need to be considered when used to capture data. Passive sensors measure reflectance coming directly from natural light, thus the data captured by these sensors are highly affected by environmental conditions. Examples of passive sensors are Analytical Spectral Devices (ASD) FieldSpec spectroradiometer [48,49]; red-green-blue (RGB) [50,51], multi- [52] and hyper-spectral [53], and thermal cameras [54]. Active sensors, however, use their own light source and therefore the resulting reflectance is much less affected by the environment, with crop circle [42] and light detection and ranging (LiDAR) [55] being typical examples. Regardless of which type is deployed, sensors must be well calibrated and raw data should be normalized before analysis for quality assurance. Individual or multiple sensors can be handheld and mounted on vehicles or platforms, depending on the experimental setup and availability [56].

The workflow of deploying sensors for phenotyping crop experiments usually involves three main steps: 1) data capture; 2) raw data processing and storage; and 3) validation and comprehensive data analysis. Raw data are captured by sensors and processed by computer software algorithms to derive digital plant parameters such as vegetation indices (VIs) or structural properties. Once validated by compatible ground truths from conventional observations, these digital parameters can be used as proxies of crop traits for subsequent analysis. For instance, one of the most common vegetative indices is the normalized difference vegetation index (NDVI), which is derived from red and near-infrared (NIR) spectral bands and is widely used as a representation of biomass, grain yield, and crop N status [57]. The 2D and 3D structural models can be reconstructed from RGB, multispectral, and thermal imagery to derive important agronomic traits for various crops under different environments such as flowering time of rice [58] and wheat [43]; crop biomass of field peas [42] and wheat [41]; plant height and biomass of rice [59] and barley [60]; seed characteristics of lentils [61], rice [62], and field peas [63]; architectural and physiological properties of apple trees [64]; height and morphological characteristics of blueberries [65]; canopy temperature of black poplars [66]; bunch architecture of grapevines [67]; and ripeness estimation [68] and fruit counts [69] of mangos. Recent advances in computer algorithms and machine learning have significantly improved the throughput of raw data processing and analysis, where the processing pipelines have enabled data capture, analysis, and extraction of multiple patterns and features simultaneously [70]. Machine learning in sensor-

and image-based phenotyping has been applied successfully for germination assessment of tomato seeds [71], head count [72,73], yield prediction in wheat [74], and prediction of seed longevity in oilseed rape from chemical compositions [75].

Since thorough discussions on the development and application of HTP tools for agriculture research alone are not the main purpose of this review; readers can find detailed information about sensors and platforms, image processing and storage, data analysis approaches from numerous excellent reviews, and references cited therein [22,26,27,36,76–79].

2.2. Why Genebank Phenomics?

There are multiple factors, both subjective and objective, that make genebank phenomics feasible and strategic, i.e., the availability of cost-efficient HTP technology; the nature of routine operations; the pressure to efficiently exploit genetic resources for crop improvement and the conservation of genetic diversity. Phenotyping is the most expensive yet indispensable component of any plant research and crop improvement program to understand the genetic basis and interaction between genotypes and environments. The use of HTP tools and methods discussed above by genebanks will shorten the time requirement, increase throughput, improve consistency, reduce the overall cost of phenotyping projects, and improve selection accuracy in breeding programs, especially for large-scale trials [80].

Genebanks complete essential seed regeneration as routine practices to maintain the viability, quality, and quantity of accessions, e.g., when the quantity and viability of specific accessions fall below a standard threshold [7,81]. Characterization of germplasm for a range of phenotypic traits is undertaken during the regeneration process, however, traits are manually recorded, can be subjective, and are time-consuming to collect, limiting the amount of data able to be captured, resulting in a wasted opportunity for a comprehensive characterization of genetic materials, with the flow-on effect of restricting their subsequent utilization. Mining superior agronomic alleles for breeding is crucial for improving crop yield and resilience, with the availability of comprehensive phenotypic data for genebank germplasm enabling researchers and breeders to more accurately identify desired accessions for breeding projects [82].

The application of low-cost HTP methods to assess the true value of genetic resources, accurate estimation of their agronomic phenotypic traits for a complete phenotypic representation of collections will significantly improve the gains of pre-breeding or breeding programs with marginal extra expenses. This is particularly useful for studying complex traits such as grain yield. Multiple secondary traits captured by HTP tools that correlate well with target traits (i.e., grain yield) can be used as surrogates in yield selection models to improve prediction accuracy. For instance, Rutkoski et al. [29] showed that the use of canopy temperature and NDVI measured by aerial thermal and hyperspectral sensors substantially improved genomic and pedigree yield predictions of 557 wheat lines across five growing environments. Interestingly, the authors also pointed out that genetic value for grain yield can be accurately estimated by using these secondary phenotypic traits in absence of pedigree and genomic data. The phenotypic profiling of genebanks' accessions can, therefore, provide direct support for phenomic selection or choice of parents in breeding programs.

Genomic selection has been proven as an excellent tool to estimate genomic breeding values and is now widely used as a routine selection method in crop breeding [83,84]. However, since its successful introduction over the last two decades, there has been a significantly faster loss of genetic variance in breeding programs compared to conventional phenotypic selection [85]. To slow down the loss of genetic diversity through genomic selection in plant breeding, a physiological breeding approach combining multiple integrative traits captured by HTP tools in conjunction with genomic selection methods, with a heavier weight placed on phenotypic components, could be an alternative [86]. The advantage of phenomic selection has been demonstrated by Rutkoski et al. [87], where the authors claimed that using an optimized breeding scheme with phenotypic selection for quantitative analysis of stem rust resistance in wheat would result in equal genetic gains as genomic selection, but higher genetic variance. This phenotypic selection approach is further supported by a recent study of Rincent et al. [88], where distinctive endophenotypes, such as transcripts, small RNAs, or metabolites, could be used as phenomic markers for the selection process. The authors found that the matrices of near infrared spectroscopy absorbance between 400 and 2500 nm of winter wheat grains and leaf tissue could provide better yield prediction than molecular-based markers. Thus, using these low-cost, high-throughput endophenotypic markers significantly improved genetic gains, while better conserving allelic diversity of breeding populations.

Finally, safeguarding genetic resources ex-situ for integrity, diversity, and allelic variability for future use is the mandatory task of every genebank whose materials fuel breeding programs, underpinning food security efforts and bringing billions of dollars in benefit [17]. For instance, by 1997, the world economy had benefited approximately $115 billion annually from using wild materials from genebanks to develop environmentally resilient and resistant crops [89]. Selections for desirable agronomic traits are the driving force of plant domestication and crop improvement. However, extensive breeding selections lead to loss of genetic variants, narrowing a crop's genetic base and an overall erosion of crop diversity through breeding programs [40,90]. Alarmingly, there is also mounting evidence that indicates that allelic variance of genebank accessions might be lost over time through seed regeneration due to genetic drift and inbreeding, while its storage size and maintenance costs will keep increasing [91]. Genebank accessions are collected from various geographical locations, thus original phenotypic variance could be lost during ex-situ conservation and seed regenerations [92]. While DNA fingerprinting is the most effective method to verify the genetic integrity of regenerated materials, the associated genotyping cost is still too high for large scale genotyping of thousands of accessions per year [62]. Thus, a complete phenotypic assessment of accessions through periodic seed regenerations could be a counter measure to ensure that original phenotypic features are preserved. Furthermore, those accessions possessing desirable agronomic characteristics can be recommended for immediate use, whereas those which do not have attributes of immediate interest can be conserved for further evaluation under different and specific environmental conditions, or potentially be discarded.

3. Phenomic Characterization and Evaluation of Genebank Accessions

Missing or incomplete passport, characterization, and evaluation data is one of the main reasons for the underutilization of genebank germplasm. For decades, the crop descriptor lists developed by Bioversity International have been routinely used to standardize genebank data collection and to facilitate the exchange of information between genebanks and end-users. This data is also used by genebanks to catalogue morphological and physiological characteristics of various crop species for germplasm validation processes [8]. In this section, we discuss the potential of deploying HTP technology to routinely collect quantitative data of specific traits in line with these descriptor lists and the possibilities of collecting additional data with a marginal cost that can enriches genebank collections. Highly heritable morphological and physiological features can provide invaluable information for strategic selection schemes used by plant breeders to speed up the development of new, high yielding environmentally adaptive cultivars [93]. Table 1 details systematic HTP approaches of genebank germplasm for morphological and physiological traits in different environments.

Table 1. Examples of phenotypic traits can be exploited from genebank germplasm by using sensors and phenotyping platforms.

Traits	Description	Sensors and Capture mode	Species	Environment	References
Morphology					
Plant architecture	Number of tillers of wheat plants	Automated RGB[1] imaging platform, LemnaTec 3D Scanalyzer	*Triticum aestivum*	Greenhouse	[94]
	Node and internode length of tomato seedlings	RGB imagery	*Solanum lycopersicum*	Greenhouse	[95]
	Characterization of plant architecture by 3D scanning reconstruction	Blue-laser scanner	*Solanum lycopersicum, Nicotiana benthamiana, Sorghum bicolor*	Greenhouse	[96]
	Canopy structure (tiller and leaf number, leaf length and angle, leaf elongation rate)	RGB imagery	*Triticum aestivum*	Greenhouse	[50]
	A phenotyping platform, PANorama, measuring architectural properties (panicle, branch, leaf) of various crop species.	RGB imaging unit	*Oryza sativa* *Zea mays* *Solanum lycopersicum*	Laboratory	[97]
Plant height	Sorghum plant height estimates	Ultrasonic, LiDAR[2]-Lite, Kinect camera, imaging array, UAV[3] RGB imagery	*Sorghum bicolor*	Field	[98]
	Wheat plant height estimation	UGV[4] and UAV RGB imagery	*Triticum aestivum*	Field	[51,99]
	Rice plant height estimation	UAV RGB imagery	*Oryza sativa*	Field	[59]
	Barley plant height measurement	UAV RGB imagery	*Hordeum vulgare*	Field	[60]
	Maize plant height estimates	UAV RGB imagery	*Zea mays*	Field	[100,101]
Leaf properties	Characterization of local leaf vein in legume leaves	Color scanner	*Vigna angularis* *Phaseolus vulgaris* *Glycine max*	Laboratory	[102]
	Leaf morphological properties and height of maize, measured by various digital phenotyping methods	3D scanner, Multi-view stereo cameras, FastTrack 3D digitizer	*Zea mays*	Laboratory	[103]
Inflorescence and fruit	An automated imaging system for monitoring the growths of maize ear and silks	RGB imaging platform	*Zea mays*	Greenhouse	[104]
	Analysis of panicle architecture and spikelet numbers in rice by an imaging tool, P-TRAP[5]	RGB imagery	*Oryza sativa*	Field, laboratory	[105,106]
	Rice panicle phenotyping using Panicle-SEG[6] algorithm	RGB imagery	*Oryza sativa*	Field	[107]
	Characterization of maize tassel traits by RGB imaging and machine vision	RGB imaging sensors	*Zea mays*	Field	[108,109]
	Analysis of oat panicle development	RGB imaging platform	*Avena sativa*	Greenhouse	[110]
	Fruit recognition and counting by an imaging robot, SPYSEE	RGB imaging robot	*Capsicum annuum*	Greenhouse	[111]
	Morphological characterization of wheat spike (grain number, size and angle, stem node)	Computed Tomography imagery	*Triticum aestivum*	Laboratory	[112]
	Automatic quantification of wheat heads	Automated RGB imagery platform, Field Scanalyzer	*Triticum aestivum*	Field	[73]
	Morphometric properties of wheat spikes	RGB imagery	*Triticum aestivum*	Laboratory	[113]

Table 1. *Cont.*

Traits	Description	Sensors and Capture mode	Species	Environment	References
Seed characteristics					
	Seed quality of field pea (color, shape, and size) analyzed by multi-spectral imaging	Built-in multi-spectral camera, EyeFoss	*Pisum sativum*	Laboratory	[114]
	Evaluating of lentil seed size by multi-spectral imaging	Built-in multi-spectral camera, EyeFoss	*Lens culinaris*	Laboratory	[61]
	Screening method to evaluate seed properties	Nuclear magnetic resonance	*Avena spp.*	Laboratory	[115]
	Rice seed shape analyzed by an image processing pipeline	Color scanner	*Oryza sativa*	Laboratory	[116]
	Analysis of maize ear, cob and kernel properties	Color scanner	*Zea mays*	Laboratory	[117]
	Estimation of ear characteristics and kernel weight in maize	RGB imagery	*Zea mays*	Field	[118]
	Morphological characteristics of wheat kernels	Color scanner and RGB imagery	*Triticum aestivum*	Laboratory	[119]
	Phenotypic classification of rice seed accessions	Multispectral imagery, VideometerLab	*Oryza sativa*	Laboratory	[62]
	Shape description of pili seed by imaging technology	Multispectral imagery, VideometerLab	*Canarium ovatum*	Laboratory	[120]
	Automated morphological characterization of rapeseed and barley seeds	Automated RGB imaging unit, phenoSeeder	*Brassica napus* *Hordeum vulgare*	Laboratory	[121]
	Automated phenotyping of oat seed properties	NIR spectroscopy, Single-Seed Analyzer	*Avena sativa*	Laboratory	[122]
Phenology					
Emergence count	Rice seedling counts	High resolution UAV RGB imagery	*Oryza sativa*	Field	[123]
	Cotton seedling detection and count	Ground-based video recording	*Gossypium hirsutum*	Field	[124]
	Germination rate estimation in tomato by color imagery	RGB imagery	*Solanum lycopersicum*	Laboratory	[71]
	Determination of plant density at emergence in wheat	High resolution UAV RGB imagery	*Triticum aestivum*	Field	[125]
Ground cover	Ground cover estimates	High resolution UAV RGB imagery	*Sorghum bicolor* *Gossypium hirsutum* *Saccharum spp.*	Field	[126]
Flowering	Automated observations of wheat flowering	CCD[7] digital camera	*Triticum aestivum*	Field	[127]
	Heading and flowering detection in wheat	Automated RGB imaging platform	*Triticum aestivum*	Field	[43]
	Automated flowering observation in rice from a time-series RGB images	Automated RGB imaging platform	*Oryza sativa*	Field	[58,128]
	Estimation of flowering time in maize	UAV RGB imagery	*Zea mays*	Field	[129]
Physiology					
Early vigor	Early vigor of field pea seedlings	Automated RGB imaging platform, LemnaTec 3D Scanalyzer and handheld active sensor, crop circle	*Pisum sativum*	Greenhouse, field	[42]
	Wheat vigor and canopy height quantification	UGV and UAV RGB imagery	*Triticum aestivum*	Field	[51]
	Monitoring plant and canopy growth dynamics	RGB or multispectral imagery, D3P[8]	*Triticum aestivum*	Greenhouse	[130]
Lodging	Lodging score estimation in barley by aerial imagery	High resolution UAV RGB imagery	*Hordeum vulgare*	Field	[131]
	Estimation of crop lodging in wheat by aerial imagery	High resolution UAV RGB imagery	*Triticum aestivum*	Field	[132]
	Rice lodging scores estimated by UNet model derived from aerial imagery	High resolution UAV RGB and multispectral imagery	*Oryza sativa*	Field	[133]

Table 1. *Cont.*

Traits	Description	Sensors and Capture mode	Species	Environment	References
Photosynthesis and respiration	Photosynthetic capacities in tobacco	Handheld hyperspectral sensor, FieldSpec.	*Nicotiana tabacum*	Field	[134]
	Leaf photosynthesis in maize	Handheld hyperspectral sensor, FieldSpec.	*Zea mays*	Field	[135]
	Leaf photosynthesis and relevant physiological parameters in wheat	Handheld hyperspectral sensor, FieldSpec.	*Triticum aestivum*	Field	[48]
Water soluble carbohydrates	Estimates of stem water soluble carbohydrates at different growth stages in wheat	Handheld hyperspectral sensor, FieldSpec.	*Triticum aestivum*	Field	[49]
	Predicting the quality of ryegrass (sugar)	Hyperspectral imaging platform	*Lolium perenne*	Field	[53]
Canopy temperature	Wheat canopy temperature measurement	Airborne thermography and wireless infra-red thermometers	*Triticum aestivum*	Field	[54]
	Maize canopy temperature measurement	UAV thermal and RGB imagery	*Zea mays*	Field	[136]
	Canopy temperature and vegetation indices of wheat	Airborne thermal and hyperspectral imagery	*Triticum aestivum*	Field	[29]
Stay green	Stay-green associates with low water-soluble carbohydrates in oat	Handheld active sensor GreenSeeker	*Avena sativa*	Field	[137]
	Characterization of maize green leaf area dynamics	UAV multispectral imagery	*Zea mays*	Field	[138]
	Senescence rate in wheat	UAV multispectral imagery	*Triticum aestivum*	Field	[52]
Biomass and yield	Biomass, ground cover and canopy height estimates	UGV LiDAR	*Triticum aestivum*	Field	[55,139]
	Estimation of shoot biomass by color imagery	RGB imaging platform, LemnaTec 3D Scanalyzer	*Triticum aestivum*	Greenhouse	[140]
	Grain yield prediction by canopy hyperspectral reflectance	Airborne hyperspectral imagery	*Triticum aestivum*	Field	[30,141]
	Wheat ear counts	Handheld thermal imagery	*Triticum aestivum*	Field	[142]
	Head counts in sorghum	High resolution UAV RGB imagery	*Sorghum bicolor*	Field	[143]
	Counting of wheat spikes	Handheld and UGV RGB imagery	*Triticum aestivum*	Field	[72,144]
	Wheat biomass and yield, nitrogen related traits	Automated RGB imaging platform, LemnaTec 3D Scanalyzer	*Triticum aestivum*	Greenhouse	[41]

[1] RGB, red green blue; [2] LiDAR, Light Detection and Ranging; [3] UAV, Unmanned Aerial Vehicle; [4] UGV, Unmanned Ground Vehicle; [5] P-TRAP, Panicle TRAit Phenotyping; [6] Panicle-SEG, Panicle segmentation algorithm; [7] CCD, charge-coupled device; [8] D3P, Digital Plant Phenotyping Platform.

3.1. Morphology

The collection of data on the morphology of accessions is a critical part of routine curatorial activities of any genebank. These data describe overall plant architecture, height, leaf shape, and angle. Conventionally, these data are visually assessed and manually recorded by curators, which are sometimes subjective in nature and prone to human errors. This labor-intensive and time-consuming notetaking can be replaced by robust HTP technology (Table 1). Morphological characteristics of various crops such as number of tillers (wheat) [94]; node and internode length (tomato) [95]; panicle, branch, and leaf number (rice, maize, tomato) [97]; and leaf shape (legumes) [102] can be easily acquired by cost-effective RGB imagery tools. The HTP technology can be flexibly applied for trait capture under various growing conditions including field and greenhouse environments. For example, plant height is an important botanical trait that is defined as the shortest distance from ground level to the upper boundary of photosynthetic tissues [145]. It is a useful indicator of crop growth rate, biomass, yield potential, and lodging resistance [46, 132]. Studies on wheat have shown that lodging can cause yield losses up to 80% [146]. Thus, strategic exploitation of genebank germplasm for novel alleles is crucial for the development of lodging resistant cultivars. Several HTP methods using sensors such as ultrasonic sensors, LiDAR or RGB cameras can be used to measure plant height in the greenhouse and field [98]. However, the method using a combination of LiDAR and RGB camera mounted on a ground- or aerial based vehicle appeared more feasible with a similarly high level of accuracy [147]. Using this method, plant height can be modelled and estimated by the principles of the structure from motion photogrammetry, where the difference between digital terrain model (DTM) and digital surface model (DSM) is the average height of plants within the plots [99]. Quantitative measurement of lodging can be derived from the differences between DSM before and after lodging events, which has been demonstrated in barley [131], wheat [132], and rice [133].

3.2. Inflorescence and Fruit

Inflorescence and fruit are important and distinctive botanical features of crops used to identify and classify genebank accessions. In physiological breeding, highly heritable traits in cereals such as spike length, spike weight, and floret number per spike are indicators of agronomic values, yield, and adaptation for selection schemes [146]. These traits can be quantitatively measured on large-scale seed regeneration trials by using cost-effective HTP technology (Table 1). For instance, Grillo et al. [148] developed a method to differentiate wheat landraces by glume size, shape, color, and texture using a color scanner. Likewise, Makanza et al. [118] designed a simple low-cost RGB imaging method to quantitatively measure seed size, number, and weight of intact maize cobs in the field. Most recently, Genaev et al. [113] described a simple RGB imaging setup which can precisely quantify morphological features such as spike shape and awnedness of the wheat spike. This work suggests that the deployment of HTP methods can help curators digitally characterize a wide range of traits related to the inflorescence and fruit, and once incorporated into genebank databases, readily provide this quantitative data for subsequent genetic analysis and breeding purposes by end-users.

3.3. Seed Characteristics

Seed traits such as shape, size, and coat color are crucial criteria for determining commodity market values and are highly controlled by genetics. Genebanks routinely characterize seeds based on general morphology and use the data for both in-house quality assurance and end-user purposes. Currently, most genebanks manually collect this data following seed regeneration cycles based on visual assessment of traits, an approach that can be subjective and potentially lead to inaccurate results. Image-based phenotyping methods using RGB, multispectral, and hyperspectral cameras could be a cost-effective and accurate substitute for manual phenotyping since shape, size, and coat color are easily reconstructed and analyzed using reflectance spectra from the seed surface, and are known to be related to chemical properties (Table 1) [149]. These HTP tools have been applied for seed quality, purity, viability, vigor testing, and variety identification on various crop species [150,151]. Potentially, this cost-effective HTP technology can be used to develop seed descriptor states for crop species [120,152] and as routine methods for managing genebank accessions as they are included into the collection from new acquisitions or seed regeneration events to avoid physical contamination and maintain genetic integrity [62], as well as genomic selection for seed traits [153].

3.4. Phenology

Understanding the timing of key physiological growth stages such as germination, flowering, and their variations is critical for crop production and breeding of new high yielding and environmentally adaptive varieties. Therefore, the documentation of crop phenological traits such as germination, flowering, and maturity are routinely recorded during genebank seed regeneration cycles. Research has shown that image-based phenotyping can be used to effectively measure these qualitative traits as a replacement to conventionally visual-based methods. For instance, HTP technology has been used for phenotype emergence [123,124], heading, and flowering [43,129] of various crop species (Table 1). This also suggests that these trait data can be systematically and simultaneously captured together with other traits by using HTP technology, reducing the cost and increasing the opportunity to explore the genetic potential of individual accessions.

3.5. Physiological and Agronomic Traits

Most genebanks choose to skip the collection of physiological and agronomic traits (known as evaluation data) from their standard curatorial procedures due to either funding shortage or the labor intensity required. However, this information together with genebank passport data is critical for prioritizing and enhancing the utilization of valuable germplasm for the selection of parents used for breeding [35,154]. Core collections can be generated using phenotypic data of useful agronomic traits [155]. A plethora of reports and publications from multiple international research groups have indicated that specially designed HTP platforms can comprehensively capture multiple trait data simultaneously that can be subsequently exploited not only by curators but also by plant scientists and breeders (Table 1).

This is particularly helpful when dissecting the genetic basis of polygenic traits such as grain yield or adaptive traits. For instance, grain yield is a critically important trait for selection in physiological breeding and can be effectively captured along with other descriptive traits using HTP technology during standard curatorial procedures at genebanks. However, grain yield is a genetically complex trait that can only be improved by simultaneously enhancing other secondary morphological and physiological traits such as plant architecture, lodging resistance, photosynthetic capacity, canopy temperature, and harvest index. This approach has been proposed for major agricultural crop species such as wheat [86,156], rice [157], and pulses [158]. Therefore, the use of HTP technology has a distinctive advantage over the conventional manual collection method, where the former can capture multiple secondary trait data quantitatively at the same time. This will reduce time, labor, and phenotyping cost with the benefit of a comprehensive data set, fully describing crop growth and yield, which is critically valuable to breeding programs. Distinctive secondary traits can also be directly used to breed for adaptation or indirectly in forward genetics for molecular cloning and gene identification. For example, stay green is an adaptive trait that provides better drought tolerance and nitrogen use efficiency in crops [56,159]. Research has shown that stay-green is a part of the drought adaptation mechanism that increases yield stability and lodging resistance in sorghum and other cereals, where it can lead to prolonged grain filling duration and improve yield [146,160]. Interestingly, other reports show that canopy temperature (CT)—an indicator of evaporative cooling from the canopy surface and an adaptive trait for high yielding and drought tolerance—is associated with stay green and deeper roots [161]. Thus, a HTP approach using a combination of sensors can capture stay green and CT traits together with other traits such as NDVI, height, biomass, and ground cover, as well as being used for selection in breeding programs [162,163].

4. Challenges

4.1. Lack of Resources

Despite the enormous potential to phenotype and characterize genebank germplasm to enhance genetic gain in plant breeding, there are several constraints that genebanks must fully address before being able to move forward. The first and perhaps the biggest challenge is the availability of resources for a long-term phenotyping scheme [7]. Although HTP phenotyping of genebank germplasm will provide valuable information for end-users, the associated cost for purchasing, establishing, and operating of sensors, phenotyping platforms, analysis, validation, and making available phenotypic data in a searchable online platform, respectively, is not a trivial task, and might not be affordable by every genebank [164,165]. Hidden costs such as equipment and database maintenance, software licensing, and upgrades need to be considered. Therefore, genebank managers need to carefully consider the balance between investment, labor cost, and achievement of goals before initiating HTP projects. For instance, low-cost simple HTP tools such as PhenoBox [166] can be developed for effective phenotyping of seed regeneration in the greenhouse without the need for complicated and costly automated phenotyping platforms reported by Nguyen et al. [41,42]. More importantly, in contrast to short-term research projects, genebanks are long-term investments with large numbers of accessions that require well planned, consistent phenotyping programs to be in place. Adequate planning and resources must be made available for effective phenotyping to be undertaken over the long-term if research and breeding programs are to achieve the increased plant production required in the future.

4.2. Technical Difficulties in Data Management and Analysis

Data capture, standardization, quality assurance, and analysis are technical challenges related to genebank phenotyping. HTP technologies generate a large volume of 'big data' in a short period of time using standardized protocols. However, a high level of infrastructure investment and a multidisciplinary approach for the appropriate storage, back up, data management, and analysis is required [167,168]. These data must be thoroughly validated before they can be used. In contrast to genomic data, plant phenotypes are non-constant, plastic, and change over time, as they are the results of instantaneous interactions between genotypes and the environments [169].

Furthermore, phenotypic data of field seed regenerations, mainly collected by passive sensors and cameras, are highly influenced by spatial and temporal climatic conditions and must be processed through sophisticated computational algorithms before the data can be made readily available for genebanks scientists [26]. Therefore, if data collection is not standardized for unpredictable weather and changes in agronomical practices, over the years data analysis will become difficult due to the disparity between different data sets, rendering the HTP efforts useless [23]. To cope with fluctuating climatic conditions, Xu [170] introduced the strategic 'envirotyping' approach, where local environmental data such as soil, weather, biotic factors, and crop management practices are documented as metadata together with plant phenotypic data.

All data must be integrated into a well-structured and publicly searchable database for end-users [171]. For instance, the Genebank Information System (GBIS) of the Leibniz Institute of Plant Genetics and Crop Plant Research (IPK), Gatersleben, Germany, currently houses approximately 151,000 crop accessions and is comprehensively managed across passport data, seed/line management, taxonomy, phenotypic characterization, and evaluation data [172]. To ensure phenotypic data generated from HTP are fully described and annotated, the plant phenotyping community has recommended a convention on the minimum information about plant phenotyping experiments (MIAPPE), where all experimental conditions are well described and published together with phenotypic data [173]. Wilkinson et al. [174] introduced the FAIR principles (findability, accessibility, interoperability, and reusability) for the management of scholarly data, where their application will enhance the handling, sharing, reuse, and interpretation of data and metadata. In addition to passport and phenotypic data, images demonstrating morphological features, which are not easily analyzed and represented by numerical data, should be included [175]. Clearly, the systematic collection of phenotype and metadata and its stewardship will assist genebank scientists to fully describe the datasets and conditions where seed regenerations are conducted and enable the interpretation of the phenotypic plasticity in statistical models.

4.3. Users' Awareness

Finally, the communication of genebank data sets to the research and breeding community is critical for increasing the successful utilization of genebank germplasm. Currently, only around 10% of genebank germplasm is used in plant breeding due to a range of technical reasons including lack of good quality phenotype and genotype data on the germplasm, as well as accession's low level of adaptation to changing environments and genetic drift [11]. Although accurate phenotypic data are valuable to plant breeders to identify outstanding performers as potential parents, phenotyping is still the most expensive part of any breeding programs [176]. Breeders are often reluctant to take risks screening a large amount of diverse genebank germplasm without any certainty of their genetic potential for beneficial agronomic traits due to the cost and significant challenge for them to identify valuable germplasm or novel alleles from a 'sea of seeds' [177]. In this context, readily available phenotypic and genomic data that enriches genebank passport data will enhance the utilization and overall value of germplasm stored in genebanks.

5. Future Perspectives

5.1. Systematic HTP Phenotyping of Routine Genebank Seed Regenerations

Despite the challenges posed by the deployment of sensors and image-based HTP protocols, systematic collection of data from genebank seed regeneration cycles can effectively derive multiple trait data for downstream research. One of the advanced features of HTP is that multiple sensors can be deployed at the same time to simultaneously and non-destructively capture many independent observations that will allow for more targeted prioritization of accessions from large genetic resources collections for downstream studies. Targeted beneficial endophenotypes of individual genebank accessions can be directly used for the low-cost phenomic selection in breeding process or prioritize germplasm for higher value in the selection of crossing parents [88]. A proposed strategic phenomic approach for the collection of multiple trait data, management of genebank collections, and increasing utilization of data and seed by end users through the adoption of HTP technologies is shown in Figure 1.

Routine seed regeneration protocols of genebanks are often conducted in small, unreplicated plots or even single rows in the field or pots in greenhouses. Seed regeneration blocks should be replicated with a reasonable number of individuals whenever possible to facilitate statistical analysis and ensure sufficient number of seeds are used to maintain the genetic diversity and integrity of accessions (Figure 1) [178,179]. A large amount of morphological, agronomic, physiological (Table 1), and environmental data [170] can simultaneously be collected from routine seed regeneration cycles over subsequent years using HTP (Figure 1). Even though these phenotypic data are highly incomplete [23], meaningful inferences can still be achieved using appropriate analysis methods, such as identifying novel alleles [25]. Measurement of grain yield from small seed regeneration plots is generally not meaningful and is sometimes impractical to measure when thousands of lines are being regenerated in a single sowing event [30]. A practical and cost-effective phenotyping approach could be used to measure secondary correlated traits such as early vigor, height, canopy properties, and biomass during the growth phase which are components contributing to grain yield. Moreover, these phenotypic data can be instantly used in conjunction with genetic data generated by advanced genotyping technologies such as diversity array technology combined with next generation sequencing (DArT-Seq) and appropriate data quenching for direct phenomic and genomic selection from landrace accessions (Figure 1) [180–182].

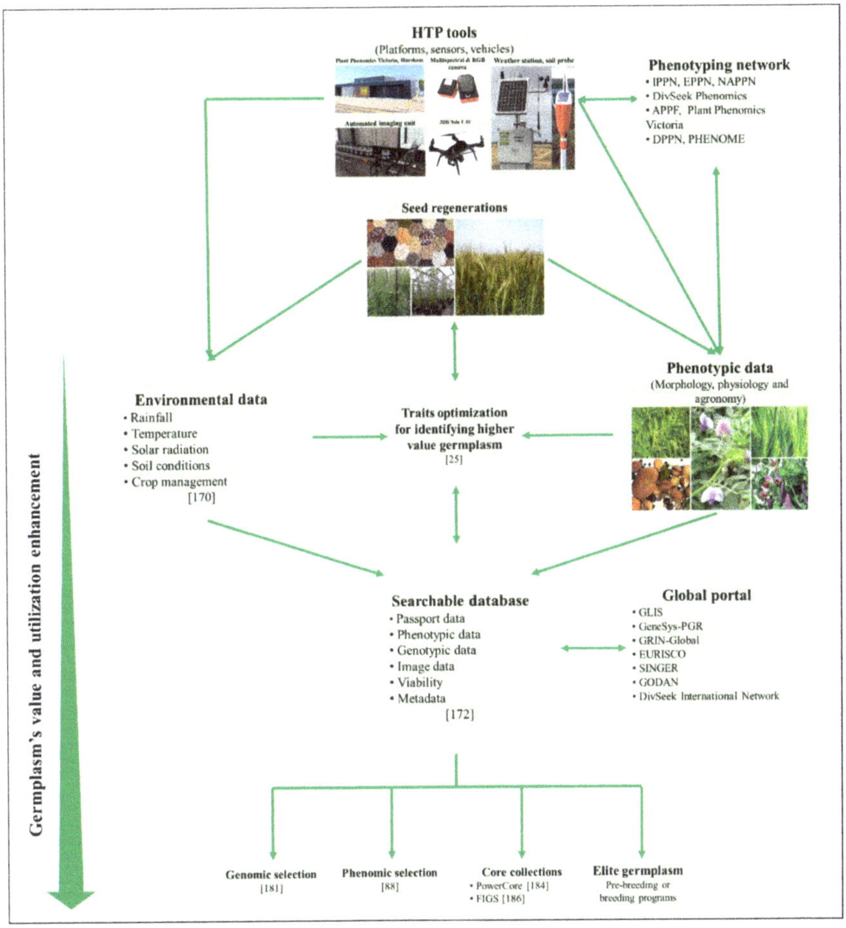

Figure 1. A proposed strategic phenomic approach to improve the value and utilization of genetic resources. IPPN, international plant phenotyping network; EPPN, European plant phenotyping network; NAPPN, North American plant phenotyping network; APPF, Australian plant phenotyping Facility; DPPN, German plant phenotyping network; PHENOME, French plant phenomic infrastructure; GLIS, global information system; GeneSys-PGR, global portal on crop genetic Resources; GRIN-Global, global germplasm resource information network; EURISCO, European plant genetic resources search catalogue; SINGER, system-wide information network for genetic resources; GODAN, global open data for agriculture.

This strategic phenomic approach has been deploying at genebanks elsewhere. For example, the Australian grains genebank (AGG), Horsham, Victoria was established in 2014 and currently houses approximately 195,000 accessions of 918 crop species such as wheat, barley, canola, field pea, chickpea, lentil, sorghum, maize, cowpea, mungbean, and millets. The number of accessions has increased by around 2750 per annum (Figure 2) [9]. Annually, AGG regenerates more than 3500 accessions of genetically diverse crop species and wild relatives in the field and greenhouse, subject to viability, quantity in stock and user demand (Figure 2)[9]. This routine activity requires large inputs of labor and resources costing in excess of A$500,000 per annum. Due to the large number of accessions being regenerated annually, using conventional phenotyping methods to obtain a complete phenotypic data set is not possible. Several field HTP platforms can acquire multiple crop traits such as plant height, biomass, leaf area index, and canopy temperature across thousands of seed regeneration plots at the same time [54,99,136]. The AGG is currently applying different HTP platforms such as automated phenotyping of Plant Phenomics Victoria, Horsham [41,42], laboratory-based phenotyping of spikes and airborne platforms (Figure 2) to capture more useful morphological, agronomic, and physiological traits from seed regeneration cycles. Once validated and analyzed, these data will be made publicly available with passport data which will help end-users prioritize higher value germplasm for targeting traits in subsequent studies (Figure 1).

Figure 2. Australian grains genebank (AGG) storage facilities and its application of HTP technology for phenotyping routine seed regenerations in laboratory, greenhouse and field.

5.2. A Combination of Genebanks' Data Mining Approaches

To enhance value and utilization of germplasm, it is crucial that traits are identified and linked with genebank accessions (Figure 1). Several methods have been proposed for mining genebank data such as using published data sources and users' feedback, core and mini core collections. phenotyping and genotyping approaches [15]. Overall, these methods aim to identify genebank accessions containing agronomic traits of interest. Core collections can be constructed by using phenotypic (Figure 1) [88,183], genomic [13], and geographical information of accessions for certain crop traits. A variety of software packages are available to develop core collections solely using phenotypic traits, for instance, Chung et al. [184] analyzed 11 quantitative and 28 qualitative phenotypic traits from 10,368 characterized rice accessions and derived a core collection of 107 entries by using PowerCore software. Similarly, Dutta et al. [185] constructed a core set of 2,208 accessions from 22,469 accessions of wheat and its wild relatives used 34 highly heritable phenotypic traits.

Accessions can also be grouped based on the geographical information data as specified in the focused identification of germplasm strategies approach (FIGS) [186,187]. The underlying principle of the FIGS approach is that crops will likely evolve under environmental selection pressures and develop their adaptation in response to extreme climatic conditions. Thus, the method uses detailed eco-geographical location and weather conditions where accessions were collected to precisely predict their adaptive traits to abiotic and biotic conditions. The FIGS approach has been successfully used to identify several core sets such wheat stem rust [188], drought in faba bean [187], powdery mildew in wheat [189], and Russian wheat aphids [190]. Using these data mining approaches will narrow down the number of accessions for further analysis while the allelic variance is still well maintained in the subsets.

Several approaches can also be used in sequence to increase the chance of identifying targeted accessions as reported by Haupt and Schmid [191] where two core collections of 183 and 366 soybean accessions were chosen from the original collection of more than 17,000 accessions by using a combination of FIGS approach and SNP genotypic markers. Given the advanced features of HTP, capturing multiple traits nondestructively at the same time, core collections can be easily developed and accessions containing promising traits of interest will be chosen for further studies.

5.3. A Collaborative Network for Data Collection, Analyzing and Sharing

To improve HTP practices, genebanks should work closely with universities, research institutes and industries to standardize seed regeneration procedures, phenotyping protocols, and calibration of sensors so that the resulting phenotypic data are comparable across genebanks and able to be fully exploited by end-users. Numerous initiatives have been implemented at the national and international scale, which aim to bring academia and industry together to address common phenotyping questions and integrate the plant phenotyping community (Figure 1) [171]. For instance, world-class plant phenotyping infrastructures have been established in Australia to enhance the capability, capacity, and scientific rigor in support of national plant phenomics studies and applications. These include the Australian plant phenotyping facility with three nodes located at the University of Adelaide, the Australian National University, and the Commonwealth Scientific and Industrial Research Organization, Canberra, respectively. Moreover, the Plant Phenomics Victoria is home to two nodes located in Horsham and Bundoora, Victoria (Figure 1). At the global scale, the International Plant Phenotyping Network [171] is an organization representing plant phenotyping centers, that formulates multidisciplinary working groups and enables the communication between stakeholders through conferences and training workshops so that up to date information about new HTP infrastructures and methodologies for various crop phenotypes can be effectively shared (Figure 1). This networking collaboration is essential to foster the advancement in plant phenotyping technologies including affordable phenotyping, sensors, and platforms, targeting traits and data analysis pipelines and data management.

To make valuable information pertaining to germplasm available for global users, a cooperative platform for data collection, analysis and sharing is urgently required. Several international initiatives such as DivSeek, breeding API, research data alliance (RDA), and global open data for agriculture (GODAN) have been developed, all of which have aimed to facilitate the integration and sharing of evaluation and characterization data so as to improve the value and utilization of germplasm [192]. For instance, the DivSeek international network is a global, community-driven initiative that facilitates the cooperation and interactions among its members through working groups [17]. Genebanks, phenotyping scientists, and breeders can develop and share methodologies, tools, and best practice phenotyping technologies to evaluate genetic resources, which improve the generation, integration, and sharing of phenotypic data [193]. Moreover, the introduction of the global information platforms such as the global portal on crop genetic resources (GeneSys-PGR) has enabled breeders and genebank users to use free online search engines to explore and request germplasm accessions conserved in genebanks worldwide [177]. The global information system is an international portal that links all current plant genetic resources systems by using unique digital object identifiers (DOIs) for individual accessions [192]. By using DOIs and linkage through these portals, invaluable phenotypic evaluation and characterization of germplasm can be effectively shared with the global user network.

Individual institutions can setup different collaboration protocols for sharing and exchanging phenotypic and genotypic data (Figure 1) [172]. For example, the International Maize and Wheat Improvement Center (CIMMYT), El Batán, State of Mexico, Mexico distributes seed all over the world and receive data in return from experimental trials that provide valuable information to assess genotype-by-environment interaction [194]. A more similar stringent protocol can be introduced to enforce the current clause in the standard material transfer agreement of the seed distribution of any genebanks, where the end-users are obliged to give back basic phenotypic data of genetic resources which they have used such as trial location, phenology, biomass, and grain yield. This information would clearly enrich the genebanks' databases and the GeneSys-PGR, which can be used as reference guides by end-users for future use. However, phenotypic and genotypic data should be linked and shared with other national and international databases of plant genetic resources through the use of DOIs and the global portals discussed above.

Although genebank scientists make use of invaluable knowledge and techniques from other research disciplines such as plant physiologist, breeders, agronomists, seed physiologists, and computer scientists, their independent translational studies are indispensable to fully utilize HTP technology to phenotype genebanks' accessions [34]. For instance, HTP methods can theoretically be applied to quantitatively analyze the 3D canopy structure of wheat by multi-view stereo and structure from motion algorithms [50]. This protocol is not yet ready for large-scale phenotyping of genebanks accessions as translational research must be conducted by genebanks to optimize the existing protocol and determine if its throughput is applicable for large-scale phenotyping of various crop species. Similarly, more studies should be dedicated to verifying and developing feasible seed testing methods by using multi- and hyper-spectral imagery for handling genebank accessions of various crop species [62].

6. Conclusions

The application of HTP technology for large-scale phenotypic characterization and validation of genebank germplasm is essential if they are to fulfill their biorepository role (i.e., in the preservation and support of further experimentation and plant breeding) [195]. With a comprehensive phenomics approach combining pedigree, genomic, and phenotyping data [17], the true value of genebank genetic resources is evident. Therefore, they should more strategically and efficiently utilized by breeding programs that should double our current rate of genetic gain to feed the growing world population under the changing conditions expected into the future.

Author Contributions: G.N.N. conceived the topic and wrote the manuscript. S.L.N. provided critical comments, edits, and acquired the funding. Both authors contributed to editing, revision and approval of the submitted manuscript. Both authors have read and agreed to the published version of the manuscript. All authors have read and agreed to the published version of the manuscript.

Funding: This work was funded by the Grains Research and Development Corporation, grant number 9176106 to S.L.N., through the Australian Grains Genebank project—Phase 3, 2017-2022.

Acknowledgments: Authors wish to thank Dr. Lance Maphosa for his critical reading and comments on the manuscript. We thank Christine Best, Katherine Dunsford and Susan Robson for assistance in taking phenotyping photos in the laboratory and the field.

Conflicts of Interest: The authors declare no conflict of interest.

References

1. Ray, D.K.; Mueller, N.D.; West, P.C.; Foley, J.A. Yield trends are insufficient to double global crop production by 2050. *PLoS ONE* **2013**, *8*, e66428. [CrossRef] [PubMed]
2. Lesk, C.; Rowhani, P.; Ramankutty, N. Influence of extreme weather disasters on global crop production. *Nature* **2016**, *529*, 84–87. [CrossRef]
3. Legrève, A.; Duveiller, E. Prevailing potential diseases and pest epidemics under a changing climate. In *Climate Change and Crop Production*; Reynolds, M.P., Ed.; CABI Press: Oxfordshire, UK, 2010; Volume 13, pp. 50–70.
4. Tanksley, S.D.; McCouch, S.R. Seed banks and molecular maps: Unlocking genetic potential from the wild. *Science* **1997**, *277*, 1063–1066. [CrossRef] [PubMed]

5. De Carvalho, M.A.P.; Bebeli, P.J.; Bettencourt, E.; Costa, G.; Dias, S.; Dos Santos, T.M.; Slaski, J.J. Cereal landraces genetic resources in worldwide GeneBanks. A review. *Agron. Sustain. Dev.* **2013**, *33*, 177–203. [CrossRef]
6. Roa, C.; Hamilton, R.S.; Wenzl, P.; Powell, W. Plant genetic resources: Needs, rights, and opportunities. *Trends Plant Sci.* **2016**, *21*, 633–636. [CrossRef] [PubMed]
7. González, M.Y.; Weise, S.; Zhao, Y.; Philipp, N.; Arend, D.; Börner, A.; Oppermann, M.; Graner, A.; Reif, J.C.; Schulthess, A.W. Unbalanced historical phenotypic data from seed regeneration of a barley ex situ collection. *Sci. Data* **2018**, *5*, 1–10. [CrossRef] [PubMed]
8. Gotor, E.; Alercia, A.; Rao, V.R.; Watts, J.; Caracciolo, F. The scientific information activity of Bioversity International: The descriptor lists. *Genet. Resour. Crop Evol.* **2008**, *55*, 757–772. [CrossRef]
9. Norton, S.L.; Khoury, C.K.; Sosa, C.C.; Castañeda-Álvarez, N.P.; Achicanoy, H.A.; Sotelo, S. Priorities for enhancing the ex situ conservation and use of Australian crop wild relatives. *Aust. J. Bot.* **2017**, *65*, 638–645. [CrossRef]
10. FAO. *The Second Report on the State of the World's Plant Genetic Resources for Food and Agriculture*; Food and Agriculture Organization of the United Nations: Rome, Italy, 2010.
11. Tadesse, W.; Sanchez-Garcia, M.; Assefa, S.G.; Amri, A.; Bishaw, Z.; Ogbonnaya, F.C.; Baum, M. Genetic gains in wheat breeding and its role in feeding the world. *CBGG* **2019**, *1*, e190005. [CrossRef]
12. Li, H.; Rasheed, A.; Hickey, L.T.; He, Z. Fast-forwarding genetic gain. *Trends Plant Sci.* **2018**, *23*, 184–186. [CrossRef]
13. Wambugu, P.W.; Ndjiondjop, M.-N.; Henry, R.J. Role of genomics in promoting the utilization of plant genetic resources in genebanks. *Brief. Funct. Genom.* **2018**, *17*, 198–206. [CrossRef] [PubMed]
14. Díez, M.J.; De la Rosa, L.; Martín, I.; Guasch, L.; Cartea, M.E.; Mallor, C.; Casals, J.; Simó, J.; Rivera, A.; Anastasio, G.; et al. Plant genebanks: Present situation and proposals for their improvement. the case of the Spanish network. *Front. Plant Sci.* **2018**, *9*. [CrossRef]
15. Anglin, N.L.; Amri, A.; Kehel, Z.; Ellis, D. A case of need: Linking traits to genebank accessions. *Biopreservation Biobanking* **2018**, *16*, 337–349. [CrossRef] [PubMed]
16. Finkers, R.; Chibon, P.-Y.; Van Treuren, R.; Visser, R.; Van Hintum, T. Genebanks and genomics: How to interconnect data from both communities? *Plant Genet. Resour.* **2014**, *13*, 90–93. [CrossRef]
17. McCouch, S.; Baute, G.J.; Bradeen, J.; Bramel, P.; Bretting, P.K.; Buckler, E.; Burke, J.M.; Charest, D.; Cloutier, S.; Cole, G.; et al. Feeding the future. *Nature* **2013**, *499*, 23–24. Available online: https://www.nature.com/articles/499023a#supplementary-information (accessed on 24 April 2020). [CrossRef] [PubMed]
18. McCouch, S.R.; McNally, K.L.; Wang, W.; Sackville, H.R. Genomics of gene banks: A case study in rice. *Am. J. Bot.* **2012**, *99*, 407–423. [CrossRef]
19. Yu, X.; Li, X.; Guo, T.; Zhu, C.; Wu, Y.; Mitchell, S.E.; Roozeboom, K.L.; Wang, D.; Wang, M.L.; Pederson, G.A.; et al. Genomic prediction contributing to a promising global strategy to turbocharge gene banks. *Nat. Plants* **2016**, *2*, 1–7. [CrossRef]
20. Kilian, B.; Graner, A. NGS technologies for analyzing germplasm diversity in genebanks. *Brief. Funct. Genom.* **2012**, *11*, 38–50. [CrossRef]
21. Philipp, N.; Weise, S.; Oppermann, M.; Börner, A.; Graner, A.; Keilwagen, J.; Kilian, B.; Zhao, Y.; Reif, J.C.; Schulthess, A.W.; et al. Leveraging the use of historical data gathered during seed regeneration of an ex-situ genebank collection of wheat. *Front. Plant Sci.* **2018**, *9*, 609. [CrossRef]
22. Furbank, R.T.; Tester, M. Phenomics—Technologies to relieve the phenotyping bottleneck. *Trends Plant Sci.* **2011**, *16*, 635–644. [CrossRef]
23. Krajewski, P.; Chen, D.; Ćwiek, H.; Van Dijk, A.D.J.; Fiorani, F.; Kersey, P.; Klukas, C.; Lange, M.; Markiewicz, A.; Nap, J.P.; et al. Towards recommendations for metadata and data handling in plant phenotyping. *J. Exp. Bot.* **2015**, *66*, 5417–5427. [CrossRef] [PubMed]
24. González, M.Y.; Philipp, N.; Schulthess, A.W.; Weise, S.; Zhao, Y.; Börner, A.; Oppermann, M.; Graner, A.; Reif, J.C. Unlocking historical phenotypic data from an ex situ collection to enhance the informed utilization of genetic resources of barley (Hordeum sp.). *Theor. Appl. Genet.* **2018**, *131*, 2009–2019. [CrossRef]
25. Keilwagen, J.; Kilian, B.; Özkan, H.; Babben, S.; Perovic, D.; Mayer, K.F.X.; Walther, A.; Poskar, C.H.; Ordon, F.; Eversole, K.; et al. Separating the wheat from the chaff—A strategy to utilize plant genetic resources from ex situ genebanks. *Sci. Rep.* **2014**, *4*, 5231. [CrossRef] [PubMed]

26. Araus, J.L.; Cairns, J.E. Field high-throughput phenotyping: The new crop breeding frontier. *Trends Plant Sci.* **2014**, *19*, 52–61. [CrossRef] [PubMed]
27. Araus, J.L.; Kefauver, S.C. Breeding to adapt agriculture to climate change: Affordable phenotyping solutions. *Curr. Opin. Plant Biol.* **2018**, *45*, 237–247. [CrossRef]
28. White, J.W.; Andrade-Sanchez, P.; Gore, M.A.; Bronson, K.F.; Coffelt, T.A.; Conley, M.M.; Feldmann, K.A.; French, A.N.; Heun, J.T.; Hunsaker, D.J.; et al. Field-based phenomics for plant genetics research. *Field Crops Res.* **2012**, *133*, 101–112. [CrossRef]
29. Rutkoski, J.; Poland, J.; Mondal, S.; Autrique, E.; Pérez, L.G.; Crossa, J.; Reynolds, M.; Singh, R. Canopy temperature and vegetation indices from high-throughput phenotyping improve accuracy of pedigree and genomic selection for grain yield in wheat. *G3 Genes Genom. Genet.* **2016**, *6*, 2799–2808. [CrossRef]
30. Krause, M.R.; González-Pérez, L.; Crossa, J.; Pérez-Rodríguez, P.; Montesinos-López, O.; Singh, R.P.; Dreisigacker, S.; Poland, J.; Rutkoski, J.; Sorrells, M.; et al. Hyperspectral reflectance-derived relationship matrices for genomic prediction of grain yield in wheat. *G3 Genes Genom. Genet.* **2019**, *9*, 1231–1247. [CrossRef]
31. Milner, S.G.; Jost, M.; Taketa, S.; Mazón, E.R.; Himmelbach, A.; Oppermann, M.; Weise, S.; Knüpffer, H.; Basterrechea, M.; König, P.; et al. Genebank genomics highlights the diversity of a global barley collection. *Nat. Genet.* **2019**, *51*, 319–326. [CrossRef]
32. Van, K.; Kim, D.H.; Shin, J.H.; Lee, S.-H. Genomics of plant genetic resources: Past, present and future. *Plant Genet. Resour.* **2011**, *9*, 155–158. [CrossRef]
33. Mir, R.R.; Reynolds, M.; Pinto, F.; Khan, M.A.; Bhat, M.A. High-throughput phenotyping for crop improvement in the genomics era. *Plant Sci.* **2019**, *282*, 60–72. [CrossRef] [PubMed]
34. Reynolds, M.; Borrell, A.; Braun, H.; Edmeades, G.; Flavell, R.; Gwyn, J.; Jordan, D.; Pixley, K.; Rebetzke, G. Translational research for climate resilient, higher yielding crops. *CBGG* **2019**, *1*, e190016. [CrossRef]
35. Reynolds, M.; Tattaris, M.; Cossani, C.M.; Ellis, M.; Yamaguchi-Shinozaki, K.; Pierre, C.S. Exploring genetic resources to increase adaptation of wheat to climate change. In *Advances in Wheat Genetics: From Genome to Field*; Ogihara, Y., Takumi, S., Handa, H., Eds.; Springer: Tokyo, Japan, 2015; pp. 355–368.
36. Tardieu, F.; Cabrera-Bosquet, L.; Pridmore, T.; Bennett, M. Plant phenomics, from sensors to knowledge. *Curr. Biol.* **2017**, *27*, R770–R783. [CrossRef] [PubMed]
37. Fahlgren, N.; Gehan, M.A.; Baxter, I. Lights, camera, action: High-throughput plant phenotyping is ready for a close-up. *Curr. Opin. Plant Biol.* **2015**, *24*, 93–99. [CrossRef]
38. Homolová, L.; Malenovský, Z.; Clevers, J.G.P.W.; García-Santos, G.; Schaepman, M.E. Review of optical-based remote sensing for plant trait mapping. *Ecol. Complex.* **2013**, *15*, 1–16. [CrossRef]
39. Araus, J.L.; Kefauver, S.C.; Zaman-Allah, M.; Olsen, M.S.; Cairns, J.E. Translating high-throughput phenotyping into genetic gain. *Trends Plant Sci.* **2018**, *23*, 451–466. [CrossRef]
40. Rebetzke, G.J.; Jimenez-Berni, J.; Fischer, R.A.; Deery, D.M.; Smith, D.J. Review: High-throughput phenotyping to enhance the use of crop genetic resources. *Plant Sci.* **2019**, *282*, 40–48. [CrossRef]
41. Nguyen, G.N.; Maharjan, P.; Maphosa, L.; Vakani, J.; Thoday-Kennedy, E.; Kant, S. A robust automated image-based phenotyping method for rapid vegetative screening of wheat germplasm for nitrogen use efficiency. *Front. Plant Sci.* **2019**, *10*. [CrossRef]
42. Nguyen, G.N.; Norton, S.L.; Rosewarne, G.M.; James, L.E.; Slater, A.T. Automated phenotyping for early vigour of field pea seedlings in controlled environment by colour imaging technology. *PLoS ONE* **2018**, *13*, e0207788. [CrossRef]
43. Sadeghi-Tehran, P.; Sabermanesh, K.; Virlet, N.; Hawkesford, M.J. Automated method to determine two critical growth stages of wheat: Heading and flowering. *Front. Plant Sci.* **2017**, *8*. [CrossRef]
44. Andrade-Sanchez, P.; Gore, M.A.; Heun, J.T.; Thorp, K.R.; Carmo-Silva, A.E.; French, A.N.; Salvucci, M.E.; White, J.W. Development and evaluation of a field-based high-throughput phenotyping platform. *Funct. Plant Biol.* **2013**, *41*, 68–79. [CrossRef] [PubMed]
45. Qiu, Q.; Sun, N.; Bai, H.; Wang, N.; Fan, Z.; Wang, Y.; Meng, Z.; Li, B.; Cong, Y. Field-based high-throughput phenotyping for maize plant using 3D LiDAR point cloud generated with a "Phenomobile". *Front. Plant Sci.* **2019**, *10*. [CrossRef] [PubMed]
46. Chapman, S.C.; Merz, T.; Chan, A.; Jackway, P.; Hrabar, S.; Dreccer, M.F.; Holland, E.; Zheng, B.; Ling, T.J.; Jimenez-Berni, J.; et al. Pheno-Copter: A low-altitude, autonomous remote-sensing robotic helicopter for high-throughput field-based phenotyping. *Agronomy* **2014**, *4*, 279–301. [CrossRef]

47. Hu, P.; Chapman, S.C.; Wang, X.; Potgieter, A.; Duan, T.; Jordan, D.; Guo, Y.; Zheng, B. Estimation of plant height using a high throughput phenotyping platform based on unmanned aerial vehicle and self-calibration: Example for sorghum breeding. *Eur. J. Agron.* **2018**, *95*, 24–32. [CrossRef]
48. Silva-Perez, V.; Molero, G.; Serbin, S.P.; Condon, A.G.; Reynolds, M.P.; Furbank, R.T.; Evans, J.R. Hyperspectral reflectance as a tool to measure biochemical and physiological traits in wheat. *J. Exp. Bot.* **2017**, *69*, 483–496. [CrossRef]
49. Dreccer, M.F.; Barnes, L.R.; Meder, R. Quantitative dynamics of stem water soluble carbohydrates in wheat can be monitored in the field using hyperspectral reflectance. *Field Crops Res.* **2014**, *159*, 70–80. [CrossRef]
50. Duan, T.; Chapman, S.C.; Holland, E.; Rebetzke, G.J.; Guo, Y.; Zheng, B. Dynamic quantification of canopy structure to characterize early plant vigour in wheat genotypes. *J. Exp. Bot.* **2016**, *67*, 4523–4534. [CrossRef]
51. Khan, Z.; Chopin, J.; Cai, J.; Eichi, V.-R.; Haefele, S.; Miklavcic, S. Quantitative estimation of wheat phenotyping traits using ground and aerial imagery. *Remote Sens.* **2018**, *10*, 950. [CrossRef]
52. Hassan, M.; Yang, M.; Rasheed, A.; Jin, X.; Xia, X.; Xiao, Y.; He, Z. Time-series multispectral indices from unmanned aerial vehicle imagery reveal senescence rate in bread wheat. *Remote Sens.* **2018**, *10*, 809. [CrossRef]
53. Shorten, P.R.; Leath, S.R.; Schmidt, J.; Ghamkhar, K. Predicting the quality of ryegrass using hyperspectral imaging. *Plant Methods* **2019**, *15*, 63. [CrossRef]
54. Deery, D.M.; Rebetzke, G.J.; Jimenez-Berni, J.A.; Bovill, W.D.; James, R.A.; Condon, A.G.; Furbank, R.T.; Chapman, S.C.; Fischer, R.A. Evaluation of the phenotypic repeatability of canopy temperature in wheat using continuous-terrestrial and airborne measurements. *Front. Plant Sci.* **2019**, *10*. [CrossRef] [PubMed]
55. Jimenez-Berni, J.A.; Deery, D.M.; Rozas-Larraondo, P.; Condon, A.G.; Rebetzke, G.J.; James, R.A.; Bovill, W.D.; Furbank, R.T.; Sirault, X.R.R. High throughput determination of plant height, ground cover, and above-ground biomass in wheat with LiDAR. *Front. Plant Sci.* **2018**, *9*. [CrossRef] [PubMed]
56. Nguyen, G.N.; Kant, S. Improving nitrogen use efficiency in plants: Effective phenotyping in conjunction with agronomic and genetic approaches. *Funct. Plant Biol.* **2018**, *45*, 606–619. [CrossRef] [PubMed]
57. Nguyen, G.N.; Panozzo, J.; Spangenberg, G.; Kant, S. Phenotyping approaches to evaluate nitrogen-use efficiency related traits of diverse wheat varieties under field conditions. *Crop Pasture Sci.* **2016**, *67*, 1139–1148. [CrossRef]
58. Guo, W.; Fukatsu, T.; Ninomiya, S. Automated characterization of flowering dynamics in rice using field-acquired time-series RGB images. *Plant Methods* **2015**, *11*, 7. [CrossRef]
59. Bendig, J.; Willkomm, M.; Tilly, N.; Gnyp, M.; Bennertz, S.; Qiang, C.; Miao, Y.; Lenz-Wiedemann, V.; Bareth, G. Very high resolution crop surface models (CSMs) from UAV-based stereo images for rice growth monitoring in Northeast China. *Int. Arch. Photogramm. Remote Sens. Spat. Inf. Sci.* **2013**, *40*, 45–50. [CrossRef]
60. Bendig, J.; Yu, K.; Aasen, H.; Bolten, A.; Bennertz, S.; Broscheit, J.; Gnyp, M.L.; Bareth, G. Combining UAV-based plant height from crop surface models, visible, and near infrared vegetation indices for biomass monitoring in barley. *Int. J. Appl. Earth Obs. Geoinf.* **2015**, *39*, 79–87. [CrossRef]
61. LeMasurier, L.S.; Panozzo, J.F.; Walker, C.K. A digital image analysis method for assessment of lentil size traits. *J. Food Eng.* **2014**, *128*, 72–78. [CrossRef]
62. Hansen, M.A.E.; Hay, F.R.; Carstensen, J.M. A virtual seed file: The use of multispectral image analysis in the management of genebank seed accessions. *Plant Genet. Resour.* **2015**, *14*, 238–241. [CrossRef]
63. McDonald, L.S.; Salisbury, P.A.; Ford, R.; Panozzo, J.F. Quantifying the colour loss of green field pea (*Pisum sativum* L.) due to bleaching. *PLoS ONE* **2019**, *14*, e0221523. [CrossRef]
64. Coupel-Ledru, A.; Pallas, B.; Delalande, M.; Boudon, F.; Carrié, E.; Martinez, S.; Regnard, J.-L.; Costes, E. Multi-scale high-throughput phenotyping of apple architectural and functional traits in orchard reveals genotypic variability under contrasted watering regimes. *Hortic. Res.* **2019**, *6*, 52. [CrossRef] [PubMed]
65. Patrick, A.; Li, C. High throughput phenotyping of blueberry bush morphological traits using unmanned aerial systems. *Remote Sens.* **2017**, *9*, 1250. [CrossRef]
66. Ludovisi, R.; Tauro, F.; Salvati, R.; Khoury, S.; Mugnozza Scarascia, G.; Harfouche, A. UAV-based thermal imaging for high-throughput field phenotyping of black poplar response to drought. *Front. Plant Sci.* **2017**, *8*. [CrossRef] [PubMed]
67. Rist, F.; Gabriel, D.; Mack, J.; Steinhage, V.; Töpfer, R.; Herzog, K. Combination of an automated 3D field phenotyping workflow and predictive modelling for high-throughput and non-invasive phenotyping of grape bunches. *Remote Sens.* **2019**, *11*, 2953. [CrossRef]

68. Gutiérrez, S.; Wendel, A.; Underwood, J. Spectral filter design based on in-field hyperspectral imaging and machine learning for mango ripeness estimation. *Comput. Electron. Agric.* **2019**, *164*, 104890. [CrossRef]
69. Xiong, J.; Liu, Z.; Chen, S.; Liu, B.; Zheng, Z.; Zhong, Z.; Yang, Z.; Peng, H. Visual detection of green mangoes by an unmanned aerial vehicle in orchards based on a deep learning method. *Biosyst. Eng.* **2020**, *194*, 261–272. [CrossRef]
70. Mochida, K.; Koda, S.; Inoue, K.; Hirayama, T.; Tanaka, S.; Nishii, R.; Melgani, F. Computer vision-based phenotyping for improvement of plant productivity: A machine learning perspective. *GigaScience* **2018**. [CrossRef]
71. Škrubej, U.; Rozman, Č.; Stajnko, D. Assessment of germination rate of the tomato seeds using image processing and machine learning. *Eur. J. Hortic. Sci.* **2015**, *80*, 68–75. [CrossRef]
72. Hasan, M.M.; Chopin, J.P.; Laga, H.; Miklavcic, S.J. Detection and analysis of wheat spikes using Convolutional Neural Networks. *Plant Methods* **2018**, *14*, 100. [CrossRef]
73. Sadeghi-Tehran, P.; Virlet, N.; Ampe, E.M.; Reyns, P.; Hawkesford, M.J. DeepCount: In-field automatic quantification of wheat spikes using simple linear iterative clustering and deep convolutional neural networks. *Front. Plant Sci.* **2019**, *10*. [CrossRef]
74. Pantazi, X.E.; Moshou, D.; Alexandridis, T.; Whetton, R.L.; Mouazen, A.M. Wheat yield prediction using machine learning and advanced sensing techniques. *Comput. Electron. Agric.* **2016**, *121*, 57–65. [CrossRef]
75. Nagel, M.; Holstein, K.; Willner, E.; Börner, A. Machine learning links seed composition, glucosinolates and viability of oilseed rape after 31 years of long-term storage. *Seed Sci. Res.* **2018**, *28*, 340–348. [CrossRef]
76. Yang, G.; Liu, J.; Zhao, C.; Li, Z.; Huang, Y.; Yu, H.; Xu, B.; Yang, X.; Zhu, D.; Zhang, X.; et al. Unmanned aerial vehicle remote sensing for field-based crop phenotyping: Current status and perspectives. *Front. Plant Sci.* **2017**, *8*, 1111. [CrossRef] [PubMed]
77. Furbank, R.T.; Jimenez-Berni, J.A.; George-Jaeggli, B.; Potgieter, A.B.; Deery, D.M. Field crop phenomics: Enabling breeding for radiation use efficiency and biomass in cereal crops. *New Phytol.* **2019**, *223*, 1714–1727. [CrossRef]
78. Lobet, G.; Draye, X.; Périlleux, C. An online database for plant image analysis software tools. *Plant Methods* **2013**, *9*, 38. [CrossRef]
79. Brown, T.B.; Cheng, R.; Sirault, X.R.R.; Rungrat, T.; Murray, K.D.; Trtilek, M.; Furbank, R.T.; Badger, M.; Pogson, B.J.; Borevitz, J.O. TraitCapture: Genomic and environment modelling of plant phenomic data. *Curr. Opin. Plant Biol.* **2014**, *18*, 73–79. [CrossRef]
80. Cobb, J.N.; DeClerck, G.; Greenberg, A.; Clark, R.; McCouch, S. Next-generation phenotyping: Requirements and strategies for enhancing our understanding of genotype–phenotype relationships and its relevance to crop improvement. *Theor. Appl. Genet.* **2013**, *126*, 867–887. [CrossRef]
81. FAO. *Genebank Standards for Plant Genetic Resources for Food and Agriculture*; Food and Agriculture Organization of the United Nations: Rome, Italy, 2014.
82. Longin, C.F.H.; Reif, J.C. Redesigning the exploitation of wheat genetic resources. *Trends Plant Sci.* **2014**, *19*, 631–636. [CrossRef]
83. Crossa, J.; Pérez-Rodríguez, P.; Cuevas, J.; Montesinos-López, O.; Jarquín, D.; De los Campos, G.; Burgueño, J.; González-Camacho, J.M.; Pérez-Elizalde, S.; Beyene, Y.; et al. Genomic selection in plant breeding: Methods, models, and perspectives. *Trends Plant Sci.* **2017**, *22*, 961–975. [CrossRef]
84. Wang, X.; Xu, Y.; Hu, Z.; Xu, C. Genomic selection methods for crop improvement: Current status and prospects. *Crop J.* **2018**, *6*, 330–340. [CrossRef]
85. Jannink, J.L. Dynamics of long-term genomic selection. *Genet. Sel. Evol.* **2010**, *42*, 35. [CrossRef] [PubMed]
86. Reynolds, M.; Langridge, P. Physiological breeding. *Curr. Opin. Plant Biol.* **2016**, *31*, 162–171. [CrossRef] [PubMed]
87. Rutkoski, J.; Singh, R.P.; Huerta-Espino, J.; Bhavani, S.; Poland, J.; Jannink, J.L.; Sorrells, M.E. Genetic gain from phenotypic and genomic selection for quantitative resistance to stem rust of wheat. *Plant Genome* **2015**, *8*, 1–10. [CrossRef]
88. Rincent, R.; Charpentier, J.-P.; Faivre-Rampant, P.; Paux, E.; Le Gouis, J.; Bastien, C.; Segura, V. Phenomic selection is a low-cost and high-throughput method based on indirect predictions: Proof of concept on wheat and poplar. *G3 Genes Genom. Genet.* **2018**. [CrossRef] [PubMed]
89. Pimentel, D.; Wilson, C.; McCullum, C.; Huang, R.; Dwen, P.; Flack, J.; Tran, Q.; Saltman, T.; Cliff, B. Economic and environmental benefits of biodiversity. *BioScience* **1997**, *47*, 747–757. [CrossRef]

90. Van De Wouw, M.; Kik, C.; Van Hintum, T.; Van Treuren, R.; Visser, B. Genetic erosion in crops: Concept, research results and challenges. *Plant Genet. Resour.* **2010**, *8*, 1–15. [CrossRef]
91. Schoen, D.J.; Brown, A.H.D. The conservation of wild plant species in seed banks: Attention to both taxonomic coverage and population biology will improve the role of seed banks as conservation tools. *BioScience* **2001**, *51*, 960–966. [CrossRef]
92. Solberg, S.O.; Yndgaard, F.; Palmè, A. Morphological and phenological consequences of ex situ conservation of natural populations of red clover (*Trifolium pratense* L.). *Plant Genet. Resour.* **2015**, *15*, 97–108. [CrossRef]
93. Tester, M.; Langridge, P. Breeding technologies to increase crop production in a changing world. *Science* **2010**, *327*, 818–822. [CrossRef]
94. Boyle, R.D.; Corke, F.M.K.; Doonan, J.H. Automated estimation of tiller number in wheat by ribbon detection. *Mach. Vis. Appl.* **2016**, *27*, 637–646. [CrossRef]
95. Yamamoto, K.; Guo, W.; Ninomiya, S. Node detection and internode length estimation of tomato seedlings based on image analysis and machine learning. *Sensors* **2016**, *16*, 1044. [CrossRef]
96. Conn, A.; Pedmale, U.V.; Chory, J.; Stevens, C.F.; Navlakha, S. A statistical description of plant shoot architecture. *Curr. Biol.* **2017**, *27*, 2078–2088.e2073. [CrossRef] [PubMed]
97. Crowell, S.; Falcão, A.X.; Shah, A.; Wilson, Z.; Greenberg, A.J.; McCouch, S.R. High-resolution inflorescence phenotyping using a novel image-analysis pipeline, PANorama. *Plant Physiol.* **2014**, *165*, 479–495. [CrossRef] [PubMed]
98. Wang, X.; Singh, D.; Marla, S.; Morris, G.; Poland, J. Field-based high-throughput phenotyping of plant height in sorghum using different sensing technologies. *Plant Methods* **2018**, *14*, 53. [CrossRef] [PubMed]
99. Holman, F.; Riche, A.; Michalski, A.; Castle, M.; Wooster, M.; Hawkesford, M. High throughput field phenotyping of wheat plant height and growth rate in field plot trials using UAV based remote sensing. *Remote Sens.* **2016**, *8*, 1031. [CrossRef]
100. Han, L.; Yang, G.; Dai, H.; Yang, H.; Xu, B.; Feng, H.; Li, Z.; Yang, X. Fuzzy clustering of maize plant-height patterns using time series of UAV remote-sensing images and variety traits. *Front. Plant Sci.* **2019**, *10*. [CrossRef]
101. Bendig, J.; Bolten, A.; Bareth, G. UAV-based imaging for multi-temporal, very high resolution crop surface models to monitor crop growth variability. *Photogramm. Fernerkund. Geoinf.* **2013**, *2013*, 551–562. [CrossRef]
102. Larese, M.G.; Granitto, P.M. Finding local leaf vein patterns for legume characterization and classification. *Mach. Vis. Appl.* **2016**, *27*, 709–720. [CrossRef]
103. Wang, Y.; Wen, W.; Wu, S.; Wang, C.; Yu, Z.; Guo, X.; Zhao, C. Maize plant phenotyping: Comparing 3D laser scanning, multi-view stereo reconstruction, and 3D digitizing estimates. *Remote Sens.* **2018**, *11*, 63. [CrossRef]
104. Brichet, N.; Fournier, C.; Turc, O.; Strauss, O.; Artzet, S.; Pradal, C.; Welcker, C.; Tardieu, F.; Cabrera-Bosquet, L. A robot-assisted imaging pipeline for tracking the growths of maize ear and silks in a high-throughput phenotyping platform. *Plant Methods* **2017**, *13*, 1–12. [CrossRef]
105. AL-Tam, F.; Adam, H.; Anjos, A.d.; Lorieux, M.; Larmande, P.; Ghesquière, A.; Jouannic, S.; Shahbazkia, H.R. P-TRAP: A panicle trait phenotyping tool. *BMC Plant Biol.* **2013**, *13*, 122. [CrossRef] [PubMed]
106. Rebolledo, M.C.; Peña, A.L.; Duitama, J.; Cruz, D.F.; Dingkuhn, M.; Grenier, C.; Tohme, J. Combining image analysis, genome wide association studies and different field trials to reveal stable genetic regions related to panicle architecture and the number of spikelets per panicle in rice. *Front. Plant Sci.* **2016**, *7*. [CrossRef] [PubMed]
107. Xiong, X.; Duan, L.; Liu, L.; Tu, H.; Yang, P.; Wu, D.; Chen, G.; Xiong, L.; Yang, W.; Liu, Q. Panicle-SEG: A robust image segmentation method for rice panicles in the field based on deep learning and superpixel optimization. *Plant Methods* **2017**, *13*, 104. [CrossRef] [PubMed]
108. Lu, H.; Cao, Z.; Xiao, Y.; Fang, Z.; Zhu, Y.; Xian, K. Fine-grained maize tassel trait characterization with multi-view representations. *Comput. Electron. Agric.* **2015**, *118*, 143–158. [CrossRef]
109. Lu, H.; Cao, Z.; Xiao, Y.; Li, Y.; Zhu, Y. Region-based colour modelling for joint crop and maize tassel segmentation. *Biosyst. Eng.* **2016**, *147*, 139–150. [CrossRef]
110. Boyle, R.; Corke, F.; Howarth, C. Image-based estimation of oat panicle development using local texture patterns. *Funct. Plant Biol.* **2015**, *42*, 433–443. [CrossRef]
111. Song, Y.; Glasbey, C.; Horgan, G.; Polder, G.; Dieleman, J.; Van der Heijden, G. Automatic fruit recognition and counting from multiple images. *Biosyst. Eng.* **2014**, *118*, 203–215. [CrossRef]

112. Xiong, B.; Wang, B.; Xiong, S.; Lin, C.; Yuan, X. 3D morphological processing for wheat spike phenotypes using computed tomography images. *Remote Sens.* **2019**, *11*, 1110. [CrossRef]
113. Genaev, M.A.; Komyshev, E.G.; Smirnov, N.V.; Kruchinina, Y.V.; Goncharov, N.P.; Afonnikov, D.A. Morphometry of the wheat spike by analyzing 2D images. *Agronomy* **2019**, *9*, 390. [CrossRef]
114. McDonald, L.S.; Panozzo, J.F.; Salisbury, P.A.; Ford, R. Discriminant analysis of defective and non-defective field pea (*Pisum sativum* L.) into broad market grades based on digital image features. *PLoS ONE* **2016**, *11*, e0155523. [CrossRef]
115. Rolletschek, H.; Fuchs, J.; Friedel, S.; Börner, A.; Todt, H.; Jakob, P.M.; Borisjuk, L. A novel noninvasive procedure for high-throughput screening of major seed traits. *Plant Biotechnol. J.* **2015**, *13*, 188–199. [CrossRef] [PubMed]
116. Tanabata, T.; Shibaya, T.; Hori, K.; Ebana, K.; Yano, M. SmartGrain: High-throughput phenotyping software for measuring seed shape through image analysis. *Plant Physiol.* **2012**, *160*, 1871–1880. [CrossRef] [PubMed]
117. Miller, N.D.; Haase, N.J.; Lee, J.; Kaeppler, S.M.; De Leon, N.; Spalding, E.P. A robust, high-throughput method for computing maize ear, cob, and kernel attributes automatically from images. *Plant J.* **2017**, *89*, 169–178. [CrossRef]
118. Makanza, R.; Zaman-Allah, M.; Cairns, J.E.; Eyre, J.; Burgueño, J.; Pacheco, Á.; Diepenbrock, C.; Magorokosho, C.; Tarekegne, A.; Olsen, M.; et al. High-throughput method for ear phenotyping and kernel weight estimation in maize using ear digital imaging. *Plant Methods* **2018**, *14*, 49. [CrossRef] [PubMed]
119. Martín-Gómez, J.J.; Rewicz, A.; Goriewa-Duba, K.; Wiwart, M.; Tocino, Á.; Cervantes, E. Morphological description and classification of wheat kernels based on geometric models. *Agronomy* **2019**, *9*, 399. [CrossRef]
120. Gentallan, R.P.; Altoveros, N.C.; Borromeo, T.H.; Endonela, L.E.; Hay, F.R.; Lalusin, A.G.; Reaño, C.E.; Yoshioka, Y. An objective method of shape descriptor state establishment using elliptic Fourier analysis (EFA). *Plant Genet. Resour.* **2019**, *17*, 480–487. [CrossRef]
121. Jahnke, S.; Roussel, J.; Hombach, T.; Kochs, J.; Fischbach, A.; Huber, G.; Scharr, H. phenoSeeder-A robot system for automated handling and phenotyping of individual seeds. *Plant Physiol.* **2016**, *172*, 1358–1370. [CrossRef]
122. Clohessy, J.W.; Pauli, D.; Kreher, K.M.; Buckler, E.S.; Armstrong, P.R.; Wu, T.; Hoekenga, O.A.; Jannink, J.-L.; Sorrells, M.E.; Gore, M.A. A low-cost automated system for high-throughput phenotyping of single oat seeds. *Plant Phenome J.* **2018**, *1*. [CrossRef]
123. Wu, J.; Yang, G.; Yang, X.; Xu, B.; Han, L.; Zhu, Y. Automatic counting of *in situ* rice seedlings from UAV images based on a deep fully convolutional neural network. *Remote Sens.* **2019**, *11*, 691. [CrossRef]
124. Jiang, Y.; Li, C.; Paterson, A.H.; Robertson, J.S. DeepSeedling: Deep convolutional network and Kalman filter for plant seedling detection and counting in the field. *Plant Methods* **2019**, *15*, 141. [CrossRef]
125. Jin, X.; Liu, S.; Baret, F.; Hemerlé, M.; Comar, A. Estimates of plant density of wheat crops at emergence from very low altitude UAV imagery. *Remote Sens. Environ.* **2017**, *198*, 105–114. [CrossRef]
126. Duan, T.; Zheng, B.; Guo, W.; Ninomiya, S.; Guo, Y.; Chapman, S.C. Comparison of ground cover estimates from experiment plots in cotton, sorghum and sugarcane based on images and ortho-mosaics captured by UAV. *Funct. Plant Biol.* **2017**, *44*, 169–183. [CrossRef] [PubMed]
127. Zhu, Y.; Cao, Z.; Lu, H.; Li, Y.; Xiao, Y. In-field automatic observation of wheat heading stage using computer vision. *Biosyst. Eng.* **2016**, *143*, 28–41. [CrossRef]
128. Desai, S.V.; Balasubramanian, V.N.; Fukatsu, T.; Ninomiya, S.; Guo, W. Automatic estimation of heading date of paddy rice using deep learning. *Plant Methods* **2019**, *15*, 76. [CrossRef] [PubMed]
129. Wu, G.; Miller, N.D.; De Leon, N.; Kaeppler, S.M.; Spalding, E.P. Predicting *Zea mays* flowering time, yield, and kernel dimensions by analyzing aerial images. *Front. Plant Sci.* **2019**, *10*. [CrossRef]
130. Liu, S.; Martre, P.; Buis, S.; Abichou, M.; Andrieu, B.; Baret, F. Estimation of plant and canopy architectural traits using the digital plant phenotyping platform. *Plant Physiol.* **2019**, *181*, 881–890. [CrossRef]
131. Wilke, N.; Siegmann, B.; Klingbeil, L.; Burkart, A.; Kraska, T.; Muller, O.; Van Doorn, A.; Heinemann, S.; Rascher, U. Quantifying lodging percentage and lodging severity using a UAV-based canopy height model combined with an objective threshold approach. *Remote Sens.* **2019**, *11*, 515. [CrossRef]
132. Singh, D.; Wang, X.; Kumar, U.; Gao, L.; Noor, M.; Imtiaz, M.; Singh, R.P.; Poland, J. High-throughput phenotyping enabled genetic dissection of crop lodging in wheat. *Front. Plant Sci.* **2019**, *10*. [CrossRef]
133. Zhao, X.; Yuan, Y.; Song, M.; Ding, Y.; Lin, F.; Liang, D.; Zhang, D. Use of unmanned aerial vehicle imagery and deep learning UNet to extract rice lodging. *Sensors* **2019**, *19*, 3859. [CrossRef]

134. Fu, P.; Meacham-Hensold, K.; Guan, K.; Bernacchi, C.J. Hyperspectral leaf reflectance as proxy for photosynthetic capacities: An ensemble approach based on multiple machine learning algorithms. *Front. Plant Sci.* **2019**, *10*. [CrossRef]
135. Yendrek, C.R.; Tomaz, T.; Montes, C.M.; Cao, Y.; Morse, A.M.; Brown, P.J.; McIntyre, L.M.; Leakey, A.D.B.; Ainsworth, E.A. High-throughput phenotyping of maize leaf physiological and biochemical traits using hyperspectral reflectance. *Plant Physiol.* **2017**, *173*, 614–626. [CrossRef]
136. Zhang, L.; Niu, Y.; Zhang, H.; Han, W.; Li, G.; Tang, J.; Peng, X. Maize canopy temperature extracted from UAV thermal and RGB imagery and its application in water stress monitoring. *Front. Plant Sci.* **2019**, *10*. [CrossRef] [PubMed]
137. Sadras, V.O.; Mahadevan, M.; Zwer, P.K. Stay-green associates with low water soluble carbohydrates at flowering in oat. *Field Crops Res.* **2019**, *230*, 132–138. [CrossRef]
138. Blancon, J.; Dutartre, D.; Tixier, M.-H.; Weiss, M.; Comar, A.; Praud, S.; Baret, F. A High-throughput model-assisted method for phenotyping maize green leaf area index dynamics using unmanned aerial vehicle imagery. *Front. Plant Sci.* **2019**, *10*. [CrossRef] [PubMed]
139. Walter, J.D.C.; Edwards, J.; McDonald, G.; Kuchel, H. Estimating biomass and canopy height with LiDAR for field crop breeding. *Front. Plant Sci.* **2019**, *10*. [CrossRef] [PubMed]
140. Golzarian, M.R.; Frick, R.A.; Rajendran, K.; Berger, B.; Roy, S.; Tester, M.; Lun, D.S. Accurate inference of shoot biomass from high-throughput images of cereal plants. *Plant Methods* **2011**, *7*, 2. [CrossRef] [PubMed]
141. Montesinos-López, O.A.; Montesinos-López, A.; Crossa, J.; De los Campos, G.; Alvarado, G.; Suchismita, M.; Rutkoski, J.; González-Pérez, L.; Burgueño, J. Predicting grain yield using canopy hyperspectral reflectance in wheat breeding data. *Plant Methods* **2017**, *13*, 4. [CrossRef]
142. Fernandez-Gallego, J.; Kefauver, S.; Gutiérrez, N.A.; Nieto-Taladriz, M.T.; Araus, J. *Automatic Wheat Ear Counting In-Field Conditions: Simulation and Implication of Lower Resolution Images*; Proc. SPIE, Remote Sensing for Agriculture, Ecosystems, and Hydrology XX, 107830M: Berlin, Germany, 2018; Volume 10783. [CrossRef]
143. Guo, W.; Zheng, B.; Potgieter, A.B.; Diot, J.; Watanabe, K.; Noshita, K.; Jordan, D.R.; Wang, X.; Watson, J.; Ninomiya, S.; et al. Aerial imagery analysis-quantifying appearance and number of sorghum heads for applications in breeding and agronomy. *Front. Plant Sci.* **2018**, *9*, 1544. [CrossRef]
144. Madec, S.; Jin, X.; Lu, H.; De Solan, B.; Liu, S.; Duyme, F.; Heritier, E.; Baret, F. Ear density estimation from high resolution RGB imagery using deep learning technique. *Agric. For. Meteorol.* **2019**, *264*, 225–234. [CrossRef]
145. Pérez-Harguindeguy, N.; Díaz, S.; Garnier, E.; Lavorel, S.; Poorter, H.; Jaureguiberry, P.; Bret-Harte, M.S.; Cornwell, W.K.; Craine, J.M.; Gurvich, D.E.; et al. New handbook for standardised measurement of plant functional traits worldwide. *Aust. J. Bot.* **2013**, *61*, 167–234. [CrossRef]
146. Reynolds, M.; Foulkes, M.J.; Slafer, G.A.; Berry, P.; Parry, M.A.J.; Snape, J.W.; Angus, W.J. Raising yield potential in wheat. *J. Exp. Bot.* **2009**, *60*, 1899–1918. [CrossRef] [PubMed]
147. Yuan, W.; Li, J.; Bhatta, M.; Shi, Y.; Baenziger, P.; Ge, Y. Wheat height estimation using LiDAR in comparison to ultrasonic sensor and UAS. *Sensors* **2018**, *18*, 3731. [CrossRef] [PubMed]
148. Grillo, O.; Blangiforti, S.; Venora, G. Wheat landraces identification through glumes image analysis. *Comput. Electron. Agric.* **2017**, *141*, 223–231. [CrossRef]
149. Cervantes, E.; Martín, J.J.; Saadaoui, E. Updated methods for seed shape analysis. *Scientifica* **2016**, *2016*. [CrossRef]
150. ElMasry, G.; Mandour, N.; Al-Rejaie, S.; Belin, E.; Rousseau, D. Recent applications of multispectral imaging in seed phenotyping and quality monitoring—An overview. *Sensors* **2019**, *19*, 1090. [CrossRef]
151. Feng, L.; Zhu, S.; Liu, F.; He, Y.; Bao, Y.; Zhang, C. Hyperspectral imaging for seed quality and safety inspection: A review. *Plant Methods* **2019**, *15*, 91. [CrossRef]
152. Gentallan, R.P.; Altoveros, N.C.; Borromeo, T.H.; Endonela, L.E.; Hay, F.R.; Lalusin, A.G.; Reaño, C.E.; Yoshioka, Y. Systematic establishment of colour descriptor states through image-based phenotyping. *Plant Genet. Resour.* **2018**, *17*, 91–94. [CrossRef]
153. Nielsen, N.H.; Jahoor, A.; Jensen, J.D.; Orabi, J.; Cericola, F.; Edriss, V.; Jensen, J. Genomic prediction of seed quality traits using advanced barley breeding lines. *PLoS ONE* **2016**, *11*, e0164494. [CrossRef]
154. Kumar, J.; Choudhary, A.K.; Gupta, D.S.; Kumar, S. Towards exploitation of adaptive traits for climate-resilient smart pulses. *Int. J. Mol. Sci.* **2019**, *20*, 2971. [CrossRef]

155. Van Hintum, T.J.; Brown, A.H.D.; Spillane, C. *Core Collections of Plant Genetic Resources*; International Plant Genetic Resources Institute: Rome, Italy, 2000.
156. Reynolds, M.; Foulkes, J.; Furbank, R.; Griffiths, S.; King, J.; Murchie, E.; Parry, M.; Slafer, G. Achieving yield gains in wheat. *Plant Cell Environ.* **2012**, *35*, 1799–1823. [CrossRef]
157. Khush, G.S. Strategies for increasing the yield potential of cereals: Case of rice as an example. *Plant Breed.* **2013**, *132*, 433–436. [CrossRef]
158. Shunmugam, A.S.K.; Kannan, U.; Jiang, Y.; Daba, K.A.; Gorim, L.Y. Physiology based approaches for breeding of next-generation food legumes. *Plants* **2018**, *7*, 72. [CrossRef] [PubMed]
159. Borrell, A.K.; Mullet, J.E.; George-Jaeggli, B.; Van Oosterom, E.J.; Hammer, G.L.; Klein, P.E.; Jordan, D.R. Drought adaptation of stay-green sorghum is associated with canopy development, leaf anatomy, root growth, and water uptake. *J. Exp. Bot.* **2014**, *65*, 6251–6263. [CrossRef] [PubMed]
160. Kamal, N.M.; Gorafi, Y.S.A.; Abdelrahman, M.; Abdellatef, E.; Tsujimoto, H. Stay-green trait: A prospective approach for yield potential, and drought and heat stress adaptation in globally important cereals. *Int. J. Mol. Sci.* **2019**, *20*, 5837. [CrossRef]
161. Kumari, M.; Pudake, R.N.; Singh, V.P.; Joshi, A.K. Association of staygreen trait with canopy temperature depression and yield traits under terminal heat stress in wheat (*Triticum aestivum* L.). *Euphytica* **2013**, *190*, 87–97. [CrossRef]
162. Rebetzke, G.J.; Jimenez-Berni, J.A.; Bovill, W.D.; Deery, D.M.; James, R.A. High-throughput phenotyping technologies allow accurate selection of stay-green. *J. Exp. Bot.* **2016**, *67*, 4919–4924. [CrossRef]
163. Li, X.; Ingvordsen, C.H.; Weiss, M.; Rebetzke, G.J.; Condon, A.G.; James, R.A.; Richards, R.A. Deeper roots associated with cooler canopies, higher normalized difference vegetation index, and greater yield in three wheat populations grown on stored soil water. *J. Exp. Bot.* **2019**, *70*, 4963–4974. [CrossRef]
164. Reynolds, D.; Baret, F.; Welcker, C.; Bostrom, A.; Ball, J.; Cellini, F.; Lorence, A.; Chawade, A.; Khafif, M.; Noshita, K.; et al. What is cost-efficient phenotyping? Optimizing costs for different scenarios. *Plant Sci.* **2019**, *282*, 14–22. [CrossRef]
165. Roitsch, T.; Cabrera-Bosquet, L.; Fournier, A.; Ghamkhar, K.; Jiménez-Berni, J.; Pinto, F.; Ober, E.S. Review: New sensors and data-driven approaches—A path to next generation phenomics. *Plant Sci.* **2019**, *282*, 2–10. [CrossRef]
166. Czedik-Eysenberg, A.; Seitner, S.; Güldener, U.; Koemeda, S.; Jez, J.; Colombini, M.; Djamei, A. The 'PhenoBox', a flexible, automated, open-source plant phenotyping solution. *New Phytol.* **2018**, *219*, 808–823. [CrossRef]
167. Coppens, F.; Wuyts, N.; Inzé, D.; Dhondt, S. Unlocking the potential of plant phenotyping data through integration and data-driven approaches. *Curr. Opin. Syst. Biol.* **2017**, *4*, 58–63. [CrossRef]
168. Bolger, A.M.; Poorter, H.; Dumschott, K.; Bolger, M.E.; Arend, D.; Osorio, S.; Gundlach, H.; Mayer, K.F.X.; Lange, M.; Scholz, U.; et al. Computational aspects underlying genome to phenome analysis in plants. *Plant J.* **2019**, *97*, 182–198. [CrossRef]
169. Poorter, H.; Fiorani, F.; Pieruschka, R.; Wojciechowski, T.; Van der Putten, W.H.; Kleyer, M.; Schurr, U.; Postma, J. Pampered inside, pestered outside? Differences and similarities between plants growing in controlled conditions and in the field. *New Phytol.* **2016**, *212*, 838–855. [CrossRef] [PubMed]
170. Xu, Y. Envirotyping for deciphering environmental impacts on crop plants. *Theor. Appl. Genet.* **2016**, *129*, 653–673. [CrossRef] [PubMed]
171. Pieruschka, R.; Schurr, U. Plant phenotyping: Past, present, and future. *Plant Phenomics* **2019**, *2019*, 7507131. [CrossRef]
172. Oppermann, M.; Weise, S.; Dittmann, C.; Knüpffer, H. GBIS: The information system of the German genebank. *Database* **2015**, *2015*. [CrossRef]
173. Ćwiek-Kupczyńska, H.; Altmann, T.; Arend, D.; Arnaud, E.; Chen, D.; Cornut, G.; Fiorani, F.; Frohmberg, W.; Junker, A.; Klukas, C.; et al. Measures for interoperability of phenotypic data: Minimum information requirements and formatting. *Plant Methods* **2016**, *12*, 44. [CrossRef]
174. Wilkinson, M.D.; Dumontier, M.; Aalbersberg, I.J.; Appleton, G.; Axton, M.; Baak, A.; Blomberg, N.; Boiten, J.-W.; da Silva Santos, L.B.; Bourne, P.E.; et al. The FAIR guiding principles for scientific data management and stewardship. *Sci. Data* **2016**, *3*. [CrossRef]

175. Colmsee, C.; Joachim Keller, E.R.; Zanke, C.; Senula, A.; Funke, T.; Oppermann, M.; Weise, S.; Scholz, U. The garlic and shallot core collection image database of IPK presenting two vegetatively maintained crops in the Federal ex situ genebank for agricultural and horticultural crops at Gatersleben, Germany. *Genet. Resour. Crop Evol.* **2012**, *59*, 1407–1415. [CrossRef]
176. Cobb, J.N.; Juma, R.U.; Biswas, P.S.; Arbelaez, J.D.; Rutkoski, J.; Atlin, G.; Hagen, T.; Quinn, M.; Ng, E.H. Enhancing the rate of genetic gain in public-sector plant breeding programs: Lessons from the breeder's equation. *Theor. Appl. Genet.* **2019**, *132*, 627–645. [CrossRef]
177. Wang, C.; Hu, S.; Gardner, C.; Lübberstedt, T. Emerging avenues for utilization of exotic germplasm. *Trends Plant Sci.* **2017**, *22*, 624–637. [CrossRef] [PubMed]
178. Richards, C.M.; Lockwood, D.R.; Volk, G.M.; Walters, C. Modeling demographics and genetic diversity in ex situ collections during seed storage and regeneration. *Crop Sci.* **2010**, *50*, 2440–2447. [CrossRef]
179. Hoban, S. New guidance for ex situ gene conservation: Sampling realistic population systems and accounting for collection attrition. *Biol. Conserv.* **2019**, *235*, 199–208. [CrossRef]
180. Bustos-Korts, D.; Boer, M.P.; Malosetti, M.; Chapman, S.; Chenu, K.; Zheng, B.; Van Eeuwijk, F.A. Combining crop growth modeling and statistical genetic modeling to evaluate phenotyping strategies. *Front. Plant Sci.* **2019**, *10*. [CrossRef]
181. Crossa, J.; Jarquín, D.; Franco, J.; Pérez-Rodríguez, P.; Burgueño, J.; Saint-Pierre, C.; Vikram, P.; Sansaloni, C.; Petroli, C.; Akdemir, D.; et al. Genomic prediction of gene bank wheat landraces. *G3 Genes Genom. Genet.* **2016**, *6*, 1819–1834. [CrossRef]
182. Gorjanc, G.; Jenko, J.; Hearne, S.J.; Hickey, J.M. Initiating maize pre-breeding programs using genomic selection to harness polygenic variation from landrace populations. *BMC Genom.* **2016**, *17*, 30. [CrossRef]
183. Girma, G.; Bhattacharjee, R.; Lopez-Montes, A.; Gueye, B.; Ofodile, S.; Franco, J.; Abberton, M. Re-defining the yam (*Dioscorea* spp.) core collection using morphological traits. *Plant Genet. Resour.* **2017**, *16*, 193–200. [CrossRef]
184. Chung, H.K.; Kim, K.W.; Chung, J.W.; Lee, J.R.; Lee, S.Y.; Dixit, A.; Kang, H.K.; Zhao, W.; McNally, K.L.; Hamilton, R.S.; et al. Development of a core set from a large rice collection using a modified heuristic algorithm to retain maximum diversity. *J. Integr. Plant Biol.* **2009**, *51*, 1116–1125. [CrossRef]
185. Dutta, M.; Phogat, B.S.; Kumar, S.; Kumar, N.; Kumari, J.; Pandey, A.C.; Singh, T.P.; Tyagi, R.K.; Jacob, S.R.; Srinivasan, K.; et al. Development of core set of wheat (*Triticum* spp.) germplasm conserved in the national genebank in India. In *Advances in Wheat Genetics: From Genome to Field*; Ogihara, Y., Takumi, S., Handa, H., Eds.; Springer: Tokyo, Japan, 2015; pp. 33–45.
186. Sanders, R. *A New Approach to Mining Agricultural Gene Banks–To Speed the Pace of Research Innovation for Food Security*; International Center for Agricultural Research in the Dry Areas: Beirut, Lebanon, 2013.
187. Khazaei, H.; Street, K.; Bari, A.; Mackay, M.; Stoddard, F.L. The FIGS (Focused Identification of Germplasm Strategy) approach identifies traits related to drought adaptation in *Vicia faba* genetic resources. *PLoS ONE* **2013**, *8*, e63107. [CrossRef]
188. Endresen, D.T.F.; Street, K.; Mackay, M.; Bari, A.; Amri, A.; De Pauw, E.; Nazari, K.; Yahyaoui, A. Sources of resistance to stem rust (Ug99) in bread wheat and Durum wheat identified using focused identification of germplasm strategy. *Crop Sci.* **2012**, *52*, 764–773. [CrossRef]
189. Bhullar, N.K.; Street, K.; Mackay, M.; Yahiaoui, N.; Keller, B. Unlocking wheat genetic resources for the molecular identification of previously undescribed functional alleles at the *Pm3* resistance locus. *Proc. Natl. Acad. Sci. USA* **2009**, *106*, 9519–9524. [CrossRef] [PubMed]
190. El Bouhssini, M.; Street, K.; Amri, A.; Mackay, M.; Ogbonnaya, F.C.; Omran, A.; Abdalla, O.; Baum, M.; Dabbous, A.; Rihawi, F.; et al. Sources of resistance in bread wheat to Russian wheat aphid *(Diuraphis noxia)* in Syria identified using the Focused Identification of Germplasm Strategy (FIGS). *Plant Breed.* **2011**, *130*, 96–97. [CrossRef]
191. Haupt, M.; Schmid, K. Combining focused identification of germplasm and core collection strategies to identify genebank accessions for central European soybean breeding. *BioRxiv* **2019**. [CrossRef] [PubMed]
192. Halewood, M.; Chiurugwi, T.; Sackville Hamilton, R.; Kurtz, B.; Marden, E.; Welch, E.; Michiels, F.; Mozafari, J.; Sabran, M.; Patron, N.; et al. Plant genetic resources for food and agriculture: Opportunities and challenges emerging from the science and information technology revolution. *New Phytol.* **2018**, *217*, 1407–1419. [CrossRef] [PubMed]

193. Fu, Y.-B. The vulnerability of plant genetic resources conserved ex situ. *Crop Sci.* **2017**, *57*, 2314–2328. [CrossRef]
194. Gourdji, S.M.; Mathews, K.L.; Reynolds, M.; Crossa, J.; Lobell, D.B. An assessment of wheat yield sensitivity and breeding gains in hot environments. *Proc. R. Soc. B Biol. Sci.* **2013**, *280*, 20122190. [CrossRef]
195. Walters, C.; Volk, G.M.; Richards, C.M. Genebanks in the post-genomic age: Emerging roles and anticipated uses. *Biodiversity* **2008**, *9*, 68–71. [CrossRef]

plants

MDPI

Article

Germplasm Acquisition and Distribution by CGIAR Genebanks

Michael Halewood [1,*], Nelissa Jamora [2], Isabel Lopez Noriega [1], Noelle L. Anglin [3], Peter Wenzl [1], Thomas Payne [4], Marie-Noelle Ndjiondjop [5], Luigi Guarino [2], P. Lava Kumar [6], Mariana Yazbek [7], Alice Muchugi [8], Vania Azevedo [9], Marimagne Tchamba [6], Chris S. Jones [10], Ramaiah Venuprasad [11], Nicolas Roux [1], Edwin Rojas [3] and Charlotte Lusty [2]

1 Alliance of Bioversity International and the International Center for Tropical Agriculture (Alliance of Bioversity and CIAT), Via dei Tre Denari 472/a, 00057 Maccarese (Fiumicino) Rome, Italy; i.lopez@cgiar.org (I.L.N.); p.wenzl@cgiar.org (P.W.); n.roux@cgiar.org (N.R.)
2 Global Crop Diversity Trust (Crop Trust), Platz der Vereinten Nationen 7, 53113 Bonn, Germany; nelissa.jamora@croptrust.org (N.J.); luigi.guarino@croptrust.org (L.G.); charlotte.lusty@croptrust.org (C.L.)
3 International Potato Center (CIP), Av. La Molina 1895, La Molina Apartado 1558, Lima 12, Peru; n.anglin@cgiar.org (N.L.A.); e.rojas@cgiar.org (E.R.)
4 International Maize and Wheat Improvement Center (CIMMYT), Apdo. Postal 6-641, 06600 Mexico, D.F., Mexico; t.payne@cgiar.org
5 Africa Rice Center (AfricaRice), Station de M'bé, Bouaké, 01 BP 2511 Bouaké, Cote d'Ivoire; m.ndjiondjop@cgiar.org
6 International Institute for Tropical Agriculture (IITA), PMB 5320, Ibadan 200001, Oyo State, Nigeria; l.kumar@cgiar.org (P.L.K.); m.tchamba@cgiar.org (M.T.)
7 International Center for Agricultural Research in the Dry Areas (ICARDA), P.O. Box 114/5055, Beirut, Lebanon; m.yazbek@cgiar.org
8 World Agroforestry (ICRAF), Box 30677, Nairobi 00100, Kenya; a.muchugi@cgiar.org
9 International Crops Research Institute for the Semi-Arid Tropics (ICRISAT), Patancheru 502 324, Telangana State, India; v.azevedo@cgiar.org
10 International Livestock Research Institute (ILRI), Box 30709, Nairobi 00100, Kenya; c.s.jones@cgiar.org
11 International Rice Research Institute (IRRI), Los Baños 4030, Laguna, Philippines; v.ramaiah@irri.org
* Correspondence: m.halewood@cgiar.org

Received: 31 July 2020; Accepted: 10 September 2020; Published: 1 October 2020

Abstract: The international collections of plant genetic resources for food and agriculture (PGRFA) hosted by 11 CGIAR Centers are important components of the United Nations Food and Agriculture Organization's global system of conservation and use of PGRFA. They also play an important supportive role in realizing Target 2.5 of the Sustainable Development Goals. This paper analyzes CGIAR genebanks' trends in acquiring and distributing PGRFA over the last 35 years, with a particular focus on the last decade. The paper highlights a number of factors influencing the Centers' acquisition of new PGRFA to include in the international collections, including increased capacity to analyze gaps in those collections and precisely target new collecting missions, availability of financial resources, and the state of international and national access and benefit-sharing laws and phytosanitary regulations. Factors contributing to Centers' distributions of PGRFA included the extent of accession-level information, users' capacity to identify the materials they want, and policies. The genebanks' rates of both acquisition and distribution increased over the last decade. The paper ends on a cautionary note concerning the potential of unresolved tensions regarding access and benefit sharing and digital genomic sequence information to undermine international cooperation to conserve and use PGRFA.

Keywords: plant genetic resources for food and agriculture; genebanks; access and benefit sharing; multilateral system; CGIAR

1. Introduction

Over the last three decades, under the auspices of the United Nations (UN) Food and Agriculture Organization (FAO), the international community has repeatedly committed itself to developing and maintaining a global system on plant genetic resources for food and agriculture (global system) (http://www.fao.org/agriculture/crops/thematic-sitemap/theme/seeds-pgr/gpa-old/gsystem/en/). This global system includes specialized international bodies that monitor the status of the conservation and use of plant genetic resources for food and agriculture (PGRFA), develop normative instruments when necessary, and support the implementation and use of those instruments. Furthermore, in 2015, the Sustainable Development Goals were adopted, including Target 2.5 concerning the sustainable management of genetic diversity, and the following target indicator focusing on PGRFA in particular: "[n]umber of plant genetic resources for food and agriculture secured in medium or long term conservation facilities" (http://www.fao.org/sustainable-development-goals/indicators/251a/en/)

Through their management of international PGRFA collections, the CGIAR Centers make important contributions to both the global system and SDG Target 2.5. Their contributions include assembling and conserving PGRFA, adding value to those materials through extensive characterization, evaluation, documentation, and health testing, and supplying samples that are free of quarantine pests and diseases to researchers, plant breeders, farmers, national and community genebanks, and seed companies around the world. The international collections hosted by the CGIAR Centers include over 760,000 accessions of crops, forages, and trees that were originally obtained from 207 countries, as well as pre-bred materials.

Over the last ten years, the CGIAR Centers' genebanks have distributed more than 1.1 million PGRFA samples to recipients in 163 countries. (Data source: Online Reporting Tool (ORT), https://grants.croptrust.org). These transfers represent approximately 23% of all PGRFA transferred following the rules of the multilateral system of access and benefit sharing (multilateral system) created by the International Treaty on Plant Genetic Resources for Food and Agriculture (Plant Treaty). The multilateral system is the internationally sanctioned mechanism for PGRFA exchanges under the global system. The CGIAR breeding programs were the source of an additional 66% (approximately 3.3 million samples) of the PGRFA transferred through the multilateral system. The remaining 11% of materials exchanged through the multilateral system were transferred by organizations and individuals outside the CGIAR. (Source: Plant Treaty Secretariat).

Given their central position within the multilateral system, the CGIAR genebanks' patterns of international acquisition and distribution of PGRFA over time are potentially significant proxies for the overall status and functioning of the global system in general, and institutions governing access to genetic resources and benefit sharing in particular. The CGIAR genebank managers previously participated in a study of factors affecting acquisitions by the CGIAR genebanks from 1984 to 2009 [1]. The study found that the following factors contributed to a significant drop in genebanks' rate of PGRFA acquisitions from the mid-1990s to 2009: decreased levels of international support for collecting expeditions, overstretched staff, inability to characterize and evaluate the materials already collected, and challenges associated with targeting gaps in existing collections. It established that the most consistent overarching factor was "the highly politicized nature of access and benefit sharing issues at the international, national, and local levels, combined with low levels of legal certainty". The study concluded on a millennial note, looking forward to the resolution of outstanding international tensions over access and benefit-sharing issues, and the full implementation of the multilateral system, with the result that more PGRFA would be made available to include in the international collections maintained by the CGIAR genebanks on behalf of the international community.

The research presented in this paper was initiated with the primary objective of revisiting the conclusions of the earlier study. We structured our research around the following questions: have the CGIAR genebanks' rates of acquisition and distribution of germplasm changed in the last 10 years

(2010–2019) as compared to the previous decade (2000–2009)? If there were significant changes, what factors—either external or internal to the CGIAR —contributed to those changes? Finally, recalling one of the main findings of the earlier study, how have international policy frameworks in particular affected the genebanks' acquisitions and distributions of PGRFA? The data and methods we used to investigate these research questions, and our principle findings are set out and discussed below.

Before proceeding, we note that the earlier study which served as initial inspiration for the research presented here focused almost exclusively on CGIAR genebanks' acquisitions of PGRFA [1]. In the early stages of our research planning, we decided to expand the scope of our investigation to also include the genebanks' rates of germplasm distribution and contributing factors.

It is important to underscore that this paper focuses almost exclusively on acquisitions and distributions of PGRFA by the CGIAR Centers' genebanks, and not by the Centers' breeding programs (other than when the breeding programs access materials from, or donate them to, the genebanks). One reason for focusing on genebanks is that data concerning the genebanks acquisitions and distributions are easier to assemble as a result of the historical, CGIAR system-wide coordination of genebank activities. It would take considerably more time and resources to compile time-sequenced data for breeders' acquisitions of germplasm in particular. Another reason for focusing exclusively on genebanks is that the mandates and contributions to the global system of the genebanks and breeders, while closely linked, are different and therefore amenable to separate studies. It is the responsibility of genebanks to assemble and maintain globally relevant collections of PGRFA, maintain the genetic integrity of conserved materials and make them available, in the form received by the Center, to recipients world-wide. The CGIAR breeding programs, on the other hand, develop new and improved materials which they distribute globally. Both make enormously important, but different, contributions to the global system. While it would be extremely interesting to examine the breeding programs' experiences in this regard, it is beyond the scope of the research presented here.

2. Materials and Methods

The CGIAR Centers hosting international PGRFA collections are Africa Rice Center, International Center for Agricultural Research in the Dry Areas (ICARDA), International Crops Research Institute for the Semi-Arid Tropics (ICRISAT), International Institute for Tropical Agriculture (IITA), International Potato Centers (CIP), International Rice Research Institute (IRRI), International Livestock Research Institute (ILRI), Alliance of Bioversity International and the International Center for Tropical Agriculture (Alliance of Bioversity and CIAT), International Maize and Wheat Improvement Center (CIMMYT), and the World Agroforestry (ICRAF). To assemble data on the genebanks' acquisitions and distributions from 2013 to 2019 inclusive, records were compiled from the CGIAR Genebank Research Program 2012–2016 and the CGIAR Genebank Platform 2017–2021. These records are stored on the Online Reporting Tool (ORT) maintained by the Global Crop Diversity Trust (Crop Trust), coordinator of the CGIAR Genebank Platform. The genebanks' managers reviewed those records, and provided additional information, subdividing acquisitions from each year into three mutually exclusive categories: from Centers' own breeding programs, from new collecting expeditions, and donations from other organizations (from outside the CGIAR) Additional information regarding the countries and crops from which Centers acquired PGRFA was assembled through reference to Centers' collecting expedition reports in the ORT and each genebank's own records. For both acquisitions and distributions by the genebanks for 2010–2012 inclusive, each genebank provided the requisite annual acquisition and distribution data, including the subdivision of acquired materials into three categories of sources, provider countries, and crops.

Each Center genebank also provided aggregate totals of their annual acquisitions and distributions for 2005–2009 inclusive. While some of these data had been compiled for CGIAR reports to the Plant Treaty's Governing Body and the FAO Commission on Genetic Resources for Food and Agriculture (CGRFA), this had not been done according to calendar years, so could not be used for this study.

These data were combined with pre-existing data on annual acquisitions and distributions by the Centers covering the period of 1980–2004 (from the 2012 study referenced above).

Genebank managers were also asked to respond to a survey (see Appendix A) providing reflections on trends in acquisitions and distributions from 2010 to 2019, contributing factors, and experiences acquiring (or attempting to acquire) new materials during the last five years. Staff from the Centers' genebanks, the Crop Trust, and the Genebank Platform Policy Module subsequently worked together in group teleconferences to review and synthesize key findings from the assembled data and survey.

The respondents/informants provided expert knowledge concerning genebanks' performance targets and quality management standards, methods for identifying and prioritizing gaps in collections to be addressed through new collecting expeditions, Centers' efforts to ensure healthy, quarantine organism-free genetic materials, and international crop conservation strategies. Literature reviews were conducted to gain insights into the experiences of genebanks outside the CGIAR. To validate data and key findings, several consultations with CGIAR genebank managers and staff were conducted.

3. Findings

3.1. Acquisitions

There was a dramatic rise in acquisitions of new PGRFA by the genebanks between 2010 and 2018, compared to the previous 10 years, though still lower than peaks reached in the 1980s and 1990s, as illustrated in Figures 1 and 2 below. The increase over the last 10 years reached its height in 2012, when the genebanks received almost 14,000 samples of distinct PGRFA to include as new accessions in the international collections. Of course, not all of the materials that the genebanks receive are ultimately accessioned. If materials received are not viable, or are redundant with respect to the materials already in the collection, they are not included in the collections. For the convenience of using a short form, the tables and figures presented below make reference to "accessions"; in fact, the data refer to materials received with the intention of including them as accessions, assuming they are viable and not redundant. In 2019, the number of newly acquired materials by the Centers to include in the international collections dropped back down to the lower levels that characterized the mid-1990s to 2009.

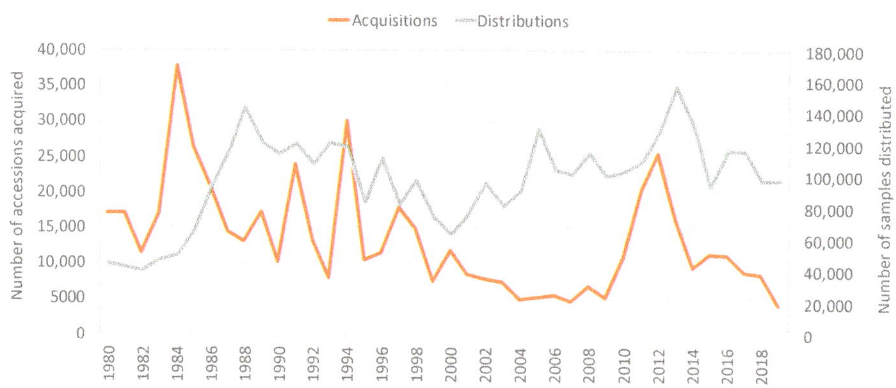

Figure 1. Acquisitions and distributions by all CGIAR Centers 1980–2019.

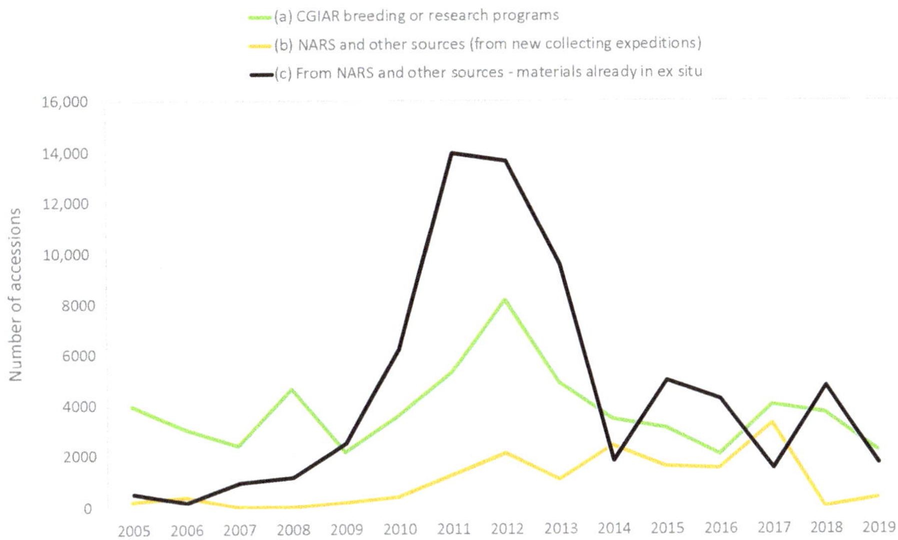

Figure 2. Details of CGIAR genebank acquisitions 2005–2019.

In total, over the course of ten years, from 2010 to 2019 inclusive, the CGIAR genebanks acquired 116,921 samples of distinct PGRFA to include in their Article 15 collections.

Approximately 65% of the materials acquired by the genebanks came from providers in 142 different countries. The remaining 35% came from the Centers' own breeding programs, as discussed below. A total of 84% of those countries are developing countries or countries with economies in transition as defined in the International Monetary Fund's World Economic Outlook Database (October 2018, accessible at: https://www.imf.org/external/pubs/ft/weo/2018/02/weodata/groups.htm). Approximately 18% of the materials from countries came from new collecting expeditions; the other 82% was material that was already in ex situ conditions prior to being sent to the Centers.

All of the materials from providers in countries were either received under the standard material transfer agreement (SMTA) adopted for exchanges of materials under the Plant Treaty's multilateral system (which allows the Center to conserve, use, and pass on the materials with the same SMTA) or under other agreements whereby the providers gave the Centers permission to subsequently distribute the material concerned with the SMTA. In this context, it is interesting to note that 31 of the countries from which the materials were made available are not currently contracting parties to the Plant Treaty, yet they were still willing to provide the materials, and have them subsequently redistributed by the CGIAR Centers, under the conditions established by the Plant Treaty. We did not analyzed whether some other countries which are now Plant Treaty members made materials available before becoming members.

Approximately one third of the PGRFA acquired by the genebanks from providers in countries between 2010 and 2019 were associated with a project coordinated by the Crop Trust from 2007 to 2012 called "Securing the Biological Basis of Agriculture" (hereinafter referred to as the "Regeneration project"), funded by the Bill and Melinda Gates Foundation and the Grains Research and Development Corporation. That project provided financial and technical support for organizations around the world to regenerate unique ex situ PGRFA that were at risk of being lost, to send a copy of the regenerated materials to an internationally recognized genebank, to send a copy for safety back-up in the Svalbard Global Seed Vault, and to make the materials available through the multilateral system of access and benefit sharing. Activities with 84 national partners in 54 countries resulted in the regeneration of approximately 73,000 threatened accessions, of which more than half were duplicated in CGIAR genebanks with permission to make them available through the multilateral system.

As of 2019, another 2256 samples of 1508 unique accessions collected from 25 countries were sent to CGIAR Center genebanks (ICARDA, ICRISAT, IRRI, IITA, CIP) by the Millennium Seed Bank (MSB), associated with the project called "Adapting Agriculture to Climate Change: Collecting, Protecting and Preparing Crop Wild Relatives" (hereinafter the "CWR project") coordinated by the Crop Trust from 2011 to 2021 [2]. The CWR project, with funding from the Norwegian government, provided financial and technical support for project partners to target and collect wild species related to crops, to create a safety back-up, and make collected material available through the multilateral system. It is critically important to have safety duplication for PGRFA accessions hosted by organizations that can ensure the required conditions for long-term storage. Details concerning the Centers, crops, providing countries, and related programs are set out in Table 1 below.

Table 1. Providers of materials acquired by CGIAR genebanks 2010–2019 (excluding CGIAR breeding programs)—Notes: associated with Regeneration project (+), CWR project (*). We separate the two genebanks hosted by the Alliance of Bioversity and CIAT.

CGIAR Genebanks (Crop Mandate)	From New Collecting Mission	From Existing Ex Situ Source	
AfricaRice (rice)	Benin Burundi Cameroon DR Congo Gambia Senegal Tanzania Uganda	Burkina Faso + Guinea + Kenya Mali + Myanmar	
Alliance-Bioversity (banana)	Cook Islands Indonesia Samoa	Cameroon + China DR Congo Germany India + Indonesia + Japan Myanmar	Nigeria Philippines + PNG + Thailand Uganda + USA Vietnam +
Alliance-CIAT (beans, cassava, forages)	Costa Rica * El Salvador *	Azerbaijan + Colombia Costa Rica + Dominican Republic + El Salvador + Malaysia	Myanmar + Nicaragua + Peru Puerto Rico + UK USA
ICARDA (wheat, barley, forages, grasspea, pea, lentils, chickpea, faba bean)	Armenia * Cyprus * Georgia Greece Jordan Kazakhstan Lebanon * Russian Federation Tajikistan	Afghanistan Albania Algeria Argentina Armenia * Australia Austria Azerbaijan * Belarus Bhutan Bolivia Bosnia and Herzegovina	Kyrgyzstan Latvia Lebanon * Libya Lithuania Macedonia Mexico Moldova Mongolia + Montenegro Morocco Nepal

Table 1. *Cont.*

CGIAR Genebanks (Crop Mandate)	From New Collecting Mission	From Existing Ex Situ Source	
		Brazil	Netherlands
		Bulgaria	New Zealand
		Canada	Pakistan
		Chile	Palestinian Authority
		China	Peru
		Colombia	Poland
		Croatia	Portugal *
		Cyprus *	Romania
		Czech Republic	Russian Federation
		Czechoslovakia	Saudi Arabia
		Denmark	Serbia
		Ecuador	Slovakia
		Egypt	Spain
		Finland	Sudan
		France	Sweden
		Georgia *	Switzerland
		Germany	Syria
		Greece	Tajikistan
		Hungary	Tunisia
		India	Turkey
		Iran	Turkmenistan
		Iraq	Ukraine
		Italy *	UK
		Japan	Uruguay
		Jordan	USA
		Kazakhstan	Uzbekistan
		Korea, Republic of	Yugoslavia
CIMMYT (wheat, maize)		Afghanistan	Italy
		Albania +	Kenya
		Australia	Japan
		Austria	Mexico +
		Azerbaijan +	Mongolia +
		Belarus +	Nepal +
		Brazil	Nicaragua +
		Bulgaria +	Nigeria
		Canada	Paraguay
		China	DPR Korea +
		Ecuador +	Peru +
		England	Philippines +
		Ethiopia	S. Africa
		France	Syria
		Georgia +	Tanzania +
		Honduras +	Turkey
		Hungary	UAE
		India	Ukraine +
		Indonesia +	Uruguay +
		Iran	USA
		Ireland	Zambia +
		Israel +	Zimbabwe
CIP (Andean roots and tubers, sweet potato, potato)	Peru *	China	Rwanda +
		Ecuador +	Uganda +
		Indonesia +	USA
		Peru +	
ICRAF (trees)		China	Senegal
		Kenya	Tanzania
		Mali	Uganda

Table 1. *Cont.*

CGIAR Genebanks (Crop Mandate)	From New Collecting Mission	From Existing Ex Situ Source	
ICRISAT			
(finger millet,	Burkina Faso	Azerbaijan +	Nepal +
pearl millet,	Ghana	Benin +	Niger +
sorghum,	India	Bulgaria +	Nigeria +
small millets,	Kenya	Burkina Faso +	Senegal +
pigeon pea,	Niger	Burundi +	Sudan +
chickpea)	Nigeria	Canada	Tanzania
	Uganda	China	Togo +
	Zimbabwe	Georgia +	Uganda +
		Ghana +	Uzbekistan +
		India +	Yemen +
		Kenya +	Zambia +
		Mali +	Zimbabwe +
IITA			
(yam, maize,	Cameroon	Afghanistan +	Mali +
cowpea,		Australia +	Mexico +
cassava, banana,		Azerbaijan +	Mozambique +
Bambara		Benin +	Namibia +
groundnut,		Botswana +	Niger +
misc. legumes,		Brazil +	Nigeria +
cocoyam)		Burkina Faso +	Norway
		Burundi +	Oman +
		Cape Verde +	Paraguay +
		Central African Republic +	Philippines +
		Chad +	Puerto Rico +
		Chile +	Russia +
		China +	Rwanda
		Colombia +	Senegal +
		Costa Rica +	Sierra Leone +
		Cote d'Ivoire +	Somalia +
		DR Congo +	South Africa +
		Egypt +	Sudan +
		France	Suriname +
		Gabon +	Swaziland +
		Georgia +	Taiwan +
		Ghana +	Tanzania +
		Guatemala +	Thailand +
		Guinea +	Togo +
		Hungary +	Turkey +
		India +	Uganda +
		Indonesia +	UK +
		Iran +	USA +
		Italy +	Vietnam +
		Kenya +	Yemen +
		Lesotho +	Zambia +
		Madagascar +	Zimbabwe +
		Malawi +	
ILRI			
(forages and fodder)		Brazil	USA
IRRI			
(rice)		Bangladesh	Malaysia +
		Brazil	Myanmar +
		Cambodia	Nepal +
		China	Pakistan +
		DPR Korea +	Philippines +
		France	Tanzania
		India	Turkey
		Indonesia +	Uganda
		Iran	UK
		Lao PDR +	USA
		Madagascar +	Vietnam +

As mentioned above, approximately 35% of the materials acquired by the genebanks between 2010 and 2019 came from the Centers' own plant breeding programs, mostly from the CIMMYT wheat breeding program. Every genebank has a policy and process, both of which are periodically reviewed, for strategically acquiring materials from breeders' collections for incorporation into the genebank for long-term conservation with the express aim of ensuring that incoming materials are likely either to represent a highly demanded material or diversity that is not already contained in the collection.

The availability of funds—both for providers and for the Centers as recipients—was one of the most frequently mentioned factors by the genebank managers affecting the ability of CGIAR Centers to acquire new materials to include in the genebanks.

The Centers representatives confirm that in many of the instances where the genebanks were able to acquire new materials, it was critically important to be able to provide financial and technical support for providers' activities related to the collection, regeneration, phytosanitary cleaning and inspection, and shipping of accessions. In some cases, the Centers provided this support, in other instances—most notably, the Regeneration and CWR projects—the support came from other organizations.

Generally speaking, the financial costs of preparing and sending materials are not particularly high. Costs arise because of the lack of capacity and means to multiply, dry, test, and clean seed that is either collected or conserved and is only available in small quantities with unknown viability. The transaction costs are greatly higher when vegetatively propagated germplasms, including roots, tubers, bananas, or trees, are involved because the movement across international borders involves stringent phytosanitary restrictions that demand expensive disease cleaning and testing upon both shipping and receipt along with extensive periods of time in quarantine based on the risk assessed by the National Plant Protection Organizations (NPPO). In such cases, the Centers need to make substantial investments in strengthening partners' capacity to test and clean materials.

Once materials arrive at the Center, they need to be processed through post-entry quarantine, health tested, and cleaned of infectious diseases, multiplied, tested for viability, and dried before being introduced into the collection and made available for distribution. In the case of clonally propagated crops, the steps can be very expensive and time-consuming, taking generally four years or longer for a new accession to become available (N. L. Anglin pers comm). Post-entry quarantine (PEQ) procedures such as growing the first generation in a quarantine greenhouse—a requirement for all newly acquired materials by the genebanks—are also expensive. Occasionally, the NPPO requires new acquisitions to remain in PEQ from four months up to two years to assess risks.

Direct costs for new collecting missions are also relatively modest. Of course, the costs per samples of materials collected greatly differs by crop type, wild or cultivated form, and geographical location. For example, similar projects, with similar budgets, working with national agricultural research system (NARS) partners to conduct collecting missions resulted in very different numbers of acquisitions; samples of 307 landraces and 94 wild relatives and forages were gathered in Tajikistan and Lebanon for the same cost as samples of 106 bananas collected in Papua New Guinea, Samoa, and the Cook Islands. The per accession costs will differ between seed and clonal crop collections by an order of magnitude more when taking into account the costs of incorporating the materials into the collections.

In addition to direct costs associated with the management and transfer of biological materials, there are transaction costs associated with getting requisite permissions to provide/access materials from both ex situ and in situ conditions. Transactional costs associated with these activities are often substantially increased in situations where the national policies and laws are unclear, or non-existent. In such cases, it can take extended rounds of communications over long periods of time with many different levels of national R&D partners, lawyers, and competent authorities before final decisions to provide materials can be made, or, as sometimes occurs, no final determination is ever communicated. Unlike some national genebanks or networks of collections [3], CGIAR does not have a centralized service that takes responsibility to help Centers comply with requisite processes for organizing collecting missions/partnerships, and obtaining requisite permissions.

Indeed, along with availability of funds, the CGIAR genebank managers emphasized that "restrictive or unclear laws or policies" were leading variables influencing their ability to acquire new PGRFA to include in the international collections. They note that the Plant Treaty's multilateral system has contributed stability and a sound legal basis for providing and receiving germplasm, and that their ability to acquire materials through new collecting is evidence of cooperation/coordination between national authorities responsible for implementing the Plant Treaty and those responsible for regulating access to genetic resources outside the multilateral system (including those implementing the CBD or Nagoya Protocol). However, the Centers are concerned that unresolved disputes concerning the enhancement of the Plant Treaty's multilateral system of access and benefit sharing, and digital sequence information (DSI) in particular (both under the Plant Treaty and the Nagoya Protocol), are holding back some countries (and providers within countries) from making more PGRFA available through the multilateral system. If international tensions over these issues remain unresolved for too long, they could further undermine the Centers' ability to access, generate, use, and distribute PGRFA and associated information.

The Center genebanks confirm that, over time, they tend to obtain materials to include in their collections from the same set of countries or subregions, where they have established connections. By corollary, there are some countries and subregions from which they rarely, if ever, obtain materials. A number of the genebanks confirm that they rarely make overtures to organizations and or countries which have strongly signaled in the past that they are unwilling to make new materials available. Their general perception is that, despite the coming into force of the Nagoya Protocol and the existence of the Plant Treaty's multilateral system of access and benefit sharing, some of these same countries have not substantially altered their approach to making materials available upon request for inclusion in the international genebanks' collections.

The initial stimulus, and subsequent financial and technical support for PGRFA providers, from these internationally coordinated projects were clearly critically important factors contributing to the extraordinary increase in PGRFA that was made available for the CGIAR Centers to include in the international collections and thus the multilateral system.

The Centers genebanks' representatives also report that the strength of the Centers' long-term relationships with providers and provider countries is equally important. Most materials are made available to Centers as part of projects with providers. Centers rarely obtain new materials to include in the genebank as a result of "cold-calling" would-be providers with simple requests for materials and no other form of engagement. Out of appreciation of these factors, the Centers most recent collective efforts under the CGIAR Genebank Platform (suspended temporarily due to the COVID-19 pandemic) to catalyze and support new collecting expeditions involve support for "two-way flows" of germplasm from the genebanks to the providers, identified on the basis of a jointly conducted analysis of potentially useful germplasm to respond to local needs, and from the providers to the genebanks, and financial and technical support for institutional capacity building for partner organizations in the country concerned. This is consistent with practices of other organizations seeking to acquire PGRFA to include in public genebanks through new collecting activities [4].

Of course, these factors affecting the ability of the genebanks to acquire new materials need to be considered within the broader context of what additional PGRFA needs to be collected, backed up, and conserved as part of the global system, either by CGIAR genebanks or other organizations hosting globally available PGRFA collections. Methods for conducting gap analyses and strategies for coordination with other organizations are considered below.

From 2007 to 2017, CGIAR Centers acquired PGRFA through collecting expeditions from 14 countries. Collecting in six of those countries was supported by the CWR project. During the same period of time, the Center for Genetic Resources in the Netherlands (CGN, with 23,000 accessions) received materials through collecting expeditions in five countries, the Leibniz Institute of Plant Genetics and Crop Plant Research (IPK, with 105,000 accessions) received materials from collecting in six countries, and the National Plant Germplasm System of the United States Department of Agriculture (NPGS-USDA, with 500,000 accessions) organized collecting expeditions in at least 20 foreign countries [5–11]. Accessions from collecting expeditions in other countries account for 8% of CGN total germplasm acquisitions [12]. At the beginning of the century, such materials represented 20% of material acquired by NPGS-USDA [3]. In the last decade, collecting expeditions have contributed 11% of all the acquisitions by CGIAR genebanks. The fact that, relative to the cumulative size of their collections, the CGIAR Centers have engaged in relatively fewer collecting activities during this period than these other organizations may be attributable to a combination of the following factors. The diversity of some CGIAR mandate crops is relatively well represented in ex situ collections when compared to the crops that IPK, CGN, and NPGS-USDA have been prioritizing in their collecting activities, i.e., vegetables (e.g., lettuce, Allium, Brassica, chicory, spinach, asparagus, and carrot), fruit and nut trees (e.g., apple, pear, pomegranate, pistachio, walnut, and hazelnut), berries (Fragaria, Rubus, and Ribes), and temperate grasses (Poa, Festuca, Agrostis, Koeleria, and Puccinellia). IPK, CGN, and NPGS-USDA have been acquiring most of their new materials from the Transcaucasia and Central Asia regions and Europe [10,11], which are arguably more open to allowing new collecting missions than other countries and regions in the world. IPK, CGN, and NPGS-USDA may also have had more, and more regular, financial resources to dedicate to supporting new collecting activities. We acknowledge that we are only scratching the surface of potential comparisons between CGIAR and other genebanks around the world; we hope to be able to deepen such analyses in the future.

3.2. Distributions

Over the course of the last 10 years, the CGIAR genebanks have distributed on average 115,000 samples of germplasm per year around the world. While there are significant fluctuations in Centers' distributions from year to year, the overall rate of Centers' distribution from 2010 to 2019 is higher than the previous decade at 95,000 samples per year (2000 to 2009 annual average), as portrayed in Figure 1 above. Comparing the total distributions between time periods 2000–2014 and 2015–2019, four Centers increased their annual distributions in the latter half of the last decade, four remained generally the same, and three decreased.

There is considerable fluctuation, from year to year, in the ratio of materials the CGIAR genebanks send to recipients within the CGIAR (mainly breeders) and to recipients outside the CGIAR, as can be seen in Figure 3. Since 2017, the Centers genebanks have been distributing proportionately more materials to recipients outside the CGIAR. Some Centers do not have crop breeding programs (e.g., Bioversity, ILRI), so almost all of their distributions are to recipients outside the CGIAR. In 2019, the Centers and crops with a high proportion of materials distributed to breeders (within their own Center or in other Centers within the CGIAR) are AfricaRice (rice); ICRISAT (pigeon pea, chickpea); ICARDA (grasspea, barley); CIP (sweet potato); IRRI (rice); IITA (cassava); and CIMMYT (wheat).

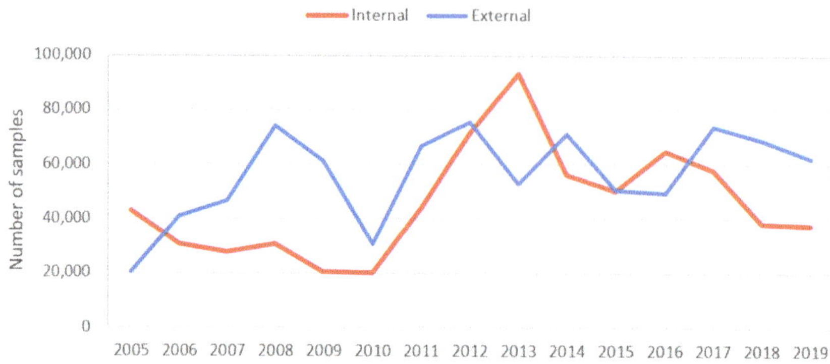

Figure 3. Centers' distributions, 2010–2019, broken down by (i) transfers within the CGIAR (internal), and (ii) transfers to recipients outside the CGIAR (external).

Between 2017 and 2019, approximately 80% of PGRFA samples distributed by the genebanks to recipients outside the CGIAR were to a combination of NARS partners, national genebanks, advanced research institutes (ARIs), and universities (see Figure 4 for more details). The proportion of samples distributed to farmers, farmer organizations, and non-governmental organizations (NGOs) during this time is still relatively small (8%), approximately the same rate as noted earlier for 2015 [13]. CIP and ICRAF were the two centers whose proportionate distribution of materials to farmers and NGOs was largest, between 2017 and 2019 inclusive, as seen in Figure 4b. Much of the material distributed by CIP during this period was part of their repatriation program in which desired potato landraces are matched to farmers' descriptions and returned to the farmers that have lost them due to normal attrition, environmental impacts (drought, hail, diseases), or other reasons, to help maintain and/or increase their diversity in farms. CIP also gives material to farmers in the repatriation program to support their needs in responding to climate change or to help improve yield.

In terms of absolute numbers, the largest provider of germplasm materials was CIMMYT, distributing 28% of the materials cumulatively distributed by the genebanks between 2010 and 2019, followed by IRRI and ICRISAT, distributing approximately 26% and 12%, respectively.

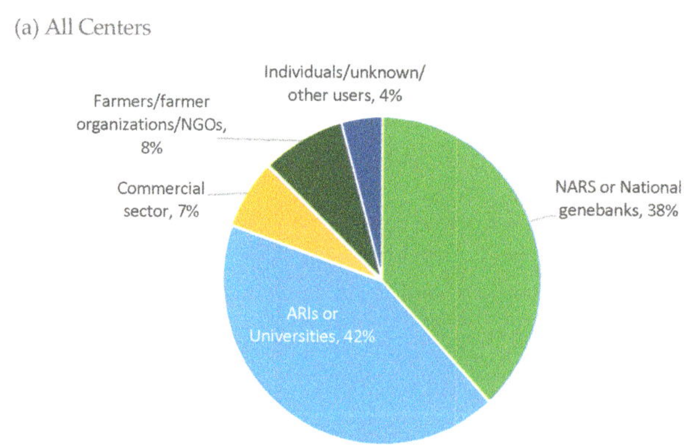

Figure 4. *Cont.*

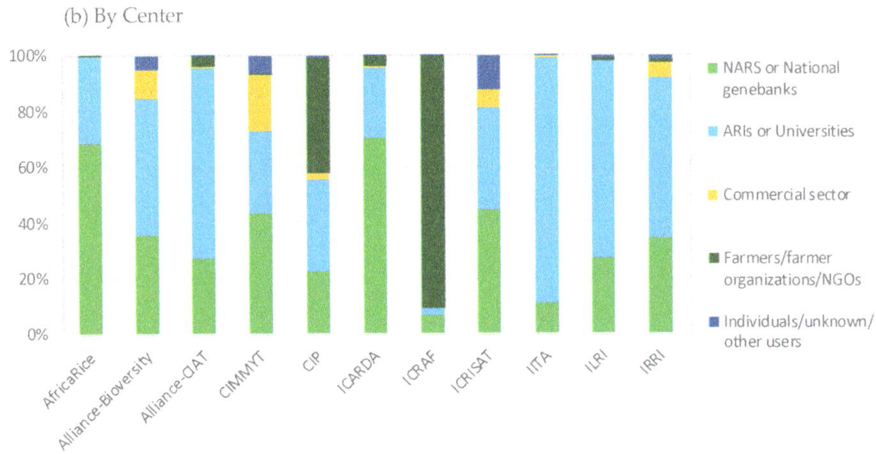

Figure 4. Recipients of germplasm distributed by CGIAR genebanks, 2017–2019.

Landraces are the most frequently requested materials (50% between 2017 and 2019), followed by breeding materials (24%), and wild relatives (13%) (see Figure 5 for more details).

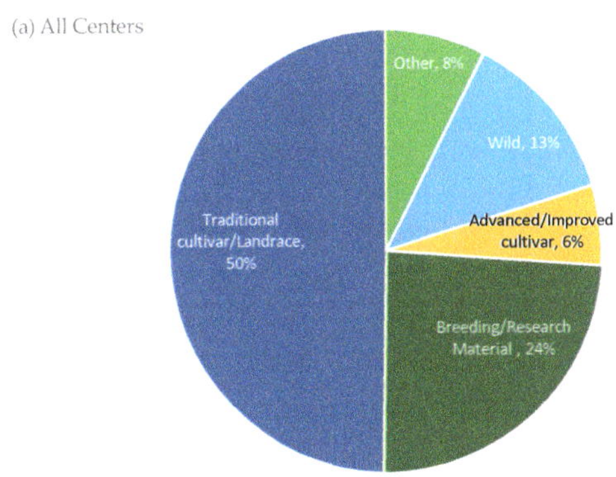

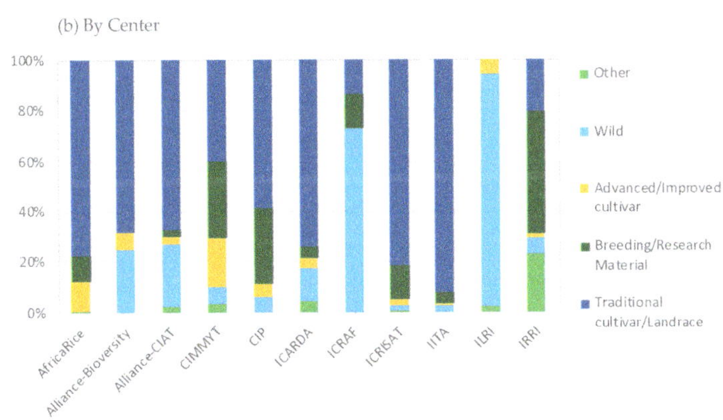

Figure 5. Types of germplasm distributed by CGIAR genebanks, 2017–2019.

The countries which received the highest numbers of samples from CGIAR genebanks, between 2017 and 2019, are set out in Table 2. These data do not include transfers within or between CCIAR Centers. Four of these countries are not contracting parties to the Plant Treaty.

Table 2. 20 countries which received the most samples from the CGIAR genebanks, 2017–2019 (not including intra- and inter-CGIAR Center transfers).

Country	2017	2018	2019	Total
1 India	11,439	15,154	5693	32,286
2 China	6480	4311	3846	14,637
3 United States	3844	3727	1543	9114
4 Italy	6834	424	1767	9025
5 Mexico	1818	3701	2667	8186
6 Morocco	2736	3780	1016	7532
7 Nigeria	2187	2471	2069	6727
8 Ethiopia	1587	1134	1792	4513
9 Australia	3600	168	611	4379
10 Colombia	519	1117	2338	3974
11 Peru	1109	634	2049	3792
12 Mali	1577	1679	459	3715
13 Japan	949	1874	409	3232
14 Sudan	62	34	3116	3212
15 Germany	2124	1011	62	3197
16 Belgium	184	2742	78	3004
17 Lebanon	-	2071	839	2910
18 Kenya	378	984	1411	2773
19 United Kingdom	429	1839	322	2590
20 Iran	643	100	1299	2042

The genebanks' annual distribution is a reflection of demand from users, often driven by the needs of the research and development projects in which they are involved, and by the profile or visibility of the genebanks among research and development, and non-governmental organizations in different countries. The Center hosting the genebank is frequently involved in the research in some way. Here, we provide some examples of projects with which distributed material was associated. The overall spike in distributions in 2013 is largely attributable to the internal CIMMYT transfers of 36,500 wheat accessions as part of CIMMYT's Seeds of Discovery (SeeD) project to CIMMYT researchers and breeders characterizing maize and wheat genetic diversity for use in breeding programs. CIMMYT's high number of distributions in 2013 also includes 8500 wheat accessions to recipients in Iran and Turkey as part of those countries' restoration efforts. In 2019, ICRAF distributed considerably more tree germplasm than in recent years as a result of requests from organizations involved in the mega project "Regreening Africa" that is being implemented in East and West Africa. Between 2017 and 2019, the ICARDA genebank's rate of distribution to recipients outside the CGIAR increased 200–300% over the previous five-year period, reflecting a combination of the Center's progress getting its genebank's operations back "up to speed" after the disruptions associated with having to relocate from Syria and a surge of new interest in the genebank's materials as a result of publicity generated when the Center retrieved materials from the Svalbard Global Seed Vault. A big increase in the ICRISAT genebank's distributions in 2014 and 2018 was associated with supporting a consortium of Indian organizations conducting chickpea and pigeon pea multilocation trials. In 2017, ICRISAT genebank's distributions were higher than usual in part because it responded to a request from Italian organizations for core collections of cereals (sorghum, pearl millet, finger, kodo, proso, barnyard, little, and foxtail millets) as part of a program to reproduce local varieties and develop new crops adaptable to Italian local conditions. The IITA genebank's increased rate of distributions in 2015 was associated with the requests for cowpea and Bambara groundnut germplasm from Nigerian universities to support their research programs on those crops.

The incidence of a newly emerging rare disease can also lead to a need to screen a large number of accessions for resistance [14,15].

The genebank managers confirm that one of the most important factors affecting demand for PGRFA is the quality and relevance of the accession-level information that the Centers compile about the materials in their collections, a finding consistent with the relevant literature [16,17]. Accession-level information helps users make informed decisions about what materials from the collections are potentially most useful for their specific purposes. It also makes it possible for the genebank to make more targeted selections of materials in response to their requests since often users do not know where to begin in choosing an appropriate accession for their needs. The genebanks highlighted the importance of trait-specific data including nutritional qualities, biotic and abiotic stresses, agronomic performance, genetic sequence information linked to traits or to provide information about relationships among accessions, and geographic information about place of collection, including climate conditions and soil type. IRRI reports there was a significant increase in the number of requests for genetic stocks of 3000 rice accessions, particularly between 2015 and 2018, after the full genome sequences were made publicly available. The genebanks have minted digital object identifiers under the Plant Treaty's global information system (GLIS-DOIs) for almost all of the accessions in the Center-hosted international collections, with a long-term view towards helping this process and linking publications and data back to accessions.

While such information is absolutely necessary to generate interest in, and demand for, materials in the genebanks, it also permits users to make more targeted requests for a narrower range of materials with each request. Otherwise, users must consider thousands of accessions from which to choose. Thus, documenting the traits and uniqueness of each accession helps create a path for utilization of the germplasm. Further, creating cores or subsets of accessions, where small groups of accessions are defined by the genebank to help narrow down the search for specific traits, has had an influence on demand and could be scaled up significantly. AfricaRice's genebank has used molecular markers to create core and mini-core sets that represent the maximum possible genetic variation contained in the African rice whole collection [18]. As a result, there has been a significant increase in requests for the mini-core set for use in rice genetic and breeding studies, and gene discovery (M.N. Ndjiondjop pers.com). The Centers' genebank managers note that as they have been able to increase the quantity of accession-level information, requests are indeed becoming more informed and better targeted. Scientists working with genebanks both outside [19,20] and with CGIAR are developing methods to assist users to identify useful materials from ex situ collections, including the ICARDA-led focused identification of germplasm strategy (FIGS) [21,22]. In 2018, IITA developed a FIGS population of drought- and heat-tolerant cowpea. This significantly increased the requests to over 5000 samples distributed to 19 recipients' countries for research purposes (T. Marimagne, pers.com)

The supply of PGRFA from the genebanks also depends upon the ready availability of a sufficient stock of pest and disease-free materials with legal certainty concerning the conditions under which the materials can be provided and received. Costs associated with multiplying, assuring plant health, and distributing samples of crops that are clonally propagated or have recalcitrant seeds are much higher than for crops of orthodox seed behavior. Under the framework of the CGIAR Genebank CRP (2012–2016) and the CGIAR Genebank Platform (2017–2021), the Crop Trust, in cooperation with the Centers, has developed performance targets and a monitoring system to assess the availability, safety duplication, documentation, and quality management of collections. Ultimately, the target is to have 90% (of accessions in the collections) immediately available and 90% safety-duplicated at two locations (for seed collections only). As of the end of 2019, 78% of all materials in the CGIAR genebank were immediately available, 60% of the seed collection was secured in safety duplication at two levels, and 77% was duplicated at the Svalbard Global Seed Vault. A total of 72% of the clonal crop collection was safety-duplicated in the form of cryopreservation or in vitro cultures in at least one location.

The COVID-19 pandemic has highlighted the strategic importance of cryopreserving clonal crops. In vitro cultures require continuous monitoring and upkeep by genebank staff in personam.

If scientists' access to these collections is limited as a result of governmental policies restricting movement, some accessions in those collections could deteriorate and be lost. If those accessions are not safety duplicated somewhere else, or if the safety duplications are in the form of in vitro collections that are similarly vulnerable to the same risk, then unique materials and potentially unique varieties no longer found in farmers' fields could disappear if they are not secured in cryopreservation. Cryopreserved back-up collections of these materials would address this risk.

The CGIAR genebanks report that, on occasion, materials they send, and materials they are meant to receive, are held up for long periods of time due to the implementation of phytosanitary regulations, occasionally to the point where materials die before they arrive at their destination.

The agreements between the CGIAR Centers and the Plant Treaty's Governing Body (signed in 2006) have created legal certainty concerning the status of the collections and the conditions under which they can be distributed. This is reflected in the fact that almost all transfers of PGRFA from the genebanks are under the Plant Treaty's SMTA (with the exception of materials sent for service agreements, or restoration or direct use by farmers in cultivation, as per the opinions of the Ad Hoc Technical Advisory Committee on the Multilateral System and SMTA) [23]. Just as the Centers received materials under the SMTA from providers in countries that are not contracting parties, the Centers' genebanks also distributed PGRFA, using the SMTA, to recipients in thirteen countries that are not Plant Treaty contracting parties (between 2012 and 2019 inclusive).

Despite these benefits to the Centers operating under the Plant Treaty's framework, the genebanks note with concern that some large seed companies, some universities, and one national agricultural research organization are unwilling to receive materials under the SMTA, which makes it impossible for the Centers to distribute materials from the genebank (and also most of the materials from the breeding programs) to them. Other genebanks noted that their ability to distribute materials from the international collections was being constrained by policies of the country in which they are located.

4. Discussion

4.1. Targeted Acquisitions within the Context of the Global System

Most of the international collections hosted by the CGIAR Centers originated as working collections to support research and breeding programs of international and national public agricultural research organizations. Their subsequent growth depended partly on taking advantage of opportunities to acquire PGRFA from a wide range of sources in an ad hoc manner, for example, being offered material:

- Previously assembled by other organizations [1];
- As a result of interactions with scientists from national programs;
- As part of the international network of base collections organized by the International Board for Plant Genetic Resources (IBPGR) [24];
- By the Centers' own breeders.

The collections grew in opportunistic "fits and starts" without resources or tools to systematically analyze their structure and coverage vis-à-vis what exists in situ, or in other collections around the world. Consequently, it can be a challenge for genebank managers to be certain that newly acquired materials are duplicates of materials they already have, or the same as materials that are conserved and made internationally available by other genebanks, or that they are truly unique. Characterization, documentation, and some cross-referencing are key to resolving this issue, but still duplicates or near duplicates are likely to be abundant within and among genebanks In recent years, the Centers' genebanks have started to take advantage of modern molecular tools such as genotyping to characterize the genetic structure of the collections and to identify genetic differences among and within accessions, for example, potato [25–27], sweet potato [28,29], cassava [30], forage grasses [31–33], Mexican wheat landraces [34], and African rice [18] IRRI generated whole genome sequences of three thousand accessions in its collection [35]. While still constrained by resource limitations, molecular-level characterization has unprecedented potential to identify redundancies not only within collections, but also across collections, both inside and outside the CGIAR along with identifying potential misclassifications, introgressions, levels of domestication, genetic origins, and putative hybrids. However, it is important to note that even with these data, it is often difficult to determine an adequate cut-off threshold for calling a material a duplicate or too genetically similar for incorporation into the collections. The cost of generating raw sequence data has dropped precipitously, but the expertise and computing power necessary to analyze the data are still expensive and time-consuming, especially when dealing with whole genomes instead of reduced-representation sequencing approaches for genotyping. There are also other biological complications to molecular data alone for resolving duplication issues. However, the CGIAR Genebank Platform has been piloting training programs with scientists from national agricultural research organizations to use genotypic information to analyze within and among accession genetic diversity for a range of crops.

All CGIAR genebanks now have acquisition policies in place and current processes are based on a more critical assessment of whether new materials add diversity to the collections or respond to a specific need or mandate. This is particularly important for clonal collections where, resources permitting, the CGIAR genebanks can use genotyping to help confirm if a sample of newly acquired material is unique (to the collection) before undertaking expensive procedures to test, clean, and reproduce the materials for introduction into the collection as new accessions [31]. Centers working with clonal crops have also developed tools to use genetic sequence information to test for the presence of viruses and bacteria, overcoming costly delays for Centers acquiring and distributing clonal PGRFA [36–38].

Under the framework of the CGIAR Genebank Platform, the Centers and the Crop Trust have developed three methods for analyzing gaps in the coverage of their own collections. First, so-called "diversity trees" have been constructed for 22 crops. The trees are developed using published literature and expert knowledge to categorize the diversity held in each crop gene pool into known variety or genotype groups or wild species, which allows the mapping of accessions into the groups and the quantitative representation (or not) of the gene pool by the collection [32,39,40]. Second, spatial analyses have been undertaken using a method to assess the ecogeographic gaps and coverage of current CGIAR crop collections. The method, which works best for collections with a high percentage of available information on the latitude–longitude of the origin of accessions, looks for relationships between geographic patterns in crop distribution with the genetic structuring, and uses these relationships to build distribution models for crop landraces [41]. Third, a method for trait-based gap analyses focuses on the analysis of the distribution of adaptive priority traits in relation to the environment using machine learning to make predictions; it works best where landraces have been associated with an environment for long enough for their traits to become associated with their environment, and presupposes well-characterized collections. Figure 6 includes a preliminarily indication of the coverage of cultivated gene pools in the CGIAR genebank collections through the use of these tools and illustrates clearly that the cultivated diversity of some crops is well represented while others are considerably less.

In addition to the three strategies listed above, some genebanks are utilizing collection-wide genotyping in order to make key decisions on any new acquisitions. This strategy is only effective for genebanks that have collected genotyping data on their entire ex situ collection with a particular marker system. After which, any new material being considered for acquisition can be genotyped with the same marker system (GBS, SNP, DArTseq, etc.) to aid in decision making. The resulting fingerprints of material being considered for acquisition are subsequently compared to the entire germplasm collection to gain genetic insights. Phylogenetic results and genetic distance measures produced from the genotyping data can clearly show which samples are unique and which are redundant to the existing germplasm collection. Once these data are available, decisions can be made on which material to introduce into the genebank, usually under the framework of maximizing genetic diversity and not introducing material that is genetically similar. This is especially a useful strategy in clonal genebanks in which introduction and virus cleaning are expensive and time-consuming.

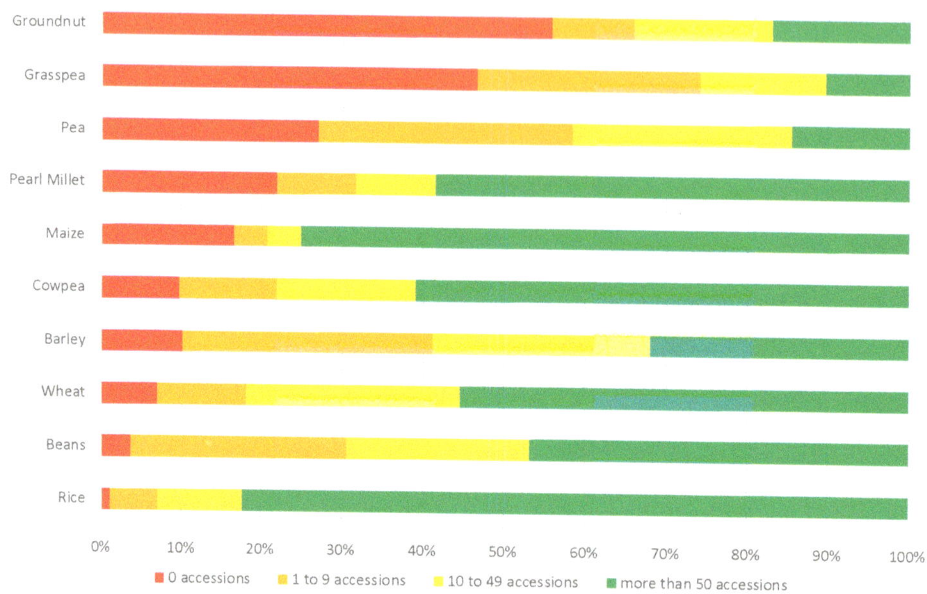

Figure 6. Preliminary assessment of the coverage of traditional landraces in international collections managed by CGIAR Centers. Note: these findings were generated by mapping accessions onto diversity trees. The colors correspond to numbers of accessions representing identified landrace end-groups or varieties making up the crop gene pool. This figure is reproduced from the 2019 Annual Report of the CGIAR Genebank Platform [40].

The Genebank Platform has recently initiated communications with national agricultural research organizations and national Plant Treaty Focal Points in 20 countries to use combinations of these tools as part of an effort to identify complementary holdings and gaps in collections, to identify priority areas for collecting (with the understanding that collected materials would be made available through the Plant Treaty's multilateral system). At the same time, they undertook to work together to identify potentially useful materials from the international collections to test in the countries concerned. While COVID-19 has had an effect on the preparations for this work, the ambition remains to support the collecting of PGRFA in target countries where there are significant gaps in ex situ conservation.

The CGIAR Centers do not have the objective to host complete collections of the diversity of their mandate crops. Rather, they seek to coordinate efforts with other collections to ensure the broadest possible coverage, long-term conservation, and availability of materials, in the global system as a whole. The Centers will pursue this approach, and the use of the tools and methods described above, in the process of defining collecting priorities and revising the current set of crop conservation strategies (available online at https://www.croptrust.org/resources/#ex-situ-conservation-strategies). The Centers (and the Crop Trust) do not organize collecting expeditions on their own. They work in collaboration with NARS partners who take responsibility for conducting the collecting, obtaining requisite permissions in compliance with national laws and regulations, and first depositing samples in their own national genebanks before transferring them to the CGIAR genebanks.

4.2. Phytosanitary Standards and Initiatives

Maintaining the health of the germplasm in the international collections is critically important to ensure against the international spread of quarantine pests and diseases. In 2019, the 11 germplasm health units (GHUs) of the 10 Centers hosting PGRFA collections collectively facilitated 2004 exchanges of materials with 141 countries. Altogether 152,469 samples of 105,961 accessions were analyzed by performing 594,909 diagnostics tests, and 13,248 samples were rejected to be replaced with healthy materials from other batches or to be subject to phytosanitary treatments to eliminate the pests concerned [40]. Most of the pests detected are target-specific species. Viral pathogens are most frequently intercepted in legume seeds and in clonal crops. Centers only distribute germplasm that is free of quarantine organisms. Untested and unclean materials are not distributed until establishment of the pest-free stocks. As of 2019, nearly 80% of the germplasm conserved in CGIAR genebanks had been tested for quarantine pests and about 78% of the collection was available for immediate distribution.

The International Plant Protection Convention (IPPC) is essential for promoting and harmonizing countries' phytosanitary regulations, and for facilitating international movement of plant genetic resources that are free of quarantine pests and diseases. However, the system still can and should be further tailored to address the particularities of the international movement and uses of plant genetic resources. Under the framework of the IPPC, contracting parties occasionally adopt International Standards for Phytosanitary Measures (ISPMs) that are tailored for different subject matters that could be a potential carrier of quarantine pests and diseases over international borders (e.g., farm equipment, commercial seed). To date, approximately 42 ISPMs have been adopted. Unfortunately, no ISPM for the regulation of international movement of plant germplasm from genebanks has been developed. The ISPM-36 on Integrated Measures for Plants for Planting and the ISPM-38 on International Movement of Seeds partially address some of the issues, but insufficiently. As a result, the national plant quarantine facilities of many countries either develop and follow their own norms or follow those prescribed through ISPMs dealing with commercial seed (or plantlets), which are not appropriate for plant germplasm shipped to and from genebanks. The commercial seed ISPMs—which anticipate tons of seed in a single shipment—prescribe testing protocols that deplete most of the very small quantities of seed transferred to and from genebanks, or create long delays which are unnecessary to address risks associated with genebank germplasm. These delays can result in materials dying in transit before they arrive at their intended destinations, or they arrive so late that they miss an entire planting season, thereby contributing to delays or cessation of planned research and or plant breeding activities. While phytosanitary issues are most challenging for recalcitrant seed or clonal crops, the genebanks report similar challenges with seed crops.

To address this situation, CGIAR germplasm health experts are working with partners from other organizations to develop a draft protocol for a comprehensive phytosanitary compliance assurance procedure that will demonstrate the best procedures in use for germplasm production and health assurance, while maintaining transparency in risk assessment and mitigation strategies to get NPPO accreditation as trusted to fast-track germplasm distribution. The initiative is referred to as the CGIAR GreenPass Phytosanitary Protocol (GreenPass). If the concept is endorsed at the level of the IPPC's Commission on Phytosanitary Measures, NPPOs could justify eliminating redundant checks by cutting some steps or reducing the processing time for material from GreenPass-accredited facilities. (These issues are addressed in more detail by Kumar et al. in this Special Edition of *Plants*.)

4.3. Genebank and Plant Genetic Resources Valuation

During the late 1990s, the genebanks received mounting criticism that materials stored in genebanks were rarely used. This was a concern because necessary investments in genebanks are difficult to obtain if potential investors do not appreciate the value of crop diversity, including the multitude of services and benefits it can provide. Several studies followed in the early 2000s which contradicted the viewpoint that genebanks were "unused" [42,43] and a large body of research documented the high rates of return from the genetic improvement of crops for yield, yield stability, quality, nutritional composition, resource use efficiency, and resistance to pests and diseases [44–49]. Most of the economic benefits have been generated from farm productivity gains which can be attributed to research and breeding programs by publicly funded institutions, such as the CGIAR, and society and consumers have especially benefited from lower food prices. Since then, however, the research on the value of international collections and the genetic diversity held in the CGIAR genebanks has not kept pace.

Recently, the CGIAR Genebank Platform and the Crop Trust supported the establishment of a Community of Practice on Genebank Impacts to revive the interest in applied economics research in this area. The work resulted in several papers which have made important contributions to earlier research on genebank valuation [50]. For example, Bernal-Galeano et al. (2020) [51] estimated a gross benefit from the "Victoria" potato variety in Uganda at USD 1.04 billion dollars, which exceeds the annual operational cost of the CIP genebank several times over. Villanueva et al. (2020) [52] estimated that a 10% increase in the genetic contribution of IRRI genebank accessions to an improved rice variety grown by rice farmers in East India is associated with a yield increase of 27%. Aberkane et al. (2020) [53], Sellitti et al. (2020) [54], and Kitonga et al. (2020) [55] focused on the use of wild, semi-natural, and landrace genetic materials in enhancing crop diversity options for breeders and farmers. The study by Ocampo-Giraldo et al. (2020) [56] highlighted the importance of combining ex situ and in situ approaches in a dynamic model of conservation. Alexandra et al. (2020) [57] narrated the formation of Pacific Community's Centre for Pacific Crops and Trees (CePaCT) in Fiji and underscored the role of a global effort to collect, conserve, and breed taro in response to disease outbreaks.

This current set of studies attests to the value of genebanks in at least two ways. First, they contribute to a better understanding of the role, function, and value of genebanks, in the light of climate change and evolving food security challenges. Several authors were able to trace the ancestry of varieties currently adopted by farmers to specific accessions in the genebanks and apportion economic gains by drawing from information on pedigrees. Second, the studies highlight the importance of long-term conservation and safety duplication of unique and diverse genetic materials, for the potential unknown use of future generations. As with Gollin (2020) [58], the findings support the need to refocus global conservation strategies on the efficient management of genetic resources, including on acquisition and conservation priorities.

4.4. Money, Politics, and Law

The findings presented above concerning the influence of political tensions, restrictive policies, and legal uncertainty are similar to the conclusion of the previous study [1] of factors affecting Centers' acquisitions between 1980 and 2009.

On the positive side, there is evidence that the Plant Treaty's multilateral system of access and benefit sharing is contributing positively to the willingness of many countries, national genebanks, and other providers to make PGRFA available and to safety-duplicate material in the CGIAR Center-hosted international collections. Perhaps the most significant evidence (which is so obvious it is often overlooked) is that all of the material in the international collections ultimately came from countries and, to date, 146 countries and the EU have voluntarily ratified the Plant Treaty, which invites the Centers to manage those collections under the Plant Treaty framework, and to make those materials available under the SMTA. As of November 2019, at least 135,000 accessions maintained in CGIAR genebanks (approximately 17% of the collections) were originally obtained from countries that were not Plant Treaty contracting parties including, but not limited to, the following countries: Mexico, China, Nigeria, Colombia, Thailand, Russia, Vietnam, Azerbaijan, Republic of South Africa, Uzbekistan, and Kazakhstan. It is partly for this reason that the CGIAR Centers feel duty bound to continue making materials available under the SMTA to recipients in countries that are not Plant Treaty members. Some contracting parties, particularly those who have been criticized in recent years for not making more PGRFA directly available through the multilateral system, complain that there is insufficient recognition of the fact that much of the genetic resources that the CGIAR Centers' genebanks distribute through the multilateral system (with the exception of accessioned breeders materials) originally came from them. Information about the country sources of materials (provenance) held in the Center-hosted international collections is available through Genesys, and Centers have published papers highlighting the origins of materials in the collection [31,53]. However, more can and should be done—through more popular, less expert-oriented mechanisms—to celebrate countries' contributions to the international collections and the multilateral system. Indeed, during the Plant

Treaty Governing Body meeting in October 2019, CGIAR undertook to work with the Plant Treaty Secretariat to publicize this information more broadly.

More recent evidence of the positive influence of the Plant Treaty on providers' willingness to include materials in the international collections is the surge of materials received by the Centers' genebanks between 2010 and 2019. All of those materials were either provided under the SMTA or with permission for Centers to subsequently make those materials available under the SMTA. It seems likely that many of those providers would not have been willing (or permitted) to provide materials for inclusion in the international collections in the absence of the Plant Treaty, the multilateral system, and the SMTA. Indeed, some genebanks have been informed by national partners that while those partners can provide PGRFA of Annex 1 materials, they cannot provide PGRFA of non-Annex 1 materials, because their national rules only apply to materials formally included within the scope of the multilateral system.

The fact that the Centers were able to acquire materials—particularly materials through new collecting expeditions—is evidence that other regulatory frameworks, apart from the Plant Treaty's multilateral system, are also contributing to the willingness/ability of providers to provide access to those materials. A lot of in situ PGRFA are not automatically included in the multilateral system, so collection must be approved by a competent authority subject to national law (or in the absence of a law, some other standard observed by the authority) to allow the material to be collected, deposited in the national genebank, and later sent under the SMTA to the CGIAR genebanks. (In this context, it is relevant to note that some countries have explicitly adopted the policy of not wanting to regulate access in this manner). The CGIAR genebanks never collect on their own. They work in partnership with national organizations who manage the interface with national authorities, logistics, and the actual collecting. Transfer within a country of material from one ABS system to another requires coordination and cooperation between the competent authorities and stakeholders.

It is certainly the case that the existence of the SMTA (because it is standard) made it possible for the Centers to process agreements to receive and manage those same materials. It would have been impossible for the Centers to negotiate unique transfer agreements with sui generis benefit sharing, dispute resolution, scope of use, and other conditions in each case, and then to put systems in place to manage materials under a plethora of different conditions.

It is important to consider the significance of the fact that most of the materials made available to the genebanks between 2010 and 2019 were associated with the Regeneration and CWR projects. First, it highlights the fact that conserving and providing PGRFA require financial and technical support and coordination, and in the absence of such support, many potential actors in the global system are unable to play their anticipated roles including securing the unique materials in their collections through safety duplication. In this context, it is important to emphasize that the funds and/or other non-monetary benefits provided to project partners in the Regeneration and CWR projects were modest, designed to subsidize/cover the costs of regeneration and or collecting, health testing and shipping, and in some cases training to do those things and for the Centers to receive the materials. Extra funding is needed to support collecting new PGRFA. Those costs, transaction costs in particular, would decrease if countries had well-defined systems for processing requests for germplasm, and clear signals from competent authorities that such requests were acceptable to consider. It is interesting to note that the USDA has a dedicated Plant Exploration Program managed by a Plant Exchange Office that is responsible for organizing collecting expeditions, reaching out to national competent authorities on behalf of the genebanks included within the national system [3]. There is no such service or function within the CGIAR. Second, the Regeneration and CWR projects highlight the continued importance, and need for, scientifically informed priority setting and international coordination to generate a shared sense of purpose and to motivate the wide range of actors, spread across the world, who ended up being engaged in those projects. The rounds of projects supported by the Plant Treaty's Benefit-sharing Fund (BSF) have played a similar role, catalyzing interest and engagement of a broad range of actors operating from the levels of individual farms to international organization, and providing financial and technical support for project participants to, among other things, collect, multiply, health-test, and share PGRFA through the multilateral system. It would be interesting, but it is beyond the scope of this paper, to identify the range of materials included in the multilateral system in general as a result of the BSF projects.

This is not to say that all of these internationally coordinated projects worked perfectly, or were immune from criticisms, or that many contracting parties, organizations, and people involved in the conservation and use of PGRFA do not have alternative visions or priorities. The point is that while the Plant Treaty's multilateral system of access and benefit sharing provides an internationally sanctioned, clearly defined legal platform of exchanging plant genetic resources, it is clear that additional financial and technical support, and inspiring visions for dispersed actors to work together, are also necessary to take advantage of the multilateral system. Other actors (including potentially the Plant Treaty's own Governing Body, or, more realistically, a specialist group it establishes) could develop a vision and internationally coordinated program that similarly stimulate and support activities that take advantage of the multilateral system to access, improve, and share PGRFA, and safeguard threatened PGRFA in furtherance of Sustainable Development Goals (e.g., pooling and evaluating a genetically diverse range of PGRFA for traits adapted to local climate changes).

Meanwhile, the CGIAR Centers' genebanks continue to distribute hundreds of thousands of PGRFA samples under the Plant Treaty's SMTA. Indeed, in the decade between 2010 and 2019 (inclusive), the Centers' genebanks distributed 21% more PGRFA samples than they did over the previous decade. These numbers attest to the utility of the multilateral system and increased reliance upon it as a means of accessing genetic materials.

The Plant Treaty's positive impact on the availability of PGRFA more generally (i.e., without involvement of the Center's genebanks) is further evidenced by the fact that, since 2014, sixteen additional countries became members of the Plant Treaty. Among them, is the USA, with the result that approximately 500,000 additional PGRFA accessions are available through the multilateral system of access and benefit sharing. In addition, during the same period, one additional international organization, the International Center for Biosaline Agriculture (ICBA), agreed to make PGRFA available under the terms of and conditions of the Plant Treaty.

On the negative side, however, there is also evidence that a number of Plant Treaty contracting parties continue to be reluctant to implement the multilateral system. Overall, national level implementation of the multilateral system is still relatively low. This is perhaps most clearly manifested in the fact that only 44 out of 146 countries that are contracting parties have confirmed what PGRFA within their borders are actually included in the multilateral system. (see "Material Available in the Multilateral System" on the Plant Treaty website at http://www.fao.org/plant-treaty/areas-of-work/the-multilateral-system/collections/en/). This may not be a "black letter law" obligation under the Treaty framework, but it is a commonly acknowledged prerequisite for the multilateral system to practically function [59]. Furthermore, some of the countries that have provided notice have confirmed only a fraction of the collections maintained in their national genebanks or other national public organizations. There are even fewer notices to the Governing Body concerning PGRFA that are voluntarily included by natural and legal persons. Few countries have reported to the Governing Body that they have adopted new policies or administrative measures to implement the multilateral system. On one hand, it is not required by the Plant Treaty that a country develop new policies or laws to implement the multilateral system; on the other hand, many countries report that in the absence of new policy instruments approved by the national government that clarify who has the right to provide materials under the multilateral system, they are unable to do so [59]. In such cases, the absence of new policy measures does indeed reflect a lack of national level implementation.

The relatively slow rate of national level implementation by a number of countries can be partially accounted for by the fact that developing country contracting parties are dissatisfied by the fact that, to date, there has been only one payment to the Plant Treaty's Benefit-Sharing Fund by a commercial user of materials from the multilateral system. This dissatisfaction contributed to the Fifth Session of the

Governing Body in 2013 launching a process to enhance the functioning of the multilateral system. That process continued until 2019, when it was suspended, with no new agreement reached, and high levels of unresolved political tension between contracting parties.

International tension and disagreement concerning access and benefit-sharing issues in other fora have also spilled over into meetings under the Plant Treaty framework. The Nagoya Protocol came into force in 2010. It was designed to address (mostly developing) countries' concerns that the Convention on Biological Diversity (CBD) did not sufficiently promote its benefit-sharing objective. However, by 2013, the Conference of the Parties to the CBD in Mexico was overtaken by tensions over benefit sharing from commercial use of DSI, which is not addressed by the Nagoya Protocol (or at least not in a way that the international community agrees upon). Since then, the issue has dominated the agendas of the CBD, Plant Treaty, FAO CGRFA, and others. Under these circumstances, some parties who are not content with levels of monetary benefit sharing from the use of PGRFA and/or genomic sequence information may be reluctant to make more materials available through the multilateral system by depositing them in the international collections hosted by the CGIAR Centers until there is some resolution to the DSI and ABS issue.

The previous study [1] of factors affecting the Centers' germplasm acquisitions from 1980 to 2009 adopted an implicitly millennial framework, looking forward to a new era when international tensions over access and benefit sharing were resolved, and national laws implementing international agreements provided requisite legal certainty for PGRFA providers and recipients alike. In 2010, it seemed reasonable to expect that this state of affairs could be achieved within a few years. Since that time, tensions over access and benefit sharing have increased, and the possibility of arriving at a set of final international agreement(s) on the issue appears to have receded still further into the distance.

5. Conclusions

Over the course of the decade 2010–2019, the CGIAR Center genebanks received a surge of PGRFA from providers around the world, with permission to make those materials available through the Plant Treaty's multilateral system of access and benefit sharing. Most of those newly deposited materials were associated with an internationally coordinated project that provided financial and technical support for the providers to regenerate and safety-duplicate unique, at-risk PGRFA in ex situ collections around the world.

The Regeneration and CWR projects, the BSF project cycles, and the CGIAR genebanks' experiences over the last ten years highlight the critical importance of internationally coordinated projects to motivate and instill otherwise dispersed and disconnected actors with a sufficiently shared sense of purpose, to make materials available through the multilateral system. Those projects and experiences also underscore the importance of providing modest levels of financial support to cover both the providers' and the recipient genebanks' costs associated with collecting, multiplying, health-testing, sending, and receiving PGRFA, and of providing technical support/training to providers/partners.

The number of collecting expeditions organized by CGIAR Centers in this period was higher than in the previous decade, but lower than some national genebanks which play an important role in the global system of conservation, sustainable use, and exchange of plant genetic resources. Increasingly sophisticated and globally coordinated gap analyses are making it possible to identify where gaps in collections exist. Gap analyses presented in this paper highlight that ex situ collections managed by the CGIAR represent well the cultivated gene pool of some crops, while there are significant gaps for other crops. Given the need to keep costs to a minimum, and the possibility of sharing responsibilities with other global system actors, CGIAR Centers must redouble their efforts to address the results of gap analyses in concert with other organizations that host internationally available collections.

While financial support, scientific leadership, and international coordination are indispensable, so too are supportive policies. The CGIAR genebanks' experiences highlight the importance of supportive policies, and conversely, the negative impacts of restrictive or unclear policies and laws, on their ability to acquire new materials to include in the international collections, and to distribute those materials to recipients around the world. In retrospect, it is perhaps surprising that over the same period of time, the international community launched and attempted to renegotiate the multilateral system of access and benefit sharing—with all the uncertainty that attends an international negotiation—and the Centers' genebanks enjoyed a surge of new materials being made available to include in the international collections. During the same period of time, they continued to distribute an extraordinary diversity of PGRFA to recipients around the world (even more than in the previous decade), a fact which reflects the persistent need/demand for access to those materials for agricultural research and development, the deep rooted nature of the CGIAR collections within the global system, and the positive influence of the Centers' Article 15 agreements with the Governing Body of the Plant Treaty.

However, with the suspension of the process to enhance the multilateral system, and widespread tensions concerning the governance of digital genomic sequence information, international disagreement over access and benefit sharing is becoming still more geopolitically polarized. There is a significant risk that this increased polarization could further undermine the willingness of a range of actors to make materials available through the multilateral system in general, and to the CGIAR genebanks in particular. The Centers highlight the importance of resolving those tensions to "head off" unintended potential negative impacts on the CGIAR's mission, the global system, and the SDGs. Rapid loss of biodiversity, climate change, the COVID-19 pandemic, rising populations, depleted soils, and a range of other challenges make the conservation, availability, and use of PGRFA more important than ever. It is essential that the Plant Treaty (and Nagoya Protocol and IPPC) is implemented in ways that support all actors in the global system to fulfill their roles.

Author Contributions: Conceptualization, M.H., N.J., I.L.N.; methodology, M.H., N.J., I.L.N.; validation, M.H., N.J., I.L.N., C.L., N.L.A., P.W., T.P., M.-N.N., L.G., P.L.K., M.Y., A.M., V.A., M.T., C.S.J., R.V., N.R., E.R.; formal analysis, M.H., N.J., I.L.N., C.L., N.L.A., P.W., T.P., M.-N.N., L.G., P.L.K., M.Y., A.M., V.A., M.T., C.S.J., R.V., N.R.; investigation, M.H., N.J., I.L.N.; data curation, N.J., N.L.A., P.W., T.P., M.-N.N., M.Y., A.M., V.A., M.T., C.S.J., R.V., N.R., E.R.; writing—original draft preparation, M.H., N.L., I.L.N.; writing—review and editing, M.H., N.J., I.L.N., C.L., T.P., N.L.A., P.W., L.G., P.L.K., A.M., C.S.J., M.-N.N., R.V.; visualization, M.H., N.J., I.L.N.; supervision, M.H., N.J.; project administration, M.H., N.J.; funding acquisition, M.H., I.L.N. All authors have read and agreed to the published version of the manuscript.

Funding: This research was funded by CGIAR Genebank Platform.

Acknowledgments: The authors thank Ana Bedmar and Gaia Gullotta for their valuable contributions to literature surveys and data compilation and analysis.

Conflicts of Interest: N.L.A., P.W., T.P., M.-N.N., M.Y., A.M., V.A., M.T., C.S.J., R.V., N.R., P.L.K., E.R. work in the management of CGIAR Centers' genebanks (and/or germplasm health units) that are the subject of this paper. N.J., C.L., L.G. work at the Crop Trust. M.H., I.L.N. are co-coordinators of the Genebank Platform Policy Module.

Appendix A

Table A1. Survey on Acquisitions and Distributions.

Q1. Center
Q2. Name of respondent (You may have more than one respondent filling in the same survey, but please fill in just one survey per Center)
Q3. Over the course of the last 10 years (2010–2019), the overall rate of acquisition of new PGRFA by your genebank from sources outside the CGIAR has...?
Answer Choices Declined Increased remained steady

Table A1. *Cont.*

Q4. Which of the following factors have affected the your genebanks' overall rate of acquisition of PGRFA from sources outside the CGIAR?
Answer Choices most relevant diversity is already collected no space left in the genebank for additional accessions no money for additional collecting missions no capacity/money to evaluate or characterize what is already in the genebank, so there is no point adding more restrictive or unclear policies or laws make it difficult to get permission to access new PGRFA unwillingness of countries to allow collecting missions or to share their ex situ materials provider's country cannot issue a certificate in compliance that satisfies the phytosanitary rules of the country hosting the my genebank other (please specify)
Q5. Please add additional information to further clarify your answer from above. If you chose more than one factor, please rank them in terms of their importance.
Q6. In the last 10 years (2010 - 2019), the overall rate of distribution of PGRFA from your genebank has...?
Answer Choices Declined Increased remained steady
Q7. Which of the following factors have affected the rate of distributions from your genebank?
Answer Choices number of requests for materials a general tendency towards more targeted requests (i.e., requestors are more specific about the range of materials they are seeking, so you end up sending less material per request) your responses to requests are more targeted (i.e., you spend more time in the past, with more information about your collection, determining which particular materials are best suited to the needs of the requestor, so you end up sending less material per request) you do not have sufficient resources to regenerate enough material to send materials in response to all requests requested materials do not meet requisite phytosanitary standards restrictive policies or laws or conditions make it difficult for the genebank to distribute PGRFA status of information available about materials in the genebank (e.g., characterization, evaluation, subsets, etc.) other (please specify)
Q8. Please add additional information to further clarify your answer from above. If you chose more than one factor, please rank them in terms of their impact (either positive or negative) on your distributions.
Q9. Are you sometimes asked for PGRFA that you do not have?
Answer Choices Yes No If 'yes', please provide a brief description of the materials concerned and why they are being asked for.
Q10. Please describe the circumstances that you believe contributed to particularly steep spikes or dips in your genebank's rate of acquisition of PGRFA in any year or years between 2010–2019 (e.g., internationally coordinated projects, organizations looking to transfer collections, joint projects with national research organizations, etc.)
Q11. Please describe the circumstances that you believe contributed to particularly steep spikes or dips in your genebank's rate of distribution of PGRFA in any year or years between 2010–2019 (e.g., sources of demand linked to particular projects in particular countries, etc.)
Q12. How many requests for PGRFA to include in your collection(s) has your genebank made to organizations outside the CGIAR in the last 5 years? (for materials in in situ and ex situ conditions)
Answer Choices 0 1–3 4–6 7–9 other (please indicate number)

Table A1. *Cont.*

Q13. Regarding the requests you made described in question above
Answer Choices (a) how many were explicitly rejected? (b) how many were ignored (i.e., simply no answer)? (c) how many were accepted, but materials are not yet acquired by your Center? (d) how many were accepted and the materials have actually been acquired by your Center? if (a) to (d) do not describe what happened, please describe the outcome
Q14. Was there a difference in the kinds of responses you received depending on whether you were seeking to acquire material from in situ conditions (therefore involving new collecting missions) or materials that were already in ex situ collections? Please explain.
Q15. Was there a difference in the kinds of responses you received depending on the types of organizations to whom you addressed your request? Please explain.
Q16. Would you have preferred to make more requests to acquire more PGRFA over the last 5 years?
Answer Choices Yes No If 'yes', why didn't you make more requests?
Q17. The Plant Treaty has:
Answer Choices made it much harder for my genebank to acquire PGRFA made it a little harder for my genebank to acquire PGRFA has not had any appreciable impact on my genebank's ability to acquire PGRFA has made it a little easier for my genebank to acquire PGRFA has made it much easier for my genebank to acquire PGRFA Briefly explain your response.
Q18. The Nagoya Protocol (to the Convention on Biological Diversity) has:
Answer Choices made it much harder for my genebank to acquire PGRFA made it a little harder for my genebank to acquire PGRFA has not had any appreciable impact on my genebank's ability to acquire PGRFA has made it a little easier for my genebank to acquire PGRFA has made it much easier for my genebank to acquire PGRFA Briefly explain your response.
Q19. The IPPC and national phytosanitary rules have:
Answer Choices made it much harder for my genebank to acquire PGRFA made it a little harder for my genebank to acquire PGRFA has not had any appreciable impact on my genebank's ability to acquire PGRFA has made it a little easier for my genebank to acquire PGRFA has made it much easier for my genebank to acquire PGRFA Briefly explain your response.
Q20. Are you concerned that unresolved international negotiations concerning digital genomic sequence information (DSI), and/or the suspension of negotiations to enhance the Plant Treaty's multilateral system of access and benefit-sharing, could, in the future, have a negative impact on the following:
CGIAR Centers' genebanks ability to access to PGRFA to include in international collections? CGIAR Centers' ability to access, generate, use and share digital genomic sequence information (DSI)? Centers' management of their Article 15 collections (including distributing PGRFA and related information)? Willingness of some organizations to enter into partnership with CGIAR Centers? If you answered 'yes' to any of the above, please explain
Q21. If you already have evidence of negative impacts as a result of the unresolved issues regarding DSI and the multilateral system enhancement, please provide details here.

References

1. Halewood, M.; Sood, R.; Hamilton, R.S.; Amri, A.; Van den Houwe, I.; Roux, N.; Tay, D. Changing rates of acquisition of plant genetic resources by international gene banks: Setting the scene to monitor an impact of the International Treaty. In *Crop Genetic Resources as a Global Commons—Challenges in International Law and Governance*; Halewood, S., Noriega, M.L., Louafi, I., Eds.; Earthscan: London, UK, 2012.
2. CWR Project. *Annual Report*; Global Crop Diversity Trust: Bonn, Germany, 2019.
3. Williams, K.A. An overview of the U.S. National Plant Germplasm System's Exploration Program. *HortScience* **2005**, *40*, 297–301. [CrossRef]
4. Brink, M.; van Hintum, T. Genebank Operation in the Arena of Access and Benefit-Sharing Policies. *Front. Plant Sci.* **2020**, *10*, 1712. [CrossRef] [PubMed]
5. Williams, K.A. *NPGS Plant Explorations and Exchanges (Presentation)*; Plant Germplasm Operating Committee: Sturgeon Bay, WI, USA, 2009.
6. Hellier, B.C. Collecting in Central Asia and the Caucasus: U.S. National plant germplasm system plant explorations. *HortScience* **2011**, *46*, 1438–1439. [CrossRef]
7. Postman, J.; Stover, E.; Aradhya, M.; Meyer, P.; Williams, K. Recent NPGS coordinated expeditions in the trans-caucasus region to collect wild relatives of temperate fruit and nut crops. *Acta Hortic.* **2012**, *948*, 191–198. [CrossRef]
8. National Germplasm Resources Laboratory. *2014 Report of the Plant Exchange Office to the Plant Germplasm Operations Committee, the Regional Technical Advisory Committees, and the Crop Germplasm Committees*; National Germplasm Resources Laboratory: Beltsville, MD, USA, January 2015. Available online: https://www.nimss.org/system/ProjectAttachment/files/000/000/053/original/2014%20Beltsville%20NGRL.pdf (accessed on 30 September 2020).
9. National Germplasm Resources Laboratory. *Summary Annual Reports of the Project Acquisition of Plant Genetic Resources through Domestic and International Plant Explorations and Associated Capacity-Building Partnerships*; USDA: Beltsville, MD, USA, 2018.
10. Weise, S. *Accessions conserved and available at the Leibniz Institute of Plant Genetics and Crop Plant Research (IPK) originating at collecting expeditions organized by IPK, or with IPK participation from 2000 to 2018*; Data facilitated through personal communication; IPK: Gatersleben, Germany, May 2020.
11. CGN. *CGN Collecting Missions from 1955 to 2019, 2020.* Available online: https://missions.cgn.wur.nl/ (accessed on 1 July 2020).
12. Kik, C. *The Dutch National Genebank CGN: A General Introduction*; NCARE: Baq'a, Jordan, 9 September 2017.
13. Westengen, O.T.; Skarbø, K.; Mulesa, T.H.; Berg, T. Access to genes: Linkages between genebanks and farmers' seed systems. *Food Secur.* **2018**, *10*, 9–25. [CrossRef]
14. Vasudevan, K.; Cruz, C.M.V.; Gruissem, W.; Bhullar, N.K. Large scale germplasm screening for identification of novel rice blast resistance sources. *Front. Plant Sci.* **2014**, *5*. [CrossRef]
15. Draz, I.S.; Abou-Elseoud, M.S.; Kamara, A.-E.M.; Alaa-Eldein, O.A.-E.; El-Bebany, A.F. Screening of wheat genotypes for leaf rust resistance along with grain yield. *Ann. Agric. Sci.* **2015**, *60*, 29–39. [CrossRef]
16. Halewood, M.; Chiurugwi, T.; Hamilton, R.S.; Kurtz, B.; Marden, E.; Welch, E.W.; Michiels, F.; Mozafari, J.; Sabran, M.; Patron, N.J.; et al. Plant genetic resources for food and agriculture: Opportunities and challenges emerging from the science and information technology revolution. *New Phytol.* **2018**, *217*, 1407–1419. [CrossRef]
17. Tanksley, S.D.; McCouch, S.R. Seed banks and molecular maps: Unlocking genetic potential from the wild. *Science* **1997**, *277*, 1063–1066. [CrossRef]
18. Ndjiondjop, M.N.; Semagn, K.; Gouda, A.C.; Kpeki, S.B.; Tia, D.D.; Sow, M.; Goungoulou, A.; Sie, M.; Perrier, X.; Ghesquiere, A.; et al. Genetic variation and population structure of *Oryza glaberrima* and development of a mini-core collection using DArTseq. *Front. Plant Sci.* **2017**, *8*. [CrossRef]
19. Yan, W.; Rutger, J.N.; Bryant, R.J.; Bockelman, H.E.; Fjellstrom, R.; Chen, M.-H.; Tai, T.H.; McClung, A.M. Development and Evaluation of a Core Subset of the USDA Rice Germplasm Collection. *Crop. Sci.* **2007**, *47*, 869–876. [CrossRef]
20. Haupt, M.; Schmid, K.J. Combining focused identification of germplasm and core collection strategies to identify genebank accessions for central European soybean breeding. *Plant Cell Environ.* **2020**, *43*, 1421–1436. [CrossRef] [PubMed]

21. Azough, Z.; Kehel, Z.; Benomar, A.; Bellafkih, M.; Amri, A. Predictive Characterization of ICARDA Genebank Barley Accessions Using FIGS and Machine Learning. *Intell. Environ.* **2019**, 121–129. [CrossRef]
22. Khazaei, H.; Street, K.; Bari, A.; Mackay, M.; Stoddard, F.L. The FIGS (focused identification of germplasm strategy) approach identifies traits related to drought adaptation in Vicia faba genetic resources. *PLoS ONE* **2013**, *8*, e63107. [CrossRef] [PubMed]
23. FAO. *Opinions and Advice of the Ad Hoc Technical Advisory Committee on the Multilateral System and the Standard Material Transfer Agreement*; FAO: Rome, Italy, 2015.
24. Thormann, I.; Engels, J.M.M.; Halewood, M. Are the old International Board for Plant Genetic Resources (IBPGR) base collections available through the Plant Treaty's multilateral system of access and benefit sharing? A review. *Genet. Resour. Crop. Evol.* **2018**, *66*, 291–310. [CrossRef]
25. Gutaker, R.; Weiß, C.L.; Ellis, D.; Anglin, N.L.; Knapp, S.; Fernández-Alonso, J.L.; Prat, S.; Burbano, H. The origins and adaptation of European potatoes reconstructed from historical genomes. *Nat. Ecol. Evol.* **2019**, *3*, 1093–1101. [CrossRef]
26. Ellis, D.; Chávez, O.; Coombs, J.; Soto, J.V.; Gomez, R.; Douches, D.S.; Panta, A.; Silvestre, R.; Anglin, N.L. Genetic identity in genebanks: Application of the SolCAP 12K SNP array in fingerprinting and diversity analysis in the global in trust potato collection. *Genome* **2018**, *61*, 523–537. [CrossRef] [PubMed]
27. Kyriakidou, M.; Achakkagari, S.R.; López, J.H.G.; Zhu, X.; Tang, C.Y.; Tai, H.H.; Anglin, N.L.; Ellis, D.; Strömvik, M.V. Structural genome analysis in cultivated potato taxa. *Theor. Appl. Genet.* **2019**, *133*, 951–966. [CrossRef]
28. Muñoz-Rodríguez, P.; Carruthers, T.; Wood, J.R.; Williams, B.R.; Weitemier, K.; Kronmiller, B.; Ellis, D.; Anglin, N.L.; Longway, L.; Harris, S.A.; et al. Reconciling Conflicting Phylogenies in the Origin of Sweet Potato and Dispersal to Polynesia. *Curr. Boil.* **2018**, *28*, 1246–1256.e12. [CrossRef]
29. Muñoz-Rodríguez, P.; Carruthers, T.; Wood, J.R.I.; Williams, B.R.M.; Weitemier, K.; Kronmiller, B.; Goodwin, Z.; Sumadijaya, A.; Anglin, N.L.; Filer, D.; et al. A taxonomic monograph of Ipomoea integrated across phylogenetic scales. *Nat. Plants* **2019**, *5*, 1136–1144. [CrossRef]
30. Ferguson, M.E.; Shah, T.; Kulakow, P.; Ceballos, H. A global overview of cassava genetic diversity. *PLoS ONE* **2019**, *14*, e0224763. [CrossRef] [PubMed]
31. Negawo, A.T.; Jorge, A.; Hanson, J.; Teshome, A.; Muktar, M.S.; Azevedo, A.L.S.; Lédo, F.J.; Machado, J.C.; Jones, C. Molecular markers as a tool for germplasm acquisition to enhance the genetic diversity of a Napier grass (Cenchrus purpureus syn. Pennisetum purpureum) collection. *Trop. Grasslands-Forrajes Trop.* **2018**, *6*, 58–69. [CrossRef]
32. Muktar, M.S.; Teshome, A.; Hanson, J.; Negawo, A.T.; Habte, E.; Entfellner, J.-B.D.; Lee, K.-W.; Jones, C. Author Correction: Genotyping by sequencing provides new insights into the diversity of Napier grass (Cenchrus purpureus) and reveals variation in genome-wide LD patterns between collections. *Sci. Rep.* **2020**, *10*, 1–2. [CrossRef]
33. Negawo, A.T.; Assefa, Y.; Hanson, J.; Abdena, A.; Muktar, M.S.; Habte, E.; Sartie, A.M.; Jones, C. Genotyping-By-Sequencing Reveals Population Structure and Genetic Diversity of a Buffelgrass (*Cenchrus ciliaris* L.) Collection. *Diversity* **2020**, *12*, 88. [CrossRef]
34. Vikram, P.; Franco, J.; Burgueño, J.; Li, H.; Sehgal, D.; Pierre, C.S.; Ortiz, C.; Sneller, C.; Tattaris, M.; Guzman, C.G.; et al. Unlocking the genetic diversity of Creole wheats. *Sci. Rep.* **2016**, *6*, 23092. [CrossRef] [PubMed]
35. Wang, W.; Mauleon, R.; Hu, Z.; Chebotarov, D.; Tai, S.; Wu, Z.; Li, M.; Zheng, T.; Fuentes, R.R.; Zhang, F.; et al. Genomic variation in 3010 diverse accessions of Asian cultivated rice. *Nature* **2018**, *557*, 43–49. [CrossRef]
36. Villamor, D.E.; Ho, T.; Al Rwahnih, M.; Martin, R.R.; Tzanetakis, I. High Throughput Sequencing For Plant Virus Detection and Discovery. *Phytopatology* **2019**, *109*, 716–725. [CrossRef]
37. Kreuze, J.F.; Pérez, A.; Untiveros, M.; Quispe-Huamanquispe, D.G.; Fuentes, S.; Barker, I.; Simon, R. Complete viral genome sequence and discovery of novel viruses by deep sequencing of small RNAs: A generic method for diagnosis, discovery and sequencing of viruses. *Virology* **2009**, *388*, 1–7. [CrossRef]
38. Silva, G.; Bömer, M.; Nkere, C.; Kumar, P.L.; Seal, S.E. Rapid and specific detection of Yam mosaic virus by reverse-transcription recombinase polymerase amplification. *J. Virol. Methods* **2015**, *222*, 138–144. [CrossRef]
39. Van Treuren, R.; Engels, J.M.M.; Hoekstra, R.; Van Hintum, T. Optimization of the composition of crop collections for ex situ conservation. *Plant Genet. Resour.* **2009**, *7*, 185–193. [CrossRef]

40. CGIAR Genebank Platform. *2019 Genebank Platform Annual Report*; Global Crop Diversity Trust: Bonn, Germany, 2020.
41. Ramirez-Villegas, J.; Khoury, C.K.; Achicanoy, H.A.; Mendez, A.C.; Diaz, M.V.; Sosa, C.C.; Debouck, D.G.; Kehel, Z.; Guarino, L. A gap analysis modelling framework to prioritize collecting for ex situ conservation of crop landraces. *Divers. Distrib.* **2020**, *26*, 730–742.
42. Rubenstein, K.D.; Smale, M.; Widrlechner, M.P. Demand for genetic resources and the U.S. National Plant Germplasm System. *Crop. Sci.* **2006**, *46*, 1021–1031.
43. Garming, H.; Roux, N.; van den Houwe, I. *The Impact of the Musa International Transit Centre—Review of Its Services and Cost-Effectiveness, and Recommendations for Rationalization of Its Operations*; Bioversity International: Montpellier, France, 2010.
44. Crop Trust. *Securing Crop Diversity for Sustainable Development*; Global Crop Diversity Trust: Bonn, Germany, 2015.
45. REvenson, R.E.; Gollin, D. Genetic resources, international organizations, and improvement in rice varieties. *Econ. Dev. Cult. Chang.* **1997**, *45*, 471–500.
46. Gollin, D.; Smale, M.; Skovmand, B. Searching an ex situ collection of wheat genetic resources. *Am. J. Agric. Econ.* **2000**, *82*, 812–827.
47. Zohrabian, A.; Traxler, G.; Caudill, S.; Smale, M. Valuing pre-commercial genetic resources: A maximum entropy approach. *Am. J. Agric. Econ.* **2003**, *85*, 429–436.
48. Johnson, N.L.; Pachico, D.; Voysest, O. The distribution of benefits from public international germplasm banks: The case of beans in Latin America. *Agric. Econ.* **2003**, *29*, 277–286.
49. Koo, B.; Pardey, P.G.; Wright, B.D. *Saving Seeds: The Economics of Conserving Crop. Genetic Resources Ex Situ in the Future Harvest Centres of the CGIAR*; CABI: Wallingford, UK, 2004.
50. Smale, M.; Jamora, N. Valuing genebanks. *Food Secur.* **2020**. [CrossRef]
51. Bernal-Galeano, V.; Norton, G.; Ellis, D.; Anglin, N.L.; Hareau, G.; Smale, M.; Jamora, N.; Alwang, J.; Pradel, W. Andean potato diversity conserved in the International Potato Center genebank helps develop agriculture in Uganda: The example. *Food Secur.* **2020**, 1–15. [CrossRef]
52. Villanueva, D.; Smale, M.; Jamora, N.; Capilit, G.L.; Hamilton, R.S. The contribution of the International Rice Genebank to varietal improvement and crop productivity in Eastern India. *Food Secur.* **2020**, 1–15. [CrossRef]
53. Aberkane, H.; Payne, T.; Kishi, M.; Smale, M.; Amri, A.; Jamora, N. Transferring diversity of goat grass to farmers' fields through the development of synthetic hexaploid wheat. *Food Secur.* **2020**. [CrossRef]
54. Sellitti, S.; Vaiknoras, K.; Smale, M.; Jamora, N.; Andrade, R.; Wenzl, P.; Labarta, R. The contribution of the CIAT genebank to the development of iron-biofortified bean varieties and well-being of farm households in Rwanda. *Food Secur.* **2020**, 1–17. [CrossRef]
55. Kitonga, K.; Jamora, N.; Smale, M.; Muchugi, A. Use and benefits of tree germplasm from the World Agroforestry genebank for smallholder farmers in Kenya. *Food Secur.* **2020**, 1–11. [CrossRef]
56. Ocampo-Giraldo, V.; Camacho-Villa, C.; Costich, D.E.; Martínez, V.A.V.; Smale, M.; Jamora, N. Dynamic conservation of genetic resources: Rematriation of the maize landrace Jala. *Food Secur.* **2020**, 1–14. [CrossRef]
57. Alexandra, S.; Jamora, N.; Smale, M.; Ghanem, M.E. The tale of taro leaf blight: A global effort to safeguard the genetic diversity of taro in the Pacific. *Food Secur.* **2020**, 1–12. [CrossRef]
58. Gollin, D. Conserving genetic resources for agriculture: Economic implications of emerging science. *Food Secur.* **2020**, 1–9. [CrossRef]
59. Joint Capacity Building Programme. *Decision-Making Tool for National Implementation of the Plant. Treaty's Multilateral System of Access and Benefit-Sharing*; Bioversity International: Maccarese (Fiumicino) Rome, Italy, 2018.

plants

MDPI

Review

Phytosanitary Interventions for Safe Global Germplasm Exchange and the Prevention of Transboundary Pest Spread: The Role of CGIAR Germplasm Health Units

P. Lava Kumar [1,*], Maritza Cuervo [2], J. F. Kreuze [3], Giovanna Muller [3], Gururaj Kulkarni [4], Safaa G. Kumari [5], Sebastien Massart [6], Monica Mezzalama [7,†], Amos Alakonya [7], Alice Muchugi [8], Ignazio Graziosi [8,‡], Marie-Noelle Ndjiondjop [9], Rajan Sharma [10] and Alemayehu Teressa Negawo [11]

1. International Institute of Tropical Agriculture (IITA), Oyo Road, PMB 5320, Ibadan 200001, Nigeria
2. The Alliance of Bioversity International and International Center for Tropical Agriculture (CIAT), Palmira 763537, Cali, Colombia; m.cuervo@cgiar.org
3. International Potato Center (CIP), Avenida La Molina 1895, Lima 15023, Peru; j.kreuze@cgiar.org (J.F.K.); g.muller@cgiar.org (G.M.)
4. International Rice Research Institute (IRRI), Los Banos 4031, Philippines; g.kulkarni@irri.org
5. International Center for Agricultural Research in the Dry Areas (ICARDA), Terbol Station, Zahle 1801, Lebanon; s.kumari@cgiar.org
6. The Alliance of Bioversity International-CIAT & University of Liège, Gembloux Agro-BioTech, Passage des déportés, 2, 5030 Gembloux, Belgium; s.massart@cgiar.org
7. International Maize and Wheat Improvement Center (CIMMYT), México-Veracruz, El Batán Km. 45, Texcoco 56237, Mexico; monica.mezzalama@unito.it (M.M.); A.Alakonya@cgiar.org (A.A.)
8. World Agroforestry Center (ICRAF), United Nations Avenue, Gigiri P.O. Box 30677, Nairobi 00100, Kenya; a.muchugi@cgiar.org (A.M.); i.graziosi@hotmail.it (I.G.)
9. Africa Rice Center (AfricaRice), 01 BP 2551, Bouake 99326, Côte d'Ivoire; M.Ndjiondjop@cgiar.org
10. International Crops Research Institute for the Semi-Arid Tropics (ICRISAT), Patancheru 502324, Hyderabad, India; r.sharma@cgiar.org
11. International Livestock Research Institute (ILRI), Addis Ababa P.O. Box 5689, Ethiopia; a.teressa@cgiar.org

* Correspondence: L.kumar@cgiar.org
† Current address: AGROINNOVA, University of Turin, Italy. Largo Paolo Braccini 2, Grugliasco, Italy.
‡ Current address: Polo GGB, Terni, Italy.

Citation: Kumar, P.L.; Cuervo, M.; Kreuze, J.F.; Muller, G.; Kulkarni, G.; Kumari, S.G.; Massart, S.; Mezzalama, M.; Alakonya, A.; Muchugi, A.; et al. Phytosanitary Interventions for Safe Global Germplasm Exchange and the Prevention of Transboundary Pest Spread: The Role of CGIAR Germplasm Health Units. *Plants* **2021**, *10*, 328. https://doi.org/10.3390/plants10020328

Academic Editor: Andreas W. Ebert and Johannes M. M. Engels
Received: 29 November 2020
Accepted: 28 January 2021
Published: 9 February 2021

Abstract: The inherent ability of seeds (orthodox, intermediate, and recalcitrant seeds and vegetative propagules) to serve as carriers of pests and pathogens (hereafter referred to as pests) and the risk of transboundary spread along with the seed movement present a high-risk factor for international germplasm distribution activities. Quarantine and phytosanitary procedures have been established by many countries around the world to minimize seed-borne pest spread by screening export and import consignments of germplasm. The effectiveness of these time-consuming and cost-intensive procedures depends on the knowledge of pest distribution, availability of diagnostic tools for seed health testing, qualified operators, procedures for inspection, and seed phytosanitation. This review describes a unique multidisciplinary approach used by the CGIAR Germplasm Health Units (GHUs) in ensuring phytosanitary protection for the safe conservation and global movement of germplasm from the 11 CGIAR genebanks and breeding programs that acquire and distribute germplasm to and from all parts of the world for agricultural research and food security. We also present the challenges, lessons learned, and recommendations stemming from the experience of GHUs, which collaborate with the national quarantine systems to export and distribute about 100,000 germplasm samples annually to partners located in about 90 to 100 countries. Furthermore, we describe how GHUs adjust their procedures to stay in alignment with evolving phytosanitary regulations and pest risk scenarios. In conclusion, we state the benefits of globally coordinated phytosanitary networks for the prevention of the intercontinental spread of pests that are transmissible through plant propagation materials.

Keywords: CGIAR; crop genetic resources; diagnostics; germplasm; crop breeding; pathogen; pest; Plant Treaty; phytosanitary regulations; transboundary pests; invasive species; prevention; quarantine; seed; seed health; virus indexing

1. Introduction

1.1. International Germplasm Transfers for Food Security and Biodiversity Conservation

The international exchange of genetic resources, such as botanic seeds and vegetative propagules, has played a crucial role in agricultural and food diversification to an extent that about 68% of national food supplies are derived from crops with a foreign origin [1]. At the forefront of these international exchanges are the CGIAR genebanks, breeding and seed system programs that have made vital contributions for over five decades by assembling germplasm from all over the world for conservation, and adding value to those materials by characterizing, breeding, and making them available to users around the world [2,3]. (*Note: Germplasm used to denote plant propagation material, both true seed (orthodox, intermediate, and recalcitrant seeds) and vegetative propagules, from genebanks and breeding programs.*) Established in 1971, the CGIAR is part of the global agricultural research system, which makes critically important contributions to the United Nations Sustainable Development Goals (SDGs) in alleviating poverty and hunger and improving food and nutrition security and in the conservation of biodiversity [4].

The 11 CGIAR genebanks conserve over 760,467 accessions of cereals, grain legumes, forages, tree species, root and tuber crops, and bananas. These represent >174 genera and over 1000 species obtained from 207 countries, which are conserved in 35 collections around the world as seeds, in vitro material, and living plants in fields or screenhouses (Supplementary Table S1) [5]. Between 2007–2016, the CGIAR centers distributed 3.91 million samples, with about 30% from genebanks and 70% from crop breeding programs, to 163 countries [3,5]. These distributions from the CGIAR programs account for almost 89% of the total annual international germplasm exchanges (*note: 'exchange' and 'transfers' used as a common term to denote germplasm exports or imports between countries.*), under the International Treaty on Plant Genetic Resources for Food and Agriculture (ITPGRFA or the Plant Treaty) [2,3,6–8]. Between 2010 to 2019, the CGIAR genebanks acquired 116,921 distinct accessions, about 35% of which were acquired through the centers' own breeding programs and 65% were acquired from collection missions or through national programs in 142 countries [7]. During the same period, the CGIAR genebanks distributed, on average, 115,000 samples of germplasm per year, and above 80% of the recipients were in developing countries [5,7]. A detailed analysis of the CGIAR genebanks' acquisition and distribution of germplasm in the last decade is presented by Halewood et al. [7]. The demand for a global movement of plant genetic resources (PGR) from the international genebanks and breeding programs is increasing due to worldwide efforts to develop nutrient-rich high-yielding varieties, which are resilient to biotic and abiotic stresses and better adapted to a changing climate, through various programs, such as the CGIAR's 'Crops to End Hunger' initiative [9]. Import and export of germplasm and other biological resources are influenced by several international and national policies, treaties, and legal frameworks [3]. The ITPGRFA and its multilateral system of access and benefit-sharing and the Convention on Biological Diversity (CBD) agreements guide the CGIAR centers' policies on germplasm acquisition, conservation, regeneration, use, and distribution [3]. The availability of pest- and disease-free germplasm is an important requirement for international distribution from genebanks and breeding programs.

1.2. Pathogen and Pest Threats to International Germplasm Transfers

It is well-known that plants and seeds can harbor pathogens and pests (hereafter referred to as pest (*note: 'pest' used to denote any species, strain or biotype of plant, animal or pathogenic agent injurious to plants or plant products.*), including bacteria, fungi, phytoplasmas, viroids, viruses, insects, nematodes, and other harmful biotic agents, and that the transfer of germplasm carries a simultaneous risk of moving pests between geographies and introducing them into territories where they are not known to exist [10,11]. International seed transfers have been recognized as important pathways for the transboundary spread of pests through human activities associated with collection and distribution [10]. The threat may become severe, if more virulent strains or races of the pathogens are introduced [12]. Even pests with a low seed transmission rate, especially viruses, may lead to the development of an epiphytotic proportion of the disease in a field, if the other conditions (e.g., occurrence of insect vectors and susceptible hosts) and climate are favorable [13].

The introduction of economically important alien pests, a term used for non-indigenous pests introduced into new territory, from their centers of origin into new environments, has been reported in many different parts of the world [14]. Considering that every plant serves as a host for several insects and microbes of both a beneficial and harmful nature, every introduction of plant material is expected to result in the introducing exotic organisms. For instance, European farmers introduced wheat and its pathogens, *Mycosphaerella graminicola* and *Phaeosphaeria nodorum*, into the Americas, Australia, and South Africa in the past 500 years [13,14]. Some examples of introduced pests causing epidemics and pandemics with disastrous consequences for food production, livelihoods, and environmental biodiversity include the Irish potato famine in the 1840s caused by *Phytophthora infestans*, which was introduced from Central America into Ireland [11,15]. Some recent examples of devastating outbreaks caused by transboundary pest introductions into regions where the CGIAR operates include the maize lethal necrosis (MLN) epidemic in East Africa caused by maize chlorotic mottle virus (MCMV), which was introduced from East Asia [16]; the fall armyworm (*Spodoptera frugiperda*) outbreak in Africa and Asia, caused by the likely introduction of insect pest from the Americas [17]; the cassava mosaic disease outbreak in East Asia, caused by the Sri Lankan cassava mosaic virus (SLCMV), which was introduced from South Asia [18]; the banana bunchy top virus (BBTV) outbreak in the sub-region of Western Africa due the spread of the virus through planting material from the sub-region of Central Africa [19]; the expansion of banana wilt caused by the *Fusarium oxysporum* Tropical Race 4, which was introduced from Southeast Asia into South (India) and West Asia (Jordon and Israel), Mozambique and Colombia [20]; the outbreak of potato cyst nematode, *Globodera pallida*, in Kenya by the likely introduction of the pest from Europe [21]; the wheat blast outbreak in Bangladesh [22] and, more recently, in Zambia [23], caused by the *Magnaporthe oryzae* pathotype *Triticum*, introduced from South America; the spread of *Candidatus* Liberibacter solanacearum haplotype A and its vector, *Bactericera cockerelli* (potato psyllid), which is responsible for the potato Zebra chip disease and potato purple top disease, which are likely to have spread from the Central American region to Ecuador in South America [24]. The occurrence of these new pests in the territories was recognized during obvious disease outbreaks. A few examples of transboundary pests that have caused economically significant disease outbreaks through their spread to sub-Saharan Africa are indicated in Table 1 and Figure 1.

Table 1. Some of the most economically important disease outbreaks, caused by pests introduced into sub-Saharan Africa (SSA).

Disease	Pest	Hosts	First Detection and Spread in SSA	Spread Pathway
Asian soybean rust	*Phakopsora pachyrhizi* (fungus)	Soybean, >150 legume species	First detected in Zambia in the 1980s; spread to most of sub-Saharan Africa	Possibly spread through air-borne urediniospores, which blew from Western India to East Africa
Banana bunchy top	*Banana bunchy top virus* (virus)	Banana/plantain	First reported in Egypt in 1902; and then in the Democratic Republic of Congo; spread to most of Central Africa, and adjoining countries in Southern and Western sub-Saharan Africa	Introduced from South-Pacific or Asia through infected suckers; further spread through exchange of infected planting material and aphid (*Pentalonia nigronervosa*) vectors
Banana bacterial wilt	*Xanthomonas campestris* pv. *Musacearum* (bacterium)	Banana/plantain	Burundi, Rwanda, Democratic Republic of Congo, Uganda, Kenya, Tanzania	Possibly introduced from Ethiopia to Uganda and then across the Great Lakes region through planting material, infected plant parts, or contaminated tools
Cassava green mite	*Monoychellus tanajoa* (Insect)	Cassava	First detected in Uganda in 1971; presently widespread in Africa	Introduced from South America into Africa; path of spread not known
Cassava mealybug	*Phenacoccus manihoti* (Insect)	Cassava	First detected in the Democratic Republic of Congo in 1973; presently widespread in Africa	Possibly introduced from South America to Africa through plant parts, or containers
Cassava brown streak	*Cassava brown streak ipomoviruses* (Virus)	Cassava	Native species in Malawi, Mozambique, and Tanzania in Southern Africa; first recorded in East Africa in 2004 in an epidemic around the Great Lakes region Burundi, Comoros, DRC, Kenya, Rwanda, Uganda, and Zambia	Spread through exchange of infected stem cuttings and whitefly (*Bemisia tabaci*) vectors
Fall armyworm	*Spodoptera frugiperda* (Insect)	Maize and other crops	First detected in Nigeria in 2016; known to occur in all African countries	Path of spread not known; likely through plants, plant parts, or cargo containers.
Fruit fly	*Bactrocera dorsalis* (Insect)	Mango and other crops	First reported in Mauritius in 1996; and on the mainland in Kenya in 2003; presently widespread in Africa and offshore islands	Pathway of spread not known; likely through plants, plant parts, or cargo containers
Maize lethal necrosis	*Maize chlorotic mottle virus*	Maize, sorghum, pearl millet	First detected in Kenya in 2011; then in Burundi, DRC, Ethiopia, Rwanda, Tanzania, and Uganda	Possibly spread from Southeast Asia
Panama disease—tropical race 4	*Fusarium oxysporum* f. sp. *Cubense* Tropical Race 4 (Fungus)	Banana/plantain	First detected in 2010 in Mozambique; no reports of further spread in Africa	Introduced from South East Asia; possibly introduction through contaminated soil or planting material

Table 1. *Cont.*

Disease	Pest	Hosts	First Detection and Spread in SSA	Spread Pathway
Papaya mealybug	*Paracoccus marginatus* (Insect)	Papaya and several hosts	First detected in Ghana in 2009; spread to most African countries	Possibly spread from Southeast Asia /South Pacific islands through plants, plant parts, or other sources
Potato cyst nematode	*Globodera pallida* (Nematode)	Potato	First detected in Algeria in 2011; subsequently in Kenya in 2018; and then in neighboring countries in East Africa	Possibly spread from European region with seed potato
Taro blight	*Phytophthora colocasiae* (Oomycete)	Taro	First reported in Nigeria in 2011; Spread to most countries in sub-Saharan Africa	Possibly spread from South Pacific islands through infected corms
Tomato leaf miner	*Tuta absoluta* (Insect)	Tomato and other solanaceous crops	First reported from Benin in 2005; spread to most countries in Africa	Wind-borne spread of insects from the northern region
Wheat blast	*Magnaporthe oryzae* pathotype *Triticum* (Fungus)	Wheat	First reported in Zambia in 2020	Possibly spread through contaminated seed/grain

Source: CABI Invasive Species Database: https://www.cabi.org/isc/ (accessed on 29 November 2020).

Figure 1. Examples of a few transboundary pests that have caused significant economic damage to food production and posed a major risk to germplasm transfers in sub-Saharan Arica. (**A**) Papaya mealybug (*Paracoccus marginatus*), (**B**) Taro blight (*Phytophthora colocasiae*), (**C**) Panama disease (*Fusarium oxysporum* Tropical Race 4), (**D**) Cassava brown streak (cassava brown streak ipomoviruses), (**E**) Maize lethal necrosis (Maize chlorotic mottle virus), (**F**) Fall armyworm (*Spodoptera frugiperda*), (**G**) Potato cyst nematode (*Globodera pallida*), and (**H**) Banana bunchy top (Banana bunchy top virus)

Many alien pests that could have caused significant damage were intercepted by quarantine authorities at the port of entry, preventing their introduction and establishment. For instance, in the USA, 7% of the 725,000 pest interception reports at the port of entry between 1984 to 2000 were attributed to plant propagative material [25]. In India,

45 viruses of quarantine importance were intercepted in imported plant germplasm between 1985–2016 [26]. The number of first reports of crop pests in new hosts and/or new regions has increased in recent years, driven by agricultural intensification, international trade, and climate change [13,27]. An analysis of 1300 known invasive pests and pathogens estimated their potential cost to global agriculture at over US$ 540 billion per year, if they continue to spread [28,29]. National capacities to prevent and manage alien pests are sub-optimal in much of Africa, as well as in parts of the Latin America, the Middle East, Central Asia, and Indochina, predisposing global biodiversity hotspots in these regions to the risk of exotic pest invasion [30]. Increasing risk of pest introduction and limited capacity to act against invasions in most countries warrant robust strategies to prevent transboundary pest spread, especially through propagation material, as such strategies decrease the chance of pest introduction, establishment, and further spread.

1.3. Transboundary Pest Risk to Germplasm Distribution and Premises for the Establishment of CGIAR Germplasm Health Programs

The nature of the CGIAR germplasm acquisition and distribution operations presents a high-risk scenario for transboundary pest spread, consequently, the safety of germplasm movement has been a major concern. The main reasons for this are the diverse origin of accessions, which are acquired from different geographies that converge at a research station for regeneration and characterization, and their distribution to diverse locations [7,15]. For instance, in the past 10 years, CGIAR genebanks have distributed 854,000 samples to 150 countries, at an average of 105,000 samples per year, catering for about 2000 requests from 100 countries [5]. The risk of exotic pest introduction through an accession into a new territory and its spread in the case of favorable environmental conditions is high. Similarly, endemic pests in regeneration sites can migrate to new territories along with the germplasm. Poor phytosanitary management also has detrimental effects on the survival of accessions during regeneration and evaluation, and it lowers the viability of the germplasm in the conservation facility, and it could lead to a loss of diversity in the collections and genetic erosion [31].

Without the proper phytosanitary measures of international agencies, such as the CGIAR, germplasm distribution increases the possibility of pest dissemination in areas previously considered to be disease-free [32,33]. Unintended pest spread through germplasm is a significant concern for the CGIAR genebanks and breeding programs, the majority of which distribute germplasm to developing countries and biodiversity hotspots that lack a sufficient quarantine capacity to prevent pest entry or respond to pest outbreaks [30]. Recognizing these pest risks, the CGIAR centers have setup Germplasm Health Units (GHUs), with the objectives of (i) averting the spread of quarantine pests with CGIAR germplasm transfers, (ii) preventing pest outbreaks, (iii) safeguarding biodiversity, and (iv) strengthening the development of phytosanitary capacities. The GHUs ensure compliance with the Food and Agriculture Organization (FAO)-International Plant Protection Convention (IPPC) procedures, which have the force of a legal treaty, and are enforced by the National Plant Protection Organization (NPPO or quarantine agency) to regulate pest spread through transfers of germplasm [34]. The movement of germplasm internationally is subjected to the same rules, with the directive that germplasm should be free of regulated pests for safe transfers across international boundaries [35]. Therefore, it is crucial to test the health of germplasm accessions before distribution and before it is used planting material. As the Center's liaison, GHUs engage with the NPPO of the host and recipient countries to organize import permits, conduct inspections of regeneration fields, conduct germplasm health testing and phytosanitation, and prepare germplasm for exportation or importation in accordance with the International Standards for Phytosanitary Measures (ISPMs) of the IPPC and other recommended actions, including the FAO-International Board for Plant Genetic Resources (IBPGR) technical guidelines for the safe movement of germplasm [36].

In this paper, we describe the mission and functions of the CGIAR GHUs and how transdisciplinary approaches are adopted for the phytosanitary protection of germplasm at all stages of the value chain, from acquisition for conservation in genebanks to regeneration for accession increase, breeding, safety duplication and regional and international distribution for the safe global movement of germplasm from the 11 CGIAR genebanks and breeding programs. We present the emerging phytosanitary challenges and the increased risk of transboundary pests and pathogens to the international exchange of germplasm, followed by lessons learned and recommendations stemming from experience of the international network of GHUs, which operate across all continents in collaboration with the NPPOs and plant health organizations. We conclude by stating the benefits of globally coordinated phytosanitary networks to prevent the transboundary spread of diseases through plant propagation materials.

2. Historical Evolution of GHU and Its Core Functions

2.1. Development of Institutional Capacity for the Prevention of Transboundary Pest Spread through Germplasm

The first set of plant quarantine procedures, as a legal measure, was established in 1873 in Germany to regulate potato tuber imports from the USA in order to prevent the spread of the Colorado potato beetle, *Leptinotarsa decemlineata*. The first international agreement on measures to control the pests through regulation of the movement of plants was established in 1878 as an "International Convention on Measures to be taken against *Phylloxera vastatrix* (present name *Viteus vitifoliae*)", an insect pest that was introduced with the vine cuttings imported from the USA to France in 1865 [14]. This agreement between seven European countries, which came into force on 3 November 1881, has specified procedures for the certification of plant material for export and import in order to control grapevine phylloxera [14]. Many countries subsequently followed suit by imposing quarantine regulations to contain the spread of pests through plants and plant products, which led to the establishment of the International Convention for the Protection of Plants in 1929 by the International Institution for Agriculture in Rome. The International Plant Protection Convention (IPPC), adopted by the sixth Conference of the FAO in 1951, which came into force on 3 April 1952, was eventually established as a multilateral treaty on plant protection and replaced previous agreements [34]. The IPPC and the 1994 World Trade Organization (WTO) agreement on the Application of Sanitary and Phytosanitary (SPS) Measures (the SPS Agreement) shaped international plant quarantine policy and standards for regulatory measures implemented by member countries for the protection of plants, animals, and human life [34]. However, the implementation of quarantine procedures differed all over the world, with many developing and underdeveloped countries unable to implement adequate measures, due to poor resources and a lack adequate physical and technical capacity [28,37].

At the time of the establishment of the early CGIAR centers—the IRRI, CIMMYT, CIAT, and IITA—in the 1960s, the subsequent formation of the CGIAR and the establishment of new centers, such as CIP, ICARDA, and ICRISAT, in 1971, many NPPO institutions were under development. In the early stage of the CGIAR, international germplasm exchange programs faced multiple challenges due to a lack of adequate baseline knowledge on pests affecting its mandate crops and weak quarantine infrastructure in many countries in which they were operating [38]. For instance, IITA depended on an intermediary quarantine station in the Netherlands for the importation of cassava from outside Africa in the 1970s and 1980s [15]. Similarly, the INIBAP (International Network for the Improvement of Banana and Plantain) Transit Centre (ITC) was established in 1985 by INIBAP at the Universiteit Leuven in Belgium as a transit center for *Musa* collection. The "INIBAP" was replaced by "International" when INIBAP and IPGRI (International Plant Genetic Resources Institute) were merged to establish Bioversity International at the end of 2006 [39].

To address the pest risks associated with the CGIAR germplasm exchange activities, a task force was established in 1975 by the IBPGR, which led to the publication of "Plant Health and Quarantine in the International Transfer of Genetic Resources" [40], outlining the control actions required to address the seed health challenges encountered by International Agricultural Research Centers (IARCs). This was followed by a series of consultations in the following decade, and an informal recommendation by the Regional

Plant Protection Organizations (RPPOs) in 1988, which fostered an FAO and IBPGR joint program to facilitate the safe exchange of germplasm, including the drafting of technical guidelines for the safe movement of germplasm for major crops [36]. A CGIAR commissioned study in 1989 on "Plant Quarantine and the International Transfer of Germplasm" recognized a lack of accurate, up-to-date information on pests and poor accessibility to updated information by national quarantine officials to be among the main problems, and the study recommended that CGIAR centers adopt standardized phytosanitary procedures for germplasm transfers [37]. This led to the foundation of an "inter-center collaboration on germplasm health and exchange", and the first meeting of the pathologists and virologists in charge of germplasm health from the CIMMYT, CIP, IBPGR, ICARDA, ICRISAT, IITA, and IRRI, which was convened by the IBPGR in Rome on October 1990 as a formal meeting of the germplasm health program [41]. In 1993, following the recommendations of the Sixth International Plant Protection Congress held in Montreal in August 1993, the CIMMYT, CIP, ICARDA, ICRISAT, IITA, and IRRI established GHUs as independent units within the centers to undertake research and facilitate the safe exchange of PGR from genebanks and breeding programs. Similar programs were subsequently established in the remaining five centers. These programs, differently named the Germplasm Health Unit (AfricaRice, Bioversity, CIAT, ICRAF, IITA, and ILRI), Seed Health Unit (CIMMYT, ICARDA, and IRRI), Health Quarantine Unit (CIP), and Plant Quarantine Unit (ICRISAT), serve the same mission of preventing phytosanitary risks associated with CGIAR germplasm activities and ensuring the safe transfer of germplasm.

2.2. GHUs as CGIAR Gateway for Safe Germplasm Exchange

The GHU mission is to maintain the multidisciplinary capacities required for health testing, ensuring the implementation of phytosanitary procedures to eliminate pests, facilitate the production of pest-free germplasm, and make "go/no-go" decisions on germplasm transfers from the centers based on phytosanitary statuses. The six strategic objectives of GHUs are: (i) to ensure that the transboundary movement of germplasm and non-seed biological materials complies with the regulatory guidelines of the importing and exporting countries and that the materials are free of quarantine pests; (ii) to develop and adopt phytosanitary procedures to generate pest-free germplasm; (iii) to develop diagnostic tools for seed health monitoring and pest surveillance; (iv) to conduct pest risk assessments of germplasm activities, including conservation, seed increase, and transfers; (v) to contribute to the development of phytosanitary capacity; and (vi) to organize a GHU Community of Practice to form a network of centers in transboundary pest prevention.

From the moment of their establishment, these centers began to develop expertise on the safe exchange of germplasm for its mandate crops by mobilizing interdisciplinary capacities for seed and plant health testing, phytosanitation, and therapy procedures in order to generate pest-free planting material or salvage germplasm after eliminating the risk of contaminating pests [36,41]. This evolution was extremely challenging due to the inadequate knowledge on pests affecting crop species of interest in operational territories. This has often led to the task of simultaneously conducting research on the identification and characterization of pests, developing diagnostics for pest detection, and establishing procedures for germplasm phytosanitation [15]. In this process, the CGIAR centers have worked closely with the NPPOs of the host countries to standardize safe germplasm exchange procedures, which were eventually transformed into a formal collaboration between the center and the host country NPPOs, leading to special accords for CGIAR germplasm transfers. For instance, in 1978, the Indian Council of Agricultural Research (ICAR) accorded permission to set up an "Export Certification Quarantine Laboratory" (Plant Quarantine Unit) at the ICRISAT headquarters in Patancheru, Hyderabad, India [38]. The need for adaptive changes remains a consistent requirement for coping with the changing phytosanitary risks in the world. For instance, the MLN outbreak in East Africa led to the redrafting of safe maize exchange procedures [42], and the characterization of casual viruses in the 2000s of cassava brown streak disease, described in the 1920s, led to the drafting of a new protocol for cassava virus indexing and the production of virus-free cassava planting material [43], which was followed by a reindexing of the cassava collection held in the IITA Genetic Resources Center at Ibadan, Nigeria. GHU programs have demonstrated a high commitment to the minimization of the pest risks associated with germplasm transfers as many national programs, especially in sub-Saharan Africa, lack sufficient facilities to carry out the required testing and/or phytosanitation of the accessions before their release for propagation use [30]. GHUs perform all their activities in close collaboration with NPPOs, RPPOs, and several national and international plant health programs. The centers' policy mandate for all germplasm, outgoing and incoming, is channeled and cleared through GHUs to ensure safe import or export, and GHUs have, over the years, developed into the centers' gateway for the safe international exchange of germplasm.

In 2017, GHUs were aligned with the Germplasm Health (GH) component of the CGIAR Genebank Platform [5]. This offered a unique opportunity for strengthening collaboration among GHUs, catalyzing a harmonized CGIAR GHU strategy, adopting a common Quality Management System (QMS) to ensure uniform standards across the centers, and the implementing cross-center R4D initiatives to address recalcitrant and emerging phytosanitary challenges. These recent developments, in close collaboration with NPPO partners, resulted in GHUs becoming a global network for transboundary pest prevention and effectively addressing the emerging needs of CGIAR programs.

3. Procedures for Germplasm Health Testing and Safe International Transfers

3.1. Multistage Phytosanitary Controls for Pest Prevention

The multidisciplinary and multistage process of GHUs for ensuring the phytosanitary safety of bioresources has five stages [15]: (i) germplasm health testing for pests using a range of diagnostic methods, including conventional bioassays, culturing methods, serological detection using enzyme-linked immunosorbent assay (ELISA), and nucleic acid-based detection (nucleic acid hybridization techniques, or various formats of DNA and RNA amplification, including polymerase chain reaction (PCR) and isothermal amplification methods, such as loop-mediated isothermal amplification (LAMP), and recombinase polymerase amplification (RPA), or sequence-independent high-throughput sequencing (HTS) and bioinformatics virus detection [44]); (ii) physical inspection to eliminate infected and physically damaged true seeds and vegetative propagules [12]; (iii) pest risk mitigation during germplasm regeneration using the most optimum procedures, including inspection of plants during the active growth stage, the use of pesticides and weed management in the field, nursery, and screenhouse production sites, and the use of virus-free planting material for clonally propagated germplasm [36]; (iv) phytosanitation (treatment) of germplasm, as a curative procedure to eliminate pests and salvage germplasm [36]; and (v) documentation for traceability and regulatory compliance, which includes an import permit issued by the NPPO of the import country that enlists phytosanitary conditions for import qualification; a phytosanitary certificate issued by the NPPO of the export country, ensuring that the germplasm complies to conditions listed in the import permit; and a health statement issued by GHUs, with a description of the germplasm origin, propagation, and health assessment information [36].

Generally, CGIAR centers acquire (imports) or distribute (exports) small quantities of germplasm as a few grams of botanic seeds or a small number of vegetative propagules of accession as in vitro plantlets, tubers, corms, or stem cuttings. All countries regulate incoming genetic resources according to the national and international laws and regulations, which are designed to prevent the risk of pest introduction. It is necessary for the CGIAR centers to align their phytosanitary compliance procedures with the host country NPPO, which is the statutory organization that sets policies, laws, and regulations to oversee plant material transfers according to the IPPC framework and agreements. The import and export of germplasm is a collaborative endeavor between the GHU and the NPPO of the country of export and import, the germplasm provider, and the germplasm material recipient (Figure 2). The GHUs submit imported germplasm to the NPPO of the host country for post-entry inspection to ensure compliance with the importation conditions, including checks for quarantine pests in the post-entry testing facility or regeneration of germplasm in a quarantine isolation area, prior to the release of safe material to the requester. A similar procedure is used to export germplasm from the centers. The GHUs conduct specific checks, as indicated on the import permit, and dispatch the material with or without seed treatment, depending on the importer's requirements.

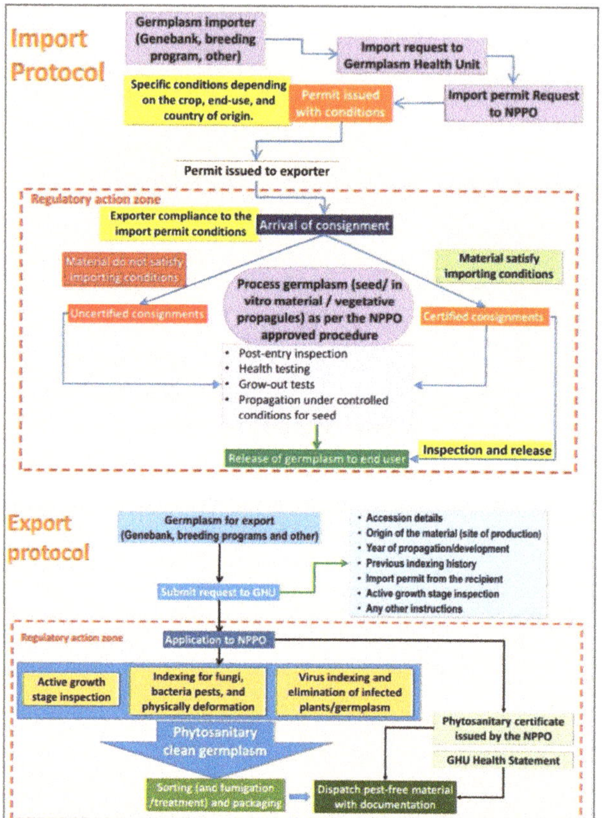

Figure 2. Schema of the germplasm import and export protocol of the CGIAR centers, which is executed in collaboration with the host country national plant protection organization (NPPO). The activities conducted under NPPO monitoring are marked as the 'regulatory action zone'. Permit = an official document issued by the NPPO authorizing the centers to transfer materials; country of origin = country in which the germplasm is regenerated.

3.2. Criteria for Pest Monitoring

A wide variety of pests is reported for each crop species, some of which are ubiquitous with the host species distribution in all geographies, and some pests are restricted to a few geographies [45]. Monitoring germplasm for pests depends on the knowledge of the pests affecting a crop species, particularly in the country regenerating the accessions, the crop propagation method (true seed or vegetatively propagation), and the ability of a pest to spread through germplasm. On the basis of the pest biology of different crop species and the economic significance and risk associated with crops and production systems, the NPPOs of each country have established national pest lists that categorize pests as "unregulated" or "regulated pests", with a further division of regulated pests into "quarantine pests" and "regulated non-quarantine pests" [46]. Pests whose introduction into an area can result in severe destruction are classified as quarantine pests. Regions can be free of quarantine pests, or such pests may exist but not be widely distributed (e.g., cassava brown streak virus, *Fusarium oxysporum* f. sp. *cubense* Race 4, and *Candidatus* Liberibacter solanacearum). Quarantine pests are strictly controlled through official monitoring measures, which are enforced by the NPPO. However, regulated non-quarantine pests are widely distributed, and their presence in germplasm causes planting material losses or initiates new disease cycles (e.g., cucumber mosaic virus). Unregulated pests include endophytes, saprophytes, and other pests of no significance.

The pest categorization and country-specific lists of quarantine and regulated non-quarantine pests are established by the NPPO. They are dynamic lists, which are updated regularly [47]. Countries and regions also use alternative classifications to designate regulated and unregulated pests. For instance, the European Plant Protection Organization (EPPO) uses "A1 pests" and "A2 pests" based on the complete absence or presence of designated pests in the EPPO region, respectively, and the Nigerian Agricultural Quarantine Services (NAQS) classifies pests under three categories: Category A (quarantine pest) for pathogens that are not present in Nigeria and/or in any country in West Africa; Category B (restricted regionally occurring pest) for pathogens that have a restricted local distribution in Nigeria and/or West Africa, against which field inspection and/or seed health testing methods can provide adequate protection; and Category C (regulated non-quarantine pests) for internationally widespread pathogens that may affect seed quality [15].

National plant pest lists provide information on pests likely to be associated with a plant species in the country of origin, and national regulated pest lists provide information on pests that need to be controlled using rigorous quarantine measures. Both these types of pest lists form the basis for setting the conditions of germplasm transfers between countries. The IPPC has established standards and frameworks for preparing a regulated pest list [48], and pest risk assessment (PRA) procedures for establishing scientifically justified regulations for the prevention of regulated pest incursions [48,49]. However, such analyses are often limited to commercially traded crops (e.g., chickpea, groundnut, maize, potato, rice, sorghum, soybean, and wheat), and the information on the pest occurrence, economic significance, distribution, and epidemiology is scanty or non-existent for several minor and orphan crops, as well as wild relatives. In addition, changes to pest nomenclature, due to taxonomic revision, which necessitates additional efforts to revise the pest lists, further complicates compliance procedures. Many countries do not update pest lists regularly, and the fact that importation or exportation conditions are based on outdated pest lists poses a challenge for regulatory compliance. Considering the variations and gaps in the country-specific lists due to limited knowledge of the pests affecting crop species, GHUs have taken a standardized approach to conducting checks for all quarantine and regulated non-quarantine pests reported for each species (detailed in Section 4). These procedures for the minimization of the risk of the spread of known and unknown pests through germplasm are in line with the FAO-IBPGR Technical Guidelines for the Safe Exchange of Germplasm [36], ISPMs, and other best practices [50].

4. Germplasm Health Testing and Pest Elimination

Seed health testing and pest detection is a first-line approach in managing seed-borne and seed-transmitted pests. In the case, of true seed crops, some pests infecting host crops are seed-borne (e.g., *Fusarium oxysporum* in cowpea), some are seed-transmitted (e.g., bean common mosaic virus in cowpea and common bean), and some are either seed-borne or seed-transmitted (e.g., *Phyllachora maydis,* which is responsible for tar spot affecting maize) [10,51]. Seed-borne and seed-transmitted pests are a concern for germplasm conservation and exchange, and procedures are therefore used to eliminate pests including the use of seed treatment methods or the regeneration and harvesting of seed from healthy plants. However, most pests affecting vegetatively propagated crops, especially intracellular pests, such as phytoplasmas, viruses, and viroids, can spread through vegetative propagules, and eliminating them requires the use of complex procedures. The GHUs routinely check for about 320 pests that are endemic in germplasm production sites, including bacteria, fungi, insects, nematodes, oomycetes, phytoplasmas, viruses, and viroids (Supplementary Table S2). The testing also covers other pests listed in the import permit of the country that receives germplasm. According to the crop mandate of the center, each GHU is specialized in enabling the production of quality germplasm in accordance with the best procedures available for the diagnosis and detection of pests, treatment for phytosanitation, and international transfers. GHUs apply similar procedures for genebanks and breeding programs, although genebank materials are more diverse, including wild species, landraces, and new acquisitions from new collection missions, which may demand complex/time-consuming procedures, owing to the different species biology and pest risks. The breeding program materials mostly comprise staple cereals, grain and oil seed legumes, roots, tubers, and banana crops. In general, managing the phytosanitary risks associated with true seed crops is relatively easy and effective, as not all the pathogens and pests are seed-transmitted, or seed-borne. Moreover, it is relatively easy to control or eliminate infections of seed-transmitted, and seed-borne pests from seed using chemical or heat treatments, thus salvaging germplasm. In the case of clonally propagated crops, however, systemically infectious pathogens, especially viruses and viroids, are difficult to eliminate without applying complex procedures, which are expensive and time-consuming. Brief details on the procedures employed to generate pest-free germplasm by crop group are summarized here.

4.1. True Seed Crops

4.1.1. Cereals

Cereal germplasm from breeding programs and genebanks is inspected for both seed-borne and seed-transmitted pests. General procedures for testing, detection, diagnosis, and seed treatment for the elimination of seed-borne pests are used [51,52], including the International Seed Trade Association (ISTA) methods, where applicable [53]. The general phytosanitary procedures used for true seed phytosanitation include, (i) active-growth stage inspection at the flowering/pre-harvest stage to check for the presence of any regulated pests and seed-transmitted pests; (ii) dry seed examination using a desk magnifier (2x) to remove the admixtures of plant debris, sclerotia, galls, insects, smut sori, and discolored and moldy seeds; (iii) seed-washing and a sedimentation test to detect the spores that could not be detected either in dry seed examination or incubation tests; (iv) standard blotter tests to detect the presence of fungi; (v) an agar test (selective media) to detect the bacterial pathogens using specific media; (vi) a seed soaking test to detect the presence of nematodes; (vii) a seed treatment involving a fungicidal treatment to remove saprophytic fungi and seed-borne pathogens; and (viii) seed fumigation using aluminum phosphide (or methyl bromide for sorghum seeds, as per the requirement of Indian NPPO at ICRISAT, India) [54]. The tests performed for some important seed-borne and seed-transmitted diseases of various CGIAR mandate crops are summarized below.

Barley: The most important seed-borne fungi are smut (*Ustilago nuda*), covered smut (*Ustilago hordei*), spot blotch (*Bipolaris sorokiniana*), head blight (*Fusarium graminearum*), barley leaf stripe (*Pyrenophora tritici-repentis*), ergot (*Claviceps purpurea*), a bacterium responsible for basal glum rot (*Pseudomonas syringae* pv. *atrofaciens*); a virus (barley stripe mosaic virus (BSMV)), a seed gall nematode (*Anguina tritici*), and an insect, the khapra beetle (*Trogoderma granarium*). Standard phytosanitary procedures are used to test and generate pest-free germplasm for import and export, including considerations of the additional conditions laid down by the NPPO of the import and export countries.

Maize: The main risks associated with maize germplasm exportation are associated with pathogens, such as *Pantoea stewartii* pv. *stewartii* maize dwarf mosaic virus, maize chlorotic mottle virus, sugarcane mosaic virus, and wheat streak mosaic virus, which have a restricted geographical distribution. These pathogens are proven to be seed-borne and seed-transmitted, although some of them have a low transmission rate of <1%. Many other maize pathogens are listed in the requirements of the country importing the germplasm, and the measures taken to guarantee that seeds are pathogen-free cover a wide range of possible threats by applying strict phytosanitary procedures in the multiplication field plots and exhaustive laboratory seed testing using conventional, serological, and molecular methods and seed treatments.

Rice: Many pests and pathogens have been identified as posing a risk to rice germplasm. GHUs use various procedures, as summarized above, for seed-borne pests, including bacteria (*Psudomonas* spp., *Xanthomonas* spp.), fungi (*Magnaporthe oryzae, Tilletia barclyayana*, etc.), oomycetes (*Sclerophthora macrospora*), phytoplasma (*Candidatus* phytoplasma 16srIII-L), virus (rice yellow mottle virus) and nematode (*Aphelenchoides besseyi*) on seeds.

Sorghum and millets: Some of the important sorghum seed-borne diseases are ergot (*Claviceps sorghi*), anthracnose (*Colletotrichum graminicola*), leaf blight (*Exserohilum turcicum*), downy mildew (*Peronosclerospora sorghi*), loose kernel smut (*Sporisorium cruentum*), long smut (*S. ehrenbergii*), head smut (*S. reilianum*), covered kernel smut (*S. sorghi*), bacterial blight (*Ralstonia andropogoni*), bacterial leaf streak (*Xanthomonas vesicola* pv. *holcicola*), and bacterial leaf spot (*Pseudomonas syringae* pv. *syringae*). Ergot (*Claviceps fusiformis*), and smut (*Moesziomyces penicillariae*) are the major seed-borne diseases of pearl millet. There are also some reports of downy mildew (*Sclerospora graminicola*) being seed-borne in nature. *Melanopsichium eleusinis, Pyricularia grisea*, and *Bipolaris* sp., are the important pathogens of small millet, for which salvaging treatment is used to recover pest-free seeds.

Wheat: The main risks associated with the germplasm exportation of bread and durum wheat are associated with pathogens, such as Karnal bunt (*Tilletia indica*), common bunt (*T. tritici* and *T. laevis*), *Alternaria triticina*, *Xanthomonas translucens* pv. *undulosa*, BSMV, and wheat streak mosaic virus (WSMV), which have a restricted geographical distribution. Nevertheless, many more wheat pathogens are listed in the requirements of the country importing the germplasm, and the measures taken to guarantee that the seeds are pathogen-free cover a wide range of possible threats by applying strict phytosanitary procedures in the multiplication field plots and exhaustive laboratory seed testing and seed treatments. Germplasm that is imported is subject to the NPPO regulations and inspected very carefully for wheat blast (*Magnaporthe oryzae* pathotype *Triticum*), dwarf bunt (*T. controversa*), and flag smut in wheat (*Urocystis agropyri*). In addition to the fungal pathogens, inspections are also carried for seed-borne insect pests, such as *T. granarium* (the khapra beetle), in seed exports and imports.

4.1.2. Grain and Oil Seed Legumes

Legume germplasm is more prone to pest attack, and many of these pests are known to spread through seeds [45]. A list of regulatory pests and pathogens frequently tested in the legume germplasm regeneration sites of CGIAR is given in Supplementary Table S2. The stringent phytosanitary and seed health testing procedures, such as those described for cereals, are also applied for legumes to prevent the transfer of fungal, bacterial, and viral diseases through legume germplasm. In general, germplasm and breeding lines for international transfers are regenerated under screenhouse conditions to avoid viral infections, and the germinated plants are inspected for viral symptoms and indexed by ELISA or PCR-based methods to ensure that plants are free from viruses prior to seed harvesting. Grow-out tests under screenhouse conditions are performed to assess seed-transmitted viruses, which is a standard practice for legumes. Although it is a time-consuming procedure, but it offers a reliable detection that eliminates the risk of viruses. Some of the important seed-borne and seed-transmitted pests, for which observations are conducted for export and import quarantine, are listed by crop below.

Bean: About 23 seed-borne bacterial, fungal, and viral pathogens are reported to be important for beans, including common blight (*Xanthomonas campestris* pv. *phaseoli*), charcoal rot (*Macrophomina phaseolina*), and anthracnose (*Colletotrichum truncatum*), along with three seed-transmitted viruses (alfaalfa mosaic virus (AMV), bean common mosaic virus (BCMV), and peanut mottle virus (PeMoV)).

Cowpea, bambara groundnut and other Vigna *species*: Several fungi and bacterial pathogens of cowpea are seed-borne, including cowpea bacterial blight (*Xanthomonas axonopodis* pv. *vignicola*), web blight (*Rhizoctonia solani*), and brown blotch (*Colletotrichum capsici*). About 10 viruses are reported to be seed-transmitted in cowpea. The most frequent viruses of interest in seed transmission are cucumber mosaic virus (CMV), cowpea yellow mosaic virus (CYMV), cowpea mottle virus (CmeV), southern bean mosaic virus (SBMV), and cowpea mild mottle virus (CPMMV). Cowpea seeds are subjected to fumigation with phostoxin (55% aluminum phosphide) to eliminate insect pests and treated with fungicide to eliminate seed-borne pathogens.

Chickpea and pigeonpea: Important seed-borne diseases of these two grain legumes are blight (*Ascochyta rabiei*), grey mold (*Botrytis cinerea*), wilt (*Fusarium oxysporum* f. sp. *ciceri*), and stem blight (*Phomopsis longicolla*) in chickpea; blight (*Botryodiplodia theobromae*) and wilt (*Fusarium oxysporum* f. sp. *udum*) in pigeonpea.

Faba bean: Twenty fungal species belonging to 13 genera were recognized as seed-borne risk (*Aspergillus*, *Penicillium*, *Alternaria*, *Botrytis*, *Cephalosporium*, *Cladosporium*, *Epicoccum*, *Fusarium*, *Rhizoctonia*, *Rhizopus*, *Stemphylium*, *Trichothecium*, and *Verticillium*), along with four seed-transmitted viruses [broad bean stain virus (BBSV), bean yellow mosaic virus (BYMV), broad bean mottle virus (BBMV), and pea seed-borne mosaic virus (PSbMV)]. Broomrape (*Orobanche* and *Phelipanche* spp.), root parasitic weeds, are also considered to pose a threat and measures are taken to avoid germplasm multiplication in the broomrape infested fields.

Groundnut: Dry root rot (M. phaseolina/Rhizoctonia bataticola), root rot (Rhizoctonia solani), pod rot (Sclerotium rolfsii), Sphaceloma arachidis (groundnut scab), *Ralstonia solanacearum* (African strains), seed bruchid (Stator pruininus), Testa nematode (*Aphelenchoides arachidis*), peanut mottle virus (PMV), peanut stripe virus (PStV), peanut clump virus (PCV), Indian peanut clump virus (IPCV), peanut stunt virus (PSV), and tobacco streak virus (TSV) are the important quarantine pests for groundnut.

Lentil: The important fungal seed-borne diseases of lentil include *Ascochyta lentis* (ascochyta blight), and *Fusarium oxysporum* f. sp. *lentis* (fusarium wilt), botrytis grey mold (*Botrytis fabae* and *B. cinerea*), Stemphylium blight (*Stemphylium botryosum*), phoma blight (*Phoma medicaginis* var. *medicaginis*), and anthracnose (*C. lindemuthianu* and *C. truncatum*); stem nematode (*Ditylenchus dipsaci*), and seed-transmitted viruses, include, AMV BYMV, PSbMV, CMV, and BBSV.

Soybean: The seed-borne fungal and bacterial pathogens of soybean are soybean bacterial pustule (*X. axonopodis* pv. *glycinea*), brown spots (*Septoria glycinea*), frogeye leaf spots (*Cercospora sojina*), yellow leaf spots (*P. manshurica*), charcoal rot (*M. phaseolina*), and anthracnose (*C. truncatum*). Many seed-transmitted viruses are also reported in soybean, including BCMV, CMV, CYMV, CmeV, CPMMV, and SBMVin West Africa. Rigorous tests are also conducted for other viruses depending on the country of origin. Seeds are fumigated with phostoxin (55% aluminum phosphide) to eliminate insect pests and fungicide treatments are given to eliminate seed-borne pathogens.

4.2. Vegetatively Propagated Crops

Banana (and plantain), cassava, potato, sweetpotato, and yam are the major vegetatively propagated crops (VPCs) exchanged by the CGIAR programs [55]. Vegetatively propagation poses the greatest risk of the introduction of pests through planting material, which can carry any infections from previous seasons to the next cropping cycle and thus accumulate pathogens, especially viruses, over generations of cultivation. Many transboundary pest introductions have been linked with the transfer of vegetative propagules: the spread of BBTV to Africa and its further spread in the continent [19,56]; in the case of potato, the necrotic strains of potato virus Y (PVY) in Brazil and aggressive strains of potato late blight in Africa and Asia and potato cyst nematode (*Globodera palladi*) in East Africa; in the case of cassava, the regional spread of cassava brown streak virus (CBSV), which is attributed to contaminated stem propagation; and in Asia, the spread of

Sri Lankan cassava mosaic virus (SLCMV) from South Asia to East Asia [57]. Therefore, many countries regulate vegetative germplasm importation, and the FAO-IPGRI technical guidelines recommend that only in vitro plants that have been tested for pathogens should be moved between countries [36]. The pollen or true seed of these crops are also exchanged for breeding purposes under adequate phytosanitary controls. By limiting international movement to sterile in vitro plants, the only concern that remains is intracellular obligate pathogens, such as viruses, viroids, and phytoplasmas. Depending on the country, some viruses are regulated by quarantine procedures (e.g., BBTV, CBSV, and PVY), and several other viruses are unregulated (e.g., sweet potato mild mosaic virus). Nonetheless, the standard procedure used by GHUs includes the generation of virus-free in vitro plants as per the FAO/IBPGR technical guidelines for the conservation and distribution of these crops [36]. All the material exported and imported are tested for viruses, and other pests under NPPO guidance, and only material free of viruses, and other pests, is released to the end-users. Unlike cereals and legumes, the phytosanitation, and testing procedures for clonally propagated crops differ according to the crop species, as explained below.

Banana: Banana is a perennial herbaceous plant, traditionally propagated using suckers (side shoots generated from underground corms), and is thus often carries both soil-borne insects and fungi, in addition to shoot-invading viruses, fungi, and bacterial agents. Several banana pathogens, like *Fusarium oxysporum* f. sp. *cubense* tropical race 4 (Panama disease), banana Xanthomonas wilt (*Xanthomonas campestris pv. musacearum*), and several viruses, such as BBTV and banana bract mosaic virus (BBrMV) have restricted geographic distribution. Guaranteeing the movement of pathogen-free germplasm is an important task to minimize the risk of these regulated quarantine pest introduction into new countries. Pathogens that are often symptomless in germplasm (e.g., in vitro plants, corms, and suckers), such as viruses, pose a special risk to the movement of vegetative germplasm. The Bioversity International-CIAT Alliance (1617 accessions), and the IITA (393 accessions) germplasm collections are managed as in vitro cultures. It has been shown that bacterial and fungal contaminants in banana shoot tip culture can be eradicated by isolating small explants, e.g., 1 mm meristems, and culturing them in vitro, but the virus infection still presents an important risk. To mitigate these risks, the Conservation Thematic Group of MusaNet, an international network for *Musa* genetic resources coordinated by Bioversity International, has recently edited a new version of technical guidelines to minimize the risk of pest introductions, through the movement of germplasm [58]. These guidelines followed a recommendation issued on the basis of an analysis of the phytosanitary procedures carried out by GHUs [56]. As per the new guidelines, at least four plants for each accession are grown for six months in a greenhouse. Leaf sampling is carried out from the limb and midrib of the three youngest leaves after 3 and 6 months for the comprehensive detection of the five most important viruses by PCR/RT-PCR: BBrMV, BBTV, banana streak virus (BSV), banana mild mosaic virus (BanMMV), and cucumber mosaic virus (CMV). Comprehensive indexing using electron microscopy is also conducted to search for any viral particle. Sanitation of the virus-infected banana accession is a complex process requiring a combination of meristem culturing, thermotherapy, and chemotherapy. Despite numerous efforts and the continuous optimization of the protocols, the success rate of banana sanitation is around 70%. An accession indexed negative is added to an in vitro banana collection for further safe propagation and distribution. All precautions are taken to avoid any further infection to in vitro plants that could arise if the plant is transferred to the field or greenhouse before distribution.

One of the major challenges for banana germplasm exchange was posed by the finding of an integration of the BSV genome, termed the eBSV (endogenous BSV), in the *M. balbisiana* genome, which contributes to the B genome. The eBSV can spontaneously release infectious particles, especially following in vitro culturing and interspecific crosses [19]. The presence of infectious eBSVs within B genomes has emerged as a main constraint for health indexing and safe *Musa* germplasm transfers. Plants apparently negative to BSV could spontaneously become positive with the expression of the eBSV sequence. The discovery of this phenomenon in bananas in the 1990s halted banana germplasm distribution from CGIAR centers. However, the Inter-African Phytosanitary Council, the Regional Plant Protection Organization of Africa, made a provision allowing the distribution within Africa of virus-free banana and plantain that may carry eBSV [19]. Based on this regulation, the IITA genebank and breeding programs distribute virus-free banana germplasm within Africa, with the informed consent of the recipients. However, the advancement of technology and knowledge on viruses integrated in host genomes provide a way to overcome this natural bottleneck to germplasm distribution. First, diagnostic techniques were established to distinguish eBSV and episomal virus particles for virus indexing purposes; secondly, molecular markers were established to identify *Musa* accessions with activatable eBSV; and lastly, a decision model was developed to enable the distribution of *Musa* germplasm with eBSV sequences based on the consent of the importer [58].

Cassava: Cassava is cultivated for tuberous roots and is traditionally propagated using stem cuttings. The crop is conserved in field collections and in vitro. The in vitro collection of 6500 accessions of cassava at the CIAT in Colombia, and 3700 accessions at the IITA in Nigeria are the largest cassava ex situ collections. The germplasm is exchanged as in vitro plants and botanic seed. Viruses and phytoplasmas pose a major threat to cassava distribution as in vitro plants. A diverse range of viruses infect cassava in Latin America, Africa, and Asia (Supplementary Table S1) [59,60]. The sanitary testing of the cassava collection held by the CIAT in Colombia checks for viruses prevalent in the region: Cassava common mosaic virus (CsCMV), cassava virus X (CsXV), and four other viruses that are associated with the cassava frogskin disease: the cassava frogskin associated virus (CsFSaV), cassava polero-like virus (CsPLV), cassava new alphaflexivirus (CsNAV), and cassava torrado-like virus (CsTLV) [60]. The sanitary testing of the in vitro African cassava collection conserved in the IITA, Nigeria, mainly checks for viruses prevalent in Africa: African cassava mosaic virus (ACMV), a complex of East African cassava mosaic viruses (EACMVs), and its strains, cassava brown streak ipomoviruses (CBSIVs), and 16Sr Phytoplasmas [57]. The cassava collection in Asia mainly focuses on viruses (Indian cassava mosaic virus (ICMV), and SLCMV), and phytoplasmas prevalent in the region. Both the CIAT and IITA GHUs have the diagnostic capability to test for all viruses known to infect cassava.

Several procedures have been established for virus and phytoplasma detection to generate virus-free planting material from meristem cultures in vitro, with or without thermotherapy, chemotherapy, or cryotherapy. The basic procedure includes a heat treatment applied to of stem cuttings with a length of about 30 at 28 and 38 °C for 6 h in the dark and 18 h in the light in an incubator [43]. Apical shoots from stem cuttings, after sanitation with 3% sodium hypochlorite, are used for meristem excision and in vitro plant development. About 2- to 4-month-old in vitro plants are virus indexed by PCR or RT-PCR to detect and eliminate virus-infected plants, and the remaining plants are re-indexed second time after 3 to 4 months to ascertain their health status. The virus-free plants are used as a mother stock for conservation as 'clean stock', and further propagation and use. It takes about 6 to 12 months to generate a virus-free stock of cassava germplasm. These procedures are known to be robust and 90% efficient in eliminating viruses. Only virus-free in vitro plants are transferred for propagation purposes. Cassava germplasm distributed as botanic seed poses little risk of virus spread, as none of the viruses reported to infect cassava have been detected in seedlings. Nonetheless, the cassava botanic seeds are surface sterilized with insecticides and pesticides, and they are germinated in screenhouses for physical inspection, before the seedlings are released to the end-users. Occasionally, cassava germplasm is transferred as stem cuttings generated from virus-free in vitro plants under insect-proof screenhouse conditions, after stem treatment with a slurry of insecticide and fungicide cocktail to eliminate microorganisms and arthropod pests.

Potato: Potato is propagated through tubers. Besides viruses, viroids, and phytoplasmas, field-produced tubers can transmit a long list of bacterial, fungal, and nematode diseases, including brown rot (*Ralstonia solanacearum phylotype IIB*), softrot (*Pectobacterium* and *Dickeya* spp.) ringrot (*Clavibacter michiganensis* subsp. *sepedonicus*), wart (*Synchytrium endobioticum*), common scab (*Streptomyces* sp.), powdery scab (*Spongospora subterranea* f. sp. *subterranea*), late blight (*Phytophthora infestans*), nematodes, and insects. Potato is known to be infected by more than 50 different viruses, but only about a handful of them (PVY, PVX, PVS, PVA, and potato leaf roll virus) are significant pathogens globally; however, some unique local viruses can be of major concern [61]. The CIP maintains one of the world's largest in-vitro genebank collections with over 7209 accessions of potato, many of which are of a local origin. The CIP uses a ISO/IEC17025 accredited process for ensuring that in vitro plants are free of all pathogens, both known and unknown. The process includes a combination of an antibacterial treatment before and during in vitro introduction, followed by virus indexing, which combines ELISA (9 viruses), RT-PCR (1 virus and 1 viroid), and a biological indicator host infection for 11 species. This indexing is repeated twice, before and after thermo-therapy and meristem tip culturing. Due to the extremely contagious and stable nature of PSTVd all plants are tested for this viroid before even starting the process of introduction and are destroyed immediately if they are found to be positive. The protocol for virus cleaning at the CIP is 91% efficient, and the most difficult viruses to clean are PVS and PVT. Only material that has been certified to be free of any pathogens after this process is permitted to be moved internationally. Any germplasm received from other institutions or countries will be tested in a similar way, before it can be made available for further distribution. Despite the rigorous indexing process, additional diagnostic tests are performed for pathogens when demanded by the importing country. Breeders occasionally move true seed and pollen internationally, and this is generally considered to be safer than moving plants or organs around. For true seed and pollen, the procedure includes testing both parents (female and male) at the pre-flowering stage for seed-transmitted viruses (APLV, APMMV, AVB-O, AMV, PVT, and PYV) and PSTVd, for which they must be negative. The seeds and pollen must be free of any insect pests and treated at −70 °C for seven days, if contamination is suspected. The seeds are surface sterilized with 2.5% sodium hypochlorite for 10 min to kill any seed-borne pathogens.

Sweetpotato: Sweetpotato is traditionally multiplied through stem (vine) cuttings. Like other VPCs, viruses are the principal concern for vegetatively propagation. More than 30 viruses have been reported to infect sweetpotato, but only a handful of them are of global significance and include crinivirus, sweet potato chlorotic stunt virus (SPCSV), the potyviruses, sweet potato feathery mottle virus (SPFMV), sweet potato virus C (SPVC), -G (SPVG), and -2 (SPV2), and several related begomoviruses (Supplementary Table S2) [62,63]. The CIP genebank holds 8,054 accessions of sweetpotato, and like potato, CIP has a ISO/IEC17025 accredited process for ensuring that in vitro cultivated sweetpotato are free of all known pathogens. The process is similar to that described for potato, except that testing is conducted for sweetpotato-specific viruses by PCR (begomoviruses) and ELISA (10 viruses), and that the biological indicator host range is replaced by a single universal indicator host, *Ipomoea setosa*, on which sweetpotato accessions are grafted. The efficiency of the current virus clean-up protocols for sweetpotato is 69%, and the difficult viruses to eliminate are begomoviruses, SPFMV, and SPVG. As for potatoes, only in vitro plants that have been confirmed to be free of known pathogens are distributed internationally. Sweetpotato pollen or seed is not commonly moved internationally by the CIP or partners, but the procedure would be similar to that for potato. Only two sweetpotato viruses, a begomovirus (sweet potato leaf curl virus), and a mastrevirus (sweet potato symptomless virus 1), have been reported to be seed-transmitted.

Yam: Unlike other clonal crops, multiple species of yam (*Dioscorea* spp.), which originated from different parts of the world, have been domesticated for the production of underground edible tubers, and the crop is traditionally propagated using tubers [64]. Out of about ten popularly grown species, *D. rotundata* (white yam) of West African origin, and *D. alata* (water yam) of Asiatic origin are widely cultivated in the world. The IITA genebank in Nigeria holds the world's largest in vitro yam collection [64]. The global yam germplasm collection comprises about 5839 accessions of about 10 species, including *D.*

rotundata, *D. alata*, *D. bulbifera*, *D. cayennensis*, *D. dumetorum*, *D. esculenta*, and a few other species. Yam is known to be infected by over 15 viruses [57,65], including yam mosaic virus (YMV), CMV, yam mild mosaic virus (YMV), several badnaviruses, generically referred to as yam bacilliform viruses (YBVs), Japanese yam mosaic virus, Chinese yam necrotic mosaic virus, yam asymptomatic virus 1, yam virus Y, yam chlorotic necrosis virus (YCNV), yam chlorotic mosaic virus (YCMV), Dioscorea mosaic-associated virus, and air potato ampelovirus 1. Of these, YMV is known to cause the most economic damage in *D. rotundata* and *D. cayanensis* worldwide, whereas YCMV and YCN are important for Chinese and Japanese yam in Asia. The other viruses detected in yam either cause mild mottling or no symptoms at all. Many yam viruses are not regulated, although the IITA uses protocols to generate virus-free in vitro plants for conservation and distribution [66]. Virus elimination is achieved by selecting asymptomatic plants for thermotherapy and regeneration of in vitro plants from meristem explants. The in vitro plants are subjected to virus indexing using PCR-based methods, after three months of post-flask growth, to ascertain their health status. Plants that test negative are bulk propagated for conservation and distribution. The integrated viral sequences of badnaviruses have been detected in yam genome sequences, but unlike in eBSV in bananas, the endogenous badnavirus sequences in yam are defective and not known to generate infectious particles. They were therefore not regarded to have any phytosanitary significance. Compared to other VPCs conserved by the CGIAR, yam phytosanitation is a lengthy process and usually takes between 12 to 24 months. Of the various associated factors, the slow in vitro meristem growth is a major bottleneck for the production of virus-free plants. True seed of yam is exchanged by breeding programs after seed treatment with a fungicide and an insecticide. Tubers generated from virus-free yam plants are also used for international exchange, especially within the West African sub-region, by breeding and seed system initiatives.

4.3. Trees

Germplasm conservation approach. The World Agroforestry Centre (ICRAF) characterizes and conserves 20,000 accessions of over 200 species of agroforestry trees, both indigenous and exotic species for agroforestry and restoration programs in Africa, Asia, and Latin America. Germplasm is conserved as seeds in genebanks or adult trees in field genebanks, where provenances are grown and evaluated in common gardens for domestication and cultivar testing [67]. Due to the wide range of plant growing conditions and geographic locations, this conservation approach implies that tree germplasm is exposed to a plethora of seed-borne pathogenic organisms, including both native and non-native microorganisms and insects. Furthermore, the vast taxonomic and geographic diversity of collected germplasm is susceptible to being impacted by poorly characterized, or unknown pests and diseases [68]. This scenario poses a major challenge in the implementation of phytosanitary measures for the detection and mitigation of pests and diseases of tree germplasm. Plant health interventions were put in place through collaborative research with the NPPO. High priority tree taxa were selected based on the strategic importance of the crop (food, timber, and non-timber product, and biodiversity conservation), and major biotic threats were determined based on laboratory testing of seed and vegetative tissue and field monitoring. Prioritized species include relevant food and multipurpose trees native to Africa, such as baobab (*Adansonia digitata*) and marula (*Sclerocarya birrea*), or exotic trees, such as Southern silky oak (*Grevillea robusta*) and eucalypts (*Eucalyptus* spp.). Special efforts have been made to characterize the taxonomy and pathogenicity of emerging fungal diseases of an unknown origin within the Botryosphaeriaceae family, which have the potential of impacting multiple tree species in Africa [69]. Standard operating procedures (SOPs) for assessing the germplasm health of both seeds and trees in the open field were developed and include: (a) seed health testing for the detection of seed-borne pathogens; (b) field assessment of incidence and damage caused by tree cankers, (c) detection and identification of tree canker fungi; and (d) field assessment of incidences and damage caused by fruit and seed insect pests. These procedures are to be used routinely as phytosanitary measures to evaluate germplasm prior to exchange or collection and for monitoring plant health in the field.

4.4. Forages

The CIAT and ILRI genebanks hold 22,694, and 18,662 accessions of forages (cereals and legumes), which include over 1400 species, the vast majority of which are wild species [5]. Most forage accessions are distributed as seeds, except for a few that rarely produce seeds, such as Napier (or elephant grass, *Pennisetum purpureum*), which are maintained by vegetatively propagation in field genebanks. The regeneration of forage germplasm faces a great variety of growth habits and development times, due to the large number of genera, as well as the considerable number of species, belonging to the same genus. All of this makes it difficult to use standardized health testing measures on a large scale. For this reason, most of these materials are regenerated in the open field and occasionally under a screenhouse, making use of the integrated pest and disease control and management of diseases expensive. Except for a few widely used species, the knowledge on pests and diseases is limited for rigorous health testing. Most countries lack sufficient records of the pests and diseases affecting forages, which complicates international germplasm transfers. As in the case of tree germplasm, strict integrated phytosanitary control measures are used to maintain germplasm health, but much work is still needed to generate baseline knowledge on seed-borne and seed-transmitted risks in order to develop robust procedures for safe exchange.

5. GHU Support for CGIAR Programs

5.1. Enabling Safe Germplasm Transfers

The GHU operations are demand-driven and accommodate the evolving needs of the genebanks and breeding programs. The centers export and receive thousands of germplasm samples from genebanks, breeding programs, and elite seeds of released cultivars for evaluation and use by national and international partners [5–7]. Over 80% of the CGIAR centers' germplasm exports are executed mainly from twelve countries, Belgium, Colombia, Cote d'Ivoire, Ethiopia, India, Kenya, Lebanon, Mexico, Morocco, Nigeria, Peru, and Philippines. All of these countries host the headquarters, main genebanks, and breeding programs of the centers, except for Belgium and Morocco. The remaining CGIAR centers' international transfers and/or regeneration activities are operated from about 12 to 16 countries, including Benin, Cameroon, Mali, Malawi, Tanzania, Turkey, Uganda, Vietnam, Zambia, and Zimbabwe, which also host the CGIAR breeding programs, and ex situ and in situ collections of genebanks. Germplasm exports from the CGIAR centers located in these countries cater for between 90 and 130 countries per year, in all continents (Figure 3).

Uniform standards are applied to all export and import events for successful pest-free germplasm transfers. For instance, in 2018 and 2019, GHUs facilitated 1300 and 2600 events of germplasm transfers from genebanks and breeding programs, respectively, to 150 countries (Figure 3) (Supplementary Figure S1). This onerous task involves the production and extensive testing to ascertain the health status of germplasm released to end-users. In 2018 and 2019, GHUs tested and removed 7% of the 335,928 genebank samples, including those for import, export, and regeneration, and 3% of the 118,044 breeding samples for import and export, were found to be infected with pests (data not shown) [5]. In this process, a total of 2.47 million diagnostic reactions were employed to analyze the 453,972 samples in the two years at an average annual cost of about US$ 12 million. The proportion of infected samples detected varies by crop, and the pests most frequently detected are endemic in regeneration sites. An example set of some pest-infested legume seeds are shown in Figure 4, and the percentages of rejections during the phytosanitary testing of crop germplasm in 2019 are shown in Figure 5. The infected samples are returned to the phytosanitary treatment cycle or replaced with healthy stock, and the infected samples are then subjected to incineration. The data and knowledge from extensive phytosanitary surveillance of germplasm helped GHUs to make improvements to their procedures and protocols, some of which have played a vital role in confirming the first occurrence of regulated pests in new territories [15–17]. The technical resources and skill set of GHUs also support the centers' and partner initiatives in combating emerging pests, suppling of reference material for diagnosis and phenotyping, developing national program capacity, and improving awareness and advocacy associated with transboundary pest prevention and control.

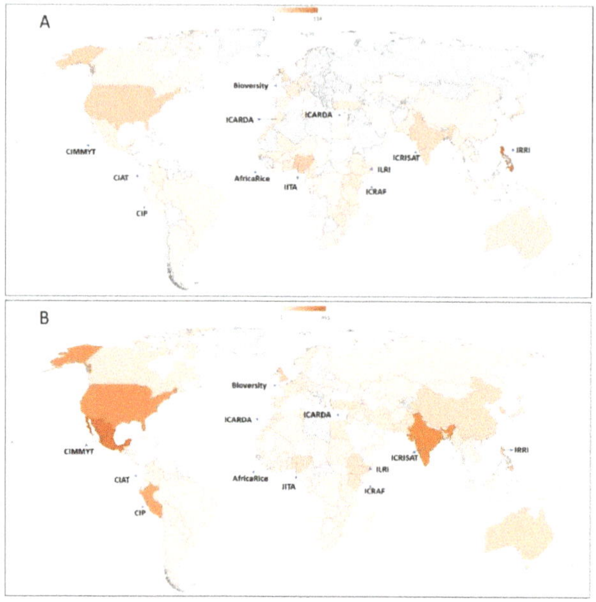

Figure 3. Countries from which the CGIAR centers received germplasm (**A**), and countries that received germplasm from the CGIAR centers (**B**) in 2018–19. The data combine transfers from the genebanks and crop breeding programs. The intensity of the orange shade indicates the number of transfer instances.

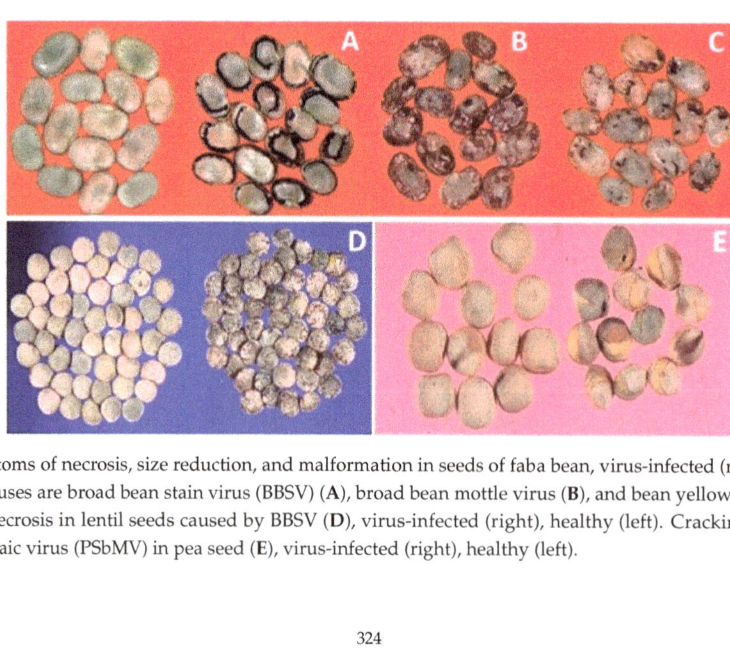

Figure 4. Symptoms of necrosis, size reduction, and malformation in seeds of faba bean, virus-infected (right) and healthy (left): causal viruses are broad bean stain virus (BBSV) (**A**), broad bean mottle virus (**B**), and bean yellow mosaic virus (**C**). Symptoms of necrosis in lentil seeds caused by BBSV (**D**), virus-infected (right), healthy (left). Cracking caused by pea seed-borne mosaic virus (PSbMV) in pea seed (**E**), virus-infected (right), healthy (left).

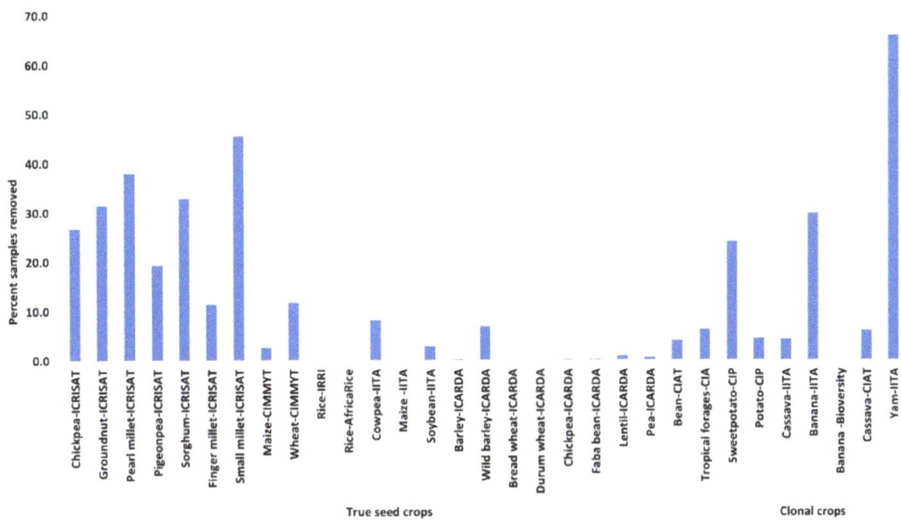

Figure 5. Percentages of samples removed, due to pest interception during phytosanitary processing.

5.2. Partnerships Enabling GHU Functions

Excellent partnerships and special arrangements, both formal and informal, between the GHUs and NPPO of the host countries, and regional plant protection organizations (e.g., Inter-African Phytosanitary Council in Africa; and Comunidad Andina de Naciones (CAN) in South America) have a significant role in enabling the successful exchange of germplasm. Such arrangements enabled the establishment of special facilities for CGIAR germplasm processing. For instance, the National Bureau of Plant Genetic Resources (NBPGR), which is responsible for the quarantine monitoring of germplasm exchanges from India, established its Regional Station at Rajendranagar, Hyderabad, India, in 1986 as the sole plant quarantine authority for clearing the germplasm and breeding material of ICRISAT's mandate crops [54]. Similarly, special arrangements exist between the CIAT and the Colombian Institute of Agriculture (ICA) in Colombia; CIP and Servicio Nacional de Sanidad Agraria (SENASA) in Peru; the CIMMYT and El Servicio Nacional de Sanidad, Inocuidad y Calidad Agroalimentaria (SENASICA) in Mexico; the IRRI and the Bureau of Plant Industry in Philippines; the ICARDA and the Plant Protection and Plant Quarantine Department of Lebanon; the IITA and the Nigerian Agricultural Quarantine Services (NAQS) in Nigeria; the ICRAF, IITA, CIP, and CIMMYT and Kenya Plant Health Inspectorate (KEPHIS) in Kenya, to name a few. Close partnerships also exist between non-NPPO agencies, for example, the relationship between the Alliance of Bioversity International-CIAT and the University of Liege (Belgium), and the ICRAF and the Kenya Forestry Research Institute (Kenya), which enable the centers' GHU activities. These arrangements between various partners recognize GHUs as part of the national diagnostic facilities, undertake collaborative research on phytosanitary issues and pest surveillance, jointly organize national and regional phytosanitary capacity development events on emerging disease control, and undertake awareness and advocacy activities to improve phytosanitary practices, policies, and regulations.

5.3. GHUs in Capacity Development

The GHUs play very important roles in the training of personnel, institutional capacity building, and raising awareness at a global level. The GHUs organize at least 10 workshops each year for staff from national and regional organizations on various phytosanitary themes, including diagnostics, seed health testing, and seed treatment. Since 2017, GHUs have organized "The International Phytosanitary Awareness Week" in coordination with the NPPOs, RPPOs, and IPPC [70]. These activities have the objective of informing interested parties about GHUs and their role in the secure distribution of germplasm, as well as the importance of pathogens and pests in crop production. The week-long event also aims to secure links between institutions through the consolidation of a "Community of Practice", fostering collaboration and exchange in such areas as joint projects, technical-scientific support, and capacity building, among other areas, with the goal of providing a 'front line' free-flow of information to assist in decision-making and facing upcoming challenges. These meetings bring together representatives of institutions from the public sector and academia, as well as staff, to take part in a variety of activities based on a main theme, which is different every year. For instance, the 2020 theme focused on *"phytosanitary safety for transboundary pest prevention"* to mark the United Nations International Year of Plant Health 2020 (IYPH 2020) [71]. These activities increase awareness of the importance of phytosanitation locally and globally and create a scientific forum for raising awareness of the phytosanitary challenges and organizational responsibilities associated with ensuring the distribution of healthy seed and sustainable agricultural production that will contribute to the global fight against hunger and malnutrition.

6. Challenges and Opportunities

6.1. Evaluation and Reevaluation of Germplasm Health

The comprehensive phytosanitary testing procedure used to declare the health status of an accession and its suitability for safe distribution or conservation is a tedious and time-consuming task (about 6 to 24 months for clonally propagated crops and 3 to 6 months for true seed crops). Untested accessions or accessions that fail health tests are marked as "unavailable for distribution". For instance, about 75% of the true seed accessions of staple crops held in ex situ collections of the CGIAR genebanks have been health certified and are available for distribution, compared to about 55% of clonally propagated accessions [5]. However, the discovery of new pests in a crop species sometimes necessitates a reevaluation for the detection of the newly reported pests and recertification of the health status of accessions and their availability for distribution. For instance, the discovery of new viruses and a phytoplasma with the cassava frogskin disease etiology in Colombia [60], led to a reevaluation and health certification of the in vitro cassava collection held at the CIAT in Colombia. Similarly, the discovery of causal viruses of cassava brown streak disease (CBSD) in 2002, and the CBSD outbreak in the Greatlakes region of Eastern Africa in 2008 [57] led to a precautionary evaluation of the in vitro and field collection of cassava conserved at the IITA Ibadan station in Nigeria. The outbreak of MLN in Eastern Africa in 2011 affected the distribution of maize germplasm in Eastern Africa, which was resumed after the establishment of procedures for the reliable detection of maize chlorotic mottle virus (MCMV) in the seed lots and phytosanitary procedures for the safe production of maize seed [42]. While instances of this type are infrequent, they result in a significant impact on germplasm distribution and result in additional costs. At the same time, these instances also aided GHUs in gaining experience in quickly adapting to new pest situations and establishing optimal protocols for phytosanitation and diagnostics to restore phytosanitary protection of germplasm and field crops. It also helped GHUs to engage with national, regional, and continental efforts to control epidemics caused by introduced transboundary pests or the emergence of new strains or species of an established threat.

6.2. Variable Standards and Different Phytosanitary Demands

GHUs work on different crops, wild relatives, and pests in various countries. Therefore, the variety of needs arises from the different phytosanitary statuses of the crops in each geography, the technologies available for detection, diagnosis, and phytosanitation, and the standards adopted by the NPPO in the country of operation [45]. For instance, the import conditions for cassava between Nigeria and Ghana are different from that between Nigeria and Vietnam, due to different pest risks. The standards for germplasm distribution from genebanks are not always well established. The NPPO adopts ISPMs designed for commercial consignments of plants and plant products, with specific modifications of their own for dealing with small sample sizes distributed from genebanks. Due to the better knowledge of pest risks, the standards for some crops are relatively well defined (e.g., banana, bean, cassava, cowpea, chickpea, groundnut, maize, potato, rice, wheat, and other crops) [47]. However, the vast taxonomic range, geographic diversity, and limited knowledge of the pest risks to crop wild relatives, trees, and forages pose significant challenges in the implementation of appropriate testing standards for pest detection. To overcome some of these challenges, GHUs began developing harmonized Quality Management Systems (QMS), termed the GHU-QMS, to achieve uniform standards across GHUs. The GHUs of the CIAT, CIMMYT, and CIP are ISO/IEC17025 accredited for the quality assurance of seed health testing methods. As of 2019, 139 SOPs have been developed, with 7 to 30 per GHU, depending on the center and country. These procedures were introduced with the aim of having GHUs conform uniform standards by the end of 2021.

The implementation of phytosanitary measures and policies for tree germplasm critically lacks in Africa. The extraordinary taxonomic and geographic diversity of the tree germplasm collected, and the availability of field genebanks, in addition to seed banks, show an opportunity for boosting the detection and characterization of emerging pathogens in line with the "sentinel plant" [72]. This should fuel fruitful collaborations with the NPPOs and IPPC and contribute to the much-needed updating of the lists of quarantine pest and diseases of tree species.

6.3. Changes in Pest Dynamics

The changes occurring in the dynamics of pests have a significant impact on germplasm transfers from the centers. Several economically important pest outbreaks in the last decade were attributed to introduced pests, as explained in the previous sections [14]. The perception of pest risk is also influenced by the severe destruction caused by unrelated pest outbreaks. For instance, the olive decline caused by introduced *Xylella fastidiosa* in Italy, citrus greening caused by *Candidatus liberibacter* spp., in the USA, and several other examples, including the Covid-19 pandemic, have a significant influence on the regulatory procedures and decision-making relating to germplasm transfers [13]. In addition, the discovery of new virus species using novel diagnostics technologies is adding to the burden of risks from the pests that are already known [73,74]. A study estimated that many alien pests introduced into countries are yet to be detected [13], a status termed "pseudo-absence", which implies the potential occurrence of a pest in the geography, but apparently considered there to be no alien pests because none had been found. This familiar but unquantified risk of "known-unknowns" and "unknown knowns" is a major threat for international germplasm exchange programs, which relies on pest occurrence knowledge in the country of the germplasm origin.

Over the years, GHUs have adjusted to changing pest dynamics, including undetermined pest risks and have taken adaptive measures to sustain operations. Following the MLN outbreak in East Africa, the CIMMYT GHU team established sampling and treatment procedures to sustain maize germplasm transfers. Similarly, the IITA-CIAT established cassava virus elimination protocols to maintain germplasm transfers between continents, including the use of transit centers for intermediary evaluation before delivering material to a final destination. Recently, the ICRAF GHU documented the invasive pests of native African tree germplasm, conserved as a resource for updating pest lists [69]. Procedures for the health testing of true seed crops from seedling to harvest, and a seed health test offers robust measures for the detection of both known and new pathogens. Clonal crops are more complicated, especially cryptic and latent viruses, which do not induce any symptoms and avoid detection. To overcome these challenges, GHUs have adopted HTS technologies and the bioinformatic reconstruction of viral sequences, which make it conceptually feasible to detect any viral agent by HTS of the nucleic acids from a host and the identification of viral sequences of known or unknown agents in the generated sequences [73,74]. These developments will strongly impact the way virus diagnostics is performed in the coming years. A pilot project focusing on the application of HTS technologies to improv the virus indexing of clonal crop germplasm accessions has been initiated for bananas, cassavas, potatoes, sweetpotato, and yams at the Bioversity International, the CIAT, the CIP, and the IITA.

6.4. Keeping up with Evolving Technologies

New technologies are evolving all the time for the phytosanitation and more accurate and rapid detection of existing and newly diagnosed pests. GHUs maintain a balance in adopting the best technologies that offer cost and time efficiency, meet regulatory requirements, and comply with ISO/QMS systems. The GHU operating system supports the use of a well-standardized procedure, so long as the procedure remains effective and offers reliable results for decision-making. The development of new standardized procedures is expensive, time-consuming, and requires extensive testing under various scenarios to determine the robustness, reliability, and suitability of the new method for the intended purpose. GHUs aim at keeping up to date and staying relevant, while avoiding change for the sake of change. As an example, the GHUs use, HTS-based diagnostic methods in the phytosanitary context is limited to virus indexing of mother stocks, while PCR and ELISA-based methods remain as 'gold standard' for virus indexing.

GHUs have identified a need to intensify efforts towards developing nucleic acid-based detection protocols for several pests that are difficult to detect through routinely used conventional tests, such as the blotter technique. Efforts are also required to standardize protocols for non-invasive techniques for detecting seed-borne pests (e.g., Videometer spectral imaging for detecting fungal pathogens and soft X-ray analysis to detect hidden seed infestation by pests) [75]. Similarly, new and safe solutions for crop protection and seed treatments are needed, as some fungicides and insecticide treatments are banned or restricted for use on specific crops in some countries. Due to the high volume of samples processed annually, the adoption of mobile digital data collection devices is necessary to facilitate the processing of materials, which would notably improve the traceability of the process and real-time data collection and analyses.

6.5. Insufficient Phytosanitary Standards for Germplasm Transfers from Genebanks and Breeding Programs

Specific phytosanitary standards for the international exchange of germplasm have not been developed. The FAO Genebanks Standards [35] lack adequate details on the procedures for the import and export germplasm from international genebanks. Therefore, the NPPOs either develop and follow their norms or follow those prescribed through ISPMs, which were established to address the SPS regulations governing the trade of plant and plant products, as part of the WTO treaty [34]. To date, the 43 ISPMs have been developed by the IPPC are aligned with the SPS requirements concerning commercial trade and large volumes of consignments. These regulations are inadequate for the purposes of the international transfer of germplasm. The ISPM 36 on the "international movement of plants for propagation" [76], and the ISPM 38 on the "international movement of seed" [77] address few issues, but are mainly designed for commercial shipment volumes. The ad hoc norms for germplasm exchange from genebanks and breeding programs introduce different requirements, depending on the country, making germplasm transfers a challenging endeavor. In addition, the existence of conflicting regulatory frameworks in different countries due to outdated regulations, outdated pest lists, or their absence constrain the exchange of germplasm. All of this causes delays in clearance, leading to germplasm having a loss of viability, before it arrives at its destinations, or a late arrival, resulting in the loss of an entire planting season.

Other challenges emerge from unforeseen changes to policies in the countries of operations. Policy changes are most often triggered by (i) new pest outbreaks, (ii) the risk perception of invasive pests spreading into territories, (iii) the introduction of new/amended procedures, and (iv) changes to administrative and implementation protocols. GHUs have adopted the flexibility needed to make necessary adjustments in order to align with policy requirements in countries of operations and thus enable germplasm distributions. In some cases, the policies do not match the biological complexities and restrict germplasm distributions. For instance, the genomes of some viruses are integrated into the host genomes (e.g., endogenous badnavirus sequences in banana and yam genomes). In essence, integrated viral genomes are an inseparable part of the host. The existing regulations do not consider these complexities, and all the germplasm with integrated virus genomes was consequently withheld from international transfers, and this amounts to over 50% of the *Musa* collection held in the CGIAR genebanks. In 2015, GHUs of IITA-Bioversity, together with the MusaNet Working Group on Genetic Resources, developed a new approach to the transfer of germplasm with integrated viral sequences (see Section 4 for details) [58]. The NPPOs have approved international transfers of *Musa* germplasm organized in accordance with this protocol, making a significant proportion of banana germplasm available again for distribution.

The current phytosanitary policies are also insufficient to address the germplasm "safety duplication" efforts (also referred to as black box conservation) in the Svalbard Global Seed Vault and/or in other third-party countries [5]. The safety duplication involves transfers of both health certified and untested accessions to another country (third-party) in sealed envelopes or as in vitro plants exclusively for conservation and repatriation to the "country of origin" when required. The NPPO requirements, however, are difficult to fulfill, as the procedures stipulate mandatory health declarations, and the entry of untested germplasm is prohibited. However, ad hoc bilateral arrangements have been established between the source of origin countries and third-party countries to facilitate safety duplication as an interim arrangement. This system is working although not always smoothly due to ambiguities arising from the different understandings of the NPPOs. GHUs are working with regulatory agencies to establish a standard policy to streamline the procedure for this important genebank activity.

To cover some of the phytosanitary policy challenges associated with germplasm exchange, GHUs have initiated the development of the "CGIAR GreenPass Phytosanitary Protocol (GreenPass)" [78], as a comprehensive procedure for the assurance of phytosanitary compliance. This protocol will detail the best procedures in use for germplasm regeneration and health assurance, while maintaining transparency in risk assessment and mitigation strategies to obtain NPPO accreditation in order to fast track germplasm distribution. It is hoped that the IPPC and other stakeholders' endorsement of this initiative will eliminate redundant checks or reduce the processing time of material from GreenPass-accredited facilities.

7. Conclusions

The CGIAR germplasm health program has over 50 years of experience [41]. GHUs have served as a vital conduit of the globally coordinated CGIAR crop research programs, which tested 1000s of germplasms and new breeding lines in multiple field sites and mega environments for the identification of lines that have superior yields, high nutrition and are resilient to biotic and abiotic stresses. The seeds of those accessions were made widely available for crop productivity improvement, leading to a broad social, economic, and environmental impact [79–81]. For instance, the International Wheat Improvement Network (IWIN) organized approximately 700 field sites in over 90 countries to develop around 1000 high-yielding, disease-resistant lines targeted at major agro-ecologies, which are delivered annually as international public goods (IPGs) [81]. To date, GHUs continue to facilitate crucial germplasm transfers to the largest number of stakeholders around the world vital to deliver IPGs with a positive impact on the SDGs associated with (i) nutrition and food security; (ii) poverty reduction; (iii) environmental health and biodiversity; and (iv) climate adaptation and greenhouse gas reduction.

The efforts of GHUs in thoroughly testing germplasm accessions for known pests, before their release for international transfer, have averted the inadvertent spread of quarantine pests. This is of great significance, as most CGIAR centers operate in countries where some of the most dreaded pests are prevalent (e.g., cassava brown streak virus, Karnal bunt, maize lethal necrosis, rice blight, and wheat blast, to name a few) [82]. Years of experience indicate that adaptability is a vital requirement for sustaining operations in an era of constant changes driven by pest outbreaks, agricultural intensification, climate variability, phytosanitary policies, and regulations [83]. A study on the patterns of invasion and spread pathways of 1517 invasive species reported that horticulture and the nursery trade are the dominant pathways for the incursion of invasive alien species [84]. The increasing international exchanges and the globalization of the world present a high risk that introduced pests will be established and expand quickly [82]. Safe and efficient germplasm transfer forms a critical preventive pest control approach for the CGIAR programs under the IPPC treaty and national laws. It is also safe to assume that the drivers responsible for transboundary pest outbreaks are difficult to contain, and high levels of vigilance will be required to monitor the pest dynamics in order to sustain the CGIAR operations. This requires regular updating of the existing protocols for hitherto unknown pests, enhanced collaboration with phytosanitary organizations and academia to obtain the most advanced information on pest detection and epidemiology, and adequate funding support, which is necessary for continuous adaptation to new pest challenges. It is imperative for GHUs to leverage technological advances in diagnostics, ICTs, remote sensing, and modeling to predict and monitor pest dynamics at a global level in order to understand their dispersal mechanisms and impact on the genebank and breeding programs in the short, medium, and long term.

GHUs high-level capacity, experience, track record, and global distribution in the developing world enable them to play an important role as centers of excellence in supporting national and regional pest and disease surveillance and rapid response. A strong case exists for positioning GHUs as part of a global network of phytosanitary hubs for the research, diagnoses, and control of established and emerging pests as part of the One CGIAR program, which is set to be operational in 2022 [85].

Supplementary Materials: The following are available online at https://www.mdpi.com/2223-7747/10/2/328/s1, Table S1: Details of crops and country locations of the CGIAR GHUs; Table S2: List of crops and pests routinely assessed by the CGIAR GHUs. Figure S1: Countries receiving germplasm from genebanks (A) and breeding programs (B); and the countries from which germplasm was received by the CGIAR genebanks (C) and CGIAR breeding programs (D), during 2018–19. The intensity of blue reflects the number of transfer events.

Author Contributions: The first author conceived the article, guided its development, and prepared the first draft; and all authors contributed equally to the preparation of the article. All authors have read and agreed to the published version of the manuscript.

Funding: The writing of this article and open access publication fees were funded through the CGIAR Genebank Platform.

Institutional Review Board Statement: Not applicable for studies not involving humans.

Informed Consent Statement: Not applicable.

Data Availability Statement: Not applicable.

Acknowledgments: We thank the editors of this special issue, Andreas Ebert and Jan Engels, for inviting us to write this article and for their critical review and valuable suggestions for improvement. We also thank Nelissa Jamora of Crop Trust (Bonn, Germany) and Alabi Tunrayo of IITA for their help with the preparation of geographic maps used in this manuscript.

Conflicts of Interest: The authors declare no conflict of interest.

References

1. Khoury, C.K.; Achicanoy, H.A.; Bjorkman, A.D.; Navarro-Racines, C.; Guarino, L.; Flores-Palacios, X.; Engels, J.M.M.; Wiersema, J.H.; Dempewolf, H.; Sotelo, S.; et al. Origins of food crops connect countries worldwide. *Proc. R. Soc. B Boil. Sci.* **2016**, *283*, 20160792. [CrossRef]
2. Galluzz, G.; Halewood, M.; Noriega, I.L.; Vernooy, R. Twenty-five years of international exchanges of plant genetic re-sources facilitated by the CGIAR genebanks: A case study on global interdependence. *Biodivers Conserv.* **2016**, *25*, 1421–1446. [CrossRef]
3. Noriega, I.L.; Halewood, M.; Abberton, M.; Amri, A.; Angarawai, I.I.; Anglin, N.; Blümmel, M.; Bouman, B.; Campos, H.; Costich, D.; et al. CGIAR Operations under the Plant Treaty Framework. *Crop. Sci.* **2019**, *59*, 819–832. [CrossRef]
4. Alston, J.M.; Pardey, P.G.; Rao, X. *The Payoff to Investing in CGIAR Research*; SoAR Foundation: Arlington, VA, USA, October 2020; Available online: https://www.instepp.umn.edu/products/payoff-investing-cgiar-research (accessed on 15 November 2020).
5. GBP Ann Rep. CGIAR Genebank Platform Annual Report 2019. 27p. Available online: https://www.genebanks.org/resources/annual-reports/ (accessed on 28 June 2020).
6. Byerlee, D.; Dubin, H.J. Crop improvement in the CGIAR as a global success story of open access and international col-laboration. *Int. J. Commons.* **2010**, *4*, 453–480.
7. Halewood, M.; Jamora, N.; Noriega, I.L.; Anglin, N.L.; Wenzl, P.; Payne, T.; Ndjiondjop, M.-N.; Guarino, L.; Kumar, P.L.; Yazbek, M.; et al. Germplasm Acquisition and Distribution by CGIAR Genebanks. *Plants* **2020**, *9*, 1296. [CrossRef] [PubMed]
8. CGIAR. Report from CGIAR Consortium to the FAO Commission on Genetic Resources for Food and Agriculture—2018. CGRFA/WG-PGR-9/18/Inf.15. Available online: http://www.fao.org/3/CA0105EN/ca0105en.pdf (accessed on 24 May 2020).
9. CtEH. Initiative on "Crops to End Hunger" Strategy and Options for CGIAR Support to Plant Breeding. Available online: https://storage.googleapis.com/cgiarorg/2018/11/SC7-B_Breeding-Initiative-1.pdf (accessed on 25 August 2020).
10. Elmer, W.H. Seeds as vehicles for pathogen introduction. *Biol. Invasions* **2001**, *3*, 263–271. [CrossRef]
11. Klinkowski, M. Catastrophic Plant Diseases. *Annu. Rev. Phytopathol.* **1970**, *8*, 37–60. [CrossRef]
12. Miew, T.W.; Misra, K.K. (Eds.) *Rice Seed Health Testing*; IRRI: Los Banos, Philippines, 1994; 122p.
13. Bebber, D.P.; Holmes, T.; Gurr, S.J. The global spread of crop pests and pathogens. *Glob. Ecol. Biogeogr.* **2014**, *23*, 1398–1407. [CrossRef]
14. Simler, A.B.; Williamson, M.A.; Schwartz, M.W.; Rizzo, D.M. Amplifying plant disease risk through assisted migration. *Conserv. Lett.* **2019**, *12*, 12605. [CrossRef]
15. Kumar, P.L.; Legg, J.P.; Ayodele, M.; Mahuku, G.; Ortega-Beltran, A.; Bandyopadhyay, R. Pest and disease surveillance, diagnostics and germplasm health in crop protection in sub-Saharan Africa. In *Critical Issues in Plant Health: 50 Years of Research in African Agriculture*; Neuenschwander, P., Tamò, M., Eds.; Burleigh Dodds Science Publishing Limited: Cambridge, UK, 2019; pp. 41–73.
16. Mahuku, G.; Lockhart, B.E.; Wanjala, B.; Jones, M.W.; Kimunye, J.N.; Stewart, L.R.; Cassone, B.J.; Sevgan, S.; Nyasani, J.O.; Kusia, E.; et al. Maize Lethal Necrosis (MLN), an Emerging Threat to Maize-Based Food Security in Sub-Saharan Africa. *Phytopathology* **2015**, *105*, 956–965. [CrossRef] [PubMed]
17. Goergen, G.; Kumar, P.L.; Sankung, S.B.; Togola, A.; Tamò, M. First Report of Outbreaks of the Fall Armyworm *Spodoptera frugiperda* (J E Smith) (Lepidoptera, Noctuidae), a New Alien Invasive Pest in West and Central Africa. *PLoS ONE* **2016**, *11*, e0165632. [CrossRef] [PubMed]
18. Minato, N.; Sok, S.; Chen, S.; Delaquis, E.; Phirun, I.; Le, V.X.; Burra, D.D.; Newby, J.C.; Wyckhuys, K.A.G.; De Haan, S. Surveillance for Sri Lankan cassava mosaic virus (SLCMV) in Cambodia and Vietnam one year after its initial detection in a single plantation in 2015. *PLoS ONE* **2019**, *14*, e0212780. [CrossRef]
19. Kumar, P.L.; Selvarajan, R.; Iskra-Caruana, M.-L.; Chabannes, M.; Hanna, R. Biology, Etiology, and Control of Virus Diseases of Banana and Plantain. *Adv. Virus Res.* **2015**, *91*, 229–269. [CrossRef] [PubMed]
20. Maymon, M.; Sela, N.; Shpatz, U.; Galpaz, N.; Freeman, S. The origin and current situation of *Fusarium oxysporum* f. sp. cubense tropical race 4 in Israel and the Middle East. *Sci. Rep.* **2020**, *10*, 1–11. [CrossRef] [PubMed]
21. Mburu, H.; Cortada, L.; Haukeland, S.; Ronno, W.; Nyongesa, M.; Kinyua, Z.; Bargul, J.L.; Coyne, D. Potato Cyst Nematodes: A New Threat to Potato Production in East Africa. *Front. Plant Sci.* **2020**, *11*, 670. [CrossRef]
22. Islam, T.; Croll, D.; Gladieux, P.; Soanes, D.M.; Persoons, A.; Bhattacharjee, P.; Hossain, M.S.; Gupta, D.R.; Rahman, M.; Mahboob, M.G.; et al. Emergence of wheat blast in Bangladesh was caused by a South American lineage of *Magnaporthe oryzae*. *BMC Biol.* **2016**, *14*. [CrossRef] [PubMed]
23. Tembo, B.; Mulenga, R.M.; Sichilima, S.; M'Siska, K.K.; Mwale, M.; Chikoti, P.C.; Singh, P.K.; He, X.; Pedley, K.F.; Peterson, G.L.; et al. Detection and characterization of fungus (*Magnaporthe oryzae* pathotype Triticum) causing wheat blast disease on rain-fed grown wheat (*Triticum aestivum* L.) in Zambia. *PLoS ONE* **2020**, *15*, e0238724. [CrossRef]
24. Caicedo, J.D.; Simbaña, L.L.; Calderón, D.A.; Lalangui, K.P.; Rivera-Vargas, L.I. First report of '*Candidatus* Liberibacter solanacearum' in Ecuador and in South America. *Australas. Plant Dis. Notes* **2020**, *15*, 6. [CrossRef]
25. McCullough, D.G.; Work, T.T.; Cavey, J.F.; Liebhold, A.M.; Marshall, D. Interceptions of Nonindigenous Plant Pests at US Ports of Entry and Border Crossings Over a 17-year Period. *Biol. Invasions* **2006**, *8*, 611–630. [CrossRef]
26. Chalam, C.; Parakh, D.B.; Maurya, A.K. Role of viral diagnostics in quarantine for plant genetic resources and prepared-ness. *Indian J. Plant Genet. Resour.* **2017**, *30*, 271–285. [CrossRef]
27. Santini, A.; Liebhold, A.; Migliorini, D.; Woodward, S. Tracing the role of human civilization in the globalization of plant pathogens. *ISME J.* **2018**, *12*, 647–652. [CrossRef]

28. Paini, D.R.; Shappard, A.W.; Cook, A.W.; De Barro, P.J.; Woomer, S.P.; Thomas, M.B. Global threat to agriculture from invasive species. *Proc. Natl. Acad. Sci. USA* **2016**, *113*, 7575–7579. [CrossRef]
29. Plant Health—State of Research. In the State of the World Plants. 2017. Available online: https://stateoftheworldsplants.com/2017/plant-health.html (accessed on 18 November 2020).
30. Early, R.; Bradley, B.A.; Dukes, J.S.; Lawler, J.L.; Olden, J.D.; Blumenthal, D.M.; Gonzalez, P.; Grosholz, E.D.; Ibañez, I.; Miller, L.P.; et al. Global threats from invasive alien species in the twenty-first century and national response capacities. *Nat. Commun.* **2016**, *7*, 12485. [CrossRef] [PubMed]
31. Thormann, I.; Engels, J.M.M. Genetic Diversity and Erosion—A Global Perspective. In *Genetic Diversity and Erosion in Plants. Sustainable Development and Biodiversity*; Ahuja, M., Jain, S., Eds.; Springer: Cham, Switzerland, 2015; Volume 7.
32. Stace-Smith, R. Role of plant breeders in disease dissemination of virus diseases. *HortScience* **1985**, *20*, 834–837.
33. Khetarpal, R.K.; Singh, S.; Parakh, D.B.; Maurya, A.K.; Chalam, V.C. Viruses intercepted in exotic germplasm during 1991–2000 in quarantine. *Indian J. Plant Genet. Resour.* **2001**, *14*, 127–129.
34. MacLeod, A.; Pautasso, M.; Jeger, M.J.; Haines-Young, R. Evolution of the international regulation of plant pests and challenges for future plant health. *Food Secur.* **2010**, *2*, 49–70. [CrossRef]
35. FAO. *Genebank Standards for Plant Genetic Resources for Food and Agriculture*; FAO: Rome, Italy, 2013; 181p.
36. Frison, E.A.; Putter, C.A.J. FAO/IBPGR Technical Guidelines for the Safe Movement of Germplasm. In Proceedings of the 5th International Congress of Plant Pathology, Kyoto, Japan, 20–27 August 1988; p. 279.
37. Plucknett, D.L.; Smith, N.J.H. *Plant Quarantine and the International Transfer of Germplasm*; CGIAR Study Paper Number 25; The World Bank: Washington, DC, USA, 1988; 65p.
38. Chiarappa, L. Plant protection at FAO-40 years of service international plant pest management. *FAO Plant Prot. Bull.* **1985**, *33*, 131–138.
39. ITC. The International Transit Center (ITC). Available online: https://www.promusa.org/ITC (accessed on 23 November 2020).
40. Hewitt, W.B.; Chiarappa, L. *Plant Health and Quarantine in International Transfer of Genetic Resources*; CRC Press: Boca Raton, FL, USA, 1977; 378p.
41. Frison, E.A. *Proceedings of the Inter-Centre Meeting on Germplasm Health and Movement*; IBPGR: Rome, Italy, 1991.
42. Mezzalama, M.; Das, B.; Prasanna, B.M. *MLN Pathogen Diagnosis, MLN-Free Seed Production and Safe Exchange to Non-Endemic Countries*; CIMMYT brochure: Mexico City, Mexico, 2015; 10p.
43. IITA. *Cassava In Vitro Processing and Gene Banking*; IITA: Ibadan, Nigeria, 2009; 38p.
44. Boonham, N.; Kreuze, J.; Winter, S.; van der Vlugt, R.; Bergervoet, J.; Tomlinson, J.; Mumford, R. Methods in virus diagnostics: From ELISA to next generation sequencing. *Virus Res.* **2014**, *186*, 20–31. [CrossRef]
45. CABI. Crop Protection Compendium. 2020. Available online: https://www.cabi.org/cpc/ (accessed on 5 October 2020).
46. FAO. Glossary of Phytosanitary Terms. In *International Standard for Phytosanitary Measures No. 5*; FAO on behalf of the Secretariat of the International Plant Protection Convention (IPPC): Rome, Italy, 2019; 35p.
47. FAO. Guidelines on List of Regulated Pests. In *International Standard for Phytosanitary Measures No. 19*; FAO on behalf of the Secretariat of the International Plant Protection Convention (IPPC): Rome, Italy, 2005; 61p.
48. FAO. Pest Risk Analysis for Quarantine Pests. In *International Standard for Phytosanitary Measures No. 11*; FAO on behalf of the Secretariat of the International Plant Protection Convention (IPPC): Rome, Italy, 2019; 40p.
49. Picard, C.; Afonso, T.; Benko-Beloglavec, A.; Karadjova, O.; Matthews-Berry, S.; Paunovic, S.A.; Pietsch, M.; Reed, P.; Van Der Gaag, D.J.; Ward, M. Recommended regulated non-quarantine pests (RNQP s), associated thresholds and risk management measures in the European and Mediterranean region. *EPPO Bull.* **2018**, *48*, 552–568. [CrossRef]
50. Rao, K.N.; Hanson, J.; Dulloo, E.M.; Ghosh, K.; Nowell, D.; Larinde, M. *Handbooks for Gene Banks No. 8 Manual of Seed Handling in Genebanks*; Biodiversity International: Rome, Italy, 2006; pp. 50–82.
51. Aveling, T.A.S. Global standards in seed health testing. In *Global Perspectives on the Health of Seeds and Plant Propagation Material*; Gullino, M.L., Munkvold, G., Eds.; Springer: London, UK, 2014; pp. 17–28.
52. Albrechtsen, S.E. *Testing Methods for Seed-Transmitted Viruses: Principles and Protocols*; CABI Publication: Wallingford, UK, 2006; 268p.
53. International Seeds Testing Association (ISTA). *International Rules for Seed Testing*; Secretariat, Zürichstrasse 50; CH-8303 Bassersdorf: Luzern, Switzerland, 2009.
54. Thakur, R.P.; Gunjotikar, G.A.; Rao, V.P. *Safe Movement of ICRISAT's Seed Crops Germplasm*; Information Bulletin no. 81; Patancheru 502 324; International Crops Research Institute for the Semi-Arid Tropics: Andhra Pradesh, India, 2010; 244p, ISBN 978-92-9066-524-3.
55. Panis, B.; Nagel, M.; Houwe, I.V.D. Challenges and Prospects for the Conservation of Crop Genetic Resources in Field Genebanks, in In Vitro Collections and/or in Liquid Nitrogen. *Plants* **2020**, *9*, 1634. [CrossRef]
56. De Clerck, C.; Crew, K.; Houwe, I.V.D.; McMichael, L.; Berhal, C.; Lassois, L.; Jijakli, M.H.; Roux, N.; Thomas, J.E.; Sebastien, M. Lessons learned from the virus indexing of *Musa* germplasm: Insights from a multiyear collaboration. *Ann. Appl. Biol.* **2017**, *171*, 15–27. [CrossRef]
57. Legg, J.P.; Iita, T.; Kumar, P.L.; Mahuku, G.; Wosula, E.; Stavolone, L.; Terry, E.; Bosque-Pérez, N.; Iita, N.; Lab, U.N.M. *Viruses Affecting African Crops and Their Vectors*; Burleigh Dodds Science Publishing: Cambridge, UK, 2019; pp. 95–135.
58. Thomas, J.E.; Sharman, M.; Lassois, L.; Declercq, C.; Iskara-Caruana, M.-L. *Technical Guidelines for the Safe Exchange of Musa germplasm*; Bioversity International: Montpellier, France, 2015; 65p.
59. Legg, J.P.; Kumar, P.L.; Makeshkumar, T.; Tripathi, L.; Ferguson, M.; Kanju, E.; Ntawuruhunga, P.; Cuellar, W. Cassava virus diseases: Biology, epidemiology, and management. *Adv. Virus Res.* **2015**, *91*, 85–142. [PubMed]

60. Carvajal-Yepes, M.; Olaya, C.; Lozano, I.; Cuervo, M.; Castaño, M.; Cuellar, W.J. Unraveling complex viral infections in cassava (*Manihot esculenta* Crantz) from Colombia. *Virus Res.* **2014**, *186*, 76–86. [CrossRef]
61. Kreuze, J.; Souza-Dias, J.A.C.; Jeevalatha, A.; Figueira, A.R.; Valkonen, J.; Jones, R.A.C. Viral Diseases in Potato. In *The Potato Crop*; Springer: Cham, Switzerland, 2019; pp. 389–430.
62. Loebenstein, G.; Fuentes, S.; Cohen, J.; Salazar, L.F. Sweetpotato. In *Virus and Virus-Like Diseases of Major Crops in Developing Countries*; Loebenstein, G., Thottappilly, G., Eds.; Kluwer Academic Publishers: Dordrecht, The Netherlands, 2003; pp. 223–248.
63. Kreuze, J.; Fuentes, S. Sweetpotato viruses. In *Encyclopedia of Virology*, 3rd ed.; Mahy, B.W.J., Van Regenmortel, M.H.V., Eds.; Elsevier: Amsterdam, The Netherlands, 2008; pp. 659–669.
64. Mahalakshmi, V.; Ng, Q.; Atalobhor, J.; Ogunsola, D.; Lawson, M.; Ortiz, R. Development of a West African yam *Dioscorea* spp. core collection. *Genet. Resour. Crop. Evol.* **2007**, *54*, 1817–1825. [CrossRef]
65. Mignouna, B.D.; Kumar, P.L.; Coyne, D.; Bandyopadhyay, R.; Ortega-Beltran, A.; Bhattacharjee, R.; De Koeyer, D. *Identifying and Managing Plant Health Risks for Key African Crops: Yam, Taro and Cocoyam*; Burleigh Dodds Science Publishing Limited: Cambridgeshire, UK, 2019; pp. 213–228.
66. Balogun, M.; Maroya, N.; Augusto, J.; Ajayi, A.; Kumar, L.; Aighewi, B.; Asiedu, R. Relative efficiency of positive selection and tissue culture for generating pathogen-free planting materials of yam (*Dioscorea* spp.). *Czech J. Genet. Plant Breed.* **2017**, *53*, 9–16. [CrossRef]
67. Loo, J.; Souvannavong, O.; Dawson, I.K. Seeing the trees as well as the forest: The importance of managing forest genetic resources. *For. Ecol. Manag.* **2014**, *333*, 1–8. [CrossRef]
68. Graziosi, I.; Tembo, M.; Kuate, J.; Muchugi, A. Pests and diseases of trees in Africa: A growing continental emergency. *PLANTS PEOPLE Planet* **2019**, *2*, 14–28. [CrossRef]
69. Cherotich, S.; Njuguna, J.; Muchugi, A.; Mutamia, J.; Otaye, D.; Graziosi, I.; Kinyanjui, Z. Botryosphaeriaceae associated with baobab (*Adansonia digitata* L.) and marula (*Sclerocarya birrea* A. Rich.) in agroforestry systems in Kenya. *Afr. J. Plant Sci.* **2020**, *14*, 411–419.
70. Germplasm Health Units. Available online: https://www.genebanks.org/the-platform/germplasm-health/ (accessed on 25 November 2020).
71. Montuori, M. Recognising the importance of plant health in today's world. *South Afr. J. Sci.* **2020**, *116*, 11–12. [CrossRef]
72. Eschen, R.; O'Hanlon, R.; Santini, A.; Vannini, A.; Roques, A.; Kirichenko, N.; Kenis, M. Safeguarding global plant health: The rise of sentinels. *J. Pest Sci.* **2018**, *92*, 29–36. [CrossRef]
73. Kreuze, J.F.; Perez, A.; Untiveros, M.; Quispe, D.; Fuentes, S.; Barker, I.; Simon, R. Complete viral genome sequence and dis-covery of novel viruses by deep sequencing of small RNAs: A generic method for diagnosis, discovery and sequencing of viruses. *Virology* **2009**, *388*, 1–7. [CrossRef] [PubMed]
74. Massart, S.; Candresse, T.; Gil, J.; Lacomme, C.; Predajna, L.; Ravnikar, M.; Reynard, J.-S.; Rumbou, A.; Saldarelli, P.; Škorić, D.; et al. A Framework for the Evaluation of Biosecurity, Commercial, Regulatory, and Scientific Impacts of Plant Viruses and Viroids Identified by NGS Technologies. *Front. Microbiol.* **2017**, *8*, 45. [CrossRef] [PubMed]
75. Feng, L.; Zhu, S.; Liu, F.; He, Y.; Bao, Y.; Zhang, A.C. Hyperspectral imaging for seed quality and safety inspection: A review. *Plant Methods* **2019**, *15*, 1–25. [CrossRef]
76. FAO. Integrated Measures for Plants for Planting. In *International Standard for Phytosanitary Measures No. 36*; FAO on behalf of the Secretariat of the International Plant Protection Convention (IPPC): Rome, Italy, 2019; 22p.
77. FAO. International Movement of Seeds. In *International Standard for Phytosanitary Measures No. 38*; FAO on behalf of the Secretariat of the International Plant Protection Convention (IPPC): Rome, Italy, 2019; 22p.
78. GreenPass Puts Germplasm Distribution in the Fast Lane. Available online: https://www.genebanks.org/news-activities/news/greenpass/ (accessed on 19 November 2020).
79. Saving the World. *Nat. Plants* **2017**, *3*, 17069. [CrossRef] [PubMed]
80. Pingali, P.L. Green Revolution: Impacts, limits, and the path ahead. *Proc. Natl. Acad. Sci. USA* **2012**, *109*, 12302–12308. [CrossRef] [PubMed]
81. Reynolds, M.P.; Braun, H.J.; Cavalieri, A.J.; Chapotin, S.; Davies, W.J.; Ellul, P.; Feuillet, C.; Govaerts, B.; Kropff, M.J.; Lucas, H.; et al. Improving global integration of crop research. *Science* **2017**, *357*, 359–360. [CrossRef] [PubMed]
82. Xing, Y.; Nopsa, J.F.H.; Andersen, K.F.; Andrade-Piedra, J.L.; Beed, F.D.; Blomme, G.; Carvajal-Yepes, M.; Coyne, D.L.; Cuellar, W.J.; Forbes, G.A.; et al. Global Cropland Connectivity: A Risk Factor for Invasion and Saturation by Emerging Pathogens and Pests. *Bioscience* **2020**, *70*, 744–758. [CrossRef] [PubMed]
83. Anderson, P.K.; Cunningham, A.A.; Patel, N.G.; Morales, F.J.; Epstein, P.R.; Daszak, P. Emerging infectious diseases of plants: Pathogen pollution, climate change and agrotechnology drivers. *Trends Ecol. Evol.* **2004**, *19*, 535–544. [CrossRef] [PubMed]
84. Turbelin, A.J.; Malamud, B.D.; Francis, R.A. Mapping the global state of invasive alien species: Patterns of invasion and policy responses. *Glob. Ecol. Biogeogr.* **2016**, *26*, 78–92. [CrossRef]
85. OneCGIAR. Available online: https://www.cgiar.org/food-security-impact/one-cgiar/ (accessed on 25 November 2020).

plants

MDPI

Review

The Role of Vegetable Genetic Resources in Nutrition Security and Vegetable Breeding

Andreas W. Ebert

World Vegetable Center, 60 Yi-Min Liao, Shanhua, Tainan 74151, Taiwan; ebert.andreas6@gmail.com

Received: 8 May 2020; Accepted: 9 June 2020; Published: 11 June 2020

Abstract: Malnutrition, comprising undernutrition, micronutrient deficiency, and overnutrition, is more widespread than hunger per se and affects most nations around the globe. The diversity and the quality of food produced and consumed are decisive factors when addressing the triple burden of malnutrition. In this context, fruit, vegetables, and nuts are increasingly moving into the focus of the nutrition community. Agricultural policies and investments in agriculture are predominantly focused on staple food production, neglecting the economic and nutritional potential of fruit and vegetables. While global vegetables are well represented in genebanks around the globe, this is much less the case for traditional vegetables. Collecting efforts in hotspots of vegetable diversity in Africa and Asia are required to conserve this germplasm before it is being replaced by modern varieties. Home gardens, community seedbanks, and variety introduction through vegetable seed kits are ways how genebanks can link with the farming community to strengthen the informal seed sector. This in turn may result in more diverse production systems and increased consumption of fruit and vegetables. In the formal seed sector, vegetable breeders need access to a wide diversity of genetic resources, predominantly farmers' varieties, landraces, and crop wild relatives. Genomics-assisted breeding is increasingly facilitating the introgression of favorable genes and quantitative trait loci (QTLs) with complex inheritance patterns from wild species into cultigens. This will lead to wider use of crop wild relatives in the development of resilient cultivars.

Keywords: malnutrition; food security; vegetables; genetic resources; ex situ conservation; home gardens; community seedbanks; variety introduction; vegetable breeding

---

1. Introduction

The Green Revolution had positive impacts on poverty reduction and lower food prices [1]. These effects were mostly driven by crop germplasm improvement programs of the CGIAR centers, resulting in impressive yield increases per hectare. The CGIAR is a global agricultural research partnership for a food secure future that was founded in the 1970s. The new, improved varieties were taken up by national agricultural agencies for adoption and broad-scale dissemination among the farming communities. From 1960 to 2000, yields across all developing countries increased 208% for wheat, 109% for rice, 157% for maize, 78% for potatoes, and 36% for cassava [2]. Estimates of the impact of crop germplasm improvement alone indicate average productivity gains, across all regions of the world, of 1.0% per annum for wheat, 0.8% for rice, 0.7% for maize, 0.6% for millets, and 0.5% for sorghum [3].

The historic success of the Green Revolution in terms of yield gains, together with lower food prices, ensured adequate quantities of staple cereal grain, thereby drastically reducing the problem of famine. However, after years of steady decline, the trend in world hunger reverted in 2015 and is now slowly increasing. About 820 million people or almost 11% of the global population suffered from chronic undernourishment in 2018 [4]. This underscores the immense challenge of achieving the Zero Hunger target of the 2030 Agenda for Sustainable Development adopted by the United Nations (UN) in 2015, which comprises 17 Sustainable Development Goals (SDGs). As stated by the United Nations

Development Program [5], the SDGs were adopted as a "call for global action to end poverty, protect the planet, and ensure that all people enjoy peace and prosperity by 2030". SDG 2—Zero Hunger calls for the eradication of hunger and all forms of malnutrition, with targets for doubling agricultural productivity and incomes of small-scale food producers (SDG 2.3), ensuring sustainable food systems (SDG 2.4), and maintaining genetic diversity (SDG 2.5).

Global demand for agricultural crops is predicted to increase until 2050 due to population growth, greater per capita purchasing power which translates into higher meat and dairy consumption, and biofuel use [6–9]. Analyses undertaken by Tilmann et al. [8] revealed that per capita demand for crop calories or protein is, since 1960, similarly dependent on per capita real income within and among seven economic groups. Those groups ranged from highest (group A) to lowest (group G) national average per capita inflation-adjusted gross domestic product (GDP). By 2050, global crop production should be 60% to 110% higher than that of 2005/2007 to feed the expected world population of more than nine billion [8–10]. Depenbusch and Klasen [11] estimated a 61% increase in calorie requirements between 2010 and 2100 if weight per age–sex group would remain stable. Considering the current trend toward increased human height and weight at a given height, expressed in terms of the average body mass index (BMI), this might add another 18.7 percentage points to the projected 61% increase in future national and global caloric requirements. Thus, an additional significant increase in food production would be required to satisfy demand.

Malnutrition, comprising undernutrition, micronutrient deficiency, and overnutrition, is more widespread than hunger per se and affects most nations around the globe. According to the 2018 Global Nutrition Report [12], children under five years of age face multiple burdens: stunting (low height for a child's weight) affects 150.8 million, wasting (low weight for a child's height) affects 50.5 million, and overweight (BMI at or above the 85th percentile and below the 95th percentile for children and teens of the same age and sex [13]) affects 38.3 million. Each year, 20 million babies are born of low birth weight (less than 2500 g) and are 20 times more likely to die in infancy than heavier babies. In 2016, 131 million children 5–9 years old, 207 million adolescents, and two billion adults were overweight [4]. About one-third of overweight adolescents and adults, and 44% of overweight children aged 5–9 were obese. Thus, overweight and obesity are at an alarming rate globally. Obesity is defined as a BMI at or above the 95th percentile for children and teens of the same age and sex [13]. In adults, BMI values of 30.0 or higher fall within the obese range.

Malnutrition is the single most important risk factor for disease. Diet-related diseases such as diabetes, cardiovascular disease, hypertension, stroke, cancer, and obesity are escalating at a global level. In 2016, an estimated 40.5 million (71%) of the 56.9 million deaths worldwide were caused by non-communicable diseases (NCDs) [14]. Approximately 32.2 million NCD deaths (80%) were attributable to cancers, cardiovascular diseases, chronic respiratory diseases, and diabetes, whereas the remaining 8.3 million (20%) were caused by other NCDs. These figures illustrate the seriousness of diet-related diseases for the healthcare sector. NCDs now pose a greater risk to morbidity and mortality than unsafe sex, alcohol, and drug and tobacco use combined [15].

To address SDG 2, it becomes obvious that there is a clear need to reorient agricultural production from a mere increase in food quantity toward delivery of more diverse and more nutritious food produced in a sustainable manner [15,16]. As incomes rise and food consumption patterns change, overnutrition from imbalanced diets also becomes a matter for concern, both in developed and in developing countries. The diversity and quality of food produced and consumed is a decisive factor when addressing the triple burden of malnutrition, i.e., undernutrition, micronutrient deficiency, and overnutrition.

Although fruit and vegetables are usually mentioned jointly when addressing malnutrition, this paper focuses mainly on the role of vegetables for nutrition security, their global production, and the current ex situ conservation of both global and traditional vegetables. The need for additional conservation efforts in hotspots of vegetable diversity to safeguard valuable germplasm for current and future breeding efforts is addressed. Ways in which genebanks can link with the farming communities to strengthen the informal seed sector and to introduce genetic diversity at the farm and household level are briefly described. In the formal seed sector, vegetable breeders need access to a wide diversity of genetic resources, predominantly farmers' varieties, landraces, and crop wild relatives (CWR). The use of conventional and modern breeding tools to make efficient use of CWR for the development of resilient cultivars is briefly discussed.

2. The Importance of Vegetables for Nutrition Security

The rise of modern agriculture is generally considered to have led to a decline in the number of plant species upon which humans depend for food [17,18]. This decline particularly affected the wild, semidomesticated, and cultivated traditional vegetables and fruits, spices, and other food plants that supplemented staple crops with micronutrients, essential for a healthy diet. Those crops/species also strengthened food security historically during failures of the main crops [19]. More recent analyses of food consumption trends, regionally and globally, revealed increases in the diversity of plants contributing to diets locally in developing countries [20]. These changes are driven by rising income, urbanization, trade liberalization, transnational food corporations, and food industry marketing and retailing through supermarkets. Consumers are now offered more diversity and convenience but at the same time cheaper, less healthy, and highly processed food with high content of fats and sugars has been made more easily accessible and affordable and is, therefore, in high demand [21]. These developments led to a Westernization transition of local diets with a preference for energy-dense foods, such as animal products, plant oils, and sugars over cereals, pulses, and vegetables, with a general preference for more global crop plants over locally produced traditional crops [22,23].

A recent study examined changes in the richness, abundance, and composition of crop commodities in the food supplies of 142 countries (comprising 98% of the global population from 1961 to 2009 [24]. During this 50-year period, national food supplies expanded quantitatively in terms of food calories, protein, fat, and weight, and significant quantities of food consumed were derived from energy-dense food sources, known to contribute to malnutrition. Because of these global food consumption trends, national food supplies became more similar in composition globally, thanks to a constant flow of a range of truly global cereal and oil crops, while the supply of local traditional crops declined. The growing reliance on the supply of those global food crops is leading to stronger interdependence among countries in their food supplies, underlying genetic resources, and nutritional priorities.

It is increasingly recognized that the global food system must shift its focus from food quantity toward dietary quality and health and environmental outcomes [15,16]. Fruit, vegetables, and nuts are increasingly entering into the focus of the nutrition community for their potential in combating the triple burden of malnutrition [25]. Traditional, underutilized crops, especially those which are locally available and culturally acceptable, are ideally placed to play a much greater role in contributing to improved nutrition and health [26], in line with the strategy proposed by the British Royal Society: "The preferred strategy to eliminate hidden hunger will always involve strategies to increase the diversity of diet with increased access to fruit and vegetables" [27]. The World Health Organization (WHO) recommends a population-wide daily intake of 400 g of edible fruit and vegetables for the prevention of NCDs, as well as for the prevention and alleviation of several micronutrient deficiencies [28]. This WHO recommendation translates to roughly five portions per day. It should be noted that potatoes, sweet potatoes, cassava, and other starchy roots are not considered as fruits or vegetables.

People able to enjoy more diverse diets, in general, also have better nutrition and health. A recent study analyzing data of a health survey in Great Britain revealed that there is a robust inverse association between fruit and vegetable consumption and mortality [29]. A study of Helen Keller International's integrated household food production program targeting women in Burkina Faso found significant improvements in several child nutrition indicators, including reductions in anemia, wasting, and diarrhea among young children [30]. The underlying agriculture production activities of this integrated program included input distribution (seeds, saplings, chicks, and gardening tools) and agricultural training. Household food production activities focused primarily on micronutrient-rich vegetables and fruits, eggs, and chicks.

Increasing production and consumption of fruit and vegetables is an obvious pathway to improve dietary diversity and enhance nutrition security, especially in the case of diets that are dominated by high-energy foods with low levels of micronutrients [29]. Such a move, coupled with a transition to diets higher in plant-based protein, will also help protect valuable habitats such as the Amazon rainforest and help meet the SDGs [31]. However, several studies suggested that current and projected fruit and vegetable production levels will fail to meet healthy consumption levels [31,32]. Following age-specific recommendations, only 40 countries representing 36% of the global population had adequate availability of fruit and vegetables by 2015 [33]. In many food-insecure countries in sub-Saharan Africa and South Asia, average fruit and vegetable consumption is well below WHO-recommended levels, with 10 countries not even meeting 30% of the recommended intake levels [34].

Similarly, projections to 2050 indicate that between 0.8 and 1.9 billion people living in countries in sub-Saharan Africa might need to contend with average fruit and vegetable availability below 400 g/person per day [33]. This estimate does not yet take food waste into consideration. Under a high-food-waste scenario, these authors predict that 139 countries representing 5.6 billion people will not be able to provide enough fruit and vegetables to the population by 2050—an increase of 1.5 billion people compared with a no-waste scenario. Reducing waste in vegetable and fruit value chains may also reduce the negative impact of their production on the environment and could keep consumer prices more accessible [35].

2.1. The Commodity Group "Vegetables"

Vegetables comprise a wide range of genera and species and are an important component of a healthy diet. They ensure nutrition security through the provision of vitamins, antioxidants, minerals, fiber, amino acids, and other health-promoting compounds [36], while enhancing diversity, flavor, and taste of many otherwise bland staple dishes. Due to their wide range of uses, it is difficult to assign crops to the commodity group "vegetables". Some legume crops, mainly grown for their dry seeds, are important vegetable crops when harvested and consumed at the immature stage as seeds, green pods, and/or leaves, as is common in Asia and Africa. This applies, for example, to crops like vegetable soybean (*Glycine*), mungbean (*Vigna radiata*), Azuki beans (*Vigna angularis*), cowpea (*Vigna unguiculata*), yard-long bean (*Vigna unguiculata* subsp. *sesquipedalis*), black gram (*Vigna mungo*), common bean (*Phaseolus*), winged bean (*Psophocarpus*), and garden pea (*Pisum*). Although the starchy root of cassava (*Manihot esculenta*) constitutes the primary use of this crop, cassava leaves are a common leafy vegetable in many African countries; hence, this crop also falls partially under vegetables.

The most dominant vegetables, also called global vegetables as they are grown in many countries around the globe, are tomatoes, cucurbits (pumpkins, squashes, cucumbers, and gherkins), alliums (onion, garlic, shallot), and chilies (sweet and hot pepper; *Capsicum* spp.). Other major vegetable crops based on farmgate value of global production, but not always truly global vegetables are spinach, brassicas (cabbages, broccoli, rape), vegetable legumes, eggplants, lettuce and chicory, carrots and turnips, and asparagus [37]. Production statistics usually do not list indigenous or traditional vegetables as these are often produced in home or family gardens or collected from the wild for family consumption, and they are, in general, only offered in local markets. The term "indigenous vegetables" primarily refers to plants grown in their centers of origin or diversity [38], but also encompasses plant species introduced from other geographical areas that adapted well, naturalized, and evolved in the new environment [39]. Indigenous vegetables are often more nutrient-dense than global vegetables [40], require low levels of external inputs, and cope well with abiotic and biotic stresses if grown on a small scale and in mixed cropping systems as is the case in their centers of origin. However, data on nutritional profiles of indigenous vegetables in raw and cooked forms are scarce. Among the traditional vegetables with high potential to be mainstreamed into urban markets and consumer diets are leafy amaranth (*Amaranthus* spp.), African eggplant (*Solanum macrocarpon*), jute mallow (*Corchorus olitorius*), roselle (*Hibiscus sabdariffa*), spider plant (*Cleome gynandra*), Ethiopian mustard (*Brassica carinata*), okra (*Abelmoschus esculentus*), and vegetable cowpea (*Vigna unguiculata*) [41].

2.2. Global Vegetable Production

The statistics of the Food and Agriculture Organization of the United Nations (FAO) cover 25 primary vegetable products (http://www.fao.org/faostat/en/#data/QC). Global primary vegetable production reached 1.09 billion tons in 2018, about 37% of global cereal production (2.96 billion tons) [42]. Asia is by far the largest producer of primary vegetables, responsible for three-quarters of global production (Figure 1). During the past 10 years (2008–2018), there was a 24% increase in global commercial vegetable production, mainly attributable to a significant production increase in Africa (32%) and Asia (28.3%) (Figure 2). The estimated farmgate value of annual global vegetable production reached 543 billion United States dollars (US$) in 2012–2013, about 65% of all food cereals combined, estimated at 837 billion US$ [37].

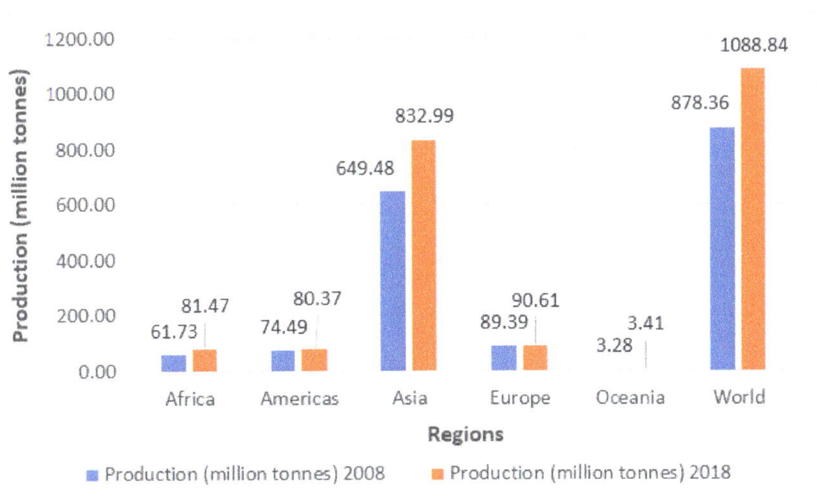

Figure 1. Comparison of global production of primary vegetables in 2008 and 2018, by major regions (Source: Statistics Division of FAO). Available online: http://www.fao.org/faostat/en/#data/QC/ (accessed on 14 March 2020).

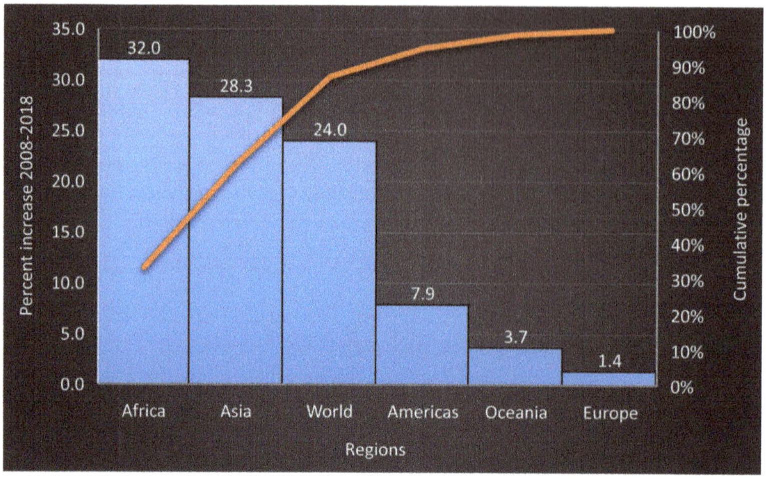

Figure 2. Percentage production increase of primary vegetables from 2008 to 2018 (calculated from production figures retrieved from Statistics Division of FAO). Available online: http://www.fao.org/faostat/en/#data/QC/ (accessed on 14 March 2020).

2.3. Research and Development Efforts Focusing on Vegetables

Although the farmgate value of annual global vegetable production is impressive when compared to global cereal production, the potential of vegetables to generate positive economic and nutritional impacts is still limited due to a relatively low level of support provided by national governments and development agencies to public sector vegetable research and development (R&D) [43]. Agricultural policies and investments in agriculture by the public and private sectors are still mainly focused on staple food production (cereals, roots, and tubers) that provide the primary source of calories, especially for low-income consumers. Fruit and vegetables, rich in micronutrients and vitamins, are not given appropriate support [44,45]. About 45% of private sector agricultural research and development investments into biotechnology and the crop seed industry are dedicated to a single crop—maize—and this investment is mainly taking place in industrialized countries [46].

With a few exceptions, private and public sector investments in vegetable research are mainly focusing on the development of hybrid cultivars of predominantly global vegetables, such as tomatoes, chilies, lettuce, cucumbers, and onions, while indigenous or traditional vegetables are being neglected. Data for 70 countries compiled by Schreinemachers et al. [37] clearly indicate the underinvestment in vegetable research and development compared to cereals. While there are 4–5 publicly funded cereal researchers per one million inhabitants in all country groups, low- and lower–middle-income countries, on average, have only one researcher working on fruit or vegetables [38]. When research investment in fruit and vegetables is placed in relation to production value of the different commodity groups, investment in fruit and vegetables is about the same for all country groups. However, higher-income countries allocate much higher research investment resources per dollar of cereal output than per dollar of vegetable output.

Investments in fruit and vegetable production should primarily target regions where projected supply is insufficient, such as sub-Saharan Africa, parts of Asia, and the Pacific. As fruit and vegetable production is mostly in the hands of small-scale producers in these regions, investments in this sector would have substantial potential to increase incomes and food security in these regions, particularly in Africa [33], thus contributing to the attainment of SDG 1—no poverty and SDG 2—zero hunger.

More research is also needed with respect to how nutrient profiles of global and traditional vegetables change in relation to cultivar choice, growing conditions, processing, and cooking methods. In addition, it is essential to study the extent of nutrient uptake and absorption by the human body.

A study of the factors determining micronutrient bioaccessibility in leafy vegetables revealed that the pectin content of the leaves impaired carotenoid bioaccessibility [47]. Leafy vegetables rich in condensed tannins, such as drumstick tree (*Moringa oleifera*), had exceptionally low content of pectin and were characterized by high micronutrient bioaccessibility. Therefore, selection and development of cultivars with high micronutrient and low pectin content is a good approach to improve absorption of micronutrients by the human gut.

Fruit and vegetables are also quite perishable, with some estimates suggesting that these commodities contribute more than 40% of total food losses and waste [48]. Investments in R&D of vegetables and fruit value chains, focusing on new and improved processing, storage, and distribution technologies, could lead to a significant reduction of food losses and waste. Moreover, value addition through the development of drying technologies and novel extraction methods transforming fruit and vegetable waste (such as pomace, peels, and seeds) into new dehydrated and nutraceutical by-products could lead to a further reduction of food waste [49].

In general, it is important that the research and development community not only focuses on the farm segment of the food system but increasingly widens the understanding and supports the transformation of the entire system. Research strategies must include vegetable processing, the logistics of storage and transport to urban markets, and wholesale of inputs and outputs in the food system [21]. R&D investment efforts in these off-farm segments which account for 40–70% of value added and costs of food merit increased attention and could lead to lower marketing margins in value chains, thus improving efficiency.

As low consumption is also a problem where fruit and vegetable availability is not a constraint, it is important to address necessary changes in consumer behavior, apart from increasing production and availability of such crops. The provision of quantified food-based dietary guidelines is indicated to give the population some orientation, and such guidelines are missing in many countries, especially in Africa [50]. Apart from behavior change communication aimed at encouraging people to adopt healthier food choices that include fruit and vegetables, other key approaches are vegetable home gardens (see below) and school gardens. The latter simultaneously strengthen demand and supply and create awareness of the importance of the inclusion of fruit and vegetables in household and school meals to improve nutrition and health [37].

2.4. *Diversifying Production Systems and the Role of Home and Household Gardens for Nutrition Security*

The growing demand for fruit and vegetables can be met through diversification of staple crop systems by including fruit and vegetable crops or through intensification of specialized fruit and vegetable systems, especially in peri-urban zones [51]. Integrating small-scale farmers with formal and informal market outlets was shown to strongly encourage those farmers to adopt diversified vegetable production and other sustainable intensification practices in Kenya, as these result in economic benefits to farmers [52].

In view of global climate change, diversifying agricultural and horticultural production will help increase resilience of farming systems for both biotic and abiotic stresses. By the end of the 21st century, crop calorific yields in single cropping systems in sub-Saharan Africa are predicted to reach only 40–55% of the crop calorific yields obtained in sequential cropping systems [53]. Multispecies cropping systems constitute practical applications of ecological principles based on biodiversity, plant interactions, and other natural regulation mechanisms [54]. Such systems offer potential advantages in productivity, yield stability, ecological sustainability, and resilience to disruptions caused by climate change and other natural events but are sometimes considered harder to manage effectively than monocultures. A wider use of neglected and undervalued fruit and vegetable crops and semi-domesticated species, either intercropped with main staples in cereal-based systems or as stand-alone crops, would provide multiple options to build temporal and spatial heterogeneity into uniform cropping systems. Such an approach will enhance resilience to biotic and abiotic stress factors which are exacerbated by global climate change and will ultimately lead to a more sustainable supply chain of diverse and nutritious food [55].

Pilot studies conducted in Kenya and Vietnam revealed that diversification of smallholder farms with underutilized and traditional vegetable crops such as African nightshade (*Solanum americanum*), cowpea (*Vigna unguiculata*) crotalaria (*Crotalaria brevidens*), French beans (*Phaseolus vulgaris*), groundnuts (*Arachis hypogea*), kale (*Brassica oleracea* var. *acephala*), pumpkin (*Cucurbita maxima*), purple amaranth (*Amaranthus blitum*), spider plant (*Cleome gynandra*), mustard greens (*Brassica juncea*), orange-fleshed sweet potato (*Ipomoea batatas*), and water spinach (*Ipomoea aquatica*) could cover vitamin A requirements of 10–31 extra people per hectare and enhance income by 25% to 185% [56]. However, a trade-off of these diversification interventions was reduced leisure time.

Apart from such specialized, intensive production systems, home or household garden interventions have a proven effect on nutritional outcomes. Widely practiced forms of gardening consist of the mixed cropping of different species of fruits, vegetables, herbs, spices, and other useful plants as a supplementary source of food and income, and such gardens are common in Asia [57]. A well-planned and well-managed garden is expected to supply the household with a diverse range and year-round harvest of fruit and vegetables for household consumption with possible surplus sold on local markets. Such interventions potentially contribute to eight of the 17 Sustainable Development Goals [58]. The key benefits of home gardening were summarized by Landon-Lane [59] as follows:

- Improved food security,
- Increased availability of food and better nutrition through food diversity,
- Income and enhanced rural employment through additional or off-season production,
- Decreased risk through diversification,
- Environmental benefits from recycling water and waste nutrients, controlling shade, dust, and erosion, and maintaining or increasing local biodiversity.

Home gardens are typically targeted at women as they are, in general, in control of meal choice and preparation, and this may lead to women's empowerment as shown in Bangladesh [60]. Experiments with home gardens in Hyderabad, southern India, including about two dozen vegetable species, showed that a small area of 6 m × 6 m can provide much of the vitamin A and C requirement for a family of four during the entire year [34]. Apart from the provision of essential vitamins, many of the vegetable crops included in home garden kits are known to be naturally nutrient-dense [40,61,62].

3. Vegetable Genetic Resources Conservation and Linkages with the Farming Community

As genetic erosion continues in situ and on-farm due to expansion of human settlements, climate change, and replacement of landraces by high-yielding hybrid cultivars, additional collecting and conservation efforts are mandatory with a major focus on crop wild relatives and poorly represented landraces of major and minor vegetable groups. Special attention is needed to conserve the genetic diversity of indigenous and underutilized vegetable crops which, in general, are poorly conserved.

3.1. Ex Situ Conservation of Vegetable Genetic Resources and Collecting Needs

The Second Report on the State of the World's Plant Genetic Resources for Food and Agriculture (SoWPGR-2) indicates that about 7.4 million accessions are currently maintained ex situ, globally, 1.4 million more than were reported in the first SoW report [63]. Information in the World Information and Early Warning System (WIEWS) on the Plant Genetic Resources for Food and Agriculture (PGRFA) database indicates that about 45% of all the accessions in the world's genebanks are cereals, followed by food legumes with 15% of global holdings [63]. About one million accessions of crops used fully or partially as vegetables are conserved ex situ [64]. In a narrow sense of crops exclusively used as vegetables, about 518,000 accessions of vegetables representing 7% of the globally held 7.4 million accessions of PGRFA are maintained ex situ [63]. Among vegetable commodities, tomatoes (84,289 accessions), chilies (73,572), brassicas (25,566), cucurbits, excluding melons and cantaloupe (39,583), alliums (29,898), okra (22,428), and eggplant (21,616) are well represented in ex situ collections at the global level [64].

The World Vegetable Center (WorldVeg) with its headquarters in Taiwan holds about 63,500 accessions of vegetable germplasm comprising 170 genera and 456 species from 158 countries of origin [41], including some of the world's largest vegetable collections held by a single institution, such as chilies, tomato, and eggplant, as well as about 12,000 accessions of indigenous vegetables [40] (https://avrdc.org/our-work/managing-germplasm/). The WorldVeg germplasm collection can either be searched in its own database AVGRIS (http://seed.worldveg.org/search/passport) containing 71,889 passport records, or in the global database Genesys (https://www.genesys-pgr.org/) with 59,954 accession records.

WorldVeg works with Kew Royal Botanic Gardens, United Kingdom (UK) and partners in the Global Crop Wild Relative project to rescue the diversity of eggplant wild relatives from Africa. Through the project, WorldVeg obtained the seeds of 217 accessions of 18 species of eggplant wild relatives [41]. These accessions are currently being multiplied and characterized by WorldVeg genebank staff. Wild relatives of cucurbits, vegetable cowpea, and okra are other priority species in need to be collected in Africa and Asia through partnerships with local players to safeguard landraces and wild relatives of vegetable crops in both continents.

WorldVeg collaborates with National Plant Genetic Resources Centers in Eastern and Southern Africa to identify hotspots of vegetable diversity and to upload vegetable biodiversity data to public databases. In total, 126 high-potential traditional African vegetables relevant for people's diets in different regions of Africa were identified in collaboration with the World Agroforestry Center and the Inland Norway University of Applied Sciences [41]. Hotspots of vegetable diversity were identified in the coastal regions of West Africa, in Cameroon, South Sudan, Ethiopia, Tanzania, Madagascar, and Eswatini (formerly known as Swaziland). This vegetable diversity is poorly represented in genebanks and requires urgent conservation action.

The large group of gourds—bitter gourd (*Momordica charantia*), sponge gourd (*Luffa aegyptiaca* and *L. acutangula*), bottle gourd (*Lagenaria siceraria*), wax/ash gourd (*Benincasa hispida*), ivy gourd (*Coccinia grandis*), snake gourd (*Trichosanthes cucumerina*), and spiny gourd (*Momordica dioica*)—requires additional collecting with a major focus on landraces from centers of origin and diversity, many of which are threatened by the introduction of hybrid cultivars by seed companies, a process which recently started in many countries in South and Southeast Asia. Collecting led by national genebanks should focus primarily on Bangladesh, Myanmar, Vietnam, and India, countries rich in landrace diversity of gourds and other underutilized traditional crops. Such efforts could be undertaken in the context of FAO's Regional Initiative on the Zero Hunger Challenge for Asia and the Pacific, which is targeting the hidden treasures embodied in neglected and underutilized species (NUS), calling them Future Smart Food [65]. These foods are considered smart as they can bolster dietary diversification, improve micronutrient intake, require fewer inputs such as chemical fertilizers, enhance soil health, and often provide resilience to climate change and adverse farming conditions. Based on national NUS scoping studies, eight countries in South and Southeast Asia already prioritized up to six promising NUS as candidates for future smart food.

3.2. *Linking Genebanks with the Farming Communities*

To combat malnutrition, there is a clear need for supportive policies to advance ex situ and on-farm/in situ conservation and documentation of underutilized traditional crops and to forge stronger links between genebanks and the farming communities to strengthen the production and consumption of a wide range of diverse vegetables and fruit trees. In formal seed systems, genebanks usually supply genetic diversity primarily to plant breeders and research organizations who act as intermediaries between them and the farming communities based on a multi-step system of selection, breeding, testing, marketing, and adoption [66]. Today, genebanks are serving an expanded range of actors and institutions, beyond the conventional R&D set-up as outlined in the revised international Genebank Standards published in 2014: "Genebanks should promote the availability of genetic resources for uses including research, breeding, education, farming, and repatriation" [67] (p. 54). Research undertaken by

Westengen et al. [68] showed that farmers, farmer organizations, and non-governmental organizations (NGOs) comprise a considerable user group of genebank material, having received about the same percentage (8%) of seed samples distributed by international genebanks in 2015 as distributed to the commercial seed sector. They identified six potential direct genebank–farmer linkages [68]: (1) reintroduction, (2) emergency seed interventions, (3) community seed banks, (4) participatory plant breeding, (5) variety introduction, and (6) integrative seed system approaches. In addition to the common role of genebanks to provide diverse crop germplasm to breeders and, thus, feeding the formal seed sector, such alternative and complementary strategies would strengthen the informal seed sector by enhancing farmers' access to crop diversity. Two of these genebank–farmer linkages, i.e., community seed banks and variety introduction, are described below.

3.2.1. Community Seedbanks for Locally Important Crop Diversity

Community seedbanks (CSB) can be described as structures in which organized groups of farmers are responsible for the different stages of the management of seed or vegetative propagules, such as selection, conservation, multiplication, exchange, and improvement [69]. CSBs were successfully established all over the world in the last few decades, often supported and funded by NGOs. The main functions of CSBs include (a) conserving and reintroducing predominantly local germplasm, (b) providing easy access to seeds for members of the community, and (c) enhancing seed and food sovereignty [70]. If responsibly managed and supported, such CSBs can support the local community effectively and serve as a reliable seed source of locally important germplasm. Even in Europe, community seed banks are being established at a rapid pace, with at least 130 initiatives reported in 2017 [71].

Key elements in ensuring sustainability for community seedbanks include capacity development to ensure management quality, sustainable mechanisms, i.e., voluntary (in-kind) contributions by farmers to reduce dependence on external funding, enabling legal and political framework conditions to ensure a safe legal basis for operation, enabling social structures among those involved, satisfactory physical infrastructure, and systematic planning processes with effective operational mechanisms.

Community seedbanks and on-farm conservation efforts of crop diversity are only successful if they are fully embedded within the farmers' livelihood strategies and support the production of nutritious food, the generation of income, and other benefits such as ecosystem services, and socio-cultural and religious practices. Successful examples were documented in a community-based agrobiodiversity management project across Latin America and Southeast Asia [24].

A project on community-based seed production of traditional vegetables in the Philippines provided technical support to farmers for the conservation and multiplication of traditional vegetables to ensure the availability of high-quality seed of promising lines for home and school gardens and commercial production [72]. The project introduced several varieties of five traditional vegetable crops from the WorldVeg genebank: jute mallow (*Corchorus olitorius*; six varieties); cucumber (*Cucumis sativus*; two varieties); bottle gourd (*Lagenaria siceraria*; four varieties); eggplant (*Solanum melongena*; six varieties); ridged gourd (*Luffa acutangula*; three varieties). These crops were complemented with farmer-saved seed of other popular local traditional vegetable crops, such as Lima bean (*Phaseolus lunatus*), butterfly pea (*Clitoria ternatea*), and vegetable hummingbird (*Sesbania grandiflora*). The project made a significant contribution to halting the on-going threat of genetic erosion of local landraces and semi-wild vegetable crops. It also empowered farmers, especially women, to save, use, exchange, and sell their seeds to sustain the diversity of crops grown on-farm and promote greater diversity in diets.

To create the necessary legal foundation for CSBs, it is recommended that national governments integrate farmers' rights to save, use, exchange, and sell farmer-saved seeds in national seed legislations [73]. In this way, CSBs represent a form to implement farmers' rights. CSBs also play an intermediary role between genebanks and local seed systems as they multiply the relatively small seed samples usually distributed by genebanks to seed lots that are big enough for the needs of farmers [68].

3.2.2. Variety Introduction of Agricultural Crops in General and Vegetable Crops in Particular

Variety introduction efforts are sometimes triggered by the need to provide farmers with a range of crop species and varieties to help them cope with environmental limitations in a changing climate by matching varieties to diverse production conditions and weather extremes [74]. To strengthen food security in sub-Saharan Africa (SSA), van Etten [74] proposed a crowd-sourcing approach to seed-based innovation, starting with the distribution of many small seed packets of different crops and varieties consisting of farmer varieties, landraces, and modern cultivars. For their distribution, existing networks such as retail stores, NGOs, and churches could be used. After having grown out the seed samples, farmers should provide basic agronomic data on the different varieties using their mobile phones. Farmers could also be requested to provide feedback on the performance of their own varieties compared to the newly introduced varieties. If local varieties prove to be superior to the introduced varieties, they could, if farmers agree, be included in the next evaluation round. This would then lead to a constant exchange of diverse varieties among the farming communities.

Such an approach was initiated in 2009 by Bioversity International through a project called Seeds for Needs (S4N) with project sites in 13 countries in Africa, Asia, and Latin America [68]. The project used crowdsourcing to evaluate both landraces and advanced cultivars and showed that landraces have great potential by offering valuable options to farmers having to deal with climate-related risks [75].

In Spain, the National Center for Plant Genetic Resources (CRF) of the National Institute for Agricultural and Food Research and Technology (INIA) recently launched an initiative to engage farmers, farmer associations, and relevant companies in the primary evaluation of germplasm that is conserved by the national ex situ collection network [76]. This initiative is linked to the First Action Plan of the National Program for the Sustainable Conservation and Use of Genetic Resources for Food and Agriculture (Order APA/63/2019). CRF will make selected varieties of local crop germplasm available to the farming community for cultivation under diverse growing conditions. In return, farmers will share information on the performance of the varieties with CRF, providing data on yield, incidence of diseases, insect pests and weeds, organoleptic quality, and other relevant data. This initiative is expected to strengthen on-farm conservation of valuable crop germplasm and to enhance agrobiodiversity, in line with SDG 2 (Zero Hunger).

The World Vegetable Center came to similar findings with variety introduction in Central Asia and the Caucasus (CAC). Starting from 2005, national research institutes in the CAC region received over several years a total of 2103 breeding lines and landraces of different vegetable crops from the World Vegetable Center [77]. About 45% of materials sent by WorldVeg were genebank accessions, mostly landraces and farmers' varieties [78]. The institutes undertook selection and adaptation trials with the materials received, and these efforts led to the registration and official release of 91 vegetable varieties as of 2017. Of these 91 new cultivars, 32 (35%) were developed from WorldVeg genebank accessions, 57 came from breeding lines, and only two resulted from actual crosses made using WorldVeg breeding lines. Meanwhile, another 10 varieties are already submitted for registration and subsequent release. The success of research institutes in releasing genebank accessions as varieties, without further cross-breeding, can be attributed to at least four factors [77]: (1) a relatively large number of accessions were tested for their performance under local conditions; (2) extensive selection trials were conducted to purify lines and stabilize traits by selecting for disease resistance and yield performance for at least three years; (3) disease pressure is relatively low in the CAC region due to favorable local climatic conditions (cold winters, hot and dry summers); (4) none of the institutes had well-functioning breeding pipelines at the start of this variety introduction process. Selection trials were, therefore, an effective way to create new cultivars.

The WorldVeg genebank of predominantly traditional vegetables in Arusha, Tanzania used a similar approach of variety introduction into East Africa via vegetable seed kit distribution. Between 2013 and 2017, this genebank distributed more than 42,500 seed kits totaling 183,193 seed samples to smallholder farmers in Tanzania, Kenya, and Uganda [79]. The seed kits usually comprised about four seed samples of different vegetable crops/varieties, enough to grow out in a home garden. About 32%

of the seed distributed came from promising accessions of the genebank, while 68% were WorldVeg breeding lines. All materials distributed, including breeding lines, were maintained by the genebank. The WorldVeg genebank in Arusha used five important criteria when composing the seed kits [80]:

(1) Only those accessions and breeding lines were selected for the vegetable seed kits which had undergone screening for yield, disease resistance, and consumer preference under local conditions in Tanzania.
(2) Distributed accessions and breeding lines were open-pollinated so that farmers could save seed for the next cropping cycle or for sharing extra seed with other famers in the community.
(3) Seed distribution channels were international NGOs, farmer groups, and local government units and WorldVeg development projects. Seed kit distribution was not intended as emergency seed aid. The distribution through projects was related to home garden projects intended for improving dietary diversity and diversifying incomes.
(4) Seed kits were distributed together with capacity building in vegetable production and seed saving provided by NGOs or project staff. Seed kits also contained instructions for good agricultural practices for the successful cultivation of the crops and information on the nutritional value of the crops included in the seed kits.
(5) Seed kits were distributed only once to the same household. A regular supply of seed was not envisaged to avoid damaging local seed enterprises which later picked up seed production and sale of some traditional vegetables.

There is a clear trend in some East African countries (Kenya, Tanzania, Uganda) that seed companies become interested in multiplying and selling traditional African vegetable crops due to high demand from the farming community and consumers. To enhance vegetable diversity grown by smallholder farmers across SSA, the WorldVeg genebank in Arusha strongly engages with partners of the formal, as well as the informal, seed sectors.

4. Vegetable Genetic Resources as Building Blocks for Vegetable Breeding

Successful breeding, in general, requires access to a wide diversity of plant genetic resources, predominantly farmers' varieties, landraces, and crop wild relatives (CWR), the building blocks of intra- and interspecific crop diversity. Such plant genetic resources represent a treasure trove of genes for vegetable and legume crop improvement [80], enabling the delivery of more nutritious quality food in sufficient quantity for the world population.

Vegetable breeding should address the needs of both growers and consumers. Vegetable growers appreciate cultivars with high yield, uniformity, good market acceptance, multiple disease and pest resistance, and abiotic stress tolerance. Consumers like to buy vegetables with good appearance, shelf life, quality, taste, and nutritional value. Increased phytonutrient density in vegetable crops could help overcome micronutrient malnutrition and improve human health. The tomato breeding program at the World Vegetable Center includes enhanced phytonutrient content as a breeding objective, and it developed high-yielding and multiple disease-resistant lines with increased content of beta-carotene, lycopene, flavonoids, or anthocyanin in different fruit types [81]. WorldVeg breeders used the *Beta* allele from the wild species *Solanum hirsutum* which shifts tomato carotenoid from lycopene almost entirely to beta-carotene and results in orange fruit color. So-called "golden tomatoes" developed by WorldVeg breeders through conventional breeding techniques contain 3–6 times more provitamin A carotenoids than standard tomatoes, and one golden fresh market tomato can provide a person's full daily vitamin A requirements. WorldVeg *Beta* breeding lines in fresh market and cherry tomato fruit types were officially released as cultivars in Mali, Taiwan, and Bangladesh [81]. However, adoption is low so far as consumers are unaware or reluctant to accept the unfamiliar orange fruit color. Many major genes affecting nutrient content in tomato are known, and there is significant scope to enhance phytonutrient content in this crop through conventional breeding [81], but few breeding programs pursue better nutrition in their breeding objectives due to lack of market incentives and not strongly enough articulated consumer demands.

Sustainable intensification of horticultural crops requires the development of new varieties with stable yields under climate change scenarios and adaptive capacity to diverse agro-ecosystems. The narrow genetic base of many vegetable crop cultivars is a major challenge for breeders aiming to develop improved varieties with multiple disease and insect pest resistance and tolerance to abiotic stresses such as heat, salinity, and drought, as well as increased input-use efficiency. Those breeding objectives require making use of interspecific crop diversity. Zhang et al. [82] were able to develop interspecific bridge lines among *Cucurbita pepo*, *C. moschata*, and *C. maxima*. With the development of these lines, it was possible to overcome the crossing barriers of interspecific hybridization and to eliminate the sexual obstacles of subsequent generations. This important breakthrough created a platform for breeders to transfer favorable traits among these species freely without the introgression of undesired characters from a wild species during the breeding process.

In contrast to public breeding programs such as those implemented by WorldVeg, for example, private seed companies primarily focus on the development of hybrid cultivars by exploiting heterosis effects and building multiple biotic stress resistance factors, as well as tolerance against abiotic stresses into a single commercial cultivar. This process ensures that growers must purchase fresh seeds for each growing cycle and the control of the parents prevents other seed companies from reproducing the hybrids. At the global level, the share of hybrid seed production is growing at a fast pace of 8–10% per annum in most major vegetable crops [83]. The global vegetable seed market was valued at US$ 9.163 billion in 2018 and is projected to increase at a compound annual growth rate of 9.4% during the period from 2019 to 2024 [84]. North America is among the largest markets for vegetable seed production and consumption, followed by the Asia-Pacific region and Europe. Tomato, cabbage, sweet pepper, and lettuce are key players in the global seed market with a share of more than 30%.

Often, private seed companies make use of public breeding products when developing hybrid cultivars. A typical case is chili pepper (*Capsicum* spp.) variety development in India. Hybrid cultivars account for about 25% of the total area under chili pepper cultivation, about 25% of the area is under improved open-pollinated varieties (OPVs), and the remaining 50% area is still grown with local landraces [85]. The current chili pepper hybrid seed market in India is estimated at about 50 t per year, with an estimated turnover worth 16 million US$, and this was made possible thanks to the use of WorldVeg cytoplasmic male sterility (CMS) breeding lines which reduce the cost of hybrid seed production by 50% compared to conventional hybrid seed development using manual emasculation. Conservative estimates suggest that hybrids involving WorldVeg germplasm and improved breeding lines as one of the parents were cultivated on more than 30,000 ha during 2012–2013 in different regions of India [85].

The potential of wild species as a source of genetic variation to bring about crop improvement was recognized early in the 20th century but is not yet widely used in crop breeding. An exception is perhaps the model crop tomato, which, in terms of production volume, is the most important vegetable crop grown worldwide. For this crop, an enormous amount of biotic and abiotic stress tolerance traits were already studied in the pool of wild relatives and extensively used in tomato breeding. Virtually all significant resistance genes to tomato diseases were sourced from wild relatives. An overview of such disease resistance genes introduced from wild species into cultivated tomato was provided by Ebert and Schafleitner [86]. However, it is essential to strengthen similar research in other major and minor vegetable crops as well.

Initially, molecular breeding complemented conventional breeding methods through marker-assisted selection (MAS) or marker-assisted backcrossing (MABC) [83]. Molecular markers intricately linked to the trait of interest can be detected and used in gene pyramiding, thus facilitating introgression of desirable, mostly monogenic traits from exotic germplasm into elite cultivars. Relatively little work was done with respect to traits that are governed by quantitative trait loci (QTLs). Traits such as yield, quality, and stress response show complex inheritance patterns that result from the segregation of numerous interacting QTLs, the expression of which is modified by the environment [87].

New breeding approaches such as "introgressiomics" allow the creation of highly diverse plant materials and populations carrying introgressions of genome segments from mostly wild crop relatives into the genetic background of crops [88]. Introgressiomics can lead to the development of chromosome substitution lines (CLs), introgression lines (ILs), and multi-parent advanced inter-cross (MAGIC) populations. Such materials can be directly used in breeding pipelines and will facilitate the development of new resilient cultivars.

ILs contain the full genome of a given crop, except for a small chromosomal segment of a wild donor parent [87]. ILs are obtained through repeated backcrossing of the hybrid to the recurrent parent. Molecular markers help tracking the introgressed fragments, thus supporting the selection of beneficial materials for subsequent backcross cycles. A final step in the development of ILs is selfing or obtaining doubled haploids to fix the introgressed fragment in a homozygous state [89]. A further advantage of ILs is the ability to intercross favorable traits that are present in different ILs for the pyramiding of desirable alleles such as yield QTLs in tomato [90].

As early as 1995, Eshed and Zamir [91] developed a novel tomato population consisting of 50 ILs originating from a cross between the drought-tolerant, green-fruited wild tomato species *Lycopersicon pennellii*, and the elite tomato inbred line M82. Each of the lines contains a single homozygous restriction fragment length polymorphism-defined *L. pennellii* chromosome segment and, together, the lines provide complete coverage of the genome and a set of lines nearly isogenic to M82. These nearly isogenic lines of the IL population provide increased sensitivity for QTL mapping compared to whole-genome segregating populations, and they were extensively used during the last two decades to map QTLs for diverse tomato traits [92].

Only recently, the first eggplant introgression line population was developed using the drought-tolerant wild species *S. incanum* as a donor parent [93]. Sixty-eight candidate genes involved in drought tolerance were identified in the set of 25 fixed ILs. Apart from drought tolerance, *S. incanum* is also known to have a high content of bioactive phenolic compounds [94], as well as resistance to some diseases [95]. Open-field and screenhouse evaluations of the eggplant IL population mentioned above revealed that desirable traits such as lack of prickles and yield did not undergo considerable changes in most ILs despite the introgression of relatively large fragments from the wild exotic parent [96]. Ten stable QTLs distributed across seven chromosomes were detected, and three of the fruit-related QTLs appeared to be syntenic to other ones previously reported in eggplant populations. The other seven stable QTLs are new ones demonstrating that eggplant ILs are highly relevant for eggplant breeding under different environments and climatic conditions.

Genomics-assisted breeding is increasingly facilitating the introgression of favorable genes and QTLs with complex inheritance patterns from wild species into cultigens. The long-term conservation of genetic resources of landraces and crop wild relatives, their full characterization and evaluation, and their availability and accessibility will be instrumental for their successful use in public and private breeding programs. Such mobilization of the biodiversity available in the wider crop genepools will allow breeders to develop varieties that bear multiple disease and insect resistance and are able to adapt to rapidly changing environmental conditions, thus boosting agricultural production and ensuring food and nutrition security.

Moving such germplasm, once identified, across borders also requires robust phytosanitary capacity and practices and appropriate distribution capacities of genebanks, especially in the case of crops with recalcitrant seeds or those that are predominantly vegetatively propagated. Effective use of crop wild relatives and landraces also requires strong research capacity and participatory pre-breeding approaches. To enhance adoption, farmers should be actively involved in the definition of breeding objectives and the selection process for the development of new varieties, either through conventional or molecular breeding methods.

In contrast to the common vegetables grown globally, traditional vegetables, especially those which originated in Africa and the Asia-Pacific region receive much less attention from the research, conservation, and breeding community, although they have the potential to play a much greater role in more nutrition-oriented agriculture [97]. Neglect by research and breeders applies to many underutilized fruit and vegetable species. These include those crops with perennial growth forms such as trees, e.g., drumstick tree (*Moringa oleifera* [55]), and shrubs with edible leaves which are suited for agroforestry systems that may have an increasing role to play in sustainable vegetable production systems in developing countries under climate change scenarios.

Compared to major staple food crops, relatively little investment was made in breeding traditional, underutilized crop varieties [98]. Hence, the latter typically do not meet modern standards for uniformity and other characteristics required in the marketplace, and they tend to be less competitive than globally grown and traded crop cultivars. Farmers' varieties, landraces, and CWR were hitherto increasingly valued and exploited for genes that provide increased biotic resistance, tolerance to abiotic stress, yield, and quality [99–101]. However, use of agricultural biodiversity should not be restricted to exploiting valuable genes for use in breeding programs if our aim is to create more robust and resilient production systems. Underutilized food sources, including fruit and vegetables, legume crops, and root and tuber crops, have the potential to make a substantial contribution to food and nutrition security, protect against internal and external market disruptions and climate uncertainties, and lead to better ecosystem functions and services, thus enhancing crop and farming sustainability [102].

5. Conclusions

Vegetable crops are key sources of essential micronutrients required for good health. They contribute variety, flavor, taste, and nutritional quality to human diets. Increasing production and consumption of vegetables constitutes a direct and affordable way to deliver better health and overcome malnutrition. Vegetable production has the potential to generate more income and employment than any other segment of the agricultural economy. Vegetables can be grown on small areas of land, close to the consumers in urban and peri-urban settings, and they do not necessarily need advanced technologies to grow them. To realize those benefits, governments and donors need to give more weight and support to the ex situ, on-farm, and in situ conservation of genetic resources (farmers' varieties, landraces, and CWR) of global, as well as traditional, vegetables. The effective utilization of those genetic resources in breeding programs and the testing and deployment of newly developed varieties with tolerance to abiotic stresses and resistance against multiple diseases and insect pests in farmers' fields will ultimately benefit the farming community and consumers. By doing so, a significant reduction or, even better, a complete elimination of the obvious and persistent gap between WHO-recommended and actual intake levels of (fruit and) vegetables would make a significant contribution to the achievement of Sustainable Development Goals related to food and nutrition security and good health, in particular SDG 2.3 aiming at doubling agricultural productivity and incomes of small-scale food producers, SDG 2.4 ensuring sustainable food systems, and SDG 2.5 maintaining genetic diversity.

Funding: This research received no external funding.

Acknowledgments: The author wishes to thank Johannes M.M. Engels for his critical review and comments on a previous version of this manuscript.

Conflicts of Interest: The author declares no conflict of interest.

References

1. Pingali, P.L. Green revolution: Impacts, limits, and the path ahead. *Proc. Natl. Acad. Sci. USA* **2012**, *109*, 12302–12308. [CrossRef] [PubMed]
2. FAO-Food and Agriculture Organization of the United Nations. *The State of Food and Agriculture 2003–2004*; FAO: Rome, Italy, 2004; p. 222.

3. Evenson, R.E.; Gollin, D. Assessing the impact of the Green Revolution, 1960 to 2000. *Science* **2003**, *300*, 758–762. [CrossRef] [PubMed]
4. FAO; IFAD; UNICEF; WFP; WHO. *The State of Food Security and Nutrition in the World 2019. Safeguarding Against Economic Slowdowns and Downturns*; FAO: Rome, Italy, 2019; p. 239.
5. United Nations Development Programme (UNDP). Sustainable Development Goals. Available online: https://www.undp.org/content/undp/en/home/sustainable-development-goals.html (accessed on 1 March 2020).
6. Godfray, H.C.J.; Beddington, J.R.; Crute, I.R.; Haddad, L.; Lawrence, D.; Muir, J.F.; Pretty, J.; Robinson, S.; Thomas, S.M.; Toulmin, C. Food security: The challenge of feeding 9 billion people. *Science* **2010**, *327*, 812–818. [CrossRef] [PubMed]
7. Foley, J.A.; Ramankutty, N.; Brauman, K.A.; Cassidy, E.S.; Gerber, J.S.; Johnston, M.; Mueller, N.D.; O'Connell, C.; Ray, D.K.; West, P.C.; et al. Solutions for a cultivated planet. *Nature* **2011**, *478*, 337–342. [CrossRef] [PubMed]
8. Tilman, D.; Balzer, C.; Hill, J.; Befort, B.L. Global food demand and the sustainable intensification of agriculture. *Proc. Natl. Acad. Sci. USA* **2011**, *108*, 20260–20264. [CrossRef] [PubMed]
9. Valin, H.; Sands, R.D.; Van der Mensbrugghe, D.; Nelson, G.C.; Ahammad, H.; Blanc, E.; Bodirsky, B.; Fujimori, S.; Hasegawa, T.; Havlik, P.; et al. The future of food demand: Understanding differences in global economic models. *Agric. Econ.* **2014**, *45*, 51–67. [CrossRef]
10. Alexandratos, N.; Bruinsma, J. *World Agriculture towards 2030/2050: The 2012 Revision*; ESA Working Paper No. 12-03; FAO: Rome, Italy, 2012; p. 147.
11. Depenbusch, L.; Klasen, S. The effect of bigger human bodies on the future global calorie requirements. *PLoS ONE* **2019**, *14*, e0223188. [CrossRef]
12. Development Initiatives. *2018 Global Nutrition Report: Shining a Light to Spur Action on Nutrition*; Development Initiatives: Bristol, UK, 2018; Available online: https://globalnutritionreport.org/reports/global-nutrition-report-2018/ (accessed on 2 March 2020).
13. Centers for Disease Control and Prevention (CDC). Overweight & Obesity. Available online: https://www.cdc.gov/obesity/index.html (accessed on 2 June 2020).
14. Bennett, J.E.; Stevens, G.A.; Mathers, C.D.; Bonita, R.; Rehm, J.; Kruk, M.E.; Riley, L.M.; Dain, K.; Kengne, A.P.; Chalkidou, K.; et al. NCD countdown 2030: Worldwide trends in non-communicable disease mortality and progress towards sustainable development goal target 3.4. *Lancet* **2018**, *392*, 1072–1088. [CrossRef]
15. Willett, W.; Rockström, J.; Loken, B.; Springmann, M.; Lang, T.; Vermeulen, S.; Garnett, T.; Tilman, D.; DeClerck, F.; Wood, A.; et al. Food in the anthropocene: The EAT-lancet commission on healthy diets from sustainable food systems. *Lancet* **2019**, *393*, 447–492. [CrossRef]
16. Fischer, G.C.; Garnett, T. *Plates, Pyramids, and Planets: Developments in National Healthy and Sustainable Dietary Guidelines: A State of Play Assessment*; Food and Agriculture Organization of the United Nations: Rome, Italy, 2016.
17. Gepts, P. Plant genetic resources conservation and utilization: The accomplishments and future of a societal insurance policy. *Crop Sci.* **2006**, *46*, 2278–2292. [CrossRef]
18. Harlan, J.R. Crops and man. *Am. Soc. Agron. Crop Sci. Soc. Am.* **1992**, *16*, 63–262.
19. Doughty, J. Dangers of reducing the range of food choice in developing countries. *Ecol. Food Nutr.* **1979**, *8*, 275–283. [CrossRef]
20. Kearney, J. Food consumption trends and drivers. *Philos. Trans. R. Soc. B Biol. Sci.* **2010**, *365*, 2793–2807. [CrossRef] [PubMed]
21. Reardon, T.; Echeverria, R.; Berdegué, J.; Minten, B.; Liverpool-Tasie, S.; Tschirley, D.; Zilberman, D. Rapid transformation of food systems in developing regions: Highlighting the role of agricultural research & innovations. *Agric. Syst.* **2019**, *172*, 47–59. [CrossRef]
22. Pingali, P. Westernization of Asian diets and the transformation of food systems: Implications for research and policy. *Food Policy* **2007**, *32*, 281–298. [CrossRef]
23. Popkin, B.M.; Adair, L.S.; Ng, S.W. Global nutrition transition and the pandemic of obesity in developing countries. *Nutr. Rev.* **2012**, *70*, 3–21. [CrossRef] [PubMed]
24. Khoury, C.K.; Bjorkman, A.D.; Dempewolf, H.; Ramirez-Villegas, J.; Guarino, L.; Jarvis, A.; Rieseberg, L.H.; Struik, P.C. Increasing homogeneity in global food supplies and the implications for food security. *Proc. Nat. Acad. Sci. USA* **2014**, *111*, 4001–4006. [CrossRef]

25. Padulosi, S.; Sthapit, B.; Lamers, H.; Kennedy, G.; Hunter, D. Horticultural biodiversity to attain sustainable food and nutrition security. *Acta Hortic.* **2018**, *1205*, 21–34. [CrossRef]
26. Kahane, R.; Hodgkin, T.; Jaenicke, H.; Hoogendoorn, C.; Hermann, M.; Hughes, J.D.A.; Padulosi, S.; Looney, N. Agrobiodiversity for food security, health and income. *Agron. Sustain. Dev.* **2013**, *33*, 671–693. [CrossRef]
27. The Royal Society. *Reaping the Benefits: Science and the Sustainable Intensification of Global Agriculture*; The Royal Society: London, UK, 2009; p. 72.
28. World Health Organization (WHO). Promoting Fruit and Vegetable Consumption. Available online: http://www.euro.who.int/en/health-topics/disease-prevention/nutrition/activities/technical-support-to-member-states/promoting-fruit-and-vegetable-consumption (accessed on 2 March 2020).
29. Oyebode, O.; Gordon-Dseagu, V.; Walker, A.; Mindell, J.S. Fruit and vegetable consumption and all-cause, cancer and CVD mortality: Analysis of Health Survey for England data. *J. Epidemiol. Community Health* **2014**, *68*, 856–862. [CrossRef]
30. Olney, D.K.; Pedehombga, A.; Ruel, M.T.; Dillon, A. A 2-year integrated agriculture and nutrition and health behavior change communication program targeted to women in Burkina Faso reduces anemia, wasting, and diarrhea in children 3–12.9 months of age at baseline: A cluster-randomized controlled trial. *J. Nutr.* **2015**, *145*, 1317–1324. [CrossRef] [PubMed]
31. Bahadur, K.K.C.; Dias, G.M.; Veeramani, A.; Swanton, C.J.; Fraser, D.; Steinke, D.; Lee, E.; Wittman, H.; Farber, J.M.; Dunfield, K.; et al. When too much isn't enough: Does current food production meet global nutritional needs? *PLoS ONE* **2018**, *13*, e0205683. [CrossRef]
32. Siegel, K.R.; Ali, M.K.; Srinivasiah, A.; Nugent, R.A.; Narayan, K.V. Do we produce enough fruits and vegetables to meet global health need? *PLoS ONE* **2014**, *9*, e104059. [CrossRef] [PubMed]
33. Mason-D'Croz, D.; Sulser, T.B.; Wiebe, K.; Rosegrant, M.W.; Lowder, S.K.; Nin-Pratt, A.; Willenbockel, D.; Robinson, S.; Zhu, T.; Cenacchi, N.; et al. Agricultural investments and hunger in Africa modeling potential contributions to SDG2–zero hunger. *World Dev.* **2019**, *116*, 38–53. [CrossRef] [PubMed]
34. Keatinge, J.D.H.; Yang, R.Y.; Hughes, J.D.A.; Easdown, W.J.; Holmer, R. The importance of vegetables in ensuring both food and nutritional security in attainment of the millennium development goals. *Food Secur.* **2011**, *3*, 491–501. [CrossRef]
35. Ellison, B.; Muth, M.K.; Golan, E. Opportunities and challenges in conducting economic research on food loss and waste. *Appl. Econ. Perspect. Policy* **2019**, *41*, 1–19. [CrossRef]
36. Tenkouano, A. The nutritional and economic potential of vegetables. In *State of the World's Food and Agriculture*; Worldwatch Institute, WW Norton & Company, Inc.: New York, NY, USA, 2011; pp. 27–38.
37. Schreinemachers, P.; Simmons, E.B.; Wopereis, M.C. Tapping the economic and nutritional power of vegetables. *Glob. Food Secur.* **2018**, *16*, 36–45. [CrossRef]
38. Harris, D.R. Vavilov's concept of centres of origin of cultivated plants: Its genesis and its influence on the study of agricultural origins. *Biol. J. Linn. Soc.* **1990**, *39*, 7–16. [CrossRef]
39. Lin, L.J.; Hsiao, Y.Y.; Kuo, C.G. *Discovering Indigenous Treasures: Promising Indigenous Vegetables from Around the World*; AVRDC-The World Vegetable Center: Shanhua, Taiwan, 2009; p. 317.
40. Keatinge, J.D.H.; Holmer, R.J.; Ebert, A.W.; Hughes, J.D.A. Less visible but yet vital for human health: Nutrient-dense indigenous vegetables and their need for urgent promotion in balanced diets. In *Promotion of Underutilized Indigenous Food Resources for Food Security and Nutrition in Asia and the Pacific*; Durst, P., Bayasgalanbat, N., Eds.; Food and Agriculture Organization of the United Nations: Bangkok, Thailand, 2014; pp. 36–45.
41. World Vegetable Center. *Annual Report 2019*; World Vegetable Center: Shanhua, Taiwan, 2020; p. 72.
42. FAO-Food and Agriculture Organization of the United Nations. FAOSTAT-Crops. 2020. Available online: http://www.fao.org/faostat/en/#data/QC/ (accessed on 14 March 2020).
43. Alston, J.M.; Pardey, P.G. Public funding for research into specialty crops. *HortScience* **2008**, *43*, 1461–1470. [CrossRef]
44. Haddad, L.; Hawkes, C.; Webb, P.; Thomas, S.; Beddington, J.; Waage, J.; Flynn, D. A new global research agenda for food. *Nature* **2016**, *540*, 30–32. [CrossRef]
45. Pingali, P. Agricultural policy and nutrition outcomes-getting beyond the preoccupation with staple grains. *Food Secur.* **2015**, *7*, 583–591. [CrossRef]

46. Fischer, R.A.; Byerlee, D.; Edmeades, G.O. *Crop Yields and Global Food Security: Will Yield Increase Continue to Feed the World?* Australian Centre for International Agricultural Research: Canberra, Australia, 2014; Volume 22, p. 634.
47. Sriwichai, W.; Berger, J.; Picq, C.; Avallone, S. Determining factors of lipophilic micronutrient bioaccessibility in several leafy vegetables. *J. Agric. Food Chem.* **2016**, *64*, 1695–1701. [CrossRef]
48. Lipinski, B.; Hanson, C.; Lomax, J.; Kitinoja, L.; Waite, R.; Searchinger, T. Reducing Food Loss and Waste. World Resources Institute Working Paper. 2013. Available online: http://www.worldresourcesreport.org (accessed on 6 April 2020).
49. Salim, N.S.M.; Singh, A.; Raghavan, V. Potential utilization of fruit and vegetable wastes for food through drying or extraction techniques. *Nov. Tech. Nutr. Food Sci.* **2017**, *1*, 1–12. [CrossRef]
50. Herforth, A.; Arimond, M.; Álvarez-Sánchez, C.; Coates, J.; Christianson, K.; Muehlhoff, E. A global review of food-based dietary guidelines. *Adv. Nutr.* **2019**, *10*, 590–605. [CrossRef]
51. Beed, F.; Dubois, T.; Yang, R.Y. Nutrition implications of urban and peri-urban agriculture. *Agric. Dev.* **2015**, *26*, 7–11.
52. Kurgat, B.K.; Ngenoh, E.; Bett, H.K.; Stöber, S.; Mwonga, S.; Lotze-Campen, H.; Rosenstock, T.S. Drivers of sustainable intensification in Kenyan rural and peri-urban vegetable production. *Int. J. Agric. Sustain.* **2018**, *16*, 385–398. [CrossRef]
53. Waha, K.; Müller, C.; Bondeau, A.; Dietrich, J.P.; Kurukulasuriya, P.; Heinke, J.; Lotze-Campen, H. Adaptation to climate change through the choice of cropping system and sowing date in sub-Saharan Africa. *Glob. Environ. Chang.* **2013**, *23*, 130–143. [CrossRef]
54. Malézieux, E.; Crozat, Y.; Dupraz, C.; Laurans, M.; Makowski, D.; Ozier-Lafontaine, H.; Rapidel, B.; De Tourdonnet, S.; Valantin-Morison, M. Mixing plant species in cropping systems: Concepts, tools and models. A review. *Agron. Sustain. Dev.* **2009**, *29*, 43–62. [CrossRef]
55. Ebert, A.W. Potential of underutilized traditional vegetables and legume crops to contribute to food and nutritional security, income and more sustainable production systems. *Sustainability* **2014**, *6*, 319–335. [CrossRef]
56. Timler, C.; Alvarez, S.; DeClerck, F.; Remans, R.; Raneri, J.; Carmona, N.E.; Mashingaidze, N.; Chatterjee, S.A.; Chiang, T.W.; Termote, C.; et al. Exploring solution spaces for nutrition-sensitive agriculture in Kenya and Vietnam. *Agric. Syst.* **2020**, *180*, 102774. [CrossRef]
57. World Vegetable Center. *The World Vegetable Center's Approach to Household Gardening for Nutrition*; World Vegetable Center: Shanhua, Taiwan, 2016; p. 35.
58. Schreinemachers, P.; Brown, S.; Roothaert, R.; Sobgui, M.C.; Toure, S.H. Research to impact: The World Vegetable Center's household garden model. *Acta Hortic.* **2018**, *1205*, 305–314. [CrossRef]
59. Landon-Lane, C. *Livelihoods Grow in Gardens-Diversifying Rural Income through Home Garden*, 2nd ed.; FAO: Rome, Italy, 2011; p. 80.
60. Patalagsa, M.A.; Schreinemachers, P.; Begum, S.; Begum, S. Sowing seeds of empowerment: Effect of women's home garden training in Bangladesh. *Agric. Food Secur.* **2015**, *4*, 24. [CrossRef]
61. Hughes, J.D.A.; Keatinge, J.D.H. The nourished Millennium: How vegetables put global goals for healthy, balanced diets within reach. In *High Value Vegetables in Southeast Asia: Production, Supply and Demand; Proceedings of the SEAVEG 2012 Regional Symposium*; Holmer, R., Linwattana, G., Nath, P., Keatinge, J.D.H., Eds.; AVRDC-The World Vegetable Center: Tainan, Taiwan, 2013; pp. 11–26.
62. Yang, R.-Y.; Keding, G.B. Nutritional contributions of important African indigenous vegetables. In *African Indigenous Vegetables in Urban Agriculture*; Shackleton, C.M., Pasquini, M., Drescher, A.W., Eds.; Earthscan: London, UK, 2009; pp. 105–143.
63. FAO-Food and Agriculture Organization of the United Nations. *The Second Report on the State of the World's Plant Genetic Resources for Food and Agriculture*; FAO: Rome, Italy, 2010; p. 399.
64. Ebert, A.W. Chapter 16-ex situ conservation of plant genetic resources of major vegetables. In *Conservation of Tropical Plant Species*; Normah, M.N., Chin, H.F., Reed, B.M., Eds.; Springer Science Business Media: New York, NY, USA, 2013; pp. 373–417. [CrossRef]
65. FAO-Food and Agriculture Organization of the United Nations. *Future Smart Food-Rediscovering Hidden Treasures of Neglected and Underutilized Species for Zero Hunger in Asia, Executive Summary*; FAO: Bangkok, Thailand, 2018; p. 36.

66. Challinor, A.J.; Koehler, A.-K.; Ramirez-Villegas, J.; Whitfield, S.; Das, B. Current warming will reduce yields unless maize breeding and seed systems adapt immediately. *Nat. Clim. Chang.* **2016**, *6*, 954–958. [CrossRef]
67. FAO-Food and Agriculture Organization of the United Nations. *Genebank Standards for Plant Genetic Resources for Food and Agriculture*; FAO: Rome, Italy, 2014; p. 182.
68. Westengen, O.T.; Skarbø, K.; Mulesa, T.H.; Berg, T. Access to genes: Linkages between genebanks and farmers' seed systems. *Food Secur.* **2018**, *10*, 9–25. [CrossRef]
69. Vernooy, R.; Sthapit, B.; Galluzzi, G.; Shrestha, P. The multiple functions and services of community seedbanks. *Resources* **2014**, *3*, 636–656. [CrossRef]
70. Vernooy, R.; Shrestha, P.; Sthapit, B. *Community Seed Banks: Origins, Evolution and Prospects*; Routledge: Abingdon, UK, 2015.
71. Andersen, R.; Shrestha, P.; Otieno, G.; Nishikawa, Y.; Kasasa, P.; Mushita, A. *Community Seed Banks: Sharing Experiences from North and South*; Diversifood: Paris, France, 2018; p. 44.
72. Ebert, A.W.; de los Santos, E.B.; San Buenaventura, A.; Imperial, R.M. Community-based seed production of traditional vegetables to enhance nutrition security and empower farmers in the Philippines. *Acta Hortic.* **2015**, *1102*, 135–142. [CrossRef]
73. Clancy, E.; Vernooy, R. *Realizing Farmers' Rights through Community-Based Agricultural Biodiversity Management*; Bioversity International: Rome, Italy, 2016; p. 8.
74. Van Etten, J. Crowdsourcing crop improvement in sub-Saharan Africa: A proposal for a scalable and inclusive approach to food security. *IDS Bull.* **2011**, *42*, 102–110. [CrossRef]
75. Bioversity International. *The Crowdsourcing Approach: Seeds for Needs*; Bioversity International: Rome, Italy, 2016; p. 2.
76. Instituto Nacional de Investigación y Tecnología Agraria y Alimentaria (INIA). Iniciativa de colaboración en Conservación de Recursos Fitogenéticos con el CRF: Sinergias con la Conservación en Finca de Variedades locales. Available online: http://wwwsp.inia.es/Comunicacion/Notasdeprensa/Lists/Notas%20Prensa/Attachments/310/NP%20CRF%20Sinergias%20conservaci%C3%B3n%20.pdf?utm_source=Agro.biodiver.se+subscribers&utm_campaign=3c430ad69f-RSS_EMAIL_CAMPAIGN&utm_medium=email&utm_term=0_949cf01306-3c430ad69f-44963817 (accessed on 2 May 2020).
77. Schreinemachers, P.; Mavlyanova, R. *Utilization of WorldVeg Vegetable Germplasm in Central Asia and the Caucasus*; World Vegetable Center: Shanhua, Taiwan, 2019; p. 27.
78. Mavlyanova, R. *Personal Communication*; World Vegetable Center, Central Asia and the Caucasus: Tashkent, Uzbekistan, 2020.
79. Stoilova, T.; van Zonneveld, M.; Roothaert, R.; Schreinemachers, P. Connecting genebanks to farmers in east Africa through the distribution of vegetable seed kits. *Plant Genet. Resour.* **2019**, *17*, 306–309. [CrossRef]
80. Dwivedi, S.L.; Ceccarelli, S.; Blair, M.W.; Upadhyaya, H.D.; Are, A.K.; Ortiz, R. Landrace germplasm for improving yield and abiotic stress adaptation. *Trends Plant Sci.* **2016**, *21*, 31–42. [CrossRef]
81. Hanson, P.M.; Yang, R.Y. Genetic improvement of tomato (*Solanum lycopersicum* L.) for phytonutrient content at AVRDC-the world vegetable center. *Ekin J. Crop Breed. Genet.* **2016**, *2*, 1–10.
82. Zhang, Q.I.; Yu, E.; Medina, A. Development of advanced interspecific-bridge lines among *Cucurbita pepo*, *C. maxima*, and *C. moschata*. *HortScience* **2012**, *47*, 452–458. [CrossRef]
83. Da Silva Dias, J.C. Guiding strategies for breeding vegetable cultivars. *Agric. Sci.* **2014**, *5*, 9–32. [CrossRef]
84. Business Wire. Global Vegetable Seed Market 2019–2024: Growth, Trends and Forecasts-Research and Markets. Available online: https://www.businesswire.com/news/home/20190430005782/en/Global-Vegetable-Seed-Market-2019-2024-Growth-Trends (accessed on 4 May 2020).
85. Reddy, M.K.; Srivastava, A.; Lin, S.W.; Kumar, R.; Shieh, H.C.; Ebert, A.W.; Chawda, N.; Kumar, S. Exploitation of AVRDC's chili pepper (*Capsicum spp.*) germplasm in India. *J. Taiwan Soc. Hortic. Sci.* **2015**, *61*, 1–9.
86. Ebert, A.W.; Schafleitner, R. Chapter 9. Utilization of wild relatives in the breeding of tomato and other major vegetables. In *Crop Wild Relatives and Climate Change*, 1st ed.; Redden, R., Yadav, S.S., Maxted, N., Dulloo, M.E., Guarino, L., Smith, P., Eds.; John Wiley & Sons Inc.: New York, NY, USA, 2015; pp. 141–172.
87. Zamir, D. Improving plant breeding with exotic genetic libraries. *Nat. Rev. Genet.* **2001**, *2*, 983–989. [CrossRef] [PubMed]
88. Prohens, J.; Gramazio, P.; Plazas, M.; Dempewolf, H.; Kilian, B.; Díez, M.J.; Fita, A.; Herraiz, F.J.; Rodríguez-Burruezo, A.; Soler, S.; et al. Introgressiomics: A new approach for using crop wild relatives in breeding for adaptation to climate change. *Euphytica* **2017**, *213*, 158. [CrossRef]

89. Herzog, E.; Falke, K.C.; Presterl, T.; Scheuermann, D.; Ouzunova, M.; Frisch, M. Selection strategies for the development of maize introgression populations. *PLoS ONE* **2014**, *9*, e92429. [CrossRef]
90. Gur, A.; Zamir, D. Mendelizing all components of a pyramid of three yield QTL in tomato. *Front. Plant Sci.* **2015**, *6*, 1096. [CrossRef]
91. Eshed, Y.; Zamir, D. An introgression line population of *Lycopersicon pennellii* in the cultivated tomato enables the identification and fine mapping of yield-associated QTL. *Genetics* **1995**, *141*, 1147–1162.
92. Lippman, Z.B.; Semel, Y.; Zamir, D. An integrated view of quantitative trait variation using tomato interspecific introgression lines. *Curr. Opin. Genet. Dev.* **2007**, *17*, 545–552. [CrossRef]
93. Gramazio, P.; Prohens, J.; Plazas, M.; Mangino, G.; Herraiz, F.J.; Vilanova, S. Development and genetic characterization of advanced backcross materials and an introgression line population of *Solanum incanum* in a *S. melongena* background. *Front. Plant Sci.* **2017**, *8*, 1477. [CrossRef]
94. Prohens, J.; Whitaker, B.D.; Plazas, M.; Vilanova, S.; Hurtado, M.; Blasco, M.; Gramazio, P.; Stommel, J.R. Genetic diversity in morphological characters and phenolic acids content resulting from an interspecific cross between eggplant, *Solanum melongena*, and its wild ancestor (*S. incanum*). *Ann. Appl. Biol.* **2013**, *162*, 242–257. [CrossRef]
95. Taher, D.; Solberg, S.Ø.; Prohens, J.; Chou, Y.Y.; Rakha, M.; Wu, T.H. World vegetable center eggplant collection: Origin, composition, seed dissemination and utilization in breeding. *Front. Plant Sci.* **2017**, *8*, 1484. [CrossRef] [PubMed]
96. Mangino, G.; Plazas, M.; Vilanova, S.; Prohens, J.; Gramazio, P. Performance of a set of eggplant (*Solanum melongena*) lines with introgressions from its wild relative, *S. incanum* under open field and screenhouse conditions and detection of QTLs. *Agronomy* **2020**, *10*, 467. [CrossRef]
97. Meldrum, G.; Padulosi, S.; Lochetti, G.; Robitaille, R.; Diulgheroff, S. Issues and prospects for the sustainable use and conservation of cultivated vegetable diversity for more nutrition-sensitive agriculture. *Agriculture* **2018**, *8*, 112. [CrossRef]
98. Stamp, P.; Messmer, R.; Walter, A. Competitive underutilized crops will depend on the state funding of breeding programmes: An opinion on the example of Europe. *Plant Breed.* **2012**, *131*, 461–464. [CrossRef]
99. Frison, E.A.; Cherfas, J.; Hodgkin, T. Agricultural biodiversity is essential for a sustainable improvement in food and nutrition security. *Sustainability* **2011**, *3*, 238–253. [CrossRef]
100. Jackson, L.E.; Pascual, U.; Hodgkin, T. Utilizing and conserving agrobiodiversity in agricultural landscapes. *Agric. Ecosyst. Environ.* **2007**, *121*, 196–210. [CrossRef]
101. McCouch, S.; Baute, G.J.; Bradeen, J.; Bradeen, J.; Bramel, P.; Bretting, P.K.; Buckler, E.; Burke, J.M.; Charest, D.; Cloutier, S. Feeding the future. *Nature* **2013**, *499*, 23–24. [CrossRef]
102. Keatinge, J.D.H.; Waliyar, F.; Jamnadass, R.H.; Moustafa, A.; Andrade, M.; Drechsel, P.; Hughes, J.D.A.; Palchamy, K.; Luther, K. Re-learning old lessons for the future of food-by bread alone no longer: Diversifying diets with fruit and vegetables. *Crop Sci.* **2010**, *50*, 51–62. [CrossRef]

plants

MDPI

Article

Macro- and Micronutrients from Traditional Food Plants Could Improve Nutrition and Reduce Non-Communicable Diseases of Islanders on Atolls in the South Pacific

Graham Lyons [1,*], Geoff Dean [2], Routan Tongaiaba [3], Siosiua Halavatau [4], Kabuati Nakabuta [3], Matio Lonalona [5] and Gibson Susumu [6]

1. School of Agriculture, Food and Wine, University of Adelaide, Waite Campus, Urrbrae, South Australia 5064, Australia
2. Tasmanian Institute of Agriculture, University of Tasmania, Prospect, Tasmania 7250, Australia; geoffrey.dean@utas.edu.au
3. Agriculture and Livestock Division, Ministry of Environment, Lands and Agriculture Development, Tanaea, South Tarawa, Kiribati; rtongaiaba@gmail.com (R.T.); sao@melad.gov.ki (K.N.)
4. UN FAO Intergovernmental Technical Panel on Soils, Tonga; halavatauj@gmail.com
5. Department of Agriculture, Ministry of Natural Resources, Energy and Environment, Vaiaku, Funafuti, Tuvalu; matiolnln@gmail.com
6. Gibson Susumu: Sustainable Agriculture Programme, The Pacific Community (SPC), Suva, Fiji; GibsonS@spc.int

* Correspondence: graham.lyons@adelaide.edu.au; Tel.: +61-8-83136533

Received: 1 June 2020; Accepted: 21 July 2020; Published: 24 July 2020

Abstract: Pacific Islanders have paid dearly for abandoning traditional diets, with diabetes and other non-communicable diseases (NCD) widespread. Starchy root crops like sweet potato, taro, and cassava are difficult to grow on the potassium-deficient soils of atolls, and high energy, low nutrient imported foods and drinks are popular. Nutritious, leafy food plants adapted to alkaline, salty, coral soils could form part of a food system strategy to reduce NCD rates. This project targeted four atolls south of Tarawa, Kiribati, and was later extended to Tuvalu. Mineral levels in diverse, local leafy food plants were compared to reveal genotype–environment interactions. Food plants varied in ability to accumulate minerals in leaves and in tolerance of mineral-deficient soils. Awareness activities which included agriculture, health, and education officers targeted atoll communities. Agriculture staff grew planting material in nurseries and provided it to farmers. Rejuvenation of abandoned giant swamp taro pits to form diversified nutritious food gardens was encouraged. Factsheets promoted the most suitable species from 24 analyzed, with multiple samples of each. These included *Cnidoscolus aconitifolius* (chaya), *Pseuderanthemum whartonianum* (ofenga), *Polyscias scutellaria* (hedge panax), and *Portulaca oleracea* (purslane). The promoted plants have been shown in other studies to have anti-NCD effects. Inclusion of the findings in school curricula and practical application in the form of demonstration school food gardens, as well as increased uptake by farmers, are needed. Further research is needed on bioavailability of minerals in plants containing phytates and tannins.

Keywords: atolls; leafy vegetables; non-communicable diseases (NCD); nutrition security; mineral nutrients; natural biofortification

1. Introduction

1.1. Epidemic of Non-Communicable Diseases (NCDs)

Since the 1940s the consumption of high-energy, low-nutrient foods, including white flour, sugar and polished rice by Pacific Islanders, combined with reduced exercise, has resulted in alarming rates of obesity, heart disease, diabetes, and certain cancers [1,2]. Indeed, around 70% of deaths in Pacific Island countries (PICs) are due to non-communicable diseases (NCDs) [3–5]. Apart from the tragic personal cost, premature death and disability undermines national economic productivity. These diseases occurred at very low rates when traditional diets and lifestyles predominated [6]. In addition, many PICs are affected by the "double burden" of NCDs and under-nutrition; for example, high rates of iron-deficiency anemia in Papua New Guinea, Fiji, Solomon Islands, and Tuvalu [7,8]. Pacific Islanders have paid dearly for forsaking traditional diets. A recent study emphasizes the need, during nutrition transitions, for public health initiatives to promote traditional diets high in vegetables, fruits, and lean protein and agricultural initiatives to promote farm diversity [9].

In addition to the health benefits of traditional diets, local food crop (including wild food) biodiversity strengthens the resilience of food systems to climate events through increasing crop species richness, thus improving food and nutrition security [10–14]. It is also economically advantageous. Growing foods such as leafy greens, breadfruit, pumpkin, and bananas to improve nutrition helps to reduce trade deficits associated with high consumption of imported foods in the Pacific. In Kiribati and Tuvalu (see Figure 1), imported food comprises about 65% of food eaten [15].

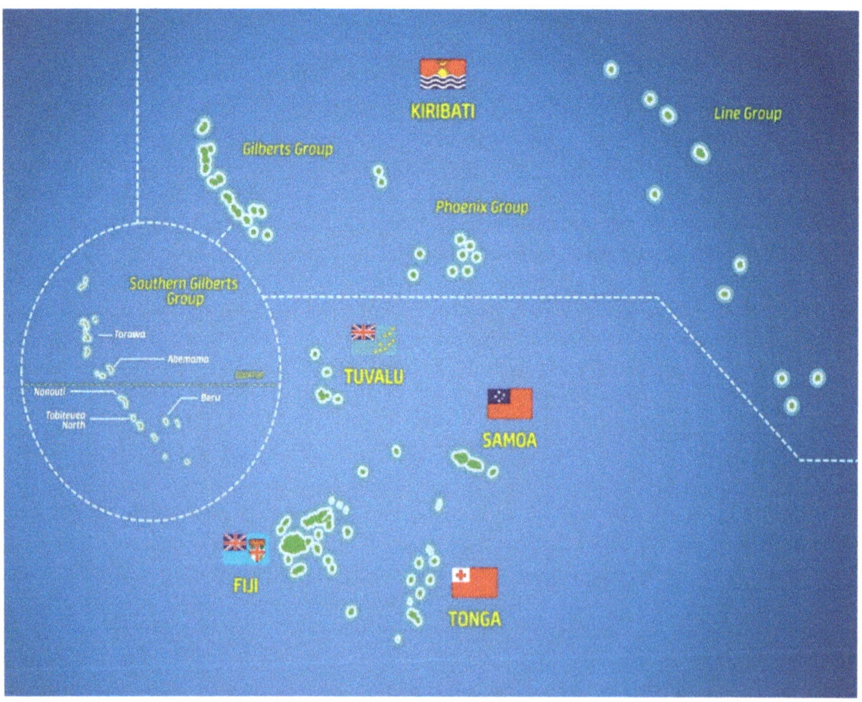

Figure 1. Map of part of the south-western Pacific Ocean, featuring Kiribati, Tuvalu, and Fiji, and (inset) the Southern Gilbert Islands. This project focused on Abemama, Tabiteuea North, Nonouti, and Beru, and (for value-chain activity) Abaiang, just north of Tarawa.

1.2. The Special Case of Atolls

Kiribati (population 114,000) and Tuvalu (11,000) are small Pacific nations where around half the people live on the main atolls of South Tarawa and Funafuti, respectively.

Atoll soils are formed almost entirely from coral (predominantly calcium carbonate with some magnesium). They are coarse textured with no clay, so water flows straight through them. Moreover, droughts are common in this part of the world. The soil is often salty, highly alkaline (pH (H_2O) 8.6–9.2), and low in nutrients such as potassium, iron, and manganese, and, unless well composted, low in organic matter [16]. Furthermore, inorganic fertilizers and chemical pesticides are prohibited on many atolls as they could pollute valuable underground fresh water. Improving soil health through targeted composting, along with growing and eating nutritious crops on atolls should lead to improved diet, nutrition, and health.

1.3. Why Leafy Plants?

On many atolls, in particular those of the Southern Gilbert Islands (part of Kiribati, see Figure 1), which often experience drought, starchy root crops can be difficult to grow, resulting in low tuber/storage root yield. This is associated with potassium deficiency and lack of sufficient water during the weeks after planting. Potassium is needed to ensure adequate storage root initiation, and the high potassium content of tubers/storage roots depletes soil potassium with each harvest [17,18].

On the other hand, hardy leafy food plants can yield well under these conditions. Many different types of leafy vegetables and leaves of other plants/food crops are grown and eaten in the Pacific region (e.g., edible ferns, kangkong, amaranth, drumstick, and leaves of starchy root crops like taro, sweet potato, and cassava) [10,19–23]. When available, indigenous vegetables are usually inexpensive and thus affordable to most people in both urban and rural areas. Despite this, they are often overlooked and regarded as "low status foods" [1,2]. However, they are important for human health, being nutritious and rich in protein, minerals, vitamins (e.g., A, B, C, K), beneficial phytocompounds, and fiber [10,22–32]. A study in Africa found that "orphan" (unimproved) leafy vegetables were popular with farmers if they were full-season varieties with high leaf yield, and resistant to pests, diseases, and abiotic stress (e.g., drought, heat, salinity). Retailers and consumers valued good appearance, long shelf-life, affordability, and high nutritional value [33].

Iron provides an example of an important micronutrient found in leafy vegetables. Lack of iron can cause iron-deficiency anemia, common in women, inducing fatigue and weakness, and in children, affecting growth, energy levels, and learning ability. Purslane, pumpkin leaves, kangkong, yellow beach pea, and chaya are all good sources of iron [23,29].

Phytocompounds such as flavonoids, anthocyanins, polyphenols, and carotenoids are beneficial to humans as antioxidants and anti-inflammatory agents in reducing the risk of diabetes, heart disease, and cancers. Examples include glucosinolates in drumstick leaves and anthocyanins in purple sweet potato leaves. Certain carotenoids, notably beta- and alpha-carotene, are converted to vitamin A when eaten, especially if consumed with some oil (e.g., coconut cream) [34]. Others, notably lutein (which is usually abundant in leafy vegetables) and zeaxanthin are important for eye health and can reduce the risk of cataracts [35]. Importantly, given the current NCD pandemic, there is growing evidence for specific activity of certain plants against diabetes and cardiovascular disease, e.g., drumstick [36–38] and chaya [39,40].

1.4. Project Objective and Strategy

The objective of this project (2014–2019) was to support and enhance an awareness program aimed at increasing production of nutritious leafy plants to reduce rates of NCDs in Kiribati and Tuvalu. This was achieved through:

- A survey of mineral nutrients in local leafy food plants collected in Kiribati and Tuvalu.

- Assessment of plants for overall nutritional content, taste, and adaptation, in particular their tolerance of drought, soil salinity, and soil alkalinity, which characterize atolls, especially those of the Southern Gilbert Islands.
- Production of a factsheet series to promote the most suitable leafy food plants for atolls.
- Collaboration with the UN International Fund for Agricultural Development (IFAD)'s Outer Islands Food and Water Project, together with agriculture, health, and education officers in Kiribati and Tuvalu, to facilitate supply of planting material and participate in workshops and school talks.

2. Results and Discussion

2.1. G x E Study

During the 2014 scoping study for the current project, we found 11 of the 12 leafy vegetables featured in the earlier factsheets growing on South Tarawa and Funafuti. This was surprising in view of the almost universal inhospitable coralline atoll soils, compared with, for example, soils of Solomon Islands. Most were growing in gardens and hedges close to homes; however, they were usually used for animal feed or as ornamentals. Clearly, raising awareness is an important program component, which includes school food gardens and curriculum development, farmer field schools, village workshops, and media promotion. See further discussion on this below.

Plant production is limited mostly by soil plant-available mineral content. Comparing mineral concentrations in genetically diverse plants provides insights into the plant–environment interactions that control mineral nutrient accumulation [41]. This not only enables identification of nutritious food plants for humans and animals, but also can lead to improvement in sustainable yield.

Mineral concentrations in leaves varied widely with species, with five-fold variations common between species grown on the same soil (Tables 1 and 2). Variation was less marked between sites (environment effect) for most minerals than is usually found, due to the relatively uniform coralline soils. For example, most minerals in leaves of chaya (*Cnidoscolus aconitifolius*) varied by no more than two-and-a-half-fold across seven sites, the exceptions being manganese and zinc (Table 3).

Table 1. Concentrations of macro- and micronutrients in leaves (dry weight basis) of different food plant species grown together on the same soil type at Vaiaku, Funafuti, Tuvalu in August 2014.

Species	Nutrient							
	Fe	Mn	B	Cu	Zn	Mg	K	N
	mg/kg	mg/kg	mg/kg	mg/kg	mg/kg	%	%	%
Brazil spinach (*Alternanthera sissoo*)	30	9	33	11	97	1.48	3.6	3.5
Chaya (*Cnidoscolus aconitifolius*)	76	19	19	9	42	0.55	1.64	5.1
Drumstick tree (*Moringa oleifera*)	52	12	21	7	39	0.61	1.09	5.2
Hedge panax (*Polyscias scutellaria*)	47	29	26	7	71	0.58	3.2	2.9
Lettuce tree (*Pisonia grandis*)	50	29	43	21	20	0.61	2.1	4.2
Ofenga (*Pseuderanthemum whartonianum*)	45	21	25	28	62	1.72	3.4	3.0
Purslane (*Portulaca oleracea*)	68	5	50	14	103	2.2	3.1	3.3
Variation (-fold)	2.5	5.8	2.8	5.6	5.2	4.0	3.3	1.8

Notes: Concentrations are on a dry weight (DW) basis throughout the manuscript. N % × 4.4 provides an estimate of crude protein %. Ca was uniformly high (range 1.61–2.20%). S was moderate in six species (0.21–0.38%) but high in drumstick tree (1.13%). Note that the data in Tables 1–3 are analyses conducted on single representative sub-samples of pooled samples for each species.

Table 2. Concentrations of macro- and micronutrients in leaves (dry weight basis) of different food plant species grown together on the same soil type at Tanaea, South Tarawa, Kiribati in August 2014.

Species	Nutrient							
	Fe	Mn	Cu	Zn	Ca	Mg	S	N
	mg/kg	mg/kg	mg/kg	mg/kg	%	%	%	%
Cassava (*Manihot esculenta*)	35	37	10	88	1.30	0.67	0.34	5.0
Chilli (*Capsicum frutescens*)	32	25	8	63	3.89	1.80	0.65	3.3
Drumstick tree (*Moringa oleifera*)	65	20	5	32	1.58	0.74	1.16	5.4
Lettuce tree (*Pisonia grandis*)	42	31	17	16	2.34	1.00	0.32	3.3
Ofenga (green) (*Pseuderanthemum whartonianum*)	26	24	7	33	2.20	2.70	0.31	2.1
Ofenga (red)	30	22	10	25	1.72	1.30	0.24	3.2
Taro (*Colocasia esculenta*)	34	35	12	29	3.30	0.63	0.21	3.8
Variation (-fold)	2.5	1.9	3.4	5.5	3.0	4.3	5.5	2.6

Table 3. Variation in selected minerals in leaves of *Cnidoscolus aconitifolius* (chaya) growing at seven locations in Tuvalu (sites 1 and 2) and Kiribati (sites 3–7) from 2014 to 2017. This study illustrates variation due mostly to differences in plant-available levels of these nutrients in soil. Most minerals (Zn, Mg, N) varied by less than three-fold.

Site	Nutrient			
	Mn	Zn	Mg	N
	mg/kg	mg/kg	%	%
1	17	50	0.50	5.0
2	4	43	0.90	4.9
3	12	35	0.56	5.2
4	19	42	0.55	5.1
5	10	27	0.71	4.7
6	4	50	0.56	5.9
7	32	79	1.11	4.2
Mean	14	47	0.70	5.0
Variation (-fold)	8.0	2.9	2.2	1.4

2.2. *Natural Biofortifiers with Variability in Micronutrient Efficiency*

Table 4 features the leafy plants which consistently accumulated the highest levels of minerals. These species could be described as natural biofortifiers of the corresponding nutrient. Leaves of pumpkin (*Cucurbita pepo*), purslane (*Portulaca oleracea*), and chilli (*Capsicum frutescens*) contained relatively high concentrations of most of the minerals, and thus, at least with respect to minerals, could be regarded as the most nutritious overall. Other studies also report the high nutritive value of these plants [42–44].

Species such as hedge panax and birdsnest fern were not observed with leaf chlorosis during this study, regardless of high soil pH. They are not exceptional Fe accumulators, e.g., birdsnest fern collected on Papaelise Island, Funafuti contained only 13 mg/kg DW of Fe, but looked healthy, and this compares with cassava, with a critical level for Fe of around 50 mg/kg [45]. Nevertheless, it is likely that plants such as birdsnest fern are efficient for Fe [46], and probably also for other nutrients, e.g., Mn, Cu, K in short supply in coralline soils. These plants seem to be able to function normally, especially with respect to photosynthesis, even when the nutrient is present at low plant-available levels in the soil. This is a different trait (involving different genes) to the ability to take up and accumulate high levels of a nutrient. Birdsnest fern is also an exceptional accumulator of K. Most cassava varieties, on the other hand, suffered from chlorosis, stunted growth, and lack of sizeable storage roots on the southern atolls. However, cassava and purslane are adept at extracting Zn from the soil and accumulating it in leaves.

Table 4. Selected mineral nutrients and the leafy vegetable species found (using opportunistic GxE analysis) in this study to be the most effective accumulators of these minerals in leaves. Samples were collected from various locations in Kiribati and Tuvalu. The values in brackets are representative concentrations of the relevant mineral for each species in this region.

Nutrient (Units)	Best Accumulators (Concentration in Leaf)
Iron (mg/kg)	Purslane (79), yellow beach pea (72), pumpkin (69), kangkong (68), chaya (65)
Manganese (mg/kg)	Giant swamp taro (94), cassava (34), taro (34), chilli (27)
Boron (mg/kg)	Chilli (60), drumstick (48), birdsnest fern (41), sweet potato (41)
Copper (mg/kg)	Tree lettuce (21), pumpkin (13), chilli (12), ofenga (11)
Zinc (mg/kg)	Purslane (119), cassava (107), pumpkin (97), hedge panax (81)
Calcium (%)	Chilli (3.8), bele (3.4), ofenga (2.7), hedge panax (2.5)
Magnesium	Purslane (2.5), ofenga (2.0)
Potassium (%)	Pumpkin (4.3), birdsnest fern (4.1), taro (3.0), kangkong (2.9)
Phosphorus (%)	Pumpkin (0.74), cassava (0.54), sweet potato (0.52)
Sulphur (%)	Drumstick (1.1), chilli (0.6), sweet potato (0.55)
Nitrogen (%)	Pumpkin (5.1), cassava (5.0), chaya (5.0), drumstick (4.7)
Selenium (μg/kg)	Drumstick (400)
Multiple nutrients	Pumpkin, purslane, chilli

Notes: Selenium is a micronutrient for humans and animals but not for higher plants; μg = micrograms; N % × 4.4 provides an estimate of crude protein % in leaves.

2.3. Factsheets

In addition to the introductory factsheet, 12 species factsheets were produced, which feature the most atoll suitable nutritious leafy vegetables identified during the project. Several of these species have also been recognized for their nutritional value in other studies (Bailey, 1992; French, 2010; SPC, 2012). The featured plants are *Amaranthus viridis* (amaranth), *Cnidoscolus aconitifolius* (chaya, tree spinach), *Moringa oleifera* (drumstick tree), *Polyscias scutellaria* (hedge panax), *Pseuderanthemum whartonianum* (ofenga, Carruthers' falseface), *Vigna marina* (yellow beach pea, beach cowpea), *Ipomoea aquatica* (kangkong), *Cucurbita pepo* (pumpkin), *Sechium edule* (choko), *Abelmoschus manihot* (bele, aibika, slippery cabbage), *Capsicum frutescens* (chilli), *Portulaca oleracea* (purslane, pigweed). Factsheet 13 discusses nutritional aspects of composting methods suitable for atolls.

Although not featured in the factsheets due to budgetary constraints (and their overall mineral levels were a little below the selected species) other nutritious leafy vegetables included *Asplenium nidus* (birdsnest fern), *Pisonia grandis* (big lettuce tree), and the leaves of these starchy root crops: *Ipomoea batatas* (sweet potato), *Manihot esculenta* (cassava), *Xanthosoma sagittifolium* (cocoyam), and *Colocasia esculenta* (taro). The nutritional value of these species has been noted in earlier studies [10,19,22,28].

Yellow beach pea was included more for its importance as a well-adapted legume on atolls than for its eating quality. It is an efficient N-fixer, with extensive root nodulation observed whenever sampled in Kiribati and Tuvalu, and is salt- and drought-tolerant. It grows better on strongly alkaline soils than *Mucuna*, *Pueraria*, *Centrosema*, *Gliricidia*, *Erythrina*, and *Sesbania*, can smother weeds, and its relatively high N, Fe, and Zn content make it a valuable green manure and compost component. Its seed pods are good to eat and highly nutritious when green, although all but the youngest of its leaves are chewy due to their high fiber content. A similar creeping legume is the Mauna Loa bean (*Canavalia cathartica*), which also thrives near beaches of Tuvalu and Kiribati, and has purple flowers, larger pods, and is a more vigorous tree-climber than *Vigna marina*.

2.4. Medicinal Effects

Although this project focused on the food/nutritional value of leafy green vegetables, traditionally in many countries they are also used for specific medical applications. For example, chaya (which originated in Mexico and Mesoamerica) protects the heart, liver, and kidneys from toxin damage [47,48]; drumstick (India and Pakistan) has anti-bacterial effects [49,50]; and bele (Papua New Guinea and

Solomon Islands) is used for bone repair and treating osteoporosis [51]. Hedge panax, drumstick, chaya, bele, and purslane are galactogogues that can stimulate lactation [10,52–54]. Indeed, purslane's generic name Portulaca means "to carry milk". This plant is so ubiquitous and prolific globally that it is usually regarded as a weed. It is renowned for its high n-3 fatty acid content [14], and in this study was found to be the best accumulator of Mg, Fe and Zn of all the plants analyzed. Pumpkin thrives in composted atoll soils and is already grown widely in both countries. Ofenga, in particular the red-leaf form, is better known in Kiribati for its embalming ability than as food.

High-protein species drumstick and chaya were fostered in Kiribati and Tuvalu under the Pacific Regional Agricultural Programme (PRAP) in the 1980s, and have adapted well to harsh atoll conditions [21]. Drumstick is high in *b*-carotene, sulphur, and selenium [14]. This species had the highest *b*-carotene level (427 mg/kg) of all plants analyzed in the earlier ACIAR project in the Pacific and northern Australia, and its protein is considered to be high in quality, with a similar pattern of essential amino acids as soybean [1]. It regularly accumulates around 12 times the concentration of selenium and around four times the concentration of sulphur compared to most other plants grown on the same soil. At the Vaiaku, Funafuti site (Table 1), drumstick leaves contained 25 times the Se concentration of the mean of the other plants growing there. Similar differences have also been observed in Africa [55]. This trait would be especially valuable in Sub-Saharan Africa, where these minerals are deficient in many soils [56–58]. Their deficiency is considered by some researchers to increase risk of HIV/AIDS [59].

Chaya is also renowned for its nutritional and medicinal effects. Like drumstick, it is an excellent source of high-quality plant protein and carotenoids and is renowned for its liver- and kidney-health enhancing effects [48].

Bele is not as climate- or insect-resilient as chaya, but has been included as its flavor is highly regarded, it is noted for health properties, including high levels of the important carotenoids lutein and *b*-carotene [1], and it grows well on composted soils with sufficient rainfall. It is the most popular leafy vegetable in Solomon Islands and Papua New Guinea.

Especially important, given the high NCD (particularly diabetes) rates in the Pacific and northern Australia, are the anti-diabetes and anti-cardiovascular disease effects of most of the plants featured in the factsheets, demonstrated in scientific studies. Studies with evidence for this include the following: drumstick [36–38,50,60], amaranth [61–63], bele [64–66], chilli [67–69], purslane [70–72], kangkong [73–75], ofenga [76–78], hedge panax [79,80], chaya [39,40], pumpkin [81,82], and choko [83,84]. Their inclusion in the diet in sufficient quantity is likely to reduce the risk of diabetes and cardiovascular disease, not only by reducing glycemic load when they are included with high-carbohydrate meals, but also because of specific anti-diabetes effects.

2.5. Mineral Deficiencies of Atoll Food Plants

The plant leaf mineral data revealed that the most common mineral deficiencies in both countries were K and Mn. For example, 51% of the plant samples had K < 15,000 mg/kg, and 46% had Mn < 15 mg/kg. Phosphorus was mostly in the adequate range of 2500–4000 mg/kg, with 23% at marginal levels. Copper was marginal (<4 mg/kg) in 19% of leaf samples. Nitrogen deficiency was rare, and 33% of leaf samples had >4% N, testament to effective long-term composting, along with the inclusion of legumes, cassava, chaya, and drumstick (which are all inherently high in N) in the sample collection. Sites which had been composted for several years were higher in N (both nitrate and ammonium), available P, K, Mg, B, Cu, Fe, Mn, and Zn.

Iron deficiency, which is usually associated with alkaline soils, was not widespread, with most leaf samples > 30 mg/kg. Critical Fe levels are species-specific; as noted earlier, hedge panax can function normally with less Fe than can cassava. Likewise, most plants had sufficient Zn and B, with levels mostly in the range 30–70 mg/kg, and only 11% of leaf samples were <20 mg/kg in either.

Leaf Na levels were, unsurprisingly, relatively high, mostly >2000 mg/kg in Kiribati and >5000 mg/kg in Tuvalu, but symptoms of Na, Cl, or NaCl toxicity were not observed, even in cocoyam with leaf Na of 31,000 mg/kg on South Tarawa. High soil and plant levels of Ca and Mg counteract Na toxicity [85,86].

2.6. Giant Swamp Taro Food Garden

The value of home gardens comprising diverse, nutritious, traditional food crops to supplement the diet of subsistence households is well documented [1,11,87–90]. The cultivation of the giant swamp taro (*Cyrtosperma merkusii*, called babai in Kiribati and pulaka in Tuvalu) is traditional on atolls. Pits are dug by hand down to the water table, which in many Kiribati atolls is only 1–2 m below the surface. Many of these pits are now neglected but they provide a strong connection to both culture and fresh ground water.

In an adaptation of this system, kangkong can be grown in the water with the swamp taro. Hence the drought-tolerance requirement is waived for this species. Its vigorous growth means that, unless harvested regularly, it can intrude upon the giant swamp taro, and may need to be grown in a separate pit. The other crops can be grown on constructed terraces formed from the pit walls, and drumstick, ofenga, hedge panax, and beach cowpea are planted around the pit at ground level. Other crops, such as bananas, pawpaw, sweet potato, and annual vegetables can be included as well (Figure 2). These babai food gardens (BFG) (or pulaka food gardens (PFG) in Tuvalu) represent a mini food system that can, once established, contribute significantly to a family's nutrition. For example, 100 g fresh weight of purslane leaves/stem will provide around 20% of an individual's daily Zn requirement. The size of the garden can be as small as 40 m^2 or as large as 0.3 ha. Even in crowded places, such as Betio on South Tarawa, there is usually room to plant a drumstick tree or two, which would soon provide a sustainable daily supply of leaves for a family.

On Nonouti, which has perhaps the toughest environment of the four atolls in the study, especially for drought frequency and duration, 27 BFG were commenced during the project, and at July 2019, 12 appeared well established and sustainable, a reasonable success rate for a new concept. Participating households usually have a BFG as well as food plants growing nearby on the land surface and around the house. These include sweet potato, ofenga, chaya, lemon grass, chilli, Brazilian spinach, pumpkin, banana, pawpaw, coconut, and breadfruit.

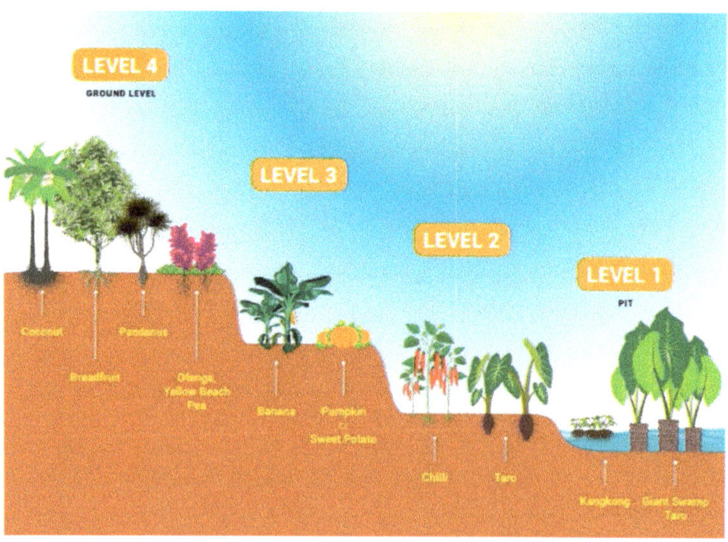

Figure 2. Layout of a babai/pulaka food garden. Other nutritious food crops can be substituted or added if desired.

2.7. How to Eat these Nutritious Vegetables

It is recommended to eat around three handfuls (around 100 g) of leafy vegetables each day. Some green leaves can be eaten uncooked, e.g., purslane, kangkong, and chilli, which preserves most vitamins, but it is usually preferable to cook them. It is important to wash the leaves in clean water first, to remove dirt and possible pathogenic microbes. Optimum cooking methods are steaming, simmering in a little water, baking, or stir frying in a little oil (ideally virgin coconut oil or coconut cream) for minimal time to limit nutrient loss. The cooking water should not be discarded, but instead used for soup. A simple method suitable for these vegetables is to chop them into small pieces (except drumstick, in which case strip the leaflets from the wiry petioles), simmer in water for 10–15 min, add coconut cream, and simmer for a further 10–15 min. Other ingredients can be added to enhance the flavor if desired. More elaborate recipes are included in several of the factsheets. A recipe book that includes most of the featured nutritious leafy plants was produced, written in Kiribati language, by the IFAD community awareness team.

The bioavailability of Fe, Zn, and other minerals will be reduced by the presence of phytate, tannins, and polyphenols, e.g., in drumstick, chaya, and purslane [91–94]. The effects of these so-called antinutrients can be reduced by various cooking methods. In chaya, for example, boiling significantly reduced phytates, oxalates, and tannins [94] and virtually eliminated cyanogenic glycosides [93].

2.8. Awareness Program and Planting Material Provision

In order to achieve impact, the project collaborated with multiple government ministries (Agriculture, Health, Education, Works), churches, NGOs, Island Councils, and communities on the target atolls. In Kiribati, about 1500 farmers attended information and training sessions on growing, handling, cooking, and preserving locally grown foods. Our findings agreed with an FAO study in Samoa which found that the main external factors which influenced people's decisions about food were availability, accessibility, cultural obligations, and family income [95].

Increasing awareness and generating interest must be met with availability of supply, whether planting material for home gardens or on a larger scale for farmers to produce for markets and tourism outlets [2]. Suitable planting material of the selected species was supplied via ALD nurseries on each atoll. The ALD Tanaea HQ, an infertile site with multi-micronutrient deficiencies, was transformed, using an improved watering system and composting, into an important germplasm source and model nutritious food garden. In Tuvalu, secondary bush was cleared, a water tank and irrigation system installed on Funafala island, Funafuti atoll and nutritious food crops grown to supply nearby Vaiaku, the main population center. This highlights the need for more resources to be devoted to conservation of diverse leafy food plants, starchy root crops, and fruits in the Pacific region [2,89].

The Pacific Community's Centre for Pacific Crops and Trees (CePaCT), Suva plays a key role in germplasm conservation and distribution. Conservation of traditional crops is especially important in countries such as Papua New Guinea and Solomon Islands, where the natural environment is threatened by logging, mining, and oil palm establishment [89].

Value chain research is essential: the producer needs to be convinced that production of green leafy vegetables is worth the effort. Strategies are needed to deliver health benefits to consumers and economic benefits to local horticultural producers and other value chain participants [13].

Improving nutrition is usually seen as the task of health agencies, but it is apparent that a cross-sectoral and multi-agency food system approach is needed [96]. The NCD pandemic can be addressed by increasing diversity on the farm and extending this diversity (of which nutritious leafy vegetables form an important part) to the diet. Involvement of children in promotion of nutritious local foods is integral; in many countries their importance in influencing lifestyle factors, especially diet, is becoming recognized. For example, schools can include food gardens featuring the most nutritious local plants, provide more nutrition education, and students can transfer knowledge back to their villages [1]. In the current study, around 1700 students (to date) have attended awareness and training programs. Further research is needed on how to optimize awareness and promotion. This is crucial for Kiribati and Tuvalu, where leafy plants were not major components of traditional diets. Studies examining the bioavailability of minerals in these plants are also needed.

3. Materials and Methods

Adaptability of leafy food plants to tough atoll conditions was clear from observation on the four atolls and on South Tarawa and Funafuti, precluding the need for formal trials. A survey was conducted to identify the most nutritious leafy food plants, in terms of minerals and protein that grow in Kiribati and Tuvalu. Particular attention was paid to species that thrive in the atoll environment. Leaf tissue samples were collected in Kiribati and Tuvalu from 2014 to 2018 (n = 140), and with the inclusion of leaf mineral data from the previous Pacific-Northern Australia nutritious leafy vegetable project (ACIAR PC/2010/063) [1] (n = 274), a total of 414 samples (with 65 food plant species, of which 24 were found growing in Kiribati and Tuvalu, and also 50 species used for herbal medicines (18) or compost (32); usually multiple samples of the same species growing at different sites were analyzed) informed the factsheets produced during the current project. In Kiribati, samples were collected on the islands of South Tarawa, Abemama, Tabiteuea North, Nonouti, and Beru. In Tuvalu, samples were collected on Vaiaku, Funafala, and Papaelise.

As with the earlier project, an opportunistic genotype–environment (GxE) strategy was employed. This included sampling of single leafy vegetable species growing at different sites as well as sampling multiple species growing at the same site. Note that due to daily time and budgetary constraints, the data presented in Tables 1–3 (in Section 2) are for single samples. The locations reported in the tables were chosen to typify the leaf mineral concentrations found throughout the study for these species.

This enabled the effects of environment (mostly soil type) and genetics (plant species/variety) to be separated, thus allowing an assessment of the ability of each species/variety to take up and accumulate essential minerals in their leaves. The minerals studied were the macronutrients nitrogen, phosphorus, potassium, calcium, magnesium, and sulphur, in addition to sodium, often present in "macro concentrations" but required in micro amounts; along with the micronutrients iron, manganese, boron, copper, and zinc. All of these macro- and micronutrients, with the possible exception of boron, are required by humans and animals as well as by plants. A sub-sample was analyzed for selenium, an essential micronutrient for humans and animals, but which is not required by higher plants. The analyses also enabled detection of any mineral deficiencies in the plants sampled.

Each leaf tissue sample comprised around two handfuls of relatively young leaves: not the youngest or older leaves, but, e.g., in sweet potato, the 5th to 9th youngest leaves, sampled from several representative plants, avoiding plants with disease (e.g., virus, scab) symptoms. The exception was pumpkin where just the tips (up to 25 cm) were sampled. If the plants were dusty (e.g., if the plants were growing near a road), they were washed in clean water. Samples were dried in a microwave oven or perspex covered trays soon after collection and placed in labelled plastic ziplock bags. Soil and compost samples were also collected from numerous sites in Kiribati and Tuvalu, and these will inform a future article.

The samples were brought to Australia under a Federal Department of Agriculture permit and irradiated. They were then acid digested and N analyzed by the combustion method (using an Elementar instrument, and limit of detection (LOD) calculated as 10 x the standard deviation of the calibration blank). Protein % was estimated by multiplying nitrogen % by 4.4. The other minerals listed above were analyzed by inductively coupled plasma atomic emission spectrometry (ICPOES) (using a radial CIROS instrument, and LOD as above), while Se was analyzed using ICP mass spectrometry (ICPMS) (using ICPMS method-*7B2G*, and LOD as above). Appropriate quality control measures were applied, including regular duplicate samples and analyses of aluminum and titanium to detect dust/soil contamination, which inflates Fe concentration.

In the previous study, carotenoids were also analyzed, and to minimize enzymatic degradation, samples were dried rapidly in a microwave oven as soon as practicable after collection, usually the same night. The carotenoids *beta*-carotene (the major pro-vitamin A carotenoid), lutein (usually the most abundant carotenoid in leaves), and *alpha*-carotene were analyzed by size-exclusion high-performance liquid chromatography (HPLC) at the Mares laboratory, Waite Campus, University of Adelaide [1]. Budgetary constraints precluded carotenoid analyses in the current study; however, these methodological details are included as carotenoid data are included for several of the species featured in the factsheets, the main study output.

The criteria for atoll suitable leafy plants, as listed under Project objective above, were (1) highly nutritious, (2) taste good, (3) tolerant of alkalinity (i.e., soil pH (H_2O) > 8.5), (4) tolerant of salt and drought (with the exception of *Ipomoea aquatica*, kangkong, which grows in fresh water), and (5) easy to grow, prepare, and cook.

For the factsheets, photographs, characteristics, uses, availability, propagation and growing methods, disease and pest threats, and advice on harvesting and storage were included for each species. Leaf mineral and carotenoid data (if the species was included in the previous factsheets) were presented in a table which included the featured leafy vegetable sampled at a representative site, compared with other leafy vegetables growing at the same site. English cabbage was also included, as a moderately nutritious yardstick, using average values of samples purchased at markets in the South Pacific.

Factsheets 1, 2, 4, 5, 6, 8, 9, 10, and 11 were adapted from those published during the ACIAR PC/2010/063 project, Feasibility study on increasing the consumption of nutritionally-rich leafy vegetables by indigenous communities in Samoa, Solomon Islands and Northern Australia (www.remoteindigenousgardens.net\T1\textgreater{}2013/08\T1\textgreater{}new-resources-fact-sheets) [23].

The factsheets (500 sets) were graphically designed, printed, and laminated to improve durability by SPC, Suva, Fiji and distributed in Kiribati and Tuvalu, and also published online: www.researchgate.net\T1\textgreater{}publication\T1\textgreater{}327261351_Tackling_NCDs_from_the_ground_up_Nutritious_ leafy_vegetables_to_improve_nutrition_security_on_atolls [29]. Later, they were translated into the Kiribati and Tuvalu languages. A recipe book featuring the selected plants was compiled and distributed to communities.

The project also trialed and promoted starchy root crops (taro, sweet potato, cassava) and non-leafy vegetables, including beans, tomato, cucumber, capsicum, eggplant, and watermelon, and these components are reported elsewhere.

Author Contributions: Conceptualization, G.L., G.D. and S.H.; Funding acquisition, G.D., S.H. and G.S.; Investigation, G.L., G.D., R.T., S.H., K.N. and M.L.; Project administration, R.T., S.H., K.N., M.L. and G.S.; Supervision, G.L., G.D., R.T., S.H., K.N., M.L. and G.S.; Visualization, G.L.; Writing—Original draft, G.L.; Writing—Review and editing, G.D., R.T., S.H., K.N., M.L. and G.S. All authors have read and agreed to the published version of the manuscript.

Funding: This research was funded by the Australian Centre for International Agricultural Research (ACIAR), grant number SMCN-2014-089.

Acknowledgments: ACIAR Program Managers Robert Edis and James Quilty, and IFAD, our main collaborator. Rosalind Kiata for her journalistic expertise in the promotion campaign. Walter Wasile, Suva for the graphics. Mary Taylor for her assistance with the scoping study for this project in 2014. This article is dedicated to Lois Englberger, an inspirational leader in improving human nutrition in the Pacific region, who founded the *Go Local* movement from her base at The Island Food Community of Pohnpei.

Conflicts of Interest: The authors declare no conflict of interest.

References

1. Lyons, G.; Goebel, R.; Tikai, P.; Stanley, K.-J.; Taylor, M. Promoting nutritious leafy vegetables in the Pacific and northern Australia. *Acta Hortic.* **2015**, *1102*, 253–260. [CrossRef]
2. Taylor, M.; Kambuou, R.; Lyons, G.; Hunter, D.; Morgan, E.; Quartermain, A.; Robert, N.; Paofa, J.; Lorens, A.; Tuia, V. Realizing the potential of indigenous vegetables through improved germplasm information and seed systems. *Acta Hortic.* **2015**, *1102*, 29–42. [CrossRef]
3. Bakineti, T.; Kiata, R. Hunger Pangs Lead to Health Complaints. News-Outer Island Food and Water Project. 2017. Available online: www.oifwpkiribati.org.ki/news (accessed on 15 December 2019).

4. SPC (The Pacific Community). *NCD Statistics for the Pacific Islands Countries and Territories*; SPC (The Pacific Community): Noumea, New Caledonia, 2010.
5. World Bank. NCD Roadmap Report. In *A Background Document on Preventing and Controlling NCDs in the Pacific. Circulated in June 2014 for Consideration by the Joint Forum Economic and Pacific Health Ministers' Meeting, July 2014*; World Bank: Washington DC, USA, 2014.
6. WHO. Pacific Islanders pay heavy price for abandoning traditional diet. *Bull. World Health Organ.* **2004**, *88*, 484–485.
7. Grieve, H.; Busch-Hallen, J.; Mellor, K. *Undernutrition in Pacific Island Countries: An Issue Requiring Further Attention*; Women's and Children's Health Knowledge Hub. Policy Brief; May 2013. Cited in Development Bulletin 78, August 2017. Urban Development in the Pacific; Thomas, P., Keen, M., Eds.; Australian National University: Canberra, Australia.
8. Goris, J.; Zomerdijk, N.; Temple, V. Nutritional status and dietary diversity of Kamea in the Gulf Province, Papua New Guinea. *Asia Pac. J. Clin. Nutr.* **2017**, *26*, 665–670. [PubMed]
9. O'Meara, L.; Williams, S.; Hickes, D.; Brown, P. Predictors of dietary diversity of indigenous food-producing households in rural Fiji. *Nutrients* **2019**, *11*, 1629. [CrossRef] [PubMed]
10. Taki, B. *Underutilized and wild indigenous food plants of Makira Ulawa Province. Record No. 51532*; SPC (The Pacific Community): Noumea, New Caledonia, 2010.
11. Ebert, A. Potential of underutilized traditional vegetables and legume crops to contribute to food and nutrition security, income and more sustainable production systems. *Sustainability* **2014**, *6*, 319–335. [CrossRef]
12. Bell, J.; Taylor, M. *Building Climate-Resilient Food Systems for Pacific Islands*; WorldFish Program Report; WorldFish: Penang, Malaysia, 2015; p. 15.
13. Taylor, M.; Jaenicke, H.; Hunter, D.; McGregor, A.; Lyons, G. Diversity for sustaining livelihoods: Examples, constraints and lessons learnt. *Acta Hortic.* **2015**, *1101*, 16. [CrossRef]
14. Bowden-Kerby, A. Agriculture and Marine Resources in a Changing Climate: Finding a Way Forward for the Line Islands, Kiribati. Technical Report. Corals for Conservation. 2017. Available online: https://www.researchgate.net/publication/318836638 (accessed on 17 December 2019).
15. SPC (The Pacific Community). *Towards a Food Secure Pacific. Framework for Action on Food Security in the Pacific 2011–2015*; SPC Land Resources Division: Suva, Fiji, 2011.
16. Halavatau, S. *Growing Root Crops on Atolls*; SPC: Suva, Fiji, 2018.
17. O'Sullivan, J.; Asher, C.; Blamey, F. *Nutritional Disorders of Sweet Potato. ACIAR Monograph No. 48*; Australian Centre for International Research: Canberra, Australia, 1997.
18. Duque, L.; Villordon, A. Root branching and nutrient efficiency: Status and way forward in root and tuber crops. *Front. Plant Sci.* **2019**, *10*, 237. [CrossRef]
19. Bailey, J.M. *The Leaves We Eat. SPC Handbook No. 31*; South Pacific Commission: Noumea, New Caledonia, 1992.
20. Thaman, R. Batiri Kei Baravi: The Ethnobotany of Pacific Island Coastal Plants. In *Atoll Research Bulletin No. 361*; National Museum of Natural History: Washington, DC, USA, 1992.
21. Webb, A.P. *Tree and Perennial Crop Introduction and Extension in Kiribati 1991–1994*; Kiribati Ministry of the Environment and Natural Resources Development: Tarawa, Kiribati, 1994.
22. French, B.R. *Leafy Greens and Vegetables in Solomon Islands*; Food Plants International: Devonport, Australia, 2010; Available online: www.b4fn.org/resources/publications/ (accessed on 8 April 2020).
23. Goebel, R.; Taylor, M.; Lyons, G. *Feasibility Study on Increasing the Consumption of Nutritionally-Rich Leafy Vegetables by Indigenous Communities in Samoa, Solomon Islands and Northern Australia*; Factsheet Series; Australian Centre for International Agricultural Research (ACIAR): Canberra, Australia, 2013. Available online: www.remoteindigenousgardens.net\T1\textgreater{}2013/08\T1\textgreater{}new-resources-fact-sheets (accessed on 10 January 2020).
24. Oomen, H.; Grubben, G. *Tropical Leaf Vegetables in Human Nutrition*; Royal Tropical Institute: Amsterdam, The Netherlands, 1978.
25. Dignan, C.; Burlingame, B.; Kumar, S.; Aalbersberg, W. *The Pacific Islands Food Composition Tables*, 2nd ed.; Food and Agriculture Organisation of the United Nations (FAO): Rome, Italy, 2004.
26. FAO. *African Leafy Vegetables in Kenya: Local Biodiversity in Production, Marketing and Consumption*; FAO: Rome, Italy, 2011.

27. Keatinge, J.D.; Yang, R.-Y.; Hughes, J.; Easdown, W.; Holmer, R. The importance of vegetables in ensuring both food and nutritional security in attainment of the Millennium development Goals. *Food Secur.* **2011**, *3*, 491–501. [CrossRef]
28. SPC (The Pacific Community). *Green Leaves Leaflet No. 8*; SPC: Noumea, New Caledonia, 2012.
29. Lyons, G.; Dean, G.; Goebel, R.; Taylor, M.; Kiata, R.; Tongaiaba, R. *Tackling NCDs from the Ground up: Nutritious Leafy Vegetables to Improve Nutrition Security on Pacific Atolls*; Australian Centre for International Agricultural Research (ACIAR): Canberra, Australia, 2018. Available online: www.researchgate.net\T1\textgreater{}publication\T1\textgreater{}327261351_Tackling_NCDs_from_the_ground_up_Nutritious_leafy_vegetables_to_improve_nutrition_security_on_atolls (accessed on 10 March 2020).
30. IFAD (United Nations International Fund for Agricultural Development). Outer Islands Food and Water Project. Kiribati IFAD Newsletter Q1. 2017. Available online: www.oifwpkiribati.org.ki (accessed on 15 March 2020).
31. Islam, M.S.; Yoshimoto, M.; Yahara, S.; Okuno, S.; Ishiguro, K.; Yamakawa, O. Identification and characterization of foliar polyphenolic composition in sweetpotato (*Ipomoea batatas* L.) genotypes. *J. Agric. Food Chem.* **2002**, *50*, 3718–3722. [CrossRef]
32. Bertrand-Protat, S. *Edible Leaves of the Pacific*; The Pacific Community (SPC): Noumea, New Caledonia, 2019.
33. Sogbohossou, E.; Achigan-Dako, E.; Maundu, P.; Solberg, S.; Deguenon, E.; Mumm, R.; Hale, I.; Van Deynze, A.; Schranz, M. A roadmap for breeding orphan leafy vegetable species: A case study of Gynandropsis gynandra (Cleomaceae). *Hortic. Res.* **2018**, *5*, 2. [CrossRef] [PubMed]
34. Haskell, M. The challenge to reach nutritional adequacy for vitamin A: B-carotene bioavailability and conversion—Evidence in humans. *Am. J. Clin. Nutr.* **2012**, *96*, 1193S–1203S. [CrossRef] [PubMed]
35. Abdel-Aal, E.S.M.; Akhtar, H.; Zaheer, K.; Ali, R. Dietary sources of lutein and zeaxanthin carotenoids and their role in eye health. *Nutrients* **2013**, *5*, 1169–1185. [CrossRef] [PubMed]
36. Mbikay, M. Therapeutic potential of Moringa oleifera leaves in chronic hyperglycemia and dyslipidemia: A review. *Front. Pharm.* **2012**, *3*, 24. [CrossRef] [PubMed]
37. Matic, I.; Guidi, A.; Keenzo, M.; Mattei, M.; Galgani, A. Investigation of medicinal plants traditionally used as dietary supplements: A review on Moringa oleifera. *J. Public Health Afr.* **2018**, *9*, 841. [CrossRef]
38. Ahmad, J.; Khan, I.; Blundell, R. Moringa oleifera and glycemic control: A review of current evidence and possible mechanisms. *Phytother. Res.* **2019**, *33*, 2841–2848. [CrossRef]
39. Oyagbemi, A.; Odetola, A.; Azeez, O. Antidiabetic properties of ethanolic extract of Cnidoscolus aconitifolius on alloxan induced diabetes mellitus in rats. *Afr. J. Med. Sci.* **2010**, *39*, S171–S178.
40. Ajiboye, B.; Oyinloye, B.; Agboinghale, P.; Ojo, A. *Cnidoscolus aconitifolius* (Mill.) I.M. Johnst leaf extract prevents oxidative hepatic injury and improves muscle glucose uptake ex vivo. *J. Food Biochem.* **2019**, *43*, e13065. [CrossRef]
41. Baxter, I.; Dilkes, B. Elemental profiles reflect plant adaptations to the environment. *Science* **2012**, *336*, 1661–1663. [CrossRef]
42. Abilgos-Ramos, R.; Mamucod, H.; Corpuz, G. Chili pepper leaves as alternative source of micronutrients. *Philipp. J. Crop Sci.* **2012**, *37* (Suppl. 1), 139.
43. Sharma, D.; Rawat, I.; Goel, H. Anticancer and anti-inflammatory activity of some dietary cucurbits. *Ind. J. Exp. Biol.* **2015**, *53*, 216–221.
44. Petropoulos, S.; Fernandes, A.; Dias, M.; Vasilakoglou, I.; Petrotos, K.; Barros, L.; Ferreira, I. Nutritional value, chemical composition and cytotoxic properties of common purslane (*Portulaca oleracea* L.) in relation to harvesting stage and plant part. *Antiox* **2019**, *8*, 293. [CrossRef] [PubMed]
45. Reuter, D.; Robinson, J. *Plant Analysis: An Interpretation Manual*, 2nd ed.; CSIRO Publishing: Melbourne, Australia, 1997.
46. Marschner, H. *Mineral Nutrition of Plants*, 2nd ed.; Academic Press: Amsterdam, The Netherlands, 2002.
47. Kuti, J.; Torres, E. Potential nutritional and health benefits of tree spinach. In *Progress in New Crops*; Janick, J., Ed.; ASHS Press: Arlington, WV, USA, 1996; pp. 516–520.
48. Oyagbemi, A.; Odetola, A. Hepatoprotective and nephroprotective effects of Cnidoscolus aconitifolius in protein energy malnutrition induced liver and kidney damage. *Pharmacogn. Res.* **2013**, *5*, 260–264. [CrossRef]
49. Gopalakrishnan, L.; Doriya, K.; Kumar, D. Moringa oleifera: A review on nutritive importance and its medicinal application. *Food Sci. Hum. Wellness* **2016**, *5*, 49–56. [CrossRef]

50. Kou, X.; Li, B.; Olayanju, J.; Drake, J.; Chen, N. Nutraceutical or pharmacological potential of Moringa oleifera Lam. *Nutrients* **2018**, *10*, 343. [CrossRef] [PubMed]
51. Puel, C.; Mathey, J.; Kati-Coulibaly, S.; Davicco, M.; Lebecque, P.; Chanteranne, B.; Horcajada, M.; Coxam, V. Preventive effect of *Abelmoschus manihot* (L.) Medik. On bone loss in the ovariectomised rats. *J. Ethnopharmacol.* **2005**, *99*, 55–60. [CrossRef]
52. Walter, A.; Lebot, V. *Garden of Oceania. Monograph No. 122*; Australian Centre for International Agricultural Research (ACIAR): Canberra, Australia, 2007; ISBN 1863204709.
53. Inu, M. Pharmacognostical profile of selected medicinal plants. In *Handbook of African Medicinal Plants*; CRC Press: London, UK, 2014.
54. Sulistiawati, Y.; Suwondo, A.; Hardjanti, T.; Soejoenoes, A.; Anwar, M.C.; Susiloretni, K. Effect of Moringa oleifera on level of prolactin and breast milk production in postpartum mothers. *Belitung Nurs. J.* **2017**, *3*, 126–133. [CrossRef]
55. Kumssa, D.; Joy, E.; Young, S.; Odee, D.; Ander, E.; Broadley, M. Variation in the mineral element concentration of Moringa oleifera Lam. And M. stenopetala (Bak. F.) Cuf.: Role in human nutrition. *PLoS ONE* **2017**, *12*, e0175503. [CrossRef]
56. Chilimba, A.; Young, S.; Black, C.; Rogerson, K.; Ander, E.; Watts, M. Maize grain and soil surveys reveal suboptimal dietary selenium intake is widespread in Malawi. *Sci. Rep.* **2011**, *1*. [CrossRef]
57. Christophersen, O.A.; Lyons, G.; Haug, A.; Steinnes, E. Selenium (Ch 16). In *Heavy Metals in Soils: Trace Metals and Metalloids in Soils and Their Bioavailability*; Alloway, B.J., Ed.; Springer: Dordrecht, The Netherlands, 2013; pp. 429–463.
58. Joy, E.; Broadley, M.; Young, S.; Black, C.; Chilimba, A.; Ander, E.; Barlow, T.; Watts, M. Soil type influences crop mineral composition in Malawi. *Sci. Total Environ.* **2015**, *505*, 587–595. [CrossRef]
59. Lyons, G.; Gondwe, C.; Banuelos, G.S.; Zambrano, M.C.; Haug, A.; Christophersen, O.A. *Selenium in the Environment and Human Health*; Banuelos, G., Lin, Z.-Y., Eds.; Taylor and Francis Group: London, UK, 2014; pp. 91–92.
60. Leone, A.; Bertoli, S.; Di Lello, S.; Bassoli, A.; Ravasenghi, S.; Borgonovo, G.; Forlani, F.; Battezzati, A. Effect of Moringa oleifera leaf powder on postprandial blood glucose response: In vivo study on Saharawi people living in refugee camps. *Nutrients* **2018**, *10*, 1494. [CrossRef]
61. Girija, K.; Lakshman, K.; Udaya, C.; Sabhya, S.; Divya, T. Anti-diabetic and anti-cholesterolemic activity of methanol extracts of three species of Amaranthus. *Asian Pac. J. Trop. Biomed.* **2011**, *1*, 133–138. [CrossRef]
62. Ashok Kumar, B.; Lakshman, K.; Jayaveea, K.; Shekar, D.; Khan, S.; Thippeswami, B.; Veerapur, V. Antidiabetic, antihyperlipidemic and antioxidant activities of methanolic extract of Amaranthus viridis Linn. In alloxan induced diabetic rats. *Exp. Toxicol. Pathol.* **2012**, *64*, 75–79. [CrossRef] [PubMed]
63. Nawale, R.; Mate, G.; Wakure, B. Ethanolic extract of Amaranthus paniculatus Linn. Ameliorates diabetes-associated complications in alloxan-induced diabetic rats. *Integr. Med. Res.* **2017**, *6*, 41–46. [CrossRef]
64. Yu, J.; Xiong, N.; Guo, H. Clinical observation on diabetic nephropathy treated with alcohol of Abelmoschus manihot. *Zhongguo Zhong Xi Yi Jie He Za Zhi* **1995**, *15*, 263–265. [PubMed]
65. Chen, Y.; Gong, Z.; Cai, G.; Gao, Q.; Chen, X.; Tang, L.; Wei, R.; Zhou, J. Efficacy and safety of Flos Abelmoschus manihot (malvaceae) on type 2 diabetic nephropathy: A systematic review. *Clin. J. Integr. Med.* **2015**, *21*, 464–472. [CrossRef]
66. Kim, H.; Dusabimana, T.; Kim, S.; Je, J.; Jeong, K.; Kang, M.; Cho, K.; Kim, H.; Park, S. Supplementation of Abelmoschus manihot ameliorates diabetic nephropathy and hepatic steatosis by activating autophagy in mice. *Nutrients* **2018**, *10*, 1703. [CrossRef]
67. Yuan, L.; Qin, Y.; Wang, L. Capsaicin-containing chili improved postprandial hyperglycemia, hyperinsulinemia, and fasting lipid disorders in women with gestational diabetes mellitus and lowered the incidence of large-for-gestational-age newborns. *Clin. Nutr.* **2016**, *35*, 388–393. [CrossRef]
68. Sonati, S.; Razavi, B.; Hosseinzadeh, H. A review of the effects of *Capsicum annuum* L. and its constituent, capsaicin, in metabolic syndrome. *Iran J. Basic Med. Sci.* **2018**, *21*, 439–448.
69. Ungurianu, A.; Seremet, O.; Gagniuc, E. Preclinical and clinical results regarding the effects of a plant-based antidiabetes formulation versus well established antidiabetic molecules. *Pharmacol. Res.* **2019**, *150*, 104522. [CrossRef]

70. Wainstein, J.; Landau, Z.; Bar Dayan, Y.; Jakubowicz, D.; Grothe, T.; Perrinjaquet-Moccetti, T.; Boaz, M. Purslane extract and glucose homeostasis in adults with type 2 diabetes: A double blind, placebo-controlled clinical trial of efficacy and safety. *J. Med. Food* **2016**, *19*, 133–140. [CrossRef] [PubMed]
71. Zheng, G.; Mo, F.; Ling, C. *Portulaca oleracea* L. alleviates liver injury in streptozoticin-induced diabetic mice. *Drug Des. Dev. Ther.* **2017**, *12*, 47–55. [CrossRef] [PubMed]
72. Hu, Q.; Niu, Q.; Song, H.; Wei, S.; Wang, S.; Yao, L.; Li, Y.-P. Polysaccharides from *Portulaca oleracea* L. regulated insulin secretion in INS-1 cells through voltage-gated Na^+ channel. *Biomed. Pharmacother.* **2019**, *109*, 876–885. [CrossRef] [PubMed]
73. Malalavidhane, S.; Wickramasinghe, S.; Jansz, E. An aqueous extract of the green leafy vegetable Ipomoea aquatic is as effective as the oral hypoglycaemic drug tolbutamide in reducing the blood sugar levels of Wistar rats. *Phytother. Res.* **2001**, *15*, 635–637. [CrossRef] [PubMed]
74. Shivananjappa, M.M. Dietary supplementation with Ipomoea aquatica (whole leaf powder) attenuates maternal and fetal oxidative stress in streptozotocin-diabetic rats. *J. Diabetes* **2013**, *5*, 25–33. [CrossRef]
75. Alam, F.; Shafique, Z.; Amjad, S.; Bin Asad, M. Enzymes inhibitors from natural sources with antidiabetic activity: A review. *Phytother. Res.* **2019**, *33*, 41–54. [CrossRef]
76. Padee, P.; Nualkaew, S.; Talubmook, C.; Sakuljaitrong, S. Hypoglycemic effect of a leaf extract of *Pseuderanthemum palatiferum* (Nees) Radlk. On normal and streptozotocin-induced diabetic rats. *J. Ethnopharmacol.* **2010**, *132*, 491–496. [CrossRef]
77. Chayarop, K.; Temsiririrkkul, R.; Peungvicha, P.; Wongkrajang, Y.; Chuakul, W.; Amnuoypol, S. Antidiabetic effects and in vitro antioxidant activity of *Pseuderanthemum palatiferum* (Nees) Radlk. Ex Lindau leaf aqueous extract. *Pharm. Soc. Asia* **2011**, *38*, 13–22.
78. Nualkaew, S.; Padee, P.; Talukmoook, C. Hypoglycemic activity in diabetic rats of stigmasterol and sitosterol-3-0-B-D-glucopyranoside isolated from *Pseuderanthemum palatiferum* (Nees) Radlk. Leaf extract. *J. Med. Plants Res.* **2015**, *9*, 629–635.
79. Hanh, T.; Nguyen, H.; Nguyen, T. α-amylase and α-glucosidase inhibitory saponins from *Polyscias fruticosa* leaves. *J. Chem.* **2016**, *4*.
80. Malini, D.; Madihah, M.; Kusmoro, J.; Kamilawati, F.; Iskandar, J. Ethnobotanical study of medicinal plants in Karangwangi, District of Cianjur, West Java. *Biosaintifika J. Biol. Biol. Educ.* **2017**, *9*, 345–356.
81. Zhu, H.; Chen, G.; Meng, G.; Xu, J. Characterization of pumpkin polysaccharides and protective effects on streptozotocin-damaged islet cells. *Chin. J. Nat. Med.* **2015**, *13*, 199–207. [CrossRef]
82. Mahmoodpoor, A.; Medghalchi, M.; Nazemiyeh, H.; Asgharian, P.; Shadvar, K.; Hamishehkar, H. Effect of Cucurbita maxima on control of blood glucose in diabetic critically ill patients. *Adv. Pharm. Bull.* **2018**, *8*, 347–351. [CrossRef] [PubMed]
83. Dire, G.; Rodrigues, J.; Oliviera, J.; Vasconcelos, S.; Siqueira, P.; Duarte, R. Biological effects of a chayote extract on Wistar rats with induced diabetes: A radiopharmaceutical analysis. *Pak. J. Biol. Sci.* **2007**, *10*, 568–574. [CrossRef] [PubMed]
84. Wu, C.; Ou, T.; Chang, C.; Chang, X.; Yang, M.; Wang, C. The polyphenol extract from Sechium edule shoots inhibits lipogenesis and stimulates lipolysis via activation of AMPK signals in HepG2 cells. *J. Agric. Food Chem.* **2014**, *62*, 750–759. [CrossRef]
85. Genc, Y.; Oldach, K.; Taylor, J.; Lyons, G. Uncoupling of sodium and chloride to assist breeding for salinity tolerance in crops. *New Phytol.* **2016**, *210*, 145–156. [CrossRef]
86. Genc, Y.; Taylor, J.; Lyons, G.; Li, Y.; Cheong, J.; Appelbee, M.; Oldach, K.; Sutton, T. Bread wheat with high salinity and sodicity tolerance. *Front. Plant Sci.* **2019**, *10*, 1280. [CrossRef]
87. Englberger, L.; Kuhnlein, H.V.; Lorens, A.; Pedrus, P.; Albert, K.; Currie, J.; Pretrick, M.; Jim, R.; Kaufer, L. Pohnpei, FSM case study in a global health project documents its local food resources and successfully promotes local food for health. *Pac. Health Dial.* **2010**, *16*, 25–36.
88. Keatinge, J.D.; Chadha, J.; Hughes, J.; Easdown, W.; Holmes, R.; Tenkouano, A.; Yang, R.; Maulyanova, R.; Neave, S.; Afari-Sefa, V.; et al. Vegetable gardens and their impact on the attainment of the millennium development goals. *Biol. Agric. Hortic.* **2012**, *28*, 71–85. [CrossRef]
89. Solberg, S.; Seta-Waken, P.; Paul, T.; Palaniappan, G.; Iramu, E. Patterns in the conservation and use of traditional vegetables from the New Guinea biodiversity hotspot. *Agroecol. Sustain. Food Syst.* **2018**, *42*, 1079–1091. [CrossRef]

90. Francis, J.; Dietershagen, J. Promoting Nutritious Food Systems in the Pacific Islands: Protein Rich Atoll Vegetables and Bioactive Compounds. The Technical Centre for Agricultural and Rural Cooperation (CTA), Ede, Netherlands. 2019. Available online: www.cta.int (accessed on 16 March 2020).
91. Makkar, H.; Becker, K. Nutritional value and antinutritional components of whole and ethanol extracted Moringa oleifera leaves. *Anim. Feed Sci. Technol.* **1996**, *65*, 211–228. [CrossRef]
92. Aberoumand, A.; Deokule, S. Screening of some nutrients and anti-nutrients components in some plant foods of Iran and India. *J. Agric. Technol.* **2010**, *6*, 777–781.
93. Fan, D.; Zhou, C. Chaya (*Cnidoscolus* spp.) leaves: Antinutritional factor analysis and effect of treatment processing on cyanogenic glycosides content. *Food Sci.* **2010**, *3*, S636.9.
94. Babalola, J.; Alabi, O. Effect of processing methods on nutritional composition, phytochemicals, and anti-nutrient properties of Chaya leaf (Cnidoscolus aconitifolius). *Afr. J. Food Sci.* **2015**, *9*, 560–565.
95. FAO. *Research Report: Knowledge, Attitudes, Beliefs and Practices Related to the Consumption of Fruits and Vegetables in Samoa*; FAO: Apia, Samoa, November 2004; Available online: www.fao.org/docrep/015/an432e/an432e00.pdf (accessed on 8 April 2020).
96. Bollars, C.; Sorensen, K.; de Vries, N.; Meertens, R. Exploring health literacy in relation to noncommunicable diseases in Samoa: A qualitative study. *BMC Public Health* **2019**, *19*, 1151. [CrossRef]

plants

MDPI

Article

Assessment of the Nutritional and Medicinal Potential of Tubers from Hairy Stork's-Bill (*Erodium crassifolium* L 'Hér), a Wild Plant Species Inhabiting Arid Southeast Mediterranean Regions

Shabtai Cohen [1], Hinanit Koltai [2], Gopinath Selvaraj [2], Moran Mazuz [2], Moran Segoli [1], Amnon Bustan [1] and Ofer Guy [1,*]

[1] Ramat Negev Desert Agro-Research Center (RN-DARC), Ramat Negev Works Ltd., D.N. Halutza 8551500, Israel; sab@inter.net.il (S.C.); moran.segoli@gmail.com (M.S.); amnonbustan@gmail.com (A.B.)

[2] Department of Ornamental Horticulture and Biotechnology, Institute of Plant Sciences, Agricultural Research Organization, Rishon LeZion 7528809, Israel; hkoltai@volcani.agri.gov.il (H.K.); gopinath@volcani.agri.gov.il (G.S.); moranjacobi@gmail.com (M.M.)

* Correspondence: oferguy243@gmail.com; Tel.: +972-50-3699001

Received: 31 July 2020; Accepted: 19 August 2020; Published: 20 August 2020

Abstract: Emerging needs for diversifying human diet and to explore novel therapeutic procedures have led to increasing attempts to retrieve traditional nourishments and recruit beneficial wild plant species. Species of the genus *Erodium* (Geraniaceae) harbor medicinal indications and substances known from folklore and scientific research. Hairy stork's bill (*Erodium crassifolium* L'Hér), is a small hemicryptophyte that inhabits arid southeast Mediterranean regions. *E. crassifolium* is among the very few Geraniaceae species known to produce tubers. Traditional knowledge holds that the tubers are edible and used by Bedouin tribes. However, no scientific information was found regarding nutrition or medicinal properties of these tubers. The objectives of our project are to unravel potential nutritional and medicinal benefits of the tubers, conduct initial steps towards domestication and develop agricultural practices enhancing *E. crassifolium* tuber yield and quality. Tubers show high water content (90%), low caloric value (23 Kcal 100^{-1} g) and considerable contents of minerals and vitamins. In addition, the tubers contain significant amounts of catechins and epigallocatechin, polyphenolic compounds known for their antioxidative, anti-inflammatory and antiproliferative activities. Furthermore, in vitro experiments demonstrated significant anti-inflammatory effects on human cell cultures. *E. crassifolium* is highly responsive to environmental changes; fertigation (700 mm) increased tuber yield by 10-fold, compared to simulated wild conditions (50–200 mm). These results indicate a significant potential of *E. crassifolium* becoming a valuable crop species. Therefore, there is a need for continued efforts in domestication, including ecotype selection, breeding, development of suitable agricultural practices and further exploration of its medicinal benefits.

Keywords: anti-inflammatory activity; antioxidants; catechin; domestication; *Erodium crassifolium*; underutilized species

1. Introduction

Human diet and therapy have featured valuable plants gathered from the wild since ancient times. The agricultural civilization developed in the past 12,000 years has been founded on the domestication of many useful species [1]. However, in the modern era, global economic considerations have significantly changed agricultural approaches and scales, which brought about overexploiting of land resources, consistent diminishing of natural biodiversity and negative modifications of human diet and health [2,3]. Globalization and urbanization have accelerated the unification of the current human diet and have led to a further narrow hoard of available useful species [4]. The recent burst of noncommunicative diseases (NCD), including obesity, diabetes, cardiovascular disorders and cancer among populations of developing countries as well as lower socioeconomic classes of developed countries, is largely attributed to nutritional and health disorders derived from current human diets [5].

The emerging need for improving human diet has led to increasing interests in traditional nourishments, such as the Mediterranean diet [6] and in primordial therapeutic aids. These are usually based on diverse resources that change through locations, seasons and include numerous wild edible plant species. Attempts to retrieve old beneficial species and recruit them for new commercial use have a vast potential, considering that a very small portion of known plant species has ever been adequately studied for such purposes [7,8]. There is an increasing research activity focused on identifying and characterizing wild plant species with particular attributes to human diet and health [4,9–15]. Special attention is paid to edible plants as sources for essential mineral elements [16] and of antioxidative-active compounds [17]. In addition, there is a rising interest in plants encompassing anti-inflammatory activities [18].

Geraniaceae family includes seven genera and about 830 species distributed from temperate to tropic and arid climates. The largest genera are *Geranium* (430 species), *Pelargonium* (280 species) and *Erodium* (80 species). The family is known for the production of essential oils and ornamentals. Many Geraniaceae species are ascribed to have various medicinal values [19–23]. Additionally, Some of the species belonging to the genus *Erodium* have recognized medicinal indications from folklore and empirical data [24]. *Erodium* species are used to treat a variety of human ailments such as colds, coughs, diarrhea, hemorrhaging and are used to dress wounds [24–28].

Hairy stork's bill (HSB) (*Erodium crassifolium* L'Hér) is a Saharo-Arabian common perennial hemicryptophyte (i.e., buds are at or near the soil surface) that inhabits shrub–steppes of arid southeast Mediterranean regions. The species is distributed from northwest at Crete [29], through few Aegean Sea islands [30], the Libyan [31] and Egyptian [32,33] coasts, north Sinai Peninsula [32,34], Cyprus [35], the Negev Desert of Israel [36], until Edom mountains of Jordan and Saudi Arabian deserts, on southeast [37]. In Israel, HSB can be found in the Negev and Judean deserts, where the annual rainfall is 30–250 mm. The species is most abundant in the stony and arid loess soils and on slopes of limestone hills [36]. Vegetative phase in form of rosette (early season) and 30 cm stems and leaves (late season) starts with the first effective rains. Flowering time is February to May and the flowers are pink to purple with dark color base (Figure 1A,B) that attracts pollinating insects. The ripe ovary splits into five diaspores, each contains one seed covered by the ovary hairy wall with a sharply pointed tip (Figure 1C). The diaspores are carried away from the mother plant by strong winds and react to humidity changes by creating screw-like twists to penetrate the soil [36]. Among all Geraniaceae species, *E. crassifolium* is among the few known to produce tubers (Figure 1D). The tubers are formed on roots at depth of 5–20 cm and are typically small and spherical (1–2 cm in diameter). HSB tubers have a sweet taste and are best in late winter or early spring when they are whitish in color [36]. Traditional knowledge holds that the tubers are edible and Bedouin tribes are their primary consumers [33]. However, no information exists regarding nutritional value of these tubers. Furthermore, in spite of enduring claims associating some medicinal properties with HSB tubers, no supportive documented evidence have been found so far.

Figure 1. Hairy stork's-bill (*E. crassifolium*). (**A**) Shrub with flowers and seed pods; (**B**) flower; (**C**) seeds with typical hairy feather-like awns; (**D**) root tubers, connected to the root-shoot transition region (RSTR), the perennial plant organ.

The objective of the present study was to evaluate, for the first time, the nutritive and medicinal potential of *E. crassifolium* for becoming a useful new crop species. In addition to a moderate nutritive potential, we demonstrate here significant in vitro anti-inflammatory capacities in the tubers, assigned to well-known bioactive compounds. Furthermore, HSB displayed impressive tuber productivity when exposed to agricultural conditions, indicating a promising potential for domestication.

2. Results and Discussion

The domestication of plant species is a complex iterating process, which includes selection according to subjective preferences [38], such as color, palatability and size. In the modern era, however, with the increasing awareness to the potential nutrition benefits or hazards of a given food product, an established list of nutrition facts has become an essential step in the initial evaluation of a candidate species. Here, for the first time, a comprehensive nutrition facts list is presented for HSB tubers (Table 1). The very high water content of HSB tubers makes them highly valuable for nomads in the hot dry desert, supporting the knowledge about their traditional use by Bedouin tribes [33]. Tubers' caloric content is about half than in carrots, but they provide considerable amounts of essential minerals such as calcium, sulfur, magnesium and iron, as well as phosphorus and potassium. In addition, HSB tubers are a good source of vitamins A and C, harboring about a quarter and a third of their contents in carrot, respectively (Table 1).

Table 1. Nutritional profile of hairy stork's bill (HSB) tubers compared to that of carrots (adopted from USDA, 2020 [39]).

Nutritional Profile	Measure Units /100 g	HSB Tubers	Carrot
Water	g	90.6	88.3
Ash		0.8	0.9
Caloric value	Kcal	23	41
Carbohydrates	g	7.9	9.6
Sugars		4.3	6
Polysaccharides		3.0	2.8
Lipids		0.1	0.24
Protein		0.6	0.93
Potassium	mg	223.4	320
Calcium		74.3	33
Phosphorus		28.0	35
Sulfur		27.4	
Magnesium		20.0	12
Sodium		18.9	69
Iron		2.2	0.3
Zinc		0.4	0.24
Vitamin A	µg	<200	835
Vitamin C (ascorbic acid)	mg	2.03	5.9
A-tocopherol (vitamin E)	IU	<1	

Compared to carrots, the attractiveness of the HSB tuber in the fresh product market is quite moderate; it is crunchy, but inadequately sweet with no flavor. HSB tubers improve when served cooked, but further culinary efforts are required. As for any wild species, a long course of selection and breeding would be necessary to bring HSB tubers to a status of a common food produce.

Enduring claims associating some medicinal properties with HSB tubers, so far with no supportive documented evidence, have encouraged our curiosity. Electromechanical analysis of the water-soluble extract from HSB tubers revealed significant reducing power, identifying at least six substances or groups of antioxidants (data not shown). Fractionation of the tubers' ethanolic extract (EE) into 11 fractions (Figure 2A) and subsequent in vitro evaluations of possible anti-inflammatory capacities revealed significant activity in fractions F3 and F4, as well as in the original EE and the pooled fractions, PF (Figure 2B).

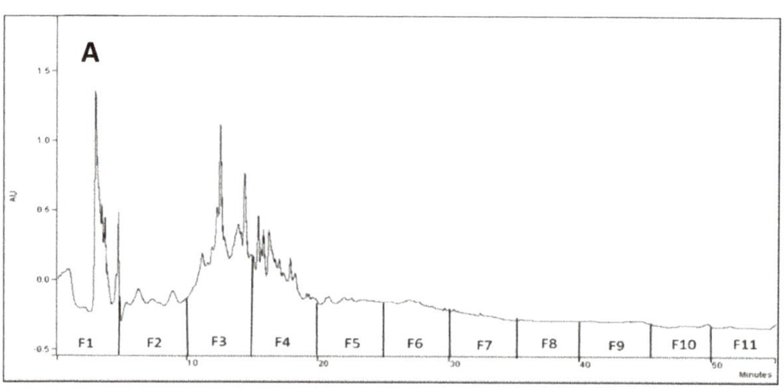

Figure 2. *Cont.*

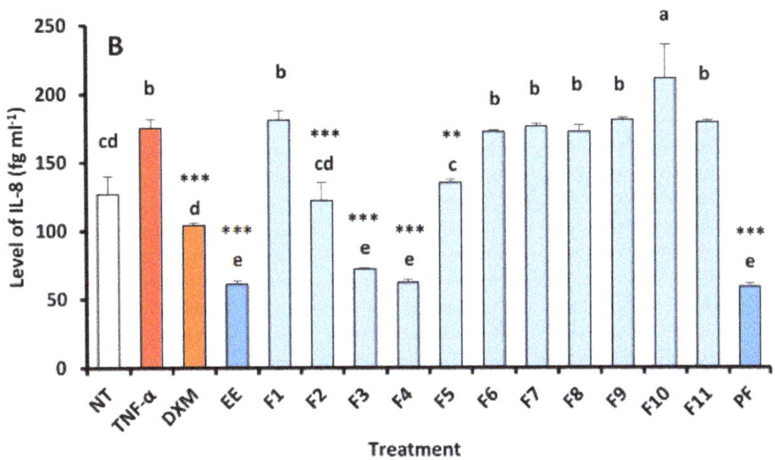

Figure 2. (**A**) HPLC profile of 70% ethanol extract (EE) of HSB tubers at 220 nm. Each fraction (F1–F11) was collected during 5 min out of the total 55 min of HPLC run; (**B**) levels of IL-8, an indicator of cell inflammatory status, in response to treatment with crude EE, EE fractions (F1–F11) and pooled fractions (PF) in in vitro trials using human cells. NT—non-treated control; TNF-α—an inflammation excitatory factor; DXM—dexamethasone (100 μM), positive control. Means of replicates were subjected to statistical analysis using Tukey–Kramer multiple comparison test. Different letters indicate significant differences between treatments; ***, ** and *, indicate $p < 0.001$, 0.01 and 0.05, respectively. Bars indicate standard error.

Biochemical analyses showed that fraction F4 comprised mainly of epigallocatechin, trans- and cis-catechin and gallic acid (Table 2), all of which are known for their robust bioactive capacities, including anti-inflammatory activity [40–46]. Interestingly, the bioactivity of fraction F4 was significantly greater than that of each compound alone at its corresponding concentration (data not shown), indicating synergic relationships in the natural extract. Furthermore, HSB tubers' extract displayed significantly stronger anti-inflammatory capacity compared to extracts of green tea (*Camellia sinensis*) or turmeric (*Curcuma longa*) [47], well-known sources of antioxidative and bioactive compounds [40,48].

Table 2. Compounds identified by GC-MS from fraction 4 (F4) of 70% ethanol extract of *E. crassifolium* tubers. RT—retention time.

Compound	RT (min)	Percentage (%) from Total Amount
Mannofuranose	28.024	17.2
α-D-xylopyranose	29.047	2.7
Gallic acid	29.766	5.7
Palmitic acid	30.853	6.4
Stearic acid	33.831	2.0
Trans-catechin	42.166	11.7
Cis-catechin	42.435	12.8
Epigallocatechin	43.007	41.5

Screening several HSB ecotypes collected from diverse locations in Israeli deserts revealed that they all share a similar range of substantial anti-inflammatory capacity; however, there were considerable differences among the ecotypes (Figure 3), indicating that the natural diversity within the species may offer a promising potential for genetic enhancement of the tubers' bioactive capacities. Accordingly, in the recent few years, seeds were collected from the wild across the Negev Desert and parts of the Rift Valley in order to broaden this potential.

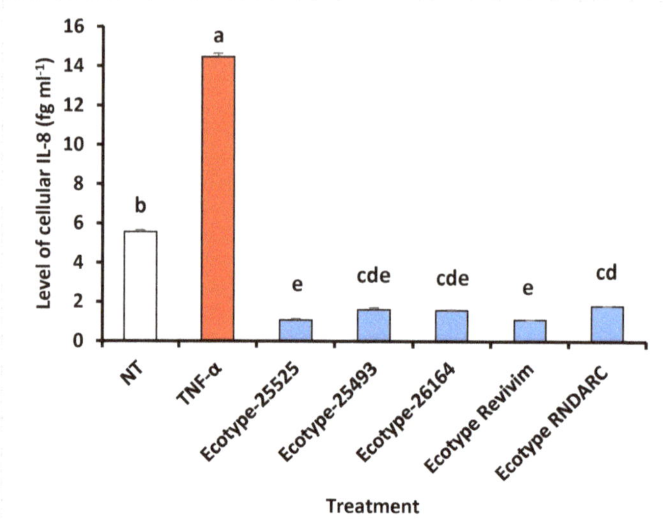

Figure 3. In vitro indications of the anti-inflammatory activity in the ethanolic extract from tubers of various *E. crassifolium* ecotypes, compared to untreated (NT) and stimulated (TNF-α) human cells. Bars indicate standard error. Different letters indicate significant differences between treatment at $p < 0.05$.

Beyond selection and breeding, optimization of the growth conditions through irrigation, soil fertilization and weeding are founding principles of agriculture [38,49,50] and positive responses of a given wild plant species to these manipulations are prerequisite to successful domestication.

In the wild, HSB is a classic opportunist desert shrub, the proliferation of which largely depends on the current water availability during a given growth season. In arid conditions, the occurrence and intensity of late autumn rain events determine the rate of seed germination and of young seedlings survival. Later on, during winter and early spring, the extents of canopy development, reproductive phase and the duration of the growing season are governed by the intermittent desert precipitation regime [51].

Simulation of various natural scenarios of the precipitation regime from November to May showed that in a relatively dry winter (50 mm), HSB exhibited low seed germination and seedlings survival rates. Conversely, many more plants persisted and grew in a rainy (200 mm) season (data not shown). Interestingly, the effects of water availability on tuber production per plant were very limited within the low precipitation range (Figure 4). However, tuber production and tuber weight increased significantly when water application increased to 700 mm, supplied consistently using drip irrigation, and no problems of germination or seedling persistence occurred. Moreover, plants produced significantly greater number (Figure 4B) of substantially larger tubers (Figure 4A). Adding fertilizer to the irrigated water had only a small influence on the number of tubers (Figure 4B), but tuber size increased by more than 50% (Figure 4A). Thus, application of a basic agricultural practice gave rise to a 10-fold increase in the mean tuber yield of an individual HSB plant (Figure 4C). Finally, the difference in plant and crop performance between wild and agricultural environments is unexpected (Figure 5); the greater germination and survival rates, fortified by enhanced tubers' growth and development have brought about the current yield potential of cultivated HSB to about 15 Mg ha^{-1}.

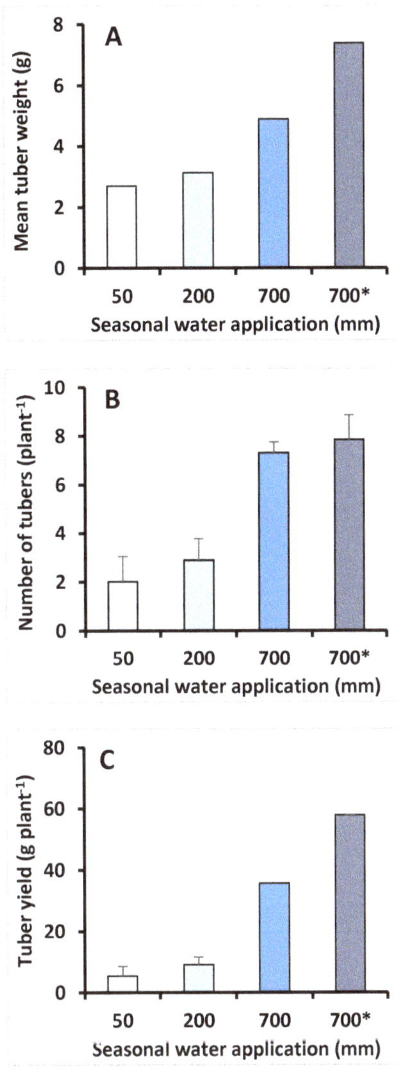

Figure 4. (**A**) Mean HSB tuber weight, (**B**) number of tubers and (**C**) mean tuber yield as affected by increased season water application at simulated natural precipitation patterns (50 and 200 mm) and at intensive agricultural environments (700 mm, drip irrigation and drip fertigation *); data were pooled from two different experiments conducted in the 2017–2018 season. Bars indicate standard deviation.

Figure 5. (A) Demonstration of an agricultural HSB cropping system at Ramat Negev with (C) representative tubers, compared to (B) the solitaire HSB phase in a wild niche, with (D) typical tubers' size, form and age.

3. Materials and Methods

3.1. Ecosystem and Plant Material

The HSB project took place at Ramat Negev Desert Agro-Research Center (RN-DARC), Israel (30°58′ N 34°42′ E), 305 m above sea level. Soil texture varies across short distances from sandy dunes to sandy-loam Loess—and consequently differ in water retention and cation exchange capacity. The natural life cycle of HSB occurs in the rainy season, germinating from November and dispersing seeds up to May. Mean annual precipitation is about 82 mm, but the amount of rain can substantially vary among years and locations, as well as the distribution and intensity of rain events. During the growing season, daily average temperatures decline from 17.8 °C in November to about 10 °C in January and then steadily rise to 23 °C in May (Figure S1).

Most of the trials with HSB employed ecotype RNDARC (Accession No. 309198), which was collected near the trial site several years ago. Additional seed sources were ecotype Revivim (Accession No. 309199), the seeds of which were collected yearly near Kibbutz Revivim and three other accessions (ecotypes 25525, 25493 and 26164), kindly received from the Israeli Gene Bank, ARO.

3.2. Field Experiments

Two separate field experiments were conducted from November 2017 to June 2018. The first examined HSB growth performance in response to distinct precipitation regimes, whereas the second experiment evaluated it under two different agricultural practices.

3.2.1. Simulated Precipitation Regimes

Treatments represent four principal scenarios of rainy seasons in the Negev Desert of Israel, differing in the intensity and frequency of rain episodes, as follows: A. low and concentrated (50 mm, comprised of two events); (B) low and scattered (50 mm, spread in ten events); (C) high and concentrated (200 mm, in four events); and, D. high and scattered (200 mm, in ten events) precipitation. Seeds were sown in loess soil at a density of eight seeds m^{-2}, in 6 m^2 plots. The experiment was organized in a random block design with four replicates. Irrigation was executed using computer-controlled sprinklers, and no fertilizer or soil amendments were used. The amount of water was monitored from germination and included rain. Plants were counted weekly. At harvest in June, tuber number and weight were determined per each surviving plant.

3.2.2. Agricultural Practice

Seeds were sown at late November 2017 in an open field on sandy soil beds. Water was supplied using two drip lines per bed (40 cm apart), with five 1.6 L h^{-1} emitters per 1 m of drip line. Two seeds were sown on either side of each emitter, resulting in a density of 16 plants m^{-2}. Fresh water (0.7 dS m^{-1}) was supplied at 6 mm day^{-1} until germination, which occurred about 2 weeks after sowing. After germination, irrigation was reduced to 4 mm day^{-1} to accomplish 700 mm until late May. An unfertilized control was tested against a fertigated treatment, which was applied using a liquid fertilizer (Shefer 4:2:6, 35% ammonium and 65% nitrate, Israel Chemicals, Ltd.) at 1.5 L m^{-3}. The experimental design was of random blocks with four replicates (plot size was 2.4 m^2). Bloom began in April and continued until June. Tubers began forming at a very early stage of plant growth and continued to form and develop throughout the season. Tuber harvest took place in June, toward seeds ripening. At harvest, tuber yield was determined; tubers were cleaned and stored at −20 °C until further examinations.

3.3. Evaluation of Tubers Nutrition Facts

To assess HSB tuber nutrition facts (Table 1), samples (2 kg each) were sent to BactoChem, Ltd., Ness Ziona, Israel, an officially certified laboratory, tightly committed to AOAC protocols.

3.4. Ethanolic Extract (EE)

HSB tubers were removed from cold storage (−20 °C) and frozen in liquid nitrogen. The frozen tubers were homogenized using an electrical blender and weighed. For each 1 g of fresh material, 4 mL of 70% ethanol were added immediately to the crushed tubers and incubated overnight at 28 °C with shaking at 180 rpm, after which the samples were centrifuged in 50-mL tubes for 5 min at 2500 rpm in an Eppendorf 5810R centrifuge with a 26 cm rotor (1820 RCF). The supernatant was transferred to new tubes and the solvent was evaporated in vacuo overnight. The remaining water content was lyophilized to powder and stored at −20 °C. From each gram of tubers, approx. 60 mg of lyophilized extract was obtained.

Just before further analysis, the lyophilized material was weighed and dissolved in 100 μL of 70% ethanol and then 900 μL of double distilled water (DDW), to obtain a 60 mg mL^{-1} sample, which was filtered through a 0.45-μm membrane.

3.5. High-Performance Liquid Chromatography (HPLC) Analysis

The filtered EE sample was separated using an Ultimate 3000 HPLC system coupled with a WPS-3000(T) Autosampler, HPG-3400 pump and DAD-300 detector. The separation was performed using a Purospher RP-18 endcapped column (250 mm × 4.6 mm I.D.; Merck KGaA, Darmstadt, Germany) with a guard column (4 mm × 4 mm I.D.). Solvent gradients were formed by varying the proportion of solvent A [water with 0.1% acetic acid (*v/v*)] to solvent B (methanol) with a flow rate of 1.0 mL min^{-1}. Solvent B was maintained initially at 10% for 5 min and then increased to

100% in 25 min. This 100% of Solvent B was maintained for 10 min, then decreased to 10% in 10 min and equilibrated for 5 min (total run time 55 min). The compound peaks were detected at three different wavelengths: 220, 240 and 280 nm. The same program was used to obtain fractions in bulk using a preparative HPLC (11250 Infinity, Agilent Technologies) using reversed-phase C18 column (Kinetex 5u EVO C18-100A—250 × 21.2 mm). After collection, the fractions were lyophilized to powder. These lyophilized fractions were resuspended in 7% ethanol and checked for their effect on IL-8 levels, as described below. Further analyses were carried out to correlate the activity and peak profile for detecting the active compound peak(s).

3.6. GC–MS Analysis

Prior to GC–MS analysis, samples underwent derivatization; 200 μL of N,O-bis(trimethylsilyl) trifluoroacetamide (BSTFA, Sigma-Aldrich, T-6381, USA) containing 1% of trimethylchlorosilane (TMCS) was added to each completely dried extract and heated to 70 °C for 20 min.

GC-MS analyses were carried out using a HP7890 gas chromatograph coupled to a HP6973 mass spectrometer (electron multiplier potential 2 KV, filament current 0.35 mA, electron energy 70 eV and the spectra were recorded over the range *m/z* 45 to 1000). An Agilent 7683 autosampler was used for sample introduction. Helium was used as a carrier gas at a constant flow of 1.1 mL s^{-1}. One μL of each sample was injected into the GC–MS using a 1:10 split ratio injection mode. An isothermal hold at 50 °C was kept for 2 min, followed by a heating gradient of 6 °C min^{-1} to 300 °C, with the final temperature held for 4 min. A 30 m, 0.25 mm ID 5% cross-linked phenylmethyl siloxane capillary column (HP-5MS) with a 0.25-μm film thickness was used for the separation, and the injection port temperature was 220 °C. The MS interface temperature was 280 °C. Peak assignments were carried out with the aid of library spectra (NIST 14.0) and compared with published data and MS data obtained from the injection of standards [(-)-Epigallocatechin, 08,108; catechin, *U*-49,040; gallic acid, 91,215] purchased from Sigma-Aldrich, Switzerland or USA.

3.7. Human Cell Culture

HaCaT is a spontaneously transformed aneuploid immortal keratinocyte cell line from adult human skin [52], widely used in scientific research [53]. HaCaT (ATCC-HB-241) normal skin cells were grown at 37 °C in a humidified 5% CO_2–95% air atmosphere. Cells were maintained in Dulbecco's modified Eagle's medium (DMEM, Biologic Industries, 01-055-1A, Beit-Haemek, Israel) with 10% fetal bovine serum (FBS, Biological Industries, 04-007-1A, Israel) and penicillin (100 units mL^{-1})—streptomycin (100 μg mL^{-1}) solution (Biologic Industries, 03-031-1B, Beit-Haemek, Israel).

Determination of Interleukin 8 (IL-8) levels in HaCaT cells

HaCaT cells were seeded into 24-well plates at 50,000 cells per well in triplicate in 500μL of media and then incubated for 24 h at 37 °C in a humidified 5% CO2–95% air atmosphere. After incubation, cell excitation was performed with recombinant human tumor necrosis factor-α (TNF-α, PeproTech, 300-01A, Cranbury, NJ, USA). Cultures in each well were treated with a final concentration of 50 ng mL^{-1} of TNFα and 50 μL plant extract. Three different controls were included in all experiments: (a) untreated cells, with neither TNF-α nor plant extracts, (b) cells treated with TNF-α alone, (c) cells treated with TNF-α and the solvent (7% ethanol). The supernatant was taken, and the level of IL-8 was measured 16 h posttreatment using the commercial Human CXCL8/IL-8 DuoSet ELISA kit (R&D Systems, DY208-05, McKinley Place MN, USA) according to the manufacturer's protocol. IL-8 is a common biomarker for inflammatory skin diseases [54,55]. The induction of detectable IL-8 levels requires a 16 h exposure to TNF-α [56]. Dexamethasone (Sigma-Aldrich, D4902, St. Louis, MO, USA) was used as a positive control [57,58].

4. Conclusions

Wild plant species may hold immense potential resources of nutritional value and therapeutic substances. In the present study, the tubers of *E. crassifolium*, an ignored desert shrub, were shown to harbor significant nutritional values and anti-inflammatory capacities. The catechins found in the tubers' ethanolic extract have well-established remedial effects in serious human ailments. Furthermore, indications exist suggesting that HSB tubers have a greater medicinal potential. While a demonstrated bioactive capacity is a prerequisite, high productivity is essential in realizing the potential of an underutilized species. The dramatic increase in HSB tubers yield in response to fertigation paves the way for this plant to become a potential industrial crop.

Supplementary Materials: The following are available online at http://www.mdpi.com/2223-7747/9/9/1069/s1, Figure S1: Mean monthly maximum, average, and minimum temperatures at RNDARC.

Author Contributions: Conceptualization, S.C., O.G. and H.K.; methodology, A.B., O.G., G.S. and M.S.; investigation, O.G., H.K., M.S., G.S. and M.M.; formal analysis, G.S., M.M., M.S. and O.G.; writing—Original draft preparation, A.B., H.K. and O.G.; writing—Review & editing, A.B., M.S., O.G. and S.C.; funding acquisition, O.G. and H.K. All authors have read and agreed to the published version of the manuscript.

Funding: This research was funded by KKL-JNF, the Israeli Ministry of Agriculture and ICA-Israel.

Conflicts of Interest: The authors declare no conflict of interest.

References

1. Zohary, D.; Hopf, M.; Weiss, E. *Domestication of Plants in the Old World*, 4th ed.; Clarendon Press: Oxford, UK, 2012.
2. Doebley, J.F.; Gaut, B.S.; Smith, B.D. The molecular genetics of crop domestication. *Cell* **2006**, *127*, 1309–1321. [CrossRef] [PubMed]
3. Hardin, G. Extensions of "the tragedy of the commons". *Science* **1998**, *280*, 682–683. [CrossRef]
4. Massawe, F.; Mayes, S.; Cheng, A. Crop diversity: An unexploited treasure trove for food security. *Trends Plant Sci.* **2016**, *21*, 365–368. [CrossRef] [PubMed]
5. Zimmet, P.Z.; Alberti, K.G.M. Introduction: Globalization and the non-communicable disease epidemic. *Obesity* **2006**, *14*, 1–3. [CrossRef] [PubMed]
6. Martínez-González, M.A.; Salas-Salvadó, J.; Estruch, R.; Corella, D.; Fitó, M.; Ros, E. Benefits of the Mediterranean diet: Insights from the PREDIMED study. *Prog. Cardiovasc. Dis.* **2015**, *58*, 50–60. [CrossRef]
7. Dogan, Y. Wild edible plants: From the past to the future. *Austin Food Sci.* **2016**, *1*, 1006.
8. Shelef, O.; Guy, O.; Solowey, E.; Kam, M.; Degen, A.A.; Rachmilevitch, S. Domestication of plants for sustainable agriculture in drylands: Experience from the Negev Desert. *Arid Land Res. Manag.* **2016**, *30*, 209–228. [CrossRef]
9. Leonti, M.; Casu, L. Traditional medicines and globalization: Current and future perspectives in ethnopharmacology. *Front. Pharmacol.* **2013**, *4*, 92. [CrossRef]
10. Heinrich, M.; Jäger, A. *Ethnopharmacolgy*; John Wiley & Sons Ltd.: Chichester, West Sussex, UK, 2015.
11. De-Cortes Sánchez-Mata, M.; Tardío, J. (Eds.) *Mediterranean Wild Edible Plants: Ethnobotany and Food Composition Tables*; Springer: Berlin/Heidelberg, Germany, 2016.
12. Benvenuti, S.; Maggini, R.; Pardossi, A. Agronomic, nutraceutical, and organoleptic performances of wild herbs of ethnobotanical tradition. *Int. J. Veg. Sci.* **2017**, *23*, 270–281. [CrossRef]
13. Carvalho, A.M.; Barata, A.M. The consumption of wild edible plants. In *Wild Plants, Mushrooms and Nuts: Functional Food Properties and Applications*; Ferreira, I.C.F.R.P., Morales, P., Barros, L., Eds.; Wiley-Blackwell: Chichester, UK, 2017; pp. 159–198.
14. Dwivedi, S.L.; Van Bueren, E.T.L.; Ceccarelli, S.; Grando, S.; Upadhyaya, H.D.; Ortiz, R. Diversifying food systems in the pursuit of sustainable food production and healthy diets. *Trends Plant Sci.* **2017**, *22*, 842–856. [CrossRef]
15. Ceccanti, C.; Landi, M.; Benvenuti, S.; Pardossi, A.; Guidi, L. Mediterranean wild edible plants: Weeds or "new functional crops"? *Molecules* **2018**, *23*, 2299. [CrossRef] [PubMed]
16. Renna, M. Wild edible plants as a source of mineral elements in the daily diet. *Progr. Nutr.* **2017**, *19*, 219–222.
17. Barros, L.; Morales, P.; Carvalho, A.M.; Ferreira, I.C. Antioxidant potential of wild food plants. In *Mediterranean Wild Edible Plants: Ethnobotany and Food Composition Tables*; De-Cortez Sánchez-Mata, M., Tardio, J., Eds.; Springer: New York, NY, USA, 2015; pp. 209–232.

18. Zhu, F.; Du, B.; Xu, B. Anti-inflammatory effects of phytochemicals from fruits, vegetables, and food legumes: A review. *Crit. Rev. Food Sci. Nutr.* **2018**, *58*, 1260–1270. [CrossRef] [PubMed]
19. Küpeli, E.; Tatli, I.I.; Akdemir, Z.S.; Yesilade, E. Estimation of antinociceptive and antiinflammatory activity of *Geranium pretense* subsp finitum and its phenolic compounds. *J. Ethnopharmacol.* **2007**, *114*, 234–240. [CrossRef] [PubMed]
20. Lu, C.H.; Li, Y.Y.; Li, L.J.; Liang, L.Y.; Shen, Y.M. Anti-inflammatory activities of fractions from *Geranium nepalense* and related polyphenols. *Drug Discov. Ther.* **2012**, *6*, 194–197. [CrossRef] [PubMed]
21. Velázquez-González, C.; Cariño-Cortes, R.; De Lucio, J.; Ortiz, M.I.; Arciniega, M.D.O.; Altamirano-Baez, D.A.; Jimenez-Angeles, L.; Bautista-Avila, M. Anti-nociceptive and anti-inflammatory activities of *Geranium bellumandits* isolated compounds. *BMC Complement. Altern. Med.* **2014**, *14*, 506. [CrossRef]
22. Bilić, V.L.; Rodríguez, J.V.; Grubešić, R.J.; Kremer, D.; Juričić, Ž. Quantitative analysis of polyphenols and antioxidant activity of Croatian populations of *Erodium cicutarium* (L.) L'Herit (Geraniaceae). *Planta Med.* **2016**, *82*, P498.
23. Shim, J.U.; Oh, P.S.; Lim, K.T. Anti-inflammatory activity of ethanol extract from *Geranium sibiricum* Linne. *J. Ethnopharmacol.* **2009**, *126*, 90–95. [CrossRef]
24. Fecka, I.; Kowalczyk, A.; Cisowski, W. Phenolic acids and depsides from some species of the *Erodium* genera. *Z. Naturforsch C* **2001**, *56*, 943–950. [CrossRef]
25. Lakhdari, W.; Dehliz, A.; Acheuk, F.; Mlik, R.; Hammi, H.; Doumandji-Mitiche, B.; Gheriani, S.; Berrekbia, M.; Guermit, K.; Chergui, S. Ethnobotanical study of some plants used in traditional medicine in the region of Oued Righ (Algerian Sahara). *J. Med. Plants Stud.* **2016**, *4*, 204–211.
26. Busso, C.A.; Lobartini, J.C. Mineral composition of *Medicago minima* and *Erodium cicutarium* under various water regimes. *Commun. Soil Sci. Plant Anal.* **2004**, *35*, 2243–2267. [CrossRef]
27. Lisbalchin, M.; Guittonneau, G.G. Preliminary investigations on the presence of alkaloids in the genus *Erodium* Lher (Geraniaceae). *Acta Bot. Gall.* **1995**, *142*, 31–35. [CrossRef]
28. Helmstadter, A. Ethnopharmacology in the work of Melville William Hilton-Simpson (1881–1938)—Historical analysis and current research opportunities. *Pharmazie* **2016**, *71*, 352–360. [PubMed]
29. Brullo, S.; Giusso del Galdo, G.; Guarino, R. Phytosociological notes on Lygeum Spartum grasslands from Crete. *Lazaroa* **2002**, *23*, 65–72.
30. Bergmeier, E.; Kypriotakis, Z.; Jahn, R.; Böhling, N.; Dimopoulos, P.; Raus, T.; Tzanoudakis, D. Flora and phytogeographical significance of the islands Chrisi, Koufonisi and nearby islets (S Aegean, Greece). *Willdenowia* **2001**, *31*, 329–356. [CrossRef]
31. El-Mokasabi, F.M. Floristic composition and traditional uses of plant species at Wadi Alkuf, Al-Jabal Al-Akhder, Libya. *Am.-Eurasian J. Agric. Environ. Sci.* **2014**, *14*, 685–697.
32. Sharawy, S.M.; Badr, A. Systematic revision of Erodium species in Egypt as reflected by variation in morphological characters and seed protein electrophoretic profiles. *Int. J. Bot.* **2008**, *4*, 225–230.
33. Bidak, L.M.; Kamal, S.A.; Halmy, M.W.; Heneidy, S.Z. Goods and services provided by native plants in desert ecosystems: Examples from the northwestern coastal desert of Egypt. *Glob. Ecol. Conserv.* **2015**, *3*, 433–447. [CrossRef]
34. Abd El-Wahab, R.H. Condition assessment of plant diversity of Gebel Maghara, North Sinai, Egypt. *Catrina* **2008**, *3*, 21–40.
35. Hand, R. Supplementary notes to the flora of Cyprus VIII. *Willdenowia* **2015**, 245–259. [CrossRef]
36. Danin, A. Flora and vegetation of Israel and adjacent areas. In *The Zoogeography of Israel*; Springer: Drodrecht, The Netherlands, 1988; pp. 18–42.
37. Osman, A.K.; Al-Ghamdi, F.; Bawadekji, A. Floristic diversity and vegetation analysis of Wadi Arar: A typical desert Wadi of the Northern Border region of Saudi Arabia. *Saudi J. Biol. Sci.* **2014**, *21*, 554–565. [CrossRef] [PubMed]
38. Abbo, S.; Gopher, A.; Lev-Yadun, S. The domestication of crop plants. In *Encyclopedia of Applied Plant Sciences*; Thomas, B., Murray, B.G., Murphy, D.J., Eds.; Academic Press: Waltham, MA, USA, 2017; Volume 3, pp. 50–54.
39. USDA. Carrot. 2020. Available online: https://fdc.nal.usda.gov/fdc-app.html#/food-details/787522/nutrients (accessed on 20 July 2020).
40. Yang, C.S.; Wang, H. Cancer preventive activities of tea catechins. *Molecules* **2016**, *21*, 1679. [CrossRef] [PubMed]

41. Mangels, D.R.; Mohler, E.R., III. Catechins as potential mediators of cardiovascular health. *Arterioscler. Thromb. Vasc. Biol.* **2017**, *37*, 757–763. [CrossRef] [PubMed]
42. Bernatoniene, J.; Kopustinskiene, D.M. The role of catechins in cellular responses to oxidative stress. *Molecules* **2018**, *23*, 965. [CrossRef] [PubMed]
43. Choubey, S.; Goyal, S.; Varughese, L.R.; Kumar, V.; Sharma, A.K.; Beniwal, V. Probing gallic acid for its broad-spectrum applications. *Mini Rev. Med. Chem.* **2018**, *18*, 1283–1293. [CrossRef]
44. Grzesik, M.; Naparło, K.; Bartosz, G.; Sadowska-Bartosz, I. Antioxidant properties of catechins: Comparison with other antioxidants. *Food Chem.* **2018**, *241*, 480–492. [CrossRef]
45. Dludla, P.V.; Nkambule, B.B.; Jack, B.; Mkandla, Z.; Mutize, T.; Silvestri, S.; Mazibuko-Mbeje, S.E. Inflammation and oxidative stress in an obese state and the protective effects of gallic acid. *Nutrients* **2019**, *11*, 23. [CrossRef]
46. Li, P.; Liu, A.; Xiong, W.; Lin, H.; Xiao, W.; Huang, J.; Liu, Z. Catechins enhance skeletal muscle performance. *Crit. Rev. Food Sci. Nutr.* **2020**, *60*, 515–528. [CrossRef]
47. Koltai, H.; Kapulnik, Y.; Fridlender, M.; Gati, E.M.; Ahmed, N.; Shemer, B.E.N. U.S. Patent Application No. 15/563,595, 2018. Erodium crassifolium L'Her plant extracts and uses thereof. Filing date: 31/03/2016.
48. Yadav, R.P.; Tarun, G. Versatility of turmeric: A review the golden spice of life. *J. Pharmacogn. Phytochem.* **2017**, *6*, 41–46.
49. Ladizinsky, G. *Plant Evolution under Domestication*; Springer: Drodrecht, The Netherlands, 1998.
50. Purugganan, M.D.; Fuller, D.Q. The nature of selection during plant domestication. *Nature* **2009**, *457*, 843–848. [CrossRef]
51. Aronson, J.; Shmida, A. Plant species diversity along a Mediterranean-desert gradient and its correlation with interannual rainfall fluctuations. *J. Arid Environ.* **1992**, *23*, 235–247. [CrossRef]
52. Boukamp, P.; Petrussevska, R.T.; Breitkreutz, D.; Hornung, J.; Markham, A.; Fusenig, N.E. Normal keratinization in a spontaneously immortalized aneuploid human keratinocyte cell line. *J. Cell Biol.* **1988**, *106*, 761–771. [CrossRef] [PubMed]
53. Schoop, V.M.; Fusenig, N.E.; Mirancea, N. Epidermal organization and differentiation of HaCaT keratinocytes in organotypic coculture with human dermal fibroblasts. *J. Investig. Dermatol.* **1999**, *112*, 343–353. [CrossRef] [PubMed]
54. Lee, H.Y.; Stieger, M.; Yawalkar, N.; Kakeda, M. Cytokines and chemokines in irritant contact dermatitis. *Mediators Inflamm.* **2013**, 916497. [CrossRef] [PubMed]
55. Turner, M.D.; Nedjai, B.; Hurst, T.; Pennington, D.J. Cytokines and chemokines: At the crossroads of cell signalling and inflammatory disease. *Biochim. Biophys. Acta* **2014**, *1843*, 2563–2582. [CrossRef]
56. Zhang, J.; Yang, X.; Wang, H.; Zhao, B.; Wu, X.; Su, L.; Xie, S.; Wang, Y.; Li, J.; Liu, J.; et al. PKCζ as a promising therapeutic target for TNFα-induced inflammatory disorders in chronic cutaneous wounds. *Int. J. Mol. Med.* **2017**, *40*, 1335–1346. [CrossRef]
57. Mizuno, K.; Morizane, S.; Takiguchi, T.; Iwatsuki, K. Dexamethasone but not tacrolimus suppresses TNF-α-induced thymic stromal lymphopoietin expression in lesional keratinocytes of atopic dermatitis model. *J. Dermatol. Sci.* **2015**, *80*, 45–53. [CrossRef]
58. Wufuer, M.; Lee, G.; Hur, W.; Jeon, B.; Kim, B.J.; Choi, T.H.; Lee, S. Skin-on-a-chip model simulating inflammation, edema and drug-based treatment. *Sci. Rep.* **2016**, *6*, 37471. [CrossRef]

plants

MDPI

Review

Punica protopunica Balf., the Forgotten Sister of the Common Pomegranate (*Punica granatum* L.): Features and Medicinal Properties—A Review

José Antonio Guerrero-Solano [1], Osmar Antonio Jaramillo-Morales [2,*], Tania Jiménez-Cabrera [1], Thania Alejandra Urrutia-Hernández [3], Alejandro Chehue-Romero [1], Elena G. Olvera-Hernández [1] and Mirandeli Bautista [1,*]

1 Academic Area of Pharmacy, Institute of Health Sciences, Autonomous University of the State of Hidalgo, San Agustin Tlaxiaca, Hidalgo 42160, Mexico; jose_guerrero@uaeh.edu.mx (J.A.G.-S.); tania_jimenez@uaeh.edu.mx (T.J.-C.); chehuea@uaeh.edu.mx (A.C.-R.); olverae@uaeh.edu.mx (E.G.O.-H.)

2 Department of Nursing and Obstetrics, Division of Life Sciences, University of Guanajuato, Ex Hacienda el Copal Km 9 Irapuato-Silao highway ap 311, Irapuato, Guanajuato 36500, Mexico

3 Academic Area of Food Chemistry, Institute of Basic Sciences and Engineering, Autonomous University of the State of Hidalgo, Pachuca-Tulancingo km 4.5 Carboneras, Pachuca de Soto, Hidalgo 42184, Mexico; thania_urrutia9356@uaeh.edu.mx

* Correspondence: oajmorales@gmail.com (O.A.J.-M.); mibautista@uaeh.edu.mx (M.B.)

Received: 1 September 2020; Accepted: 12 September 2020; Published: 16 September 2020

Abstract: *Punica protopunica* Balf. is one of only two species housed by the *Punica* genera. *Punica protopunica*. Balf., known as Socotran pomegranate, is an endemic, isolated species found only in Socotra archipelago in the northwestern Indian Ocean, and is considered to be the ancestor of pomegranate. This review stems from the fact that in many *Punica granatum* L. articles, *Punica protopunica* Balf. is mentioned, but just in an informative way, without mentioning their taxonomic and genetic relationship and their medicinal properties. It is there where the need arises to know more about this forgotten species: "the other pomegranate tree." A large part of the human population does not know of its existence, since only its "sister" has spread throughout the world. The present review deals with the taxonomy and origin of *Punica protopunica* Balf., the morphology of the tree, distribution, cultivation, vulnerability, and as well as its relationship with *Punica granatum* L. It also discusses its uses in traditional medicine, its antioxidant capacity, and the medicinal properties of this forgotten species.

Keywords: *Punica protopunica* Balf.; *Punica granatum* L.; *Punica* genera; *Lythraceae*

1. Introduction

Myrtales is an order within the classification of terrestrial green plants (Viridiplantae-Streptophyta) [1,2]. The Lythraceae family (from the order of the Myrtales) is composed of herbs, shrubs and trees that are mainly recognized for their flaky bark, crumpled petals from the bud (when emerging out of the rim of the calyx tube of the sepals), leaves oppositely paired, seeds with multi-layered outer integuments, and the fruit is usually a capsule [1–4]. The Lythraceae family comprises 31 genera including the *Punica* genera [5]. This is quite surprising as the *Punica* genera has previously been assigned to the monogenic *Punicaceae* family [6,7]. However, the results of numerous molecular and morphological investigations revealed the close relationship of the genera *Punica* with the Lythraceae family [1,8–13]. Previously it was considered a monogenic Punicaceae family that contains only one genera, *Punica* [14,15]. According to Pliny, the name *Punica* was given by the Romans, referring to the city of Cartago, in Tunis (Punic, Phoenician, Carthaginian), from where the best pomegranate (from Latin "*pome*" witch means apple and "*granate*" meaning many seeded) arrived in Europe. The genera *Punica* contains two species, *Punica granatum* L. and *Punica protopunica* Balf., [16]. Initially, *Punica granatum* L. was known as *Malum punicum*, the apple of Cartago, but later, Carl Linnaeus (1707–1778) chose the current name, with a specific epithet of *granatum*, which means granular [6]. On the other hand, *Punica protopunica* Balf. was first described by the Scottish botanist Isaac Bayley Balfour (1853–1922) during his arboreal and botanical expedition in 1880, and published in the Proceedings of the Royal Society of Edinburgh in 1882 [4,14–16]. *P. granatum* is native to the region that covers territories from a part of Iran to northern India [17,18]. Wild *P. granatum* L. types have their natural distribution in central Asia from Iran, Afghanistan, Turkmenistan, to northern India, and this region is considered the center of origin of pomegranate [6]. Later, the pomegranate was distributed to the Mediterranean, East Asia, America and South Africa, and this distribution originated the genetic diversity of the pomegranate [19], on the other hand, *P. protopunica* Balf. is endemic to the Socotra archipelago (located between the Arabian sea and the Guardafui channel in the Indian Ocean, off the coast of the Horn of Africa) [20]. In this context, the objective of this review was to compile the available information on the *P. protopunica* Balf. species (morphology, distribution, cultivation, vulnerability, uses), including its antioxidant capacity and the medicinal properties, to make it known and allow a wider use of this forgotten wild species.

2. Results

2.1. Taxonomic Positioning and Distribution

P. protopunica Balf. (taxonomic positioning in Table 1) is an endemic species, found only in the remote archipelago of Socotra, and is considered as one of the most important endemic species on the archipelago [21,22]. Socotra belongs to the Republic of Yemen, it is located at 12°19′–12°42′ N and 53°18′–54°32′ E, on the Arabian sea of the Indian Ocean [21,23]. Socotra archipelago, also known as the "Galápagos of the Indian ocean" is a group of four islands, Socotra being the most important and largest one. Socotra archipelago (isolated from the rest of the world) is an island with great biological diversity (approximately 900 plant species, 30% endemic) and it hosts unique fauna and flora [24]. People of Socotra use medicinal plants and it is known that this people has a deep respect for nature and its environment [24]. Socotra was included in 2008 to the select list of World Heritage by the United Nations Educational, Scientific and Cultural Organization (UNESCO) under the criteria of natural site [25]. Additionally, *P. protopunica* Balf. is considered from an independent evolutionary path (due to isolation from the rest of the world) [26], commonly considered as "the other pomegranate tree", it is an unknown species worldwide, in contrast to the sacred status granted by the name of its species, *protopunica*: prototype [27].

Table 1. Taxonomic positioning of *Punica protopunica* Balf.

Classification	Denomination	Common Name
Order	Myrtales	-
Family	Lythraceae	Loosestrife
Subfamily	Punicoideae	-
Genera	*Punica*	Pomegranate
Species	*Punica protopunica*	Wild pomegranate, Socotra pomegranate

Adapted from [28,29].

P. protopunica Balf. is distributed in different regions of the island, but mainly grows in the central-western part of Socotra, in humid forest regions, in the Haggeher mountains whose slopes are characterized by being made of granite and having a maximum elevation of 1503 m. It also grows on Diksam, the island's limestone plateau, which rises precipitously 1520 m above sea level. The total area covered has been calculated to be 1/15 of the total area of Socotra (3796 km^2) [27].

2.2. Morphology

Balfour, the discoverer of *P. protopunica* Balf., described it as "trees with branches, often thorny; elliptic leaves round sheath, oblique; below the oblong, obtuse flower bracts; obovate petals; joined carpels, horizontal basal tone center spiral. From Socotra, a new species that abounds and grows on the peninsula" [30]. Additionally, Balfour wrote: "In general habit, it is not unlike the pomegranate, but its leaves are larger and coarser, and it wants the delicate character of the foliage of that species. The flowers, too, are somewhat smaller, and their turbinate base is more angular. The fruit is very much less in size" [31]. *P. protopunica* Balf. has morphological differences compared to *P. granatum* L.; it has larger, narrower leaves, different foliage, continuous flowering, and smaller, pink (not red) flowers. The fruit of *P. protopunica* Balf. is round, pommel-shaped, with a maximum diameter of 3 cm and a characteristic yellow-greenish or red-brown color when ripe, is smaller, evergreen, with white seeds and less sweet than *Punica granatum* L. [20,30–33]. Table 2 shows morphological characteristics of the species.

Table 2. Morphological characteristics of *P. protopunica* Balf.

Punica protopunica Balf.	Morphological Characteristics
General habit	The tree can reach a height of 2.5 to 4.5 m. It is considered a small tree or shrub, but if it reaches more than 9.14 m it can be classified as a tree, that is, it can be considered as both main forms. There are trees that are wider than they are tall (for example, trees that grow on the slopes rocks of Socotra) and trees taller than they are wide (typical of trees growing on the Socotra limestone plateau.) Generally, the tree is equal in width to height, with an upright shrub appearance.
Bark, branches and trunk	The bark is reddish-brown when the tree is young, but changes to a grayish hue as it grows older. The branches have thorns.
Leaves	The leaves grow to a length of 3 cm, in pairs on the opposite sides of the stalk, they are perennial, their most common shape is elliptical or oblong, although there are also circular or oval and obovate leaves (a single branch can have leaves of all the shapes described). Its color is dark green, with a bright tone.
Flowers	The flowers have obovate or oval petals, although they are sometimes heart-shaped. Its color is light pink with glitter and its shape is "trumpet". Flowering occurs from December until the summer of the following year. Their physiology allows them to produce fertilization and pollination. It is a self-crossing species.
Fruits	The shape of the fruits is almost identical to that of modified tangelos (Citrus x tangelo). They retain their floral chalice tube. The peel is hard and its color when ripe is light green to greenish yellow and may or may not contain pink reflections. Inside they present a spongy pericarp with membranes (endocarp) that separate the arils into compartments. In turn, each aril contains a membrane, pulp juice and a seed.
Seeds	The seeds are inside the arils and there are hundreds of them, they are relatively light.

Adapted from [20,26,27,29,31].

2.3. Cultivation

A hardiness zone is a geographical area defined in the quality of a specific category of plant life capable of growing. The most widely used system is that of the United States Department of Agriculture (USDA), which includes 13 zones characterized by annual extreme minimum temperatures. The use of this system has spread throughout the world and has been adapted in other countries. Using this system, *P. protopunica* Balf. has USDA hardiness zone 7a through 11b; from −17.8 °C to +10 °C [20], which means that, emulating optimal conditions, it can be grown in other regions of the world.

P. protopunica Balf. can grow out of the island of Socotra, however certain requirements must be met that are necessary for its growth: (1) temperatures above 10 °C; (2) sunlight from 6 to 8 h per day and protection against the wind; (3) constant accumulation of humidity (1000 mm) with a percentage of 20% to 40%; and (4) alkaline soils with pH 7 and content of calcareous or rocky gravel. In cultivation, sowing, cutting and grafting can be methods of propagation. Botanist Alan Radcliffe-Smith from the Kew Royal Botanic Gardens Herbarium (1938–2007) successfully propagated *P. protopunica* Balf. using all three methods, although cutting and grafting did not result in fruit in the varieties [22,27].

2.4. Vulnerability and Conservation of the Species

As early as 1978, *P. protopunica* Balfwas considered a vulnerable species with fragmented populations. Later in 2004 [34] the red list of the International Union for Conservation of Nature (IUCN) specified that *P. protopunica* Balf. was a vulnerable species; It indicated that efforts should be made to protect *P. protopunica* Balf., since it is the only congener of *P. granatum* L., although, according to Miller [29,34], this information is outdated, since it has been shown that *P. protopunica* Balf. is widely found on the island of Socotra and is quite common in some regions. Its total area of occupation is approximately 100 km^2, 2/15 of the total area of Socotra. Miller also reported that *P. protopunica* Balf. has a fragmented distribution with different subpopulations and regenerates actively, however, there are large areas in which the tree does not grow, except in some areas with remaining populations without regeneration being observed. The foliage of *P. protopunica* Balf. is of no interest to livestock and the tree is not cut down for fuelwood, even in dry spells. Wood is not important neither as firewood nor for construction.

Socotra was included on the World Heritage List in 2008 as a natural site, which has had a positive impact on the conservation of its species [35]. The inclusion was under the selection criteria number 10 that establishes: "to contain the most important and significant natural habitats for in-situ conservation of biological diversity, including those containing threatened species of outstanding universal value from the point of view of science or conservation". In its operational guidelines, UNESCO requests that in order to maintain this status, the assets and properties included on the list must be protected by well-established and delimited legislation and institutional regulations, to guarantee their safeguarding. Likewise, the states must demonstrate that actions are carried out to protect them at the national, regional and local levels, and must attach the appropriate texts to the nomination with a detailed explanation of how this protection operates [36]. So the archipelago species, including *P. protopunica* Balf., are now more protected than ever, however, UNESCO has detected some factors that affect the property, such as livestock farming/grazing of domesticated animals, management plan, uncontrolled developments including ground transport infrastructure: the road network, absence of biosecurity politics to eradicate the introduction of invasive species, extreme weather events (storms and cyclones) and industrial activities among others [37,38]. There is a development and conservation program for the archipelago called: Socotra Archipelago Conservation and Development Program (SCDP). This is an initiative of the Republic of Yemen to develop and conserve the archipelago's resources. The SCDP is supported by the United Nations Development Program (UNDP) and the governments of three European countries (Italy, Poland and the Netherlands), in conjunction with international donors and private non-profit organizations. The mission of the SCDP is to join the efforts of all the aforementioned organizations and countries for the human development of the population and conservation of the biodiversity of the Socotra archipelago with a sustainable approach [39].

2.5. Relationship between *P. protopunica* Balf., and *P. granatum* L.

In 1973, Shilkina reported that Socotran pomegranate tree wood contains fiber tracheids, (water-conducting cells). She found that tracheids are the only characteristic that differentiates *P. protopunica* Balf. from the rest of botanical and arboreal varieties of the order Myrtales. She stated that this unique characteristic is not shared by even *P. granatum* L. Therefore, she urged that *P. protopunica* Balf. be not only the only member of its species, but also the only species of a new genera. It appears that, based on the anatomy of the xylem, *P. protopunica* Balf. was suggested as an ancestor of the domesticated species *P. granatum* [7]. On the other hand, in *P. protopunica* Balf., 2n = 14, therefore, the haploid number of chromosomes is n = 7, unlike n = 8 in *P. granatum* L.; this difference is considered as a primitive characteristic of *P. protopunica* Balf. from the evolutionary point of view, since n = 8 is a development factor [40].

Recently, Youssef et al., [33] analyzed genetic diversity and the relationship between *P. protopunica* Balf. and eleven accessions of *P. granatum* L. (from Egypt, México and Yemen), using the following molecular markers: (1) amplified sequence-related polymorphism (SRAP); (2) amplification polymorphism of the target region (TRAP); (3) amplified intron-directed polymorphism (ITAP); and (4) sequence analysis of the *pgWD40* gene (involved in anthocyanin biosynthesis). It was found that the relationship between accessions of *P. granatum*, grouped by regions, was approximately 90% similar, while, evidently, the degree of genetic variation was altered within of each region. However, these markers revealed the relationship between *P. protopunica* and *P. granatum* at 33% similarity. ITAP, TRAP and SRAP generated a total of 719 bands, of these, 193 were specific for *P. protopunica* Balf., and 234 bands were shared between both species. The pgWD40 gene analysis showed 100% identity between *P. granatum* L. accessions and 98% with *P. protopunica* Balf. Phylogenetic analysis of the WD40 sequences of species, including both species of the *Punica* genera confirmed the relationship between *P. protopunica* Balf. and *P. granatum* L., supporting the hypothesis that *P. protopunica* Balf. could be an ancestor of *P granatum* L.

Moreover, Muhammad et al., [41] studied the genetic association among the genotypes of the species of *Punica* (20 genotypes of *P. granatum* L., 20 of *P. protopunica* Balf.), based on morphological and biochemical characterization. The phylogenetic tree was constructed with the 40 genotype data matrix based on morphology to represent the similarity of the two species. The phylogenetic tree divided the two species in two lineages (Regions R-I and R-II). R-I holds the 20 genotypes of *P. granatum* L., while R-II enclosed all genotypes of *P. protopunica* Balf. The similarity indexes were performed for the genotype of 2 species that was 53.84% for *P. protopunica* Balf. and *P. granatum* L. In the biochemical characterization, the total seed protein profiling was carried out on slab gel electrophoresis; 10 bands were recorded in both species (molecular weight 15 KDa–180 KDa) intra locus contribution toward the genetic disagreement was 10% in *P. protopunica* Balf. and 50% in *P. granatum* L. Inter species locus contribution toward genetic diversity was 50%.

2.6. *Uses in Traditional Medicine*

Despite the great scientific advancement in medicine and pharmaceuticals, much of Yemenis actively practice traditional medicine, they use medicinal plants for their daily health care needs and have a long tradition in herbal medicine. The vegetation and flora of Socotra provide healers with "natural pharmacies" with a great variety of plants, to prepare phytomedicine, in order to alleviate a great variety of human and veterinary diseases [42]. Some authors have reported the use of *P. protopunica* Balf. fruit peel, seed and flower in traditional medicine; the extraction and consumption techniques are by decoction, boiling, infusion, maceration of ethanol and fresh juice. The above mentioned are helpful for treating diseases such as peptic ulcer, diarrhea, dysentery, sores and wounds, urinary infections, dry cough, digestive problems, skin disease, mouth and throat sore and jaundice. It is also used because of the anthelmintic and anti-diabetic properties [29,34,41,43–46].

2.7. *Bioactive Compounds of Pomegranate and Their Medicinal Properties*

2.7.1. Phenolic Content and Antioxidant Activity

Muhammad et al., in 2019 [41], evaluated the antioxidant potential of methanolic extracts of *P. protopunica* Balf. and *P. granatum* L. species, both cultivated in Swat Valley, KP, Pakistan. They found high amounts of total phenols in both species, as well as flavonoids and antioxidant activity, with *P. prototunica* Balf. showing the highest flavonoid content. Antioxidant activity was similar between species. The other study was led by Al-Huqail et al., 2018 [47]. They studied the antioxidant effect of the aqueous ethanolic extracts of the peel and seed coat of *P. granatum* L. and *P. protopunica* Balf., in vitro. The two extracts not only contained significantly different phenolic and total flavonoid contents but also different phytochemical constituents. Gas chromatography mass spectroscopy (GC-MS) analysis of the peel extracts revealed twenty-six compounds. The main ones were benzenepropanoic acid, 1H-pyrrole-2,5-dione, 1,2-benzenedicarboxylic acid, 1-(propylthio)-(CAS) ethanol (CAS) ethylalcohol methyl ester of 3-methoxypropionic and 2-propanol. *P. protopunica* Balf. seed coat extract showed the presence of 14 phytochemical constituents, the major constituents were Di-2 (2-ethylhexyl) phthalate, 1,2-benzenedicarboxylic acid, propanoic acid, 2-hydroxy-ethyl formic acid and benzoic acid. In the malondialdehyde method (MDA), hydrogen peroxide (H_2O_2) scavenging and DPPH· assays, the two seed coat extracts exhibited very high antioxidant activities, with higher activity observed for the *P. granatum* L. extract [47]. These differences in the antioxidant activity in the two species may be attributed to their different phytochemical constituents. The importance of the high concentration of phenolic compounds is that they protect cells from the damaging effect of free radicals, molecules responsible for altering biological systems, causing diseases or accelerating aging [48].

2.7.2. Antimicrobial, Antiviral and Antiprotozoal Activity

Mothana and Lindequist [49] evaluated the antimicrobial effect of twenty-five medicinal plants of the island of Socotra, including the fruit and leaves extracts of *P. protopunica* Balf. (4 mg of the dried extract), on nine types of Gram-positive and Gram-negative bacteria: *Bacillus subtilis* (ATCC 6059), *Candida maltosa* (SBUG), *Escherichia coli* (ATCC 11229), *Micrococcus flavus* (SBUG 16), *Pseudomonas aeruginosa* (ATCC 27853), *Staphylococcus aureus* (ATCC 6538), multi-resistant strains *Staphylococcus aureus*, *Staphylococcus epidermidis* 847, and *Staphylococcus haemolyticus* 535 and against a species of yeast. The methanolic extract of *P. protopunica* Balf., was found to be one of the species with the highest antimicrobial activity, especially on Gram-positive bacteria including multi-resistant strains of *Staphylococcus*, but without activity on yeast.

The antiviral activity of the methanolic and aqueous extracts of twenty-five medicinal plants including *P. protopunica* Balf. fruit and leaves, were evaluated in two in vitro models (unreported concentrations) by Mothana et al., [50], one with MDCK cells with type A influenza virus and the other with Vero cells infected with herpes simplex virus type 1 (HSV-1). HSV-1 was more sensitive than type A influenza against the extracts evaluated. The half maximal inhibitory concentration (IC_{50}) for *P. protopunica* Balf. was anti-influenza virus A = 75.7 μg/mL and anti-HSV-1 = 5.8 μg/mL. The species was not considered by the authors as a plant with potential for the development of antiviral drugs.

Additionally, Mothana et al., [45] evaluated the in vitro antiprotozoal activity of twenty Socotra plants including the methanolic extract of the fruit of *P. protopunica* Balf., (at 5 concentrations: 0.25, 1, 4, 16, and 64 μg/mL), *Plasmodium falciparum* erythrocytic schizonts were used to evaluate antiplasmodial activity. The antileishmanial activity was evaluated using a model of intracellular amastigotes of *Leishmania infantum*, and finally the antitripanosomal activity was evaluated using intracellular amastigotes of *Trypanosoma cruzi* and free trypomastigotes of *T. brucei*. The results indicated that there is selective activity of *P. protopunica* Balf., against *Plasmodium* (IC50 2.2 μg/mL), and a potential for relevant antileishmanial and antitrypanosomal activity was also found. In the same line of research, Barzinji et al. [44] investigated the antimalarial efficacy in vitro of methanolic and aqueous extracts of thirteen traditionally used plant species from Yemen, including *P. protopunica* Balf., in blood samples from positive malaria patients. The methanolic extract from *P. protopunica* Balf. (20 mg) was one of the three extracts with the highest antimalarial activity (IC 0.98 μg/mL), and also exhibited schizont maturation inhibition (SMI) of 31.25 μg/mL [49].

2.7.3. Anticancer Activity

In 2007, Mothana et al., published the study "anticancer potential of Yemeni plants used in folk medicine". Twenty-four methanolic extracts of common plants in traditional medicine in Socotra and other Yemen states were evaluated. *P. protopunica* Balf., (leaf and fruit extracts) was included. They used a microtiter plate assay based on cell staining with violet crystal to evaluate the in vitro cytotoxic potency of the extracts at different concentrations, with 5 human cancer cell lines: two urinary bladder carcinomas (5637 and RT-112), two lung cancer line (A-427 and LCLC-103H) and a breast cancer line (MCF-7). The methanolic extracts of *P. protopunica* Balf. exhibited a moderate potency of toxicity (seventh place of the extracts evaluated) in all tumor cell lines with IC_{50} values of 16.5 to 37.6 μg/mL. The methanolic extract of *Dendrosicyos socotrana* had the greatest cytotoxic effect against all the cancer cell lines analyzed [51].

2.7.4. Cytotoxicity

In some of the studies cited above, cytotoxicity tests were carried out. Cultured MRC-5 SV2 cells were used for the toxicity test for *P. protopunica* Balf. fruit and leaves extracts. Cell viability was evaluated fluorimetrically. Fluorescence was measured and cell viability results were compared versus control group (data were expressed as a percentage reduction in cell viability). The IC_{50} for *P. protopunica* Balf. was 29.5 ± 3.7 μg, which is interpreted as low toxicity by the authors [45]. A cytotoxicity assay on the proliferation of MDCK and Vero cells was also carried out using culture plates and incubated at 37 °C with 5% CO_2. Confluent monolayers were incubated with dilutions of extracts (100, 50, 25, and 12.5 μg/mL) in culture medium for 3 days. The 50% inhibitory cellular concentration (ICC_{50}) was measured by a dye absorption assay, with culture medium as a control. The ICC_{50} of the pomegranate was less than 10 μg, which was considered non-cytotoxic [50]. With the little information available, it cannot be stated that there is no toxicity of the tree and its components, however the trials point to that conclusion.

3. Discussion

Socotra Archipelago is known to be a section of what was once the Gondwana Supercontinent [52,53]. The biodiversity and geology of the island of Socotra are living and tangible proof of the historical biogeography of the supercontinent of Gondwana [54]. In the early Cretaceous, Gondwana included the continental areas of Africa, South Asia, Polynesia and Central America, which later separated to create the continents as we know them today. Subsequently, the 4 islands of the Socotra archipelago, was isolated from the Indian Ocean approximately 20 million years ago, when the African and Arab plates separated, which resulted in the formation of the Gulf of Aden [52,55]. At some point in that transformation, the Socotra archipelago was left behind, isolating all the living matter contained in that piece of land. It is known that Socotra is not an island formed by volcanic lava but is a separate portion of a continent, so the endemic species are really old and "rare" in the eyes of modern humans. The dragon's blood tree *Dracaena cinnabari* Balf. and Socotran *Adenium obesum* tree are clear examples, that look like they came out of a science-fiction novel. This is how many of the species that have been home to the island for millions of years had a different evolution, with fewer changes than in other lands, hence the genetics of *P. protopunica* Balf.

The morphological differences between *P. protopunica* and *P. granatum* are evident, but the most notable are the size of the tree, which can be more than 7 m in natural conditions for *P. granatum* L. [41], as opposed to the 4.5 m (average height) for *P. protopunica* Balf. [20]. The flowers are orange to red in *P. granatum* L. and smaller bright pink in *P. protopunica* Balf. [27,56], and the color of the fruit peel in *P. granatum* L. can be yellow, reddish yellow, or different shades of green and red (starting from pink to crimson); purple is less common and there are even unique varieties, such as black pomegranate, which acquires its coloration from immaturity and remains so until overripe [40,57]. On the other hand, as we already mentioned, *P. protopunica* Balf., has a fruit that ripens from green to greenish yellow, which can have a dark pink hue and they are in continuous bloom [23,29].

As we have seen, *P. protopunica* Balf. is endemic to Socotra and the distribution is centered on the island, however one of the studies we reviewed used specimens of *P. protopunica* Balf. from Pakistan (it was introduced to this region) [41]. It has been reported that with special care, the species can be cultivated outside the region where it grows naturally [22], making it a viable option for the species to be cultivated in other regions of the world, and taking into account vulnerability reports in Socotra and thus preserve it in a better way. In addition, efforts have been made to collect, conserve and evaluate the germplasm of *Punica* species [58]. There are collections of germplasm of wild and domesticated varieties of pomegranate in gene banks and seed banks of Albania, China, Cyprus, Egypt, France, Germany, Greece, Hungary, India, Iran, Israel, Italy, Morocco, Portugal, Spain, Tunisia, Turkey, Turkmenistan, Ukraine, USA and Uzbekistan [59–61]. However the largest collection is located in St. Petersburg, Russia. Interestingly, there was a collection considered the largest, in Garrygala Turkmenistan, but Levin reported that it was destroyed when Turkmenistan separated from Russia [40].

The domestication process of *P. granatum* L. gave rise to fruits and plants with magnum seeds, some infertile seeds and fruits, as well as fruits and seeds of different shades of color [58]. Chandra et al. (2010) [19] gave us a detailed history of the pomegranate, explaining that this was from the first fruit crops to be domesticated and planted in the years 4000 and 3000 BC., being one of the oldest edible first fruits [19,62]. It is known to have been cultivated in Egypt and consumed in India (it was an important food in Indian royalty) so early that there is a Sanskrit word for pomegranate. There are also records of its consumption in China during the Han and Sung dynasties, carried from the Middle East by merchants. It was adopted and consumed regularly in medieval Europe and spread around the world in European conquests [60].

Due to the globalization of pomegranate cultivation, there are genetic variations within the same species. More than 500 varieties of *P. granatum* L. are known, although few varieties bring their cultivation to a commercial level of production (about 50) [63]. In contrast, *P. protopunica* Balf. has a smaller wild fruit, a lesser variety of colors, and an acidic flavor that makes it an inedible fruit. There is a great polarization when referring to the two species of the *Punica* genera, on the one hand, there is the common pomegranate, widespread throughout the world, reported as a super fruit in any recipe magazine article (*P. granatum* L.). The other is the Socotran pomegranate (*P. protopunica* Balf.), which is little known, hidden from the eyes of the world and only used by the people of Socotra for medicinal purposes, the one without admirers.

P. protopunica Balf. is one of the only two species of the *Punica* genera, being considered as the "sister" of *P. granatum* L. [14]. However, according to its origin (due to its independent evolutionary line and which seems to be an ancestor of the *Punica* genera), we think that it may be, beyond the taxonomic classification, the "grandmother" of *P. granatum* L.

P. protopunica Balf. has strong ties to *P. granatum* L., and a strong relationship with the flora of the adjacent continental areas of Arabia and Northeast Africa, tropical Africa, Madagascar, India, South Asia, Polynesia and Central America that, as already mentioned, were united in the Cretaceous. It is believed that at least since the late Cretaceous, much of Socotra was emerged, considering itself one of the longest isolated landmasses on earth. Its vascular plants have an endemicity index, quite similar to that of other islands such as the Canary Islands, and it is a refuge for interesting paleoendemisms of very ancient origin, including the case of *P. protopunica* Balf. [64].

Herbalism is one of the most used treatments in traditional Yemeni medicine, predominately in rural territories, herbalism is practiced by Yemen population to all kind of ailments [65]. The uses of *P. protopunica* Balf. in traditional medicine agree with the uses that have been reported in *P. granatum* L., although *P. granatum* L. has more uses and has had more effects attributed to it. *P. granatum* L. is traditionally used for diarrhea, stomatitis, ulcers, bleeding, enemas, vaginal discharge, inflammation of the pancreas, gallbladder diseases, dysentery and stomach disorders, antiparasitic (taenicide and others), antibacterial, inflammatory diseases, astringent, abortion, burns, pain, snakebite, bronchitis, cough, and nausea [66–70]. Traditional and alternative medicine has many followers in the Republic of Yemen, because access to occidental medicine is still restricted, and Yemen is a country with difficult access [71].

Pomegranate is a source of bioactive compounds, present both in the fruit (peel and arils), and in the leaves and bark [72]. Ozgen et al. [73] affirm that pomegranate is a fruit rich in phenolic antioxidants, specifically anthocyanins, but their content varies between varieties, or sub-species. The fruit has a large number of flavonoids, it is estimated that about 0.2% to 1% of the weight of the fruit represent this group of compounds, of which about 30% of all the anthocyanidins are in the peel [74]. Pomegranate juice has a high content of polyphenols, significant amounts of ellagic acid, caffeic acid, chlorogenic acid, coumaric acid, catechins, ferulic acid and a large list of anthocyanins [75]. The literature indicates that pomegranate contains 124 different compounds and that among these phytochemicals, high molecular weight polyphenols (such as ellagitannins and punicalagin) are likely to mediate most of the fruit's protective effects against harmful agents [75]. In the peel, almost 48 phenolic compounds have been identified [76]. Some authors identified and quantified phenolic compounds with more than 50 varieties of pomegranate fruits, using high-performance liquid chromatography with electrospray ionization and mass spectrometry (HPLC-DAD-ESI/MS^n), they concluded that ellagitannins were the most abundant compounds in all the investigated samples (mesocarp, peel, arils, juices), and all the varieties had ellagitannins and anthocyanins [77,78]. All these data were taken from studies in *P. granatum* L. It can be hypothesized that these compounds are also present in *P. protopunica* Balf., however there is a large information gap in this regard. Hundreds of studies have been conducted in which the antioxidant capacity and total phenols of *P. granatum* L. are evaluated, however, for *P. protopunica* Balf., the scenario is different. It is curious that there are not enough published studies of the content of total phenols and antioxidant effect of *P. protopunica* Balf. Nor did we find studies of the nutrimental and chemical composition of the fruit. This may be due in the first place to the isolation of the species, and also due to the difficulty in obtaining government permits, as it originates from a declared a World Heritage Site. Perhaps there are unpublished works.

Many of the uses in traditional medicine have been explained and demonstrated with a great diversity of in vitro and in vivo studies. For example, in *P. granatum* L. hypoglycemic, hypocholesterolemic, hypotriglyceridemic, antihypertensive, anti-atherosclerotic, anti-inflammatory, against metabolic syndrome, various types of cancer, antimicrobial and antifungal, healing, among others effects were reported [79–94]. However, not enough reports of biological effects of the *P. protopunica* Balf., species were found. The antimicrobial, antiviral, antiparasitic and anticancer activity found in *P. protopunica* Balf. has also been found in *P. granatum* L., and many of the mechanisms of action reported by the authors are related with the high concentration of bioactive compounds of secondary metabolism of the fruit, specially ellagic acid, gallic acid, punicic acid and flavonoids as quercetin and kaempferol [45,77,82,85,87,94–102]. Finally, the consumption of *P. granatum* L., including the peel and edible parts, as well as its extracts, are considered safe in vitro and in vivo [103–109], and these results agree with the non-cytotoxicity results of *P. protopunica* Balf., however, more studies in *P. protopunica* Balf. are required to conclude that.

The data provided in this review lead us to suggest that this poorly studied species may have similar pharmacological effects like *P. granatum* L., because it belongs to the same genera and the probability that they share the same active compounds, although perhaps in different proportions. So its pharmacological properties can be better, or just the opposite.

4. Materials and Methods

4.1. Search Criteria

A search for information on the subject was made, using as inclusion criteria all the articles, books and official web pages published to date (June 2020). In this study, published peer-reviewed articles without language restrictions (unpublished data were not included) in repositories such as PubMed, ScienceDirect, Worldwide science, Springer link, Refseek, SciELO, Cochrane Library were considered as primary sources. Secondary sources of information included data from official web pages and books of taxonomic, botanical, cultivation and vulnerability information of the *P. protopunica* Balf., species with the keywords: *Punica protopunica* Balf., wild pomegranate, Socotra pomegranate, *Punica granatum*

L., *Punica* gender, Lythraceae, Myrtales, *Punica protopunica* taxonomy, health effects, biological effects, composition, chemistry, cultivation, distribution and morphology.

4.2. Data Extraction

The data used for this article were collected, classified, summarized, analyzed, compared, discussed and written. The conclusions were made accordingly. The data extracted from each study were extracted and reported using a thematic and subtopic analysis, the results were also compared with *P. granatum* L., as a reference species.

5. Conclusions

In summary, this review supports the idea that the *P. protopunica* Balf. species can be considered as a powerful source of different pharmacological activities. Taking into account that it has been widely reported that this species exhibits antiviral, antimicrobial, antiprotozoal, antioxidant and anticancer activities. It was shown that there is a relationship between *P. protopunica* and *P. granatum*. It is not entirely clear whether the species remains vulnerable, but there are efforts to conserve it. The presence of various bioactive compounds could justify their effect, however, it is necessary further studies to demonstrate its pharmacological actions and their possible adverse effects. Additionally, it is essential to consider clinical studies to establish adequate doses in humans to evaluate the bioactive compounds of *P. protopunica* Balf., whereas the use of traditional medicine and complementary or alternative medicine are increasingly used in developing and developed countries.

6. Prospects

There is a wide range of possibilities for studying this species. In biomedical sciences, more information is needed on the effect of *P. protopunica* Balf. on human health. In vitro and in vivo models in various pathologies are necessary. It is suggested to delve into the mechanisms of action of the species in each pathology studied. Studies of the chemical composition and evaluation of its bioactive compounds (tree and especially of the fruit) are necessary, and compare these results with the composition of *P. granatum* L. Clinical studies are considered necessary to corroborate the effects reported in traditional medicine, once it has been reported that there is no toxicity. It is suggested to use other compound extraction methods and the use of high, medium and low polarity solvents.

On the other hand, the scientific community has focused all its efforts on studying *P. granatum* L. and the immense variety of cultivars it presents. There are thousands of articles on *P. granatum* L., there are high-quality books and reviews, and in most of them, they only mention *P. protopunica* Balf. as a curiosity. This review represents an effort and a call for the scientific community and the population in general to know more about this wonderful species, and to the extent possible, studies in the area of biology, health and pharmacology can be carried out. By studying the benefits of this species, a culture of respect and care can be generated. It is recommended that if scientists wish to study this species, they cultivate it in its places of origin, to preserve the species and take care of the wonderful place that houses it, Socotra.

Author Contributions: Conceptualization, writing, original draft preparation and methodology, J.A.G.-S.; formal analysis, writing, T.J.-C. and T.A.U.-H.; investigation, supervision, A.C.-R.; investigation, conceptualization, E.G.O.-H.; supervision, review and editing, O.A.J.-M. and M.B. All authors have read and agreed to the published version of the manuscript.

Funding: One of the authors (J.A.G.-S.) is grateful for a scholarship N.554424 provided by the National Council of Science and Technology of Mexico (CONACyT).

Acknowledgments: We acknowledge to PROFEXCE 2020-2021 program.

Conflicts of Interest: The authors declare that there are no conflict of interest in the publication of this article.

References

1. Conti, E.; Litt, A.; Sytsma, K.J. Circumscription of Myrtales and their relationships to other rosids: Evidence from rbcL sequence data. *Am. J. Bot.* **1996**, *83*, 221–233. [CrossRef]
2. Dahlgren, R.; Thorne, R.F. The order Myrtales: Circumscription, variation, and relationships. *Ann. Mo. Bot. Gard.* **1984**, 633–699. [CrossRef]
3. Graham, S.A.; Graham, A. Ovary, fruit, and seed morphology of the Lythraceae. *Int. J. Plant Sci.* **2014**, *175*, 202–240. [CrossRef]
4. Stevens, P.F. Angiosperm Phylogeny Website Missouri. Available online: http://www.mobot.org/mobot/research/apweb/welcome.html (accessed on 8 June 2020).
5. The Plant List. Lythraceae—The Plant List. 2013. Available online: http://www.theplantlist.org/1.1/browse/A/Lythraceae/ (accessed on 20 August 2020).
6. Zeynalova, A.M.; Novruzov, E.N. Origin, taxonomy and systematics of pomegranate. *Proc. Inst. Bot. ANAS* **2017**, *37*, 21–26.
7. Shilkina, I.A. On the xylem anatomy of the genus Punica L. *Bot. Zhurnal* **1973**, *58*, 1628–1630.
8. Graham, S.A.; Hall, J.; Sytsma, K.; Shi, S.-H. Phylogenetic analysis of the Lythraceae based on four gene regions and morphology. *Int. J. Plant Sci.* **2005**, *166*, 995–1017. [CrossRef]
9. Conti, E.; Fischbach, A.; Sytsma, K.J. Tribal relationships in Onagraceae: Implications from rbcL sequence data. *Ann. Mo. Bot. Gard.* **1993**, 672–685. [CrossRef]
10. Conti, E.; Litt, A.; Wilson, P.G.; Graham, S.A.; Briggs, B.G.; Johnson, L.A.S.; Sytsma, K.J. Interfamilial relationships in Myrtales: Molecular phylogeny and patterns of morphological evolution. *Syst. Bot.* **1997**, 629–647. [CrossRef]
11. Huang, Y.; Shi, S. Phylogenetics of Lythraceae sensu lato: A Preliminary Analysis Based on Chloroplast rbcL Gene, psaA-ycf3 Spacer, and Nuclear rDNA Internal Transcribed Spacer (ITS) Sequences. *Int. J. Plant Sci.* **2002**, *163*, 215–225. [CrossRef]
12. Shi, S.; Huang, Y.; Tan, F.; He, X.; Boufford, D.E. Phylogenetic analysis of the Sonneratiaceae and its relationship to Lythraceae based on ITS sequences of nrDNA. *J. Plant Res.* **2000**, *113*, 253–258. [CrossRef]
13. Narzary, D.; Ranade, S.A.; Divakar, P.K.; Rana, T.S. Molecular differentiation and phylogenetic relationship of the genus Punica (Punicaceae) with other taxa of the order Myrtales. *Rheedea* **2016**, *26*, 37–51.
14. Stover, E.D.; Mercure, E.W. The pomegranate: A new look at the fruit of paradise. *HortScience* **2007**, *42*, 1088–1092. [CrossRef]
15. Graham, S.A. Lythraceae. In *Flowering Plants Eudicots*; Springer: Berlin/Heidelberg, Germany, 2007; Volume 9, pp. 226–246. [CrossRef]
16. International Plant Names Index (IPNI) Punica Protopunica|International Plant Names Index. Available online: https://www.ipni.org/n/554132-1 (accessed on 16 June 2020).
17. Mars, M. Pomegranate plant material: Genetic resources and breeding, a review. *Options Mediterr. Ser. A* **2000**, *42*, 55–62.
18. Morton, J.F.; Dowling, C.F. *Fruits of Warm Climates*; JF Morton: Miami, FL, USA, 1987; Volume 20534.
19. Chandra, R.; Babu, K.D.; Jadhav, V.T.; Jaime, A.; Silva, T.D. Origin, history and domestication of pomegranate. *Fruit Veg. Cereal Sci. Biotechnol.* **2010**, *2*, 1–6.
20. World of Flowering Plants Punica Protopunica (Socotran Pomegranate). Available online: https://worldoffloweringplants.com/punica-protopunica-socotran-pomegranate/ (accessed on 10 June 2020).
21. Hedberg, I.; Hedberg, O. Conservation of vegetation in Africa south of the Sahara. In Proceedings of the A Symposium Held at the 6th Plenary Meeting of the" Association Pour L'étude Taxonomique de la Flore d'Afrique Tropicale"(AETFAT), Uppsala, Sweden, 12–16 September 1966.
22. Royal Botanic Gardens Kew Punica Protopunica Balf. f|Plants of the World Online|Kew Science. Available online: http://www.plantsoftheworldonline.org/taxon/urn:lsid:ipni.org:names:554132-1 (accessed on 16 June 2020).
23. Tardelli, M.; Baldini, R. Botanical report on the island of Socotra (Yemen). *Port. Acta Biol.* **2000**, *19*, 443–454.
24. Abulhawa, T. Socotra: Enduring bonds between people and place. In *World Heritage*; UNESCO: Paris, France, 2015; pp. 16–21.
25. Van-Damme, K. *Socotra: Natural History of the Islands and Their People*; Odyssey Books & Guides: Hong Kong, China, 2006.

26. Kosenko, V.N. Palynomorphology of representatives of the family Punicaceae. *Bot. Zhurnal* **1985**, *70*, 39–41.
27. Marriner, D. Socotra Pomegranate Tree (Punica Protopunica): The Vulnerable Predicament of the "Other Pomegranate." DerdriuMarriner. 2020. Available online: https://wizzley.com/socotra-pomegranate-tree-punica-protopunica/ (accessed on 10 June 2020).
28. Integrated Taxonomic Information System (ITIS) Standard Report Page: Punica Granatum. Available online: https://www.itis.gov/servlet/SingleRpt/SingleRpt?search_topic=TSN&search_value=27278#null (accessed on 20 June 2020).
29. Miller, A. Punica Protopunica, Pomegranate Tree. The UICN Red List of Threatened Species. 2004. Available online: https://dx.doi.org/10.2305/IUCN.UK.2004.RLTS.T30404A9544416.en (accessed on 23 June 2020).
30. Biodiversity Heritage Library. *Proceedings of the Royal Society of Edinburgh*; Neill and Company: Edinburgh, Scotland, 1882; Volume XI. Available online: https://www.biodiversitylibrary.org/page/48711006 (accessed on 23 June 2020).
31. Balfour, I.B. Botany of Socotra. In *Transactions of the Royal Society of Edinburgh*; Grant, R., Ed.; Neill and Company: Edinburgh, Scotland, 1888; Volume 31, pp. 1–446.
32. Al Shawish, F.; Hamed, F.; Al-Issa, I. Evaluation of some qualitative and chemical characteristics for the most important pomegranate (Punica granatum) accessions in Yemen. *Damascus Univ. J. Agric. Sci.* **2006**, *22*, 227–241.
33. Youssef, M.; Alhammadi, A.S.; Ramírez-Prado, J.H.; Sánchez-Teyer, L.F.; Escobedo-GraciaMedrano, R.M. Remarks on genetic diversity and relationship of Punica protopunica and P. granatum assessed by molecular analyses. *Genet. Resour. Crop Evol.* **2018**, *65*, 577–590. [CrossRef]
34. Miller, A.G.; Morris, M. *Ethnoflora of the Soqotra Archipelago*; Royal Botanic Garden Edinburgh: Edinburgh, Scotland, 2004; ISBN 1872291597.
35. Van-Damme, K.; Banfield, L. Past and present human impacts on the biodiversity of Socotra Island (Yemen): Implications for future conservation. *Zool. Middle East* **2011**, *54*, 31–88. [CrossRef]
36. UNESCO World Heritage Centre. *Operational Guidelines for the Implementation of the World Heritage Convention*; UNESCO: Paris, France, 1999. Available online: https://whc.unesco.org/en/guidelines/ (accessed on 26 June 2020).
37. Pironnet, T.; Marsh, E. *Proteger a las Personas a Través de la Naturaleza*; World Wide Fund for Nature: Gland, Switzerland, 2016; ISBN 9782940529315.
38. UNESCO World Heritage Centre. Socotra Archipelago (Yemen). Available online: https://whc.unesco.org/en/soc/3891 (accessed on 27 June 2020).
39. Clearing House Mechanism of Biodiversity in Yemen Socotra Conservation and Development Programme. Available online: http://ye.chm-cbd.net/information-1/links/related-links/socotra-conservation-and-development-programme (accessed on 28 June 2020).
40. Levin, G.M. *Pomegranate*, 1st ed.; East Libr. Drive Tempe Third Millenn.: St Petersburg, Russia, 2006; pp. 1–129.
41. Muhammad, N.; Uddin, N.; Liu, M.; Ali, N. Ethno medicine, antioxidant potential and inter-specific variations of Punica L. species growing in Swat district, KP, Pakistan. *Int. J. Bot. Stud.* **2019**, *4*, 36–46.
42. Mothana, R.A. Island Soqotra: A unique source of medicinal plants with anticancer, antimicrobial, antiviral and antiprotozoal potentials. *UOFK* **2015**, *31*, 45–49.
43. Al-Dubai, A.S.; Al-Khulaidi, A.A. Medicinal and Aromatic Plants of Yemen (In Arabic) Sana'a. Obadi Center for Studies and Publishing: Sana'a, Yemen, 1996; pp. 125–130.
44. Barzinji, A.K.R.; Nasher, A.K.; Mothana, R.A.; Al-Hamadi, M.M.S. In vitro Antimalarial activity of selected Yemeni plants used in traditional medicine. International Journal of Medicinal Plants. *Photon* **2014**, *107*, 526–535.
45. Mothana, R.A.; Al-Musayeib, N.M.; Matheeussen, A.; Cos, P.; Maes, L. Assessment of the in vitro antiprotozoal and cytotoxic potential of 20 selected medicinal plants from the island of Soqotra. *Molecules* **2012**, *17*, 14349–14360. [CrossRef]
46. Schopen, A. *Traditionelle Heilmittel in Jemen*; Steiner: Wiesbaden, Germany, 1983; ISBN 3515040676.
47. Al-Huqail, A.A.; Elgaaly, G.A.; Ibrahim, M.M. Identification of bioactive phytochemical from two Punica species using GC–MS and estimation of antioxidant activity of seed extracts. *Saudi J. Biol. Sci.* **2018**, *25*, 1420–1428. [CrossRef]

48. Durazzo, A.; Lucarini, M.; Souto, E.B.; Cicala, C.; Caiazzo, E.; Izzo, A.A.; Novellino, E.; Santini, A. Polyphenols: A concise overview on the chemistry, occurrence, and human health. *Phyther. Res.* **2019**, *33*, 2221–2243. [CrossRef]
49. Mothana, R.A.A.; Lindequist, U. Antimicrobial activity of some medicinal plants of the island Soqotra. *J. Ethnopharmacol.* **2005**, *96*, 177–181. [CrossRef] [PubMed]
50. Mothana, R.A.A.; Mentel, R.; Reiss, C.; Lindequist, U. Phytochemical screening and antiviral activity of some medicinal plants from the island Soqotra. *Phyther. Res. Int. J. Devoted Pharmacol. Toxicol. Eval. Nat. Prod. Deriv.* **2006**, *20*, 298–302. [CrossRef]
51. Mothana, R.A.A.; Gruenert, R.; Lindequist, U.; Bednarski, P.J. Study of the anticancer potential of Yemeni plants used in folk medicine. *Die Pharm. Int. J. Pharm. Sci.* **2007**, *62*, 305–307.
52. Vasconcelos, R.; Pujol-Buxó, E.; Llorente, G.A.; Saeed, A.; Carranza, S. Micro-hotspots for conservation: An umbrella tree species for the unique Socotran reptile fauna. *Forests* **2020**, *11*, 353. [CrossRef]
53. Autin, J.; Bellahsen, N.; Leroy, S.; Husson, L.; Beslier, M.-O.; d'Acremont, E. The role of structural inheritance in oblique rifting: Insights from analogue models and application to the Gulf of Aden. *Tectonophysics* **2013**, *607*, 51–64. [CrossRef]
54. Macey, J.R.; Kuehl, J.V.; Larson, A.; Robinson, M.D.; Ugurtas, I.H.; Ananjeva, N.B.; Rahman, H.; Javed, H.I.; Osman, R.M.; Doumma, A. Socotra Island the forgotten fragment of Gondwana: Unmasking chameleon lizard history with complete mitochondrial genomic data. *Mol. Phylogenet. Evol.* **2008**, *49*, 1015–1018. [CrossRef] [PubMed]
55. Balini, M. The Himalayan connection of the Middle Triassic brachiopod fauna from Socotra (Yemen). *Bull. Geosci.* **2018**, *93*, 2. [CrossRef]
56. Kumar, P.A.; Sudhakaran, S.; Mohan, T.C.; Pamanna, D.; Kumar, P.R.; Shanthanna, P. Evaluation of colour enhance potential of three natural plant pigment sources (African tulip tree flower, red paprika, pomegranate peel) in goldfish (Carassius auratus). *Int. J. Fish. Aquat. Stud.* **2017**, *5*, 47–51.
57. Fahan, A. The leaf. The flower. The seed. *Plant Anat. Hakkibutz Hameuhad Publ. House Ltd. Jerus.* **1976**, 171–212.
58. Verma, N.; Mohanty, A.; Lal, A. Pomegranate genetic resources and germplasm conservation: A review. *Fruit Veg. Cereal Sci. Biotechnol.* **2010**, *4*, 120–125.
59. Frison, E.A.; Servinsky, J. *Directory of European Institutions Holding Crop Genetic Resources Collections: Vol. 2. Index*; European Cooperative Programme for Plant Genetic Resources: Rome, Italy, 1995; ISBN 9290432594.
60. Heber, D.; Schulman, R.N.; Seeram, N.P. *Pomegranates: Ancient Roots to Modern Medicine*; CRC Press: Boca Raton, FL, USA, 2006; ISBN 1420009869. [CrossRef]
61. Still, D.W. Chapter 13 Pomegranates: A Botanical Perspective. In *Pomegranates: Ancient Roots to Modern Medicine*; Seeram, N.P., Heber, D., Shulman, R.N., Eds.; CRC Press: Boca Raton, FL, USA, 2006; pp. 199–210. ISBN 1420009869.
62. Da Silva, J.A.T.; Rana, T.S.; Narzary, D.; Verma, N.; Meshram, D.T.; Ranade, S.A. Pomegranate biology and biotechnology: A review. *Sci. Hortic.* **2013**, *160*, 85–107. [CrossRef]
63. Mirjalili, S.A. Pomegranate: Biodiversity and genetic resources, a review. *Rostaniha* **2016**, *17*, 1–18.
64. Tosco, R.B.; Navarro, M.L.R.; Scholz, S. Archipiélagos macaronésicos (VIII) y otras islas del mundo. *Makaronesia Boletín Asoc. Amigos Mus. Cienc. Nat. Tenerife* **2007**, *9*, 120–129.
65. Oleski, A.; Lindequist, U.; Mothana, R.A.A.; Melzig, M.F. Screening of selected Arabian medicinal plant extracts for inhibitory activity against peptidases. *Die Pharm. Int. J. Pharm. Sci.* **2006**, *61*, 359–361.
66. Amin, G.H. Popular medical plants of Iran. *Farhang Publ. Tehran* **1991**, *61*, 51–52.
67. Lansky, E.; Shubert, S.; Neeman, I. Pharmacological and therapeutic properties of pomegranate. In Proceedings of the Symposium on Production, Processing and Marketing of Pomegranate in the Mediterranean Region: Advances in Research and Technology. Séminaires Méditerranéens (CIHEAM), Orihuela, Spain, 15–17 October 2000; pp. 231–235.
68. Ma, K.I.; Du, M.; Liao, M.; Chen, S.; Yin, G.; Liu, Q.; Wei, Q.; Qin, G. Evaluation of wound healing effect of punica granatum L Peel extract on deep second-degree burns in rats. *Trop. J. Pharm. Res.* **2015**, *14*, 73–78. [CrossRef]
69. Poyrazoğlu, E.; Gökmen, V.; Artık, N. Organic acids and phenolic compounds in pomegranates (*Punica granatum* L.) grown in Turkey. *J. Food Compos. Anal.* **2002**, *15*, 567–575. [CrossRef]

70. Tanveer, A.; Farooq, U.; Akram, K.; Hayat, Z.; Shafi, A.; Nazar, H.; Ahmad, Z. Pomegranate extracts: A natural preventive measure against spoilage and pathogenic microorganisms. *Food Rev. Int.* **2015**, *31*, 29–51. [CrossRef]
71. Chhetri, B.K.; Ali, N.A.A.; Setzer, W.N. A survey of chemical compositions and biological activities of Yemeni aromatic medicinal plants. *Medicines* **2015**, *2*, 67–92. [CrossRef]
72. Mirdehghan, S.H.; Rahemi, M. Seasonal changes of mineral nutrients and phenolics in pomegranate (*Punica granatum* L.) fruit. *Sci. Hortic.* **2007**, *111*, 120–127. [CrossRef]
73. Ozgen, M.; Durgaç, C.; Serçe, S.; Kaya, C. Chemical and antioxidant properties of pomegranate cultivars grown in the Mediterranean region of Turkey. *Food Chem.* **2008**, *111*, 703–706. [CrossRef]
74. Akhtar, S.; Ismail, T.; Fraternale, D.; Sestili, P. Pomegranate peel and peel extracts: Chemistry and food features. *Food Chem.* **2015**, *174*, 417–425. [CrossRef]
75. Gil, M.I.; Tomás-Barberán, F.A.; Hess-Pierce, B.; Holcroft, D.M.; Kader, A.A. Antioxidant activity of pomegranate juice and its relationship with phenolic composition and processing. *J. Agric. Food Chem.* **2000**, *48*, 4581–4589. [CrossRef]
76. Bhandari, P.R. Pomegranate (*Punica granatum* L). Ancient seeds for modern cure? Review of potential therapeutic applications. *Int. J. Nutr. Pharmacol. Neurol. Dis.* **2012**, *2*, 171. [CrossRef]
77. Fischer, U.A.; Carle, R.; Kammerer, D.R. Identification and quantification of phenolic compounds from pomegranate (*Punica granatum* L.) peel, mesocarp, aril and differently produced juices by HPLC-DAD–ESI/MSn. *Food Chem.* **2011**, *127*, 807–821. [CrossRef]
78. Balli, D.; Cecchi, L.; Khatib, M.; Bellumori, M.; Cairone, F.; Carradori, S.; Zengin, G.; Cesa, S.; Innocenti, M.; Mulinacci, N. Characterization of Arils Juice and Peel Decoction of Fifteen Varieties of *Punica granatum* L.: A Focus on Anthocyanins, Ellagitannins and Polysaccharides. *Antioxidants* **2020**, *9*, 238–261. [CrossRef]
79. Deng, Y.L.; Li, Y.L.; Zheng, T.T.; Hu, M.X.; Ye, T.H.; Xie, Y.; Yin, W.Y. The extract from Punica granatum (pomegranate) leaves promotes apoptosis and impairs metastasis in prostate cancer cells. *J. Sichuan Univ. Med. Sci. Ed.* **2018**, *49*, 8–12.
80. Wang, Z.; Chen, Y. Antihypertensive Effect of Pomegranate Polyphenols in Spontaneously Hypertensive Rats. *China Pharm.* **2016**, *19*, 255–258.
81. Neyrinck, A.M.; Van Hée, V.F.; Bindels, L.B.; De Backer, F.; Cani, P.D.; Delzenne, N.M. Polyphenol-rich extract of pomegranate peel alleviates tissue inflammation and hypercholesterolaemia in high-fat diet-induced obese mice: Potential implication of the gut microbiota. *Br. J. Nutr.* **2013**, *109*. [CrossRef] [PubMed]
82. Medjakovic, S.; Jungbauer, A. Pomegranate: A fruit that ameliorates metabolic syndrome. *Food Funct.* **2013**, *4*, 19–39. [CrossRef] [PubMed]
83. Ismail, T.; Sestili, P.; Akhtar, S. Pomegranate peel and fruit extracts: A review of potential anti-inflammatory and anti-infective effects. *J. Ethnopharmacol.* **2012**, *143*, 397–405. [CrossRef]
84. Stowe, C.B. The effects of pomegranate juice consumption on blood pressure and cardiovascular health. *Complement. Ther. Clin. Pract.* **2011**, *17*, 113–115. [CrossRef]
85. Betanzos-Cabrera, G.; Guerrero-Solano, J.A.; Martínez-Pérez, M.M.; Calderón-Ramos, Z.G.; Belefant-Miller, H.; Cancino-Diaz, J.C. Pomegranate juice increases levels of paraoxonase1 (PON1) expression and enzymatic activity in streptozotocin-induced diabetic mice fed with a high-fat diet. *Food Res. Int.* **2011**, *44*, 1381–1385. [CrossRef]
86. Silva, N.C.C.; Fernandes Júnior, A. Biological properties of medicinal plants: A review of their antimicrobial activity. *J. Venom. Anim. Toxins Incl. Trop. Dis.* **2010**, *16*, 402–413. [CrossRef]
87. Mohan, M.; Waghulde, H.; Kasture, S. Effect of pomegranate juice on Angiotensin II-induced hypertension in diabetic wistar rats. *Phyther. Res.* **2010**, *24*, S196–S203. [CrossRef]
88. Miguel, M.G.; Neves, M.A.; Antunes, M.D. Pomegranate (*Punica granatum* L.): A medicinal plant with myriad biological properties-A short review. *J. Med. Plant Res.* **2010**, *4*, 2836–2847.
89. Hossin, F.L.A. Effect of pomegranate (Punica granatum) peels and it's extract on obese hypercholesterolemic rats. *Pak. J. Nutr.* **2009**, *8*, 1251–1257. [CrossRef]
90. Bagri, P.; Ali, M.; Aeri, V.; Bhowmik, M.; Sultana, S. Antidiabetic effect of Punica granatum flowers: Effect on hyperlipidemia, pancreatic cells lipid peroxidation and antioxidant enzymes in experimental diabetes. *Food Chem. Toxicol.* **2009**, *47*, 50–54. [CrossRef] [PubMed]
91. Lei, F.; Zhang, X.N.; Wang, W.; Xing, D.M.; Xie, W.D.; Su, H.; Du, L.J. Evidence of anti-obesity effects of the pomegranate leaf extract in high-fat diet induced obese mice. *Int. J. Obes.* **2007**, *31*, 1023–1029. [CrossRef]

92. Rosenblat, M.; Hayek, T.; Aviram, M. Anti-oxidative effects of pomegranate juice (PJ) consumption by diabetic patients on serum and on macrophages. *Atherosclerosis* **2006**, *187*, 363–371. [CrossRef] [PubMed]
93. Aviram, M.; Rosenblat, M.; Gaitini, D.; Nitecki, S.; Hoffman, A.; Dornfeld, L.; Volkova, N.; Presser, D.; Attias, J.; Liker, H.; et al. Pomegranate juice consumption for 3 years by patients with carotid artery stenosis reduces common carotid intima-media thickness, blood pressure and LDL oxidation. *Clin. Nutr.* **2004**, *23*, 423–433. [CrossRef]
94. Aviram, M.; Dornfeld, L. Pomegranate juice consumption inhibits serum angiotensin converting enzyme activity and reduces systolic blood pressure. *Atherosclerosis* **2001**, *158*, 195–198. [CrossRef]
95. Adams, L.S.; Seeram, N.P.; Aggarwal, B.B.; Takada, Y.; Sand, D.; Heber, D. Pomegranate Juice, Total Pomegranate Ellagitannins, and Punicalagin Suppress Inflammatory Cell Signaling in Colon Cancer Cells. *J. Agric. Food Chem.* **2006**, *54*, 980–985. [CrossRef]
96. Hora, J.J.; Maydew, E.R.; Lansky, E.P.; Dwivedi, C. Chemopreventive Effects of Pomegranate Seed Oil on Skin Tumor Development in CD1 Mice. *J. Med. Food* **2003**, *6*, 157–161. [CrossRef]
97. Khan, N.; Hadi, N.; Afaq, F.; Syed, D.N.; Kweon, M.-H.; Mukhtar, H. Pomegranate fruit extract inhibits prosurvival pathways in human A549 lung carcinoma cells and tumor growth in athymic nude mice. *Carcinogenesis* **2007**, *28*, 163–173. [CrossRef]
98. Kim, N.D.; Mehta, R.; Yu, W.; Neeman, I.; Livney, T.; Amichay, A.; Poirier, D.; Nicholls, P.; Kirby, A.; Jiang, W.; et al. Chemopreventive and adjuvant therapeutic potential of pomegranate (Punica granatum) for human breast cancer. *Breast Cancer Res. Treat.* **2002**, *71*, 203–217. [CrossRef]
99. Malik, A.; Afaq, F.; Sarfaraz, S.; Adhami, V.M.; Syed, D.N.; Mukhtar, H. Pomegranate fruit juice for chemoprevention and chemotherapy of prostate cancer. *Proc. Natl. Acad. Sci. USA* **2005**, *102*, 14813–14818. [CrossRef] [PubMed]
100. Mehta, R.; Lansky, E.P. Breast cancer chemopreventive properties of pomegranate (Punica granatum) fruit extracts in a mouse mammary organ culture. *Eur. J. Cancer Prev.* **2004**, *13*. [CrossRef] [PubMed]
101. Pacheco-Palencia, L.A.; Noratto, G.; Hingorani, L.; Talcott, S.T.; Mertens-Talcott, S.U. Protective Effects of Standardized Pomegranate (*Punica granatum* L.) Polyphenolic Extract in Ultraviolet-Irradiated Human Skin Fibroblasts. *J. Agric. Food Chem.* **2008**, *56*, 8434–8441. [CrossRef] [PubMed]
102. Dell'Agli, M.; Galli, G.V.; Bulgari, M.; Basilico, N.; Romeo, S.; Bhattacharya, D.; Taramelli, D.; Bosisio, E. Ellagitannins of the fruit rind of pomegranate (Punica granatum) antagonize in vitro the host inflammatory response mechanisms involved in the onset of malaria. *Malar. J.* **2010**, *9*, 208. [CrossRef]
103. Cerdá, B.; Cerón, J.J.; Tomás-Barberán, F.A.; Espín, J.C. Repeated oral administration of high doses of the pomegranate ellagitannin punicalagin to rats for 37 days is not toxic. *J. Agric. Food Chem.* **2003**, *51*, 3493–3501. [CrossRef]
104. El Deeb, K.S.; Eid, H.H.; Ali, Z.Y.; Shams, M.M.; Elfiky, A.M. Bioassay-guided fractionation and identification of antidiabetic compounds from the rind of Punica Granatum Var. nana. *Nat. Prod. Res.* **2019**, 1–4. [CrossRef]
105. Gautam, R.K.; Sharma, S.; Sharma, K.; Gupta, G. Evaluation of Antiarthritic Activity of Butanol Fraction of Punica granatum Linn. Rind Extract against Freund's Complete Adjuvant-Induced Arthritis in Rats. *J. Environ. Pathol. Toxicol. Oncol.* **2018**, *37*. [CrossRef]
106. Kulkarni, A.P.; Mahal, H.S.; Kapoor, S.; Aradhya, S.M. In vitro studies on the binding, antioxidant, and cytotoxic actions of punicalagin. *J. Agric. Food Chem.* **2007**, *55*, 1491–1500. [CrossRef]
107. Patel, C.; Dadhaniya, P.; Hingorani, L.; Soni, M.G. Safety assessment of pomegranate fruit extract: Acute and subchronic toxicity studies. *Food Chem. Toxicol.* **2008**, *46*, 2728–2735. [CrossRef]
108. Ouachrif, A.; Khalki, H.; Chaib, S.; Mountassir, M.; Aboufatima, R.; Farouk, L.; Benharraf, A.; Chait, A. Comparative study of the anti-inflammatory and antinociceptive effects of two varieties of Punica granatum. *Pharm. Biol.* **2012**, *50*, 429–438. [CrossRef] [PubMed]
109. Vidal, A.; Fallarero, A.; Peña, B.R.; Medina, M.E.; Gra, B.; Rivera, F.; Gutierrez, Y.; Vuorela, P.M. Studies on the toxicity of *Punica granatum* L.(Punicaceae) whole fruit extracts. *J. Ethnopharmacol.* **2003**, *89*, 295–300. [CrossRef] [PubMed]

plants

MDPI

Article

Genome-Wide Association Mapping for Stripe Rust Resistance in Pakistani Spring Wheat Genotypes

Madiha Habib [1], Faisal Saeed Awan [1,*], Bushra Sadia [1] and Muhammad Anjum Zia [2]

[1] Centre of Agricultural Biochemistry and Biotechnology, University of Agriculture, Faisalabad 38000, Pakistan; madihahabib217@gmail.com (M.H.); bushra.sadia@uaf.edu.pk (B.S.)
[2] Department of Biochemistry, University of Agriculture, Faisalabad 38000, Pakistan; anjum.zia@uaf.edu.pk
* Correspondence: faisal.saeed@uaf.edu.pk; Tel.: +92-41-9201087

Received: 13 May 2020; Accepted: 15 June 2020; Published: 19 August 2020

Abstract: Stripe rust caused by the pathogen *Puccinia striiformis* f. sp. *tritici* (*Pst*) is a major threat for wheat, resulting in low yield and grain quality loss in many countries. Genetic resistance is a prevalent method to combat the disease. Mapping the resistant loci and their association with traits is highly exploited in this era. A panel of 465 Pakistani spring wheat genotypes were evaluated for their phenotypic response to stripe rust at the seedling and adult plant stages. A total of 765 single nucleotide polymorphism (SNP) markers were applied on 465 wheat genotypes to evaluate their stripe rust response against nine races during the seedling test and in three locations for the field test. Currently, twenty SNPs dispersed on twelve chromosomal regions (1A, 1B, 1D, 2A, 2B, 4A, 4B, 5B, 6A, 6B, 6D and 7B) have been identified that were associated with rust race-specific resistance at the seedling stage. Thirty SNPs dispersed on eighteen chromosomal regions (1A, 1B, 1D, 2A, 2B, 2D, 3A, 3B, 3D, 4B, 5A, 5B, 6A, 6B, 6D, 7A, 7B and 7D) are associated with adult plant resistance. SNP loci IWB3662 was linked with all three Pakistani races, and likewise IWA2344 and IWA4096 were found to be linked with three different USA races. The present research findings can be applied by wheat breeders to increase their resistant capability and yield potential of their cultivars, through marker-assisted selection.

Keywords: wheat; genome-wide association studies; association mapping; SNP; stripe rust

1. Introduction

Triticum aestivum L., commonly known as bread wheat, is the world's major food crop for about 2 billion people. It fulfills the nutritional requirements of almost 1/3 of the world's population (about 35%) by providing half of their protein and above half of their calorie requirement [1]. Wheat contains a high quantity of dietary nutrients which includes carbohydrates (70%), crude fibers (22%), proteins (12%), water (12%), fats (2%) and minerals (1.8%) [2].

Wheat can withstand a wide range of climate changes but it is also prone to different biotic and abiotic stresses. Among the biotic stresses (rust, smut, mildew, bunt etc.), rust has global economic and historic significance. Rust (leaf, stripe and stem) is the most devastating fungal disease that is threatening the overall world wheat production [3] and its cyclic rotation is considered responsible for famine in many parts of the world.

Wheat in its growing season, if faced with a cool environment, is mostly arrested by stripe (yellow) rust which is a major foliar disease caused by *Puccinia striiformis* f. sp. *tritici* (*Pst*) [4], but nowadays wheat growing in warmer regions is also prone to stripe rust epidemics [5,6]. Mostly, 0.1–5% wheat yield losses are observed due to stripe rust disease and rarely, losses expand up to 25% [4]. The yield loss due to stripe rust can exceed up to 100% for susceptible wheat landraces. Wheat infected with stripe rust causes yield loss by reducing the kernel quality, the kernel number per spike, the low-test weight and the plant height [7,8].

Pakistan is ranked among the top 10 wheat-producing countries. Globally, Pakistan ranked at seventh with 25.5 million tonnes annual wheat production [9]. In Pakistan, almost 70% of the wheat cultivation area (5.8 MH) is susceptible to stripe rust [10]. Northern and central-west areas of Pakistan are majorly threatened with stripe rust. A major yellow rust epidemic in Pakistan was observed in 1995, which caused a 20% loss in the affected areas [11]. Due to climate change, the risk factor of disease decreased but it is still at the doorstep of Pakistan. For the years 2009 and 2010, due to the high prevalence of disease in close neighbors like Uzbekistan, significant yield losses were observed. Stripe rust epidemic losses mounted to almost 360M$ in the USA for the year 2004, 100M$ in Pakistan for the year 2005 [12], 127M$ in Australia for the year 2009 [13] and 30M$ in Morocco for the year 2009 [14].

Wheat rust can be controlled by fungicides, cultural practices, and planting resistant varieties [15]. Fungicidal spray at the appropriate time can reduce the risk of loss caused by pathogens but it requires a healthy finance. Despite fungicidal spray, the areas with a high disease pressure are prevalent to substantial loss if susceptible cultivars are present [16]. In Pakistan, fungicidal control is also not a sensible choice for >8 MH area due to its high cost, spray machinery requirement, the alarming conditions of water resources, as well as air and soil pollution.

Genetic resistance is the most prevailing approach to protect the crop against the yield losses caused by diseases [17]. Resistance to stripe rust is characterized as seedling resistance (all stage resistance) and adult plant resistance (APR). All stage resistance is mostly controlled by a single resistant gene (race specific) and provides high resistance throughout the plant's development but is readily overcome due to changes in the virulence by the emergence of new rust pathogens [18]. Alternatively, wheat resistance against stripe rust can be improved by adult plant resistance (APR), or race-nonspecific resistance or partial resistance. This nonspecific resistance is efficiently controlled by minor and effective multiple loci which are quantitatively inherited [19] and mostly appears at the later growing stages of the plant's development [20]. Pyramiding four to five race-nonspecific resistance genes [21] makes the genotypes more durable at the later growing stages then the race-specific resistance. Gene pyramiding can be achieved by using a molecular marker closely linked with the APR and hence can be further used in breeding through marker-assisted selection (MAS).

Molecular mapping helps in the identification of phenotypic traits at the genomic level by using molecular markers closely linked with the required traits, and hence used in marker-assisted selection and gene pyramiding. Recently, single nucleotide polymorphisms (SNPs), due to their high-density form, shown by the iSelect array, have been proven as valuable mapping markers across the whole wheat genome. The usage of SNPs in genome-wide association studies, based on linkage disequilibrium (LD), increases the efficiency of finding linked loci to the desired traits in the diverse population. Association mapping (AM) or genome-wide association studies (GWAS) utilizes the LD to spot the association among genetic polymorphism and phenotypic variation [22,23]. LD is the principle behind AM which studies the nonrandom association of alleles at different loci [22]. AM has its importance to map QTL (quatitative trait loci) due to the availability of a faster, higher density and cheaper molecular marker. It also minimizes the time and cost by utilizing the diverse populations to determine the linkage disequilibrium between the alleles and to identify the marker–trait association, over the biparental population development [24].

The aim of the present research was to identify the genetic divergence pattern of stripe rust resistance resources in Pakistani spring wheat genotypes. The present study also focused on the identification and mapping of seedling and field resistance minor loci for wheat stripe rust resistance using single nucleotide polymorphism (SNP) markers.

2. Results

2.1. Phenotypic Variation to Stripe Rust Response

The phenotypic characterization of 465 spring wheat genotypes was done with nine *Pst* races (six races selected from the United State of America and three races from Pakistan). The seedling test was performed in a greenhouse (controlled condition) and the rust-infection score (infection type (IT)) collected is summarized in Figures 1 and 2. The frequency of resistance and susceptibility varied with all the stripe rust races. About 15, 56, 72, 106, 31 and 57 Pakistani accessions were resistant (IT score = 0–3) to the USA races PSTv-37, PSTv-198, PSTv-51, PSTv-40, PSTv-14 and PSTv-4, respectively. Most of the genotypes were susceptible (IT score = 7–9) as 384, 268, 224, 156, 200 and 220 to the USA races of stripe rust PSTv-37, PSTv-198, PSTv-51, PSTv-40, PSTv-14 and PSTv-4 respectively (Figure 1).

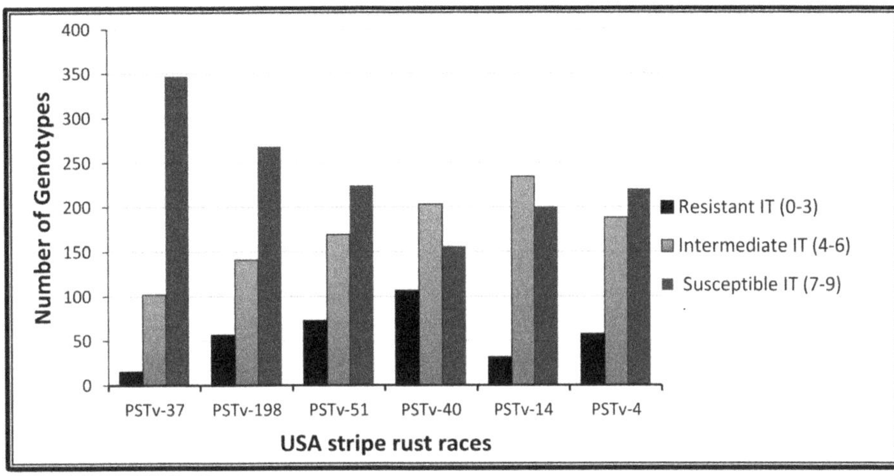

Figure 1. Infection distribution of the 465 spring wheat genotypes at the seedling stage based on their infection type (IT) score, tested with six stripe rust races from the USA.

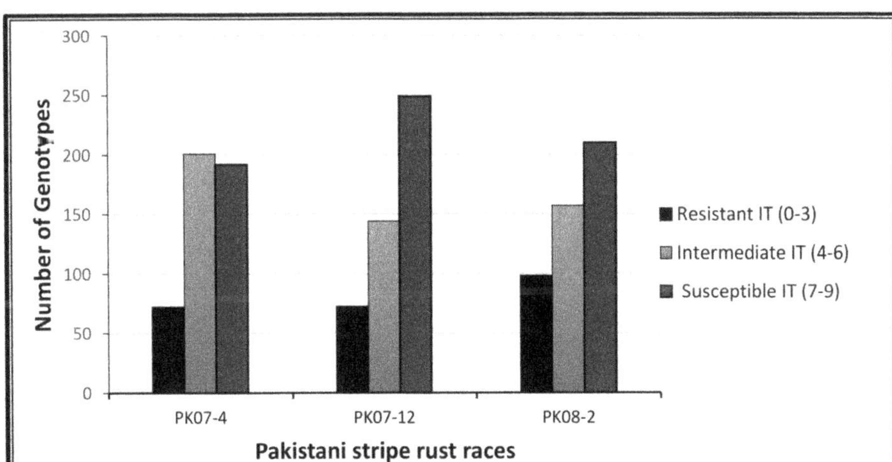

Figure 2. Infection distribution of the 465 spring wheat genotypes at the seedling stage based on their infection type (IT) score, tested with three stripe rust races from Pakistan.

Wheat genotypes showed a slightly different behavior with the Pakistani stripe rust races such that the numbers of resistant genotypes were greater when compared to the USA stripe rust races (Figure 2). The phenotypic analysis revealed 72 resistant genotypes to stripe rust races PK07-4 and PK07-12, while 98 genotypes showed a resistance response against the PK08-2 stripe rust race. Most of the genotypes were susceptible to the USA races PSTv-37 and PSTv-198 and the Pakistani race PK07-12. Most of the genotypes showed resistance to the USA races PSTv-40 and PSTv-51 and to the Pakistani race PK08-2 (Figures 1 and 2).

In a field trial, the adult stage response of the wheat genotype against the stripe rust was slightly different as compared to the controlled conditions in the greenhouse. In a field, the resistance behavior against stripe rust infection of the genotypes was observed, namely that of PAK-University of Agriculture, Faisalabad (UAF)17 (IT = 421, SEV = 414), PAK-UAF16 (IT = 402, SEV = 407), USA-Mount Vernon (MTV)18 (IT = 292, SEV = 270) and USA-Palouse conservation field station (PCFS)18 (IT = 234, SEV = 317) with both their infection type (IT) scores and disease severity (SEV) scores (Figures 3 and 4). The genotypes susceptibility frequencies were higher in the USA locations than in Pakistan, hence the response varied with the environment. In Pakistan, due to the warm environment, the disease impact was slightly less but was not negligible as compared to the cold environment near Pullman, WA, USA. The estimation of the variance component showed a highly significant behavior ($p < 0.0001$) of the genotypes and the genotype x environment interaction across all the environments. Non-significant responses of environmental variance indicate the variable climate conditions and the influence of different races in the disease development. A high range of the coefficient of determination and broad sense heritability of (H^2) 87% for the IT and 93% for the SEV indicate the reliability of the dataset for GWAS (Table 1).

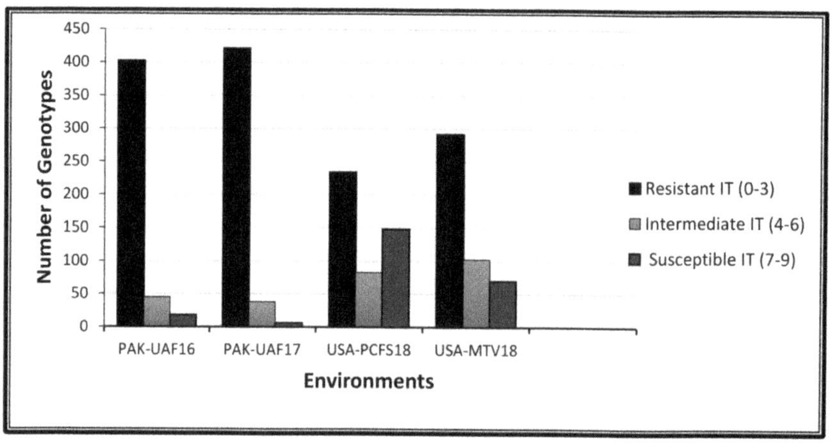

Figure 3. Infection distribution of 465 spring wheat genotypes at the adult plant stage and the stripe rust infection type (IT) score across three environments. Scoring for the adult plant response performed in three different environments (two years in Pakistan considered as one location and two locations in the USA).

2.2. Population Structure and Linkage Disequilibrium

The population structure was performed on a total 465 entire wheat genotypes by the Bayesian clustering approach. An admixture-based model was used to cluster the genotype into three subpopulations based on the ΔK (Figure 5; Supplementary Table S4) value. Population structure reduces the false positive association among the markers and traits. Subpopulation one (Q1) contains 85 individuals. Similarly, subpopulation two (Q2) contains 179 and subpopulation three (Q3) contains

201 individuals. Each subpopulation is shown by a different color in the cluster analysis. The length of each color represents the estimated contribution of each sample to the subgroups.

After the filtration of minor allelic frequency, 765 SNP markers out of 1500 SNPs were used for the linkage disequilibrium analysis (LD). Linkage disequilibrium (LD) depends on many factors, including the population structure, genetic drift, chromosomal region and natural selection. LD decay relies on the value of r^2, whose value is calculated for all chromosomes. Critical r^2 value 0.12 was identified for all 465 spring wheat genotypes by taking the 95th percentile of the coefficient square (showed by the red line in Figure 6). The mean r^2 value across the genome was found 0.03 with 50 cm distance. The highest number of pair-markers were found on the A genome (48%) followed by the B genome (46%) and the D genome (5%). Chromosome 1A had the highest number of pairs (2138) and 4D had the minimum number (nine) of pair markers. The LD decay was constructed using chromosomal distance and the critical r^2 value as the threshold to indicate the LD decay length, which attained 1.25 cm for the whole genome (Figure 6).

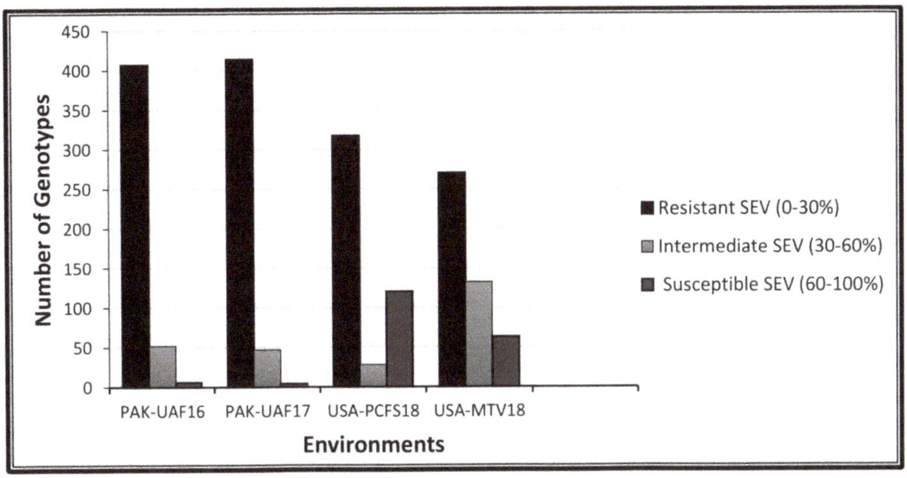

Figure 4. Infection distribution of 465 spring wheat genotypes at the adult plant stage and the stripe rust disease severity (SEV) score across three environments. Scoring for the adult plant response performed in three different environments (two years in Pakistan considered as one location and two locations in the USA).

Table 1. Mean, range and the variance components of *Puccinia striiformis* under different environments.

Parameter	PAK-IT	PAK-SEV	USA-IT	USA-SEV	ALL-IT	ALL-SEV
Range	0-8	0-80	2-8	1-100	0-8	0-100
Mean	2.14	12.32	4.17	30.85	3.15	21.53
Var(G)	1.529 **	175.66 **	3.035 **	440.97 **	0.962 **	119.39 **
Var(E)	0.006 ^{ns}	0.238 ^{ns}	0.3 ^{ns}	0.303 ^{ns}	1.47 ^{ns}	114.26 ^{ns}
Var(G*E)	0.446 **	25.54 **	2.094 **	447.7 **	2.64 **	433.91 **
Var(Total)	1.98	201.44	5.429	888.98	5.07	667.57
R^2	0.87	0.93	0.76	0.687	0.54	0.43
SD	1.418	14.37	2.38	30.93	2.21	25.8
H^2	87	93	71	66	48	46

Asterisks ** significant at $p < 0.0001$ and * significant at $p < 0.005$, **ns**: non-specific, **IT**: infection type, **SEV**: disease severity, **Var**: variance, **R^2**: coefficient of determination, **SD**: standard deviation, **H^2**: broad sense heritability.

2.3. Marker–Trait Association at Seedling Stage

Genome-wide association studies (GWAS) of SNP markers with IT scores of nine stripe rust races identify twenty SNPs associated with rust resistance at the seedling stage (Table 2). Manhattan and Q-Q plots, showing the marker–trait association of stripe rust-resistance response at the seedling stage, with all nine races IT scores are provided in the Supplementary Figures S1 and S2. Twelve chromosomal regions (1A, 1B, 1D, 2A, 2B, 4A, 4B, 5B, 6A, 6B, 6D and 7B) were found to be significantly associated ($p < 0.0001$) with the IT scores of nine rust races at the seedling stage. A total of fourteen SNP loci were identified to be linked with resistance against USA races and twelve loci were identified to be linked with stripe rust-resistance response to Pakistani races. One SNP present on chromosome 1A (IWB3662) was found to be associated with all three Pakistani rust races, namely PK07-4, PK07-12 and PK08-2. A SNP on the 1B chromosome (IWB12258) was linked with the IT score of PK08-2. At the chromosome 1D, two SNPs IWA7171 (linked to PK07-12) and IWA4344 (linked to PSTv-4) were identified. Two SNP markers, IWB50806 linked with PK07-12 and IWA7638 linked with PSTv-14, were identified on chromosome 2A. The targeted SNP IWA2344 was associated with three rust races PSTv-198, PSTv-51 and PSTv-40 at chromosome 2B. Similarly, for the three rust races, PSTv-198, PSTv-51 and PSTv-4, one SNP (IWA4097) marker on the chromosome 1B was identified. Two SNPs were identified at chromosome 2B associated with six stripe rust race. At chromosome 4A, one SNP (IWA3361) identified associated with PSTv-4. Three SNPs identified on the 4B chromosome, IWB12434, IWA2031 and IWA408, were associated with PK07-12, PSTv-14 and PK08-2, respectively. The SNP IWB5781 was associated with PSTv-4, identified at chromosome 5B. At 6A three SNPs identified IWA8022 linked with PSTv-40 and IWA3463, IWB25252 both linked with PK07-12. One SNP (IWB7615) on chromosome 6B was associated with PSTv-51. IWB12259 was linked with PK07-4 at the location 6D. Finally, two SNPs (IWA6857 and IWB10895) were associated with PK07-12 and PSTv-40 on chromosome 7B. Seven SNPs were identified at genome A, ten SNPs at genome B and three SNPs at genome D. The largest number of SNPs (six) was identified to be linked to PK07-12. The identified SNP markers associated with rust race-specific resistance can be linked to different Yr genes based on the virulence and avirulence formula of races (Supplementary Table S2 and S3).

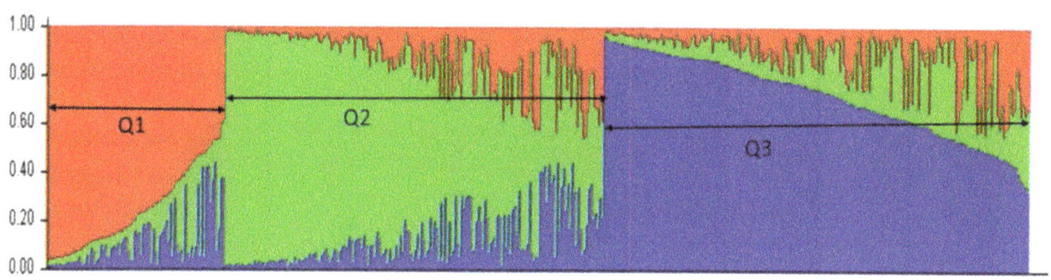

Figure 5. Estimated population structure of the 465 spring wheat genotypes (K = 3) based on the Q matrix using the single nucleotide polymorphism (SNP) markers. Q: three (K = 3) different subpopulations as Q1, Q2 and Q3.

Table 2. SNPs associated with the seedling response to nine *Pst* races in 465 genotypes of wheat ($p < 0.0001$).

QTL [a]	SNP Id. [b]	Chr. [c]	Position [d] (cm)	MAF [e]	$-\log_{10}(P)$ [f]								
					PK07-4	PK07-12	PK08-2	PSTv-37	PSTv-198	PSTv-51	PSTv-40	PSTv-14	PSTv-4
QYr.uaf-1A.1	IWB3662	1A	10.69	0.46	8.14	9.03	6.59	–	–	–	–	–	–
QYr.uaf-1B.1	IWB12258	1B	28.76	0.24	–	–	4.77	–	–	–	–	–	–
QYr.uaf-1D.1	IWA7171	1D	90.30	0.30	–	4.57	–	–	–	–	–	–	–
QYr.uaf-1D.2	IWA4344	1D	90.30	0.21	–	–	–	–	–	–	–	–	4.69
QYr.uaf-2A.1	IWB50806	2A	81.49	0.47	4.42	–	–	–	–	–	–	–	–
QYr.uaf-2A.2	IWA7638	2A	162.89	0.46	–	–	–	–	–	–	–	4.3	–
QYr.uaf-2B.1	IWA2344	2B	96.99	0.26	–	–	–	–	6.91	6.09	5.55	–	–
QYr.uaf-2B.2	IWA4097	2B	113.86	0.10	–	–	–	–	8.36	7.53	–	–	5.32
QYr.uaf-4A.1	IWA3361	4A	48.84	0.24	–	–	–	–	–	–	–	–	4.66
QYr.uaf-4B.1	IWB12434	4B	74.62	0.46	–	4.32	–	–	–	–	–	–	–
QYr.uaf-4B.2	IWA2031	4B	98.65	0.16	–	–	–	–	–	–	–	4.51	–
QYr.uaf-4B.3	IWA408	4B	114.87	0.30	–	–	4.23	–	–	–	–	–	–
QYr.uaf-5B.1	IWB5781	5B	56.60	0.47	–	–	–	–	–	–	–	–	4.19
QYr.uaf-6A.1	IWA8028	6A	77.14	0.27	–	–	–	–	–	–	4.06	–	–
QYr.uaf-6A.2	IWA3463	6A	83.04	0.34	–	4.02	–	–	–	–	–	–	–
QYr.uaf-6A.3	IWB25252	6A	95.87	0.26	–	4.28	–	–	–	–	–	–	–
QYr.uaf-6B.1	IWB7615	6B	64.82	0.43	–	–	–	–	–	4.81	–	–	–
QYr.uaf-6D.1	IWB12259	6D	155.10	0.18	4.51	–	–	–	–	–	–	–	–
QYr.uaf-7B.1	IWA6857	7B	76.31	0.08	–	5.4	–	–	–	–	–	–	–
QYr.uaf-7B.2	IWB10895	7B	103.21	0.45	–	–	–	–	–	–	4.07	–	–

[a] Putative name of the identified quantitative trait loci (QTL) from the marker–trait association, [b] SNP marker associated with all stage resistance to stripe rust against the different races, [c,d] Chromosome number and position of the associated SNP marker according to [25], [e] MAF: Minor allelic frequency, [f] *p*-value of marker–trait association.

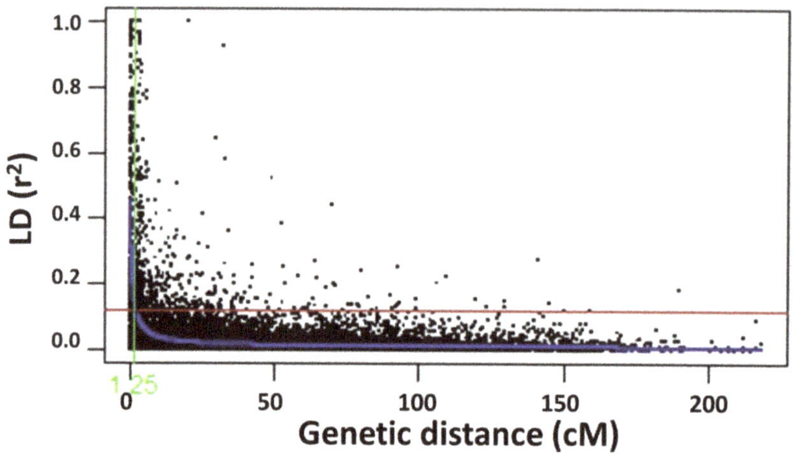

Figure 6. Scatter plot of the linkage disequilibrium (LD) decay with the critical r^2 value and the genetic chromosome distance (cm) for the whole genome. The red line shows the critical r^2 value i.e., 0.12.

2.4. *Marker–Trait Association at Adult Plant Stage*

Genome-wide association studies of 765 SNPs with the IT and SEV responses of 465 wheat genotypes against stripe rust at three different locations were performed. The association with the IT scores at different locations, namely PAK-UAF16, PAK-UAF17, USA-MTV18, USA-PCFS18 and with the BLUE (best linear unbiased estimator) value identifies twenty-two SNPs that were associated with stripe rust resistance at the adult plant stage. Significant SNPs was identified based on $p < 0.001$. A total of fifteen chromosomal regions (1A, 1B, 1D, 2A, 2B, 2D, 3A, 3B, 4B, 5A, 5B, 6A, 6D, 7A and 7B) were mapped with the identified SNPs. The highest number of SNPs with an IT score was identified in the USA environment, whereas with the SEV response, twenty-two SNPs were mapped in sixteen chromosomal regions (1A, 1B, 1D, 2A, 2B, 2D, 3A, 3B, 3D, 4B, 5A, 6A, 6B, 7A, 7B and 7D). Overall, thirty significant SNPs were identified with both IT and SEV response. The highest number of loci (eight) was mapped for the 2A chromosomes with the SNPs IWA11136, IWB12554 and IWA6798 at 9.41 cm, 143.6 cm and 150.1 cm respectively. The SNP name, chromosome, position and the probability of its association with IT and SEV, are discussed in detail in Table 3. Five chromosome regions (1D, 3D, 4B, 6D and 7A) were identified that were mapped with significant SNPs in the Pakistan environments with both the IT and SEV disease scores of the genotypes. Nine regions were identified with both the IT and SEV response in the environment USA-MTV18 and eight regions were mapped with significant SNP markers in the environment USA-PCFS18 (Table 3). In the present work, the SNP marker, IWB11136 (2A at 9.41 cm), was found to be linked with all the USA locations with both IT and SEV resistant scores of stripe rust. Manhattan and Q-Q plots showing the marker–trait association with the stripe rust resistance response IT and SEV scores, at all three locations and with the BLUE value, are provided in Supplementary Figures S3 and S4.

Table 3. SNPs associated with adult plant response (IT and SEV) at three locations with $p < 0.001$.

QTL [a]	SNP Id. [b]	Chr. [c]	Pos. [d] (cm)	MAF [e]	−log10(p) Based on IT [f] Score					−log10(p) Based on SEV [g] Score				
					PAK-UAF16	PAK-UAF17	USA-MTV18	USA-PCFS18	BLUE-IT	PAK-UAF16	PAK-UAF17	USA-MTV18	USA-PCFS18	BLUE-SEV
QYr.uaf-1A.2	IWB3662	1A	10.69	0.45	–	–	–	–	–	–	–	–	–	3.24
QYr.uaf-1A.3	IWA2541	1A	95.55	0.22	–	–	–	–	3.52	–	–	–	–	–
QYr.uaf-1A.4	IWA4080	1A	96.3	0.5	–	–	4.03	–	–	–	–	–	–	–
QYr.uaf-1B.2	IWA5278	1B	70.08	0.34	–	–	–	3.46	–	–	–	–	3.34	3.47
QYr.uaf-1D.3	IWA642	1D	67.72	0.32	–	–	–	3.16	–	–	–	–	–	–
QYr.uaf-1D.2	IWA4344	1D	90.3	0.21	3.22	–	–	–	3.27	3.08	–	–	–	–
QYr.uaf-2A.3	IWB11136	2A	9.41	0.36	–	–	4.71	3.25	3	–	–	4.25	3.03	3.23
QYr.uaf-2A.4	IWB12554	2A	143.65	0.28	–	–	–	–	–	–	–	3.04	–	–
QYr.uaf-2A.5	IWA6798	2A	150.11	0.4	–	–	–	–	–	–	–	3.29	–	–
QYr.uaf-2B.3	IWA3478	2B	130.62	0.31	–	–	5.19	–	–	–	–	5.7	–	3.48
QYr.uaf-2D.1	IWA5673	2D	82.82	0.16	–	–	3.32	–	–	–	–	3.32	–	–
QYr.uaf-3A.1	IWA288	3A	49.1	0.32	–	–	–	3.59	–	–	–	–	–	–
QYr.uaf-3A.2	IWB73771	3A	66.48	0.37	–	–	–	4.24	–	–	–	–	–	–
QYr.uaf-3A.3	IWB68593	3A	195.1	0.36	–	–	–	–	–	–	–	3.6	–	–
QYr.uaf-3B.1	IWB11085	3B	25.09	0.15	–	–	3.41	–	–	–	–	–	–	–
QYr.uaf-3B.2	IWB11270	3B	67.67	0.31	–	–	3.45	–	3.57	–	–	3.8	–	5.38
QYr.uaf-3B.3	IWB36652	3B	71.65	0.39	–	–	–	3.73	–	–	–	–	4.3	–
QYr.uaf-3D.1	IWA4725	3D	4.56	0.29	–	–	–	–	–	–	3.18	–	–	–
QYr.uaf-4B.4	IWB60835	4B	0	0.34	3.48	–	–	–	–	–	3.9	–	–	–
QYr.uaf-4B.2	IWA2031	4B	98.65	0.16	–	–	–	4.71	4.57	–	–	–	–	5.63
QYr.uaf-5A.1	IWB38719	5A	88.7	0.39	–	–	3.13	–	4.24	–	–	–	4.13	–
QYr.uaf-5A.2	IWA589	5A	123.21	0.25	–	–	–	3.12	4.87	–	–	–	8.34	7.07
QYr.uaf-5B.2	IWB9459	5B	110.56	0.25	–	–	4.2	–	–	–	–	–	–	–
QYr.uaf-6A.4	IWB29623	6A	40.6	0.32	–	–	3.23	–	–	–	–	4.25	–	–
QYr.uaf-6B.2	IWB26626	6B	113.67	0.39	–	–	–	–	–	–	–	–	–	5.32
QYr.uaf-6D.2	IWA6673	6D	9.47	0.38	3.05	–	–	–	–	–	–	–	–	–
QYr.uaf-7A.1	IWA4173	7A	218.7	0.48	–	–	–	–	3.56	3.15	–	–	–	–
QYr.uaf-7B.3	IWB22838	7B	73.79	0.22	–	–	–	3.58	–	–	–	–	6.3	–
QYr.uaf-7B.4	IWA6401	7B	77.73	0.34	–	–	–	–	–	–	–	–	–	5.12
QYr.uaf-7D.1	IWB26628	7D	144.96	0.44	–	–	–	–	–	–	–	–	–	3.33

[a] Putative name of identified quantitative trait loci (QTL) from the marker–trait association, [b] SNP marker associated with all-stage resistance to stripe rust against different races, [cd] Chromosome number and the position of the associated SNP marker according to [25], [e] MAF: Minor allelic frequency, [f] *p*-value of marker–trait association with the IT scores of all locations, [g] *p*-value of the marker–trait association with SEV scores of all locations. PAK-UAF: University of Agriculture, Faisalabad, Pakistan; MTV: Mount Vernon, USA; PCFS: Palouse conservation field station, USA; BLUE: Best linear unbiased estimator.

3. Discussion

3.1. Phenotypic Variation to Stripe Rust Response

The present study was conducted to study the spectrum of genetic diversity, as well as to map the resistance loci in 465 Pakistani spring wheat genotypes in response to stripe rust (*Puccinia striiformis* f. sp. *tritici*) at seedling and adult plant stage. Profiling with high-density SNP markers helps in the identification of genomic diversity, population structure among genotypes and marker–trait association to identify well-worn and novel resistance sources in germplasm. In previous reports, variable numbers of stripe rust races were used to compute the response against the specific race [19,26–28]. In the current study, a total of nine *Pst* races, including six from the USA and three of Pakistani origin were used to evaluate the genotypes response at the seedling stage also three different locations, tested for adult plant response (two years in Pakistan considered as one location and two locations in the USA).

In the current study, almost 50% intermediate response was observed at the seedling stage. Moreover, a high level of resistance (IT score = 0–3) response appeared at the adult plant stage and a low level of resistance phenotype appeared at the seedling stage. Similarly, a more susceptible response appeared at the seedling stage than at the adult plant stage [16,27–29]. Wheat genotypes were more susceptible to USA stripe rust races compared to Pakistani stripe rust races, where the resistant phenotype response was dominant. Pakistani wheat genotypes have long history of cultivation [30] before the emergence of modern wheat. This suggests that they have more interaction with *Pst* races, prevailing themselves as a major source of resistance to *Pst* [31].

Population structure, based on the Bayesian model, divides the 465 genotypes into three subpopulations (K = 3). Based on the genotype diversity, a different number of subpopulations were reported, as two subpopulations [16], three subpopulations [29], six subpopulations [32], seven subpopulation [31] and eight subpopulations [27]. Critical r^2 was used to estimate the extent of the LD decay with the line intersecting the smooth curve [19]. In present study, the critical r^2 value for the whole genome was 0.12 (for 765 SNPs) and we found LD decay at a confidence interval of 1.25, previously reported as 1.6 cm [31].

3.2. Genome-Wide Association Analysis with Stripe Rust Response

The major factors that influenced the association mapping were the population size, the germplasm choice and the marker density over whole genome [32]. Several association mapping studies were reported for stripe rust using different molecular markers as diversity arrays technology (DArT) [26,33,34], simple sequence repeats (SSR) [35,36] and SNP [19,27,29,37,38].

Marker–trait association at the seedling stage with nine different races identified twenty QTLs covering twelve chromosomes (1A, 1B, 1D, 2A, 2B, 4A, 4B, 5B, 6A, 6B, 6D and 7B) with $p < 0.0001$. The SNP loci *QYr.uaf-1D.2* associated with stripe rust resistance at the seedling stage was also found to be associated with rust resistance at the adult plant stage.

SNP loci *QYr.uaf-1A.1*, mapped on the short arm of 1A positioned at 10.69 cm, was found to be associated with resistance to all three Pakistani races (PK07-4, PK07-12 and PK08-2). Likewise, earlier it was reported that the position of SNP IWB3662 lay within the confidence interval of *YrEDWL-1AS*, associated with resistance to PSTv-14 and PSTv-37 in durum wheat [19] and *QYrid.ui-1A_Rio* Blanco [39]. Three SNP markers (IWB12795, IWB20633 and IWB56353) associated with the loci *QYr.tsw-1A* were positioned on the 1A chromosome had already been reported [16] in close proximity to currently identified loci that were linked with resistance to PSTv-4 in spring wheat germplasm. Stripe rust-resistant locus *QYr.uaf-1B.1* tagged with SNP IWB12258 in current research findings was found in close proximity of IWA1191 [28]. The Chromosomal position 1B reported with many *Yr*-associated genes as *Yr9, Yr10, Yr15, Yr34/Yr26, Yr29, Y64, Yr65, YrAlp* and *YrH52* [31]. Chromosome 1B locus *QYr.uaf-1B.1* was found in close proximity to *QYrcau-1BS_AQ24788-53* [40] from Chinese winter wheat and *Yr9* resistant gene [31]. In Pakistan, *Yr9* was first reported in 1994, and after that many cultivated varieties were developed with this gene, due to its linkage with other genes (*Lr26, Sr31,*

Pm8) and pleiotropic effect [41]. The successful translocation of the *Yr9* gene from rye and alongside high-yield potential, made *Yr9* a highly dominated gene in Pakistani germplasm [35]. Wheat varieties carrying this gene were highly cultivated during 1990s, but resistance breakdown after a few years made the resistant cultivars susceptible [42]. A major yellow rust epidemic was observed in 1995 with a 20% loss in the affected area in Pakistan. At that time, Pak81, also known as Veery#5 carrying *Yr9* gene, predominated. Two major cultivars, Pak81/Pirasabak 85 became susceptible during the period 1994–1995 due to the ineffectiveness of the *Yr9* gene and Inquilab 91 in 2002 due to the virulence occurrence of the *Yr27* gene [11].

Single stripe rust resistant QTL, *QYr.tsw-1D*, against the USA race Pstv-14, as reported by [16], was found in close proximity to SNP IWA7171 and IWA4344, positioned at 90.30 cm on the 1D chromosome. QTL *QYr.uaf-1D.1*, was linked with resistance to PK07-12 and PSTv-4 in Pakistan's spring wheat population. SNP loci *QYr.uaf-2B.1* (IWA2344) was positioned on the 2B at 96.99 cm, was found to be linked to all-stage resistance with the three stripe rust races PSTv-198, PSTv-51, PSTv-40, and was found in the confidence interval of *QYr.ucw-2B_UC1110* [43], *QYr.inra-2B.1_Camp* Remy [44] and *QYr.cim-2BS_Francolin* [45]. It was found that the second identified loci of chromosome 2B (*QYr.uaf-2B.2*) positioned at 113.86 cm was also associated with the three rust races PSTv-198, PSTv-51 and PSTv-4. The locus (*QYr.uaf-2B.2*) was found in the confidence interval of *QYraq.cau-2BL_Aquileja* [46] and *Yr53* [47].

Three QTLs *QYr.uaf-4B.1, QYr.uaf-4B.2, QYr.uaf-4B.3* at chromosome 4B were found in close proximity to *QYr.jic-4B_Alcedo* [48], *QYr.jic-4B_Guardian* [49], *QYr.vt-4BL_VA00W-38* [50] and *YrEDWL-4BL*, in novel durum wheat linked to the stripe rust race PSTv-14 and PSTv-51 [19]. One QTL *QYr.uaf-5B.1* on the 5B chromosome positioned at 56.6 cm, was found within the confidence interval of *QYr.ufs-5B_Cappelle-Desprez* [51]. Two SNP markers (IWA3463, IWB25252) linked the QTL *QYr.uaf-6A* to stripe rust resistance, and PK07-12 was found in close proximity to the previously reported QTL *QYr.cim-6AL_Francolin* [45]. *QYr.uaf-7B.1* and *QYr.uaf-7B.2* were found to be linked with resistance to the stripe rust races PK07-12 and PSTv-40, respectively, and were found to be linked to the *QYr.sun-7B_Kukri* [52], *Yr39* gene [53] and near to the marker IWA312 (76.1 cm) that was linked with resistance to the *Pst* races Pstv-37 and Pstv-40 [28].

3.3. Genome-Wide Association Analysis for Adult Plant Response

Genome-wide association (GWA) analysis of stripe rust using high-density SNP markers was carried out across the three locations. In total, thirty loci were identified that were significantly ($p < 0.001$) associated with the infection type (IT) and disease severity (SEV) score of stripe rust in multi-environments. These thirty loci were present in eighteen genomic regions, namely 1A, 1B, 1D, 2A, 2B, 2D, 3A, 3B, 3D, 4B, 5A, 5B, 6A, 6B, 6D, 7A, 7B and 7D. In the current study, three loci were mapped on the chromosomes 1A, 2A, 3A and 3B. Two QTLs were identified on each chromosome 1D, 4B, 5A and 7B. One QTL was identified on each chromosome 1B, 2B, 2D, 3D, 5B, 6A, 6B, 6D, 7A and 7D.

In the present research work, the QTL, *QYr.uaf-2A.3* (linked SNP IWB11136), identified to be positioned at 9.41 cm on the 2A chromosome, was significantly associated with six locations including the IT (USA-MTV18, USA-PCFS18, BLUE-IT) and SEV response (USA-MTV18, USA-PCFS18 and BLUE-SEV). The same SNP marker IWB11136 (*QYr.tsw-2A.3*) that was identified previously was also found significantly associated with all the tested locations for the stripe rust response [16]. This locus was also found within the confidence interval of *QYr.tam-2AS* [54] from the hard winter wheat TAM111 and *Yr17* genes [55]. The *Yr17* gene was developed by 2NS–2AS locus translocation (25 to 38 cm) from *T. ventricosum* (2NS), a famous wild *Triticeae* species to 2AS of wheat chromosome [55,56]. The 2NS–AS translocation was first carried out in winter bread wheat VPM1 and afterwards in California and Washington, where many winter wheat cultivars were developed such as Madsen, Hyak and Expresso (spring wheat). Furthermore, the 2B QTL *QYr.uaf-2B.3* positioned at 130.6 cm was linked with both the IT and SEV score at the USA location MTV, and was linked with the *Yr* genes *Yr53* and *Yr43* [47]. At chromosome 3B, two SNP markers (IWB11270 and IWB36652) were associated with the QTLs

QYr.uaf-3B.2 and *QTL QYr.uaf-3B.3*, and were aligned with both the IT and SEV scores of USA-MTV, USA-PCFS18 and the BLUE value, which was found in close proximity to *QYr.cim-3B_Pastor* and *QYr.inra-3Bcentr_Renan* [57,58].

Presently, two QTLs *QYr.uaf-3B.2* (IWB11270) were positioned at 67.67 cm on 3B and *QYr.uaf-5A.2* (IWA589) was positioned at 123.21 cm on chromosome 5A, which was significantly associated ($p < 0.001$) with stripe rust resistance at four different locations. Locus *QYr.uaf-3B.2* was found in the confidence interval of *QYr.cim-3B_Pastor* [58], whereas the QTL locus *QYrEDWL-5AL.2* reported in Ethiopian durum wheat was in close proximity to *QYr.uaf-5A.2* [19]. *Qyr.wpg-6B.1* (IWA7257) in winter wheat and *QYr.cim-6BL_Pastor* was found in close proximity *to QYr.uaf-6B.2* (IWB26626) positioned at 113.67 cm on the chromosome 6B [37,57]. Five SNPs loci identified in the current study were linked with resistance to either the IT or SEV response of the stripe rust at three different locations. These included *QYr.uaf-1B.2*, *QYr.uaf-1D.2*, *QYr.uaf-2B.3*, *QYr.uaf-4B.2* and *QYr.uaf-5A.1*. The identified resistance source can be utilized as the breeding line for enhancing wheat resistance against disease.

4. Materials and Methods

4.1. Collection of Genetic Material

Four hundred and sixty-five (465) genotypes of bread wheat (*Triticum aestivum* L.) were collected from the Wheat Research Institute, Ayub Agricultural Research Institute (AARI), Pakistan, Faisalabad.

4.2. Field Based Resistance Screening to Puccinia striiformis (Pst)

The panel of 465 spring wheat genotypes was tested for stripe rust response under field conditions. All genotypes were grown at the experimental area of the Centre of Agricultural Biochemistry and Biotechnology (CABB), University of Agriculture, Faisalabad (31°26′ N, 73°6′ E) for two consecutive years 2015–2016 and 2016–2017. Each entry was planted in a 1 m long row by keeping a row-to-row distance of 36 cm and the sowing was done by planting two seeds per hole and maintaining 8 cm plant-to-plant distance. The highly susceptible variety of stripe rust i.e., Morocco was used as the susceptible check in the field experiments in Pakistan to increase the disease pressure. All 465 Pakistani wheat genotypes were also sown at two different locations in the USA at the Palouse conservation field station (PCFS), Pullman, WA (46°43′59″ N; 117°10′19″ W) and Mount Vernon (MTV), WA (48°25′16″ N; 122A°20′2″ W) in the year 2018 for the evaluation of stripe rust resistance under natural disease conditions. Spring wheat Avocet susceptible (AvS) was used as the susceptible check to increase the disease pressure at both locations planted after every twenty lines in the USA. Each line of the genotype was grown up in a 0.5 m long and a 0.3 m wide row. All standard agronomic practices were followed for the crop production.

The evaluation of *Pst* at the adult plant stage was done in a field with the natural conditions of disease epidemics at all three locations (two years in Pakistan considered as one location and two locations in USA). In the field, the data scoring was done by visualizing the impact of the disease on the flag leaf of the susceptible check that the IT score, which ranged from 7–9 and the disease severity score which ranged from 70–100% [19]. One location was in Pakistan at the University of Agriculture, Faisalabad (UAF) for two consecutive years 2015–2016 (PAK-UAF16) and 2016–2017 (PAK-UAF17). The other two locations data was recorded at the USA for year 2018 at Pullman, WA (USA-PCFS18) and at Mount Vernon, WA (USA-MTV18). Infection type (IT) and disease severity (SEV) were two disease phenotype scores that were recorded for the *Puccinia striiformis* (*Pst*) infection in the field. The IT score based on the 0–9 scale is explained in Supplementary Table S1 whereas the SEV recorded as the % age area of the flag leaf covered with disease and was scored from 0 to 100%.

4.3. Greenhouse-Based Resistance Screening to *Puccinia striiformis* (*Pst*)

The seedling response of 465 wheat genotypes was evaluated with isolates of six *Pst* USA races including PSTv-37, PSTv-198, PSTv-51, PSTv-40, PSTv-14, PSTv-4 [25] and three isolates of Pakistani races PK07-4, PK07-12, PK08-2 [59] under controlled greenhouse conditions. The virulence and avirulence formulae of the stripe rust race isolates are provided in the Supplementary Tables S2 and S3. All the stripe rust races were collected from the USDA, Wheat Health, Genetics and Quality Research Unit, Pullman, WA. Four to five seeds of each genotype were planted per well in a 96 wells tray filled with the number 1 sunshine mix growing medium (Sungro Horticulture, Bellevue, WA, USA). Trays were regularly watered and kept in a rust free greenhouse at 20 °C with 50% relative humidity (RH). The inoculation was done when the seedling reached the 2-leaf stage after approximately 9–10 days of sowing. The inoculation of each race was done by mixing the rust urediniospores with talcum powder. Inoculated plants were incubated in a dark dew chamber for 24 h at 10 °C and 100% relative humidity and then moved to the greenhouse having the 8 °C day, 16 °C night temperature and 16 h photoperiod. The reaction to *Puccinia striiformis* f. sp. *tritici* was scored after 18 to 20 days of inoculation, using the 0–9 scale for the infection type (IT) [60,61]. Based on the infection type, the genotypes were grouped as 0–3 = resistant; 4–6 = intermediate; 7–9 = susceptible. The scale of the infection type (IT) disease score is discussed in Supplementary Table S1.

4.4. Statistical Analysis

The range, mean, standard deviation, coefficient of determination (R^2) were scored within and across the environments using JMP Genomics 15.1.0 (SAS Institute Inc., Cary, NC, USA, 2007). Broad sense heritability (H^2) was calculated by the variance component obtained from REML (random effects model) computed using JMP software. The BLUE (best linear unbiased estimator) value for the IT and SEV scores of all environments was calculated using the PROC MIXED procedure in SAS v9.3 (SAS Institute Inc., Cary, NC, USA, 2007) considering the genotype as a fixed effect [16].

4.5. DNA Extraction, SNP Genotyping

Wheat genotypes sown in greenhouse and young leaves used for DNA extraction using a robotic system of oKtopureTM at the Western regional small grain genotyping laboratory (WRSGGL) (Washington State University, Pullman, WA, USA) [62] for SNP genotyping against stripe rust. Targeted amplicon sequencing (TAS) for stripe rust resistant genes was done using NextSeq® 500 (Illumina, Inc., Pullman, WA, USA). Genotypic calling and removing monomorphic as well as low quality SNPs, was carried out using GenomeStudio Software v2011.1. (Illumina, Inc., Pullman, WA, USA) to call bi-allelic SNPs AA, AB and BB, for this default clustering algorithm, was used. A total 1500 SNPs were yielded and subjected to TASSEL (trait analysis by association, evolution and linkage) software v.5.2.61 [63] to remove the SNPs with minor allelic frequency MAF < 0.05 and to made kinship matrix. A total 765 high-quality SNPs were selected and projected on to a consensus map of hexaploid wheat to order them based on the chromosome position [25]: these 765 SNPs were used for association analysis.

4.6. Population Structure and Linkage Disequilibrium (LD)

Major genetic structure of the selected genotypes was determined using 150 SNP markers with inter marker distances >5 cm from each other using the Bayesian model-based clustering algorithm in STRUCTURE v2.3.4 [64]. The number of subpopulations (K) was estimated by the running simulation from burn-in 10,000 iteration to 100,000 Monte Carlo Markov Chain (MCMC) replicates. K was run from 1–10 times and 10 independent runs were set for each run. The STRUCTURE results were visualized to determine the value of K (subpopulation) based on the ad hoc criterion by using the STRUCTURE HARVESTER [65,66].

The measurement of the linkage disequilibrium between the pairs of the SNP marker was estimated using the program TASSEL (v5.2.61). The LD parameters D′ and r^2 among the loci and comparison-wise significance was computed by 1000 permutations. The critical r^2 value was determined by taking the 95th percentile of the unlinked markers [67]. The scatter plot among the r^2 and distance on chromosome, of all significant ($p < 0.001$) pairwise combinations, were used to fit the locally weighted polynomial regression curve (LOESS) to estimate the extent of the LD decay in the R environment [16] using the critical r^2 value.

4.7. Genome-Wide Association (GWA) Analysis

Integrated mixed-model (MLM) method for association mapping, which accounts for multiple levels of relatedness, was used to narrate the genetic polymorphism to important phenotypic variation in specific traits [68]. An association test was performed using both (1) the Genome Association and Prediction Integrated Tool (GAPIT) [69] and (2) fixed and random model circulating probability unification (FarmCPU) package [70] implemented in R software v.3.6.1 (https://www.r-project.org/). The MLM model utilized trait data, genotype data K (kinship) and PCA (principle component analysis) to find the marker–trait association. The model comparison was done to select the best model for the marker–trait association (MTA) with each trait as K + P (kinship and principal component) [44] and K + Q (kinship and population structure) [68]. The final results were analyzed by FarmCPU, selecting the model (K + Q) based on their respective Q-Q plots. Significant MTA was described based on the *p*-value. Markers with $p < 0.0001$ were considered significant for the seedling test and $p < 0.001$ for the field experiment. Marker–trait association was performed with all nine rust races data, scored at the seedling stage and with the field data of all the environments separately for the IT score, the disease severity score and with the best linear unbiased estimator (BLUE) value using the FarmCPU package implemented in R software v.3.6.1.

5. Conclusions

GWAS provides a good outline of the distribution and frequency of resistance genes over the whole world subpopulation. This spring wheat Pakistani germplasm was proved an efficient source of phenotypic diversification to combat stripe rust infection for both seedling and field experiments and to determine the yield QTLs related to the yield components. The genotypes possessing a higher fraction of resistance loci of stripe rust divulged themselves as a parental breeding line and hence can increase the breeding efficiency for stripe rust resistance. The present research findings can be exploited by wheat breeders to increase the resistant capability and yield potential of their cultivars by marker-assisted selection breeding.

Supplementary Materials: The following are available online at http://www.mdpi.com/2223-7747/9/9/1056/s1. Table S1: Stripe rust infection type (IT) recording scale. Table S2: Virulence and avirulence formula for the USA stripe rust races. Table S3: Virulence and avirulence formula for the Pakistan stripe rust. Table S4: ΔK value based on LnP(K) for 465 wheat genotypes. Figures S1 and S3: Manhattan representing the number of chromosome and their associated SNPs. Figures S2 and S4: Q-Q plot representing the number of chromosome and their associated SNPs.

Author Contributions: M.H. and F.S.A. were involved in planning and executing and manuscript writing; B.S., M.A.Z. contributed to the field data analysis. All authors have read and agreed to the published version of the manuscript.

Funding: This research was funded by Higher Education Commision, Pakistan 4616.

Acknowledgments: We pay our gratitude to Xianming Chen from WSU (Washington State University) for his assistance in genotyping and supporting greenhouse experiments and Deven R. See from WSU for providing computational support in analyzing genotypic data. We are also thankful to AARI, FSD for providing wheat seeds. This work was supported by the Higher Education Commission (HEC), Govt. of Pakistan.

Conflicts of Interest: The authors declare that they have no conflict of interest in the publication.

Abbreviations

Pst	*Puccinia striiformis* f. sp. *tritici*
MH	Million hectare
Yr	Yellow rust
USA	United State of America
APR	Adult plant resistance
MAS	Marker assisted selection
LD	Linkage disequilibrium
AM	Association mapping
GWAS	Genome-wide association mapping
QTL	Quantitative trait loci
SNP	Single nucleotide polymorphism
SEV	Severity
DArT	Diversity Arrays Technology
SSR	Simple sequence repeats
PCFS	Palouse conservation field station
MTV	Mount Vernon
UAF	University of Agriculture, Faisalabad
AvS	Avocet susceptible
MAF	Minor allelic frequency
TAS	Targeted amplicon sequencing
MLM	Mixed linear model
MTA	Marker–trait association
TASSEL	Trait analysis by association, evolution and linkage

References

1. Ogbonnaya, F.; Abdalla, O.; Mujeeb-Kazi, A.; Kazi, A.G.; Xu, S.S.; Gosman, N.; Lagudah, E.; Bonnett, D.; Sorrells, M.E.; Tsujimoto, H. Synthetic Hexaploids: Harnessing Species of the Primary Gene Pool for Wheat Improvement. *Plant Breed.* **2013**, *37*, 35–122. [CrossRef]
2. Farooq, J.; Khaliq, I.; Akbar, M.; Petrescu-mag, I.V.; Hussain, M. Genetic analysis of some grain yield and its attributes at high temperature stress in wheat (*Triticum aestivum* L.). *Ann. RSCB* **2015**, *19*, 71–81. [CrossRef]
3. Bhanupriya, B.; Satyanarayana, N.H.; Mukherjee, S.; Sarkar, K.K. Genetic diversity of wheat genotypes based on principal component analysis in Gangetic alluvial soil of West Bengal. *J. Crop. Weed.* **2014**, *10*, 104–107.
4. Wellings, C. Global status of stripe rust: A review of historical and current threats. *Euphytica* **2011**, *179*, 129–141. [CrossRef]
5. Mboup, M.; Leconte, M.; Gautier, A.; Wan, A.; Chen, W.; De Vallavieille-Pope, C.; Enjalbert, J. Evidence of genetic recombination in wheat yellow rust populations of a Chinese oversummering area. *Fungal Genet. Biol.* **2009**, *46*, 299–307. [CrossRef] [PubMed]
6. Hovmøller, M.; Walter, S.; Justesen, A.F. Escalating Threat of Wheat Rusts. *Science* **2010**, *329*, 369. [CrossRef]
7. Prescott, J.M.; Burnett, P.A.; Saari, E.E.; Ransom, J.K.; Bowman, J.; De Milliano, W.; Singh, R.P.; Abeyo Bekele, G. *Wheat Diseases and Pests: A Guide for Field Identification*; CIMMYT: Ciudad de México, Mexico, 1986; p. 135.
8. Ma, H.; Singh, R.P. Expression of adult resistance to yellow rust at different growth stages of wheat. *Plant Dis.* **1996**, *80*, 375–379. [CrossRef]
9. Statista. 2019. Available online: https://www.statista.com/statistics/237912/global-top-wheat-producing-countries/ (accessed on 5 November 2019).
10. Singh, R.P.; William, H.M.; Huerta-Espino, J.; Rosewarne, G. Wheat rust in Asia: Meeting the challenges with old and new technologies. In Proceedings of the 4th International Crop Science Congress, Brisbane, QLD, Australia, 26 September–1 October 2004; Volume 26.
11. Solh, M.; Nazari, K.; Tadesse, W.; Wellings, C.R. The growing threat of stripe rust worldwide. In Proceedings of the Borlaug Global Rust Initiative (BGRI) Conference, Beijing, China, 1–4 September 2012; pp. 1–4.
12. Duveiller, E.; Singh, R.P.; Nicol, J.M. The challenges of maintaining wheat productivity: Pests, diseases, and potential epidemics. *Euphytica* **2007**, *157*, 417–430. [CrossRef]

13. Murray, G.M.; Brennan, J.P. *The Current and Potential Costs from Diseases of Wheat in Australia*; Grains Research and Development Corporation: Canberra, Australia, 2009; p. 70.
14. Rahmatov, M. Sources of resistance to yellow rust and stem rust in wheat-alien introgressions. In Introductory paper at the Faculty of Landscape Planning. *Hortic. Agric. Sci.* **2013**, *3*, 1–64.
15. Chen, X. Epidemiology and control of stripe rust [Puccinia striiformis f. sp. tritici] on wheat. *Can. J. Plant Pathol.* **2005**, *27*, 314–337. [CrossRef]
16. Godoy, J.G.; Rynearson, S.; Chen, X.M.; Pumphrey, M. Genome-Wide Association Mapping of Loci for Resistance to Stripe Rust in North American Elite Spring Wheat Germplasm. *Phytopathology* **2018**, *108*, 234–245. [CrossRef] [PubMed]
17. Roelfs, A.P.; Bushnell, W.R. *The Cereal Rusts: Diseases, Distribution, Epidemiology and Control*; Academic Press: Orlando, FL, USA; London, UK, 1985; Volume 2, p. 606.
18. Kolmer, J. Tracking wheat rust on a continental scale. *Curr. Opin. Plant Biol.* **2005**, *8*, 441–449. [CrossRef] [PubMed]
19. Liu, W.; Maccaferri, M.; Rynearson, S.; Letta, T.; Zegeye, H.; Tuberosa, R.; Chen, X.; Pumphrey, M. Novel Sources of Stripe Rust Resistance Identified by Genome-Wide Association Mapping in Ethiopian Durum Wheat (*Triticum turgidum* ssp. durum). *Front. Plant Sci.* **2017**, *8*, 774. [CrossRef] [PubMed]
20. Line, R.F. Stripe rust of wheat and barley in North America: A retrospective historical review. *Annu. Rev. Phytopathol.* **2002**, *40*, 75–118. [CrossRef] [PubMed]
21. Singh, R.P.; Huerta-Espino, J.; William, H.M. Genetics and breeding for durable resistance to leaf and stripe rusts in wheat. *Turk. J. Agric.* **2005**, *29*, 121–127.
22. Flint-Garcia, S.; Thornsberry, J.M.; Buckler, E.S.; Iv, B. Structure of Linkage Disequilibrium in Plants. *Annu. Rev. Plant Biol.* **2003**, *54*, 357–374. [CrossRef]
23. Zondervan, K.T.; Cardon, L.R. The complex interplay among factors that influence allelic association. *Nat. Rev. Genet.* **2004**, *5*, 89–100. [CrossRef]
24. Yu, J.; Buckler, E.S. Genetic association mapping and genome organization of maize. *Curr. Opin. Biotechnol.* **2006**, *17*, 155–160. [CrossRef]
25. Wang, S.; Wong, D.; Forrest, K.; Allen, A.; Chao, S.; Huang, B.E.; Maccaferri, M.; Salvi, S.; Milner, S.G.; Cattivelli, L.; et al. Characterization of polyploid wheat genomic diversity using a high-density 90 000 single nucleotide polymorphism array. *Plant. Biotechnol. J.* **2014**, *12*, 787–796. [CrossRef]
26. Tadesse, W.; Ogbonnaya, F.; Jighly, A.; Nazari, K.; Rajaram, S.; Baum, M. Association Mapping of Resistance to Yellow Rust in Winter Wheat Cultivars and Elite Genotypes. *Crop. Sci.* **2014**, *54*, 607–616. [CrossRef]
27. Zegeye, H.; Rasheed, A.; Makdis, F.; Badebo, A.; Ogbonnaya, F.C. Genome-wide association mapping for seedling and adult plant resistance to stripe rust in synthetic hexaploid wheat. *PLoS ONE* **2014**, *9*, e0105593. [CrossRef] [PubMed]
28. Muleta, K.T.; Bulli, P.; Rynearson, S.; Chen, X.; Pumphrey, M. Loci associated with resistance to stripe rust (*Puccinia striiformis* f. sp. tritici) in a core collection of spring wheat (*Triticum aestivum*). *PLoS ONE* **2017**, *12*, e0179087. [CrossRef] [PubMed]
29. Liu, W.; Maccaferri, M.; Chen, X.; Laghetti, G.; Pignone, D.; Pumphrey, M.; Tuberosa, R. Genome-wide association mapping reveals a rich genetic architecture of stripe rust resistance loci in emmer wheat (*Triticum turgidum* ssp. dicoccum). *Theor. Appl. Genet.* **2017**, *130*, 2249–2270. [CrossRef]
30. Tengberg, M. Crop husbandry at Miri Qalat, Makran, SW Pakistan (4000–2000 BC). *Veg. Hist. Archeobot.* **1999**, *8*, 3–12. [CrossRef]
31. Maccaferri, M.; Zhang, J.; Bulli, P.; Abate, Z.; Chao, S.; Cantu, D.; Bossolini, E.; Chen, X.; Pumphrey, M.; Dubcovsky, J. A Genome-Wide Association Study of Resistance to Stripe Rust (*Puccinia striiformisf.* sp.tritici) in a Worldwide Collection of Hexaploid Spring Wheat (*Triticum aestivum* L.). *G3 Genes Genomes Genet.* **2015**, *5*, 449–465. [CrossRef]
32. Joukhadar, R.; El-Bouhssini, M.; Jighly, A.; Ogbonnaya, F. Genome-wide association mapping for five major pest resistances in wheat. *Mol. Breed.* **2013**, *32*, 943–960. [CrossRef]
33. Jighly, A.; Oyiga, B.C.; Makdis, F.; Nazari, K.; Youssef, O.; Tadesse, W.; Abdalla, O.; Ogbonnaya, F. Genome-wide DArT and SNP scan for QTL associated with resistance to stripe rust (*Puccinia striiformis* f. sp. tritici) in elite ICARDA wheat (*Triticum aestivum* L.) germplasm. *Theor. Appl. Genet.* **2015**, *128*, 1277–1295. [CrossRef]
34. Singh, A.K.; Knox, R.E.; Depauw, R.M.; Cuthbert, R.D.; Campbell, H.L.; Shorter, S.; Bhavani, S.; Cuthbert, R. Stripe rust and leaf rust resistance QTL mapping, epistatic interactions, and co-localization with stem rust resistance loci in spring wheat evaluated over three continents. *Theor. Appl. Genet.* **2014**, *127*, 2465–2477. [CrossRef]

35. Tabassum, S.; Ashraf, M.; Chen, X.M. Evaluation of Pakistani wheat germplasm for stripe rust resistance using molecular marker. *Sci. China Life Sci.* **2010**, *53*, 1–12. [CrossRef]
36. Akfirat, F.S.; Ertugrul, F.; Hasancebi, S.; Aydin, Y.; Akan, K.; Mert, Z.; Cakir, M.; Uncuoglu, A.A. Chromosomal location of genomic SSR markers associated with yellow rust resistance in Turkish bread wheat (*Triticum aestivum* L.). *J. Genet.* **2013**, *92*, 233–240. [CrossRef]
37. Naruoka, Y.; Garland-Campbell, K.A.; Carter, A.H.; Garland-Campbell, K.A. Genome-wide association mapping for stripe rust (*Puccinia striiformis* F. sp. tritici) in US Pacific Northwest winter wheat (*Triticum aestivum* L.). *Theor. Appl. Genet.* **2015**, *128*, 1083–1101. [CrossRef]
38. Shewaye, Y.; Taddesse, W. Association Mapping of Seedling and Adult Plant Resistance for Stripe Rust Resistance in Spring Bread Wheat (*Triticum aestivum* L.). *Adv. Crop. Sci. Technol.* **2018**, *6*, 1–10. [CrossRef]
39. Chen, J.; Chu, C.; Souza, E.J.; Guttieri, M.J.; Chen, X.; Xu, S.; Hole, D.; Zemetra, R. Genome-wide identification of QTL conferring high-temperature adult-plant (HTAP) resistance to stripe rust (*Puccinia striiformis* f. sp. tritici) in wheat. *Mol. Breed.* **2011**, *29*, 791–800. [CrossRef]
40. Quan, W.; Hou, G.; Chen, J.; Du, Z.; Lin, F.; Guo, Y.; Liu, S.; Zhang, Z. Mapping of QTL lengthening the latent period of *Puccinia striiformis* in winter wheat at the tillering growth stage. *Eur. J. Plant. Pathol.* **2013**, *136*, 715–727. [CrossRef]
41. Saari, E.E. The current stripe rust status in south Asia. In *Proceedings of Wheat Research Report of National Wheat Research Program*; National Wheat Research Program: Bhairahawa, Nepal, 1995; p. 12.
42. Yahyaoui, A.; Abdalla, O.; Morgonouv, H.K.; Torabi, M.; Cetin, L.; Saidov, L.M. Occurrence of stripe rust and effectiveness of resistance genes in Western and Central Asia and the Caucasus (WCAC). In Proceedings of the ASA-CSSA-SSSA International Annual Meetings, Seattle, WA, USA, 31 October–4 November 2004.
43. Lowe, I.; Cantu, D.; Dubcovsky, J. Durable resistance to the wheat rusts: Integrating systems biology and traditional phenotype-based research methods to guide the deployment of resistance genes. *Euphytica* **2010**, *179*, 69–79. [CrossRef] [PubMed]
44. Mallard, S.; Gaudet, D.; Aldeia, A.; Abelard, C.; Besnard, A.L.; Sourdille, P.; Dedryver, F. Genetic analysis of durable resistance to yellow rust in bread wheat. *Theor. Appl. Genet.* **2005**, *110*, 1401–1409. [CrossRef] [PubMed]
45. Lan, C.; Rosewarne, G.M.; Singh, R.P.; Herrera-Foessel, S.A.; Huerta-Espino, J.; Basnet, B.; Zhang, Y.; Yang, E. QTL characterization of resistance to leaf rust and stripe rust in the spring wheat line Francolin# 1. *Mol. Breed.* **2014**, *34*, 789–803. [CrossRef]
46. Guo, Q.; Zhang, Z.J.; Xu, Y.B.; Li, G.; Feng, J.; Zhou, Y. Quantitative Trait Loci for High-Temperature Adult-Plant and Slow-Rusting Resistance to *Puccinia striiformis* f. sp. tritici in Wheat Cultivars. *Phytopathology* **2008**, *98*, 803–809. [CrossRef]
47. Xu, L.S.; Wang, M.N.; Cheng, P.; Kang, Z.S.; Hulbert, S.H.; Chen, X.M. Molecular mapping of Yr53, a new gene for stripe rust resistance in durum wheat accession PI 480148 and its transfer to common wheat. *Theor. Appl. Genet.* **2013**, *126*, 523–533. [CrossRef]
48. Jagger, L.J.; Newell, C.; Berry, S.T.; MacCormack, R.; Boyd, L.A. The genetic characterisation of stripe rust resistance in the German wheat cultivar Alcedo. *Theor. Appl. Genet.* **2010**, *122*, 723–733. [CrossRef] [PubMed]
49. Melichar, J.P.E.; Berry, S.; Newell, C.; MacCormack, R.; Boyd, L.A. QTL identification and microphenotype characterisation of the developmentally regulated yellow rust resistance in the UK wheat cultivar Guardian. *Theor. Appl. Genet.* **2008**, *117*, 391–399. [CrossRef] [PubMed]
50. Christopher, M.D.; Liu, S.; Hall, M.D.; Marshall, D.S.; Fountain, M.O.; Johnson, J.W.; Milus, E.A.; Garland-Campbell, K.A.; Chen, X.; Griffey, C.A. Identification and Mapping of Adult Plant Stripe Rust Resistance in Soft Red Winter Wheat VA00W-38. *Crop. Sci.* **2013**, *53*, 871–879. [CrossRef]
51. Agenbag, G.M.; Pretorius, Z.A.; Boyd, L.A.; Bender, C.M.; Prins, R. Identification of adult plant resistance to stripe rust in the wheat cultivar Cappelle-Desprez. *Theor. Appl. Genet.* **2012**, *125*, 109–120. [CrossRef] [PubMed]
52. Bariana, H.; Bansal, U.; Schmidt, A.; Lehmensiek, A.; Kaur, J.; Miah, H.; Howes, N.; McIntyre, C.L. Molecular mapping of adult plant stripe rust resistance in wheat and identification of pyramided QTL genotypes. *Euphytica* **2010**, *176*, 251–260. [CrossRef]
53. Lin, F.; Chen, X.M. Genetics and molecular mapping of genes for race-specific all-stage resistance and non-race-specific high-temperature adult-plant resistance to stripe rust in spring wheat cultivar Alpowa. *Theor. Appl. Genet.* **2007**, *114*, 1277–1287. [CrossRef]

54. Basnet, B.R.; Ibrahim, A.; Chen, X.; Singh, R.P.; Mason, E.R.; Bowden, R.L. Molecular mapping of stripe rust resistance in hard red winter wheat TAM 111 adapted to the US High Plains. *Crop. Sci.* **2014**, *54*, 1361–1373. [CrossRef]
55. Helguera, M.; Khan, I.A.; Kolmer, J.; Lijavetzky, D.; Zhong-Qi, L.; Dubcovsky, J. PCR Assays for the Lr37-Yr17-Sr38 Cluster of Rust Resistance Genes and Their Use to Develop Isogenic Hard Red Spring Wheat Lines. *Crop. Sci.* **2003**, *43*, 1839–1847. [CrossRef]
56. Friebe, B.; Badaeva, E.D.; Gill, B.S.; Tuleen, N.A. Cytogenetic identification of Triticum peregrinum chromosomes added to common wheat. *Genome* **1996**, *39*, 272–276. [CrossRef]
57. Rosewarne, G.M.; Singh, R.P.; Huerta-Espino, J.; Herrera-Foessel, S.A.; Forrest, K.L.; Hayden, M.; Rebetzke, G.J. Analysis of leaf and stripe rust severities reveals pathotype changes and multiple minor QTLs associated with resistance in an Avocet × Pastor wheat population. *Theor. Appl. Genet.* **2012**, *124*, 1283–1294. [CrossRef]
58. Dedryver, F.; Paillard, S.; Mallard, S.; Robert, O.; Trottet, M.; Negre, S.; Verplancke, G.; Jahier, J. Characterization of genetic components involved in durable resistance to stripe rust in the bread wheat "Renan". *Phytopathology* **2009**, *99*, 968–973. [CrossRef]
59. Sharma-Poudyal, D.; Chen, X.M.; Wan, A.M.; Zhan, G.M.; Kang, Z.S.; Cao, S.Q.; Jin, S.L.; Morgounov, A.; Akin, B.; Mert, Z.; et al. Virulence Characterization of International Collections of the Wheat Stripe Rust Pathogen, *Puccinia striiformis* f. sp. tritici. *Plant. Dis.* **2013**, *97*, 379–386. [CrossRef] [PubMed]
60. Line, R.F.; Qayoum, A. *Virulence, Aggressiveness, Evolution and Distribution of Races of Puccinia Striiformis (The Cause of Stripe Rust of Wheat) in North America*; Technical Bulletin; United State Department of Agriculture, Agricultural Research Service: Washington, DC, USA, 1992; pp. 1968–1987.
61. McNeal, F.H.; Konzak, C.F.; Smith, E.P.; Tate, W.S.; Russell, T.S. A uniform system for recording and processing cereal research data US Dept. *Agric. Res. Serv. ARS* **1971**, *34*, 121–143.
62. Jarana, A.P.; Carandang, S.L.; Tandayu, E.S.; Cariño, G.A.; Sanchez, G.S.B.; Llantada, K.J. High-throughput sample collection and DNA extraction using PlantTrak and oKtopure systems at IRRI's [International Rice Research Institute] genotyping services laboratory. *Philipp. J. Crop. Sci. (Philipp.)* **2015**, *40*, 86.
63. Bradbury, P.J.; Zhang, Z.; Kroon, D.E.; Casstevens, T.; Ramdoss, Y.; Buckler, E.S. TASSEL: Software for association mapping of complex traits in diverse samples. *Bioinformatics* **2007**, *23*, 2633–2635. [CrossRef] [PubMed]
64. Pritchard, J.K.; Stephens, M.; Rosenberg, N.A.; Donnelly, P. Association Mapping in Structured Populations. *Am. J. Hum. Genet.* **2000**, *67*, 170–181. [CrossRef] [PubMed]
65. Evanno, G.; Regnaut, S.; Goudet, J. Detecting the number of clusters of individuals using the software structure: A simulation study. *Mol. Ecol.* **2005**, *14*, 2611–2620. [CrossRef]
66. Earl, D.; Vonholdt, B.M. STRUCTURE HARVESTER: A website and program for visualizing STRUCTURE output and implementing the Evanno method. *Conserv. Genet. Resour.* **2011**, *4*, 359–361. [CrossRef]
67. Breseghello, F.; Sorrells, M.E. Association Mapping of Kernel Size and Milling Quality in Wheat (*Triticum aestivum* L.) Cultivars. *Genetics* **2005**, *172*, 1165–1177. [CrossRef]
68. Yu, J.; Pressoir, G.; Briggs, W.H.; Bi, I.V.; Yamasaki, M.; Doebley, J.F.; McMullen, M.D.; Gaut, B.S.; Nielsen, D.M.; Holland, J.B.; et al. A unified mixed-model method for association mapping that accounts for multiple levels of relatedness. *Nat. Genet.* **2005**, *38*, 203–208. [CrossRef]
69. Lipka, A.E.; Tian, F.; Wang, Q.; Peiffer, J.; Li, M.; Bradbury, P.J.; Gore, M.A.; Buckler, E.S.; Zhang, Z. GAPIT: Genome association and prediction integrated tool. *Bioinformatics* **2012**, *28*, 2397–2399. [CrossRef]
70. Liu, X.; Huang, M.; Fan, B.; Buckler, E.S.; Zhang, Z. Iterative Usage of Fixed and Random Effect Models for Powerful and Efficient Genome-Wide Association Studies. *PLoS Genet.* **2016**, *12*, e1005767. [CrossRef] [PubMed]

plants

MDPI

Article

Assessment of Genetic Diversity for Drought, Heat and Combined Drought and Heat Stress Tolerance in Early Maturing Maize Landraces

Charles Nelimor [1,2,3], Baffour Badu-Apraku [2,*], Antonia Y. Tetteh [4] and Assanvo S. P. N'guetta [3]

1 WASCAL Graduate Research Program on Climate Change and Biodiversity, Université Felix Houphouët Boigny, Abidjan 01 BPV 34, Cote d'Ivoire; nelimor.c@edu.wascal.org
2 International Institute of Tropical Agriculture, Ibadan 200001, Nigeria
3 Department of Bioscience, Université Felix Houphouët Boigny, Abidjan 01 BPV 34, Cote d'Ivoire; nguettaewatty@gmail.com
4 Department of Biochemistry and Biotechnology, Kwame Nkrumah University of Science and Technology, University Post Office Box PMP, Kumasi 00233, Ghana; aytetteh@gmail.com
* Correspondence: b.badu-apraku@cgiar.org

Received: 27 September 2019; Accepted: 15 November 2019; Published: 17 November 2019

Abstract: Climate change is expected to aggravate the effects of drought, heat and combined drought and heat stresses. An important step in developing 'climate smart' maize varieties is to identify germplasm with good levels of tolerance to the abiotic stresses. The primary objective of this study was to identify landraces with combined high yield potential and desirable secondary traits under drought, heat and combined drought and heat stresses. Thirty-three landraces from Burkina Faso (6), Ghana (6) and Togo (21), and three drought-tolerant populations/varieties from the Maize Improvement Program at the International Institute of Tropical Agriculture were evaluated under three conditions, namely managed drought stress, heat stress and combined drought and heat stress, with optimal growing conditions as control, for two years. The phenotypic and genetic correlations between grain yield of the different treatments were very weak, suggesting the presence of independent genetic control of yield to these stresses. However, grain yield under heat and combined drought and heat stresses were highly and positively correlated, indicating that heat-tolerant genotypes would most likely tolerate combined drought and stress. Yield reduction averaged 46% under managed drought stress, 55% under heat stress, and 66% under combined drought and heat stress, which reflected hypo-additive effect of drought and heat stress on grain yield of the maize accessions. Accession GH-3505 was highly tolerant to drought, while GH-4859 and TZm-1353 were tolerant to the three stresses. These landrace accessions can be invaluable sources of genes/alleles for breeding for adaptation of maize to climate change.

Keywords: climate change; combined drought and heat stress; drought; heat; landraces; maize

---

1. Introduction

Climate change is predicted to increase global temperatures and reduce rainfall patterns, with adverse effects, particularly, on the critical stages of plant growth and development, resulting in yield losses. Productivity of maize, the major staple cereal crop in sub-Saharan Africa (SSA) is thus threatened [1,2]. Rainfall under future climate change scenarios in SSA will either occur late or stop earlier than usual [1], while temperatures in large areas have already exceeded the threshold for maize growth and productivity [2,3]. Maize is highly vulnerable to drought stress (DS) and heat stress (HS) during the reproductive stages [3,4]. Drought stress at flowering and grain-filling stages of maize causes delayed silking, increased anthesis-silking interval, and reduced kernel set [5], resulting in grain yield (GY) losses above 50% [6,7]. Under extreme conditions, DS at a few days before anthesis to the beginning of grain-filling period reduced GY of maize by as much as 90% [8]. High temperatures occurring at two weeks before flowering resulted in leaf firing, increased rate of leaf senescence [9] and premature lodging [7]. At the on-set of flowering, high temperature stress caused tassel blasting, leading to reduced pollen production and viability, reduced pollination rate, shortened grain-filling period, and reduced kernel and grain weight [9,10]. Together, yield losses due to these altered morpho-physiological traits was estimated at 45% or more [11,12].

Under HS, plants open their stomata to cool their leaves by transpiration, but when plants have to keep their stomata closed to reduce water loss during combined drought and heat stress (DSHS) conditions, the leaf temperature remain high, resulting in increased yield losses compared to the separate effects of the two stresses. For example, each degree rise in temperature above 30 °C resulted in 1% reduction in maize GY under optimal growing conditions (OGC), 1.7% under DS and up to 40% or more under DSHS [1]. Meseka et al. [7] found that while DS reduced GY by 58%, DSHS reduced GY by 77%. It was demonstrated that under extreme conditions, DSHS forced farmers to abandon their farmlands [13]. To cope with these stresses, farmers will need 'climate smart' maize varieties with low risk of failure. To develop such maize varieties, there is the need for continuous access to a wide range of alleles that are scattered in germplasm resources, particularly, the landraces.

Landraces of maize have been selected for several years for adaptation to local conditions, harbor abundant genetic diversity for important agronomic characteristics such as, phenology, growing season, resistance to diseases and insects and tolerance to abiotic stresses, including drought and/or high temperatures [14]. The availability of such genetic diversity is critical for genetic enhancement, allowing not only increased genetic gains and improved nutritional quality, but also for broadening the genetic base of elite varieties. Maize breeding efforts over the last few decades achieved remarkable success in terms of DS tolerance [3]. However, to ensure continued gains in genetic improvement, new DS tolerant source populations are needed [3]. Moreover, in contrast to DS, research on HS and DSHS has only recently began in tropical and subtropical maize [3,7,15,16]. Identifying new sources of variation for tolerance to DS, HS and DSHS will greatly complement these breeding efforts.

The primary objective of this study was to identify landraces that combine desirable secondary traits with good levels of tolerance to DS, HS and DSHS for use in maize breeding programs as sources of variation for breeding for climate change resilience.

2. Results

2.1. Variation in Weather Conditions during the Trial Periods

In 2018, the average night and day temperatures during the period when the HS and DSHS trials were carried out ranged from 17 to 25 °C and 36 to 40 °C, respectively (Table 1). During the evaluation period in 2019, average temperatures varied from 18 to 26 °C at night and from 31 to 39 °C at daytime. There were incidences of rains after grain-filling stages in May and June, with minor effects on trials. In each year, the lowest day and night temperatures were observed at the time of planting in February, and the highest in April (Table 1). In both years, the peak temperature occurred in April (Figure 1), which coincided with the flowering and grain filling stages. During this period, night temperatures varied from 18 to 27 °C in 2018 and 23 to 29 °C in 2019, while daytime temperatures ranged from 36 to 41 °C in both years.

Table 1. Monthly average temperature and rainfall recorded at Kadawa, Nigeria during the trial periods in 2018 and 2019.

	2018			2019		
Month	**Night (°C)**	**Day (°C)**	**Rainfall (mm)**	**Night (°C)**	**Day (°C)**	**Rainfall (mm)**
February	17	36	0	18	31	0
March	19	38	0	24	37	0
April	24	40	0	26	39	0
May	26	39	18	26	37	37
June	25	37	47	24	34	36

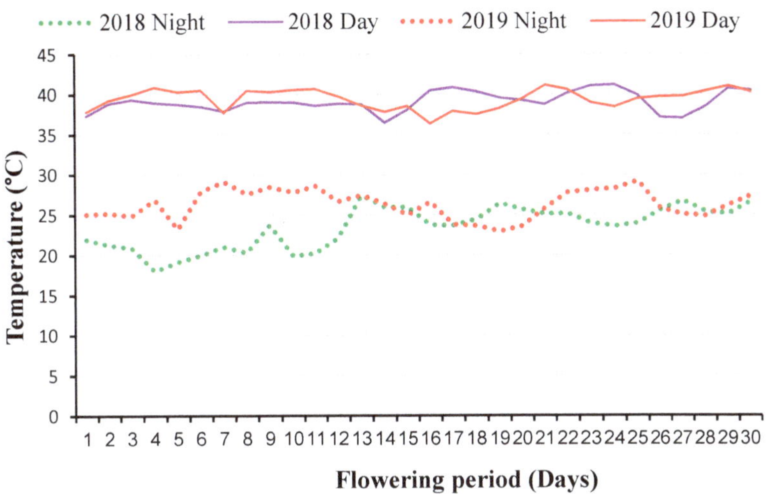

Figure 1. Average day and night temperatures recorded at Kadawa during the flowering period in April.

2.2. *Variance and Heritability Estimates*

Broad-sense heritability estimates of GY of individual trials ranged from 0.37 to 0.87 under OGC, 0.66 to 0.83 under managed drought stress (MDS), 0.72 to 0.76 under HS and 0.51 to 0.61 under DSHS (Supplementary Table S1). The combined ANOVA of the 36 early maturing maize accessions evaluated under each of the stress conditions showed highly significant ($p < 0.001$) mean squares for GY and other measured traits for year, genotypes, and genotype × year interactions. Under OGC, year, and genotypes had significant effects ($p < 0.05$ or $p < 0.001$) on all measured traits except anthesis-silking interval (ASI) and husk cover (HC) (Table 2). Results of combined analysis across two years under MDS revealed that the effects of environments and genotypes were significant ($p < 0.05$ or $p < 0.001$) for all traits except environmental effects on GY and plant aspect (PASP) (Table 2). Similarly, under HS and DSHS, the combined ANOVA showed that year and genotypes and their interaction were significant (p < 0.005 or 0.001) for GY and most other traits (Table 3). Under each evaluation condition, the genotypes contributed the largest percentage to total sum of squares when compared with other sources of variation. Consequently, repeatability values of majority of the traits were high (>0.60).

Table 2. Mean squares and repeatability values of grain yield and other traits of 36 early maturing maize accessions evaluated under optimal growing environments and managed drought stress between 2017 and 2019 in Nigeria.

Source	df	GY	AD	SD	ASI	PLHT	EHT	HC	EPP	PASP	EASP	EAROT	SG	RL	SL
Optimal Growing Conditions															
Env	2	2,654,514.1 *	115.6 *	198.3 **	11.7	29620.0 **	21,117.5 **	1.4	1.5 **	2.0 *	5.4 *	0.4 **	-	-	-
Rep (Env)	3	441,211.6	30.0 *	102.7 *	24.3 *	2160.6 *	1,390.5 **	0.125	0	0.4	1.7	0.1 **	-	-	-
Block (Env * Rep)	30	481,717.9 *	9	5.9	6.8	1156.6 **	405.1 *	0.79 *	0	1.1 *	1.3 *	0.0 *	-	-	-
Genotype	35	1,341,158.1 **	26.3 **	34.9 **	11.4 *	1342.1 **	627.5 **	2.2 **	0.1 **	2.0 **	3.4 **	0.0 *	-	-	-
Env *Genotype	70	216,762.9	10.3	5.3	4.5	556.5 *	311.6 *	0.5	0.0 *	0.4	0.6	0.0 *	-	-	-
Error	75	266,755.3	7.6	10.5	5.5	223.7	120.7	0.4	0	0.4	0.7	0	-	-	-
Repeatability		0.86	0.64	0.79	0.66	0.63	0.55	0.84	0.56	0.84	0.85	-	-	-	-
Managed Drought Stress															
Env	1	3,976,484	109.1 **	164.6 **	25.4 **	7629.7 **	9232.1 **	55.1 **	0.1 *	0.6	4.0 **	0.5 **	4.0 *	0.0 **	0.1 *
Rep (Env)	2	2,258,791	1	7.7 *	5.9 *	357.8 *	685.2 **	0.3	0	1.5 *	0.4	0.1 *	0	0	0.0 *
Block (Env * Rep)	20	1,039,093	2.1	2.6	0.7	254.4 *	100.7	0.6	0	0.5	0.3	0	0.8	0.0 *	0
Genotype	35	6,056,025.5 **	26.0 **	29.6 **	2.5 **	1957.9 **	981.2 **	96.0 **	0.1 **	2.4 **	2.7 **	0.1 **	1.4 *	0.0 **	0.0 **
Env * Genotype	35	1,840,190	4.8 **	5.6 **	1.5 *	733.2 **	342.9 **	95.8 **	0.0 *	0.7 *	0.6 *	0.0 *	0.8	0	0
Error	50	1,638,119	1.5	2.1	0.7	121.6	82.4	0.5	0	0.3	0.3	0	0.6	0	0
Repeatability		0.69	0.84	0.83	0.38	0.63	0.65	0.03	0.5	0.74	0.79	0.49	0.51	0.39	0.57

*, ** Significance at 0.01 and 0.001, respectively; df: degree of freedom; Env: environment/year; Rep: replication; GY: Grain yield; AD: Days to 50% anthesis; SD: Days to 50% silking; ASI: Anthesis-silking interval; PLHT: Plant height; EHT: Ear height; HC: Husk cover; EPP: Ears per plant; PASP: Plant aspect; EASP: EAROT: Ear rot; EASP: Ear aspect; SG: Stay green characteristic; RL: Root lodging; SL: Stalk lodging; TB: Tassel blast; LF: Leaf firing.

Table 3. Mean squares and repeatability values of grain yield and other traits of 36 early maturing maize accessions evaluated under heat stress and combined drought and heat stress during the 2018 and 2019 dry seasons at Kadawa, Nigeria.

Source	df	GY	AD	SD	ASI	PLHT	EHT	HC	EPP	PASP	EASP	SG	RL	SL	EAROT	TB	LF
Heat Stress																	
Env	1	57,677,836.7 **	604.3 **	855.7 **	21.8 *	28713.3 **	14600.7 **	64.0 **	2.8 **	42.3 **	134.2 **	29.3 **	-	0	0.7 **	0	-
Rep (Env)	2	4541,533.3 *	4.5	5.2	6.7	7009.3 **	2634.0 **	0.9	0.3 *	4.3 **	3	9.2 **	-	0.0 *	0.01	0	-
Block (Env * Rep)	20	768121.6	15.1 *	13.3 *	1.3	428.9 *	260.0 *	0.7 *	0.1 *	0.9 *	2.4	1.8 *	-	0	0.0 *	0	-
Genotype	35	1,836,764.3 **	27.7 **	31.2 **	4.3 *	951.3 **	495.2 **	0.7*	0.1 *	0.8 *	4.1 *	1.0 *	-	0	0.0 *	0	-
En * Genotype	35	907,223.9 *	5.6	4.9	1.8	114.6	96.8	0.5	0	0.3	1.9	0.5	-	0	0	0	-
Error	50	444,221	6	4.3	2.5	133.6	113.6	0.4	0	0.4	1.8	0.5	-	0	0	0	-
Repeatability		0.57	0.85	0.89	0.58	0.89	0.82	0.21	0.68	0.62	0.68	0.49	-	0.17	-	0.25	-
Combined Drought and Heat Stress																	
Env	1	6,582,075.3 **	40.1 *	242.8 **	85.6 *	14336.1 **	7764.5 **	40.1 **	0.2 *	1.6	4.7 *	0	0	0	-	0.1 **	0.1
Rep (Env)	2	1,772,416.6 **	72.4 **	148.4 **	14.3	2181.8 **	490.6 *	5.1 *	0.1	2.8 *	2.4 *	0.2	0	0	-	0.0 *	0.4
Block (Env * Rep)	20	271,106.8 *	12.2 *	12.5 *	4.3	339.0 **	254.7 **	1.4 *	0	0.6	1	1	0	0	-	0	0.1
Genotype	35	924,112.7 **	29.9 **	33.4 **	4.7	972.4 **	795.8 **	1.2*	0.1 **	1.0 *	3.3 **	1.5 *	0	0	-	0.0 **	0.1
Env * Genotype	35	405,956.3 *	7.0 *	7.9	2.9	228.3 **	171.8 **	0.7	0	0.5	1.3 *	0.6	0	0	-	0	0
Error	50	145,812.4	3.7	5.7	5.6	86	46.5	0.5	0	0.5	0.6	0.6	0	0	-	0	0
Repeatability		0.60	0.78	0.77	0.03	0.77	0.78	0.36	0.78	0.58	0.66	0.61	0.52	0.43	-	0.68	0.29

*, ** Significance at 0.01 and 0.001, respectively; df: degree of freedom; Env: environment/year; Rep: replication; GY: Grain yield; AD: Days to 50% anthesis; SD: Days to 50% silking; ASI: Anthesis-silking interval; PLHT: Plant height; EHT: Ear height; HC: Husk cover; EPP: Ears per plant; PASP: Plant aspect; EASP: EAROT: Ear rot; EASP: Ear aspect; SG: Stay green characteristic; RL: Root lodging; SL: Stalk lodging; TB: Tassel blast; LF: Leaf firing.

2.3. Phenotypic and Genetic Correlations between Treatments, and Trait Associations

The genetic correlations between GY under HS and DSHS was very high and positive (0.94) whereas that between OGC and, HS and MDS were moderate and positive, 0.61 and 0.65, respectively. Genetic correlations observed between GY of the other treatments were very weak, ranging from −0.01 between MDS and DSHS to 0.29 for OGC and DSHS (Table 4). A similar trend was observed for the phenotypic correlations between GY of the treatments (Table 4). The phenotypic correlation between flowering traits [days to anthesis (AD), days to silking (SD) and ASI] of the different treatments were positive and ranged from very low (0.07) between ASI under MDS and HS to very high (0.84) between AD under HS and DSHS (Supplementary Table S2). The association between majority of the other traits including GY were highly significant ($p < 0.001$) and positive for HS vs. DSHS.

Table 4. Genetic correlations (down-diagonal) and phenotypic correlations (top-diagonal) between grain yield of 36 maize accessions under optimal growing conditions, managed drought stress, heat stress and combined drought and heat stress conditions.

Treatment	Optimal Growing Conditions	Managed Drought Stress	Heat Stress	Combined Drought and Heat Stress
Optimal growing conditions		0.52 **	0.45 **	0.23
Managed drought stress	0.65		0.21	−0.01
Heat stress	0.61	0.28		0.69 ***
Combined drought and heat stress	0.29	−0.01	0.94	

, * Significant at 0.01, and 0.001, respectively.

2.4. Step-Wise Multiple Regression and Path Coefficient Analyses

Under MDS, the step-wise multiple regression analysis identified ear aspect (EASP) and PASP as traits with significant direct effects on GY, accounting for approximately 91% of the total variation in GY (Figure 2). of these two traits, EASP had the highest direct effect (−0.62) on yield, while PASP had the lowest direct effect on GY (−0.36). There were also two traits in the second order, including ears per plant (EPP) and ASI, each contributed indirectly to GY through EASP and PASP. Traits classified in the third order included stay green characteristic (SG), HC, SD, and plant height (PLHT). The indirect contribution of the remaining four third-order traits to yield through the second-order traits are clearly shown in Figure 2. Four traits, root lodging (RL), ear rot (EAROT), stalk lodging (SL) and ear height (EHT) were identified as the fourth-order traits with significant indirect effect on GY. While EAROT and SL contributed indirectly to GY through only PLHT and RL, EHT had indirect effects through two or more traits.

Across the HS trials, three traits (EASP, PASP and RL) were identified by the step-wise multiple regression in the first order, which explained about 86% of the total variation in GY (Figure 3). EASP had the highest direct effect on GY (−0.56), while RL was the least direct contributor to GY (−0.15). Four traits (EPP, HC, PLHT and AD) were categorized into second- order traits, each contributing indirectly to GY through one or two first order traits. The traits grouped in the third order included SD, ASI, EAROT, SG, tassel blast (TB) and EHT. Of these, ASI and EAROT had significant negative indirect effects on GY through EPP and AD. There were no fourth-order traits.

Under DSHS, only two traits, EASP and EPP were categorized as first order traits accounting for about 89% of the total variation in GY (Figure 4). While the contribution of EASP to variation in GY was negative (−0.52), that of EPP was positive (0.46). Two traits, namely SD and HC, were classified as second order traits, with each contributing indirectly to GY through the two first order traits. The traits classified into third order were AD, ASI, PASP and TB, all of which had positive indirect effects on GY. Five traits, EHT, SG, RL, SL and leaf firing (LF) were classified as fourth-order traits, each contributing to GY through two or more of the third order traits. Plant height was the only trait classified into the fifth order.

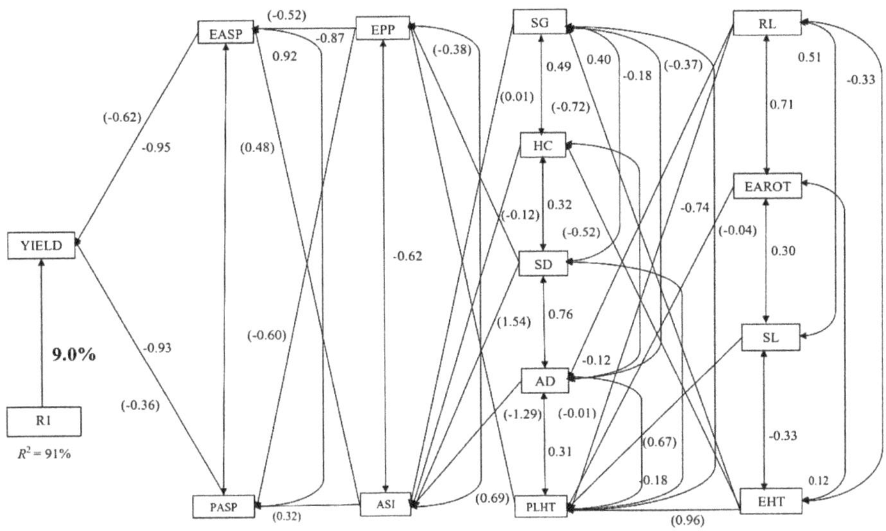

Figure 2. Path analysis diagram depicting the causal relationship of measured traits of the 36 maize accessions under managed drought stressed conditions. Note: Value written in bold is the error effects; the direct path coefficients are values in parenthesis and other values are correlation coefficients. R1 is error effects, R^2 = coefficient of determination.

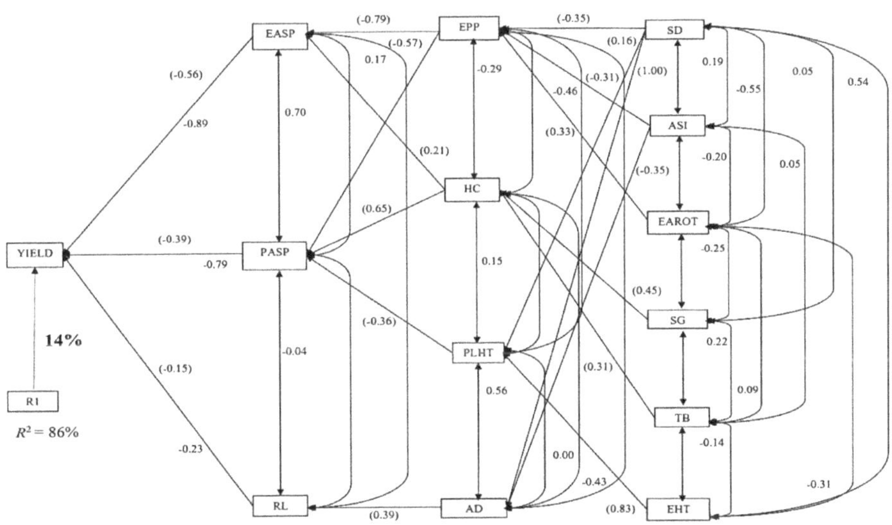

Figure 3. Path analysis diagram depicting the causal relationship of measured traits of the 36 maize accessions under heat stressed conditions. Note: Value written in bold is the error effects; the direct path coefficients are values in parenthesis and other values are correlation coefficients. R1 is error effects, R^2 = coefficient of determination.

2.5. Abiotic Stress Strongly Affected Traits and Reduced Grain Yield Levels

Under OGC, GY averaged 3205 kg/ha (Supplementary Table S3). Grain yield was reduced to 1744 kg/ha under MDS, 1443 kg/ha under HS and to 1088 kg/ha under DSHS. Anthesis was, on average, reached at 55 days under OGC and MDS and about 60 days under HS (67) and DSHS (65). As a result of reduced water availability, ASI averaged five days under MDS. Mean ASI under HS and DSHS, was three days, similar to that recorded under OGC (2 days). Plant height was reduced by approximately 25% under MDS, 14% under HS and by 7% under DSHS. Husk cover was less affected by stress conditions as indicated by the average rating value of 4 under the different evaluation conditions. A similar trend was observed for SG and PASP. Ear aspect averaged five under OGC and MDS, and six under HS and DSHS (Supplementary Table S3). While the number of ears per plant averaged 0.81, 0.57, and 0.53 under OGC, MDS, and HS, respectively, more barren plants, reflected as reduced EPP (0.40) were observed under DSHS. As a direct consequence, average yield reduction was higher under DSHS (66.04%) compared to MDS (45.60%) and HS (54.98%).

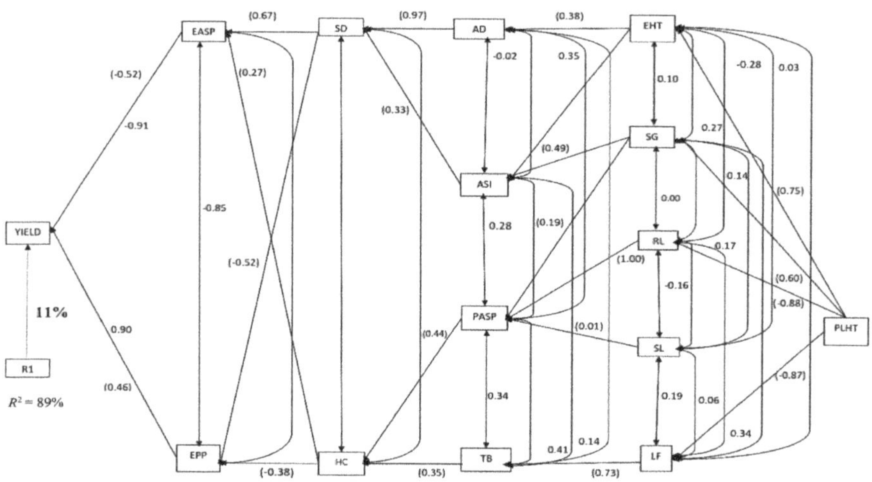

Figure 4. Path analysis diagram depicting the causal relationship of measured traits of the 36 maize accessions under combined drought and heat stressed conditions. Note: Value written in bold is the error effects; the direct path coefficients are values in parenthesis and other values are correlation coefficients. R1 is error effects, R^2 = coefficient of determination.

2.6. *Germplasm Tolerant to Abiotic Stresses*

The means of GY of top 11 accessions (top 10 landraces and best check) and worst 5 landraces identified using the base index under DS, HS and DSHS are presented in Table 5. Under MDS, the index values varied from −13.66 for TZm-1441 to 14.98 for GH-3505. Top 10 landraces with positive index values yielded above 2000 kg/ha under both MDS and OGC, except TZm-1473 that yielded less (1952 kg/ha) under MDS. Nine of the top 10 performing landraces under MDS (GH-3505, TZm-1317, TZm-1307, GH-4859, GH-5756, TZm-1273, TZm-1353, TZm-1312 and TZm-1311) had a yield advantage of between 4% to 42% over the best check (Aburohemaa). Under HS, the index values ranged from −8.04 for TZm-1319 to 13.53 for Check 1 (2011 TZE-W DT STR Synthetic), whereas under DSHS, it varied from −10.85 for TZm-1277 to 13.66 for Check 1 (2011 TZE-W DT STR Synthetic). Based on index selection, the outstanding landraces under HS included GH-4859, TZm-1353, TZm-1488, TZm-1441, TZm-1466, TZm-1473, TZm-1309, TZm-1325 and TZm-1317). Of these, TZm-1353 out-yielded the best check (2011 TZE-W DT STR Synthetic) by approximately 20% while GH-4859 produced yield comparable to 2011 TZE-W DT STR Synthetic. Under DSHS, the top 10 landraces identified by the base index included GH-4859, TZm-1473, TZm-1325, TZm-1441, TZm-1466, TZm-1273, TZm-1551, TZm-1452 and TZm-1353. Two landrace accessions (GH-4859 and TZm-1353) combined high yield potential with desirable secondary traits (reduced ASI, TB and LF, and increased PHLT, SG, and EPP as well as good PASP and EASP).

Table 5. Grain yield and other agronomic traits of maize accessions (best check, and top 10 and worst 5 landraces based on index selection) evaluated under managed drought stress, heat stress and combined drought and heat stress between 2017 and 2019.

Accession	GY (Kg/ha)	AD (Days)	SD (Days)	ASI (Days)	PLHT (cm)	PASP (Scale: 1–9)	SG (Scale: 1–9)	EPP	EASP (Scale: 1–9)	YR (%)	BI
Managed Drought Stress											
GH-3505	3901.61	56	58	3	183	3	3	0.94	3	14.75	14.98
TZm-1317	2586.92	54	57	3	180	4	3	0.79	4	34.25	8.52
TZm-1307	2336.55	54	56	2	172	5	3	0.83	5	43.38	6.57
GH-4859	2591.27	56	59	3	150	4	3	0.49	5	23.94	6.03
GH-5756	2236.72	57	60	3	163	5	3	0.80	5	50.80	5.63
TZm-1273	2251.48	56	59	3	141	5	3	0.77	5	5.70	5.50
TZm-1353	2359.25	52	56	4	153	4	4	0.64	4	44.06	5.13
TZm-1312	2462.75	57	60	4	162	5	3	0.68	5	42.47	5.10
TZm 1311	2248.39	58	62	4	173	5	3	0.75	5	23.36	4.89
TZm-1473	1951.92	49	50	2	153	5	3	0.68	5	38.71	4.67
Check 2	2139.95	52	57	5	122	5	4	0.69	5	45.29	2.33
TZm-1504	823.39	54	62	8	124	6	4	0.46	7	63.58	−6.91
TZm-1488	573.92	56	64	8	112	7	5	0.30	8	82.41	−11.89
TZm-1446	505.46	55	63	8	109	7	5	0.22	8	76.24	−12.53
TZm-1551	478.75	55	64	9	93	7	5	0.20	8	78.85	−13.21
TZm-1441	473.36	50	60	10	100	7	5	0.21	8	80.06	−13.66
MEAN	**1743.62**	**55**	**60**	**5**	**142**	**5**	**4**	**0.57**	**5**	**45.60**	
Heat Stress											
Check 1	3151.11	67	69	2	156	4	3	0.75	4	54.50	13.53
GH-4859	3066.86	70	72	2	172	4	3	0.71	4	9.99	11.67
TZm-1353	3922.93	65	69	4	186	4	3	0.65	4	6.98	10.75
TZm-1488	1913.70	67	68	1	165	4	3	0.69	5	41.34	7.42
TZm-1441	1809.11	60	63	3	148	5	3	0.62	5	23.80	4.01
TZm-1466	2155.31	69	73	4	180	5	4	0.70	5	46.38	3.57
TZm-1473	2090.47	60	63	3	138	5	4	0.66	5	34.36	3.27
TZm-1309	1560.76	67	68	2	158	5	4	0.69	6	14.28	2.57

Table 5. *Cont.*

Accession	GY (Kg/ha)	AD (Days)	SD (Days)	ASI (Days)	PLHT (cm)	PASP (Scale: 1–9)	SG (Scale: 1–9)	EPP	EASP (Scale: 1–9)	YR (%)	BI
TZm-1325	1557.57	62	64	2	134	5	4	0.71	5	24.82	2.13
TZm-1317	1655.82	65	67	2	171	5	4	0.57	6	57.92	1.43
TZm-1273	921.25	67	71	4	160	5	4	0.44	7	61.41	−4.03
TZm-1499	494.64	72	73	1	192	5	4	0.20	8	77.52	−4.24
TZm-1299	771.25	69	73	4	170	5	4	0.34	7	62.80	−4.80
GH-5756	688.64	71	72	2	168	5	5	0.39	8	84.85	−4.87
TZm-1277	1262.86	69	72	4	168	6	5	0.36	7	62.07	−4.97
TZm-1504	750.33	64	69	4	157	6	4	0.35	7	66.81	−7.16
TZm-1319	666.63	66	71	5	180	6	4	0.35	8	79.96	−8.04
MEAN	**1443.00**	**67**	**70**	**3**	**162**	**6**	**4**	**0.53**	**6**	**54.98**	
Combined Drought and Heat Stress											
Check 1	2581.69	64	65	1	175	4	3	0.95	4	62.72	13.66
GH-4859	1965.05	69	71	2	152	5	4	0.74	5	42.32	7.37
TZm-1473	1950.12	60	63	3	150	5	4	0.64	5	38.77	6.13
TZm-1325	1783.11	60	62	2	145	5	4	0.50	5	13.93	5.20
TZm-1441	2164.05	63	66	4	157	6	4	0.66	5	8.84	5.19
TZm-1466	1807.08	67	70	3	179	4	4	0.72	5	55.04	5.09
TZm-1273	1162.94	67	68	1	201	5	3	0.48	5	51.29	3.83
TZm-1551	1554.25	62	66	3	165	5	3	0.56	5	31.34	3.58
TZm-1452	1534.90	57	61	4	168	5	3	0.50	5	28.37	3.50
TZm-1353	1796.76	62	66	4	189	4	3	0.46	5	57.40	3.27
TZm-1488	1285.44	65	68	3	173	4	3	0.44	6	60.60	1.82
TZm-1316	740.47	65	68	3	188	5	6	0.23	7	80.16	−4.55
GH-3505	565.45	68	72	4	166	5	3	0.07	8	87.64	−5.46
TZm-1439	566.12	69	74	5	178	7	4	0.21	8	67.20	−6.40
TZm-1504	512.65	66	70	4	167	6	5	0.16	8	77.33	−6.61
TZm-1277	0.00	67	71	5	172	6	4	0.00	8	100	−10.85
MEAN	**1088.39**	**65**	**68**	**3**	**175**	**6**	**4**	**0.40**	**6**	**66.04**	

GY: Grain yield; AD: Days to 50% anthesis; SD: Days to 50% silking; ASI: Anthesis-silking interval; PLHT: Plant height, PASP: EPP: Ears per plant; Plant aspect; PLHT: SG: Stay green characteristics; EASP: Ear aspect; YR: Yield reduction; BI: base index. Check 1: 2011 TZE-W DT STR Synthetic; Check 2: Aburohemaa.

2.7. Grouping of Accessions under Abiotic Stresses

Phylogenetic constellation plots generated from the standardized data of grain yield, plant and ear aspect scores, anthesis-silking interval, ears per plant and stay green characteristics under MDS, HS and DSHS are presented in Figure 5. Under MDS, the accessions were classified into five major groups. The number of accessions in the clusters ranged from one in cluster I to 11 in clusters II and III. The accessions of clusters I, II and III were characterized by increased ears per plant, delayed senescence, and desirable plant and ear aspects that resulted in positive base index values (Supplementary Table S4). Under HS, the accessions were grouped into two clusters, each consisting of four sub-clusters (Figure 5). The first major cluster consisted of 15 accessions, which included the three checks. Majority of the accessions in this group had high grain yields, increased ears per plant, shorter ASI and good ear aspect (Supplementary Table S4). Consequently, the base index was positive for this cluster. The second major cluster consisted of 21 landraces that were largely barren with poor ear aspect scores and low grain yield, resulting in negative base index (BI) values averaging −3.79. Similarly, under DSHS, the 36 maize accessions were clustered into three major groups (Figure 5). Cluster I consisted of 12 accessions, which included two checks whereas clusters II and III were represented by 14 and 10 accessions, respectively. Most of the accessions in cluster I were characterized by increased ears per plant, good ear aspect, high grain yield and positive base index values. In contrast, the accessions of clusters II and III were generally barren with poor ear aspect scores, low yield resulting in negative base index values (Supplementary Table S4).

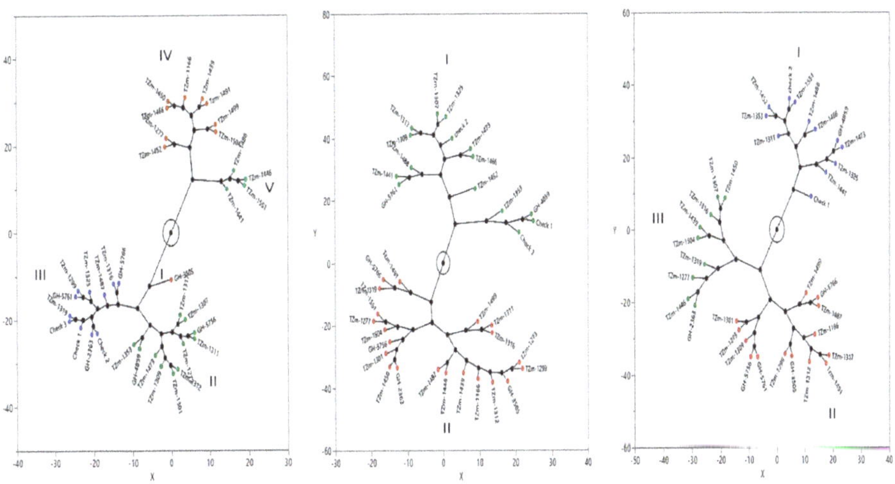

Figure 5. Phylogenetic constellation plots displaying the relationships between 33 maize landraces and three improved populations/varieties evaluated under managed drought stress (**left**), heat stress (**middle**) and combined drought and heat stress (**right**). Cluster I, II and III (**left**), and I (**middle** and **right**) are represented by drought, heat and combined drought and heat-tolerant accessions, respectively while the remaining clusters consisted of susceptible accessions.

3. Discussion

Under the prevailing and future conditions of climate change, DS, HS and DSHS stresses constitute the most important environmental factors constraining maize production in SSA [2,3]. Results of climate simulation studies showed that these stresses will most likely coincide with the flowering and grain filling stages of maize growth and development in large areas of SSA [3,12] and will result in huge yield losses on farmer's field. The value of landraces as potential donors of beneficial alleles for breeding for climate change resilience is well-recognized [14]. Our study aimed at identifying landraces with good levels of tolerance to DS, HS and DSHS for use in maize breeding programs as potential sources of beneficial alleles for developing cultivars with enhanced resilience to climate change as well as to identify key stress-adaptive secondary traits that could lead to genetic enhancement for grain yield under DS, HS and DSHS stressed environments.

The sites selected for this study were previously used for screening maize genotypes for high levels of tolerance to DS and/or HS for climate change adaptation [4,6,7]. As shown in Table 1, the HS trials were exposed to elevated temperatures, while the DSHS trials were subjected to prolonged DS at elevated temperatures. In particular, temperatures during the reproductive stages highly exceeded the optimal threshold value of 34 °C for lowland tropical maize (Figure 1), indicating that the environments used for this study were appropriate for screening the maize germplasm for tolerance to HS and DSHS.

The significant mean squares observed for GY and most other measured traits under each evaluation condition indicated that substantial genetic variation existed among the accessions, which should facilitate selection for DS, HS and DSHS tolerance and key secondary traits conferring tolerance under the research conditions. These observations corroborated the results of Gouesnard et al. [17] who suggested the presence of significant genetic variability for tolerance to abiotic stresses in tropical maize landraces. Moreover, the presence of significant means squares of genotype by environment interaction for most of the traits indicated that the accessions varied in their responses to the stresses of the different years. These findings are consistent with the results of Meseka et al. [7].

Although broad-sense heritability estimates of GY of single trials under the stresses were moderately high, these results were consistent with previous studies of maize under abiotic stress [3,6,18]. However, as indicated by Cairns et al. [3], broad-sense heritability values of single trials can be inflated because genetic variance and genotype × trial interaction variance are confounded. The high repeatability values (60%) observed for most measured traits including GY under the stressed and non-stressed environments suggested consistency in the expression of the traits under the research conditions. These results largely provided a good indication of the performance of the accessions for breeding purposes. Similar observations were reported in maize under multiple stresses [3,19].

Grain yield observed under OGC was to some extent predictive of GY under both MDS and HS conditions as indicated by the moderately positive genetic and phenotypic correlations. These observations were in agreement with the results of previous studies on abiotic stress in maize [3,7,19,20]. The GY under OGC and DSHS had weak positive genetic and phenotypic correlations (0.29 and 0.23, respectively), suggesting the presence of independent genetic factors controlling yield potential in the two conditions. These results are consistent with the findings of earlier studies by Cairns et al. [3]. The lack of both genetic and phenotypic correlations between GY under MDS and HS as well as between MDS and HS indicated that tolerance to these stresses were modulated by different genetic mechanisms. These results are in agreement with the findings of earlier studies [3,7,18], who found that tolerance to DS was genetically distinct from tolerance to DSHS. In contrast to the results of Cairns et al. [3] however, GY observed under HS and DSHS was positive and moderately high at both the phenotypic and genetic levels. This result was in agreement with the findings of Tandzi et al. [21] who demonstrated that HS tolerant maize genotypes were most likely to be tolerant to DSHS conditions. Furthermore, the significant phenotypic correlations between the same traits under the different research conditions indicated the existence of common genetic elements governing the expression of the measured traits under the different research conditions.

The stresses applied in this study had significant negative impacts on GY and other relevant traits. This result is in agreement with the reports of Cairns et al. [3] and Trachsel et al. [18]. In particular, while the days to anthesis was on the average unaffected under MDS, silking was significantly affected, resulting in delayed ASI that affected asynchrony between male and female flowers and eventually led to reduced ears per plant. Days to anthesis was on average delayed by 10 days under DSHS and by 12 days under HS whereas, silking was delayed by 13 and 11 days under HS and DSHS, respectively. These observations could be attributed to delayed emergence owing to the severe cold due to harmattan at the time of planting. The moderate reductions in plant heights observed under

HS and DSHS indicated that the plants were only affected by these stresses towards the end of the vegetative phase. Similarly, the lower average plant height observed under MDS could be attributed to the incidence of the drought stress at an early stage of plant growth and development. Indeed, the MDS was imposed during the early growth stages (25 DAP) compared to the HS and DSHS, which were imposed at 32 DAP. Traits such as EPP and EASP were most negatively affected under HS and DSHS compared to MDS. Consequently, average yield reduction under MDS was lower than that observed under HS and DSHS. Comparison of the yield loss under MDS and HS to DSHS revealed that the latter had hypo-additive effects (i.e., the effect of combined stress was higher than the individual effects but lower than their sum). These results were most probably due to the interaction of HS and DS on stomatal movements. Plants have to either close their stomata under DSHS at elevated temperature to prevent water loss or keep stomata opened to cool the leaves through transpiration [22]. Alterations in these morpho-physiological mechanisms under DSHS might have caused osmotic imbalances, resulting in the huge yield losses. These results are in agreement with the findings of previous workers who reported higher yield losses from the combined effects of DS and HS than DS and HS applied alone [7,13].

Selection based on grain yield alone under DS, HS and DSHS often limits selection gains because of its complex nature and the low heritability of the trait under stress [23]. Consequently, secondary traits that are highly heritable and are associated with GY have been widely used as selection criteria for improved yield potential under abiotic stresses. In maize for instance, significant genetic gains were reported under abiotic stresses such as low nitrogen and DS by complementing GY with key secondary traits [24]. In particular, ASI, senescence, tassel blast, ears per plant, kernel per row, tassel sterility, pollen viability, and stigma receptivity and other morpho-physiological traits, such as leaf firing were identified as key secondary traits associated with GY under DS, HS and DSHS [15,16]. In the present study, sequential multiple regression analyses revealed both ear and plant aspects, and to some extent, ears per plant and root lodging as the major determinants of yield potential, accounting for more than 85% of the observable differences in grain yield levels under the stresses. The repeatability values of these traits were moderately high. Therefore, both ear and plant aspects, ears per plant and root lodging have the potential to improve the selection efficiency for GY under the target stresses. These results partly corroborated the findings of Meseka et al. [7] who reported ear and plant aspects as well as ears per plant as the principal yield determinant traits under DS and DSHS.

An important objective of the present study was to identify landraces with outstanding performance under each of the stresses for use in maize breeding programs as sources of tolerance to the stress factors. The outstanding landraces identified by the base index under each treatment, could be invaluable sources of beneficial genes/alleles for improving DS and/or HS tolerance in elite maize germplasm. In particular, GH-3505 yielded approximately 4 tons ha^{-1} under MDS, making it interesting for use in drought-prone areas. Moreover, the accessions that yielded above the best-improved checks under each research condition should be prioritized for introgression into breeding pipelines. Of special interest for breeders will be accessions GH-4859 and TZm-1353, which combined desirable secondary traits with good levels of tolerance (positive base index values) to all the applied stresses. The fact that only two accessions were tolerant to all the target stresses was most likely due to the fact that different genetic mechanisms underlie tolerance to the three stresses applied in the present study. Furthermore, cluster analysis based on phylogenetic constellation plots largely separated tolerant accessions from their susceptible counterparts indicating that the two classes of accessions were genetically dissimilar. Such clustering patterns of maize genotypes were previously reported under DS and DSHS [7] as well as DS, HS and DSHS [21].

4. Materials and Methods

4.1. Genetic Materials

One hundred and ninety-six (196) maize landraces, representing gene pools from Burkina Faso, Ghana and Togo, were sourced from germplasm banks at International Institute of Tropical Agriculture (IITA), Nigeria and the Plant Genetics Resources Institute of Ghana. The germplasm was evaluated for phenotypic variation during the main growing season in 2017 and 2018 [25]. Thirty-three landraces, comprising six each from Burkina Faso and Ghana, and 21 from Togo, were used for this study. The landraces were selected based on the expression of the adaptive traits such as shortness and early flowering under OGC [25]. Three improved populations/varieties with enhanced tolerance to DS and/or HS, which served as checks were obtained from the Maize Improvement Program at IITA (IITA-MIP), Ibadan, Nigeria. The list of the genetic materials evaluated in this study is presented in Table 6.

Table 6. List of the 36 maize accessions that were evaluated for tolerance to drought, heat and combined drought and heat stress between 2017 and 2019 at Ikenne and Kadawa, Nigeria.

No.	Accession	Origin	No.	Accession	Origin	No.	Accession	Origin
1	TZm-1273	Togo	13	TZm-1353	Togo	25	TZm-1491	Burkina Faso
2	TZm-1277	Togo	14	TZm-1439	Togo	26	TZm-1499	Burkina Faso
3	TZm-1299	Togo	15	TZm-1441	Togo	27	TZm-1504	Burkina Faso
4	TZm-1301	Togo	16	TZm-1446	Togo	28	GH-2363	Ghana
5	TZm-1307	Togo	17	TZm-1450	Togo	29	GH-3505	Ghana
6	TZm-1309	Togo	18	TZm-1452	Togo	30	GH-4859	Ghana
7	TZm-1311	Togo	19	TZm-1466	Togo	31	GH-5756	Ghana
8	TZm-1312	Togo	20	TZm-1473	Togo	32	GH-5761	Ghana
9	TZm-1316	Togo	21	TZm-1551	Togo	33	GH-5766	Ghana
10	TZm-1317	Togo	22	TZm-1166	Burkina Faso	34	2011 TZE-W DT STR Synthetic (Check 1)	IITA-MIP
11	TZm-1319	Togo	23	TZm-1487	Burkina Faso	35	Aburohemaa (Check 2)	IITA-MIP
12	TZm-1325	Togo	24	TZm-1488	Burkina Faso	36	2015 TZE-W DT STR Syn C0 (Check 3)	IITA-MIP

IITA-MIP: Maize Improvement Program at the International Institute of Tropical Agriculture.

4.2. Trial Establishment and Agronomic Management

The 36 maize accessions (33 landraces plus 3 DS and/or HS tolerant populations/varieties) were evaluated for two years under OGC, MDS, HS and DSHS. The genetic materials, experimental design, and replications were the same for all evaluation conditions. The trials were arranged in 6 by 6-alpha lattice design (incomplete design) with two replications. A plot consisted of one row, 3 m long. Rows were spaced 0.75 m apart and the distance between hills was 0.40 m. Three seeds were planted in a hill and thinned to two per stand at two weeks after planting (WAP), resulting in a final plant population density of 66,666 plants ha^{-1}. Pre-emergence weeds were controlled by applying gramoxone and atrazine at the rates of 1.5 L gramoxone and 2.5 L atrazine in 200 L of water ha^{-1}. Subsequently, manual weeding was done to keep trials free from weeds.

In the first experiment, the accessions were evaluated at Ikenne (6°53′ N, 3°42′ E, 60 m altitude, 1200 mm annual rainfall), Nigeria, under MDS during the dry seasons of 2017/2018 and 2018/2019. The soil type at Ikenne is Eutric nitrisol [26]. The MDS at Ikenne was achieved by using a sprinkler irrigation system that provided 17 mm of water per week up to 25 days after planting (DAP). Thereafter, the irrigation was withheld until maturity, so that the maize plants depended on stored water in the soil for growth and development.

In the second experiment, the accessions were evaluated for tolerance to HS and DSHS at Kadawa (11°39′ N, 8°27′ E, 500 m altitude), Nigeria, where drought stress at elevated temperature occur between February and June. The soil type at Kadawa is characterized as Regosols, with mainly sandy to clay loam texture [27]. At Kadawa, air temperature during the dry season often range from 33 to 45 °C [7]. This enabled establishment of trials under HS and DSHS, in which water supply was carefully controlled by a furrow irrigation system. The trials under HS and DSHS were laid in adjacent blocks, separated by 15 m to prevent spill-over of irrigation water. The trials were planted on the same day in mid-February each year, and flowering and grain-filling stages occurred in April when rainfall incidence was negligible and temperatures ranged from 36 and 41 °C, allowing for exposure of the accessions to HS and DSHS (Figure 1). A furrow irrigation system was used to supply water to the crop every four days during the first 32 DAP. Thereafter, irrigation water was withdrawn from the DSHS block while the HS block continued to receive irrigation water until physiological maturity. Irrigation was resumed one week after grain filling and was applied only once per week to avoid complete loss of DSHS trials. Meteorological data were recorded during the trial period with the aid of an automated weather station.

The same set of genetic materials were initially grown under OGC during the main growing seasons of 2017 and 2018 at Ikenne and Mokwa (9°18′ N, 5°185 4′ E, altitude 457 m, 1100 mm annual rainfall), Nigeria. The soil at Mokwa is luvisol with 0.27, 0.035 and 0.48% organic C, organic N and P content [26].

4.3. Traits Measured

Data were recorded on several traits at flowering including the number of days from planting to when 50% of the plants in a plot were shedding pollen and had emerged silks under DS, HS, DSHS and OGC, respectively. Anthesis-silking interval (ASI) was computed as the difference between days to 50% silking and anthesis. In addition, the number of plants showing leaf firing and tassel blasting were counted on plot basis at the flowering and grain filling stages on HS and DSHS trials and converted to percentages. Plant and ear heights (PLHT and EHT) were measured as the distance from the base of the plant to the height of the first tassel branch and the node bearing the upper ear, respectively. Root lodging (SL) (percentage of plants leaning more than 30° from the vertical), and stalk lodging (SL) (percentage broken at or below the highest ear node) were recorded as taking a few days to harvest. At 70 DAP, Plant Aspect (PASP) was scored based on the general architectural appeal of plants in a plot (standability, vigour, plant and ear height, uniformity of plants, ear placement and size, as well as disease damage and lodging) using a scale of 1 to 9, where 1 = excellent overall phenotypic appeal; 2 = very good overall phenotypic appeal; 3 = good overall phenotypic appeal; 4 = satisfactory overall phenotypic appeal; 5 = acceptable phenotypic appeal; 6 = undesirable phenotypic appeal, 7 = poor overall phenotypic appeal, 8 = very poor phenotypic appeal and 9 = completely undesirable phenotypic appeal. Similarly, husk cover (HC) was rated on a scale of 1 to 9, where 1 = husks tightly arranged and extended beyond the ear tip and 9 = exposed ears. In addition, stay green characteristic (SG) was recorded for the MDS, HS and DSHS plots at 70 DAP on a scale of 1 to 9, where 1 = 10% dead leaf area; 2 = 20% dead leaf area; 3 = 30% dead leaf area, 4 = 40% dead leaf area; 5 = 50% dead leaf area; 6 = 60% dead leaf area; 7 = 70% dead leaf area; 8 = 80% dead leaf area and 9 represented 90%–100% dead leaf area. At harvest, number of ears per plant (EPP) was obtained by dividing the total number of ears per plot by the total number of plants harvested. Ear aspect (EASP) was recorded based on general appeal of the ears without the husks (ear size and number; uniformity of size, colour and texture; extent of grain filling, insect and disease damage) using a scale of 1 to 9, where 1 = excellent (clean, uniform, large, and fully filled ears with no disease/insect damage); 2 = very good ears with no disease/insect damage and fully filled grains, one or two irregularity in cob size; 3 = good with no disease/insect damage and fully filled grains, one or two irregularity in cob size; 4= mild insect damage, no disease, fully filled grains, one or two irregularity in cob size poor; 5= mild disease/insect damage and fully filled grains, one or two irregularity in cob size, 6 = severe disease/insect damage and fully filled grains, smaller cobs, non-uniform cob size; 7 = severe disease/insect damage, scanty grain filling, few ears, non-uniformity of cobs; 8= severe disease/insect damage, scanty grain filling, very few ears and 9 = only one or no ears. In the stressed trials, harvested ears from each plot were shelled to determine the percentage grain moisture. Grain yield in kg ha^{-1} was computed from the shelled grain weight and adjusted to 15% moisture content, whereas in the experiments under OGC, harvested ears of each plot were weighed and GY was estimated based on 80% shelling percentage and adjusted to 15% moisture content.

4.4. Data Analysis

Data from each evaluation condition (OGC, MDS, HS and DSHS) was analysed separately. Variance components were estimated using the mixed model procedure in SAS version 9.4 [28] based on the standard linear mixed model described by Vargas et al. [29]. All effects except for genotypes were considered random and the Best Linear Unbiased Predictor (BLUP) was estimated for all measured traits following the procedure of Robinson [30].

Broad-sense heritability (H^2) of GY of each trial was estimated using META-SAS v4 [29]. Within treatment repeatability of the traits [31] were calculated on accession-mean basis using the following formula:

$$R = \frac{\sigma_g^2}{\sigma_g^2 + \frac{\sigma_{ge}^2}{e} + \frac{\sigma_e}{re}}$$

where σ_g^2 is the genotypic variance, σ_{ge}^2 is the variance of genotype × environment, σ_e is the residual variance; e is the number of environments, and r is the number of replicates per environment.

The phenotypic and genetic correlations between GY of the different treatments were computed following the procedure of Cooper et al. [32] as:

$$\rho_g = \frac{\overline{\sigma_{g(jj')}}}{\overline{\sigma_{g(j)}\sigma_{g(j')}}}$$

where ρ_g is the genetic correlation, $\overline{\sigma_{g(jj')}}$ is the arithmetic mean of all pairwise genotypic covariances between environment j (say, optimum) and j' (say, drought), and $\overline{\sigma_{g(j)}\sigma_{g(j')}}$ is the arithmetic average of all pairwise geometric means between the genotypic variance components of the environment.

The sequential path coefficient analyses [33] was performed using the Statistical Package for Social Sciences, SPSS version 17.0 [34] to determine the secondary traits with significant contributions to GY under each research condition. A step-wise regression analysis was used to categorize the predictor traits into first, second and third order based on their individual contributions to total variation in GY with minimized multicollinearity [35]. The first step involved the regression of all the traits on GY and those with significant contributions to GY at $p < 0.05$ were identified as first order traits. Subsequently, traits that were not identified as first order traits were regressed on each of the first order traits to identify those with significant contributions to GY through each of the first order traits and were categorized as second-order traits. The procedure was repeated to identify traits in subsequent orders. The path coefficients were represented by the standardized b-values obtained from the regression analysis [33,35,36].

A base index (BI) that integrated superior grain yield, EPP, ASI, PASP, EASP, and SG was used to select the best and worst performing genotypes under each treatment [37]. Each trait was first standardized with standard deviation of 1 and a mean of zero to minimize the effect of the different scales prior to integrating into the BI. The BI was computed using the equation:

$$BI = [(2 \times YLD_S) + EPP - ASI - PASP - EASP - SG]$$

where YLD_S is GY under stress, *PASP* is plant aspect, *EASP* is ear aspect, *EPP* is ears per plant, *ASI* is anthesis-silking interval and *SG* is the stay-green characteristic. A positive *BI* value indicated tolerance to the applied stress while negative values indicated susceptibility [37].

Subsequently, all traits included in the *BI* were used as an input file to generate a phylogenetic constellation plot, which classified the accessions into genetic groups. The phylogenetic constellation plot was performed using JMP pro 14.10 [28] based on Ward's algorithm.

5. Conclusions

In conclusion, highly significant differences were observed among the accessions under each of the evaluation conditions thus, enabling identification of donors with good levels of tolerance to DS, HS and DSHS. Introgression of these outstanding donors into breeding materials will help maximize genetic gains under the stress conditions. In particular, accession GH-3505 was highly tolerant to DS while GH-4859 and TZm-1353 combined desirable secondary traits with good levels of tolerance to all the stresses. These accessions can immediately be used in tropical maize breeding programs to develop cultivars with enhanced tolerance to abiotic stress to sustain production and thus, food security in the face of climate change. The lack of correlation between the stress treatments and the poor overlap of genotypes performing well across all the applied treatments indicated that tolerance to these stresses were under independent genetic control, thus emphasizing the need to screen germplasm under each abiotic stress separately. However, the results also, showed that heat-tolerant accessions were most likely to tolerate combined drought and heat stress conditions. Identifying the genomic regions potentially underlying tolerance and the gene action controlling the stress-adaptive traits might further facilitate the introgression of these valuable landraces into breeding pipelines.

Supplementary Materials: The following are available online at http://www.mdpi.com/2223-7747/8/11/518/s1, Table S1: Genotypic and residual variance, and broad-sense heritability estimates of grain yield (kg/ha) of the nine individual trials. Table S2: Association of traits measured under optimal growing conditions with same traits under managed drought stress, heat stress and combined drought and heat stress, and those measured under managed drought stress with same traits under heat stress and combined drought and heat stress as well as with those measured under heat stress with the same traits under combined drought and heat stress. Table S3: Mean grain yield and other traits of 36 maize accessions evaluated under optimal growing conditions, managed drought stress, heat stress and combined drought and heat stress between 2017 and 2019, in Nigeria. Table S4: Cluster means (base index values and other secondary traits) of 36 maize accessions evaluated under optimal growing conditions, managed drought stress, heat stress and combined drought and heat stress between 2017 and 2019, in Nigeria.

Author Contributions: Conceptualization: A.S.P.N. and A.Y.T.; Methodology: C.N. and B.B.-A.; Supervision: B.B.-A., A.S.P.N. and A.Y.T.; Data analysis and manuscript draft: C.N.; Manuscript review and editing, B.B.-A., A.Y.T. and A.S.P.N.

Funding: This research was funded by the German Federal Ministry of Education, through the West African Science Service Centre on Climate Change and Adapted Land-use (WASCAL). This work was also supported by the Bill & Melinda Gates Foundation [OPP1134248] through the funding support to the Stress Tolerant Maize for Africa (STMA) Project.

Acknowledgments: We are grateful to the Genetic Resource Centre in IITA and the Plant Genetics Resources Institute of Ghana for providing maize accessions used in this study.

Conflicts of Interest: The authors declare no conflicts of interest.

References

1. Lobell, D.B.; Bänziger, M.; Magorokosho, C.; Vivek, B. Nonlinear heat effects on African maize as evidenced by historical yield trials. *Nat. Clim. Chang.* **2011**, *1*, 42–45. [CrossRef]
2. Cairns, J.E.; Hellin, J.; Sonder, K.; Araus, J.L.; MacRobert, J.F.; Thierfelder, C.; Prasanna, B.M. Adapting maize production to climate change in Sub-Saharan Africa. *Food Secur.* **2012**, *5*, 345–360. [CrossRef]
3. Cairns, J.E.; Crossa, J.; Zaidi, P.H.; Grudloyma, P.; Sanchez, C.; Araus, J.L.; Menkir, A. Identification of drought, heat, and combined drought and heat tolerant donors in maize. *Crop Sci.* **2013**, *53*, 1335–1346. [CrossRef]

4. Badu-Apraku, B.; Fakorede, M.A.B. *Improvement of Early and Extra-Early Maize for Combined Tolerance to Drought and Heat Stress in Sub-Saharan Africa*; Advances in Genetic Enhancement of Early and Extra-Early Maize for Sub-Saharan Africa; Springer: Cham, Switzerland, 2017; pp. 311–358.
5. Schussler, J.R.; Westgate, M.E. Maize kernel set at low water potential. I. Sensitivity to reduced assimilates during early kernel growth. *Crop Sci.* **1991**, *31*, 1189–1195. [CrossRef]
6. Menkir, A.; Crossa, J.; Meseka, S.; Bossey, B.; Ado, S.G.; Obeng-Antwi, K.; Alidu, H. Comparative performance of top-cross maize hybrids under managed drought stress and variable rainfed environments. *Euphytica* **2016**, *212*, 455–472. [CrossRef]
7. Meseka, S.; Menkir, A.; Bossey, B.; Mengesha, W. Performance Assessment of Drought Tolerant Maize Hybrids under Combined Drought and Heat Stress. *Agronomy* **2018**, *8*, 274. [CrossRef]
8. NeSmith, D.S.; Ritchie, J.T. Effects of soil water-deficits during tassel emergence on development and yield components of maize (*Zea mays* L.). *Field Crop. Res.* **1992**, *28*, 251–256. [CrossRef]
9. Zaidi, P.H.; Zaman-Allah, M.; Trachsel, S.; Seetharam, K.; Cairns, J.E.; Vinayan, M.T. *Phenotyping for Abiotic Stress Tolerance in Maize Heat Stress: A Field Manual*; CIMMYT: Mexico City, Mexico, 2016.
10. Cicchino, M.; Rattalino-Edreria, J.I.; Uribelarrea, M.; Otegui, M.E. Heat stress in field-grown maize: Response of physiological determinants of grain yield. *Crop Sci.* **2011**, *50*, 1438–1448. [CrossRef]
11. Bassu, S.; Brisson, N.; Durand, J.L.; Boote, K.; Lizaso, J.; Jones, J.W.; Basso, B. How do various maize crop models vary in their responses to climate change factors? *Glob. Chang. Biol.* **2014**, *20*, 2301–2320. [CrossRef]
12. Sultan, B.; Gaetani, M. Agriculture in West Africa: Climate change and impacts scenarios and potential for adaptation. *Front. Plant Sci.* **2016**, *7*, 1262. [CrossRef]
13. Heiniger, R.W. *The Impact of Early Drought on Corn Yield*; North Carolina State University: Raleigh, NC, USA, 2001.
14. Dwivedi, S.L.; Ceccarelli, S.; Blair, M.W.; Upadhyaya, H.D.; Are, A.K.; Ortiz, R. Landrace germplasm for improving yield and abiotic stress adaptation. *Trends Plant Sci.* **2016**, *21*, 31–42. [CrossRef] [PubMed]
15. Alam, M.S.; Seetharam, K.; Zaidi, P.H.; Dinesh, A.; Vinayan, M.T.; Nath, U.K. Dissecting heat stress tolerance in tropical maize (*Zea mays* L.). *Field Crop. Res.* **2017**, *204*, 110–119. [CrossRef]
16. Noor, J.J.; Vinayan, M.T.; Umar, S.; Devi, P.; Iqbal, M.; Seetharam, K.; Zaidi, P.H. Morpho-physiological traits associated with heat stress tolerance in tropical maize (*Zea mays* L.) at reproductive stage. *Aust. J. Crop Sci.* **2019**, *13*, 536. [CrossRef]
17. Gouesnard, B.; Zanetto, B.; Welcker, C. Identification of adaptation traits to drought in collections of maize landraces from southern Europe and temperate regions. *Euphytica* **2015**, *209*, 565–584. [CrossRef]
18. Trachsel, S.; Leyva, M.; Lopez, M.; Suarez, E.A.; Mendoza, A.; Montiel, N.G.; Macias, M.S.; Burgueno, J.; San Vicente, F. Identification of tropical maize germplasm with tolerance to drought, nitrogen deficiency, and combined heat and drought stress. *Crop Sci.* **2016**, *56*, 3031–3045. [CrossRef]
19. Badu-Apraku, B.; Yallou, C.G.; Obeng-Antwi, K.; Alidu, H.; Talabi, A.O.; Annor, B.; Oyekunle, M.; Akaogu, I.C.; Aderounmu, M. Yield Gains in Extra-Early Maize Cultivars of Three Breeding Eras under Multiple Environments. *Agron. Appl. Genet. Resour.* **2017**, *109*, 418–431. [CrossRef]
20. Bänziger, M.; Lafitte, H.R. Efficiency of secondary traits for improving low-nitrogen target environments. *Crop Sci.* **1997**, *37*, 1110–1117. [CrossRef]
21. Tandzi, L.N.; Bradley, G.; Mutengwa, C. Morphological responses of maize to drought, heat and combined drought and heat stresses. *J. Biol. Sci.* **2019**, *19*, 7–16. [CrossRef]
22. Mahrookashani, A.; Siebert, S.; Hüging, H.; Ewert, F. Independent and combined effects of high temperature and drought stress around anthesis on wheat. *J. Agron. Crop Sci.* **2017**, *203*, 453–463. [CrossRef]
23. Bolaños, J.; Edmeades, G.O. Eight cycles of selection for drought tolerance in lowland tropical maize. I. Responses in grain yield, biomass, and radiation utilization. *Field Crop. Res.* **1993**, *31*, 233–252. [CrossRef]
24. Edmeades, G.O.; Bolaños, J.; Chapman, S.C.; Lafiite, H.R.; Bänziger, M. Selection improves drought tolerance in tropical maize populations. I. Gains in biomass, grain yield and harvest index. *Crop Sci.* **1999**, *39*, 1306–1315. [CrossRef]
25. Nelimor, C.; Badu-Apraku, B.; Nguetta, S.P.A.; Tetteh, A.Y.; Garcia-Oliveira, A.L. Phenotypic characterization of maize landraces from Sahel and Coastal West Africa reveals marked diversity and potential for genetic improvement. *J. Crop Improv.* **2019**. [CrossRef]
26. Soil Survey Staff. *Soil Taxonomy: A Basic System of Soil Classification for Making and Interpreting Soil Surveys*, 2nd ed.; USDA-NRCS Agriculture Handb; U.S. Gov. Print. Office: Washington, DC, USA, 1999; p. 436.

27. Gabasawa, A.I.; Mohammed, H.; Yusuf, A.A. Biological nitrogen fixation and pod yield of groundnut (*Arachis hypogaea* L.) as influenced by salt affected Alfasol at Kadawa, Nigeria. *Int. J. Plant Soil Sci.* **2014**, *3*, 1479–1489. [CrossRef]
28. SAS Institute Inc. *SAS User's Guide: Statistics*; Version 9.4; SAS Institute Inc.: Cary, NC, USA, 2017.
29. Vargas, M.; Combs, E.; Alvarado, G.; Atlin, G.; Mathews, K.; Crossa, J. META: A suit of SAS programs to analyse multi-environment breeding trials. *Agron. J.* **2013**, *105*, 11–19. [CrossRef]
30. Robinson, G.K. That BLUP is a good thing: The estimation of random effects. *Stat. Sci.* **1991**, *6*, 15–32. [CrossRef]
31. Falconer, D.S.; Mackay, T.F.C. *Introduction to Quantitative Genetics*, 4th ed.; Longman: New York, NY, USA, 1996.
32. Cooper, M.; DeLacy, I.H.; Basford, K.E. Relationship among analytical methods used to analyse genotypic adaptation in multi-environment trials. In *Plant Adaptation and Crop Improvement*; Cooper, M., Hammer, G.L., Eds.; Cambridge Univ. Press: Cambridge, UK, 1996; pp. 193–224.
33. Mohammadi, S.A.; Prasanna, B.M.; Singh, N.N. Sequential path model for determining interrelationships among grain yield and related characters in maize. *Crop Sci.* **2003**, *43*, 1690–1697. [CrossRef]
34. Statistical Package for Social Sciences (SPSS) Inc. *SPSS Base 17.0 for Windows User's Guide*; SPSS Inc.: Chicago, IL, USA, 2007.
35. Badu-Apraku, B.; Akinwale, R.O.; Oyekunle, M. Efficiency of secondary traits in selecting for improved grain yield in extra-early maize under *Striga*-infested and *Striga*-free environments. *Plant Breed.* **2014**, *133*, 373–380. [CrossRef]
36. Talabi, A.O.; Badu-Apraku, B.; Fakorede, M.A.B. Genetic variances and relationship among traits of an early-maturing maize population under drought-stress and low-N environments. *Crop Sci.* **2016**, *57*, 681–692. [CrossRef]
37. Badu-Apraku, B.; Fakorede, M.A.B.; Talabi, A.O.; Oyekunle, M.; Akaogu, I.C.; Akinwale, R.O.; Annor, B.; Melaku, G.; Fasanmade, Y.; Aderounmu, M. Gene action and heterotic groups of early white quality protein maize inbreds under multiple stress environments. *Crop Sci.* **2015**, *56*, 183–199. [CrossRef]

plants

MDPI

Article

Genetic Diversity, Nitrogen Fixation, and Water Use Efficiency in a Panel of Honduran Common Bean (*Phaseolus vulgaris* L.) Landraces and Modern Genotypes

Jennifer Wilker [1], Sally Humphries [2], Juan Carlos Rosas-Sotomayor [3], Marvin Gómez Cerna [4], Davoud Torkamaneh [1], Michelle Edwards [1], Alireza Navabi [1,†] and K. Peter Pauls [1,*]

1 Department of Plant Agriculture, University of Guelph, Guelph, ON N1G 2W1, Canada; jwilker@uoguelph.ca (J.W.); dtorkama@uoguelph.ca (D.T.); edwardsm@uoguelph.ca (M.E.)
2 Department of Sociology and Anthropology, University of Guelph, Guelph, ON N1G 2W1, Canada; shumphri@uoguelph.ca
3 Departamento de Ciencia y Producción Agropecuaria, Escuela Agrícola Panamericana, Zamorano, Tegucigalpa 11101, Honduras; jcrosas@zamorano.edu
4 Fundación para la Investigación Participativa con Agricultores de Honduras, La Ceiba, Atlántida 561, Honduras; marvin.gomez@est.zamorano.edu
* Correspondence: ppauls@uoguelph.ca; Tel.: +1-519-824-4120 (ext. 54136)
† Deceased on 10 March 2019.

Received: 31 July 2020; Accepted: 11 September 2020; Published: 19 September 2020

Abstract: Common bean (*Phaseolus vulgaris* L.) provides critical nutrition and a livelihood for millions of smallholder farmers worldwide. Beans engage in symbiotic nitrogen fixation (SNF) with Rhizobia. Honduran hillside farmers farm marginal land and utilize few production inputs; therefore, bean varieties with high SNF capacity and environmental resiliency would be of benefit to them. We explored the diversity for SNF, agronomic traits, and water use efficiency (WUE) among 70 Honduran landrace, participatory bred (PPB), and conventionally bred bean varieties (HON panel) and 6 North American check varieties in 3 low-N field trials in Ontario, Canada and Honduras. Genetic diversity was measured with a 6K single nucleotide polymorphism (SNP) array, and phenotyping for agronomic, SNF, and WUE traits was carried out. STRUCTURE analysis revealed two subpopulations with admixture between the subpopulations. Nucleotide diversity was greater in the landraces than the PPB varieties across the genome, and multiple genomic regions were identified where population genetic differentiation between the landraces and PPB varieties was evident. Significant differences were found between varieties and breeding categories for agronomic traits, SNF, and WUE. Landraces had above average SNF capacity, conventional varieties showed higher yields, and PPB varieties performed well for WUE. Varieties with the best SNF capacity could be used in further participatory breeding efforts.

Keywords: nitrogen fixation; symbiosis; bean; landrace; PPB; participatory breeding; climate resilient; Honduras

1. Introduction

The common bean (*Phaseolus vulgaris* L.) is the most important food legume grown and consumed worldwide. High in protein, fiber, and essential nutrients, the nutritional profile and affordability of beans relative to other protein sources make beans a dietary staple in developing economies.

A member of the Fabaceae family, common bean is a predominantly self-pollinating species with a genome size of 587 Mbp and ploidy of 2n = 2x = 22 [1]. The center of origin for common bean is in present-day Central Mexico [2]. As a result of geographical dispersion, the ancestral bean diverged and evolved into two domesticated gene pools. Larger-seeded market classes evolved and were domesticated in South America and belong to the Andean gene pool, while smaller-seeded market classes evolved and were domesticated in Central America and belong to the Middle American gene pool [2–7].

Beans are traditionally grown as monocrops or with maize in either a relay cropping or an intercropping system in Honduras. There are two growing seasons—the rainy *Primera* (May–September) and the traditionally dry *Postrera* (October–January)—although climate change causes more fluctuation in precipitation levels and the duration of these seasons. Food insecurity is an issue in hillside communities and is a particular problem during the summer months before the *Primera* harvest. In some locations, this hungry period is termed *los junios* after the month during which food becomes scarce. More than 50% of bean production in Honduras takes place on steep hillside slopes (15–30° and greater) [8]. In addition, the country's infrastructure is poor, and less than 30% of the bean growing area is located within an hour's travel to a road, thus restricting market access [8].

Bean production in Honduras is affected by various biotic and abiotic stresses, and productivity is low and averaged 785 kg ha^{-1} in 2018 compared to yields in Canada, which averaged 2888 kg ha^{-1} in 2018 [9]. Bean diseases and insect pests comprise the primary biotic stresses of Honduran bean production. The most impactful diseases are Bean Golden Mosaic Virus (BGMV), rust (caused by *Uromyces phaseoli*), web blight (caused by *Rhizoctonia solani*), anthracnose (caused by *Colletotrichum lindemuthianum*), and angular leaf spot (ALS; caused by *Pseudoscercospora griseola*) [8]. The whitefly (*Bemisia tabaci*) is the vector for BGMV and is the most important insect pest of bean in Honduras. Weevils (*Acanthoscelides obtectus* and *Zabrotes subfasciathus*) are serious pests of stored beans, reducing marketability and damaging seed for planting. Climate change is expected to affect the impact of these biotic stresses in bean production and may lead to a shift in the complex of pests and diseases involved [10].

Extreme weather, such as high temperatures and flooding, including from hurricanes, reduces bean production. Climate models for Honduras predict higher temperatures and reduced overall rainfall but more extreme weather events increasing floods in the coming decades [11]. Another abiotic stress impacting bean production is soil health. Soils across Central America are deficient in available phosphorus (P), nitrogen (N), calcium (Ca), and organic matter, and aluminum (Al) and manganese (Mn) toxicity are exacerbated by low soil pH levels. Bean productivity is limited by soil nutrient availability, particularly N and P [12]. Nitrogen deficiency reduces grain yield because N is a structural component of various essential molecules, including chlorophyll, amino acids, nucleic acids, and lipids, required for the production of storage carbohydrates and proteins. Soils can be supplemented with nitrogen fertilizer throughout the growing season to avert yield losses; however, synthetic amendments are expensive, difficult to access, and generally not used by bean growers in Honduras. Instead, bean growers rely on organic forms of N, including that derived through symbiotic nitrogen fixation (SNF), as a nutrient source for their crop.

Beans are capable of generating their own organic nitrogen through SNF where nitrogen-fixing bacteria infect root nodules and reduce atmospheric nitrogen into forms useable by the host plant in exchange for carbohydrates [13]. SNF is a complex biological process and its efficiency is impacted by abiotic, biotic, and genetic factors, including soil nutrient levels, environmental conditions, the presence of efficient *Rhizobium* strains, and genetic constitution of the crop grown [12]. Recent studies have confirmed that SNF capacity in beans has a wide range and can reach high, yield-sustaining levels under optimal conditions [14–22]. For example, Kamfwa et al. [14] reported a range of 3.6 to 98.2% nitrogen derived from the atmosphere, and Aserse et al. [20] found that inoculated beans had comparable yields to those grown with nitrogen fertilizer.

Smallholder hillside farmers (0.5–5 ha) comprise approximately 70% of Honduran bean growers, and the remainder of production occurs in foothill and valley regions by larger-scale producers. Hillside farmers cultivate marginal land with steep slopes and low soil fertility, they tend towards subsistence production, and produce primarily for household consumption following traditional practices and planting traditional crop varieties. These smallholder farmers have limited access to markets, which has a two-fold impact, reducing the influence of market demands on what growers produce and limiting access to modern bean varieties and production inputs. These constraints notwithstanding, hillside farmers market approximately 50% of their bean harvest.

Landraces, known locally as *criollos*, comprise the majority of varieties traditionally grown by hillside farmers in Honduras [23]. Ninety-five percent of beans produced across the country are small, light red beans, which are preferred by Hondurans [8], and are also exported to El Salvador and to the United States to meet the needs of Central Americans who have emigrated there. Some black beans are grown in Honduras and are primarily exported to neighboring Guatemala where that market class is favored [8]. Landraces have local genetic adaptation, high genetic diversity, and lack formal genetic improvement [24]. The genetic heterogeneity of bean landraces lends resilience and makes them able to adapt to the changeable growing conditions of mountain hillsides and other marginal areas where they are grown. Among the preferred traits of the landraces included in this study are adaptation to cultivation at a range of altitudes, more marketable seed coat color and appealing kinesthetic properties, and yield stability in a changeable climate. In addition to traditional landraces, hillside bean farmers also grow conventionally bred and participatory bred varieties.

Conventional bean breeding in Honduras has been primarily the responsibility of the *Programa de Investigaciones en Frijol (PIF)* in the Department of Agronomy at the *Escuela Agricola Pan-Americana* (Zamorano) since the late 1980s when government funding to the agricultural research department (*Dirección de Ciencia y Tecnología Agropecuaria, DICTA*) was reduced restricting agricultural research and extension services [25]. The International Center for Tropical Agriculture (CIAT), the Bean/Cowpea Collaborative Research Support Program (CRSP), and the Regional Cooperative Bean Program (*Programa Cooperativo Regional de Frijol*-PROFRIJOL) have also been involved in variety development and/or providing funding for bean research in Honduras. Zamorano's early breeding focus was on developing conventional varieties with BGMV resistance and improved heat and drought tolerance for the lowland and valley production regions of Honduras [25]. By the late 1990s, Zamorano took leadership in bean breeding for the region, developing small red varieties for Honduras, Guatemala, El Salvador, Nicaragua, and Costa Rica, as well as black bean varieties for Guatemala and Haiti [26]. Conventionally bred commercial varieties are adapted for cultivation across a wide geographic region and have disease resistance and agronomic traits, which can bolster their yields. Adoption of commercial varieties among hillside growers is limited for a number of reasons, including darker seed coat color and other culinary traits, which reduce their marketability, as well as poor yield performance compared to landraces [23]. In the mid 1990s, Zamorano embarked on collaborative research with social scientists from CIAT to explore the social and economic factors that impact adoption of conventional varieties [23]. As a result, participatory research has become an important method used in the Zamorano bean breeding program.

The term participatory plant breeding encompasses two main methods of plant variety development, 'participatory varietal selection' and 'participatory plant breeding'. Participatory varietal selection (PVS) involves farmers locally testing varieties or advanced breeding lines provided by a formal plant breeder and making selections based on their needs [27]. Participatory plant breeding (PPB) involves farmers locally testing early stage (F_2-F_3) breeding material and can further involve the farmers actively participating in choosing parents and driving variety development by selecting progeny that meet local needs and preferences. As in other reports on the subject, the term PPB will be used to refer to both PPB and PVS in this study [28,29].

Participatory bean breeding at Zamorano has been facilitated through collaboration with CIAT-initiated *comités de investigación agrícola local* (CIALs), which are village-level farmer research teams that create a space where applied agricultural research can be carried out. For this study, we collaborated with the *Fundación para la Investigación Participativa con Agricultores de Honduras* (FIPAH) and *Programa de Reconstrucción Rural* (PRR). FIPAH supports over 100 CIALs, backstopped by regional offices across the country (https://fipah-hn.org/). PRR is an NGO that works with smallholder farmers in Santa Barbara and Lempira and supports approximately 60 CIALs (https://www.iaf.gov/grants/honduras/2017-prr/) [30]. CIAL members are trained in the scientific method, and most CIALs focus their research on obtaining higher-yielding and climate-resilient corn and bean varieties. The relationship between the CIALs and the bean breeding program at Zamorano is collegial and formal, responding to the needs of the farmers while the research is carried out with scientific rigor [23]. Traits of interest to the farmers are emphasized, and trials are performed using statistically valid designs and research methods. Landraces, conventional varieties, and germplasm from across the region are used in PPB efforts. PPB generates varieties that combine the local adaptation of landraces with improved traits from conventional genotypes such as disease resistance and higher yields. Other traits that factor into selection by farmer-researchers include seed color, appearance and size, pod length, plant architecture, even ripening, early maturity, and various culinary qualities [23]. Zamorano has developed some PPB varieties using landraces as parents in Honduras, Costa Rica, and Nicaragua [31]. Between 1994 and 2015, 24 PPB varieties were developed by Zamorano in collaboration with CIAL groups using participatory research methods, and one of these varieties, 'PM2-Don Rey', has been supported for national registration [29]. Adoption of the PPBs among CIAL members is above 60%, and PPBs are gaining wide acceptance among other farmers in communities where participatory research is carried out [29]. Extensive discussion of the development of 'Macuzalito', 'Cedron', 'Amilcar', 'Esperanceño', 'Chepe', and 'PM2-Don Rey' (representing both PPV and PPB methods of variety development) can be found in Humphries et al. [29].

Due to limited production resources and the threat of climate change, farmers in remote hillside communities would benefit from growing high-yielding common bean varieties that are climate resilient and have high nitrogen fixing capacity. To examine Honduran bean germplasm for these traits of interest, we curated a panel of Honduran bean genotypes representative of the traditional landraces and the participatory bred varieties grown by hillside bean farmers, as well as Honduran conventional and North American checks. The current study tests the hypothesis that bean landraces are a good source of germplasm with a high capacity for nitrogen fixation. The objectives of this study were to determine the genotypic and phenotypic diversity of the Honduran panel and to identify germplasm sources for breeding improved varieties suited to hillside production in Honduras.

2. Results

2.1. Analysis of Genetic Relatedness

Landrace and PPB plant material for the Honduran panel were sourced from six municipalities across west–central Honduras. The majority of genotypes came from Yoro (26) and Francisco Morazán (17), with less than 20 genotypes coming from Intibucá, Santa Bárbara, Comayagua, and Lempira combined (Figure 1). Descriptions of the genotypes can be found in Materials and Methods Section 4.1.

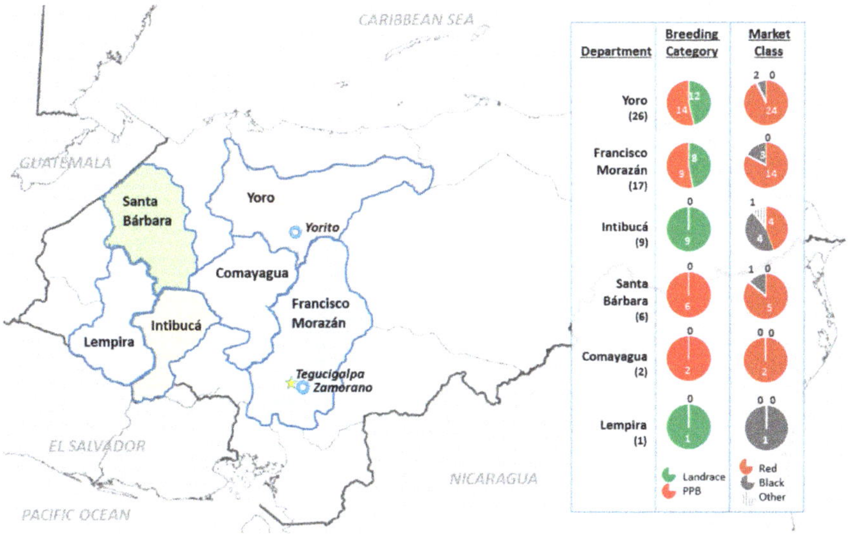

Figure 1. Map of west–central Honduras, outlining the six departments from which landrace and participatory bred (PPB) bean genotypes were sourced for the Honduran Panel. The chart at the right describes the number of landraces and PPB genotypes obtained from each department and the market classes to which those genotypes belong. The location of Yorito where the Honduran field trial was carried out, Zamorano where the *Escuela Agrícola Panamericana* is located, as well as the capital of Honduras, Tegucigalpa, are shown.

The genetic structure of the Honduran panel was explored to determine the evolutionary relatedness of the genotypes in the panel and the genetic composition of the genotypes. It is apparent from the topology of the phylogenetic tree (Figure 2A) that the landrace genotypes ('CRI') generally group into clusters of connected branches in the tree structure that are positioned in the left half of the figure, denoted as groupings I, II, and III (Figure 2A). The PPB genotypes ('PPB') grouped into separate clusters that are positioned in the right half of the figure, denoted as grouping IV (Figure 2A). Grouping I at the left of the tree, is comprised of the 'Milpero' genotypes, two landraces (HON70 and HON43), and Merlot (HON62). The Milpero landraces belong to diverse market classes, including black, small red, white, and carioca, and they included genotypes that did not flower at Elora in 2014. The remaining landrace genotype clusters were generally delineated by market class membership, with black genotypes comprising grouping II (including HON07, HON45, HON41, HON42, HON65, HON40, and HON43) and small red genotypes comprising grouping III (including HON08, HON09, HON11, HON27, HON68, HON67, HON66, HON51, HON34, HON48, HON38, HON46, HON50, and HON49). The landraces 'Concha Rosada' (HON02) and 'Rosado' (HON22) are displaced and found among the PPB branches of the tree. The North American check genotypes ('CHK'; including HON64, HON61, and HON59) formed a separate cluster that branched off between the Milpero landraces and the black landraces in grouping II. 'OAC Rosito' (HON63), clustered with the Honduran PPBs. All Honduran conventional genotypes ('CNV'; including black HON54, and small red HON77, HON52, HON80, and HON55) grouped with the PPBs (grouping IV), except 'Dorado' (HON56), which is found among the landraces (grouping III). Six PPB genotypes (including HON10, HON05, HON12, HON25, HON33, and HON72) were found within the landrace clusters of the tree (groupings II and III).

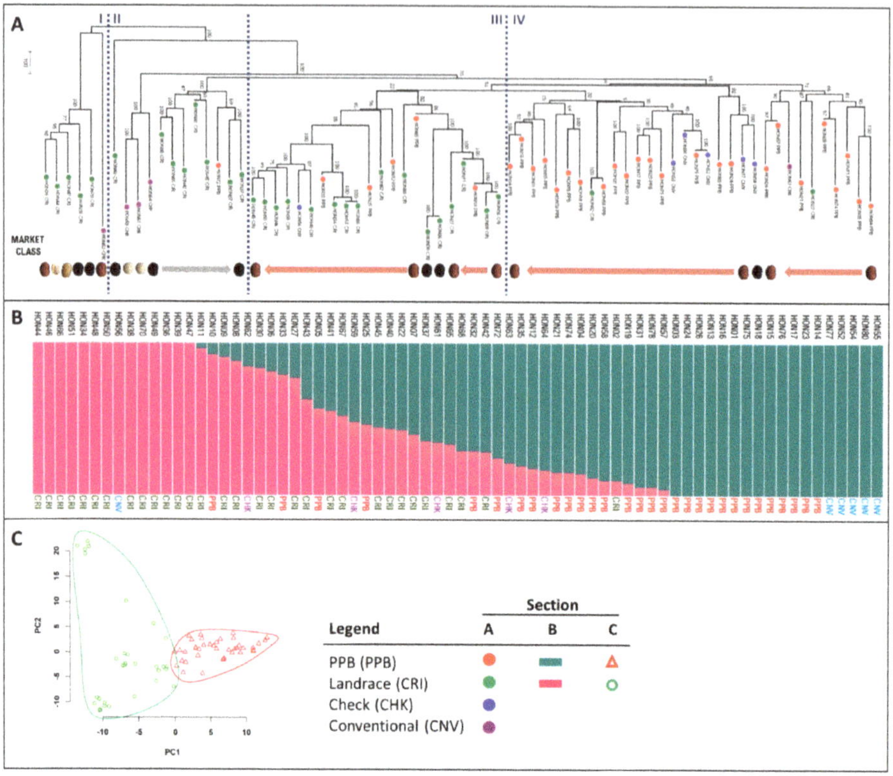

Figure 2. Analysis of genetic structure and relatedness of 72 genotypes of the Honduran panel. (**A**) Dendrogram of evolutionary genetic relatedness. Abbreviations are: participatory bred (PPB), Landrace (CRI), North American check (CHK), and Honduran conventional (CNV) genotypes. Market classes are denoted by representative beans. (**B**) Genetic structure plot using two genetic groupings ($\Delta K = 2$). **C**. Principal component analysis indicating two genetic groupings in the panel. Genotype descriptions are found in Section 4.1. (Note: In section A, grouping names I–IV are assigned to natural subsections of the tree for descriptive purposes and do not correspond to the genetic groups presented in section B.).

The genetic similarity of genotypes in the panel is depicted in a STRUCTURE plot using two subpopulations ($K = 2$) (Figure 2B). Fourteen of the landraces (including all of the Milpero types) belong to one genetic subgroup at the left of the plot and the PPB varieties belong to the other subgroup at the right of the plot, with an intermediary admixed group (Figure 2B). The Honduran conventional genotypes, except 'Dorado' (HON56), group with the PPBs. The North American check genotypes are found among the admixed genotypes, along with some PPBs and landraces. The principle component analysis of the panel also indicates the relatedness of the genotypes using two genetic groupings (Figure 2C). PC1 divides the genotypes into PPB (green triangles) and landrace (red circles) categories. Along the PC2 axis, the landraces show wide dispersion, with the Milpero group forming a small cluster near the axis at the top of the plot, and the North American check genotypes are scattered among the landraces. In contrast, PC2 generates a denser cluster of PPB genotypes, and the Honduran Conventional genotypes are at the right edge of the plot.

2.2. Nucleotide Diversity and Population Differentiation: Landrace and PPB Categories

Nucleotide diversity was measured in the two largest groupings within the Honduran panel, the landraces and the PPBs, to ascertain the genetic diversity of these groups. According to the π statistic, nucleotide diversity for the landrace category overall ($\pi = 3.20 \times 10^{-4}$) was significantly greater ($P = 0.04$, Welch two-sample *t*-test) than that found in the PPB category overall ($\pi = 2.89 \times 10^{-4}$). Additionally, according to the *D* statistic, the overall nucleotide diversity for the landrace category ($D = 0.669$) was significantly greater ($P = 0.02$, Welch two-sample *t*-test) than that found in the PPB category ($D = 0.476$). The positive Tajima's *D* value indicates that both landraces and PPBs are under balancing selection and implies that both categories are probably experiencing different selective pressure. Fifty-six subregions (>100 Mbp long) across the genome were identified where landrace π values exceeded PPB π values by more than 3 times (Table 1, Figure 3A). These regions, identified on all 11 chromosomes, may contain loci related to traits favored by selection associated with formal plant breeding (Figure 3A).

Table 1. Regions of the *P. vulgaris* (2.0) genome where high nucleotide diversity (π) was discovered in landrace genotypes compared to PPB genotypes. A literature search was performed to identify candidate genes within these regions. See Table S1 for candidate gene annotation. (Chr—Chromosome).

Region of High Diversity			π Value		Candidate Genes
Chr	**Start (Mbp)**	**End (Mbp)**	**Landrace**	**PPB**	**Number**
1	23	24	5.50×10^{-7}	1.31×10^{-7}	2
1	40	41	7.71×10^{-7}	1.83×10^{-7}	4
1	41	42	1.24×10^{-6}	2.49×10^{-7}	4
1	42	43	1.22×10^{-6}	3.98×10^{-7}	1
1	47	48	8.60×10^{-7}	9.66×10^{-8}	3
2	4	5	1.95×10^{-6}	6.51×10^{-7}	0
2	22	23	1.71×10^{-6}	1.64×10^{-7}	8
2	32	33	5.53×10^{-7}	1.22×10^{-7}	5
2	48	49	5.92×10^{-6}	1.62×10^{-6}	18
3	34	35	6.73×10^{-7}	1.97×10^{-7}	8
3	48	49	1.17×10^{-6}	1.89×10^{-7}	8
4	12	13	1.53×10^{-6}	2.62×10^{-7}	0
4	17	18	4.05×10^{-7}	6.55×10^{-8}	0
4	38	39	4.74×10^{-7}	9.66×10^{-8}	6
4	40	41	4.32×10^{-7}	1.27×10^{-7}	0
5	24	25	3.55×10^{-7}	6.55×10^{-8}	1
5	25	26	2.10×10^{-6}	1.92×10^{-7}	0
5	27	28	7.46×10^{-7}	1.31×10^{-7}	1
5	32	33	1.16×10^{-6}	3.28×10^{-7}	11
6	9	10	1.21×10^{-6}	6.55×10^{-8}	7
6	13	14	1.02×10^{-6}	1.31×10^{-7}	14
6	15	16	4.82×10^{-7}	1.55×10^{-7}	6
7	8	9	4.32×10^{-7}	9.66×10^{-8}	1
7	20	21	5.27×10^{-7}	6.55×10^{-8}	1
7	26	27	3.98×10^{-6}	1.01×10^{-6}	0
7	35	36	3.36×10^{-6}	7.38×10^{-7}	9
7	39	40	2.43×10^{-6}	3.32×10^{-7}	8
8	11	12	5.44×10^{-6}	1.74×10^{-6}	1
8	15	16	1.63×10^{-6}	5.32×10^{-7}	2
8	18	19	1.30×10^{-6}	3.98×10^{-7}	1
8	23	24	5.27×10^{-7}	6.55×10^{-8}	0
8	29	30	1.35×10^{-6}	3.32×10^{-7}	0
8	38	39	3.29×10^{-7}	3.33×10^{-8}	0

Table 1. *Cont.*

Region of High Diversity			π Value		Candidate Genes
Chr	Start (Mbp)	End (Mbp)	Landrace	PPB	Number
8	41	42	7.62×10^{-7}	2.26×10^{-7}	2
8	44	45	1.35×10^{-6}	1.62×10^{-7}	0
8	45	46	9.15×10^{-7}	2.90×10^{-7}	0
8	51	52	2.86×10^{-6}	6.33×10^{-7}	6
8	52	53	1.21×10^{-6}	9.66×10^{-8}	12
8	57	58	3.45×10^{-7}	6.55×10^{-8}	0
9	16	17	7.81×10^{-7}	1.60×10^{-7}	0
9	18	19	9.04×10^{-7}	3.33×10^{-8}	3
9	21	22	1.94×10^{-6}	4.99×10^{-7}	0
10	17	18	2.97×10^{-7}	9.32×10^{-8}	0
10	33	34	2.26×10^{-6}	7.41×10^{-7}	0
10	35	36	9.78×10^{-7}	2.62×10^{-7}	2
10	37	38	1.07×10^{-6}	1.94×10^{-7}	7
10	39	40	3.00×10^{-6}	7.34×10^{-7}	1
11	2	3	8.74×10^{-7}	1.93×10^{-7}	3
11	7	8	8.00×10^{-7}	1.55×10^{-7}	23
11	10	11	1.37×10^{-6}	3.01×10^{-7}	7
11	37	38	2.64×10^{-6}	7.84×10^{-7}	1
11	40	41	9.07×10^{-7}	3.01×10^{-7}	3
11	45	46	7.85×10^{-6}	9.80×10^{-7}	9
11	47	48	7.31×10^{-6}	1.53×10^{-6}	0
11	48	49	1.52×10^{-6}	4.99×10^{-7}	0
11	53	54	6.90×10^{-7}	2.28×10^{-7}	11

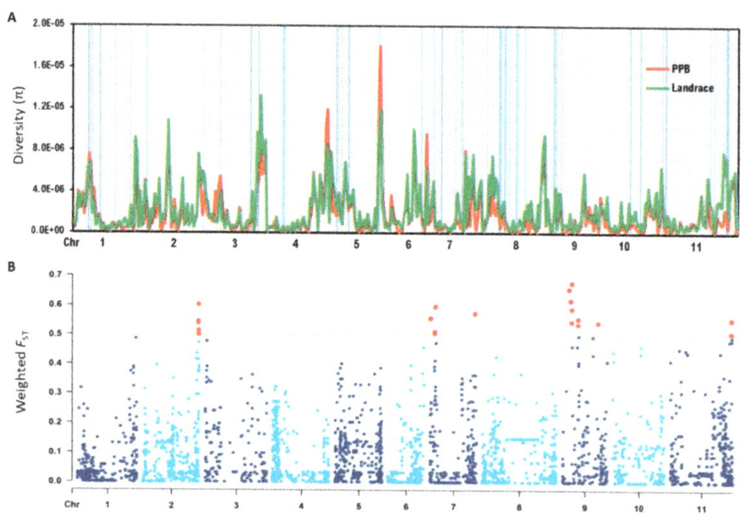

Figure 3. Nucleotide diversity and population differentiation. (**A**) Patterns of nucleotide diversity (π) across the genome between *P. vulgaris* landrace (green) and PPB (red) genotypes. Blue vertical bars show the strongly differentiated (3×) regions. (**B**) Weighted F_{ST} plot of genetic variance differentiation among landrace and PPB categories. Significant SNPs are red. Significance threshold $F_{ST} > 0.5$. SNP—single nucleotide polymorphism; Chr—Chromosome.

Calculation of population genetic differentiation (F_{ST}) between landrace and PPB beans enabled identification of loci under selection between landrace and PPB genotypes. Twenty-six single nucleotide polymorphisms (SNPs) with significant weighted F_{ST} values (>0.5) were found on Pv02, Pv07, Pv09, and Pv11 (Table 2, Figure 3B). These SNPs do not fall within the regions of high nucleotide diversity identified in the π comparison above.

Table 2. Regions of the *P. vulgaris* (2.0) genome where SNPs with significantly high weighted F_{ST} values (>0.5) were found. JBrowse was used to search for candidate genes within 100 Kb of significant SNPs. Candidate gene descriptions are listed in Table S2. (https://legumeinfo.org/genomes/jbrowse/). (Chr—chromosome).

Chr	SNP Position (Mbp)	F_{ST} Value	Number of Candidate Genes
2	48.9	0.541	8
	49.1	0.546	11
	49.1	0.603	13
	49.2	0.514	10
	49.2	0.501	
7	0.6	0.558	15
	0.7	0.558	
	4.2	0.505	13
	4.2	0.511	
	4.7	0.597	7
	38.9	0.574	10
9	5.5	0.656	8
	6.9	0.618	3
	7.7	0.546	5
	7.8	0.677	9
	7.9	0.589	
	13.5	0.536	11
	13.6	0.536	
	13.7	0.556	11
	30.6	0.543	6
11	52.4	0.553	10
	52.4	0.506	
	52.4	0.553	
	52.4	0.553	
	52.5	0.553	
	52.5	0.553	

2.3. *Identification of Candidate Genes*

Two approaches were used to identify candidate genes associated with regions of significantly high (3×) nucleotide diversity (π) in landraces and high population differentiation (F_{ST}) between landraces and PPBs, including: exploring the recent bean literature for reported quantitative trait loci (QTL) and searching the bean genome using JBrowse.

QTL associated with various traits have been reported in the literature, including those related to agronomic traits [5,32] and nitrogen fixation [15,33–36]. Our literature search revealed 10 QTL that fall within regions of significantly high landrace π values, 8 of which are associated with agronomic traits and 2 of which are associated with SNF-related traits. In a GWAS study of agronomic traits in the Middle American Diversity panel (MDP), Wilker et al. (unpublished) found QTL for days to flowering on Pv01 (23.2 Mbp), Pv02 (48.6 Mbp), and Pv06 (13.9 Mbp); QTL for days to maturity on Pv07 (35.6 Mbp) and Pv11 (40.3 Mbp); and QTL for hundred seed weight on Pv01 (23.2 Mbp), Pv05 (32.5 Mbp), and Pv11 (53.5 Mbp). Various candidate genes were found underlying these agronomic QTL and more detailed information is available in Table S1. In a separate GWAS study of agronomic traits in the MDP, Moghaddam et al. [32] found a QTL on Pv01 (42.9 Mbp) associated with growth habit, which contained an RNA polymerase-associated protein RTF1 homolog (Phvul.001G167200). For SNF-related traits, reported QTL that fall within regions of high landrace π values are associated with seed %N content and plant biomass. In a GWAS study of SNF related traits in the MDP, Wilker et al. (unpublished) found a QTL associated with seed %N content at 22.8 Mbp on Pv02. The QTL contains seven putative candidate genes, including a Ras homologous (Rho)/Rho of plants (Rop) family GTPase

(Phvul.002G106600). These genes play a role in the symbiotic interaction between the host plant and rhizobia [37]. Two separate studies investigating SNF and related traits in the Andean and the Middle American gene pools identified a QTL associated with shoot biomass at 45.1 Mbp on Pv11 [33,35]. Shoot biomass supports root symbiosis through carbohydrates generated through photosynthesis as well as serving as a sink for N generated through SNF, which is a source of N ultimately stored in the seed [35]. Beyond searching the recent literature for QTL associated with SNF and agronomic traits, we also examined the study by Schmutz et al. [5] which identified >1800 domestication candidate genes in the Middle American gene pool. Over 140 of the domestication genes identified by Schmutz et al. [5] fall within regions of high nucleotide diversity discovered in our study (see Table S1). Two of these genes have a role in symbiosis: Phvul.008G217100 is a short open reading frame (sORF) small protein of the glycerin rich protein family and is expressed during nodule ontology [38]; and Phvul.010G102300 belongs to the plant nuclear factor Y (NF Y) gene family, whose members are involved in nodule ontology, epidermal infection, and rhizobia discrimination [39].

The bean genome was explored using JBrowse in 100 kb segments centered on SNPs with significant genetic differentiation (F_{ST}) to identify putative candidate genes. All genes found within these regions are listed in Table S2. A diverse range of gene types and functions were seen, including plant defense and stress response genes, enzymes, and transcription factors. The PubMed Central database of NCBI (https://www.ncbi.nlm.nih.gov/pmc/) was used to search for published research on putative candidate genes, and some of those findings are discussed here. The region flanking the significant F_{ST} SNP on Pv02 (48.9 Mb) contains two leucine rich repeat disease resistance proteins, Phvul.002G323708 and Phvul.002G323712. This region was identified by Oladzad et al. [40] as a major QTL peak in their GWAS study of *Rhizoctonia solani* resistance in Andean beans. A second region on Pv02 (49.0 Mb) contains a disease resistance gene and one associated with nodulation. Tock et al. [41] found that the pentatricopeptide repeat superfamily protein (Phvul.002G326200) at 49.0 Mb was associated with halo blight damage, while Nova-Franco et al. [42] found that a 1-aminocyclopropane-1-carboxylate oxidase gene (Phvul.002G326600) in this region was associated with nodule senescence. A third region on Pv02 (49.2 Mb) contains a protein kinase superfamily protein (Phvul.002G328300) that Zuiderveen et al. [43] found to be significantly associated with Anthracnose resistance in a GWAS of Andean beans. The region centered at 38.9 Mb on Pv07 contains a protein kinase superfamily protein (Phvul.007G268200), which was downregulated in a slow darkening pinto bean study [44]. On Pv09, the region located at 7.8 Mb contains a GATA transcription factor (Phvul.009G035400), which belongs to a family of transcription factors that have been studied in soybean under nitrogen stress and may play a role in nitrogen metabolism [45]. In the region centered on 13.5 Mb on Pv09, a cytokinin oxidase/dehydrogenase 1 gene (Phvul.009G081800) is located that was found to be upregulated in bean root cortical cells inoculated with arbuscular mycorrhizal fungi under drought stress, compared to noninoculated roots [46].

2.4. Diversity for Symbiotic Nitrogen Fixation

The influences of genotype, environment, and the genotype by environment interaction were significant for the combined locations ANOVA for %Ndfa (Table S3); therefore, each environment was analyzed separately for this trait. At each location, significant differences were found between genotypes for %Ndfa (Tables S4–S6). At Elora 2014 (N = 49), the average nitrogen fixation capacity was 49.3% and ranged from 21.0% to 64.4%, a difference of 43.4% between the least and most effective genotypes. At Elora 2015 (N = 62), the average nitrogen fixation capacity was higher at 55.8%, yet the range for this trait was narrower with a low of 40.5% and a high of 67.3%, a difference of 26.8% between the least and most effective genotypes. At Yorito (N = 53), the average nitrogen fixation capacity was 49.0% with a range of 14.0% to 66.4%, a difference of 52.4% between the least and most effective genotypes, which was the greatest range in performance of all locations.

Further, in a separate ANOVA for each location, the genotypes were divided into breeding history categories and their means were compared. In these analyses, significant differences were found among breeding categories at two of the three trial locations. At Elora 2014, the landrace genotypes (*N* = 20, *M* = 52.5 %Ndfa) performed better than all other breeding categories, although the difference between categories was not significant (Table 3). It is evident from the Elora 2014 %Ndfa histogram (Figure S1A) that many landrace genotypes had above average nitrogen fixation performances. At Elora 2015, the Honduran conventional genotypes (*N* = 7, *M* = 59.0 %Ndfa) and the landraces (*N* = 26, *M* = 58.3 %Ndfa) exhibited the best nitrogen fixing capacities, but they were not significantly different from each other (Table 3). The average nitrogen fixing capacities of the North American check genotypes (*N* = 5, *M* = 50.0 %Ndfa) and the PPBs (*N* = 24, *M* = 53.2 %Ndfa) were similar at Elora 2015 and significantly lower than the values for the Honduran conventional and landrace genotypes (Table 4). Of particular note, Merlot (HON62) fixed the most N at Yorito (66.4%), almost 6% more than the next best genotype. This genotype, bred for Northern US production, also performed well at Elora in 2014 (64.3 %Ndfa), but had the worst performance among conventional genotypes at Elora 2015 (43.5 %Ndfa). Merlot has very dark green leaves, an indicator of plant N status, and consistently had high leaf chlorophyll content when measured with the SPAD meter in the Elora trials. (SPAD was not measured at Yorito.) As with the Elora 2014 trial, many landrace genotypes had above average nitrogen fixation performances at Elora 2015 (Figure S1B). At Yorito, the landrace genotypes (*N* = 22, *M* = 46.4 %Ndfa) showed significantly higher nitrogen fixing capacities than the PPBs (*N* = 22, *M* = 40.1 %Ndfa), whereas, the check and Honduran conventional genotypes had intermediary SNF means and did not have significantly different nitrogen fixing performance values when compared to each other nor the other breeding categories (Table 5). As we found at the other trial locations, the landraces showed above average nitrogen fixing performance at Yorito (Figure S1C).

The top five landraces with the highest SNF performance at Yorito were Vaina Rosada (60.6 %Ndfa; HON34), Cincuenteño (59.5 %Ndfa; HON48), Negro Cuarenteño (57.0 %Ndfa; HON42), Olanchano Negro (56.4 %Ndfa; HON65), and Paraísito (53.6 %Ndfa; HON49). These landraces represent both small red and black market classes and are from three different departments (Yoro, Francisco Morazán, and Intibucá). Vaina Rosada, Cincuenteño, and Paraísito are already used in participatory breeding efforts between Zamorano and FIPAH, and a number of the resulting PPB varieties were included in our panel (including HON05, HON23, HON25, HON26, HON28, HON31, HON32, and HON33). The PPB progeny of these landraces ranged in SNF capacity from 26.6 to 53.3 %N at Yorito, which is noteworthy considering SNF was not a breeding objective. Amilcar (53.3 %Ndfa; HON05) is among the top five SNF performing PPB varieties at Yorito, which also included Conan 33 (55.7 %Ndfa; HON24), Campechano (54.5 %Ndfa; HON57), San Jose (51.1 %Ndfa; HON35), and Arbolito Negro (50.8 %Ndfa; HON72). Both small red and black beans are represented in this list, and they come from three departments (Yoro, Francisco Morazán, and Santa Barbara).

Table 3. F-test of fixed effect breeding category in the GLIMMIX analysis, and the breeding category LSmeans comparisons of genotypes in the HON panel grown at Elora, 2014.

	N Derived from the Atmosphere (%)		Carbon Discrimination (Δ) (‰)		Flowering (Days)		Maturity (Days)		Yield (kg ha^{-1})	
	F	*Pr* > *F*	*F*	*Pr* > *F*	*F*	*Pr* > *F*	*F*	*Pr* > *F*	*F*	*Pr* > *F*
Breeding category	1.86	0.1441	1.51	0.2195	6.13	0.0007	3.06	0.0368	7.15	0.0004
	LSmean *	*SE*	*LSmean* *	*SE*	*LSmean* *	*SE*	*LSmean* *	*SE*	*LSmean* *	*SE*
Check	51.4 ^{a}	0.04	20.2 ^{a}	0.23	48.6 ^{ab}	1.44	114.9 ^{a}	2.0	933.7 ^{a}	53.18
Conventional	43.3 ^{a}	0.04	19.6 ^{a}	0.24	51.0 ^{ab}	1.44	108.6 ^{a}	2.82	700.9 ^{ab}	77.10
Landrace	52.5 ^{a}	0.02	20.1 ^{a}	0.14	48.2 ^{b}	0.87	109.9 ^{a}	1.19	912.8 ^{a}	28.39
PPB	46.7 ^{a}	0.02	19.9 ^{a}	0.14	51.6 ^{a}	0.83	114.0 ^{a}	1.26	721.6 ^{b}	30.51

* Means labeled with different letters within trait categories are significantly different according to ANOVA, $p = 0.05$.

Table 4. F-test of fixed effect breeding category in the GLIMMIX analysis, and the breeding category LSmeans comparisons of genotypes in the HON panel grown at Elora, 2015.

	N Derived from the Atmosphere (%)		Carbon Discrimination (Δ) (‰)		Flowering (Days)		Maturity (Days)		Yield (kg ha^{-1})	
	F	*Pr > F*	*F*	*Pr > F*	*F*	*Pr > F*	*F*	*Pr > F*	*F*	*Pr > F*
Breeding category	6.69	0.0004	0.88	0.4535	0.94	0.4225	3.27	0.0251	3.66	0.0148
	LSmean *	*SE*	*LSmean* *	*SE*	*LSmean* *	*SE*	*LSmean* *	*SE*	*LSmean* *	*SE*
Check	50.0 ^{a}	0.03	19.8^{a}	0.32	50.8 ^{a}	0.83	113.1 ^{ab}	1.65	1454.7 ^{ab}	145.59
Conventional	59.0 ^{bc}	0.02	19.8^{a}	0.29	49.6 ^{a}	0.75	109.9 ^{ab}	1.42	1613.2 ^{ab}	121.38
Landrace	58.3 ^{b}	0.01	19.5^{a}	0.24	49.3 ^{a}	0.36	109.0 ^{a}	0.63	1396.5 ^{a}	58.55
PPB	53.2 ^{ac}	0.01	19.6^{a}	0.24	49.3 ^{a}	0.36	112.1 ^{b}	0.65	1686.2 ^{b}	62.74

* Means labeled with different letters within trait categories are significantly different according to ANOVA, $p = 0.05$.

Table 5. F-test of fixed effect breeding category in the GLIMMIX analysis, and the breeding category LSmeans comparisons of genotypes in the HON panel grown at Yorito, 2014–2015.

	N Derived from the Atmosphere (%)		Carbon Discrimination (Δ) (‰)		Days to Flowering (Days)		Yield (kg ha^{-1})	
	F	*Pr > F*	*F*	*Pr > F*	*F*	*Pr > F*	*F*	*Pr > F*
Breeding category	3.72	0.0143	3.18	0.0280	3.60	0.0163	1.34	0.2647
	LSmean *	*SE*	*LSmean* *	*SE*	*LSmean* *	*SE*	*LSmean* *	*SE*
Check	49.5 ^{ab}	0.04	18.0 ^{ab}	0.25	39.0 ^{ab}	1.28	669.9 ^{a}	135.22
Conventional	40.1 ^{ab}	0.03	18.0 ^{ab}	0.22	36.8 ^{ab}	1.06	956.5 ^{a}	108.87
Landrace	46.4 ^{a}	0.01	18.2 ^{a}	0.10	36.2 ^{a}	0.53	745.2 ^{a}	52.16
PPB	40.1 ^{b}	0.01	17.7 ^{b}	0.10	38.3 ^{b}	0.54	823.2 ^{a}	52.62

* Means labeled with different letters within trait categories are significantly different according to ANOVA, $p = 0.05$.

Leaf chlorophyll content was measured at Elora in 2014 and 2015, and these values were analyzed in separate ANOVAs because of the repeated-measure nature of trait data collection. The combined ANOVA indicated significant differences for the fixed effects of genotype, environment, and the genotype by environment interaction at both the early vegetative and reproductive stages (Table S7). When this trait was analyzed by location, significant differences were found between genotypes at both locations and for both growth stages (Table S7).

2.5. Diversity for Agronomic Traits

A series of agronomic traits were measured for this study, including carbon discrimination ($\Delta^{13}C$) as an indicator of water use efficiency, plant growth stages (days to flowering and maturity), yield (kg ha^{-1}), and hundred seed weight. Significant differences were found for the fixed effects of genotype, environment, and the genotype by environment interaction for the agronomic traits carbon discrimination ($\Delta^{13}C$), days to flowering (DTF) and days to maturity (DTM), yield (kg ha^{-1}), and hundred seed weight (HSW) in a combined ANOVA (Table S3). These traits were therefore analyzed further within locations.

Significant differences were found between genotypes at all locations (Tables S4–S6) for carbon discrimination (in $\Delta^{13}C$ units) calculated according to the method of Farquhar et al. [47] from seed carbon discrimination ($\delta^{13}C$) values obtained from isotope analysis of seed samples. At Elora 2014 ($N = 48$), $\Delta^{13}C$ values ranged from 18.4‰ to 21.4‰ (Figure S2A), and at Elora 2015 ($N = 62$), the range was similar, from 17.4‰ to 21.1‰ (Figure S2B). At Yorito ($N = 53$), the $\Delta^{13}C$ values were lower, ranging from 16.4‰ to 20.0‰ (Figure S2C). When genotypes were divided into breeding categories and compared, significant differences were only found at Yorito (Table 5), where the average PPB $\Delta^{13}C$ value ($N = 22$, $M = 17.7$‰) was significantly lower than the average landrace $\Delta^{13}C$ value ($N = 22$, $M = 18.2$‰) (Table 5).

Significant differences were found between genotypes in DTF measured at Elora in 2014 and 2015 (Tables S4–S6). At Elora 2014 (N = 58), the average was 50 DTF with a range of 42–62 DTF (Figure S3A). Some Honduran genotypes (including HON9, HON13, HON18, HON28, HON32, HON33, HON36, HON39, HON44, and HON47) did not flower at that first trial location and were replaced by different genotypes at the subsequent locations. At Elora 2015 (N = 57), the average was 50 DTF with a range of 42–55 DTF (Figure S3B). When genotypes were divided into breeding categories and compared, significant differences were found at Elora 2014 only (Table 2). Overall, the landrace genotypes flowered the earliest (N = 22, M = 48 days) and they were significantly earlier than the PPB genotypes (N = 26, M = 52 days) (Table 2).

Significant differences were found among genotypes for DTM measured at Elora in 2014 and 2015 (Tables S4–S6). At Elora 2014 (N = 35), the average was 112 DTM with a range of 97–120 DTM (Figure S4A). At Elora 2015 (N = 56), the average was 111 DTM with a range of 94–115 DTM (Figure S4B). Significant differences in DTM were found only at Elora 2015 (Table 3), when genotypes were grouped by breeding history and contrasted. At Elora 2015, landraces (N = 23, M = 109 days) matured significantly earlier than PPBs (N = 22, M = 112 days).

Significant differences were found among the yields of genotypes in the Elora trials only (Tables S4 and S5). At Elora 2014 (N = 35), the average yield was 828.4 kg ha^{-1} with a range from 325.2–1124.2 kg ha^{-1} (Figure S5A). At Elora 2015 (N = 62), the average yield was 1558.6 kg ha^{-1} with a range from 600.5–2263.4 kg ha^{-1} (Figure S5B). At Yorito (N = 58), the average yield was 791.1 kg ha^{-1} with a range from 299–1471 kg ha^{-1} (Figure S5C). Significant differences were found only in the Elora trials among breeding categories (Tables 2 and 3). At Elora 2014, the North American check (N = 5, M = 933.7 kg ha^{-1}) and the Honduran landrace (N = 14, M = 912.8 kg ha^{-1}) genotypes yielded significantly more than the other categories (Table 2), and at Elora 2015, the PPBs (N = 23, M = 1686.2 kg ha^{-1}) yielded significantly more than the landraces (N = 27, M = 1396.5 kg ha^{-1}; Table 3). At Yorito, the Honduran conventional genotypes (N = 6, M = 956.5 kg ha^{-1}) returned the highest yields, followed by the PPBs (N = 24, M = 823.2 kg ha^{-1}), while the landraces (N = 24, M = 745.2 kg ha^{-1}) and the North American checks (N = 4, M = 669.9 kg ha^{-1}) had lower yields (Table 4).

Significant differences were found among hundred seed weight (HSW) calculated for samples from the Elora trials (Tables S4 and S5). At Elora in 2014 (N = 49), the average HSW was 20.2 g with a range from 13.3–29.4 g. At Elora in 2015 (N = 62), the average HSW was 21.4 g with a range from 14.5–31.9 g. No significant differences were found among genotypes grouped by breeding history (data not shown).

Plant height was measured only at the Yorito location (N = 60), and significant differences were found between genotypes for this trait (Table S6). The average height was 35.3 cm with a range of 4 cm to 47 cm. Significant differences were not found when breeding history categories were contrasted (data not shown).

2.6. Trait Correlation and Genotype by Trait Biplot Analyses

Pearson correlation analyses were performed on LSmeans for each trial environment to determine trait interactions (Table S8). In addition, trait correlations and genotype performance were visualized using genotype × trait biplots for each location (Figure 4). In the biplots, positive correlations between traits are evidenced by vectors forming acute angles, for example between SPAD and HSW at Elora in 2014 and 2015 (Figure 4B), whereas negative correlations between traits are evidenced by obtuse angles formed between vectors, such as that formed by DTF and yield at Elora in 2014 (Figure 4A). A right-angle formed between trait vectors indicates a weak or lack of association between those traits. The results of our correlation and biplot analyses support each other.

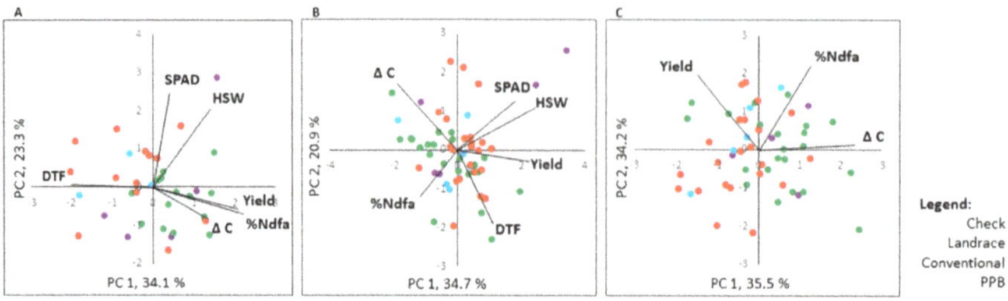

Figure 4. Biplot analysis of traits for genotypes of the Honduran panel in three location years. (**A**) Elora 2014; (**B**) Elora 2015; and (**C**) Yorito 2014–15. DTF, days to flowering; Yield, yield (kg ha^{-1}); HSW, 100 seed weight (g); Δ C, carbon discrimination; %Ndfa, percent nitrogen derived from the atmosphere; and SPAD, leaf chlorophyll content at 100% flowering.

2.6.1. %Ndfa

At Elora in 2014, %Ndfa was negatively correlated with DTF (r = −0.31) but was positively correlated with $\Delta^{13}C$ (r = 0.45) and with yield (r = 0.38) (Table S8). In the Elora 2014 biplot (Figure 4A), the landrace genotypes are clustered towards the yield and the %Ndfa vectors. The Honduran conventional genotypes are more closely associated with the DTF vector, as are the majority of the Honduran conventional genotypes (Figure 4A).

At Elora in 2015, %Ndfa was not significantly associated with any other trait (Table S8). The biplot analysis showed that DTF and $\Delta^{13}C$ are not associated with %Ndfa, and leaf chlorophyll content at flowering (SPAD) and HSW have a negative relationship with %Ndfa (Figure 4B). As in the Elora 2014 biplot, the landrace genotypes cluster towards the %Ndfa vector (Figure 4B).

At Yorito, %Ndfa was not significantly correlated with other traits (Table S8). The biplot analysis for Yorito shows landrace genotypes cluster more towards the %Ndfa and $\Delta^{13}C$ vectors, whereas the PPB genotypes cluster away from the %Ndfa vector and are more closely associated with the yield vector (Figure 4C).

2.6.2. Agronomic Traits

Leaf chlorophyll content at the early reproductive stage (SPAD) was positively correlated with hundred seed weight (HSW) at both Elora 2014 (r = 0.36) and Elora 2015 (r = 0.44) (Table S8). DTF was negatively associated with yield at Elora 2014 (r = −0.48), but no association was found between these traits at Elora 2015. DTF was negatively associated with $\Delta^{13}C$ at Elora 2015 (r = −0.37), but no association was found between these traits at Elora 2014 (Table S8). Yield was positively associated with HSW (r = 0.49) and negatively associated with $\Delta^{13}C$ (r = −0.33) at Elora 2015; however, these associations were not repeated at the other trial locations (Table S8).

2.6.3. High-Yielding and High-Fixing Genotypes

The aim of any breeding program is to generate high-yielding genotypes, and in this study an additional goal was to identify genotypes that were also high-N fixing. It is particularly useful to examine genotype performance at Yorito, where growing conditions are representative of the marginal production regions in Honduras. At Yorito, 14 genotypes had above-average yields coupled with above-average SNF performance (Figure 5). These included four Honduran conventional varieties (HON56, HON53, HON55, and HON77), four landraces (HON2, HON49, HON65, and HON66) and six PPB varieties (HON5, HON26, HON35, HON57, HON72, and HON78). Of the six PPB varieties, five were developed through participatory varietal selection, and one was developed through participatory plant breeding. These genotypes are dispersed throughout the phylogenetic tree (Figure 2A), suggesting a lack of close genetic relatedness; however, three of the high-yielding high-fixing PPB varieties share common genotypes in their pedigrees: HON5 and HON26 have a common landrace parent, Cincuenteño (HON48); HON26 shares a conventional parent, Tio Canela 75 (HON55), with HON78; and, HON5 has Amadeus 77 as a parent, which is a daughter of Tio Canela 75. There was no apparent relationship between release date and higher yields, nor was there a temporal trend for SNF performance.

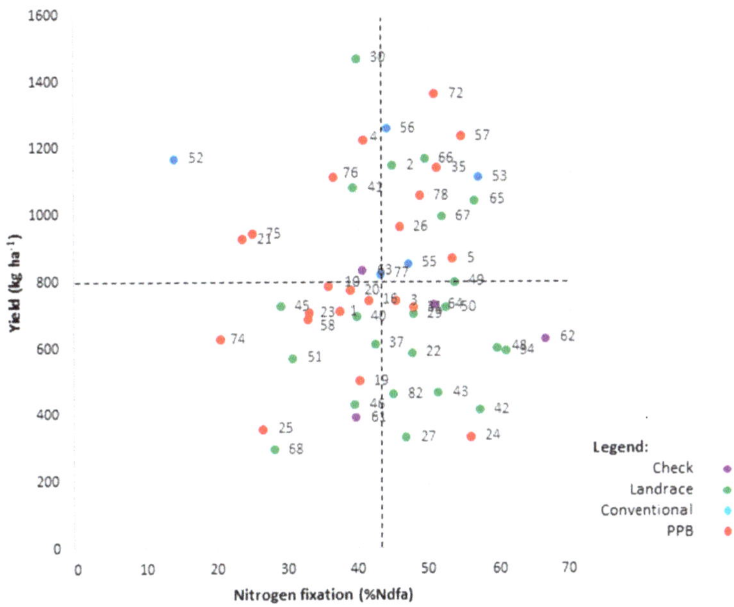

Figure 5. Genotype performance for nitrogen fixation (x axis) and yield (y axis) at Yorito. For this trial, average yield was 799 kg ha^{-1} and average nitrogen fixation was 43.3 %Ndfa, and these values are shown on the plot in dashed lines, dividing the plot into quadrants. Genotypes in the upper right quadrant are higher yielding and higher nitrogen-fixing.

3. Discussion

3.1. Genotype Origins and Pedigree Explain Honduran Panel Structure

The patterns observed in the phylogenetic tree and STRUCTURE diagrams derived from the SNP compositions of the lines in the Honduran panel largely agree with what is known about their geographic origins and their pedigrees, but there are also a few exceptions. The two large groupings in the dendrogram based on SNP similarities (groups I–III and IV) show a clear separation (with some exceptions) between landraces (CRI) and material that has been conventionally bred or is the product of participatory plant breeding (PPB). Among the landraces, the small red beans that came from Yoro, Francisco Morazán, and Intibucá, were randomly interspersed throughout group III of the phylogenetic tree with no particular pattern, based on genotype origin. In contrast, clustering of genotypes by region of origin is found among the black bean landraces (groups I and II). Group I consists of the black bean landraces, which came from Intibucá, Yoro, Francisco Morazán, and Lempira, and contains the Milpero landraces (HON36, HON47, HON44, and HON39) and 'Negro Opalaca' (HON70), which are all from Intibucá, and 'Negro' (HON43), which is found alone on the next branch is from the neighboring department, Lempira ['Merlot' (HON62) is also found in this region of the tree and is discussed below]. These landraces are the most distantly related genotypes with respect to the rest of the panel. The Milpero landraces were daylight sensitive when grown at Elora in 2014, and their photoperiod sensitivity was likely inherited from a common ancestor. All photoperiod sensitive varieties in the panel may carry the dominant *ppd* gene responsible for control of this trait [48]. The remaining black landraces from Yoro, Francisco Morazán, and Intibucá are found without any particular pattern throughout the next branch of the tree, and two are found among the small red landrace branches (HON06 and HON30). The STRUCTURE analysis shows that most of the black genotypes are admixed, suggesting a closer genetic relationship to the conventional and PPB germplasm. No small red landraces were found within the black landrace branches; however, two small red landraces, 'Concha Rosada' (HON02) and 'Rosado' (HON22), are displaced and found in group IV among the PPBs. The STRUCTURE analysis shows that Concha Rosada contains only ~10% genetic material from the landrace subpopulation, and Rosado is almost equally admixed between the landrace and PPB subpopulations. In the case of Concha Rosada, this may be explained by the fact that this landrace is widely grown and has been used as a parent in participatory breeding efforts and thus shares ancestry with many PPB varieties. Alternatively, Concha Rosada may not be a traditional landrace but instead a creolized variety derived from a formal-sector variety introduced to the Yorito region in the early 1980s [23]. The reason that Rosado is found among the PPB varieties in the tree is less apparent. Rosado has not been used as a parent for any of the PPB varieties in the panel according to the pedigree information available. Rosado recently arrived in the Yorito region, and a survey of bean farmers revealed that its origin is unknown [29]. According to M. Gomez (FIPAH), the initial population of Rosado showed phenotypic heterogeneity, and some selection has been made to create a uniform line for PPB breeding in collaboration with Zamorano. At Elora 2014, Rosado had uneven maturity, which may indicate that the seed planted that year, and the seed grown for DNA extraction, was not a fixed homogeneous line, and this heterogeneity may have resulted in misplacement of this genotype in the phylogenetic tree.

The organization of the PPB branches of the phylogenetic tree can be explained in part by common ancestry. The Honduran conventional varieties, 'Tio Canela 75' (HON55) and 'Amadeus-77' (HON52), and the landraces, 'Estica' (HON11) and 'Vaina Rosada' (HON34), have been used frequently in generating the PPB genotypes in this panel. For example, five of the seven genotypes in the left-most PPB branch (HON19, HON18, HON15, HON16, and HON14) are derived from crosses with Tio Canela 75 or Estica, or both. The STRUCTURE plot indicates that these genotypes contain a greater proportion of conventional/PPB than landrace genetic material, and therefore, they are found within the PPB branches of the tree. However, six PPB varieties (HON10, HON05, HON12, HON25, HON33, and HON72) are found among the landrace branches of the tree, which indicates that they share greater genetic similarity to their landrace parent than the other genotypes in their pedigrees. For example, FPV 921-65 (HON33) has the landrace Vaina Rosada (HON34) and Amadeus 77 in its pedigree. FPV 921-65 is found in the same branch as Vaina Rosada among the landraces in the tree, and the STRUCTURE analysis shows that FPV 921-65 has more similarity to the landrace genetic subgroup. Marcelino (HON10) and Amilcar (HON05) also contain >50% landrace genetic material (Figure 2B) and are found among the landrace branches. Fourteen PPB genotypes have between 5% and 50% admixture with the landrace subgroup. Alleles favoring agronomic characteristics for local adaptation and culinary traits contributed by landrace parents were likely prioritized under selection among PPB progeny from crosses between landrace and conventional genotypes, resulting in PPB varieties consisting of a large proportion of landrace genetic material.

Common parentage may explain the clustering of the five Honduran conventional genotypes (Amadeus-77, CENTA San Andrés, Tio Canela 75, Aifi Wuriti, and Carrizalito) in the panel. CENTA San Andrés, Amadeus-77, and Carrizalito have common ancestry. Tio Canela 75 is a parent of breeding line EAP 9510-77, which was released in the early 2000s in countries across Central America, including in El Salvador, as CENTA San Andrés, and in Honduras, as Amadeus-77. Tio Canela 75 is also a parent of EAP 9510-1, a sister line of EAP 9510-77, which was released in Honduras as Carrizalito in 2003 and as Telire in Costa Rica in 2004 [49]. Accordingly, these genotypes are closely associated in the phylogenetic tree, although some genetic differentiation has occurred between CENTA San Andres and Amadeus-77. The line 'MD 30-75' (released as Tio Canela 75) was used as a parent in generating four of the five conventional genotypes' mentioned above. This is likely a result of the effort to introduce BGMV resistance into Central American germplasm, as MD 30-75 is a highly resistant line, which carries the *bgm-1* resistance gene [50]. DEORHO (HON53) was not genotyped in our study; however, it also has MD 30-75 in its pedigree, and it is reasonable to suppose that it would also appear in this region of the dendrogram. Aifi Wuriti was the only black conventional genotype included in the panel. Interestingly, it is most closely related to three small red genotypes (Carrizalito, Conan 33, and Cedrón) rather than the small black genotypes, although Aifi Wuriti does not appear in the pedigrees of any of the small black PPB genotypes in the panel. The final Honduran conventional genotype, 'Dorado', is unusual because it is found among the landrace genotypes in the phylogenetic tree and the STRUCTURE analysis. This may be explained by the lack of a common genotype in its pedigree compared to the other conventional genotypes.

The placement of the North American check varieties (except OAC Rosito) as a separate group within the dendrogram is consistent with the unique alleles that they would be expected to have relative to the genotypes in this panel. The location of the check varieties in the landrace portion of the tree (groups I and II) may reflect the genetic diversity and wide-ranging geographic origins of the germplasm used in the University of Guelph/AAFC and Michigan State University/USDA-ARS bean breeding programs, which are aimed at introducing resistance to abiotic and biotic stresses and improving various agronomic traits. The locations of the North American checks interspersed throughout the admixed portion of the STRUCTURE plot also indicate that genetic diversity has been retained in these modern North American genotypes. OAC Rosito is a special case, since it was derived from El Salvadoran germplasm, and it is found among the PPBs in the phylogenetic analysis where it is most closely related to two PPB genotypes, Campechano (HON57) and Quebradeño (HON26). According to the STRUCTURE analysis, OAC Rosito is more genetically similar to the Honduran Conventional/PPB sub-group than the landraces. This suggests either that El Salvadoran landraces are not similar to Honduran landraces, or more likely that the landrace population from which OAC Rosito was developed was actually a creolized conventional variety. Germplasm provided by Zamorano has been used in El Salvadoran variety development since the early 2000s [26], and this could explain the genetic similarity of OAC Rosito to the PPB varieties in our panel, which were developed in collaboration with Zamorano.

3.2. Optimizing Use of Genetic Diversity of Honduran Landraces and PPB Varieties

The larger nucleotide diversity among the landraces ($\pi = 3.20 \times 10^{-4}$) than observed in the PPBs ($\pi = 2.89 \times 10^{-4}$) in the current study is consistent with general observations that landraces are more diverse than materials that are products of selection [51–54]. However, other studies that compared diversity in wild to domesticated bean accessions found wider diversification between those groups than we found between landraces and PPBs in our panel. Nanni et al. [55] reported that within the Mesoamerican gene pool, nucleotide diversity was 3.2 times higher among wild genotypes ($\pi = 17.34 \times 10^{-3}$) than domesticated genotypes ($\pi = 5.43 \times 10^{-3}$). The difference between landraces and PPBs in the current study was only 1.1 times, probably because of the small population size and because these genotype groups do not represent extremes of the diversity continuum that was sampled in the previous study. Both landraces and PPBs are selections from wild accessions, and the PPBs have probably not been as strongly selected as conventionally derived varieties.

The high level of diversity in Honduran landraces suggests they could be a source of novel alleles that could be used in breeding to improve various traits. Landraces are adapted to the environmental conditions of the locations where they were maintained, in some cases, over thousands of years. Landraces that were grown in fluctuating environments and in low-input agricultural systems may be enriched for rare alleles enabling phenotypic plasticity and inherent responsiveness to diverse abiotic and biotic stresses [56]. Landraces, in regions where they are still grown, have often been pushed to marginal production environments where their performance often exceeds that of modern cultivars [29,57,58].

Dry bean landrace germplasm across Mesoamerica is genetically diverse [59]. Soil conditions across this region are poor, and the terrain ranges from low to high altitude with steep slopes, leading to certain trait adaptations in the landraces. For example, 'Common Red Mexican', a red-seeded landrace from Mexico, has been found to be drought resistant [60], while 'Puebla 152', a black-seeded landrace also from Mexico, has superior SNF capacity [61]. Originating in the Andean region, G19833, a 'Chaucha Chuga' landrace from Perú, has tolerance to high levels of soil aluminum and low levels of phosphorus [56] and resistance to a number of bean diseases [62,63]. Our survey of the literature found genes in high landrace π regions associated with abiotic stress tolerance, phosphorus use efficiency, and nitrogen fixation (see Table S1). Conservation of landraces and mobilization of the unique genetic diversity they contain through plant breeding can help address the future need for higher yielding and climate resilient varieties.

3.3. *Regions of High Genetic Differentiation Indicate Regions Impacted by Selection*

Genetic divergence between the PPB and landrace subpopulations in the panel is indicated by regions of high genetic differentiation (F_{ST}), and these genomic regions may contain loci that have been subjected to selection pressure. We identified several regions on chromosomes Pv02, Pv07, Pv09, and Pv11 where F_{ST} values exceeded 0.5. Similarly, in a study of genetic diversity of Italian bean landraces, Lioi et al. [64] reported that genomic regions related to domestication were concentrated on Pv02, Pv07, and Pv09 for Mesoamerican types. For comparisons between wild and domesticated bean landraces, Papa et al. [65] also reported significantly larger levels of F_{ST} differentiation around genomic regions associated with domestication. While the genetic distance between the Honduran landraces and PPB genotypes included in our study is not likely as wide as that between the wild and landrace genotypes investigated by Papa et al. [65], similar trends towards genetic differentiation between landrace and PPB genotypes developed with modern breeding objectives and germplasm could be expected.

In particular, the genomic regions with large F_{ST} differences may be associated with traits that were a focus of selection in PPB breeding. However, an extensive search of the recent bean literature did not reveal any known QTLs associated with agronomic, SNF, or WUE traits that are located within the regions of large F_{ST} differentiation identified in this study. This may be because this is the first genomic survey study of Honduran material, and the distinguishing traits between landraces and PPB materials are specific to materials from that region or expressed in that location. In particular, because we are comparing two domesticated groups of genotypes, namely farmer traditional landraces and PPB varieties, the genes underlying the regions of large differentiation found in our study could be those responsible for local adaptation, culinary qualities, and favorable plant traits, rather than traits associated with domestication [5]. Additionally, the conventional germplasm used to generate PPB genotypes, either through crosses with landraces or through varietal selection, is largely limited to material in the Zamorano breeding program, which may have a specific genetic architecture.

In spite of the lack of previous QTL evidence for selection for domestication in the high F_{ST} regions, several genes in those regions that have been studied for various reasons may be associated with domestication. For example, disease resistance genes, such as those found on Pv02 for Rhizoctonia resistance (Phvul.002G323708, Phvul.002G323712; [40]), Halo blight resistance (Phvul002G326200; [41]), and Anthracnose resistance (Phvul.002G328300; [43]), and on Pv08 for Anthracnose resistance (Phvul.008G019600; [66]), and for the bean-rust interaction (Phvul.008G270500; [67]), have been associated with domestication in several crops [68,69]. Genes controlling agronomic traits have been identified in domestication studies in other crops [70–72]. Additionally, genes that affect survival in diverse growing conditions may have also been favored over the course of domestication. Two such genes are located in a region of high genetic diversity on Pv08; the ethylene-responsive transcription factor (Phvul.008G019600; [73]), an ortholog of an Arabidopsis gene known to be involved in flooding tolerance [74], and the transcription factor IIIA (Phvul.008G270400), which is upregulated in phosphorus-restricted conditions [75]. Soltani et al. [73] suggest that further studies are needed to understand the process of local adaptation and allelic selection using bean landraces and wild populations. Insight to develop climate-resilient crops can be drawn from the study of crop adaptation under natural selection and domestication [76,77].

3.4. Landraces are Superior Nitrogen Fixers

Although some genotype- and environment-influenced variability was seen in our study, our examination of symbiotic nitrogen fixation in the Honduran panel revealed a wide range of capacity for this trait. The superiority of the landraces for SNF capacity at all trial locations may be the consequence of the continual selections of these materials under conditions of low soil fertility endemic to Central America. Even today, these materials continue to be grown by small scale farmers who do not have access to fertilizer inputs. Under these conditions, bean genotypes which have developed efficient associations with nitrogen-fixing bacteria would have a larger source of nitrogen for metabolic processes and better phenotypic fitness compared to poor nitrogen-fixing genotypes. Strong nitrogen-fixers would have a competitive advantage in the low input environments and would likely be preferentially selected over time. There may be parallels between the selection pressures during landrace evolution and the selection of heirloom bean varieties, which have also been shown to have strong SNF capacity [18].

There are few studies that have investigated SNF capacity of bean landrace genotypes. Heilig et al. [33] used 'Puebla 152', a black-seeded Mexican landrace known for its nitrogen fixing capacity [78], in a cross with conventional genotype 'Zorro' to create a RIL population to study SNF. The authors found that Puebla 152 fixed between 13.0 to 45.5% of the nitrogen in samples (seed + biomass), which was slightly more than Zorro, which fixed between 5.4 to 44.4% [33]. Many landraces in our study fixed more N than Puebla 152, indicating that Honduran landraces may be a useful source of SNF capacity. The SNF performance of Zorro ranged from 47.7 to 53.0%Ndfa in our study, a mid-range performance among our check genotypes, and overall better than its performance in Heilig's study [33].

The SNF performance of the progeny of the cross between the conventional genotype Zorro and the landrace Puebla 152 may be predictive of the performance of PPB varieties that are crosses between Honduran conventional varieties and landraces. The SNF performance of Honduran PPB varieties ranged between 20.5 to 55.7%Ndfa at Yorito. Although the focus of the participatory breeding program between FIPAH and Zamorano has been to generate higher-yielding genotypes, rather than on improving SNF performance, the SNF capacity seen for the PPB varieties falls within the upper range of that found for the RILs in Heilig's study [33].

For the Honduran panel, the insights gleaned from Yorito are of particular interest because this location has growing conditions representative of small-scale growers across the region, and as much as possible, local growing practices were employed in the trial. At Yorito, there was a range in SNF performance in the landraces, and overall genotypes belonging to this breeding history group performed better than the others studied. In addition, PPB genotypes derived from crosses with the best SNF-performing landraces had strong SNF capacity. Zamorano used the methods of participatory varietal selection to develop these PPB varieties with CIALs, enabling local growers to evaluate genotype performance on their farms. For example, Amilcar and San Jose were tested by various CIALs through the regional adaptation nursery (VIDAC, Vivero de Adaptación Centroamericano) in the mid-2000s. Amilcar has the high-SNF performing landrace Cincuenteño in its pedigree. Generally, native Rhizobia inhabit tropical soils and farmers do not use Rhizobia inoculants, although Zamorano disseminates SNF knowledge through the CIALs, including effective Rhizobia inoculants and protocols for use. In addition, the breeding program has the capacity to test SNF performance in '*bancales*' where soil N levels are low and SNF-related traits, such as nodulation, can be observed. PPB breeding for enhanced SNF capacity could be expanded if grower demand and the threat of climate change and resulting raising input costs warrant it.

In addition, the range in SNF performance among the conventionally bred North American checks and Honduran conventional varieties was wide. The superior SNF capacity Merlot exhibited at Yorito would suggest it has value as a breeding parent for this trait in Honduras; however, it has a larger seed size and a dark red seed coat; traits that are less preferred by Honduran consumers and could be challenging to select against in a breeding program. Of the Honduran conventional genotypes, DEORHO (HON53) fixed the highest amount of N (56.8%), but it performed poorly in Elora fixing 34.4% (2014) and 52.6% (2015) of its N. It is not found in the pedigrees of any PPB genotype included in our panel, but DEORHO has been a popular variety in commercial growing regions of the country. It has disease resistance, high yield, and the preferred light red seed coat color, making it a good candidate for future PPB breeding efforts.

3.5. PPB Genotypes Have Superior Water Use Efficiency Values

Plants that have higher water use efficiency (WUE) are more drought tolerant, and WUE can be estimated using carbon differentiation (Δ^{13}C) values measured from plant biomass. During photosynthesis, plants discriminate against the incorporation of the heavy C isotope (^{13}C), depleting ^{13}C in plant biomass and driving lower Δ^{13}C values [79]. Plants with comparatively low biomass Δ^{13}C values can be considered more drought tolerant. WUE has been studied in beans, including landraces. A study by Munoz-Perea et al. [80] of the WUE of 16 dry bean genotypes in drought-stressed and nonstressed environments found that the two landraces included differed in their responses to drought stress, but Common Red Mexican was among the best performers under drought stress conditions. In contrast, in our study, the significantly lower Δ^{13}C values measured in Yorito for the PPB genotypes than the landraces indicates that the PPB varieties in our panel may be more resilient to drought conditions than the landraces.

The drought resistant characteristics of the PPBs were likely contributed by the conventional parents. For example, PM2-Don Rey (HON23) was the most WUE PPB genotype at Yorito, with a Δ^{13}C value of 16.36‰. PM2-Don Rey was developed through PPB methods from a cross between the landrace, Paraísito, and the Honduran conventional variety, Carrizalito. It was released as a drought-resistant variety in 2016 [81], and it is the first variety from the EAP-Zamorano-CIAL PPB collaborations to be released at the national level. A second PPB genotype, Marcelino (HON10), was developed through participatory varietal selection, and it had similar WUE (16.43‰) to PM2-Don Rey. The PPBs FPY-724-43 (HON16; 17.03‰), Cedrón (HON03; 17.41‰), and Amilcar (HON05; 17.42‰) have the next best WUEs.

Three landraces had WUE values below 17.5‰, including Concha Rosada (HON02; 16.85‰), Chapin Rojo (HON27; 16.86‰), and Chirineño (HON67; 17.30‰). Concha Rosada is of particular note because it is favored by poor farmers for its early maturity, which allows it to escape drought conditions late in the growing season [29,30]. Our study indicates that Concha Rosada not only has drought resistance through 'drought escape' but also has WUE characteristics that enable it to survive drought.

Among the conventionally bred genotypes, Carrizalito (HON77) and OAC Rosito (HON63) were the most water use efficient with low Δ^{13}C values of 17.47‰ and 17.54‰ at Yorito. Carrizalito is a commercial Honduran variety, and it was used as a parent contributing disease resistance, agronomic, and likely WUE traits, to PM2-Don Rey. Among the check genotypes, OAC Rosito, recently developed at the University of Guelph from an El Salvadoran plant introduction, had the best WUE performance. This characteristic is likely retained from its domestication in Central America, and this enabled it to outperform the other check genotypes that have been developed for production in the Great Lakes region of North America.

In the coming decades, the effects of climate change are predicted to bring drier conditions to Honduras, and drought-resistant crops will help protect yields through periods of minimal rainfall. It has been proposed that WUE can be improved through selection and breeding. In alfalfa, evaluating genotypes for Δ^{13}C and selecting for lower Δ^{13}C values has been used to improve WUE in this important forage species [79]. The current results indicate that there is variation among the Honduran PPB and conventional bean germplasm in WUE traits, and selecting for lower Δ^{13}C values could be applied to beans in Honduras to generate improved varieties that are more resilient to drought conditions.

3.6. Conventional Genotypes Have Superior Yields

Releasing varieties with higher yields is the objective of modern breeding programs, and our trial results suggest that improvements impacting yield have been made along the breeding history continuum from landraces to PPBs to conventional varieties. Considering the Honduran germplasm, the landraces were the lowest-yielding group at Yorito, followed by PPBs and the conventional varieties. This result is counter to the findings of early experiments performed by CIALs, where landraces out-yielded conventional materials [23]. However, our trial was conducted at the FIPAH office in Yorito, where soil fertility is less restrictive and the plot is flat, whereas the early CIAL trials were conducted in farmers' fields, which have low-fertility soils and sloped land; conditions for which conventional materials were not developed. The superior performance of the conventional materials in our trial at Yorito is consistent with the aim of modern breeding practices in generating higher yielding varieties. The North American checks were the highest yielding at Elora 2014, as they were bred for performance in this region, whereas the Honduran conventional and PPBs performed poorly at Elora 2014. The landraces also performed well at Elora 2014, and this may be attributed to phenotypic plasticity resulting from retention of useful nucleotide diversity enabling them to perform well in a new environment.

3.7. Utility of Panel Genotypes for Breeding

The different breeding and/or selection histories for the materials contributing to the phenotypic diversity present in the Honduran panel may provide opportunities for improving different traits in beans in the same way that a number of studies with different crops have found unique benefits from the use of landraces. In wheat, for example, cultivation of landraces in low-input systems has led to the conservation of traits that increase the duration of photosynthesis, which can lead to an increase in grain yield [82]. In a study comparing barley landrace and modern cultivar performances under stress conditions, the landraces were higher yielding and were less likely to fail outright [58]. The advantage of using landraces as parents in breeding programs has also been explored. In a study examining barley yields under drought conditions, progeny from crosses using landrace genotypes were found to be higher yielding than progeny from crosses without landraces in their pedigrees [56]. The authors concluded that breeding crops for vulnerable environments could be enhanced by identifying landrace alleles associated with yield performance and abiotic stress adaptation and employing these in breeding programs [56].

In the current study, landraces, which had superior symbiotic nitrogen fixation characteristics could be excellent sources of novel alleles for this trait. Similarly, PPB materials, which had superior WUE, and cultivated varieties, which had superior yields within their target environments, might be exploited, respectively, for these purposes. In general, all the germplasm types that were tested represent useful resources for breeding for important traits in the face of climate change and increasing production costs/demands.

The diversity for SNF capacity inherent in Honduran bean landraces, and their unique adaptation to the microclimates where they are grown, leads us to conclude that the inclusion of landrace germplasm in breeding for enhanced SNF would produce high fixing genotypes with growth and culinary characteristics already accepted by small-scale bean growers.

4. Materials and Methods

4.1. Plant Material

The Honduran Panel was assembled in 2014 at the University of Guelph in collaboration with agronomists at FIPAH. The initial panel contained 27 landraces, 26 PPB varieties, and 5 Honduran conventional checks provided by FIPAH, as well as 6 North American checks sourced from the University of Guelph bean breeding program.

The landraces consisted of traditional inbred varieties unimproved by modern plant breeding, which are grown by subsistence farmers in hillside communities. The PPB varieties were generated either through participatory varietal selection (PVS) or participatory plant breeding (PPB) through a collaboration between the bean breeding program at Zamorano and CIALs associated with FIPAH. The landraces and PPB varieties were sourced by M. Gomez (FIPAH) from six departments in west–central Honduras. The landraces were from Yoro, Intibucá, Francisco Morazán, and Lempira, and the PPB varieties were from Yoro, Francisco Morazán, Santa Bárbara, and Comayagua (Figure 1). Seed was either collected directly from farmers in their communities or sourced from central seed banks maintained by FIPAH and PRR. The five Honduran conventional checks were developed for production in lower to mid-altitude, hillside and valley commercial-production regions of the country. The six North American varieties consisted of Merlot and OAC Rosito as small red market class checks, Zorro as a black market class check, and three navy beans: OAC Mist, a high-nitrogen-fixing genotype, R99, a nonfixing mutant, and its parent line OAC Rico. All genotypes in the panel belong to race Mesoamerica [83].

In the first trial location (Elora 2014), 10 Honduran genotypes were found to be daylight sensitive and were not grown at the subsequent locations. Additionally, seed of 16 genotypes that exhibited uneven maturity in Elora 2014 were sent to Puerto Rico for seed increase over the winter of 2015. For the second trial location (Yorito, 2014–2015), 12 new genotypes (6 landraces, 5 improved, and 1 Honduran conventional check) were added to the panel. For the third trial location (Elora 2015), seed harvested from Elora 2014, from the Puerto Rican seed increase, and from Yorito were used, as available. An additional Honduran conventional variety was grown that year to fill in the experimental design. Overall, a total of 77 genotypes were tested in the Honduran panel, 50 genotypes of which were grown at all 3 locations. A summary of the genotypes included in the Honduran Panel, including trial year, market class, seed source, and pedigree information is provided in Tables 6–8 according to breeding history.

Table 6. Genotypes of the North American check and Honduran conventional breeding categories that were included in the HON panel tested at three field locations, 2014–15. The trials in which each genotype was included and whether the entry was SNP genotyped are indicated. Breeding category, market class, genealogy, and origin are provided where available.

									Origin	
HON Entry	**VARIETY**	**Elora '14**	**Yorito '14-'15**	**Elora '15**	**SNP Genotyped**	**Breeding Category**	**Market Class**	**Genealogy**	**Institution or Organization**	**Notes**
59	OAC Rico	x		x	x	Check	Navy	(Ex Rico 23/Narda)/Ex Rico 23 *See* [84]	University of Guelph	Resistant to BCMV and Anthracnose. Tolerant to white mold. Unremarkable SNF capacity. In other studies it fixed approximately 53% of N [18].
60	R99[1]	x	x	x	x	Check	Navy	*See* [85]	Agriculture Agri-Food Canada (AAFC)	Non-nodulating experimental line. Derived from OAC Rico through ethyl methan sulphonate (EMS) mutagesis [86]. Used in natural abundance method to establish a baseline nitrogen level in seed measured for ^{14}N and ^{15}N accumulation.
61	OAC Mist	x	x	x	x	Check	Navy	*See* [87]	University of Guelph	High yielding, late season. Resistant to BCMV and CBB. Generally high SNF capacity. Farid and Navabi (2015) reported that OAC Mist fixed as much as 78.5% of N [16]. Wilker et al. [18] reported that OAC Mist fixed an average of 55% N.
62	Merlot	x	x	x	x	Check	Small red	*See* [88]	United States Department of Agriculture—Agriculture Research Service (USDA-ARS)	Intense red seed color. Larger seed size than Honduran beans (mean HSW 39.2 g). Resistant to rust, BCMV, and BCMNV. Susceptible to anthracnose. Moderate SNF capacity. Wilker et al. (unpublished) found Merlot fixed as much as 64.9 % of its N.

Table 6. *Cont.*

									Origin	
HON Entry	**VARIETY**	**Elora '14**	**Yorito '14-'15**	**Elora '15**	**SNP Genotyped**	**Breeding Category**	**Market Class**	**Genealogy**	**Institution or Organization**	**Notes**
63	OAC Rosito	x	x	x	x	Check	Small red	*See* [89]	University of Guelph	Developed from a diverse landrace originating in El Salvador [89]. Dark red seed color. Similar seed size to Honduran beans (mean HSW 21.7 g). Resistant to BCMV. Susceptible to Anthracnose and CBB. The SNF capacity of OAC Rosito has not been examined previous to the current study.
64	Zorro	x	x	x	x	Check	Black	*See* [90]	Michigan State University	Resistant to rust and anthracnose and is less affected by white mold. Moderate SNF capacity. Wilker et al. [18] reported that Zorro fixed an average of 59%, and Wilker et al. (unpublished) found that Zorro fixed as much as 46.9% of its N.
80	CENTA San Andrés [3]			x	x	Conventional	Small red	EAP 9510-77, [MD 30-75/DICTA 105]	PIF/Zamorano, UPR, CENTA, El Salvador; 2003	Same breeding line as Amadeus-77 [49]. Resistant to BGYMV and BCMV. Heat tolerant and adapted for production in lower-altitude coastal areas [49].
52	Amadeus-77 [3]	x	x	x	x	Conventional	Small red	EAP 9510-77, [MD 30-75/DICTA 105]	PIF/Zamorano, UPR, DICTA, Honduras; 2003	Same breeding line as CENTA San Andres [49]. Resistant to BGYMV and BCMV. Heat tolerant and performs well in low altitude, coastal areas. Widely adopted across Central America, and in 2010, accounted for around 50% of commercial production in the region [26].

Table 6. *Cont.*

										Origin
HON Entry	**VARIETY**	**Elora '14**	**Yorito '14-'15**	**Elora '15**	**SNP Genotyped**	**Breeding Category**	**Market Class**	**Genealogy**	**Institution or Organization**	**Notes**
56	Dorado [3]	x	x	x	x	Conventional	Small red	DOR 364, [BAT 1215 x (RAB 166 x DOR 125)]	Profrijol, DICTA, Honduras; 1990	Also known as 'DOR 364'. Resistant to BGYMV and BCMV [49]. Yields well across environments and has mid-range maturity; however, it has a dark red seed coat [49].
53	DEORHO [2,3]	x	x	x		Conventional	Small red	SRC 2-18-1, [Milenio/MD 30-75]	PIF/Zamorano, UPR, DICTA, Honduras; 2007	Also known as 'DEHORO' and 'INTA Matagalpa'. Resistant to BGYMV and BCMV. Higher yielding and desirable light red seed coat color [91]. Popular with Honduran growers, accounting for 23% of the red bean acreage in 2010 [26]. DEORHO was not grown for DNA extraction and consequently was not included in the genetic analyses carried out for this study.
55	Tio Canela 75 [2,3]	x	x	x	x	Conventional	Small red	MD 30-75, [DOR 483//DOR 391/Pompadour J]	PIF/Zamorano, UPR, Honduras; 1996	Resistant to BGYMV and BCMV [49]. Yields well across environments, has mid-range maturity, and has a shiny red seed [49]. Tio Canela 75 is a parent line of Amadeus-77 and Carrizalito.
77	Carrizalito [3]		x	x	x	Conventional	Small red	EAP 9510-1, [MD 30-75/DICTA 105]	PIF/Zamorano, UPR, DICTA, Honduras; 2003	Resistant to BGYMV and BCMV. Early maturity (68–70 DAP) and upright plant architecture. High yielding variety, adapted to mid-altitude production [91].
54	Aifi Wuriti [2]	x	x	x	x	Conventional	Black	EAP 9712-13, Tio Canela 75/DICTA 105/BG12WB12//Tio Canela 75/DICTA 105/AL12	PIF/Zamorano, UPR, SNS, Haiti; 2008	Also known as 'Negro Olfirwit'. Resistant to BGYMV, BCMV, is tolerant of low soil fertility, and is early-maturing [92]. Popular in Haiti and the Dominican Republic and was successfully adopted by growers in southeast Guatemala [92].

[1] R99 was genotyped but not included in the genetic analyses (Figure 2). [2] Genotypes exhibiting uneven maturity at Elora 2014 and sent to Puerto Rico for seed increase in winter 2015. [3] Varieties developed using conventional breeding methods according to J.C. Rosas, pers. comm.

Table 7. Genotypes of the Landrace breeding category that were included in the HON panel tested at three field locations, 2014–15. The trials in which each genotype was included and whether the entry was SNP genotyped are indicated. Breeding category, market class, and origin details are provided where available.

								Origin			
HON Entry	**VARIETY**	**Elora '14**	**Yorito '14–'15**	**Elora '15**	**SNP Genotyped**	**Breeding Category**	**Market Class**	**Institution, Farmer, or Organization**	**Locality**	**Municipality**	**Department**
02	Concha Rosada [2]	x	x	x	x	Landrace	Small red	FIPAH	Yorito	Yorito	Yoro
06	Negro Pedreño	x	x	x	x	Landrace	Black	Odir Palma	La Esperanza	Yorito	Yoro
07	Negro Concha Blanca [2]	x	x	x	x	Landrace	Black	Odir Palma	La Esperanza	Yorito	Yoro
08	Balin Rojo [2]	x	x	x	x	Landrace	Small red	Edy Hernandez	La Patastera	Yorito	Yoro
09	Carmelita [1]	x			x	Landrace	Small red	Francisco Murillo	La Esperanza	Yorito	Yoro
11	Estica	x		x	x	Landrace	Small red	Irene Hernandez	La Esperanza	Yorito	Yoro
22	Rosado [2]	x	x	x	x	Landrace	Small red	Odir Palma	La Esperanza	Yorito	Yoro
27	Chapin Rojo	x	x	x	x	Landrace	Small red	Daniel Vargas	El Injerto	Comayagua	Francisco Morazán
29	Uva	x	x	x		Landrace	Black	Alonso Gutierrez	San Jose	Vallecillo	Francisco Morazán
30	Chapin Negro [2]	x	x	x	x	Landrace	Black	Ovidio Valeriano	Nocoro	Vallecillo	Francisco Morazán
34	Vaina Rosada	x	x	x	x	Landrace	Small red	CIAL San Jose	San Jose	Vallecillo	Francisco Morazán
36	Milpero Negro [1]	x			x	Landrace	Black	Bertilio Antonio Rodriguez	San Pedrito	Opalaca	Intibucá
37	Negro Vaina Blanca	x	x	x	x	Landrace	Small red	Carmen Azucenaa Giron	Guayabal	Jesus de Otoro	Intibucá
38	Mano de Piedra	x	x	x	x	Landrace	Small red	Maria Laines	Maye	Jesus de Otoro	Intibucá
39	Milpero Rojo [1]	x			x	Landrace	Small red	Bertilio Antonio Rodriguez	San Pedrito	Opalaca	Intibucá
40	Vaina Blanca	x	x	x	x	Landrace	Black	Tiburcio Dias	Pueblo Viejo	NA	NA
41	Negro Arbolito	x	x	x	x	Landrace	Black	Armando Pineda	Crucita Oriente	Jesus de Otoro	Intibucá
42	Negro Cuarenteño	x	x	x	x	Landrace	Black	Maria Laines	Maye	Jesus de Otoro	Intibucá

Table 7. *Cont.*

HON Entry	VARIETY	Elora '14	Yorito '14–'15	Elora '15	SNP Genotyped	Breeding Category	Market Class	Origin: Institution, Farmer, or Organization	Origin: Locality	Origin: Municipality	Origin: Department
43	Negro [2]	x	x	x	x	Landrace	Black	Antonio Espinosa	Iguala	Lempira	Lempira
44	Milpero Gatiador [1]	x			x	Landrace	Carioca	Evelino Sanchez	La Vegas	NA	NA
45	Ponga la Olla	x	x	x	x	Landrace	Black	Estalin Diaz	Pueblo Viejo	NA	NA
46	Madura Parejo	x	x	x	x	Landrace	Small red	Maria Lainez	Maye	Jesus de Otoro	Intibucá
47	Milpero Blanco [1]	x			x	Landrace	White	Maria Juana Gutierrez	Monte Verde	Opalaca	Intibucá
48	Cincuenteño	x	x	x	x	Landrace	Small red	FIPAH	Yorito	Yorito	Yoro
49	Paraísito	x	x	x	x	Landrace	Small red	FIPAH	Yorito	Yorito	Yoro
50	Rojo de Seda	x	x	x	x	Landrace	Small red	FIPAH	Yorito	Yorito	Yoro
51	Marciano	x	x	x	x	Landrace	Small red	FIPAH	Yorito	Yorito	Yoro
65	Olanchano Negro		x	x	x	Landrace	Black	CIAL San Jose de la Mora	San Jose de la Mora	Vallecillo	Francisco Morazán
66	Seda-Vallecillo		x	x	x	Landrace	Small red	CIAL San José de la Mora	San Jose de La Mora	Vallecillo	Francisco Morazán
67	Chirineño		x	x	x	Landrace	Small red	CIAL Chirinos	Chirinos	Cedros	Francisco Morazán
68	Roseño		x	x	x	Landrace	Small red	Adan Bustillo	La Fortuna	Victoria	Yoro
70	Negro Opalaca		x	x	x	Landrace	Black	NA	Monte Verde	San Francisco de Opalaca	Intibucá
82	Olanchano Rojo		x			Landrace	Small red	San Jose de la Mora	San Jose de la Mora	Vallecillo	Francisco Morazán

[1] Genotypes exhibiting daylight sensitivity at Elora 2014. [2] Genotypes exhibiting uneven maturity at Elora 2014 and sent to Puerto Rico for seed increase in winter 2015. [NA] Information not available.

Table 8. Genotypes of PPB breeding category that were included in the HON panel tested at three field locations, 2014–15. The trials in which each genotype was included and whether the entry was SNP genotyped are indicated. Breeding category, market class, and origin details are provided where available.

									Seed Origin			
HON Entry	VARIETY	Elora '14	Yorito '14–'15	Elora '15	SNP Genotyped	Breeding Category	Market Class	Genealogy	Institution, Farmer, or Organization	Locality	Municipality	Department
01	Macuzalito	x	x	x	x	PPB (PPB)[3]	Small red	PPB-9911-44-5-13M, [Concha Rosada//SRC 1-1-18/SRC 1-12-1]	PIF/Zamorano, FIPAH, Honduras; 2004	Yorito	Yorito	Yoro
13	FPY-722-53 [1]	x			x	PPB (PPB)	Small red	FPY-722-53, Tio Canela 75/ Estica	PIF/Zamorano, FIPAH, CIAL Santa Cruz	Santa Cruz	Yorito	Yoro
14	FPY-722-38	x	x	x	x	PPB (PPB)	Small red	FPY-722-38, Tio Canela 75/Estica	PIF/Zamorano, FIPAH, CIAL Santa Cruz	Santa Cruz	Yorito	Yoro
15	FPY-722-13 [1]	x	x		x	PPB (PPB)	Small red	FPY-722-13, Tio Canela 75/Estica	PIF/Zamorano, FIPAH, CIAL Santa Cruz	Santa Cruz	Yorito	Yoro
16	FPY-724-43 [1]	x	x	x	x	PPB (PPB)	Small red	FPY-724-43, Macuzalito/Estica	PIF/Zamorano, FIPAH, CIAL Santa Cruz	Santa Cruz	Yorito	Yoro
18	FPY-721-16 [1]	x			x	PPB (PPB)	Small red	FPY-721-16, Amadeus 77/Estica	PIF/Zamorano, FIPAH, CIAL Santa Cruz	Yorito	Yorito	Yoro
19	FPY-722-41 [2]	x	x	x	x	PPB (PPB)	Small red	FPY-722-41, Tio Canela 75/Estica	PIF/Zamorano, FIPAH, CIAL Santa Cruz, Amilcar Orellana	La Esperanza	Yorito	Yoro
23	PM2-Don Rey [2]	x	x	x	x	PPB (PPB)	Small red	IBC-302-29, Carrizalito//Carrizalito/Paraísito	PIF/Zamorano, UPR, DICTA, ASOCIAL Vallecillo, Reinaldo Funez; 2014	San Isidro	Vallecillo	Francisco Morazán
25	FPV-921-4	x	x	x	x	PPB (PPB)	Small red	FPV-921-4, Vaina Rosada/Amadeus 77	CIAL San Isidro	San Isidro	Vallecillo	Francisco Morazán
26	Quebradeño	x	x		x	PPB (PPB)	Small red	IBC-307-7, [TC75//TC75/Cincuenteño]	CIAL Quebrada	Trinidad de Quebrada	Vallecillo	Francisco Morazán

Table 8. *Cont.*

HON Entry	VARIETY	Elora '14	Yorito '14–'15	Elora '15	SNP Genotyped	Breeding Category	Market Class	Genealogy	Seed Origin			
									Institution, Farmer, or Organization	Locality	Municipality	Department
28	FPV-921-61 [1]	x				PPB (PPB)	Small red	FPV-921-61, Vaina Rosada/Amadeus 77	CIAL San Isidro	San Isidro	Vallecillo	Francisco Morazán
31	FPV-923-25 [2]	x	x	x	x	PPB (PPB)	Small red	FPV-923-25, Vaina Rosada/Conan 33	CIAL San Isidro	San Isidro	Vallecillo	Francisco Morazán
32	FPV-923-21 [1]	x			x	PPB (PPB)	Small red	FPV-923-21, Vaina Rosada/Conan 33	CIAL San Isidro	San Isidro	Vallecillo	Francisco Morazán
33	FPV 921-65 [1]	x			x	PPB (PPB)	Small red	FPV 921-65, Vaina Rosada/Amadeus 77	CIAL San Isidro	San Isidro	Vallecillo	Francisco Morazán
03	Cedrón	x	x	x	x	PPB (PVS)[4]	Small red	EAP 9508-93, [Bribri//MD 30-37//PR 9177-214-1/Tio Canela 75]	PIF/Zamorano, FIPAH, Honduras; 2005	Yorito	Yorito	Yoro
04	Chepe	x	x	x	x	PPB (PVS)	Small red	703-SM15216-11-5	PIF/Zamorano, CIAT, FIPAH, Honduras; 2012	Yorito	Yorito	Yoro
05	Amilcar [2]	x	x	x	x	PPB (PVS)	Small red	IBC 308-24, Amadeus 77//Amadeus 77/Cincuenteño	PIF/Zamorano, FIPAH, Honduras; 2012	Yorito	Yorito	Yoro
10	Marcelino [2]	x	x	x	x	PPB (PVS)	Small red	EAP 9508-41, Bribri/MD 30-37//PR 9177-214-1/Tio Canela 75	PIF/Zamorano, FIPAH, Edy Hernandez; 2012	La Patastera	Yorito	Yoro
12	Esperanceño [2]	x	x	x	x	PPB (PVS)	Small red	PR 0310-26-3-3, VAX 6	PIF/Zamorano, UPR, CIAT, FIPAH, Pablo Orellana; 2011	La Esperanza	Yorito	Yoro
17	Paisano PF	x	x	x	x	PPB (PVS)	Small red	MER-2212-28, Milenio/Amadeus 77	PIF/Zamorano, PRR, CIAL Palmicha Fatima; 2011	Palmichal Fatima	Siguatepeque	Comayagua
20	523-DFBS 15092-04-4	x	x	x	x	PPB (PVS)	Small red	523-DFBS 15092-04-4	PIF/Zamorano, CIAT, FIPAH	Yorito	Yorito	Yoro
21	519-DFBZ 15094-39-4 [2]	x	x	x	x	PPB (PVS)	Small red	519-DFBZ 15094-39-4	PIF/Zamorano, CIAT, FIPAH	Yorito	Yorito	Yoro

Table 8. *Cont.*

HON Entry	VARIETY	Elora '14	Yorito '14–'15	Elora '15	SNP Genotyped	Breeding Category	Market Class	Genealogy	Seed Origin			
									Institution, Farmer, or Organization	Locality	Municipality	Department
24	Conan 33	x	x	x	x	PPB (PVS)	Small red	PRF-9659-25B-1, [EAP 9503/RS3//Bribri/MD 30-37////EAP 9503/RS3//A429/K2///V8025/XR 16492//APN83/CNC]	PIF/Zamorano, FIPAH, Asocial Vallecillo; 2005	Trinidad de Quebrada	Vallecillo	Francisco Morazán
35	San Jose	x	x	x	x	PPB (PVS)	Small red	X0-233-171-4, VAX 3	PIF/Zamorano, UPR, CIAT, FIPAH, CIAL San José; date NA	San Jose	Vallecillo	Francisco Morazán
57	Campechano	x	x	x	x	PPB (PVS)	Small red	SX14825-7-1	PIF/Zamorano, CIAT, PRR, ASOCIALAYO; 2012	La Buena Fe	Zacapa	Santa Bárbara
58	Don Kike	x	x	x	x	PPB (PVS)	Small red	MDSX14797-6-1	PIF/Zamorano, CIAT, PRR, ASOCIALAYO; 2012	La Buena Fe	Zacapa	Santa Bárbara
72	Arbolito Negro		x	x	x	PPB (PVS)	Black	SJC 729-89, Negro Vaina Blanca/PRF 9924-50N	PIF/Zamorano, PRR, ASOCIALAYO; 2013	La Buena Fe	Zacapa	Santa Bárbara
74	Rojo Delicia		x	x	x	PPB (PVS)	Small red	703-SM15216-11-4-VR	PIF/Zamorano, CIAT, PRR, ASOCIALAYO; 2015	La Buena Fe	Zacapa	Santa Bárbara
75	Don Cristóbal		x	x	x	PPB (PVS)	Small red	SRC1-12-1-8, [DOR476//XAN155/DOR364]	PIF/Zamorano, CIAT, PRR, CIAL Laguna Seca; 2015	Laguna Seca	Taulabe	Comayagua
76	Victoria		x	x	x	PPB (PVS)	Small red	SRS 56-3, [Amadeus77/SEA5]	PIF/Zamorano, PRR, CIAL Nueva Esperanza; 2015	Nueva Esperanza	Concepción Sur	Santa Bárbara
78	Nueva Esperanza		x	x	x	PPB (PVS)	Small red	DICZA 9801, UPR9606-2-2/MD 30-37	PIF/Zamorano, PRR, CIAL Nueva Esperanza; 2005	Nueva Esperanza	Concepción Sur	Santa Bárbara

[1] Genotypes exhibiting daylight sensitivity at Elora 2014. [2] Genotypes exhibiting uneven maturity at Elora 2014 and sent to Puerto Rico for seed increase in winter 2015. [3] PPB are participatory plant bred varieties derived from a cross between a landrace and a breeding line; classified by FIPAH. [4] PVS are varieties derived from participatory varietal selection where breeding lines are selected through generations of testing by CIALs; classified by FIPAH. [NA] Information not available.

4.2. Field Experimental Design and Maintenance

Field trials were carried out at the University of Guelph Elora research station (ERS) in summer 2014 and 2015 and at Yorito, Honduras in the *Postrera* season (planted December, 2014).

4.2.1. Elora

The Elora fields were selected based on low soil nitrogen levels as measured by preplant soil tests and by site crop rotation histories that indicated that no dry bean crops had been produced in those fields for the previous decade at a minimum. In 2014, nitrates (NO_3^-) were low at 7.1 ppm, and ammonium ($NH^+{}_4$) levels were 3.2 ppm. In 2015, nitrates (NO_3^-) were very low at 4.8 ppm, and ammonium ($NH^+{}_4$) levels were 4.5 ppm.

A square lattice design (8 × 8) with two replications was used for each trial. At the ERS in 2014, 135 seeds of each genotype were grown in 4-row plots (150 cm × 190 cm, 37.5 cm between rows) with approximately 6 cm between plants. At the ERS in 2015, 60 seeds of each genotype were grown in 4-row plots (150 cm × 90 cm, 37.5 cm between rows) with approximately 5 cm between plants within rows.

Clean seed of each genotype was coated with commercially available Nodulator (Becker-Underwood) *Rhizobium leguminosarum* bv *phaseoli* peat-based inoculant prior to planting. The day before planting, inoculant powder (1/4 teaspoon, approximately 0.4 g in 2014; 1/8 teaspoon, approximately 0.2 g in 2015) was added to each seed envelope and the contents were shaken to coat the seeds. Inoculated seed was stored at the ERS at 4 °C until planting to maintain inoculant viability. The entire contents of each envelope (coated seed + loose inoculant powder) was planted. Successful inoculation was confirmed each year by observing pink (active) nodules on a few plants chosen at random throughout the trial at flowering time.

At the ERS, plots were maintained with standard practices throughout the growing season, except no-nitrogen fertilizer was used. Preplant fertilizer (0-20-20) at a rate of 200 kg ha^{-1} was applied approximately one week prior to planting. Preplant herbicide was applied to control broadleaf weeds, and pesticides were applied as needed throughout the growing season at standard rates to control leafhoppers, anthracnose, and root rot (see details in Wilker et al. [18]). Plots were manually weeded once before canopy closure each year.

At Elora 2014, the harvest was staggered according to maturity. The plots were pulled by hand at maturity and threshed at the side of the field using a Wintersteiger plot combine (Wintersteiger AG, Upper Austria, Austria) with a Classic Seed-Gauge weighing system by Harvest-Master (Juniper Systems Inc., Logan, UT, USA), and plot seed weight and moisture content were recorded. Plots that did not reach reproductive/physiological maturity were not harvested. In 2015, plot harvest took place after all plots reached maturity with an SPC20 Almaco plot combine (ALMACO, Nevada, IO, USA), which automatically recorded moisture and weight (kg ha^{-1}) for each plot at 13% moisture.

4.2.2. Yorito

The Yorito trial site was chosen based on field uniformity, access to irrigation, and proximity to the FIPAH regional office. Soil NO_3^- levels were 18 ppm ("moderate" to "high") at Yorito, and the field had been used for bean and maize production in previous seasons. The FIPAH agronomist, M. Gomez, indicated that the trial site conditions were representative of the bean production areas around Yorito.

Seed of each North American check variety were provided by the University of Guelph, and seed of all Honduran bean genotypes in the trial were sourced in Honduras. A square lattice design (8 × 8) with two replications was used for the trial. At Yorito, 100 seeds of each genotype were grown in 2-row plots (100 cm × 500 cm, 50 cm between rows) with approximately 30 cm between plants and 3 seeds sown per hole, as per the traditional planting system.

Inoculant for the trial, a mixture of *Rhizobium etli* (CIAT 632) and *R. tropici* (CIAT 899) strains, was provided by J.C. Rosas (EAP-Zamorano) and was applied according to a protocol provided by *PIF* at a rate of 500 g ha^{-1} [93]. Briefly, the peat-based inoculant powder was applied to slightly moistened seed to ensure it adhered well to the seeds, and once sown, the seeds were covered with soil to protect the inoculant from the sun. CIAT 899 has high symbiotic stability and efficient N-fixation characteristics [15].

Plots were maintained with standard practices throughout the growing season. A preplant fertilizer of 12-24-12 NPK was used at a rate of 64.81 kg ha^{-1} as a formulation without N was not available in Honduras. Carbendazim (DEROSAL, Bayer) at a rate of 400 mL ha^{-1} was used to protect the trial from angular leaf spot (*Pseudoscercospora griseola*), common bacterial blight (*Xanthomonas axonopodis* pv. *phaseoli*), and rust (*Uromyces appendiculatus*).

At Yorito, plots were harvested by hand over a number of days as each plot matured. Yield (kg ha^{-1}) was measured at harvest, the two plots of each genotype were bulked, and a subsample of seed was sent to Guelph for further analyses.

4.3. Phenotyping

4.3.1. Elora

As plots initiated the reproductive phase, DTF was recorded as the date when 50% of the plants in a plot had one flower open. Relative leaf chlorophyll content was measured twice during the growing season (when the mean number of plots had reached (1) the second trifoliate stage and (2) at 100% flowering) using an SPAD 502 Plus Chlorophyll Meter (Konica-Minolta). The meter was calibrated according to manufacturers' instructions each time the unit was powered-on (https://www.specmeters.com/assets/1/22/2900P_SPAD_502.pdf). The middle leaflet in the top-most, fully expanded trifoliate leaf was used for the measurements, and three plants were sampled at random per plot.

As plots reached physiological maturity, DTM was recorded as the date when plots were ready to harvest. Three plants were randomly sampled from mature plots, placed in large paper bags, and dried in a repurposed tobacco kiln (De Cloet Bulk Curing Systems, model TPG-360, Tillsonburg, ON, Canada) at 33 °C at the ERS for 24–48 h. Roots were cut from each plant and the above-ground biomass was weighed. Plants were then threshed using an indoor belt thresher (Agriculex SPT-1A, Guelph, ON, Canada), their seed collected, weighed, and counted. Hundred seed weights (HSW) were calculated.

4.3.2. Yorito

As plots initiated the reproductive phase, DTF was recorded as the date when 50% of the plants in a plot had one flower open. Disease ratings and agronomic scores were collected throughout the growing season; however, statistical analyses revealed no significant differences between genotypes, and these traits are not further reported here. DTM was recorded as the date when 90% of the pods in a plot had changed color.

4.4. Isotope Analysis

Seed from each plot was processed and analyzed as detailed in Wilker et al. [18]. Briefly, seed was finely ground, precisely measured, and the isotope analyses were carried out using mass spectrometry at the Agriculture and Agri-food Canada (AAFC) gas chromatography mass spectrometry facility in Lethbridge, Alberta. Samples were analyzed for δ^{15}N (‰) and δ^{13}C (‰).

To calculate the percent nitrogen derived from the atmosphere (%Ndfa), the natural abundance method was used on seed samples in this study [94]. Seed N represents the total N accumulated by a plant over the course of the growing season, and seed N values are representative of whole-plant N values [21]. Additionally, processing seed samples is more efficient than shoot tissue.

The natural abundance method uses the following equation,

$$\%Ndfa = \frac{\delta^{15}N_{reference\ plant} - \delta^{15}N_{fixing\ plant}}{\delta^{15}N_{reference\ plant} - B}$$

where $\delta^{15}N_{ref.\ plant}$ is the rate of $\delta^{15}N$ in the reference genotype (R99), $\delta^{15}N_{fixing\ plant}$ is the $\delta^{15}N$ of the N-fixing bean genotype, and B is the average $\delta^{15}N$ of beans grown in an environment where its entire N source is from fixation [95]. The B-value was obtained for this experiment as described by Farid (2015) [36]. Briefly, $\delta^{15}N$ was measured and averaged for 20 bean genotypes from both the Andean and Middle American gene pools, which were grown in a growth room in N-free media. Normalized $\delta^{15}N$ values were used for all genotypes, and an average of $\delta^{15}N$ values for R99 were used in %Ndfa calculations.

To estimate water use efficiency, $\delta^{13}C$ values obtained from GCMS seed analysis were used following the methods proposed by Farquhar et al. [47]. Because the current WUE discussion utilizes $\Delta^{13}C$ values, the raw $\delta^{13}C$ values were converted to $\Delta^{13}C$ using the following equation:

$$\Delta C = \frac{\delta^{13}C_{air} - \delta^{13}C_{plant}}{1 + \delta^{13}C_{plant}}$$

where $\delta^{13}C_{air}$ is the current free atmospheric level of approximately −8‰ and $\delta^{13}C_{plant}$ is calculated per plant seed sample using appropriate C isotope standards. For example, a plant with a $\delta^{13}C$ value of −28.2‰ yields $\Delta^{13}C = (-0.008 + 0.0282)/(1 - 0.0282) = 0.0202/0.9718 = 20.7‰$.

4.5. Genotyping

To enable discovery of the genetic structure of the HON panel, 73 genotypes (see Table 1) were genotyped for single nucleotide polymorphisms (SNPs). DNA was extracted following the same methods outlined in Wilker et al. [18]. Briefly, plants were grown in a controlled environment (16 h photoperiod, 22 °C) at the University of Guelph, and young-leaf tissue samples were harvested, freeze dried, and the DNA extracted according to manufacturer's instructions using the NucleoSpin Plant II kit (Macherey-Nagel, Dueren, Germany). DNA of adequate quality was sent to the Genome Quebec Innovation Centre (McGill University, Montreal) for SNP genotyping using the Illumina Infinium iSelect Custom Genotyping BeadChip (BARCBEAN6K_3) containing 5398 SNPs [96]. TASSEL was used to filter the SNP data (MAF > 0.01) to a set representing 72 individuals and containing 4314 polymorphic SNPs [97]. Missing data comprised less than 3% of the data, and these were subsequently imputed using Beagle v4.1 [98] as described by Torkamaneh and Belzile [99].

4.6. Population Structure

The population structure of the Honduran panel was determined using a number of methods, as follows. First, the population structure was estimated using variational Bayesian inference implemented in fastSTRUCTURE [100]. Five runs were performed for each number of populations (K) set from 1 to 10 using the 4.3K genome-wide SNPs identified in this study. A ChooseK analysis was conducted to determine the number of subpopulations that maximize the marginal likelihood. Then, a principal component analysis (PCA) was conducted in TASSEL and plotted using PCAshiny in R [101]. Finally, the evolutionary relationships among the genotypes of the panel were inferred using the Neighbor-Joining method with the genome-wide SNP data [102]. The taxa were clustered together using the bootstrap test (1000 replicates) [103]. The tree was drawn to scale, with branch lengths (next to the branches) in the same units as those of the evolutionary distances used to infer the phylogenetic tree. The evolutionary distances were computed using the Maximum Composite Likelihood method and the units correspond to the number of base substitutions per site. Evolutionary analyses were conducted in MEGA7 [104,105].

4.7. Genetic Diversity

The levels of genetic diversity in the landrace and PPB breeding history categories of the HON panel were assessed using the 4.3K SNP dataset and VCFtools [106]. The π statistic provides an indication of polymorphism within a population as measured by nucleotide diversity, and Tajima's

D (D) provides an indication of selection pressure [107,108]. Both π and D were measured in sliding windows of 1 Mb across the genome using the—window-pi and—TajimaD options in VCFtools [106], which resulted in an average of 6 SNPs per window. The pairwise π and D values were also calculated among different subpopulations. Genome-wide averages of π and D for each breeding history category were generated by taking the average across all windowed calculations. Landrace and PPB π values were compared across the genome, and regions where landrace π exceeded PPB π by more than 3 times were considered highly differentiated, and the regions that were at least 25,000 bp in length were considered significant. To investigate the level of differentiation between the landrace and PPB genotypes, the F_{ST} statistic was computed. F_{ST} was calculated using the—weir-fst-pop option in VCFtools in sliding windows of 100 bp across the genome [106]. Weighted F_{ST} values range from 0 with no genetic differentiation to 1 where fixation of alleles has occurred. F_{ST} values exceeding 0.5 were considered significant in our analysis.

4.8. Candidate Gene Investigation

A literature search was carried out to identify previously reported QTL and genes that colocalize with regions where π values were highly differentiated between the Landrace and PPB categories. JBrowse (https://legumeinfo.org/genomes/jbrowse/) was used to explore the bean genome around SNPs with significant weighted F_{ST} values in order to identify candidate genes. A 100 kb region centered on each significant marker was searched.

4.9. Statistical Analysis

Analysis of variance (ANOVA) tests were performed on the phenotypic data collected from each environment and all environments combined, using the GLIMMIX procedure in SAS (version 9.4, SAS Institute, Cary, NC, USA, 2012). In the combined analysis, genotype, environment, and the genotype-by-environment interaction were considered fixed effects, while all other effects and their interactions were considered random. In the separate environment analyses, genotypes were considered fixed effects, while all other effects and the interaction effects were considered random. The Shapiro–Wilk test was performed on the residuals in the UNIVARIATE procedure to test their normality [109]. Random and independent distributions of the residuals were visually examined by plotting the studentized residuals against the predicted values. Data that generated outlier residuals were removed from the data set. Further, single degree of freedom contrasts were conducted in GLIMMIX between breeding history categories—landraces, PPBs, and conventional and check genotypes—contrasting each category to each of the others. Repeated measures of leaf chlorophyll content (SPAD) were taken, and separate ANOVA tests were used to compare SPAD values at each time point in a combined analysis and by environment. The least squared means (LSmeans) for each trait were computed using the LSMEANS statement in the GLIMMIX procedure for each genotype.

Using the LSmeans calculated above, the pair-wise Pearson's coefficients of correlation were computed for all traits in the CORR procedure in SAS. The PRINCOMP and PRINQUAL procedures were used in SAS to generate the principal component (PC) values, to estimate the proportion of variance accounted for by each PC, and to plot PC1 against PC2 to generate genotype x trait (GT) biplots to determine genotype and trait interactions in each environment [110].

5. Conclusions

The aim of our study was to evaluate a large set of Honduran landraces and varieties generated through participatory plant breeding, as well as check conventional genotypes, to ascertain their value in future breeding efforts. We used simple genomics and phenotyping to characterize the panel. Our genetic analyses found that the panel is divided into predominantly-landrace and predominantly-PPB groupings, with Honduran conventional genotypes sharing most similarity to the PPBs. Breeding history and pedigrees account for this division. The genetic diversity analysis revealed that landraces have retained a higher level of nucleotide diversity than PPB genotypes, which we attribute to selection pressure imposed by breeding for different production environments/objectives, and the use of a small number of conventional/elite parents in breeding efforts. The nucleotide diversity inherent in landraces can be used to increase the frequency of rare alleles in breeding programs. Beyond genetic characterization, it is important to classify germplasm for trait phenotypic diversity, which could be employed in breeding, because landraces that have evolved in adverse environments contribute adaptive traits to variety development.

Two traits that contribute to climate resiliency, nitrogen fixation capacity, and water use efficiency were evaluated in our study. Genotypes with good nitrogen fixation capacity are an asset for remote hillside growers who have limited funds and limited opportunity to purchase inputs because of poor market access. Landraces were shown to have superior SNF capacity and are already favored by hillside producers. Genotypes with enhanced water use efficiency will also be an asset to hillside growers in a future with drier and more changeable growing conditions. The PPB and conventional varieties in our study show promising characteristics for drought resilience. Further evaluation of PPB varieties under drought conditions is warranted.

Honduran bean production will continue to be carried out predominantly by small-scale hillside producers. The widely grown farmer landraces are locally adapted and accepted by consumers, and future breeding efforts should deliver varieties that maintain the inherent SNF capacity of landraces, while enhancing drought resilience and producing high yields. PPB methods employed in the breeding efforts between EAP-Zamorano and FIPAH- and PRR-supported CIALs have succeeded in generating promising PPB varieties. One variety that combines these characteristics is Amilcar (HON05). It is a small red PPB variety selected among germplasm provided by Zamorano for improvement by a CIAL in Yoro. Ultimately, Amilcar has been widely accepted because of its commercial value, culinary qualities, and disease resistance, but its climate resiliency traits and yield potential are of critical importance. Landrace characterization, conservation, and employment in breeding programs will bring continued benefits.

Supplementary Materials: The following are available online at http://www.mdpi.com/2223-7747/9/9/1238/s1, **Table S1**: Candidate genes identified in regions of the *P. vulgaris* (2.0) genome where high nucleotide diversity (π) was discovered in landrace genotypes compared to PPB genotypes [5,32,33,35,37–39,46,63,67,111–115]. **Table S2**: Candidate genes identified in regions of the *P. vulgaris* (2.0) genome where SNPs with significantly high weighted FST values (>0.5) were found [40,42–44,46,116,117]. **Table S3**: F-test of fixed and Pearson's χ^2 test of random effects in the combined GLIMMIX mixed model analysis of the Honduran panel genotypes tested in multiple field locations in Ontario, Canada and Yorito, Honduras, 2014–2015. **Table S4**: F-test of fixed effect of genotype overall and by breeding history category, and the Pearson's χ^2 test of random effects in the GLIMMIX analysis of 63 genotypes tested at Elora, Ontario, Canada, 2014. **Table S5**: F-test of fixed effect of genotype overall and by breeding history category, and the Pearson's χ^2 test of random effects in the GLIMMIX analysis of 63 genotypes tested at Elora, Ontario, Canada, 2015. **Table S6**: F-test of fixed effect of genotype overall and by breeding history category, and the Pearson's χ^2 test of random effects in the GLIMMIX analysis of 63 genotypes tested at Yorito, Honduras, 2014–2015. **Table S7**: F-test of fixed effects and Pearson's χ^2 test of random effects in the GLIMMIX mixed-model analysis of leaf chlorophyll content (SPAD) at early vegetative (SPAD1) and reproductive (SPAD2) stages in the Honduran panel overall (Combined) and at Elora in 2014 and 2015. **Table S8**: Phenotypic (r_p) correlations among %Ndfa and other traits estimated in the Honduran panel grown in three locations in 2014–2015. Values shown for traits within environments are significantly correlated at the 5% level. The number of genotypes analyzed in each correlation is presented in brackets for each significant correlation. **Figure S1**: Histograms of Nitrogen derived from the atmosphere (%) values of genotypes comprising the HON panel tested at three field locations from 2014–2015. A. Elora 2014, B. Elora 2015, and C. Yorito. Breeding history category averages with standard errors are presented, followed by individual genotype LSmeans with standard errors. North American check genotypes (CK; purple), Honduran conventional genotypes (CV; blue), landraces (LR; green), and PPB varieties (PB; red). Genotype numbers correspond to those listed in Tables 6–8. **Figure S2**: Histograms of carbon discrimination (Δ) values of genotypes comprising the HON panel tested at three field locations from 2014–2015. A. Elora 2014, B. Elora 2015, and C. Yorito. Breeding history category averages with standard errors are presented, followed by individual genotype LSmeans with standard errors. North American check genotypes (CK; purple), Honduran conventional genotypes (CV; blue), landraces (LR; green), and PPB varieties (PB; red). Genotype numbers correspond to those listed in Tables 6–8. **Figure S3**: Histograms of days to flowering of genotypes comprising the HON panel tested at two field locations from 2014–2015. A. Elora 2014, B. Elora 2015. Breeding history category averages with standard errors are presented, followed by individual genotype LSmeans with standard errors. North American check genotypes (CK; purple), Honduran conventional genotypes (CV; blue), landraces (LR; green), and PPB varieties (PB; red). Genotype numbers correspond to those listed in Tables 6–8.

Figure S4: Histograms of days to maturity of genotypes comprising the HON panel tested at two field locations from 2014–2015. A. Elora 2014 and B. Elora 2015. Breeding history category averages with standard errors are presented, followed by individual genotype LSmeans with standard errors. North American check genotypes (CK; purple), Honduran conventional genotypes (CV; blue), landraces (LR; green), and PPB varieties (PB; red). Genotype numbers correspond to those listed in Tables 6–8. **Figure S5**: Histograms of yield (kg ha^{-1}) values of genotypes comprising the HON panel tested at three field locations from 2014–2015. A. Elora 2014, B. Elora 2015, and C. Yorito. Breeding history category averages with standard errors are presented, followed by individual genotype LSmeans with standard errors. North American check genotypes (CK; purple), Honduran conventional genotypes (CV; blue), landraces (LR; green), and PPB varieties (PB; red). Genotype numbers correspond to those listed in Tables 6–8. **Data Repository files**: Phenotypic, SNP, and genetic diversity datasets. Available at https://doi.org/10.5683/SP2/DJB7VY

Author Contributions: Conceptualization, J.W., A.N., S.H. and M.G.C.; methodology, J.W., K.P.P. and A.N.; software, D.T. and M.E.; validation, J.W. and M.E.; formal analysis, J.W., A.N., M.E. and D.T.; investigation, J.W.; resources, J.W., K.P.P., A.N., M.G.C. and J.C.R.-S.; data curation, J.W. and D.T.; writing—original draft preparation, J.W.; writing—review and editing, K.P.P.; visualization, J.W. and D.T.; supervision, K.P.P.; project administration, J.W.; funding acquisition, K.P.P., and S.H. All authors have read and agreed to the published version of the manuscript.

Funding: This research was funded by the Ontario Bean Growers, Agriculture Agri-Food Canada through their Cluster Program, the Ontario Ministry for Food and Rural Affairs, the Ontario Ministry for Research and Innovation, and the Food from Thought Program at the University of Guelph funded by the Canada First Research Excellence Fund. SeedChange/USC Canada, Global Affairs Canada, International Development Research Centre, World Accord, Development Fund, Norway, USAID.

Acknowledgments: GCMS analyses were performed by Brett Hill at AAFC, Lethbridge and arranged by Frederic Marsolais, AAFC, London.

Conflicts of Interest: The authors declare no conflict of interest. The funders had no role in the design of the study; in the collection, analyses, or interpretation of data; in the writing of the manuscript, or in the decision to publish the results.

References

1. Bitocchi, E.; Rau, D.; Bellucci, E.; Rodriguez, M.; Murgia, M.L.; Gioia, T.; Santo, D.; Nanni, L.; Attene, G.; Papa, R. Beans (Phaseolus ssp.) as a model for understanding crop evolution. *Front. Plant Sci.* **2017**, *8*, 1–21. [CrossRef] [PubMed]
2. Bitocchi, E.; Nanni, L.; Bellucci, E.; Rossi, M.; Giardini, A.; Zeuli, P.S.; Logozzo, G.; Stougaard, J.; McClean, P.; Attene, G.; et al. Mesoamerican origin of the common bean (*Phaseolus vulgaris* L.) is revealed by sequence data. *Proc. Natl. Acad. Sci. USA* **2012**, *109*, E788–E796. [CrossRef] [PubMed]
3. Mamidi, S.; Rossi, M.; Annam, D.; Moghaddam, S.; Lee, R.; Papa, R.; McClean, P. Investigation of the domestication of common bean (*Phaseolus vulgaris*) using multilocus sequence data. *Funct. Plant Biol.* **2011**, *38*, 953–967. [CrossRef] [PubMed]
4. Mamidi, S.; Rossi, M.; Moghaddam, S.M.; Annam, D.; Lee, R.; Papa, R.; McClean, P.E. Demographic factors shaped diversity in the two gene pools of wild common bean *Phaseolus vulgaris* L. *Heredity* **2013**, *110*, 267–276. [CrossRef] [PubMed]
5. Schmutz, J.; McClean, P.E.; Mamidi, S.; Wu, G.A.; Cannon, S.B.; Grimwood, J.; Jenkins, J.; Shu, S.; Song, Q.; Chavarro, C.; et al. A reference genome for common bean and genome-wide analysis of dual domestications. *Nat. Genet.* **2014**, *46*, 707–713. [CrossRef]
6. Gepts, P.; Osborn, T.C.; Rashka, K.; Bliss, F.A. Phaseolin-Protein Variability in Wild Forms and Landraces of the Common Bean (*Phaseolus vulgaris*): Evidence for Multiple Centers of Domestication. *Econ. Bot.* **1986**, *40*, 451–468. [CrossRef]
7. Bitocchi, E.; Bellucci, E.; Giardini, A.; Rau, D.; Rodriguez, M.; Biagetti, E.; Santilocchi, R.; Spagnoletti Zeuli, P.; Gioia, T.; Logozzo, G.; et al. Molecular analysis of the parallel domestication of the common bean (*Phaseolus vulgaris*) in Mesoamerica and the Andes. *New Phytol.* **2013**, *197*, 300–313. [CrossRef]
8. Bernsten, R.H. Bean-Growing Environments in Honduras. Available online: https://msu.edu/~{}bernsten/beanatlas/CountryPages--withGIS/Honduras/2.BeanGrowingEnvironment/hon_text_bean_growinging_environments.htm (accessed on 28 October 2019).
9. FAO. FAOSTAT Database, Dry Bean Yield. 2018. Available online: http://www.fao.org/faostat/en/#data (accessed on 22 May 2020).

10. Beebe, S.; Ramirez, J.; Jarvis, A.; Rao, I.M.; Mosquera, G.; Bueno, J.M.; Blair, M.W. Genetic Improvement of Common Beans and the Challenges of Climate Change. *Crop Adapt. Clim. Chang.* **2011**, 356–369.
11. World Bank Group. Climate Change Knowledge Portal, Honduras. Available online: https://climateknowledgeportal.worldbank.org/country/honduras/climate-data-projections (accessed on 14 May 2020).
12. Beebe, S. Common Bean Breeding in the Tropics. In *Plant Breeding Reviews, Volume 36*; Janick, J., Ed.; Wiley-Blackwell: Hoboken, NJ, USA, 2012; Volume 36, pp. 357–426, ISBN 9781118358566.
13. Lindstrom, K.; Mousavi, S.A. Effectiveness of nitrogen fixation in rhizobia. *Microb. Biotechnol.* **2020**, *13*, 1314–1335. [CrossRef]
14. Kamfwa, K.; Cichy, K.A.; Kelly, J.D. Genome-wide association analysis of symbiotic nitrogen fixation in common bean. *Theor. Appl. Genet.* **2015**, *128*, 1999–2017. [CrossRef]
15. Diaz, L.M.; Ricaurte, J.; Cajiao, C.; Galeano, C.H.; Rao, I.; Beebe, S.; Raatz, B. Phenotypic evaluation and QTL analysis of yield and symbiotic nitrogen fixation in a common bean population grown with two levels of phosphorus supply. *Mol. Breed.* **2017**, *37*, 76. [CrossRef]
16. Farid, M.; Navabi, A. N2 fixation ability of different dry bean genotypes. *Can. J. Plant Sci.* **2015**, *95*, 1243–1257. [CrossRef]
17. Heilig, J.A.; Wright, E.M.; Kelly, J.D. Symbiotic nitrogen fixation of black and navy bean under organic production systems. *Agron. J.* **2017**, *109*, 2223–2230. [CrossRef]
18. Wilker, J.; Navabi, A.; Rajcan, I.; Marsolais, F.; Hill, B.; Torkamaneh, D.; Pauls, K.P. Agronomic Performance and Nitrogen Fixation of Heirloom and Conventional Dry Bean Varieties Under Low-Nitrogen Field Conditions. *Front. Plant Sci.* **2019**, *10*, 1–21. [CrossRef] [PubMed]
19. Sanyal, D.; Osorno, J.M.; Chatterjee, A. Influence of Rhizobium inoculation on dry bean yield and symbiotic nitrogen fixation potential and symbiotic nitrogen fixation potential. *J. Plant Nutr.* **2020**, *43*, 798–810. [CrossRef]
20. Aserse, A.A.; Markos, D.; Getachew, G.; Yli-halla, M. Rhizobial inoculation improves drought tolerance, biomass and grain yields of common bean (*Phaseolus vulgaris* L.) and soybean (*Glycine max* L.) at Halaba and Boricha in Southern Ethiopia. *Arch. Agron. Soil Sci.* **2020**, *66*, 488–501. [CrossRef]
21. Barbosa, N.; Portilla, E.; Buendia, H.F.; Raatz, B.; Beebe, S.; Rao, I. Genotypic differences in symbiotic nitrogen fixation ability and seed yield of climbing bean. *Plant Soil* **2018**, *428*, 223–239. [CrossRef]
22. Reinprecht, Y.; Schram, L.; Hill, B.; Pauls, K.P. Effects of Nitrogen Application on Nitrogen Fixation in Common Bean Production. *Front. Plant Sci.* **2020**, *11*, 1172. [CrossRef]
23. Humphries, S.; Gallardo, O.; Jiménez, J.; Sierra, F. *Linking Small Farmers to the Formal Research Sector: Lessons from a Participatory Bean Breeding Programme in Honduras*; AgREN: Odisha, India, 2005.
24. Villa, T.C.C.; Maxted, N.; Scholten, M.; Ford-Lloyd, B. Defining and identifying crop landraces. *Plant Genet. Resour.* **2005**, *3*, 373–384. [CrossRef]
25. Mather, D.L.; Bernsten, R.; Rosas, J.C.; Viana Ruano, A.; Escoto, D. The economic impact of bean disease resistance research in Honduras. *Agric. Econ.* **2003**, *29*, 343–352. [CrossRef]
26. Reyes, B.A. *The Economic Impact of Improved Bean Varieties and Determinants of Market Participation: Evidence from Latin America and Angola*; Michigan State University: East Lansing, MI, USA, 2012.
27. Ceccarelli, S.; Grando, S.; Singh, M.; Michael, M.; Shikho, A.; Al Issa, M.; Al Saleh, A.; Kaleonjy, G.; Al Ghanem, S.M.; Al Hasan, A.L.; et al. A methodological study on participatory barley breeding I. Selection phase. *Euphytica* **2000**, *133*, 185–200. [CrossRef]
28. Almekinders, C.J.M.; Elings, A. Collaboration of farmers and breeders: Participatory crop improvement in perspective. *Euphytica* **2001**, *122*, 425–438. [CrossRef]
29. Humphries, S.; Rosas, J.C.; Gómez, M.; Jiménez, J.; Sierra, F.; Gallardo, O.; Avila, C.; Barahona, M. Synergies at the interface of farmer-scientist partnerships: Agricultural innovation through participatory research and plant breeding in Honduras. *Agric. Food Secur.* **2015**, *4*, 1–17. [CrossRef]
30. Gomez, M.; Rosas, J.C.; Humphries, S.; Jiménez, J.; Barahona, M.; Avila, C.; Orellana, P.; Sierra, F. Pushing back against bureaucracy: Farmers' role in decentralizing plant breeding and seed production in Honduras. In *Farmers and Plant Breeding*; Westengen, O.T., Winge, T., Eds.; Routledge: London, UK, 2020.
31. Rosas, J.C. Contribuciones del Programa de Investigaciones en Frijol en Centro América y El Caribe. *Ceiba* **2013**, *52*, 65–73. [CrossRef]

32. Moghaddam, S.M.; Mamidi, S.; Osorno, J.M.; Lee, R.; Brick, M.; Kelly, J.; Miklas, P.; Urrea, C.; Song, Q.; Cregan, P.; et al. Genome-Wide Association Study Identifies Candidate Loci Underlying Agronomic Traits in a Middle American Diversity Panel of Common Bean. *Plant Genome* **2016**, *9*, 1–21. [CrossRef]
33. Heilig, J.A.; Beaver, J.S.; Wright, E.M.; Song, Q.; Kelly, J.D. QTL analysis of symbiotic nitrogen fixation in a black bean population. *Crop Sci.* **2017**, *57*, 118–129. [CrossRef]
34. Cichy, K.A.; Porch, T.G.; Beaver, J.S.; Cregan, P.; Fourie, D.; Glahn, R.P.; Grusak, M.A.; Kamfwa, K.; Katuuramu, D.N.; McClean, P.; et al. A *Phaseolus vulgaris* Diversity Panel for Andean Bean Improvement. *Crop Sci.* **2015**, *55*, 2149–2160. [CrossRef]
35. Kamfwa, K.; Cichy, K.A.; Kelly, J.D. Identification of quantitative trait loci for symbiotic nitrogen fixation in common bean. *Theor. Appl. Genet.* **2019**, *132*, 1375–1387. [CrossRef]
36. Farid, M. *Symbiotic Nitrogen Fixation in Common Bean*; University of Guelph: Guelph, ON, Canada, 2015.
37. Flores, A.C.; Via, V.D.; Savy, V.; Villagra, U.M.; Zanetti, M.E.; Blanco, F. Comparative phylogenetic and expression analysis of small GTPases families in legume and non-legume plants. *Plant Signal. Behav.* **2018**, *13*, e1432956. [CrossRef]
38. Guillén, G.; Díaz-camino, C.; Loyola-torres, C.A.; Aparicio-fabre, R.; Hernández-lópez, A.; Díaz-sánchez, M.; Sanchez, F. Detailed analysis of putative genes encoding small proteins in legume genomes. *Front. Plant Sci.* **2013**, *4*, 208.
39. Rípodas, C.; Castaingts, M.; Clúa, J.; Blanco, F.; Zanetti, M.E. Annotation, phylogeny and expression analysis of the nuclear factor Y gene families in common bean (*Phaseolus vulgaris*). *Front. Plant Sci.* **2015**, *5*, 761.
40. Oladzad, A.; Porch, T.; Rosas, J.C.; Moghaddam, S.M.; Beaver, J.; Beebe, S.E.; Burridge, J.; Jochua, C.N.; Miguel, M.A.; Miklas, P.N.; et al. Single and multi-trait GWAS identify genetic factors associated with production traits in common bean under abiotic stress environments. *G3 Genes Genomes Genet.* **2019**, *9*, 1881–1892.
41. Tock, A.J.; Fourie, D.; Walley, P.G.; Holub, E.B.; Soler, A.; Cichy, K.A.; Pastor-Corrales, M.A.; Song, Q.; Porch, T.G.; Hart, J.P.; et al. Genome-wide linkage and association mapping of halo blight resistance in common bean to race 6 of the globally important bacterial pathogen. *Front. Plant Sci.* **2017**, *8*, 1–17. [CrossRef] [PubMed]
42. Nova-Franco, B.; Íñiguez, L.P.; Valdés-López, O.; Alvarado-Affantranger, X.; Leija, A.; Fuentes, S.I.; Ramírez, M.; Paul, S.; Reyes, J.L.; Girard, L.; et al. The micro-RNA72c-APETALA2-1 node as a key regulator of the common bean-Rhizobium etli nitrogen fixation symbiosis. *Plant Physiol.* **2015**, *168*, 273–291. [CrossRef] [PubMed]
43. Zuiderveen, G.H.; Padder, B.A.; Kamfwa, K.; Song, Q.; Kelly, J.D. Genome-Wide association study of anthracnose resistance in andean beans (*Phaseolus vulgaris*). *PLoS ONE* **2016**, *11*, e0156391. [CrossRef]
44. Duwadi, K.; Austin, R.S.; Mainali, H.R.; Bett, K.; Marsolais, F.; Dhaubhadel, S. Slow darkening of pinto bean seed coat is associated with significant metabolite and transcript differences related to proanthocyanidin biosynthesis. *BMC Genom.* **2018**, *19*, 1–13. [CrossRef]
45. Zhang, C.; Hou, Y.; Hao, Q.; Chen, H.; Chen, L.; Yuan, S.; Shan, Z.; Zhang, X.; Yang, Z.; Qiu, D.; et al. Genome-wide survey of the soybean GATA transcription factor gene family and expression analysis under low nitrogen stress. *PLoS ONE* **2015**, *10*, e0125174. [CrossRef]
46. Recchia, G.H.; Konzen, E.R.; Cassieri, F.; Caldas, D.G.G.; Tsai, S.M. Arbuscular mycorrhizal symbiosis leads to differential regulation of drought-responsive genes in tissue-specific root cells of common bean. *Front. Microbiol.* **2018**, *9*, 1–24. [CrossRef]
47. Farquhar, G.D.; Ehleringer, J.R.; Hubick, K.T. Carbon Isotope Discrimination and Photosynthesis. *Annu. Rev. Plant Physiol. Plant Mol. Biol.* **1989**, *40*, 337–503. [CrossRef]
48. Wallace, D.H.; Yourstone, K.S.; Masaya, P.N.; Zobel, R.W. Photoperiod gene control over partitioning between reproductive and vegetative growth. *Theor. Appl. Genet.* **1993**, *86*, 6–16. [CrossRef]
49. Rosas, J.C.; Beaver, J.S.; Beebe, S.; Viana, A. Nomenclatura de variedades de frijol común liberadas en Centro América y El Caribe. *Agron. Mesoam.* **2004**, *15*, 221–224. [CrossRef]
50. Singh, S.P. Chapter 2: Common Bean (*Phaseolus vulgaris* L.). In *Genetic Resources, Chromosome Engineering, and Crop Improvement: Grain Legumes, Volume 1*; Singh, R.J., Jauhar, P.P., Eds.; Taylor & Francis: Boca Raton, FL, USA, 2005; p. 32.

51. Haudry, A.; Cenci, A.; Ravel, C.; Bataillon, T.; Brunel, D.; Poncet, C.; Hochu, I.; Poirier, S.; Santoni, S.; Glémin, S.; et al. Grinding up wheat: A massive loss of nucleotide diversity since domestication. *Mol. Biol. Evol.* **2007**, *24*, 1506–1517. [CrossRef] [PubMed]
52. Maron, L.G.; Guimarães, C.T.; Kirst, M.; Albert, P.S.; Birchler, J.A.; Bradbury, P.J.; Buckler, E.S.; Coluccio, A.E.; Danilova, T.V.; Kudrna, D.; et al. Aluminum tolerance in maize is associated with higher MATE1 gene copy number. *Proc. Natl. Acad. Sci. USA* **2013**, *110*, 5241–5246. [CrossRef]
53. Bellucci, E.; Bitocchi, E.; Ferrarini, A.; Benazzo, A.; Biagetti, E.; Klie, S.; Minio, A.; Rau, D.; Rodriguez, M.; Panziera, A.; et al. Decreased nucleotide and expression diversity and modified coexpression patterns characterize domestication in the common bean. *Plant Cell* **2014**, *26*, 1901–1912. [CrossRef] [PubMed]
54. Zhou, Z.; Jiang, Y.; Wang, Z.; Gou, Z.; Lyu, J.; Li, W.; Yu, Y.; Shu, L.; Zhao, Y.; Ma, Y.; et al. Resequencing 302 wild and cultivated accessions identifies genes related to domestication and improvement in soybean. *Nat. Biotechnol.* **2015**, *33*, 408–414. [CrossRef] [PubMed]
55. Nanni, L.; Bitocchi, E.; Bellucci, E.; Rossi, M.; Rau, D.; Attene, G.; Gepts, P.; Papa, R. Nucleotide diversity of a genomic sequence similar to SHATTERPROOF (PvSHP1) in domesticated and wild common bean (*Phaseolus vulgaris* L.). *Theor. Appl. Genet.* **2011**, *123*, 1341–1357. [CrossRef] [PubMed]
56. Dwivedi, S.L.; Ceccarelli, S.; Blair, M.W.; Upadhyaya, H.D.; Are, A.K.; Ortiz, R. Landrace Germplasm for Improving Yield and Abiotic Stress Adaptation. *Trends Plant Sci.* **2016**, *21*, 31–42. [CrossRef]
57. Ceccarelli, S. Adaptation to low/high input cultivation. *Euphytica* **1996**, *92*, 203–214. [CrossRef]
58. Ceccarelli, S. Specific adaptation and breeding for marginal conditions. *Euphytica* **1994**, *77*, 205–219. [CrossRef]
59. Beebe, S.; Skroch, P.W.; Tohme, J.; Duque, M.C.; Pedraza, F.; Nienhuis, J. Structure of genetic diversity among common bean landraces of Middle American origin based on correspondence analysis of RAPD. *Crop Sci.* **2000**, *40*, 264–273. [CrossRef]
60. Muñoz-Perea, C.G.; Terán, H.; Allen, R.G.; Wright, J.L.; Westermann, D.T.; Singh, S.P. Selection for drought resistance in dry bean landraces and cultivars. *Crop Sci.* **2006**, *46*, 2111–2120. [CrossRef]
61. Bliss, F.A. Breding common bean for improved biological nitrogen fixation. *Plant Soil* **1993**, *152*, 71–79. [CrossRef]
62. Broughton, W.J.; Hernández, G.; Blair, M.; Beebe, S.; Gepts, P.; Vanderleyden, J. Beans (*Phaseolus* spp.)—Model food legumes. *Plant Soil* **2003**, *252*, 55–128. [CrossRef]
63. Hurtado-Gonzales, O.P.; Gilio, T.A.S.; Pastor-Corrales, M.A. Resistant Reaction of Andean Common Bean Lamdrace G19833, Reference Genome, to 13 Races of Uromyces appendiculatus Suggests Broad Spectrum Rust Resistance. In Proceedings of the Annual report of the Bean Improvement Cooperative, East Lansing, MI, USA, 30 October–1 November 2017; pp. 27–28.
64. Lioi, L.; Zuluaga, D.L.; Pavan, S.; Sonnante, G. Genotyping-by-sequencing reveals molecular genetic diversity in Italian common bean landraces. *Diversity* **2019**, *11*, 154. [CrossRef]
65. Papa, R.; Acosta, J.; Delgado-Salinas, A.; Gepts, P. A genome-wide analysis of differentiation between wild and domesticated *Phaseolus vulgaris* from Mesoamerica. *Theor. Appl. Genet.* **2005**, *111*, 1147–1158. [CrossRef] [PubMed]
66. Padder, B.A.; Kamfwa, K.; Awale, H.E.; Kelly, J.D. Transcriptome profiling of the *Phaseolus vulgaris*—Colletotrichum lindemuthianum pathosystem. *PLoS ONE* **2016**, *11*, e0165823. [CrossRef]
67. Ayyappan, V.; Kalavacharla, V.; Thimmapuram, J.; Bhide, K.P.; Sripathi, V.R.; Smolinski, T.G.; Manoharan, M.; Thurston, Y.; Todd, A.; Kingham, B.; et al. Genome-wide profiling of histone modifications (H3K9me2 and H4K12ac) and gene expression in rust (uromyces appendiculatus) inoculated common bean (*Phaseolus vulgaris* L.). *PLoS ONE* **2015**, *10*, e0132176. [CrossRef]
68. Ma, Y.; Liu, M.; Stiller, J.; Liu, C. A pan-transcriptome analysis shows that disease resistance genes have undergone more selection pressure during barley domestication 06 Biological Sciences 0604 Genetics. *BMC Genom.* **2019**, *20*, 1–11. [CrossRef]
69. Zheng, F.; Wu, H.; Zhang, R.; Li, S.; He, W.; Wong, F.L.; Li, G.; Zhao, S.; Lam, H.M. Molecular phylogeny and dynamic evolution of disease resistance genes in the legume family. *BMC Genom.* **2016**, *17*, 1–13. [CrossRef]
70. Varshney, R.K.; Saxena, R.K.; Upadhyaya, H.D.; Khan, A.W.; Yu, Y.; Kim, C.; Rathore, A.; Kim, D.; Kim, J.; An, S.; et al. Whole-genome resequencing of 292 pigeonpea accessions identifies genomic regions associated with domestication and agronomic traits. *Nat. Genet.* **2017**, *49*, 1082–1088. [CrossRef]

71. Varshney, R.K.; Thudi, M.; Roorkiwal, M.; He, W.; Upadhyaya, H.D.; Yang, W.; Bajaj, P.; Cubry, P.; Rathore, A.; Jian, J.; et al. Resequencing of 429 chickpea accessions from 45 countries provides insights into genome diversity, domestication and agronomic traits. *Nat. Genet.* **2019**, *51*, 857–864. [CrossRef] [PubMed]
72. Zhang, X.; Zhang, J.; He, X.; Wang, Y.; Ma, X.; Yin, D. Genome-wide association study of major agronomic traits related to domestication in peanut. *Front. Plant Sci.* **2017**, *8*, 1–10. [CrossRef] [PubMed]
73. Soltani, A.; Mafimoghaddam, S.; Oladzad-Abbasabadi, A.; Walter, K.; Kearns, P.J.; Vasquez-Guzman, J.; Mamidi, S.; Lee, R.; Shade, A.L.; Jacobs, J.L.; et al. Genetic analysis of flooding tolerance in an andean diversity panel of dry bean (*Phaseolus vulgaris* L.). *Front. Plant Sci.* **2018**, *9*. [CrossRef] [PubMed]
74. Liu, P.; Sun, F.; Gao, R.; Dong, H. RAP2.6L overexpression delays waterlogging induced premature senescence by increasing stomatal closure more than antioxidant enzyme activity. *Plant Mol. Biol.* **2012**, *79*, 609–622. [CrossRef]
75. da Silva, D.A.; Tsai, S.M.; Chiorato, A.F.; da Silva Andrade, S.C.; de Fatima Esteves, J.A.; Recchia, G.H.; Carbonell, S.A.M. Analysis of the common bean (*Phaseolus vulgaris* L.) transcriptome regarding efficiency of phosphorus use. *PLoS ONE* **2019**, *14*, e0210428. [CrossRef] [PubMed]
76. Henry, R.J.; Nevo, E. Exploring natural selection to guide breeding for agriculture. *Plant Biotechnol. J.* **2014**, *12*, 655–662. [CrossRef] [PubMed]
77. Henry, R.J. Innovations in plant genetics adapting agriculture to climate change. *Curr. Opin. Plant Biol.* **2019**. [CrossRef]
78. Chaverra, M.H.; Graham, P.H. Cultivar Variation in Traits Affecting Early Nodulation of Common Bean. *Crop Sci.* **1992**, *32*, 1432–1436. [CrossRef]
79. Moghaddam, A.; Raza, A.; Vollmann, J.; Ardakani, M.R.; Wanek, W.; Gollner, G.; Friedel, J.K. Carbon isotope discrimination and water use efficiency relationships of alfalfa genotypes under irrigated and rain-fed organic farming. *Eur. J. Agron.* **2013**, *50*, 82–89. [CrossRef]
80. Muñoz-Perea, C.G.; Allen, R.G.; Westermann, D.T.; Wright, J.L.; Singh, S.P. Water use efficiency among dry bean landraces and cultivars in drought-stressed and non-stressed environments. *Euphytica* **2007**, *155*, 393–402. [CrossRef]
81. Drought-Resistant Bean Variety Offers Hope to Central American Communities. Available online: https://www.canr.msu.edu/legumelab/uploads/files/Success_Drought_resistant_beans_Central_America.pdf (accessed on 9 September 2020).
82. Lopes, M.S.; El-Basyoni, I.; Baenziger, P.S.; Singh, S.; Royo, C.; Ozbek, K.; Aktas, H.; Ozer, E.; Ozdemir, F.; Manickavelu, A.; et al. Exploiting genetic diversity from landraces in wheat breeding for adaptation to climate change. *J. Exp. Bot.* **2015**, *66*, 3477–3486. [CrossRef] [PubMed]
83. Singh, S.P.; Gepts, P.; Debouck, D.G. Races of Common Bean (*Phaseolus vulgaris*, Fabaceae). *Econ. Bot.* **1991**, *45*, 379–396. [CrossRef]
84. Beversdorf, W. OAC Rico Field Bean. *Can. J. Plant Sci.* **1984**, *64*, 753–755. [CrossRef]
85. Park, S.J.; Buttery, B.R. Registration of Ineffective Nodulation Mutant R69 and Nonnodulation Mutant R99 Common Bean Genetic Stocks. *Crop Sci.* **2006**, *46*, 1415–1417. [CrossRef]
86. Park, S.J.; Buttery, B.R. Ethyl-methane sulphonate (EMS) induced nodulation mutants of common bean (*Phaseolus vulgaris* L.) lacking effective nodules. *Plant Soil* **1992**, *139*, 295–298. [CrossRef]
87. Khanal, R.; Rupert, T.; Navabi, A.; Smith, T.H.; Michaels, T.E.; Burt, A.J.; Pauls, K.P.; Rex, O.A.C. Cultivar Description: Mist Common Bean. *Can. J. Plant Sci.* **2017**, *97*, 192–195.
88. Hosfield, G.L.; Varner, G.V.; Uebersax, M.A.; Kelly, J.D. Registration of 'Merlot' Small Red Bean. *Crop Sci.* **2004**, *44*, 351–352. [CrossRef]
89. Aguilera, A.; Smith, T.H.; Pauls, K.P. Cultivar Description: Small Red Bean 'OAC Rosito'. Available online: https://www.gobeans.ca/downloads/OAC_Rosito-SmallRed.pdf (accessed on 15 November 2019).
90. Kelly, J.D.; Varner, G.V.; O'Boyle, P.; Long, B. Registration of 'Zorro' Black Bean. *J. Plant Regist.* **2009**, *3*, 226–230. [CrossRef]
91. Jamora, N.; Maredia, M. Database of Improved Beans and Cowpea Varietal Releases in Major Bean and Cowpea Producing Countries in Africa. *Lat. Am. US* **2010**.
92. Beaver, J.S.; Rosas, J.C.; Myers, J.; Acosta, J.; Kelly, J.D.; Nchimbi-Msolla, S.; Misangu, R.; Bokosi, J.; Temple, S.; Arnaud-Santana, E.; et al. Contributions of the Bean/Cowpea CRSP to cultivar and germplasm development in common bean. *Field Crops Res.* **2003**, *82*, 87–102. [CrossRef]
93. Rosas, J.C.; Vargas, A.G. Rhizobium Biofertilizante para el frijol común.

94. Shearer, G.; Kohl, D.H. Natural 15N abundance as a method of estimating the contribution of biologically fixed nitrogen to N2-fixing systems: Potential for non-legumes. *Plant Soil* **1988**, *110*, 317–327. [CrossRef]
95. Peoples, M.B.; Unkovich, M.J.; Herridge, D.F. Measuring Symbiotic Nitrogen Fixation in Legumes. In *Nitrogen Fixation in Crop Production*; Emerich, D.W., Krishnan, H.B., Eds.; American Society of Agronomy: Madison, WI, USA, 2009.
96. Song, Q.; Jia, G.; Hyten, D.L.; Jenkins, J.; Hwang, E.Y.; Schroeder, S.G.; Osorno, J.M.; Schmutz, J.; Jackson, S.A.; McClean, P.E.; et al. SNP assay development for linkage map construction, anchoring whole-genome sequence, and other genetic and genomic applications in common bean. *G3 Genes Genomes Genet.* **2015**, *5*, 2285–2290. [CrossRef] [PubMed]
97. Bradbury, P.J.; Zhang, Z.; Kroon, D.E.; Casstevens, T.M.; Ramdoss, Y.; Buckler, E.S. TASSEL: Software for association mapping of complex traits in diverse samples. *Bioinformatics* **2007**, *23*, 2633–2635. [CrossRef] [PubMed]
98. Browning, B.L.; Browning, S.R. Genotype Imputation with Millions of Reference Samples. *Am. J. Hum. Genet.* **2016**, *98*, 116–126. [CrossRef] [PubMed]
99. Torkamaneh, D.; Belzile, F. Scanning and Filling: Ultra-Dense SNP Genotyping Combining Genotyping-by-Sequencing, SNP Array and Whole-Genome Resequencing Data. *PLoS ONE* **2015**, *10*, e0131533. [CrossRef] [PubMed]
100. Raj, A.; Stephens, M.; Pritchard, J.K. FastSTRUCTURE: Variational inference of population structure in large SNP data sets. *Genetics* **2014**, *197*, 573–589. [CrossRef]
101. R Core Team. *R: A Language and Environment for Statistical Computing*; R Foundation for Statistical Computing: Vienna, Austria, 2017.
102. Saitou, N.; Nei, M. The neighbor-joining method: A new method for reconstructing phylogenetic trees. *Mol. Biol. Evol.* **1987**, *4*, 406–425.
103. Felsenstein, J. Confidence Limits on Phylogenies: An Approach Using the Bootstrap. *Evolution* **1985**, *39*, 783. [CrossRef]
104. Tamura, K.; Nei, M.; Kumar, S. Prospects for inferring very large phylogenies by using the neighbor-joining method. *Proc. Natl. Acad. Sci. USA* **2004**, *101*, 11030–11035. [CrossRef]
105. Kumar, S.; Stecher, G.; Tamura, K. MEGA7: Molecular Evolutionary Genetics Analysis Version 7.0 for Bigger Datasets. *Mol. Biol. Evol.* **2016**, *33*, 1870–1874. [CrossRef]
106. Danecek, P.; Auton, A.; Abecasis, G.; Albers, C.A.; Banks, E.; DePristo, M.A.; Handsaker, R.E.; Lunter, G.; Marth, G.T.; Sherry, S.T.; et al. The variant call format and VCFtools. *Bioinformatics* **2011**, *27*, 2156–2158. [CrossRef] [PubMed]
107. Nei, M.; Li, W.H. Mathematical model for studying genetic variation in terms of restriction endonucleases. *Proc. Natl. Acad. Sci. USA* **1979**, *76*, 5269–5273. [CrossRef] [PubMed]
108. Tajima, F. Statistical Method for Testing the Neutral Mutation Hypothesis by DNA Polymorphism. *Genetics* **1989**, *123*, 585–595. [PubMed]
109. Shapiro, A.S.S.; Wilk, M.B. An Analysis of Variance Test for Normality (Complete Samples). *Biometrika* **1965**, *52*, 591–611. [CrossRef]
110. Yan, W.; Rajcan, I. Biplot analysis of test sites and trait relations of soybean in Ontario. *Crop Sci.* **2002**, *42*, 11–20. [CrossRef]
111. Martin, K.; Singh, J.; Hill, J.H.; Whitham, S.A.; Cannon, S.B. Dynamic transcriptome profiling of Bean Common Mosaic Virus (BCMV) infection in Common Bean (*Phaseolus vulgaris* L.). *BMC Genom.* **2016**, *17*. [CrossRef]
112. Mohanta, T.K.; Kumar, P.; Bae, H. Genomics and evolutionary aspect of calcium signaling event in calmodulin and calmodulin-like proteins in plants. *BMC Plant Biol.* **2017**, *17*, 38. [CrossRef]
113. De Souza, L.P.; Scossa, F.; Proost, S.; Bitocchi, E.; Papa, R.; Tohge, T.; Fernie, A.R. Multi-tissue integration of transcriptomic and specialized metabolite profiling provides tools for assessing the common bean (*Phaseolus vulgaris*) metabolome. *Plant J.* **2019**, *97*, 1132–1153. [CrossRef]
114. Konzen, E.R.; Recchia, G.H.; Cassieri, F.; Gomes Caldas, D.G.; Berny Mier, Y.; Teran, J.C.; Gepts, P.; Tsai, S.M. DREB genes from common bean (*Phaseolus vulgaris* L.) show broad to specific abiotic stress responses and distinct levels of nucleotide diversity. *Int. J. Genom.* **2019**, *2019*. [CrossRef]
115. Gonzalez, A.M.; Godoy, L.; Santalla, M. Dissection of Resistance Genes to Pseudomonas syringae pv. phaseolicola in UI3 Common Bean Cultivar. *Int. J. Mol. Sci.* **2017**, *18*, 2503. [CrossRef]

116. MacQueen, A.H.; White, J.W.; Lee, R.; Osorno, J.M.; Schmutz, J.; Miklas, P.N.; Myers, J.; McClean, P.E.; Juenger, T.E. Genetic Associations in Four Decades of Multi-Environment Trials Reveal Agronomic Trait Evolution in Common Bean. *Genetics* **2020**, *215*, 267–284. [CrossRef] [PubMed]
117. Przysiecka, Ł.; Książkiewicz, M.; Wolko, B.; Naganowska, B. Structure, expression profile and phylogenetic inference of chalcone isomerase-like genes from the narrow-leafed lupin (*Lupinus angustifolius* L.) genome. *Front. Plant Sci.* **2015**, *6*, 268. [CrossRef] [PubMed]

plants

MDPI

Article

DNA Fingerprinting and Species Identification Uncovers the Genetic Diversity of Katsouni Pea in the Greek Islands Amorgos and Schinoussa

Evangelia Stavridou [1], Georgios Lagiotis [1], Lefkothea Karapetsi [1], Maslin Osathanunkul [2,3] and Panagiotis Madesis [1,*]

1 Institute of Applied Biosciences, Centre for Research and Technology, Thermi, GR-570 01 Thessaloniki, Greece; estavrid@certh.gr (E.S.); glagiotis@certh.gr (G.L.); lefki8@certh.gr (L.K.)

2 Department of Biology, Faculty of Science, Chiang Mai University, Chiang Mai 50200, Thailand; maslin.cmu@gmail.com

3 Research Center in Bioresources for Agriculture, Industry and Medicine, Chiang Mai University, Chiang Mai 50200, Thailand

* Correspondence: pmadesis@certh.gr

Received: 21 March 2020; Accepted: 5 April 2020; Published: 9 April 2020

Abstract: Pea (*P. sativum* L.), one of the most important legume crops worldwide, has been traditionally cultivated in Lesser Cyclades since ancient times. The commonly known traditional pea cultivar, 'Katsouni', is endemic to the islands of Amorgos and Schinoussa and is of great local economic importance. Despite the widespread cultivation of 'Katsouni' in both islands, it is still unknown whether the current Schinoussa and Amorgos pea populations are distinct landraces, and if they have common evolutionary origin. To assist conservation and breeding of the pea crop, the genetic diversity and phylogenetic relationships of 39 pea samples from Amorgos and 86 from Schinoussa were studied using DNA barcoding and ISSR marker analyses. The results indicate that both populations are different landraces with distinct geographical distribution and are more closely related to *P. sativum* subsp. *elatius* than the *P. abyssinicum* and *P. fulvum* species. Further characterization of the 'Katsouni' landraces for functional polymorphisms regarding pathogen resistance, revealed susceptibility to the powdery mildew (*Erysiphe pisi* DC.). This work represents the first investigation on the genetic diversity and population structure of the 'Katsouni' cultivar. Exploiting the local genetic diversity of traditional landraces is fundamental for conservation practices and crop improvement through breeding strategies.

Keywords: pea landraces; Amorgos; Schinoussa; DNA Barcoding; ISSR genotyping; HRM analysis; powdery mildew

1. Introduction

Species in the economically important *Fabaceae* family have been a staple human food for millennia and their use is closely related to human evolution [1]. Legumes, such as Spanish vetchling (*Lathyrus clymenum* L.), lentils (*Lens culinaris* M.) and beans (*Phaseolus vulgaris* L.) are an important plant-based protein source, rich in mineral nutrients, complex starch and fibers, and contain health-promoting antioxidants, such as carotenoids and phenolic compounds [2–4]. The usage of leguminous crops in traditional crop rotation systems, reduces the need for synthetic nitrogen-based fertilizers by forming symbiotic relationships with nitrogen (N)-fixing soil bacteria [5]. Such management practices are of great ecological importance and have high potential for conservation agriculture, considering legumes are functional either as growing crop or as crop residue [6].

Pea (*P. sativum* L.) is among the most important legume crops, such as chickpea (*Cicer arietinum* L.), lentil and faba bean (*Vicia faba* L.), in temperate climates and has a wide geographical distribution, with field pea being specifically adapted to a wide range of climates and altitudes. The *Pisum* species are of high commercial importance and are cultivated worldwide for dry and fresh consumption. According to the International Legume Database (ILDIS) and to the classification of Maxted and Ambrose (2001) [7], the *Pisum* genus includes three species: i) *Pisum abyssinicum*, ii) *Pisum fulvum* and iii) *Pisum sativum* L., which further includes the wild pea, *Pisum sativum* subsp. *elatius* (M. Bieb. Asch. & Graebn) and the domesticated pea, *Pisum sativum* subsp. *sativum*.

Phylogenetic analyses of various pea taxa with molecular markers indicate that hybridization between wild peas is not an extensive phenomenon [8]. The recently annotated pea genome sequence and the resequencing of data from 42 wild, landrace and cultivar *Pisum* genotypes, provided further insights into legume genome evolution [9]. It has been suggested that the common ancestor of the *Pisum* species was probably cytogenetically similar to *P. sativum* subsp. *elatius*, which evolved across the Mediterranean and Middle East [8] and gave rise to *P. sativum* subsp. *sativum* and *P. fulvum* in the northern Middle East. Regarding *P. abyssinicum*, two main hypotheses exist; it is considered the result of a domestication event from a southern *P. sativum* subsp. *elatius* ancestor [10] followed by a migration to Abyssinia, possibly through ancient human trading routes [11], indicating at least two domestication events independent of *P. sativum* subsp. *sativum* [9,12]. The alternative hypothesis about the origins of *P. abyssinicum* suggests that it derived from a hybridization event between *P. fulvum* and *P. sativum* subsp. *elatius*, which occurred in the western half region of the Fertile Crescent [13] and then a small sample was introduced to north-eastern Africa, where it evolved into the modern *P. abyssinicum* [14]. The very low genetic diversity in the Abyssinian pea suggests that the taxon has recently experienced a severe bottleneck or is a relatively young taxon [10] and the hybridization event has most likely occurred about 4000 years bp [15].

The *Pisum* genus is very diverse, showing the gamut of relatedness that reflect taxonomic identifiers, eco-geography and breeding gene pools [8,12,16,17]. Several phenotypic classification studies on pea germplasm are based on agronomical characteristics and morphological descriptors [18–21], which are unreliable for the evaluation of pea genetic resources and the identification of different cultivars in the Fabaceae family [22], especially considering the environmental effects on the expression of the genotype. Several different molecular methods have been previously employed to assess the genetic diversity in the *Pisum* genus, such as Random Amplification of Polymorphic DNA (RAPD) [23–25], Inter-Single Sequence Repeats (ISSRs) [23,24,26], Single Sequence Repeats (SSR) [27,28], Retrotransposon-Based Insertion Polymorphism (RBIP) markers [13,17,29–31] and Expressed Sequence Tags (EST)-derived genomic markers [32]. Additionally, high-throughput parallel genotyping via genome-wide next generation sequencing techniques have also been used to study the diversity of wild pea [8,33,34].

An alternative method for the simple and accurate authentication of plant species is DNA Barcoding. The *rbcL* and *matK* regions have been recommended as core DNA barcodes for plant identification [35]. In *Fabaceae* species, four coding chloroplast regions (*rpoB*, *rpoC1*, *rbcL*, and *matK*) and two non-coding nuclear regions (*ITS1* and *ITS2*) have been used as barcodes [22,36–38]. DNA barcoding has also been used to reconstruct the phylogenetic relationship of the main Mediterranean leguminous crops [39]. Furthermore, the combination of DNA barcoding with high resolution melting analysis (Bar-HRM) has, thus far, been proved an effective approach for the identification of diverse plant species, their Protected Designation of Origin (PDO) products and quantification of adulterants [40–43].

In Greece, pulses have been traditionally cultivated since the ancient times and is a staple food in the local culinary culture. A popular Greek dish (namely 'Fava') is typically prepared using different legume species, such as yellow- split peas (*P. sativum* L.) or faba beans (*Vicia faba*). However, in the island of Santorini, the authentic PDO 'Fava Santorinis', is exclusively prepared from a local grass pea variety of *L. clymenum*. In other Cycladic islands, especially in Amorgos and Schinoussa, 'Fava' is prepared from the dried peeled and split seeds of an endemic *Pisum* cultivar also known as 'Katsouni', named after the convex shape of a small sickle's lobe used for mowing the crop. 'Katsouni' is a traditional product and a crop of great economic importance for the Lesser Cyclades. Currently, is in the process to be appointed as a PDO EU mark, offering a significant income to the local farmers. 'Katsouni' is fully adapted to the local climatic conditions of the Cyclades, with dry and hot summers and mild winters. It is rich in proteins (over 22%) and can be stored after drying the seeds throughout the year.

Historic records indicate that the 'Katsouni' landrace is the result of long-term selection and evolution from prehistoric times that occurred in Amorgos. Since the mid-19th century 'Katsouni' was transferred to the deserted Schinoussa island by residents of neighbouring Amorgos, who moved to settle there, and has since been cultivated uninterruptedly. However, Schinoussa was not always an abandoned island. Archaeological excavations have revealed findings indicating great activity from the prehistoric times to the Classical and Hellenistic period [44]. In the Byzantine times, trade and commerce were essential components of the island's prosperity [45], which heralded an age of advancement, especially during the Venetian rule (13th–16th century). However, during the Ottoman rule (16th–19th century) the island was deserted as indicated by the famous botanist and traveler de Tournefort (2003) [46]. In the late Middle Ages, in 1537, Cyclades along with Schinoussa were plundered by the Ottoman pirate Haiderin Barbarossa [47] and piracy continued up to the 19th century [48]. With Schinoussa becoming a pirates' den, we hypothesized that either: (i) the current Schinoussa pea population is an independent landrace, which is possibly the result of introgression of *P. abyssinicum* (transferred by pirates) into the *P. sativum* subsp. *elatius* germplasm; (ii) the Amorgos and Schinoussa populations belong to the same widely distributed landrace, or (iii) the two landraces emerged from the split of an ancestral population.

The knowledge of genetic relationships and diversity among individual landraces is fundamental for conservation practices and the selection of appropriate parents in breeding programs. Hence, in the present study, we evaluated the application of ISSR marker analysis and DNA barcoding for the molecular characterization of local Amorgos and Schinousssa pea populations. Furthermore, considering that powdery mildew (*E. pisi* DC.) severely affects pea crops worldwide [49], and in the frame of targeting functional polymorphisms, we aimed at characterizing the two landraces for the presence of the powdery mildew resistance gene (*er1-7*) with HRM analysis. The diversity assessment of local landraces may not only provide insights in understanding pea phylogenetics and population genetics, but also broaden pea breeding strategies.

2. Results

2.1. DNA Barcoding, Sequencing and Tree Analysis

To identify potential inter- and intra-specific variation between the two pea populations of Amorgos and Schinoussa, 24 pea samples were analyzed in the present study using the *ITS2*, *trnL* and *rpoC*, and 21 for the *psbA trnH* and *matK* barcoding regions. The selection of *matK* and *psbA-trnH* was based on the unique SNPs observed in the aligned sequences among the three species and especially between *P. abyssinicum* and *P. sativum* (Figure S1). The amplification of *trnL* and *rpoC* regions was 100% successful, whilst for the *psbA-trnH*, *ITS2* and *matK* sequences the rate was 95.24%, 95.8% and 90.47%, respectively.

The positive amplicons were sequenced, and based on the BLAST results, most of the markers were able to identify the samples at the genus level, but not at the species level, when blasted against the NCBI database. The BLAST entries matched with all the three *Pisum* species (*P. sativum* subsp. *sativum*, *P. sativum* subsp. *elatius*, *P. fulvum* and *P. abyssinicum*), with over 96% similarity in identity and coverage. However, the *ITS2* marker identified all target sequences as *P. sativum* subsp. *elatius* with 100% similarity to the available sequences.

The sequences of each barcode gene were aligned and compared against indicative BLAST entries. In *psbA-trnH* we observed an intel polymorphism among our samples and the NCBI database entries. Most of our samples showed a gap in the barcoding region similar to that observed in the *elatius* subspecies sequences, but not in the *P. abyssinicum* and *P. fulvum* sequences (Figure S1). Other observed variations include the *P. abyssinicum*-specific polymorphism of a thymine (T) in place of a cytocine (C), and an unspecific SNP (G/T) present exclusively in the samples of both islands and the *P. sativum* subsp. *elatius* NCBI entries (Figure S1). Regarding the *ITS2* barcode, only the brown colored peas from both islands shared a common nucleotide variation of a guanine (G) in place of an adenine (A) (Figure S2). The *matK* barcoding showed a *P. fulvum*-specific polymorphism (C/A) and other unspecific SNPs (Figure S2). However, the observed variations in *trnL* and *rpoC* regions were not consistent among either seed coat color and/or landrace (Figure S2). Furthermore, both *matK* and *psbA-trnH* regions showed higher diversity and pairwise distance values compared to the other markers (Table S1); yet, the observed low genetic variation values across all barcoding regions (Table S1) indicate that the populations are closely related with a recent common ancestor.

To provide a basic illustration of the phylogenetic associations between the landraces, the DNA barcoding data was used to calculate the genetic distances and generate Neighbour-joining dendrograms (Figure 1 and Figure S3). The *psbA-trnH* dendrogram illustrates the clustering of the *P. abyssinicum* sequences supported by 76% bootstrap value, which are distinct from the rest of the samples at 64%, emphasizing that our samples probably do not belong to *P. abyssinicum* (Figure 1). In contrast, both the main structures of the *trnL* and *ITS2* dendrograms (Figure S3) separated the two landraces in two distinct clusters, which also corresponded to the represented geographical regions, although supported by a low bootstrap value (30–60%). The main structure of the *rpoC* dendrogram (Figure S3) presented also two clusters separating the pea taxa from Schinoussa to those of Amorgos island, supported by a higher bootstrap value (59–66%). However, the *matK* dendrogram (Figure S3) did not show any significant patterns. Taking into consideration the results from the barcoding analysis, the two populations probably belong to the *P. sativum* subsp. *elatius*.

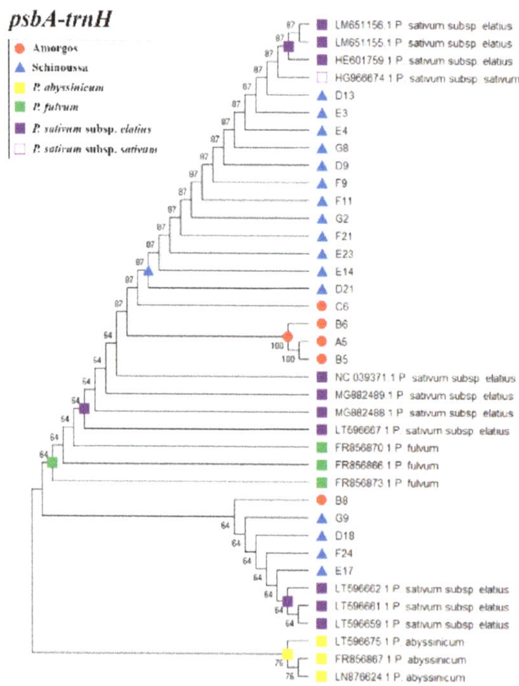

Figure 1. The *psbA-trnH* dendrogram illustrating the phylogenetic relationships between Amorgos (red circle) and Schinoussa (blue triangle) pea populations. The corresponding NCBI sequences of the *P. sativum* subsp. *elatius* (purple square), *P. sativum* subsp. *sativum* (white square), *P. abyssinicum* (yellow square) and *P. fulvum* (green square) were used as reference taxa.

2.2. ISSR Genotyping

To further investigate the genetic differences between the two populations, we used fifteen ISSR markers, out of which six were found to be polymorphic. ISSR analysis of the pea populations using the six polymorphic markers yielded 66 bands in total; five unique bands were identified for the Schinousasa population, whereas the remaining 61 bands were shared between the two populations (Table 1). The Schinoussa population presented a significantly greater polymorphism compared to Amorgos by displaying a greater number of different alleles (Na), whilst the number of effective alleles (Ne) did not show significant differences (Table 2).

Table 1. Band patterns of the Amorgos and Schinoussa populations resulted from the ISSR analysis.

Population	Number of Bands	Number of Band Frequency (> = 5%)	Number of Unique Bands
Amorgos	61	59	0
Schinoussa	66	66	5

Table 2. Mean value and standard error over loci for Amorgos and Schinoussa populations.

Population		N	Na	Ne	I	h	uh
Amorgos	Mean	39	1.742	1.403	0.373	0.244	0.250
	SE	0.000	0.073	0.043	0.031	0.022	0.023
Schinoussa	Mean	86	2.000	1.483	0.443	0.289	0.292
	SE	0.000	0.000	0.042	0.026	0.020	0.021

Mean = Mean value, SE = Standard error, N = Number of alleles, Na = Number of different alleles, Ne = Number of effective alleles = 1/(p^2 + q^2), I = Shannon's Information Index = −1* (p * Ln (p) + q * Ln(q)), h = Diversity = 1 − (p^2 + q^2), uh = Unbiased diversity = (N/(N − 1)) * h.

The genetic differentiation between landraces (PhiPT distances) was significantly different (PhiPT = 0.188; $P \leq 0.001$), indicating that the two landraces are geographically distinct. The Analysis of Molecular Variance (AMOVA) based on the PhiPT values indicated that most of the genetic diversity occurred within landraces (80%), while the variability among landraces contributed to the 20% of the observed genetic diversity (Table 3).

Table 3. AMOVA analysis of the Amorgos and Schinoussa pea populations.

Source	Df	SS	MS	Est. Var.	%
Among populations	1	133.042	133.042	2.307	20
Within populations	123	1133.326	9.214	9.214	80
Total	124	1266.368		11.521	100

Df = Degrees of freedom, SS = Sum of Squares, MS = Mean Square, Est. Var. = Estimated Variance.

The Principal Coordinates Analysis (PCoA) generated two major clusters, in which samples from Amorgos and Schinoussa were clearly separated (Figure 2A). Additionally, the clustering based on seed coat color supports that the two landraces present genetic differences (Figure 2B), given Amorgos' Black, Brown and Green peas do not overlap with the corresponding colors of Schinoussa peas, supporting the hypothesis that the Amorgos and Schinoussa pea populations are two distinct landraces. The UPGMA dendrogram illustrates that the pea samples originating from the same geographic location are closely clustered (Figure 3).

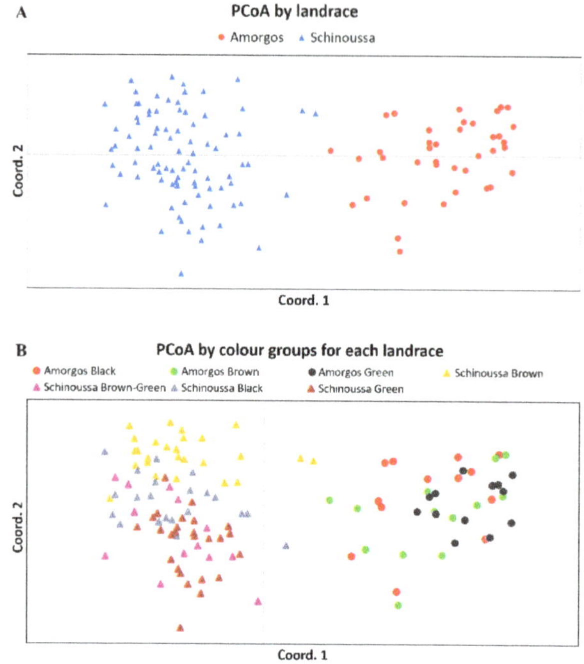

Figure 2. Principal Coordinates Analysis (PCoA) of the Amorgos and Schinoussa pea populations, clustering for: (**A**) population, and (**B**) seed coat color from each landrace. The PCoA analysis shows the separation of the two populations as distinct landraces based on their region of origin.

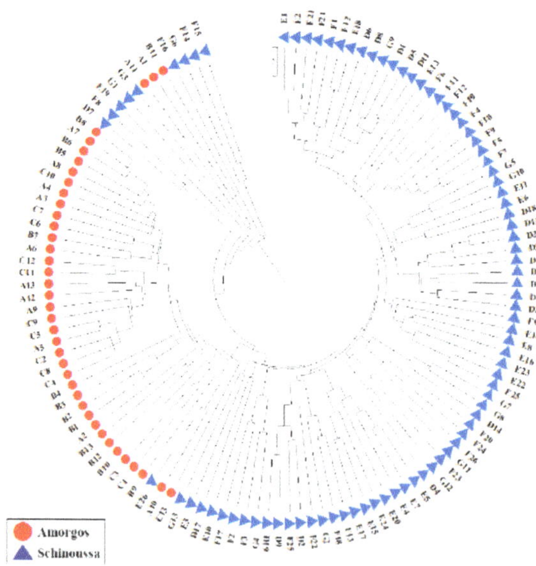

Figure 3. Dendrogram of the Amorgos and Schinousa pea populations based on the UPGMA analysis of the ISSR polymorphisms. Individuals are shape- and color-coded based on their region of origin (Red circle: Amorgos, Blue triangle: Schinoussa).

2.3. Molecular Screening for Powdery Mildew Using HRM Analysis

HRM analysis using a specific functional marker was employed to characterize the two landraces for the presence of the powdery mildew resistance gene. Regarding the specific InDel marker, all the pea samples tested were found negative to the resistant allele (*er1-7*), across both landraces (Figure 4A). This was also confirmed by sequencing of selected samples, where the resistant allele was absent and therefore the 10-bp sequence TCATGTTATT was present (Figure 4B).

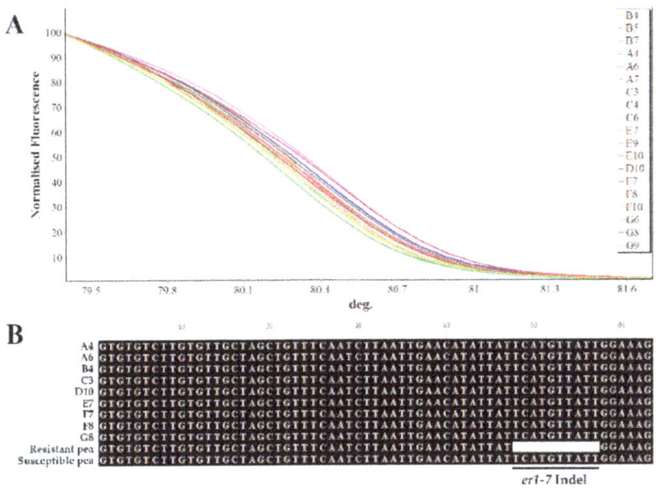

Figure 4. HRM analysis coupled with a co-dominant functional marker specific for *er1-7*. (**A**) Normalized fluorescence graph of selected pea samples per seed coat color and landrace. In the x axis deg. indicates temperature in °C. (**B**) Sequence alignment of the *er1-7* region from selected pea samples and the corresponding reference sequences obtained from Sun et al. (2016) [50].

3. Discussion

Exploiting genetic diversity from local traditional landraces is fundamental for conservation practices and breeding programs, especially under the pressure posed for adaptation to climate change worldwide. To promote the local pea landraces of the Cyclades we aimed at identifying the species and understanding the genetic relationship of the local Amorgos and Schinousssa pea populations, using DNA barcoding and ISSR marker analysis. Over the centuries, from the prehistoric times and the Bronze Age (3000–1100 B.C in Greece) to the mid-19th century, the Aegean was a field of pirate action [51,52]. Oral traditions and place names, throughout the history of Schinoussa, suggest that the island was used as a ground and shelter for pirates. One of our hypotheses was that the landrace of Schinoussa belongs to *P. abyssinicum*, however, our results strongly suggest otherwise. Although both landraces showed a geographical clustering according to the PCoA and the UPGMA analysis of the ISSR polymorphisms, both were identified based on DNA Barcoding as *P. sativum* subsp. *elatius*.

Herein, the inability of most of the barcoding markers to discriminate samples at the species level, being not variable enough to resolve phylogeny of the genus, is probably due to the conserved chloroplast sequences and the low mutation rate [53]. The inability to discriminate among the *P. abyssinicum*, *P. sativum* subsp. *elatius* and *P. fulvum* is in agreement with the view that *P. abyssinicum* is an ancient hybrid of the two species [14]. Nevertheless, despite the inability of the *psbA-trnH* spacer on identifying species due to the frequently observed intraspecific inversions [54], in this study it was shown as the most effective marker for separating the two pea landraces from *P. abyssinicum*. Additionally, although the *matK* barcode showed the lowest resolution capacity in our study, it was sensitive enough to discriminate the two pea landraces from *P. fulvum*. Thus, the most informative barcoding markers to draw conclusions concerning the species identification in the two populations is the combination of *psbA-trnH* and *matK*. Taking into consideration the phylogenetic trees and barcoding results from all the studied markers, the two landraces are more closely related to *P. sativum* subsp. *elatius*, than to the *P. abyssinicum* and *P. fulvum* species.

There is a limited number of studies on *Pisum* germplasm that mainly are focusing on SSR analysis [55]. Herein, the ISSR analysis showed that the populations were distinguished into geographical regions, as separated clusters, indicating the adaptation of these traditional landraces to relatively different agro/climatic conditions (Figures 2 and 3). This geographical isolation could have potentially resulted to the genetic drift of the two landraces. This result is in accordance with the larger molecular differences detected between pea landraces collected in Maritime areas of Spain [26]. Pea is known as self-pollinating with occasional cross-pollination which allows spontaneous hybridization [56]. As a self-pollinated crop, higher genetic diversity is expected among cultivars than within cultivars. However, the AMOVA analysis showed larger genetic diversity within cultivars (80%), which is in accordance with similar findings in other legumes such as chickpea [57]. This may be attributed to the natural interspecific crosses that can occur between *Pisum* species, serving as a source of additional genetic diversity for the selection of common pea [15]. This indicates that despite autogamy, the analysis of genetic diversity on some plants per landrace might be useful in breeding programs [28]. The large genetic diversity might be due to the long-term adaptation of the landraces to the local environment and the diverse agro-ecological systems [55], in combination with putative migration events among the regions, followed by introgression with pre-existing germplasms [28].

Aiming to identify unique traits in the two pea traditional landraces we screened for the powdery mildew resistance gene (*er1*). The er1 is the loss-of-function mutation in the disease susceptibility-related *PsMLO1* gene [58] and the most widespread across resistant pea germplasm, conferring penetration resistance [50,59]. HRM coupled with a functional marker has been previously reported to be highly efficient and cost-effective for routine large-scale screening of pea germplasm for resistance to powdery mildew [60]. This approach allowed for the accurate genotyping of both the homozygous and heterozygous resistant peas [58,60]. In this study, using the *er1-7* functional marker coupled with HRM, we identified that the two landraces of the Cycladic islands are susceptible to powdery mildew. These results indicate the need for further crop improvement of the traditional pea landraces by introducing resistance through molecular-assisted breeding strategies.

4. Materials and Methods

4.1. Plant Material and DNA Isolation

The plant material used in this work includes samples from two pea populations from the islands of Amorgos and Schinoussa (Table 4 and Figure 5). The pea samples were grouped based on the geographical region of origin and the seed coat color; groups A-C from Amorgos and groups D-G from Schinoussa (Table 4). Color grouping was based on the three main seed coat colors (black, brown and green) observed in both populations, except for Schinoussa, which had an additional brown-green hue (Table 4). Following, a subsample of seeds from each group was planted in pots containing 2:1 peat:perlite in order to obtain fresh leaf material for DNA extraction. Total genomic DNA was isolated from approximately 0.1 g of fresh leaf material for each sample following the modified CTAB protocol as described by Doyle and Doyle (1987) [61]. After extraction, the DNA samples were re-diluted in 1X TE buffer (10 mM Tris-Cl pH 8.0, 1 mM EDTA) and stored at −20 °C. DNA quantity and quality were assessed by regular spectrophotometric procedures using the UV-Vis Spectrophotometer Q5000 (Quawell Technology Inc., U.S.A.) and gel electrophoresis in 1% agarose gel.

Table 4. Samples from Amorgos and Schinoussa pea populations used in this work. The samples were grouped in sub-groups (A–G) according to region of origin and seed coat color.

Sample Group	Number of Individuals	Region	Seed Coat Color
A	13	Amorgos	Black
B	13	Amorgos	Brown
C	13	Amorgos	Green
D	21	Schinoussa	Black
E	26	Schinoussa	Brown
F	26	Schinoussa	Green
G	13	Schinoussa	Brown-Green

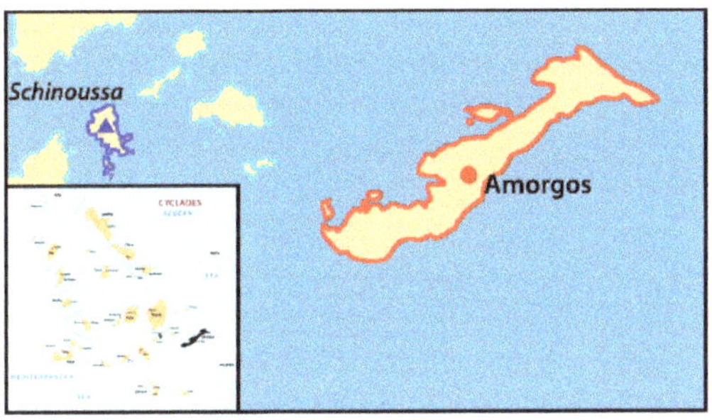

Figure 5. Map of the sampling sites in Amorgos and Schinoussa islands.

4.2. DNA Barcoding and Sequencing Analysis

For the identification of the two landraces, we performed DNA barcoding analysis using the *ITS2* [62], *trnL* [63], *rpoC*, *matK* [35], and *psbA-trnH* [64] barcoding markers. PCR amplification was performed on a Rotor-Gene 6000 real-time 5-Plex HRM PCR Thermocycler (Corbett Research, Sydney, Australia), using the Rotor-Gene Q software version 2.0.2 (Corbett Life Science, Cambridge, UK). PCR reaction mixtures with a total volume of 20 μL consisted of approximately 20 ng genomic DNA, 1× PCR buffer, 0.5 μM forward and reverse primers, 0.2 mM dNTPs, 1.5 mM SYTO™ 9 Green Fluorescent Nucleic Acid Stain (Invitrogen, Eugene, Oregon, USA), and 1 U Kapa Taq DNA polymerase (Kapa Biosystems, USA). The universal regions were amplified using the following protocol: initial denaturation at 95 °C for 4 min, followed by 35 cycles of 95 °C for 30 sec, corresponding annealing temperature (Ta) °C for 60 sec, and 72 °C for 60 sec with a final extension phase at 72 °C for 3 min.

After sequencing, the sequences of the five candidate regions were aligned with the MUSCLE algorithm and genetic distances were calculated using Molecular Evolutionary Genetics Analysis X (MEGA X; Version 10.05) based on the K2P-distance model [65] to evaluate divergence between the two populations. The Neighbour-joining clustering method was used to demonstrate the represented differences as an unrooted dendrogram using MEGA X [65]. Statistical support for each constructed tree was provided by two statistical data analysis as bootstrapping (1000 replications) and pairwise distance model.

Species identification based on the sequence similarity approach was performed with the National Center for Biotechnology Information (NCBI) database [66] by basic local alignment search tool (BLAST; setting: blastn, megablast) [67] and all regions of the three various *Pisum* species were used as query sequences. Correct identification was concluded when the best BLAST hit of the query sequence had over 96% query coverage and identity.

4.3. ISSR Genotyping and Data Analysis

For the distance-based analysis of the two populations we used 15 ISSR markers of which six (UBC811, UBC818, UBC827, UBC841, UBC873, UBC880) were selected for further analysis based on their discrimination efficiency. The total volume of PCR reaction was 25 μL containing 1X PCR buffer, 0.2 mM dNTPs, 10 μM primer, 1 U/μL Taq DNA Polymerase (Kapa Biosystems Ltd.) and 20 ng template DNA. The profile of the PCR reaction program was an initial denaturation for 4 min at 94 °C, followed by 35 cycles at 94 °C for 30 s, 40 sec annealing at the corresponding Ta °C, and 40 sec extension at 72 °C, ending with final extension phase at 72 °C for 7 min. The amplified PCR products were run on 1.5% agarose gel with 1X TAE buffer at 100 V and visualized with the UV Minibis Pro (DNR Bio-Imaging Systems, Jerusalem, Israel) instrument. Band scoring was performed using the Logger Pro 3.15 software.

DNA fragment profiles were scored in a binary fission with '0' indicating the absence and '1' indicating presence of a band. Using the binary haploid data, a pairwise individual-by-individual genetic distance matrix was constructed using the Jaccard coefficient. The percentage of polymorphic loci (P), Number of alleles (N), Number of different alleles (Na), Number of effective alleles (Ne), gene diversity (expected heterozygosity, He), Shannon's diversity index (I), Diversity (h) and unbiased genetic distances (uh) according to [68] were subsequently calculated. The hierarchical distribution of genetic diversity among and within populations was also characterized by Analysis of Molecular Variance (AMOVA) and Principal Co-ordinate Analysis (PCoA). All of the above analyses were performed using the GenAlex 6.5 software package [69]. The Unweighted Pair Group Method based on Arithmetic Averages (UPGMA) clustering analysis for analyzing the similarity estimates was performed using MEGA X [65] and expressed as a dendrogram.

4.4. Molecular Screening for Powdery Mildew Resistance Using HRM Analysis

We used a co-dominant functional marker specific for *er1-7*, the InDel111–120, associated with pea resistance to powdery mildew [50,59,60]. Representative samples were selected from each color group of both populations. The total volume of PCR reaction was 20 μL containing 1X PCR buffer, 0.2 mM dNTPs, 10 μM of the er1-7 primer, 1 U/μL Taq DNA Polymerase (Kapa Biosystems Ltd.) and 20 ng template DNA. PCR conditions where: preheating for 5 min and initial denaturation at 95 °C, followed by 35 cycles at 94 °C for 30 s transition, 30 sec annealing at 56 °C. The fluorescence was measured at the end of each extension step during the PCR cycles. The HRM was performed by an initial pre-melt conditioning of the PCR products at 95 °C for 5 sec and 50 °C for 30 sec, followed by a melt at range of 75–85 °C in increments of 0.1 °C every 2 sec. Fluorescence was measured at the end of each increment.

5. Conclusions

This work represents the first investigation focused on the molecular characterization of pea traditional accessions collected from the semi-arid region of Lesser Cyclades with the main goal to evaluate their genetic diversity and their population structure. Based on the DNA Barcoding and the ISSR marker analysis: i) both landraces have been identified as *P. sativum* subsp. *elatius* germplasm and ii) the landraces show a tendency for differentiation, which is in accordance with the geographical distribution of the genetic structure as an underlying evolutionary process. However, further research and phylogenetic analysis is required with larger population sizes for better understanding of the evolutionary processes that led to these differences, as well as for the preservation of existing diversity in *ex-situ* collections.

Supplementary Materials: The following are available online at http://www.mdpi.com/2223-7747/9/4/479/s1, Table S1: Pairwise distance and diversity analysis within and between populations of the five barcodes, Figure S1: Sequence alignments of the *psbA-trnH* region, Figure S2: Sequence alignments of the *ITS2*, *trnL*, *rpoC* and *matK* regions from selected pea samples and the corresponding Pisum reference sequences available in the NCBI, Figure S3: The *ITS2*, *trnL*, *rpoC* and *matK* dendrograms illustrating the phylogenetic relationships between Amorgos (red circle) and Schinoussa (blue triangle) pea populations.

Author Contributions: Conceptualization, P.M.; methodology, E.S., G.L., L.K., M.O.; data curation, E.S., G.L.; writing—original draft preparation, E.S.; writing—review and editing, E.S., G.L., M.O.; visualization, G.L., E.S.; supervision, P.M.; funding acquisition, P.M. All authors have read and agreed to the published version of the manuscript.

Funding: This research has been co-financed by the European Regional Development Fund of the European Union and Greek national funds through the Operational Program Competitiveness, Entrepreneurship and Innovation, under the call RESEARCH-CREATE-INNOVATE (project code: T1EDK-04448).

Acknowledgments: We would like to thank Dimitris Roukas for providing the pea landrace seeds and vital information regarding the origin of the seeds and the history behind its cultivation in the two islands. We would also like to thank the University of Chiang Mai for financial support.

Conflicts of Interest: The authors declare no conflict of interest.

References

1. Weiss, E. Current state of the art. In *Domestication of Plants in the Old World—The Origin and Spread of Domesticated Plants in Southwest Asia, Europe, and the Mediterranean Basin*; Oxford University Press: Oxford, UK, 2012; pp. 1–8.
2. Dahl, W.J.; Foster, L.M.; Tyler, R.T. Review of the health benefits of peas (Pisum sativum L.). *Br. J. Nutr.* **2012**, *108*, 3–10. [CrossRef] [PubMed]
3. Hall, C.; Hillen, C.; Robinson, J.G. Composition, nutritional value, and health benefits of pulses. *Cereal Chem.* **2017**, *94*, 11–31. [CrossRef]
4. Singh, B.; Singh, J.P.; Shevkani, K.; Singh, N.; Kaur, A. Bioactive constituents in pulses and their health benefits. *J. Food Sci. Technol.* **2017**, *54*, 858–870. [CrossRef] [PubMed]
5. Crews, T.E.; Peoples, M.B. Legume versus fertilizer sources of nitrogen: Ecological tradeoffs and human needs. *Agric. Ecosyst. Environ.* **2004**, *102*, 279–297. [CrossRef]
6. Stagnari, F.; Maggio, A.; Galieni, A.; Pisante, M. Multiple benefits of legumes for agriculture sustainability: An overview. *Chem. Biol. Technol. Agric.* **2017**, *4*, 1–13.
7. Maxted, N.; Ambrose, M. Peas (Pisum L.). In *Plant Genetic Resources of Legumes in the Mediterranean*; Kluwer Academic Publishers: Dordrecht, The Netherlands, 2001; pp. 181–190.
8. Smýkal, P.; Hradilová, I.; Trn, O.; Brus, J.; Rathore, A.; Das, R.R.; Bhattacharyya, D.; Richards, C.; Coyne, J.C.; Pirintsos, S. Genomic diversity and macroecology of the crop wild relatives of domesticated pea. *Nat. Sci. Rep.* **2017**, *7*, 1–12. [CrossRef]
9. Kreplak, J.; Madoui, M.; Cápal, P.; Novák, P.; Labadie, K.; Aubert, G.; Bayer, P.E.; Gali, K.K.; Syme, R.A.; Main, D.; et al. A reference genome for pea provides insight into legume genome evolution. *Nat. Genet.* **2019**, *51*, 1411–1422. [CrossRef]
10. Weeden, N.F. Domestication of pea (Pisum sativum L.): The case of the Abyssinian pea. *Front. Plant Sci.* **2018**, *9*, 1–11. [CrossRef]
11. Pagani, L.; Kivisild, T.; Tarekegn, A.; Ekong, R.; Plaster, C.; Gallego Romero, I.; Ayub, Q.; Mehdi, S.Q.; Thomas, M.G.; Luiselli, D.; et al. Ethiopian genetic diversity reveals linguistic stratification and complex influences on the Ethiopian gene pool. *Am. J. Hum. Genet.* **2012**, *91*, 83–96. [CrossRef]
12. Trněný, O.; Brus, J.; Hradilová, I.; Rathore, A.; Das, R.R.; Kopecký, P.; Coyne, C.J.; Reeves, P.; Richards, C.; Smýkal, P. Molecular evidence for two domestication events in the pea crop. *Genes* **2018**, *9*, 535–555. [CrossRef]
13. Jing, R.; Vershinin, A.; Grzebyta, J.; Shaw, P.; Smýkal, P.; Marshall, D.; Ambrose, M.J.; Ellis, T.H.N.; Flavell, A.J. The genetic diversity and evolution of field pea (*Pisum*) studied by high throughput retrotransposon based insertion polymorphism (RBIP) marker analysis. *BMC Evol. Biol.* **2010**, *10*, 1–20. [CrossRef] [PubMed]
14. Smýkal, P.; Kenicer, G.; Flavell, A.J.; Corander, J.; Kosterin, O.; Redden, R.J.; Ford, R.; Coyne, C.J.; Maxted, N.; Ambrose, M.J.; et al. Phylogeny, phylogeography and genetic diversity of the Pisum genus. *Plant Genet. Resour. Charact. Util.* **2011**, *9*, 4–18. [CrossRef]
15. Kosterin, O.E. Abyssinian pea (Lathyrus schaeferi Kosterin nom. nov. pro Pisum abyssinicum A. Br.) is a problematic taxon. *Acta Biol. Sib.* **2017**, *21*, 158–169. [CrossRef]
16. Ellis, T.H.N. Pisum. In *Wild Crop Relatives: Genomic and Breeding Resources Legume Crops and Forages*; Kole, C., Ed.; Springer: Berlin/Heidelberg, Germany, 2011; pp. 237–248. ISBN 9783642204463.

17. Jing, R.; Ambrose, M.A.; Knox, M.R.; Smykal, P.; Hybl, M.; Ramos, Á.; Caminero, C.; Burstin, J.; Duc, G.; van Soest, L.J.M.; et al. Genetic diversity in European Pisum germplasm collections. *Theor. Appl. Genet.* **2012**, *125*, 367–380. [CrossRef]
18. Nisar, M.; Ghafoor, A.; Ahmad, H.; Khan, M.R.; Qureshi, A.S.; Ali, H.; Islam, M. Evaluation of genetic diversity of pea germplasm through phenotypic trait analysis. *Pak. J. Bot.* **2008**, *40*, 2081–2086.
19. Smýkal, P.; Horáček, J.; Dostálová, R.; Hýbl, M. Variety discrimination in pea (Pisum sativum L.) by molecular, biochemical and morphological markers. *J. Appl. Genet.* **2008**, *49*, 155–166. [CrossRef]
20. Ouafi, L.; Alane, F.; Rahal-Bouziane, H.; Abdelguerfi, A. Agro-morphological diversity within field pea (Pisum sativum L.) genotypes. *Afr. J. Agric. Res.* **2016**, *11*, 4039–4047.
21. Singh, S.R.; Ahmed, N.; Singh, D.B.; Srivastva, K.K.; Singh, R.K.; Mir, A. Genetic variability determination in garden pea (Pisum sativum L sub sp. hortense Asch. and Graebn.) by using the multivariate analysis. *Legum. Res. Int. J.* **2017**, *40*, 416–422. [CrossRef]
22. Newmaster, S.G.; Ragupathy, S. Testing plant barcoding in a sister species complex of pantropical Acacia (Mimosoideae, Fabaceae). *Mol. Ecol. Resour.* **2009**, *9*, 172–180.
23. Tar'an, B.; Zhang, C.; Warkentin, T.; Tullu, A.; Vandenberg, A. Genetic diversity among varieties and wild species accessions of pea (Pisum sativum L.) based on molecular markers, and morphological and physiological characters. *Genome* **2005**, *48*, 257–272. [CrossRef]
24. Kapila, R.K.; Naryal, S.; Dhiman, K.C.; Singh, S.K.; Sharma, J.K. Identification of field and Garden Pea (Pisum sativum L.) varieties using RAPD and ISSR markers. *Seed Res.* **2012**, *40*, 1–9.
25. Iqbal, A.; Razzaq, A.; Hadi, F.; Nisar, M.; Ozturk, M.; Altay, V. Assessment of Genetic Diversity Among Hybrid Pea Lines (Pisum Sativum L.) as Revealed by Random Amplified Polymorphic DNA (RAPD) Markers. *Fresenius Environ. Bull.* **2018**, *27*, 6447–6453.
26. Lázaro, A.; Aguinagalde, I. Genetic variation among Spanish pea landraces revealed by Inter Simple Sequence Repeat (ISSR) markers: Its application to establish a core collection. *J. Agric. Sci.* **2006**, *144*, 53–61. [CrossRef]
27. Adhikari, D.N.; Khanna, V.; Singh, K.N. Crossability and Genetic Diversity Studies in Pea (Pisum sativum). *Curr. Investig. Agric. Curr. Res.* **2018**, *3*, 443–452.
28. Mohamed, A.; García-Martínez, S.; Loumerem, M.; Carbonell, P.; Ruiz, J.J.; Boubaker, M. Assessment of genetic diversity among local pea (Pisum sativum L.) accessions cultivated in the arid regions of Southern Tunisia using agro-morphological and SSR molecular markers. *Genet. Resour. Crop Evol.* **2019**, *66*, 1189–1203. [CrossRef]
29. Smýkal, P.; Hýbl, M.; Corander, J.; Jarkovský, J.; Flavell, A.J.; Griga, M. Genetic diversity and population structure of pea (Pisum sativum L.) varieties derived from combined retrotransposon, microsatellite and morphological marker analysis. *Theor. Appl. Genet.* **2008**, *117*, 413–424. [CrossRef] [PubMed]
30. Martin-Sanz, A.; Caminero, C.; Jing, R.; Flavell, A.J.; de la Vega, M.P. Genetic diversity among Spanish pea (Pisum sativum L.) landraces, pea cultivars and the World Pisum sp. core collection assessed by retrotransposon-based insertion polymorphisms (RBIPs). *Span. J. Agric. Res.* **2011**, *9*, 166–178. [CrossRef]
31. Solberg, S.Ø.; Brantestam, A.K.; Olsson, K.; Leino, M.W.; Weibull, J.; Yndgaard, F. Diversity in local cultivars of Pisum sativum collected from home gardens in Sweden. *Biochem. Syst. Ecol.* **2015**, *62*, 194–203. [CrossRef]
32. Jain, S.; Kumar, A.; Mamidi, S.; McPhee, K. Genetic Diversity and Population Structure Among Pea (Pisum sativum L.) Cultivars as Revealed by Simple Sequence Repeat and Novel Genic Markers. *Mol. Biotechnol.* **2014**, *56*, 925–938. [CrossRef]
33. Siol, M.; Jacquin, F.; Chabert-Martinello, M.; Smýkal, P.; Le Paslier, M.C.; Aubert, G.; Burstin, J. Patterns of genetic structure and linkage disequilibrium in a large collection of pea germplasm. *G3 Genes Genomes Genet.* **2017**, *7*, 2461–2471. [CrossRef]
34. Holdsworth, W.L.; Gazave, E.; Cheng, P.; Myers, J.R.; Gore, M.A.; Coyne, C.J.; McGee, R.J.; Mazourek, M. A community resource for exploring and utilizing genetic diversity in the USDA pea single plant plus collection. *Hortic. Res.* **2017**, *4*, 17017. [CrossRef] [PubMed]
35. CBOL Plant Working Group; Hollingsworth, P.M.; Forrest, L.L.; Spouge, J.L.; Hajibabaei, M.; Ratnasingham, S.; van der Bank, M.; Chase, M.W.; Cowan, R.S.; Erickson, D.L.; et al. A DNA barcode for land plants. *Proc. Natl. Acad. Sci. USA* **2009**, *106*, 12794–12797.
36. Gao, T.; Chen, S. Authentication of the Medicinal Plants in Fabaceae by DNA Barcoding Technique. *Planta Med.* **2009**, *75*, 417. [CrossRef]

37. Gao, T.; Sun, Z.; Yao, H.; Song, J.; Zhu, Y.; Ma, X.; Chen, S. Identification of fabaceae plants using the DNA barcode matK. *Planta Med.* **2011**, *77*, 92–94. [CrossRef]
38. Gao, T.; Yao, H.; Song, J.; Liu, C.; Zhu, Y.; Ma, X.; Pang, X.; Xu, H.; Chen, S. Identification of medicinal plants in the family Fabaceae using a potential DNA barcode ITS2. *J. Ethnopharmacol.* **2010**, *130*, 116–121. [CrossRef]
39. Madesis, P.; Ganopoulos, I.; Ralli, P.; Tsaftaris, A. Barcoding the major Mediterranean leguminous crops by combining universal chloroplast and nuclear DNA sequence targets. *Genet. Mol. Res.* **2012**, *11*, 2548–2558. [CrossRef]
40. Jaakola, L.; Suokas, M.; Häggman, H. Novel approaches based on DNA barcoding and high-resolution melting of amplicons for authenticity analyses of berry species. *Food Chem.* **2010**, *123*, 494–500. [CrossRef]
41. Ganopoulos, I.; Madesis, P.; Darzentas, N.; Argiriou, A.; Tsaftaris, A. Barcode High Resolution Melting (Bar-HRM) analysis for detection and quantification of PDO "Fava Santorinis" (Lathyrus clymenum) adulterants. *Food Chem.* **2012**, *133*, 505–512. [CrossRef]
42. Ganopoulos, I.; Madesis, P.; Tsaftaris, A. Universal ITS2 Barcoding DNA Region Coupled with High-Resolution Melting (HRM) Analysis for Seed Authentication and Adulteration Testing in Leguminous Forage and Pasture Species. *Plant Mol. Biol. Rep.* **2012**, *30*, 1322–1328. [CrossRef]
43. Bosmali, I.; Ordoudi, S.A.; Tsimidou, M.Z.; Madesis, P. Greek PDO saffron authentication studies using species specific molecular markers. *Food Res. Int.* **2017**, *100*, 899–907. [CrossRef] [PubMed]
44. Philaniotou, O. Lesser Cyclades. In *Archaeology: Agean Islands*; Vlachopoulos, A.G., Ed.; Melissa Publishing House: Athens, Greece, 2006; pp. 286–289. ISBN 960-204-272-9.
45. Chatzilazarou, D.T. An Aegean coastal settlement at the 'end of Late Antiquity': The case of Schinoussa near Naxos. In *Naxos and the Byzantine Aegean: Insular Responses to Regional Change*; Crow, J., Hill, D., Eds.; The Norwegian Institute at Athens: Athens, Greece, 2018; Volume 7, pp. 195–222. ISBN 9789608514577.
46. De Tournefort, J.P. *Relation d'un Voyage du Levant (1700-1702)*, 1st ed.; Lydaki, E., Ed.; Crete University Press: Crete, Greece, 2003; ISBN 960-524-173-0.
47. Loupis, D. Piracy in the Ottoman Naval Texts (16th–17th century). In *Pirates and Corsairs*; Kalliga, C., Malliaris, A., Eds.; Hestia Publishers: Athens, Greece, 2003; p. 294. ISBN 960-05-1106-3.
48. Ginio, E. Piracy and Redemption in the Aegean Sea during the first half of the Eighteenth century. *Turcica* **2001**, *33*, 135–147. [CrossRef]
49. Sun, S.; Deng, D.; Duan, C.; Zong, X.; Xu, D.; He, Y.; Zhu, Z. Two novel er1 alleles conferring powdery mildew (Erysiphe pisi) resistance identified in a worldwide collection of pea (Pisum sativum L.) germplasms. *Int. J. Mol. Sci.* **2019**, *20*, 5071. [CrossRef] [PubMed]
50. Sun, S.; Deng, D.; Wang, Z.; Duan, C.; Wu, X.; Wang, X.; Zong, X.; Zhu, Z. A novel er1 allele and the development and validation of its functional marker for breeding pea (Pisum sativum L.) resistance to powdery mildew. *Theor. Appl. Genet.* **2016**, *129*, 909–919. [CrossRef]
51. Barber, R.L.N. *The Cyclades in the Bronze Age*; Duckworth Publishers: London (Richmond), UK, 1987; ISBN 0715621602.
52. Treuil, R.; Darcque, P.; Poursat, J.-C.; Touchais, G. *Les Civilisations Égéennes du Néolithique et de l'Age du Bronze*; Deuxième, É., Ed.; Presses Universitaires de France: Paris, France, 1989.
53. Li, B.; Zheng, Y. Dynamic evolution and phylogenomic analysis of the chloroplast genome in Schisandraceae. *Sci. Rep.* **2018**, *8*, 1–11. [CrossRef]
54. Whitlock, B.A.; Hale, A.M.; Groff, P.A. Intraspecific inversions pose a challenge for the trnH-psbA plant DNA barcode. *PLoS ONE* **2010**, *5*, e11533. [CrossRef] [PubMed]
55. Baloch, F.S.; Alsaleh, A.; de Miera, L.E.S.; Hatipoğlu, R.; Çiftçi, V.; Karaköy, T.; Yildiz, M.; Özkan, H. DNA based iPBS-retrotransposon markers for investigating the population structure of pea (Pisum sativum) germplasm from Turkey. *Biochem. Syst. Ecol.* **2015**, *61*, 244–252. [CrossRef]
56. Bogdanova, V.S.; Mglinets, A.V.; Shatskaya, N.V.; Kosterin, O.E.; Solovyev, V.I.; Vasiliev, G.V. Cryptic divergences in the genus Pisum L. (peas), as revealed by phylogenetic analysis of plastid genomes. *Mol. Phylogenet. Evol.* **2018**, *129*, 280–290. [CrossRef]
57. Gubae, T.T.; Petros, Y. Genetic diversity of chickpea (Cicer arietinum L.) cultivars from Ethiopia by using ISSR markers. *J. Agric. Biotechnol. Sustain. Dev.* **2018**, *10*, 178–184.
58. Pavan, S.; Schiavulli, A.; Appiano, M.; Miacola, C.; Visser, R.G.F.; Bai, Y.; Lotti, C.; Ricciardi, L. Identification of a complete set of functional markers for the selection of er1 powdery mildew resistance in Pisum sativum L. *Mol. Breed.* **2013**, *31*, 247–253. [CrossRef]

59. Sun, S.; He, Y.; Dai, C.; Duan, C.; Zhu, Z. Two major er1 alleles confer powdery mildew resistance in three pea cultivars bred in Yunnan Province, China. *Crop J.* **2016**, *4*, 353–359. [CrossRef]
60. Ganopoulos, I.; Mylona, P.; Mellidou, I.; Kalivas, A.; Bosmali, I.; Kontzidou, S.; Osathanunkul, M.; Madesis, P. Microsatellite genotyping and molecular screening of pea (Pisum sativum L.) germplasm with high-resolution melting analysis for resistance to powdery mildew. *Plant Gene* **2018**, *15*, 1–5. [CrossRef]
61. Doyle, J.J.; Doyle, J.L. A Rapid DNA Isolation Procedure for Small Quantities of Fresh Leaf Tissue. *Phytochem. Bull* **1987**, *19*, 11–15.
62. Yao, H.; Song, J.; Liu, C.; Luo, K.; Han, J.; Li, Y.; Pang, X.; Xu, H.; Zhu, Y.; Xiao, P.; et al. Use of ITS2 region as the universal DNA barcode for plants and animals. *PLoS ONE* **2010**, *5*, e13102. [CrossRef] [PubMed]
63. Taberlet, P.; Coissac, E.; Pompanon, F.; Gielly, L.; Miquel, C.; Valentini, A.; Vermat, T.; Corthier, G.; Brochmann, C.; Willerslev, E. Power and limitations of the chloroplast trnL (UAA) intron for plant DNA barcoding. *Nucleic Acids Res.* **2007**, *35*, e14. [CrossRef] [PubMed]
64. Tate, J.A.; Simpson, B.B. Paraphyly of Tarasa (Malvaceae) and Diverse Origins of the Polyploid Species. *Syst. Bot.* **2003**, *28*, 723–737.
65. Kumar, S.; Stecher, G.; Li, M.; Knyaz, C.; Tamura, K. MEGA X: Molecular evolutionary genetics analysis across computing platforms. *Mol. Biol. Evol.* **2018**, *35*, 1547–1549. [CrossRef]
66. Benson, D.A.; Clark, K.; Karsch-Mizrachi, I.; Lipman, D.J.; Ostell, J.; Sayers, E.W. GenBank. *Nucleic Acids Res.* **2015**, *43*, D30–D35. [CrossRef]
67. Altschul, S.F.; Gish, W.; Miller, W.; Myers, E.W.; Lipman, D.J. Basic local alignment search tool. *J. Mol. Biol.* **1990**, *215*, 403–410. [CrossRef]
68. Nei, M. Estimation of average heterozygosity and genetic distance from a small number of individuals. *Genetics* **1978**, *89*, 583–590.
69. Peakall, R.; Smouse, P.E. GenALEx 6.5: Genetic analysis in Excel. Population genetic software for teaching and research-an update. *Bioinformatics* **2012**, *28*, 2537–2539. [CrossRef]

plants

MDPI

Article

Evaluation of Wild Potato Germplasm for Tuber Starch Content and Nitrogen Utilization Efficiency

Silvia Bachmann-Pfabe * and Klaus J. Dehmer

Leibniz Institute of Plant Genetics and Crop Plant Research (IPK), Genebank Department, Satellite Collections North, Gross Luesewitz Potato Collections, Parkweg 3a, 18190 Gross Luesewitz, Germany; dehmer@ipk-gatersleben.de

* Correspondence: pfabe@ipk-gatersleben.de; Tel.: +49-38209-82-314

Received: 10 June 2020; Accepted: 1 July 2020; Published: 2 July 2020

Abstract: Potato wild relatives provide a considerable source of variation for important traits in cultivated potato (*Solanum tuberosum* L.) breeding. This study evaluates the variation of tuber starch content and nitrogen utilization efficiency (NutE) in wild potato germplasm. For the experiments regarding starch content, 28 accessions of ten different tuber-bearing wild *Solanum*-species were chosen, and in vitro plantlets were raised from seeds. Twenty plantlets (= genotypes) per accession were then cultivated in the greenhouse until natural senescence and tuber starch content was determined. The average tuber starch content across all genotypes tested was 21.7% of fresh mass. Contents above 28% of fresh mass were found in 50 genotypes, belonging to the species *S. chacoense*, *S. commersonii*, *S. jamesii*, and *S. pinnatisectum*. Subsequently, 22 wild genotypes revealing high tuber starch contents and four modern varieties of cultivated potato were studied as in vitro plantlets under optimal and low N supply (30 and 7.5 mmol L^{-1} N). Low N supply lead to a genotype-dependent reduction of shoot dry mass between 13 and 46%. The majority of the wild types also reduced root dry mass by 26 to 62%, while others maintained root growth and even exceeded the NutE of the varieties under low N supply. Thus, wild potato germplasm appears superior to cultivars in terms of tuber starch contents and N utilization efficiency, which should be investigated in further studies.

Keywords: genetic resources; *Solanum chacoense*; stress tolerance

1. Introduction

Besides being one of the most important food crops worldwide, potato (*Solanum tuberosum* L.) plays an important role in industry due to its starchy tubers. Starch is used in bakery products, thickening products, soups and noodles but also for the production of paper, textiles, building materials, pharmaceutical products, chemicals and biodegradable packaging materials [1]. Compared to other starches, potato starch has superior characteristics because it is easily isolated, of high purity and of large granule size, needs low temperatures for gelatinization and produces gels with highest viscosity [2]. Depending on genotype and growing environment, a fresh potato tuber contains about 20% of dry mass, 60–80% of which is starch [3]. Dedicated starch varieties may even reach a starch content of up to 23% of fresh mass [4]. The nitrogen (N) fertilizer regime influences starch yield by positively affecting canopy development and photosynthesis efficiency, dry matter partitioning to the tubers, tuber bulking and tuber yield formation [5,6]. Furthermore, field N availability influences starch quality parameters, such as granule size, viscosity and breakdown [7]. Consistently, Maltas et al. [8] reported a highly significant effect of different N fertilizer rates on total tuber yield, the percentage of large tubers and starch concentration under field conditions in cv. Bintje and Laura.

Depending on environmental conditions and genotype, the potato crop has been found to remove 90 to 190 kg N ha^{-1} [9]. However, the shallow, less branched and less dense root system of potato does not allow the exploration of a large soil volume or to retrieve nitrogen from deeper soil layers, and hence, potato demands a high level of readily available soil N at the right period of growth [5,9–12]. In a review, Iwama [13] reported that most of the potato roots are present in the upper 30 cm of the soil and only a small fraction extends to 100 cm. In combination with the fact that potatoes are often cultivated on coarse and sandy soils and under irrigation, these areas face an increasing potential for nitrate leaching and contamination of groundwater [14].

Thus, improving the N use efficiency in potato production is not only of economic, but also of environmental concern, and different measures such as split application of N or foliar application of urea are being discussed in order to reduce N leaching [5]. Improving the N uptake and use efficiency of the potato crop itself is also an important approach, and many studies evaluated the N use efficiency in cultivated potatoes [8,15–21]. In contrast, only few studies evaluated the N use efficiency and/or tuber starch contents of native Andean cultivars or wild potato germplasm [22–25]. The wild relatives of the cultivated potato could be an important source of variation for root length and morphology, tuber starch content and N use efficiency. The secondary and tertiary genepool of potato has intensively been studied as a source of disease resistance [26–28], and was used, amongst others, to improve foliar late blight and nematode resistance of *S. tuberosum* [29,30]. In terms of N use efficiency, Errebhi et al. [22] compared 39 wild potato accessions of 23 species with three cultivated varieties under high and zero N in the field. They found some wild potato genotypes which were able to take up significantly more applied N than their cultivated relatives. Genotypes of *S. microdontum* and *S. chacoense* were identified as the ones with the highest N uptake efficiency (NupE) [22]. Selected native Andean cultivars indicated a similar nitrogen use efficiency to commercial cultivars, but showed, despite different environmental conditions, a highly consistent performance across a two-year field study [24]. To our knowledge, the most comprehensive study of tuber starch content and quality in exotic germplasm, was provided by Jansen et al. [25]. Accessions of 46 wild and cultivated potato species showed a high variation in starch contents ranging between 3.8 and 39.6% of fresh mass. Highest starch contents were predominately found in genotypes of species *S. pinnatisectum* and *S. chacoense* [25].

Based on the above-mentioned findings, our study aimed to (I) update and assess the variability of the tuber starch contents in wild potato germplasm and to (II) study the nitrogen use efficiency of genotypes with high tuber starch contents in relation to modern cultivars.

2. Results

2.1. Variation of Tuber Starch Contents in Wild Potato Germplasm after Greenhouse Cultivation

In 2013, altogether 28 different wild potato accessions (= populations) representing ten different species were cultivated in the greenhouse to evaluate their tuber starch contents. For each accession 20 different genotypes were cultivated as in vitro plantlets, however, the results only include genotypes which produced sufficient tubers for starch analysis (in total 506 genotypes, Table 1). On average, of all the 506 genotypes analyzed, the starch content in the tubers amounted to 21.7% of fresh mass (FM). The lowest average starch content with 14.2% of FM was measured for accession Gross Luesewitz Potato Collections (GLKS) 31559 (*S. stenotomum*), while accession GLKS 30211 (*S. commersonii*) showed the highest starch content with on average 30.0% of FM. Interestingly, all genotypes of accession GLKS 30211 showed high tuber starch contents ranging between 26.4 and 33.3% of FM, indicating a rather low variation within this population (Table 1). From *S. chacoense,* all the 15 accessions had an average starch content of 22.5% of FM and showed a rather low variation between the populations (CV = 8.71%), and a higher within population variation (CV: 11.5–23.9%, Table 1). Ten accessions of *S. chaoense* encompassed genotypes with a tuber starch content higher than 28% of FM, a target value which was considered as selection criterion for high-starch genotypes in this study. Regarding *S. pinnatisectum*, the tubers of the five accessions studied, had an average starch content of 22.2% of FM, but showed starch contents up to 36.6% of FM. Altogether 11 genotypes of three accessions produced starch contents higher than 28% (Table 1).

Table 1. Wild potato accessions of the Gross Luesewitz Potato Collections (GLKS) used in the study ranked according to their tuber starch content. Given are the number of genotypes tested per accession, the mean starch content (%) in the tubers, its minimum and maximum values and the coefficient of variation (CV %).

Accession GLKS	*Solanum*-Species	No. Genotypes	Starch Content (% of FM) Mean	Min	Max	CV %
30211	*S. commersonii*	13	30.0	26.4	33.3	8.60
30916	*S. chacoense*	20	26.8	19.5	31.6	11.5
31595	*S. pinnatisectum*	15	26.8	18.9	30.6	11.5
30475	*S. jamesii*	17	25.7	12.7	32.6	18.7
30177	*S. chacoense*	20	24.8	15.4	31.4	16.6
30159	*S. chacoense*	19	24.3	19.0	30.9	12.0
30154	*S. chacoense*	20	23.2	11.0	31.8	17.9
30160	*S. chacoense*	19	23.2	15.8	32.7	16.8
31600	*S. pinnatisectum*	18	23.2	13.0	36.6	24.7
30156	*S. chacoense*	20	23.1	15.3	29.1	17.6
30191	*S. chacoense*	20	22.5	15.6	28.2	15.7
30197	*S. chacoense*	19	22.5	17.9	28.0	13.8
30181	*S. chacoense*	20	22.3	15.4	29.8	19.9
30995	*S. chacoense*	18	22.1	15.6	29.0	17.1
31025	*S. chacoense*	20	21.6	18.6	25.9	11.7
31610	*S. pinnatisectum*	19	21.5	14.7	31.9	24.2
30665	*S. chacoense*	19	21.1	13.0	27.2	16.0
30135	*S. chacoense*	20	20.4	13.8	33.0	23.9
31602	*S. pinnatisectum*	16	20.1	12.6	26.2	19.6
30688	*S. microdontum*	20	20.0	14.1	26.2	16.6
30148	*S. chacoense*	18	19.7	11.0	25.9	23.2
31605	*S. pinnatisectum*	7	19.6	16.1	22.7	11.4
30134	*S. chacoense*	20	19.3	14.0	25.6	18.0
32852	*S. hondelmannii*	18	19.2	13.7	24.4	17.2
31583	*S. tarijense*	19	16.9	13.5	20.7	11.8
34995	*S. tuberosum* subsp. *andigena*	16	16.6	12.6	21.6	17.9
30944	*S. sparsipilum*	20	16.0	11.4	22.2	18.8
31559	*S. stenotomum*	16	14.2	7.1	22.4	23.0

For subsequent N efficiency experiments, genotypes with starch contents higher than 28% of FM were to be used. In *S. chacoense*, altogether 23 genotypes were identified revealing a tuber starch content above 28% of FM. Overall, 15 genotypes, belonging to accessions GLKS 30135, GLKS 30154, GLKS 30156, GLKS 30159, GLKS 30160, GLKS 30177, GLKS 30181, GLKS 30916, and GLKS 30995 were selected for the N experiments and re-cultivated in 2014 to validate their starch contents (Table 2). In *S. pinnatisectum*, 11 genotypes showed tuber starch contents above 28% of FM, and genotype GLKS 31600_10 with a starch content of 36.6% was selected for up-coming experiments and re-tested in 2014. Additionally, two genotypes with the highest starch contents of *S. microdontum* (GLKS 30688_04, GLKS 30688_12) and *S. stenotomum* (GLKS 31559_11, GLKS 31559_14) as well as one genotype of *S. tuberosum* subsp. *andigena* (GLKS 34995_18) were selected for the N efficiency experiments in order to cover a broader spectrum of *Solanum* species, even if they had starch contents below 28% of FM (Table 2). The tuber starch contents determined in 2013 and after re-testing of selected genotypes in 2014 correlated well ($r = 0.72$, $p \leq 0.01$), confirming the high-starch properties of the majority of the selected accessions.

Table 2. Selected accessions and genotypes of the Gross Luesewitz Potato Collections used for the N efficiency experiments as well as their respective tuber starch contents (%) in 2013 and 2014.

Accession GLKS	Genotype No.	*Solanum*-Species	Starch Content (% of FM) 2013	2014	Mean
30135	05	*S. chacoense*	28.0	25.3	26.7
30135	19	*S. chacoense*	33.0	29.8	31.4
30154	09	*S. chacoense*	31.8	25.2	28.5
30156	16	*S. chacoense*	29.1	26.7	27.9
30159	05	*S. chacoense*	30.9	36.9	33.9
30160	13	*S. chacoense*	29.5	26.2	27.9
30160	15	*S. chacoense*	32.7	-	32.7
30177	01	*S. chacoense*	24.2	29.0	26.6
30177	02	*S. chacoense*	30.9	27.3	29.1
30177	15	*S. chacoense*	30.1	26.0	28.1
30177	17	*S. chacoense*	28.3	32.0	30.2
30177	20	*S. chacoense*	31.4	-	31.4
30181	06	*S. chacoense*	28.6	-	28.6
30181	18	*S. chacoense*	29.8	26.2	28.0
30688	04	*S. microdontum*	26.2	31.8	29.0
30688	12	*S. microdontum*	25.3	23.9	24.6
30916	08	*S. chacoense*	31.6	29.7	30.7
30995	18	*S. chacoense*	29.0	28.7	28.9
31559	11	*S. stenotomum*	15.2	11.4	13.3
31559	14	*S. stenotomum*	22.1	16.6	19.4
31600	10	*S. pinnatisectum*	36.6	31.1	33.9
34995	18	*S. tuberosum* subsp. *andigena*	18.0	22.6	20.3

2.2. Dry Yield of Shoots, Roots and Root-DM:Shoot-DM Ratio in the N Experiments

Shoot and root DM as well as the root-DM:shoot-DM ratio were predominately affected by the genotype and, to a lesser extent, by the factor treatment. For these traits, the factor genotype explained up to 68% of the variation in the data, while the factor treatment explained between 6 and 22%. The genotype × treatment interaction explained 11 and 15% of the total variation for root DM and root-DM:shoot-DM ratio, respectively, but played only a minor role for shoot DM (3.58%, Table 3).

In the high N (30 mmol L^{-1}) treatment, the shoot DM of the genotypes varied between 214 and 682 mg $vessel^{-1}$ (Table 4). The lowest shoot DM was observed for the genotypes GLKS 31600_10, GLKS 30177_17 and cv. Kiebitz, while the highest biomass was achieved by the genotypes GLKS 30177_20, GLKS 30181_06 and GLKS 30160_15. These genotypes even exceeded the performance of cv. Tomba, which showed the highest shoot DM amongst the standard varieties. Under low N supply (7.5 mmol L^{-1}), shoot DM ranged between 156 and 549 mg $vessel^{-1}$. Shoot DM decreased under low N supply on average by 115 mg $vessel^{-1}$ (23%) and the shoot biomass reduction was significant for all the genotypes tested, except for cv. Kiebitz and cv. Eurobravo (Table 4). The strongest reduction (>30%) was observed for genotypes GLKS 30135_19, GLKS 30995_18, GLKS 30177_02 and GLKS 31559_11. A moderate shoot DM reduction (15 to 20%) at simultaneously high yields in the control was observed for genotypes GLKS 30135_05, GLKS 30160_13 and GLKS 30177_20. Again, cv. Tomba produced the highest shoot DM amongst the standard varieties under low N supply. However, several wild potato genotypes performed as well or even exceeded the shoot DM of cv. Tomba under reduced N conditions (Table 4).

Table 3. Evaluated traits and variance explained (%) by the factors genotype, treatment and interaction. Given are the sum of squares (SSQ) resulting from two-factor analysis of variance and F-test.

Trait	Total SSQ	Genotype			Treatment			Genotype × Treatment			Residuals	
		SSQ	%	Sign.	SSQ	%	Sign.	SSQ	%	Sign.	SSQ	%
DM shoot mg vessel^{-1}	3,164,366	2,167,098	68.5	***	697,852	22.1	***	113,436	3.58	***	185,980	5.88
DM root mg vessel^{-1}	1,229,096	777,076	62.9	***	260,060	21.1	***	134,179	10.9	***	63,202	5.12
N uptake shoot [a]	77.24	3.03	3.93	***	71.54	92.6	***	1.15	1.49	***	1.51	1.95
N uptake root [a]	66.63	17.72	28.0	***	40.00	62.8	***	3.40	5.33	***	2.54	3.99
N uptake total [a]	68.96	2.93	4.25	***	64.08	92.9	***	0.87	1.26	***	1.08	1.56
NutE [a]	43.57	9.20	21.1	***	32.56	74.7	***	0.79	1.82	***	1.01	2.32
Root-DM:Shoot-DM ratio [a]	31.26	21.15	67.7	***	1.87	6.00	***	4.81	15.4	***	3.43	11.0
Root-N:Shoot-N ratio [a]	30.69	18.29	59.6	***	4.56	14.9	***	4.15	13.5	***	3.69	12.0

[a] log transformation of the data prior to ANOVA; SSQ = sum of squares; % = percent share of total sum of squares, Sign. = significance of F-test (* $p \leq 0.05$; ** $p \leq 0.01$; *** $p \leq 0.001$).

Table 4. Mean dry mass and N uptake of shoots and roots (mg vessel^{-1}) of different wild potato genotypes cultivated under optimal and reduced N supply for 21 days in a climate chamber. Results of ANOVA and post hoc comparison of means.

Genotype	*Solanum*-Species	Shoot DM			Root DM			Shoot N			Root N			N Uptake Total		
		Control	Reduced	*p*	Control	Reduced	*p*	Control	Reduced	*p*	Control	Reduced	*p*	Control	Reduced	*p*
30135_05	*S. chacoense*	552 a	462 a	**	135 b	51.7 b	***	19.5 b	7.58 a	***	3.01 b	0.85 b	***	22.5 b	8.48 b	***
30135_19	*S. chacoense*	604 a	411 a	***	177 b	78.0 b	***	24.6 a	6.79 b	***	5.43 a	1.40 b	***	30.1 a	8.20 b	***
30154_09	*S. chacoense*	496 a	431 a	*	304 a	143 a	***	20.0 b	6.57 b	***	6.54 b	2.21 a	***	26.5 a	8.78 b	***
30156_16	*S. chacoense*	575 a	459 a	***	240 a	159 a	***	23.8 a	6.47 b	***	4.61 a	2.20 a	***	28.5 a	8.67 b	***
30159_05	*S. chacoense*	528 a	371 b	***	192 b	133 a	***	20.7 b	5.72 b	***	4.86 a	2.22 a	***	25.6 b	7.95 b	***
30160_13	*S. chacoense*	600 a	506 a	**	337 b	151 a	***	19.3 b	6.30 b	***	7.52 b	2.40 a	***	26.8 a	8.73 b	***
30160_15	*S. chacoense*	652 b	513 a	***	277 a	204 b	***	24.4 a	6.31 b	***	6.86 b	3.11 b	***	31.3 a	9.43 a	***
30177_01	*S. chacoense*	436 b	311 b	***	250 a	153 a	***	19.7 b	5.07 b	***	5.91 b	2.45 a	***	25.7 b	7.58 b	***
30177_02	*S. chacoense*	495 a	297 b	***	157 b	62.5 b	***	22.2 a	5.99 b	***	6.17 b	2.43 a	***	28.3 a	8.44 b	***
30177_15	*S. chacoense*	620 a	485 a	***	226 a	203 b	ns	24.7 a	6.95 b	***	4.56 a	2.66 b	***	29.2 a	9.61 a	***
30177_17	*S. chacoense*	280 b	196 b	**	156 b	74.5 b	***	17.3 b	5.19 b	***	4.55 a	1.76 a	***	22.1 b	6.96 b	***
30177_20	*S. chacoense*	682 b	549 b	***	318 b	219 b	***	25.5 a	6.85 b	***	5.00 a	2.99 b	***	30.6 a	9.96 a	***
30181_06	*S. chacoense*	680 b	508 a	***	291 a	117 a	***	23.0 a	7.45 b	***	5.29 a	1.85 a	***	28.3 a	9.84 b	***
30181_18	*S. chacoense*	397 b	335 b	*	76.0 b	61.5 b	ns	18.8 b	6.99 b	***	3.43 a	1.35 b	***	22.3 b	8.35 b	***
30688_04	*S. microdontum*	491 b	377 b	***	128 b	118 a	ns	23.5 a	7.17 b	***	4.50 a	2.08 a	***	28.1 a	9.27 b	***
30688_12	*S. microdontum*	497 a	384 a	***	139 b	116 a	ns	24.2 a	6.58 b	***	4.07 a	2.05 a	***	28.2 a	8.64 b	***
30916_08	*S. chacoense*	536 a	403 a	***	160 b	104 b	**	22.8 a	5.80 b	***	3.65 a	1.83 a	***	26.4 a	7.64 b	***
30995_18	*S. chacoense*	458 b	306 b	***	103 b	51.5 b	**	22.5 a	5.84 b	***	2.87 b	1.04 b	***	25.4 b	6.90 b	***

Table 4. *Cont.*

Genotype	*Solanum*-Species	Shoot DM Control	Shoot DM Reduced	*p*	Root DM Control	Root DM Reduced	*p*	Shoot N Control	Shoot N Reduced	*p*	Root N Control	Root N Reduced	*p*	N Uptake Total Control	N Uptake Total Reduced	*p*
31559_11	*S. stenotomum*	498 a	266 b	***	195 b	95.0 b	***	23.6 a	6.40 b	***	4.59 a	1.51 a	***	28.2 a	7.91 b	***
31559_14	*S. stenotomum*	408 b	323 b	**	77.3 b	89.7 b	ns	22.6 a	7.01 b	***	3.19 b	1.34 b	***	25.8 b	8.36 b	***
31600_10	*S. pinnatisectum*	214 b	156 b	*	30.0 b	40.7 b	ns	11.3 b	5.49 b	***	1.48 b	1.44 a	ns	12.8 b	6.93 b	***
34995_18	*S. tuberosum* subsp. *andigena*	403 b	313 b	**	173 b	118 a	**	23.1 a	6.56 b	***	5.37 a	1.79 a	***	28.5 a	8.39 b	***
Eurobravo	*S. tuberosum* subsp. *tuberosum*	452 b	400 a	ns	157 b	102 b	**	22.6 a	7.48 a	***	4.70 a	1.68 a	***	27.3 a	9.17 b	***
Kiebitz	*S. tuberosum* subsp. *tuberosum*	302 b	268 b	ns	66.0 b	50.2 b	ns	19.9 b	7.58 a	***	2.96 b	0.97 b	***	22.9 b	8.56 b	***
Maxi	*S. tuberosum* subsp. *tuberosum*	456 b	374 b	*	164 b	95.0 b	***	23.6 a	7.74 a	***	4.00 a	1.38 b	***	27.6 a	9.12 b	***
Tomba	*S. tuberosum* subsp. *tuberosum*	**567 a**	**452 a**	***	**270 a**	**153 a**	***	**27.1 a**	**9.34 a**	***	**4.31 a**	**1.91 a**	***	**31.4 a**	**11.3 a**	***
mean		496	381		185	115		21.9	6.66		4.59	1.88		26.6	8.56	

The letters within one column indicate whether there is a significant difference to the best cultivar Tomba ("b") or not ("a", Dunnett's test $p \leq 0.05$), underlined are mean values significantly higher than that of cv. Tomba; asterisks indicate a significant difference between the treatments within one genotype (pairwise comparisons, Tukey adjustment; *** $p \leq 0.001$, ** $p \leq 0.01$, * $p \leq 0.05$, ns = not significant).

The root DM varied between 30 and 337 mg vessel^{-1} in the control treatment. The lowest root growth (<80 mg vessel^{-1}) was observed for genotypes GLKS 31600_10, cv. Kiebitz, GLKS 30181_18 and GLKS 31559_14, while the genotypes GLKS 30154_09, GLKS 30160_13 and GLKS 30177_20 produced more than 300 mg vessel^{-1}. Under low N conditions, the root DM varied between 41 and 219 mg vessel^{-1}. The different genotypes either maintained or decreased root DM due to N deficit. The decrease was highest (about 60%) for the genotypes GLKS 30135_05, GLKS 30177_02, GLKS 30181_06, while the genotypes GLKS 30177_15, GLKS 30181_18, GLKS 30688_04, GLKS 30688_12, GLKS 31559_14, GLKS 31600_10 and cv. Kiebitz maintained root mass. Most interestingly, genotype GLKS 30177_15 produced a high root mass in the control and maintained it under reduced N conditions. Amongst the cultivars, cv. Tomba produced the highest root mass in both treatments. However, several wild potato genotypes produced a significantly higher root biomass under control (GLKS 30160_13, GLKS 30177_20) or low N conditions (GLKS 30160_15, GLKS 30177_15, GLKS 30177_20) compared to cv. Tomba (Table 4).

Relating the root to the shoot biomass improves the understanding of genotype-specific reactions to low N supply. The root-DM:shoot-DM ratio varied strongly between the wild types, ranging from 0.14 to 0.61 in the control and 0.10 to 0.48 in the low N treatment, but was generally comparable to that of the standard varieties (Table 5). High values above 0.50 were calculated for genotypes GLKS 30154_09, GLKS 30177_01, GLKS 30160_13 and GLKS 30177_17, indicating a strong root growth in relation to the shoot. Genotypes reacted differently to low N supply, either by maintaining (13), reducing (11) or increasing (2) root-DM:shoot-DM ratio. The genotypes GLKS_31559_14 and GLKS_31600_10 showed the lowest root-DM:shoot-DM ratio in the control, but increased root growth at the expense of the shoot under low N.

Table 5. Mean root-DM:shoot-DM ratio as well as partitioning of N taken up by the different wild potato genotypes (root-N:shoot-N ratio) cultivated under optimal and reduced N conditions in a climate chamber for 21 days. Results of ANOVA and post hoc comparison of means.

Genotype	*Solanum*-Species	Root-DM:Shoot-DM Ratio Control	Root-DM:Shoot-DM Ratio Reduced	*p*	Root-N:Shoot-N Ratio Control	Root-N:Shoot-N Ratio Reduced	*p*
30135_05	*S. chacoense*	0.25 b	0.10 b	***	0.15 a	0.11 b	*
30135_19	*S. chacoense*	0.29 b	0.19 b	***	0.22 a	0.21 a	ns
30154_09	*S. chacoense*	0.61 a	0.33 a	***	0.33 b	0.34 b	ns
30156_16	*S. chacoense*	0.42 a	0.35 a	ns	0.19 a	0.34 b	***
30159_05	*S. chacoense*	0.36 a	0.36 a	ns	0.23 b	0.39 b	***
30160_13	*S. chacoense*	0.55 a	0.30 a	***	0.39 b	0.38 b	ns
30160_15	*S. chacoense*	0.42 a	0.40 a	ns	0.28 b	0.49 b	***
30177_01	*S. chacoense*	0.57 a	0.48 b	ns	0.30 b	0.48 b	***
30177_02	*S. chacoense*	0.32 b	0.20 b	***	0.28 b	0.41 b	**
30177_15	*S. chacoense*	0.36 a	0.42 a	ns	0.19 a	0.38 b	***
30177_17	*S. chacoense*	0.56 a	0.37 a	***	0.26 b	0.34 b	*
30177_20	*S. chacoense*	0.46 a	0.40 a	ns	0.20 a	0.44 b	***
30181_06	*S. chacoense*	0.43 a	0.23 b	***	0.23 b	0.25 a	ns
30181_18	*S. chacoense*	0.18 b	0.18 b	ns	0.18 a	0.19 a	ns
30688_04	*S. microdontum*	0.26 b	0.31 a	ns	0.19 a	0.29 b	**
30688_12	*S. microdontum*	0.28 b	0.30 a	ns	0.17 a	0.31 b	***
30916_08	*S. chacoense*	0.30 b	0.26 a	ns	0.16 a	0.32 b	***
30995_18	*S. chacoense*	0.22 b	0.17 b	*	0.13 a	0.18 a	*
31559_11	*S. stenotomum*	0.39 a	0.36 a	ns	0.19 a	0.24 a	ns
31559_14	*S. stenotomum*	0.19 b	0.27 a	*	0.14 a	0.19 a	*
31600_10	*S. pinnatisectum*	0.14 b	0.26 a	***	0.13 a	0.26 a	***
34995_18	*S. tuberosum* subsp. *andigena*	0.43 a	0.38 a	ns	0.23 b	0.27 a	ns
Eurobravo	*S. tuberosum* subsp. *tuberosum*	0.35 a	0.26 a	*	0.21 a	0.22 a	ns
Kiebitz	*S. tuberosum* subsp. *tuberosum*	0.22 b	0.19 b	ns	0.15 a	0.13 b	ns
Maxi	*S. tuberosum* subsp. *tuberosum*	0.36 a	0.25 a	*	0.17 a	0.18 a	ns
Tomba	*S. tuberosum* subsp. *tuberosum*	**0.47 a**	**0.34 a**	*	**0.16 a**	**0.20 a**	*
mean		0.36	0.29		0.21	0.29	

The letters within one column indicate whether there is a significant difference to the best cultivar Tomba ("b") or not ("a", Dunnett's test $p \leq 0.05$), underlined are mean values significantly higher than that of cv. Tomba; asterisks indicate a significant difference between the treatments within one genotype (pairwise comparisons, Tukey adjustment; *** $p \leq 0.001$, ** $p \leq 0.01$, * $p \leq 0.05$, ns = not significant).

2.3. N Uptake, N Partitioning and N Efficiency Parameters in the N Experiments

Interestingly, the genotypic variation for the shoot and total N uptake was low and explained only about 4% of the total variation. In contrast, the factor genotype explained between 20 and 60% of the total variation for the traits root N uptake, NutE and N partitioning (root-N:shoot-N ratio). Additionally, a clear genotype × treatment interaction was observed for the root-N:shoot-N ratio (Table 3). In the control treatment, the average shoot N uptake was 22.0 mg N vessel^{-1} and varied only by ± 3.18 mg vessel^{-1} (CV = 14.5%). In the low N treatment, shoot N uptake was on average reduced by 15.3 mg N vessel^{-1} (69%), and the reduction was significant for all the genotypes tested (Table 4). Amongst the standard varieties, cv. Tomba had the highest N uptake in the control, and many wild potato genotypes achieved shoot N uptakes as high as cv. Tomba. However, under low N supply, the wild potato genotypes generally lag behind cv. Tomba, except for genotype GLKS 30135_05.

In both treatments, root N uptake accounted for approximately one quarter of the shoot N uptake but showed a stronger variation between the genotypes. In the control, root N uptake was on average 4.59 mg vessel^{-1} and varied by ± 1.36 mg vessel^{-1} (CV = 29.6%). Under low N supply, root N uptake decreased on average by 59% to 1.88 mg vessel^{-1}. The reduction was significant for all the genotypes tested, except for GLKS 31600_10. Again, cv. Tomba was the best cultivar in terms of root N uptake in both treatments, but, in contrast to shoot N uptake, root N uptake of many wild potato genotypes even exceed that of cv. Tomba under control as well as under reduced N conditions (Table 4).

Total N uptake (shoot + root N uptake) generally ranged from 22.1 to 31.4 mg vessel^{-1} (except GLKS 31600_10 with only 12.8) in the control, and from 6.90 to 11.3 mg vessel^{-1} in the reduced N treatment. Relating the total N uptake of the plant to the total amount of N supplied revealed an average NupE of 102% in the control and 132% in the low N treatment. This indicates that the plants took up all the N provided via the nutrient solution and that an additional amount of N was introduced into the system via the ten shoot tips. The NutE, expressed as the amount of total biomass (shoot + root dry mass) produced per unit N taken up, increased, on average, from 25 units in the control to 57 units in the low N treatment. NutE differed moderately between the genotypes in the control treatment where it ranged from 16 to 35 units. In the reduced N treatment, NutE varied strongly from 28 to 78 units. Although all genotypes significantly increased their NutE under low N conditions, many wild potato genotypes exceeded the NutE of the standard varieties. For example, in the low N treatment 13 of the wild potato genotypes achieved a similar and eight genotypes a significantly better NutE than the best cultivar Tomba (Figure 1). The N partitioning, as the ratio of N taken up by the root and N taken up by the shoots (root-N:shoot-N ratio), gives insight into the distribution of N within the plant. In the control, most of the wild types showed values similar to that of the standard varieties, while nine wild types showed significantly higher values than cv. Tomba. Under reduced N conditions, the majority of the wild types increased the root-N:shoot-N ratio, while it remained constant for all the standard varieties. This indicates a stronger partitioning of N to the root for wild potato genotypes, in particular under N deficit.

To evaluate the stress performance of the different genotypes, two commonly used indices (stress susceptibility index SSI, stress tolerance index STI) were applied based on total plant dry mass (shoot DM + root DM). In Figure 2 the accessions and genotypes are ranked according to their SSI and their STI from the best to the weakest genotype. The SSI, as a measure of the yield stability under stress conditions, identified cv. Kiebitz, GLKS 31559_14 and GLKS 30181_18 as the three most stable genotypes across both environments, while the reaction to N stress was most pronounced for GLKS 30177_17, GLKS 30177_02 and GLKS 31559_11. On the other hand, the STI identified GLKS 30177_20, GLKS 30160_15 and GLKS 30160_13 as the most promising genotypes, because they produced high total yields under control as well as under stress conditions, whereas cv. Kiebitz and GLKS 31600_10 ranked least, due to relatively low yields in both treatments (Figure 2).

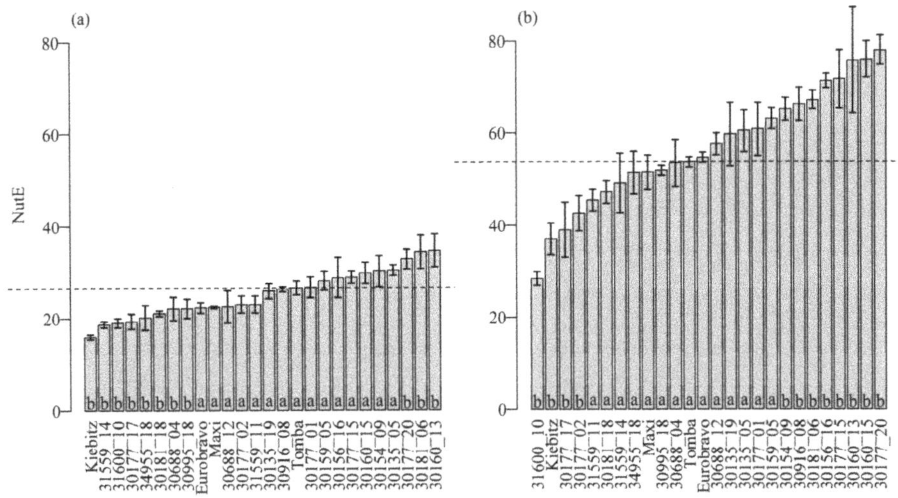

Figure 1. Mean nitrogen utilization efficiency (NutE) of 22 different wild potato genotypes and the four standard varieties under optimal (**a**) and reduced N conditions (**b**) after 21 days of cultivation in a climate chamber. Error bars indicate standard deviation, the dotted line indicates the mean NutE of the best cv. Tomba. The different letters at the bottom of the bars indicate whether there is a significant difference to the best cultivar Tomba ("b") or not ("a", Dunnett's test $p \leq 0.05$).

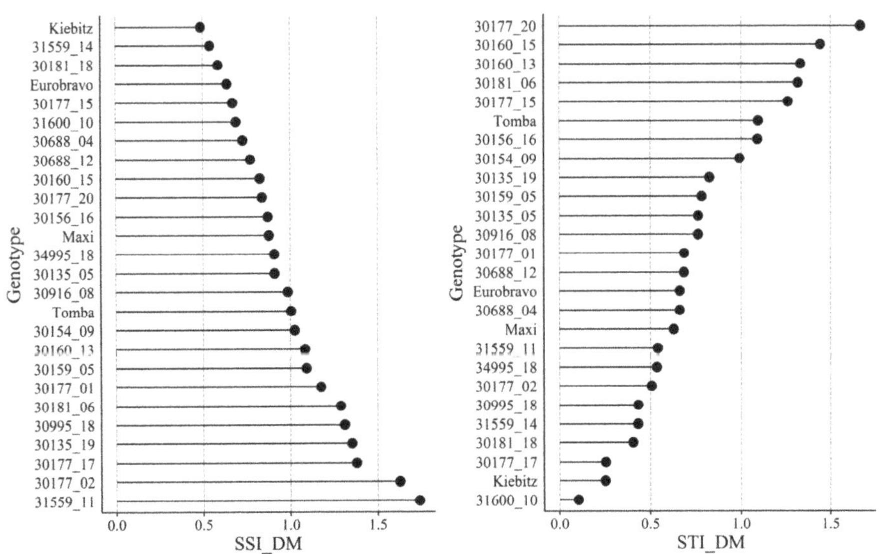

Figure 2. Stress susceptibility index (SSI) and stress tolerance index (STI) of different wild potato genotypes and four cultivars based on their total dry mass production (DM) under optimal and reduced N supply during 21 days of cultivation in a climate chamber. Genotypes are ranked according to their performance, with the best being on top of the graph.

3. Discussion

We assessed the tuber starch content in wild potato accessions and studied the reaction of selected wild types to N deficiency. After greenhouse cultivation, the average starch content of all tested wild potato genotypes amounted to 21.7% of FM, but ranged from minimal 7.1 to maximal 36% of FM. This indicates considerable variation for this trait in the wild relatives of cultivated potato. Furthermore, in 50 out of the 506 genotypes, tuber starch content was higher than 28%, which clearly exceeds the starch contents of modern cultivars. Based on our study, germplasm with high tuber starch content is predominately found in *S. chacoense*, *S. commersonii*, *S. jamesii* and *S. pinnatisectum*. Our results are in line with Jansen et al. [25], where tuber starch contents ranged between 3.8 and 39.6% of FM in wild species cultivated in the greenhouse. Here, genotypes of the species *S. chacoense* and *S. pinnatisectum* revealed the highest starch contents. In comparison, 14 modern starch potato varieties cultivated in pot experiments revealed starch contents between 13.9 and 21.9% of FM [20]. Under field conditions, tuber starch contents ranged from 7 to 20% of FM in a set of 300 potato cultivars, breeding clones, landraces and diploid clones [31]. This points out the potential of wild potato germplasm to increase tuber starch contents in cultivars. However, next to the tuber starch content, also the starch yield, as a result of starch content multiplied by tuber yield, plays an important role for industrial starch production. Whether the high starch contents in the wild species will be maintained when tuber size increases due to breeding, has still to be examined. Studies of Schönhals [31] showed that tuber yield is negatively correlated with tuber starch content, while no significant correlation between tuber yield and starch content was found by Bombik et al. [32].

An increase in tuber starch content and starch yield should be linked with a high resource use efficiency. This holds particularly true for the element N, because potato cultivation bears high risks of N leaching due to its high demand of readily available N in soil and the small root system of the crop [5,6,14]. Based on our results mentioned above, 22 wild potato genotypes with high tuber starch contents and four commercial cultivars were studied for their N use efficiency by cultivating them as in vitro plantlets in 500 mL vessels filled with a nutrient solution containing 30 or 7.5 mmol L^{-1} N for 21 days in a climate chamber. This system allowed us to screen a high number of plants under low space requirement and highly controlled conditions. Several other reports underline the potential of in vitro cultures for the evaluation of potato germplasm with respect to abiotic and biotic stress or rooting characteristics, because it provides conditions independent from weather conditions, pathogens, N leaching or immobilization events [16,33–35]. Our results revealed a high variation in shoot and root DM development between the wild potato genotypes. N deficiency significantly reduced shoot DM for all wild types and the cultivars (except for cv. Kiebitz and cv. Eurobravo), with the extent of shoot DM reduction being genotype dependent. Most interestingly, genotypes GLKS 30135_05, GLKS_30160_13 and GLKS 30177_20 of *S. chacoense* showed a moderate shoot DM reduction due to low N supply whilst producing a comparably high shoot biomass under high N conditions. The root DM varied considerably and the wild types GLKS 30160_13, GLKS 30160_15, GLKS 30177_15 and GLKS 30177_20 clearly exceeded the root growth of the best cultivar Tomba under high and/or low N supply, indicating that wild potato germplasm could considerably contribute to enhance root growth of *S. tuberosum* cultivars. Under N deficit, the genotypes followed different strategies in terms of root development. A significant reduction of root DM was observed for the majority of the genotypes, while seven genotypes maintained root biomass. This was also reflected in the root-DM:shoot-DM ratio which was either maintained (13 genotypes), reduced (11 genotypes) or increased (two genotypes). To sustain or even increase root biomass at the expense of the shoots is a well-known reaction of plants to nutrient deficiency and helps to maintain the nutrient uptake from soil or nutrient solution by exploring a larger (soil) volume [36,37]. In contrast, other wild types seem to preferably invest into shoot growth, probably in order to maintain photosynthetic activity. Different strategies to cope with low N as observed in our experiment are also known from cultivated potatoes. A higher root:total mass ratio under N deficiency was reported for the majority of 17 modern starch and table potato varieties during the course of 18 days of in vitro cultivation [16]. On the other hand, the authors also identified some genotypes which missed the ability to stimulate root growth at the expense of the shoots under N deficiency. This is in accordance with previous studies under climate chamber conditions where some cultivars reduced root FM with increasing N stress, while others showed an increased root development upon N reduction and even maintained root growth at very low N levels [17]. From their studies, Schum et al. [16,17] observed that genotypes with high biomass production and fast nitrogen uptake under high N supply did not enhance root growth under low N and clearly reduced biomass production. On the other hand, genotypes with comparatively slow growth under high N supply increased root mass under low N supply [16,17]. Besides, it should be considered that other well-known responses of plants to nutrient deficiency stress such as changes in the root architecture, increase of root length, root surface area, root volume or number of root hairs [33], is not necessarily reflected in changes of the total root biomass as measured in our experiment. In relation to the total N applied, the plants took up almost all the N available in both treatments (except for GLKS 31600_10). This explains the low genotypic variation for total N uptake (Table 3). In some cases, N uptake of the genotypes even exceeded the amount of N provided via the nutrient solution, probably due to the additional N introduced into the system via the transferred shoot tips. Therefore, it is difficult to evaluate the NupE of the different accessions. Nevertheless, on the basis of the uniform N uptakes of the genotypes, the results give insight into genotype-dependent N partitioning and provide a clear picture in terms of NutE.

Most of the N taken up by the different genotypes was partitioned to the shoots under high N supply in our study. Under low N supply, a clear shift towards the roots was observed for many genotypes (Table 5). This confirms previous results, where generally a higher percentage of N was translocated to the roots or tubers under nutrient deficiency [6,36]. The increase in root-N:shoot-N ratio became especially evident for many wild potato genotypes, while this was less pronounced for the standard varieties.

NutE was on average 25.3 units in the control and increased to 57.3 units under N deficiency and the genotypic variation in NutE was particularly high in the low N treatment (Figure 1). This is in line with several studies [16,17,20,24]. In our experiment, many wild potato genotypes exceeded the NutE of the standard varieties under low N supply. Outstanding genotypes were GLKS 30154_09, GLKS 30916_08, GLKS 30181_06, GLKS 30156_16, GLKS 30177_15, GLKS 30160_13, GLKS 30160_15 and GLKS 30177_20 of *S. chacoense*. This indicates that these genotypes need considerably less N to produce the same amount of biomass. A high NutE is often related to a good translocation of N from the root to the shoot and/or a reallocation from older leaves to the younger leaves in order to maintain the photosynthetic activity and eventually to the reproductive organs [8,36]. The superior performance of *S. chacoense* genotypes was also found in field studies where, amongst 39 wild potato accessions, genotypes of *S. microdontum* and *S. chacoense* revealed the highest total biomass (tubers + roots + shoots + fruits), a high NupE and N recovery from soil, even exceeding the performance of the control varieties cv. Russet Norkotah and cv. Red Norland [22]. The authors attributed the higher N recovery by the wild species to the deeper penetrating, denser, and more branched root system that is advantageous for nutrient uptake. However, it generally has to be considered that wild potato species form only small tubers, in some cases, produce stolons rather than tubers under long day conditions. Whether the high N recovery rate of the wild type will be maintained after crossing to cultivars needs detailed examination. Hybrids of *S. chacoense* and a haploid *S. tuberosum* line (USW551) were studied in the field with high and zero N supply by Errebhi et al. [23]. Here, hybrids showed highest N use efficiency and produced a total biomass (tubers + roots/stolons + shoots + fruits) higher or similar than that of commercial varieties, but tuber yield was low [23].

For a final assessment of the overall performance of the different genotypes under N deficiency, we studied the two stress indices SSI and STI based on the total DM. Zhao et al. [38] studied different indices to evaluate low N tolerance in maize, and advised to use several indices and not to rely on only one. The SSI for example, proposed by Fischer and Maurer [39] for evaluating the yield stability under stressed and non-stressed environments, does not consider the yield of a respective genotype in relation to the other genotypes tested under control conditions [38]. Cv. Kiebitz and GLKS 31559_14, for example, exhibited the lowest SSI and could thus be considered as the ones with the lowest N stress susceptibility. That is confirmed by no significant changes in shoot or root mass under low N as compared to high N. However, these genotypes produced a low total biomass during the three-week in vitro culture compared to the other genotypes tested even in the high N treatment. This might indicate that genotypes with a slow biomass development, and by this a probably rather low internal N demand, react less sensitive to a reduction in N supply than fast growing types with a strong biomass development. By calculating the STI, these genotype-specific growth rates were considered, and here genotypes with a high biomass development under both control and stress conditions rank best. Under this premise, cv. Kiebitz and GLKS 31559_14 were rather intolerant to N stress, while GLKS 30177_20, GLKS_30160_15, GLKS_30160_13 and GLKS 30181_06, GLKS_30177_15 and cv. Tomba were more tolerant. Finally, genotypes being among the best under both indices will be very interesting candidates for further research and pre-breeding. Here, we consider GLKS 30177_20, GLKS 30177_15 and GLKS 30160_15 of *S. chacoense* as the most relevant genotypes, because they combine high shoot and root biomasses in both treatments with a moderate reduction in shoot and root biomass under low N supply. Furthermore, the best performers revealed the highest share of root biomass in relation to total biomass, maintained root-DM:shoot-DM ratio under low N, but partitioned more N to the roots than other genotypes and revealed a high internal N utilization efficiency. *S. chacoense* is a well-known source of pest and disease resistance, resistance to cold-induced sweetening and abiotic stresses such as drought tolerance, but its tubers contain high levels of toxic steroidal glycoalkaloids [40,41]. As a diploid species (2n = 2x = 24, EBN 2), hybridization with tetraploid *S. tuberosum* (2n = 4x = 48, EBN 4) is possible after a ploidy reduction in the *S. tuberosum* parent to the diploid level, followed by backcrossing [42], but it is also an interesting future candidate for diploid breeding programs [43].

Apart from a sole comparison of different genotypes, our study highlights the variation of N use efficiency between genotypes within one population. The five genotypes of GLKS 30177 reacted differently to N stress. While GLKS 30177_15 and GLKS 30177_20 belong to the best performing genotypes, GLKS 30177_02 and GLKS 30177_17 exhibited a medium to low shoot DM, a strong reduction in root DM and root-DM:shoot-DM ratio under low N, low NutE and a low stability under stress (SSI). Instead, GLKS 30177_01 can be considered as an intermediate type. These results underline the high diversity of the different genotypes within a wild potato accession which is maintained as a population in gene banks. Furthermore, it highlights the importance to study, describe and maintain individual wild potato genotypes in order to promote the use of wild potato germplasm in breeding and research [44]. Therefore, the tested genotypes in this study are maintained clonally via in vitro propagation at the Gross Luesewitz Potato Collections.

Genotypes GLKS 34995_18, GLKS 31559_14 and GLKS 31559_11 of *S. tuberosum* subsp. *andigena* and *S. stenotomum*, belonging to the cultivated part of series Tuberosa and being most related to *S. tuberosum* according to the taxonomy of Hawkes [45], showed no outstanding performance in the respective experiment. Tuber starch contents as well as shoot and root biomass or N uptakes and efficiencies were on an intermediate to low level. Although revealing the highest tuber starch content, the genotype of *S. pinnatisectum* (GLKS 31600_10) showed a rather weak performance under the given experimental conditions. This was indicated by the lowest shoot and root biomass as well as the lowest N uptakes compared to the other wild potato genotypes. Since this was the sole genotype of this species, we can only speculate whether this is a generally low-yielding species or if the experimental conditions were unfavorable. However, its reactions to N deficit clearly distinguished it from the other genotypes. Although not statistically significant, absolute root mass increased by about 35% under N deficit and the proportion of root biomass in relation to total biomass increased most at the expense of the shoot mass. In relation to its exceptionally high tuber starch contents, it is worth studying the root parameters of this species in further experiments.

4. Materials and Methods

4.1. Plant Material

In total, a set of 28 accessions was selected from the Gross Luesewitz Potato Collections (GLKS, Gross Luesewitz, Germany) of the Leibniz Institute of Plant Genetics and Crop Plant Research (Table 6). The set comprised 15 accessions of the species *S. chacoense* Bitter, five accessions of *S. pinnatisectum* Dunal, and one each of *S. tuberosum* subsp. *andigena* Hawkes, *S. commersonii* Dunal, *S. hondelmannii* Hawkes and Hjerting, *S. jamesii* Torrey, *S. microdontum* Bitter, *S. sparsipilum* (Bitt.) Juz. and Bukasov, *S. stenotomum* Juz. and Bukasov, and *S. tarijense* Hawkes (Table 6). These species were selected because they are known for other interesting traits such as disease resistance and/or because they originate from regions with high temperature and low rainfall, and by this may additionally provide tolerance to heat and drought. The latter is especially expected from *S. chacoense*, which originates from the Chaco-Region, a hot and semi-arid region in southern America. Detailed passport data of the wild potato accessions maintained at the IPK Potato Collections in Gross Luesewitz are also available via the genebank information system (GBIS). For in vitro establishment, 50 seeds of the respective accession were pretreated in gibberellic acid (500 ppm) for 24 h at room temperature to improve germination. After that, seeds were treated with 5% NaClO solution to sterilize them and placed in a test tube (one seed per tube) containing about 6.0 mL of a solid culture medium under sterile conditions. The solid culture medium was composed as described by Murashige and Skoog [46]. The seeds were then placed in a climate chamber at 20 °C and 12 h of light (150–250 µmol m^{-2} s^{-1}). After approximately four weeks, 20 well developed genotypes per accession were chosen for further experiments and multiplied. For multiplication, the plantlet of a respective genotype was cut in up to four nodal sections which were then transferred to new tubes with solid culture medium and grown in a climate chamber as described above. Prior to their cultivation in the respective experiments, genotypes were tested for quarantine diseases like virus (X, Y, L, S, M, A), potato spindle tuber viroid (PSTVd), bacteria (*Clavibacter michiganensis* ssp. *sepedonicus*, *Ralstonia solanacearum*) and Andean viruses (Andean Potato Latent Virus (APLV-Col, APLV-Col 2, APLV-Hu), Andean Potato Mottle Virus (APMoV-B, APMoV-H), Potato Black Ringspot Virus (PBRSV), Aracacha Virus B, Oca strain (AVB-O), Potato Virus T (PVT), Potato Virus V (PVV), Potato Yellowing Virus (PYV)).

4.2. Evaluation of Tuber Starch Contents

In 2013, 20 genotypes of each accession and three plantlets per genotype were transferred into pots (16 × 16 cm, 16 cm deep) filled with a turf-based planting substrate (95% white turf, 5% sand, 1.5 kg NPK (14% N, 16% P_2O_5, 18% K_2O, micro nutrients), Einheitserde GmbH, Uetersen, Germany) and cultivated in the greenhouse. One accession was finally represented by 60 pots (20 genotypes and three plants per genotype). Plants were irrigated daily with rain- or tap water according to their needs. After natural senescence (three to four months after planting), irrigation was stopped, the aboveground plant biomass was removed, and the tubers were harvested separately for each genotype. Due to limited greenhouse capacities, starch content evaluations of accessions GLKS 30211, GLKS 30475, GLKS 31583, GLKS 31595 and GLKS 31559 were performed in 2014 in the same way as described above. Accordingly, accessions GLKS 31559, GLKS 32852 and GLKS 34995 were repeated in 2014 because too many genotypes were lost during greenhouse cultivation or too few tubers were produced. Furthermore, tubers of genotypes with high starch contents were re-cultivated in 2014 in order to validate the results.

Table 6. Overview of the accessions analyzed for starch content and nitrogen use efficiency as well as passport data of the accessions and taxonomic classification (according to Hawkes [45]).

GLKS-Accession	*Solanum*-Species, Ploidy, EBN ^	Series *	Origin °
30211	*S. commersonii* Dunal (2x, 1)	COM	URY
30475	*S. jamesii* Torrey (2x, 1)	PIN	USA
31595, 31600, 31602, 31605, 31610	*S. pinnatisectum* Dunal (2x, 1)	PIN	MEX
31559	*S. stenotomum* Juz. and Bukasov (2x, 2)	TUBc	BOL
34995	*S. tuberosum* subsp. *andigena* Hawkes (4x, 4)	TUBc	UNK
32852	*S. hondelmannii* Hawkes and Hjerting (2x, na)	TUBw	BOL
30688	*S. microdontum* Bitter (2x, 3x, 2)	TUBw	ARG
30944	*S. sparsipilum* (Bitt.) Juz. and Bukasov (2x, 2)	TUBw	BOL
30134, 30135, 30148, 30154, 30156, 30159, 30160, 30177, 30181, 30191, 30197, 30665, 30916, 30995, 31025	*S. chacoense* Bitter (2x, 2)	YNG	ARG
31583	*S. tarijense* Hawkes (2x, 2)	YNG	UNK

^ EBN = Endosperm Balance Number; * COM = Commersonia, PIN = Pinnatisecta, TUBc = Tuberosa cultivated, TUBw = Tuberosa wild, YNG = Yungasensa; ° ARG = Argentina, BOL = Bolivia, MEX = Mexico, URY = Uruguay, USA = United States of America, UNK = unknown.

4.3. Plant Material and Experimental Setup of N Experiments

For the N efficiency studies, genotypes were selected which had a tuber starch content higher than 28% of FM in 2013 and grew reliably in vitro and in the greenhouse. These comprised 16 genotypes from nine different accessions of *S. chacoense* and one genotype of *S. pinnatisectum*. Additionally, to cover a broader spectrum of *Solanum* species, two genotypes of *S. microdontum* and *S. stenotomum* as well as one genotype of *S. tuberosum* subsp. *andigena* were added (see Table 2). The selected genotypes were multiplied in vitro as described above. Finally, after having produced 40 plantlets per selected genotype, shoots tips of approximately 1.5 to 2.0 cm length were transferred to the testing system.

An in vitro method for early evaluation of nitrogen use efficiency traits as described in Schum et al. [17] was applied. In brief, 500 mL glass cultivation vessels were filled with 62 mL of a nutrient solution based on Murashige and Skoog [46] (Table 7). For the N experiments, two N levels were applied, containing 0.420 g L^{-1} N (control) and 0.105 g L^{-1} N (reduced N), respectively, being equivalent to 30 mmol L^{-1} and 7.5 mmol L^{-1} N. Ten shoot tips of one genotype were cultivated in one vessel for 21 days, and fixed via a perforated stainless steel plate. The transfer of the shoots to the experimental system was carried out under sterile conditions and the vessels were closed with a cellulose ring to enable gas exchange plus a glass lid to prevent contamination. They were placed in a climate chamber with 12 h of light and a constant temperature of 20 °C in a complete randomized design. All treatment × genotype combinations were repeated four times. Due to the high number of accessions and genotypes to be multiplied and tested, combined with the unequal growth rate of the different accessions, several consecutive experiments were conducted. For comparison, four modern varieties (cultivars, cv.) were used, kindly provided by the breeders; cv. Kiebitz (Norika, Germany), cv. Maxi (Bayerische Pflanzenzuchtgesellschaft, Germany), cv. Eurobravo and Tomba (Europlant Pflanzenzucht, Germany).

Table 7. Composition of the nutrient solution used in the control and reduced N treatment of the N efficiency experiments.

Chemical	Unit	Control	Reduced	Nutrients	Control	Reduced
NH_4NO_3	g/L	0.825	0.206	N	0.420	0.105
KNO_3	g/L	0.950	0.238	K	0.784	0.784
KCl	g/L	0.701	1.226	Cl	0.545	0.795
$CaCl_2 \times 6H_2O$	g/L	0.655		Ca	0.120	
$MgSO_4 \times 7H_2O$	g/L	0.370		Mg	0.036	
KH_2PO_4	g/L	0.170		P	0.039	
$FeSO_4 \times 7H_2O$	g/L	0.028		Fe	0.006	
Na × EDTA	g/L	0.037				
$MnSO_4 \times H_2O$	mg/L	17.10		Mn	5.558	
$ZnSO_4 \times 7\,H_2O$	mg/L	8.600		Zn	1.955	
H_3BO_3	mg/L	6.200		B	1.084	
$CuSO_4 \times 5\,H_2O$	mg/L	0.025		Cu	0.006	
$CoCl_2 \times 6H_2O$	mg/L	0.025		Co	0.012	
$Na_2MoO_4 \times 2H_2O$	mg/L	0.250		Mo	0.119	
Organic stock sol.	mL/L	1.0		S	0.056	
Sucrose	g/L	30				

4.4. *Laboratory Analyses and Calculations*

4.4.1. Tuber Starch Contents

Ten tubers of similar size (approx. 2 cm long) were selected per genotype, washed and analyzed immediately for their starch content via the underwater weight method using a balance (KERN PES 6200-2M, Kern & Sohn GmbH, Balingen, Germany) equipped with a cage to sink the tubers in water. The specific gravity (SG) was calculated based on the weight in air divided by difference of weight in air and weight under water. The starch content was calculated based on studies by Lunden [47] and as described in Meise et al. [20]:

$$\text{Starch (\% of FM)} = -211.89 + 209.06 * \text{SG}$$

4.4.2. Yield, Nitrogen Uptake and Stress Indices

Harvested shoots and roots formed per in vitro vessel were weighed to determine fresh mass (FM). Dry mass (DM) was determined after drying the shoot and root biomass in an oven at 60 °C and weighing. The dry plant material was then ground in a mixer mill (Retsch, Tissue Lyser, Quiagen GmbH, Duesseldorf, Germany) for one minute at a frequency of 30 s^{-1} using 3 mm steel beads. After that, the dried and ground plant material was analyzed for its N content using an elementar analyzer (Eurovector EA 3000 b, HEKAtech GmbH, Wegberg, Germany). Shoot and root N uptake were calculated by multiplying the measured N content with the shoot or root dry mass. Total N uptake was calculated by summing up shoot and root N uptake. The root-DM:shoot-DM ratio was calculated by dividing the root DM by the shoot DM. Similarly, the N uptake into the root was divided by the shoot N uptake to reflect the partitioning of N in the plant, and was denoted as root-N:shoot-N ratio. The N uptake efficiency (NupE, %) was calculated by dividing the total N uptake by the amount of N supplied in the respective treatment:

$$\text{NupE (\%)} = \frac{\text{total N uptake (mg per vessel)}}{\text{total N supplied (mg per vessel)}}$$

The relation of the total N taken up by the plant to the total dry mass produced is denoted as N utilization efficiency (NutE) and was calculated as follows:

$$\text{NutE (arb. u.)} = \frac{\text{total dry mass (mg per vessel)}}{\text{total N uptake (mg per vessel)}}$$

Furthermore, the stress susceptibility index (SSI) was calculated based on total dry mass (SSI_{DM}) as introduced by Fischer and Maurer [39]:

$$\text{SSI} = \frac{(1 - \text{Ps}/\text{Pc})}{(1 - \text{meanPs}/\text{meanPc})}$$

where Ps is the parameter (DM, N uptake) determined under stress conditions and Pc is the parameter determined under control conditions. This is related to the mean of all genotypes tested under stress conditions (mean Ps) divided by the mean of all genotypes under control conditions (mean Pc). Additionally, the stress tolerance index was calculated as proposed by Fernandez [48]:

$$\text{STI} = \frac{\text{Pc} * \text{Ps}}{(\text{meanPc})^2}$$

4.5. Statistical Analyses

All statistical analyses were performed using the software R (version 3.3.2, R Foundation for Statistical Computing, Vienna, Austria) [49]. For tuber starch content, one-way analysis of variance (ANOVA) was applied to test for within- and between-accession variations. Finally, mean, minimum and maximum values as well as coefficient of variation (CV) were calculated for each accession. Pearson correlation coefficient was calculated using the RcmdrMisc package [50].

For dry mass and N uptake traits, two-way ANOVA was used to test the effect of genotype and treatment as well as their interaction on the respective trait. Because the climate chamber allows the cultivation of the plantlets under highly controlled and standardized conditions, all consecutive experiments were analyzed together in one model. A linear model using the "lmer" procedure of the package "lme4" [51] was applied, with genotype, treatment and genotype × treatment as fixed effects. Assumptions such as normality of residuals and homogeneity of variances were tested prior to ANOVA using q-q-plots, the Shapiro–Wilk normality test and the Levene's test (package "cars" [52]). If assumptions were not met, data were log transformed. If significant factor effects were identified ($p \leq 0.05$), post hoc comparison of means was performed using the Tukey test of the "lsmeans" package [53] to compare all means. In the results, significant differences between the control and reduced N are indicated by asterisks. The Dunnett's test against the control of the package "multcomp" [54] was used to identify means differing significantly from the best cultivar. In the results, means significantly different (higher or lower) from cv. Tomba are indicated by the lowercase letter "b". Additionally, means significantly higher than cv. Tomba are underlined.

The original data of the experiments are provided at the e!DAL repository [55] under the DOI 10.5447/ipk/2020/19.

5. Conclusions

We assessed the tuber starch content of 506 wild potato genotypes under greenhouse conditions. Of them, 50 revealed tuber starch contents above 28% of FM, clearly exceeding the starch contents of commercial cultivars. Amongst the wild types with high starch content, three were superior in terms of N utilization efficiency (NutE) as indicated by the in vitro screening in a climate chamber under high and low N levels for 21 days. GLKS 30177_15, GLKS 30177_20 and GLKS 30160_15 of species *S. chacoense* produced the highest shoot and root biomass under N stress and showed only a moderate reduction of the total biomass under low N compared to the high N treatment. NutE of these genotypes was high and exceeded that of most other wild types and the standard varieties. Combining two common stress tolerance indices (SSI and STI) proved to be a helpful tool for the identification of genotypes with a high and stable biomass production under stress compared to non-stress conditions. Based on our study, the identified genotypes of *S. chacoense* are a promising source for further research projects aiming to improve starch contents and N use efficiency in cultivated potato. Most wild potato genotypes of this study are maintained in vitro and are available at the IPK Gross Luesewitz Potato Collections.

Author Contributions: Data analysis, manuscript conceptualization, writing and editing: S.B.-P.; project administration, methodology, review of the manuscript: K.J.D. All authors have read and agreed to the published version of the manuscript.

Funding: This project was funded by the FNR Agency for Renewable Resources (22023411) on behalf of the German Federal Ministry of Food and Agriculture.

Acknowledgments: We thank K. Löschner for excellent technical assistance during the greenhouse and climate chamber studies, and C. Brandt and U. Behrendt for fruitful discussions and proof reading of the manuscript. We also thank our partners from the Federal Research Centre for Cultivated Plants (JKI) for their advice and support in applying the in vitro testing system as well as the breeders for kindly providing modern cultivars for the studies. The publication of this article was funded by the Open Access Fund of the Leibniz Association.

Conflicts of Interest: The authors declare no conflict of interest. The funders had no role in the design of the study; in the collection, analyses, or interpretation of data; in the writing of the manuscript, or in the decision to publish the results.

References

1. Kaur, B.; Ariffin, F.; Bhat, R.; Karim, A.A. Progress in starch modification in the last decade. *Food Hydrocoll.* **2012**, *26*, 398–404. [CrossRef]
2. Tomasik, P. Specific physical and chemical properties of potato starch. Invited Review. *Food* **2009**, *3*, 45–56.
3. Robertson, T.M.; Alzaabi, A.Z.; Robertson, M.D.; Fielding, B.A. Starchy carbohydrates in a healthy diet: The role of the humble potato. Review. *Nutrients* **2018**, *10*, 1764. [CrossRef] [PubMed]
4. Bruns, H. Landessortenversuch Stärkekartoffeln (Variety Trials Starch Potatoes). Chamber of Agriculture Lower Saxony, Meppen. 2015. Available online: https://docplayer.org/48527547-Landessortenversuch-staerkekartoffeln-2015.html (accessed on 7 May 2020).
5. Koch, M.; Naumann, M.; Pawelzik, E.; Gransee, A.; Thiel, H. The importance of nutrient management for potato production. Part I: Plant nutrition and yield. *Potato Res.* **2019**, *63*, 97–119. [CrossRef]
6. Zebarth, B.J.; Rosen, C.J. Research perspective on nitrogen BMP development for potato. *Am. J. Potato Res.* **2007**, *84*, 3–18. [CrossRef]
7. Ebúrneo, J.M.; Garcia, E.L.; Santos, T.P.R.; de Souza, E.D.F.C.; Soratto, R.P.; Fernandes, A.M.; Leonel, M. Influence of nitrogen fertilization on the characteristics of potato starch. *Aust. J. Crop. Sci.* **2018**, *12*, 365–373. [CrossRef]
8. Maltas, A.; Dupuis, B.; Sinaj, S. Yield and quality response of two potato cultivars to nitrogen fertilization. *Potato. Res.* **2018**, *61*, 97–114. [CrossRef]
9. Muleta, H.D.; Aga, M.C. Role of nitrogen on potato production: A review. *J. Plant Sci.* **2019**, *7*, 36–42.
10. Schum, A.; Jansen, G. Physiological response to nitrogen deficiency stress of in vitro grown potato genotypes. *Acta Hortic.* **2012**, *961*, 465–469. [CrossRef]
11. Fixen, P.E.; Bruulsema, T.W. Potato management challenges created by phosphorus chemistry and plant roots. *Am. J. Potato Res.* **2014**, *91*, 121–131. [CrossRef]
12. Kelling, K.A.; Rosen, C.J.; Stark, J.C.; Essah, S.Y.C. Potato phosphorus management and utilization for today and tomorrow. *Am. J. Potato Res.* **2014**, *91*, 199–210. [CrossRef]
13. Iwama, K. Physiology of the potato: New insights into root system and repercussion for crop management. *Potato Res.* **2008**, *51*, 333–353. [CrossRef]
14. Davenport, J.R.; Milburn, P.H.; Rosen, C.J.; Thornton, R.E. Environmental impacts of potato nutrient management. *Am. J. Potato Res.* **2005**, *82*, 321–328. [CrossRef]

15. Xie, X.; Li, X.-Q.; Zebarth, B.J.; Niu, S.; Tang, R.; Tai, H.H.; Bizimungu, B.; Wu, W.; Haroon, M. Rapid screening of potato cultivars tolerant to nitrogen deficiency using a hydroponic system. *Am. J. Potato Res.* **2018**, *95*, 157–163. [CrossRef]
16. Schum, A.; Meise, P.; Jansen, G.; Seddig, S.; Ordon, F. Evaluation of nitrogen efficiency associated traits of starch potato cultivars under in vitro conditions. *Plant Cell Tiss. Organ. Cult.* **2017**, *130*, 651–665. [CrossRef]
17. Schum, A.; Jansen, G. In vitro method for early evaluation of nitrogen use efficiency associated traits in potato. *J. Appl. Bot. Food Qual.* **2014**, *87*, 256–264.
18. Zebarth, B.J.; Tai, G.; Tarn, R.; de Jong, H.; Milburn, P.H. Nitrogen use efficiency characteristics of commercial potato cultivars. *Can. J. Plant Sci.* **2003**, *84*, 589–598. [CrossRef]
19. Meise, P.; Seddig, S.; Uptmoor, R.; Ordon, F.; Schum, A. Impact of nitrogen supply on leaf water relations and physiological traits in a set of potato (*Solanum tuberosum* L.) cultivars under drought stress. *J. Agro. Crop Sci.* **2018**, *204*, 359–374. [CrossRef]
20. Meise, P.; Seddig, S.; Uptmoor, R.; Ordon, F.; Schum, A. Assessment of yield and yield components of starch potato cultivars (*Solanum tuberosum* L.) under nitrogen deficiency and drought stress conditions. *Potato Res* **2019**, *62*, 193–220. [CrossRef]
21. Tiemens-Hulscher, M.; Lammerts van Bueren, E.T.; Struik, P.C. Identification of genotypic variation for nitrogen responses in potato (*Solanum tuberosum*) under low nitrogen input circumstances. In Proceedings of the 1st International IFOAM-conference on Organic Animal and Plant Breeding, Santa Fe, NM, USA, 25–28 August 2009; International Federation of Organic Agriculture Movement: Bonn, Germany, 2009; pp. 354–361.
22. Errebhi, M.; Rosen, C.J.; Lauer, F.I.; Martin, M.W.; Bamberg, J.B.; Birong, D.E. Screening of exotic potato germplasm for nitrogen uptake and biomass production. *Am. J. Potato Res.* **1998**, *75*, 93–100.
23. Errebhi, M.; Rosen, C.J.; Lauer, F.I.; Martin, M.W.; Bamberg, J.B. Evaluation of tuber-bearing *Solanum* species for nitrogen use efficiency and biomass partitioning. *Am. J. Potato Res.* **1999**, *76*, 143–151. [CrossRef]
24. Zebarth, B.J.; Tarn, T.R.; de Jong, H.; Murphy, A. Nitrogen use efficiency characteristics of Andigena and Diploid Potato Selections. *Am. J. Potato Res.* **2008**, *85*, 210–218. [CrossRef]
25. Jansen, G.; Flamme, W.; Schüler, K.; Vandrey, M. Tuber and starch quality of wild and cultivated potato species and cultivars. *Potato Res.* **2001**, *44*, 137–146. [CrossRef]
26. Khiutti, A.; Spooner, D.M.; Jansky, S.H.; Halterman, D.A. Testing taxonomic predictivity of foliar and tuber resistance to *Phytophthora infestans* in wild relatives of potato. *Phytopathology* **2015**, *105*, 1198–1205. [CrossRef] [PubMed]
27. López, M.; Riegel, R.; Lizana, C.; Behn, A. Identification of virus and nematode resistance genes in the Chilota Potato Genebank of the Universidad Austral de Chile. *Chilean J. Agric. Res.* **2015**, *75*, 320–327. [CrossRef]
28. Bachmann-Pfabe, S.; Hammann, T.; Kruse, J.; Dehmer, K.J. Screening of wild potato genetic resources for combined resistance to late blight on tubers and pale potato cyst nematodes. *Euphytica* **2019**, *215*, 48. [CrossRef]
29. Rodewald, J.; Trognitz, B. *Solanum* resistance genes against *Phytophthora infestans* and their corresponding avirulence genes. *Mol. Plant Pathol.* **2013**, *14*, 740–757. [CrossRef]
30. Dalamu, B.V.; Umamaheshwari, R.; Sharma, R.; Kaushik, S.K.; Joseph, T.A.; Singh, B.P.; Gebhardt, C. Potato cyst nematode (PCN) resistance: Genes, genotypes and markers—An update. *SABRAO J. Breed. Genet.* **2012**, *44*, 202–228.
31. Schönhals, E.M. Identifying novel diagnostic SNP markers for potato (*Solanum tuberosum* L.) tuber starch and yield by association mapping. Ph.D. Thesis, University of Cologne, Cologne, Germany, 25 May 2014.
32. Bombik, A.; Rymuza, K.; Olszewski, T. Variation and correlation of starch potato utility features and tuber quality traits. *Acta Agrophys.* **2019**, *26*, 29–42. [CrossRef]
33. Christensen, C.T.; Zotarelli, L.; Haynes, K.G.; Colee, J. Rooting characteristics of *Solanum chacoense* and *Solanum tuberosum* in vitro. *Am. J. Potato Res.* **2017**, *94*, 588–598. [CrossRef]
34. Khrais, T.; Leclerc, Y.; Donnelly, D.J. Relative salinity tolerance of potato cultivars assessed by in vitro screening. *Am. J. Potato Res.* **1998**, *75*, 207–2010. [CrossRef]
35. Khan, M.A.; Munive, S.; Bonbierbale, M. Early generation in vitro assay to identify potato populations and clones tolerant to heat. *Plant Cell Tiss. Organ. Cult.* **2016**, *121*, 45–52. [CrossRef]
36. Lammerts van Bueren, E.T.; Struik, P.C. Diverse concepts of breeding for nitrogen use efficiency. A review. *Agron. Sustain. Dev.* **2017**, *37*, 50. [CrossRef]

37. Cambui, C.A.; Svennerstam, H.; Gruffman, L.; Nordin, A.; Ganeteg, U.; Näsholm, T. Patterns of plant biomass partitioning depend on nitrogen source. *PLoS ONE* **2011**, *6*, e19211. [CrossRef] [PubMed]
38. Zhao, Z.; He, K.; Feng, Z.; Li, Y.; Chang, L.; Zhang, X.; Xu, S.; Liu, J.; Xue, J. Evaluation of yield-based low nitrogen tolerance indices for screening maize (*Zea mays* L.) inbred lines. *Agronomy* **2019**, *9*, 240. [CrossRef]
39. Fischer, R.A.; Maurer, R. Drought resistance in spring what cultivars. 1. Grain yield responses. *Aust. J. Agric. Res.* **1978**, *29*, 897–912. [CrossRef]
40. Jansky, S. Breeding for disease resistance in potato. In *Plant Breeding Reviews*, 1st ed.; Janick, J., Ed.; Wiley Online Library: Hoboken, NJ, USA, 2010; Volume 33, pp. 69–155.
41. Zhao, Q.; Zhao, B.; Zhang, Q.; Yu, B.; Cheng, L.; Jin, R.; Wang, Y.; Zhang, J.; Wang, D.; Zhang, F. Screening for chip-processing potato line from introgression of wild species germplasm with post-harvest storage and chip qualities. *Am. J. Potato Res.* **2013**, *90*, 425–439. [CrossRef]
42. Lambert, E.S.; Pinto, C.A.B.P. Agronomic performance of potato interspecific hybrids. *Crop Breed. Appl. Biot.* **2002**, *2*, 179–188. [CrossRef]
43. Leisner, C.P.; Hamilton, J.P.; Crisovan, E.; Manrique-Carpintero, N.C.; Marand, A.P.; Newton, L.; Pham, G.M.; Jiang, J.; Douches, D.S.; Jansky, S.; et al. Genome sequence of M6, a diploid inbred clone of the high-glycoalkaloid-producing tuber-bearing potato species *Solanum chacoense*, reveals residual heterozygosity. *Plant J.* **2018**, *94*, 562–570. [CrossRef]
44. Bethke, P.C.; Haltermann, D.A.; Jansky, S. Are we getting better at using wild potato species in light of new tools? *Crop Sci.* **2017**, *57*, 1241–1258. [CrossRef]
45. Hawkes, J.G. *The potato. Evolution, Biodiversity and Genetic Resources*, 1st ed.; Belhaven Press: London, UK, 1990.
46. Murashige, T.; Skoog, F. A revised medium for rapid growth and bio essays with tobacco tissue cultures. *Physiol. Plant.* **1962**, *15*, 473–497. [CrossRef]
47. Lunden, A.P. Undersokelser overforholdet mellom poteternes specifikke vekt og deres torrstoff—og stivelsesindhold [Studies on the relation between the specific gravity of potatoes and their content of dry matter and starch]. *Forsk Forsok i Landbr. Res. Nor. Agric.* **1956**, *7*, 81–107.
48. Fernandez, G.C.J. Effective selection criteria for assessing stress tolerance. In Proceedings of the International Symposium on Adaptation of Vegetables and other Food Crops in Temperature and Water Stress, Tainan, Taiwan, 13–18 August 1992; Kuo, C.G., Ed.; Asian Vegetable Research and Development Center: Tainan, Taiwan, 1992; pp. 257–270.
49. R Development Core Team. A language and environment for statistical computing. R Foundation for Statistical Computing, Vienna, Austria. Available online: http://www.R-project.org/ (accessed on 15 June 2016).
50. Fox, J. RmcdrMisc: R commander miscellaneous functions. R package version 1.0-5, 2016. Available online: https://rdrr.io/cran/RcmdrMisc/ (accessed on 11 September 2019).
51. Bates, D.; Maechler, M.; Bolker, B.; Walker, S. Fitting linear mixed-effects models using lme4. *J. Stat. Softw.* **2015**, *67*, 1–48. [CrossRef]
52. Fox, J.; Weisberg, S. *An {R} Companion to Applied Regression*, 3rd ed.; Sage: Thousand Oaks, LA, USA, 2019; Available online: https://r-forge.r-project.org/projects/car/ (accessed on 13 June 2019).
53. Lenth, R.V. The R package lsmeans. *J. Stat. Softw.* **2016**, *69*, 1–33. [CrossRef]
54. Hothorn, T.; Bretz, F.; Westfall, P. Simultaneous inference in general parametric models. *Biometrical J.* **2008**, *50*, 346–363. [CrossRef] [PubMed]
55. Arend, D.; Junker, A.; Scholz, U.; Schüler, D.; Wylie, J.; Lange, M. PGP repository: A plant phenomics and genomics data publication infrastructure. *Database* **2016**, *2016*, 1–10. [CrossRef] [PubMed]

plants

Article

Genetic Diversity, Population Structure and Marker-Trait Association for 100-Seed Weight in International Safflower Panel Using SilicoDArT Marker Information

Fawad Ali [1,2], Muhammad Azhar Nadeem [3], Muzaffer Barut [2,4], Ephrem Habyarimana [5], Hassan Javed Chaudhary [1], Iftikhar Hussain Khalil [6], Ahmad Alsaleh [7], Rüştü Hatipoğlu [4], Tolga Karaköy [3], Cemal Kurt [4], Muhammad Aasim [3], Muhammad Sameeullah [2], Ndiko Ludidi [8], Seung Hwan Yang [9], Gyuhwa Chung [9,*] and Faheem Shehzad Baloch [2,*]

1 Department of Plant Sciences, Quaid-I-Azam University, Islamabad 45320, Pakistan; fawadali365@gmail.com (F.A.); hassaan@qau.edu.pk (H.J.C.)
2 Department of Field Crops, Faculty of Agriculture and Natural Science, Bolu Abant Izzet Baysal University, 14030 Bolu, Turkey; mbarut@cu.edu.tr (M.B.); sameepbg@gmail.com (M.S.)
3 Faculty of Agricultural Sciences and Technologies, Sivas University of Science and Technology, 58140 Sivas, Turkey; azharjoiya22@gmail.com (M.A.N.); tkarakoy@sivas.edu.tr (T.K.); mshazim@gmail.com (M.A.)
4 Department of Field Crops, Faculty of Agriculture, Çukurova University, 1000 Adana, Turkey; rhatip@mail.cu.edu.tr (R.H.); ckurt@cu.edu.tr (C.K.)
5 CREA Research Center for Cereal and Industrial Crops, 40128 Bologna, Italy; ephrem.habyarimana@crea.gov.it
6 Department of Plant Breeding and Genetics, The University of Agriculture, Peshawar 25130, Pakistan; drihkhalil@gmail.com
7 Science and Technology Application and Research Center (BİLTEM), Yozgat Bozok University, 66900 Yozgat, Turkey; ahmad.alsaleh@bozok.edu.tr
8 Department of Biotechnology and Center of Excellence in Food Security, University of the Western Cape, Robert Sobukwe Road, Bellville 7530, South Africa; nludidi@uwc.ac.za
9 Department of Biotechnology, Chonnam National University, Chonnam 59626, Korea; ymichigan@chonnam.ac.kr
* Correspondence: chung@chonnam.ac.kr (G.C.); balochfaheem13@gmail.com (F.S.B.)

Received: 8 April 2020; Accepted: 19 May 2020; Published: 21 May 2020

Abstract: Safflower is an important oilseed crop mainly grown in the arid and semi-arid regions of the world. The aim of this study was to explore phenotypic and genetic diversity, population structure, and marker-trait association for 100-seed weight in 94 safflower accessions originating from 26 countries using silicoDArT markers. Analysis of variance revealed statistically significant genotypic effects ($p < 0.01$), while Turkey samples resulted in higher 100-seed weight compared to Pakistan samples. A Constellation plot divided the studied germplasm into two populations on the basis of their 100-seed weight. Various mean genetic diversity parameters including observed number of alleles (1.99), effective number of alleles (1.54), Shannon's information index (0.48), expected heterozygosity (0.32), and unbiased expected heterozygosity (0.32) for the entire population exhibited sufficient genetic diversity using 12232 silicoDArT markers. Analysis of molecular variance (AMOVA) revealed that most of the variations (91%) in world safflower panel are due to differences within country groups. A model-based structure grouped the 94 safflower accessions into populations A, B, C and an admixture population upon membership coefficient. Neighbor joining analysis grouped the safflower accessions into two populations (A and B). Principal coordinate analysis (PCoA) also clustered the safflower accessions on the basis of geographical origin. Three accessions; Egypt-5, Egypt-2, and India-2 revealed the highest genetic distance and hence might be recommended as candidate parental lines for safflower breeding programs. The mixed linear model i.e., the Q + K model, demonstrated that two DArTseq markers (DArT-45483051 and DArT-15672391) had significant association ($p < 0.01$) for 100-seed weight. We envisage that identified DArTseq markers associated with 100-seed weight will be helpful to develop high-yielding cultivars of safflower through marker-assisted breeding in the near future.

Keywords: *Carthamus tinctorius*; genotyping by sequencing; germplasm characterization; GWAS; oilseed crop

1. Introduction

Safflower (*Carthamus tinctorius* L.) belongs to the family *Compositae* and it is a self-pollinated crop with a genome size of about 1.4 GB [1]. Safflower is cultivated in 20 different countries of the world on a total cultivated area of 1,140,002 hectares that produces approximately 948,516 tons [2]. It is grown as an important industrial crop for different purposes, which include extraction of edible oil, production of dyes, and several uses in the pharmaceutical industry [3–5]. Safflower has better adaptation to stress conditions such as salinity and drought, although it produces oil in lower quantity than other oilseed crops [6]. Safflower also gained importance because it has the capability of biofuel production [7]. *Carthamus* species have been utilized since the pre-historic period as its archeological remains were found at sites in Syria since 7500 BC. Safflower was distributed from its center of domestication (i.e., Syria) to linked regions comprising Egypt, the Aegean region and southern Europe [8].

Safflower is considered an underutilized crop in comparison to other oilseed crops such as soybean, rapeseed and sunflower [3]. Key factors contributing to its underutilized status include lower oil content and seed yield, insect pest susceptibility, and lower resistance to diseases, which decrease safflower productivity and quality [9]. Narrow genetic diversity of local and traditional varieties necessitate exploration of the genetic diversity of the available germplasm by collecting accessions from different geographies worldwide. Such collections will provide information that will aid safflower conservation and utilization in the future [10]. Domestication of safflower has bottlenecked its genetic diversity, which greatly reduced its adaptive potential against different environmental stresses [10,11]. Exploration of genetic diversity is regarded as an important tool yielding a good source of trait variations [12,13]. Genetic diversity present in the germplasm contributes to execution of successful breeding programs [14]. Germplasm possessing diverse traits may be helpful in breeding programs for the development of elite cultivars [15,16]. Characterization of crop germplasm also helps in food security through the identification of novel genetic variations [11–14]. The Food and Agriculture Organization described the decrease of crop genetic diversity as one of the important factors that negatively impact food security and the environment [17]. The availability of limited information regarding genetic diversity and lack of the efficient genomic tools are considered as hampering factors to the improvement of safflower agronomic traits in breeding programs. Great emphasis on the molecular characterization and development of efficient molecular markers in safflower is required to accelerate safflower breeding activities [18–20]. Safflower genetic diversity using different molecular markers has been estimated [2,10,18,21–27].

Next generation sequencing technologies, such as genotyping by sequencing (GBS) and multiplex sequencing, aid in the generation of massive genetic data for various applications [28]. Application of the current polymerase chain reaction (PCR)-based marker technologies—aiming at whole genome analysis for association studies, construction of genetic maps, assessment of the collected germplasm for large scale molecular evaluation, and genome wide selection of the desirable alleles—are not attainable due to consumable and labor costs [28]. The application of DNA hybridization-based technologies like some SNP technologies and Diversity Arrays Technology are more suitable for such purposes. Hassani et al. [29] implemented DArTseq technology to assess genetic diversity in 89 safflower accessions originating from different countries of the world. They applied 1136 silicoDArT markers along with 2295 SNPs in their investigation.

Linkage analysis, also known as QTL mapping, helps in the identification of genomic regions controlling complex plant traits. QTL mapping is a time-consuming technique that needs mapping populations to be developed from bi-parents. QTL mapping captures less allelic variation utilizing bi-parental populations due to the very low rate of occurrence of recombination events and low mapping resolution [30]. Association mapping is a more efficient and faster technique, which provides higher resolution of complex plant traits in comparison to QTL mapping. Association mapping emerged as a promising technique to avoid limitations present in QTL mapping [31,32]. Relationships between plant traits and genetic polymorphisms observed in a heterogeneous assembly of distinct individuals, utilizing naturally occurring recombination events, aid in fine scale mapping of traits. Ambreen et al. [21] and Ebrahimi et al. [33] identified marker-trait associations in safflower, utilizing SSR and AFLP marker systems, respectively. Yield in any crop is the most important trait that is polygenic and difficult to measure and improve as it is highly influenced by other contributing traits. Thus, indirect selection for yield is highly preferred in comparison to direct selection focusing on the yield-contributing traits [34]. This study aimed at the establishment of the usefulness of silicoDArT markers to investigate phenotypic and genetic diversity, population structure and marker-trait associations for 100-seed weight. To attain these aims, we implemented a total of 12232 silicoDArT markers detected by a DArTseq approach of genotyping by sequencing in a safflower panel collected from 26 countries.

2. Results

2.1. Phenotypic Data Evaluation

During this study, 100-seed weight at both locations (Pakistan and Turkey) was recorded with the help of an electronic seed counter and weighing balance by taking undamaged and fully matured seeds. Analysis of variance (ANOVA) for both locations revealed highly significant differences among the studied safflower accessions for 100-seed weight (Table 1). Minimum, maximum, and mean values for 100-seed weight reflected sufficient phenotypic variation in the studied safflower panel (Table 2). Overall 100-seed weight ranged from 2.17 to 5.32 g with an average of 3.33 g. This reflects the presence of genetic variability and suggests that the safflower accessions studied here are suitable for association analysis. Mean 100-seed weight in Pakistan was comparatively lower than 100-seed weight obtained from Turkey (Table 2). Safflower accessions Egypt-5, China-3 and China-5 recorded superior 100-seed weight at both locations (Supplementary Table S1). Frequency distribution revealed a normal distribution of 100-seed weight at both locations (Pakistan and Turkey) and better mean performance of the safflower accessions for 100-seed weight in Turkey compared to Pakistan (Figure 1).

Table 1. Analysis of variance for 100-seed weight of safflower germplasm.

			Pakistan		
Source	**DF**	**SS**	**Mean Square**	**F Value**	**Pr > F**
Block	5	0.66	0.13	3.25	0.1108
Check	1	4.58	4.58	112.09	0.0001 **
Accessions	86	42.38	0.49	12.04	0.0053 **
Error	5	0.20	0.04		
			Turkey		
Source	**DF**	**SS**	**Mean Square**	**F Value**	**Pr > F**
Block	5	0.82	0.16	6.99	0.0261
Check	1	2.13	2.13	90.86	0.0002 **
Accessions	86	30.14	0.35	14.93	0.0032 **
Error	5	0.11	0.02		

** Significant at $p < 0.01$.

Table 2. Minimum, maximum, mean, and standard deviation (StD) of the 100-seed weight in international safflower panel.

100-Seed Weight (g)	Minimum	Maximum	Mean	Std. Deviation
Pakistan	1.88	5.29	3.26	0.74
Turkey	2.16	5.32	3.33	0.59
Overall	2.17	5.31	3.29	0.59

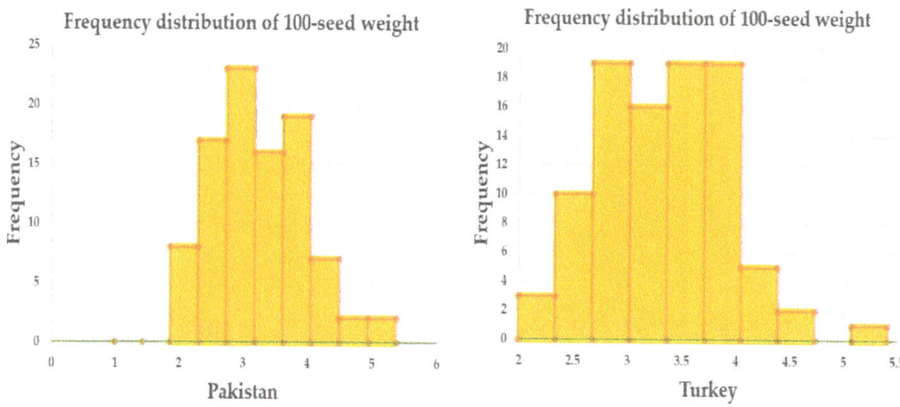

Figure 1. Frequency distribution chart for 100-seed weight for both studied locations (Pakistan and Turkey).

The implemented constellation plot clearly divided international safflower panel into two populations on the basis of their 100-seed weight. A total of 46 and 48 accessions clustered in population A and population B, respectively (Figure 2).

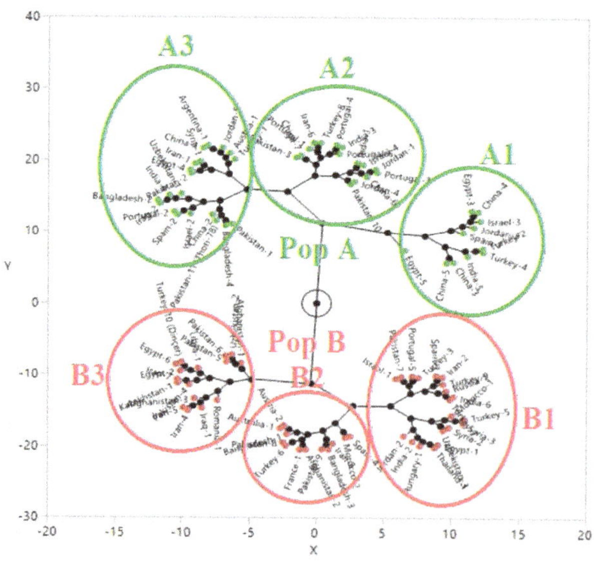

Figure 2. Constellation plot for 100-seed weight in international safflower panel.

2.2. DArTseq Profiling by GBS

DArTseq profiling of 94 safflower accessions resulted in a total of 29,048 silicoDArT markers. This dataset was filtered by accounting markers having less than 5% missing data, polymorphism information content (PIC) value of 0.10 to 0.50, call rate greater than 0.81, and 100% reproducibility, to retain 12,232 high quality markers for further analysis. Figure 3 shows the distribution of PIC values of the filtered silicoDArT marker dataset. The whole safflower panel revealed maximum and minimum PIC values of 0.50 and 0.10 respectively, with an average of 0.31. Highest and lowest call rate values of 1.00% and 0.81%, with an average of 0.93%, were obtained through the whole safflower panel.

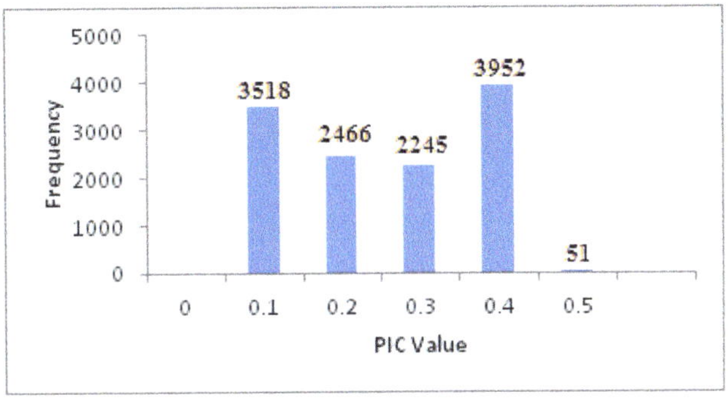

Figure 3. Frequency distribution of PIC values of 12232 silicoDArT markers.

2.3. Genetic Diversity and Population Structure Analysis in Safflower Panel

Various diversity parameters such as observed number of alleles (1.99), effective number of alleles (1.54), Shannon's information index (0.48), expected heterozygosity (0.32), and unbiased expected heterozygosity (0.32) reflected a good level of genetic variation in the studied germplasm (Table 3, Supplementary Table S2). Maximum genetic distance (0.76) was found between Egypt-2 and India-2 accessions, while mean genetic distance for the entire safflower population was 0.50 (Supplementary Table S3). Diversity indices were investigated on country basis, and Pakistan and Turkey revealed the existence of maximum percentage of polymorphic loci and high diversity parameters from the rest of countries (Table 3).

Table 3. Diversity indices calculated to investigate genetic diversity for whole safflower panel and accessions grouped according to country of origin panel with silicoDArT markers.

Population/Country	Polymorphic Loci (%)	Na	Ne	I	He	uHe	Mean GD	GD Range
Overall population	-	1.99	1.54	0.48	0.32	0.32	0.50	0.14–0.76
Afghanistan	74.97	1.53	1.45	0.41	0.28	0.34	0.46	-
Austria	49.96	1.25	1.35	0.30	0.21	0.28	0.48	-
Bangladesh	87.37	1.74	1.57	0.48	0.33	0.37	0.44	-
China	98.44	1.98	1.66	0.56	0.38	0.41	0.46	-
Egypt	96.73	1.94	1.63	0.54	0.36	0.40	0.41	-
India	96.73	1.95	1.65	0.55	0.37	0.41	0.48	-
Iran	98.44	1.96	1.65	0.55	0.37	0.40	0.45	-
Iraq	49.90	1.24	1.35	0.30	0.21	0.28	0.42	-
Israel	87.37	1.73	1.57	0.48	0.33	0.38	0.44	-
Jordan	93.53	1.90	1.63	0.53	0.36	0.40	0.27	-
Morocco	49.90	1.23	1.34	0.30	0.20	0.26	0.42	-
Pakistan	99.81	1.98	1.69	0.58	0.39	0.42	0.44	-
Portugal	96.73	1.93	1.63	0.51	0.36	0.40	0.42	-
Spain	87.37	1.81	1.59	0.49	0.34	0.39	0.38	-
Syria	74.97	1.61	1.50	0.42	0.28	0.34	0.38	-
Turkey	99.82	1.99	1.68	0.58	0.39	0.42	0.53	-
Uzbekistan	74.97	1.62	1.52	0.43	0.29	0.35	0.48	-

Na: Observed number of alleles, Ne: Number of effective alleles, I: Shannon's information index, He: Expected heterozygosity, uHe: Unbiased expected heterozygosity, GD: Jaccard Genetic distance.

Pairwise kinship coefficients ranged from −1.45 to 1.24 for the entire safflower panel. A total of 51.17% kinship values ranged from −0.40 to 0.4. 99% of the kinship coefficient values, which ranged from 0.60 to 1.00; however, 0.21% of the kinship coefficients ranged from 1.10 to 1.30, respectively (Figure 4). Analysis of molecular variance (AMOVA) revealed the division of the total variation into two stratum; i.e., among countries (9%) and within country group (91%) (Table 4). The ΔK peak at K = 3 in the structure analysis revealed that the genetic structure of the 94 safflower accessions is divided into three groups (Figure 5).

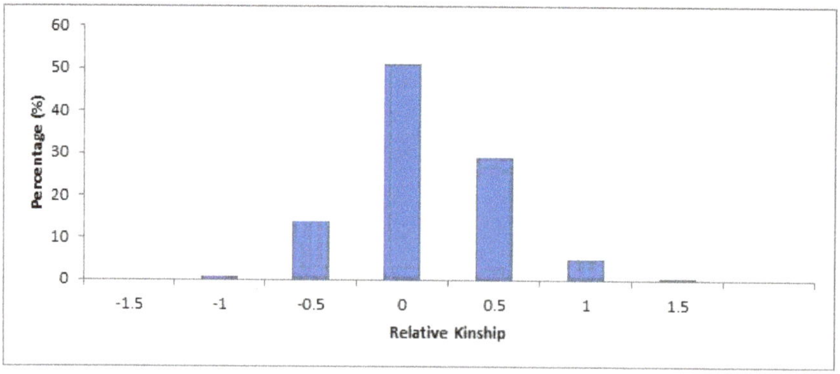

Figure 4. The proportion of pairwise kinship coefficients in international safflower panel.

Table 4. Analysis of molecular variance among countries and within country groups of safflower germplasm.

Source of Variation	Df	SS	MS	Est. Var.	% Variations
Among Countries	25	56719.33	3336.43	225.00	9
Within Country	68	165646.96	2179.56	2179.56	91
Total	93	222366.29	-	2404.57	100

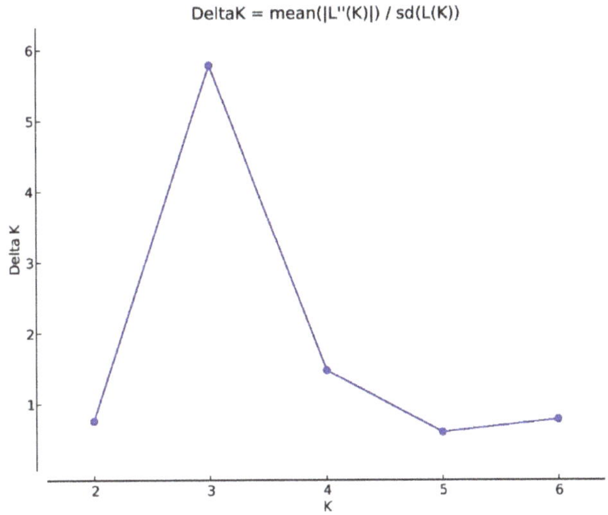

Figure 5. Delta K for the entire safflower population indicating the presence of three subpopulations at K = 3.

The Bayesian clustering model grouped the international safflower panel into three main populations implemented in STRUCTURE software on the basis of membership coefficient: 17 accessions in population A, 6 accessions in population B, and 21 accessions in population C. The remaining 50 accessions were clustered as admixture population (Figure 6). Clustering of the safflower accessions within the same population revealed membership coefficients of either 80% or greater than 80%. The Neighbor Joining analysis divided the 94 safflower accessions into two populations (A and B), each containing 47 accessions (Figure 7). PCoA was also performed and results showed that collection countries played some role in the clustering and an admixture of accessions was found as well, which had a similar structure clustering (Figure 8).

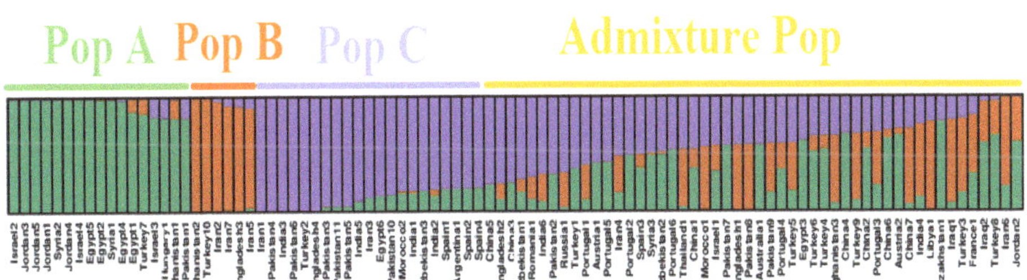

Figure 6. Structure-based clustering of the 94 safflower accessions using silicoDArT molecular markers.

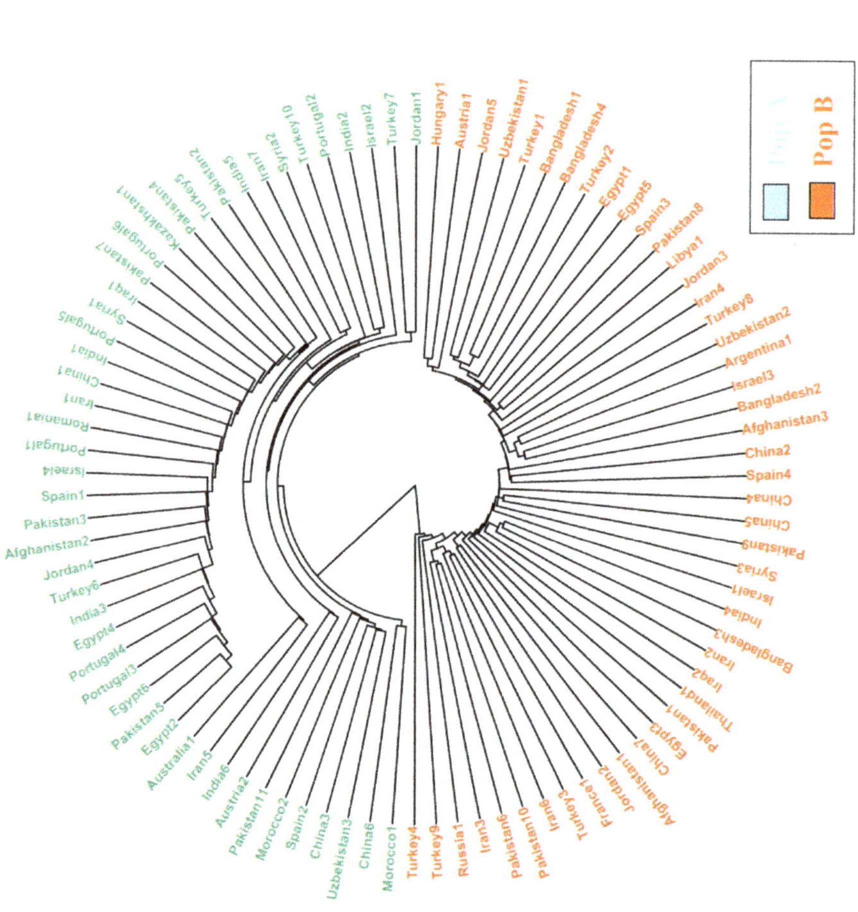

Figure 7. Neighbor joining-based clustering of the 94 safflower accessions using silicoDArT molecular markers.

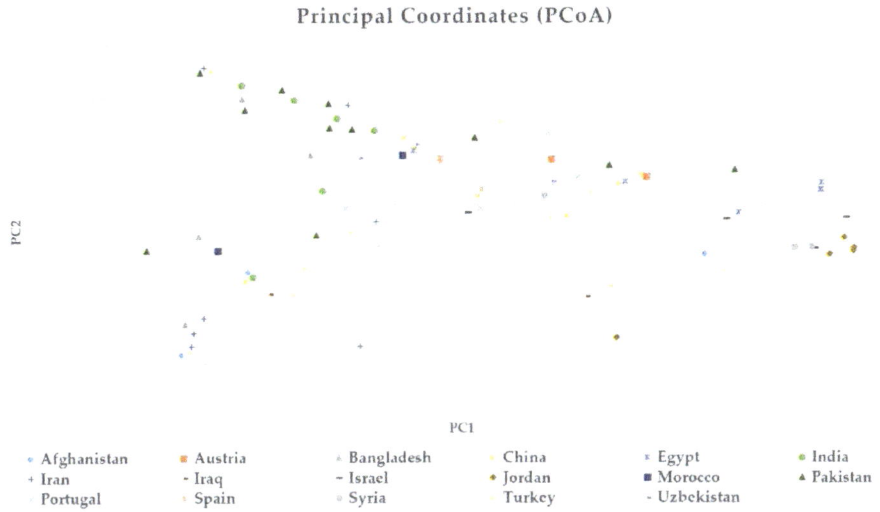

Figure 8. Principal coordinate analysis (PCoA) of the 94 safflower accessions using silicoDArT molecular markers.

2.4. *Marker-Trait Associations and Putative Candiadate Genes Identification for 100-Seed Weight*

The MLM (Q + K) model was performed to assess marker-trait associations for 100-seed weight in the international safflower panel. DArT-45483051 and DArT-15672391 showed marker-trait association for 100-seed weight by revealing statistically significant association (p-value; 1.17E-04 and 1.15E-04), respectively (Figure 9). DArT-45483051 and DArT-15672391 markers were present on supposedly chromosome 2 and 3, respectively. DArT-45483051 and DArT-15672391 markers contributed a total of 17.40% and 18.60% variation for 100-seed weight, respectively (Table 5). BLAST search against NCBI for DArT-45483051 resulted in AT1G01040 as a putative candidate gene. A search using the DArT-15672391 marker resulted in the retrieval of AT5G58040 as a putative gene for this marker.

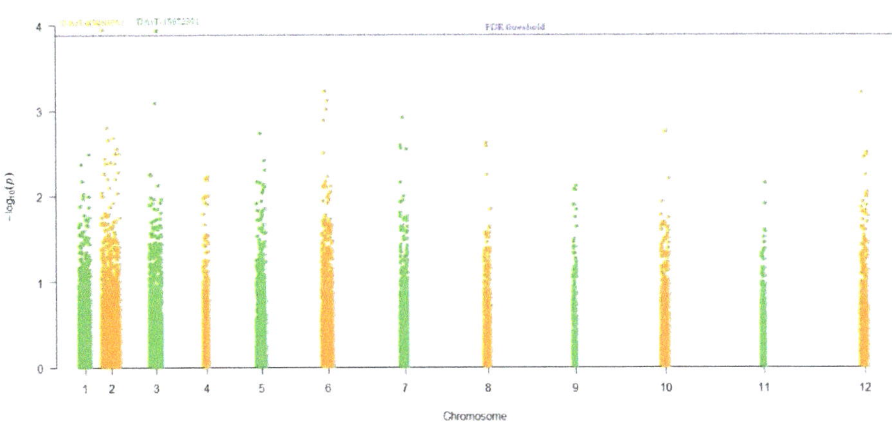

Figure 9. Pseudo manhattan plot for 100-seed weight in world safflower panel. DArT-45483051 and DArT-15672391 were considered statistically (FDR thresholds p = 0.01) associated with this trait.

Table 5. Marker-trait associations of the 100-seed weight with its associated markers in international safflower panel of 94 accessions.

Trait	Markers	*p*-Value	R^2	Putative Gene
100-seed weight	DArT-45483051	1.17E-04	17.40%	*AT1G01040*
	DArT-15672391	1.15E-04	18.60%	*AT5G58040*

3. Discussion

The studied safflower panel revealed statistically significant differences for 100-seed weight. Analysis of variance (ANOVA) confirmed the statistically significant genotypic effects in both locations (Table 1). These results were found to be in line with El-Lattief et al. [35] as they also found statistically significant genotypic effects for various agronomic traits of safflower, including 1000-seed weight. Frequency distribution for 100-seed weight at both locations (Pakistan and Turkey) was also calculated for a better understanding of the distribution of data. There was a normal frequency distribution of 100-seed weight at both locations. Frequency distribution revealed that only one safflower accession (Egypt-5) resulted in average 100-seed weight of > 5.00 g at both locations (Figure 1). Overall mean (3.29 g) and range of 100-seed weight in this study were higher than that reported in a previous study [36]. Mean and range of 100-seed weight obtained from Turkey were slightly higher than those obtained from Pakistan. Köse et al. [37] ascribe variations in seed weight to be a result of the reaction of genotypes to different environments. Our ANOVA also confirmed that there is a statistically significant genotypic effect on 100-seed weight. Constellation plot clustered the whole germplasm into two main populations, i.e., 46 and 48 accessions present in population A and B based on their 100-seed weight (Figure 2). Further division of accessions into subpopulations occurred. i.e., A1, A2, A3, B1, B2, and B3. Subpopulation A1 clustered only those 12 safflower accessions having 100-seed weight >4 g. Subpopulation A2 clustered safflower accessions having 3.70 to 4.00 g 100-seed weight. Accessions having 100-seed weight range 3.24 to 3.59 g clustered together, making subpopulation A3. Subpopulation B1 clustered safflower accessions having 100-seed weight ranging from 3.13 to 3.29 g. Accessions having 100-seed weight in a range of 2.64 to 2.85 g clustered together to make subpopulation B2. All accessions having 100-seed weight <2.60 g were present in subpopulation B3. As it is obvious from the above discussed results, 12 accessions present in the A1 subpopulation have higher 100-seed weight than the rest of the accessions. Therefore, these accessions can be used for breeding activities of safflower for high yield.

DArTseq technology gained the attention of scientists globally due to low cost and high throughput nature. DArTseq technology has been used to explore the genetic diversity and population structure of different crops with a large number of entries and complex genomes [38,39]. Hassani et al. [29] used DArTseq technology to explore genetic variations in a world panel of 89 safflower genotypes of diverse origin. The safflower panel utilized in their investigation is different from our panel except one accession, i.e., Afghanistan-1. During this study we also aimed to explore the phenotypic and genetic diversity, population structure and marker-trait association in an international safflower panel using silicoDArT markers. Hassani et al. [29] used 1136 silicoDArT and 2295 SNP markers, while we used a higher number of markers (12232) for the molecular characterization. Moreover, Hassani et al. [29] used germplasm from 12 countries, while we included germplasm from 26 countries to explore population structure more extensively.

Polymorphism information content (PIC) value is a measure of polymorphism which provides information regarding the genetic diversity or DNA segment in a studied population, and indicates the allele's evolutionary pressure and mutations that occurred at a locus over a time period. The range of the PIC value (0.10 to 0.50) obtained in this study suggests the existence of a high level of genetic variation that might be derived utilizing a large number of good quality markers in a diversified safflower panel. An average PIC value of 0.31 across all the silicoDArT markers was obtained during this study. PIC values were distributed asymmetrically and were skewed towards the lower values.

More than 50% of the implemented silicoDArT markers revealed a PIC value of more than 0.30, which indicates the informativeness and usefulness of these markers for genetic diversity, population structure, and marker-trait association in safflower.

Diversity parameters including observed number of alleles (Na), effective number of alleles (Ne), Shannon's information index (I), expected heterozygosity (He), and unbiased expected heterozygosity (uHe) for the entire population of 94 safflower accessions, which were 1.99, 1.54, 0.48, 0.32, and 0.32, respectively. Previous use of different gel-based marker systems obtained lower diversity metrics values than our current results from the silicoDArT marker system [2,18,40,41]. The most prominent reason for getting good diversity results is likely due to higher capability of the silicoDArT marker system in comparison with other gel-based marker systems. Differences in the experimental materials might also be another reason of revealing higher polymorphism in this study. Furthermore, results of diversity indices on the basis of collection countries revealed the highest polymorphism and genetic diversity for safflower genotypes from Pakistan and Turkey, while the lowest polymorphism and genetic diversity was obtained for safflower accessions originating from Iraq and Morocco. In a similar way, the highest mean genetic distance was observed for accessions originating from Turkey, and followed by India, Austria and Uzbekistan. The lowest mean genetic distance was observed for accessions originating from Jordan, followed by Spain (Table 3).

The Jaccard coefficients of genetic distance resulted in a mean value of 0.50 for the entire population of 94 safflower accessions. A maximum genetic distance was proposed between safflower accessions Egypt-2 and India-2, followed by Egypt-5 and India-2 with genetic distance values of 0.76 and 0.76, respectively. The highest genetic similarity was recorded between safflower accessions Spain-1 and Spain-2, with a genetic distance value of 0.14. The presence of higher genetic similarity between safflower accessions is possibly because of their origin from common parents (Supplementary Table S3). Safflower accessions containing desirable plant traits can be integrated in different breeding programs to aid in the development of superior cultivars [2]. The most diverse safflower accessions identified (Egypt-2, India-2, and Egypt-5) during the current evaluation can be recommended as candidate parental lines for future safflower breeding activities. The inferences obtained from kinship coefficient estimations with silicoDArT markers are robust to population structure. Negative kinship coefficients were also observed, suggesting an unrelated relationship between the safflower accessions. The close relatives can be inferred fairly reliably based on the estimated kinship coefficients. Thus, it is suggested that most of the safflower accessions were less related, having kinship coefficients of either 0 or below 0 (Figure 4). Analysis of molecular variance (AMOVA) revealed the division of the total variation into two strata, i.e., among countries and within country. A total of 91% of the genetic diversity was present within country group (Table 4). This is supported by Hassani et al. [29], where the majority of genetic variation among accessions within populations was obtained. The presence of a higher level of genetic variation within populations can be attributed to gene flow, which depends on the informal seed exchanges between farmers of different ecological zones [42].

3.1. Genetic Structure and Diversity in Safflower Panel

The three clustering algorithms important to genetic diversity and population structure analysis (model-based structure, Neighbor Joining, and PCoA) were implemented and revealed that the safflower accessions were successfully grouped by the silicoDArT markers based on geographical regions. Among the three clustering algorithms, more preference was given to the model-based structure algorithms. The reason for giving such a high preference to the structure is that this algorithm revealed more robustness in the previous works [43,44]. Structure algorithm divided the whole germplasm panel into four genetic populations: population A, B, C, and an admixture population. These different populations will aid in the selection of the parental accessions, which can used to design and conduct various crossing combinations for safflower genetic improvement (Figure 6).

Grouping of the safflower accessions obtained from structure analysis was based on the geographical origins of germplasm. Population A clustered safflower accessions originating from

Israel, Jordan, Syria, Egypt, Turkey, Hungary, Afghanistan, and Pakistan. Out of 17 of these accessions, 13 belong to the geographical locations situated in the Mediterranean region. Safflower accessions from the Mediterranean region clustered together and revealed their genetic similarity, and possibly shared a similar parentage. Clustering of the safflower accessions from Mediterranean countries proposes this region as a center of safflower domestication, with Syria having a predominant role [8]. Population B clustered safflower accessions from the South Asian countries, i.e., Afghanistan, Iran and Bangladesh, but not Turkey. Population C revealed safflower accessions from Iran, Pakistan, India, Turkey, Bangladesh, Egypt, Morocco, Uzbekistan, Spain, and Argentina. All four populations clustered accessions from Turkey. Our results are strongly supported by Hassani et al. [29] as they observed that safflower accessions from the Asian continent like Pakistan, Iran, Turkey, and India were found genetically similar and grouped into the same cluster. Clustering of the safflower accessions originating from the Mediterranean region to other localities suggested the distribution of safflower accessions from the Mediterranean region to other geographies. Turkey signifies a high level of biodiversity and differentiation center for safflower among the continents, thus reflecting a key role in the connection of different continents (Asia and Europe) with each other [15], and might possibly have played a role in the distribution of safflower from its domestication center. Our current findings are also supported by previously conducted archeological and molecular characterization studies using safflower accessions with its wild progenitors. These studies concluded the domestication of safflower in Fertile Crescent and its distribution to other parts of the world, i.e., Europe, Africa, the Middle and Far East [45,46]. All safflower accessions with a membership coefficient below 80% were classified as admixture populations. Accessions from countries such as Afghanistan, Australia, Austria, Bangladesh, China, Egypt, France, India, Iran, Israel, Iraq, Jordan, Kazakhstan, Libya, Morocco, Pakistan, Portugal, Romania, Russia, Syria, Thailand, Turkey, and Uzbekistan were clustered in the admixture population. The representation of safflower accessions from such a wide range of countries in the admixture population might be due to their use in international breeding programs. Other important factors like mutation, migration and selection by humans might also be responsible for the occurrence of admixture populations in safflower accessions of different origin [21,27].

Neighbor joining analysis divided the studied germplasm into two populations based on their geographical origin (Figure 7). Structure-based clustering of the 94 safflower accessions was also greatly supported by the principal coordinate analysis (PCoA) with silicoDArT markers information. PCoA resulted in clustering of safflower accessions on the basis of their geographical origins (Figure 8). Clustering results by PCoA were also supported by the structure, as admixture of accessions was also observed in the PCoA clustering as well.

3.2. Marker-Trait Associations for 100-Seed Weight

Seed yield is a complex trait with many underlying factors contributing to it. Comprehensive understanding of the relationship between seed yield and other contributing traits is crucial to the process of selection and ultimately to crop improvement [47]. In safflower, 100-seed weight is regarded as an important yield trait. Seed weight is proposed as an important trait to the process of selection in safflower breeding programs [48–50]. Furthermore, Chaudhary [48] observed the positive effects of 1000-seed weight on seed yield in safflower. Identification of loci influencing important plant morpho-agronomic traits is a prerequisite to marker-assisted breeding for enhancement of crop productivity. Our current investigation involved the identification of two silicoDArT markers (DArT-45483051 and DArT-15672391) associated with 100-seed weight (Table 5). Earlier studies reported different loci/markers linked with 100-seed weight. Ambreen et al. [21] reported two loci (NGSaf_306 and NGSaf_309) associated with 100-seed weight utilizing SSR markers. Mirzahashemi et al. [51] identified one QTL (qThsw5) associated with 100-seed weight.

The BLAST search against DArT-45483051 marker resulted in AT1G01040, a putative candidate gene. AT1G01040 is highly associated with seed embryo health and embryo shape at seed maturity and ovule development [52]. Therefore, the marker DArT-45483051 associated with 100-seed weight can be suggested for marker-assisted breeding of safflower for this trait. The BLAST search against DArT-15672391 retrieved a gene (AT5G58040) that encodes the pre-mRNA polyadenylation factor FIP1. Recently, Téllez-Robledo et al. [53] explored the role of polyadenylation factor FIP1 for plant development and root response to abiotic stresses. Polyadenylation factor FIP1 plays an important role in the plant embryo cotyledonary stage of development. Paez-Garcia et al. [54] established that better root growth ultimately contributes to higher crop yield. Based on the role of polyadenylation factor FIP1 during the plant embryo cotyledonary stage, AT5G58040 should be considered as having a potentially important role in seed weight. The studied germplasm reflected a wide range of phenotypic variations for 100-seed weight. Moreover, various genetic diversity indices also confirmed the existence of higher polymorphism in the evaluated germplasm. Characterization of germplasm provides us with an opportunity to unlock the novel genetic variations that can be utilized for breeding purposes [55–57]. This is a pioneer study concerning the investigation of marker-trait association for 100-seed weight for safflower using GBS analysis. We believe that these identified markers can be utilized in safflower marker-assisted breeding in order to develop cultivars with improved yield.

4. Materials and Methods

4.1. Plant Materials and Phenotypic Evaluation

A total of 94 safflower accessions originating from 26 countries were used as plant materials in this study. Seeds of the evaluated germplasm were provided by the United States Department of Agriculture (USDA) (Supplementary Table S4). The experimental materials were sown at two diverse locations, i.e., Pakistan and Turkey. The dirst experiment was conducted at the National Agricultural Research Center (Pakistan), whereas the second experiment was conducted at the research and experimental area of Bolu Abant Izzet Baysal University (Turkey) during 2016–2017 and 2018, respectively. Field experiments were performed by implementing an augmented block design. Seeds of each safflower accession were planted in elementary plots with a row length, inter-row and intra-row spacing of 3 m, 50 cm, and 10 cm respectively. A total of 10 plants for each accession were maintained for the phenotypic characterization. Thori-78 and Dinçer were included as check cultivars. Di-ammonium phosphate (DAP) and ammonium sulfate were applied as a source of fertilizer, while standard cultural practices were performed at both locations. Safflower accessions were harvested at their proper maturity at both locations. Measurement of 100-seed weight was done with the help of an electronic seed counter by taking undamaged and fully matured seeds randomly in triplicate.

4.2. Genomic DNA Isolation

To extract the genomic DNA from each accession, fresh, healthy and young leaves were harvested and kept frozen in the laboratory at −80 °C. DNA isolation of each safflower accession was performed utilizing the bulk of leaves from 10 individuals. The individuals used for the purpose of DNA isolation were from plants from the original seeds from the gene bank. DNA isolation was performed according to CTAB protocol [58] and a specific protocol suggested by Diversity Arrays Technology [59]. DNA concentration was estimated with agarose gel (0.80%) and was then confirmed with NanoDrop (DeNovix DS-11 FX, USA). For the genotyping by sequencing (GBS) analysis, DNA was diluted and a 50 ng.μl^{-1} DNA concentration was maintained. The prepared DNA samples were sent to Diversity Array Technology Pty, Ltd., Bruce, Australia, for DArTseq analyses of GBS [60].

4.3. DArTseq-Generated SilicoDArT Marker Analysis

DArTseq technology is a complexity reduction method and next generation sequencing platform [61]. DArTseq facilitated the selection of the genome fractions containing active genes associated with agronomically important plant traits [62]. Digestion/ligation reactions were used for the processing of DNA samples following the method described by Kilian et al. [63]. Mixed fragments

(PstI–MseI) were amplified by performing 30 rounds of PCR cycles. Details of silicoDArT markers analysis can be found in earlier studies [63,64].

4.4. Statistical Analysis

4.4.1. Phenotypic Data Analysis

Online software developed by Rathore et al. [65] for statistical inferences of augmented block design was used. Analysis of variance (ANOVA) for both locations was calculated through SAS 9.3 version [66]. Data recorded on 100-seed weight of both field experiments was averaged and used to calculate parameters like minimum, maximum, mean, standard deviation, and frequency distribution utilizing statistical software XLSTAT (Addinsoft, 2018) [67]. The cluster constellation plot for 94 safflower accessions was constructed through JMP 14.1.0 statistical software (2018, SAS Institute Inc., Cary, NC, USA).

4.4.2. DArTseq Markers Analysis

All images were analyzed from the DArTseq platform using DArTsoft v.7.4.7 (DArT P/L, Canberra, Australia). SilicoDArT are dominant markers that were detected through DArTseq and scored using the binary fashion, where 1 or 0 represent presence or absence of the restriction fragment in the genomic representation of each sample, respectively [12,68]. Screening of the markers was done with various parameters including call rate, polymorphism information content (PIC) and reproducibility being considered. Markers with PIC, reproducibility and call rate lower than 0.10, 100% and 0.80% were ignored during bioinformatics analyses to avoid false inferences.

4.4.3. Genetic Diversity Analyses

The proportion of shared alleles that were obtained from silicoDArT markers were used to compute the genetic distances among the safflower accessions using Jaccard's coefficients of genetic distance. Important diversity metrics such as observed number of alleles (Na), effective number of alleles (Ne), Shannon's Information Index (I), expected heterozygosity (He), and unbiased expected heterozygosity (uHe) were estimated for the entire population using GenAlEx 6.5 [69]. The kinship coefficients between safflower accessions were calculated with hierfstat R package to investigate the pairwise relationships of the 94 safflower accessions. Analysis of molecular variance (AMOVA) was computed with GenAlEx 6.5 [69] considering total variation into two strata, i.e., among countries and within countries.

Population structure of the studied safflower accessions was evaluated with STRUCTURE software (version 2.3.4; [70]). The most suitable number of clusters (K subpopulations) ranging from 1 to 10 was determined applying STRUCTURE software following the protocol of Evanno et al. [71]. For each K value and for each run, 10 independent runs were set. The initial burn-in period was set to 500 with 500,000 MCMC (Markov Chain Monte Carlo) iterations with no prior information on the origin of individuals. The most probable number of subpopulations was investigated by following the methodology suggested by Evanno et al. [71]. Each accession was assigned to a specific population on the basis of a membership coefficient. The PCoA was performed with GenAlEx 6.5 [69], while the Neighbor Joining tree was constructed with hierfstat R package in R statistical software. The populations obtained from the Neighbor Joining and PCoA were named and colored with the same clusters pattern identified with model-based Structure algorithm for coherence purposes.

4.4.4. Functional Analysis for Putative Candidate Gene Identification

To investigate the putative candidate genes, sequences of identified silicoDArT markers were used to perform BLAST searches against the National Center for Biotechnology Information (NCBI) [72], and the Phytozome V.12.1 [73] database. Moreover, detailed information about putative candidate genes was obtained from the TAIR database [74].

4.4.5. Genome-Wide Association Mapping for 100-Seed Weight

A Mixed linear model (MLM, Q + K) approach was applied to inspect marker-trait associations (MTAs) via TASSEL 5.0.5 [75]. The population and family structure were corrected utilizing Q-metrics (Q) and kinship (K) during association analysis, as suggested by Nadeem et al. [76,77]. Scaled identity was utilized to detect kinship matrix by the descent method applied in TASSEL 5.0.5 [75]. In the results of association analysis, the p value signifies the relatedness of a marker with the associated trait, and R^2 reflects the proportion of phenotypic variation resulting from a significant marker [78]. SilicoDArT markers with Bonferroni and FDR thresholds $p = 0.01$ were taken as significantly associated with the 100-seed weight. A Pseudo-Manhattan plot was developed using the qq-man R Package in the R 4.0.0 statistical software [79].

5. Conclusions

The current evaluation revealed a good level of phenotypic and genetic diversity in the studied safflower panel from the silicoDArT markers information. Analysis of variance (ANOVA) revealed the highly significant genotypic effect for 100-seed weight. Frequency distribution resulted in a normal distribution for 100-seed weight across the two locations (Pakistan and Turkey). The mean and range of 100-seed weight obtained from Turkey was higher than that from Pakistan. Analysis of molecular variance (AMOVA) revealed the division of total variations into two stratum i.e., among countries and within country. A total of 91% of the genetic variation was present within country and low variation (9%) was observed among the countries. Findings of AMOVA were also supported by the results of 100-seed weight and genetic distance. A good range of variations in 100-seed weight and genetic distance calculated at countries basis confirmed that mostly variations resulted in this study are because of diverse individuals within countries. Safflower accessions Egypt-5, Egypt-2, and India-2 showed the highest genetic distance among the studied panel and hence might be recommended as candidate parental lines for safflower breeding programs. Moreover, Egypt-5 is the only accession among the studied international safflower panel that reflected 100-seed weight of > 5.00 g at both locations and the highest genetic distance (0.76 with silicoDArT markers). Model-based structure analysis, Neighbor joining analysis and Principal coordinate analysis (PCoA) clustered the safflower accessions on the basis of their geographical origin. This is a pioneer study uncovering the marker-trait association for 100-seed weight in safflower. A total of two silicoDArT markers (DArT-45483051 and DArT-15672391) showed statistically significant association for 100-seed weight and these markers can be used in marker-assisted breeding to develop safflower cultivars with improved yield.

Supplementary Materials: The following are available online at http://www.mdpi.com/2223-7747/9/5/652/s1, Supplementary Table S1: Mean data for 100-seed weight across two locations (Pakistan and Turkey), Supplementary Table S2: Calculated diversity parameters for the whole safflower germplasm, Supplementary Table S3: Genetic distance between the 94 safflower accessions, Supplementary Table S4: List of 94 safflower accessions evaluated to explore genetic diversity and population structure with silicoDArT molecular markers.

Author Contributions: Conceptualization, F.S.B., H.J.C.; methodology, F.S.B., M.A.N.; software, M.A.N., E.H.; validation, F.A., M.A.N.; formal analysis, E.H., M.A.N., M.S.; investigation, F.A., M.A.N., M.B., A.A.; resources, H.J.C., G.C., S.H.Y.; data curation, F.A., M.A.N., M.B.; writing—original draft preparation, F.A., M.A.N.; writing—review and editing, M.A.N., R.H., S.H.Y., C.K., A.A., N.L., M.A., T.K., I.H.K.; supervision, F.S.B. All authors have read and agreed to the published version of the manuscript.

Funding: This research received no external funding

Acknowledgments: The authors are very thankful to TUBİTAK (The Scientific and Technological Research Council of Turkey) for their support to Fawad Ali under the TUBITAK-2216 fellowship program.

Conflicts of Interest: The authors declare no conflict of interest.

References

1. Kumari, S.; Choudhary, R.C.; Kumara, S.R.; Saharan, V.; Joshi, A.; Munot, J. Assessment of genetic diversity in safflower (Carthamus tinctorius L.) genotypes through morphological and SSR marker. *J. Pharmacogn. Phytochem.* **2017**, *6*, 2723–2731.
2. Ali, F.; Yılmaz, A.; Nadeem, M.A.; Habyarimana, E.; Subaşı, I.; Nawaz, M.A.; Chaudhary, H.J.; Shahid, M.Q.; Ercişli, S.; Zia, M.A.B.; et al. Mobile genomic element diversity in world collection of safflower (Carthamus tinctorius L.) panel using iPBS-retrotransposon markers. *PLoS ONE* **2019**, *14*, e0211985. [CrossRef] [PubMed]
3. Ali, F.; Yilmaz, A.; Chaudhary, H.J.; Nadeem, M.A.; Rabbani, M.A.; Arslan, Y.; Nawaz, M.A.; Habyarimana, E.; Baloch, F.S. Investigation of Morpho-Agronomic Performance and Selection Indices in the International Safflower Panel for Breeding Perspectives. *Turk. J. Agric. For.* **2020**, *44*, 103–120. [CrossRef]
4. Ali, F.; Nadeem, M.A.; Habyarimana, E.; Yilmaz, A.; Nawaz, M.A.; Khalil, I.H.; Ercisli, S.; Chung, G.; Chaudhary, H.J.; Baloch, F.S. Molecular characterization of genetic diversity and similarity centers of safflower accessions with ISSR markers. *Braz. J. Bot.* **2020**, *43*, 109–121. [CrossRef]
5. Dajue, L.; Mündel, H.H. *Safflower. Carthamus Tinctorius L. Promoting the Conservation and Use of Underutilized and Neglected Crops*, 7th ed.; Institute of Plant Genetics and Crop Plant Research: Rome, Italy, 1996; pp. 1–83.
6. Ferwerda, J.; Oilseed Crops. Weiss, E.A., Ed.; Elsevier: Harlow, Essex, UK, 1983; ISBN 0-582-46338-6.
7. Dordas, C.A.; Sioulas, C. Dry matter and nitrogen accumulation, partitioning, and retranslocation in safflower (Carthamus tinctorius L.) as affected by nitrogen fertilization. *Field Crops Res.* **2009**, *110*, 35–43. [CrossRef]
8. Marinova, E.; Riehl, S. Carthamus species in the ancient Near East and south-eastern Europe: Archaeobotanical evidence for their distribution and use as a source of oil. *Veg. Hist. Archaeobot.* **2009**, *18*, 341–349. [CrossRef]
9. Zeinali, E. *Safflower (Characteristics, Production and Utilization)*; Gorgan University of Agricultural Science and Natural Resources: Gorgan, Iran, 1999; p. 137.
10. Yang, Y.X.; Wu, W.; Zheng, Y.L.; Chen, L.; Liu, R.J.; Huang, C.Y. Genetic diversity and relationships among safflower (Carthamus tinctorius L.) analyzed by inter-simple sequence repeats (ISSRs). *Genet. Resour. Crop. Evol.* **2007**, *54*, 1043–1051. [CrossRef]
11. Mayerhofer, M.; Mayerhofer, R.; Topinka, D.; Christianson, J.; Good, A.G. Introgression potential between safflower (Carthamus tinctorius) and wild relatives of the genus Carthamus. *BMC Plant Biol.* **2011**, *11*, 47. [CrossRef]
12. Baloch, F.S.; Alsaleh, A.; Shahid, M.Q.; Çiftçi, V.; de Miera, L.E.S.; Aasim, M.; Nadeem, M.A.; Aktaş, H.; Özkan, H.; Hatipoğlu, R. A whole genome DArTseq and SNP analysis for genetic diversity assessment in durum wheat from central fertile crescent. *PLoS ONE* **2017**, *12*, e0167821. [CrossRef]
13. Nadeem, M.A.; Habyarimana, E.; Çiftçi, V.; Nawaz, M.A.; Karaköy, T.; Comertpay, G.; Shahid, M.Q.; Hatipoğlu, R.; Yeken, M.Z.; Ali, F.; et al. Characterization of genetic diversity in Turkish common bean gene pool using phenotypic and whole-genome DArTseq-generated silicoDArT marker information. *PLoS ONE* **2018**, *13*, e0205363. [CrossRef]
14. Mary, S.; Gopalan, A. Dissection of genetic attributes yield traits of fodder cowpea in F3 and F4. *J. Appl. Sci. Res.* **2006**, *2*, 805–808.
15. Arystanbekkyzy, M.; Nadeem, M.A.; Aktas, H.; Yeken, M.Z.; Zencirci, N.; Nawaz, M.A.; Ali, F.; Haider, M.S.; Tunc, K.; Chung, G.; et al. Phylogenetic and taxonomic relationship of turkish wild and cultivated emmer (triticum turgidum ssp. dicoccoides) revealed by iPBSretrotransposons markers. *Int. J. Agric. Biol.* **2018**, *10*. [CrossRef]
16. Yaldiz, G.; Camlica, M.; Nadeem, M.A.; Nawaz, M.A.; Baloch, F.S. Genetic diversity assessment in Nicotiana tabacum L. with iPBS-retrotransposons. *Turk. J. Agric. For.* **2018**, *42*, 154–164. [CrossRef]
17. Castañeda-Álvarez, N.P.; Khoury, C.K.; Achicanoy, H.A.; Bernau, V.; Dempewolf, H.; Eastwood, R.J.; Guarino, L.; Harker, R.H.; Jarvis, A.; Maxted, N. Global conservation priorities for crop wild relatives. *Nat. Plants* **2016**, *2*, 16022. [CrossRef]
18. Johnson, R.C.; Kisha, T.; Evans, M. Characterizing safflower germplasm with AFLP molecular markers. *Crop Sci.* **2007**, *47*, 1728–1736. [CrossRef]
19. Amiri, R.; Azdi, S.; Ghanadha, M.; Abd, M. Detection of DNA polymorphism in landrace populations of safflower in Iran using RAPD-PCR technique. *Iran. J. Agric. Sci.* **2001**, *32*, 737–745.

20. Sehgal, D.; Raina, S.N.; Devarumath, R.M.; Sasanuma, T.; Sasakuma, T. Nuclear DNA assay in solving issues related to ancestry of the domesticated diploid safflower (*Carthamus tinctorius* L.) and the polyploid (*Carthamus*) taxa, and phylogenetic and genomic relationships in the genus *Carthamus* L. (Asteraceae). *Mol. Phylogenet. Evol.* **2009**, *53*, 631–644. [CrossRef]
21. Ambreen, H.; Kumar, S.; Kumar, A.; Agarwal, M.; Jagannath, A.; Goel, S. Association mapping for important agronomic traits in safflower (*Carthamus tinctorius* L.) core collection using microsatellite markers. *Front. Plant Sci.* **2018**, *9*, 402. [CrossRef]
22. Ambreen, H.; Kumar, S.; Variath, M.T.; Joshi, G.; Bali, S.; Agarwal, M.; Kumar, A.; Jagannath, A.; Goel, S. Development of genomic microsatellite markers in *Carthamus tinctorius* L.(safflower) using next generation sequencing and assessment of their cross-species transferability and utility for diversity analysis. *PLoS ONE* **2015**, *10*, e0135443. [CrossRef]
23. Amini, F.; Saeidi, G.; Arzani, A. Study of genetic diversity in safflower genotypes using agro-morphological traits and RAPD markers. *Euphytica* **2008**, *163*, 21–30. [CrossRef]
24. Khan, M.A.; von Witzke-Ehbrecht, S.; Maass, B.L.; Becker, H.C. Relationships among different geographical groups, agro-morphology, fatty acid composition and RAPD marker diversity in safflower (*Carthamus tinctorius*). *Genet. Resour. Crop Evol.* **2009**, *56*, 19–30. [CrossRef]
25. Kumar, S.; Ambreen, H.; Murali, T.; Bali, S.; Agarwal, M.; Kumar, A.; Goel, S.; Jagannath, A. Assessment of genetic diversity and population structure in a global reference collection of 531 accessions of *Carthamus tinctorius* L.(Safflower) using AFLP markers. *Plant Mol. Biol. Rep.* **2015**, *33*, 1299–1313. [CrossRef]
26. Lee, G.A.; Sung, J.S.; Lee, S.Y.; Chung, J.W.; Yi, J.Y.; Kim, Y.G.; Lee, M.C. Genetic assessment of safflower (*Carthamus tinctorius* L.) collection with microsatellite markers acquired via pyrosequencing method. *Mol. Ecol. Resour.* **2014**, *14*, 69–78. [CrossRef]
27. Pearl, S.A.; Burke, J.M. Genetic diversity in *Carthamus tinctorius* (Asteraceae; safflower), an underutilized oilseed crop. *Am. J. Bot.* **2014**, *101*, 1640–1650. [CrossRef]
28. Raman, H.; Raman, R.; Nelson, M.N.; Aslam, M.; Rajasekaran, R.; Wratten, N.; Cowling, W.A.; Kilian, A.; Sharpe, A.G.; Schondelmaier, J. Diversity array technology markers: Genetic diversity analyses and linkage map construction in rapeseed (Brassica napus L.). *DNA Res.* **2011**, *19*, 51–65. [CrossRef]
29. Hassani, S.M.R.; Talebi, R.; Pourdad, S.S.; Naji, A.M.; Fayaz, F. In-depth genome diversity, population structure and linkage disequilibrium analysis of worldwide diverse safflower (*Carthamus tinctorius* L.) accessions using NGS data generated by DArTseq technology. *Mol. Biol. Rep.* **2020**, *47*, 2123–2135. [CrossRef] [PubMed]
30. Flint-Garcia, S.A.; Thuillet, A.C.; Yu, J.; Pressoir, G.; Romero, S.M.; Mitchell, S.E.; Doebley, J.; Kresovich, S.; Goodman, M.M.; Buckler, E.S. Maize association population: A high-resolution platform for quantitative trait locus dissection. *Plant J.* **2005**, *44*, 1054–1064. [CrossRef] [PubMed]
31. Yu, J.; Buckler, E.S. Genetic association mapping and genome organization of maize. *Curr. Opin. Biotech.* **2006**, *17*, 155–160. [CrossRef]
32. Abdurakhmonov, I.Y.; Abdukarimov, A. Application of association mapping to understanding the genetic diversity of plant germplasm resources. *Int. J. Plant Genom.* **2008**. [CrossRef] [PubMed]
33. Ebrahimi, F.; Majidi, M.M.; Arzani, A.; Mohammadi-Nejad, G. Association analysis of molecular markers with traits under drought stress in safflower. *Crop Pasture Sci.* **2017**, *68*, 167–175. [CrossRef]
34. Mahmood, T.; Rahman, M.H.; Stringam, G.R.; Yeh, F.; Good, A. Molecular markers for yield components in *Brassica juncea*—Do these assist in breeding for high seed yield? *Euphytica* **2005**, *144*, 157–167. [CrossRef]
35. El-Lattief, E.A. Evaluation of 25 safflower genotypes for seed and oil yields under arid environment in upper Egypt. *Asian J. Crop Sci.* **2012**, *4*, 72–79. [CrossRef]
36. Rahim, A.D.A. Dimension, geometric, agricultural and quality characteristics of safflower seeds. *Turk. J. Field Crops* **2014**, *19*, 7–12.
37. Köse, A.; Koşar, F.; Bilir, Ö. Agricultural performances of some safflower (*Carthamus tinctorius* L.) lines developed by single plant selection method. *J. Agric. Sci.* **2018**, *24*, 1–11. [CrossRef]
38. Chen, T.; Tantasawat, P.A.; Wang, W.; Gao, X.; Zhang, L. Population structure of Chinese southwest wheat germplasms resistant to stripe rust and powdery mildew using the DArT-seq technique. *Ciênc. Rural.* **2018**, *48*, 1–9. [CrossRef]

39. Nadeem, M.A.; Nawaz, M.A.; Shahid, M.Q.; Doğan, Y.; Comertpay, G.; Yıldız, M.; Hatipoğlu, R.; Ahmad, F.; Alsaleh, A.; Labhane, N.; et al. DNA molecular markers in plant breeding: Current status and recent advancements in genomic selection and genome editing. *Biotechnol. Biotechnol. Equip.* **2018**, *32*, 261–285. [CrossRef]
40. Panahi, B.; Neghab, M.G. Genetic characterization of Iranian safflower (*Carthamus tinctorius*) using inter simple sequence repeats (ISSR) markers. *Physiol. Mol. Biol. Plants* **2013**, *19*, 239–243. [CrossRef]
41. Sung, J.S.; Cho, G.T.; Lee, G.A.; Baek, H.J.; Huh, M.K. Phylogenetic Relationships and Genetic Diversity in Collected Resources of *Carthamus tinctorius* by Random Amplified Polymorphic DNA Markers. *J. Life Sci.* **2010**, *20*, 1764–1771. [CrossRef]
42. Hirano, R.; Kikuchi, A.; Kawase, M.; Watanabe, K.N. Evaluation of genetic diversity of bread wheat landraces from Pakistan by AFLP and implications for a future collection strategy. *Genet. Resour. Crop Evol.* **2008**, *55*, 1007–1015. [CrossRef]
43. Bouchet, S.; Pot, D.; Deu, M.; Rami, J.F.; Billot, C.; Perrier, X.; Rivallan, R.; Gardes, L.; Xia, L.; Wenzl, P. Genetic structure, linkage disequilibrium and signature of selection in sorghum: Lessons from physically anchored DArT markers. *PLoS ONE* **2012**, *7*, e33470. [CrossRef]
44. Newell, M.A.; Cook, D.; Hofmann, H.; Jannink, J.L. An algorithm for deciding the number of clusters and validation using simulated data with application to exploring crop population structure. *Ann. Appl. Stat.* **2013**, *7*, 1898–1916. [CrossRef]
45. Chapman, M.A.; Hvala, J.; Strever, J.; Burke, J.M. Population genetic analysis of safflower (*Carthamus tinctorius*; Asteraceae) reveals a Near Eastern origin and five centers of diversity. *Am. J. Bot.* **2010**, *97*, 831–840. [CrossRef]
46. Pearl, S.A.; Bowers, J.E.; Reyes-Chin-Wo, S.; Michelmore, R.W.; Burke, J.M. Genetic analysis of safflower domestication. *BMC Plant Biol.* **2014**, *14*, 43. [CrossRef]
47. Yol, E.; Uzun, B. Geographical patterns of sesame accessions grown under Mediterranean environmental conditions, and establishment of a core collection. *Crop Sci.* **2012**, *52*, 2206–2214. [CrossRef]
48. Chaudhary, S.K. Path analysis for seed yield in safflower (*Carthamus tinctorius* L.) in acid soil under mid altitude conditions. *Int. J. Trop. Agric.* **1990**, *8*, 129–132.
49. Pascual-Villalobos, M.J.; Alburquerque, N. Genetic variation of a safflower germplasm collection grown as a winter crop in southern Spain. *Euphytica* **1996**, *92*, 327–332. [CrossRef]
50. Omidi, T.A.H. Correlation between traits and path analysis for grain and oil yield in spring safflower. *Sesame Safflower Newsl.* **2000**, *15*, 78–82.
51. Mirzahashemi, M.; Mohammadi-Nejad, G.; Golkar, P. A qtl linkage map of safflower for yield under drought stress at reproductive stage. *Iran. J. Genet. Plant Breed.* **2015**, *4*, 20–27.
52. Zumajo-Cardona, C.; Ambrose, B.A. Phylogenetic analyses of key developmental genes provide insight into the complex evolution of seeds. *Mol. Phylogenet. Evol.* **2020**, *147*, 106778. [CrossRef]
53. Téllez-Robledo, B.; Manzano, C.; Saez, A.; Navarro-Neila, S.; Silva-Navas, J.; de Lorenzo, L.; González-García, M.P.; Toribio, R.; Hunt, A.G.; Baigorri, R.; et al. The polyadenylation factor FIP1 is important for plant development and root responses to abiotic stresses. *Plant J.* **2019**, *99*, 1203–1219. [CrossRef]
54. Paez-Garcia, A.; Motes, C.M.; Scheible, W.R.; Chen, R.; Blancaflor, E.B.; Monteros, M.J. Root traits and phenotyping strategies for plant improvement. *Plants* **2015**, *4*, 334–355. [CrossRef]
55. Yildiz, M.; Kocak, M.; Nadeem, M.A.; Cavagnaro, P.; Barboza, K.; Baloch, F.S.; Argün, D.; Keleş, D. Genetic diversity analysis in the Turkish pepper germplasm using iPBS retrotransposonbased markers. *Turk. J. Agric. For.* **2020**, *44*, 1–14. [CrossRef]
56. Nadeem, M.A.; Karaköy, T.; Yeken, M.Z.; Habyarimana, E.; Hatipoğlu, R.; Çiftçi, V.; Nawaz, M.A.; Sönmez, F.; Shahid, M.Q.; Yang, S.H.; et al. Phenotypic characterization of 183 Turkish common bean accessions for agronomic, trading, and consumer-preferred plant characteristics for breeding purposes. *Agronomy* **2020**, *10*, 272. [CrossRef]
57. Barut, M.; Nadeem, M.A.; Karaköy, T.; Baloch, F.S. DNA fingerprinting and genetic diversity analysis of world quinoa germplasm using iPBS-retrotransposon marker system. *Turk. J. Agric. For.* **2020**, *44*. [CrossRef]
58. Doyle, J.J.; Doyle, J.L. Isolation of plant DNA from fresh tissue. *Focus* **1990**, *12*, 39–40.
59. Available online: https://www.diversityarrays.com/orderinstructions/plant-dna-extractionprotocolfor-dart/ (accessed on 20 December 2019).

60. Available online: http://www.diversityarrays.com/ (accessed on 20 December 2019).
61. Elshire, R.J.; Glaubitz, J.C.; Sun, Q.; Poland, J.A.; Kawamoto, K.; Buckler, E.S.; Mitchell, S.E. A robust, simple genotyping-by-sequencing (GBS) approach for high diversity species. *PLoS ONE* **2011**, *6*, e19379. [CrossRef]
62. Raman, H.; Raman, R.; Kilian, A.; Detering, F.; Carling, J.; Coombes, N.; Diffey, S.; Kadkol, G.; Edwards, D.; McCully, M. Genome-wide delineation of natural variation for pod shatter resistance in *Brassica napus*. *PLoS ONE* **2014**, *9*, e101673. [CrossRef]
63. Kilian, A.; Wenzl, P.; Huttner, E.; Carling, J.; Xia, L.; Blois, H.; Caig, V.; Heller-Uszynska, K.; Jaccoud, D.; Hopper, C. Diversity arrays technology: A generic genome profiling technology on open platforms. In *Data Production and Analysis in Population Genomics*; François, P., Aurélie, B., Eds.; Springer: Totowa, NJ, USA, 2012; pp. 67–89.
64. Li, H.; Vikram, P.; Singh, R.P.; Kilian, A.; Carling, J.; Song, J.; Burgueno-Ferreira, J.A.; Bhavani, S.; Huerta-Espino, J.; Payne, T. A high density GBS map of bread wheat and its application for dissecting complex disease resistance traits. *BMC Genomics* **2015**, *16*, 216. [CrossRef]
65. Rathore, A.; Parsad, R.; Gupta, V. Computer aided construction and analysis of augmented designs. *J. Indian Soc. Agric. Stat* **2004**, *57*, 320–344.
66. *Statistical Analysis Software/Stat Software*, Version 9.3, 2nd ed.; SAS Institute: Cary, NC, USA, 2011.
67. Available online: www.xlstat.com (accessed on 3 April 2020).
68. Cömertpay, G.; Baloch, F.S.; Kilian, B.; Ülger, A.C.; Özkan, H. Diversity assessment of Turkish maize landraces based on fluorescent labelled SSR markers. *Plant Mol. Biol. Rep.* **2012**, *30*, 261–274. [CrossRef]
69. Peakall, R.; Smouse, P.E. GENALEX 6: Genetic analysis in Excel. Population genetic software for teaching and research. *Mol. Ecol. Notes* **2006**, *6*, 288–295. [CrossRef]
70. Pritchard, J.K.; Stephens, M.; Donnelly, P. Inference of population structure using multilocus genotype data. *Genetics* **2000**, *155*, 945–959.
71. Evanno, G.; Regnaut, S.; Goudet, J. Detecting the number of clusters of individuals using the software STRUCTURE: A simulation study. *Mol. Ecol.* **2005**, *14*, 2611–2620. [CrossRef]
72. Available online: http://www.ncbi.nlm.nih.gov/ (accessed on 4 May 2020).
73. Available online: http://phytozome.jgi.doe.gov/pz/portal.html (accessed on 4 May 2020).
74. Available online: http://www.arabidopsis.org (accessed on 4 May 2020).
75. Bradbury, P.J.; Zhang, Z.; Kroon, D.E.; Casstevens, T.M.; Ramdoss, Y.; Buckler, E.S. Tassel: Software for association mapping of complex traits in diverse samples. *Bioinform* **2007**, *23*, 2633–2635. [CrossRef]
76. Nadeem, M.A.; Gündogdu, M.; Ercisli, S.; Karaköy, T.; Saracoglu, O.; Habyarimana, E.; Lin, X.; Hatipoglu, R.; Nawaz, M.A.; Sameeullah, M.; et al. Uncovering phenotypic diversity and DArTseq marker loci associated with antioxidant activity in common bean. *Genes* **2020**, *11*, 36. [CrossRef]
77. Zia, M.A.B.; Demirel, U.; Nadeem, M.A.; Caliskan, M.E. Genome-wide association study identifies various loci underlying agronomic and morphological traits in diversified potato panel. *Physiol. Mol. Biol. Plants.* **2020**, *26*, 1003–1020. [CrossRef]
78. Jin, F.X.; Ji, S.D.; Xie, X.B.; Kang, J.W.; Ju, H.G.; Ahn, S.N. Detection of epistatic interaction of two qtls, gw8. 1 and gw9. 1, underlying grain weight using nearly isogenic lines in rice. *Breed. Sci.* **2011**, *61*, 69–75. [CrossRef]
79. Turner, S.D. qqman: An R package for visualizing GWAS results using QQ and Manhattan plots. *Biorxiv* **2014**, *005165*, 1–2.

plants

Article

The Cypriot Indigenous Grapevine Germplasm Is a Multi-Clonal Varietal Mixture

Apostolis Grigoriou [1], Georgios Tsaniklidis [2], Marianna Hagidimitriou [3] and Nikolaos Nikoloudakis [1,*]

1 Department of Agricultural Sciences, Biotechnology and Food Science, Cyprus University of Technology, Limassol CY-3603, Cyprus; apo_gregoriou@yahoo.gr
2 Institute of Olive Tree, Subtropical Plants and Viticulture, Hellenic Agricultural Organization 'Demeter' (NAGREF), 71003 Heraklio, Greece; tsaniklidis@nagref-her.gr
3 Department of Biotechnology, Agricultural University of Athens, 10561 Athens, Greece; marianna@aua.gr
* Correspondence: n.nikoloudakis@cut.ac.cy

Received: 16 May 2020; Accepted: 13 August 2020; Published: 14 August 2020

Abstract: Cypriot vineyards are considered as one among the earliest niches of viticulture and a pivotal hub for the domestication and dissemination of grapevine. The millennial presence of *Vitis* spp. in this Eastern Mediterranean island has given rise to a plethora of biotypes that have not been adequately characterized, despite their unique attributes and stress tolerance. This ancient germplasm also has an additional value since it survived the phylloxera outbreak; hence, it possesses a large amount of genetic diversity that has been unnoticed. In order to provide useful insights to the lineage of Cypriot vineyards, a two-year-spanning collection of centennial grapevine cultivars mostly regarded to belong to four indigenous variety clusters ("Mavro", "Xynisteri", "Maratheftiko", and "Veriko") was initiated. There were 164 accessions across the broader Commandaria wine zone sampled and characterized using a universal microsatellite primer set. Genetic analysis indicated that considered indigenous Cypriot germplasm has a polyclonal structure with a high level of heterozygosity. Moreover, several lineages or unexplored varieties may exist, since a larger than considered number of discrete genotypes was discovered. Furthermore, it was established that grapevine lineages in Cyprus were shaped across eras via clonal, as well as, sexual propagation. The special attributes of the Cypriot landscape are discussed.

Keywords: Cyprus; domestication; microsatellites; *Vitis vinifera* subsp. *sativa*; *Vitis vinifera* subsp. *Sylvestris*

1. Introduction

Grapes are the most emblematic deciduous woody vines. The imprint of viticulture to agriculture, society, culture, and even religion rituals has been unparalleled, and it can be stated without any reluctance that viticulture is intertwined with human history throughout eras since antiquity. At present, the total global grape production is 75 million tons and the surface of vineyards spans at around 7.5 million ha (FAO). In terms of revenue, the global wine market alone has perceived a stable expansion over the years and has developed as a multi-billion industry [1].

Taxonomically, grapes belong to the Vitaceae family which consists of 14 genera and about 900 species [2]. Based on seed morphological data and fossils, it has been determined that the Vitaceae cluster arose nearly 60 million years ago in the North American region [3]. Nowadays, cultivated species (classified as *Vitis vinifera* L. or *Vitis vinifera* subsp. *vinifera*) have a widespread distribution and can be found across all continents. Nonetheless, *Vitis* spp. progenitors can be exclusively found at specific areas; *V. vinifera* subsp. *sylvestris* (regarded as the Eurasian *Vitis* wild form) is found in North Africa and Eurasia, while *Vitis labrusca* is restricted to Eastern North America. The distinction of wild ancestors from cultivated forms can be challenging, since at least in the broader Mediterranean zone, a wide variety of feral forms and hybrids are present. These comprise cultivation escapees, as well as, seed-disseminated weedy types growing in local niches that occur primarily in surroundings areas of vineyards [4].

Currently, *V. vinifera* subsp. *sylvestris* is considered to be the wild relative of cultivated grapes (*Vitis vinifera* subsp. *vinifera*) and several botanical characteristics are used in order to differentiate the two taxa. One of the most important characters is the flower, as wild grapevines are dioecious, in contrast to virtually all cultivated genotypes that are hermaphrodites [5]. Due to the dioecious manner of the wild grapevine species and the high rates of heterozygosity, asexual propagation was frequently adopted in the grapevine lineage and domestication, in order to preserve desirable traits [6]. Selection of elite traits resulted to a domestication syndrome; thus, generating notable morphological variability in leaves and seeds, as well as a shift from dioecy to self-pollinating hermaphroditism and an increase in berry size and sugar content [7].

Increasing indications converge that the domestication of grapevines was disseminated alongside the advent of wine production from the center of domestication placed at the Near East and the Southern Caspian Belt [8]. The diffusion of grapevines from the principal center of domestication into adjacent areas of Northern Africa and Europe trailed through three main pathways. Firstly, towards Mesopotamia, the broader Eastern Mediterranean Basin and the Southmost Balkans (Pre-Bronze Age). Then, moving via Sicily to Western Europe, and lastly, domesticated grapevine forms reached to Central Europe 1000 years before the common era [9]. Concurrently, introgression of regional forms and secondary domestication events cannot be excluded [10–12]. This millennial ongoing process via asexual propagation, as well as trait introgression through inter- and intra-specific crossings resulted in a plethora of discrete and/or overlapping genotypes.

However, there were milestone eras and events that influenced the plethora of genetic grapevine diversity. Early Muslim conquests in the 7th century, and occupation of the primary as well as secondary domestication regions, introduced an extended period of gradual decline of wine production. This in succession could have forced a subsequent extinction of genotypes used for vinification and a turnover towards table grapes production [13]. A more recent substantial reduction of genetic resources in both wild and cultivated grapevines occurred when grape phylloxera outbroken in Europe at the late 1800s [14]. As a result, several American indigenous *Vitis* species were introduced as rootstocks, or used for incorporating disease resistant traits to hybrids [15]. Hence, varieties that were used as scions were mostly preserved but non-grafted genotypes were likely perished. Nowadays, a third wave of genetic erosion occurs via habitat fragmentation of wild and feral *Vitis* forms, supported by an augmented interest of global wine producers for only a handful of elite varieties [16].

Cyprus was on the map of ancient trade routes and, as a result, on the crossroads of the westward dissemination of grapevines to Europe. Cyprus and viticulture have been inextricably related since antiquity [17]. Regardless of its millennial vinification history [18], there is also one paramount reason why Cyprus is a grapevine hotspot of conserved diversity. The phylloxera pest occurs in all Eurasia, but Cyprus is regarded as a protected zone [19]. This constitutes Cyprus as an invaluable genetic resource reservoir of pivotal significance, since the present genotypes are living fossils and epochal remnants of grapevine domestication and distribution. Furthermore, wild grapevines can still be detected in fragmented populations or sporadically across the country; mainly located near small streams or forests. Despite its significance, up till now there has not been any widespread attempt to properly record the diversity of domestic grapevines.

Moreover, a precise number of varieties and clones is not clearly demarcated and there is quite uncertainty regarding the actual number since synonyms and homonyms can occur. Furthermore, ampelographic data are not largely available. Pierre Galet referred that "among the vines of Cyprus, there are (some) fifteen indigenous varieties": "Xynisteri", "Mavro", "Maratheftiko", "Lefkas", "Opthalmo", "Promara", "Spourtiko", "Flouriko", "Yiannoudhi", "Katomylitiko", "Kanella", "Morokanella", "Michalias", "Skouro mavro", "Rodhino rose", "Rozoudi rose", and "Maroucho black" [20]. On the other hand, only 12 native cultivars were genotyped by Hvarleva and co-workers [21]. Still, the number of indigenous registered varieties in the national plant variety catalogue is further reduced, probably due to insufficient description.

One of the most emblematic (and globally recognized) dessert wines is Commandaria. Commandaria is an amber-colored traditional wine produced in Cyprus, and it was the first type of wine receiving the controlled appellation of origin certification amid Cypriot wines [22]. One of the aspects that gives Commandaria a unique status comes from the fact that it is one of infrequent cases of wines in production that essentially follow the same principles practiced for millennia. According to Cypriot regulation, Commandaria production is only permitted in an explicit zone of fourteen villages, situated at the slopes of the Troodos Mountain at a spanning elevation from 500 to 900 m.a.s.l. The contemporary vineyard zone covers circa 9000 ha with non-grafted plants. Moreover, Commandaria can be only produced via the vinification of sun-dried grape berries of the two local varieties; Mavro and Xynisteri [23]. Xynisteri is the major white grape Cypriot variety and accounts for 30% of Cypriot vineyards, while Mavro is the prominent black grape variety and occupies almost 40% of Cyprus vineyards. Besides Mavro and Xynisteri varieties that are used in Commandaria, Maratheftiko is also gaining recognition as the most auspicious variety to develop elegant and quality wines. Maratheftiko is a red grapevine variety, and even though it is a cultivated *Vitis* form, it lacks hermaphrodite flowers and is frequently co-planted with other cultivars in order to attain fruit-set and development [24].

Several aspects can affect the outcome of vinification. Among these, primarily the variety (or genotype) and *terroir* are among the most complex and debated issues in viniculture and oenology. The effect of the genotype and the existence of different clones within varieties is often overlooked in studies, hence making it difficult to drive meaningful conclusions (the possible diversity due to genetic background is not considered). Still, clonal selection is extensively practiced in viticulture, signifying that somatic alterations have a substantial unaccounted outcome on berry and wine attributes [25]. Since Cypriot genotypes have been locally cultivated for millennia, it is expected that genetic diversity within variety clusters exist, but largely remain uncharted and unaccredited.

Since a nationwide molecular characterization of the Cypriot germplasm has not been conducted so far, the size of genetic diversity in Cypriot grapevines must be largely underestimated (as indicated by the low and uncertain number of varieties reported, compared to other countries having similar size and geography). Nowadays newly established vineyards are propagated clones of the same genotype; hence, focusing on antique grapevines could in fact be more effective in demarking genetic diversity. The hypothesis when commencing this project was that relic genotypes that are currently considered as a variety (due to the lack of robust ampelographic description and identification) could in fact be landraces (populations), discrete clones, or different varieties that may have a potential in enriching the Cypriot grapevine germplasm. Towards attaining meaningful insights for the Cypriot grapevine landscape, 164 centennial genotypes (putatively belonging to four local Cypriot varieties; Mavro, Xynisteri, Maratheftiko, and Veriko) were sampled across the Commandaria zone and genotyped using 11 nuclear SSRs in order to reveal their genetic variability. This study is the starting point towards the prominence and safeguarding of local indigenous germplasm, and the highlighting of historical knowledge and heritage concerning the millennia viniculture of Cyprus.

2. Materials and Methods

2.1. Plant Material

A total of 164 centennial grapevine accessions putatively attributed to the four mentioned populations, Maratheftiko, Mavro, Veriko, and Xynisteri (Table S1), were collected across Cypriot vineyards (Figure 1). Sampling was conducted across the seven wine-zone districts in the Commandaria region and the wine villages of Cyprus, during a spanning period of two years (2018–2019). Vines were carefully chosen through the support of local industry members and vineyard owners to certify a record of vine age, while distinctive phenotypic characters were considered according to the growth stage (Supplementary Data).

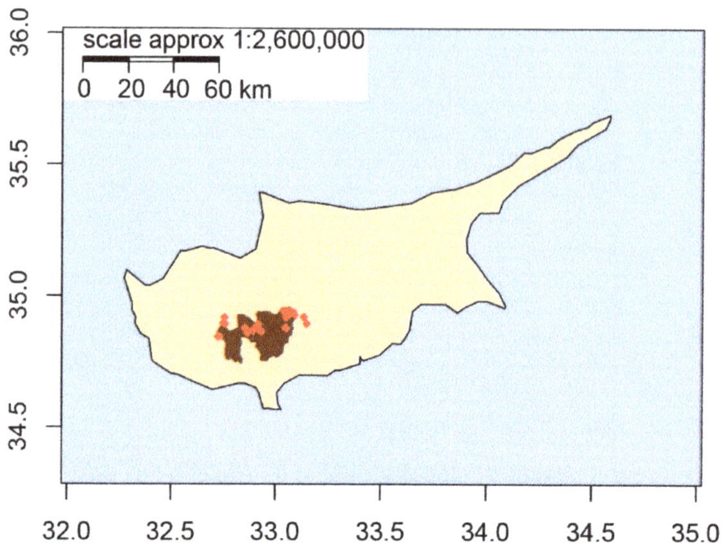

Figure 1. Map of Cyprus. Shaded area (brown) indicates the Commandaria zone and red rhombi mark the collection sites.

2.2. DNA Extraction Protocol

For DNA extraction, leaves were excised and kept between humid paper towels on ice, until storage at −80 °C. Leaf tissue (100–200 mg) was weighed in 2-mL rounded Eppendorf tubes, flash frozen in liquid nitrogen and freeze-dried overnight. Two stainless steel balls (3 mm) were added in each Eppendorf, samples were quickly frozen in liquid nitrogen, and tissues were crushed for 30 sec (at full speed) using a mixer mill MM 200 (Retsch). Samples were kept at −80°C until DNA extraction. In preliminary tests, it was established that DNA extraction using standard DNA extraction procedures (CTAB or commercial DNA extraction kits) proved problematic (for multiplex PCR) due to the highly recalcitrant nature of vine tissues. Hence, before DNA extraction, crushed tissues were pre-washed twice with sorbitol wash buffer (Sorbitol wash buffer: 100 mM Tris-HCl pH 8.0, 0.35 M Sorbitol, 5 mM EDTA pH 8, 0.1% (w/v) Polyvinylpyrrolidone PVP-40, 0.1% β-mercaptoethanol) to remove interfering metabolites [26]. After the wash of the inhibitors, cell lysis was achieved using a high-salt CTAB extraction buffer (2% CTAB (hexadecyltrimethylammonium bromide), 100 mM TrisHCl pH = 8, 20 mM EDTA, 1.4 M NaCl, 0.1% (w/v) Polyvinylpyrrolidone PVP-40, 0.1% β-mercaptoethanol). Impurities were further removed using equal volumes of chloroform: Isoamyl alcohol (24:1) and DNA were precipitated with saturated NaCl (1/10th of the lysis buffer volume) and ice-cold ethanol (twice of the lysis buffer volume). The pellet was washed twice with 100 ul of 70% cold ethanol, air dried, and dissolved in 100 μL of dd H_2O. DNA concentration and purity were estimated using nanodrop spectrophotometry.

2.3. Microsatellite Genotyping

Eleven SSR markers were selected for the analysis of the Cypriot germplasm (Table S2). Nine of them are suggested within the European projects Genres081 and GrapeGen06, proposing the use of a common set of microsatellite markers [27]. All forward primers employed were designed with a M13(-21) tail at the 5′-end and carried a fluorescent dye label (FAM^{TM}, ROX^{TM}, or TAMRA^{TM}).

This enabled the run of a one-tube, single-reaction nested PCR, as previously described [28]. Based on expected allele sizes from databases (Italian Vitis Database, http://www.vitisdb.it/; Vitis International Variety Catalogue VIVC, http://www.vivc.de/) and pilot single locus PCR reactions, three different panels were selected (Table S2), in order to ensure non-overlapping allele sizes across loci. Furthermore, even though green fluorescent dye (JOE^{TM}) was initially tested for multiplexing, it was not finally employed, since it was determined that it interfered with red (ROX^{TM}) and blue (FAM^{TM}) spectra, causing minor peak pull-ups that could result to erroneous scoring.

For PCR reactions, the mix contained 50 ng of template DNA, 10 pmol of a labeled M13 tailed forward primer, 10 pmol of the reverse and 2.5 pmol of the forward primer, 0.2 mM dNTPs, 0.5 U KAPA Taq DNA Polymerase (Kapa Biosystems), and a 2.5 mM final concentration of $MgCl_2$ in a 12 μL final reaction volume. Conditions for the PCR amplification were: 94 °C (5 min for initial denaturation), followed by 39 cycles at 94/56/72 °C (60 s), and a final extension at 72 °C for 30 min.

Amplification products were verified using a standard 2% agarose electrophoresis and diluted at a 1:40 ratio with dd H_2O. One μL of the dilutions was added to 10 μL deionized formamide and 0.2 μL of DNA size standard (GeneScan 500-LIZ, Applied Biosystems, Foster City, CA, USA), before denaturing at 95 °C (5 min). Allele fragments were separated (in three discrete panels) by capillary electrophoresis using an Applied Biosystems 3130® Genetic Analyzer (Applied Biosystems, Foster City, CA, USA). Two control varieties (Cabernet Sauvignon and Merlot) and four Cypriot genotypes were replicated in every run, to normalize microsatellite allele sizes and associate genotypes across runs and databases. Allele scoring was performed by two researchers independently, and tandem software was utilized to verify/correct bins [29].

2.4. Genetic Relationships and Analysis of Population Structure

Microsatellite data curation and formatting was performed via the MS Excel add-in GENALEX v. 6.501 [30]. All genotypes were included for calculating allelic frequencies across loci studied. In order to assess the discriminating power among unique genetic profiles, a genotype accumulation curve was constructed. Additionally, the number of multi-locus genotypes (MLGs), as well as the number of expected MLG (eMLGs) were calculated. Moreover, genotypic diversity was assessed with several indexes (H: Shannon–Wiener Index of MLG diversity, G: Stoddart and Taylor's index of MLG diversity, lambda Simpson's index, E.5: Evenness of the alleles, and Hexp: Nei's unbiased gene diversity).

The same dataset was similarly used to test for linkage disequilibrium and Hardy–Weinberg equilibrium (HWE) in the grapevine accessions. Genetic relationships between individuals (MLGs) were assessed using the dissimilarity distance calculation and visualized as a minimum spanning network (MSN) and a discriminant analysis of principal components (DAPC). All of the above-mentioned statistics/analyses were performed using the Poppr (V. 2.8.5) package [31] and the RStudio suite (V 1.2.5033; R V 3.6.2).

A phylogenetic tree was also constructed using the binary template (converted from allele size) using the R package polysat [32]. An approximate likelihood-ratio test (aLRT) for branch support was achieved by means of the SH-like parameter as previously described [33]. The newick -formatted tree was displayed and manipulated using the iTOL v4 server [34].

Finally, a Bayesian statistic employing method for estimating genetic kinship was performed using Structure 2.3.4 [35]. The admixture model was selected and 20 independent repeats per K value (extending from 1 to 20) were run. Each run involved 100,000 iterations burning period and a post burning simulation of 1,000,000. Validation of the most probable number of clusters K and visualization was achieved using the Clumpak server (http://clumpak.tau.ac.il/).

3. Results

3.1. Genetic Affiliations across the Cypriot Germplasm

From the broader Commandaria zone of Cyprus 164 vines putatively attributed to four varieties (Maratheftiko, Mavro, Veriko, and Xynisteri) were genotyped using a primer set of 11 SSR markers. The genotype accumulation curve (Figure S1) indicated that the 11 microsatellite loci were acceptable in order to delineate all the multi-locus genotypes (MLGs) present among cultivars (Table 1). The probability of identity (PI) of two samples to have the same genotype was also calculated for the dataset, and it was concluded that the cumulative capacity of the 11 loci resulted to a PI value of 6.2×10^{-7} to 7.8×10^{-10} across populations.

Table 1. Genetic variability across Cypriot grapevine populations. Total number of discrete genotypes and genetic diversity indices (clone corrected).

	Pop	*MLG (N)*	*eMLG*	*H*	*Hexp*	*Ia* *	*rbarD* *	*Ho*	*He*
1	Xynisteri	32 (74)	10	3.466	0.489	1.03	0.118	0.700	0.430
2	Mavro	41 (67)	10	3.714	0.667	4.141	0.427	0.923	0.626
3	Maratheftiko	8 (19)	8	2.079	0.567	−0.395	−0.148	0.981	0.526
4	Veriko	2 (4)	2	0.693	0.652	NA	NA	0.932	0.474
5	Total	83 (164)	10	4.419	0.732	4.143	0.42	0.884	0.514

MLG = number of observed multi-locus genotypes, *N* = total number of samples, *eMLG* = expected multi-locus genotype, *H* = Shannon–Wiener index of MLG diversity, *Hexp* = Nei's 1978 expected heterozygosity, *Ia* = the index of association, *rbarD* = the standardized index of association * $p < 0.001$, *Ho* = observed heterozygosity (calculated by GeneAlEx), *He* = expected heterozygosity (calculated by GeneAlEx).

All loci were found polymorphic (Table S3), and in total 102 alleles were detected, varying from six at loci VrZAG112 and VVMD25 up to 16 at locus VVMD28; with a mean of 9.27 alleles per locus (Table 2). Even though many discrete alleles were detected across loci, a few alleles were predominant, while the rest were noticed at lower frequencies (Table S4). After clone correction filtering (removal of redundant genotypes), several diversity indices were established across loci (Table 2).

Table 2. Allele summary statistics for the 11 analyzed loci of Cypriot grapevine populations.

Locus	*Alleles*	*1-D*	*Hexp*	*Evenness*
VVMD27	7	0.75	0.75	0.8
VrZAG79	13	0.82	0.82	0.7
VrZAG67	12	0.71	0.71	0.64
VrZAG62	8	0.81	0.81	0.89
VrZAG112	6	0.5	0.5	0.52
VVS2	7	0.68	0.68	0.68
VVMD5	8	0.79	0.8	0.86
VVMD7	8	0.75	0.75	0.75
VVMD28	16	0.83	0.83	0.68
VVMD25	6	0.58	0.58	0.65
VVMD32	11	0.8	0.8	0.8
mean	9.27	0.73	0.73	0.73

Alleles = number of observed alleles, *1-D* = Simpson index, *Hexp* = Nei's 1978 gene diversity, *Evenness* = distribution of genotype abundances.

The Simpson index varied from 0.5 for locus VrZAG112, up to 0.83 for locus VVMD28 (mean 0.73); in general, being proportional to the number of alleles detected among discrete loci. Still, across loci the high percentages of heterozygosity and evenness of alleles, elevated the efficiency of these markers to reveal the genetic polymorphism. Indices indicated the high potential for this marker set to define variability in the Cypriot panel of grapevines.

Multi-locus genotype analysis across populations recorded 32 MLGs for Xynisteri, 41 for Mavro, eight for Maratheftiko, and two for Veriko (Table 1). The most frequent MLG across populations was MLG 76 (16 vines) and MLG 81 (14 vines) that were detected for Xynisteri (Table S3). Correspondingly, several accessions were also found to be common in Mavro, (MLG 38 in 14 vines) and Maratheftiko population (MLG 8 in 7 vines), but at a lesser extent.

Even though genotypic abundancy can be primarily inferred from the number of MLGs detected, unequal sample size can cause a partial bias; hence, an estimate of the quantity of genotypes that could be anticipated at the major common sample size established on rarefaction (eMLG) could be more suitable. The expected number of eMLGs was also computed and was found comparable among Xynisteri and Mavro populations (10; Table 1). Still, a somewhat higher amount of genetic diversity was established in the Mavro cluster ($H = 3.714$, $Hexp = 0.667$), even though many accessions shared a genotype.

In order to determine the mode of reproduction across these centennial varieties, a test of linkage disequilibrium was conducted for clone corrected genotypes. Consequently, disequilibrium indices were calculated (*Ia* and *rbarD*); it seems that populations in Cyprus were mostly clonally propagated. Moreover, we explored the probability that loci were under Hardy–Weinberg equilibrium (Table 1). It was established that several loci were in HWE ($p < 0.5$). This suggests that even though at a reduced rate, sexual propagation has also occurred in the lineage of Cypriot grapevine varieties.

3.2. Population Structure of the Main Cypriot Grapevine Varieties

Estimates of population differentiation across loci (Figure S2) were elevated suggesting that a great amount of genetic diversity still exists within each considered cluster. In order to infer the number of groups of genetically related accessions, a multivariate approach partitioning between- and within-group components was used. The discriminant analysis of principal components (DAPC; Figure 2A) revealed that a clear genetic distinction across the Cypriot populations studied is evident. Moreover, it was established that significant within-variability also exists. The minimum spanning network (MSN) analysis (Figure 2B) depicted a structure having limited reticulation, indicative of clonal propagation where somatic mutations can result in novel genotypes that are highly affiliated to the core. Still, at low levels, reticulation was evident among genotypes. In general, the four main clusters of genotypes distinguished reflected the discrete genetic background of the four Cypriot strains studied. Interestingly, linear and reticulate relations were depicted within the germplasm of Xynisteri and Mavro populations, while the Maratheftiko cluster had a lower kinship to the core of genotypes.

The population structure among the Cypriot collection was also assessed using the Bayesian procedure STRUCTURE (Figure 3). The optimal for the ad hoc number, based on the second order rate of probability of the likelihood function respecting to Delta K, was attained for K = 2. At low cluster complexities, this analysis, indicated two discrete genetic groups (Delta K = 2229.2). The majority of putatively Mavro, Maratheftiko (black berry varieties), and Veriko (rose berry varieties) accessions were found highly constrained (all having a mean proportion of membership higher than 0.9, and were clustered in the first genetic category (Table S5). On the other hand, genotypes regarded as Xynisteri (white berry variety) had lower affinity to the first group and were affiliated with the second genetic ancestry. Within each population there were instances of admixture genotypes, characteristic of sexual propagation. As the level of K complexity increased (higher K clusters), a finer and a more detailed structure was portrayed (also differentiating Veriko genotypes; data not shown).

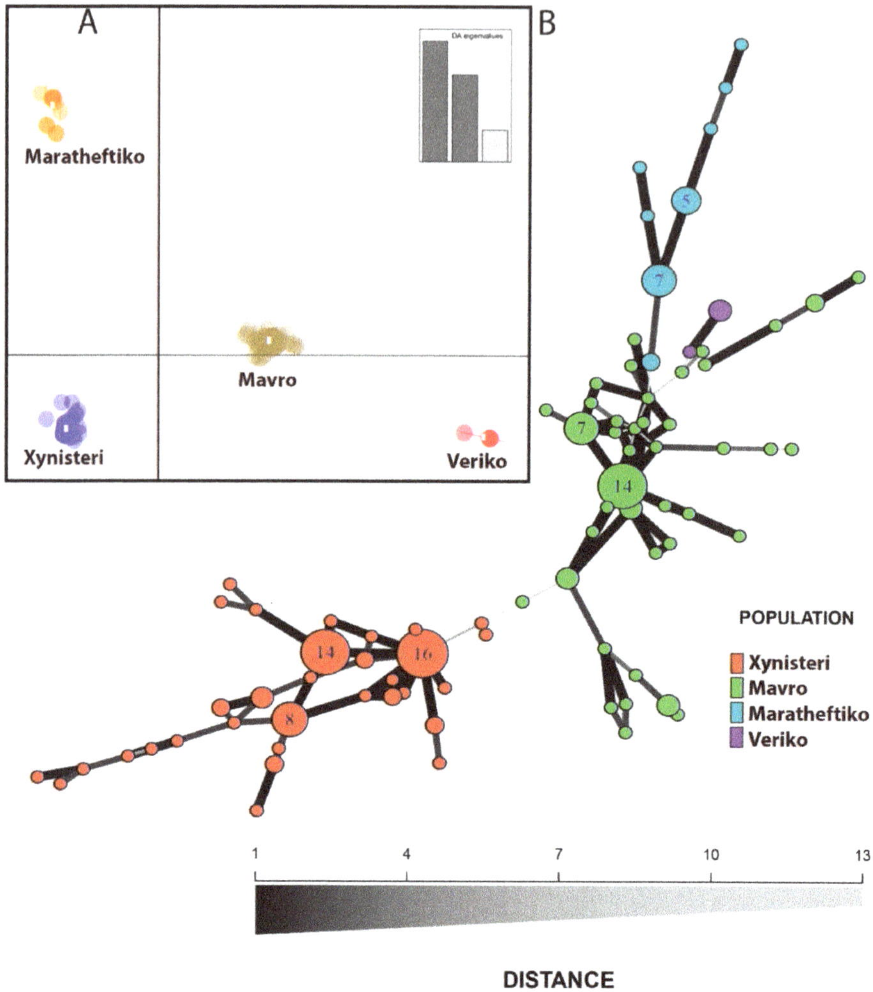

Figure 2. (**A**) Discriminant analysis of principal components (DAPC) depicting a clear cut-off across populations (DAPC cross-validation values indicated that the proportion of successful prediction (larger than 0.8) was optimum for 10 PCA axes). (**B**) Minimum spanning network (MSN) of the Cypriot grapevine varieties studied. Linear and reticulated affiliations are evident across Xynisteri (red), Mavro (green), Veriko (blue), and Maratheftiko (purple) clusters. Nodal size is proportional to the number of accessions sharing an MLG.

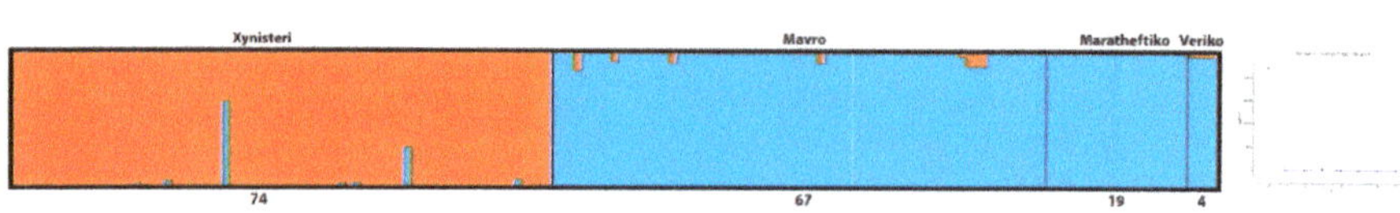

Figure 3. Structure analysis for Cypriot grapevine germplasm at K = 2 cluster and delta K probability for Cypriot genotypes. Numbers correspond to the analyzed samples.

Finally, a phylogenetic tree was constructed in order to also visualize the genetic relationships across genotypes using a hierarchical clustering approach with branch support (Figure S3). The dendrogram produced was analogous to the previous analyses, though with a notable distinction. The Maratheftiko cluster was not positioned as an outgroup as in Figure 2, but had a greater affinity to Mavro group, as it was placed among Mavro accessions. Moreover, Veriko accessions were highly affiliated to Mavro genotypes. Still, there were a few instances where several genotypes were sporadically clustered to different than expected groups, showing that misnaming is possible, or that these genotypes could in fact be different than registered varieties.

4. Discussion

The main scope of the current study was to evaluate the level of existing genetic diversity across Cypriot vineyards in the broader Commandaria area, and possibly identify novel genotypes that could provide evolutionary insights of the Cypriot grapevine germplasm structure. It was also intended to delineate the extent and imprint of genotype variability in one of the world's oldest niches of viticulture and pivotal landmark of grapevine domestication. Despite the fact that Cyprus is proximate to the grapevine center of domestication and a vital region of *Vitis* sp. westward dissemination, a very low number of native varieties has been reported so far [20,21]. Furthermore, it is speculated that the lack of robust ampelographic descriptors permits the unnoticed misuse of several "alike" forms as the same variety. As a result, there is the possibility that grapevine variability in the form of landraces, clones, or novel varieties remains largely unnoticed till today.

Towards that objective, two-year genetic resources collecting across the vineyards of Cyprus in the broader Commandaria zone was initiated and 164 centennial grapevines were obtained. The samples were genetically characterized using a universal set of microsatellite primers at eleven loci. Genotypic fingerprinting using microsatellite markers can deliver a valuable tool that allows for inter-accessions assessment of discrete collections, as well as, the pedigree analysis of hybrids and cultivar certification [36–38].

Among the 164 Cypriot accessions (registered and putatively assigned to the four clusters), a total of 83 multi-locus genotypes (MLGs) were identified (Table 1). Interestingly, 64 MLGs were unique and thus were detected only in one individual. As a consequence, 19 MLGs were found to be common (at different frequencies) among the remaining accessions. Hence, it was established that centennial grapevines of Cyprus are characterized by extensive genetic diversity and a small fraction of the sampled grapevine collection was composed of redundant germplasm. In a recent study, Drori and co-workers [13] collected and characterized grapevine genetic resources in Israel (a proximate to Cyprus region, having similar size and edaphoclimatic conditions). It was also concluded that a large proportion of the *V. vinifera* subsp. *sativa* and *V. vinifera* subsp. *sylvestris* genotypes had a unique genetic profile (about 40%) and did not correspond to known varieties/genotypes.

In viticulture, it is known that grapevine cultivars frequently consist of discrete clones, sharing mutual phenotypic traits and grouped as a variety cluster [39]. If, however, clones belonging to a cluster have traits discrete enough they are considered as different varieties. Nevertheless, several genetically affiliated varieties are very similar morphologically and hence hard to distinguish based on visual observation [40]. Conversely, clones of varieties can considerably differ in several characteristics without a significant change in genetic profiles [41]. Microsatellites can be proven a valuable tool delineating the above-mentioned dilemma, since such markers are extremely informative for attaining the level of heterozygosity across and within grapevine varieties. Across grapevine genotypes, a mean of nine alleles per locus [42] and a maximum of 23 alleles has been reported [41]. In the present study, 11 microsatellite loci were used, and a comparable level of allelic diversity was attained. An average of 9.27 alleles was obtained while in the case of the most polymorphic marker (VVMD28), 16 discrete alleles were recorded (Table 2). Moreover, several genotypes that were putatively assigned to the four clusters had extensive allelic discrepancies; hence, the possibility they are misidentified as the aforementioned cultivars and are in fact discrete varieties cannot be uncritically ruled out.

In genetic studies, when dealing with clonal taxa, analyses can be typically conducted using (or not) clonal correction. Since allele frequencies can be affected by the number of redundant genotypes, clone correction is often advised in order to better depict genetic relationships and kinships, as well as removing potential bias [31]. In the current survey, clone correction was performed in order to robustly estimate genetic indices (Tables 1 and 2). A substantial level of heterozygosity was revealed within the Cypriot varieties (*Hexp* = 0.732). The within cultivar heterozygosity is infrequently reported across studies, but in several cases, it seems that a moderate-to-high proportion of heterozygosity exists. Heterozygosity levels across grapevine clones generally vary from 0.47 for "Tannat" [43] to 0.87 for "Pinot Noir" and "Riesling" [44] cultivars, having a mean of 0.77 [40]. Despite the fact that data across studies are somewhat difficult to compare, since there are discrepancies in the number of loci and accessions employed, several common conclusions can still be attained [16]. In general, even though reports employ grapevine datasets of a few dozens to more than thousand accessions (focusing on eight to eleven loci), average to elevated heterozygosity values are reported [42,45–49]. Thus, it seems that diversity is present within cultivars, as well as across cultivars.

Since Cypriot grapevines have been cultivated for millennia, a high amount of diversity is expected among clones or varieties. Pinot is another characteristic cultivar where several clones are considered, and a great amount of heterogeneity exists. Unfortunately, the lack of precise ampelographic data on Cypriot germplasm complicates the cut-off line that defines the optimal taxonomic status. As a result, the existence of unidentified/misidentified varieties, hybrids, or feral forms considered as a "true-to-type" variety is a possibility. In the current study the predominant MLGs of Xynisteri cluster (MLG 76, MLG 81, and MLG 79) differed in one out of 11 loci, hence probably representing different clones of the same variety. This was also the case for the Mavro (MLG 38 and MLG 36) and Maratheftiko (MLG 7 and MLG 8) clusters, which were also identical in 10 out of 11 loci. Conversely, in the study of Hvarleva et al. [21], genotypes referring to discrete varieties differed at almost half of the analyzed loci in bilateral comparisons. In an equivalent study utilizing clones of Pinot, which is generally considered as one of the oldest varieties, significant discrepancies of SSR fragment lengths within clones of "Pinot gris" (loci VVS2, VMCNG1E1, VMC8A7, VMC7G3, VrZAG79, and VrZAG 25) and Pinot noir (loci VVS2, VVIM10, VMCNG1E1, VMC1F10, VMC2H4, VMV8A7, VMC7G3, VVMD28, and VrZAG 25) were established [50]. Still, in the case of Cypriot germplasm, a clear demarcation of what constitutes a variety, clone, or a landrace (population) is not an easy task, especially due to the lack of robust ampelographic data. This can be depicted in the DAPC (Figure 2A) that clearly demarks four clusters. Still, the within-group variation seems relatively small and a confident conclusion of the cut-off line separating clones from varieties cannot be established. In fact, one variety may consist of a smaller or larger number of similar and more or less related clones. Nevertheless, the findings presented here showcase beyond any doubt that a much larger than currently considered genetic variation is present across Cypriot vineyards and that several grapevines varieties are in fact misidentified by farmers.

The high amount of heterozygosity within Cypriot-cultivated grapevines could also be the result of a recurrent hybridization and parallel domestication events due to the long cultivation history and the proximity to the hotspot of grapevine domestication. In that direction, in our dataset several tri-allelic profiles across markers and genotypes were observed for markers VVMD32 and VVMD5, indicative of hybridization or chimerism. This phenomenon in clonally propagated grapevines such as monumental cultivars could participate in clonal variation and hamper proper variety identification or pedigree analysis [51]. Zarouri and co-workers [52] have also reported that amplification of multiple alleles per locus in one accession is possible. Feral forms and wild grapevines are still present in Cyprus, and sporadically, individual plants have been marked in remote regions near riverbeds.

Unfortunately, microsatellite profiles of Cypriot grapevine genotypes (using a universal primer set) are not available in public databases such as VIVC. Furthermore, partial comparison of the current dataset to the one previously described [21], revealed discrepancies in allele sizes or shifts. Such disparities are often reported in microsatellite markers studies, since inter-laboratory protocols can influence the outcome; the use of different DNA polymerase types, PCR elongation time, template concentration, or different fluorochromes can affect the addition of an adenine nucleotide at the 3′ end, resulting in allele size shifts due to the differences in molecular weight or even spectra pull-ups [53]. Hence, comparison across inter-laboratory studies must be cautiously regarded. Additionally, screening of the profiles attained in the current study to the VIVC database did not reveal similarities. The novelty of these centennial genotypes is further supported by the fact that foreign cultivars were only fairly recently (1970s) introduced into Cyprus [21].

Nonetheless, the genetic relationships of the current dataset largely correlates to previous studies that included Cypriot genotypes [21] even though a different primer set was used. Mavro and Maratheftiko populations seem to be genetically affiliated compared to Xynisteri that was the most genetically distant form, from the core cluster of Cypriot genotypes. Maratheftiko, or locally known as "Vamvakada" (due to a white coating at the back of its leaves resembling cotton; vamvaki is the Greek word for cotton) is an irregular occurrence of grape variety. Maratheftiko is one of the rare cases of grapevine cultivated forms that lacks hermaphrodite flowers. More specifically, its flowers present a well-developed pistil, but the stamens are reflexed; hence, Maratheftiko is incapable of self-pollination. This is the main problem hampering its cultivation in Cyprus despite being a drought tolerant variety and having exceptional vinification capabilities. Maratheftiko is considered among the most promising Cypriot varieties across local wineries since it produces red wines with an intense full body, having soft tannins and unique aromas when harvested and vinified correctly [54]. In the current survey it was clustered as an outgroup, having a limited genetic affiliation to the core of other Cypriot genotypes (Figure 2B). The Hardy–Weinberg equilibrium (HWE) indicates that at least in the case of the Maratheftiko cluster, widespread hybridizations have occurred throughout its lineage. In that direction, previous studies employing SNP markers have described a Levant domestication of the vinifera cluster and have displayed signs of introgression from local *sylvestris* forms as grapevines disseminated to Europe [12]. The extent to which local wild forms contributed to the establishment of the grapevine germplasm globally remains a disputable subject [9].

Still, the indices of association and significant *rbarD* values indicated that Cypriot genotypes were also clonally propagated since a considerable disequilibrium among loci was attained signifying conscious selection of elite genotypes during domestication. The clonal propagation of Cypriot genotypes can also be depicted from the network analysis since a limited reticulation can be observed among clusters (Figure 2), even though at slower rates compared to sexual propagation, clonal propagation can give rise to genetic variation [55]. Clonal polymorphism within perennial species has been mainly attributed to naturally occurring mutations throughout grapevine growth [39].

The "wild" characteristics of the Cypriot germplasm is also supported by vinification features of these genotypes. Xynisteri ('Xyno' is the Greek word for sour) is the predominant white variety and has a sharp taste due to the high level of acids [54]. Other traits indicating that Xynisteri retains some wild type features is the exceptional resilience against drought and elevated temperature [56], which are frequent features across wild *V. vinifera* spp. *sylvestris* forms [57]. In fact, even though domestication has resulted into higher yield, larger berries, and a higher sugar content, this selection was at the expense of biotic and abiotic stress capacity [9]. Recent studies have stressed the significance of wild forms for breeding purposes against mildew pathogens [58,59] or abiotic resilience and berry quality [60]. Past pedigree records and passport data (http://www.vivc.de/) reveal that Xynisteri has been broadly used in hybrid crosses in Israel during the 1930s, establishing a series of Hebron varieties used mainly as rootstocks. Nowadays, Xynisteri is largely tested outside the narrow range of Cyprus, at high temperate-arid regions of Australia, highlighting the ready-to-use potential of Cypriot germplasm for grapevine sustainability purposes [56].

In terms of conservation, several projects for the preservation and highlighting of local genetic resources have been directed in grapevine growing countries [16]. As a result, a significant number of minor varieties have been collected and characterized. Since several elite foreign varieties were only fairly recently (1970s) introduced into Cyprus [21], the Cypriot germplasm largely preserves its original structure throughout antiquity and is a remnant of the westward grapevine domestication and dissemination. In conclusion, the genetic characterization of such genetic resources is of paramount importance not only having a local interest but extendable to global viticulture. Furthermore, in the current study it was portrayed that the existence of a plethora of discrete genotypes is in fact an indication that more Cypriot varieties can exist and the need for proper ampelographic characterization is imperative.

Supplementary Materials: All relevant data are appended as supporting information. The following are available online at http://www.mdpi.com/2223-7747/9/8/1034/s1. Supplementary data: Morphological characters used for the discrimination of varieties/ Leaf diversity across accessions. Figure S1: Genotype accumulation curve depicting the efficiency of SSRs in delineating the Cypriot Vitis spp. Genotypes. Figure S2: Genetic variability estimates from clone corrected genotypes. Figure S3: Phylogenetic tree of Cypriot grapevine genotypes. Table S1: Sampling locations and germplasm information across collection sites within the Commandaria zone. Table S2: Primers and panels used for the amplification of alleles across 11 loci. Table S3: Alleles size and MLGs across the 164 Cypriot genotypes sampled and analyzed. Table S4: Frequency of alleles detected for every locus in the Cypriot grapevine collection. Table S5: Ancestry probability of STRUCTURE analysis across the 164 Cypriot genotypes.

Author Contributions: A.G. collected the plant material, performed the investigation process, and developed fingerprinting data. G.T. contributed in data acquisition, validation, and reviewed the manuscript. M.H. contributed in data acquisition, validation and analysis, contributed to the discussion of results, and reviewed the manuscript. N.N. conceptualized and supervised the study, curated data and analysis, wrote the final draft. All authors have read and agreed to the published version of the manuscript.

Funding: This research received no external funding.

Acknowledgments: The authors would like to thank the Vlassides, Tsiakkas, and Santa Irene wineries for their help with providing the germplasm studied and the anonymous reviewers for their constructive comments.

Conflicts of Interest: The authors declare no conflict of interest.

References

1. Anderson, K.; Wittwer, G.U.K. and Global Wine Markets by 2025, and Implications of Brexit. *J. Wine Econ.* **2017**, *12*, 221–251. [CrossRef]
2. Wen, J. Vitaceae. In *The Families and Genera of Vascular Plants*; Kubitzki, K., Ed.; Springer: Berlin, Germany, 2007; Volume 9, pp. 466–478.
3. Chen, I.; Manchester, S.R. Seed morphology of modern and fossil Ampelocissus (Vitaceae) and implications for phytogeography. *Am. J. Bot.* **2007**, *94*, 1534–1553. [CrossRef] [PubMed]
4. Zohary, D.; Hopf, M.; Weiss, E. *Domestication of Plants in the Old World: The Origin and Spread of domesticated Plants in Southwest Asia, Europe, and The Mediterranean Basin*; Oxford University Press on Demand: Oxford, UK, 2012; ISBN 0199549060.
5. Murphy, C. Grapes: Origins and Development BT. In *Encyclopedia of Global Archaeology*; Smith, C., Ed.; Springer: New York, NY, USA, 2014; pp. 3107–3111. ISBN 978-1-4419-0465-2.
6. Cunha, J.; Ibáñez, J.; Teixeira-Santos, M.; Brazão, J.; Fevereiro, P.; Martínez-Zapater, J.M.; Eiras-Dias, J.E. Genetic Relationships Among Portuguese Cultivated and Wild Vitis vinifera L. Germplasm. *Front. Plant Sci.* **2020**, *11*, 1–14. [CrossRef] [PubMed]
7. This, P.; Lacombe, T.; Thomas, M.R. Historical origins and genetic diversity of wine grapes. *Trends Genet.* **2006**, *22*, 511–519. [CrossRef]
8. McGovern, P.E. Ancient Wine: The Search for the Origins of Viniculture. *Econ. Bot.* **2004**, *58*, 488.
9. Riaz, S.; De Lorenzis, G.; Velasco, D.; Koehmstedt, A.; Maghradze, D.; Bobokashvili, Z.; Musayev, M.; Zdunic, G.; Laucou, V.; Andrew Walker, M.; et al. Genetic diversity analysis of cultivated and wild grapevine (*Vitis vinifera* L.) accessions around the Mediterranean basin and Central Asia. *BMC Plant Biol.* **2018**, *18*, 1–14. [CrossRef]
10. Arroyo-García, R.; Ruiz-García, L.; Bolling, L.; Ocete, R.; López, M.A.; Arnold, C.; Ergul, A.; Söylemezoğlu, G.; Uzun, H.I.; Cabello, F.; et al. Multiple origins of cultivated grapevine (*Vitis vinifera* L. ssp. sativa) based on chloroplast DNA polymorphisms. *Mol. Ecol.* **2006**, *15*, 3707–3714. [CrossRef]
11. Imazio, S.; Labra, M.; Grassi, F.; Scienza, A.; Failla, O. Chloroplast microsatellites to investigate the origin of grapevine. *Genet. Resour. Crop Evol.* **2006**, *53*, 1003–1011. [CrossRef]

12. Myles, S.; Boyko, A.R.; Owens, C.L.; Brown, P.J.; Grassi, F.; Aradhya, M.K.; Prins, B.; Reynolds, A.; Chia, J.M.; Ware, D.; et al. Genetic structure and domestication history of the grape. *Proc. Natl. Acad. Sci. USA* **2011**, *108*, 3530–3535. [CrossRef]
13. Drori, E.; Rahimi, O.; Marrano, A.; Henig, Y.; Brauner, H.; Salmon-Divon, M.; Netzer, Y.; Prazzoli, M.L.; Stanevsky, M.; Failla, O.; et al. Collection and characterization of grapevine genetic resources (*Vitis vinifera*) in the Holy Land, towards the renewal of ancient winemaking practices. *Sci. Rep.* **2017**, *7*, 44463. [CrossRef]
14. Riaz, S.; Pap, D.; Uretsky, J.; Laucou, V.; Boursiquot, J.M.; Kocsis, L.; Andrew Walker, M. Genetic diversity and parentage analysis of grape rootstocks. *Theor. Appl. Genet.* **2019**, *132*, 1847–1860. [CrossRef] [PubMed]
15. Olmo, H.P. Grapes. *Evol. Crop Plants* **1976**, *114*, 294–298.
16. Emanuelli, F.; Lorenzi, S.; Grzeskowiak, L.; Catalano, V.; Stefanini, M.; Troggio, M.; Myles, S.; Martinez-Zapater, J.M.; Zyprian, E.; Moreira, F.M.; et al. Genetic diversity and population structure assessed by SSR and SNP markers in a large germplasm collection of grape. *BMC Plant Biol.* **2013**, *13*, 1–17. [CrossRef] [PubMed]
17. Vrontis, D.; Thrassou, A. The renaissance of Commandaria: A strategic branding prescriptive analysis Demetris Vrontis * and Alkis Thrassou. *J. Glob. Bus. Adv.* **2011**, *4*, 302–316. [CrossRef]
18. Johnson, H.; Robinson, J. *The World Atlas of Wine.*; Mitchell Beazley: London, UK, 2001.
19. EFSA Panel on Plant Health (PLH). Scientific Opinion on the risk to plant health posed by Daktulosphaira vitifoliae (Fitch) in the EU territory, with the identification and evaluation of risk reduction options. *EFSA J.* **2014**, *12*, 3678. [CrossRef]
20. Galet, P. *Vines Wines Cyprus. 4000 Years Tradit*; Food and Agriculture Organization of the United Nations: Roma, Italy, 1993; pp. 61–73.
21. Hvarleva, T.; Hadjinicoli, A.; Atanassov, I.; Atanassov, A.; Ioannou, N. Genotyping Vitis vinifera L. cultivars of Cyprus by microsatellite analysis. *Vitis J. Grapevine Res.* **2005**, *44*, 93–97.
22. Ioannou-Papayianni, E.; Kokkinofta, R.I.; Theocharis, C.R. Authenticity of cypriot sweet wine commandaria using FT-IR and chemometrics. *J. Food Sci.* **2011**, *76*, 420–427. [CrossRef]
23. Constantinou, S.; Gómez-Caravaca, A.M.; Goulas, V.; Segura-Carretero, A.; Koundouras, S.; Manganaris, G.A. The impact of postharvest dehydration methods on qualitative attributes and chemical composition of 'Xynisteri' grape (*Vitis vinifera*) must. *Postharvest Biol. Technol.* **2018**, *135*, 114–122. [CrossRef]
24. Chrysargyris, A.; Xylia, P.; Antoniou, O.; Tzortzakis, N. Climate change due to heat and drought stress can alter the physiology of maratheftiko local cyprian grapevine variety. *J. Water Clim. Chang.* **2018**, *9*, 1–13. [CrossRef]
25. Anesi, A.; Stocchero, M.; Dal Santo, S.; Commisso, M.; Zenoni, S.; Ceoldo, S.; Tornielli, G.B.; Siebert, T.E.; Herderich, M.; Pezzotti, M.; et al. Towards a scientific interpretation of the terroir concept: Plasticity of the grape berry metabolome. *BMC Plant Biol.* **2015**, *15*, 1–17. [CrossRef]
26. Inglis, P.W.; Marilia de Castro, R.P.; Resende, L.V.; Grattapaglia, D. Fast and inexpensive protocols for consistent extraction of high quality DNA and RNA from challenging plant and fungal samples for high-throughput SNP genotyping and sequencing applications. *PLoS ONE* **2018**, *13*, e0206085. [CrossRef] [PubMed]
27. Töpfer, R.; Sudharma, K.N.; Kecke, S.; Marx, G.; Eibach, R.; Maghradze, D.; Maul, E. The Vitis International Variety Catalogue (VIVC): New design and more information. *Bull. L'OIV.* **2009**, *82*, 45–55.
28. Schuelke, M. An economic method for the fluorescent labeling of PCR fragments. *Nat. Biotechnol.* **2000**, *18*, 233–234. [CrossRef] [PubMed]
29. Matschiner, M.; Salzburger, W. TANDEM: Integrating automated allele binning into genetics and genomics workflows. *Bioinformatics* **2009**, *25*, 1982–1983. [CrossRef]
30. Peakall, R.O.D.; Smouse, P.E. GENALEX 6: Genetic analysis in Excel. Population genetic software for teaching and research. *Mol. Ecol. Notes* **2006**, *6*, 288–295. [CrossRef]
31. Kamvar, Z.N.; Brooks, J.C.; Grünwald, N.J. Novel R tools for analysis of genome-wide population genetic data with emphasis on clonality. *Front. Genet.* **2015**, *6*, 208. [CrossRef]
32. Clark, L.V.; Jasieniuk, M. Polysat: An R package for polyploid microsatellite analysis. *Mol. Ecol. Resour.* **2011**, *11*, 562–566. [CrossRef]
33. Trifinopoulos, J.; Nguyen, L.-T.; von Haeseler, A.; Minh, B.Q. W-IQ-TREE: A fast online phylogenetic tool for maximum likelihood analysis. *Nucleic Acids Res.* **2016**, *44*, 1–4. [CrossRef]

34. Letunic, I.; Bork, P. Interactive Tree Of Life (iTOL) v4: Recent updates and new developments. *Nucleic Acids Res.* **2019**, *47*, W256–W259. [CrossRef]
35. Pritchard, J.K.; Stephens, M.; Donnelly, P. Inference of population structure using multilocus genotype data. *Genetics* **2000**, *155*, 945–959.
36. Laucou, V.; Launay, A.; Bacilieri, R.; Lacombe, T.; Adam-Blondon, A.-F.; Bérard, A.; Chauveau, A.; De Andrés, M.T.; Hausmann, L.; Ibáñez, J.; et al. Extended diversity analysis of cultivated grapevine Vitis vinifera with 10K genome-wide SNPs. *PLoS ONE* **2018**, *13*, e0192540. [CrossRef] [PubMed]
37. Lacombe, T.; Boursiquot, J.-M.; Laucou, V.; Di Vecchi-Staraz, M.; Péros, J.-P.; This, P. Large-scale parentage analysis in an extended set of grapevine cultivars (*Vitis vinifera* L.). *Theor. Appl. Genet.* **2013**, *126*, 401–414. [CrossRef] [PubMed]
38. Laucou, V.; Lacombe, T.; Dechesne, F.; Siret, R.; Bruno, J.-P.; Dessup, M.; Dessup, T.; Ortigosa, P.; Parra, P.; Roux, C.; et al. High throughput analysis of grape genetic diversity as a tool for germplasm collection management. *Theor. Appl. Genet.* **2011**, *122*, 1233–1245. [CrossRef]
39. Pelsy, F. Molecular and cellular mechanisms of diversity within grapevine varieties. *Heredity* **2010**, *104*, 331–340. [CrossRef] [PubMed]
40. Aradhya, M.K.; Dangl, G.S.; Prins, B.H.; Boursiquot, J.-M.; Walker, M.A.; Meredith, C.P.; Simon, C.J. Genetic structure and differentiation in cultivated grape, *Vitis vinifera* L. *Genet. Res.* **2003**, *81*, 179–192. [CrossRef] [PubMed]
41. This, P.; Jung, A.; Boccacci, P.; Borrego, J.; Botta, R.; Costantini, L.; Crespan, M.; Dangl, G.S.; Eisenheld, C.; Ferreira-Monteiro, F.; et al. Development of a standard set of microsatellite reference alleles for identification of grape cultivars. *Theor. Appl. Genet.* **2004**, *109*, 1448–1458. [CrossRef]
42. Pelsy, F. Untranslated leader region polymorphism of Tvv1, a retrotransposon family, is a novel marker useful for analyzing genetic diversity and relatedness in the genus Vitis. *Theor. Appl. Genet.* **2007**, *116*, 15–27. [CrossRef]
43. González Techera, A.; Jubany, S.; Ponce De León, I.; Boido, E.; Dellacassa, E.; Carrau, F.M.; Hinrichsen, P.; Gaggero, C. Molecular diversity within clones of cv. Tannat (*Vitis vinifera*). *Vitis J. Grapevine Res.* **2004**, *43*, 179–185.
44. Pelsy, F.; Hocquigny, S.; Moncada, X.; Barbeau, G.; Forget, D.; Hinrichsen, P.; Merdinoglu, D. An extensive study of the genetic diversity within seven French wine grape variety collections. *Theor. Appl. Genet.* **2010**, *120*, 1219–1231. [CrossRef]
45. Ibáñez, J.; Vélez, M.D.; Teresa de Andrés, M.; Borrego, J. Molecular markers for establishing distinctness in vegetatively propagated crops: A case study in grapevine. *Theor. Appl. Genet.* **2009**, *119*, 1213–1222. [CrossRef]
46. Muzzalupo, I.; Vendramin, G.G.; Chiappetta, A. Genetic biodiversity of Italian olives (*Olea europaea*) germplasm analyzed by SSR markers. *Sci. World J.* **2014**, *2014*, 296590. [CrossRef] [PubMed]
47. Cipriani, G.; Spadotto, A.; Jurman, I.; Gaspero, G.D.; Crespan, M.; Meneghetti, S.; Frare, E.; Vignani, R.; Cresti, M.; Morgante, M.; et al. The SSR-based molecular profile of 1005 grapevine (*Vitis vinifera* L.) accessions uncovers new synonymy and parentages, and reveals a large admixture amongst varieties of different geographic origin. *Theor. Appl. Genet.* **2010**, *121*, 1569–1585. [CrossRef] [PubMed]
48. Grassi, F.; Labra, M.; Imazio, S.; Spada, A.; Sgorbati, S.; Scienza, A.; Sala, F. Evidence of a secondary grapevine domestication centre detected by SSR analysis. *Theor. Appl. Genet.* **2003**, *107*, 1315–1320. [CrossRef] [PubMed]
49. De Michele, R.; La Bella, F.; Gristina, A.S.; Fontana, I.; Pacifico, D.; Garfi, G.; Motisi, M.; Crucitti, D.; Abbate, L.; Carimi, F. Phylogenetic relationship among wild and cultivated Grapevine in Sicily: A hotspot in the Middle of the Mediterranean Basin. *Front. Plant Sci.* **2019**, *10*, 1506. [CrossRef] [PubMed]
50. Jahnke, G.; Májer, J.; Varga, P.; SzÖke, B. Analysis of clones of Pinots grown in Hungary by SSR markers. *Sci. Hortic.* **2011**, *129*, 32–37. [CrossRef]
51. Franks, T.; Botta, R.; Thomas, M.R.; Franks, J. Chimerism in grapevines: Implications for cultivar identity, ancestry and genetic improvement. *Theor. Appl. Genet.* **2002**, *104*, 192–199. [CrossRef]
52. Zarouri, B.; Vargas, A.M.; Gaforio, L.; Aller, M.; de Andrés, M.T.; Cabezas, J.A. Whole-genome genotyping of grape using a panel of microsatellite multiplex PCRs. *Tree Genet. Genomes* **2015**, *11*, 17. [CrossRef]
53. Sutton, J.T.; Robertson, B.C.; Jamieson, I.G. Dye shift: A neglected source of genotyping error in molecular ecology. *Mol. Ecol. Resour.* **2011**, *11*, 514–520. [CrossRef]

54. Copper, A.W.; Johnson, T.E.; Danner, L.; Bastian, S.E.P.; Collins, C. Preliminary sensory and chemical profiling of Cypriot wines made from indigenous grape varieties Xynisteri, Maratheftiko and Giannoudhi and acceptability to Australian consumers. *Oeno One* **2019**, *53*, 229–248. [CrossRef]
55. Anestiadou, K.; Nikoloudakis, N.; Hagidimitriou, M.; Katsiotis, A. Monumental olive trees of Cyprus contributed to the establishment of the contemporary olive germplasm. *PLoS ONE* **2017**, *12*, e0187697. [CrossRef]
56. Tzortzakis, N.; Chrysargyris, A.; Aziz, A. Adaptive response of a native mediterranean grapevine cultivar upon short-term exposure to drought and heat stress in the context of climate change. *Agronomy* **2020**, *10*, 249. [CrossRef]
57. Azri, W.; Cosette, P.; Guillou, C.; Rabhi, M.; Nasr, Z.; Mliki, A. Physiological and proteomic responses to drought stress in leaves of two wild grapevines (*Vitis sylvestris*): A comparative study. *Plant Growth Regul.* **2020**, *91*, 37–52. [CrossRef]
58. Duan, D.; Halter, D.; Baltenweck, R.; Tisch, C.; Tröster, V.; Kortekamp, A.; Hugueney, P.; Nick, P. Genetic diversity of stilbene metabolism in *Vitis sylvestris*. *J. Exp. Bot.* **2015**, *66*, 3243–3257. [CrossRef] [PubMed]
59. Toffolatti, S.L.; Maddalena, G.; Salomoni, D.; Maghradze, D.; Bianco, P.A.; Failla, O. Evidence of resistance to the downy mildew agent Plasmopara viticola in the Georgian Vitis vinifera germplasm. *Vitis J. Grapevine Res.* **2016**, *55*, 121–128. [CrossRef]
60. Revilla, E.; Carrasco, D.; Benito, A.; Arroyo-García, R. Anthocyanin composition of several wild grape accessions. *Am. J. Enol. Vitic.* **2010**, *61*, 536–543. [CrossRef]

plants

MDPI

Article

Genetic Structure of Wild Germplasm of Macadamia: Species Assignment, Diversity and Phylogeographic Relationships

Thuy Mai *, Mobashwer Alam, Craig Hardner, Robert Henry and Bruce Topp

Queensland Alliance for Agriculture and Food Innovation, The University of Queensland, Brisbane, QLD 4072, Australia; m.alam@uq.edu.au (M.A.); c.hardner@uq.edu.au (C.H.); robert.henry@uq.edu.au (R.H.); b.topp@uq.edu.au (B.T.)

* Correspondence: thi.mai3@uqconnect.edu.au

Received: 15 May 2020; Accepted: 30 May 2020; Published: 3 June 2020

Abstract: Macadamia is an Australian native rainforest tree that has been domesticated and traded internationally for its premium nuts. Common cultivars rely upon a limited gene pool that has exploited only two of the four species. Introducing a more diverse germplasm will broaden the genetic base for future crop improvement and better adaptation for changing environments. This study investigated the genetic structure of 302 accessions of wild germplasm using 2872 SNP and 8415 silicoDArT markers. Structure analysis and principal coordinate analysis (PCoA) assigned the 302 accessions into four distinct groups: (i) *Macadamia integrifolia*, (ii) *M. tetraphylla*, and (iii) *M. jansenii* and *M. ternifolia*, and (iv) admixtures or hybrids. Assignment of the four species matched well with previous characterisations, except for one *M. integrifolia* and four *M. tetraphylla* accessions. Using SNP markers, 94 previously unidentified accessions were assigned into the four distinct groups. Finally, 287 accessions were identified as pure examples of one of the four species and 15 as hybrids of *M. integrifolia* and *M. tetraphylla*. The admixed accessions showed the highest genetic diversity followed by *M. integrifolia*, while *M. ternifolia* and *M. jansenii* accessions were the least diverse. Mantel test analysis showed a significant correlation between genetic and geographic distance for *M. integrifolia* (r = 0.51, $p = 0.05$) and a positive but not significant correlation for *M. tetraphylla* (r = 0.45, $p = 0.06$). This study provides a population genetics overview of macadamia germplasm as a background for a conservation strategy and provides directions for future macadamia breeding.

Keywords: genetic diversity; DArT markers; macadamia; dendrogram; principal coordinate analysis; population structure; population genetics; wild species

1. Introduction

The genus *Macadamia* belongs to the Proteaceae family and is composed of four species: *Macadamia integrifolia* Maiden and Betche, *M. jansenii* C.L. Gross and P.H. Weston, *M. ternifolia* F. Muell and *M. tetraphylla* L.A.S. Johnson. The natural distribution of the four species is in the subtropical rainforest from south-east Queensland (QLD) to north-east New South Wales (NSW), Australia [1,2]. *M. integrifolia* and *M. ternifolia* occur in south-east QLD, while *M. tetraphylla* is distributed mainly in northern NSW, with some extension into south-east QLD [3]. Overlapping zones exist between *M. integrifolia* and *M. tetraphylla* and between *M. integrifolia* and *M. ternifolia*, with natural hybridisation occurring in these zones [4]. The fourth species, *M. jansenii*, has been found only in a single location in Bulburin (QLD) that is 180 km north from the nearest population of *M. integrifolia* [5]. Of these four species, *M. integrifolia* and *M. tetraphylla* produce edible nuts, and hence, most of the commercial cultivars belong to either of these two species or their hybrids [6]. The other two species, *M. jansenii* and *M. ternifolia*, have not been used in directed breeding, due to their bitter inedible nuts containing high levels of cyanogenic glycoside [7–9]. Most current commercial varieties appear to be two to four generations from the wild [1]. The majority of global macadamia production relies upon the cultivars from the Hawaiian breeding program, which is comprised mostly of a limited gene pool of *M. integrifolia* [10,11]. However, wild genetic resources have the potential to provide parents with desirable traits, including small tree size, nuts with thinner shells, resistance/ tolerance to biotic and abiotic factors, etc. [12] Exploring the genetic potential of the wild germplasm will facilitate further exploitation of these resources in trait improvement.

The study of genetic diversity and population structure can determine the potential of wild germplasm in future breeding. The genetic structure is formed over time due to the multiple actions of migration, selection, mutation, and genetic drift [13], as well as the mode and method of reproduction. A diverse species has the opportunity for selection of the fittest alleles while low diversity leads to the risk of extinction [14]. Broadening the genetic base of breeding material requires the identification of diverse parents for crossing with the cultivated crop [14]. Understanding the genetic relationship among the parents is essential to avoid inbreeding depression, particularly for the improvement of complex traits. Therefore, knowledge of the genetic divergence is a prerequisite to maximise heterosis in the breeding progeny [15]. The history of world macadamia breeding is very short, and the existing cultivars are only a few generations from the wild germplasm growing naturally in the rain forest. Recently, a chloroplast genome sequencing project on wild and cultivated germplasm indicated that all major Hawaiian cultivars share a single chlorotype probably derived from a small sample form single location [16]. As current macadamia plantations are mostly dependent on Hawaiian cultivars, this limited sampling suggests there is an opportunity for future genetic improvement by exploiting the diverse genetic resource. In addition, the genetic base can be expanded in breeding material by selecting diverse parents from wild germplasm for crossing with the cultivated crop.

Molecular markers are considered as the most suitable tool to estimate genetic diversity, due to their polymorphic nature and independence to environmental effects [17]. Several molecular marker systems have been developed to study in macadamia and only a few of them were used for the genetic characterisation of wild germplasm. An isozyme-based study was conducted by Aradhya et al. [18] to identify the genetic relationships among 40 cultivars (35 *M. integrifolia*, three *M. tetraphylla*, and two hybrids) and three *M. ternifolia* accessions. Mast et al. [19] studied the relationships among the four *Macadamia* species and their wild relatives, using three cpDNAs (*matK*, *atpB*, and *ndhF*), three nDNA (*waxy* loci *1* and *2*, and *PHYA*) genomic regions. However, this study used only one accession per species. The genetic structure of a large number of wild germplasm accessions was studied by Peace [4] using low throughput RAF (randomly amplified DNA fingerprinting, dominant) and RAMiFi (randomly amplified microsatellite fingerprinting, co-dominant) markers. All these marker systems have limitations, including a low total number of markers, low marker density, and low genome coverage, and hence, are seldom used in genomic studies. SSRs (simple sequence repeat markers or microsatellite) are considered as one of the best marker systems for genetic studies, with many advantages, such as stability, PCR-based amplification, and relatively low cost [17]. SSRs were used in the genetic diversity study of wild *M. integrifolia* [20] and wild *M. tetraphylla* populations [21,22]. However, there is a limited number of SSR primers available for macadamia, particularly those that successfully amplify across species [23], and as such, may not be effective for a large-scale genetic study of all four wild species.

The rapid advancement of next generation sequencing technology (NGS) enables the discovery of high-throughput and cost-effective molecular marker systems. Using NGS technology, Diversity Array Technology (DArT) developed a marker system that facilities affordable whole-genome level genetic characterisation. DArT has been successfully applied for the genetic diversity, population structure and genetic mapping studies of many crop species [24–26]. Recently, Alam et al. [27] used 11,526 silicoDArTs and 3956 SNPs to study the genetic diversity and population structure of 80 macadamia cultivars. O'Connor et al. [28] reported the genetic diversity, population structure and linkage disequilibrium of 295 seedling progenies from 29 selected parents, using 16,171 silicoDArTs and 4113 SNPs. These studies suggest that DArTseq markers could be applied for genomic studies in the wild germplasm of macadamia.

In this study, for the first time, we used high-throughput DArTseq platforms for the genetic characterisation of a large number of wild accessions of macadamia. The aims were to: (1) assess the population structure of wild macadamia germplasm, (2) explore the genetic diversity among the accessions within species, and (3) determine the relationship between genetic and geographic distance within *M. integrifolia* and *M. tetraphylla*.

2. Results

2.1. Quality of DArTseq Markers

DArTseq platforms generated 13,221 SNP and 47,811 silicoDArT markers. The call rates of SNP markers varied from 0.20 to 1.00, with an average of 0.62 (Supplementary Materials Table S1). Of the 13,221 SNPs, the call rate of 4184 markers (32%) was > 0.80 (Figure 1a). The reproducibility of SNPs varied from 0.86 to 1.00, and most of them (98%) were over 0.95 (Figure 1a). The call rate of silicoDArT markers varied from 0.81 to 1.00 (Table S1). The average call rate was very high (>0.99), with 94% of markers having a call rate over 0.95 (Figure 1a). The range of reproducibility was 0.95 to 1.00, in which 98% of the markers had very high value (0.99) of reproducibility. Mean one ratio was higher in SNPs (0.32) than in silicoDArT markers (0.08) (Table S1). Most silicoDArTs (82%) had one ratio below 0.05, compared with only 28% of SNPs (Figure 1b). Considering the quality parameters: call rate (>0.80), reproducibility (>0.95) and one ratio (>0.05), 2872 SNPs and 8415 silicoDArTs were retained for further analysis (Tables S2 and S3). The remaining markers had PIC values from 0 to 0.5 for both SNPs and silicoDArTs (Figure 1c). Mean PIC was 0.26 for silicoDArTs and 0.22 for SNPs. Only 120 of the SNPs (4%) had low PIC (<0.05), compared with 825 (9.8%) of silicoDArTs (Figure 1c).

2.2. Population Assignment

The K and Q values from STRUCTURE analysis were used for the assignment of individual accessions in each species/hybrid group. Population clusters in PCoA validated the species representation.

The ΔK from the STRUCTURE analysis of SNP markers was significant when K = 2, 3 and 5, with a peak at K = 3 (Figure 2a). The optimal peak at K = 3 suggested that the 302 accessions in the germplasm were derived from three distinct clusters, as represented by different colours in the structure analysis (Figure 2b, K = 3), here named Cluster I (blue), Cluster II (green) and Cluster III (red). Cluster I was composed of 18 accessions, including eight previously labelled as *M. ternifolia*, two as *M. jansenii*, and eight as undefined species (Table S4). These undefined eight accessions, originally labelled as mixed/hybrid populations, were collected from the natural distribution of *M. ternifolia*. Considering this distribution and their morphological appearance (Thuy Mai, pers. observations), these accessions were classified as *M. ternifolia*. Cluster II contained 99 predefined *M. integrifolia* and 36 accessions of previously undefined species. Cluster III was comprised of 94 predefined *M. tetraphylla* and 38 undefined species. There were 17 accessions, including one predefined *M. integrifolia*, four predefined *M. tetraphylla*, and 12 accessions from planted/unknown/mixed populations that showed the genetic admixture (hybrid) of clusters. For example, the accession "M034", which was previously labelled as *M. integrifolia*, consisted of 6% of Cluster II (predominantly *M. integrifolia*), 16% of Cluster III (predominantly *M. tetraphylla*), but 78% of Cluster I (predominantly *M. jansenii* and *M. ternifolia*). Two previously labelled *M. tetraphylla* accessions ("M265" and "M266") were identified as admixtures of Cluster II & III. The accession "M265" was composed of 61% of Cluster II and 39% of Cluster III, and accession "M266" was composed of 52% of Cluster III.

The pattern displayed for K = 2 (Figure 2b) grouped the accessions of three species, *M. jansenii*, *M. ternifolia* and *M. tetraphylla*, into one cluster, and separated *M. integrifolia* accessions in another cluster. The pattern displayed for K = 5 (Figure 2b) still did not separate the accessions of the two species *M. jansenii* and *M. ternifolia* but divided the accessions of *M. integrifolia* into smaller sub-clusters.

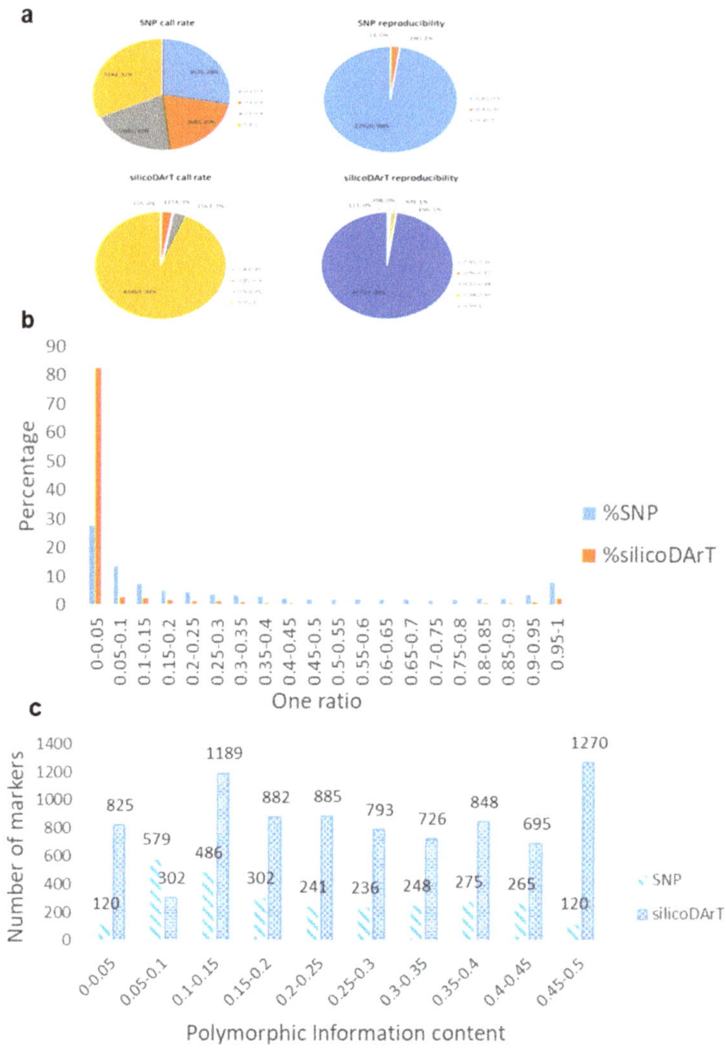

Figure 1. Distribution of SNP and silicoDArT marker data for several quality parameters. (**a**) call rate and reproducibility; (**b**) one ratio and (**c**) polymorphic information content (PIC) value.

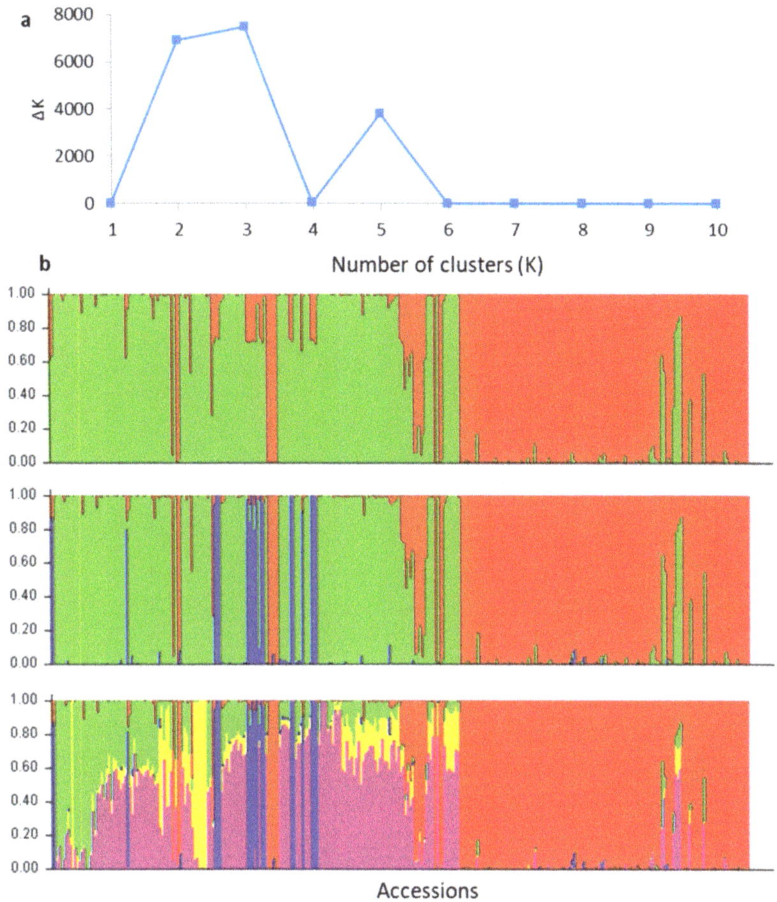

Figure 2. Population structure of 302 accessions based on 2872 SNPs, as inferred by STRUCTURE. (**a**) Best value of K based on Evanno's ΔK; (**b**) Individual membership proportions (Q value) in two, three and five clusters, with each cluster represented by a colour block. Each vertical line represents one accession. The accessions were sorted on the x-axis by latitude from north to south.

Principal coordinate analyses (PCoA) of SNP markers via distance matrix with data standardization identified three distinct groups of the four species (Figure 3). This result was consistent with the result of STRUCTURE analysis at K = 3. The first two coordinates of PCoA explained 61.09% of total variation in SNPs. The accessions of two species *M. ternifolia* and *M. jansenii* formed Cluster I. Cluster II was formed by the accessions of *M. integrifolia* and Cluster III included the accessions of *M. tetraphylla*. The accession "M034", which was assigned as admixture in STRUCTURE analysis, was clustered in the *M. jansenii*/*M. ternifolia* group and shows a close relationship with two *M. jansenii* accessions. The accession "M160", which was also assigned as admixture in STRUCTURE analysis composing 78% of Cluster III, was clustered in the *M. tetraphylla* group.

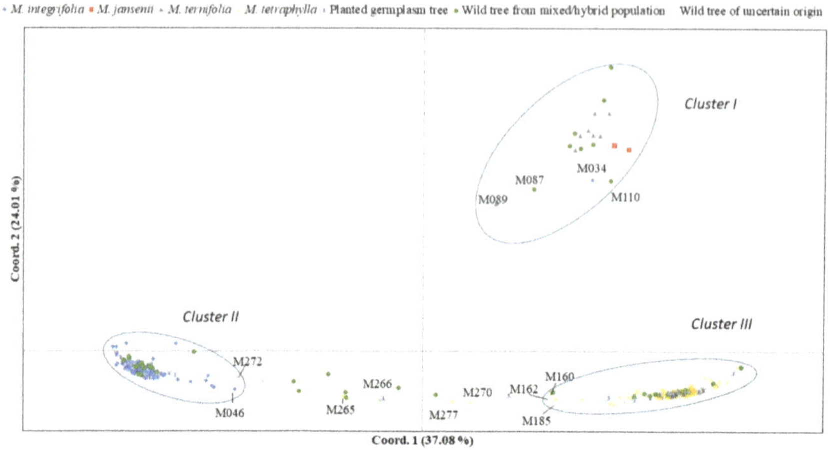

Figure 3. Principal coordinate analysis (PCoA) of the 302 accessions based on 2872 SNP markers, showing the three distinct groups and the admixtures. The first two coordinates of PCoA explained 34.06% of the total variation.

Finally, based on the STRUCTURE analysis and PCoA, we assigned 302 wild accessions into 287 pure accessions, representing the four distinct species, and 15 admixtures. Pure accessions are composed of 135 *M. integrifolia*, 133 *M. tetraphylla*, and 19 *M. ternifolia*/*M. jansenii* (Table S4).

2.3. Genetic Diversity

We estimated the genetic diversity parameters among 302 accessions, using both SNP and silicoDArT markers (Table 1). For SNP markers, the number of effective alleles (*Ne*) ranged from 1.08 to 1.34, with the lowest in *M. ternifolia*/*M. jansenii* group and the highest in admixture. Shannon's index (*I*) ranged from 0.11 (*M. ternifolia*/*M. jansenii*) to 0.33 (admixture), with a mean of 0.23. In all clusters, the observed heterozygosity (*Ho*) was smaller than the expected heterozygosity (*He*). *He* was highest in admixture (0.21) and lowest in *M. ternifolia*/ *M. jansenii* (0.07), with a mean of 0.15. Interestingly, accessions from the *M. integrifolia* group showed the highest percentage (86.53%) of polymorphism (%P), followed by admixture (74.93%), *M. tetraphylla* (71.5%), *M. ternifolia*/*M. jansenii* (32%) groups. Similar results were also observed for silicoDArT markers (Table 1).

Nei's genetic distance (D), based on SNP markers (Table 2), ranged from 0.06 between admixtures and *M. integrifolia* to 0.27 between *M. integrifolia* and *M. ternifolia*/*M. jansenii* groups. *M. tetraphylla* accessions shows lower genetic distance with *M. integrifolia* (D = 0.2) than that of *M. ternifolia*/*M. jansenii* accessions (D = 0.23). Hence, the admixture accessions showed almost similar genetic distance with *M. integrifolia* (0.06) and *M. tetraphylla* (0.07) germplasm. Although the estimated value of genetic distance using silicoDArT markers was lower than that of SNPs, the genetic relationship between species was similar in both marker systems (Table 2).

The analysis of molecular variance (AMOVA) showed a higher proportion of variance detected within species than among clusters (Table 3). For SNPs, the percentage of genetic variation within species (55%) was higher than that among species (45%). A similar pattern of genetic variation was observed using silicoDArT markers (Table 3).

Table 1. Genetic diversity parameters for Macadamia accessions based on SNP and silicoDArT markers. N = number of accessions, Na = number of different alleles, Ne = number of effective alleles, I = Shannon's information index, Ho= observed heterozygosity, He = expected heterozygosity, %P = percentage of polymorphic loci, SE = standard error.

Group		N	Na	Ne	Ho	He	I	%P
SNPs								
M. ternifolia/M. jansenii (Cluster I)	Mean	19	1.28	1.08	0.04	0.07	0.11	32.00
	SE		0.01	0.01	0.00	0.00	0.00	
M. integrifolia (Cluster II)	Mean	135	1.87	1.28	0.12	0.18	0.29	86.53
	SE		0.01	0.01	0.00	0.00	0.00	
M. tetraphylla (Cluster III)	Mean	133	1.72	1.20	0.08	0.12	0.20	72.32
	SE		0.01	0.01	0.00	0.00	0.00	
Admixture	Mean	15	1.75	1.34	0.18	0.21	0.33	74.93
	SE		0.01	0.01	0.00	0.00	0.00	
Total	Mean	302	1.66	1.22	0.11	0.15	0.23	47.20
	SE		0.01	0.00	0.00	0.00	0.00	19.30
silicoDArTs								
M. ternifolia/M. jansenii (Cluster I)	Mean	19	0.58	1.11	-	0.07	0.11	24.01
	SE		0.01	0.00	-	0.00	0.00	
M. integrifolia (Cluster II)	Mean	135	1.66	1.35	-	0.21	0.33	79.94
	SE		0.01	0.00	-	0.00	0.00	
M. tetraphylla (Cluster III)	Mean	133	1.47	1.26	-	0.16	0.24	70.86
	SE		0.01	0.00	-	0.00	0.00	
Admixture	Mean	15	1.53	1.36	-	0.22	0.34	70.18
	SE		0.01	0.00	-	0.00	0.00	
Total	Mean	302	1.31	1.27	-	0.16	0.26	61.25
	SE		0.01	0.00	-	0.00	0.00	12.61

Table 2. Pairwise population matrix of Nei's genetic distance (D) among the clusters of Macadamia wild germplasm, using SNP and silicoDArT markers.

Group	*M. integrifolia*	*M. tetraphylla*	*M. ternifolia/M. jansenii*
SNPs			
M. tetraphylla	0.20		
M. ternifolia/M. jansenii	0.27	0.23	
Admixture	0.06	0.07	0.22
silicoDArTs			
M. tetraphylla	0.17		
M. ternifolia/M. jansenii	0.17	0.16	
Admixture	0.08	0.05	0.16

Table 3. Summary statistic of AMOVA analysis in Macadamia germplasm using SNP and silicoDArT markers. df = degrees of freedom, SS = sum of squared observations, MS = mean of squared observations, Est. Var = estimated variance, % Var. = percentage of total variance. PhiPT = var. among groups (species)/ (var. among groups + var. within group).

Source	df	SS	MS	Est. Var.	% Var.	PhiPT Statistic	*p*-Value
SNPs							
Among species	3	175,814	58,604	951	45%		
Within species	298	344,429	1155	1155	55%		
Total	301	520,243		2107	100%	0.45	0.001
silicoDArTs							
Among species	3	74,996	24,999	401	34%		
Within species	298	235,037	789	789	66%		
Total	301	310,033		1190	100%	0.34	0.001

2.4. Phylogeographic Relationships among the Accessions of *M. integrifolia* and *M. tetraphylla*

We identified the genetic relationships within each species of *M. integrifolia* and *M. tetraphylla* presented in the dendrogram (Figure 4). Most accessions from the same locality grouped together, although some accessions from a locality clustered with accessions from other localities. For example, within *M. integrifolia*, the accession "M010" from Gundiah and four accessions from Nambour grouped with Beenleigh accessions (Figure 4a). Two accessions, "M027" and "M037", from Gympie grouped with accessions from Gundiah and Nambour, respectively. Within *M. tetraphylla*, the accession "M263" from Lismore grouped with Murwillumbah, while the accessions "M208" and "M209" from Murwillumbah grouped with Lismore (Figure 4b). Some of the accessions showed an unexpectedly variable branch length compared to other accessions from the same cluster. Accessions with the longest branches represent the most diverged accessions within the population. For example, accession "M148" had ~50% longer branch than other Beenleigh accessions. Similarly, accessions "M036", "M048", "M136" and "M008" of *M. integrifolia*, and "M227", and "M267" of *M. tetraphylla* had significantly longer branches.

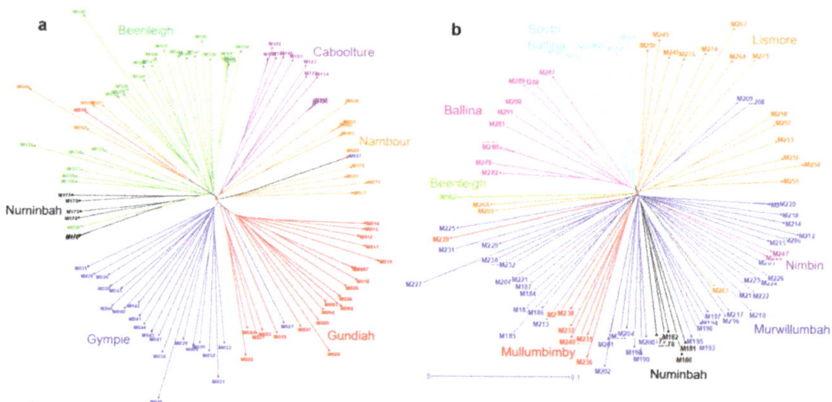

Figure 4. Unweighted neighbour-joining dendrograms using 2872 SNP markers, showing the genetic relationships among (**a**) 99 *M. integrifolia* accessions and (**b**) 94 *M. tetraphylla* accessions. Localities are represented by different colours.

To explore the genetic basis of the geographic relationship we calculated the correlation between the genetic distance and geographic distances of the localities for each species. The pairwise genetic distance among the accessions of six localities of *M. integrifolia* (Table S5) varied from 0.019 to 0.055, with an average of 0.034. The genetic distance between Gundiah and Gympie was the closest (0.019), while it was the farthest (0.055) between the accessions of Numinbah and Caboolture. The pairwise genetic distances among the accessions of *M. tetraphylla* localities (Table S6) were higher than that of *M. integrifolia*. The range of variation in *M. tetraphylla* was 0.013 to 0.151, with a mean of 0.063. Accessions from locality Murwillumbah and Lismore showed the closest genetic distance (0.013), while the highest (0.151) was observed between the accessions of Beenleigh and Nimbin.

The Mantel test analysis showed a significant correlation between genetic distance and geographic distance ($r = 0.51$, $p = 0.05$) among *M. integrifolia* localities (Figure 5a). Meanwhile, the correlation between genetic and geographic distance among *M. tetraphylla* localities was positive but not significant ($r = 0.45$, $p = 0.06$) (Figure 5b).

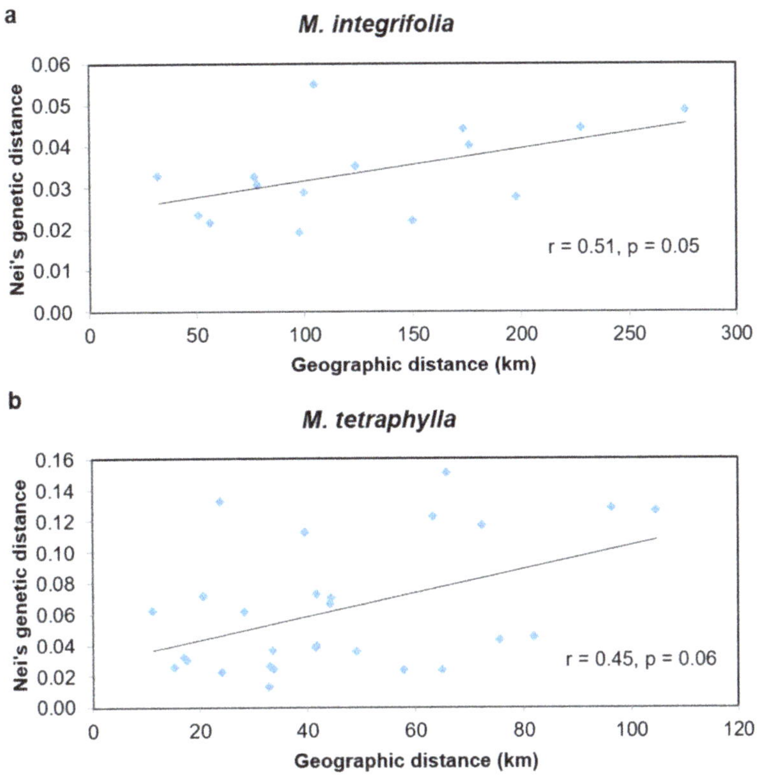

Figure 5. Correlation between genetic and geographical distance among (**a**) six localities of *M. integrifolia*, and (**b**) eight localities of *M. tetraphylla*, based on a Mantel test at 999 random permutations.

3. Discussion

3.1. Species Assignment of Wild Macadamia Germplasm

Population structure and PCoA of DArTseq based SNP markers facilitated the assignment of individuals in corresponding species or hybrid groups. Our species classification of each accession matches well with the field note and previous DNA study [4]. Results from this study confirmed that most of the previous phenotypic characterisations were successful for species' identification of wild germplasm.

All but one *M. integrifolia* accessions were clustered together in the same group. "M034", an accession from the "Mooloo" region of the "Gympie" locality, was identified as *M. ternifolia/M jansenii*, although it was previously recorded as *M. integrifolia*. A co-investigation of wild germplasm using 15 SSR markers identified that the same accession "M034" is a clone of another accession (X-CANB896104) from Canberra Botanic Garden. Interestingly, accession "X-CANB896104" is recorded as a cutting from wild *M. ternifolia* from Mary Cairncross Park, Maleny (Cathy Nock., pers communication). Further phenotypic characterisation confirmed that "M034" is a *M. ternifolia*. Among the 99 accessions of *M. tetraphylla*, four accessions were assigned as hybrid of *M. tetraphylla* and *M. integrifolia*. Accessions "M265" and "M266" shared almost 50% from each species, whereas "M270" and "M277" had ~30% from *M. integrifolia* genotype and ~70% *M. tetraphylla* (Figure 3, Table S3).

This study clearly demonstrated that the accessions of *M. tetraphylla* and *M. integrifolia* formed two distinct populations. The accessions of *M. jansenii* and *M. ternifolia* clustered together, although they were collected from geographically distant locations. The close relationship between *M. jansenii* and *M. ternifolia* is supported by the previous molecular studies [29,30]. These findings indicate that the accessions of both *M. jansenii* and *M. ternifolia* may have the same genetic lineage, or that one may be an ancestor of the other. Possibly, these two small groups of populations may have separated due to past climatic extremes and adapted as small groups in two distinct locations. It is to be noted that there may be some sampling effect of the small number of accessions of *M. jansenii* (n = 2) and *M. ternifolia* (n = 16) in our study. Though *M. jansenii* and *M. ternifolia* differ in a few simple traits including leaf tip, leaf serration and flower colour, they are morphologically similar, both with small tree size, and small and bitter nuts [9,31]. However, there may be further debate on the species differentiation of these two small populations. Investigation on evolutionary genetics and time divergence on four *Macadamia* species can be used to confirm the speciation of wild macadamia germplasm.

The species status of previously unidentified accessions, including those of unknown origins, planted germplasm and mixed/hybrid populations were resolved. Out of 94 unidentified accessions, 36 were assigned as *M. integrifolia*, 39 as *M. tetraphylla*, 8 as *M. ternifolia* and 11 as hybrids/admixtures of *M. integrifolia* and *M. tetraphylla* (Table S5). The genotypic classification was consistent with our field observation on phenotypic characteristics of representative species (Thuy Mai, pers communication) The species composition of many modern and heritage cultivars is uncertain [1], but this finding supports the potential of SNP markers to resolve their species status.

Some accessions we identified as *M. tetraphylla* had been collected from further north than the accepted distribution of this species. These were the accessions "M056", and "M057" from population 16 (Palmwoods, Nambour QLD), accession "M054" from population 106 (Mapleton Kenilworth, Nambour QLD) and five accessions from population 36 (Mount Glorious, Caboolture QLD). However, these populations were noted at the time of collection as planted or uncertain populations, and it seems highly likely that their locations were the result of human activity.

3.2. Genetic Diversity of the Four Macadamia Species

In this study, we developed new knowledge on genetic diversity in wild accessions by using high-throughput silicoDArT and SNPs marker. In both types of markers, the average expected heterozygosity (*He*) across the wild germplasm and clusters forming "pure" species was lower than in previous genetic studies of macadamia cultivars [21–23,32]. However, the cluster of hybrid accessions from wild *M. integrifolia* and *M. tetraphylla*, which is composed of a small number of accessions (n = 15), showed greater gene diversity than previous studies on cultivars and pure wild species in the current study. This result indicates that crossing between two species can be conducted, to increase genetic diversity in the future breeding program.

Our results suggested that *M. integrifolia* and *M. tetraphylla* contained two- to three-fold greater diversity than the *M. jansenii* and *M. ternifolia* cluster. Population size has a significant effect on genetic diversity [33]. Generally, smaller population size leads to lower genetic diversity [34]. Extinction and contraction of species' distribution during successive ice ages has resulted in reduced population size and resultant diversity bottlenecks in other Australian flora, such as *Acacia, Banksia, Eucalyptus* etc. [35]. Certainly, *M. jansenii* formed a very small population, with less than 100 individuals comprising the whole species [5,9]. In this study, only a small number of *M. jansenii* (n = 2) and *M. ternifolia* (n = 17) have been tested. The lower diversity of *M. jansenii* and *M. ternifolia* in our study may be the result of small populations. An investigation with larger sample numbers from diverse distributions of *M. jansenii* and *M. ternifolia* should be conducted, to define their genetic diversity more completely and accurately.

3.3. Phylogeographic Relationship among the Accessions of M. integrifolia and M. tetraphylla

Geographical distance is one of the major contributing factors in species differentiation. Knowledge of the genetic structure of a species over its geographic distribution is important to develop an understanding of the evolutionary processes [36]. In this study, the neighbour-joining tree, based on dissimilarity matrix (Figure 6), showed that the accessions from the same geographical area appeared to be grouped together. However, some accessions (e.g., "M010", "M027", "M037", "M263", "M208", "M209") were found to be clustered with accessions of different geographical regions (Figure 6). This result was supported by a previous study on chloroplast genome sequence, where Nock et al. [16] reported the relocation of some accessions within the *M. integrifolia* germplasm. Since the gene flow for both species is restricted within a short distance (~50 km) [1], the impact of environmental parameters, such as water, gravity, and animals like rodents [37], or a result of human activity on the seed transportation [16], could be considered as the cause of relocation.

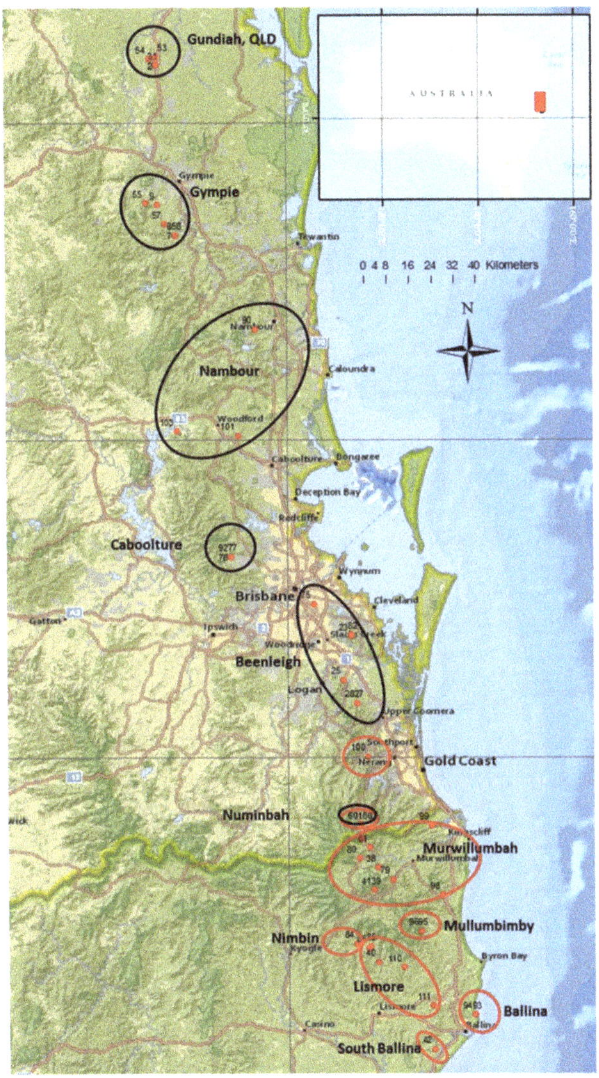

Figure 6. Map showing origins of representative wild-germplasm accessions of *M. integrifolia* and *M. tetraphylla*. The black circles represent six localities of *M. integrifolia* population, and the red circles represent eight localities of *M. tetraphylla* populations.

Phylogeographic analysis of *M. integrifolia* accessions revealed a significant positive correlation between genetic and geographic distance (r = 0.51, $p = 0.05$). This correlation was higher than a previous study using RAF markers [4] (r = 0.16, $p = 0.016$), and in almond germplasm (r = 0.173, $p = 0.226$) [38]. Based on chloroplast genome study, Nock et al. [16] also found a latitudinal population structure among these accessions. We also observed a positive relationship between genetic and geographic distance among the accessions of wild *M. tetraphylla*. Although, this correlation (r = 0.45) is non-significant, it is higher than that found by Peace [4] (r = 0.13). The non-significant phylogeographic relationship in *M. tetraphylla* suggested that geographic distance may not be the main factor influencing the genetic distance between populations of this species, although geographical boundaries, low gene flow and genetic drift are typically key factors explaining genetic differentiation in fragmented populations [39].

The phylogenetic trees (Figure 4) of *M. integrifolia* and *M. tetraphylla* provided an indication of ancestral lineage of the accessions of each species. Interestingly, the root of most of the accessions of both species originated from the population of Numinbah (Figure 4). It is to be noted that Numinbah is an overlapping region of both species, and its surrounding regions are the sources of mixed/hybrid population. We hypothesise that there is a possibility of early divergence of these two species at Numinbah. Later, smooth leaved *M. integrifolia* may have adapted to the north and serrated leaved *M. tetraphylla* may have adapted to the southern regions. Further genomic investigation with more accessions from Numinbah can explain the origin of these two species.

4. Materials and Methods

4.1. Collection of Plant Samples

The germplasm field trials were established in 2000 and 2001 in Tiaro, Queensland (QLD) and Alstonville, New South Wales (NSW), Australia. Wild accessions were collected from multiple geographical regions (Figure 6), covering the natural distribution of the four species. Ramets of each accession were propagated clonally as rooted cuttings. During the collecting trip, the germplasm collector (S. Faulkner) classified the populations as "wild", "planted" (trees were cultivated from the local wild trees), "hybrid/mixed", or "uncertain". The accession(s) from each population may belong to more than one of these classifications; therefore, the original description for each accession was noted. Accessions of the rare species *M. jansenii* were added to the field trial in July 2011. From these collections, we studied 302 accessions, including 100 *M. integrifolia*, two *M. jansenii*, eight *M. ternifolia*, 98 *M. tetraphylla*, and 94 accessions of undefined species (from mixed/hybrid populations, uncertain origins or planted germplasm). These accessions originated from 75 populations (one to seven accessions per population, averaging 4.1) across 52 regions (one to three populations per region) from 14 localities (one to nine regions per locality) (Table S7). The accessions were ordered by latitude from north to south and coded from "M001" to "M302".

4.2. DNA Extraction and Genotyping

Newly flushed young leaves were collected from the ex situ trials at Tiaro and Alstonville in December 2017 and placed in zipped plastic bags inside a cool box with ice blocks. The materials were stored in a cold room at 4 °C before transfer to DArT Pty Ltd. (Canberra, Australian Capital Territory, Australia), to perform DNA extraction. A total of 2–3 mg of leaf tissue from each sample was sub-sampled into a 1.1 mL microtube containing a disposable steel ball bearing. Leaf samples were crushed using a QIAGEN Tissue Lyser (Qiagen, Germany). A Freedom EVO robotic Tecan 100 (Tecan Group, Switzerland) was used for DNA extraction following an existing protocol recommended by DArT (https://www.diversityarrays.com/). DNA samples were incubated with loading dye at 37 °C for two hours and then checked for quality control on 0.8% agarose electrophoresis gel for 30 min at 100 V.

Accessions were genotyped for dominant silicoDArT and co-dominant SNP markers following an established protocol developed by Kilian et al. [40] An appropriate method (*Pst*I + *Hha*I) of complexity reduction was selected to detect DArTseq-based markers using next generation sequencing (NGS) technology. The silicoDArT and SNP markers were scored as a binary data format in which the score was "1" for presence, "0" for absence and "-", for failure to score of a marker in the genomic representation of each sample. The full details of methodology of developing DArTseq-based markers for macadamia were previously described by Alam et al. [27] and O'Connor et al. [28]

4.3. Analysis of Processed Marker Data

DArTseq platforms generated SNP and silicoDArT markers. DArTsoft v7.4 software was used to automatically identify and score the polymorphic markers, using a proprietary marker calling algorithms. The quality of the markers was tested for call rate (%), reproducibility (%), one ratio, and polymorphic information content (PIC). The call rate determines the success of reading the marker sequence across the samples, and was estimated from the percentage of samples for which the score was either "0" (absence of marker) or "1" (presence of marker). The scoring of reproducibility involves the proportion of 138 technical replicates for which the marker score exhibited consistency. One ratio was determined as the proportion of samples for which genotypes were scored as "1". PIC is the degree of diversity of the marker in the population and shows the usefulness of the marker for linkage analysis. Quality control and filtering were applied to both SNP and silicoDArT markers, including call rate (>80%), reproducibility (>95%), and one ratio (>0.05).

4.4. Analysis of Population Structure and Genetic Diversity

The population structure of the 302 accessions of four species was identified based on SNP markers by using the Bayesian model-based program STRUCTURE v2.3.4 [41–44]. A burn-in length of 10,000 cycles and Markov chain Monte Carlo (MCMC) of 50,000 runs were set for the structure analysis. Cluster values (K) ranging from one to 10 were performed, each with ten different iterations. Results from STRUCTURE were uploaded to Structure Harvester [45], through http://taylor0.biology.ucla.edu/structureHarvester/, to determine the optimum number of populations (best value of K) using ΔK value, as described by Evanno et al. [46] The individual ancestry proportion (Q value) at the best K value was determined for each accession. Based on Q values, the accessions were identified as pure or admixed species: accessions with Q value more than 0.8 were considered as "pure" and accessions with Q value less than 0.8 were assigned as "admixture" [14,47].

The genetic diversity and principal coordinate analysis (PCoA) were performed using GenAlEx v6.5.2 software [48]. Both SNP and silicoDArT markers were used to calculate the diversity parameters, including the mean number of alleles per locus (Na), number of effective alleles (Ne), observed and expected heterozygosity (Ho and He), Shannon's information index of diversity (I) and the percentage of polymorphic loci (%P). Pairwise population matrix of Nei's genetic distance [49] and the analysis of molecular variance (AMOVA) were also estimated.

4.5. Phylogeographic Relationships Analysis

M. integrifolia and *M. tetraphylla* are the two major species used in current breeding programs. In this study, we have identified a large number of accessions of *M. integrifolia* and *M. tetraphylla*. Therefore, a detailed study was undertaken on the genetic relationships and geographical diversity of those species. Only accessions identified as "pure" representatives of each species that were confirmed by structure analysis, including 99 *M. integrifolia* accessions from six localities (Figure 6, black circles), and the 94 *M. tetraphylla* accessions from eight localities (Figure 6, red circles), were used. The "pure" accessions collected from planted wild germplasm or unknown origin were not included. DARwin v6.0 software [50] was used to estimate pairwise Jaccard's genetic dissimilarity indices using 2872 SNP markers. A dendrogram was constructed by clustering accessions, based on a dissimilarity matrix using the unweighted neighbour-joining method. Clade strength in the dendrogram was tested using 100 bootstraps.

The correlation between genetic distance and geographic distance was estimated. The Nei's genetic distance matrix and the pairwise geographic distance among the localities were calculated using GenAlEx v6.5.2 [48]. The Mantel test was used to determine the correlation between genetic and geographic distance, using 999 random permutations.

5. Conclusions

This is the first study to investigate the genetic structure of a large collection of wild macadamia germplasm using thousands of high-throughput molecular markers. A total of 302 wild accessions were characterised, using 2872 SNP and 8415 silicoDArT markers. Our population structure and principal co-ordinate analyses identified three distinct populations, in which *M. jansenii* and *M. ternifolia* formed a single cluster. The Nei's genetic distance analysis clearly demonstrated that *M. jansenii* and *M. ternifolia* are related and showed greater heterozygosity in *M. ternifolia* than in *M. jansenii*. However, the limited number of accessions available in this study from *M. jansenii* and *M. ternifolia* limits the strength of our conclusion on the diversity and population structure of these species. We suggest that further analysis with more accessions from these two species should be conducted, to increase understanding of genetic diversity and clarify their classification as distinct species. We observed the significant correlation between genetic and geographic distance among *M. integrifolia* populations. Additionally, we were able to confirm the species identity of unknown wild accessions and suggested the use of these markers to resolve the unclear species composition of domesticated cultivars.

Supplementary Materials: The following are available online at http://www.mdpi.com/2223-7747/9/6/714/s1, Table S1. Summary of markers; Table S2. List of SNP markers; Table S3. List of silicoDArT markers; Table S4. Q value at K = 3, Table S5. Genetic distance of *M. integrifolia*; Table S6. Genetic distance of *M. tetraphylla*; Table S7. 302 accessions information.

Author Contributions: Conceptualization, B.T. and T.M.; Methodology, T.M. and M.A.; Software, T.M. and M.A.; Validation, T.M.; Formal Analysis, T.M.; Resources, T.M.; Data Curation, T.M. and M.A.; Writing—Original Draft Preparation, T.M.; Writing—Review and Editing, M.A., C.H., R.H., B.T. and T.M.; Visualization, T.M.; Supervision, B.T., M.A., R.H. and C.H.; Project Administration, B.T. and M.A.; Funding Acquisition, B.T. All authors have read and agreed to the published version of the manuscript.

Funding: This research has been funded by Hort Innovation, using the Macadamia research and development levy and contributions from the Australian Government. Hort Innovation is the grower owned, not-for-profit research and development corporation for Australian horticulture. PhD scholarship for TM was provided by Vietnam International Education Development (The Key Biotechnology Program—Ministry of Agriculture and Rural Development) and The University of Queensland.

Acknowledgments: The authors acknowledge Julia Playford for initiating the germplasm collection. Thanks to Agnelo Furtado (QAAFI), Jodi Neal (DAF), Katie O'Connor (QAAFI), Cathy Nock (Southern Cross University), Diversity Arrays Technology staff, Macadamia team and DAF staff, for their guidance and comments.

Conflicts of Interest: The authors declare no conflict of interest.

Data Availability Statement: The SNP and silicoDArT markers generated and analysed during this study are obtainable from The University of Queensland's Institutional Data Access/Ethics Committee, but restrictions apply to the availability of these data. The dataset "DArTseq markers of wild macadamia species" is available at https://doi.org/10.14264/uql.2018.395, for researchers who meet the criteria for access to confidential data. Contact data@library.uq.edu.au.

References

1. Hardner, C.M.; Peace, C.; Lowe, A.J.; Neal, J.; Pisanu, P.; Powell, M.; Schmidt, A.; Spain, C.; Williams, K. Genetic resources and domestication of macadamia. In *Horticultural Reviews*; Janick, J., Ed.; Wiley-Blackwell: Hoboken, NJ, USA, 2009; Volume 35, pp. 1–125.
2. Topp, B.L.; Nock, C.J.; Hardner, C.M.; Alam, M.; O'Connor, K.M. Macadamia (*Macadamia* spp.) Breeding. In *Advances in Plant Breeding Strategies: Nut and Beverage Crops*; Al-Khayri, J., Jain, S.D.J., Eds.; Springer International Publishing: Cham, Switzerland, 2019; pp. 221–251.
3. Powell, M.; Accad, A.; Shapcott, A. Where they are, why they are there, and where they are going: Using niche models to assess impacts of disturbance on the distribution of three endemic rare subtropical rainforest trees of *Macadamia* (Proteaceae) species. *Aust. J. Bot.* **2014**, *62*, 322–334. [CrossRef]
4. Peace, C.P. Genetic Characterisation of Macadamia with DNA Markers. Ph.D. Thesis, University of Queensland, Brisbane, Australia, 2004.
5. Shapcott, A.; Powell, M. Demographic structure, genetic diversity and habitat distribution of the endangered, Australian rainforest tree help facilitate an introduction program. *Aust. J. Bot.* **2011**, *59*, 215–225. [CrossRef]

6. Hardner, C.; Topp, B. Macadamia integrifolia and *M. tetraphylla*. In *Trees for Life in Oceania: Conservation and Utilisation of Genetic Diversity*; Lee, T., John, D., Clarke, B., Eds.; Australian Centre for International Agricultural Research: Canberra, Australia, 2018; pp. 147–151.
7. Dahler, J.M.; McConchie, C.; Turnbull, C.G.N. Quantification of cyanogenic glycosides in seedlings of three *Macadamia* (Proteaceae) species. *Aust. J. Bot.* **1995**, *43*, 619–628. [CrossRef]
8. Vivian-Smith, A. Cyanogenesis and the Fruit Growth of Three *Macadamia* Species. Bachelor's Thesis, University of Queensland, Brisbane, Australia, 1995.
9. Gross, C.L.; Weston, P.H. *Macadamia jansenii* (Proteaceae), a new species from central Queensland. *Aust. Syst. Bot.* **1992**, *5*, 725–728. [CrossRef]
10. Shigeura, G.T.; Ooka, H. Macadamia nuts in Hawaii, history and production. In *Research Series, Hawaii Institute of Tropical Agriculture and Human Resources*; University of Hawaii: Honolulu, HI, USA, 1984; pp. 6–22.
11. Hamilton, R.A.; Fukunaga, E.T. *Growing Macadamia Nuts in Hawaii*; Hawaii Agricultural Experiment Station: Honolulu, HI, USA, 1959; Volume 121, p. 59.
12. Topp, B.; Hardner, C.M.; Neal, J.; Kelly, A.; Russell, D.; McConchie, C.; O'hare, P. Overview of the Australian macadamia industry breeding program. *Acta Horticult.* **2016**, *1127*, 45–50. [CrossRef]
13. Yang, S.; Xue, S.; Kang, W.; Qian, Z.; Yi, Z. Genetic diversity and population structure of *Miscanthus lutarioriparius*, an endemic plant of China. *PLoS ONE* **2019**, *14*, e0211471. [CrossRef]
14. Singh, B.P.; Singh, B.; Mishra, S.; Kumar, V.; Singh, N.K. Genetic diversity and population structure in Indian wild rice accessions. *Aust. J. Crop Sci.* **2016**, *10*, 144–151.
15. Steiger, D.L.; Moore, P.H.; Zee, F.; Liu, Z.; Ming, R. Genetic relationships of macadamia cultivars and species revealed by AFLP markers. *Euphytica* **2003**, *132*, 269–277. [CrossRef]
16. Nock, C.J.; Hardner, C.M.; Montenegro, J.D.; Ahmad Termizi, A.A.; Hayashi, S.; Playford, J.; Edwards, D.; Batley, J. Wild origins of macadamia domestication identified through intraspecific chloroplast genome sequencing. *Front. Plant Sci.* **2019**, *10*, 1–15. [CrossRef]
17. Saeed, S.; Barozai, M.Y. A review on genetic diversity on wild plants by using different genetic markers. *Pure Appl. Biol.* **2012**, *1*, 68–71. [CrossRef]
18. Aradhya, M.; Yee, L.; Zee, F.; Manshardt, R. Genetic variability in *Macadamia*. *Genet. Resour. Crop Evol.* **1998**, *45*, 19–32. [CrossRef]
19. Mast, A.R.; Willis, C.L.; Jones, E.H.; Downs, K.M.; Weston, P.H. A smaller *Macadamia* from a more vagile tribe: Inference of phylogenetic relationships, divergence times, and diaspore evolution in *Macadamia* and relatives (tribe Macadamieae; Proteaceae). *Am. J. Bot.* **2008**, *95*, 843–870. [CrossRef] [PubMed]
20. Neal, J.M. *The Impact of Habitat Fragmentation on Wild Macadamia Integrifolia Maiden and Betche (Proteaceae) Population Viability*; University of New England: Armidale, Australia, 2007.
21. O'Connor, K.; Powell, M.; Nock, C.; Shapcott, A. Crop to wild gene flow and genetic diversity in a vulnerable *Macadamia* (Proteaceae) species in New South Wales, Australia. *Biol. Conserv.* **2015**, *191*, 504–511. [CrossRef]
22. Spain, C.S.; Lowe, A.J. Genetic consequences of subtropical rainforest fragmentation on *Macadamia tetraphylla* (Proteaceae). *Silvae Genet.* **2011**, *60*, 241–249. [CrossRef]
23. Nock, C.J.; Elphinstone, M.S.; Ablett, G.; Kawamata, A.; Hancock, W.; Hardner, C.M.; King, G.J. Whole genome shotgun sequences for microsatellite discovery and application in cultivated and wild *Macadamia* (Proteaceae). *Appl. Plant Sci.* **2014**, *2*, apps.1300089. [CrossRef] [PubMed]
24. Roorkiwal, M.; von Wettberg, E.J.; Upadhyaya, H.D.; Warschefsky, E.; Rathore, A.; Varshney, R.K. Exploring germplasm diversity to understand the domestication process in *Cicer* spp. using SNP and DArT markers. *PLoS ONE* **2014**, *9*, e102016. [CrossRef] [PubMed]
25. Baloch, F.S.; Alsaleh, A.; Shahid, M.Q.; Çiftçi, V.; de Miera, L.E.S.; Aasim, M.; Nadeem, M.A.; Aktaş, H.; Özkan, H.; Hatipoğlu, R. A whole genome DArTseq and SNP analysis for genetic diversity assessment in durum wheat from Central Fertile Crescent. *PLoS ONE* **2017**, *12*, e0167821. [CrossRef] [PubMed]
26. Grzebelus, D.; Iorizzo, M.; Senalik, D.; Ellison, S.; Cavagnaro, P.; Macko-Podgorni, A.; Heller-Uszynska, K.; Kilian, A.; Nothnagel, T.; Allender, C.; et al. Diversity, genetic mapping, and signatures of domestication in the carrot (*Daucus carota* L.) genome, as revealed by Diversity Arrays Technology (DArT) markers. *Mol. Breed.* **2014**, *33*, 625–637. [CrossRef]
27. Alam, M.; Neal, J.; O'Connor, K.; Kilian, A.; Topp, B. Ultra-high-throughput DArTseq-based silicoDArT and SNP markers for genomic studies in macadamia. *PLoS ONE* **2018**, *13*, e0203465. [CrossRef]

28. O'Connor, K.; Kilian, A.; Hayes, B.; Hardner, C.; Nock, C.; Baten, A.; Alam, M.; Topp, B. Population structure, genetic diversity and linkage disequilibrium in a macadamia breeding population using SNP and silicoDArT markers. *Tree Genet. Genomes* **2019**, *15*, 24. [CrossRef]
29. Sharp, D.; Playford, J. Native macadamia germplasm resources. In *Australian Macadamia Society Annual Conference Report*; Australian Macadamia Society: Lismore, Australia, 1997; p. 20.
30. Peace, C.; Vithanage, V.; Turnbull, C.; Carroll, B.J. Characterising macadamia germplasm with codominant radiolabelled DNA fingerprinting (RAF) markers. *Acta Horticult.* **2002**, *575*, 371–380. [CrossRef]
31. Storey, W.B. The ternifolia group of *Macadamia* species. *Pac. Sci.* **1965**, *19*, 507–514.
32. Peace, C.; Vithanage, V.; Neal, J.; Turnbull, C.; Carroll, B. A comparison of molecular markers for genetic analysis of macadamia. *J. Horticult. Sci. Biotechnol.* **2004**, *79*, 965–970. [CrossRef]
33. Sun, M. Effects of population size, mating system, and evolutionary origin on genetic diversity in *Spiranthes sinensis* and *S. hongkongensis*. *Conserv. Biol.* **1996**, *10*, 785–795. [CrossRef]
34. Shapcott, A.; Lamont, R.W.; O'Connor, K.M.; James, H.; Conroy, G.C. Population genetics of *Philotheca sporadica* (Rutaceae) to advise an offset translocation program. *Conserv. Genet.* **2015**, *16*, 687. [CrossRef]
35. Sniderman, J.K.; Jordan, G.J.; Cowling, R.M. Fossil evidence for a hyperdiverse sclerophyll flora under a non–Mediterranean-type climate. *Proc. Natl. Acad. Sci. USA* **2013**, *110*, 3423–3428. [CrossRef]
36. Brunet, J.; Larson-Rabin, Z.; Stewart, C.M. The distribution of genetic diversity within and among populations of the rocky mountain Columbine: The impact of gene flow, pollinators, and mating system. *Int. J. Plant Sci.* **2012**, *173*, 484–494. [CrossRef]
37. Pisanu, P.C. Survivorship of the Threatened Subtropical Rainforest Tree *Macadamia tetraphylla* L. Johnson (Proteaceae) in Small Habitat Fragments. Ph.D. Thesis, University of New England, Armidale, Australia, 2001.
38. Halász, J.; Kodad, O.; Galiba, G.; Skola, I.; Ercisli, S.; Ledbetter, C.; Hegedűs, A. Genetic variability is preserved among strongly differentiated and geographically diverse almond germplasm: An assessment by simple sequence repeat markers. *Tree Genet. Genomes* **2019**, *15*, 1–13. [CrossRef]
39. Nkatha, G.M.; Alice, M.; Anmin, Y.; Jianchu, X.; Aizhong, L. Genetic diversity analysis reveals genetic differentiation and strong population structure in *Calotropis* plants. *Sci. Rep.* **2018**, *8*, 1–10.
40. Kilian, A.; Wenzl, P.; Huttner, E.; Carling, J.; Xia, L.; Blois, H.; Caig, V.; Heller-Uszynska, K.; Jaccoud, D.; Hopper, C. Diversity arrays technology: A generic genome profiling technology on open platforms. In *Data Production and Analysis in Population Genomics: Methods and Protocols*; Humana Press: Totowa, NJ, USA, 2012; pp. 67–89.
41. Pritchard, J.; Stephens, M.; Donnelly, P. Inference of population structure using multilocus genotype data. *Genetics* **2000**, *155*, 945–959.
42. Falush, D.; Stephens, M.; Pritchard, J.K. Inference of population structure using multilocus genotype data: Linked loci and correlated allele frequencies. *Genetics* **2003**, *164*, 1567–1587.
43. Falush, D.; Stephens, M.; Pritchard, J.K. Inference of population structure using multilocus genotype data: Dominant markers and null alleles. *Mol. Ecol. Notes* **2007**, *7*, 574–578. [CrossRef]
44. Hubisz, M.J.; Falush, D.; Stephens, M.; Pritchard, J.K. Inferring weak population structure with the assistance of sample group information. *Mol. Ecol. Resour.* **2009**, *9*, 1322–1332. [CrossRef]
45. Earl, D.; von Holdt, B. STRUCTURE HARVESTER: A website and program for visualizing STRUCTURE output and implementing the Evanno method. *Conserv. Genet. Resour.* **2012**, *4*, 359–361. [CrossRef]
46. Evanno, G.; Regnaut, S.; Goudet, J. Detecting the number of clusters of individuals using the software structure: A simulation study. *Mol. Ecol.* **2005**, *14*, 2611–2620. [CrossRef]
47. Bernard, A.; Barreneche, T.; Lheureux, F.; Dirlewanger, E. Analysis of genetic diversity and structure in a worldwide walnut (*Juglans regia* L.) germplasm using SSR markers. *PLoS ONE* **2018**, *13*, e0208021. [CrossRef]
48. Peakall, R.; Smouse, P. GenAlEx 6.5: Genetic analysis in Excel. Population genetic software for teaching and research—An update. *Bioinformatics* **2012**, *28*, 2537–2539. [CrossRef]
49. Nei, M. Genetic distance between populations. *Am. Nat.* **1972**, *106*, 283–292. [CrossRef]
50. Perrier, X.; Jacquemoud-Collet, J.P. DARwin Software. 2006. Available online: http://darwin.cirad.fr/darwin (accessed on 3 June 2020).

plants

Article

Genetic Diversity and Population Structure of *Rhododendron rex* Subsp. *rex* Inferred from Microsatellite Markers and Chloroplast DNA Sequences

Xue Zhang [1,2,3,4,†], Yuan-Huan Liu [1,†], Yue-Hua Wang [1,2,3] and Shi-Kang Shen [1,2,3,*]

1. School of Life Sciences, Yunnan University, Kunming 650091, China; zhxue@nxu.edu.cn (X.Z.); lyh49662479@163.com (Y.-H.L.); wangyh58212@126.com (Y.-H.W.)
2. School of Ecology and Environmental Sciences & Yunnan Key Laboratory for Plateau Mountain Ecology and Restoration of Degraded Environments, Yunnan University, Kunming 650091, China
3. Yunnan Key Laboratory of Plant Reproductive Adaptation and Evolutionary Ecology, Yunnan University, Kunming 650091, China
4. Breeding Base for State Key Laboratory of Land Degradation and Ecological Restoration in Northwest China, Key Laboratory for Restoration and Reconstruction of Degraded Ecosystem in Northwest China of Ministry of Education, Ningxia University, Yinchuan 750021, China

* Correspondence: ssk168@ynu.edu.cn; Tel.: +86-871-65031412

† These authors contribute equals to this work.

Received: 7 February 2020; Accepted: 5 March 2020; Published: 7 March 2020

Abstract: Genetic diversity is vital to the sustainable utilization and conservation of plant species. *Rhododendron rex* subsp. *rex* Lévl. is an endangered species endemic to the southwest of China. Although the natural populations of this species are facing continuous decline due to the high frequency of anthropogenic disturbance, the genetic information of *R. rex* subsp. *rex* is not yet elucidated. In the present study, 10 pairs of microsatellite markers (nSSRs) and three pairs of chloroplast DNA (cpDNAs) were used in the elucidation of the genetic diversity, population structure, and demographic history of 11 *R. rex* subsp. *rex* populations. A total of 236 alleles and 12 haplotypes were found. A moderate genetic diversity within populations (H_E = 0.540 for nSSRs, Hd = 0.788 for cpDNA markers), high historical and low contemporary gene flows, and moderate genetic differentiation (nSSR: F_{ST} = 0.165***; cpDNA: F_{ST} = 0.841***) were detected among the *R. rex* subsp. *rex* populations. Genetic and geographic distances showed significant correlation ($p < 0.05$) determined by the Mantel test. The species exhibited a conspicuous phylogeographical structure among the populations. Using the Bayesian skyline plot and species distribution models, we found that *R. rex* subsp. *rex* underwent a population demography contraction approximately 50,000–100,000 years ago. However, the species did not experience a recent population expansion event. Thus, habitat loss and destruction, which result in a population decline and species inbreeding depression, should be considered in the management and conservation of *R. rex* subsp. *rex*.

Keywords: *Rhododendron*; conservation strategies; genetic differentiation; gene flow; populations contraction

1. Introduction

Rhododendron is the largest woody plant genus in Ericaceae, containing more than 1000 recognized species, of which 567 species representing six subgenera are known from China [1]. Wild *Rhododendron* species are the major components of alpine and subalpine vegetation and widely distributed in America, Europe, and Asia, which have tropical to polar climates [2,3]. Therefore, these species are potential genetic resources for the development of new cultivars that can adapt to diverse environmental conditions [4]. In addition, plants in the genus *Rhododendron* L. produce numerous chemical constituents and are recognized as an important source of bioactive phytochemicals [5]. Some *Rhododendron* species are used as traditional medicine in China, India, Europe, and North America against various diseases, such as inflammation, pain, skin ailments, common cold, and gastrointestinal disorders [5]. However, as an important natural resource for human daily life and ecosystem composition, most *Rhododendron* species are facing risk of extinction due to the high frequency of anthropogenic disturbance [6]. Thus, research on the population genetic information of *Rhododendron* species is undoubtedly beneficial for germplasm protection and sustainable utilization [6–9].

Inferring genetic information is recognized as the undisputed basis for the sustainable exploitation and conservation of plant diversity [10,11]. Different molecular markers are used in assessing genetic information and identifying distinct plant populations for management and conservation [12–14]. Microsatellite markers (SSRs) are used in revealing the genetic characteristics and related influence factors of plant species at individual and population levels due to their desirable advantages [13,15]. Chloroplast DNA (cpDNA), which is transmitted only through seeds in most angiosperms, is exceptionally conserved in gene content and organization, providing sufficient information for genome-wide evolutionary studies [16]. cpDNA can reveal a more highly geographical structure than a nuclear genome [17] and is generally used in the detection of phylogeographical patterns in plant species [18,19]. Thus, nSSRs and cpDNA were extensively and successfully documented to study the genetic diversity, variation, and population demographic of plant species [17,20–22].

Habitat loss and destruction are global problems that continue to threaten global biodiversity [23,24]. Firstly, habitat destruction and loss can cause a decline in the distribution range and population and limit the natural regeneration of a species. Secondly, habitat destruction and loss can increase selfing rates and decrease pollen diversity, thereby affecting a species's reproductive success [23,25]. Finally, habitat destruction and loss increase genetic drift and inbreeding and reduce gene flow in the fragmented populations of species and substantially decrease species genetic diversity and adaptation to the changing environment. Some studies suggested that woody plants are less likely to lose genetic diversity after habitat fragmentation and destruction than herbaceous species [26]; however, recent reports showed that habitat loss and fragmentation are associated with increased level of inbreeding, reduced gene flow, genetic variation, plant progeny quality, and genetic extinction debt in woody species [24,27]. Thus, understanding the current genetic information of endangered woody plants subjected to habitat loss and destruction is necessary for effective conservation and management.

Rhododendron species are not only popular woody ornamental plants but also play an important role in alpine and subalpine ecosystems. In addition, *R. rex* is an important wild germplasm source of the genus *Rhodendron* in China and an endangered plant endemic to the southwest of China [1]. Three subspecies (*R. rex* subsp. *rex*, *R. rex* subsp. *gratum*, and *R. rex* subsp. *fictolacteum*) are recognized in the *R. rex* complex. Recently, the wild populations of *R. rex* subsp. *rex* are facing continuous decline due to the high frequency of anthropogenic disturbance and forest destruction. Genetic information is important to the management and sustainable exploitation of species, particularly those threatened by habitat loss and destruction. However, the genetic diversity and structure of the wild populations of *R. rex* subsp. *rex* remain unexplored. In the present study, the genetic diversity and differentiation, population structure, and demographic history of 11 *R. rex* subsp. *rex* populations are inferred using 14 pairs of microsatellite markers and three cpDNA sequences. The following central questions are addressed: (1) What is the level of genetic diversity in *R. rex* subsp. *rex*? How does they apportion among/within the populations? (2) How is the genetic structure of the remnant population? Are they affected by historical and contemporary gene flows? (3) How is the phylogenetic relationship of haplotypes? Are they reflected by the demographic history in *R. rex* subsp. *rex*? This result is used to design optimum management strategies for *R. rex* subsp. *rex* conservation.

2. Materials and Methods

2.1. Plant Material Sampling

We collected 212 individuals of *R. rex* subsp. *rex* from 11 natural populations. Four of these populations (BJS, DLT, BCL, and JZS) with 63 individuals were distributed in Yunnan province, whereas seven populations (QLB1, QLB2, QLB3, GDX, LJS, LZS, and YS) with 149 individuals were distributed in Sichuan province, China (Table 1). Our sampling locations covered all the herbarium sampling sites and documented sites of *R. rex* subsp. *rex*. During field sampling, sampled site, sampled individuals, and altitude were recorded (Figure 1 and Table 1). Fresh leaves were collected from individuals of *R. rex* subsp. *rex* separated by a minimum distance of 15 m and then dried in silica gel immediately. The samplings were stored at −4 °C until DNA extraction.

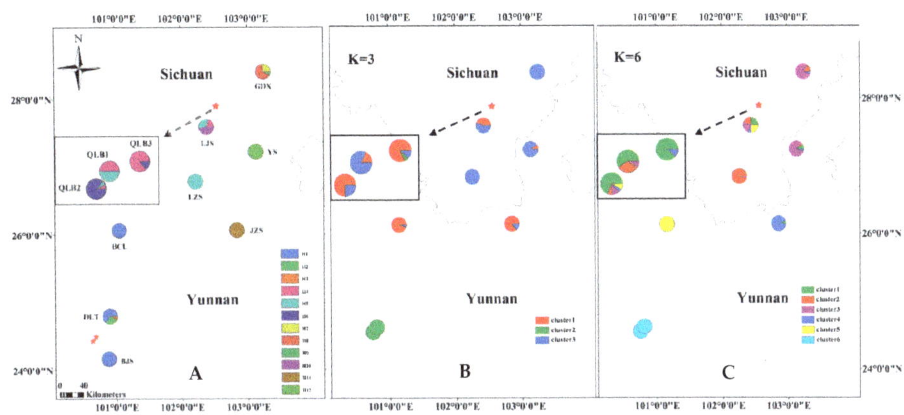

Figure 1. Distribution of chloroplast DNA (cpDNA) haplotypes (**A**); map of the geographic distribution of nuclear microsatellite clusters when the assumed cluster numbers are (**B**) $K = 3$ and (**C**) $K = 6$ in 11 populations of *Rhododendron rex subsp. rex*.

Table 1. Details of sample locations, sample size (*N*), haplotype diversity (*Hd*), and nucleotide diversity (Pi) surveyed for cpDNA sequences of *R. rex* subsp. *rex*. SSR—microsatellite marker.

Location	Population Code	Latitude	Longitude	Altitude (m)	*N* (cpDNA/SSR)	Haplotypes (No.)	cpDNA	
							Hd	*Pi*
Yunnan	BJS	24°24′31″	100°38′15″	2670	6/6	H1	0	0
	DLT	24°28′57″	100°41′47″	2660	14/15	H1, H2, H3	0.538	0.00031
	BCL	26°3′26″	101°03′11″	2950	15/21	H1	0	0
	YS	27°13′09″	103°07′43″	2887	16/23	H12	0	0
	JZS	26°04′07″	102°49′56″	3250	16/21	H11	0	0
Sichuan	QLB1	27°53′46″	102°30′56″	3250	14/22	H4, H5	0	0
	QLB2	27°53′19″	102°30′36″	3303	14/23	H4, H5, H6	0.264	0.00028
	QLB3	27°54′0.4″	102°31′44″	3332	14/17	H4, H6	0.264	0.00028
	GDX	28°24′29″	103°14′33″	2966	15/20	H7, H8, H9	0.514	0.00115
	LJS	27°35′19″	102°23′34″	2833	15/20	H4, H5, H10	0	0
	LZS	26°47′48″	102°12′30″	3335	16/24	H5	0	0
Total	11				155/212	H1–H12	0.788	0.0018

2.2. DNA Extraction, PCR Amplification, and Sequencing

We extracted genomic DNA of *R. rex* subsp. *rex* from the silica-dried leaves through a modified cetyltrimethyl ammonium bromide (CTAB) method [28]. Purified DNA was amplified by three universal cpDNA sequences (*rbc*L, *mat*K, and *psb*A-*trn*H). A total of 14 SSR markers were selected from recently developed nuclear microsatellites in *Rhododendron* subg. *Hymenanthes* according to their clarity and reproducibility (Table S1) [29–31]. PCR amplification was performed in accordance with methods of Zhang et al. [1]. Forward SSR primers were labeled with a fluorescent dye (FAM, TAMRA, or HEX) and visualized by an ABI 3730xl automated sequencer at Sangon Biotech Services Company Ltd. (Shanghai, China). Fragment sizes were read with the GeneMapper version 4.0. CERVUS [32] was used in eliminating four loci as existing null alleles ($F_{(\mathrm{Null})} > 0.4$) [33]. PCR products by three cpDNA intergenic spacers were sequenced in both directions by Sangon Biotech Services Company Ltd. (Shanghai, China).

2.3. Data Analysis

2.3.1. Data Analysis of Microsatellite Markers

The dataset was edited and formatted with GenAlEx ver. 6.3 [34]. We used Genepop ver. 4.1.4 to test the Hardy–Weinberg equilibrium (HWE) for each locus and population [35]. The universal genetic diversity parameters were calculated using GenAlEx ver. 6.3 [34] and POPGENE ver. 1.32 [35]. Then, rarefied allelic richness (*Ra*), total diversity (H_T), and the level of gene differentiation (G_{ST}) among *R. rex* subsp. *rex* populations were estimated by FSTAT ver. 2.9.3 [13,36]. Analysis of the molecular variance (AMOVA) was implemented in the estimation of genetic variation by using Arlequin ver. 3.11 [37,38], and F_{ST} values with 10^3 permutations were calculated for the assessment of genetic differentiation between the pairwise populations of *R. rex* subsp. *rex*.

The historical gene flow (*Nm*) between the pairs of *R. rex* subsp. *rex* populations was calculated using Wright's principles using formula $Nm = (1 - F_{ST})/4F_{ST}$ [39]. In addition, pollen to seed gene flow ratio (m_p/m_s) was calculated using the Ennos formula [40]. To estimate contemporary migration patterns, we estimated the contemporary inter-population migrations in *R. rex* subsp. *rex* using the BayesAss version 3.0 by 3×10^6 Markov chain Monte Carlo (MCMC) iterations, with a burn-in of 10^6 iterations and a sampling frequency of 2000 by setting delta at 0.15 (the default value) [41–43].

Isolation by distance was examined in GenAlEx ver. 6.3 on the basis of the correlation of a geographic distance for pairwise populations with $F_{ST}/(1 - F_{ST})$ value [34]. Population structure was accessed through unweighted pair group mean analysis (UPGMA) and principal coordinate analysis. TFPGA ver. 1.3 with 5000 permutations [44] and GenAlEx ver. 6.3 [34] were used, respectively. The Bayesian clustering analysis with an admixture model to understand the population structure of *R. rex* subsp. *rex* using STRUCTURE ver. 2.2 was also explored [22,45]. *K*-values in the model ranged from two to 15 with 20 independent variables for each set with a burn-in of 1×10^5 iterations and 1×10^5 subsequent Markov chain Monte Carlo steps [45]. The final best-fit number of the clusters was determined by ΔK values in STRUCTURE HARVESTER ver. 0.6.8 [46,47].

By performing a heterozygosity excess test, we explored the demographic history of the populations. We used two different models, namely, stepwise mutation and two-phased models, to construct the recent bottleneck statistic in BOTTLENECK ver. 1.2.02 (Sign and Wilcoxon tests) [48]. We further analyzed the genetic bottleneck with Garza–Williamson index (GWI) in Arlequin ver. 3.11 [38]. GWI lower than the critical *Mc* value of 0.68 indicated a reduction in population size [1,38,49].

2.3.2. Data Analysis of cpDNA Sequences

We used SeqMan II [50] and Bioedit ver. 7.0.4.1 [51] to treat the raw sequencing data and manually edited and assembled these sequences [22]. Three cpDNA intergenic spacers of *R. rex* subsp. *rex* were combined using PAUP 4.0 [52].

The haplotypes and variable sites of combined cpDNA sequences were calculated by DnaSP ver. 5.0 [53]. *Nei's* nucleotide diversity (*Pi*) and haplotype diversity (*Hd*) indices of *R. rex* subsp. *rex* were tested within a population and among populations. The haplotype distribution in each sampled population was plotted by ArcGIS ver. 10.2. In addition, we calculated H_T and within-population gene diversity (H_S) with Permut ver. 1.0 [22]. The two values of population differentiation G_{ST} and N_{ST}

were computed in accordance with the methods described by Pons and Petit [54] and with the same software. AMOVA of cpDNA sequences was constructed with Arlequin ver. 3.11 [37,38].

A genealogical haplotype network was constructed by Network ver. 4.2.0.1 for the estimation of the relationship per haplotype, and an indel was treated as a single mutational event [55]. The phylogenetic relationships of the as-obtained haplotypes of *R. rex* subsp. *rex* were inferred by Bayesian methods and neighbor-joining method in MrBayes ver. 3.1.2 [56]. *Nerium oleander* (EU916729.1, GQ997664.1 and AY899942.1) was selected as the outgroup species.

The evolutionary rates of seed plants (1.01×10^{-9}) were used for each Beast analysis in BEAST ver. 1.6.1 [57–59] with 10^7 iterations and a burn-in of 10^6 under the Hasegawa–Kishino–Yano (HKY) model and a strict clock. The most suitable model (HKY) was determined by Mega ver. 6.06 [60]. The results were visualized using the software FigTree ver. 1.4.2. The signatures of demographic changes in *R. rex* subsp. *rex* populations were assessed. We calculated pairwise mismatch distribution, neutrality tests (Tajima's D and Fu's F_S), the sum of squared deviations and the raggedness index, and their *p*-values using DnaSP ver. 5.0 [53] and Arlequin ver. 3.11 [38].

2.3.3. Analysis of Species Distribution Model

Species distribution models were constructed for the identification of the species' potential distribution during the last glacial maximum (LGM; ~21–18 ka) at present and in the future (model rcp45 for the years 2050, model rcp85 for the years 2070) by MAXENT v. 3.3.3k [61]. For each time period, models were run for 25 replicates, and default parameters were used. A total of 28 points comprised our 11 sampled sites and 17 records compiled in the Chinese Virtual Herbarium (www.cvh.org.cn), and 19 bioclimatic variables were obtained from the WorldClim database [62].

3. Results

3.1. SSR Data

We identified 169 alleles at 10 polymorphic loci among 11 *R. rex* subsp. *rex* populations, ranging from eight (R-40, R-49) to 30 (R-30), with an average of 16.9 alleles per locus (). All loci and populations conformed to HWE ($p > 0.05$; Table S3). At the locus level, genetic diversity and variation exhibited certain dissimilarities. However, no remarkable difference was detected between populations (Table 2). N_P varied from 2 (BJS) to 12 (DLT and JZS), *Ra* varied from 3.071 (BJS) to 4.231 (JZS), A_E varied from 2.011 (BJS) to 3.954 (YS), and *I* varied from 0.740 (BJS) to 1.319 (YS). The minimum values of H_O (0.300) and H_E (0.399) occurred in population BJS. The mean value of fixation indices (*Fis* = 0.171; Table 2) was positive for *R. rex* subsp. *rex* populations, suggesting a slightly increased level of inbreeding.

Table 2. Genetic diversity of populations in *R. rex* subsp. *rex*.

Population	N_P	*Ru*	N_A	A_E	*I*	H_O	H_E	*Fis*	*PPB* (%)
BCL	10	3.574	5.800	3.215	1.061	0.429	0.474	0.119	100.00%
BJS	2	3.071	3.100	2.011	0.740	0.300	0.399	0.331	90.00%
DLT	12	4.178	6.100	3.804	1.281	0.513	0.578	0.148	90.00%
GDX	3	3.681	5.800	3.479	1.183	0.452	0.547	0.200	100.00%
JZS	12	4.231	6.100	3.085	1.252	0.401	0.605	0.357	100.00%
LJS	8	3.841	6.700	3.169	1.228	0.478	0.561	0.167	100.00%
LZS	7	3.676	6.200	3.114	1.230	0.558	0.585	0.068	90.00%
QLB1	4	3.689	5.800	3.213	1.183	0.515	0.556	0.098	100.00%
QLB2	3	3.618	6.200	2.994	1.086	0.417	0.486	0.165	100.00%
QLB3	5	3.718	5.900	3.118	1.187	0.498	0.541	0.111	90.00%
YS	11	3.937	6.900	3.954	1.319	0.547	0.605	0.119	100.00%
Mean	7	3.747	5.873	3.196	1.159	0.464	0.540	0.171	96.36%

Note: N_A, mean number of alleles; A_E, number of effective alleles; *I*, Shannon's information index; H_O, observed heterozygosity; H_E, expected heterozygosity; N_P, number of private alleles; *Ra*: rarefied allelic richness; *Fis*, fixation index; *PPB* (%), percentage of polymorphic loci.

AMOVA indicated that 83.53% genetic variation occurred within populations, whereas 16.47% variation was estimated among the populations (Table 3). Genetic differentiation was observed among populations (F_{ST} = 0.165, 0.15 < F_{ST} < 0.25).

Table 3. Analysis of molecular variance (AMOVA) based on 14 microsatellites and three cpDNA sequences in *R. rex* subsp. *rex*. d.f.: degrees of freedom.

	Source of Variation	d.f.	Sum of Squares	Variance Components	Percentage of Variation (%)	
SSR data	Among populations	10	237.748	0.548	16.47	F_{ST} = 0.165 ***
	Within populations	413	1148.398	2.781	83.53	
	Total	423	1386.146	3.329		
cpDNA sequences	Among populations	10	276.023	1.940	84.07	F_{ST} = 0.841 ***
	Within populations	144	52.919	0.367	15.93	
	Total	154	328.942	2.314		

Note: *** $p < 0.001$, most significant difference.

A high level of historical gene flow of pairwise populations was detected in *R. rex* subsp. *rex* (Table 4). The minimum gene flow was generated from populations BJS and QLB2 (0.307), whereas the maximum gene flow was generated from populations QLB2 and QLB3 (7.452). The relative contribution of m_p/m_s was 24.775, indicating that pollen dispersal played an important role in the gene flow of *R. rex* subsp. *rex*. Except for other populations migrating to LJS, a non-significant level of inter-population contemporary migration rate between the populations of *R. rex* subsp. *rex* was detected (M < 0.05, Table 5).

Table 4. Historical gene flows between 11 populations of *R. rex* subsp. *rex*.

Population	BCL	BJS	DLT	GDX	JZS	LJS	LZS	QLB1	QLB2	QLB3	YS
BCL	0										
BJS	0.311	0									
DLT	0.469	2.024	0								
GDX	0.734	0.504	0.817	0							
JZS	1.257	0.463	0.699	1.439	0						
LJS	2.013	0.439	0.715	2.62	2.208	0					
LZS	1.582	0.481	0.781	1.721	1.454	2.86	0				
QLB1	1.895	0.421	0.674	1.636	2.146	10.591	2.97	0			
QLB2	2.363	0.307	0.462	1.059	1.649	3.782	1.882	4.021	0		
QLB3	2.029	0.365	0.572	1.093	1.582	2.785	2.099	3.461	7.452	0	
YS	1.100	0.517	0.833	2.642	1.525	3.207	1.621	1.862	1.479	1.617	0

Table 5. Contemporary migration rate between populations of *R. rex* subsp. *rex* by BayesAss with 95% confidence intervals.

Population->	BCL	BJS	DLT	GDX	JZS	LJS	LZS	QLB1	QLB2	QLB3	YS
BCL	0.695	0.029	0.029	0.028	0.027	0.055	0.028	0.028	0.027	0.028	0.028
BJS	0.029	0.696	0.028	0.028	0.028	0.051	0.028	0.028	0.029	0.027	0.029
DLT	0.029	0.027	0.695	0.028	0.028	0.053	0.027	0.027	0.029	0.028	0.029
GDX	0.028	0.027	0.027	0.697	0.029	0.052	0.027	0.028	0.028	0.028	0.028
JZS	0.027	0.028	0.028	0.026	0.695	0.057	0.027	0.028	0.028	0.029	0.027
LJS	0.030	0.028	0.027	0.029	0.028	0.719	0.027	0.028	0.029	0.027	0.029
LZS	0.028	0.028	0.029	0.028	0.028	0.057	0.694	0.027	0.026	0.028	0.027
QLB1	0.029	0.027	0.027	0.028	0.027	0.055	0.027	0.694	0.028	0.029	0.028
QLB2	0.029	0.028	0.026	0.029	0.027	0.055	0.028	0.029	0.695	0.027	0.027
QLB3	0.028	0.029	0.028	0.028	0.027	0.055	0.027	0.028	0.028	0.695	0.028
YS	0.028	0.028	0.026	0.028	0.027	0.056	0.028	0.026	0.029	0.028	0.695

Note: population->: population migration into the other populations.

The optimal K value was 3 with ΔK of 63.924, and the second fit value was 6 with ΔK of 16.473 according to STUCTURE analysis (Figure S1A,B). At K of 3, the populations GDX, YS, and JZS were similar, BJS and DLT were related, and the remaining populations BCL, LJS, LZS, QLB1, QLB2, and QLB3 comprised one group (Figure 1B; Figure S1A). At $K = 6$, the populations BCL and LZS were further distinguished from LJS, QLB1, QLB2, and QLB3, and JZS was further distinguished from GDX and YS (Figure 1C; Figure S1A). This result was in accordance with the conclusions of UPGMA (Figure S1C) and principal component analysis (PCA) at the population level (Figure 2A). In conclusion, the populations of *R. rex* subsp. *rex* should be grouped into three groups according to the genetic structure analysis results by using SSR data. A significant correlation between genetic and geographic distances was determined by Mantel test ($p < 0.050$; Figure 2A).

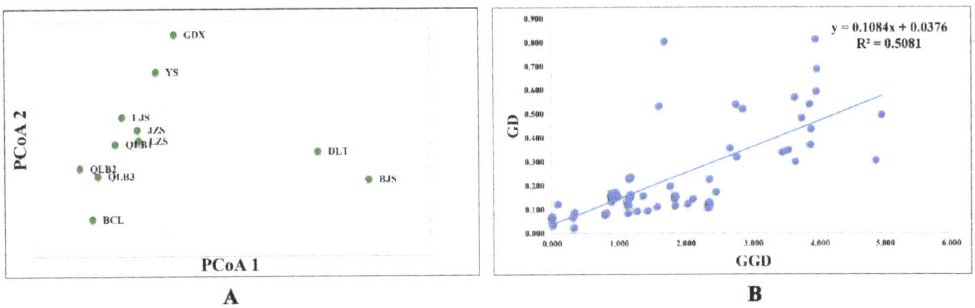

Figure 2. Principal coordinate analysis (**A**) and the plot of geographical distance against genetic distance (**B**) for *R. rex* subsp. *rex* by SSR data analysis.

As shown in Table 6, most of the probabilities of the Wilcoxon and Sign tests under both models in *R. rex* subsp. *rex* populations were non-significant ($p > 0.05$). In addition, the allele distribution per population was presented as a normal L-shaped distribution. The above-mentioned results indicated that the *R. rex* subsp. *rex* populations conformed to the mutation–drift equilibrium. However, GWI values were lower than the critical *Mc* indices (0.68), implying that the populations of *R. rex* subsp. *rex* underwent a demographic reduction in history.

Table 6. Bottleneck analysis of 11 populations in *R. rex* subsp. *rex*.

Population	Two Phased Model (T.P.M)		Step Mutation Model (S.M.M)		Mode Shift	Garza–Williamson Index
	Sign Test	Wilcoxon Test	Sign Test	Wilcoxon Test		
BCL	0.614	0.539	0.170	0.813	L	0.336
BJS	0.211	0.410	0.068	0.545	L	0.399
DLT	0.399	0.652	0.183	0.839	L	0.329
GDX	0.158	0.862	0.002 **	0.998	L	0.492
JZS	0.176	0.813	0.183	0.958	L	0.361
LJS	0.178	0.862	0.169	0.958	L	0.278
LZS	0.074	0.947	0.003 **	0.995	L	0.297
QLB1	0.370	0.423	0.181	0.947	L	0.284
QLB2	0.371	0.461	0.058	0.984	L	0.323
QLB3	0.065	0.862	0.074	0.984	L	0.333
YS	0.612	0.461	0.389	0.722	L	0.349

Note: ** $p < 0.01$, significant difference.

3.2. *cpDNA Sequence*

The three cpDNAs, *mat*K, *psb*A-*trn*H, and *rbc*L were 828, 398, and 658 bp in length, respectively (GenBank accession numbers: MN228019-MN228483). The 1884-bp combined cpDNA sequences of *R. rex* subsp. *rex* had 18 polymorphic sites and 12 haplotypes (H1–H12) (Table 1). The detailed haplotype distribution per population is displayed in Figure 1.

The populations DLT (0.538 and 0.00031) and GDX (0.514 and 0.00115) exhibited additional maximum values of *Hd* and *Pi* per site, followed by QLB2 (*Hd* = 0.264, *Pi* = 0.00028) and QLB3 (*Hd* = 0.264, *Pi* = 0.00028), whereas no diversity was found in the seven remaining populations (Table 1). In summary, the total *Hd* and *Pi* for *R. rex* subsp. *rex* were 0.78768 and 0.00180, respectively. The value of H_T (0.909) was higher than that of H_S (0.337), and the value of N_{ST} (0. 0.774) was significantly higher than that of G_{ST} (0.629; $p < 0.05$; Table S4). These results indicated the remarkable phylogeographic structure among the populations of *R. rex* subsp. *rex*. AMOVA indicated that 84.07% genetic variation was partitioned among populations, whereas 15.93% was partitioned within populations (Table 3). This result was inconsistent with the result of nSSRs data. Moreover, significant genetic differentiation was observed among *R. rex* subsp. *rex* populations ($F_{ST} = 0.841, p < 0.001$).

The phylogenetic relationships of 12 cpDNA haplotypes are shown in Figure 3A. H7, H8, H9, and H12 were grouped into one clade. Among the remaining haplotypes, H1, H2, and H3 were grouped into one clade, while H4, H5, H10, and H11 were grouped into another clade. H6 was differentiated from the remaining others. However, the result of the haplotype network diagram shown that H6 distributed more closely to H5, and the 12 cpDNA haplotypes should be divided into three groups (Figure 3B).

A

B

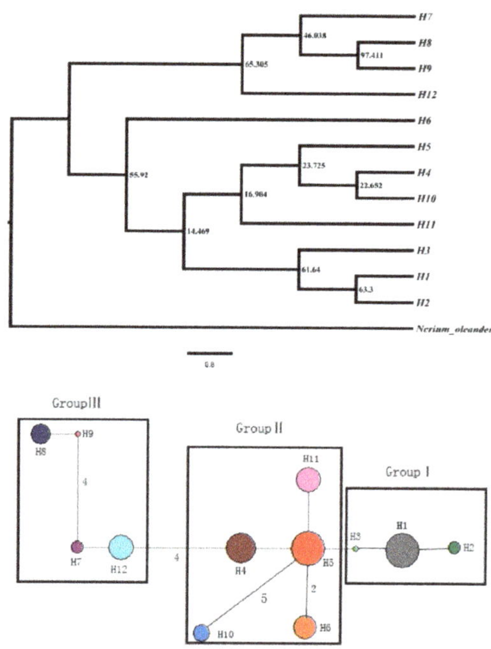

Figure 3. Bayesian tree (**A**) and the network of haplotypes (**B**) based on combined cpDNA sequences. (**A**) The numbers on branches indicate the posterior probability; (**B**) the size of the circles corresponds to the frequency of each haplotype, and the vertical branches indicate mutational steps.

Only the Fu and Li'*D* yielded significantly positive values ($p < 0.05$; Table S5) according to the neutrality test. This result indicated that no recent population expansion in *R. rex* subsp. *rex* occurred, and this was supported by the effects of mismatch distributions shown in the multimodal graph (Figure 4A). Based on the Bayesian analysis, the skyline plot indicated that the historical demographic of *R. rex* subsp. *rex* populations experienced a contraction event approximately 50,000–100,000 years ago and had no recent expansion (Figure 4B).

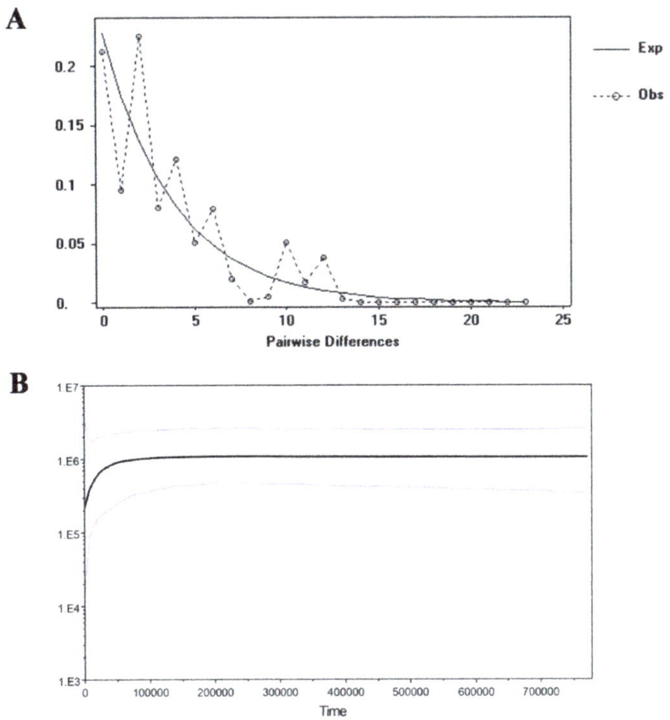

Figure 4. Mismatch distribution (**A**) and Bayesian skyline plot based on combined cpDNA sequences (**B**). (**A**) The solid lines show expected values, whereas the dashed lines represent observed values under a model of sudden population expansion. (**B**) The black line indicates effective population size fluctuation throughout.

3.3. Species Distribution Model

According to predictions of *R. rex* subsp. *rex*'s past, present, and future potential distributions, the predicted current distributions showed a clear range contraction relative to the LGM distributions. Moreover, the moderate habitat suitability (>0.31) was slightly removed to the northeastern direction (Figure 5A,B). The potential distribution with moderate to high habitat suitability (>0.31) for the years 2050 and 2070 was slightly extended compared with the present-day model (Figure 5C,D).

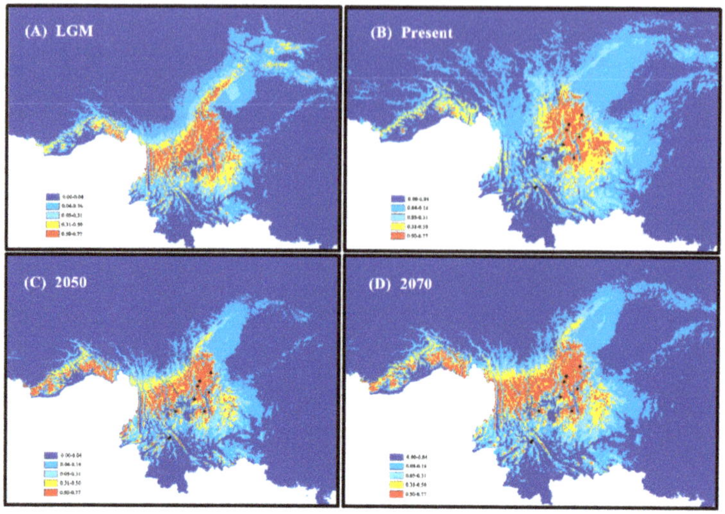

Figure 5. Distribution dynamics of *R. rex* subsp. *rex* using MAXENT. Predicted distributions are shown for (**A**) the last glacial maximum (LGM), (**B**) the present, (**C**) the year 2050, and (**D**) the year 2070. Color-coded keys represent different habitat suitability.

4. Discussion

4.1. Genetic Diversity in R. rex Subsp. rex Populations

The genetic diversity of species reflects its long-term evolution and adaptation demographic history [63]. Based on nSSR data, *R. rex* subsp. *rex* has lower genetic diversity (H_E = 0.540) than the other species of *Rhododendron*, such as *R. protistum* var. *giganteum* (H_E = 0.602) [9], *R. jinggangshanicum* (H_E = 0.642) [64], *R. simsii* (H_E = 0.754) [65], *R. ripense* (H_E = 0.800) [66], and *R. brachycarpum* (H_E = 0.815) [67], but has higher genetic diversity than *R. ferrugineum* (H_E = 0.500) [68]. Meanwhile, the genetic diversity of *R. rex* subsp. *rex* is evidently higher than that of the "narrow" species (H_E = 0.420) and lower than that of the "widespread" species (H_E = 0.620) [69]. Inconsistent with the results of microsatellite markers, the genetic diversity of *R. rex* subsp. *rex* (Pi = 0.00180) assessed by cpDNA shows a higher tendency toward high genetic diversity than the genetic diversities of 20 species of *Rhododendron* sect. *Brachycalyx* (insular species, Pi = 0.00040; continental species, Pi = 0.00160) in East Asia [70], but a lower tendency than the genetic diversity of bird-dispersed arctic–alpine plant *Vaccinium vitis-idaea* in Ericaceae (Pi = 0.00240) [71]. The value of H_T estimated in *R. rex* subsp. *rex* (0.909) is higher than the mean value of H_T (0.6747) in 170 plant species according to cpDNA [19,72]. Therefore, *R. rex* subsp. *rex* possesses a relatively moderate genetic diversity compared with the other species in *Rhododendron*. In general, life span, reproductive mode, and breeding system are the important factors in genetic diversity [6,22,69]. As in other outcrossing and long-lived species in *Rhododendron*, high historical gene flow among ancestral population mitigates the loss of genetic diversity and further results in a moderate or high genetic diversity in remnant populations [69,73]. Thus, the current levels of genetic diversity in *R. rex* subsp. *rex* may be attributed to the species' long-lived habit, which is similar to other perennial woody plants [9,22].

4.2. Genetic Differentiation and Structure among R. rex Subsp. rex Populations

The F_{ST} value of *R. rex* subsp. *rex* indicated that a moderate genetic differentiation among populations occurred. A total of 83.75% genetic variation occurred within *R. rex* subsp. *rex* populations with regard to nSSR markers, whereas 83.53% variation was partitioned among populations with regard to cpDNA sequences. This discordance should be affected by dispersal mechanisms among populations of plant species in *Rhododendron* [74,75]. Insect visitors are the primary pollen dispersal vectors for *Rhododendron* species. Various insect vectors evolved longer dispersal distance for pollen, whereas the seeds dispersed by wind traveled less than 10 m albeit in open landscapes [74]. Meanwhile, this different consequence might be related to the type and evolutionary rates of different genome sequences [76]. In general, the evolutionary rate of nuclear genomes transmitted by parents was higher than that of maternally inherited chloroplast genomes [77]. Therefore, cpDNA variations reflected a past change, whereas nSSR variations inferred recent events in the population demographics of *R. rex* subsp. *rex*.

On the basis of genetic structure analysis by SSR data, the populations of *R. rex* subsp. *rex* were grouped into three groups, and the correlation between genetic and geographic distances was significant ($p < 0.05$). Phylogenetic trees and genealogical haplotype networks based on cpDNA sequences showed that three reciprocally lineages were detected. This species possessed unique genetic lineages and endemic cpDNA haplotypes in its separate refuge populations. The closely related haplotypes H1, H2, and H3 were distributed in populations BJS, BCL, and DLT; H4, H5, H6, H10, and H11 occurred primarily in populations LJS, JZS, LZS, QLB1, QLB2, and QLB3; H7, H8, H9, and H12 were only detected in populations GDX and YS.

Habitat dislocation and overexploitation accelerate the generation of genetic differentiation among populations [9,13]. In the sampled regions, large-scale land reclamation and unreasonable forest destruction can be observed, which resulted in habitat loss and fragmented distribution in *R. rex* subsp. *rex* natural populations. In addition, gene flow is a fundamental micro-evolutionary force, influencing genetic differentiation among populations [78,79]. The contemporary gene flow of *R. rex* subsp. *rex* is lower than that of the related species of *R. protistum* var. *giganteum* [9], which plays an important role in the formation of genetic structure and differentiation among *R. rex* subsp. *rex* populations. Moreover, breeding system is an important factor for the genetic differentiation and structure of a species [6,22,69]. Although both selfing and outcrossing are detected in *Rhododendron* species [80,81], the breeding system in *R. rex* subsp. *rex* is yet to be explored. Based on the positive value of fixation indices (F_{is}) and the phenomenon of all populations deviated from HWE in the present study, we can reasonably speculate that inbreeding is present in populations of *R. rex* subsp. *rex*. Hence, the mating system and its influences on genetic differentiation and structure must be further elucidated in *R. rex* subsp. *rex*.

4.3. Population Demographic History of the R. rex Subsp. rex

Exploring the historical demography of a species can facilitate our knowledge of its ancient evolutionary environment [58]. Quaternary glaciers profoundly affected the distribution and genetic variation of plant species. Tremendous global climatic oscillations during quaternary glaciations with several glacial–interglacial cycles caused the expansion and contraction of plant distribution [82]. Most plants were subjected to population demographic stability or expansion throughout the LGM [83,84]. The Bayesian skyline plot of cpDNA showed that *R. rex* subsp. *rex* experienced a notable reduction approximately 50,000–100,000 years ago. This supposition is supported by the GWI values, which are lower than the critical *Mc* indices (0.68). Microsatellite-based bottleneck analysis indicated that no recent bottleneck event occurred in the natural populations of *R. rex* subsp. *rex*. Therefore, the population demographic contraction detected in the *R. rex* subsp. *rex* might have been the result of climate oscillations, and the finding is consistent with the results obtained from other species, such as *Cycas simplicipinna* [63]. Typically, rapid population expansion occurred in the post-glacial period because temperatures increased to warm conditions [85]. However, based on neutrality and mismatch distribution tests, no recent population expansion occurred in *R. rex* subsp. *rex*. We speculate that the populations of *R. rex* subsp. *rex* might have survived in situ rather than migrating long distances to suitable habitats and that evolutionary adaptation might have occurred in the cold environment. The current existing populations of *R. rex* subsp. *rex* were limited in distribution at 2400–3400 m elevation, and this condition might partly support our speculation.

In addition, the complex topology of physical environmental condition in southwest China might cause geographical barriers between population migrations. This scenario was also found in the population demography of *Leucomeris decora* [86].

5. Conclusions

The present study firstly investigated the genetic diversity, population structure, and demographic history of 11 remnant *R. rex* subsp. *rex* populations. A moderate genetic diversity, a high genetic differentiation, and a conspicuous geographical structure were detected in *R. rex* subsp. *rex*. The species possessed unique genetic lineages and endemic cpDNA haplotypes in its separate refuge populations. In addition, we found that *R. rex* subsp. *rex* experienced a population contraction approximately 50,000–100,000 years ago based on the comprehensive analysis of demographic history. Furthermore, no recent population expansion occurred in this species. Hence, the conservation of *R. rex* subsp. *rex* should focus on habitat destruction and loss, resulting in a population decline and inbreeding depression within populations. Furthermore, all the remnant adult trees of *R. rex* subsp. *rex* should receive priority protection for the maintenance of its genetic diversity. This research exhibited tremendous ecological value for the future conservation and sustainable utilization of *R. rex* subsp. *rex* and other similar plants, which are subjected to climate oscillation, inbreeding depression, overexploration, and habitat destruction.

Supplementary Materials: The following are available online at http://www.mdpi.com/2223-7747/9/3/338/s1: Figure S1. Bayesian inference of the number of clusters when $K = 3$ and $K = 8$ (A), delta K values obtained (B), and an unweighted pair-group method with arithmetic averages (UPGMA) phenogram of *R. rex* subsp. *rex* (C) based on nSSR; Table S1. The information of 14 microsatellite primers for the *R. rex* subsp. *rex*; Table S2. Summary of the 10 microsatellite loci used to the 11 populations of *R. rex* subsp. *rex*; Table S3. *p*-Value of Hardy–Weinberg equilibrium test for 11 populations of *R. rex* subsp. *rex*; Table S4. Genetic diversity and differentiation parameters for the combined cpDNA in 11 populations of *R. rex* subsp. *rex*; Table S5. Parameters of neutrality tests based on cpDNA of *R. rex* subsp. *rex*.

Author Contributions: S.-K.S., Y.-H.L., X.Z., and Y.-H.W. initiated and designed the research. S.-K.S. obtained funding for this study. S.-K.S., Y.-H.L., and X.Z. collected the materials and performed the experiments. S.-K.S., Y.-H.L., X.Z., and Y.-H.W. wrote and revised the paper. All authors read and approved the version to be published. All authors have read and agreed to the published version of the manuscript.

Funding: This study was financially supported by the National Key Research and Development Project of China (2017YFC0505204), the National Natural Science Foundation of China (31560224, 31870529), the Young Academic and Technical Leader Raising Foundation of Yunnan Province (2018HB035), the Open Fund of Yunnan Key Laboratory for Plateau Mountain Ecology and Restoration of Degraded Environments (2018DG005), and the Program for Excellent Young Talents, Yunnan University.

Acknowledgments: The authors thank Xiu-Yan Feng from the Kunming Institute of Botany, Chinese Academy of Science for her construction suggestions. We also thank Fang-Li Liu, Xiong-Li Zhou, and Yue Zhang for their assistance with the field sampling.

Conflicts of Interest: The authors declare no conflicts of interest.

References

1. Zhang, Y.; Zhang, X.; Wang, Y.H.; Shen, S.K. *De Novo* Assembly of transcriptome and development of novel EST-SSR markers in *Rhododendron rex* Lévl. through illumina sequencing. *Front. Plant Sci.* **2017**, *8*, 1664. [CrossRef]
2. Yoichi, W.; Jin, X.F.; Peng, C.I.; Tamaki, I.; Tomaru, N. Contrasting diversification history between insular and continental species of three-leaved azaleas (*Rhododendron* sect. *Brachycalyx*) in East Asia. *J. Biogeogr.* **2017**, *44*, 1065–1076. [CrossRef]
3. Xing, W.; Liao, J.; Cai, M.; Xia, Q.; Liu, Y.; Zeng, W.; Jin, X. *De novo* assembly of transcriptome from *Rhododendron latoucheae* Franch. using illumina sequencing and development of new EST-SSR markers for genetic diversity analysis in *Rhododendron*. *Tree Genet. Genomes* **2017**, *13*, 53. [CrossRef]
4. Wei, H.; Dhanaraj, A.L.; Rowland, L.J.; Fu, Y.; Krebs, S.L.; Arora, R. Comparative analysis of expressed sequence tags from cold-acclimated and non-acclimated leaves of *Rhododendron catawbiense* Michx. *Planta* **2005**, *221*, 406–416. [CrossRef]

5. Qiang, Y.; Zhou, B.; Gao, K. Chemical constituents of plants from the genus. *Rhododendron* **2011**, *8*, 792–815.
6. Wu, F.Q.; Shen, S.K.; Zhang, X.J.; Wang, Y.H.; Sun, W.B. Genetic diversity and population structure of an extremely endangered species: The world's largest *Rhododendron*. *AoB Plants* **2015**, *7*, 10696–10700. [CrossRef] [PubMed]
7. Scheiber, S.M.; Robacker, C.D.; Lindstrom, O.M. Evaluation of deciduous azaleas for cold hardiness potential in the southeastern united states. *J. Environ. Hortic.* **2000**, *18*, 230–233.
8. Scariot, V.; De, K.E.; Handa, T.; De, R.J. Comparative study of the discriminating capacity and effectiveness of AFLP, STMS and EST markers in assessing genetic relationships among evergreen azaleas. *Plant Breed.* **2010**, *126*, 207–212. [CrossRef]
9. Wu, F.Q.; Shen, S.K.; Zhang, X.; Yang, G.S.; Wang, Y.H. Inferences of genetic structure and demographic history of *Rhododendron protistum* var. *giganteum*—The world's largest *Rhododendron* using microsatellite markers. *Flora* **2017**, *233*, 1–6. [CrossRef]
10. Frankham, R. Genetics and extinction. *Biol. Conserv.* **2005**, *126*, 131–140. [CrossRef]
11. Divakara, B.N.; Upadhyaya, H.D.; Wani, S.P.; Gowda, C.L.L. Biology and genetic improvement of *Jatropha curcas* L.: A review. *Appl. Energy* **2010**, *87*, 732–742. [CrossRef]
12. Han, Y.C.; Teng, C.Z.; Chang, F.H.; Robert, G.W.; Zhou, M.Q.; Hu, Z.L.; Song, Y.C. Analyses of genetic relationships in *Nelumbo nucifera* using nuclear ribosomal ITS sequence data, ISSR and RAPD markers. *Aquat. Bot.* **2007**, *87*, 141–146. [CrossRef]
13. Zhang, J.J.; Ye, Q.G.; Gao, P.X.; Yao, X.H. Genetic footprints of habitat fragmentation in the extant populations of *Sinojackia* (Styraceaceae): Implications for conservation. *Bot. J. Linn. Soc.* **2012**, *170*, 232–242. [CrossRef]
14. Mutegi, S.M.; Muchugi, A.; Carsan, S.; Kariba, R.; Jamnadass, R.; Oballa, P.; Brunner, A.M.; Runo, S. Genetic diversity of the African poplar (*Populus ilicifolia*) populations in Kenya. *Tree Genet. Genomes* **2016**, *12*, 1–7. [CrossRef]
15. Wambulwa, M.C.; Meegahakumbura, M.K.; Chalo, R.; Kamunya, S.; Muchugi, A.; Xu, J.C.; Liu, J.; Li, D.Z.; Gao, L.M. Nuclear microsatellites reveal the genetic architecture and breeding history of tea germplasm of east Africa. *Tree Genet. Genomes* **2016**, *12*, 1–10. [CrossRef]
16. Pervaiz, T.; Sun, X.; Zhang, Y.; Tao, R.; Zhang, J.; Fang, J. Association between chloroplast and mitochondrial dna sequences in Chinese *Prunus* genotypes (*Prunus persica*, *Prunus domestica*, and *Prunus avium*). *BMC Plant Biol.* **2015**, *15*, 4. [CrossRef]
17. Yuan, Q.J.; Zhang, Z.Y.; Peng, H.; Ge, S. Chloroplast phylogeography of *dipentodon* (dipentodontaceae) in southwest China and northern Vietnam. *Mol. Ecol.* **2008**, *17*, 1054–1065. [CrossRef]
18. Hewitt, G. The genetic legacy of the Quaternary ice ages. *Nature* **2000**, *405*, 907–913. [CrossRef]
19. Petit, R.J.; Duminil, J.; Fineschi, S.; Hampe, A.; Salvini, D.; Vendramin, G.G. Comparative organization of chloroplast, mitochondrial and nuclear diversity in plant populations. *Mol. Ecol.* **2005**, *14*, 689–701. [CrossRef]
20. Garcia-Lor., A.; Luro, F.; Ollitrault, P.; Navarro, L. Genetic diversity and population structure analysis of mandarin germplasm by nuclear, chloroplastic and mitochondrial markers. *Tree Genet. Genomes* **2015**, *11*, 1–15. [CrossRef]
21. Mingeot, D.; Husson, C.; Mertens, P.; Watillon, B.; Bertin, P.; Druart, P. Genetic diversity and genetic structure of black alder (*Alnus glutinosa* [L.] Gaertn) in the Belgium-Luxembourg-France cross-border area. *Tree Genet. Genomes* **2016**, *12*, 1–12. [CrossRef]
22. Zhang, X.; Shen, S.K.; Wu, F.Q.; Wang, Y.H. Inferring genetic variation and demographic history of *Michelia yunnanensis* Franch. (Magnoliaceae) from chloroplast DNA sequences and microsatellite markers. *Front. Plant Sci.* **2017**, *8*, 583. [CrossRef] [PubMed]
23. Breed, M.; Ottewell, K.; Gardner, M.; Marklund, M.H.K.; Dormontt, E.E.; Lowe, A.J. Mating patterns and pollinator mobility are critical traits in forest fragmentation genetics. *Heredity* **2015**, *115*, 108–114. [CrossRef] [PubMed]
24. Aguilar, R.; Cristóbal-Pérez, E.J.; Balvino-Olvera, F.J.; Aguilar-Aguilar, M.; Aguirre-Acosta, N.; Ashworth, L.; Lobo, J.A.; Martén-Rodríguez, S.; Fuchs, E.J.; Sanchez-Montoya, G.; et al. Habitat fragmentation reduces plant progeny quality: A global synthesis. *Ecol. Lett.* **2019**, *22*, 1163–1173. [CrossRef]
25. Sáyago, R.; Quesada, M.; Aguilar, R.; Ashworth, L.; Lopezaraiza-Mikel, M.; Martén-Rodríguez, S. Consequences of habitat fragmentation on the reproductive success of two Tillandsia species with contrasting life history strategies. *AoB Plants* **2018**, 10. [CrossRef]

26. Hamrick, J.L. Response of forest trees to global environmental changes. *For. Ecol. Manag.* **2004**, *197*, 323–335. [CrossRef]
27. Vranckx, G.; Jacquemyn, H.; Muys, B.; Honnay, O. Meta-analysis of susceptibility of woody plants to loss of genetic diversity through habitat fragmentation. *Conserv. Biol.* **2011**, *26*, 228–237. [CrossRef]
28. Doyle, J. DNA protocols for plants—CTAB total DNA isolation. In *Molecular Techniques in Taxonomy*; Hewitt, G.M., Ed.; Springer: Berlin, Germany, 1991.
29. Wang, X.; Huang, Y.; Long, C.L. Isolation and characterization of twenty-four microsatellite loci or *Rhododendron decorum* Franch (Ericaceae). *Hortscience* **2009**, *44*, 2028–2030. [CrossRef]
30. Wang, N.; Qin, Z.C.; Yang, J.B.; Zhang, J. Development and characterization of 15 microsatellite loci for *Rhododendron delavayi* Franch. (Ericaceae). *Hortscience* **2010**, *45*, 457–459. [CrossRef]
31. Li, L.F.; Yin, D.X.; Song, N. Genomic and EST microsatellites for *Rhododendron aureum* (Ericaceae) and cross-amplification in other congeneric species. *Am. J. Bot.* **2011**, *98*, 250–252. [CrossRef]
32. Dakin, E.E.; Avise, J.C. Microsatellite null alleles in parentage analysis. *Heredity* **2004**, *93*, 504–509. [CrossRef] [PubMed]
33. Fan, L.; Zheng, H.; Milne, R.I.; Zhang, L.; Mao, K. Strong population bottleneck and repeated demographic expansions of *Populus adenopoda* (salicaceae) in subtropical China. *Ann. Bot.* **2018**, *121*, 665–679. [CrossRef] [PubMed]
34. Peakall, R.; Smouse, P.E. GENALEX 6: Genetic analysis in excel. Population genetic software for teaching and research. *Mol. Ecol. Notes* **2006**, *6*, 288–295. [CrossRef]
35. Yeh, F.C.; Yang, R.; Boyle, T.B.; Ye, Z.; Mao, J.X. *POPGENE, the User-Friendly Shareware for Population Genetic Analysis*; University of Alberta: Edmonton, AB, Canada, 1997; Available online: http://www.ualberta.ca/~{}fyeh/ (accessed on 5 February 2020).
36. Goudet, J. FSTAT (Version 1.2): A computer program to calculate F-statistics. *J. Hered.* **1995**, *86*, 485–486. [CrossRef]
37. Excoffier, L.; Smouse, P.E.; Quattro, J.M. Analysis of molecular variance inferred from metric distances among DNA haplotypes: Application to human mitochondrial DNA restriction data. *Genetics* **1992**, *131*, 479–491.
38. Excoffier, L.; Laval, G.; Schneider, S. Arlequin (Version 3.0): An integrated software package for population genetics data analysis. *Evol. Bioinform.* **2005**, *1*, 47. [CrossRef]
39. Wright, S. Evolution in Mendelian populations. *Genetics* **1931**, *16*, 97.
40. Ennos, R.A. Estimating the relative rates of pollen and seed migration among plant populations. *Heredity* **1994**, *72*, 250–259. [CrossRef]
41. Abbott, R.J.; Brochmann, C. History and evolution of the arctic flora: In the footsteps of eric hultén. *Mol. Ecol.* **2003**, *12*, 299–313. [CrossRef]
42. Wilson, G.A.; Rannala, B. Bayesian inference of recent migration rates using multilocus genotypes. *Genetics* **2003**, *163*, 1177–1191.
43. Meirmans, P.G. Nonconvergence in Bayesian estimation of migration rates. *Mol. Ecol. Resour.* **2014**, *14*, 726–733. [CrossRef] [PubMed]
44. Miller, M.P. Tools for population genetic analyses (TFPGA) 1.3: A Windows program for the analysis of allozyme and molecular population genetic data. *Comput. Softw.* **1997**, *4*, 157.
45. Pritchard, J.K.; Stephens, M.; Donnelly, P. Inference of population structure using multilocus genotype data. *Genetics* **2000**, *155*, 945–959. [PubMed]
46. Evanno, G.; Regnaut, S.; Goudet, J. Detecting the number of clusters of individuals using the software STRUCTURE: A simulation study. *Mol. Ecol.* **2005**, *14*, 2611–2620. [CrossRef] [PubMed]
47. Earl, D.A. STRUCTURE HARVESTER: A website and program for visualizing STRUCTURE output and implementing the Evanno method. *Conserv. Genet. Resour.* **2012**, *4*, 359–361. [CrossRef]
48. Piry, S.; Luikart, G.; Cornuet, J. BOTTLENECK: A computer program for detecting recent reductions in the effective size using allele frequency data. *J. Hered.* **1999**, *90*, 502–503. [CrossRef]
49. Garza, J.C.; Williamson, E.G. Detection of reduction in population size using data from microsatellite loci. *Mol. Ecol.* **2001**, *10*, 305–318. [CrossRef]
50. Swindell, S.R.; Plasterer, T.N. *Sequence Data Analysis Guidebook*; Humana Press: New York, NY, USA, 1997; pp. 75–89.
51. Hall, T.A. BioEdit: A user-friendly biological sequence alignment editor and analysis program for Windows 95/98/NT. *Nucleic Acids Symp. Ser.* **1999**, *41*, 95–98.

52. Swofford, D.L. *PAUP*: Phylogenetic Analysis Using Parsimony (and Other Methods)*; Version 4.0b10; Sinauer Associates, Inc.: Sunderland, MA, USA, 2002.
53. Librado, P.; Rozas, J. DnaSP v5: A software for comprehensive analysis of DNA polymorphism data. *Bioinformatics* **2009**, *25*, 1451–1452. [CrossRef]
54. Pons, O.; Petit, R.J. Measuring and testing genetic differentiation with ordered versus unordered alleles. *Genetics* **1996**, *144*, 1237–1245.
55. Bandelt, H.J.; Forster, P.; Röhl, A. Median-joining networks for inferring intraspecifc phylogenies. *Mol. Biol. Evol.* **1999**, *16*, 37–48. [CrossRef] [PubMed]
56. Ronquist, F.; Huelsenbeck, J.P. MrBayes 3: Bayesian phylogenetic inference under mixed models. *Bioinformatics* **2003**, *19*, 1572–1574. Available online: http://tree.bio.ed.ac.uk/software/figtree/ (accessed on 5 February 2020). [CrossRef] [PubMed]
57. Batalha-Filho, H.; Cabanne, G.S.; Miyaki, C.Y. Phylogeography of an Atlantic forest passerine reveals demographic stability through the last glacial maximum. *Mol. Phylogenet. Evol.* **2012**, *65*, 892–902. [CrossRef] [PubMed]
58. Graur, D.; Li, W.H. *Fundamentals of Molecular Evolution*; Sinauer Associates: Sunderland, MA, USA, 2000.
59. Drummond, A.J.; Rambaut, A. BEAST: Bayesian evolutionary analysis by sampling trees. *BMC Evol. Biol.* **2007**, *7*, 214. [CrossRef]
60. Tamura, K.; Stecher, G.; Peterson, D.; Filipski, A.; Kumar, S. MEGA6: Molecular evolutionary genetics analysis version 6.0. *Mol. Boil. Evol.* **2013**, *30*, 2725–2729. [CrossRef]
61. Phillips, S.J.; Anderson, R.P.; Schapire, R.E. Maximum entropy modeling of species geographic distributions. *Ecol. Model.* **2006**, *190*, 231–259. [CrossRef]
62. Hijmans, R.J.; Cameron, S.E.; Parra, J.L.; Jones, P.G.; Jarvis, A. Very high resolution interpolated climate surfaces for global land areas. *Int. J. Clim.* **2005**, *25*, 1965–1978. Available online: http://www.worldclim.org (accessed on 5 February 2020). [CrossRef]
63. Feng, X.Y.; Wang, Y.; Gong, X. Genetic diversity, genetic structure and demographic history of *Cycas simplicipinna* (Cycadaceae) assessed by DNA sequences and SSR markers. *BMC Plant Biol.* **2014**, *14*, 187. [CrossRef]
64. Li, M.W.; Chen, S.F.; Shi, S.; Zhang, Z.; Liao, W.B.; Wu, W.; Zhou, R.C.; Fan, Q. High genetic diversity and weak population structure of *Rhododendron jinggangshanicum*, a threatened endemic species in Mount Jinggangshan of China. *Biochem. Syst. Ecol.* **2015**, *58*, 178–186. [CrossRef]
65. Tan, X.X.; Li, Y.; Ge, X.J. Development and characterization of eight polymorphic microsatellites for *Rhododendron simsii* Planch (Ericaceae). *Conserv. Genet.* **2009**, *10*, 1553–1555. [CrossRef]
66. Kondo, T.; Nakagishi, N.; Isagi, Y. Shaping of genetic structure along Pleistocene and modern river systems in the hydrochorous riparian azalea, *Rhododendron ripense* (Ericaceae). *Am. J. Bot.* **2009**, *96*, 1532–1543. [CrossRef] [PubMed]
67. Hirao, A.S. Kinship between parents reduces offspring fitness in a natural population of *Rhododendron brachycarpum*. *Ann. Bot.* **2010**, *105*, 637–646. [CrossRef]
68. Bruni, I.; De Mattia, F.; Labra, M.; Grassi, F.; Fluch, S.; Berenyi, M.; Ferrari, C. Genetic variability of relict *Rhododendron ferrugineum* L. populations in the Northern Apennines with some inferences for a conservation strategy. *Plant Biosyst.* **2012**, *146*, 24–32. [CrossRef]
69. Nybom, H. Comparison of different nuclear DNA markers for estimating intraspecific genetic diversity in plants. *Mol. Ecol.* **2004**, *13*, 1143–1155. [CrossRef] [PubMed]
70. Yoichi, W.; Sakaguchi, S.; Ueno, S.; Tomaru, N.; Uehara, K. Development and characterization of EST-SSR markers for the genus *Rhododendron* section *Brachycalyx* (Ericaceae). *Plant Species Biol.* **2017**, *32*, 455–459. [CrossRef]
71. Ikeda, H.; Yoneta, Y.; Higashi, H.; Eidesen, P.B.; Barkalov, V.; Yakubov, V.; Brochmann, C.; Setoguchi, H. Persistent history of the bird-dispersed arctic-alpine plant *Vaccinium vitis-idaea* L. (Ericaceae) in Japan. *J. Plant Res.* **2015**, *128*, 437–444. [CrossRef] [PubMed]
72. Hamrick, J.L.; Godt, M.J.W. Allozyme diversity in plant species. In *Plant Population Genetics, Breeding, and Genetic Resources*; Brown, A.H.D., Clegg, M.T., Kahler, A.L., Weir, B.S., Eds.; Sinauer: Sunderland, MA, USA, 1989; pp. 43–63.

73. Setoguchi, H.; Mitsui, Y.; Ikeda, H.; Nomura, N.; Tamura, A. Genetic structure of the critically endangered plant *Tricyrtis ishiiana* (Convallariaceae) in relict populations of Japan. *Conserv. Genet.* **2011**, *12*, 491–501. [CrossRef]
74. Escaravage, N.; Wagner, J. Pollination effectiveness and pollen dispersal in a *Rhododendron ferrugineum* (Ericaceae) population. *Plant Biol.* **2004**, *6*, 606–615. [CrossRef]
75. Stephenson, C.M.; Kohn, D.; Park, K.; Atkinson, R.J.; Edwards, C.; Travis, J.M.J. Testing mechanistic models of seed dispersal for the invasive *Rhododendron ponticum* (L.). *Perspect. Plant Ecol. Evol. Syst.* **2007**, *9*, 15–28. [CrossRef]
76. Han, Q.X.; Higashi, H.; Mitsui, Y.; Setoguchi, H. Distinct phylogeographic structures of wild radish (*Raphanus sativus* L. var. raphanistroides Makino) in Japan. *PLoS ONE* **2015**, *10*, e0135132. [CrossRef]
77. Cheng, B.B.; Zheng, Y.Q.; Sun, Q.W. Genetic diversity and population structure of *Taxus cuspidata* in the Changbai mountains assessed by chloroplast DNA sequences and microsatellite markers. *Biochem. Syst. Ecol.* **2015**, *63*, 157–164. [CrossRef]
78. Yao, X.H.; Ye, Q.G.; Kang, M.; Huang, H.W. Microsatellites analysis reveals interpopulation differention and gene flow in endangered tree *Changiostyrax dolichocarpa* (Styracaceae) with fragmented distribution in central China. *New Phytol.* **2007**, *176*, 472–480. [CrossRef] [PubMed]
79. Côté, C.L.; Gagnaire, P.A.; Bourret, V.; Verreault, G.; Castonguay, M.; Bernatchez, L. Population genetics of the American eel (*Anguilla rostrata*): $F_{st} = 0$ and North Atlantic Oscillation effects on demographic fluctuations of a panmictic species. *Mol. Ecol.* **2013**, *22*, 1763–1776. [CrossRef] [PubMed]
80. Escaravage, N.; Pornon, A.; Doche, B.; Till-Bottraud, I. Breeding system in an alpine species: *Rhododendron ferrugineum* (Ericaceae) in the French northern Alps. *Can. J. Bot.* **1997**, *75*, 736–743. [CrossRef]
81. Hahn, C.Z.; Michalski, S.G.; Durka, W. Gene flow in, and mating system of, *Rhododendron simsii* in a nature reserve in subtropical china. *Nord. J. Bot.* **2017**, *35*, 1–7. [CrossRef]
82. Harrison, S.P.; Yu, G.; Takahara, H.; Prentice, I.C. Diversity of temperate plants in East Asia. *Nature* **2001**, *413*, 129–130. [CrossRef]
83. Bisconti, R.; Canestrelli, D.; Colangelo, P.; Nascetti, G. Multiple lines of evidence for demographic and range expansion of a temperate species (*Hyla sarda*) during the last glaciation. *Mol. Ecol.* **2011**, *20*, 5313–5327. [CrossRef]
84. Pinheiro, F.; De Barros, F.; Palma-Silva, C.; Fay, M.F.; Lexer, C.; Cozzolino, S. Phylogeography and genetic differentiation along the distributional range of the orchid *Epidendrum fulgens*: A Neotropical coastal species not restricted to glacial refugia. *J. Biogeogr.* **2011**, *38*, 1923–1935. [CrossRef]
85. Gao, Y.D.; Zhang, Y.; Gao, X.F.; Zhu, Z.M. Pleistocene glaciations, demographic expansion and subsequent isolation promoted morphological heterogeneity: A phylogeographic study of the alpine *Rosa sericea* complex (Rosaceae). *Sci. Rep.* **2015**, *5*, 11698. [CrossRef]
86. Zhao, Y.J.; Gong, X. Genetic structure of the endangered *Leucomeris decora* (Asteraceae) in China inferred from chloroplast and nuclear DNA markers. *Conserv. Genet.* **2014**, *13*, 271–281. [CrossRef]

plants

MDPI

Article

Genetic Distinctiveness Highlights the Conservation Value of a Sicilian Manna Ash Germplasm Collection Assigned to *Fraxinus angustifolia* (Oleaceae)

Loredana Abbate [1,†], Francesco Mercati [1,*,†], Giuseppe Di Noto [2], Myriam Heuertz [3], Francesco Carimi [1], Sergio Fatta del Bosco [1] and Rosario Schicchi [2]

1 Institute of Biosciences and Bioresources (IBBR), National Research Council, Corso Calatafimi 414, 90129 Palermo, Italy; loredana.abbate@ibbr.cnr.it (L.A.); francesco.carimi@ibbr.cnr.it (F.C.); sergio.fatta@ibbr.cnr.it (S.F.d.B.)

2 Department of Agricultural, Food and Forestry Sciences (SAAF), University of Palermo, Via Archirafi 38, 90123 Palermo, Italy; pippo.dinoto@virgilio.it (G.D.N.); rosario.schicchi@unipa.it (R.S.)

3 Institut National de Recherche Pour l'agriculture, l'alimentation et l'environnement (INRAE), Univ. Bordeaux, BIOGECO, 69 route d'Arcachon, F-33610 Cestas, France; myriam.heuertz@inra.fr

* Correspondence: francesco.mercati@ibbr.cnr.it; Tel.: +39-091-6574578

† These authors contributed equally to this work.

Received: 24 July 2020; Accepted: 12 August 2020; Published: 14 August 2020

Abstract: The cosmopolitan genus *Fraxinus* comprises about 40 species occupying several habitats in the Northern Hemisphere. With some species hybridizing and sharing genetic variants, questions remain on the species assignment of germplasm within the genus *Fraxinus* despite numerous species-specific assessments. A multidisciplinary approach was employed to provide a definitive insight into the genetics of an endangered *Fraxinus* "manna ash" collection, located in a rich plant biodiversity hotspot of the Madonie Mountains (Sicily). Although the collection size was small, genetic diversity, assessed by chloroplast (cpSSR) and nuclear (nSSR) microsatellites (SSR—Simple Sequence Repeats), allowed identifying three different chloroplast haplotypes, with one (H5) dominant, and several polymorphic loci, able to discriminate most of the local accessions studied. Molecular data were linked to cytofluorimetric and phenotypic evaluations and, contrary to popular belief that manna ash is *Fraxinus ornus* L., the germplasm currently used for manna production belongs to *Fraxinus angustifolia* Vahl. Interestingly, joint analysis of our genetic panel with a large European dataset of *Fraxinus* spp. suggested the presence of a possible glacial refuge in Sicily, confirming its importance as biodiversity source. Our results will be helpful for the design of long-term conservation programs for genetic resources, such as in situ and ex situ conservation, seed collection and tree reintroduction.

Keywords: *Fraxinus* spp.; manna; local varieties nSSR; cpSSR; cytometry; morphological traits

1. Introduction

The genus *Fraxinus* (Oleaceae) comprises 45–65 tree species and is represented across large areas of Europe, Asia and North America [1,2]. *Fraxinus* species show considerable variability in their flowering biology, ecological requirements and distribution ranges as a result of dispersal and vicariance processes as well as adaptive evolution underlying diversification in the genus [3].

In Europe, three ash species are present: the common ash, *Fraxinus excelsior* L., the flowering ash, *Fraxinus ornus* L., and the narrow-leaved ash, *Fraxinus angustifolia* Vahl [4]. *F. excelsior*, a polygamous species with male, female and hermaphrodite individuals, [5,6], is found throughout the continent except in the Mediterranean region, while *F. ornus* is androdiecious [3] and grows in relatively high altitude areas of the eastern Mediterranean basin [7,8]. Finally, *F. angustifolia*, a close relative to *F. excelsior*, shows hermaphrodite flowers and is very common around the Mediterranean basin and to the west of the Black Sea in the Danube basin. Due to the level of local adaptation [9], different botanical names were assigned to *F. angustifolia*. Three groups can be identified at subspecies level, structured by geographical regions [7,10]: *F. angustifolia* ssp. *oxycarpa* (Bieb. ex Willd.) Franco and Rocha Afonso observed in the in East Central and Southeastern Europe, including the Balkans; *F. angustifolia* ssp. *syriaca* (Boiss.) Yalt. in the east of Europe, from Turkey to Pakistan; and *F. angustifolia* ssp. *angustifolia* in southwestern Europe. While *F. excelsior* grows on a wide range of soil types, preferring nutrient-rich substrates, *F. angustifolia* grows near surface and ground waters, and thus its habitat is more restricted [11,12]. Indeed, the distribution of this species in the Mediterranean region is irregular and limited to smaller and isolated populations on drier sites at higher altitudes or on wetland sites [13]. The levels and distribution of genetic diversity in *F. excelsior* and *F. angustifolia* reflect the differential climatic and ecological affinities of both species, as well as the signature of ancient and contemporary hybridization [4,14].

Sicily boasts several botanical highlights. Both trees and endemic herbs survived through the last glacial maximum, highlighting the role of Sicily as a refuge area for thermophilous European plants during glacial times [15]. In addition Sicily, the biggest region of Italy, has been identified as one of 52 putative glacial refuges based on phylogeographic data from trees and herbs, and lies within one of ten regional hotspots of plant species biodiversity in the Mediterranean basin [16].

The presence of the genus *Fraxinus* in Sicily is restricted mainly to scattered groves and natural populations of *F. angustifolia* and *F. ornus* spread in the hilly area of the Madonie and Nebrodi Regional Natural Parks, at altitudes between 100 and 900 m a.s.l. In this area, the relevance of *Fraxinus* spp. is extremely high, from economical, botanical, naturalistic and historical points of view [17,18]. In fact, the *F. ornus* cultivation, brought to Sicily at the time of the Arab invasion (ninth-eleventh century) and known as manna ash, is grown for the production of manna, a crystal coagulate exuding after cutting 5–10 cm long slits with a billhook along and inside the whole thickness of the bark. From these slits a purplish and bitter liquid flows out. After contact with the air, this liquid turns white and becomes sweet, rapidly coagulating and forming crystal-like layers, called manna. Then manna is harvested and placed in specific trays for drying. Its production varies from 0.2 to 1.5 kg per tree per year, depending on ash cultivar [18]. However specific pedoclimatic conditions (persistent dry, hot periods and steady ventilation during summertime) are mandatory for the bioaccumulation of osmotic sugars in the phloem of *Fraxinus* and to obtain the complete drying of the sap that exudes from cuttings. The most abundant constituent of manna is mannitol (around 50% of total chemical compounds), a hexavalent alcohol known also as "manna sugar"; monosaccharide, oligosaccharide, mineral and volatile constituents are also well represented [18,19]. Due to its characteristics, the manna is used in cosmetic, pharmaceutical and confectionary industries [20]. In Sicily, the presence of *Fraxinus* groves has given rise to a traditional agroforestry landscape of unique value and relevant economic importance until the late 1950s. In the last 50–60 years, after the introduction of synthetic mannitol, the integrity and the extension of the manna ash landscapes have been dramatically reduced. Today, because of habitat fragmentation, naturally poor regeneration and land use change, the maintenance of the crop is threatened. The present landscape of Sicilian manna *Fraxinus* is limited to a fragmented and isolated area of few hundreds of hectares in the territory around Castelbuono and Pollina (Palermo-Italy), in which natural populations, single and scattered trees are distributed. Large differences in several phenotypic traits, such as leaf morphology, inflorescence features and infructescence characteristics, are observed among the different plants. However, based on their distinctive morphological features, several cultivars (or varieties) were distinguished and identified with dialect names, since the nineteenth century [17,20–22]. The declining status of the manna ash germplasm has been confirmed by the last survey on the residual consistency of the crop in the Madonie Mountains [17,18]. In less than 150 years, about 50% of the germplasm of the manna ash cultivars has been lost; therefore, both the germplasm of the cultivated varieties and the ash cultivation are threatened with extinction. Previous studies, based on morphological and chemical analysis, confirmed the presence of 16 cultivars, 13 of which putatively belonging to *F. angustifolia* and the remaining three putatively belonging to *F. ornus* [17]. However, the value of these approaches to characterize manna cultivars and varieties is limited due to their drawbacks, such as the influence of environment on trait expression, epistatic interactions and pleiotropic effects. Therefore, there is an urgent need to focus on the unknown genetic background, structure, relationships and diversity of the *Fraxinus* gene pool distributed in Sicily, to provide useful information aimed both to prevent biodiversity loss, safeguarding this endangered plant populations, and develop future breeding programs.

Microsatellite (SSR, simple sequence repeat) markers are codominant, highly polymorphic, reproducible, well distributed throughout the genome and suitable for automated analysis, and are; therefore, widely used markers for plant variety characterization [23–28]. SSRs have proved to be ideal tools, in conjunction with analysis of morphological characteristics, for analyzing genetic diversity and structure and evolutionary relationships between populations of rare, isolated and endangered species. Chloroplast DNA sequences are suitable tools to highlight the postglacial colonization routes [4,29,30] because they are usually non-recombinant in angiosperms and they are transmitted through seeds only [31,32], implying that the effective population size is more reduced for chloroplast DNA than for nuclear DNA. Nuclear and plastid DNA microsatellites have been extensively used to describe the phylogeography of *Fraxinus* spp. and clarify the hybridization process among ash species [4,14,33,34].

Nuclear DNA content has an important role in systematics, and it is a useful tool in biodiversity evaluation [35,36]. Flow cytometry is a quick and effective method to assess genome size (i.e., the amount of DNA in cell nuclei) [37,38]. Studies of genome size variation can shed light on the molecular mechanisms underlying this type of variation [39–41], which is a relevant indicator of genetic divergence and speciation processes [42,43].

In this study, we used morphological traits, genetic profiles (nuclear and plastid SSR markers) and genome size estimates (nuclear DNA content) to characterize the biodiversity within and among native accessions of *Fraxinus* in Sicily, belonging to a historical manna ash collection that, from previous studies [18], seems to be represented by varieties mainly characterized as *F. angustifolia*. We also present the first assessment of evolutionary relationships between the *F. angustifolia* germplasm from Sicily and *F. angustifolia* germplasm from throughout Europe. Knowledge of the extent of genetic diversity and characterization of population genetic structure of native *Fraxinus* spp. in Sicily will have important implications for the conservation of this biodiversity, to provide useful knowledge for the management of the manna ash germplasm collection, in-situ/ex-situ conservation and to save this precious cultural heritage.

This analysis is a first step towards selecting breeding material and establishing conservation strategies to sustain and potentially increase the overall production and productivity of manna ash and the continual use of this agroforestry tree resource through varietal improvement and suitable agronomic practices under southern Mediterranean conditions.

2. Materials and Methods

2.1. Plant Material

Leaf samples of native manna ash (*Fraxinus* spp.) germplasm located in Madonie National Park (Palermo, Sicily; 37°53′ N 14°01′ E/37.883° N 14.017° E) were collected. Individual trees were selected according to specific criteria based on morphological features, historical notes and their use for manna production [17,18,21,22,44]. The list of accessions studied, and their dialectical name, is indicated in Table 1 and Table S1. At least three individual trees (replicates) were collected for each local variety studied, except for monumental trees. In total thirty-four (34) trees (accessions) were analyzed. Four *F. ornus* trees, collected in the same area, were also added in the study and used as reference. Young and fully expanded leaves were immediately frozen in liquid nitrogen and conserved at −80 °C until used for DNA extraction. This sample set, comprising 38 accessions (Table S1) belonging to manna ash varieties and *F. ornus* (FOR) collected in Sicily, is named sample set #1 from here on.

To full investigate genetic relationships and population genetic structure among the analyzed manna ash accessions and natural populations of *Fraxinus* spp., our sample set (#1) was added to SSR profiles already available. To avoid redundancy in the results, this second sample set (#2) included only samples showing unique genetic profiles isolated from sample set #1, and SSR unique profiles of 475 individual ash trees described in Gérard et al. (2013) [14], belonging to *F. angustifolia* (148; FAN), *F. excelsior* (274; FEX) and hybrids between *F. angustifolia* and *F. excelsior* (53; HYB) collected in the natural, partially overlapping, distribution ranges of the two species across Europe (Table S2). This second sample set is named sample set #2 from here on.

2.2. Morphological Trait Variation

For each accession from set #1, including four *F. ornus* trees, a set of morphological traits were evaluated (Table S1). The most discriminant morphological features among *Fraxinus* species (e.g., the samara stalk length, the number of samaras/raceme, the colour of apical buds) were recorded; three replicates for each trait were evaluated.

2.3. Genome Size and Ploidy Level

The genome sizes of collected accessions (Table 1) were estimated by flow cytometry using the *F. ornus* (2C-value = 1.98 pg) as reference standard. Nuclei isolated from a single mature leaf were analyzed in three technical replicates for each accession of set #1. The analysis was carried out with a Partec PAS flow cytometer (Partec, http://www.partec.de/), equipped with a mercury lamp. Fully expanded leaves (1 cm^2) were chopped in a glass Petri dish with 1 mL nuclei extraction buffer OTTO1 [45] and 3 drops of Tween 20. After 3 min, 1 mL of OTTO2 [45] supplemented with DAPI (4 µg/mL) was added. The solution was filtered through a 30 µm Cell-Trics disposable filter (Partec). The relative fluorescence intensity of stained nuclei was measured on a linear scale, and 4000–5000 nuclei for each sample were analyzed. DNA content histograms were generated using the Partec software package (Partec-FlowMax®, Münster, Germany).

2.4. DNA Extraction and Genotyping

Genomic DNA was extracted from 100 mg of powdered, frozen, young leaf tissue of each individual from sample set #1 using the QiagenDNeasy Plant Mini Kit (Qiagen, Hilden, Germany). The purity and quantity of the DNA extracts were assessed with a NanoDrop 1000 spectrophotometer (Thermo Scientific, Waltham, MA, USA). Molecular investigations were carried out by amplifying 6 plastid (chloroplast) microsatellite (SSR, simple sequence repeat) markers (cpSSR) (ccmp2, ccmp3, ccpm4, ccmp6, ccmp7, and ccmp10 [46]), 3 plastid DNA regions (atpB-rbcL, CPFRAX6 and matK [33]) and 11 nuclear microsatellite nSSR (FEMSATL4, FEMSATL11, FEMSATL12, FEMSATL16, FEMSATL19 [47], M230 [48], EST-SSR326, EST-SSR427, EST-SSR431, EST-SSR520, EST-SSR528 [49]) markers, respectively. PCR reactions were performed following the procedures reported in Garfi et al. (2013) [50], using primers fluorescently labelled with FAM, VIC, NED and PET. The fragments were separated by capillary electrophoresis using an ABI PRISM 3130 Genetic Analyzer (Applied Biosystems) to detect length polymorphism only. Fragments were sized and binned into alleles using Gene Mapper v. 4.1 software (Thermo Fisher Scientific, Waltham, MA, USA), (Applied Biosystems) (Table S2).

2.5. Data Analysis

Plastid DNA haplotypes were defined based on the concatenation of fragment length polymorphism at cpSSRs and amplified plastid DNA regions. To assign the species of origin of the sampled materials, the obtained haplotypes were compared visually with previously published cpSSR profiles from the three European ash species reported in Heuertz et al. (2006) [4].

Genetic relationships between nSSR profiles of both sample sets (set #1; set #2) were estimated using Bruvo's distance [51] in the poppr package [52] in R Core Team (2020; http://www.R-project.org). A dendrogram was computed from each distance matrix using the UPGMA (Unweighted Pair Group Method with Arithmetic Mean) implemented in the adegenet package [53] in R. Bootstrap analysis was performed based on 1000 replicate samples to assess the robustness of the inferred evolutionary relationships in the dendrogram.

After establishing that Sicilian manna ash is *F. angustifolia* based on the listed analysis, we assessed the genetic differentiation between Sicilian manna ash and the collection of European *F. angustifolia*, *F. excelsior* and their hybrids using Wright's fixation index (*Fst*) [54] and Nei's (1973) distance [55] computed through hierfstat [56] in R. The genetic variability across *F. angustifolia* (FAN accessions belonging to set #2 was estimated using the observed (*Ho*) and expected (*He*) heterozygosities [55], Shannon's index (*I*) and the inbreeding coefficient (*Fis*) for sampling locations in GenAlEx 6.5 [57]. Each locus and population was tested for Hardy-Weinberg equilibrium deviation with the exact test through Genepop v. 3.4 [58], with the default parameters (dememorization = 10,000, number of batches = 100, and number of iterations/batch = 5000). To compare the genetic diversity of population studied with different sample sizes the allelic richness (*AR*) within sampling locations was also evaluated though SPAGeDi software [59]. Population pairwise fixation index (*Fst*) values were then computed for only *F. angustifolia* accessions, grouped based on their sampling location or country and a neighbor-joining (NJ) tree using was developed using the adegenet package [53].

Finally, to evaluate the genetic structure of *F. angustifolia* distributed across Europe (belonging to set #2 a discriminant analysis of principal components (DAPCs) was employed. Samples belonging to *F. angustifolia* were grouped based on their geographic area of origin (in five main groups: Balkans, Eastern Europe, France, Italy and Portugal) and DAPC analysis, implemented in the adegenet package [53], was carried out to infer population subdivision of the germplasm studied. The number of principal components (PCs) retained was evaluated using the cross-validation procedure. The *K*-means algorithm "find.clusters" was used to independently verify the assignment of individuals to clusters.

3. Results

Ploidy level, DNA content and morphological trait variation of the manna ash collection DNA content of the 34 accessions belonging to the manna ash collection and of four *F. ornus* individual trees (set #1) was estimated based on the relative intensity of fluorescence using the known *F. ornus* genome size (1.98) as reference [60]. DNA content estimates were congruent with all individuals of set #1 being diploids ($2n = 2 \times = 46$) [61]. Flow cytometry-derived genome sizes (pg/2C) for manna ash varieties ranged from 1.539 to 1.544 (Table 1).

Broad variability in several phenotypic traits related to leaf morphology, inflorescence and infructescence features were observed among the different local varieties belonging to the Sicilian manna ash germplasm collection (Table S1). Leaf shape, number of leaflets, leaflet margin, petiole length and rachis wings were highly variable (Table S1). Also the inflorescences showed a great range of variation. Differences between varieties were mainly observed in the structure of reproductive organs, including the terminal clustering of multiple carpels, the partial basal fusion of individual carpels and the attachment and orientation of the seeds (Table S1). The mean values of some morphological traits showed most high variability between the manna ash collection and the *F. ornus* reference trees (Table 1). Specifically, the largest phenotypic variability was found in the number of samaras/raceme (5.8 vs. 46.0), followed by samara stalk length (1.1 vs. 0.4) and apical leaf length/width ratio (0.28 vs. 0.52) (Table 1). Moreover, flowering time (November–January for manna ash vs. May–June for *F. ornus*) and the colour of apical buds (red in manna ash vs. white in *F. ornus*) were clearly distinct for manna ash and *F. ornus* (Table 1).

Table 1. Genome sizes and morphological traits of studied samples, belonging to manna ash collection. *F. ornus* was used as reference. In bracket the number of samples analyzed belonging to each local variety/species. Three replicates for each trait and the genome sizes evaluation were analyzed.

Type	Local Varieties	Ploidy Level	2C DNA Value *	Apical Leaf (Length/Width Ratio) *	Samara Stalk Length (cm) *	Number of Samaras/Raceme *	Flowering Time	Apical Buds Colour
Manna ash	Abbassa cappeddu (3)	2n	1.540 ± 0.01	0.21 ± 0.01	0.81 ± 0.07	5.51 ± 0.20	November–January	Red
	Baciciu (4)	2n	1.544 ± 0.03	0.42 ± 0.02	1.12 ± 0.07	3.50 ± 1.02	November–January	Red
	Cavolo (3)	2n	1.542 ± 0.02	0.23 ± 0.09	2.11 ± 0.06	5.50 ± 0.07	November–January	Red
	Macigna (3)	2n	1.541 ± 0.01	0.42 ± 0.03	0.79 ± 0.02	6.01 ± 0.70	November–January	Red
	Nivuru (3)	2n	1.541 ± 0.04	0.35 ± 0.01	3.04 ± 0.01	7.10 ± 0.62	November–January	Red
	Nsiriddu (3)	2n	1.544 ± 0.03	0.24 ± 0.08	1.15 ± 0.05	5.09 ± 0.07	November–January	Red
	Russu (3)	2n	1.542 ± 0.04	0.32 ± 0.07	0.68 ± 0.07	5.52 ± 0.27	November–January	Red
	Sarvaggio (3)	2n	1.539 ± 0.02	0.31 ± 0.05	0.77 ± 0.02	5.01 ± 0.19	November–January	Red
	Verdello (7)	2n	1.542 ± 0.05	0.21 ± 0.01	0.76 ± 0.08	6.53 ± 0.35	November–January	Red
	Frassino monumentale (1)	2n	1.541 ± 0.01	0.23 ± 0.03	0.92 ± 0.03	7.05 ± 0.10	November–January	Red
	Nsiriddu monumentale (1)	2n	1.544 ± 0.01	0.22 ± 0.01	0.91 ± 0.05	7.03 ± 0.32	November–January	Red
	mean	-	*1.542* **	*0.28* **	*1.18* **	*5.80* **	-	-
F. ornus (4)		2n	1.98 ± 0.01	0.52 ± 0.02	0.41 ± 0.03	46.02 ± 2.05	May–June	White

* Mean values. ** Significantly different at the 0.01 probability level between local varieties and *F. ornus*.

3.1. Genetic Variation of Plastid DNA Microsatellites (cpSSRs) and Amplified Regions in the Manna Ash Collection

Five (ccmp3, ccmp4, ccmp7, atpB/rbcL and matK) out of nine plastid DNA markers were monomorphic in set #1, displaying fragment lengths of 97, 140, 117, 157 and 253 bp, respectively. The other loci showed low but significant levels of polymorphism (Table 2): Three size variants, with amplification fragment sizes of 364, 365 and 366 bp, were observed at CPFRAX6, whereas two distinct size variants separated by one and four nucleotides were showed at ccmp6 and ccmp2, respectively. Finally, ccmp10 displayed three specific size variants, with amplicons of 103, 104 and 106 bp (Table 2), respectively. The size variants combined into a total of 4 haplotypes (Table 2). Except for "Verdello", the samples belonging to each local variety showed one specific haplotype. The four haplotypes detected in set #1 represented three of the twenty-two previously characterized cpSSR-based haplotypes [4]: Manna ash accessions carried either a sub-variant of H5 (92% of accessions) or haplotype H10 (8% of accessions), these haplotypes having previously been observed in *F. angustifolia* and in *F. excelsior*; the *F. ornus* reference trees carried H19, a haplotype private to *F. ornus* (Table 2).

Table 2. Haplotypes detected based on fragment length polymorphism using six chloroplast microsatellites (cpSSR) and three plastid DNA regions in the germplasm belonging to the studied manna ash collection. *F. ornus* was used as reference. In bracket the number of samples analyzed belonging to each local variety/species.

Samples/Local Varieties	N_h [#]	ccmp2	ccmp3	ccmp4	ccmp6	ccmp7	ccmp10	CPFRAX6	atpB/rbcL	matK	Haplotypes *
Abbassa cappeddu (3)	1	194	97	140	98	117	104	365	157	253	H10
Baciciu (4)	1	194	97	140	98	117	103	365	157	253	H5
Cavolo (3)	1	194	97	140	98	117	103	365	157	253	H5
Macigna (3)	1	194	97	140	98	117	103	365	157	253	H5
Nivuru (3)	1	194	97	140	98	117	103	365	157	253	H5
Nsiriddu (3)	1	194	97	140	98	117	103	366	157	253	H5 [a]
Russu (3)	1	194	97	140	98	117	103	365	157	253	H5
Sarvaggio (3)	1	194	97	140	98	117	103	365	157	253	H5
		194	97	140	98	117	103	365	157	253	H5
Verdello (7)	3	194	97	140	98	117	103	366	157	253	H5 [a]
		194	97	140	98	117	104	365	157	253	H10
Frassino monumentale (1)	1	194	97	140	98	117	103	365	157	253	H5
Nsiriddu monumentale (1)	1	194	97	140	98	117	103	365	157	253	H5
F. ornus (4)	1	190	97	140	99	117	106	364	157	253	H19

[#] Nh, number of haplotypes. * Haplotype classification following Heuertz et al. (2006) [4], without CPFRAX6, atpB/rbcL and matK profiles.[a] H5 sub-variation identified using also CPFRAX6, atpB/rbcL and matK profiles.

3.2. Genetic Diversity of Manna Ash Germplasm

The genetic diversity of the manna ash germplasm and *F. ornus* genotypes, the so called set #1, was investigated using 11 nSSR (Table S2). Phylogenetic analysis based on Bruvo's distance and the UPGMA algorithm generated a dendrogram that comprised four main clusters across set #1 (Figure 1). A total of 20 unique SSR profiles were detected (16 from manna ash collection and 4 for *F. ornus* accessions, respectively), with "Frassino monumentale" as the most distant sample among the manna ash genotypes. Except for varieties "Baciciu", "Macigna" and "Nsiriddu", all accessions were assigned to their expected local variety (Figure 1), showing "Verdello" as the local variety with the highest genetic variability. In addition, the accessions analyzed belonging to "Sarvaggio", "Nivuru", "Russo" and "Abbassa cappeddu" can be considered clones, respectively, showing the same genetic profile within the same variety. The four reference trees of *F. ornus* grouped together in a cluster that behaved like an outgroup.

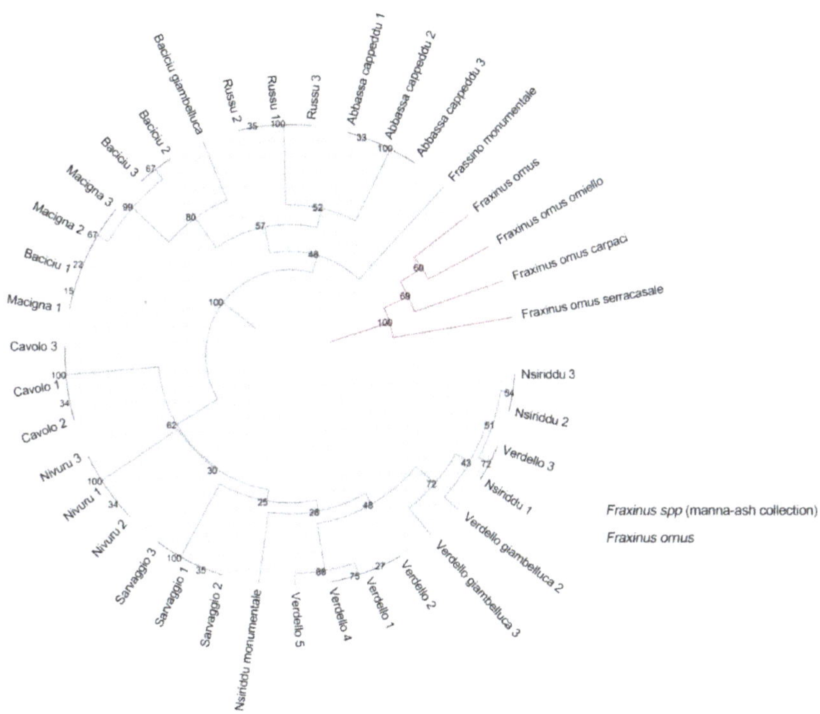

Figure 1. Genetic relationship developed using the profiles obtained from eleven nSSR (Table S2) among studied local varieties (34 samples) collected in Sicily, belonging to the manna ash germplasm. Four genotypes belonging to *F. ornus* were used as references. The dendrogram was developed using the UPGMA and Bruvo's distance [51].

3.3. Genetic Diversity of Fraxinus spp. in Europe

In order to compare the genetic relationships and diversity between the manna ash collection (sample set #1) and the germplasm of *F. angustifolia* and *F. excelsior* throughout the partially overlapping natural distribution ranges of both species across Europe, we combined our SSR profiles of manna ash with the genotypic profiles described in Gérard et al. (2013) [14] into a second dataset called sample set #2, which included 495 unique profiles at 11 SSRs. UPGMA analysis based on Bruvo's distance [51] identified groups based on species (Figure 2; Table S2). Indeed, three main clusters were highlighted, with two branches represented almost exclusively by *F. angustifolia* and *F. excelsior* samples, respectively. As expected, samples belonging to *F. ornus* clustered as the most distant genotypes, behaving as outgroup, while the individuals morphologically classified as hybrids were distributed across the two main groups. Interestingly, *F. angustifolia* samples coming from Southern Italian regions from Gérard et al. (2013) [14] data clustered together with genotypes of our manna ash collection (Figure 2), highlighting a distinctive, basal position of Southern Italian *F. angustifolia* germplasm in comparison with the range-wide collection of the species.

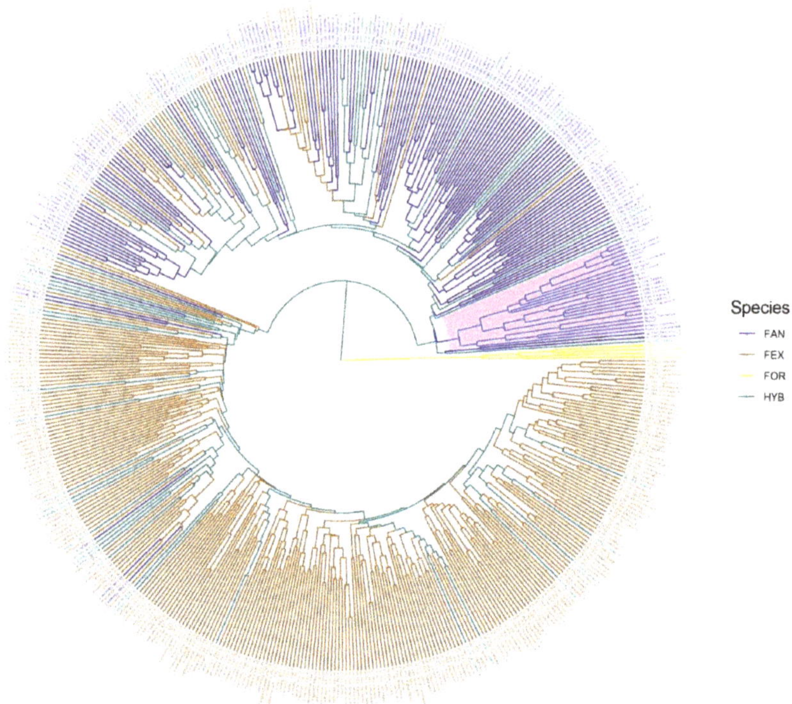

Figure 2. Genetic relationships among *Fraxinus* spp. accessions-based UPGMA and Bruvo's distance, developed using unique nSSR profiles of sample belonging to set #2, including *F. angustifolia* (FAN), *Fraxinus excelsior* (FEX), and *F. angustifolia* x *F. excelsior* (HYB). *F. ornus* (FOR) was used as outgroup. Manna ash genetic profiles collected in the South of Italy are highlighted (in pink) in the phylogenetic tree.

Interestingly, within FAN, samples collected in Southern Italy (Sicily and Calabria) showed lower genetic distance (Nei, 1973 [55]) and lower *Fst* values with respect to the other FAN populations collected in Europe, whereas comparison to FEX populations resulted in higher values (Table 3). Hybrids of both ash species were closer both to FAN (*Fst* = 0.030; *Nei* = 0.159) and FEX (*Fst* = 0.028; *Nei* = 0.110) than to FAN (Italy) (*Fst* = 0.107; *Nei* = 0.536) (Table 3).

Table 3. *Fst* values (below diagonal) and *Nei* (1973) genetic distances (above diagonal) evaluated through nSSR on sample set #2. Each parameter was calculated for plants belonging to *F. angustifolia* collected in Italy—FAN (Italy), *F. angustifolia* sampled across Europe—FAN, *F. excelsior*—FEX, and hybrids—HYB (*F. angustifolia* × *F. excelsior*).

Group.	FAN (Italy)	FAN	HYB	FEX
FAN (Italy)	0.000	0.284	0.536	0.874
FAN	0.066	0.000	0.159	0.437
HYB	0.107	0.030	0.000	0.110
FEX	0.163	0.084	0.028	0.000

Different genetic parameters (I, He, Ho, Fis, AR and Fst) were also estimated for *F. angustifolia* sampling locations belonging to set #2 (Table 4). Hungary and France2 locations had, respectively, the lowest (0.504) and highest (0.680) values for genetic diversity (He). Overall, genetic diversity parameters (I, He, Ho) displayed similar values across the locations studied (Table 4). Sixteen out of twenty-one locations showed an excess of heterozygotes, showing a negative inbreeding coefficient (*Fis*). However, except Hungary population (−0.607), *Fis* was close to zero therefore all groups could be considered in equilibrium (Table 4).

Table 4. Summary of genetic variation statistics at 11 nSSR loci on FAN samples belonging to set #2. Individuals were grouped based on sampling location (see Table S2).

Group ID.	*n.*	*I*	*Ho*	*He*	*Fis*	*AR*
Bulgaria	8	0.964	0.615	0.525	−0.208	2.73
Croatia1	8	1.156	0.591	0.563	−0.025	3.05
Croatia2	8	1.000	0.614	0.523	−0.135	2.71
France1	7	1.259	0.645	0.632	−0.030	3.32
France2	7	1.320	0.803	0.680	−0.216 *	3.42
France3	8	1.316	0.614	0.658	0.039	3.32
France4	8	1.171	0.659	0.607	−0.074	3.03
France5	8	1.094	0.489	0.572	0.204 **	2.93
France6	8	1.256	0.648	0.626	−0.042	3.22
France7	8	0.953	0.659	0.513	−0.269	2.63
Hungary	7	0.801	0.805	0.506	−0.607 ***	2.28
Italy1	16	1.170	0.591	0.553	0.040 ***	2.96
Italy2	8	1.032	0.623	0.541	−0.112 ***	2.85
Macedonia	8	1.325	0.750	0.651	−0.152	3.39
Montenegro	8	1.215	0.659	0.607	−0.087	3.13
Portugal1	6	1.185	0.621	0.624	0.017	3.17
Portugal2	8	1.029	0.534	0.529	0.038	2.86
Serbia1	8	1.323	0.623	0.623	−0.022	3.40
Serbia2	6	1.193	0.667	0.580	−0.159	3.26
Turkey	3	1.089	0.727	0.601	−0.241	3.55
Ukraine	8	1.040	0.610	0.553	−0.109	2.88

n. = number of samples for each group; *I* = Shannon's index; *Ho* = observed heterozygosity; *He* = expected; *Fis* = inbreeding coefficient; *AR* = allelic richness; * $p < 0.05$; ** $p < 0.01$; *** $p < 0.001$.

Overall allelic richness (*AR*) ranged from 1.173 (SSR431) to 3.746 (Fem12) for SSR431 and Fem12 markers, respectively (Table S3). *AR* was higher in trees collected in Turkey (3.55) than the other populations. Macedonia, Montenegro, Portugal1, five out of seven French groups and the two population collected in Serbia showed similar *AR* values (ranging from 3.05 to 3.42), while population belonging to Hungary had the lowest value (2.28; Table 4).

The *Fst* values were also used to assess the genetic relationships across the population belonging to the *F. angustifolia* collection here studied (Figure 3; Table S4). Three main clusters were observed in the NJ-tree: six out of seven populations collected in France were gathered together with the two Italian groups and Portugal2 population; except the samples collected in Croatia, the other Balkan populations clustered with samples belonging to Turkey and Hungary; and, finally, the remaining populations of France (France7), Balkans (Croatia1 and 2) and Portugal (Portugal1) were linked to Bulgaria and Ukraine populations (Figure 3). On the other hand, comparing the *Fst* recorded on samples grouped by country, the greatest distance was observed between the pair Hungary–Bulgaria; except Croatia, the populations belonging to Balkan clustered close in the NJ-tree. Portugal and France were genetically separated, with the last population near the branch that gathered Turkey and Italy (Figure S1).

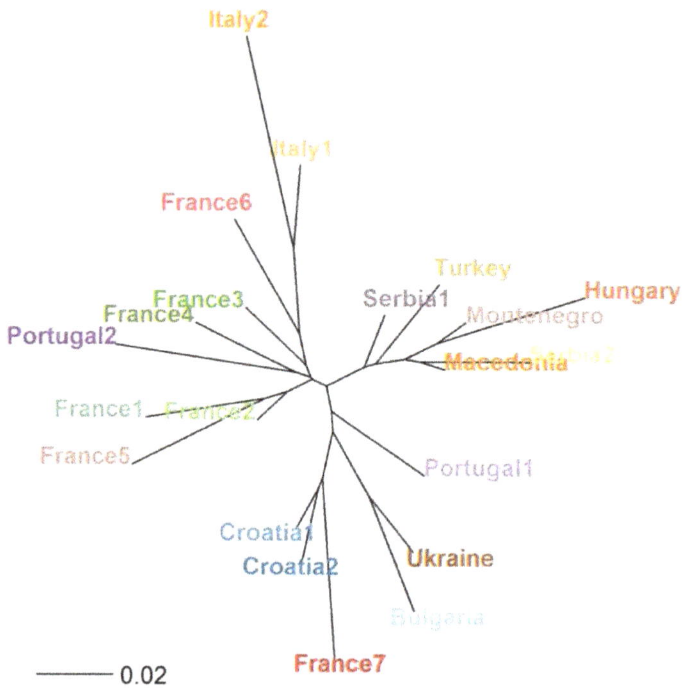

Figure 3. Neighbor-joining (NJ) tree based on population pairwise fixation index. Genetic distances were computed among populations grouped based on sampling locations.

DAPC identified five clusters corresponding to the five groups selected (Figure 4). Clusters including the samples belonging to France and Portugal were overlapping, as well as clusters containing Balkans and Eastern European genotypes. On the contrary all samples collected in Italy (from Calabria and Sicily) were separated from the other groups. Except for some samples, the plot of the first two principal components distinguished clearly three groups represented by genotypes from Eastern, Western and Mediterranean Europe, respectively (Figure 4).

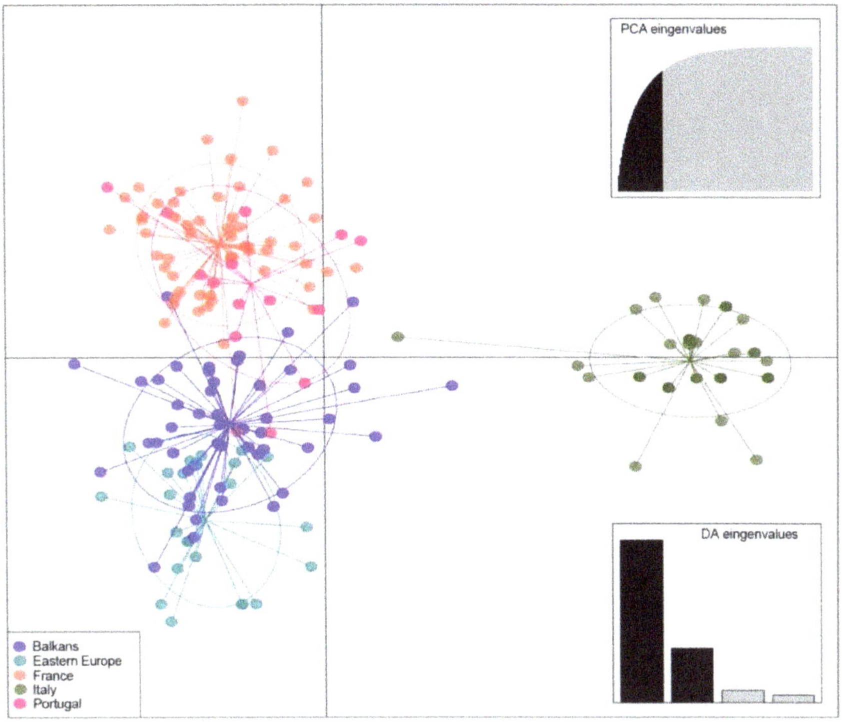

Figure 4. DAPC clustering of European germplasm of *F. angustifolia* studied, using the first two principal components (*Y*-axis and *X*-axis, respectively). The samples are grouped in five main groups: Balkans (blue; Croatia, Macedonia, Montenegro, Serbia, and Turkey), Eastern Europe (cyan; Bulgaria, Hungary, and Ukraine), France (orange), Italy (green) and Portugal (pink).

4. Discussion

The genetics of endangered *Fraxinus* spp. populations is of great interest for both conservation and evolutionary aspects. Using a multidisciplinary approach, including molecular analysis, we demonstrated in this study that an important and residual manna ash germplasm collection located in the Madonie Mountains (Sicily, Italy), belongs to *F. angustifolia*, the narrow-leaved ash. We also showed that this population contained high genetic diversity and, together with a population from Calabria, was strongly differentiated from *F. angustifolia* populations from other locations in Europe (Balkans, Eastern Europe, France, and Portugal). The assessment of the genetic diversity is crucial to plan activities aimed to conservation of this important and endangered forest tree genetic resource.

Notwithstanding the limited number of individuals studied, plants belonging to the manna ash collection analyzed showed a wide variation in several phenotypic traits, such as leaf characteristics, inflorescence and infructescence features, plant vigor and morphology, showing a continuous and narrow range of values for individual traits. Leaf morphology is the strongest differentiating character within the local collection, although this trait is not successfully used to characterize the varieties, being strongly influenced by environmental factors. Indeed, it is well known that plants react to water stress, which is particularly strong in the Mediterranean area, by modifying their leaf traits. Considering that the trees collected come from a heterogeneous mountain environment in which neighboring individuals compete for the same ground and light resources, the observed phenotypic variations might be attributed to the impact of environmental factors on growth habit. Regarding fruit morphology, the shape and the length of the fruits of "manna" varieties were quite heterogeneous, as well as the colour or samaras changing from light green of most cultivars to veined brown of "Nsiriddu". These traits; thus, showed a weak distinguishing power [62]. Nevertheless distinctive morphological traits separated all manna ash varieties from *F. ornus*. Specifically, in the local varieties the samara stalk length was greater (nearly three times) than in *F. ornus*, while the apical leaf length/width ratio and the number of samaras/raceme of manna ash samples were approximately two and a half and eight times lower, respectively, separating the two groups compared. Variety identification based on morphological data can be verified using molecular analysis, because the environmental effects, epistatic interactions and pleiotropic effects can interfere with morphological traits evaluation, as in many forest trees [63].

The genome size showed a small change in the 2C nuclear DNA content within the samples belonging to the manna ash collection (from 1.539 to 1.544 pg). The existence of intraspecific variation in genome size has been reported in several plant species [64–66]. However, due to the little shift here reported, the genome size, as well as the ploidy level, can be considered stable in our collection, as commonly observed within the species [67]. The variations recorded can be explained by technical issues, as reported for some angiosperms [68,69]. The 2C DNA values observed in the manna ash collection unambiguously assign the manna germplasm to *F. angustifolia* (1.540 pg [60]) and are different from both *F. excelsior* (1.68 pg; [60]), and *F. ornus* (1.98 pg; [60]), showing a larger genome size due to probably a high number of repetitive elements and/or several ribosomal gene repetitions [60].

Molecular analyses confirmed the differences highlighted through morphological and cytological approaches. In agreement with the low chloroplast DNA mutation rate detected in the Oleaceae [70], cpSSRs and plastid DNA regions identified a low genetic diversity in the manna ash collection, identifying three haplotypes, two sub-variants of H5 (covering 92% of samples) and H10 (the remaining 8%), respectively. The haplotype detected in the reference trees (H19) was previously found only in *F. ornus*, specifically in trees from Italy and Corsica. In agreement with the increased botanical similarity of *F. angustifolia* and *F. excelsior* and a partially overlapping phenology [6,71,72], an extensive sharing of cpDNA haplotypes between *F. angustifolia* and *F. excelsior* has been found previously but without common profiles between them and *F. ornus* [4]. Indeed, both H5 and H10 variants have been shown to be shared haplotypes between *F. angustifolia* and *F. excelsior* although in different geographic regions [4,73]. The majority of manna ash accessions carried haplotype H05, a haplotype observed in *F. angustifolia* throughout central and southern Italy and in *F. excelsior* in a restricted area of the Eastern Alps; this particular distribution could be related to two possible different and independent glacial refuges of ash species [73]. The other haplotype (H10) observed in "Abbassa cappeddu" variety and in some trees belonging to "Verdello" was detected in *F. angustifolia* trees collected in Portugal and Corsica [4] and only in the Czech Republic for *F. excelsior*.

Microsatellites are reported as very effective marker in terms of high information content and discrimination power owing to high allelic variation and allowing clear identification of populations or varieties in several plant species [74–77]. Nuclear SSR (nSSR) profiles showed a noteworthy rate of polymorphism and, except for "Baciciu" and "Cavolo", were able to group each accession with its own variety. The most represented local variety named "Verdello" was grouped in a main cluster, together with tree belonging to "Nsiriddu", showing a high variability, in agreement to the profiles obtained in natural populations of *F. angustifolia* in Greece [78]. Among studied varieties, "Verdello" and "Nsiriddu" showed an upright growth habit and are considered the most productive ones [18], yielding high quality manna with very similar chemical and organoleptic characteristics.

Although only a slight genetic differentiation between *F. angustifolia* and *F. excelsior* was reported [79,80], our findings highlighted a clear separation between trees belonging to two species collected throughout their natural distribution across Europe. Significant patterns of distance were found among species, with *F. ornus* clearly behaving like outgroup. Interestingly, a private branch grouping only *F. angustifolia* trees collected in the South of Italy, including the profiles recorded for manna ash collection, was highlighted. In agreement to cluster analysis, the pairwise fixation index (*Fst*) highlighted a clear distinctness for the populations from Southern Italy, close to Turkish and French populations. In addition both cluster and *Fst* analysis allowed to separate the two Italian populations highlight the feature belonging to manna ash collection. Our results are consistent with *Fraxinus* spp. distribution, indeed common ash (*F. excelsior*) is found mainly in the north regions throughout Europe [4,81] and has expanded its range in the south probably during cooler climate episodes, maintaining the current relic populations in temperate locations such as the Elburz Mountains (Turkey), Calabria (Italy), and in Sicily, more specifically in the Nebrodi Mountains, where a small population (200 plants) of *F. excelsior* spp. *siciliensis* [82,83] is located in three different sites (Caronia, Longi and Alcara Li Fusi), showing specific features (e.g., reduced size and blooming at the same time of leaf emission; [84] due to the high isolation. On the contrary, *F. angustifolia* is mostly restricted to the Mediterranean region [81] and the favorable environmental conditions allowed this species to spread in the Sicilian Nebrodi Mountains, giving rise at the end of 19th century more than 16 varieties distinguished for both morphological traits and manna production [21,85]. The remarkable diversity and strong genetic structure in the Sicilian collection of *F. angustifolia* highlighted by both cp and nSSRs probably is related to the geographic isolation in which they have been found for about 10,000 years starting from the end of the last glaciation. This trend of speciation in *Fraxinus*, related to specific geographic zones, is in agreement to previous work [71], and would be driven by geological and climatic modifications [86] as reported for other species [87]. Phylogenetic analysis showed also a distribution of hybrids (HYB) within both *F. excelsior* and *F. angustifolia* groups, reflecting the sharing of haplotypes between the two species [4]. More interestingly, in the last group the hybrids and trees belonging to common ash were linked to *F. angustifolia* trees collected in Balkan and Eastern Europe, in agreement to the species distribution [4], whereas the samples from Southern Italy were the most genetically distant plants. Our finding can be explained by the marked different flowering time that characterizes the two species in Sicily making the interspecific cross unfavorable.

Finally, grouping the *F. angustifolia* trees studied on the grounds of their geographic area, discriminant analysis based on nuclear microsatellites underlined a gradient from west to east in *Y*-axis. This evidence is consistent with the classification of *F. angustifolia* that can be grouped into three geographic subspecies: (i) *F. angustifolia* Vahl ssp. *angustifolia* (Portugal and the Western Mediterranean), (ii) ssp. *oxycarpa* (M. Bieb. ex Willd.) Franco and Rocha Afonso (Northeast Spain to Turkey), and (iii) ssp. *syriaca* (Boiss.) Yalt (Turkey and Asia Minor) [6–8]. Additionally, samples belonging to Calabria and Sicily were separated from the others. The clear diversity harbored by *F. angustifolia* populations from South Italy, allow to hypothesize a possible glacial refuge of species in this area, according to a temperature increasing during the late glacial maximum (18,000 years bp) in the Mediterranean eastern compared to western one, as already reported for the Turkish populations [4,88].

5. Conclusions

In summary, using a multidisciplinary approach the present study was focused to characterize in depth a historical manna ash collection represented by local varieties collected in the Madonie area (Sicily, Italy), in order to safeguard an important cultural heritage. Our finding definitively clarifies that the local varieties actually used for manna production belong to *F. angustifolia*. This study provides useful information for germplasm management, finalized to improve the production and productivity of agroforestry species investigated. In addition, the evidences here reported suggest the presence of an additional glacial refuge for *F. angustifolia* in Italy, confirming the importance of Sicily as source of biodiversity. Furthermore, our study could represent an invitation for botanists to expand the historical knowledge of the collection. Due to the importance of ash germplasm studied, all local varieties were propagated in two repository fields, at the Professional Institute for Agriculture and the Environment (I.P.A.A.) "Luigi Failla Tedaldi" (Castelbuono, Italy) and at the private Schicchi's repository located in the Croce-Foresta district (Castelbuono, Italy) (info at rosario.schicchi@unipa.it) making available this genetic resources for future breeding programs and to develop future national/international collaborations.

Supplementary Materials: The following are available online at http://www.mdpi.com/2223-7747/9/8/1035/s1, Table S1: Morphological traits of studied samples, belonging to manna ash collection and samples belonging to *F. ornus*, used as reference. In brackets the number of samples analyzed belonging to each local variety/species. Three replicates for each trait were analyzed. Table S2: Nuclear SSR profiles of samples belonging to (A) set #1 and (B) set #2. Table S3: Allelic richness (*AR*) evaluated for FAN accessions grouped based on sampling location. In the table geographic coordinates (Long and Lat) and AR values for each nSSR were indicated. Table S4: Pairwise population Fst values. Figure S1: Neighbor-joining (NJ) tree based on population pairwise fixation index. Genetic distances were computed among populations grouped based on country of origin.

Author Contributions: Conceptualization, L.A., F.M., S.F.d.B. and R.S.; Data curation, L.A., F.M. and M.H.; Formal analysis, L.A. and F.M.; Funding acquisition, R.S.; Investigation, L.A. and F.M.; Resources, G.D.N. and R.S.; Supervision, F.M.; Writing-original draft, L.A., F.M., M.H. and S.F.d.B.; Writing-review & editing, L.A., F.M., G.D.N., M.H., F.C., S.F.d.B. and R.S. All authors have read and agreed to the published version of the manuscript.

Funding: This research received no external funding.

Acknowledgments: The authors are grateful to Maria Luisa Virga (Assessorato dell'Agricoltura, dello Sviluppo Rurale e della Pesca Mediterranea-Dipartimento Regionale dell'Agricoltura-Regione Sicilia) for encouraging the meeting between the research groups.

Conflicts of Interest: The authors declare no conflict of interest.

References

1. Wallander, E.; Albert, V.A. Phylogeny and classification of Oleaceae based on rps16 and trnL-F sequence data. *Am. J. Bot.* **2000**, *87*, 1827–1841. [CrossRef] [PubMed]
2. Hinsinger, D.D.; Gaudeul, M.; Couloux, A.; Bousquet, J.; Frascaria-Lacoste, N. The phylogeography of Eurasian Fraxinus species reveals ancient transcontinental reticulation. *Mol. Phylogenet. Evol.* **2014**, *77*, 223–237. [CrossRef] [PubMed]
3. Wallander, E. Systematics of Fraxinus (Oleaceae) and evolution of dioecy. *Plant Syst. Evol.* **2008**, *273*, 25–49. [CrossRef]
4. Heuertz, M.; Carnevale, S.; Fineschi, S.; Sebastiani, F.; Hausman, J.F.; Paule, L.; Vendramin, G.G. Chloroplast DNA phylogeography of European ashes, Fraxinus sp. (Oleaceae): Roles of hybridization and life history traits. *Mol. Ecol.* **2006**, *15*, 2131–2140. [CrossRef] [PubMed]
5. Picard, J.-F. Contribution à l'étude de la biologie florale et de la fructification du Frêne commun (*Fraxinus excelsior* L.). *Rev. For. Fr.* **1982**, 97. [CrossRef]
6. Wallander, E. Evolution of Wind-Pollination in Fraxinus (Oleaceae): An Ecophylogenetic Approach. Ph.D. Thesis, Göteborg University, Göteborg, Sweden, 2001.
7. Tutin, T.G.; Heywood, V.H.; Burges, N.A.; Moore, D.M.; Valentine, D.H.; Walters, S.M.; Webb, D.E.; Chater, A.O.; Defilipps, R.A.; Richardson, I.B.K.; et al. *Flora Europaea: Plantaginaceae to Compositae (and Rubiaceae)*; Cambridge University Press: Cambridge, UK, 1972.
8. Huntley, B.; Birks, H.J.B.; Harry, J.B. *Atlas of Past and Present Pollen Maps for Europe, 0–13,000 Years Ago*; Cambridge University Press: Cambridge, UK, 1983; ISBN 978-0-521-23735-2.
9. Temunović, M.; Franjić, J.; Satovic, Z.; Grgurev, M.; Frascaria-Lacoste, N.; Fernández-Manjarrés, J.F. Environmental Heterogeneity Explains the Genetic Structure of Continental and Mediterranean Populations of Fraxinus angustifolia Vahl. *PLoS ONE* **2012**, *7*, e42764. [CrossRef]
10. FRAXIGEN, 2005 Ash Species in Europe: Biological Characteristics and Practical Guidelines for Sustainable Use—ProQuest. Available online: https://search.proquest.com/openview/e333985a9ac22793988179029545b7da/1?pq-origsite=gscholarandcbl=29663 (accessed on 17 June 2020).
11. Thomas, P.A. Biological Flora of the British Isles: Fraxinus excelsior. *J. Ecol.* **2016**, *104*, 1158–1209. [CrossRef]
12. Caudullo, G.; Houston Durrant, T. Fraxinus angustifolia in Europe: Distribution, habitat, usage and threats. In *European Atlas of Forest Tree Species 2016*; San-Miguel-Ayanz, J., de Rigo, D., Caudullo, G., Houston Durrant, T., Mauri, A., Eds.; Publications Office of the European Union: Luxembourg, 2016; p. 97.
13. Bogdan, S.; Kajba, D.; Franjic, J.; Idzojtic, M.; Skvorc, Z.; Trupčević, I.K. Genetic variation in quantitative traits within and among Croatian narrow-leaved ash (Fraxinus angustifolia Vahl) populations assessed in an open-pollinated progeny test. *Period. Boil.* **2007**, *109*, 1–11.

14. Gérard, P.R.; Temunović, M.; Sannier, J.; Bertolino, P.; Dufour, J.; Frascaria-Lacoste, N.; Fernández-Manjarrés, J.F. Chilled but not frosty: Understanding the role of climate in the hybridization between the Mediterranean Fraxinus angustifolia Vahl and the temperate *Fraxinus excelsior* L. (Oleaceae) ash trees. *J. Biogeogr.* **2013**, *40*, 835–846. [CrossRef]
15. Noti, R.; van Leeuwen, J.F.N.; Colombaroli, D.; Vescovi, E.; Pasta, S.; La Mantia, T.; Tinner, W. Mid- and late-Holocene vegetation and fire history at Biviere di Gela, a coastal lake in southern Sicily, Italy. *Veg. Hist. Archaeobot.* **2009**, *18*, 371–387. [CrossRef]
16. Médail, F.; Diadema, K. Glacial refugia influence plant diversity patterns in the Mediterranean Basin. *J. Biogeogr.* **2009**, *36*, 1333–1345. [CrossRef]
17. Mazzola, P.; Schicchi, R.; Spadaro, V. Ricerche finalizzate al recupero e conservazione del germoplasma dei frassini da manna sulle Madonie (Sicilia settentrionale). In Proceedings of the 101 Congresso Società Botanica Italiana, Caserta, Italy, 27–29 September 2006; p. 137.
18. Schicchi, R.; Camarda, L.; Spadaro, V.; Pitonzo, R. Caratterizzazione chimica della manna estratta nelle Madonie (Sicilia) da cultivar di Fraxinus angustifolia e di Fraxinus ornus (Oleaceae). *Quad. Bot. Amb. Appl.* **2006**, *17*, 151–174.
19. Giardinieri, A.; Schicchi, R.; Geraci, A.; Rosselli, S.; Maggi, F.; Fiorini, D.; Ricciutelli, M.; Loizzo, M.R.; Bruno, M.; Pacetti, D. Fixed oil from seeds of narrow-leaved ash (F. angustifolia subsp. angustifolia): Chemical profile, antioxidant and antiproliferative activities. *Food Res. Int.* **2019**, *119*, 369–377. [CrossRef]
20. Guarcello, R.; Gaglio, R.; Todaro, A.; Alfonzo, A.; Schicchi, R.; Cirlincione, F.; Moschetti, G.; Francesca, N. Insights Into the Cultivable Microbial Ecology of "Manna" Ash Products Extracted From Fraxinus angustifolia (Oleaceae) Trees in Sicily, Italy. *Front. Microbiol.* **2019**, *10*, 984. [CrossRef]
21. Minà Palumbo, F. Monografia sulla coltivazione del frassino. *Agric. Lal.* **1876**, *1875*, 19–25.
22. Falci, R. Il frassino da manna in Sicilia. *Boll. R. Orto Bot. Giard. Col. Palermo* **1910**, *9*, 145–168.
23. Weber, J.L. Informativeness of human (dC-dA) n (dG-dT) n polymorphisms. *Genomics* **1990**, *7*, 524–530. [CrossRef]
24. Morgante, M.; Olivieri, A.M. PCR-amplified microsatellites as markers in plant genetics. *Plant J.* **1993**, *3*, 175–182. [CrossRef]
25. Bacchi, M.; Leone, M.; Mercati, F.; Preiti, G.; Sunseri, F.; Monti, M. Agronomic Evaluation and Genetic Characterization of Different Accessions in Lentil (Lens culinaris Medik.). *Ital. J. Agron.* **2010**, *5*, 303. [CrossRef]
26. Jiao, Y.; Jia, H.; Li, X.; Chai, M.; Jia, H.; Chen, Z.; Wang, G.; Chai, C.; van de Weg, E.; Gao, Z. Development of simple sequence repeat (SSR) markers from a genome survey of Chinese bayberry (Myrica rubra). *BMC Genom.* **2012**, *13*, 201. [CrossRef]
27. Mercati, F.; Longo, C.; Poma, D.; Araniti, F.; Lupini, A.; Mammano, M.M.; Fiore, M.C.; Abenavoli, M.R.; Sunseri, F. Genetic variation of an Italian long shelf-life tomato (*Solanum lycopersicon* L.) collection by using SSR and morphological fruit traits. *Genet. Resour. Crop Evol.* **2015**, *62*, 721–732. [CrossRef]
28. Mercati, F.; Catarcione, G.; Paolacci, A.R.; Abenavoli, M.R.; Sunseri, F.; Ciaffi, M. Genetic diversity and population structure of an Italian landrace of runner bean (*Phaseolus coccineus* L.): Inferences for its safeguard and on-farm conservation. *Genetica* **2015**, *143*, 473–485. [CrossRef] [PubMed]
29. King, R.A.; Ferris, C. Chloroplast DNA phylogeography of *Alnus glutinosa* (L.) Gaertn. *Mol. Ecol.* **1998**, *7*, 1151–1161. [CrossRef]
30. Palmé, A.E. Chloroplast DNA variation, postglacial recolonization and hybridization in hazel, Corylus avellana. *Mol. Ecol.* **2002**, *11*, 1769–1779. [CrossRef] [PubMed]
31. Rajora, O.P.; Dancik, B.P. Chloroplast DNA inheritance in Populus. *Theor. Appl. Genet.* **1992**, *84*, 280–285. [CrossRef] [PubMed]
32. Dumolin, S.; Demesure, B.; Petit, R.J. Inheritance of chloroplast and mitochondrial genomes in pedunculate oak investigated with an efficient PCR method. *Theor. Appl. Genet.* **1995**, *91*, 1253–1256. [CrossRef]
33. Harbourne, M.E.; Douglas, G.C.; Waldren, S.; Hodkinson, T.R. Characterization and primer development for amplification of chloroplast microsatellite regions of Fraxinus excelsior. *J. Plant Res.* **2005**, *118*, 339–341. [CrossRef]
34. Erichsen, E.O.; Budde, K.B.; Sagheb-Talebi, K.; Bagnoli, F.; Vendramin, G.G.; Hansen, O.K. Hyrcanian forests—Stable rear-edge populations harbouring high genetic diversity of Fraxinus excelsior, a common European tree species. *Divers. Distrib.* **2018**, *24*, 1521–1533. [CrossRef]

35. Kellogg, E.A. Relationships of cereal crops and other grasses. *PNAS* **1998**, *95*, 2005–2010. [CrossRef]
36. Leitch, I.J.; Chase, M.W.; Bennett, M.D. Phylogenetic Analysis of DNA C-values Provides Evidence for a Small Ancestral Genome Size in Flowering Plants. *Ann. Bot.* **1998**, *82*, 85–94. [CrossRef]
37. Doležel, J.; Bartoš, J. Plant DNA Flow Cytometry and Estimation of Nuclear Genome Size. *Ann. Bot.* **2005**, *95*, 99–110. [CrossRef] [PubMed]
38. Doležel, J.; Greilhuber, J.; Suda, J. Estimation of nuclear DNA content in plants using flow cytometry. *Nat. Protoc.* **2007**, *2*, 2233–2244. [CrossRef] [PubMed]
39. Petrov, D.A.; Sangster, T.A.; Johnston, J.S.; Hartl, D.L.; Shaw, K.L. Evidence for DNA Loss as a Determinant of Genome Size. *Science* **2000**, *287*, 1060–1062. [CrossRef] [PubMed]
40. Bennetzen, J.L.; Ma, J.; Devos, K.M. Mechanisms of Recent Genome Size Variation in Flowering Plants. *Ann. Bot.* **2005**, *95*, 127–132. [CrossRef] [PubMed]
41. Harkess, A.; Mercati, F.; Abbate, L.; McKain, M.; Pires, J.C.; Sala, T.; Sunseri, F.; Falavigna, A.; Leebens-Mack, J. Retrotransposon Proliferation Coincident with the Evolution of Dioecy in Asparagus. *G3 Genes Genomes Genet.* **2016**, *6*, 2679–2685. [CrossRef]
42. Murray, B.G. When does Intraspecific C-value Variation become Taxonomically Significant? *Ann. Bot.* **2005**, *95*, 119–125. [CrossRef]
43. Garnatje, T.; Garcia, S.; Vilatersana, R.; Vallès, J. Genome Size Variation in the Genus Carthamus (Asteraceae, Cardueae): Systematic Implications and Additive Changes During Allopolyploidization. *Ann. Bot.* **2006**, *97*, 461–467. [CrossRef]
44. Raimondo, F.M.; Bonomo, R.; Lentini, F. Il frassino da manna in Sicilia, pianta officinale da rivalutare. *Giorn. Bot. Ital.* **1980**, *114*, 141–142.
45. Otto, F.J. Preparation and Staining of Cells for High-Resolution DNA Analysis. In *Flow Cytometry and Cell Sorting*; Radbruch, A., Ed.; Springer: Berlin/Heidelberg, Germany, 1992; pp. 65–68. ISBN 978-3-662-02785-1.
46. Weising, K.; Gardner, R.C. A set of conserved PCR primers for the analysis of simple sequence repeat polymorphisms in chloroplast genomes of dicotyledonous angiosperms. *Genome* **1999**, *42*, 9–19. [CrossRef]
47. Lefort, F.; Brachet, S.; Frascaria-Lacoste, N.; Edwards, K.J.; Douglas, G.C. Identification and characterization of microsatellite loci in ash (*Fraxinus excelsior* L.) and their conservation in the olive family (Oleaceae). *Mol. Ecol.* **1999**, *8*, 1088–1089. [CrossRef]
48. Brachet, S.; Jubier, M.F.; Richard, M.; Jung-Muller, B.; Frascaria-Lacoste, N. Rapid identification of microsatellite loci using 5′ anchored PRC in the common ash Fraxinus excelsior. *Mol. Ecol.* **1999**, *8*, 160–163.
49. Aggarwal, R.K.; Allainguillaume, J.; Bajay, M.M.; Barthwal, S.; Bertolino, P.; Chauhan, P.; Consuegra, S.; Croxford, A.; Dalton, D.L.; Belder, E.D.; et al. Permanent Genetic Resources added to Molecular Ecology Resources Database 1 August 2010–30 September 2010. *Mol. Ecol. Resour.* **2011**, *11*, 219–222. [CrossRef] [PubMed]
50. Garfi, G.; Mercati, F.; Fontana, I.; Collesano, G.; Pasta, S.; Vendramin, G.G.; De Michele, R.; Carimi, F. Habitat features and genetic integrity of wild grapevine *Vitis vinifera* L. subsp. sylvestris (C.C. Gmel.) Hegi populations: A case study from Sicily. *Flora Morphol. Distrib. Funct. Ecol. Plants* **2013**, *208*, 538–548. [CrossRef]
51. Bruvo, R.; Michiels, N.K.; D'souza, T.G.; Schulenburg, H. A simple method for the calculation of microsatellite genotype distances irrespective of ploidy level. *Mol. Ecol.* **2004**, *13*, 2101–2106. [CrossRef] [PubMed]
52. Kamvar, Z.N.; Tabima, J.F.; Grünwald, N.J. Poppr: An R package for genetic analysis of populations with clonal, partially clonal, and/or sexual reproduction. *PeerJ* **2014**, *2*, e281. [CrossRef]
53. Jombart, T.; Ahmed, I. Adegenet 1.3-1: New tools for the analysis of genome-wide SNP data. *Bioinformatics* **2011**, *27*, 3070–3071. [CrossRef]
54. Wright, S. The Interpretation of Population Structure by F-Statistics with Special Regard to Systems of Mating. *Evolution* **1965**, *19*, 395–420. [CrossRef]
55. Nei, M. Analysis of Gene Diversity in Subdivided Populations. *PNAS* **1973**, *70*, 3321–3323. [CrossRef]
56. Goudet, J. Hierfstat, a package for r to compute and test hierarchical F-statistics. *Mol. Ecol. Notes* **2005**, *5*, 184–186. [CrossRef]
57. Peakall, R.; Smouse, P.E. genalex 6: Genetic analysis in Excel. Population genetic software for teaching and research—An update. *Bioinformatics* **2012**, *28*, 2537–2539. [CrossRef]

58. Raymond, M.; Rousset, F. An Exact Test for Population Differentiation. *Evolution* **1995**, *49*, 1280–1283. [CrossRef] [PubMed]
59. Hardy, O.J.; Vekemans, X. Spagedi: A versatile computer program to analyse spatial genetic structure at the individual or population levels. *Mol. Ecol. Notes* **2002**, *2*, 618–620. [CrossRef]
60. Siljak-Yakovlev, S.; Temunović, M.; Robin, O.; Raquin, C.; Frascaria-Lacoste, N. Molecular–cytogenetic studies of ribosomal RNA genes and heterochromatin in three European Fraxinus species. *Tree Genet. Genomes* **2014**, *10*, 231–239. [CrossRef]
61. Ivanova, D.; Dimitrova, D.; Vladimirov, V. Chromosome numbers of selected woody species from the Bulgarian. *Phytol. Balc.* **2006**, *12*, 79–84.
62. Jarni, K.; Westergren, M.; Kraigher, H.; Brus, R. Morphological variability of Fraxinus angustifolia Vahl in the north-western Balkans. *Acta Soc. Bot. Pol.* **2011**, *80*, 245–252. [CrossRef]
63. Ziegenhagen, B.; Fladung, M. DNA Markers for Identification and Evaluation of Genetic Resources in Forest Trees: Case Studies in Abies, Picea and Populus. In *Molecular Marker Systems in Plant Breeding and Crop Improvement*; Lörz, H., Wenzel, G., Eds.; Springer: Berlin/Heidelberg, Germany, 2005; pp. 413–429. ISBN 978-3-540-26538-2.
64. Schmuths, H.; Meister, A.; Horres, R.; Bachmann, K. Genome Size Variation among Accessions of Arabidopsis thaliana. *Ann. Bot.* **2004**, *93*, 317–321. [CrossRef]
65. Wang, Y.; Bigelow, C.A.; Jiang, Y. Ploidy Level and DNA Content of Perennial Ryegrass Germplasm as Determined by Flow Cytometry. *HortScience* **2009**, *44*, 2049–2052. [CrossRef]
66. Díez, C.M.; Gaut, B.S.; Meca, E.; Scheinvar, E.; Montes-Hernandez, S.; Eguiarte, L.E.; Tenaillon, M.I. Genome size variation in wild and cultivated maize along altitudinal gradients. *New Phytol.* **2013**, *199*, 264–276. [CrossRef]
67. Eilam, T.; Anikster, Y.; Millet, E.; Manisterski, J.; Sagi-Assif, O.; Feldman, M. Genome size and genome evolution in diploid Triticeae species. *Genome* **2007**, *50*, 1029–1037. [CrossRef]
68. Greilhuber, J. Intraspecific Variation in Genome Size in Angiosperms: Identifying its Existence. *Ann. Bot.* **2005**, *95*, 91–98. [CrossRef]
69. Loureiro, J.; Rodriguez, E.; Doležel, J.; Santos, C. Flow Cytometric and Microscopic Analysis of the Effect of Tannic Acid on Plant Nuclei and Estimation of DNA Content. *Ann. Bot.* **2006**, *98*, 515–527. [CrossRef] [PubMed]
70. Besnard, G.; Khadari, B.; Baradat, P.; Bervillé, A. Olea europaea (Oleaceae) phylogeography based on chloroplast DNA polymorphism. *Theor. Appl. Genet.* **2002**, *104*, 1353–1361. [CrossRef] [PubMed]
71. Jeandroz, S.; Roy, A.; Bousquet, J. Phylogeny and Phylogeography of the Circumpolar GenusFraxinus(OLEACEAE) Based on Internal Transcribed Spacer Sequences of Nuclear Ribosomal DNA. *Mol. Phylogenet. Evol.* **1997**, *7*, 241–251. [CrossRef] [PubMed]
72. Morand-Prieur, M.E.; Vedel, F.; Raquin, C.; Brachet, S.; Sihachakr, D.; Frascaria-Lacoste, N. Maternal inheritance of a chloroplast microsatellite marker in controlled hybrids between Fraxinus excelsior and Fraxinus angustifolia. *Mol. Ecol.* **2002**, *11*, 613–617. [CrossRef] [PubMed]
73. Brewer, S. Recolonisation Postglaciaire de Quelques Taxons Tempérés en Europe: Une Approche Spatiale et Temporelle. Ph.D. Thesis, Cornell University, Marseille, France, 2002.
74. Thomson, M.J.; Septiningsih, E.M.; Suwardjo, F.; Santoso, T.J.; Silitonga, T.S.; McCouch, S.R. Genetic diversity analysis of traditional and improved Indonesian rice (*Oryza sativa* L.) germplasm using microsatellite markers. *Theor. Appl. Genet.* **2007**, *114*, 559–568. [CrossRef] [PubMed]
75. Blair, M.W.; Díaz, L.M.; Buendía, H.F.; Duque, M.C. Genetic diversity, seed size associations and population structure of a core collection of common beans (*Phaseolus vulgaris* L.). *Theor. Appl. Genet.* **2009**, *119*, 955–972. [CrossRef]
76. Carimi, F.; Mercati, F.; De Michele, R.; Fiore, M.C.; Riccardi, P.; Sunseri, F. Intra-varietal genetic diversity of the grapevine (*Vitis vinifera* L.) cultivar 'Nero d'Avola' as revealed by microsatellite markers. *Genet. Resour. Crop Evol.* **2011**, *58*, 967. [CrossRef]
77. Laidò, G.; Mangini, G.; Taranto, F.; Gadaleta, A.; Blanco, A.; Cattivelli, L.; Marone, D.; Mastrangelo, A.M.; Papa, R.; De Vita, P. Genetic Diversity and Population Structure of Tetraploid Wheats (*Triticum turgidum* L.) Estimated by SSR, DArT and Pedigree Data. *PLoS ONE* **2013**, *8*, e67280. [CrossRef]

78. Papi, R.M.; Spanos, K.A.; Kyriakidis, D.A. Genetic variation of Fraxinus angustifolia natural populations in Greece based on nuclear and chloroplast microsatellite markers. *Eur. J. For. Res.* **2012**, *131*, 1151–1161. [CrossRef]
79. Fernandez-Manjarres, J.F.; Gerard, P.R.; Dufour, J.; Raquin, C.; Frascaria-Lacoste, N. Differential patterns of morphological and molecular hybridization between Fraxinus excelsior L. and Fraxinus angustifolia Vahl (Oleaceae) in eastern and western France. *Mol. Ecol.* **2006**, *15*, 3245–3257. [CrossRef]
80. Gerard, P.R.; Fernandez-Manjarres, J.F.; Frascaria-Lacoste, N. Temporal cline in a hybrid zone population between Fraxinus excelsior L. and Fraxinus angustifolia Vahl. *Mol. Ecol.* **2006**, *15*, 3655–3667. [CrossRef] [PubMed]
81. Hinsinger, D.D.; Basak, J.; Gaudeul, M.; Cruaud, C.; Bertolino, P.; Frascaria-Lacoste, N.; Bousquet, J. The Phylogeny and Biogeographic History of Ashes (Fraxinus, Oleaceae) Highlight the Roles of Migration and Vicariance in the Diversification of Temperate Trees. *PLoS ONE* **2013**, *8*, e80431. [CrossRef] [PubMed]
82. Ilardi, V.; Raimondo, F.M. The genus *Fraxinus*, L. (Oleaceae) in Sicily. *Flora Mediterr.* **1999**, *9*, 305–318.
83. Schicchi, R.; Marino, P. Taxa relitti della flora forestale siciliana e problemi di conservazione. Relict taxa of the Sicilian forest flora and conservation issues. *Biogeographia* **2011**, *30*, 141–150.
84. Schicchi, R.; Camarda, L.; Di Stefano, V.; Spadaro, V.; Pitonzo, R. Analisi qualitativa dei costituenti cumarinici presenti in campioni di manna prodotti da esemplari di Fraxinus angustifolia (Oleaceae) delle Madonie. In Proceedings of the Atti Del 102 Congresso Della Societa Botanica Italiana, Palermo, Italy, 26–29 September 2007; p. 388.
85. Minà Palumbo, F. Notizie sui frassini di Sicilia e sulla coltivazione dell'Amolleo in Castelbuono. *Stamp. Leg. F. Ruffino Palermo* **1847**, 17.
86. Donoghue, M.J.; Bell, C.D.; Li, J. Phylogenetic Patterns in Northern Hemisphere Plant Geography. *Int. J. Plant Sci.* **2001**, *162*, 41–52. [CrossRef]
87. Magri, D.; Fineschi, S.; Bellarosa, R.; Buonamici, A.; Sebastiani, F.; Schirone, B.; Simeone, M.C.; Vendramin, G.G. The distribution of Quercus suber chloroplast haplotypes matches the palaeogeographical history of the western Mediterranean. *Mol. Ecol.* **2007**, *16*, 5259–5266. [CrossRef]
88. Hayes, A.; Kucera, M.; Kallel, N.; Sbaffi, L.; Rohling, E.J. Glacial Mediterranean sea surface temperatures based on planktonic foraminiferal assemblages. *Quat. Sci. Rev.* **2005**, *24*, 999–1016. [CrossRef]

plants

MDPI

Article

Patterns of Genetic Diversity in Highly Invasive Species: Cogongrass (*Imperata cylindrica*) Expansion in the Invaded Range of the Southern United States (US)

Rima D. Lucardi [1,*], Lisa E. Wallace [2] and Gary N. Ervin [3]

1 USDA Forest Service, Southern Research Station, Forestry Sciences Laboratory, 320 East Green Street, Athens, GA 30602, USA
2 Department of Biological Sciences, Old Dominion University, Norfolk, VA 23529, USA; lewallac@odu.edu
3 Department of Biological Sciences, Mississippi State University, Mississippi State, MS 39762, USA; gervin@biology.msstate.edu
* Correspondence: rima.lucardi@usda.gov; Tel.: +1-(706)559-4278

Received: 11 March 2020; Accepted: 25 March 2020; Published: 31 March 2020

Abstract: The spatial expansions of invasive organisms in the novel range are generally expected to follow an isolation-by-distance relationship (IBD) if the invasion is biologically driven; however, many invasions are facilitated anthropogenically. This research focused on the extant expansion patterns of cogongrass (*Imperata cylindrica*). Cogongrass is a widespread invasive species throughout the southern United States (US). Patterns of infestation vary among US states. Cogongrass is pyrogenic, and its invasion threatens softwood (*Pinus* spp.) plantations, a substantial economic market for this US region. Over 600 individuals were sampled from seven invaded US states, using amplified fragment length polymorphisms (AFLPs) to assess genetic diversity and population structure. We suspected that differences in historical management efforts among US states influenced differences in genetic diversity and structure. We detected two genetic lineages at the highest level of analysis. One genetic lineage was locally restricted, whereas the other was found throughout the study region. Admixed individuals were found in all US states and consistently co-occurred with the dominant lineage, suggesting that secondary contact and hybridization may have facilitated expansion. The widespread prevalence of only one of the two detected genetic lineages suggests a primary genetic lineage responsible for on-going population expansion in the US.

Keywords: AFLP; genetic diversity; invasive; Poaceae; population genetics; range expansion

1. Introduction

Biological invasions continue to increase during this era of increased global connectivity [1], while research has sought to understand the biological mechanisms which contribute to novel invasion success or the failure to establish or expand beyond incipient populations [2]. Ultimately, the success of a novel plant invasion can be constrained by its biological and evolutionary limits, such as its evolutionary history, geographic origin, propagule pressure and multiple introductions, or a species' ability to adapt to novel environments [3–6]. However, biology is not the only factor in successful plant invasions; the consequences of anthropogenic activity and vectoring are also influential [5,7,8]. It is known that changes in the structure and/or function of the recipient habitat may provide or limit opportunities for invasion success and can be possibly aided by human-assisted dispersal [9–13]. Disturbances can create new habitat resources (i.e., increased availability of canopy openings, or increased frequency of edge habitat) or facilitate the successful dispersal of invasive plant propagules (creation of "stepping-stone" patches, novel dispersal routes and/or mechanisms) in the spread and establishment of invasive plants [14–16]. Habitat-modifying activities by humans are common in the regular maintenance and management of rights-of-way, roads, and in agricultural and silvicultural practices in the United States (US). For many weedy plant species, human activities influence inter-patch connectivity by creating new habitats and by enhancing dispersal [17–20]. Long-distance dispersal can increase the overall rates of spread, but also has the potential to influence the population's genetic structure by facilitating intraspecific hybridization among independent novel introductions or through the homogenization of a regional population of a particular plant invader.

Invasions, such as cogongrass (*Imperata cylindrica*), are of great concern to the agriculture of softwood timber plantations, primarily pines (*Pinus* spp.). Previous work on this species has shown that anthropogenic land use (e.g., the presence and maintenance of transportation corridors or forestry management practices) is an important driver of this species' distribution, and therefore of range expansion [21,22]. Cogongrass (*Imperata cylindrica*) has invaded seven southern US states and has negatively impacted both the economy and the ecology of these states and the agronomy of the region [23,24]. This invasive grass is a federally listed noxious weed [25] and is considered a global weed of substantial consequence [26,27]. In addition to its invasion across the US softwood timber-growing region, cogongrass is widespread throughout the tropical and sub-tropical regions of the Old World [26,28]. In the US, this plant invasion has benefitted from multiple introductions from previously isolated genetic lineages [29–31]. Previous research observed two genetic lineages of parental material introduced from East Asia, along with documented occurrences of multiple introductions that probably facilitated the establishment of cogongrass in the southern US [31]. The continued spread of cogongrass into neighboring states is a significant concern [23,24,32]. Cogongrass is reported to be an obligate-outcrosser that produces thousands of viable seeds per inflorescence [23,33]. Despite its impressive sexual reproductive capacity, the range expansion of this species has been primarily attributed to the human-aided movement of rhizomatous fragments [24,30]. Again, previous research found cogongrass populations to be surprisingly genetically diverse for a clonally reproductive grass; therefore, this reproductive flexibility of both sexual and asexual approaches has contributed to its fitness and its successful establishment and invasion over the last 100 years [31]. Previous research also allows us to exclude interspecific hybridization with a congeneric directly contributing to current range expansion [34–36] at this time. Human-aided spread, both purposeful and inadvertent, is the probable vector of erratic range expansion, especially during the mid-20th century when propagules were transported from Mississippi and Alabama into Florida and elsewhere [29,30]. Consequences of cogongrass, in both the native and invaded ranges, include reductions in biodiversity, monotypic stands, and timber loss; furthermore, cogongrass is pyrogenic, meaning this grass species is highly-flammable, contributing to alterations in fire regimes to more frequent and/or intense fire events [23,24,26,33]. In managed timber plantations, young *Pinus* spp. are especially susceptible to cogongrass fire events; after a fire, cogongrass rhizomes are generally the first to re-sprout and then dominate the landscape, excluding other plants and animals from recolonizing [24,26,34].

Cogongrass was sampled from, and in cooperation with, state and federal agencies in the following US states experiencing extant infestation and practicing any form of current management: Alabama, Florida, Georgia, Louisiana, Mississippi, South Carolina and Texas. An anonymous genome-wide scan provided by amplified fragment length polymorphism (AFLPs) markers is considered adequate for this analysis. The same AFLP markers have been previously utilized to assess genetic diversity and population structure in invasive US cogongrass populations [31,36–40]; however, previous published research was limited in spatial scale. In this analysis, the US state-level genetic diversity and population structure across the infested range were evaluated. Furthermore, this evaluation sought if anthropogenic activities influenced the extant patterns of genetic diversity and population structure during novel range expansion. One example of a major anthropogenic influence is the American Recovery and Reinvestment Act (ARRA) of 2008, which allocated funding from the US federal government to state agencies for the purpose of improving the control and management of this invasive grass. Alabama received $6.3-mil (USD) and Georgia received $1.8-mil (USD) earmarked for cogongrass eradication. Mississippi received $1.2-mil (USD) and South Carolina received $700,000 (USD) for general invasive plant control and management. These widely variable financial allocations, and each state's use of the funding, may have contributed to differential patterns of genetic diversity or structure among infested US states. Furthermore, we expected genetic variance among US states due to the historical treatment and management of cogongrass in duration, chemical or physical management, and historical efforts. Though state political borders are unlikely barriers to dispersal for propagule movement within the southern US, the influence of differential funding availability and state-level management practices on genetic diversity and structure are variable due to jurisdictional boundaries; thus, we expected cogongrass in states at the expanding fronts of the invasion (South Carolina and Texas) to be less genetically diverse than in the states that received direct introductions and where invasion has been present for the most amount of time (Mississippi and Alabama). Alternatively, states that received the most funding (e.g., Alabama and Georgia) may be less genetically diverse than states that received substantially less funding (e.g., South Carolina).

2. Results

2.1. Genetic Diversity

The AFLP analysis resulted in 2057 polymorphic loci from 676 cogongrass individuals. We observed a good reproducibility for this method (SE = 0.004; 95% CI; <1 mismatch per individual per locus) among the samples analyzed. The state-level average percentage of polymorphic loci was 23% (SE ± 7%), ranging from 2% (GA) to 56% (SC). The mean Shannon's Information Index (I) was 0.030 (SE ± 0.001). Nei's gene diversity for all states analyzed ranged from 0.006 (GA) to 0.042 (SC) with a mean of 0.023 (Table 1). Heterozygosity values also trended similarly to Shannon's (I) and Nei's gene diversity: the mean heterozygosity (both H_e and UH_e) was 0.016 (SE ± 0.001). Spearman's correlation coefficient (ρ) values between the sample size and genetic diversity were calculated for (1) all states, and (2) for states with more than 50 sampled individuals (AL, FL, LA, MS and SC). All relationships between the sample size and genetic diversity were not significant ($P > 0.05$), suggesting that unevenly sampling states did not unnecessarily bias results or interpretation.

Table 1. State-level sampling and location information with genetic and genotypic diversity estimates.

State	Counties Sampled	Other Location Information	Number of Individuals (n)	Shannon's Information Index (I)	Nei's Gene Diversity	Number of Genotypes	Effective Number of Genotypes
Alabama (AL)	Baldwin, Hale, Lee, Mobile, Washington, Sumter	Talladega NF, Frank Boykin WMA	208	0.033 ± 0.002	0.023	92	0.935
Florida (FL)	Alachua, Duval, Indian River, Miami-Dade, Osceola, Sarasota	Disney Wilderness Preserve (TNC), Miami-Dade municipal parks	129	0.033 ± 0.002	0.028	49	0.816
Georgia (GA)	Baker, Crawford, Decatur, Mitchell, Thomas, Worth	Georgia Forestry Commission	13	0.009 ± 0.001	0.006	2	0.154
Louisiana (LA)	St. Tammany, Washington	Benscreek WMA	62	0.024 ± 0.002	0.029	14	0.64
Mississippi (MS)	Greene, Harrison, Jasper, Jones, Scott, Smith, Wayne	Desoto NF, Bienville NF	180	0.038 ± 0.003	0.023	117	0.94
South Carolina (SC)	Berkeley, Greenwood, Saluda, Union	Frances Marion NF, Sumter NF	74	0.063 ± 0.002	0.042	41	0.888
Texas (TX)	Tyler	Texas Forest Service	10	0.012 ± 0.002	0.011	6	0.778
Overall			676			321	
Mean (±SE)				0.030 ± 0.001	0.023 ± 0.001		0.736 ± 0.105

An overall reduction was observed in the clonal diversity analysis from 676 sampled individuals to 321 unique multi-locus genotypes (Table 1 and Figure 1). The mean genotype diversity (range: 0 to 1) for all states was 0.736 (SE ± 0.105). We observed the lowest genotype diversity in GA at 0.154, reducing the effective number of genotypes to 1.154. We observed the highest genotype diversity in MS and AL (>0.90), with FL and SC close behind (0.816 and 0.888, respectively; Table 1 and Figure 1).

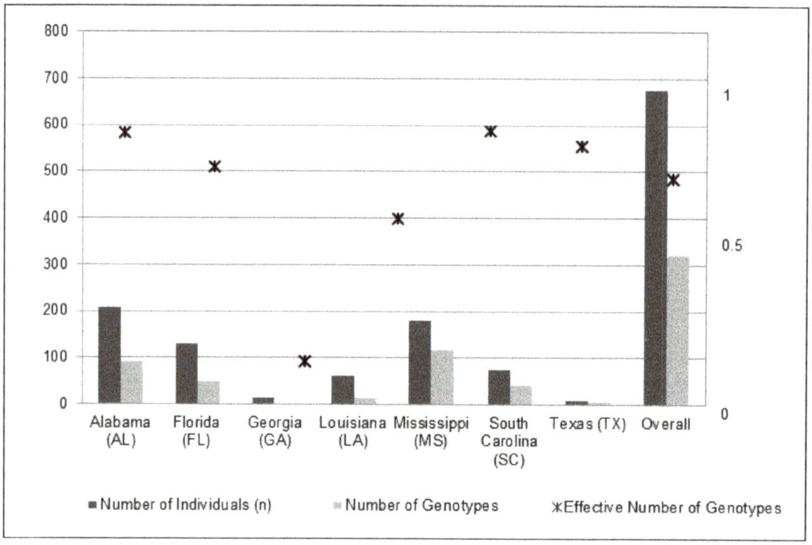

Figure 1. Genotypic diversity relative to the number of individual samples in each state and overall. The effective number of genotypes for the US state of Georgia was <1; therefore, no symbol is present for that state.

2.2. Population Structure

Of the 676 individuals sampled, 485 were assigned with >90% posterior probability to a single lineage; 73 to the MS-type lineage (red, Figure 1) and 412 to the AL-type lineage (blue, Figure 2). Admixture was present in both clusters (mean $\alpha = 0.165$), and 191 individuals could not be confidently assigned to either cluster with strong confidence. A Bayesian cluster analysis was conducted in the program STRUCTURE, which supported an inference of two distinct genetic clusters ($K = 2$; mean LnP(D) = −107,561; Figure 3) extant in the region examined: a MS-type lineage and an AL-type lineage (Figure 2). Individuals assigned to the MS-type lineage are only present in central MS, based on our sampling, and there is a single outlier individual in SC. It is in this geographic locale of central MS that both lineages are present and co-occur at the patch level, with varying proportions of admixed individuals. In all other states in the region, the AL-type lineage is dominant and co-occurs with admixed individuals, except in GA where only the AL-type lineage was found. A cluster analysis at the patch- and state-level in GA is consistent with the genetic diversity, supporting a low heterogeneity, and all individuals in GA were assigned to the AL-type lineage. Additional Bayesian cluster analyses were conducted on all individuals excluding the individuals strongly assigned to the MS-type lineage. No population substructure was observed in the AL-type and in ambiguous individuals, and this remained consistent with the initial analysis.

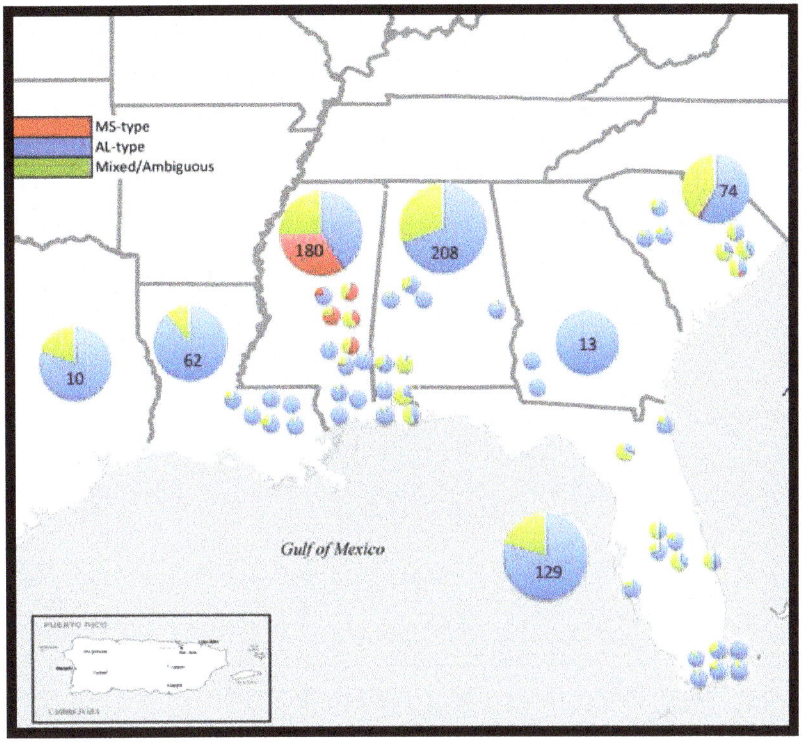

Figure 2. Map of the southern US with the proportion of individuals assigned to MS-type (red), AL-type (blue) or ambiguous (green) based on a 90% threshold of assignment from STRUCTURE.

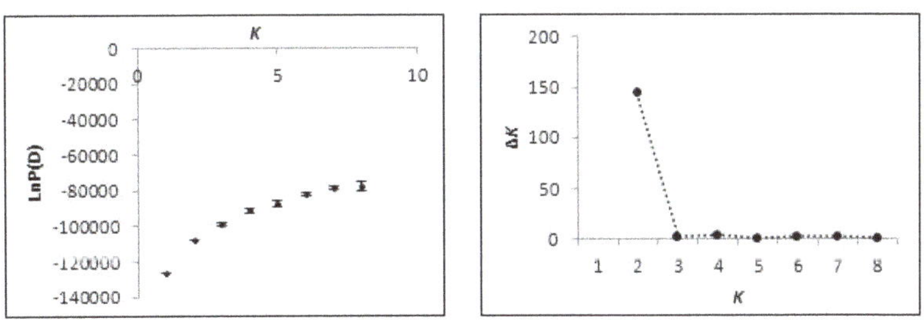

Figure 3. Summary evaluation of the posterior probability values [LnP(D)] used for the determination of the most likely number of clusters (*K*) from STRUCTURE simulations.

Significant population pairwise F_{ST} values were observed between the two genetic groups within MS patches ($F_{ST} = 0.330$, $P < 0.05$; Table 2). Please note that the MS-central group was significantly dissimilar from all other groups tested ($F_{ST} > 0.3$); therefore, a pairwise F_{ST} analysis separated MS populations into two groups: 1) coastal Mississippi (MS-Coast) populations from the other 2) Mississippi lineage (MS-Central) (Table 2). The SC population was the least genetically differentiated from the FL populations ($F_{ST} = 0.060$), while the AL and SC populations were the least genetically differentiated from the TX population ($F_{ST} = 0.094$, 0.083, respectively; Table 2). Pairwise population F_{ST} values were similar between SC and most other states analyzed ($F_{ST} < 0.2$), excepting MS-central ($F_{ST} = 0.314$). Pairwise F_{ST} between TX and AL ($F_{ST} = 0.090$) was very similar to pairwise F_{ST} values between TX and SC, and between TX and FL (both F_{ST} = 0.083). The greatest genetic dissimilarity existed between TX and GA (F_{ST} = 0.553).

Table 2. Seven groups analyzed for population pairwise F_{ST}[1], where MS-type (central) was separated from coast (AL-type) for the state-level analysis.

MS-Coast	AL	LA	SC	TX	GA	FL	
0.132	*						**AL**
0.096	**0.175**	*					**LA**
0.098	**0.100**	**0.120**	*				**SC**
0.217	**0.090**	**0.264**	**0.083**	*			**TX**
0.150	**0.277**	**0.050**	**0.158**	**0.553**	*		**GA**
0.146	**0.094**	**0.158**	**0.060**	**0.083**	**0.281**	*	**FL**
0.330	**0.320**	**0.444**	**0.314**	**0.348**	**0.478**	**0.385**	**MS-Central**

[1] Significant values in bold, *P* = 0.05; * indicates a value of 0, when pairwise comparison of the same group.

A principal coordinate analysis (PCoA) was consistent with the population structure as inferred by the Bayesian cluster analysis in STRUCTURE. The first two axes explained 61% of individual genetic variance (Figure 4). Two clusters were observed in the PCoA plot: one fairly contained cluster and the other being broad and loosely organized. Individuals sampled from MS were present in both clusters, where individuals sampled from central MS (MS-Central) appear localized to the bottom right quadrant (red diamonds, Figure 4) and individuals sampled from the MS Gulf Coast (MS-Coast) cluster with individuals from AL, FL and other states. Individuals sampled from AL are broadly distributed throughout the larger cluster, indicating a high degree of individual genetic variation. Individuals from FL also presented a similar pattern. Sampled individuals from TX (all collected from a single site) did not form a tight cluster, but grouped with samples from AL, FL, LA and SC. Georgia; these individuals from these states, however, form a very tight cluster within the broad cluster (Figure 4), overlapping within a subset of the AL, LA, MS and SC individuals, potentially indicating genetic relationships among geographically disparate populations.

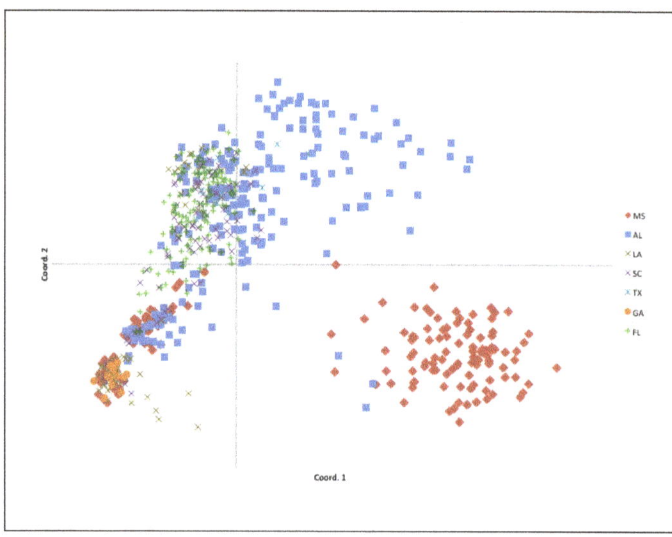

Figure 4. Principle coordinates analysis (PCoA) of individual genetic covariance with data standardization (N = 676, 61% of variation accounted for by the first two axes). All seven US states are represented in this ordination.

The genetic differentiation was evaluated between the MS-type cluster and all other individuals based on inferences from STRUCTURE and the PCoA. Ambiguous individuals were lumped with the AL-type in an analysis of molecular variation (AMOVA; Table 3); this grouping was derived because of the low value of admixture from STRUCTURE (mean $\alpha = 0.165$) and the high proportion of assignment to the AL-type in ambiguous individuals (<0.90, but >0.50). We observed a significant genetic differentiation ($F_{ST} = 0.363$, $P < 0.001$) between the groups as defined above.

Table 3. Analysis of molecular variance (AMOVA) where two groups were inferred from the STRUCTURE analysis.

Source of Variation	d.f.	Sum of Squares	Percentage of Variation	*P*-Value
Among Groups	1	1325.57	27.43	<0.001
Among populations within groups	6	1283.18	8.88	0.110
Within populations	668	12,151.01	63.68	<0.001
Total	675	14,776.76		
$F_{ST} = 0.363$ ($P < 0.001$), $F_{SC} = 0.122$ ($P < 0.001$), $F_{CT} = 0.274$ ($P = 0.11$)				

3. Discussion

The detection of two distinct genetic lineages throughout the invaded region is consistent with other published research on the cogongrass invasion in the US [29–31,36]. This analysis, however, demonstrated that one of those two genetic lineages (the dominant AL-type) has been responsible for the majority of the spatial spread and persistence in the US. The dominant AL-type lineage is more heterogeneous, widespread, and is genetically distinct ($F_{ST} = 0.363$, $P < 0.001$) from the MS-type lineage, even when inclusive of ambiguously assigned individuals. A two-lineage scenario provided the only evidence of strong population structuring throughout the invaded US range. The majority of individuals sampled for this work were assigned with a 90% or greater probability to the AL-type lineage (Figure 2). The other detected genetic group (the MS-type) appears to be geographically constrained to central MS (Figure 2). This pattern may be due to the reduced competitive or dispersal ability of this lineage, lack of anthropogenic activity promoting its spread or other potentially heritable fitness factors restricting its ability to spread from this region.

The present analyses further demonstrated that, although the two genetic groups remain significantly differentiated at the scale of this study, admixture is present within individuals at the local, patch scale (mean $\alpha = 0.165$). Admixture is also suggested to have increased upon regional evaluation as compared to previous research constrained to MS and AL (where mean $\alpha = 0.08$) [31]. Similarly, recent admixture in the invasive weed, *Silene vulgaris*, was detected in the novel range (North America), and demonstrated a relationship between fitness benefits and increased heterozygosity in invasive populations [41]. Such results suggest the possibility that genetic mixing contributed to the successful spread of cogongrass across the southern US; however, measures of intraspecific phenotypic variation, and its relationship to genetic variation and/or hybridization, are needed to demonstrate whether significant genetic admixture has contributed to cogongrass' success [42,43]. Typically, multiple introductions increase the genetic diversity and probability of success in a wide array of invasive, colonizing organisms [44–48].

Although we found a significant genetic distance to exist among the states (Table 2), the overwhelming evidence favored a regional population dominated by the AL-type lineage, with a highly-restricted MS-type found mainly restricted in the central portion of Mississippi (Figure 2). There also appears to be an element of biological influence on the population structure, as the majority of the genetic variation in cogongrass is partitioned within sampled locations (64%). Similar to previous findings at smaller scales (in terms of sample size and geography), the majority of genetic variation being partitioned within patches continues to be unexpected for a species thought to expand locally via clonal propagation. The regional degree of genetic differentiation in cogongrass ($F_{ST} = 0.363$, $P < 0.001$)

is similar to that of early successional plant species (mean F_{ST} = 0.37) [49]. Even considering the fact that the majority of genetic differentiation is partitioned within local patches, average heterozygosity for state-level cogongrass sampling (H_e and UH_e = 0.016 ± 0.001) is still lower than such values in long-lived perennial plant species (H_e = 0.68) or those capable of selfing (H_e = 0.41) [49].

We expected that differential funding and historical management efforts among the seven US states would contribute to contemporary state-level genetic diversity differences across the region. Differing state-level management practices, efforts, and starting-acreages of infestation affected the relative genetic diversity (Table 1). The state of Alabama (AL) received the greatest allocation of cogongrass-specific funding from the American Recovery and Reinvestment Act (ARRA), $6.3-mil (USD). Despite receiving the largest allocation of federal funding, AL cogongrass populations on public and private lands were not much reduced since 2008. Though no published data has resulted since then, we do know that AL chose to outsource the ARRA funds to a private invasive plant control consulting firm responsible for the detection, treatment and repeated visits/re-application on infested sites. Since a follow-up chemical application is necessary in any cogongrass management plan or strategy [24], Alabama's utilization of ARRA funds has been viewed as one of the greatest modern failures in cogongrass control efforts. Alabama continues to be one of the most infested states in the US, superseded only by Florida. When compared to the neighboring state of Georgia (GA), which received $1.8-mil (USD) from the ARRA specifically for cogongrass eradication, the state of GA resulted in the lowest genetic and genotypic diversity (Figure 2), along with the least amount of acreage infested excluding the infestation site in Texas. Additional evidence as to why GA is so different from other states in terms of the distribution and diversity of cogongrass infestation is that the Georgia Forestry Commission began treating cogongrass in the late 1960s (Art Miller personal correspondence, GFC *retired*) with chemical herbicide and mechanical removal decades before all other southern US states. Furthermore, the Georgia Forestry Commission currently utilizes a chemical-application strategy that has been most effective [24], along with semi-annual revisits to document infested sites. Therefore, the state of Georgia has been able to stem the majority of cogongrass infestation to the southwestern corner of the state, which borders Gulf Coastal Alabama. The states of GA and TX resulted in the lowest overall genetic diversity. We considered that the smaller sample sizes in these two states may have contributed to this pattern, 13 and 10 individuals, respectively; however, Spearman's correlation tests did not find significant relationships between sample sizes and genetic/genotypic diversity outcomes (Figure 1, Table 1). The genotypic diversity in GA was slightly over 15%, whereas in TX it was over 77%. From our analysis, this suggested that cogongrass sites sampled in GA persist with both a low genetic and genotypic diversity. Furthermore, all individuals in GA were strongly assigned to the AL-type, also suggesting a superior ability to persist despite the fitness benefits expected to accompany a high genotypic diversity contributing to increased adaptability [3]. This pattern observed in GA, which is not mirrored in TX, indicates that the success of an invader can be based on the fitness of one to a few adapted genotypes [50,51].

Cogongrass possesses a broad global distribution [23,24], indicating a generalist phenotype, preferentially colonizing and positively associated with disturbance [21], and originating from potentially multiple points in Asia [29,30]. We consider cogongrass a suitable candidate for rapid adaptation and evolution in the invaded range due anthropogenic activities [52]. Furthermore, the chemical control of invasive plant populations is a human-induced selective pressure, unevenly applied and distributed across the landscape. These human activities directly and indirectly affect the population and genetic structure of secondarily invading populations, and will continue to do so.

The regional expansion of cogongrass since its initial introduction in 1919 has benefitted from the direct and purposeful anthropogenic transport of propagules, multiple introductions and secondary contact [29–31]. The majority of propagule spread had been previously attributed to the transport and establishment of rhizome fragments; however, the lack of a strong decay in isolation-by-distance (IBD) genetic-geographic relationships suggested this is not the case [31,36,40]. These data suggest that the cogongrass expansion throughout the South has probably benefitted from reproductive flexibility, the on-going movement of viable propagules (both seeds and rhizome fragments) and potential adaptive persistence under active management by chemical control. Given the clear regional dominance by the AL-type lineage of plants, further investigation into the mechanisms responsible for the successful invasion of cogongrass might benefit from a functional genetic comparison between the AL- and MS-lineage's phenotypes and phenotypic responses, as well as intraspecific global references.

4. Materials and Methods

4.1. Study Area and Sampling

Live cogongrass leaf tissues were collected during 2008–2009 from seven states in the southern region of the US: Alabama (AL, n = 208, 10 patches), Florida (FL, n = 129, 14 patches), Georgia (GA, n = 13, 2 patches), Louisiana (LA, n = 62, 6 patches), Mississippi (MS, n = 180, 11 patches), South Carolina (SC, n = 74, 7 patches) and Texas (TX, n = 10, single patch site; Figure 2, Table 1, N = 676). Each tiller was assumed as being representative of an "individual" or a ramet in the patch/population (while acknowledging that individual patches may have arisen from only one to a few colonizing propagules resulting in few genets comprising a population). Leaf tissues were collected in the field from cogongrass populations. Patches were identified as contiguous sites of cogongrass, which often occur as circular patches in open areas or as long, narrow patches along roadside rights-of-way. Leaf tissues were placed into individually-labeled plastic bags and then stored in a cooler with 1–2 cups of ice (or ice substitute) per large cooler during collection and transport (for maximum 24–36 h, at ambient vehicle temperature). Silica gel containing color indicator was poured into individual bags to dry the leaf tissue in the lab upon unpacking of leaf tissues. Drying leaf tissues in silica gel were dried for at least 1 week at room temperature.

Cogongrass (*I. cylindrica*) is a listed Federal Noxious Weed; all sampling was conducted with approval by the U.S. Dept. of Agriculture, Animal and Plant Health Inspection Service, Plant Pest Quarantine (Permit#: P526P-12-00211, P526-080721-005). Additional permissions were required for access to specific areas, including a collection agreement with The Nature Conservancy (TNC) and Miami-Dade County Parks and Recreation (Permit #145). Most sampling was conducted on public land including National Forests (USDA Forest Service), interstate/highway rights-of-way and on private land with permission from state forestry agencies and associated landowners, as appropriate.

4.2. Molecular Analysis

Approximately 1-cm^2 of dried individual leaf tissue was aseptically transferred into a 2-mL microcentrifuge tube for each individual. The extraction of total DNA from leaf tissues, primers, reagents and PCR conditions were described previously in Lucardi et al. [31,36] based on Vos et al. [53]. Lack of reproducibility is considered one of the negative aspects of dominant markers, such as AFLPs; therefore, we accompanied all runs with positive and negative controls.

Fragment data were visualized in GeneMarker® (SoftGenetics, LLC, State College, PA, USA) and exported into a general text format. Detected fragments were sorted on migration size (basepairs) and objectively scored utilizing an independently developed procedure (Lucardi and Walker, unpublished methodology) using both Excel 2007 (Microsoft Corporation, Redmond, WA, USA) and PASW v.18.0 (SPSS, IBM Corporation, Armonk, NY, USA) as specified in Lucardi et al. [31,36].

Data conversions of presence-absence matrices were conducted in AFLPdat source script [54] in R v2.15.1. The genetic diversity was assessed using the expected heterozygosity (biased (H_e) and unbiased (UH_e)) and Shannon's Information/Diversity Index (I), serving as a coefficient of similarity, for each state for state-level diversity estimates (GenAlEx 6.3) [55]. Population genetic diversity metrics, such as Shannon's Information Index [I], were estimated from allele frequencies inferred from dominant data and are subject to Hardy-Weinberg equilibrium assumptions. These assumptions can reduce the accuracy in allele frequency estimations from dominant data, but reliable results for a comparative study can be achieved with adequate population sampling and a sufficient number of primer sets, which generate a large number of detected polymorphic loci [39,56,57].

Because of cogongrass' capacity to clonally reproduce *via* rhizomes, we conducted a clonal analysis using the "Clones" function within the AFLPdat package [54]. The standard error among positive control replicates (0.004) functioned as the error parameter for detecting identical multi-locus genotypes. The number of unique multi-locus genotypes per population contributes toward a more accurate assessment of the actual genetic diversity. Uneven sample sizes may bias the interpretation of results due to statistical dependence between variables. We used the Spearman's correlation coefficient to determine if relationships are present between sample size and genetic diversity. We tested the relationships between sample sizes and frequency-based estimates using Spearman's correlation method ("cor.s") in R v.2.15.1.

We evaluated the population structure with population pairwise F_{ST} (Arlequin v.3.5 [58]) and the Bayesian cluster analysis program, STRUCTURE v.2.3.3 [59], to infer the most likely number of clusters (or K) based on posterior log-likelihood probability values from each simulation. We utilized the Evanno et al. [60] method to detect the population structure when $K \geq 2$. Multiple simulations were conducted for $K = 1$ through 7, based on the seven sampled US states, with the following parameters: admixture ancestry model (to infer α), burn-in 10,000 and 50,000 Markov Chain Monte Carlo (MCMC) [58]. The ad hoc statistic, ΔK, was plotted to determine the mode. Individuals were assigned to a single lineage if they contained a probability of membership of at least 0.90 (90% threshold) to one of the clusters; all other individuals that were not strongly assigned to one of the lineages were assigned as ambiguous. We further assessed the population structure with a principal coordinates analysis (PCoA, GenAlEx v.6.3) and analysis of molecular variation (AMOVA, [61]) performed in Arlequin v.3.5 [58] using genetic distances, testing the structure based on inferences made from multi-locus genetic information analyzed in a cluster analysis.

Author Contributions: R.D.L., L.E.W. and G.N.E. initiated the initial research design; R.D.L. collected field samples (with assistance), conducted the molecular research and data analyses, R.D.L. wrote the manuscript. L.E.W. and G.N.E. significantly contributed to the writing and interpretation. All authors have read and agreed to the published version of the manuscript.

Funding: This research was funded by in part by the United States Geological Survey, Biological Resources Discipline (#04HQAG0135) and the United States Department of Agriculture (2006-03613 and 2008-35320-18679) to GNE.

Acknowledgments: We are very grateful to the Georgia Forestry Commission (M. McClure and C. Bates), the Texas Forest Service (M. Murphrey), our cooperators at Auburn University (N. Loewenstein), C. Matson and D. Tharp (The Nature Conservancy), S.C. Hughes, and K.A. Bradley (previously of the Institute for Regional Conservation, FL) for their assistance with sampling and field identifications. We also thank Drs. Brian Counterman and Diana Outlaw (Mississippi State University), Charles T. Bryson (retired, USDA ARS), and Jacob Walker (Arkansas Research Center) for reviewing an early draft of this manuscript.

Conflicts of Interest: The authors declare no conflict of interest.

Data Accessibility: Sampling coordinates are detailed in the Appendix section of Lucardi 2012, available freely here: [http://sun.library.msstate.edu/ETD-db/theses/available/etd-10192012-122638/unrestricted/LUCARDI_Lib_v2.3_291012.pdf].

References

1. Meyerson, L.A.; Mooney, H.A. Invasive alien species in an era of globalization. *Front. Ecol. Environ.* **2007**, *5*, 199–208. [CrossRef]
2. Schulz, A.N.; Lucardi, R.D.; Marsico, T.D. Successful Invasions and Failed Biocontrol: The Role of Antagonistic Species Interactions. *BioScience* **2019**, *69*, 711–724. [CrossRef]
3. Lee, C.E. Evolutionary genetics of invasive species. *Trends Ecol. Evol.* **2002**, *17*, 386–391. [CrossRef]
4. Dlugosch, K.M.; Parker, I.M. Founding events in species invasions: Genetic variation, adaptive evolution, and the role of multiple introductions. *Mol. Ecol.* **2008**, *17*, 431–449. [CrossRef] [PubMed]
5. Ward, S.M.; Gaskin, J.F.; Wilson, L.M. Ecological Genetics of Plant Invasion: What Do We Know? *Invasive Plant Sci. Manag.* **2008**, *1*, 98–109. [CrossRef]

6. Catford, J.A.; Jansson, R.; Nilsson, C. Reducing redundancy in invasion ecology by integrating hypotheses into a single theoretical framework. *Divers. Distrib.* **2009**, *15*, 22–40. [CrossRef]
7. Hulme, P.E. Trade, transport and trouble: Managing invasive species pathways in an era of globalization. *J. Appl. Ecol.* **2009**, *46*, 10–18. [CrossRef]
8. Holzmueller, E.J.; Jose, S. Invasion success of cogongrass, an alien C4 perennial grass, in the southeastern United States: Exploration of the ecological basis. *Boil. Invasions* **2010**, *13*, 435–442. [CrossRef]
9. Vitousek, P.; Mooney, H.A.; Lubchenco, J.; Melillo, J.M. Human Domination of Earth's Ecosystems. *Science* **1997**, *277*, 494–499. [CrossRef]
10. Hobbs, R.J. Land-use changes and invasions. In *Invasive Species in a Changing World*; Mooney, H.A., Hobbs, R.J., Eds.; Island Press: Washington, DC, USA, 2000.
11. Harrison, S.; Hohn, C.; Ratay, S. Distribution of Exotic Plants along Roads in a Peninsular Nature Reserve. *Boil. Invasions* **2002**, *4*, 425–430. [CrossRef]
12. Shuster, W.D.; Herms, C.P.; Frey, M.N.; Doohan, D.J.; Cardina, J. Comparison of survey methods for an invasive plant at the subwatershed level. *Boil. Invasions* **2005**, *7*, 393–403. [CrossRef]
13. Christen, D.; Matlack, G.R. The role of roadsides in plant invasions: A demographic approach. *Conserv. Boil.* **2006**, *20*, 385–391. [CrossRef] [PubMed]
14. With, K.A. The Landscape Ecology of Invasive Spread. *Conserv. Boil.* **2002**, *16*, 1192–1203. [CrossRef]
15. Rouget, M.; Richardson, D.M. Inferring Process from Pattern in Plant Invasions: A Semimechanistic Model Incorporating Propagule Pressure and Environmental Factors. *Am. Nat.* **2003**, *162*, 713–724. [CrossRef]
16. Lázaro-Lobo, A.; Ervin, G.N. A global examination on the differential impacts of roadsides on native vs. exotic and weedy plant species. *Glob. Ecol. Conserv.* **2019**, *17*, e00555. [CrossRef]
17. Hodkinson, D.J.; Thompson, K. Plant Dispersal: The Role of Man. *J. Appl. Ecol.* **1997**, *34*, 1484–1496. [CrossRef]
18. Tikka, P.M.; Högmander, H.; Koski, P.S. Road and railway verges serve as dispersal corridors for grassland plants. *Landsc. Ecol.* **2001**, *16*, 659–666. [CrossRef]
19. Watkins, R.Z.; Chen, J.; Pickens, J.; Brosofske, K.D. Effects of Forest Roads on Understory Plants in a Managed Hardwood Landscape. *Conserv. Boil.* **2003**, *17*, 411–419. [CrossRef]
20. Lockwood, J.L.; Hoopes, M.F.; Marchetti, M.P. *Invasion Ecology*; Blackwell: Malden, MA, USA, 2007.
21. Holly, D.C. Multi-Scale Evaluation of Mechanisms Associated with the Establishment of a Model Invasive Species in Mississippi: *Imperata cylindrica*. Ph.D. Thesis, Mississippi State University, Starkville, MS, USA, 2008; 140p.
22. Ervin, G.N.; Holly, D.C. Examining Local Transferability of Predictive Species Distribution Models for Invasive Plants: An Example with Cogongrass (*Imperata cylindrica*). *Invasive Plant Sci. Manag.* **2011**, *4*, 390–401. [CrossRef]
23. Bryson, C.T.; Carter, R. Cogongrass, *Imperata cylindrica*, in the United States. *Weed Technol.* **1993**, *7*, 1005–1009. [CrossRef]
24. MacDonald, G.E. Cogongrass (*Imperata cylindrica*)—Biology, Ecology, and Management. *Crit. Rev. Plant Sci.* **2004**, *23*, 367–380. [CrossRef]
25. The PLANTS Database. Available online: http://plants.usda.gov (accessed on 23 March 2020).
26. Hubbard, C.E.; Gray, A.P.; Brown, D.; Whyte, R.O. *Imperata Cylindrica: Taxonomy, Distribution, Economic Significance, and Control*; Imperial Agricultural Bureaux Joint Publication: Oxford, UK, 1944; Volume 7, 63p.
27. Holm, L.G.; Donald, P.; Pancho, J.V.; Herberger, J.P. *The World's Worst Weeds: Distribution and Biology*; The University Press of Hawaii: Honolulu, HI, USA, 1977; p. 609.
28. Gabel, M.L. A Biosystematic Study of the Genus *Imperata* (Gramineae: Andropogoneae). Ph.D. Thesis, Iowa State University, Ames, IA, USA, 1982; 94p.
29. Tabor, P. Cogon grass, *Imperata cylindrica* (L.) Beauir in the South-eastern United States. *Agron. J.* **1949**, *41*, 270. [CrossRef]
30. Tabor, P. Cogongrass in Mobile County, Alabama. *Agron. J.* **1952**, *44*, 50. [CrossRef]
31. Lucardi, R.D.; Wallace, L.E.; Ervin, G.N. Invasion Success in Cogongrass (*Imperata cylindrica*): A Population Genetic Approach Exploring Genetic Diversity and Historical Introductions. *Invasive Plant Sci. Manag.* **2014**, *7*, 59–75. [CrossRef]
32. Dickens, R. Cogongrass in Alabama after Sixty Years. *Weed Sci.* **1974**, *22*, 177–179. [CrossRef]

33. Dozier, H.; Gaffney, J.F.; McDonald, S.K.; Johnson, E.R.; Shilling, D.G. Cogongrass in the United States: History, Ecology, Impacts, and Management. *Weed Technol.* **1998**, *12*, 737–743. [CrossRef]
34. Howard, J. Imperata brasiliensis, *I. cylindrical.* In *Fire Effects Information System*; U.S. Department of Agriculture, Forest Service, Rocky Mountain Research Station, Fire Science Laboratory (Producer): Bozeman, MT, USA, 2005; Available online: https://www.fs.fed.us/database/feis/plants/graminoid/impspp/all.html (accessed on 29 March 2020).
35. Hall, D.W. Is cogon grass really an exotic? *Wildland Weeds* **1998**, *1*, 14–15.
36. Lucardi, R.D.; Wallace, L.E.; Ervin, G.N. Evaluating hybridization as a potential facilitator of successful cogongrass (*Imperata cylindrica*) invasion in Florida, USA. *Boil. Invasions* **2014**, *16*, 2147–2161. [CrossRef]
37. Amsellem, L.; Noyer, J.L.; LeBourgeois, T.; Hossaert-McKey, M. Comparison of genetic diversity of the invasive weed *Rubus alcefolius* Poir. (Rosaceae) in its native range and in areas of introduction, using amplified fragment length polymorphism (AFLP) markers. *Mol. Ecol.* **2000**, *9*, 443–455. [CrossRef]
38. Campbell, D.; Duchesne, P.; Bernatchez, L. AFLP utility for population assignment studies: Analytical investigation and empirical comparison with microsatellites. *Mol. Ecol.* **2003**, *12*, 1979–1991. [CrossRef]
39. Meudt, H.; Clarke, A. Almost Forgotten or Latest Practice? AFLP applications, analyses and advances. *Trends Plant Sci.* **2007**, *12*, 106–117. [CrossRef]
40. Capo-Chichi, L.J.A.; Faircloth, W.H.; Williamson, A.G.; Patterson, M.G.; Miller, J.H.; Van Santen, E. Invasion Dynamics and Genotypic Diversity of Cogongrass (*Imperata cylindrica*) at the Point of Introduction in the Southeastern United States. *Invasive Plant Sci. Manag.* **2008**, *1*, 133–141. [CrossRef]
41. Keller, S.R.; Fields, P.D.; Berardi, A.E.; Taylor, D.R. Recent admixture generates heterozygosity-fitness correlations during the range expansion of an invading species. *J. Evol. Boil.* **2014**, *27*, 616–627. [CrossRef]
42. Ellstrand, N.C.; Schierenbeck, K.A. Hybridization as a stimulus for the evolution of invasiveness in plants? *Proc. Natl. Acad. Sci. USA* **2000**, *97*, 7043–7050. [CrossRef] [PubMed]
43. Molofsky, J.; Keller, S.R.; Lavergne, S.; Kaproth, M.A.; Eppinga, M.B. Human-aided admixture may fuel ecosystem transformation during biological invasions: Theoretical and experimental evidence. *Ecol. Evol.* **2014**, *4*, 899–910. [CrossRef] [PubMed]
44. Blackburn, T.M.; Duncan, R.P. Determinants of establishment success in introduced birds. *Nature* **2001**, *414*, 195–197. [CrossRef] [PubMed]
45. Genton, B.J.; Shykoff, J.A.; Giraud, T. High genetic diversity in French invasive populations of common ragweed, *Ambrosia artemisiifolia*, as a result of multiple sources of introduction. *Mol. Ecol.* **2005**, *14*, 4275–4285. [CrossRef]
46. Walls, R. Hybridization and Plasticity Contribute to Divergence among Coastal and Wetland Populations of Invasive Hybrid Japanese Knotweed s.l. (*Fallopia* spp.). *Chesap. Sci.* **2009**, *33*, 902–918. [CrossRef]
47. Mothapo, N.P.; Wossler, T.C. Behavioural and chemical evidence for multiple colonisation of the Argentine ant, *Linepithema humile*, in the Western Cape, South Africa. *BMC Ecol.* **2011**, *11*, 6. [CrossRef]
48. Garbelotto, M.; Guglielmo, F.; Mascheretti, S.; Croucher, P.J.P.; Gonthier, P. Population genetic analyses provide insights on the introduction pathway and spread patterns of the North American forest pathogen *Heterobasidion irregulare* in Italy. *Mol. Ecol.* **2013**, *22*, 4855–4869. [CrossRef]
49. Nybom, H. Comparison of different nuclear DNA markers for estimating intraspecific genetic diversity in plants. *Mol. Ecol.* **2004**, *13*, 1143–1155. [CrossRef] [PubMed]
50. Sakai, A.K.; Allendorf, F.W.; Holt, J.S.; Lodge, D.M.; Molofsky, J.; With, K.A.; Baughman, S.; Cabin, R.J.; Cohen, J.E.; Ellstrand, N.C.; et al. The Population Biology of Invasive Species. *Annu. Rev. Ecol. Syst.* **2001**, *32*, 305–332. [CrossRef]
51. Vellend, M.; Drummond, E.B.M.; Tomimatsu, H. Effects of genotype identity and diversity on the invasiveness and invisibility of plant populations. *Oecologia* **2010**, *162*, 371–381. [CrossRef] [PubMed]
52. Hufbauer, R.A.; Facon, B.; Ravigné, V.; Turgeon, J.; Foucaud, J.; Lee, C.E.; Rey, O.; Estoup, A. Anthropogenically induced adaptation to invade (AIAI): Contemporary adaptation to human-altered habitats within the native range can promote invasions. *Evol. Appl.* **2011**, *5*, 89–101. [CrossRef] [PubMed]
53. Vos, P.; Hogers, R.; Bleeker, M.; Reijans, M.; Van De Lee, T.; Hornes, M.; Friters, A.; Pot, J.; Paleman, J.; Kuiper, M.; et al. AFLP: A new technique for DNA fingerprinting. *Nucleic Acids Res.* **1995**, *23*, 4407–4414. [CrossRef] [PubMed]
54. Ehrich, D. aflpdat: A collection of r functions for convenient handling of AFLP data. *Mol. Ecol. Notes* **2006**, *6*, 603–604. [CrossRef]

55. Peakall, R.; Smouse, P.E. genalex 6: Genetic analysis in Excel. Population genetic software for teaching and research. *Mol. Ecol. Notes* **2006**, *6*, 288–295. [CrossRef]
56. Mariette, S.; Le Corre, V.; Austerlitz, F.; Kremer, A. Sampling within the genome for measuring within-population diversity: Trade-offs between markers. *Mol. Ecol.* **2002**, *11*, 1145–1156. [CrossRef]
57. Bonin, A.; Ehrich, D.; Manel, S. Statistical analysis of amplified fragment length polymorphism data: A toolbox for molecular ecologists and evolutionists. *Mol. Ecol.* **2007**, *16*, 3737–3758. [CrossRef]
58. Excoffier, L.; Lischer, H.E.L. Arlequin suite ver 3.5: A new series of programs to perform population genetics analyses under Linux and Windows. *Mol. Ecol. Resour.* **2010**, *10*, 564–567. [CrossRef]
59. Pritchard, J.K.; Stephens, M.; Donnelly, P. Inference of population structure using multilocus genotype data. *Genetics* **2000**, *155*, 945–959. Available online: http://pritch.bsd.uchicago.edu/structure.html (accessed on 29 March 2020). [PubMed]
60. Evanno, G.; Regnaut, S.; Goudet, J. Detecting the number of clusters of individuals using the software structure: A simulation study. *Mol. Ecol.* **2005**, *14*, 2611–2620. [CrossRef] [PubMed]
61. Excoffier, L.; Smouse, P.E.; Quattro, J.M. Analysis of Molecular Variance Inferred from Metric Distances among DNA Haplotypes: Application to Human Mitochondrial DNA Restriction Data. *Genetics* **1992**, *131*, 479–491. [PubMed]

plants MDPI

Review

A Dual Strategy of Breeding for Drought Tolerance and Introducing Drought-Tolerant, Underutilized Crops into Production Systems to Enhance Their Resilience to Water Deficiency

Amparo Rosero [1,*], Leiter Granda [2], Jhon A. Berdugo-Cely [1], Olga Šamajová [3], Jozef Šamaj [3] and Radim Cerkal [2]

1 Corporación Colombiana de Investigación Agropecuaria–AGROSAVIA, Centro de Investigación Turipaná, Km 13 vía Montería, 250047 Cereté, Colombia; jberdugo@agrosavia.co
2 Department of Crop Science, Breeding and Plant Medicine, Mendel University in Brno, Zemedelska 1, 613 00 Brno, Czech Republic; leiter.granda@gmail.com (L.G.); radim.cerkal@mendelu.cz (R.C.)
3 Department of Cell Biology, Centre of the Region Haná for Biotechnological and Agricultural Research, Faculty of Science, Palacký University, Šlechtitelů 27, 783 71 Olomouc, Czech Republic; olga.samajova@upol.cz (O.Š.); jozef.samaj@upol.cz (J.Š.)
* Correspondence: erosero@agrosavia.co

Received: 31 July 2020; Accepted: 22 September 2020; Published: 24 September 2020

Abstract: Water scarcity is the primary constraint on crop productivity in arid and semiarid tropical areas suffering from climate alterations; in accordance, agricultural systems have to be optimized. Several concepts and strategies should be considered to improve crop yield and quality, particularly in vulnerable regions where such environmental changes cause a risk of food insecurity. In this work, we review two strategies aiming to increase drought stress tolerance: (i) the use of natural genes that have evolved over time and are preserved in crop wild relatives and landraces for drought tolerance breeding using conventional and molecular methods and (ii) exploiting the reservoir of neglected and underutilized species to identify those that are known to be more drought-tolerant than conventional staple crops while possessing other desired agronomic and nutritive characteristics, as well as introducing them into existing cropping systems to make them more resilient to water deficiency conditions. In the past, the existence of drought tolerance genes in crop wild relatives and landraces was either unknown or difficult to exploit using traditional breeding techniques to secure potential long-term solutions. Today, with the advances in genomics and phenomics, there are a number of new tools available that facilitate the discovery of drought resistance genes in crop wild relatives and landraces and their relatively easy transfer into advanced breeding lines, thus accelerating breeding progress and creating resilient varieties that can withstand prolonged drought periods. Among those tools are marker-assisted selection (MAS), genomic selection (GS), and targeted gene editing (clustered regularly interspaced short palindromic repeat (CRISPR) technology). The integration of these two major strategies, the advances in conventional and molecular breeding for the drought tolerance of conventional staple crops, and the introduction of drought-tolerant neglected and underutilized species into existing production systems has the potential to enhance the resilience of agricultural production under conditions of water scarcity.

Keywords: crop diversity; drought tolerance; genetic approaches; neglected and underutilized species

---

1. Introduction

Crops are dependent on rainfall, and so water scarcity is the primary productivity constraint in arid and semiarid tropical areas [1]. In these areas, water deficiency can last for periods longer than four months. Additionally, when El Niño Southern Oscillation (ENSO) occurs, the amount of water available during the rainy season significantly drops, while rainfall is concentrated within a period of a few months. As a meteorological event, drought is a period in which the potential evaporation exceeds the rainfall. Agricultural drought is the result of water flow imbalance between the environmental demands of evapotranspiration and water transport in the soil-root system [2]. In this context, drought tolerance is described as the ability of a plant to live, grow, and reproduce successfully with a limited water supply or during periodic conditions of water deficit [3]. Regarding crops, a drought response is defined as a change in yield as a consequence of impaired plant development [4].

The challenge to produce under water scarcity conditions requires integrated actions and strategies to remodel the crop genetic background and the cropping systems. For it, wild relatives and landraces and neglected and underutilized species (NUS) are a rich source of genetic diversity. Thus, crop wild relatives and landraces due to their local adaptations are a vast resource of genetic diversity for developing more productive, nutritious, and resilient crop varieties [5–7]. In maize, landraces from dry habitats have been used successfully in breeding for water-limited environments, and wild species that are relatives of cultivated crops have been on the agenda as possible donors for drought tolerance [8]. Similarly, several NUS are drought and heat stress-tolerant, resistant to pest and diseases, and adapted to semi-arid and arid environments [9–11]. However, NUS are considered those species to which little attention is paid or that are entirely ignored by agricultural researchers, plant breeders, and policymakers [12]. Some NUS show potential to be introduced in cropping systems for crop diversification, e.g., quinoa has been accepted as an alternative crop in the northern latitudes of Europe [13]. The wider use of NUS would increase the agricultural biodiversity to buffer against crop vulnerability to water scarcity and would provide the quality of food and diverse food sources to address both food and nutritional security [14].

This review focuses on two strategies to increase drought stress tolerance: (i) the use of natural genes that have evolved over time and are preserved in crop wild relatives and landraces and (ii) exploiting the reservoir of neglected and underutilized species to identify those that are known to be more drought-tolerant than conventional staple crops for introducing them into existing cropping systems to make them more resilient to water deficiency conditions. We also highlight the use of phenomics and genomics as methods and approaches to characterize, identify, and use the desired traits related to drought tolerance (Figure 1).

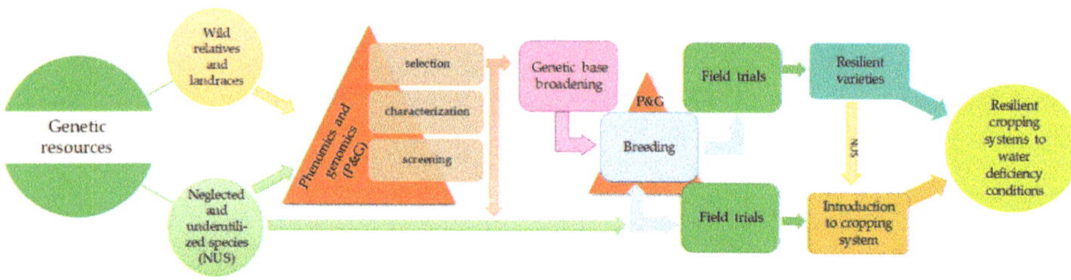

Figure 1. Dual strategy of breeding for drought tolerance and the introduction of underutilized crops to make more resilient cropping systems to water deficiency conditions.

2. General Overview of Physiological Responses of Plants to Drought Stress Conditions

In arid environments, crops are exposed to extreme water-limiting conditions, which have become more extreme in recent decades, leading to reductions in yield or even total yield loss. Drought conditions trigger a progressive process in plants that begins with an early priming and preconditioning stage, followed by an intermediate stage in preparation for acclimation and a late stage of new homeostasis with reduced growth (Figure 2) [15]. Signal transduction pathways connecting the recognition of environmental stress factors and the initiation of plant responses often involve several intracellular changes, including variations in Ca^{2+} concentration, reactive oxygen species (ROS) accumulation, and cytosolic K^+. Several proteins in the plasmalemma and tonoplast recognize these intracellular messengers, acting not only during signaling sensing but, also, in response to stress conditions [16].

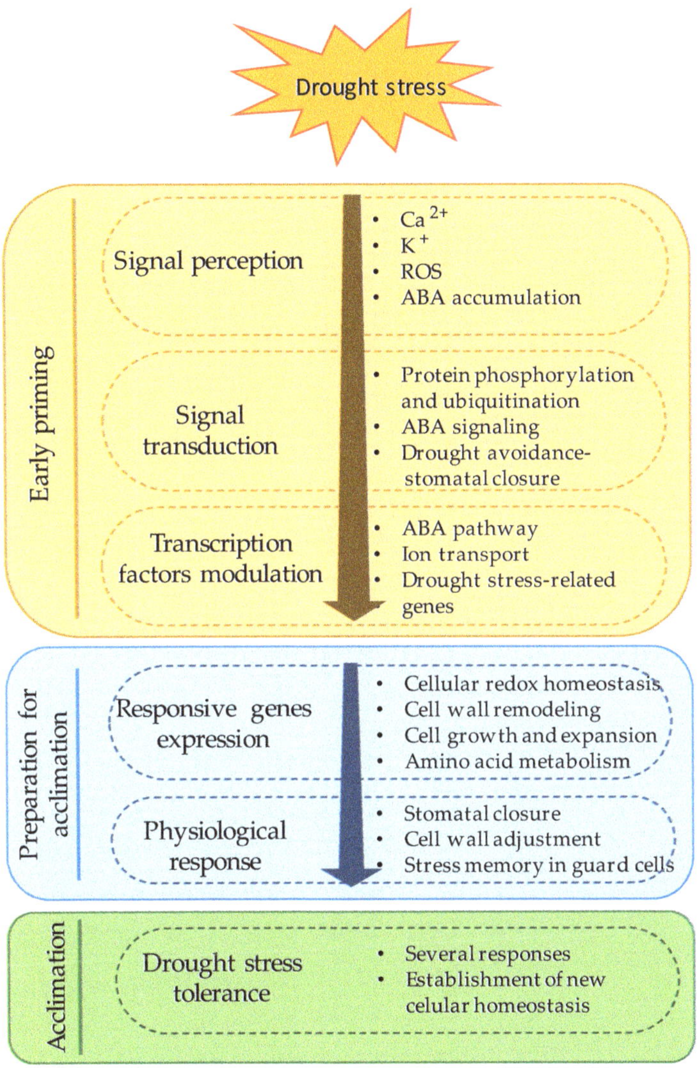

Figure 2. General description of physiological responses of plants to drought stress conditions. ROS: reactive oxygen species and ABA: abscisic acid.

The cascade of morphophysiological responses to drought stress is primarily controlled by the early accumulation of abscisic acid (ABA), ion transport, and the induction of the associated signaling pathway genes [17]. Signal transduction that occurs in response to the early accumulation of ABA during drought stress is mediated by protein phosphorylation and ubiquitination. This post-translational modification of specific proteins triggers a cascade of physiological responses that include a decrease in stomatal conductance as an early avoidance response to drought stress, resulting in rapid stomatal closure [18–20]; furthermore, the regulation of several transcription factors (TFs) involved in osmotic stress and the increased expression of expansion genes involved in cell wall adjustments are a preparatory step towards drought acclimation [15,17,21]. TFs, as major regulators of plant responses to drought stress, affect the adaptation of plants to drought stress through their involvement in the transcriptional regulation of ABA- and drought stress-related gene expressions [21,22]. An elevated resistance to drought can also be achieved by the increased expression of cuticular wax biosynthesis genes leading to an enhanced cuticular wax accumulation in both leaves and stems [23].

ABA-dependent kinases related to stomatal closure are sucrose nonfermenting 1 (SNF1)-related protein kinase 2.6 (SnRK2.6, also known as OST1), which has the overall control of the stomata, and SnRK2.2 and SnRK2.3, which are only involved in the implementation of stress memory in guard cells during the subsequent dehydration process [24]. The OST1-dependent phosphorylation of the plasma membrane intrinsic protein 2;1 (PIP2;1) aquaporin produces an increase in the guard cell permeability to water and, possibly, hydrogen peroxide to trigger stomatal closure [25].

Drought-sensitive plants accumulate significantly more reactive oxygen species (ROS) and reactive nitrogen species (RNS) than tolerant genotypes [26,27]. Cellular redox homeostasis is disturbed as a consequence of extra ROS generation under drought stress. ABA induces the production of nitric oxide (NO) in guard cells, which, together with RNS, is a secondary messenger that modifies the enzyme activity and gene regulation [28,29]. At a specific level of NO and H_2O_2 treatments, the destruction of the mesophyll cell ultrastructure caused by drought stress is attenuated, increasing leaf chlorophyll, chlorophyll fluorescence values, and soluble carbohydrate and protein contents. Thus, endogenous NO and H_2O_2 may play crucial roles in rooting and the photosynthetic performance under drought conditions [30].

ABA is produced in the cytosol by the plastid-carotenoid pathway via the cleavage of xanthophyll precursors [20,31], and an impairment of the plastid-carotenoid pathway produces photo-oxidation and ABA-deficiency [31]. Genes involved in carotene biogenesis are not only rate-limiting for ABA synthesis but are also involved in plant responses to drought-stress conditions [20].

3. Use of Crop Diversity in Plant Breeding for Drought-Tolerance Traits

Valuable genes from natural inter- and intraspecific diversity can be used to take advantage of several mechanisms of survival and coadaptation in plants produced by natural selection [32]. Some of these genes are conserved by farmers (in landraces) or are present in crop wild relatives, and the narrow genetic base of modern cultivars is becoming a major bottleneck for crop improvement efforts; therefore, crop wild relatives have been extremely valuable in adapting crop varieties to changing climatic conditions [33,34]. *Helianthus anomalus*, a diploid annual sunflower species of hybrid origin that is endemic to active desert dunes, was successfully used in sunflower breeding with tolerance to drought stress [35]. In rice, *Oryza glaberrima* has been used in interspecific backcrossing to improve the drought resistance in *Oryza sativa* [36]. Similarly, the wild emmer wheat (*Triticum dicoccoides*) is highly tolerant to drought compared with its domesticated counterpart [37]. Additional examples of drought-tolerant varieties of major stables obtained through conventional breeding are presented in Table 1.

Table 1. Progress in the improvement of drought tolerance (DT) in major crops, number of drought-tolerant varieties or donors selected using conventional breeding, and the use of crop wild relatives as sources of drought-tolerant genes.

Crop	No. DT Varieties or Donors	Method	Wild Species as Possible Sources for DT Introgression
Maize	41	Thirty-five varieties obtained by conventional breeding ([38,39] results published in the official internet web pages of CYMMYT, KARI, and WEMA projects)	Information not available

Table 1. *Cont.*

Crop	No. DT Varieties or Donors	Method	Wild Species as Possible Sources for DT Introgression
Potato	22	Conventional breeding [40–44]	*Solanum juzepczukii* B., *S. cardiophyllum*, and *S. gandarillasii*, *S. tarijense* [42]
Rice	16	Conventional breeding [45–47]	*Oryza longistaminata*, *O. rufipogon* [48], *O. meridionalis*, and *O. nivara* [49], *O. glaberrima* [36]
Sugarcane	24	Conventional breeding [50–53]	*Saccharum spontaneum* [54], and *S. robustum* [53]
Wheat	2	Conventional breeding [55]	*Aegilops kotsehyi*, *A. variabilis*, *A. speltoides*, *A. umbellulata*, *A. squarrosa* [54] and *A. tauschii* [56]
Cassava	2	Conventional breeding [57]	*Manihot glaziovii* [58], *M. pseudoglaziovii*, *M. stipularis*, *M. caerulenscens* [59], *M. carthaginensis*, and *M. dichotoma* [60]

There is evidence of success in some crops that have been obtained by genetic introgression. Although, in some cases, it could be a time-demanding method, the introduction of new high-throughput "omics" technologies (some are described below) improves the efficiency for drought-tolerance traits in intra- and interspecific introgressions. In rice, a transcriptomic analysis established that drought-tolerant genotypes (drought-tolerant donor parent and progeny) were functionally enriched in oxidoreductase and lyase activities compared with a cultivated variety [61]. Thus, many traits related to mechanisms of drought tolerance have evolved over time and are present in wild relatives. The process of introducing genetic diversity from wild species into cultivars requires a significant amount of time, resources, and human capacity. The evident success of this strategy cannot be estimated only in terms of released varieties; for example, its contribution as additional genetic diversity in some crops should also be considered. However, in many species, wild relatives and landraces are poorly represented in gene banks, largely unavailable, and, therefore, underutilized [6]. For this reason, the ex situ conservation of genetic resources, especially wild relatives, should be a priority to guarantee their future availability [5,6]; however, this conservation is constrained for technical and funding challenges. Therefore, crop prioritization should be done for target wild relative conservations; ex situ conservations should be guaranteed especially for crops with importance to global food supplies and production systems worldwide [5]. Alternatively, the establishment of in situ conservation reserves could be used for major and minor crops to support the custodians of agrobiodiversity, the local communities [6]. Crops that grow in regions that currently are under high threats of water scarcity should consider seeking gene sources from wild relatives; for that, the characterizations of the wild relatives and landraces enable the detection of traits that can be used and introduced in improved varieties to provide greater adaptation and resilience to such restrictive environmental conditions.

4. Introduction of Neglected and Underutilized Species into Cropping Systems

The introduction of neglected and underutilized species (NUS) into the current cropping systems could help reduce food scarcity and diversify the homogeneous crops systems. This as a contribution to improve the human diet that currently has an overreliance on very limited numbers of major crops, mainly as sources of energy-dense foods but poor-quality nutrition [62,63]. As research on drought tolerance has mainly focused on major staple crops, the potential of some NUS being naturally more drought-tolerant than most staple crops has been overlooked. Moreover, NUS can be used in strategies to diversify cropping systems by introducing new species, thus increasing the genetic diversity and resilience of production systems [63]. The domestication and breeding of new crops is a long-term solution for drought constraints [64]; however, this should be considered and could be carried out through (i) understanding the physiological, genetic, and molecular basis of natural mechanisms and the plasticity that allows their adaptations to drought stress, (ii) integrating new knowledge in breeding and field crop management in priority species, and (iii) articulating strategies and actionable recommendations to encourage their cultivation and make technologies and innovations widely available [64]. Crops that are naturally adapted to arid and semi-arid regions of the world exhibit several drought-tolerance mechanisms. Halophyte plants, such as quinoa (*Chenopodium quinoa*), have adaptations to drought stress that include an increase in Na^+ and K^+ transporters in cell and vacuole membranes, the rapid alteration of the plasma H^+-ATPase activity, high contents of antioxidant compounds, vesicles for salt secretion, and lower stomatal density, among others [65,66]. Although, sweet potatoes (*Ipomoea batatas*) cannot be categorized as a NUS, in certain countries, this species is not cultivated and could have high potential, since it shows the ability to grow and produce under adverse conditions. In this species, the role of phytoene synthase (IbPSY1) regulated by the orange protein (IbOr) in abiotic stress tolerance has been confirmed [67,68]. Both proteins are related to carotenoid biosynthesis and accumulation. In buckwheat, FeBREB1 and FtMYB10 are the functional genes associated with the drought response, and the proteomic profile showed an overexpression of oxidoreductase activity, oxidation–reduction processes, xyloglucan:xyloglucosyl transferase activity, and apoplasts [69,70]. Thus, some NUS grow in marginal lands and extreme conditions (drought, salinity, heat, etc.) using versatile and adaptive mechanisms.

Several NUS are hardy, resilient, and long used for food by traditional communities, particularly in the primary regions of the diversity of each crop. Although the ongoing globalization of food systems worldwide has led to a uniformization of crops grown globally at the detriment of tradition, it also has contributed to crop introductions in countries from the primary regions of diversity of the crops [71]. Therefore, countries use introduced crops from regions of diversity other than their own ("foreign crops"), confirming that crop introduction is a process that has occurred throughout the history of agricultural crops. However, this process has been affected by crop homogeneity of the global food supply, which has limited the current agriculture to a focus on eleven species [62] (these species, consequently, have been the main focus of the research activities). Thus, research focused on NUS needs to be encouraged and is required to dissect their value and promote their use as alternative crops to create more resilient cropping systems. Thanks to dedicated research, previously neglected and underutilized species such as quinoa (*Chenopodium quinoa*), buckwheat (*Fagopyrum* sp.), millet (*Pennisetum glaucum*), cowpea (*Vigna unguiculata*), and sweet potato (*Ipomoea batatas*), among others, were shown to have adaptative capacities to water deficiency and have been successfully introduced as new commercial crops into production systems (Table 2). Global efforts relating to these NUS have allowed them to be explored and introduced into agricultural systems in different regions worldwide, confirming that this is a key strategy for crop diversification, nutritional enhancement, and adaptation to changing climates for future needs.

Table 2. Neglected and underutilized species successfully introduced to several countries as alternative crops for drought tolerance.

Crop	Crop Origin	Status	Countries with Registered Varieties
Quinoa (*Chenopodium quinoa* Willd.)	Andean region [72]	Introduced	The Netherlands (5), Denmark (4), France (1), Canada(4) USA (3), Australia (2), Germany (1), Ukraine (2) [73,74]

Table 2. *Cont.*

Crop	Crop Origin	Status	Countries with Registered Varieties
Amaranth (*Amaranthus hypochondriacus*, *Amaranthus cruentus* L., and *Amaranthus caudatus* L.)	High tropical and subtropical lands of America [75]	Introduced	Russian (7 *Ah*, 8 *Ac*, 6 *Acr*), Germany (1 *Ah*, 1 *Acr*), Slovakia (1 *Ac*, 2 *Acr*), The Netherlands (1 *Ac*), Romania (1 *Ac*, 1 *Acr*), Ukraine (1 *Ac*), Poland (2 *Acr*), New Zealand (1 *Acr*), Czech Republic (1 *Acr*) [74]
Millet (*Pennisetum glaucum* L.)	Africa [76]	Introduced	Brazil (13), Russia (5), USA (3), Ukraine (1), Mexico (3), Australia (1) [74]
Buckwheat (*Fagopyrum* sp.)	China [77]	Introduced	Ukraine (19), Denmark (10), USA (4,) Moldova (3), Canada (2), Australia (1) [74]
Cowpea (*Vigna unguiculata*)	Southern Africa [78]	Introduced	Brazil (13), Australia (8), China (7), Turkey (7), Moldova (6), Korea (4), Romania (3), Bulgaria (2), Poland (1), Portugal (1) [74]
Sweetpotato (*Ipomoea batatas*)	Central America and north of South America [79]	Introduced	Switzerland (3), Israel (11,) Romania (2), Slovenia (8), Ukraine (4), USA (89), South Africa (29), China (42) [74]
Andean Lupin (*Lupinus mutabilis*)	Andean region [80]	Introduced	The Netherlands (1), Czech Republic (1), Germany (1) [74]

In parentheses is the number of registered varieties in each country. *Ah*: *Amaranthus hypochondriacus*, *Ac*: *Amaranthus caudatus*, and *Acr*: *Amaranthus cruentus*.

Exploiting the potential of NUS provides a highly diversified agricultural production system capable of sustaining food and nutritional security in water-deficient environments [81]. Species such as bambara groundnut (*Vigna subterranea*), taro (*Colocasia esculenta*), teff (*Eragrostis Tef*), yam (*Dioscorea esculenta*), moringa (*Moringa oleifera*), fonio (*Digitaria exilis*), safflower (*Carthamus tinctorius*), cañahua (*Chenopodium pallidicaule*), and tepary bean (*Phaseoulus acutifolius*), among others, could be alternative crops for several regions worldwide due to their natural adaptation to arid or semi-arid regions (center of origin or diversity) [71] to contribute to crop diversification and cropping systems that are more resilient to water-deficient conditions. However, more studies should be performed to understand the natural mechanisms and plasticity that allow their adaptation to current climate alterations, to identify priority species, to design their own field crop management, and to articulate strategies and actionable recommendations to encourage their cultivation and improvement.

5. Methods and Approaches to Improve Crop Tolerance to Drought Stress

Several methods and approaches could be used to characterize, identify, and apply desired traits related to drought tolerance in crops and their relatives and uncover their potential to promote more resilient cropping systems. This review is focused on both phenotyping and genomic approaches.

5.1. Phenotyping Methods for Drought-Tolerance Trait Evaluations

Plant growth and development change because of physiological alterations in response to water deficiency. These morphological traits are related to changes in metabolic patterns in source or sink organs [82]. Morphoanatomical and physiological adaptations can be determined by measuring certain traits, primarily those related to constitutive early vigor, starch storage, growth maintenance, and desiccation tolerance. These traits are important components of crop yield but can be expressed differently among genotypes. Consequently, levels of drought tolerance are expected to differ because of the phenotypic plasticity of the genotypes. In this context, phenotyping methods could complement and improve the efficiency and accuracy of field measurements that are, subsequently, to select desirable genotypes. For example, imaging techniques are suitable to measure the response to abiotic stress [83]. A decrease in stomatal conductance produces an increase in leaf or canopy temperature, which can easily be detected by thermal imaging, and changes in the assimilation rate, stomatal conductance, and intrinsic water use efficiency can be estimated [84,85]. Hyperspectral and near-infrared imaging are also used to correlate these parameters, which can easily be monitored without destructive sampling [86]. Fluorescence imaging is used to estimate the phenotypic parameters of the photosynthetic status, quantum yield, nonphotochemical quenching, and leaf health [83].

Morphology and color distributions can serve as indicators of developmental processes and stress responses of plants, and such alterations can be effectively detected by RGB color imaging. This is an effective, quantitative, and low-cost method to determine variations in several morphological traits and other yield-related parameters, such as growth, biomass, and architecture, among others, employing routines developed to convert pixels from red/green/blue images [87]. The use of RGB imaging for plant phenotyping is improving the quality of morphoagronomic characterizations, because it uses quantitative parameters as opposed to conventional qualitative parameters. Plant growth, morphology, and physiology can be automatically monitored via nondestructive analysis using RGB imaging software that is readily available. Thus, RGB is currently the most extensive imaging technology used [88]. Additionally, traits detected by imaging can be correlated with yield, resulting in cost-effective models that can identify target traits for breeding, including drought-tolerance traits and genotypes with higher plasticity.

In roots, hydrotropism is a phenotypic strategy in response to the water supply [89]. Similar to aboveground tissues, several mechanisms are related to morphological and physiological adaptations in response to drought stress. In soybeans, the plants with more lateral roots, a thicker lateral root system, and forks showed higher yields under water-deficient conditions in clay or sandy soils compared to those of the susceptible varieties [90]. Several phenotyping systems have been developed to evaluate the response in roots to drought conditions. Those systems use high-throughput analysis to determine the stem diameter, median and maximum root width, root top, and bottom angles (using Digital Imagen of Root Traits (DIRT) [91]), properties of the primary and lateral roots (employing RootGraph [92]), and other parameters. Although image analysis has improved with time, the extraction of roots from the soil produces damage that can affect the evaluation of the actual plant response. The 3D reconstruction systems that are currently under development perform nondestructive evaluations of the root system, as they do for aboveground organs, to address this problem [93].

All the previously described imaging methods could be useful tools to characterize wild species, cultivars, landraces, and orphan crops with mechanisms to tolerate drought stress conditions. The identification of these desired traits contributes to the efficient use of genotypes with higher tolerances to drought stress, expressed primarily in yields under water-deficient conditions.

5.2. *Potential of Genomic Approaches to Improve Crop Tolerances to Drought Stress*

For the development of genetic materials with drought tolerance, selection has focused on genotypes with high yield levels cultivated under dry conditions. For example, in wheat and maize, selections have been based on evaluating the plant phenotypes and physiological responses to drought stress [94]. Selection based on plant responses to drought stress is affected by low heritability, genetic interaction, environment-genotype interactions, and polygenic effects; therefore, the selection is slow, as massive phenotypic screening is required [95]. In this context, future breeding programs focusing on drought tolerance require the combination of plant breeding, genomics, statistics, experimental design, and genetic diversity management strategies. The combinations of these approaches can offer new opportunities in the genetic dissection of important quantitative traits (e.g., drought tolerance) through the identification of quantitative trait loci (QTL), the implementation of marker-assisted selection (MAS) in breeding programs, and the cloning of these QTLs/genes and their editing using genetic-engineering strategies [94,96]. All approaches are studied using data from "omics" strategies, such as genomics, transcriptomics, proteomics, and metabolomics, which include large amounts of information; therefore, bioinformatics approaches are required for its analysis; hence, the importance of this area of computation in the analyses that integrate genotypic and phenotypic information, such as QTL mapping, single-nucleotide polymorphism (SNP), and gene discovery, as well as genomic selection, among others. Currently, the generation and use of bioinformatics tools, the open or closed code source (paid programs), and the use of molecular databases such as NCBI (National Center for Biotechnology Information) allows the analysis and interpretation of this information at different levels and varying degrees of complexity [97,98].

Genetic engineering has allowed the manipulation of the plant genome to study the gene structure and the function of candidate genes [95]. However, for drought stress, this strategy has not allowed the development of drought-tolerant varieties [94,99]. For example, maize (the highest-yielding cereal crop worldwide) is susceptible to drought during the flowering period, and the materials generated with genetic modifications to improve their drought tolerance presented a reduction in their yield traits. However, the overexpression of the trehalose-6-phosphate phosphatase (TPP) gene in maize plants has allowed the identification of materials that show an increase in yield of a similar proportion to the material generated without genetic modifications cultivated in mild and severe drought conditions [100]. Nonetheless, in Arabidopsis, the genetic basis of drought tolerance has been studied, and related homologous genes and their products have been identified and manipulated using genetic engineering, which demonstrated the feasibility for agriculturally important crop species. For example, ABA is a key player in drought tolerance and avoidance, as described above (Section 2). ABA regulates stomata closure in cases of water deficiency and plant desiccation and activates stress response genes by regulating diverse transcription factors. Therefore, if the primary aim is to increase plant sensitivity to ABA, this plant hormone is one of the main targets for genetic breeding regarding plant drought tolerance [96].

Several approaches are proposed to achieve plant tolerance to drought in transgenic crops. One approach is based on the manipulation of the ABA pathway, either by downregulating or by upregulating genes and the corresponding proteins involved in ABA signaling, biosynthesis, or degradation [7]. Farnesyltransferase ERA1 participates in ABA signaling, and plants transformed with antisense ERA1 constructed under a drought-inducible promoter showed higher tolerance to drought [101]. Another approach relies on the regulation of transcriptional activities of several genes by controlling the functions of the ABA-dependent transcription factors MYB2, MYC2, CBF, and AREB or the ABA-independent transcription factors such as DREB2, DREB3, and ZHDH [96,102,103]. ABA-independent transcription factors regulate the expression of stress-responsive genes; DREBs were first identified in Arabidopsis, but their role in stress tolerance has been demonstrated in several crops (e.g., maize, wheat, rice, rye, barley, soybean, and tomato). Hence, because DREBs have universal roles in abiotic stress responses in plants [104], they are ideal candidates for the genetic manipulation of the drought response in transgenic plant lines. Finally, but not less important, some signaling pathways that include reactive oxygen species (ROS), sucrose nonfermenting 1-related protein kinase (SnRKs), mitogen-activated protein kinases (MAPKs), calcium-dependent protein kinases (CDPKs) and phosphatases are involved in the regulation of plant responses to drought [105–107]; these pathways are also valuable targets in the genetic engineering of drought tolerance. The experimental results of genetic manipulation in crops are positive. In maize, the increased expression of the orthologous maize transcription factor (ZmNF-YBs) increased the tolerance to drought based on the responses of various stress-related parameters, including the chlorophyll content, stomatal conductance, leaf temperature, reduced wilting, and the maintenance of photosynthesis. These adaptations contributed to higher grain yields in water-limited environments [108]. The overexpression of phytochrome-interacting factor 3 (PIF3) increases the tolerance to dehydration and salt stress, and in ZmPIFs transgenic plants, the relative water and chlorophyll contents and chlorophyll fluorescence increases, in addition to a significant increase in cell membrane stability under stress conditions [109,110].

Drought tolerance is a quantitative trait controlled by many genes, so the genetic manipulation to generate new cultivars with drought tolerance using the single genetic interventions strategy is difficult [111,112]. However, the understanding of the genetic bases of drought tolerance has increased [101], and many genes associated with this trait have been identified [113] and used in the implementation of gene-editing systems [114], gene silencing [115] and overexpression methods [116] to generate materials with possible drought tolerance [100]. Some examples of success using these strategies to produce new materials with possible drought tolerance in the main crops are presented in Table 3. As with the phenotypic selections, the genomic analysis for drought tolerance has been studied, measuring traits related to yield in dry conditions [113], because there are few reports about regions of the genome associated with specific drought-response components. Additionally, in species such as wheat, the QTLs associated with drought tolerance remain large, and their uses in breeding programs are limited [95].

Table 3. Examples of the contribution of genomic approaches in the breeding of major crops for drought tolerance (DT).

Crop	Genotypes or Varieties Names	Method	DT Source	Reference
Maize	PH4CV-T, PH6WC-T, Chang7-2-T, and Zheng58-T	Overexpression	VPP gene	[117]
Potato	Cultivar Sante	Overexpression	STANN1 mRNA	[118]
Rice	U7, U14	Overexpression	OsOAT gene	[119]
Sugarcane	ZmRab17:AtDREB2A CA	Overexpression	AtDREB2A CA transcription factor	[120]
Wheat	Transgenic Durum Wheat cv. Maali	Overexpression	TdPIP2 gene	[121]
Cassava	South China 124 (SC124) cassava variety	Silencing	HSP90 protein	[122]
Soybean	Transgenic soybean plants	Overexpression	GmFDL19 transcription factor	[123]

Genomic analyses for drought tolerance have allowed the identification of genes, transcription factors, miRNAs, hormones, and proteins involved in this type of stress [96], including loci related to ABA pathway signaling [124], leaf senescence [125], and other drought-related traits [96]. For example, for related ABA signaling, one of these QTLs contained a candidate gene coding for 9-cis-epoxycarotenoid dioxygenase 2, which is an essential enzyme during ABA synthesis that is expressed under drought stress [124]. QTLs related to plant growth and physiological parameters in wheat under water-deficit conditions were identified from a genetic map of 3200 SNP markers and 783 loci, and strong effects of these QTLs related to drought conditions were detected, providing evidence of their potential in plant breeding. Nevertheless, low heritability values showed that these traits are highly influenced by the environment [87]. Similarly, in barley, two primary QTLs associated with drought stress and leaf senescence were identified. Water deficiency negatively affected the biomass yield, leaf color, and electron transport rate values, whereas the osmolality, free proline content, and total content of soluble sugars increased under drought stress [125]. In rice, asymmetric root growth and an increase in the root growth angle were observed upon the introduction of the DEEPER ROOTING 1 gene (DRO1), which is a quantitative trait locus in rice that controls the root growth angle; therefore, the resulting line avoided the occurrence of drought by increasing deep rooting and maintained a high yield performance under drought conditions [126]. However, drought tolerance is a highly complex trait, so the cultivar development for this target has been achieved by classical breeding [95].

Due to the interest in drought tolerance, genomic tools such as genomic selection (GS) and marker-assisted selection (MAS) are required in new breeding programs [127]. To implement MAS, first, the identification of molecular markers and genes/QTLs that explain the phenotypic variance of drought tolerance is required, and due to drought stress in plants, many changes in gene expressions are observed. For this, the identification of candidate genes that are expressed in drought-stress conditions is a major strategy, and genomic technologies such as microarray and transcriptomic analyses have been useful in the identification of these genes [101,128]. These technologies include new strategies in the identification of candidate genes associated with important agronomic traits in species with genome sequencing that have been developed as novel in-silico platforms for gene discovery and that help scientists to identify candidate genes through the knowledge available in genetic databases and public information, allowing candidate gene prioritization [129]. Although the advance of genomic technology has allowed the identification of QTLs for drought tolerance, this information has been underutilized in the generation of new cultivars with drought tolerance [113]. To improve drought tolerance, QTLs associated with this trait have been introgressed into breeding populations and selected genotypes using MAS. For example, in rice, alleles of QTLs related to drought stress were transferred into different genetic backgrounds, and their effects were identified; moreover, the phenotypic evaluations confirmed the success of MAS for this trait [130].

Many favorable alleles for drought tolerance are present in crop wild relatives; thus, new alleles from those wild species could be included in breeding programs. These alleles should come mainly from genotypes adapted to target environments and traits. For this, it is necessary to include and evaluate these materials in all breeding programs, particularly in crops that show low levels of genetic diversity due to bottlenecks caused by selection and inbreeding [131]. In crops such as wheat and barley, QTLs related to drought have been identified and introgressed from wild relative species [132], allowing the inclusion of new favorable alleles in the breeding populations. Another approach to identifying regions that explain the phenotypic variation of traits (e.g., drought tolerance) in plants is the Genome-Wide Association Study (GWAS), where it is possible to associate a specific single-nucleotide polymorphism (SNP)/gene with a phenotypic variation of any trait [133]. The GWAS combines phenotypic and genomic information to identify statistical associations in order to explain the phenotypic variations of the target trait. This approach uses all allelic variations available and the possibility to identify marker-trait associations. On the other hand, resolution mapping depends on the linkage disequilibrium (LD) extension in the mapped population. These populations can be found in gene banks, core collections, and breeding populations, allowing the analysis of all historical recombination events in the used population. Additionally, the GWAS considers the relationship between the samples and their genetic structure in order to minimize false positives [134]. For drought stress, the GWAS has been used to identity SNPs related to high-temperature tolerance and their yield effects in crops such as wheat [135], cotton [136], rice [137], and maize [138], among other species.

Considering the above, new approaches in genomics strategies such as GS combined with high-throughput phenotyping are starting to be used to achieve different important target traits; this is because this approach allowed the identification of better lines through prediction models for complex traits such as stress (e.g., drought tolerance), employing the genomic estimated breeding value (GEBV) at the individual level [95]. GS is a form of MAS but shows new characteristics for identifying promising materials in comparison to conventional MAS that uses some markers/genes previously identified, with significant effects in the genetic government of the target traits, to improve their level [95]. In recent years, the most popular technologies to identify molecular markers in a partial representation of the genome (use of enzyme restriction) are genotyping strategies based on next-generation sequencing such as genotyping by sequencing (GBS), RAD-seq, diversity arrays technology (DArT), and the complexity reduction of polymorphic sequences (CRoPS), among others [139,140].

Genomic selection implemented prediction models that allowed selecting genotypes without phenotypic evaluation. In this sense, alleles with high and low effects are analyzed; the breeding of complex traits such as drought tolerance is well-supported by this approach. GS strategies in the plant breeding of maize and barley have reduced the selection time by almost half per cycle compared to the phenotypic selection [141]. GS in plant breeding has been used to improve different important traits such as grain yield in maize [142], soybean cyst nematode resistance [143], amylase activity in barley [144], and grain yield and plant height in rice [145], among others. To this end, genomic selection offers a new method to accelerate the breeding process [146]. The use of GS to improve drought tolerance has been reported in maize [147,148] and chickpea [149], establishing a low-to-medium prediction accuracy for yield and secondary traits related to drought stress. The implementation of GS has shown that the prediction accuracy is affected by the breeding population types, training population size, the complexity of the trait, and the number of markers used. Additionally, the inclusion of marker-trait associations identified by QTL and GWAS analyses in prediction models increases the prediction accuracy [150,151]. Thus, it is necessary to continue with the study of the genetic architecture of complex traits by GWAS and QTLs analyses.

Currently, it is possible to generate new alleles in known genes through second-generation gene-editing methods, the most popular being the zinc-finger nucleases (ZFN), transcription activator-like effector nucleases (TALEN), and the clustered regularly interspaced short palindromic repeat (CRISPR)/CRISPR-associated nuclease protein (CRISPR/Cas) system [152,153]. The CRISPR/Cas9 system is a technique to edit genes that function as an endonuclease that induces double-strand breaks (DSB) at specific genome sites, followed by DNA repair, which includes, in the gene sequences, some insertions and/or deletions [154]. This system for genome editing has been used in species such as Arabidopsis, tomato, rice, maize, and wheat [154–157], among others. This gene-editing system is known to be simple to implement, has design flexibility, is low-cost, and is highly efficient [158]. The CRISPR/Cas9 system for drought stress has been used to edit the genes ARGOS8 in maize [159]; SlMAPK3 and SlNPR1 in tomato [160,161]; MIR169a and OST2 in Arabidopsis [162]; and OsDERF1, OsPMS3, OsEPSPS, OsMSH1, and OsMYB5 in rice [163], among others.

Genetic information on drought tolerance pathways in crops has proven to be a promising approach to improve crops using genetic engineering, marker-assisted selection, and genomic selection, thus demonstrating the feasibility of integrating this approach into drought tolerance challenges in cropping systems.

6. Concluding Remarks

In this review, we described two strategies to improve drought-stress tolerance in crops: (i) the use natural genes for drought stress tolerance that have evolved over time and are present in crop wild relatives and landraces and (ii) exploiting the potential of neglected and underutilized species and introducing them into cropping systems to make them more resilient to water deficiency conditions. For both, the richness of genetic diversity represents an invaluable reserve for breeding, crop diversification, nutritional enhancement, and adaptation to changing climates, which should be recognized and conserved for future needs. The mechanisms of the drought tolerance of crop gene pools and neglected and underutilized species guarantee food security in environments where they grow naturally and/or are cultivated. However, despite the recent studies in this field, much information remains unknown. Future studies should continue and integrate several approaches (including both phenomics and genomics) to explore, characterize, identify, and use desired traits, contributing to the development of crops and cropping systems with tolerances to drought stress. An understanding of the morphoanatomical, physiological, and genetic mechanisms involved in the responses to drought stress in crop wild relatives and neglected and underutilized species is fundamental to recognize their potential for crop breeding or crop diversification. Therefore, the integration of crop breeding supported by phenomics and genomics is key for improving the tolerance to drought stress in crops, as has been reflected in species such as corn and other mentioned species. Likewise, for neglected and underutilized species such as quinoa, whose development as a crop in recent years can be considered as the result of the integration of these strategies and introduced into existing cropping systems.

This is a practical direction for future research to make more resilient cropping systems to water deficiency conditions.

Author Contributions: All authors listed have made a substantial, direct, and intellectual contribution to this work; further. All authors have read and agreed to the published version of the manuscript.

Funding: The Internal Grant Agency, FA, Mendel University in Brno (No. IP 40/2015), NAAR (No. QK1910197) (both in the Czech Republic), and the Ministerio de Agricultura y Desarrollo Rural (MADR) of Colombia (Agreement No. 1828) funded this study.

Acknowledgments: To William Gomez for support and time for discussions.

Conflicts of Interest: The authors declare no conflict of interest.

References

1. Simelton, E.S.; Fraser, E.D.G.; Termansen, M.; Benton, T.G.; Gosling, S.N.; South, A.; Arnell, N.W.; Challinor, A.J.; Dougill, A.J.; Forster, P.M.D.F. The socioeconomics of food crop production and climate change vulnerability: A global scale quantitative analysis of how grain crops are sensitive to drought. *Food Secur.* **2012**, *4*, 163–179. [CrossRef]
2. Lipiec, J.; Doussan, C.; Nosalewicz, A.; Kondracka, K. Effect of drought and heat stresses on plant growth and yield: A review. *Int. Agrophysics* **2013**, *27*, 463–477. [CrossRef]
3. Turner, N. Drought resistance and adaptation to water deficits in crop plants. In *Stress Physiology in Crop Plants*; Mussell, H., Staples, R.C., Eds.; John Wiley & Sons: New York, NY, USA, 1979; pp. 343–372, ISBN 978-0-4710-3809-2.
4. Passioura, J.B. Drought and drought tolerance. *Plant Growth Regul.* **1996**, *20*, 79–83. [CrossRef]
5. Castañeda-Álvarez, N.P.; Khoury, C.K.; Achicanoy, H.A.; Bernau, V.; Dempewolf, H.; Eastwood, R.J.; Guarino, L.; Harker, R.H.; Jarvis, A.; Maxted, N.; et al. Global conservation priorities for crop wild relatives. *Nat. Plants* **2016**, *2*, 16022. [CrossRef]
6. Vincent, H.; Amri, A.; Castañeda-Álvarez, N.P.; Dempewolf, H.; Dulloo, E.; Guarino, L.; Hole, D.G.; Mba, C.; Toledo, A.; Maxted, N. Modeling of crop wild relative species identifies areas globally for in situ conservation. *Commun. Biol.* **2019**, *2*, 136. [CrossRef] [PubMed]
7. Villa, T.C.C.; Maxted, N.; Scholten, M.; Ford-Lloyd, B. Defining and identifying crop landraces. *Plant Genet. Resour.* **2005**, *3*, 373–384. [CrossRef]
8. Xu, Y.; Skinner, D.J.; Wu, H.; Palacios-Rojas, N.; Araus, J.L.; Yan, J.; Gao, S.; Warburton, M.L.; Crouch, J.H. Advances in maize genomics and their value for enhancing genetic gains from breeding. *Int. J. Plant Genom.* **2009**, *2009*, 1–30. [CrossRef]
9. Chivenge, P.; Mabhaudhi, T.; Modi, A.T.; Mafongoya, P. The potential role of neglected and underutilised crop species as future crops under water scarce conditions in sub-Saharan Africa. *Int. J. Environ. Res. Public Health* **2015**, *12*, 5685–5711. [CrossRef]
10. Mabhaudhi, T.; Chimonyo, V.; Modi, A.T. Status of underutilised crops in South Africa: Opportunities for developing research capacity. *Sustainibility* **2017**, *9*, 1569. [CrossRef]
11. Modi, A.T.; Mabhaudhi, T. Water use and drought tolerance of selected traditional and indigenous crops. In *Final Report of Water Research Commission Project K5/1771//4*; Water Research Commission: Pretoria, South Africa, 2013; ISBN 978-1-4312-0434-2.
12. Padulosi, S.; Thompson, J.; Rudebjer, P. *Fighting Poverty, Hunger and Malnutrition with Neglected and Underutilized Species (NUS): Needs, Challenges and the Way Forward: Neglected and Underutilized Species*; Bioversity International: Rome, Italy, 2013; ISBN 978-92-9043-941-7.
13. Bazile, D.; Pulvento, C.; Verniau, A.; Al-Nusairi, M.S.; Ba, D.; Breidy, J.; Hassan, L.; Mohammed, M.I.; Mambetov, O.; Otambekova, M.; et al. Worldwide evaluations of quinoa: Preliminary results from post international year of quinoa FAO projects in nine countries. *Front. Plant Sci.* **2016**, *7*, 850. [CrossRef] [PubMed]
14. Mustafa, M.A.; Mayes, S.; Massawe, F. Crop diversification through a wider use of underutilised crops: A strategy to ensure food and nutrition security in the face of climate change. In *Sustainable Solutions for Food Security*; Springer Science and Business Media LLC.: Berlin, Germany, 2019; pp. 125–149.

15. Harb, A.; Krishnan, A.; Ambavaram, M.M.; Pereira, A. Molecular and physiological analysis of drought stress in arabidopsis reveals early responses leading to acclimation in plant growth. *Plant Physiol.* **2010**, *154*, 1254–1271. [CrossRef] [PubMed]
16. Shabala, S.; Bose, J.; Fuglsang, A.T.; Pottosin, I. On a quest for stress tolerance genes: Membrane transporters in sensing and adapting to hostile soils. *J. Exp. Bot.* **2015**, *67*, 1015–1031. [CrossRef] [PubMed]
17. Osakabe, Y.; Osakabe, K.; Shinozaki, K.; Tran, L.-S.P. Response of plants to water stress. *Front. Plant Sci.* **2014**, *5*, 86. [CrossRef] [PubMed]
18. Ding, S.; Zhang, B.; Qin, F. Arabidopsis RZFP34/CHYR1, a ubiquitin E3 ligase, regulates stomatal movement and drought tolerance via SnRK2.6-mediated phosphorylation. *Plant Cell* **2015**, *27*, 3228–3244. [CrossRef] [PubMed]
19. Wang, Y.; Noguchi, K.; Ono, N.; Inoue, S.-I.; Terashima, I.; Kinoshita, T. Overexpression of plasma membrane H+-ATPase in guard cells promotes light-induced stomatal opening and enhances plant growth. *Proc. Natl. Acad. Sci. USA* **2013**, *111*, 533–538. [CrossRef] [PubMed]
20. Daszkowska-Golec, A. The role of abscisic acid in drought stress: How ABA helps plants to cope with drought stress. In *Drought Stress Tolerance in Plants*; Springer Science and Business Media LLC: Berlin, Germany, 2016; Volume 2, pp. 123–151.
21. Yang, J.; Zhang, G.; An, J.; Li, Q.; Chen, Y.; Zhao, X.; Wu, J.; Wang, Y.; Hao, Q.; Wang, W.; et al. Expansin gene TaEXPA2 positively regulates drought tolerance in transgenic wheat (*Triticum aestivum* L.). *Plant Sci.* **2020**, *298*, 110596. [CrossRef]
22. Fang, Q.; Wang, X.; Wang, H.; Tang, X.; Liu, C.; Yin, H.; Ye, S.; Jiang, Y.; Duan, Y.; Luo, K. The poplar R2R3 MYB transcription factor PtrMYB94 coordinates with abscisic acid signaling to improve drought tolerance in plants. *Tree Physiol.* **2019**, *40*, 46–59. [CrossRef]
23. Zhong, M.-S.; Jiang, H.; Cao, Y.; Wang, Y.-X.; You, C.-X.; Li, Y.-Y.; Hao, Y.-J. MdCER2 conferred to wax accumulation and increased drought tolerance in plants. *Plant Physiol. Biochem.* **2020**, *149*, 277–285. [CrossRef]
24. Virlouvet, L.; Fromm, M. Physiological and transcriptional memory in guard cells during repetitive dehydration stress. *New Phytol.* **2014**, *205*, 596–607. [CrossRef]
25. Grondin, A.; Rodrigues, O.; Verdoucq, L.; Merlot, S.; Leonhardt, N.; Maurel, C. Aquaporins contribute to ABA-triggered stomatal closure through OST1-mediated phosphorylation. *Plant Cell* **2015**, *27*, 1945–1954. [CrossRef]
26. Yang, L.; Fountain, J.C.; Wang, H.; Ni, X.; Ji, P.; Lee, R.D.; Kemerait, R.C.; Scully, B.T.; Guo, B. Stress sensitivity is associated with differential accumulation of reactive oxygen and nitrogen species in maize genotypes with contrasting levels of drought tolerance. *Int. J. Mol. Sci.* **2015**, *16*, 24791–24819. [CrossRef]
27. Laxa, M.; Liebthal, M.; Telman, W.; Chibani, K.; Dietz, K.-J. The role of the plant antioxidant system in drought tolerance. *Antioxidants* **2019**, *8*, 94. [CrossRef] [PubMed]
28. Smékalová, V.; Doskočilová, A.; Komis, G.; Šamaj, J. Crosstalk between secondary messengers, hormones and MAPK modules during abiotic stress signalling in plants. *Biotechnol. Adv.* **2014**, *32*, 2–11. [CrossRef] [PubMed]
29. Ullah, A.; Manghwar, H.; Shaban, M.; Khan, A.H.; Akbar, A.; Ali, U.; Ali, E.; Fahad, S. Phytohormones enhanced drought tolerance in plants: A coping strategy. *Environ. Sci. Pollut. Res.* **2018**, *25*, 33103–33118. [CrossRef]
30. Liao, W.-B.; Huang, G.-B.; Yu, J.; Zhang, M. Nitric oxide and hydrogen peroxide alleviate drought stress in marigold explants and promote its adventitious root development. *Plant Physiol. Biochem.* **2012**, *58*, 6–15. [CrossRef] [PubMed]
31. Fang, J.; Chai, C.; Qian, Q.; Li, C.; Tang, J.; Sun, L.; Huang, Z.; Guo, X.; Sun, C.; Liu, M.; et al. Mutations of genes in synthesis of the carotenoid precursors of ABA lead to pre-harvest sprouting and photo-oxidation in rice. *Plant J.* **2008**, *54*, 177–189. [CrossRef]
32. Ford-Lloyd, B.V.; Schmidt, M.; Armstrong, S.J.; Barazani, O.; Engels, J.; Hadas, R.; Hammer, K.; Kell, S.P.; Kang, D.; Khoshbakht, K.; et al. Crop Wild Relatives—Undervalued, underutilized and under threat? *Bioscience* **2011**, *61*, 559–565. [CrossRef]
33. Dempewolf, H.; Baute, G.; Anderson, J.; Kilian, B.; Smith, C.; Guarino, L. Past and future use of wild relatives in crop breeding. *Crop. Sci.* **2017**, *57*, 1070–1082. [CrossRef]

34. Mammadov, J.; Buyyarapu, R.; Guttikonda, S.K.; Parliament, K.; Abdurakhmonov, I.Y.; Kumpatla, S.P. Wild relatives of maize, rice, cotton, and soybean: Treasure troves for tolerance to biotic and abiotic stresses. *Front. Plant Sci.* **2018**, *9*, 886. [CrossRef]
35. Ludwig, F.; Rosenthal, D.M.; Johnston, J.A.; Kane, N.C.; Gross, B.L.; Lexer, C.; Dudley, S.A.; Rieseberg, L.H.; Donovan, L.A. Selection on leaf ecophysiological traits in a desert hybrid helianthus species and early-generation hybrids. *Evolution* **2004**, *58*, 2682–2692. [CrossRef]
36. Ndjiondjop, M.N.; Manneh, B.; Cissoko, M.; Dramé, N.; Kakaï, R.G.; Bocco, R.; Baimey, H.; Wopereis, M. Drought resistance in an interspecific backcross population of rice (Oryza spp.) derived from the cross WAB56-104 (O. sativa)×CG14 (O. glaberrima). *Plant Sci.* **2010**, *179*, 364–373. [CrossRef]
37. Budak, H.; Kantar, M.; Kurtoglu, K.Y. Drought tolerance in modern and wild wheat. *Sci. World J.* **2013**, *2013*, 1–16. [CrossRef] [PubMed]
38. Alpha, Y.; Kureh, I.; Menkir, A.; Kartung, P.; Tarfa, B.; Amaza, P. Participatory on-farm evaluation of the performance of drought-tolerant maize varieties in the Guinea savannas of Nigeria. *Int. J. Food Agric. Environ.* **2006**, *4*, 192–196.
39. Xu, J.; Yuan, Y.; Xu, Y.; Zhang, G.; Guo, X.-S.; Wu, F.; Wang, Q.; Rong, T.; Pan, G.; Cao, M.; et al. Identification of candidate genes for drought tolerance by whole-genome resequencing in maize. *BMC Plant Biol.* **2014**, *14*, 83. [CrossRef] [PubMed]
40. Al Sharari, S.; Alsadon, A.; Alharbi, A. Evaluation of drought tolerance of potato cultivars under greenhouse conditions. *Acta Hortic.* **2007**, *747*, 67–74. [CrossRef]
41. International Potato Center (CIP). *Annual Report CIP 1993–1994*; CIP: Lima, Peru, 1994; p. 192. ISSN 0256-6311.
42. Monneveux, P.; Ramirez, D.A.; Pino, M.-T. Drought tolerance in potato (*S. tuberosum* L.). *Plant Sci.* **2013**, *205*, 76–86. [CrossRef] [PubMed]
43. Rolando, J.; Ramirez, D.A.; Yactayo, W.; Monneveux, P.; Quiroz, R. Leaf greenness as a drought tolerance related trait in potato (*Solanum tuberosum* L.). *Environ. Exp. Bot.* **2015**, *110*, 27–35. [CrossRef]
44. Sharma, N.; Rawal, S.; Kadian, M.; Arya, S.; Bonierbale, M.; Singh, B. Evaluation of advanced potato clones for drought tolerance in arid zone in Rajasthan, India. *Potato J.* **2014**, *41*, 189–193.
45. Reyes, L. Making rice less thirsty. *Rice Today* **2009**, *8*, 12–15.
46. Singh, A.; Singh, A.K.; Singh, V.; Singh, N.; Singh, V.N.; Shamim, M.; Vikram, P.; Singh, S. Genetic variability among traits associated with grain yield of rice (*Oryza sativa* L.) exposed to drought at flowering stage. *Afr. J. Agric. Res.* **2014**, *9*, 1252–1264.
47. Zheng, X.-G.; Chen, L.; Lou, Q.-J.; Xia, H.; Li, M.-S.; Luo, L. Changes in DNA methylation pattern at two seedling stages in water saving and drought-resistant rice variety after drought stress domestication. *Rice Sci.* **2014**, *21*, 262–270. [CrossRef]
48. Liu, L.; Lafitte, R.; Guan, D. Wild Oryza species as potential sources of drought-adaptive traits. *Euphytica* **2004**, *138*, 149–161. [CrossRef]
49. Thanh, P.T.; Sripichitt, P.; Chanprame, S.; Peyachoknagu, S. Transfer of drought resistant character from wild rice (*Oryza meridionalis* and *Oryza nivara*) to cultivated rice (*Oryza sativa* L.) by backcrossing and immature embryo culture. *Kasetsar J.* **2006**, *40*, 582–594.
50. Cia, M.; Guimaraes, A.; Medici, L.O.; Chabregas, S.; Azevedo, R. Antioxidant responses to water deficit by drought-tolerant and -sensitive sugarcane varieties. *Ann. Appl. Biol.* **2012**, *161*, 313–324. [CrossRef]
51. Da Costa, M.L.M.; Amorim, L.L.B.; Onofre, A.V.C.; De Melo, L.J.O.T.; De Oliveira, M.B.M.; De Carvalho, R.; Benko-Iseppon, A.M. Assessment of genetic diversity in contrasting sugarcane varieties using inter-simple sequence repeat (ISSR) markers. *Am. J. Plant Sci.* **2011**, *2*, 425–432. [CrossRef]
52. Ishaq, M.N.; Olaoye, G.; Akinsanya, T.O. Screening sugar-cane germplasm for drought tolerance in Nigeria. *Plant Genetic Resources Newsletter* **2008**, *154*, 48–54.
53. Srivastava, M.K.; Li, C.N.; Li, Y.R. Development of sequence characterized amplified region (SCAR) marker for identifying drought tolerant sugarcane genotypes. *Aust. J. Crop. Sci.* **2012**, *6*, 763–767.
54. Mohanan, K. Breeding for special purposes. In *Essentials of Plant Breeding*; Mohanan, K.V., Ed.; Prentice-Hall of India Pvt Limited: New Delhi, India, 2010; p. 107. ISBN 978-812-033-968-2.
55. Finkel, E. Richard Richards profile: Making every drop count in the buildup to a blue revolution. *Science* **2009**, *323*, 1004–1005. [CrossRef]
56. Sohail, Q.; Inoue, T.; Tanaka, H.; Eltayeb, A.E.; Matsuoka, Y.; Tsujimoto, H. Applicability of Aegilops tauschii drought tolerance traits to breeding of hexaploid wheat. *Breed. Sci.* **2011**, *61*, 347–357. [CrossRef]

57. Turyagyenda, L.F.; Kizito, E.B.; Ferguson, M.; Baguma, Y.; Agaba, M.; Harvey, J.; Osiru, D.S.O. Physiological and molecular characterization of drought responses and identification of candidate tolerance genes in cassava. *AoB Plants* **2013**, *5*, plt007. [CrossRef]
58. Nassar, N.; Abreu, L.; Teodoro, D.; Graciano-Ribeiro, D. Drought tolerant stem anatomy characteristics in Manihot esculenta (Euphorbiaceae) and a wild relative. *Genet. Mol. Res.* **2010**, *9*, 1023–1031. [CrossRef] [PubMed]
59. Narina, S.; Jasti, M.; Buyyarapu, R.; Bhattacharjee, R. Manihot. In *Wild Crop Relatives Genomic and Breeding Resources Industrial Crops*; Kole, C., Ed.; Springer: Berlin/Heidelberg, Germany, 2011; pp. 133–156, ISBN 978-3-642-21102-7.
60. Chavez, R.; Reyes, R.; Roca, W. In vitro culture for the conservation of wild Manihot species. In *Review of Advances in Plant Biotechnology 1985–1988*; Mujeeb-Kazi, A., Sitch, L., Eds.; CYMMYT and IRRI: Ciudad de México, Mexico, 1989; pp. 301–307, ISBN 978-968-612-734-8.
61. Huang, L.; Zhang, F.; Wang, W.; Zhou, Y.; Fu, B.; Li, Z. Comparative transcriptome sequencing of tolerant rice introgression line and its parents in response to drought stress. *BMC Genom.* **2014**, *15*, 1–16. [CrossRef] [PubMed]
62. Khoury, C.K.; Bjorkman, A.; Dempewolf, H.; Ramirez-Villegas, J.; Guarino, L.; Jarvis, A.; Rieseberg, L.H.; Struik, P.C. Increasing homogeneity in global food supplies and the implications for food security. *Proc. Natl. Acad. Sci. USA* **2014**, *111*, 4001–4006. [CrossRef] [PubMed]
63. Mayes, S.; Massawe, F.; Alderson, P.G.; Roberts, J.A.; Azam-Ali, S.N.; Hermann, M. The potential for underutilized crops to improve security of food production. *J. Exp. Bot.* **2011**, *63*, 1075–1079. [CrossRef] [PubMed]
64. Boote, K.J.; Ibrahim, A.M.H.; Lafitte, R.; McCulley, R.; Messina, C.D.; Murray, S.C.; Specht, J.E.; Taylor, S.; Westgate, M.E.; Glasener, K.; et al. Position statement on crop adaptation to climate change. *Crop. Sci.* **2011**, *51*, 2337–2343. [CrossRef]
65. Bose, J.; Rodrigo-Moreno, A.; Lai, D.; Xie, Y.; Shen, W.; Shabala, S. Rapid regulation of the plasma membrane H+-ATPase activity is essential to salinity tolerance in two halophyte species, Atriplex lentiformis and Chenopodium quinoa. *Ann. Bot.* **2014**, *115*, 481–494. [CrossRef]
66. Yang, A.; Akhtar, S.S.; Amjad, M.; Iqbal, S.; Jacobsen, S.-E. Growth and physiological responses of quinoa to drought and temperature stress. *J. Agron. Crop. Sci.* **2016**, *202*, 445–453. [CrossRef]
67. Shao, H.; Yong, B.; Xu, P.; Zheng, H.; Liao, R.; Wang, X.; Li, X.; Zhang, L.; Shen, J. Phytoene synthase gene (PSY) from sweet potato (Ipomoea batatas Lam) enhances tolerance to abiotic stress. *Braz. Arch. Biol. Technol.* **2018**, *61*, e18160558. [CrossRef]
68. Park, S.; Kim, H.S.; Jung, Y.J.; Kim, S.H.; Ji, C.Y.; Wang, Z.; Jeong, J.C.; Lee, H.-S.; Lee, S.Y.; Kwak, S.-S. Orange protein has a role in phytoene synthase stabilization in sweetpotato. *Sci. Rep.* **2016**, *6*, 33563. [CrossRef]
69. Fang, Z.; Xu, X.; Gao, J.; Wang, P.; Liu, Z.; Feng, B. Characterization of FeDREB1 promoter involved in cold- and drought-inducible expression from common buckwheat (*Fagopyrum esculentum*). *Genet. Mol. Res.* **2015**, *14*, 7990–8000. [CrossRef]
70. Wu, Q.; Zhao, G.; Bai, X.; Zhao, W.; Xiang, D.; Wan, Y.; Wu, X.; Sun, Y.; Tan, M.; Peng, L.; et al. Characterization of the transcriptional profiles in common buckwheat (*Fagopyrum esculentum*) under PEG-mediated drought stress. *Electron. J. Biotechnol.* **2019**, *39*, 42–51. [CrossRef]
71. Khoury, C.K.; Achicanoy, H.A.; Bjorkman, A.; Navarro-Racines, C.; Guarino, L.; Flores-Palacios, X.; Engels, J.M.M.; Wiersema, J.H.; Dempewolf, H.; Sotelo, S.; et al. Origins of food crops connect countries worldwide. *Proc. R. Soc. B: Biol. Sci.* **2016**, *283*, 20160792. [CrossRef]
72. Risi, C.; Galwey, N.W. The Chenopodium grains of the Andes: Inca crops for modern agriculture. *Adv. Appl. Biol.* **1984**, *10*, 145–216.
73. Jacobsen, S.-E. The scope for adaptation of quinoa in Northern Latitudes of Europe. *J. Agron. Crop. Sci.* **2017**, *203*, 603–613. [CrossRef]
74. International Union for the Protection of New Varieties of Plants (UPOV). 2020 PLUTO: Plant Variety Database. Available online: https://www.upov.int/pluto/en/ (accessed on 2 September 2020).
75. Di Fabio, A.; Parraga, G. *Origin, Production and Utilization of Pseudocereals*; Wiley: Hoboken, NJ, USA, 2017; pp. 1–27.
76. Oumar, I.; Mariac, C.; Pham, J.-L.; Vigouroux, Y. Phylogeny and origin of pearl millet (Pennisetum glaucum [L.] R. Br) as revealed by microsatellite loci. *Theor. Appl. Genet.* **2008**, *117*, 489–497. [CrossRef]

77. Zhou, M.; Tang, Y.; Deng, X.; Ruan, C.; Kreft, I.; Tang, Y.; Wu, Y. Overview of buckwheat resources in the world. In *Buckwheat Germplasm in the World*; Zhou, M., Kreft, I., Suvorova, G., Tang, Y., Wu, Y., Eds.; Academic Press: London, UK, 2018; pp. 1–7. ISBN 978-0-12-811006-5.
78. Padulosi, S.; Ng, N.Q. Origin, taxonomy, and morphology of *Vigna unguiculata* (L.) Walp. In *Advances in Cowpea Research*; Singh, B.B., Mohan Raj, D.R., Dashiell, K.E., Jackai, L.E., Eds.; Sayce Publishing: Devon, UK, 1997; pp. 1–12. ISBN 978-131-110X.
79. Muñoz-Rodríguez, P.; Carruthers, T.; Wood, J.R.; Williams, B.R.; Weitemier, K.; Kronmiller, B.; Ellis, D.; Anglin, N.L.; Longway, L.; Harris, S.A.; et al. Reconciling conflicting phylogenies in the origin of sweet potato and dispersal to polynesia. *Curr. Biol.* **2018**, *28*, 1246–1256.e12. [CrossRef] [PubMed]
80. Atchison, G.W.; Nevado, B.; Eastwood, R.J.; Contreras-Ortiz, N.; Reynel, C.; Madriñán, S.; Filatov, D.A.; Hughes, C.E. Lost crops of the Incas: Origins of domestication of the Andean pulse crop tarwi, Lupinus mutabilis. *Am. J. Bot.* **2016**, *103*, 1592–1606. [CrossRef] [PubMed]
81. Singh, M.; Malhotra, N.; Sharma, K. Buckwheat (*Fagopyrum* sp.) genetic resources: What can they contribute towards nutritional security of changing world? *Genet. Resour. Crop. Evol.* **2020**, *67*, 1–20. [CrossRef]
82. Cuellar-Ortiz, S.M.; Arrieta-Montiel, M.D.L.P.; Acosta-Gallegos, J.; Covarrubias, A.A. Relationship between carbohydrate partitioning and drought resistance in common bean. *Plant Cell Environ.* **2008**, *31*, 1399–1409. [CrossRef]
83. Li, L.; Zhang, Q.; Huang, D. A review of imaging techniques for plant phenotyping. *Sensors* **2014**, *14*, 20078–20111. [CrossRef]
84. McAusland, L.; Davey, P.A.; Kanwal, N.; Baker, N.R.; Lawson, T. A novel system for spatial and temporal imaging of intrinsic plant water use efficiency. *J. Exp. Bot.* **2013**, *64*, 4993–5007. [CrossRef] [PubMed]
85. Takahashi, S.; Monda, K.; Negi, J.; Konishi, F.; Ishikawa, S.; Hashimoto-Sugimoto, M.; Gotô, N.; Iba, K. Natural variation in stomatal responses to environmental changes among arabidopsis thaliana ecotypes. *PLoS ONE* **2015**, *10*, e0117449. [CrossRef] [PubMed]
86. Kim, D.M.; Zhang, H.; Zhou, H.; Du, T.; Wu, Q.; Mockler, T.C.; Berezin, M. Highly sensitive image-derived indices of water-stressed plants using hyperspectral imaging in SWIR and histogram analysis. *Sci. Rep.* **2015**, *5*, 15919. [CrossRef] [PubMed]
87. Parent, B.; Shahinnia, F.; Maphosa, L.; Berger, B.; Rabie, H.; Chalmers, K.; Kovalchuk, A.; Langridge, P.; Fleury, D. Combining field performance with controlled environment plant imaging to identify the genetic control of growth and transpiration underlying yield response to water-deficit stress in wheat. *J. Exp. Bot.* **2015**, *66*, 5481–5492. [CrossRef] [PubMed]
88. Humplík, J.F.; Lazár, D.; Husičková, A.; Spíchal, L. Automated phenotyping of plant shoots using imaging methods for analysis of plant stress responses—A review. *Plant Methods* **2015**, *11*, 1–10. [CrossRef] [PubMed]
89. Farooq, M.A.; Wahid, A.; Kobayashi, N.; Fujita, D.; Basra, S.M.A. Plant drought stress: Effects, mechanisms and management. *Agron. Sustain. Dev.* **2009**, *29*, 185–212. [CrossRef]
90. Prince, S.; Mutava, R.N.; Nguyen, N.; Pathan, S.M.; Shannon, G.J.; Murphy, M.; Zhang, Z.; Kim, Y.H.; Valliyodan, B.; Nguyen, H.T. Evaluation of high yielding soybean germplasm under water limitation. *J. Integr. Plant Biol.* **2015**, *58*, 475–491. [CrossRef]
91. Das, A.; Schneider, H.M.; Burridge, J.D.; Ascanio, A.K.M.; Wojciechowski, T.; Topp, C.N.; Lynch, J.P.; Weitz, J.S.; Bucksch, A. Digital imaging of root traits (DIRT): A high-throughput computing and collaboration platform for field-based root phenomics. *Plant Methods* **2015**, *11*, 51. [CrossRef]
92. Cai, J.; Zeng, Z.; Connor, J.N.; Huang, C.Y.; Melino, V.J.; Kumar, P.; Miklavcic, S.J. RootGraph: A graphic optimization tool for automated image analysis of plant roots. *J. Exp. Bot.* **2015**, *66*, 6551–6562. [CrossRef]
93. Kumar, P.; Cai, J.; Miklavcic, S.J. A complete system for 3D reconstruction of roots for phenotypic analysis. *Adv. Exp. Med. Biol.* **2014**, *823*, 249–270. [CrossRef]
94. Sallam, A.; Alqudah, A.M.; Dawood, M.F.A.; Baenziger, P.S.; Börner, A. Drought stress tolerance in wheat and barley: Advances in physiology, breeding and genetics research. *Int. J. Mol. Sci.* **2019**, *20*, 3137. [CrossRef] [PubMed]
95. Mohammadi, R. Breeding for increased drought tolerance in wheat: A review. *Crop. Pasture Sci.* **2018**, *69*, 223–241. [CrossRef]
96. Khan, A.; Sovero, V.; Gemenet, D.C. Genome-assisted breeding for drought resistance. *Curr. Genom.* **2016**, *17*, 330–342. [CrossRef] [PubMed]

97. Upadhyay, J.; Joshi, R.; Singh, B.; Bohra, A.; Vijayan, R.; Bhatt, M.; Bisht, S.P.S.; Wani, S.H. *Application of Bioinformatics in Understanding of Plant Stress Tolerance*; Springer Science and Business Media LLC: Berlin, Germany, 2017; pp. 347–374.
98. Sharma, V.; Sarkar, I.N. Bioinformatics opportunities for identification and study of medicinal plants. *Briefings Bioinform.* **2012**, *14*, 238–250. [CrossRef] [PubMed]
99. Blum, A. Genomics for drought resistance—Getting down to earth. *Funct. Plant Biol.* **2014**, *41*, 1191–1198. [CrossRef]
100. Nuccio, M.L.; Wu, J.; Mowers, R.; Zhou, H.-P.; Meghji, M.; Primavesi, L.F.; Paul, M.J.; Chen, X.; Gao, Y.; Haque, E.; et al. Expression of trehalose-6-phosphate phosphatase in maize ears improves yield in well-watered and drought conditions. *Nat. Biotechnol.* **2015**, *33*, 862–869. [CrossRef] [PubMed]
101. Daszkowska-Golec, A.; Skubacz, A.; Sitko, K.; Słota, M.; Kurowska, M.; Szarejko, I. Mutation in barley ERA1 (Enhanced Response to ABA1) gene confers better photosynthesis efficiency in response to drought as revealed by transcriptomic and physiological analysis. *Environ. Exp. Bot.* **2018**, *148*, 12–26. [CrossRef]
102. Joshi, R.; Wani, S.H.; Singh, B.; Bohra, A.; Dar, Z.A.; Lone, A.; Pareek, A.; Singla-Pareek, S.L. Transcription factors and plants response to drought stress: Current understanding and future directions. *Front. Plant Sci.* **2016**, *7*, 1029. [CrossRef]
103. Zhang, L.; Zhao, G.; Xia, C.; Jia, J.; Liu, X.; Kong, X. A wheat R2R3-MYB gene, TaMYB30-B, improves drought stress tolerance in transgenic Arabidopsis. *J. Exp. Bot.* **2012**, *63*, 5873–5885. [CrossRef]
104. Jangale, B.L.; Chaudhari, R.S.; Azeez, A.; Sane, P.V.; Sane, A.P.; Krishna, B. Independent and combined abiotic stresses affect the physiology and expression patterns of DREB genes differently in stress-susceptible and resistant genotypes of banana. *Physiol. Plant.* **2018**, *165*, 303–318. [CrossRef]
105. Bundó, M.; Coca, M.A. Calcium-dependent protein kinase OsCPK10 mediates both drought tolerance and blast disease resistance in rice plants. *J. Exp. Bot.* **2017**, *68*, 2963–2975. [CrossRef] [PubMed]
106. Mitula, F.; Tajdel, M.; Cieśla, A.; Kasprowicz-Maluśki, A.; Kulik, A.; Babula-Skowronska, D.; Michalak, M.; Dobrowolska, G.; Sadowski, J.; Ludwików, A. Arabidopsis ABA-activated kinase MAPKKK18 is regulated by protein phosphatase 2C ABI1 and the ubiquitin-proteasome pathway. *Plant Cell Physiol.* **2015**, *56*, 2351–2367. [CrossRef] [PubMed]
107. Samajova, O.; Plihal, O.; Al-Yousif, M.; Hirt, H.; Šamaj, J. Improvement of stress tolerance in plants by genetic manipulation of mitogen-activated protein kinases. *Biotechnol. Adv.* **2013**, *31*, 118–128. [CrossRef] [PubMed]
108. Wang, B.; Li, Z.; Ran, Q.; Li, P.; Peng, Z.; Zhang, J. ZmNF-YB16 overexpression improves drought resistance and yield by enhancing photosynthesis and the antioxidant capacity of maize plants. *Front. Plant Sci.* **2018**, *9*, 709. [CrossRef] [PubMed]
109. Wu, G.; Zhao, Y.; Shen, R.; Wang, B.; Xie, Y.; Ma, X.; Zheng, Z.; Wang, H. Characterization of maize phytochrome-interacting factors in light signaling and photomorphogenesis. *Plant Physiol.* **2019**, *181*, 789–803. [CrossRef]
110. Gao, Y.; Jiang, W.; Dai, Y.; Xiao, N.; Zhang, C.; Li, H.; Lu, Y.; Wu, M.; Tao, X.; Deng, D.; et al. A maize phytochrome-interacting factor 3 improves drought and salt stress tolerance in rice. *Plant Mol. Biol.* **2015**, *87*, 413–428. [CrossRef]
111. Khan, S.; Anwar, S.; Yu, S.; Sun, M.; Yang, Z.; Gao, Z.-Q. Development of drought-tolerant transgenic wheat: Achievements and limitations. *Int. J. Mol. Sci.* **2019**, *20*, 3350. [CrossRef]
112. Yang, S.; Vanderbeld, B.; Wan, J.; Huang, Y. Narrowing down the targets: Towards successful genetic engineering of drought-tolerant crops. *Mol. Plant* **2010**, *3*, 469–490. [CrossRef]
113. Gupta, P.K.; Balyan, H.S.; Gahlaut, V. QTL analysis for drought tolerance in wheat: Present status and future possibilities. *Agronomy* **2017**, *7*, 5. [CrossRef]
114. Shan, Q.; Wang, Y.; Li, J.; Zhang, Y.; Chen, K.; Liang, Z.; Zhang, K.; Liu, J.; Xi, J.J.; Qiu, J.; et al. Targeted genome modification of crop plants using a CRISPR-Cas system. *Nat. Biotechnol.* **2013**, *31*, 686–688. [CrossRef]
115. Yin, F.; Gao, J.; Liu, M.; Qin, C.; Zhang, W.; Yang, A.; Xia, M.; Zhang, Z.; Shen, Y.; Lin, H.; et al. Genome-wide analysis of Water-stress-responsive microRNA expression profile in tobacco roots. *Funct. Integr. Genom.* **2014**, *14*, 319–332. [CrossRef] [PubMed]
116. Saad, A.S.I.; Li, X.; Li, H.-P.; Huang, T.; Gao, C.-S.; Guo, M.-W.; Cheng, W.; Zhao, G.-Y.; Liao, Y.-C. A rice stress-responsive NAC gene enhances tolerance of transgenic wheat to drought and salt stresses. *Plant Sci.* **2013**, *203*, 33–40. [CrossRef] [PubMed]

117. Jia, T.-J.; Jing-Jing, L.; Wang, L.-F.; Cao, Y.-Y.; Ma, J.; Wang, H.; Zhang, D.-F.; Li, H.-Y. Evaluation of drought tolerance in ZmVPP1-overexpressing transgenic inbred maize lines and their hybrids. *J. Integr. Agric.* **2020**, *19*, 2177–2187. [CrossRef]
118. Szalonek, M.; Sierpien, B.; Rymaszewski, W.; Gieczewska, K.; Garstka, M.; Lichocka, M.; Sass, L.; Paul, K.; Vass, I.; Vanková, R.; et al. Potato annexin STANN1 promotes drought tolerance and mitigates light stress in transgenic solanum tuberosum L. plants. *PLoS ONE* **2015**, *10*, e0132683. [CrossRef]
119. You, J.; Hu, H.; Xiong, L. An ornithine δ-aminotransferase gene OsOAT confers drought and oxidative stress tolerance in rice. *Plant Sci.* **2012**, *197*, 59–69. [CrossRef] [PubMed]
120. De Souza, W.R.; Oliveira, N.G.; Vinecky, F.; Ribeiro, A.P.; Basso, M.F.; Casari, R.A.D.C.N.; Cunha, B.A.D.B.; Duarte, K.E.; Santiago, T.R.; Martins, P.K.; et al. Field evaluation of At DREB 2A CA overexpressing sugarcane for drought tolerance. *J. Agron. Crop. Sci.* **2019**, *205*, 545–553. [CrossRef]
121. Ayadi, M.; Brini, F.; Masmoudi, K. Overexpression of a wheat aquaporin gene, TdPIP2;1, enhances salt and drought tolerance in transgenic durum wheat cv. maali. *Int. J. Mol. Sci.* **2019**, *20*, 2389. [CrossRef]
122. Wei, Y.; Liu, W.; Hu, W.; Yan, Y.; Shi, H. The chaperone MeHSP90 recruits MeWRKY20 and MeCatalase1 to regulate drought stress resistance in cassava. *New Phytol.* **2020**, *226*, 476–491. [CrossRef]
123. Li, Y.; Chen, Q.; Nan, H.; Li, X.; Lu, S.; Zhao, X.; Liu, B.; Guo, C.; Kong, F.; Cao, D. Overexpression of GmFDL19 enhances tolerance to drought and salt stresses in soybean. *PLoS ONE* **2017**, *12*, e0179554. [CrossRef]
124. Fan, Y.; Shabala, S.; Ma, Y.; Xu, R.; Zhou, M. Using QTL mapping to investigate the relationships between abiotic stress tolerance (drought and salinity) and agronomic and physiological traits. *BMC Genom.* **2015**, *16*, 43. [CrossRef]
125. Wehner, G.; Balko, C.; Enders, M.M.; Humbeck, K.; Ordon, F. Identification of genomic regions involved in tolerance to drought stress and drought stress induced leaf senescence in juvenile barley. *BMC Plant Biol.* **2015**, *15*, 125. [CrossRef] [PubMed]
126. Uga, Y.; Sugimoto, K.; Ogawa, S.; Rane, J.; Ishitani, M.; Hara, N.; Kitomi, Y.; Inukai, Y.; Ono, K.; Kanno, N.; et al. Control of root system architecture by DEEPER ROOTING 1 increases rice yield under drought conditions. *Nat. Genet.* **2013**, *45*, 1097–1102. [CrossRef] [PubMed]
127. Technow, F.; Messina, C.D.; Totir, L.R.; Cooper, M. Integrating crop growth models with whole genome prediction through approximate bayesian computation. *PLoS ONE* **2015**, *10*, e0130855. [CrossRef] [PubMed]
128. Ghorbani, R.; Alemzadeh, A.; Razi, H. Microarray analysis of transcriptional responses to salt and drought stress in Arabidopsis thaliana. *Heliyon* **2019**, *5*, e02614. [CrossRef]
129. Hassani-Pak, K.; Singh, A.; Brandizi, M.; Hearnshaw, J.; Amberkar, S.; Phillips, A.L.; Doonan, J.H.; Rawlings, C.J. KnetMiner: A comprehensive approach for supporting evidence-based gene discovery and complex trait analysis across species. *bioRxiv* **2020**, *4*, 017004.
130. Noraziyah, A.A.S.; Swamy, B.P.M.; Wickneswari, R.; Cruz, M.T.S.; Raman, A.; Kumar, A. Marker assisted pyramiding of drought yield QTLs into a popular Malaysian rice cultivar, MR219. *BMC Genet.* **2016**, *17*, 30. [CrossRef]
131. Lopes, M.S.; El-Basyoni, I.S.; Baenziger, P.S.; Singh, S.; Royo, C.; Ozbek, K.; Aktas, H.; Ozer, E.; Ozdemir, F.; Manickavelu, A.; et al. Exploiting genetic diversity from landraces in wheat breeding for adaptation to climate change. *J. Exp. Bot.* **2015**, *66*, 3477–3486. [CrossRef]
132. Merchuk-Ovnat, L.; Barak, V.; Fahima, T.; Ordon, F.; Lidzbarsky, G.A.; Krugman, T.; Saranga, Y. Ancestral QTL alleles from wild emmer wheat improve drought resistance and productivity in modern wheat cultivars. *Front. Plant Sci.* **2016**, *7*, 452. [CrossRef]
133. Alqudah, A.M.; Sallam, A.; Baenziger, P.S.; Börner, A. GWAS: Fast-forwarding gene identification and characterization in temperate Cereals: Lessons from Barley—A review. *J. Adv. Res.* **2020**, *22*, 119–135. [CrossRef]
134. Gupta, M.; Chawla, V.; Garg, P.; Yadav, N.; Munjal, R.; Sharma, B. Genetic analysis of yield and heat stress related traits in wheat (*Triticum aestivum* L. em. Thell) using microsatellite markers. *J. Appl. Nat. Sci.* **2015**, *7*, 739–744. [CrossRef]
135. Qaseem, M.F.; Qureshi, R.; Muqaddasi, Q.H.; Shaheen, H.; Kousar, R.; Röder, M.S. Genome-wide association mapping in bread wheat subjected to independent and combined high temperature and drought stress. *PLoS ONE* **2018**, *13*, e0199121. [CrossRef]

136. Hou, S.; Zhu, G.; Li, Y.; Li, W.; Fu, J.; Niu, E.; Li, L.; Zhang, D.; Guo, W. Genome-wide association studies reveal genetic variation and candidate genes of drought stress related traits in cotton (*Gossypium hirsutum* L.). *Front. Plant Sci.* **2018**, *9*, 9. [CrossRef] [PubMed]
137. Kadam, N.N.; Struik, P.C.; Rebolledo, M.C.; Yin, X.; Jagadish, S.V.K. Genome-wide association reveals novel genomic loci controlling rice grain yield and its component traits under water-deficit stress during the reproductive stage. *J. Exp. Bot.* **2018**, *69*, 4017–4032. [CrossRef] [PubMed]
138. Yuan, Y.; Cairns, J.E.; Babu, R.; Gowda, M.; Makumbi, D.; Magorokosho, C.; Zhang, A.; Liu, Y.; Wang, N.; Hao, Z.; et al. Genome-wide association mapping and genomic prediction analyses reveal the genetic architecture of grain yield and flowering time under drought and heat stress conditions in maize. *Front. Plant Sci.* **2019**, *9*, 9. [CrossRef]
139. Gurgul, A.; Miksza-Cybulska, A.; Szmatoła, T.; Jasielczuk, I.; Piestrzynska-Kajtoch, A.; Fornal, A.; Semik-Gurgul, E.; Bugno-Poniewierska, M. Genotyping-by-sequencing performance in selected livestock species. *Genomics* **2019**, *111*, 186–195. [CrossRef]
140. Ray, S.; Satya, P. Next generation sequencing technologies for next generation plant breeding. *Front. Plant Sci.* **2014**, *5*, 367. [CrossRef] [PubMed]
141. Wang, X.; Xu, Y.; Hu, Z.; Yang, Z. Genomic selection methods for crop improvement: Current status and prospects. *Crop. J.* **2018**, *6*, 330–340. [CrossRef]
142. Zhao, Y.; Gowda, M.; Liu, W.; Würschum, T.; Maurer, H.P.; Longin, F.H.; Ranc, N.; Reif, J.C. Accuracy of genomic selection in European maize elite breeding populations. *Theor. Appl. Genet.* **2011**, *124*, 769–776. [CrossRef] [PubMed]
143. Bao, Y.; Vuong, T.; Meinhardt, C.; Tiffin, P.; Denny, R.; Chen, S.; Nguyen, H.T.; Orf, J.H.; Young, N.D. Potential of association mapping and genomic selection to explore PI 88788 derived soybean cyst nematode resistance. *Plant Genome* **2014**, *7*, 1–13. [CrossRef]
144. Lorenzana, R.E.; Bernardo, R. Accuracy of genotypic value predictions for marker-based selection in biparental plant populations. *Theor. Appl. Genet.* **2009**, *120*, 151–161. [CrossRef]
145. Spindel, J.; Begum, H.; Akdemir, D.; Virk, P.; Collard, B.C.; Redoña, E.; Atlin, G.; Jannink, J.-L.; McCouch, S.R. Genomic selection and association mapping in rice (*Oryza sativa*): Effect of trait genetic architecture, training population composition, marker number and statistical model on accuracy of rice genomic selection in elite, tropical rice breeding lines. *PLoS Genet.* **2015**, *11*, e1004982. [CrossRef]
146. Sallam, A.H.; Endelman, J.B.; Jannink, J.-L.; Smith, K.P. Assessing genomic selection prediction accuracy in a dynamic barley breeding population. *Plant Genome* **2015**, *8*, 1–15. [CrossRef]
147. Cerrudo, D.; Cao, S.; Yuan, Y.; Martínez, C.; Suarez, E.A.; Babu, R.; Zhang, X.; Trachsel, S. Genomic selection outperforms marker assisted selection for grain yield and physiological traits in a maize doubled haploid population across water treatments. *Front. Plant Sci.* **2018**, *9*, 366. [CrossRef] [PubMed]
148. Shikha, M.; Kanika, A.; Rao, A.R.; Mallikarjuna, M.G.; Gupta, H.S.; Nepolean, T. Genomic selection for drought tolerance using genome-wide SNPs in maize. *Front. Plant Sci.* **2017**, *8*, 550. [CrossRef] [PubMed]
149. Li, Y.; Ruperao, P.; Batley, J.; Edwards, D.; Khan, T.; Colmer, T.D.; Pang, J.; Siddique, K.H.M.; Sutton, T. Investigating drought tolerance in chickpea using genome-wide association mapping and genomic selection based on whole-genome resequencing data. *Front. Plant Sci.* **2018**, *9*, 190. [CrossRef]
150. Zhang, A.; Wang, H.; Beyene, Y.; Semagn, K.; Liu, Y.; Cao, S.; Cui, Z.; Ruan, Y.; Burgueño, J.; Vicente, F.S.; et al. Effect of trait heritability, training population size and marker density on genomic prediction accuracy estimation in 22 bi-parental tropical maize populations. *Front. Plant Sci.* **2017**, *8*, 8. [CrossRef]
151. Zhang, X.; Pérez-Rodríguez, P.; Burgueño, J.; Olsen, M.; Buckler, E.; Atlin, G.; Prasanna, B.M.; Vargas, M.; Vicente, F.M.S.; Crossa, J. Rapid cycling genomic selection in a multiparental tropical maize population. *G3 Genes Genomes Genet.* **2017**, *7*, 2315–2326. [CrossRef]
152. Čermák, T.; Baltes, N.J.; Cegan, R.; Zhang, Y.; Voytas, D.F. High-frequency, precise modification of the tomato genome. *Genome Biol.* **2015**, *16*, 232. [CrossRef]
153. Li, J.-F.; Norville, J.E.; Aach, J.; McCormack, M.; Zhang, D.; Bush, J.; Church, G.M.; Sheen, J. Multiplex and homologous recombination–mediated genome editing in Arabidopsis and Nicotiana benthamiana using guide RNA and Cas9. *Nat. Biotechnol.* **2013**, *31*, 688–691. [CrossRef]
154. Ito, Y.; Nishizawa, A.; Endo, M.; Mikami, M.; Toki, S. CRISPR/Cas9-mediated mutagenesis of the RIN locus that regulates tomato fruit ripening. *Biochem. Biophys. Res. Commun.* **2015**, *467*, 76–82. [CrossRef]

155. Feng, Z.; Mao, Y.; Xu, N.; Zhang, B.; Wei, P.; Yang, N.-L.; Wang, Z.; Zhang, Z.; Zheng, R.; Yang, L.; et al. Multigeneration analysis reveals the inheritance, specificity, and patterns of CRISPR/Cas-induced gene modifications in Arabidopsis. *Proc. Natl. Acad. Sci. USA* **2014**, *111*, 4632–4637. [CrossRef] [PubMed]
156. Du, H.; Zeng, X.; Zhao, M.; Cui, X.; Wang, Q.; Yang, H.; Cheng, H.; Yu, D. Efficient targeted mutagenesis in soybean by TALENs and CRISPR/Cas9. *J. Biotechnol.* **2016**, *217*, 90–97. [CrossRef]
157. Sun, Y.; Zhang, X.; Wu, C.; He, Y.; Ma, Y.; Hou, H.; Guo, X.; Du, W.; Zhao, Y.; Xia, L. Engineering herbicide-resistant rice plants through CRISPR/Cas9-mediated homologous recombination of acetolactate synthase. *Mol. Plant* **2016**, *9*, 628–631. [CrossRef] [PubMed]
158. Ma, X.; Zhang, Q.; Zhu, Q.; Liu, W.; Chen, Y.; Qiu, R.; Wang, B.; Yang, Z.; Li, H.; Lin, Y.; et al. A robust CRISPR/Cas9 system for convenient, high-efficiency multiplex genome editing in monocot and dicot plants. *Mol. Plant* **2015**, *8*, 1274–1284. [CrossRef] [PubMed]
159. Shi, J.; Gao, H.; Wang, H.; Lafitte, H.R.; Archibald, R.L.; Yang, M.; Hakimi, S.M.; Mo, H.; Habben, J.E. ARGOS8 variants generated by CRISPR-Cas9 improve maize grain yield under field drought stress conditions. *Plant Biotechnol. J.* **2016**, *15*, 207–216. [CrossRef] [PubMed]
160. Wang, L.; Chen, L.; Li, R.; Zhao, R.; Yang, M.; Sheng, J.; Shen, L. Reduced drought tolerance by CRISPR/Cas9-mediated SlMAPK3 mutagenesis in tomato plants. *J. Agric. Food Chem.* **2017**, *65*, 8674–8682. [CrossRef]
161. Li, R.; Liu, C.; Zhao, R.; Wang, L.; Chen, L.; Yu, W.; Zhang, S.; Sheng, J.; Shen, L. CRISPR/Cas9-mediated SlNPR1 mutagenesis reduces tomato plant drought tolerance. *BMC Plant Biol.* **2019**, *19*, 38. [CrossRef]
162. Osakabe, Y.; Watanabe, T.; Sugano, S.S.; Ueta, R.; Ishihara, R.; Shinozaki, K.; Osakabe, K. Optimization of CRISPR/Cas9 genome editing to modify abiotic stress responses in plants. *Sci. Rep.* **2016**, *6*, 26685. [CrossRef]
163. Zhang, H.; Zhang, J.; Wei, P.; Zhang, B.; Gou, F.; Feng, Z.; Mao, Y.; Yang, L.; Zhang, H.; Xu, N.; et al. The CRISPR/Cas9 system produces specific and homozygous targeted gene editing in rice in one generation. *Plant Biotechnol. J.* **2014**, *12*, 797–807. [CrossRef]

plants

MDPI

Article

The Role of Genetic Resources in Breeding for Climate Change: The Case of Public Breeding Programmes in Eighteen Developing Countries

Gea Galluzzi [1,*], Aseffa Seyoum [2], Michael Halewood [1], Isabel López Noriega [1] and Eric W. Welch [2]

1 Bioversity International, Via dei Tre Denari 472/a, Maccarese (Fiumicino), 00057 Rome, Italy; m.halewood@cgiar.org (M.H.); i.lopez@cgiar.org (I.L.N.)

2 Center for Science, Technology, and Environmental Policy Studies, School of Public Affairs, Arizona State University, 411 N Central Ave, Phoenix, AZ 85004, USA; aseyoumw@yahoo.com (A.S.); EricWelch@asu.edu (E.W.W.)

* Correspondence: geagalluzzi@gmail.com; Tel.: +39-348-403-0812

Received: 29 July 2020; Accepted: 21 August 2020; Published: 31 August 2020

Abstract: The role of plant breeding in adapting crops to climate changes that affect food production in developing countries is recognized as extremely important and urgent, alongside other agronomic, socio-economic and policy adaptation pathways. To enhance plant breeders' capacity to respond to climate challenges, it is acknowledged that they need to be able to access and use as much genetic diversity as they can get. Through an analysis of data from a global survey, we explore if and how public breeders in selected developing countries are responding to climate challenges through a renewed or innovative use of plant genetic resources, particularly in terms of types of material incorporated into their breeding work as well as sources of such germplasm. It also looks at the possible limitations breeders encounter in their efforts towards exploring diversity for adaptation. Breeders are clearly considering climate challenges. In general, their efforts are aimed at intensifying their breeding work on traits that they were already working on before climate change was so widely discussed. Similarly, the kinds of germplasm they use, and the sources from which they obtain it, do not appear to have changed significantly over the course of recent years. The main challenges breeders faced in accessing germplasm were linked to administrative/legal factors, particularly related to obtaining genetic resources across national borders. They also underscore technical challenges such as a lack of appropriate technologies to exploit germplasm sets such as crop wild relatives and landraces. Addressing these limitations will be crucial to fully enhance the role of public sector breeders in helping to adapt vulnerable agricultural systems to the challenges of climate change.

Keywords: genetic resources; plant breeding; climate change adaptation; genebanks; policy; developing countries

1. Introduction

The International Panel on Climate Change (IPCC) [1] defines climate change as a change in the mean value and/or variability of the climate's properties, that persists for an extended period, usually for decades or longer. Agriculture is extremely vulnerable to climate change: increasing temperatures and declining precipitation over semi-arid regions are likely to reduce yields for a number of primary crops in the next two decades [2–7]; the intensity and distribution of pests and disease outbreaks may also become increasingly unpredictable, with serious impacts on agricultural productivity [8–10]. While the overall impacts of climate change on agricultural systems are expected to be negative [11], the effects may vary both in type and magnitude across geographical areas [12,13]: developing countries are likely to be most affected, not only because of their predominantly low input, rain-fed cropping systems which rely on somewhat regular weather patterns, but also because of their more rapid population growth, which determines greater pressures on agricultural production and more serious food insecurity risks [2,14–16].

Adaptation to climate change is the practice and process of adjusting to climate-induced adverse effects [2,14,17]. In agriculture, adaptation to climate change is about farmers' and other stakeholders' responses to environmental disturbances that affect their cropping systems. This may involve a broad array of alterations from livelihood and agronomic strategies to policy changes [18,19]. Among those strategies that focus on the production system itself, climate change adaptation strategies are commonly divided in two pathways: agronomic management (which tends to be short term) and genetic improvement (longer term) [9,13]. Agronomic management strategies encompass changing cultivation practices (timing or location of cropping activities, techniques of land preparation, weed/pest/disease management [18]) and the adoption of new varieties or shifting to alternative crops or crop combinations [9,13,20–25]. Genetic improvement, on the contrary, involves the development and adoption of new, better adapted varieties of the crop of interest. While farmers have always conducted genetic improvement within their informal seed networks [26–30], a crucial role has been played over the last century by public (and, increasingly, private) plant breeding [31,32]. Indeed, "modern", scientific plant breeding has contributed to massive and rapid yield increases in many crops, as well as to increased tolerance to a variety of biotic and abiotic stresses; however, with more severe and frequent challenges from aggravated climate change, plant breeders are being called to place extra efforts in improving and accelerating the tools and working strategies they use, to timely provide farmers with adapted varieties [33–36], particularly in more climate vulnerable developing countries.

An essential building block for any innovation in plant breeding to occur is access to, and use of genetic diversity from existing wild or domesticated species [37–40], particularly when the genetic target dealt with is complex, as in the case of adaptive traits for climate stress responses. In developing new cultivars, breeders can start from a range of different germplasm types and germplasm sources. Among the former, landraces and crop wild relatives remain perhaps the largest reservoir of genetic diversity, including traits of tolerance or resistance to environmental stressors, even those associated with climate change [30,36,41–54]. On the other hand, advanced breeding or elite lines which have undergone pre-breeding efforts, may harbour less genetic diversity but be more "ready to use" materials for breeders, thanks to the useful information accumulated on their structure and properties [49,55–59]. In terms of sources, breeders can tap into the collections of national and international genebanks and germplasm research centres, which host a wide range of different materials for different crops [55,57,60], or search for the diversity they may need directly in natural or farmers' fields, particularly in areas of origin or domestication. Recent years have witnessed the establishment of national laws and regulations that have changed the rules of germplasm access, distribution and sharing, potentially affecting the use of genetic materials by breeders [57,61–65]. Additional factors, including human, technical and financial resources available in any breeding programme, as well as linkages and collaborations with other national and international institutions, are likely to contribute to shaping the way breeders use genetic resources in the context of climate change adaptation [66,67].

It is against the above context that we undertook a survey of plant breeders in 19 developing countries to see how they are perceiving climate change's impacts on their breeding objectives, and if these changes had knock-on effects on the kinds of genetic resources they use and where they access them. In particular, the objectives of the study are: (1) assessing breeders' perception of climate change and the traits they give priority to in breeding; (2) exploring trends in the use of genetic materials in breeding programs over the course of the last two to five years (in terms of types of materials used and sources of access); (3) assessing the association between changes in the way breeders use genetic resources and their perception of climate change priorities; and (4) examining if and how regulatory, technical, financial or other issues influence the extent to which breeders are capable of innovating with genetic resources.

Section 2 presents the results of the study, Section 3 provides the discussion and conclusions; data sources, data collection procedures and methods of analysis are presented in Section 4. We chose to focus on public sector breeders only, based on the consideration that in the developing world, private breeding continues to be modest for most crops, while public support to breeding is possibly still the major avenue for the development of new varieties [66], and for focusing on traits (such as environmental stability and sustainability) or crops which may be under-researched in private sector breeding [68].

2. Results

Complete responses were received from 200 breeders in 18 countries, who answered in one out of three languages (English, French and Spanish), as summarized in Table 1.

Table 1. Sample frame of the survey.

Language	Country	Respondents
English	China	38
	Ethiopia	15
	India	33
	Jordan	6
	Kenya	5
	Nepal	20
	Philippines	5
	Rwanda	5
	Uganda	12
	Zambia	1
French	Burkina Faso	7
	Cote d'Ivoire	10
	Morocco	9
Spanish	Bolivia	3
	Brazil	12
	Costa Rica	3
	Guatemala	2
	Peru	14
Total		200

Around 57% of the respondents were from national agricultural research systems, while about 24% belonged to academic institutions. Although we had targeted breeders which to the best of our knowledge belonged to public institutions, we received a few responses from the private sector, which anyhow represented only 5%. Respondents from national non-governmental or non-profit organizations totaled 6%. The remaining 8% of respondents worked in other organizations including local, community or farmers' organizations as well as international non-governmental and non-profit organizations. Regarding gender, 146 breeders were male, while 32 were female. At the time of the survey, the average age of respondents was 49 years, and the average time they had worked as breeders was just over 17 years, of which, on average, 15 years were spent working in their current organization. Over half of the breeders (113) had a doctorate degree, while over half of the others (46) had a master's degree.

2.1. *Perception of Extreme Weather Patterns*

The plant breeders were asked about their perception of extreme weather events occurring in their region of work over the last five years. The results show that most breeders agreed on irregular drought periods and irregular rainfall patterns as being the most seriously increasing weather phenomena (see Figure 1). A considerable number of plant breeders also observed an increasing tendency toward a late onset of cropping seasons, a pattern that was strongly correlated to perceptions of irregular rainfall

(p-value = 0.007864) as well as with irregular and excessive droughts (p-value = 0.00000002616 and 0.002548 respectively) (data not shown).

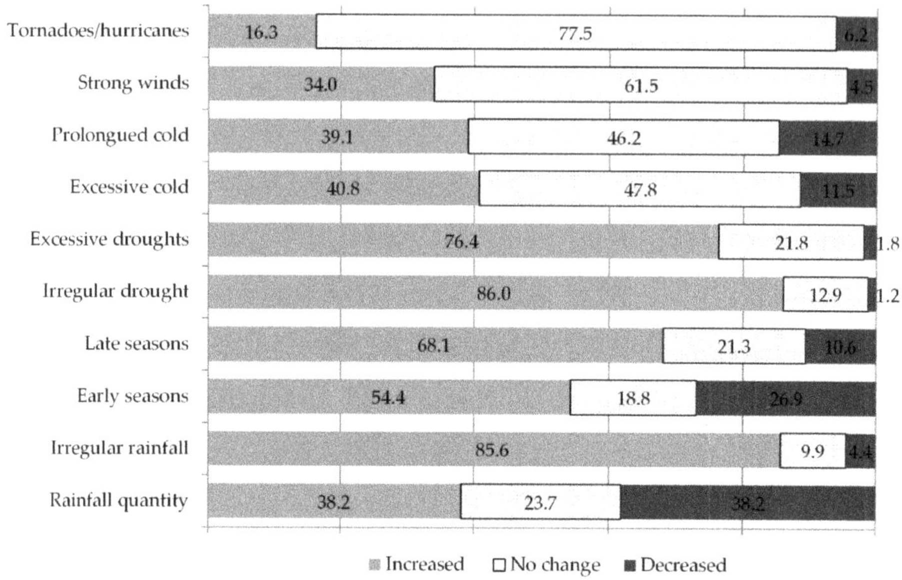

Figure 1. Plant breeders' perceptions about weather patterns observed in their regions. Numbers on or next to the bars represent percentages over the total responses.

The crops which breeders work on, do not seem to significantly influence their perception of climate change. To detect if there were significant differences across regions in terms of climate change perceptions, we subdivided our responses into six geographical regions, namely Central America (with five responses), East Africa (38), East Asia (43), North Africa and the Middle East (15), South America (29), South Asia (53) and West Africa (17). Ordinal means per region revealed that the most serious problem affecting all of the areas was increasing irregularity in rainfall and drought as well as from increasing drought periods. Inter-regional differences in perceived weather patterns were not statistically significant (based on an ANOVA analyses of variance) but may provide an indication of prevailing trends. Breeders in West Africa reported to be dealing with increasing unpredictability in the onset and end of rainy and dry seasons; in East and West Africa an increase in cold weather periods was reported, while Central America was the only region where a slight decrease in overall rainfall quantity was described (in all other regions, rainfall appears to be changing little or slightly increasing). Central America also scored highest in terms of increases in strong winds and hurricanes. Late onset of seasons, more than early, was reported as an issue in many regions, namely in Central America, East Africa, East Asia, North Africa and the Middle East, and South Asia.

2.2. Changes in the Traits on Which Breeders Are Working

The results presented in Figure 2 describe the changes in the importance of the traits on which breeders are working. Breeding for resistance to pests and diseases is the priority that has become most important over the past five years (87%), followed closely by breeding for drought tolerance (82%), shorter growth cycles (63%) and tolerance to high temperatures (54%). In another set of questions concerning the traits that have always been most important regardless of recent changes, breeders flagged the same traits in the same order of importance. Similarly, breeding for low temperature as well as for water logging tolerance have not consistently increased in importance, and, indeed, their overall relevance even before the last five years was ranked quite low by respondents. Over 85% of breeders listed other important traits in addition to these priority ones, including quality traits (63%), traits related to tolerance to low soil fertility (20%) and yield (11%). In summary, recent years appear to have witnessed an increase in the urgency with which breeders are working on traits that already had high priority in their longer-term breeding work. The "other" traits category was chosen by a number of respondents, but 66% of these did not specify in the open text space what they meant. Those who did, flagged the importance of working for improving the quality of the crops' final product (15% of the total "other" responses), tolerance to limiting soil conditions (acidity, salinity, low nitrogen, aluminium, 7%) and yield improvements (5%).

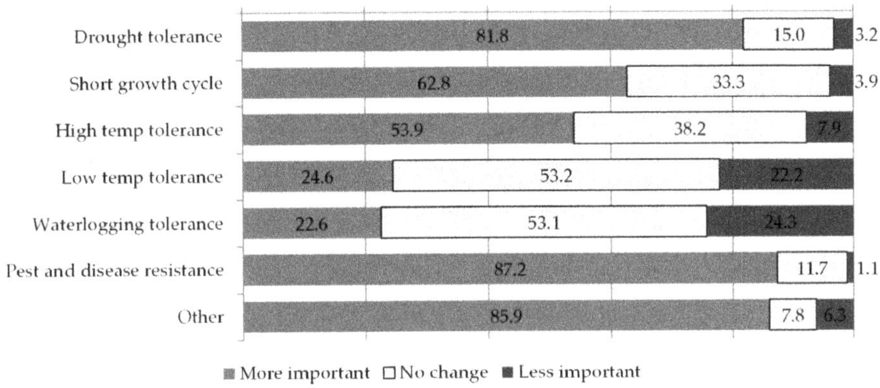

Figure 2. Change in the importance of traits in breeders' work. Numbers on or next to the bars represent percentages over the total responses.

While our questions on changing trait importance required choosing among a list of single priority traits, the results we received reflect the complexity of breeders' work, which entails focusing on different but somehow related traits at the same time: focusing more on pest/disease resistance and drought tolerance was related to an increase in the importance of all other breeding objectives; strong associations were also found among breeding for tolerance to low temperatures and waterlogging and between breeding for high and low temperature tolerance, as if a more encompassing breeding objective was obtaining a general improvement in temperature stress response mechanisms. The many positive associations reported between the changing importance of different traits confirm the complex, multi-trait nature of a crop's adaptation to biotic and abiotic stresses. As expected, changes in the importance of traits during recent years showed a strong association with breeders' perception towards changes in some of the climate patterns (Table 2): plant breeders who observe an increase in irregular and excessive drought also increase their focus towards drought resistance and shorter growth cycles; those focusing more on pest and disease resistance are those who report a more prevalent shift in seasonality and drought occurrences; those detecting excessive and prolonged cold periods (although not common among the respondents) tend to breed more for traits of tolerance to low temperatures. In summary, the recent focus of breeders has not shifted significantly from the most important traits they had been working on routinely, but it does closely mirror their perceptions of climatic events and trends.

Table 2. Correlation between the change in importance of traits in breeding and perceived changes in climate.

	Change in Breeders' Priority Traits					
Perceived changes in climate	Drought Tolerance	Short Growth Cycle	High Temp Tolerance	Low Temp Tolerance	Waterlogging Tolerance	Pest/Disease Resistance
Rainfall quantity	−0.07	−0.11	−0.02	0.05	0.03	0.01
Irregular rainfall	0.11	0.12	0.03	−0.03	−0.09	0.11
Early seasons	−0.08	0.07	0.11	0.10	−0.12	0.13
Late seasons	0.12	0.11	0.14 *	−0.16	−0.05	0.23 ***
Irregular drought	0.18 **	0.24 ***	0.10	−0.05	−0.06	0.15 *
Excessive drought	0.20 **	0.08	0.08	0.06	−0.02	0.18 **
Excessive cold	0.13	0.04	0.11	0.31 ***	0.15	0.08
Prolonged cold	0.03	0.03	0.02	0.27 ***	0.07	0.10
Strong winds	−0.06	0.03	−0.05	0.05	−0.06	0.03
Tornadoes/hurricanes	0.03	0.03	-0.01	0.01	0.02	−0.02

* Significant at the 0.1 probability level. ** Significant at the 0.05 probability level. *** Significant at the 0.01 probability level.

The majority of breeders reported that farmers' preferences for resistance to pests and diseases, drought tolerance and short-cycle varieties had increased the most, followed by high temperature tolerance; significant positive associations were detected between breeders' increasing attention to pest/disease resistance, drought tolerance, high and low temperature tolerance and farmers' increased preference for each of these traits. A caveat must be made here: farmers' preferences, as described in this article, were reported by breeders in the survey and not gathered among farmers directly. We cannot therefore be certain that the breeders' perceptions are not self-made or just intuited. In a different section of our survey we asked breeders if their institutions had specific partnerships with relevant field-level organizations that allowed them to interact with farmers and determine their needs. The majority of responses were positive; most breeders stated that their programme interacted with local, community or farmers' organizations (over 78%) as well as with national agricultural research organizations (72%). This points to an encouraging scenario of collaboration and dialogue between researchers and grassroots/community-level organizations to ensure that breeders' work is aligned with farmers' needs (see Table 3).

Table 3. Association between breeders' priorities and breeders' perceived farmers' preferences for traits.

	Change in Breeders' Priority Traits					
Change in famers' priority traits	Drought Tolerance	Short Growth Cycle	High Temp Tolerance	Low Temp Tolerance	Waterlogging Tolerance	Pest/Disease Resistance
Drought tolerance	0.35 ***	0.15 **	0.11	−0.10	−0.05	−0.01
Short growth cycle	0.22 ***	0.53 ***	0.08	0.04	−0.08	0.09
High temp tolerance	−0.06	0.09	0.57 ***	−0.04	−0.02	0.17 **
Low temp tolerance	−0.02	−0.08	−0.07	0.47 ***	−0.02	0.06
Waterlogging tolerance	−0.04	−0.07	−0.08	0.02	0.41 ***	0.02
Pest and disease resistance	0.01	0.09	0.03	0.13 *	0.11	0.20 ***

* Significant at the 0.1 probability level. ** Significant at the 0.05 probability level. *** Significant at the 0.01 probability level.

2.3. *Changes in the Genetic Materials Used in Breeding Programs*

Having noted a positive association between breeders' perceptions of climate change and of farmers' needs, on the one hand, and the traits breeders' target, on the other hand, we analysed if an association existed between breeders' perceptions of the prevailing climate patterns and the changes in the types of genetic resources they use. First, breeders were asked to describe the relative proportion of crop wild relatives, landraces and advanced/elite lines in their routine work, and then to describe any recent (two to five years) increase or decrease of each germplasm type. Advanced or elite lines were reported as the prevalent material type in breeders' routine work (53%), followed by landraces (25%), other kinds of materials (15%) and crop wild relatives (7%). The use of crop wild relatives showed a positive association with the use of landraces, suggesting that plant breeders who use a greater proportion of the one material type are also likely to be using more of the other. On the contrary, a trade-off between the use of elite lines and less advanced materials exists, indicating that the choice of advanced materials comes with a more likely reduction in the use of landraces and crop wild relatives (data not shown). In terms of recent change, the only category of germplasm that was reported to have increased among the majority of breeders (58%) was that of advanced/elite lines. Conversely, for landraces and crop wild relatives, the majority of responses indicated no significant change in their use (42% and 49% respectively) (see Figure 3).

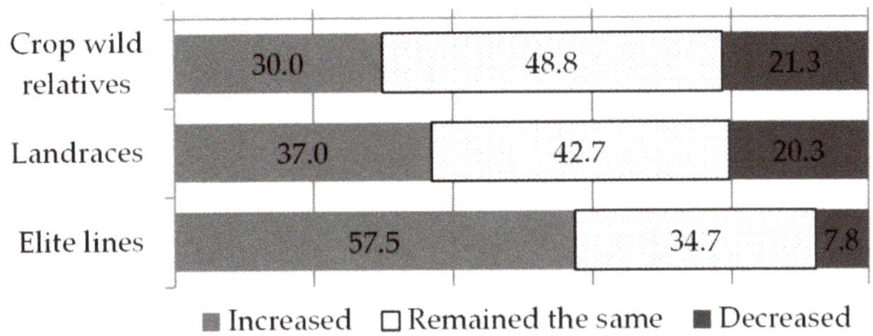

Figure 3. Change in the proportion of germplasm types used by breeders. Numbers on or next to the bars represent percentages over the total responses.

The crops that breeders reported to be working on were grouped into categories: cereals and pseudo-cereals (114 responses), grain legumes (27 responses), roots and tubers (25 responses), fruits (16 responses) and other crops (18 responses). No significant differences were found between crop categories in terms of changes in the types of materials used.

A few elements of breeders' perception of climate change were related to recent changes in the types of materials they use. The significance of these relationships is limited to landraces and elite lines since no association exists with crop wild relatives (which, as we have seen, are more scarcely used). When breeders observe increasing rainfall quantity and irregularity, or colder weather, they tend to use more landraces in their breeding work. On the other hand, those plant breeders observing later onset of seasons, tend to use smaller proportions of landraces. Elite lines are prevalent among breeders concerned with increasing drought irregularity (see Table 4).

Table 4. Association between breeders' perception of climate change and changes in their use of germplasm.

	Crop Wild Relatives	Landraces	Advanced/Elite Lines
Rainfall quantity	0.12	0.15 *	−0.06
Irregular distribution of rainfall	−0.01	0.15 **	−0.02
Early seasons	−0.02	0.04	−0.19 **
Late seasons	−0.09	−0.17 **	0.09
Irregular drought	−0.08	−0.01	0.22 ***
Excessive droughts	−0.00	0.07	0.07
Excessive cold	0.01	0.22 ***	−0.09
Prolonged cold	−0.02	0.22 ***	−0.08
Strong winds	0.09	0.01	−0.12
Tornadoes/hurricanes	0.02	0.10	−0.01

* Significant at the 0.1 probability level. ** Significant at the 0.05 probability level. *** Significant at the 0.01 probability level.

These associations are hard to interpret or even match in literature, since all these climate challenges can potentially be met by working with both advanced materials as well as with landraces and even crop wild relatives. However, using more landraces positively correlates with a higher number of climate challenges, suggesting greater climate-related efforts or successes among breeders who explore potentially more variable reservoirs of genetic diversity.

2.4. Changing Sources of Breeding Materials

The majority of breeders reported to be mostly using ex situ material (82%, if we combine all internal material and that coming from genebanks and universities, excluding community genebanks). Thirty-five percent of the ex situ germplasm comes from the breeder's own organization, and 23% comes from the CGIAR. The share of materials from the national genebank in the breeders' own country is rather low, probably due to the fact that most respondents belonged to the NARO, which often also includes or manages the national genebank, and hence referred to this source as their "own organisation". A less consistent, but still significant, proportion of materials also comes from farmers' fields and natural areas (11%) and community genebanks (5%), which we consider to be a conservation strategy closer to on farm than to ex situ [69,70] (see Figure 4).

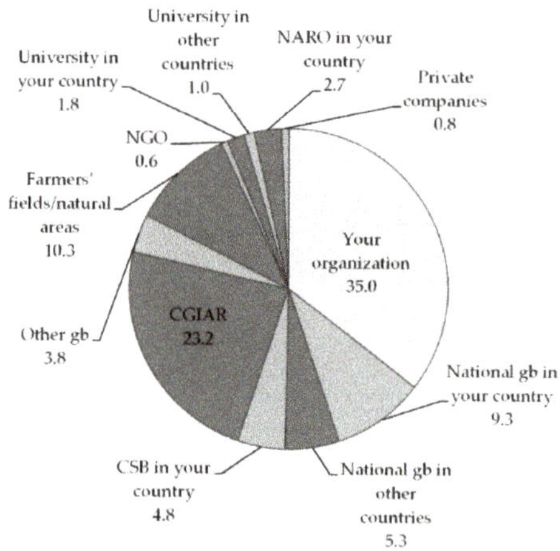

Figure 4. Sources of germplasm used by breeders. CGIAR, Consultative Group on International Agricultural Research; CSB, community seed bank; gb, genebank; NARO, National Agriculture Research Organization; NGO, non-governmental organization.

Association analyses between changes in sources of germplasm and material types highlighted a few relevant relationships. Breeders who have been using more crop wild relatives over recent years, although few in number, increasingly sourced them from a variety of ex situ collections (both public, including the CGIAR, and private, and mostly outside their countries' borders) rather than in situ (i.e., in farmers' or natural fields). Increasing landrace use is indeed the only trend that is associated to greater on farm sourcing; other typical sources for those breeders using more landraces are a variety of (public) institutions, all within the national territory. Increased use of elite lines was strongly associated with foreign or international sources—that is, genebanks and universities, and the CGIAR (see Table 5).

These relationships mirror what was already happening in breeders' routine work, previous to recent years (data not shown), and are therefore not indicative of any recent change in behaviour or innovation.

Table 5. Association between changing material types and sources of breeding material.

		Germplasm Types	
	Crop Wild Relatives	**Landraces**	**Advanced/Elite Lines**
Sources of genetic materials: Your own collection or the collection/genebank in your organization	0.10	0.19 **	0.03
Farmers' fields or natural areas	0.09	0.28 ***	−0.07
National genebanks in your country	0.13	0.29 ***	−0.03
National genebanks in other countries	0.28 ***	0.03	0.17 *
Farmer community genebanks in your country	−0.06	0.32 ***	0.08
CGIAR	0.19 *	0.04	0.23 **
Other genebanks	0.05	−0.03	0.13
Non-governmental organizations	0.23 *	−0.09	0.05
University researchers in your country	0.15	0.11	0.04
University researchers in other countries	0.08	−0.01	0.23 **
Researchers in national agricultural research organizations	0.14	0.22 **	0.12
Private companies	0.25 **	−0.05	0.17

* Significant at the 0.1 probability level. ** Significant at the 0.05 probability level. *** Significant at the 0.01 probability level.

2.5. Policy, Financial and Other Limitations

To analyse whether other factors are related to breeders' behaviour with respect to germplasm use, we asked breeders about if and how (always/often, sometimes, rarely/never) they are subject to limitations that affect their capacity to access and use diverse sets of germplasm. The limitations that breeders were offered to choose among were of a legal, administrative, financial and technical nature (see Figure 5). The types of restrictions that the majority of breeders reported to be always or often affected by, were of an administrative and policy nature, the most serious being burdensome procedures imposed by providers (60%) and national rules or laws (58%). These were followed by restrictions to the further transmission of the material received (50%), difficulties related to the use of material transfer agreements (48%), intellectual property regimes (47%), international rules (45%), internal administrative procedures (43%) and an unwillingness by providers to share materials (43%). Restrictions of technical (breeders' ability to use the germplasm) and financial (payments for obtaining germplasm) nature were less frequently reported.

Some of these limitations were significantly related to breeders' efforts to incorporate a greater proportion of specific germplasm types into their work (Table 6). Using more advanced materials, which was a widespread trend in our sample population, was related to experiencing administrative limitations both by providers and by breeders' own organizations. The few breeders who are increasing their use of landraces and crop wild relatives experienced other limitations, including financial (for both landraces and crop wild relatives), technical capacity (crop wild relatives) and political (landraces) issues. The positive and significant association between the difficulty of using landraces and the "costs of shipment" variable is rather surprising, since respondents declared to be mostly and increasingly sourcing landraces from national or internal sources to their organization. It could be that breeders who are using more landraces are close to exhausting their "routine" sources of variation and are eager to obtain more samples of this germplasm category, which are unavailable within their closest collections. It may be that in seeking beyond these, they incur in shipping costs which they are unable to cover.

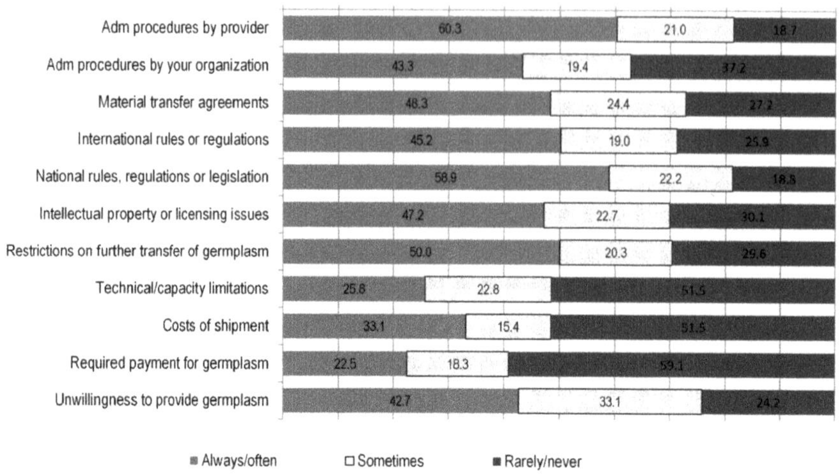

Figure 5. Administrative (adm)/legal and technical factors affecting breeders' work. Numbers on or next to the bars represent percentages over the total responses.

Table 6. Association between limiting factors and changes in the use of germplasm types.

		Germplasm Types		
		Crop Wild Relatives	**Landraces**	**Elite Lines**
Limiting factors	Administrative procedures set by providers	0.03	0.10	−0.13 *
	Administrative procedures set by your organization	0.07	−0.01	−0.14 *
	Material transfer agreements	−0.01	−0.01	−0.02
	International rules or regulations	0.05	0.03	−0.08
	National rules, regulations or legislation	−0.04	−0.02	−0.09
	Intellectual property or licensing issues	−0.08	−0.11	−0.06
	Restrictions on the further transfer of the germplasm to third parties	−0.05	0.09	−0.00
	Technical/capacity limitations	−0.23 ***	−0.00	−0.10
	Costs of shipment	−0.18 **	0.22 ***	−0.02
	Required payment for germplasm	−0.14	0.12	−0.03
	Unwillingness of other organizations to provide germplasm	−0.10	−0.18 **	0.04

* Significant at the 0.1 probability level. ** Significant at the 0.05 probability level. *** Significant at the 0.01 probability level.

When prompted about the effect of access and benefit sharing policies on their work, most breeders reported that they were scarcely influenced by these policies (25%), although another 21% said their situation had worsened; few breeders (12%) responded that they did not know if and how these policies had affected their work. These percentages are in a way surprising if compared to the importance breeders had given to legal and administrative barriers in the previously analysed question, but may suggest that there are either other non-ABS barriers which we did not capture, or that there is not a widespread awareness and knowledge about national and international ABS frameworks and their inter-relations. Regarding in particular the effect of the International Treaty on Plant Genetic Resources for Food and Agriculture (ITPGRFA), 37% of the breeders stated that it had made little or no difference to their work, and 22% answered that they did not know. Among the surveyed countries, China (from where we received numerous responses) and Bolivia were not Parties to the International Treaty at the time of the survey (China is still not). When asked about the lack of specific technologies or tools, limited access to molecular tools and approaches (from sequencing to genomics and proteomics instruments, together with the capacity to use them) was the most urgent factor (for 68% of respondents). Infrastructure for phenotyping, controlled trials, micro-propagation and characterization followed (24%). The availability of genetic materials or information about them was deemed to be a critical limitation by only around 6% of respondents. Breeders were also asked if the budget available for their breeding programme as well as international donor funding had changed. The majority of breeders reported that both had increased (57% and 39% respectively). Positive changes in financial availability were related to a greater use of crop wild relatives in breeding, which makes sense in light of the fact that the few breeders working more with wild material had reported financial limitations.

3. Discussion

Overall, our results suggest that the surveyed plant breeders are well aware of increasingly urgent climate patterns in their target regions, and that their response is to increase their focus on traits that were already the highest priority in earlier years. For the vast majority of breeders, regardless of geographical or crop focus, these traits included pest and disease resistance and drought tolerance, followed by shorter growth cycles, which they believe to also be the traits most desired by farmers. Our results confirm that recent climate changes have exacerbated the breeders' sense of urgency in addressing biotic and abiotic challenges that were already a high priority [71,72]: indeed, research has suggested that dry regions will become drier and wet regions wetter in response to global warming, a trend labelled as the "rich get richer" mechanism [73]. This tendency would naturally lead breeders to devote even more efforts to traits that have always received the highest importance in their work, such as drought tolerance, particularly in those developing countries which rely on mostly rain-fed cropping systems and are hence more heavily affected [71,74]. The prevailing tendency among the surveyed breeders was to increase their use of elite lines versus landraces and crop wild relatives. One might have expected instead to see increased reliance on more genetically diverse materials as potential sources of genetic traits adapted to changing climate changes, as highlighted in the introduction. However, increased use of landraces, albeit not widespread in our sample population, correlated with the highest number of climate challenges, suggesting that those relatively few breeders who are working more with this material type, are able to tackle specific climate change needs. However, financial, technical and policy related disincentives appear to be very influential in limiting a more widespread increase in the use of landraces, and even more so crop wild relatives. Of course, it is also possible that some of the advanced lines that breeders are increasingly using derive from introgression of traits from crop wild relatives or traditional varieties as a result of pre-breeding conducted elsewhere, thus guaranteeing the incorporation of new, potentially adaptive diversity anyway.

As far as sources of material are concerned, breeders prioritised materials coming from their own organization and secondly the international network of CGIAR Centers. A similar order of prevalence in germplasm sources has been observed in other studies. A survey among wheat breeders in developed and developing countries found that top priority was given by the majority of breeders to lines readily available within their own programmes; in developing countries, CGIAR lines and released varieties were the second most heavily used type of germplasm [75], as also emerging here. Indeed, it is widely acknowledged that the CGIAR has played a critical role in the provision of germplasm to developing countries since its inception, through its genebanks, breeding programmes and international nurseries [76–80]. Germplasm originally received from foreign or international programmes such as the CGIAR gradually becomes internalised into national programmes, likely making the dependence on continued international sourcing less prominent over time, compared to in-house sourcing. Indeed, international or foreign germplasm is often re-distributed by the original recipient to additional colleagues, particularly in developing countries [59]. In addition, some national programmes have consistently improved their capability of carrying out their own crosses and developing their own improved lines, with more targeted use of CGIAR improved material.

Perhaps the most interesting result of our survey is that sixty percent of the breeders confirmed that their work is always/often affected by burdensome procedures imposed by providers, and by national rules or laws regulating access and further transfer of materials. We recorded similarly high response rates with respect to the deleterious impacts of international rules and regulations, intellect property and licensing, even their own organization's administrative procedures. Overall, this points to the lack of a supportive policy environment as a greater limiting factor than their own technical capacity to use genetic resources, their ability to pay for them or budgetary limitations. We were not clearly able to pinpoint the effect of specific international ABS rules (either those under the CBD/Nagoya or the ITPGRFA) since breeders' responses to our questions on if/how these frameworks had affected their work (in general or with specific materials) suggested a limited effect or one of which they were not clearly aware.

Increasing the use of different material types is affected by policy and technical limitations in different ways. It is elite lines and landraces that are most affected by policy/legal issues, the former by administrative procedures imposed by providers, the latter by an unwillingness by providers to grant access to germplasm. Since breeders' increasing use of elite lines is related to using external, foreign sources of germplasm, there may indeed be more barriers for introducing these materials, including on the phytosanitary side: it has been noted that certain countries have adopted phytosanitary policies that have led to lower acceptance rates of international genetic materials, or that some countries may not have the capacity to carry out all the analyses that their phytosanitary policies require, resulting in decreasing requests [57]. The fact that landraces are the material type specifically related to the response on providers' "unwillingness" to share material, may be due to the long-lasting heritage of traditional and cultural knowledge and the strong sense of ownership that is associated to landraces, often instigated by civil society organizations or local and national governments [81]. Ownership and sovereignty issues become even more relevant in light of the fact that many of the breeders working with landraces are also using more materials collected from on farm sources. While on farm sourcing yields materials subject to ongoing evolutionary forces, hence potentially more adapted to climate change [57,82], it also makes it more likely to raise cultural identity and resource ownership issues. In addition to on farm sources, breeders who are using more landraces are also accessing more materials from ex situ institutions, all of which fall within the breeders' own institution or her/his national borders. The fact that they don't turn to the CGIAR as much, despite the centres' facilitated access system under the conditions of the International Treaty on Plant Genetic Resources [60,83], may be due to the higher transaction costs (increased landrace use positively correlated to increased shipping costs) and possible delays due to international phytosanitary issues [57]. Furthermore, in a scenario of increasingly cumbersome international exchanges, personal contacts and closer relationships of trust may be more effective in smoothening germplasm transactions, particularly those which involve culturally-relevant materials [57].

Technical and financial challenges outweighed political/legal or ownership issues when it came respondents' ability to access and use crop wild relatives. While there are a number of success stories around the introduction of useful traits from the highly variable pools of crop wild relatives [41,46–50], the transfer of alleles from wild populations tends to be slow, genetically tricky and expensive compared to when using advanced or elite lines [12,55,84]. Other limitations, such as the lack of genetic materials or information about them, only affected a minority of breeders working with wild relatives, a minority which seeks to mine an even greater variety of ex situ sources, including foreign ex situ institutions and collections from the private sector. The lack of significant in situ sourcing of crop wild relatives goes against the recommendation of using materials subject to natural climate change selective pressures and may be an issue if we also consider the far from complete CWR representation in genebanks [10,85]. However, the possibility of sourcing materials from the wild requires a knowledge of any existing in situ conservation strategies [86] as well as targeted collection missions, which many countries don't have the finances or the technical tools to carry out. The efficiency of in situ sourcing could also be improved by greater availability and capacity to use eco-geographical and climate modelling tools, which can significantly narrow down the collection areas to be surveyed while maximising the chance to find adapted materials [87–90].

While most of the surveyed breeders are continuing to pursue similar strategies working with genetic resources under increasing climate change awareness, our results offer some interesting insights from the minority of breeders who are diversifying their germplasm materials and sources. Overcoming the barriers they experience may encourage a broader diffusion of those diversity-based strategies that literature describes as essential in responding to climate change [36–50,91]. Though some argue this potential is overstated [92,93], introducing more advanced genomic/phenotyping tools into any breeding programme has the potential to improve the power and speed of exploring large pools of diversity [94,95]. Although many molecular and genomic tools are becoming increasingly affordable, they still require substantial investment and training to be fully integrated into breeding programmes worldwide, particularly in developing countries [96,97]. The widespread diffusion of high-throughput field phenotyping infrastructure and capacities appears to be even slower [66,98,99]. The technical, technological and digital divide affecting researchers in some countries has increasingly important implications in terms of access to and use of genetic resources in breeding for climate relevant objectives. For example the opportunities to link historical climate change data with genomic analysis of germplasm stored in long-established collections, could provide an additional tool for breeders to select wild varieties with potential adaptive traits, particularly in centres of crop diversity [52]. Eco-geographical modelling tools could aid the identification of sites where to conduct multi-environment trials allowing the evaluation of germplasm across a range of climate-relevant target environments [100,101]. Financial, technology transfer and capacity building bottlenecks remain very real, particularly for certain (minor) crops and contexts that offer limited incentives for private sector involvement, for instance through public-private partnerships or consortia. Thus, increasing support from national governments and networking with foreign or international institutions will be crucial.

Above all technical trends and limitations, the persistence of high-level policy and legal bottlenecks affecting access to and use of PGRFA would deserve a more detailed and dedicated investigation than the one made possible by our survey data. Much has been written on how to smoothen national and international rules and regulations, to achieve a more mutually supportive implementation of the numerous agreements on access to and use of germplasm (and most particularly of the Convention on Biological Diversity—CBD, its Nagoya Protocol, and the International Treaty on PGRFA). It would appear that substantial work is still required, since the many actors now involved in national policy development and implementation are uncertain how to do this in practice. This scenario risks negatively affecting researchers and crop breeders, who instead could and should be playing an important role, hand in hand with farmers, in adapting agriculture to future climate challenges.

4. Materials and Methods

Data for this study were collected in 2013 through a web-based survey designed by a research team from the Centre for Science, Technology and Environment Policy Studies of Arizona State University and Bioversity International. The study targets were plant breeders in national programmes dealing with food security crops in their countries. Target countries were 19 developing countries across Asia, Africa and Latin America (namely Bhutan, Bolivia, Brazil, Burkina Faso, China, Costa Rica, Ethiopia, Guatemala, India, Ivory Coast, Jordan, Kenya, Morocco, Nepal, Peru, Philippines, Rwanda, Uganda and Zambia). Many of these are countries where Bioversity International has been running projects and has contacts within the plant genetic resources community. In addition, the selection also took into account countries' human development index and their climate vulnerability.

The sample frame for the target population was prepared by collecting contacts of potential respondents from electronic sources, including country reports from the Global Partnership Initiative for Plant Breeding Capacity Building, the African Crop Science Society and a list of national and international workshop participants, and through Bioversity International's network of national project partners. Complete contact information for approximately 1092 potential respondents was collected within the 19 countries. The team checked for repetition of names or incorrect email addresses (for example, through pinging) before administering the survey. The survey was pre-tested by the study team members. The final version was translated to English, French and Spanish, written into the Sawtooth software© and administered online. Potential respondents received an email describing the project and requesting their participation in the survey, with an appropriate consent statement. The letter provided information about the survey and its purposes as well as the link to the instrument, a unique identification and an individual password. The survey contained 65 questions covering a wide range of issues (technical, policy, financial) regarding breeders' work; the questions we are analysing in this paper focused on breeders' perceptions of climate change and of farmers' climate-related needs over the past two-to-five years, the priority traits they work on, the types of germplasm they use and the sources of access for such germplasm. For priority traits, types and sources of germplasm, breeders were asked both about their business-as-usual behaviour and about any changes occurring over the last two-to-five years. Other questions analysed here concerned the institution breeders worked in, the resource allocation for their programme, as well as gender, age, academic qualifications and breeding experience. Except for age and years of work as breeders, all the other questions were multiple choice, with several pre-defined responses. Finally, we also looked into breeders' responses to an open ended question on what tools they would need to improve their breeding work and efficiency. All categorical responses were coded numerically for subsequent analyses, which employed basic descriptive statistics such as mean, frequency and association analyses. All data analysis was performed in R [102].

Author Contributions: Conceptualization, G.G., A.S., M.H., I.L.N. and E.W.W.; Formal analysis, G.G. and A.S.; Methodology, G.G., A.S. and E.W.W.; Software, G.G. and A.S.; Supervision, M.H., I.L.N. and E.W.W.; Writing-original draft, G.G. and A.S.; Writing-review & editing, G.G., A.S., M.H. and I.L.N. All authors have read and agreed to the published version of the manuscript.

Funding: This work was implemented as part of the CGIAR Research Program on Climate Change, Agriculture and Food Security (CCAFS), which is carried out with support from the CGIAR Trust Fund and through bilateral funding agreements. For details please visit https://ccafs.cgiar.org/donors. The views expressed in this document cannot be taken to reflect the official opinions of these organizations. Support was also received from the CGIAR Genebank Platform and from the Directorate-General for International Cooperation, Ministry of Foreign Affairs of the Netherlands through the grant "Strengthening national capacities to implement the International Treaty on Plant Genetic Resources for Food and Agriculture" (Grant contract number: 21820/DSO0113359).

Acknowledgments: The authors would like to thank all breeders who took the time to answer the survey, providing us with the basis for the present analysis.

Conflicts of Interest: The authors declare no conflict of interest. The funders had no role in the design of the study; in the collection, analyses, or interpretation of data; in the writing of the manuscript, or in the decision to publish the results.

References

1. IPCC. Summary for policymakers. In *Climate Change 2014: Impacts, Adaptation, and Vulnerability. Part A: Global and Sectoral Aspects. Contribution of Working Group II to the Fifth Assessment Report of the Intergovernmental Panel on Climate Change*; Field, C.B., Barros, V.R., Dokken, D.J., Mach, K.J., Mastrandrea, M.D., Bilir, T.E., Chatterjee, M., Ebi, K.L., Estrada, Y.O., Genova, R.C., et al., Eds.; Cambridge University Press: Cambridge, UK; New York, NY, USA, 2014; pp. 1–32.
2. Lobell, D.B.; Burke, M.B.; Tebaldi, C.; Mastrandrea, M.D.; Falcon, W.P.; Naylor, R.L. Prioritizing Climate Change Adaptation Need for Food Security in 2030. *Science* **2008**, *319*, 607–610. [CrossRef] [PubMed]
3. Thornton, P.K.; Ericksen, P.J.; Herrero, M.; Challinor, A.J. Climate variability and vulnerability to climate change: A review. *Glob. Chang. Biol.* **2014**, *20*, 3313–3328. [CrossRef] [PubMed]
4. Wassmann, R.; Jagadish, S.V.K.; Heuer, S.; Ismail, A.; Redona, E.; Serraj, R.; Singh, R.K.; Howell, G.; Pathak, H.; Sumfleth, K. Chapter 2. Climate Change Affecting Rice Production: The Physiological and Agronomic Basis for Possible Adaptation Strategies. *Adv. Agron.* **2009**, *101*, 59–122.
5. Vadez, V.; Kholova, J.; Choudhary, S.; Zindy, P.; Terrier, M.; Krishnamurthy, L.; Ratna Kumar, P.; Turner, N.C. Responses to Increased Moisture Stress and Extremes: Whole Plant Response to Drought under Climate Change. In *Crop Adaptation to Climate Change*; Yadav, S.S., Redden, R.J., Hatfield, J.L., Lotze-Campen, H., Hall, A.E., Eds.; Wiley-Blackwell: Chichester, UK, 2011; pp. 186–197.

6. Schafleitner, R.; Ramirez, J.; Jarvis, A.; Evers, D.; Gutierrez, R.; Scurrah, M. Adaptation of the Potato Crop to Changing Climates. In *Crop Adaptation to Climate Change*; Yadav, S.S., Redden, R.J., Hatfield, J.L., Lotze-Campen, H., Hall, A.E., Eds.; Wiley-Blackwell: Chichester, UK, 2011; pp. 287–297.
7. Singh, R.P.; Vara Prasad, P.V.; Sharma, A.K.; Raja Reddy, K. Impacts of High-Temperature Stress and Potential Opportunities for Breeding. In *Crop Adaptation to Climate Change*; Yadav, S.S., Redden, R.J., Hatfield, J.L., Lotze-Campen, H., Hall, A.E., Eds.; Wiley-Blackwell: Chichester, UK, 2011; pp. 166–185.
8. Oszako, T.; Nowakowska, J.A. Climate change and food security: Challenges for plant health, plant breeding and genetic resources. *Folia For. Pol.* **2015**, *57*, 194–197. [CrossRef]
9. Chapman, S.C.; Chakraborty, S.; Dreccer, M.F.; Howden, S.M. Plant adaptation to climate change—opportunities and priorities in breeding. *Crop Pasture Sci.* **2012**, *63*, 251–268. [CrossRef]
10. Maxted, N.; Kell, S. *Establishment of a Global Network for the in Situ Conservation of Crop Wild Relatives: Status and Needs*; FAO Commission on Genetic Resources for Food and Agriculture: Rome, Italy, 2009; p. 266.
11. Nelson, G.C.; Rosegrant, M.W.; Koo, J.; Robertson, R.; Sulser, T.; Zhu, T.; Ringler, C.; Msangi, S.; Palazzo, A.; Batka, M.; et al. *Climate Change Impact on Agriculture and Costs of Adaptation. Food Policy Report*; International Food Policy Research Institute: Washington, DC, USA, 2009; Volume 21, p. 20.
12. Mba, C.; Guimaraes, E.P.; Ghosh, K. Re-orienting crop improvement for the changing climatic conditions of the 21st century. *Agric. Food Secur.* **2012**, *1*, 7. [CrossRef]
13. Turner, N.C.; Meyer, R. Synthesis of Regional Impacts and Global Agricultural Adjustments. In *Crop Adaptation to Climate Change*; Yadav, S.S., Redden, R.J., Hatfield, J.L., Lotze-Campen, H., Hall, A.E., Eds.; Wiley-Blackwell: Chichester, UK, 2011; pp. 156–165.
14. Smit, B.; Burton, I.; Klein, R.J.T.; Wandel, J. An Anatomy of Adaptation to Climate Change and Variability. *Clim. Chang.* **2000**, *45*, 223–251. [CrossRef]
15. Battisti, D.S.; Naylor, R.L. Historical Warnings of Future Food Insecurity with Unprecedented Seasonal Heat. *Science* **2009**, *323*, 240–244. [CrossRef]
16. Lotze-Campden, H. Climate Change, Population Growth, and Crop Production: An Overview. In *Crop Adaptation to Climate Change*; Yadav, S.S., Redden, R.J., Hatfield, J.L., Lotze-Campen, H., Hall, A.E., Eds.; Wiley-Blackwell: Chichester, UK, 2011; pp. 1–11.
17. Vinet, L.; Zhedanov, A. A 'missing' family of classical orthogonal polynomials. In *Climate Change 2014: Impacts, Adaptation, and Vulnerability. Part A: Global and Sectoral Aspects. Contribution of Working Group II to the Fifth Assessment Report of the Intergovernmental Panel on Climate Change*; Cambridge University Press: Cambridge, UK; New York, NY, USA, 2010; pp. 1–32.
18. Howden, S.M.; Soussana, J.-F.; Tubiello, F.N.; Chhetri, N.; Dunlop, M.; Meinke, H. Adapting agriculture to climate change. *Proc. Natl. Acad. Sci. USA* **2007**, *104*, 19691–19696. [CrossRef]
19. Redden, R.J.; Yadav, S.S.; Hatfield, J.L.; Prasanna, B.M.; Vasal, S.K.; Lafarge, T. The Potential of Climate Change Adjustment in Crops: A Synthesis. In *Crop Adaptation to Climate Change*; Yadav, S.S., Redden, R.J., Hatfield, J.L., Lotze-Campen, H., Hall, A.E., Eds.; Wiley-Blackwell: Chichester, UK, 2011; pp. 482–494.
20. Deresa, T.T.; Hassan, R.M.; Ringler, C.; Alemu, T.; Yesuf, M. Determinants of Farmers' Choice of Adaptation Methods to Climate Change in the Nile Basin of Ethiopia. *Glob. Environ. Chang.* **2009**, *19*, 248–255. [CrossRef]
21. Kurukulasuriya, P.; Mendelsohn, R. Crop switching as a strategy for adapting to climate change. *Afr. J. Agric. Resour. Econ.* **2008**, *2*, 105–126.
22. Tucker, C.M.; Eakin, H.; Castellanos, E.J.; Tucker, C.M. Perceptions of risk and adaptation: Coffee producers, market shocks, and extreme weather in Central America and Mexico. *Glob. Environ. Chang.* **2010**, *20*, 23–32. [CrossRef]
23. D'Alpoim Guedes, J.; Bocinsky, R.K. Climate change stimulated agricultural innovation and exchange across Asia. *Sci. Adv.* **2018**, *4*, eaar4491. [CrossRef]
24. Gowda, C.L.L.; Saxena, K.B.; Srivastava, R.K.; Upadhyaya, H.D.; Silim, S.N. Pigeonpea: From an Orphan to A Leader in Food Legumes. In *Biodiversity in Agriculture: Domestication, Evolution, and Sustainability*; Cambridge University Press: New York, NY, USA, 2011; pp. 362–373.
25. Uprety, D.C.; Sirohi, G.S. Comparative Study on the Effect of Water Stress on the Photosynthesis and Water Relations of Triticale, Rye and Wheat. *J. Agron. Crop Sci.* **1987**, *159*, 349–355. [CrossRef]
26. Harlan, J.R.; De Wet, J.M.J.; Price, E.G. Comparative evolution in cereals. *Evolution* **1973**, *27*, 311–325. [CrossRef] [PubMed]

27. Zohary, D. Unconscious selection and the evolution of domesticated plants. *Econ. Bot.* **2004**, *58*, 5–10. [CrossRef]
28. Ceccarelli, S. Efficiency of Plant Breeding. *Crop Sci.* **2015**, *55*, 87–97. [CrossRef]
29. Vigouroux, Y.; Mariac, C.; De Mita, S.; Pham, J.-L.; Gérard, B.; Kapran, I.; Sagnard, F.; Deu, M.; Chantereau, J.; Ali, A.; et al. Selection for Earlier Flowering Crop Associated with Climatic Variations in the Sahel. *PLoS ONE* **2011**, *6*, e19563. [CrossRef]
30. Mercer, K.L.; Perales, H.R. Evolutionary response of landraces to climate change in centers of crop diversity. *Evol. Appl.* **2010**, *3*, 480–493. [CrossRef]
31. Lafarge, T.; Peng, S.; Hasegawa, T.; William, P.; Quick, S.V.; Jagadish, K.; Wassmann, R. Genetic Adjustment to Changing Climates: Rice. In *Crop Adaptation to Climate Change*; Yadav, S.S., Redden, R.J., Hatfield, J.L., Lotze-Campen, H., Hall, A.E., Eds.; Wiley-Blackwell: Chichester, UK, 2011; pp. 298–313.
32. Trethowan, R.M.; Mahmood, T. Genetic Options for Improving Productivity of Wheat in Water-Limited and Temperature-Stressed Environments. In *Crop Adaptation to Climate Change*; Yadav, S.S., Redden, R.J., Hatfield, J.L., Lotze-Campen, H., Hall, A.E., Eds.; Wiley-Blackwell: Chichester, UK, 2011; pp. 218–237.
33. Haussmann, B.I.G.; Fred Rattunde, H.; Weltzien-Rattunde, E.; Traoré, P.S.C.; vom Brocke, K.; Parzies, H.K. Breeding Strategies for Adaptation of Pearl Millet and Sorghum to Climate Variability and Change in West Africa. *J. Agron. Crop Sci.* **2012**, *198*, 327–339. [CrossRef]
34. Rosegrant, M.W.; Cline, S.A. Global Food Security: Challenges and Policies. *Science* **2003**, *302*, 1917–1919. [CrossRef] [PubMed]
35. Tester, M.; Langridge, P. Breeding Technologies to Increase Crop Production in a Changing World. *Science* **2010**, *327*, 818–822. [CrossRef] [PubMed]
36. McCouch, S.; Baute, G.J.; Bradeen, J.; Bramel, P.; Bretting, P.K.; Buckler, E.; Burke, J.M.; Charest, D.; Cloutier, S.; Cole, G.; et al. Feeding the future. *Nature* **2013**, *499*, 23. [CrossRef] [PubMed]
37. FAO. *Coping with Climate Change—The Roles of Genetic Resources for Food and Agriculture. Rome*; Food and Agriculture Organization of the United Nations: Rome, Italy, 2015; ISBN 9789251084410.
38. Michael, T.P.; VanBuren, R. Progress, challenges and the future of crop genomes. *Curr. Opin. Plant Biol.* **2015**, *24*, 71–81. [CrossRef] [PubMed]
39. Brozynska, M.; Furtado, A.; Henry, R.J. Genomics of crop wild relatives: Expanding the gene pool for crop improvement. *Plant Biotechnol. J.* **2016**, *14*, 1070–1085. [CrossRef]
40. Zhang, H.; Mittal, N.; Leamy, L.J.; Barazani, O.; Song, B.-H. Back into the wild—Apply untapped genetic diversity of wild relatives for crop improvement. *Evol. Appl.* **2017**, *10*, 5–24. [CrossRef]
41. Lopes, M.S.; El-Basyoni, I.; Baenziger, P.S.; Singh, S.; Royo, C.; Ozbek, K.; Aktas, H.; Ozer, E.; Ozdemir, F.; Manickavelu, A.; et al. Exploiting genetic diversity from landraces in wheat breeding for adaptation to climate change. *J. Exp. Bot.* **2015**, *66*, 3477–3486. [CrossRef]
42. Ceccarelli, S. Evolution, plant breeding and biodiversity. *J. Agric. Environ. Int. Dev.* **2009**, *103*, 131–145.
43. Ceccarelli, S.; Valkoun, J.; Erskine, W.; Weigand, S.; Miller, R.; van Leur, J. Plant Genetic Resources and Plant Improvement as Tools to Develop Sustainable Agriculture. *Exp. Agric.* **1992**, *28*, 89–98. [CrossRef]
44. Ceccarelli, S.; Grando, S.; Maatougui, M.; Michael, M.; Slash, M.; Haghparast, R.; Rahmanian, M.; Taheri, A.; Al-Yassin, A.; Benbelkacem, A.; et al. Plant breeding and climate changes. *J. Agric. Sci.* **2010**, *148*, 627–637. [CrossRef]
45. Esquinas-Alcázar, J. Protecting crop genetic diversity for food security: Political, ethical and technical challenges. *Nat. Rev. Genet.* **2005**, *6*, 946. [CrossRef] [PubMed]
46. Dempewolf, H.; Eastwood, R.J.; Guarino, L.; Khoury, C.K.; Müller, J.V.; Toll, J. Adapting Agriculture to Climate Change: A Global Initiative to Collect, Conserve, and Use Crop Wild Relatives. *Agroecol. Sustain. Food Syst.* **2014**, *38*, 369–377. [CrossRef]
47. Tanksley, S.D.; McCouch, S.R. Seed Banks and Molecular Maps: Unlocking Genetic Potential from the Wild. *Science* **1997**, *277*, 1063–1066. [CrossRef] [PubMed]
48. Ayala, M.; Guzmán, C.; Alvarez, J.B.; Peña, R.J. Characterization of genetic diversity of puroindoline genes in Mexican wheat landraces. *Euphytica* **2013**, *190*, 53–63. [CrossRef]
49. Machida-Hirano, R. Diversity of potato genetic resources. *Breed. Sci.* **2015**, *65*, 26–40. [CrossRef]
50. Lakew, B.; Henry, R.J.; Eglinton, J.; Baum, M.; Ceccarelli, S.; Grando, S. SSR analysis of introgression of drought tolerance from the genome of Hordeum spontaneum into cultivated barley (Hordeum vulgare ssp vulgare). *Euphytica* **2013**, *191*, 231–243. [CrossRef]

51. Bennett, E. Adaptation in Wild and Cultivated Plant Populations. In *Genetic Resources in Plants—Their Exploration and Cultivation IBP Handbook No 11*; Frankel, O.H., Bennett, E., Eds.; Blackwell Scientific Publications: Oxford, UK, 1970; pp. 115–129.
52. Gutaker, R.M.; Groen, S.C.; Bellis, E.S.; Choi, J.Y.; Pires, I.S.; Bocinsky, R.K.; Slayton, E.R.; Wilkins, O.; Castillo, C.C.; Negrão, S.; et al. Genomic history and ecology of the geographic spread of rice. *Nat. Plants* **2020**, *6*, 492–502. [CrossRef]
53. Upadhyaya, H.D.; Dwivedi, S.L.; Ambrose, M.; Ellis, N.; Berger, J.; Smýkal, P.; Debouck, D.; Duc, G.; Dumet, D.; Flavell, A.; et al. Legume genetic resources: Management, diversity assessment, and utilization in crop improvement. *Euphytica* **2011**, *180*, 27–47. [CrossRef]
54. Li, L.; Redden, R.J.; Zong, X.; Berger, J.D.; Bennett, S.J. Ecogeographic analysis of pea collection sites from China to determine potential sites with abiotic stresses. *Genet. Resour. Crop Evol.* **2013**, *60*, 1801–1815. [CrossRef]
55. Fowler, C.; Smale, M.; Gaiji, S. *The Demand for Crop Genetic Resources from International Collections*; International Food Policy Research Institute: Washington, DC, USA, 2003.
56. Bamberg, J.; del Rio, A.; Huaman, Z.; Vega, S.; Martin, M.; Salas, A.; Pavek, J.; Kiru, S.; Fernandez, C.; Spooner, D. A decade of collecting and research on wild potatoes of the Southwest USA. *Am. J. Potato Res.* **2003**, *80*, 159–172. [CrossRef]
57. López Noriega, I.; Halewood, M.; Galluzzi, G.; Vernooy, R.; Bertacchini, E.; Gauchan, D.; Welch, E. How Policies Affect the Use of Plant Genetic Resources: The Experience of the CGIAR. *Resources* **2013**, *2*, 231–269. [CrossRef]
58. Henry, R.J. Genomics strategies for germplasm characterization and the development of climate resilient crops. *Front. Plant Sci.* **2014**, *5*, 68. [CrossRef] [PubMed]
59. Smale, M.; Day Rubenstein, K. The Demand for Crop Genetic Resources: International Use of the US National Plant Germplasm System. *World Dev.* **2002**, *30*, 1639–1655. [CrossRef]
60. Galluzzi, G.; Halewood, M.; Lopez Noriega, I.; Vernooy, R. Twenty-five years of international exchanges of plant genetic resources facilitated by the CGIAR genebanks: A case study on global interdependence. *Biodivers. Conserv.* **2016**, *25*, 1421–1446. [CrossRef]
61. Seyoum, A.; Welch, E.W. Ex Post Use Restriction and Benefit-sharing Provisions for Access to Non-plant Genetic Materials for Public Research. *Appl. Econ. Perspect. Policy* **2015**, *37*, 667–691. [CrossRef]
62. Halewood, M. International efforts to pool and conserve crop genetic resources in times of Radical Legal Change. In *Intellectual Property Rights—Initiative for Policy Dialogue*; Cimoli, M., Dosi, G., Maskus, K.E., Okediji, R.L., Reichman, J.H., Eds.; Oxford University Press: Oxford, UK, 2014.
63. Halewood, M.; López Noriega, I.; Louafi, S. *Crop Genetic Resources as a Global Commons: Challenges in International Governance and Law*; Routledge: Abingdon, UK, 2013.
64. Chiarolla, C.; Jungcurt, S. *Outstanding Issues on Access and Benefit Sharing under the Multilateral System of the International Treaty on Plant Genetic Resources for Food and Agriculture*; The Berne Declaration/Development Fund: Zurich, Switzerland; Oslo, Norway, 2011.
65. Vernooy, R.; Ruiz, M. Conclusions: Race to the Bottom versus Slow Walk to the Top. In *The Custodians of Biodiversity: Sharing Access to and Benefits of Genetic Resources*; Ruiz, M., Vernooy, R., Eds.; Earthscan; International Development Research Center of Canada (IDRC): London, UK; Ottawa, ON, Canada, 2012; pp. 163–180.
66. Morris, M.; Edmeades, G.; Pehu, E. The global need for plant breeding capacity: What roles for the public and private sectors? *HortScience* **2006**, *41*, 30–39. [CrossRef]
67. Gepts, P.; Hancock, J. The future of plant breeding. *Crop Sci.* **2006**, *46*, 1630–1634. [CrossRef]
68. Thirtle, C.; Srinivasan, C.; Heisey, P. Public Sector Plant Breeding In A Privatizing World. *United States Dep. Agric. Econ. Res. Serv. Agric. Inf. Bull.* **2001**, *772*, 22.
69. Vernooy, R.; Sthapit, B.; Galluzzi, G.; Shrestha, P. The Multiple Functions and Services of Community Seedbanks. *Resources* **2014**, *3*, 636–656. [CrossRef]
70. Vernooy, R.; Shrestha, P.; Sthapit, B. *Community Seed Banks*; Earthscan: London, UK, 2015; ISBN 9780415708050.
71. Ceccarelli, S.; Grando, S.; Baum, M.; Udupa, S.M. Breeding for Drought Resistance in a Changing Climate. In *Challenges and Strategies of Dryland Agriculture*; CSSA Special Publication SV—32; Crop Science Society of America and American Society of Agronomy: Madison, WI, USA, 2004; pp. 167–190. ISBN 978-0-89118-611-3.

72. Garrett, K.A.; Dendy, S.P.; Frank, E.E.; Rouse, M.N.; Travers, S.E. Climate Change Effects on Plant Disease: Genomes to Ecosystems. *Annu. Rev. Phytopathol.* **2006**, *44*, 489–509. [CrossRef]
73. Durack, P.J.; Wijffels, S.E.; Matear, R.J. Ocean Salinities Reveal Strong Global Water Cycle Intensification During 1950 to 2000. *Science* **2012**, *336*, 455–458. [CrossRef] [PubMed]
74. Ceccarelli, S.; Grando, S.; Baum, M. Participatory Plant Breeding in Water-limited Environments. *Exp. Agric.* **2007**, *43*, 411–435. [CrossRef]
75. Rejesus, R.M.; Smale, M.; van Ginkel, M. Wheat breeders' perspectives on genetic diversity and germplasm use: Findings from an international survey. *Plant Var. Seeds* **1996**, *9*, 129–147.
76. Evenson, R.E.; Gollin, D. Genetic resources, international organisations, and improvement in rice varieties. *Econ. Dev. Cult. Chang.* **1997**, *45*, 471–500. [CrossRef]
77. Evenson, R.E.; Gollin, D. Assessing the impact of the Green Revolution, 1960 to 2000. *Science* **2003**, *300*, 758–762. [CrossRef]
78. Lantican, M.A.; Braun, H.J.; Payne, T.S.; Singh, R.P.; Sonder, K.; Baum, M.; van Ginkel, M.; Erenstein, O. *Impacts of International Wheat Improvement Research, 1994–2014*; International Maize and Wheat Improvement Center (CIMMYT): Mexico City, Mexico, 2016.
79. Renkow, M.; Byerlee, D. The impacts of CGIAR research: A review of recent evidence. *Food Policy* **2010**, *35*, 391–402. [CrossRef]
80. Varshney, R.; Ojiewo, C.; Monyo, E. A decade of Tropical Legumes projects: Development and adoption of improved varieties, creation of market-demand to benefit smallholder farmers and empowerment of national programmes in sub-Saharan Africa and South Asia. *Plant Breed.* **2019**, *138*, 379–388. [CrossRef]
81. Villa, T.C.C.; Maxted, N.; Scholten, M.; Ford-Lloyd, B. Defining and identifying crop landraces. *Plant Genet. Resour.* **2005**, *3*, 373–384. [CrossRef]
82. Bellon, M.; van Etten, J. Climate change and on-farm conservation of crop landraces in centres of diversity. In *Plant Genetic Resources and Climate Change*; CABI: Wallingford, UK, 2014; pp. 137–150, ISBN 978-1-78064-197-3.
83. Fowler, C.; Smale, M.; Gaiji, S. Unequal Exchange? Recent Transfers of Agricultural Resources and their Implications for Developing Countries. *Dev. Policy Rev.* **2001**, *19*, 181–204. [CrossRef]
84. Baenziger, P.S.; Salah, I.; Little, R.S.; Santra, D.K.; Regassa, T.; Wang, M.Y. Structuring an efficient organic wheat breeding program. *Sustainability* **2011**, *3*, 1190–1205. [CrossRef]
85. Castañeda-Álvarez, N.P.; Khoury, C.K.; Achicanoy, H.A.; Bernau, V.; Dempewolf, H.; Eastwood, R.J.; Guarino, L.; Harker, R.H.; Jarvis, A.; Maxted, N.; et al. Global conservation priorities for crop wild relatives. *Nat. Plants* **2016**, *2*, 16022. [CrossRef] [PubMed]
86. Jarvis, A.; Lane, A.; Hijmans, R. The effect of climate change on crop wild relatives. *Agric. Ecosyst. Environ.* **2008**, *126*, 13–23. [CrossRef]
87. Parra-Quijano, M.; Iriondo, J.M.; Torres, E. Ecogeographical land characterization maps as a tool for assessing plant adaptation and their implications in agrobiodiversity studies. *Genet. Resour. Crop Evol.* **2011**, *59*, 205–217. [CrossRef]
88. Kantar, M.B.; Sosa, C.C.; Khoury, C.K.; Castañeda-Álvarez, N.P.; Achicanoy, H.A.; Bernau, V.; Kane, N.C.; Marek, L.; Seiler, G.; Rieseberg, L.H. Ecogeography and utility to plant breeding of the crop wild relatives of sunflower (Helianthus annuus L.). *Front. Plant Sci.* **2015**, *6*, 841. [CrossRef] [PubMed]
89. Van Zonneveld, M.; Thomas, E.; Galluzzi, G.; Scheldeman, X. Mapping the ecogeographic distribution of biodiversity and GIS tools for plant germplasm collectors. In *Collecting Plant Genetic Diversity: Technical Guidelines—2011 Update*; Guarino, L., Ramanatha Rao, V., Goldberg, E., Eds.; CAB International on behalf of Bioversity International, the Food and Agriculture Organization of the United Nations (FAO), the World Conservation Union (IUCN) and the United Nations Environment Programme (UNEP): Wallingford, UK, 2011; p. 26.
90. Berger, J.D.; Mackay, M.C.; Street, K.A.; Konopka, J.; Adhikari, K.; Clarke, H.J.; Sandhu, J.S.; Nayyar, H. Emerging Opportunities for Agriculture: Investigating Plant Adaptation by Characterizing Germplasm Collection Habitats. In *Global Issues, Paddock Action: Proceedings of the 14th Australian Agronomy Conference, Adelaide, Australia, 21–25 September 2008*; The Regional Institute: Adelaide, Australia, 2008; p. 5395.
91. Meilleur, B.A.; Hodgkin, T. In situ conservation of crop wild relatives. *Biodivers. Conserv.* **2004**, *13*, 663–684. [CrossRef]